Main groups — Transition metals — Main groups

1 1A	2 2A	3 3B	4 4B	5 5B	6 6B	7 7B	8	9 8B	10	11 1B	12 2B	13 3A	14 4A	15 5A	16 6A	17 7A	18 8A
1 **H** 1.00794																	2 **He** 4.00260
3 **Li** 6.941	4 **Be** 9.01218											5 **B** 10.81	6 **C** 12.011	7 **N** 14.0067	8 **O** 15.9994	9 **F** 18.998403	10 **Ne** 20.1797
11 **Na** 22.98977	12 **Mg** 24.305											13 **Al** 26.98154	14 **Si** 28.0855	15 **P** 30.97376	16 **S** 32.065	17 **Cl** 35.453	18 **Ar** 39.948
19 **K** 39.0983	20 **Ca** 40.078	21 **Sc** 44.9559	22 **Ti** 47.867	23 **V** 50.9415	24 **Cr** 51.996	25 **Mn** 54.9380	26 **Fe** 55.845	27 **Co** 58.9332	28 **Ni** 58.6934	29 **Cu** 63.546	30 **Zn** 65.38	31 **Ga** 69.723	32 **Ge** 72.64	33 **As** 74.9216	34 **Se** 78.96	35 **Br** 79.904	36 **Kr** 83.798
37 **Rb** 85.4678	38 **Sr** 87.62	39 **Y** 88.9059	40 **Zr** 91.224	41 **Nb** 92.9064	42 **Mo** 95.96	43 **Tc** (98)	44 **Ru** 101.07	45 **Rh** 102.9055	46 **Pd** 106.42	47 **Ag** 107.8682	48 **Cd** 112.41	49 **In** 114.82	50 **Sn** 118.710	51 **Sb** 121.760	52 **Te** 127.60	53 **I** 126.9045	54 **Xe** 131.29
55 **Cs** 132.9055	56 **Ba** 137.33	57 ***La** 138.9055	72 **Hf** 178.49	73 **Ta** 180.9479	74 **W** 183.84	75 **Re** 186.207	76 **Os** 190.23	77 **Ir** 192.217	78 **Pt** 195.08	79 **Au** 196.9666	80 **Hg** 200.59	81 **Tl** 204.3833	82 **Pb** 207.2	83 **Bi** 208.9804	84 **Po** (209)	85 **At** (210)	86 **Rn** (222)
87 **Fr** (223)	88 **Ra** 226.0254	89 **†Ac** 227.0278	104 **Rf** (267)	105 **Db** (268)	106 **Sg** (271)	107 **Bh** (272)	108 **Hs** (270)	109 **Mt** (276)	110 **Ds** (281)	111 **Rg** (280)	112 * (285)	113 **Uut** (284)	114 **Uuq** (289)	115 **Uup** (288)	116 **Uuh** (293)		118 **Uuo** (294)

*Lanthanide series	58 **Ce** 140.116	59 **Pr** 140.9077	60 **Nd** 144.242	61 **Pm** (145)	62 **Sm** 150.36	63 **Eu** 151.964	64 **Gd** 157.25	65 **Tb** 158.9254	66 **Dy** 162.500	67 **Ho** 164.9303	68 **Er** 167.259	69 **Tm** 168.9342	70 **Yb** 173.05	71 **Lu** 174.9668
†Actinide series	90 **Th** 232.0381	91 **Pa** 231.0359	92 **U** 238.0289	93 **Np** 237.048	94 **Pu** (244)	95 **Am** (243)	96 **Cm** (247)	97 **Bk** (247)	98 **Cf** (251)	99 **Es** (252)	100 **Fm** (257)	101 **Md** (258)	102 **No** (259)	103 **Lr** (262)

*Element 112 has a proposed name of Copernicium which is, at the time of this publication, under review by IUPAC.

Source: Atomic Weights of the Elements 2007, IUPAC. See http://www.chem.qmul.ac.uk/iupac/AtWt

Values in parentheses are mass numbers of the longest-lived known isotopes.

Greek Alphabet

A	α	alpha	N	ν	nu
B	β	beta	Ξ	ξ	xi
Γ	γ	gamma	O	o	omicron
Δ	δ	delta	Π	π	pi
E	ε	epsilon	P	ρ	rho
Z	ζ	zeta	Σ	σ	sigma
H	η	eta	T	τ	tau
Θ	θ	theta	Υ	υ	upsilon
I	ι	iota	Φ	ϕ	phi
K	κ	kappa	X	χ	chi
Λ	λ	lambda	Ψ	ψ	psi
M	μ	mu	Ω	ω	omega

Names and Symbols for the Elements

Z	*Symbol*	*Name*	*Z*	*Symbol*	*Name*	*Z*	*Symbol*	*Name*
1	H	Hydrogen	40	Zr	Zirconium	79	Au	Gold (Aurum)
2	He	Helium	41	Nb	Niobium	80	Hg	Mercury (Hydrargyrum)
3	Li	Lithium	42	Mo	Molybdenum	81	Tl	Thallium
4	Be	Beryllium	43	Tc	Technetium	82	Pb	Lead (Plumbum)
5	B	Boron	44	Ru	Ruthenium	83	Bi	Bismuth
6	C	Carbon	45	Rh	Rhodium	84	Po	Polonium
7	N	Nitrogen	46	Pd	Palladium	85	At	Astatine
8	O	Oxygen	47	Ag	Silver (Argentum)	86	Rn	Radon
9	F	Fluorine	48	Cd	Cadmium	87	Fr	Francium
10	Ne	Neon	49	In	Indium	88	Ra	Radium
11	Na	Sodium (Natrium)	50	Sn	Tin (Stannum)	89	Ac	Actinium
12	Mg	Magnesium	51	Sb	Antimony (Stibium)	90	Th	Thorium
13	Al	Aluminum	52	Te	Tellurium	91	Pa	Protactinium
14	Si	Silicon	53	I	Iodine	92	U	Uranium
15	P	Phosphorus	54	Xe	Xenon	93	Np	Neptunium
16	S	Sulfur	55	Cs	Cesium	94	Pu	Plutonium
17	Cl	Chlorine	56	Ba	Barium	95	Am	Americium
18	Ar	Argon	57	La	Lanthanum	96	Cm	Curium
19	K	Potassium (Kalium)	58	Ce	Cerium	97	Bk	Berkelium
20	Ca	Calcium	59	Pr	Praseodymium	98	Cf	Californium
21	Sc	Scandium	60	Nd	Neodymium	99	Es	Einsteinium
22	Ti	Titanium	61	Pm	Promethium	100	Fm	Fermium
23	V	Vanadium	62	Sm	Samarium	101	Md	Mendelevium
24	Cr	Chromium	63	Eu	Europium	102	No	Nobelium
25	Mn	Manganese	64	Gd	Gadolinium	103	Lr	Lawrencium
26	Fe	Iron (Ferrum)	65	Tb	Terbium	104	Rf	Rutherfordium
27	Co	Cobalt	66	Dy	Dysprosium	105	Db	Dubnium
28	Ni	Nickel	67	Ho	Holmium	106	Sg	Seaborgium
29	Cu	Copper (Cuprum)	68	Er	Erbium	107	Bh	Bohrium
30	Zn	Zinc	69	Tm	Thulium	108	Hs	Hassium
31	Ga	Gallium	70	Yb	Ytterbium	109	Mt	Meitnerium
32	Ge	Germanium	71	Lu	Lutetium	110	Dy	Darmstadtium
33	As	Arsenic	72	Hf	Hafnium	111	Rg	Roentgenium
34	Se	Selenium	73	Ta	Tantalum	112*	Cn	Copernicium
35	Br	Bromine	74	W	Tungsten (Wolfram)	113	Uut	Ununtrium
36	Kr	Krypton	75	Re	Rhenium	114	Uuq	Ununquadium
37	Rb	Rubidium	76	Os	Osmium	115	Uup	Ununpentium
38	Sr	Strontium	77	Ir	Iridium	116	Uuh	Ununhexium
39	Y	Yttrium	78	Pt	Platinum	118	Uuo	Ununoctium

The names in parentheses are the sources of the symbols, but are not used when referring to the elements. Iron, copper, silver, tin, gold, and lead (and sometimes antimony) anions are named by using the name in parentheses. For example, $[Fe(CN)_6]^{3-}$ is called hexacyanoferrate (III).

* Element 112 has a proposed name of Copernicium which is, at the time of this publication, under review by IUPAC.

Fourth Edition

INORGANIC CHEMISTRY

Gary L. Miessler
St. Olaf College Northfield, Minnesota

Donald A. Tarr
St. Olaf College Northfield, Minnesota

Prentice Hall
Boston Columbus Indianapolis New York San Francisco Upper Saddle River
Amsterdam Cape Town Dubai London Madrid Milan Munich Paris Montréal Toronto
Delhi Mexico City São Paulo Sydney Hong Kong Seoul Singapore Taipei Tokyo

Publisher: Dan Kaveney
Editor in Chief, Chemistry and Geosciences: Nicole Folchetti
Assistant Editors: Carol DuPont/Jessica Neumann
Marketing Manager: Erin Gardner
Editorial Assistant: Fran Falk
Marketing Assistant: Nicola Houston
Managing Editor, Chemistry and Geosciences: Gina M. Cheselka
Project Manager: Wendy A. Perez
Senior Manufacturing and Operations Manager: Nick Sklitsis
Operations Specialist: Maura Zaldivar
Composition/Full Service: GEX Publishing Services
Art Editor: Ronda Whitson
Art Studio: Stacy B. Smith
Art Director: Jayne Conte
Cover Designer: Jodi Notowitz
Photo Researcher: Elaine Soares

Photo Credits: p. 82, Figure 4.1, Rachel Miessler; p. 122, Problem 4.8a, Naomi Miessler; p. 122, Problem 4.8f Andrey Prokhorov/iStockphoto; p. 122, Problem 4.8g Murray Creek/murraycreek.net; p. 122, Problem 4.9f Naomi Miessler; p. 123, Problem 4.16a JackF/Shutter Stock Images; p. 123, Problem 4.16b ARTEKI/Shutter Stock Images; p. 123, Problem 4.16c Kheng Guan Toh/Shutter Stock Images; p. 123, Problem 4.16d JackF/Shutter Stock Images; p. 123, Problem 4.16e JackF/Shutter Stock Images.

Printed in the United States
10 9 8 7 6 5 4 3 2 1

Prentice Hall
is an imprint of

ISBN 13: 978-0-13-615383-2
ISBN 10: 0-13-615383-6

The passing of my valued colleague Don Tarr has left a void, both in the collegial work we conducted together for more than two decades and in the areas of expertise in which he made so many contributions. Don was a valued friend and dedicated community servant who was appreciated more broadly than he ever knew. This edition is dedicated to his memory.

BRIEF CONTENTS

CONTENTS

PREFACE

The pace of developments in inorganic chemistry makes it an increasing challenge to provide a textbook that is contemporary and meets the needs of those who will use it. Many constructive suggestions have been provided by students, faculty, reviewers, and other users. These are greatly appreciated and have been adopted to the extent that space, time, and the scope of the book allow. The main emphasis in preparing this edition has been to bring it up-to-date as much as possible while providing clear explanations and a variety of helpful features. In addition, with broader electronic access to research journals and search tools such as Web of Science and SciFinder Scholar, there has been an increased emphasis on end-of-chapter problems that can make use of the original literature and electronic searches. Features in the fourth edition that distinguish it from the third include the following:

- New and expanded discussions have been incorporated in many chapters to reflect topics of recent interest, for example, receptor–guest complexes (Chapter 6), quantum dots (Chapter 7), graphene and nanotubes (Chapter 8), metal–organic frameworks (Chapter 9), carbide and cumulene ligands (Chapter 13), olefin metathesis (Chapter 14), quintuple bonds (Chapter 15), platinum-based antitumor agents (Chapter 16), and greenhouse gases (Chapter 16).
- The sections on VSEPR and the ligand close-packing model in Chapter 3 have been expanded to provide a wider variety of examples.
- In keeping with a focus on symmetry, Chapter 4 has been expanded by including applications of symmetry to Raman spectra, illustrations of the symmetry of translational and rotational motion, and a broader array of end-of-chapter problems.
- At the encouragement of many users, the number of end-of-chapter problems has been increased by more than 25 percent, with most new problems based on the recent inorganic literature. The total number of problems now exceeds 500. The number of problems involving molecular modeling software has been increased.
- The number of in-chapter examples and exercises has also been increased, with about 10 percent more exercises than in the third edition.
- Reference information has been added on angular functions for f orbitals (Appendix B.8). In addition, the atomic weights of the elements provided in the periodic table inside the front cover have been updated to include the most recent IUPAC recommendations, and the values of physical constants inside the back cover have been revised to use the most recent values cited on the NIST Web site.
- Partly for fun, some new diagrams have been included of particularly interesting structures, for example, a carbon peapod (Chapters 1 and 8), a "double concave hydrocarbon buckycatcher" (Chapter 6), nanotube unzipping and the high-pressure O_8 allotrope (Chapter 8), and a molecular Ferris wheel and a diplatinum complex held together by a 16-carbon tether (Chapter 13).
- This edition employs the use of a second color to better highlight and clarify key features of the art and chemistry within the body of the text.

Sections of text have been revised to improve organization and clarity, for example, on Slater's rules (Chapter 2), applications of symmetry to molecular orbitals (Chapter 5), and bonding in coordination complexes (Chapter 10). In addition, numerous smaller revisions have been made in response to readers' suggestions. Keeping a text within its scope and maintaining a reasonable length are challenging, and difficult decisions about content are inevitable. I hope that the text will serve readers well, even if some may find that a favorite topic is missing or not covered in sufficient detail. Overall, this edition is the broadest in scope of its four editions to date.

Finally, Paul Fischer of Macalester College has provided valuable advice on this edition and will bring creative ideas and scholarship to future editions, joining me as a coauthor. We welcome comments on this edition and suggestions for topics to be included in the next.

SUPPLEMENTS

For the Instructor

INSTRUCTOR IMAGES FOR DOWNLOAD (ISBN: 0-13-614291-5) All art, photos, and tables from the textbook have been digitized and converted into JPG files for instructor use. Instructors may insert these into their own lecture presentation notes to suit classroom needs and enhance lectures.

For the Student

SOLUTIONS MANUAL (ISBN: 0-13-612867-X) by Gary Miessler & Donald Tarr, Saint Olaf College. Includes fully worked-out solutions to all end-of-chapter problems in the text.

ACKNOWLEDGMENTS

Thanks, first, to my wife Becky and daughters Naomi and Rachel for their understanding, support, and love through long hours over many months. Special thanks to Rachel and Naomi for contributing their photographic expertise to Chapter 4. I also appreciate all that Carol DuPont and Jessica Neumann, my editors at Pearson Prentice Hall, and Kelly Morrison at GEX Publishing Services have contributed, especially their patience when waiting for materials from me.

Finally, I greatly appreciate the helpful suggestions of the reviewers and other faculty listed below and of the many St. Olaf students who have pointed out needed improvements; in many ways, students have been the best reviewers of all. While constraints on the scope and length of the text have meant that not all suggestions could be included, I am truly thankful that so many individuals have been willing to be helpful, and the ideas that have been offered will be considered again for the next edition.

Reviewers of the Fourth Edition of **Inorganic Chemistry**

Nancy Deluca
University of Massachusetts–Lowell

James J. Dechter
University of Central Oklahoma

Stephanie K. Hurst
Northern Arizona University

Derek P. Gates
University of British Colombia

Reviewers of Previous Editions of **Inorganic Chemistry**

John Arnold
University of California–Berkeley

Ronald Bailey
Rensselaer Polytechnic University

Robert Balahura
University of Guelph

Craig Barnes
University of Tennessee–Knoxville

Daniel Bedgood
Arizona State University

Simon Bott
University of Houston

Joe Bruno
Wesleyan University

Charles Dismukes
Princeton University

Kate Doan
Kenyon College

Charles Drain
Hunter College

Jim Finholt
Carleton College

Daniel Haworth
Marquette University

Laura Pence
University of Hartford

Greg Peters
University of Memphis

Cortland Pierpont
University of Colorado

Robert Pike
College of William and Mary

Jeffrey Rack
Ohio University

Gregory Robinson
University of Georgia

Lothar Stahl
University of North Dakota

Karen Stephens
Whitworth College

Robert Stockland
Bucknell University

Dennis Strommen
Idaho State University

Patrick Sullivan
Iowa State University

Duane Swank
Pacific Lutheran University

Michael Johnson
University of Georgia

Jerome Kiester
University of Buffalo

Katrina Miranda
University of Arizona

Michael Moran
West Chester University

Wyatt Murphy
Seton Hall University

Mary-Ann Pearsall
Drew University

William Tolman
University of Minnesota

Robert Troy
Central Connecticut State University

Edward Vitz
Kutztown University

Richard Watt
University of New Mexico

Tim Zauche
University of Wisconsin–Platteville

Chris Ziegler
University of Akron

Gary L. Miessler
St. Olaf College
Northfield, Minnesota

1

Introduction to Inorganic Chemistry

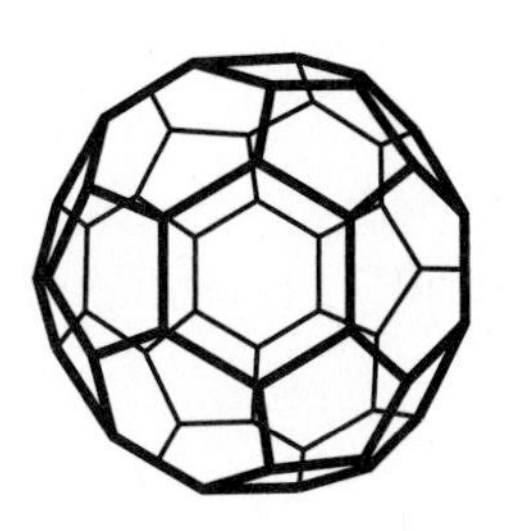

1.1 WHAT IS INORGANIC CHEMISTRY?

If organic chemistry is defined as the chemistry of hydrocarbon compounds and their derivatives, inorganic chemistry can be described broadly as the chemistry of "everything else." This includes all the remaining elements in the periodic table, as well as carbon, which plays a major role in many inorganic compounds. Organometallic chemistry, a very large and rapidly growing field, bridges both areas by considering compounds containing direct metal–carbon bonds, and it includes catalysis of many organic reactions. Bioinorganic chemistry bridges biochemistry and inorganic chemistry, and environmental chemistry includes the study of both inorganic and organic compounds. As can be imagined, the inorganic realm is vast, providing essentially limitless areas for investigation.

1.2 CONTRASTS WITH ORGANIC CHEMISTRY

Some comparisons between organic and inorganic compounds are in order. In both areas, single, double, and triple covalent bonds are found, as shown in Figure 1.1; for inorganic compounds, these include direct metal–metal bonds and metal–carbon bonds. However, although the maximum number of bonds between two carbon atoms is three, there are many compounds that contain quadruple bonds between metal atoms. In addition to the sigma and pi bonds common in organic chemistry, quadruply bonded metal atoms contain a delta (δ) bond (Figure 1.2); a combination of one sigma bond, two pi bonds, and one delta bond makes up the quadruple bond. The delta bond is possible in these cases because the metal atoms have *d* orbitals to use in bonding, whereas carbon has only *s* and *p* orbitals available.

More recently, compounds with "fivefold" bonds between transition metals have been reported, accompanied by discussion as to whether these bonds merit the designation "quintuple." An example is shown in Figure 1.3.

In organic compounds, hydrogen is nearly always bonded to a single carbon. In inorganic compounds, hydrogen is frequently encountered as a bridging atom between two or more other atoms. Bridging hydrogen atoms can also occur in metal cluster compounds. In these clusters, hydrogen atoms form bridges across edges or faces of polyhedra of metal atoms. Alkyl groups may also act as bridges in inorganic compounds, a function rarely encountered in organic chemistry except in reaction

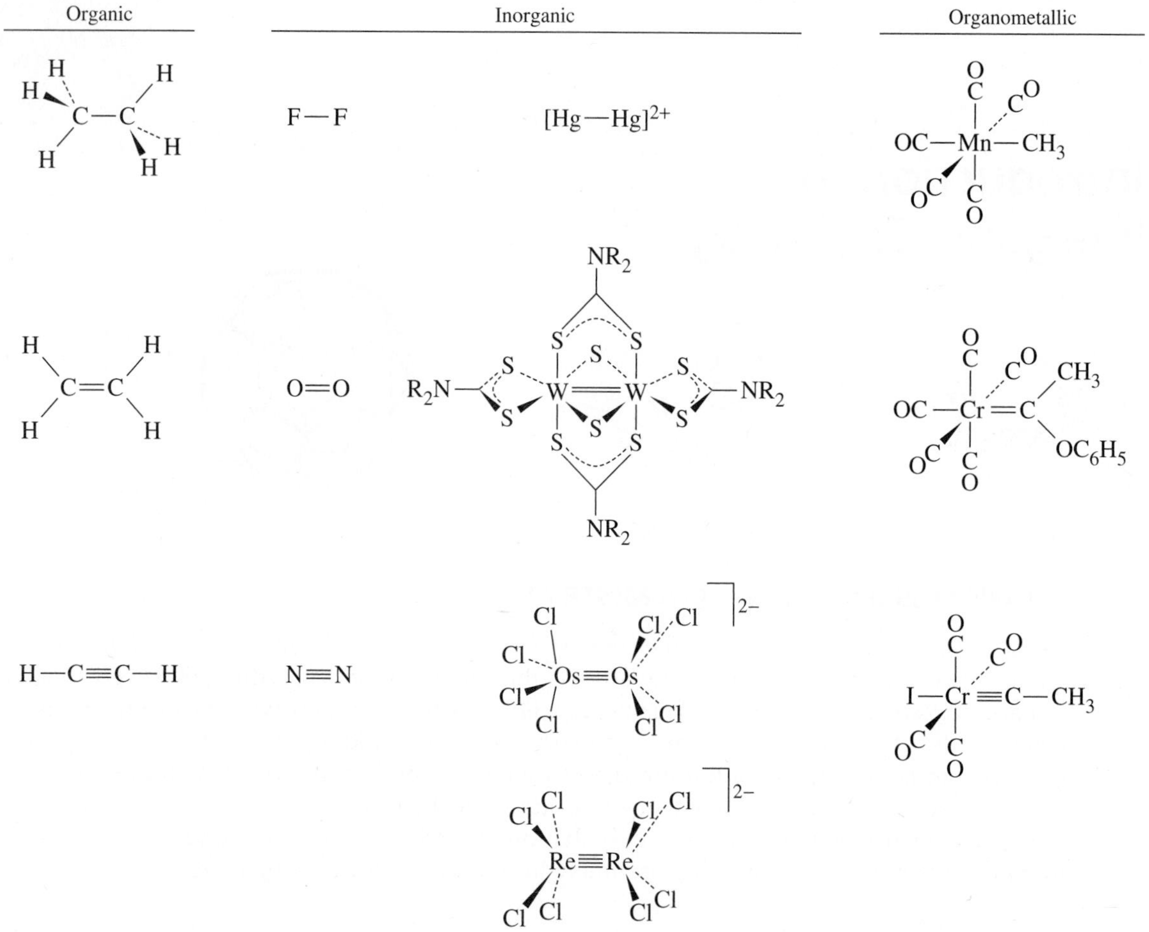

Organic	Inorganic		Organometallic
H_3C-CH_3	F—F	$[Hg-Hg]^{2+}$	$(OC)_5Mn-CH_3$
$H_2C=CH_2$	O=O	$[W_2(S_2CNR_2)_4(\mu\text{-}S)_2]$	$(OC)_5Cr=C(CH_3)(OC_6H_5)$
H—C≡C—H	N≡N	$[Cl_4Os\equiv OsCl_4]^{2-}$	$I-Cr(CO)_4\equiv C-CH_3$
		$[Cl_4Re\equiv ReCl_4]^{2-}$	

FIGURE 1.1 Single and Multiple Bonds in Organic and Inorganic Molecules.

intermediates. Examples of terminal and bridging hydrogen atoms and alkyl groups in inorganic compounds are shown in Figure 1.4.

Some of the most striking differences between the chemistry of carbon and that of many other elements are in coordination number and geometry. Although carbon is usually limited to a maximum coordination number of four (a maximum of four atoms bonded to carbon, as in CH_4), numerous inorganic compounds have central atoms with coordination numbers of five, six, seven, and higher; the most common

FIGURE 1.2 Examples of Bonding Interactions.

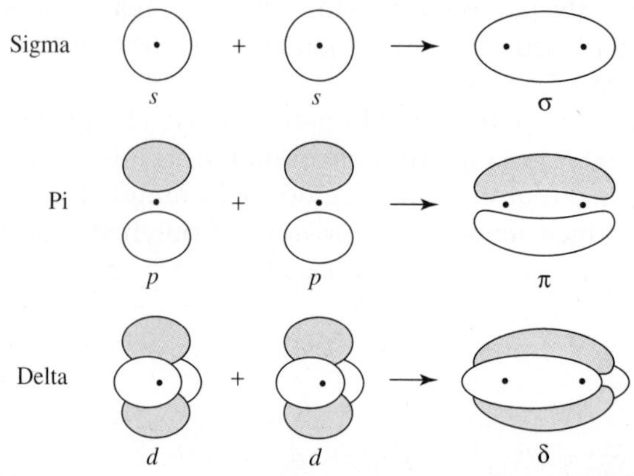

Sigma: $s + s \rightarrow \sigma$

Pi: $p + p \rightarrow \pi$

Delta: $d + d \rightarrow \delta$

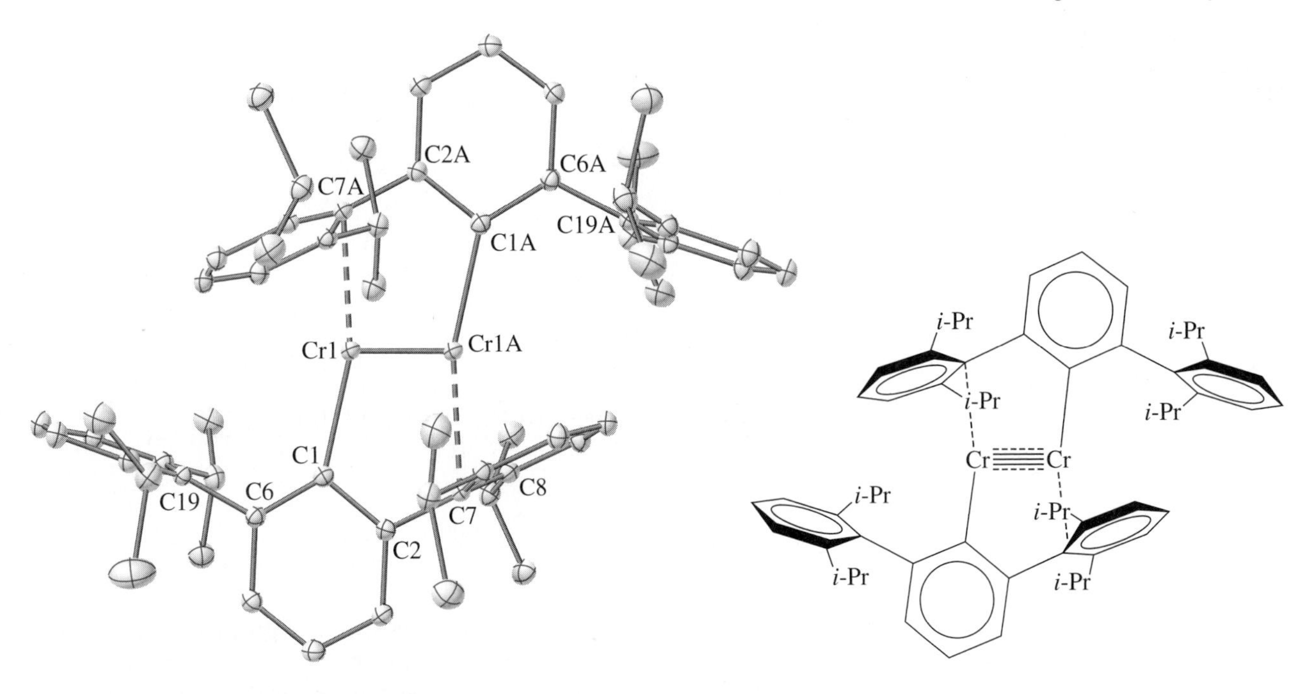

FIGURE 1.3 Example of Fivefold Bonding.

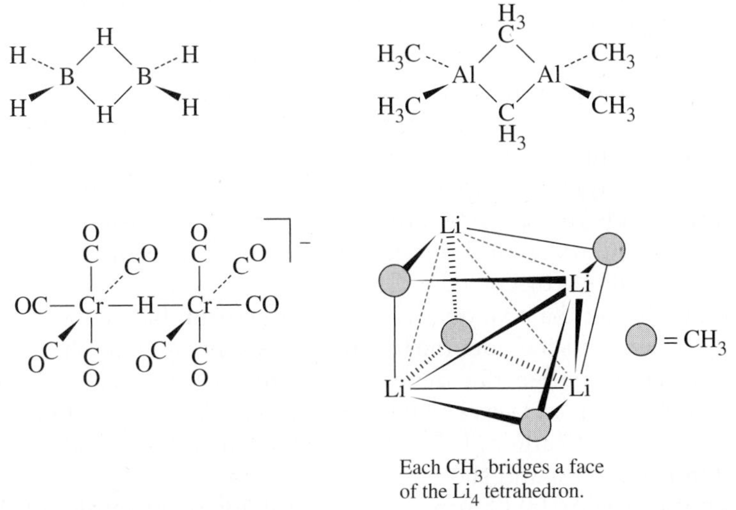

Each CH_3 bridges a face of the Li_4 tetrahedron.

= CH_3

FIGURE 1.4 Examples of Inorganic Compounds Containing Terminal and Bridging Hydrogens and Alkyl Groups.

coordination geometry for transition metals is an octahedral arrangement around a central atom, as shown for $[TiF_6]^{3-}$ in Figure 1.5. Furthermore, inorganic compounds present coordination geometries different from those found for carbon. For example, although 4-coordinate carbon is nearly always tetrahedral, both tetrahedral and square-planar shapes occur for 4-coordinate compounds of both metals and nonmetals. When metals are in the center, with anions or neutral molecules bonded to them (frequently through N, O, or S), these are called *coordination complexes;* when carbon is the element directly bonded to metal atoms or ions, they are also organometallic complexes.

The tetrahedral geometry usually found in 4-coordinate compounds of carbon also occurs in a different form in some inorganic molecules. Methane contains four hydrogens in a regular tetrahedron around carbon. Elemental phosphorus is tetratomic (P_4) and tetrahedral, but with no central atom. Examples of some of the geometries found for inorganic compounds are shown in Figure 1.5.

FIGURE 1.5 Examples of Geometries of Inorganic Compounds.

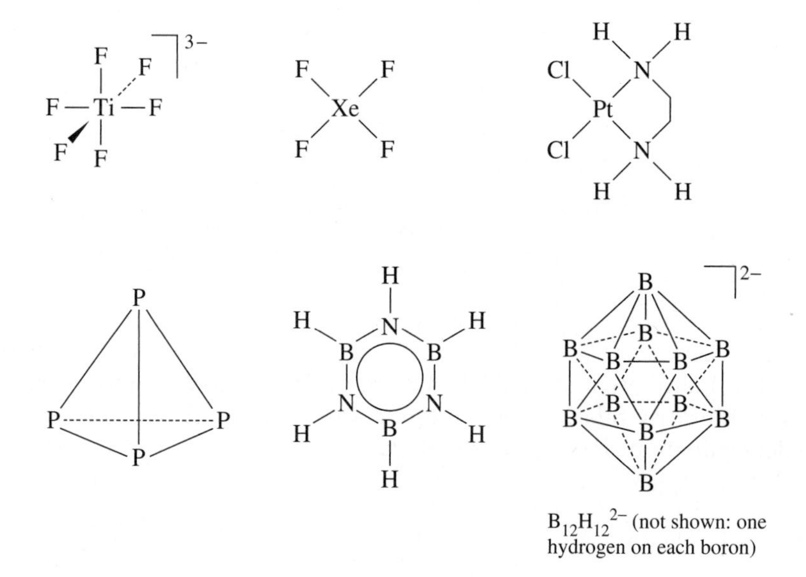

$B_{12}H_{12}^{2-}$ (not shown: one hydrogen on each boron)

Aromatic rings are common in organic chemistry, and aryl groups can also form sigma bonds to metals. However, aromatic rings can also bond to metals in a dramatically different fashion using their pi orbitals, as shown in Figure 1.6. The result is a metal atom bonded above the center of the ring, almost as if suspended in space. In many cases, metal atoms are sandwiched between two aromatic rings. Multiple-decker sandwiches of metals and aromatic rings are also known.

Carbon plays an unusual role in a number of metal cluster compounds in which a carbon atom is at the center of a polyhedron of metal atoms. Examples of carbon-centered clusters with five, six, or more surrounding metals are known; two of these are shown in Figure 1.7. The contrast between the role that carbon plays in these clusters and its usual role in organic compounds is striking, and attempting to explain how carbon can form bonds to the surrounding metal atoms in clusters has provided an interesting challenge to theoretical inorganic chemists. A molecular orbital picture of bonding in these clusters is discussed in Chapter 15.

In addition, during the past quarter century the realm of chemistry of elemental carbon has flourished. This includes the fullerenes, most notably the cluster C_{60}, labeled "buckminsterfullerene" after the developer of the geodesic dome. Many other fullerenes are now known and serve as the cores of a variety of derivatives. In addition, numerous other forms of carbon, for example carbon nanotubes and graphene, have also attracted much interest and show potential for a range of applications in fields as diverse as nanoelectronics, body armor, and drug delivery. Examples of these newer forms of carbon are shown in Figure 1.8.

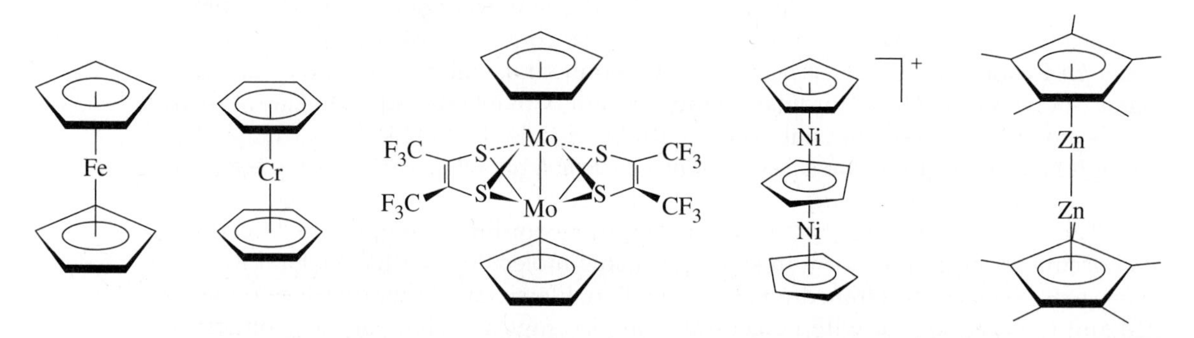

FIGURE 1.6 Inorganic Compounds Containing Pi-Bonded Aromatic Rings.

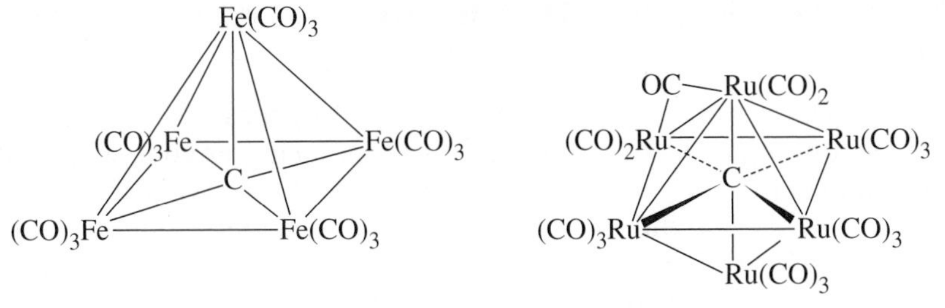

FIGURE 1.7 Carbon-Centered Metal Clusters.

There are no sharp dividing lines between subfields in chemistry. Many of the subjects in this book, such as acid–base chemistry and organometallic reactions, are of vital interest to many organic chemists. Others such as oxidation-reduction reactions, spectra, and solubility relations also interest analytical chemists. Subjects related to structure determination, spectra, conductivity, and theories of bonding appeal to physical chemists. Finally, the use of organometallic catalysts provides a connection to petroleum and polymer chemistry, and the presence of coordination compounds such as hemoglobin and metal-containing enzymes provides a similar tie to biochemistry. This list is not intended to describe a fragmented field of study, but rather to show some of the interconnections between inorganic chemistry and other fields of chemistry.

The remainder of this chapter is devoted to a short history of the origins of inorganic chemistry, intended to provide a sense of connection to the past and to place some aspects of inorganic chemistry within the context of larger historical events. In later chapters, brief histories are given with the same intention. Although time and space do not allow for much attention to history, it would be misleading to give the impression that any field of chemistry has sprung full-blown from any one person's work or has appeared suddenly. Although certain events, such as a new theory or a new type of compound or reaction, can later be identified as marking a dramatic change of direction in inorganic chemistry, new ideas are built on past achievements. In some cases, experimental observations from the past become understandable in the light of new theoretical developments. In others, the theory is already in place, ready for the new compounds or phenomena that it will explain.

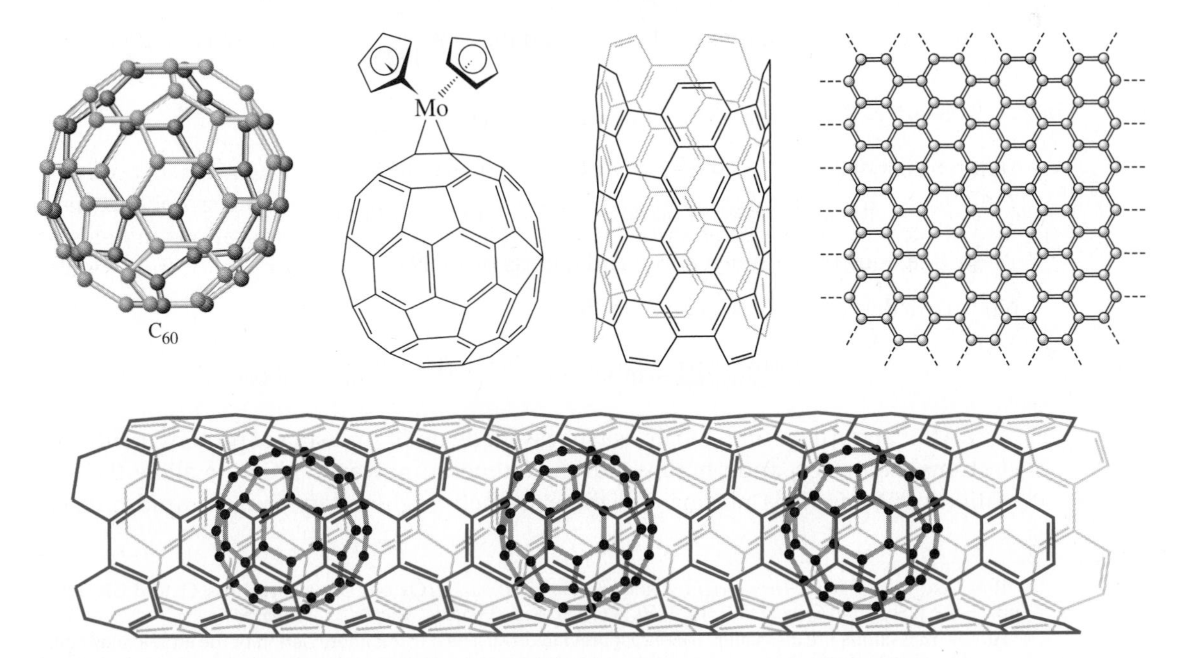

FIGURE 1.8 The Fullerene C_{60}, a Fullerene Compound, a Carbon Nanotube, Graphene, and a Carbon Peapod.

1.3 GENESIS OF THE ELEMENTS: THE BIG BANG AND FORMATION OF THE EARTH

We begin our discussion of inorganic chemistry with the genesis of the elements and the creation of the universe. Among the difficult tasks facing anyone who attempts to explain the origin of the universe are the inevitable questions: "What about the time just before the creation? Where did the starting material, whether energy or matter, come from?" The whole idea of an origin at a specific time means that there was nothing before that instant. By its very nature, no theory attempting to explain the origin of the universe can be expected to extend infinitely back in time.

Current opinion favors the big bang theory[1] over other theories of the origin of the universe, although many controversial points have yet to be explained. Other theories, such as the steady-state or oscillating theories, have their advocates, and the creation of the universe is certain to remain a source of controversy and study.

According to the big bang theory, the universe began approximately 1.37×10^{10} years ago with an extreme concentration of energy in a very small space. In fact, extrapolation back to the time of origin requires zero volume and infinite temperature. How this can be envisioned remains a challenge. What is almost universally agreed on is that the universe is expanding rapidly from an initial event, during which neutrons were formed that decayed quickly (half-life = 11.3 min) into protons, electrons, and antineutrinos:

$$\mathrm{n} \longrightarrow \mathrm{p} + \mathrm{e}^- + \overline{v}_\mathrm{e}$$

or

$${}^1_0\mathrm{n} \longrightarrow {}^1_1\mathrm{H} + {}^{\;\,0}_{-1}\mathrm{e} + \overline{v}_e$$

In this and subsequent equations,

${}^1_1\mathrm{H} = \mathrm{p} =$ a proton of +1 charge and mass of 1.007 atomic mass unit (u) [2]

$\gamma =$ a gamma ray (high-energy photon) with zero mass

${}^{\;\,0}_{-1}\mathrm{e} = \mathrm{e}^- =$ an electron of −1 charge and mass of $\frac{1}{1823}$ u (also known as a β particle)

${}^0_1\mathrm{e} = \mathrm{e}^+ =$ a positron with +1 charge and mass of $\frac{1}{1823}$ u

$v_\mathrm{e} =$ a neutrino with no charge and a very small mass

$\overline{v}_\mathrm{e} =$ an antineutrino with no charge and a very small mass

${}^1_0\mathrm{n} =$ a neutron with no charge and a mass of 1.009 u

Nuclei are described by the convention

$${}^{\text{mass number}}_{\text{atomic number}}\text{symbol} \quad \text{or} \quad {}^{\text{protons plus neutrons}}_{\text{protons}}\text{symbol}$$

After about 1 second, the universe was made up of a plasma of protons, neutrons, electrons, neutrinos, and photons, but the temperature was too high to allow the formation of atoms. This plasma and the extremely high energy caused rapid nuclear

[1] P. A. Cox, *The Elements, Their Origin, Abundance and Distribution*, Oxford University Press, Oxford UK, 1990, pp. 66–92; J. Selbin, *J. Chem. Educ.*, **1973**, *50*, 306, 380; A. A. Penzias, *Science*, 1979, **105**, 549.

[2] Atomic mass units are also called *daltons*, abbreviated Da. The current abbreviation of the atomic mass unit is *u*, although the former *amu* is also still common. More accurate masses are given inside the back cover of this text.

reactions. As the temperature dropped to about 10^9 K, the following reactions occurred within a matter of minutes:

$$^{1}_{1}H + ^{1}_{0}n \longrightarrow ^{2}_{1}H + \gamma$$

$$^{2}_{1}H + ^{2}_{1}H \longrightarrow ^{3}_{1}H + ^{1}_{1}H$$

$$^{2}_{1}H + ^{2}_{1}H \longrightarrow ^{3}_{2}He + ^{1}_{0}n$$

$$^{3}_{2}He + ^{1}_{0}n \longrightarrow ^{4}_{2}He + \gamma$$

The first is the limiting reaction, because the reverse reaction is also fast. The interplay of the rates of these reactions gives an atomic ratio of He/H = 1/10, which is the abundance observed in young stars.

By this time, the temperature had dropped enough to allow the positive particles to capture electrons to form atoms. Because atoms interact less strongly with electromagnetic radiation than do individual subatomic particles, the atoms could now interact with each other more or less independently from the radiation. The atoms began to condense into stars, and the radiation moved with the expanding universe. This expansion caused a red shift, leaving the background radiation with wavelengths in the millimeter range, characteristic of a temperature of 2.7 K. This radiation was observed in 1965 by Penzias and Wilson and is supporting evidence for the big bang theory.

Within one half-life of the neutron (11.3 min), half the matter of the universe consisted of protons, and the temperature was near 5×10^8 K. The nuclei formed in the first 30 to 60 minutes were those of deuterium (^{2}H), ^{3}He, ^{4}He, and ^{5}He. (Helium 5 has a very short half-life of 2×10^{-21} seconds and decays back to helium 4, effectively limiting the mass number of the nuclei formed by these reactions to 4.) The following reactions show how these nuclei can be formed in a process called *hydrogen burning:*

$$^{1}_{1}H + ^{1}_{1}H \longrightarrow ^{2}_{1}H + ^{0}_{1}e + v_e$$

$$^{2}_{1}H + ^{1}_{1}H \longrightarrow ^{3}_{2}He + \gamma$$

$$^{3}_{2}He + ^{3}_{2}He \longrightarrow ^{4}_{2}He + 2\,^{1}_{1}H$$

The expanding material from these first reactions began to gather together into galactic clusters and then into more dense stars, where the pressure of gravity kept the temperature high and promoted further reactions. The combination of hydrogen and helium with many protons and neutrons led rapidly to the formation of heavier elements. In stars with internal temperatures of 10^7 to 10^8 K, the reactions forming ^{2}H, ^{3}He, and ^{4}He continued, along with reactions that produced heavier nuclei. The following *helium-burning* reactions are among those known to take place under these conditions:

$$2\,^{4}_{2}He \longrightarrow ^{8}_{4}Be + \gamma$$

$$^{4}_{2}He + ^{8}_{4}Be \longrightarrow ^{12}_{6}C + \gamma$$

$$^{12}_{6}C + ^{1}_{1}H \longrightarrow ^{13}_{7}N \longrightarrow ^{13}_{6}C + ^{0}_{1}e + v_e$$

In more massive stars, with temperatures of 6×10^8 K or higher, the carbon-nitrogen cycle is possible:

$$^{12}_{6}C + ^{1}_{1}H \longrightarrow ^{13}_{7}N + \gamma$$

$$^{13}_{7}N \longrightarrow ^{12}_{6}C + ^{0}_{1}e + v_e$$

$$^{12}_{6}C + ^{1}_{1}H \longrightarrow ^{13}_{7}N + \gamma$$

$$^{13}_{7}N + ^{1}_{1}H \longrightarrow ^{15}_{8}O + \gamma$$

$$^{15}_{8}O \longrightarrow ^{15}_{7}N + ^{0}_{1}e + v_e$$

$$^{15}_{7}N + ^{1}_{1}H \longrightarrow ^{4}_{2}He + ^{12}_{6}C$$

The net result of this cycle is the formation of helium from hydrogen, with gamma rays, positrons, and neutrinos as by-products. In addition, even heavier elements are formed:

$$^{12}_{6}C + ^{12}_{6}C \longrightarrow ^{20}_{10}Ne + ^{4}_{2}He$$

$$2\ ^{16}_{8}O \longrightarrow ^{28}_{14}Si + ^{4}_{2}He$$

$$2\ ^{16}_{8}O \longrightarrow ^{31}_{16}S + ^{1}_{0}n$$

At still higher temperatures, further reactions take place:

$$\gamma + ^{28}_{14}Si \longrightarrow ^{24}_{12}Mg + ^{4}_{2}He$$

$$^{28}_{14}Si + ^{4}_{2}He \longrightarrow ^{32}_{16}S + \gamma$$

$$^{32}_{16}S + ^{4}_{2}He \longrightarrow ^{36}_{18}Ar + \gamma$$

Even more massive elements can be formed, with the actual amounts depending on a complex relationship between their inherent stability and the temperature and lifetime of the star. The curve of inherent stability of nuclei reaches a maximum at $^{56}_{26}Fe$, accounting for the high relative abundance of iron in the universe. If these reactions continued indefinitely, the result should be nearly complete dominance of elements near iron over the other elements. However, as parts of the universe cooled, the reactions slowed or stopped. Consequently, both lighter and heavier elements are common. Formation of elements of higher atomic number takes place by the addition of neutrons to a nucleus, followed by electron emission decay. In environments of low neutron density, this addition of neutrons is relatively slow, one neutron at a time; in the high neutron density environment of a nova, 10 to 15 neutrons may be added in a very short time, and the resulting nucleus is neutron rich:

$$^{56}_{26}Fe + 13\ ^{1}_{0}n \longrightarrow ^{69}_{26}Fe \longrightarrow ^{69}_{27}Co + ^{0}_{-1}e$$

The very heavy elements are also formed by reactions such as this. After the addition of the neutrons, beta decay (loss of electrons from the nucleus as a neutron is converted to a proton plus an electron) leads to nuclei with larger atomic numbers. Figure 1.9 shows the cosmic abundances of some of the elements.

Gravitational attraction combined with rotation gradually formed the expanding cloud of material into relatively flat spiral galaxies containing millions of stars each. Complex interactions led to black holes and other types of stars, some of which exploded as supernovas and scattered their material widely. Further gradual accretion of some of this material into planets followed. At the lower temperatures found in planets, the buildup of heavy elements stopped, and decay of unstable radioactive isotopes of the elements became the predominant nuclear reactions.

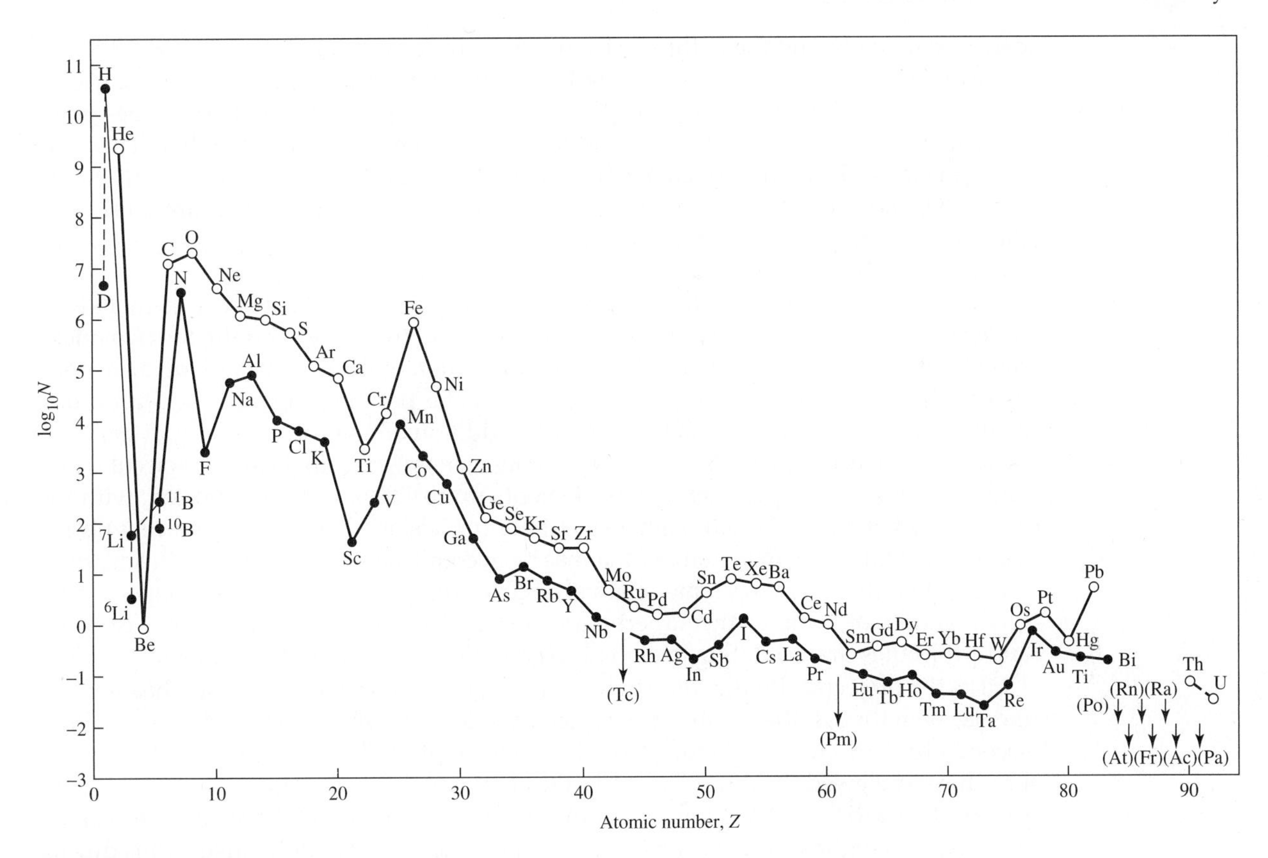

FIGURE 1.9 Cosmic Abundances of the Elements.

(Reprinted with permission from N. N. Greenwood and A. Earnshaw, *Chemistry of the Elements*, Butterworth-Heinemann, Oxford UK, 1997, p. 4.)

1.4 NUCLEAR REACTIONS AND RADIOACTIVITY

Some nuclei that formed were stable, never undergoing further reactions. Others have lifetimes ranging from 10^{16} years to 10^{-16} seconds. The usual method of describing nuclear decay is in terms of the **half-life**, or the time needed for half the nuclei to react. Because decay follows first-order kinetics, the half-life is a well-defined value not dependent on the amount of nuclei present. In addition to the overall curve of nuclear stability, which has its most stable region near atomic number $Z = 26$ (iron), combinations of protons and neutrons at each atomic number exhibit different stabilities. In some elements, such as fluorine (^{19}F), there is only one stable **isotope**, a specific combination of protons and neutrons. In others, such as chlorine, there are two or more stable isotopes. ^{35}Cl has a natural abundance of 75.77 percent, and ^{37}Cl has a natural abundance of 24.23 percent. Both are stable, as are all the natural isotopes of the lighter elements. Tin, with 10 stable isotopes, has the most. The radioactive isotopes of these elements have short half-lives and have had more than enough time to decay to more stable elements. ^{3}H, ^{14}C, and a few other radioactive nuclei are continually being formed by cosmic rays and have a low constant concentration.

Heavier elements ($Z = 40$ or higher) may also have radioactive isotopes with longer half-lives. As a result, some of these radioactive isotopes have not had time to decay completely, and the natural substances are radioactive. Further discussion of isotopic abundances and radioactivity can be found in larger or more specialized sources.[3]

As atomic mass increases, the ratio of neutrons to protons in stable isotopes gradually increases from 1:1 to 1.6:1 for $^{238}_{92}U$. Nuclei also have energy levels similar to the electron energy levels described in Chapter 2, which result in stable nuclei with 2, 8, 20, 28, 50, 82, and 126 protons or neutrons. In nature, the most stable nuclei are those with the numbers of both protons and neutrons matching one of these numbers; $^{4}_{2}He$, $^{16}_{8}O$, $^{40}_{20}Ca$, and $^{208}_{82}Pb$ are examples.

Elements not present in nature can be formed by bombardment of one element with nuclei of another; if the atoms are carefully chosen, and the energy is right, the two nuclei can merge to form one nucleus and then eject a portion of the nucleus to form a new element. This procedure has been used to extend the periodic table beyond uranium. Neptunium and plutonium can be formed by addition of neutrons to uranium followed by release of electrons (β particles). Still heavier elements require heavier projectiles and higher energies. Using this approach, elements up to 118 have been reported, with the exception of element 117, although element 112 (provisionally named *copernicium*) is the element with highest mass number that has been confirmed by the IUPAC.[4] Calculations indicate that there may be some relatively stable isotopes, with half-lives longer than a few seconds, of some of the superheavy elements if the appropriate target isotopes and projectiles are used. Suggestions include ^{248}Cm, ^{250}Cm, and ^{244}Pu as targets and ^{48}Ca as the projectile. Predictions such as this have fueled the search for still heavier elements, even though their stability is typically so low that they must be detected within seconds of their creation, before they decompose into lighter elements. Hoffman and Lee[5] have reviewed the efforts to study the chemistry of these new elements. The subtitle of their article, "One Atom at a Time," describes the difficulty of such studies. In one case, α-daughter decay chains of ^{265}Sg were detected from only three atoms during 5000 experiments, but this was sufficient to show that Sg(VI) is similar to W(VI) and Mo(VI) in forming neutral or negative species in HNO_3—HF solution, but not like U(VI), which forms $[UO_2]^{2+}$ under these conditions. Element 108, hassium, formed by bombarding ^{248}Cm with high-energy atoms of ^{26}Mg, was found to form an oxide similar to that of osmium on the basis of six oxide molecules carried from the reaction site to a detector by a stream of helium.[6] This may be the most massive atom on which "chemistry" has been performed to date.

1.5 DISTRIBUTION OF ELEMENTS ON EARTH

Theories that attempt to explain the formation of the specific structures of the Earth are at least as numerous as those for the formation of the universe. Although the details of these theories differ, there is general agreement that the Earth was much hotter during its early life, and that the materials fractionated into gaseous, liquid, and solid states at that time. As the surface cooled, the lighter materials in the crust solidified and still float on a molten inner layer, according to the plate tectonics explanation of geology. There is also general agreement that the Earth has a core of iron and nickel, which is solid at the center and liquid above that. The outer half of the Earth's radius is composed of silicate minerals in the mantle; silicate, oxide, and sulfide minerals in the crust;

---

[3] N. N. Greenwood and A. Earnshaw, *Chemistry of the Elements*, 2nd ed., Butterworth-Heinemann, Oxford, 1997; J. Silk, *The Big Bang: The Creation and Evolution of the Universe*, W. H. Freeman, San Francisco, 1980.

[4] R. C. Barber, H. W. Gäggeler, P. J. Karol, H. Nakahara, E.Vardaci, and E. Vogt, *Pure Appl. Chem.*, **2009**, *81*, 1331. Information on reports of other elements with high mass numbers can be found at the IUPAC site, *http://www.iupac.org*.

[5] D. C. Hoffman and D. M. Lee, *J. Chem. Educ.*, **1999**, *76*, 331.

[6] *Chem. Eng. News*, June 4, 2001, p. 47.

and a wide variety of materials at the surface, including abundant water and the gases of the atmosphere.

The different types of forces apparent in the early planet Earth can now be seen indirectly in the distribution of minerals and elements. In locations where liquid magma broke through the crust, compounds that are readily soluble in such molten rock were carried along and deposited as ores. Fractionation of the minerals then depended on their melting points and solubilities in the magma. In other locations, water was the source of the formation of ore bodies. At these sites, water leached minerals from the surrounding area and later evaporated, leaving the minerals behind. The solubilities of the minerals in either magma or water depend on the elements, their oxidation states, and the other elements with which they are combined. A rough division of the elements can be made according to their ease of reduction to the element and their combination with oxygen and sulfur. **Siderophiles**, (iron-loving elements) concentrate in the metallic core; **lithophiles**, (rock-loving elements) combine primarily with oxygen and the halides and are more abundant in the crust; and **chalcophiles** (Greek, *khalkos*, copper) combine more readily with sulfur, selenium, and arsenic and are also found in the crust. **Atmophiles** are present as gases. These divisions are shown in the periodic table in Figure 1.10.

As an example of the action of water, we can explain the formation of bauxite (hydrated Al_2O_3) deposits by the leaching away of the more soluble salts from aluminosilicate deposits. The silicate portion is soluble enough in water that it can be leached away, leaving a higher concentration of less soluble Al_2O_3. This is shown in the following reaction, in which H_4SiO_4 is a generic representation for a number of soluble silicate species.

$$\underset{\text{Aluminosilicate}}{4KAlSi_3O_8(s)} + 4\,CO_2 + 22\,H_2O \longrightarrow 4\,K^+ + 4\,HCO_3^- + \underset{\text{Higher concentration of Al}}{Al_4Si_4O_{10}(OH)_8(s)} + \underset{\text{Silicate (leached away)}}{8\,H_4SiO_4(aq)}$$

This mechanism provides at least a partial explanation for the presence of bauxite deposits in tropical areas or in areas that were once tropical, with large amounts of rainfall in the past.

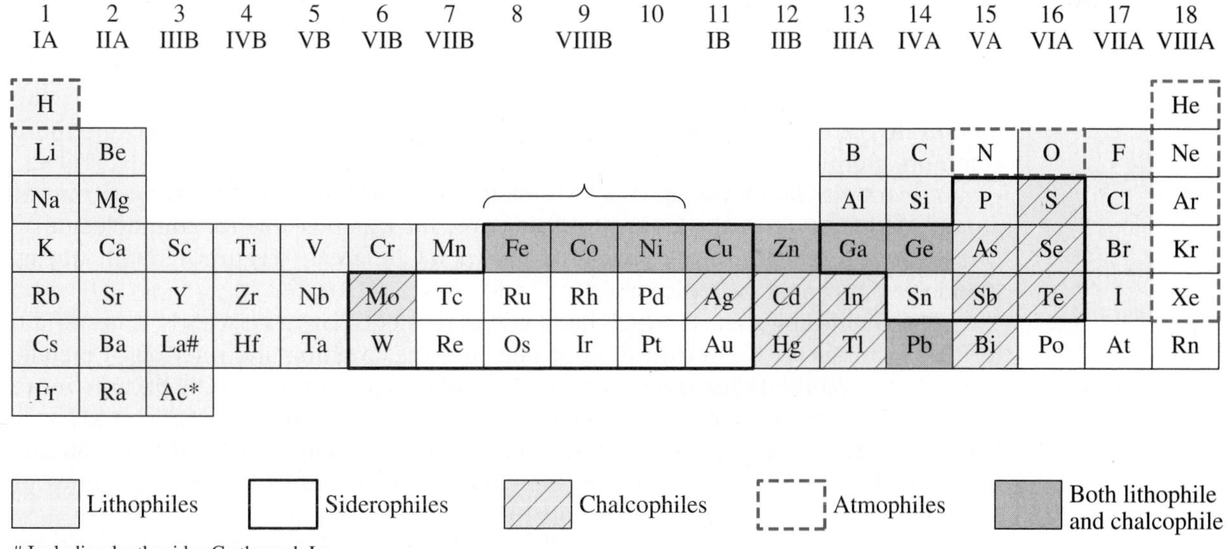

FIGURE 1.10 Geochemical Classification of the Elements.

(Adapted with permission from P. A. Cox, *The Elements, Their Origin, Abundance, and Distribution*, Oxford University Press, Oxford, 1990, p. 13.)

Further explanations of these geological processes must be left to more specialized sources.[7] Such explanations are based on concepts treated later in this text. For example, modern acid–base theory helps explain the different solubilities of minerals in water or molten rock and their resulting deposits in specific locations. The divisions illustrated in Figure 1.10 can be partly explained by this theory, which is discussed in Chapter 6 and used in later chapters.

1.6 THE HISTORY OF INORGANIC CHEMISTRY

Even before alchemy became a subject of study, many chemical reactions were used and the products applied to daily life. For example, the first metals used were probably gold and copper, which can be found in the metallic state. Copper can also be readily formed by the reduction of malachite—basic copper carbonate, $Cu_2(CO_3)(OH)_2$—in charcoal fires. Silver, tin, antimony, and lead were also known as early as 3000 BCE. Iron appeared in classical Greece and in other areas around the Mediterranean Sea by 1500 BCE. At about the same time, colored glasses and ceramic glazes were introduced, largely composed of silicon dioxide (SiO_2, the major component of sand) and other metallic oxides, which had been melted and allowed to cool to amorphous solids.

Alchemists were active in China, Egypt, and other centers of civilization early in the first centuries CE. Although much effort went into attempts to "transmute" base metals into gold, the treatises of these alchemists also described many other chemical reactions and operations. Distillation, sublimation, crystallization, and other techniques were developed and used in their studies. Because of the political and social changes of the time, alchemy shifted into the Arab world and later—about 1000 to 1500 CE—reappeared in Europe. Gunpowder was used in Chinese fireworks as early as 1150, and alchemy was also widespread in China and India at that time. Alchemists appeared in art, literature, and science until at least 1600, by which time chemistry was beginning to take shape as a science. Roger Bacon (1214–1294), recognized as one of the first great experimental scientists, also wrote extensively about alchemy.

By the seventeenth century, the common strong acids—nitric, sulfuric, and hydrochloric—were known, and more systematic descriptions of common salts and their reactions were being accumulated. The combination of acids and bases to form salts was appreciated by some chemists. As experimental techniques improved, the quantitative study of chemical reactions and the properties of gases became more common, atomic and molecular weights were determined more accurately, and the groundwork was laid for what later became the periodic table. By 1869, the concepts of atoms and molecules were well established, and it was possible for Mendeleev and Meyer to describe different forms of the periodic table. Figure 1.11 illustrates Mendeleev's original periodic table.[8]

The chemical industry, which had been in existence since very early times in the form of factories for the purification of salts and the smelting and refining of metals, expanded as methods for the preparation of relatively pure materials became more common. In 1896, Becquerel discovered radioactivity, and another area of study was opened. Studies of subatomic particles, spectra, and electricity finally led to the atomic theory of Bohr in 1913, which was soon modified by the quantum mechanics of Schrödinger and Heisenberg in 1926 and 1927.

---

[7] J. E. Fergusson, *Inorganic Chemistry and the Earth*, Pergamon Press, Elmsford, NY, 1982; J. E. Fergusson, *The Heavy Elements*, Pergamon Press, Elmsford, NY, 1990.

[8] The original table was published in *Zeitschrift für Chemie*, **1869**, *12*, 405. It can be found in English translation, together with a page from the German article, at http://web.lemoyne.edu/~giunta/mendeleev.html. See M. Laing, *J. Chem. Educ.*, **2008**, *85*, 63 for illustrations of Mendeleev's various versions of the periodic table.

			Ti = 50	Zr = 90	? = 180
			V = 51	Nb = 94	Ta = 182
			Cr = 52	Mo = 96	W = 186
			Mn = 53	Rh = 104.4	Pt = 197.4
			Fe = 56	Ru = 104.2	Ir = 198
			Ni = Co = 59	Pd = 106.6	Os = 199
H = 1			Cu = 63.4	Ag = 108	Hg = 200
	Be = 9.4	Mg = 24	Zn = 65.2	Cd = 112	
	B = 11	Al = 27.4	? = 68	Ur = 116	Au = 197?
	C = 12	Si = 28	? = 70	Sn = 118	
	N = 14	P = 31	As = 75	Sb = 122	Bi = 210?
	O = 16	S = 32	Se = 79.4	Te = 128?	
	F = 19	Cl = 35.5	Br = 80	J = 127	
Li = 7	Na = 23	K = 39	Rb = 85.4	Cs = 133	Tl = 204
		Ca = 40	Sr = 87.6	Ba = 137	Pb = 207
		? = 45	Ce = 92		
		?Er = 56	La = 94		
		?Yt = 60	Di = 95		
		?In = 75.6	Th = 118 ?		

FIGURE 1.11 Mendeleev's 1869 Periodic Table. Two years later, Mendeleev revised his table into a form similar to a modern short-form periodic table, with eight groups across.

Inorganic chemistry as a field of study was extremely important during the early years of the exploration and development of mineral resources. Qualitative analysis methods were developed to help identify minerals and, combined with quantitative methods, to assess their purity and value. As the Industrial Revolution progressed, so did the chemical industry. By the early twentieth century, plants for the production of ammonia, nitric acid, sulfuric acid, sodium hydroxide, and many other inorganic chemicals produced on a large scale were common.

In spite of the work of Werner and Jørgensen on coordination chemistry near the beginning of the twentieth century, and the discovery of a number of organometallic compounds, the popularity of inorganic chemistry as a field of study gradually declined during most of the first half of the century. The need for inorganic chemists to work on military projects during World War II rejuvenated interest in the field. As work was done on many projects (not least of which was the Manhattan Project, in which scientists developed the fission bomb that later led to the development of the fusion bomb), new areas of research appeared, and new theories were proposed that prompted further experimental work. A great expansion of inorganic chemistry started in the 1940s, sparked by the enthusiasm and ideas generated during World War II.

In the 1950s, an earlier method used to describe the spectra of metal ions surrounded by negatively charged ions in crystals (**crystal field theory**)[9] was extended by the use of molecular orbital theory[10] to develop **ligand field theory** for use in coordination compounds, in which metal ions are surrounded by ions or molecules that donate electron pairs. This theory, discussed in Chapter 10, gave a more complete picture of the bonding in these compounds. The field developed rapidly as a result of this theoretical framework, availability of new instruments, and the generally reawakened interest in inorganic chemistry.

In 1955, Ziegler[11] and associates and Natta[12] discovered organometallic compounds that could catalyze the polymerization of ethylene at lower temperatures and pressures than the common industrial method used up to that time. In addition, the polyethylene formed was more likely to be made up of linear, rather than branched,

[9] H. A. Bethe, *Ann. Physik*, **1929**, *3*, 133.
[10] J. S. Griffith and L. E. Orgel, *Q. Rev. Chem. Soc.*, **1957**, *XI*, 381.
[11] K. Ziegler, E. Holzkamp, H. Breil, and H. Martin, *Angew. Chem.*, **1955**, *67*, 541.
[12] G. Natta, *J. Polym. Sci.*, **1955**, *16*, 143.

molecules and, as a consequence, was stronger and more durable. Other catalysts were soon developed, and their study contributed to the rapid expansion of organometallic chemistry, still one of the fastest growing areas of chemistry today.

The study of biological materials containing metal atoms has also progressed rapidly. Again, the development of new experimental methods allowed more thorough study of these compounds, and the related theoretical work provided connections to other areas of study. Attempts to make *model* compounds that have chemical and biological activity similar to the natural compounds have also led to many new synthetic techniques. Two of the many biological molecules that contain metals are shown in Figure 1.12. Although these molecules have very different roles, they share similar ring systems.

One current area that bridges organometallic chemistry and bioinorganic chemistry is the conversion of nitrogen to ammonia:

$$N_2 + 3\,H_2 \longrightarrow 2\,NH_3$$

This reaction is one of the most important industrial processes, with over 100 million tons of ammonia produced annually worldwide, primarily for fertilizer. However, in spite of metal oxide catalysts introduced in the Haber–Bosch process in 1913, and improved since then, it is also a reaction that requires temperatures near 400 °C and 200 atm pressure and that still results in a yield of only 15 percent ammonia. Bacteria, however, manage to fix nitrogen (convert it to ammonia and then to nitrite and nitrate) at 0.8 atm at room temperature in nodules on the roots of legumes. The nitrogenase enzyme that catalyzes this reaction is a complex iron–molybdenum–sulfur protein. The structure of the active sites has been determined by X-ray crystallography.[13] This

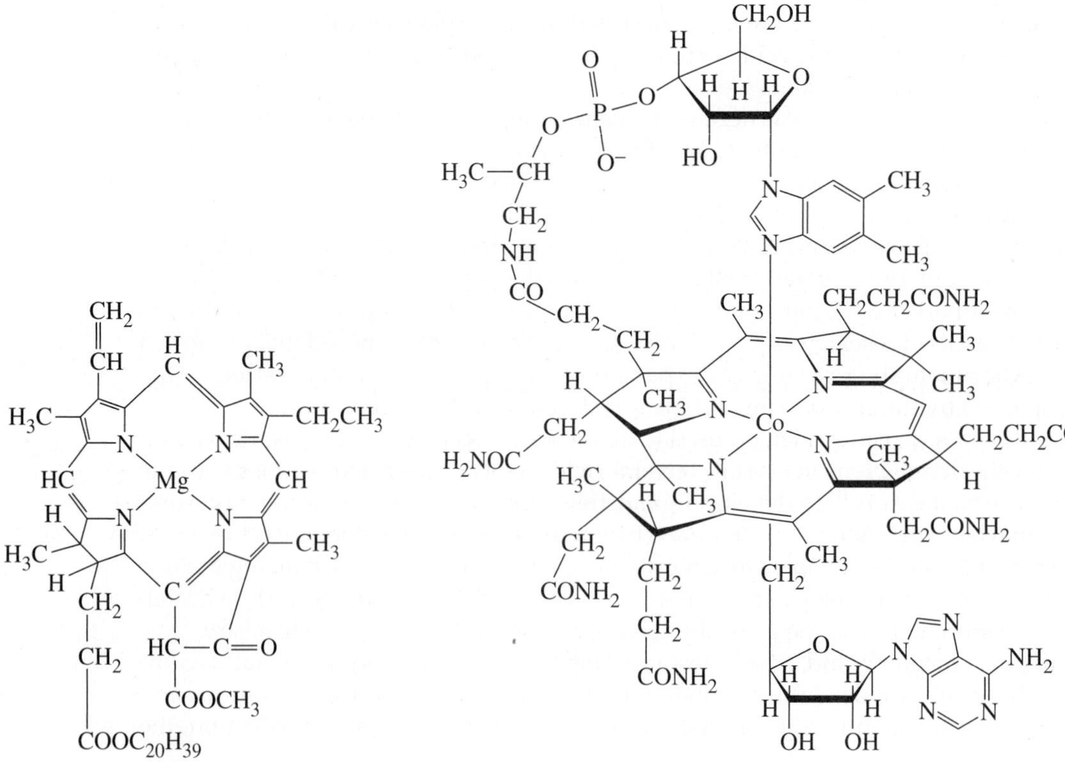

(a) (b)

FIGURE 1.12 Biological Molecules Containing Metal Ions. (a) Chlorophyll *a*, the active agent in photosynthesis. (b) Vitamin B_{12} coenzyme, a naturally occurring organometallic compound.

[13] M. K. Chan, J. Kin, and D. C. Rees, *Science*, **1993**, *260*, 792.

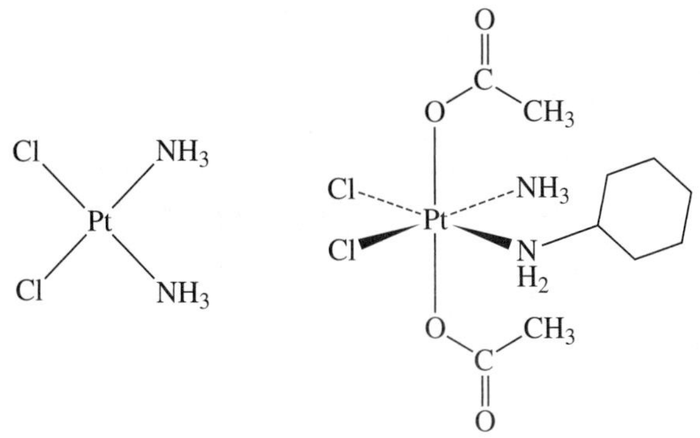

FIGURE 1.13 Cisplatin and Satraplatin.

problem and others linking biological reactions to inorganic chemistry are described in Chapter 16.

Inorganic chemistry also has a variety of medical applications. Notable among these is the development of platinum-containing antitumor agents, the first of which was the *cis* isomer of $Pt(NH_3)_2Cl_2$, cisplatin. First approved for clinical use approximately 30 years ago, cisplatin has served as the prototype for a variety of anticancer agents; for example, satraplatin, the first orally available platinum anticancer drug to reach clinical trials.[14] These two compounds are shown in Figure 1.13; they are also discussed in Chapter 16.

With this brief survey of the marvelously complex field of inorganic chemistry, we now turn to the details in the remainder of this book. The topics included provide a broad introduction to the field. However, even a cursory examination of a chemical library or one of the many inorganic journals shows some important aspects of inorganic chemistry that must be omitted in a short textbook. The references cited in the text suggest resources for further study, including historical sources, texts, and reference works that can provide useful additional material.

General References

For those interested in further discussion of the physics of the big bang and related cosmology, a nonmathematical treatment is in S. W. Hawking's *A Brief History of Time*, Bantam, New York, 1988. The title of P. A. Cox, *The Elements, Their Origin, Abundance, and Distribution*, Oxford University Press, Oxford, 1990, describes its contents exactly. The inorganic chemistry of minerals, their extraction, and their environmental impact at a level understandable to anyone with some background in chemistry can be found in J. E. Fergusson's *Inorganic Chemistry and the Earth*, Pergamon Press, Elmsford, NY, 1982. Among the many general reference works available, three of the most useful and complete are N. N. Greenwood and A. Earnshaw's *Chemistry of the Elements*, 2nd ed., Butterworth-Heinemann, Oxford, 1997; F. A. Cotton, G. Wilkinson, C. A. Murillo, and M. Bochman's *Advanced Inorganic Chemistry*, 6th ed., John Wiley & Sons, New York, 1999; and A. F. Wells's *Structural Inorganic Chemistry*, 5th ed., Oxford University Press, New York, 1984. An interesting study of inorganic reactions from a different perspective can be found in G. Wulfsberg's *Principles of Descriptive Inorganic Chemistry*, Brooks/Cole, Belmont, CA, 1987.

[14] For a review of modes of interaction of cisplatin and related drugs, see P. C. A. Bruijnincx and P. J. Sadler, *Curr. Opin. Chem. Bio.*, **2008**, *12*, 197.

2 Atomic Structure

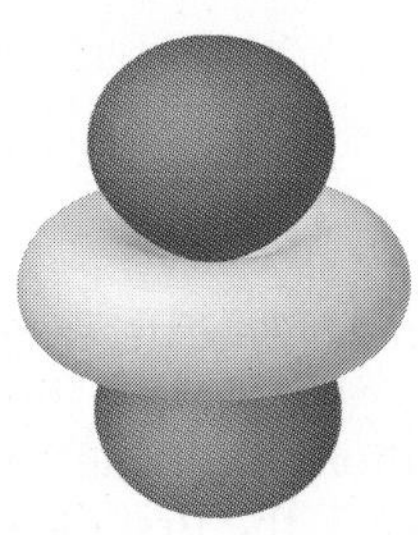

The theories of atomic and molecular structure depend on quantum mechanics to describe atoms and molecules in mathematical terms. Fortunately, it is possible to gain a practical understanding of the principles of atomic and molecular structure with only a moderate amount of mathematics rather than the mathematical sophistication involved in quantum mechanics. This chapter presents the fundamentals needed to explain atomic and molecular structures in qualitative or semiquantitative terms.

2.1 HISTORICAL DEVELOPMENT OF ATOMIC THEORY

Although the Greek philosophers Democritus (460–370 BCE) and Epicurus (341–270 BCE) presented views of nature that included atoms, many centuries passed before experimental studies could establish the quantitative relationships needed for a coherent atomic theory. In 1808, John Dalton published *A New System of Chemical Philosophy*,[1] in which he proposed that

> . . . the ultimate particles of all homogeneous bodies are perfectly alike in weight, figure, etc. In other words, every particle of water is like every other particle of water; every particle of hydrogen is like every other particle of hydrogen, etc.[2]

and that atoms combine in simple numerical ratios to form compounds. The terminology he used has since been modified, but he clearly presented the ideas of atoms and molecules, described many observations about heat (or *caloric*, as it was called), and made quantitative observations of the masses and volumes of substances combining to form new compounds. Because of confusion about elemental molecules such as H_2 and O_2, which he assumed to be monatomic H and O, he did not find the correct formula for water. Dalton said that

> When two measures of hydrogen and one of oxygen gas are mixed, and fired by the electric spark, the whole is converted into steam, and if the pressure be great, this steam becomes water. It is most probable then that there is the same number of particles in two measures of hydrogen as in one of oxygen.[3]

[1] John Dalton, *A New System of Chemical Philosophy*, 1808; reprinted with an introduction by Alexander Joseph, Peter Owen Limited, London, 1965.
[2] Ibid., p. 113.
[3] Ibid., p. 133

In fact, he then changed his mind about the number of molecules in equal volumes of different gases:

> At the time I formed the theory of mixed gases, I had a confused idea, as many have, I suppose, at this time, that the particles of elastic fluids are all of the same size; that a given volume of oxygenous gas contains just as many particles as the same volume of hydrogenous; or if not, that we had no data from which the question could be solved. . . . I [later] became convinced. . . . that every species of pure elastic fluid has its particles globular and all of a size; but that no two species agree in the size of their particles, the pressure and temperature being the same.[4]

Only a few years later, Avogadro used data from Gay-Lussac to argue that equal volumes of gas at equal temperatures and pressures contain the same number of molecules, but uncertainties about the nature of sulfur, phosphorus, arsenic, and mercury vapors delayed acceptance of this idea. Widespread confusion about atomic weights and molecular formulas contributed to the delay; in 1861, Kekulé gave 19 different possible formulas for acetic acid![5] In the 1850s, Cannizzaro revived the argument of Avogadro and argued that everyone should use the same set of atomic weights rather than the many different sets then being used. At a meeting in Karlsruhe in 1860, Cannizzaro distributed a pamphlet describing his views.[6] His proposal was eventually accepted, and a consistent set of atomic weights and formulas gradually evolved. In 1869, Mendeleev[7] and Meyer[8] independently proposed periodic tables nearly like those used today, and from that time, the development of atomic theory progressed rapidly.

2.1.1 The Periodic Table

The idea of arranging the elements into a periodic table had been considered by many chemists, but either the data to support the idea were insufficient, or the classification schemes were incomplete. Mendeleev and Meyer organized the elements in order of atomic weight and then identified families of elements with similar properties. By arranging these families in rows or columns, and by considering similarities in chemical behavior, as well as atomic weight, Mendeleev found vacancies in the table and was able to predict the properties of several elements—gallium, scandium, germanium, and polonium—that had not yet been discovered. When his predictions proved accurate, the concept of a periodic table was quickly established (see Figure 1.11). The discovery of additional elements not known in Mendeleev's time and the synthesis of heavy elements have led to the more complete, modern periodic table, shown inside the front cover of this text.

In the modern periodic table, a horizontal row of elements is called a **period**, and a vertical column is a **group** or **family**. The traditional designations of groups in the United States differ from those used in Europe. The International Union of Pure and Applied Chemistry (IUPAC) has recommended that the groups be numbered 1 through 18, a recommendation that has generated considerable controversy. In this text, we will use the IUPAC group numbers with the traditional American numbers in parentheses. Some sections of the periodic table have traditional names, as shown in Figure 2.1.

[4] Ibid., pp. 144–145.
[5] J.R. Partington, *A Short History of Chemistry*, 3rd ed., Macmillan, London, 1957; reprinted, 1960, Harper & Row, New York, p. 255.
[6] Ibid., pp. 256–258.
[7] D. I. Mendeleev, *J. Russ. Phys. Chem. Soc.*, **1869**, *i*, 60.
[8] L. Meyer, *Justus Liebigs Ann. Chem.*, **1870**, *Suppl. vii*, 354.

FIGURE 2.1 Names for Parts of the Periodic Table.

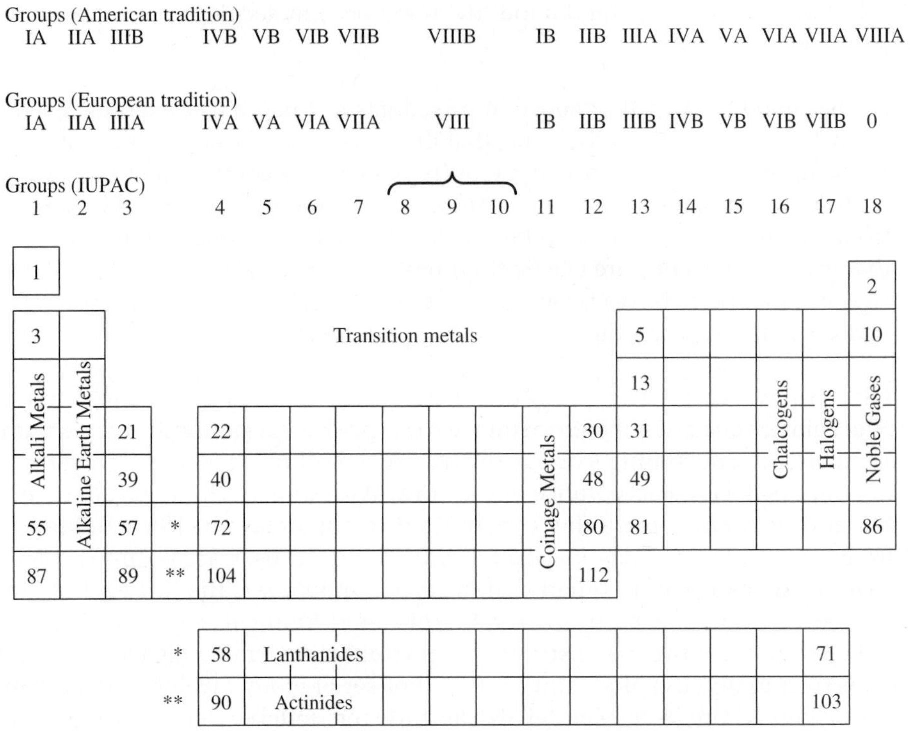

Groups (American tradition)
IA IIA IIIB IVB VB VIB VIIB VIIIB IB IIB IIIA IVA VA VIA VIIA VIIIA

Groups (European tradition)
IA IIA IIIA IVA VA VIA VIIA VIII IB IIB IIIB IVB VB VIB VIIB 0

Groups (IUPAC)
1 2 3 4 5 6 7 8 9 10 11 12 13 14 15 16 17 18

1
2
3 Transition metals 5 10
13
21 22 30 31
39 40 48 49
55 57 * 72 80 81 86
87 89 ** 104 112
Alkali Metals
Alkaline Earth Metals
Coinage Metals
Chalcogens
Halogens
Noble Gases

* 58 Lanthanides 71
** 90 Actinides 103

2.1.2 Discovery of Subatomic Particles and the Bohr Atom

During the 50 years after the periodic tables of Mendeleev and Meyer were proposed, experimental advances came rapidly. Some of these discoveries are shown in Table 2.1.

Parallel discoveries in atomic spectra showed that each element emits light of specific energies when excited by an electric discharge or heat. In 1885, Balmer showed that the energies of visible light emitted by the hydrogen atom are given by the equation

$$E = R_H\left(\frac{1}{2^2} - \frac{1}{n_h^2}\right)$$

where

n_h = integer, with $n_h > 2$

R_H = Rydberg constant for hydrogen $= 1.097 \times 10^7\ \mathrm{m}^{-1} = 2.179 \times 10^{-18}\ \mathrm{J}$

TABLE 2.1 Discoveries in Atomic Structure

1896	A. H. Becquerel	Discovered radioactivity of uranium
1897	J. J. Thomson	Showed that electrons have a negative charge, with charge/mass = 1.76×10^{11} C/kg
1909	R. A. Millikan	Measured the electronic charge as 1.60×10^{-19} C, therefore the mass of the electron is 9.11×10^{-31} kg, $\frac{1}{1836}$ the mass of the H atom
1911	E. Rutherford	Established the nuclear model of the atom; a very small, heavy nucleus surrounded by mostly empty space
1913	H. G. J. Moseley	Determined nuclear charges by X-ray emission, establishing atomic numbers as more fundamental than atomic masses

And the energy is related to the wavelength, frequency, and wave number of the light, as given by the equation

$$E = hv = \frac{hc}{\lambda} = hc\bar{v}$$

where[9]

$$\begin{aligned} h &= \text{Planck constant} = 6.626 \times 10^{-34}\ \text{J s} \\ v &= \text{frequency of the light, in s}^{-1} \\ c &= \text{speed of light} = 2.998 \times 10^{8}\ \text{m s}^{-1} \\ \lambda &= \text{wavelength of the light, frequently in nm} \\ \bar{v} &= \text{wavenumber of the light, usually in cm}^{-1} \end{aligned}$$

The Balmer equation was later made more general, as spectral lines in the ultraviolet and infrared regions of the spectrum were discovered, by replacing 2^2 by n_l^2, with the condition that $n_l < n_h$. These quantities, n_i, are called **quantum numbers**. (These are the **principal quantum numbers**; other quantum numbers are discussed in Section 2.2.2.) The origin of this energy was unknown until Niels Bohr's quantum theory of the atom,[10] first published in 1913 and refined over the following 10 years. This theory assumed that negative electrons in atoms move in stable circular orbits around the positive nucleus with no absorption or emission of energy. However, electrons may absorb light of certain specific energies and be excited to orbits of higher energy; they may also emit light of specific energies and fall to orbits of lower energy. The energy of the light emitted or absorbed can be found, according to the Bohr model of the hydrogen atom, from the equation

$$E = R_H\left(\frac{1}{n_l^2} - \frac{1}{n_h^2}\right)$$

where

$$\begin{aligned} R &= \frac{2\pi^2 \mu Z^2 e^4}{(4\pi\varepsilon_0)^2 h^2} \\ \mu &= \text{reduced mass of the electron–nucleus combination} \\ \frac{1}{\mu} &= \frac{1}{m_e} + \frac{1}{m_{nucleus}} \\ m_e &= \text{mass of the electron} \\ m_{nucleus} &= \text{mass of the nucleus} \\ Z &= \text{charge of the nucleus} \\ e &= \text{electronic charge} \\ h &= \text{Planck constant} \\ n_h &= \text{quantum number describing the higher energy state} \\ n_l &= \text{quantum number describing the lower energy state} \\ 4\pi\varepsilon_0 &= \text{permittivity of a vacuum} \end{aligned}$$

This equation shows that the Rydberg constant depends on the mass of the nucleus as well as on the fundamental constants.

Examples of the transitions observed for the hydrogen atom and the energy levels responsible are shown in Figure 2.2. As the electrons drop from level n_h to n_l (h for higher

[9] More accurate values for the constants and energy conversion factors are given inside the back cover of this book.
[10] N. Bohr, *Philos. Mag.*, **1913**, *26*, 1.

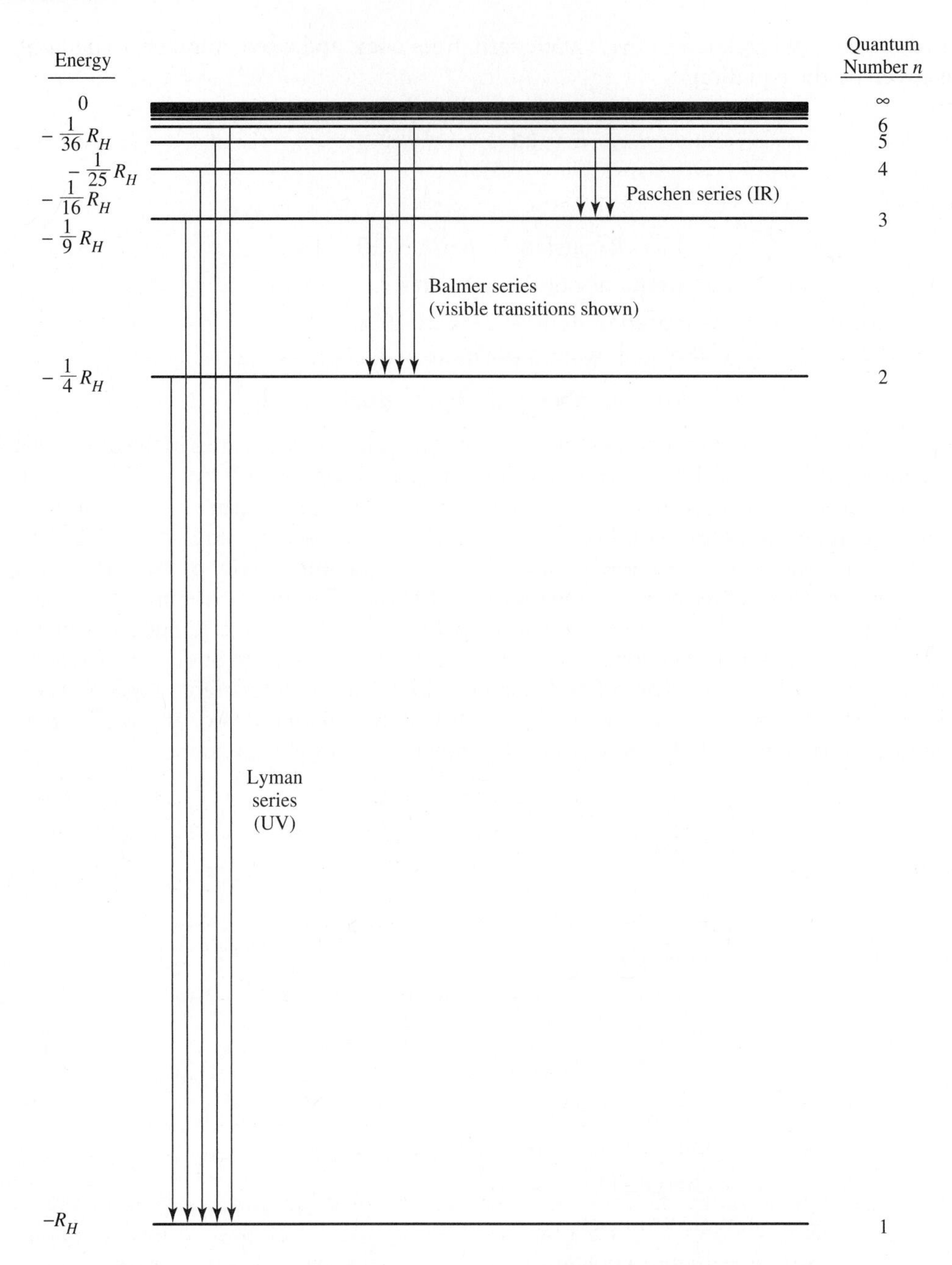

FIGURE 2.2 Hydrogen Atom Energy Levels.

level, l for lower level), energy is released in the form of electromagnetic radiation. Conversely, if radiation of the correct energy is absorbed by an atom, electrons are raised from level n_l to level n_h. The inverse-square dependence of energy on n results in energy levels that are far apart in energy at small n and become much closer in energy at larger n. In the upper limit, as n approaches infinity, the energy approaches a limit of zero. Individual electrons can have more energy, but above this point, they are no longer part of the atom; an infinite quantum number means that the nucleus and the electron are separate entities.

▶ **Exercise 2.1** Determine the energy of the transition from $n_h = 3$ to $n_l = 2$ for the hydrogen atom, in both joules and cm^{-1} (a common unit in spectroscopy, often used as an energy unit, since $\overline{\nu}$ is proportional to E). This transition results in a red line in the visible emission spectrum of hydrogen. (Solutions to the exercises are given in Appendix A.)

When applied to hydrogen, Bohr's theory worked well; when atoms with more electrons were considered, the theory failed. Complications such as elliptical rather than circular orbits were introduced in an attempt to fit the data to Bohr's theory.[11] The developing experimental science of atomic spectroscopy provided extensive data for testing Bohr's theory and its modifications, and it forced the theorists to work hard to explain the spectroscopists' observations. In spite of their efforts, the Bohr theory eventually proved unsatisfactory; the energy levels shown in Figure 2.2 are valid only for the hydrogen atom. An important characteristic of the electron, its wave nature, still needed to be considered.

According to the de Broglie equation,[12] proposed in the 1920s, all moving particles have wave properties described by the equation

$$\lambda = \frac{h}{mv}$$

where

$$\begin{aligned} \lambda &= \text{wavelength of the particle} \\ h &= \text{Planck constant} \\ m &= \text{mass of the particle} \\ v &= \text{velocity of the particle} \end{aligned}$$

Particles massive enough to be visible have very short wavelengths, too small to be measured. Electrons, on the other hand, have observable wave properties because of their very small mass.

Electrons moving in circles around the nucleus, as in Bohr's theory, can be thought of as forming standing waves that can be described by the de Broglie equation. However, we no longer believe that it is possible to describe the motion of an electron in an atom so precisely. This is a consequence of another fundamental principle of modern physics, **Heisenberg's uncertainty principle**,[13] which states that there is a relationship between the inherent uncertainties in the location and momentum of an electron. The x component of this uncertainty is described as

$$\Delta x \, \Delta p_x \geq \frac{h}{4\pi}$$

where

$$\begin{aligned} \Delta x &= \text{uncertainty in the position of the electron} \\ \Delta p_x &= \text{uncertainty in the momentum of the electron} \end{aligned}$$

The energy of spectral lines can be measured with great precision (as an example, the Rydberg constant is known to 11 significant figures). This in turn allows precise determination of the energy of electrons in atoms. This precision in energy also implies precision in momentum (Δp_x is small); therefore, according to Heisenberg, there is a large uncertainty in the location of the electron (Δx is large). These concepts mean that we cannot treat electrons as simple particles with their motion described precisely, but we must instead consider the wave properties of electrons, characterized by a degree of uncertainty in their location. In other words, instead of being able to describe precise **orbits** of electrons, as in the Bohr theory, we can only describe **orbitals**, regions that describe the probable location of electrons. The **probability** of finding the electron at a particular point in space, also called the **electron density**, can be calculated—at least in principle.

---

[11] G. Herzberg, *Atomic Spectra and Atomic Structure*, 2nd ed., Dover Publications, New York, 1994, p. 18.

[12] L. de Broglie, *Philos. Mag.* **1924**, *47*, 446; *Ann. Phys.*, **1925**, *3*, 22.

[13] W. Heisenberg, *Z. Phys.*, **1927**, *43*, 172.

2.2 THE SCHRÖDINGER EQUATION

In 1926 and 1927, Schrödinger[14] and Heisenberg[13] published papers on wave mechanics, descriptions of the wave properties of electrons in atoms, that used very different mathematical techniques. In spite of the different approaches, it was soon shown that their theories were equivalent. Schrödinger's differential equations are more commonly used to introduce the theory, and we will follow that practice.

The Schrödinger equation describes the wave properties of an electron in terms of its position, mass, total energy, and potential energy. The equation is based on the **wave function**, Ψ, which describes an electron wave in space; in other words, it describes an atomic orbital. In its simplest notation, the equation is

$$H\Psi = E\Psi$$

where

$$\begin{aligned} H &= \text{the Hamiltonian operator} \\ E &= \text{energy of the electron} \\ \Psi &= \text{the wave function} \end{aligned}$$

The **Hamiltonian operator**, frequently called simply the *Hamiltonian*, includes derivatives that **operate** on the wave function.[15] When the Hamiltonian is carried out, the result is a constant (the energy) times Ψ. The operation can be performed on any wave function describing an atomic orbital. Different orbitals have different wave functions and different values of E. This is another way of describing quantization in that each orbital, characterized by its own function Ψ, has a characteristic energy.

In the form used for calculating energy levels, the Hamiltonian operator is

$$H = \frac{-h^2}{8\pi^2 m}\left(\frac{\partial^2}{\partial x^2} + \frac{\partial^2}{\partial y^2} + \frac{\partial^2}{\partial z^2}\right) - \frac{Ze^2}{4\pi\,\varepsilon_0\sqrt{x^2 + y^2 + z^2}}$$

This part of the operator describes the *kinetic energy* of the electron

This part of the operator describes the *potential energy* of the electron, the result of electrostatic attraction between the electron and the nucleus. It is commonly designated as V.

where

$$\begin{aligned} h &= \text{Planck constant} \\ m &= \text{mass of the particle (electron)} \\ e &= \text{charge of the electron} \\ \sqrt{x^2 + y^2 + z^2} = r &= \text{distance from the nucleus} \\ Z &= \text{charge of the nucleus} \\ 4\pi\varepsilon_0 &= \text{permittivity of a vacuum} \end{aligned}$$

When this operator is applied to a wave function Ψ,

$$\left[\frac{-h^2}{8\pi^2 m}\left(\frac{\partial^2}{\partial x^2} + \frac{\partial^2}{\partial y^2} + \frac{\partial^2}{\partial z^2}\right) + V(x, y, z)\right]\Psi(x, y, z) = E\Psi(x, y, z)$$

---

[14] E. Schrödinger, *Ann. Phys.* (Leipzig), **1926**, *79*, 361, 489, 734; **1926**, *80*, 437; **1926**, *81*, 109; *Naturwissenshaften*, **1926**, *14*, 664; *Phys. Rev.*, **1926**. *28*, 1049.

[15] An *operator* is an instruction or set of instructions that states what to do with the function that follows it. It may be a simple instruction such as "multiply the following function by 6," or it may be much more complicated than the Hamiltonian. The Hamiltonian operator is sometimes written $\hat{H}$ with the ^ (hat) symbol designating an operator.

where $$V = \frac{-Ze^2}{4\pi\varepsilon_0 r} = \frac{-Ze^2}{4\pi\varepsilon_0\sqrt{x^2 + y^2 + z^2}}$$

The potential energy V is a result of electrostatic attraction between the electron and the nucleus. Attractive forces, like those between a positive nucleus and a negative electron, are defined by convention to have a negative potential energy. An electron near the nucleus (small r) is strongly attracted to the nucleus and has a large negative potential energy. Electrons farther from the nucleus have potential energies that are small and negative. For an electron at infinite distance from the nucleus ($r = \infty$), the attraction between the nucleus and the electron is zero, and the potential energy is zero.

Because every atomic orbital is described by a unique Ψ, there is no limit to the number of solutions of the Schrödinger equation for an atom. Each Ψ describes the wave properties of a given electron in a particular orbital. The probability of finding an electron at a given point in space is proportional to Ψ^2. A number of conditions are required for a physically realistic solution for Ψ:

1. The wave function Ψ must be single-valued. — There cannot be two probabilities for an electron at any position in space.
2. The wave function Ψ and its first derivatives must be continuous. — The probability must be defined at all positions in space and cannot change abruptly from one point to the next.
3. The wave function Ψ must approach zero as r approaches infinity. — For large distances from the nucleus, the probability must grow smaller and smaller (the atom must be finite).
4. The integral $\int_{\text{all space}} \Psi_A \Psi_A^* \, d\tau = 1$ — The total probability of an electron being *somewhere* in space = 1. This is called **normalizing** the wave function.[16]
5. The integral $\int_{\text{all space}} \Psi_A \Psi_B^* \, d\tau = 0$ — All orbitals in an atom must be orthogonal to each other. In some cases, this means that the axes of orbitals must be perpendicular, as with the p_x, p_y, and p_z orbitals.

2.2.1 The Particle in a Box

A simple example of the wave equation, the one-dimensional particle in a box, shows how these conditions are used. We will give an outline of the method; details are available elsewhere.[17] The "box" is shown in Figure 2.3. The potential energy $V(x)$ inside the box, between $x = 0$ and $x = a$, is defined to be zero. Outside the box, the potential energy is infinite. This means that the particle is completely trapped in the box and would require an infinite amount of energy to leave the box. However, there are no forces acting on it within the box.

The wave equation for locations within the box is

$$\frac{-h^2}{8\pi^2 m}\left(\frac{\partial^2\Psi(x)}{\partial x^2}\right) = E\Psi(x), \quad \text{because } V(x) = 0$$

---

[16] Because the wave functions may have imaginary values (containing $\sqrt{-1}$), Ψ^* (where Ψ^* designates the complex conjugate of Ψ) is used to make the integral real. In many cases, the wave functions themselves are real, and this integral becomes $\int_{\text{all space}} \Psi_A^2 \, d\tau$.

[17] G. M. Barrow, *Physical Chemistry*, 6th ed., McGraw-Hill, New York, 1996, pp. 65, 430, calls this the "particle on a line" problem. Many other physical chemistry texts also include solutions.

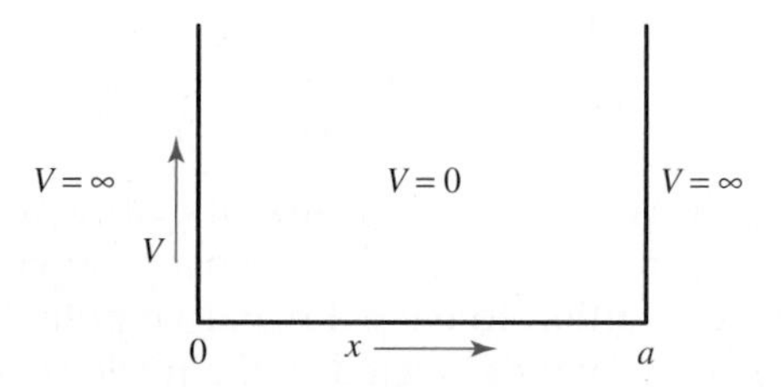

FIGURE 2.3 Potential Energy Well for the Particle in a Box.

Sine and cosine functions have the properties that we associate with waves—a well-defined wavelength and amplitude—and we may therefore propose that the wave characteristics of our particle may be described by a combination of sine and cosine functions. A general solution to describe the possible waves in the box would then be

$$\Psi = A \sin rx + B \cos sx$$

where A, B, r, and s are constants. Substitution into the wave equation allows solution for r and s (see Problem 6 at the end of the chapter):

$$r = s = \sqrt{2mE}\,\frac{2\pi}{h}$$

Because Ψ must be continuous and must equal zero at $x < 0$ and $x > a$, because the particle is confined to the box, Ψ must go to zero at $x = 0$ and $x = a$. Because $\cos sx = 1$ for $x = 0$, Ψ can equal zero in the general solution above only if $B = 0$. This reduces the expression for Ψ to

$$\Psi = A \sin rx$$

At $x = a$, Ψ must also equal zero; therefore, $\sin ra = 0$, which is possible only if ra is an integral multiple of π:

$$ra = \pm n\pi \quad \text{or} \quad r = \frac{\pm n\pi}{a}$$

where n = any integer $\neq 0$.[18] Because both positive and negative values yield the same results, substituting the positive value for r into the solution for r gives

$$r = \frac{n\pi}{a} = \sqrt{2mE}\,\frac{2\pi}{h}$$

This expression may be solved for E:

$$E = \frac{n^2 h^2}{8ma^2}$$

These are the energy levels predicted by the particle-in-a-box model for any particle in a one-dimensional box of length a. The energy levels are quantized according to **quantum numbers** $n = 1, 2, 3, \ldots$

Substituting $r = n\pi/a$ into the wave function gives

$$\Psi = A \sin \frac{n\pi x}{a}$$

And applying the normalizing requirement $\int \Psi\Psi^* \, d\tau = 1$ gives

$$A = \sqrt{\frac{2}{a}}$$

[18] If $n = 0$, then $r = 0$ and $\Psi = 0$ at all points. The probability of finding the electron is $\int \Psi\Psi^* \, dx = 0$, and there is no electron at all.

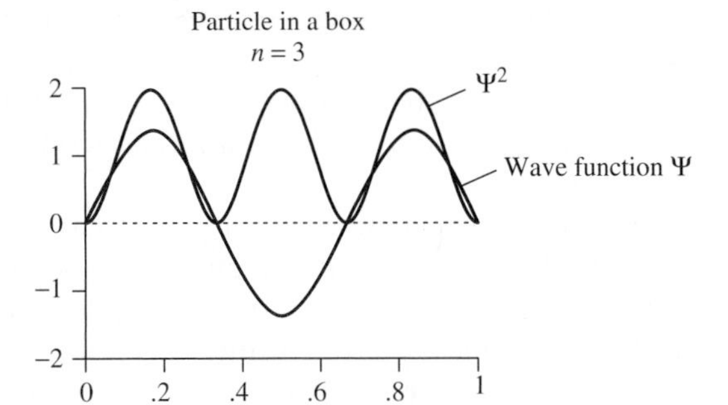

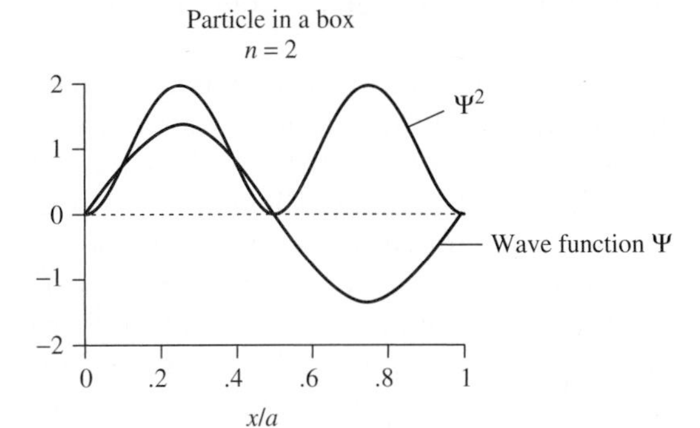

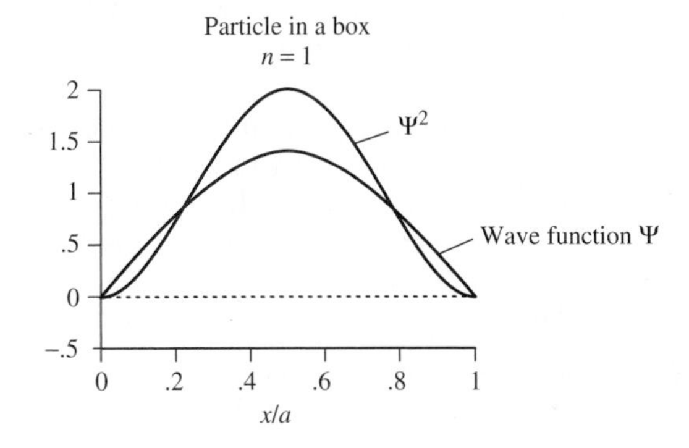

FIGURE 2.4 Wave Functions and Their Squares for the Particle in a Box with n = 1, 2, and 3.

The total solution is then

$$\Psi = \sqrt{\frac{2}{a}} \sin \frac{n\pi x}{a}$$

The resulting wave functions and their squares for the first three states—the ground state and first two excited states—are plotted in Figure 2.4.

The squared wave functions are the probability densities, and they show the difference between classical and quantum mechanical behavior. Classical mechanics predicts that the electron has equal probability of being at any point in the box. The wave nature of the electron gives it the extremes of high and low probability at different locations in the box.

2.2.2 Quantum Numbers and Atomic Wave Functions

The particle-in-a-box example shows how a wave function operates in one dimension. Mathematically, atomic orbitals are discrete solutions of the three-dimensional Schrödinger equations. The same methods used for the one-dimensional box can be expanded to three dimensions for atoms. These orbital equations include three quantum numbers, n, l, and m_l. A fourth quantum number, m_s, a result of relativistic corrections to the Schrödinger equation, completes the description by accounting for the magnetic moment of the electron. The quantum numbers are summarized in Table 2.2. Tables 2.3 and 2.4 describe wave functions.

The fourth quantum number explains several experimental observations. Two of these observations are that lines in alkali metal emission spectra are doubled, and that a beam of alkali metal atoms splits into two parts if it passes through a magnetic field. Both of these can be explained by attributing a magnetic moment to the electron: it behaves like a tiny bar magnet. This is usually described as the *spin* of the electron, because a spinning electrically charged particle also has a magnetic moment; but this should not be taken as an accurate description; it is a purely quantum mechanical property.

The quantum number n is primarily responsible for determining the overall energy of an atomic orbital; the other quantum numbers have smaller effects on the energy. The quantum number l determines the angular momentum and shape of an orbital and has a smaller effect on the energy. The quantum number m_l determines the orientation of the angular momentum vector in a magnetic field, or the position of the orbital in space, as shown in Table 2.3. The quantum number m_s determines the orientation of the electron's magnetic moment in a magnetic field, either in the direction of the field $\left(+\frac{1}{2}\right)$ or opposed to it $\left(-\frac{1}{2}\right)$. When no field is present, all m_l values—all three p orbitals or all five d orbitals—have the same energy, and both m_s values have the same energy. Together, the quantum numbers n, l, and m_l define an atomic orbital; the quantum number m_s describes the electron spin within the orbital.

One feature that should be mentioned is the appearance of $i\left(=\sqrt{-1}\right)$ in the p and d orbital wave equations in Table 2.3. Because it is much more convenient to work with real functions than complex functions, we usually take advantage of another property of the wave equation. For differential equations of this type, any linear combination of solutions to the equation—sums or differences of the functions, with each multiplied by any coefficient—is also a solution to the equation. The combinations usually chosen for

TABLE 2.2 Quantum Numbers and Their Properties

Symbol	Name	Values	Role
n	Principal	1, 2, 3, . . .	Determines the major part of the energy
l	Angular momentum[19]	0, 1, 2, . . ., $n - 1$	Describes angular dependence and contributes to the energy
m_l	Magnetic	0, ±1, ±2, . . ., ±l	Describes orientation in space (angular momentum in the z direction)
m_s	Spin	$\pm\frac{1}{2}$	Describes orientation of the electron spin (magnetic moment) in space

Orbitals with different l values are known by the following labels, derived from early terms for different families of spectroscopic lines:

l	0	1	2	3	4	5, . . .
Label	*s*	*p*	*d*	*f*	*g*	continuing alphabetically

[19] Also called the azimuthal quantum number

TABLE 2.3 Hydrogen Atom Wave Functions: Angular Functions

Angular Factors					Real Wave Functions			
Related to Angular Momentum				Functions of θ	In Polar Coordinates	In Cartesian Coordinates	Shapes	Label
l	m_l	Φ	Θ		$\Theta\Phi(\theta, \phi)$	$\Theta\Phi(x, y, z)$		
$0(s)$	0	$\frac{1}{\sqrt{2\pi}}$	$\frac{1}{\sqrt{2}}$		$\frac{1}{\sqrt{2\pi}}$	$\frac{1}{2\sqrt{\pi}}$	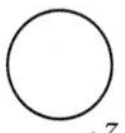	s
$1(p)$	0	$\frac{1}{\sqrt{2\pi}}$	$\frac{\sqrt{6}}{2}\cos\theta$		$\frac{1}{2}\sqrt{\frac{3}{\pi}}\cos\theta$	$\frac{1}{2}\sqrt{\frac{3}{\pi}}\frac{z}{r}$	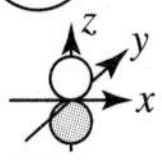	p_z
	+1	$\frac{1}{\sqrt{2\pi}}e^{i\phi}$	$\frac{\sqrt{3}}{2}\sin\theta$		$\frac{1}{2}\sqrt{\frac{3}{\pi}}\sin\theta\cos\phi$	$\frac{1}{2}\sqrt{\frac{3}{\pi}}\frac{x}{r}$	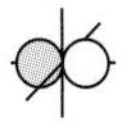	p_x
	−1	$\frac{1}{\sqrt{2\pi}}e^{-i\phi}$	$\frac{\sqrt{3}}{2}\sin\theta$		$\frac{1}{2}\sqrt{\frac{3}{\pi}}\sin\theta\sin\phi$	$\frac{1}{2}\sqrt{\frac{3}{\pi}}\frac{y}{r}$	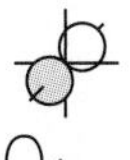	p_y
$2(d)$	0	$\frac{1}{\sqrt{2\pi}}$	$\frac{1}{2}\sqrt{\frac{5}{2}}(3\cos^2\theta - 1)$		$\frac{1}{4}\sqrt{\frac{5}{\pi}}(3\cos^2\theta - 1)$	$\frac{1}{4}\sqrt{\frac{5}{\pi}}\frac{(2z^2 - x^2 - y^2)}{r^2}$		d_{z^2}
	+1	$\frac{1}{\sqrt{2\pi}}e^{i\phi}$	$\frac{\sqrt{15}}{2}\cos\theta\sin\theta$		$\frac{1}{2}\sqrt{\frac{15}{\pi}}\cos\theta\sin\theta\cos\phi$	$\frac{1}{2}\sqrt{\frac{15}{\pi}}\frac{xz}{r^2}$		d_{xz}
	−1	$\frac{1}{\sqrt{2\pi}}e^{-i\phi}$	$\frac{\sqrt{15}}{2}\cos\theta\sin\theta$		$\frac{1}{2}\sqrt{\frac{15}{\pi}}\cos\theta\sin\theta\sin\phi$	$\frac{1}{2}\sqrt{\frac{15}{\pi}}\frac{yz}{r^2}$	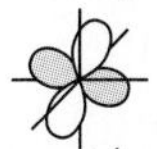	d_{yz}
	+2	$\frac{1}{\sqrt{2\pi}}e^{2i\phi}$	$\frac{\sqrt{15}}{4}\sin^2\theta$		$\frac{1}{4}\sqrt{\frac{15}{\pi}}\sin^2\theta\cos 2\phi$	$\frac{1}{4}\sqrt{\frac{15}{\pi}}\frac{(x^2 - y^2)}{r^2}$		$d_{x^2-y^2}$
	−2	$\frac{1}{\sqrt{2\pi}}e^{-2i\phi}$	$\frac{\sqrt{15}}{4}\sin^2\theta$		$\frac{1}{4}\sqrt{\frac{15}{\pi}}\sin^2\theta\sin 2\phi$	$\frac{1}{4}\sqrt{\frac{15}{\pi}}\frac{xy}{r^2}$	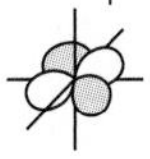	d_{xy}

Source: Adapted from G. M. Barrow, *Physical Chemistry*, 5th ed., McGraw-Hill, New York, 1988, p. 450, with permission.

NOTE: The relations $(e^{i\phi} - e^{-i\phi})/(2i) = \sin\phi$ and $(e^{i\phi} + e^{-i\phi})/2 = \cos\phi$ can be used to convert the exponential imaginary functions to real trigonometric functions, combining the two orbitals with $m_l = \pm 1$ to give two orbitals with $\sin\phi$ and $\cos\phi$. In a similar fashion, the orbitals with $m_l = \pm 2$ result in real functions with $\cos^2\phi$ and $\sin^2\phi$. These functions have then been converted to Cartesian form by using the functions $x = r\sin\theta\cos\phi$, $y = r\sin\theta\sin\phi$, and $z = r\cos\theta$.

TABLE 2.4 Hydrogen Atom Wave Functions: Radial Functions

			Radial Functions $R(r)$, with $\sigma = Zr/a_0$
Orbital	n	l	$R(r)$
$1s$	1	0	$R_{1s} = 2\left[\frac{Z}{a_0}\right]^{3/2} e^{-\sigma}$
$2s$	2	0	$R_{2s} = \left[\frac{Z}{2a_0}\right]^{3/2}(2 - \sigma)e^{-\sigma/2}$
$2p$		1	$R_{2p} = \frac{1}{\sqrt{3}}\left[\frac{Z}{2a_0}\right]^{3/2}\sigma e^{-\sigma/2}$
$3s$	3	0	$R_{3s} = \frac{2}{27}\left[\frac{Z}{3a_0}\right]^{3/2}(27 - 18\sigma + 2\sigma^2)e^{-\sigma/3}$
$3p$		1	$R_{3p} = \frac{1}{81\sqrt{3}}\left[\frac{2Z}{a_0}\right]^{3/2}(6 - \sigma)\sigma e^{-\sigma/3}$
$3d$		2	$R_{3d} = \frac{1}{81\sqrt{15}}\left[\frac{2Z}{a_0}\right]^{3/2}\sigma^2 e^{-\sigma/3}$

the p orbitals are the sum and difference of the p orbitals having $m_l = +1$ and –1, normalized by multiplying by the constants $\frac{1}{\sqrt{2}}$ and $\frac{i}{\sqrt{2}}$, respectively:

$$\Psi_{2px} = \frac{1}{\sqrt{2}}(\Psi_{+1} + \Psi_{-1}) = \frac{1}{2}\sqrt{\frac{3}{\pi}}\left[R(r)\right] \sin\theta\cos\phi$$

$$\Psi_{2py} = \frac{i}{\sqrt{2}}(\Psi_{+1} - \Psi_{-1}) = \frac{1}{2}\sqrt{\frac{3}{\pi}}\left[R(r)\right] \sin\theta\sin\phi$$

The same procedure used on the d orbital functions for $m_l = \pm 1$ and ± 2 gives the functions in the column headed $\Theta\Phi(\theta, \phi)$ in Table 2.3, which are the familiar d orbitals. The d_{z^2} orbital ($m_l = 0$) actually uses the function $2z^2 - x^2 - y^2$, which we shorten to z^2 for convenience.[20] These functions are now real functions, so $\Psi = \Psi^*$ and $\Psi\Psi^* = \Psi^2$.

A more detailed look at the Schrödinger equation shows the mathematical origin of atomic orbitals. In three dimensions, Ψ may be expressed in terms of Cartesian coordinates (x, y, z) or in terms of spherical coordinates (r, θ, ϕ). Spherical coordinates, as shown in Figure 2.5, are especially useful in that r represents the distance from the nucleus. The spherical coordinate θ is the angle from the z axis, varying from 0 to π, and ϕ is the angle from the x axis, varying from 0 to 2π. It is possible to convert between Cartesian and spherical coordinates using the following expressions:

$$x = r\sin\theta\cos\phi$$

$$y = r\sin\theta\sin\phi$$

$$z = r\cos\theta$$

In spherical coordinates, the three sides of the volume element are $r\,d\theta$, $r\sin\theta\,d\phi$, and dr. The product of the three sides is $r^2\sin\theta\,d\theta\,d\phi\,dr$, equivalent to $dx\,dy\,dz$. The volume of the thin shell between r and $r + dr$ is $4\pi r^2\,dr$, which is the integral over ϕ from 0 to π and over θ from 0 to 2π. This integral is useful in describing the electron density as a function of distance from the nucleus.

Ψ can be factored into a radial component and two angular components. The **radial function R** describes electron density at different distances from the nucleus; the

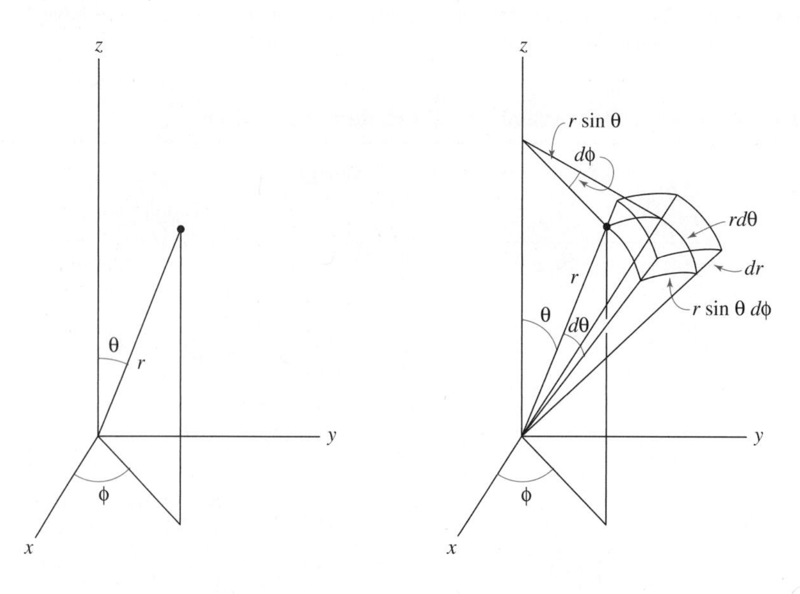

FIGURE 2.5 Spherical Coordinates and Volume Element for a Spherical Shell in Spherical Coordinates.

[20] We should really call this the $d_{2z^2-x^2-y^2}$ orbital!

angular functions Θ and Φ describe the shape of the orbital and its orientation in space. The two angular factors are sometimes combined into one factor, called Y:

$$\Psi(r, \theta, \phi) = R(r)\Theta(\theta)\Phi(\phi) = R(r)Y(\theta, \phi)$$

R is a function only of r; Y is a function of θ and ϕ, and it gives the distinctive shapes of s, p, d, and other orbitals. R, Θ, and Φ are shown separately in Tables 2.3 and 2.4.

ANGULAR FUNCTIONS The angular functions Θ and Φ determine how the probability changes from point to point at a given distance from the center of the atom; in other words, they give the shape of the orbitals and their orientation in space. The angular functions Θ and Φ are determined by the quantum numbers l and m_l. The shapes of s, p, and d orbitals are shown in Table 2.3 and Figure 2.6.

In the center of Table 2.3 are the shapes for the Θ portion; when the Φ portion is included, with values of $\phi = 0$ to 2π, the three-dimensional shapes in the far-right column are formed. In the diagrams of orbitals in Table 2.3, the orbital lobes are shaded where the wave function is negative. *The different shadings of the lobes represent different signs of the wave function* Ψ. The probabilities are the same for locations with positive and negative signs for Ψ, but it is useful to distinguish regions of opposite signs for bonding purposes, as we will see in Chapter 5.

RADIAL FUNCTIONS The radial factor $R(r)$ (Table 2.4) is determined by the quantum numbers n and l, the principal and angular momentum quantum numbers.

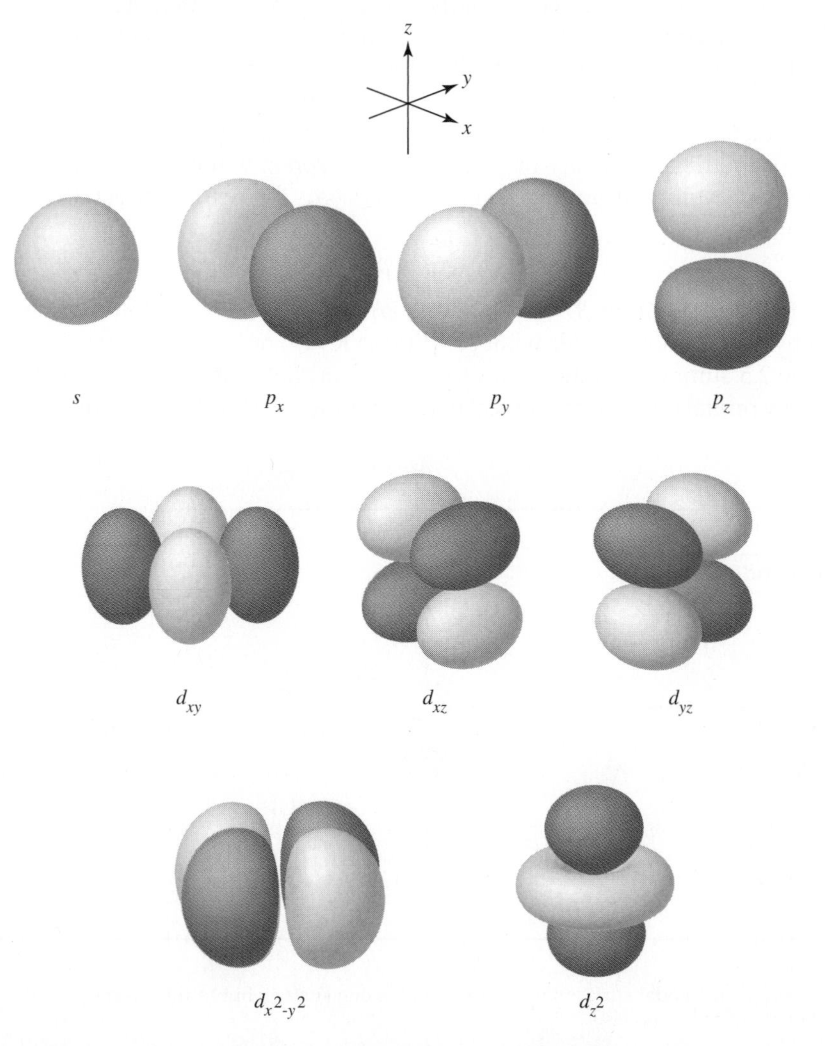

FIGURE 2.6 Selected Atomic Orbitals. (Adapted with permission from G. O. Spessard and G. L. Miessler, *Organometallic Chemistry*, Prentice Hall, Upper Saddle River, NJ, 1997, p. 11, Fig. 2-1.)

The **radial probability function** is $4\pi r^2 R^2$. This function describes the probability of finding the electron at a given distance from the nucleus, summed over all angles, with the $4\pi r^2$ factor the result of integrating over all angles. The radial wave functions and radial probability functions are plotted for the $n = 1$, 2, and 3 orbitals in Figure 2.7. Both $R(r)$ and $4\pi r^2 R^2$ are scaled with a_0, the Bohr radius, to give reasonable units on the axes of the graphs. The Bohr radius, $a_0 = 52.9$ pm, is a common unit in quantum mechanics. It is the value of r at the maximum of Ψ^2 for a hydrogen 1*s* orbital, and it is also the radius of a 1*s* orbital according to the Bohr model.

In all the radial probability plots, the electron density, or probability of finding the electron, falls off rapidly as the distance from the nucleus increases. It falls off most quickly for the 1*s* orbital; by $r = 5a_0$, the probability is approaching zero. By contrast, the 3*d* orbital has a maximum at $r = 9a_0$ and does not approach zero until approximately $r = 20a_0$. All the orbitals, including the *s* orbitals, have zero probability at the center of the nucleus, because $4\pi r^2 R^2 = 0$ at $r = 0$. The radial probability functions are a combination of $4\pi r^2$, which increases rapidly with r, and R^2, which may have maxima and minima, but generally decreases exponentially with r. The product of these two factors gives the characteristic probabilities seen in the plots. Because chemical reactions depend on the shape and extent of orbitals at large distances from the nucleus, the radial probability functions help show which orbitals are most likely to be involved in reactions.

NODAL SURFACES At large distances from the nucleus, the electron density, or probability of finding the electron, falls off rapidly. The 2*s* orbital also has a **nodal surface**, a surface with zero electron density, in this case a sphere with $r = 2a_0$ where the probability is zero. Nodes appear naturally as a result of the wave nature of the electron; they occur in the functions that result from solving the wave equation for Ψ. A node is a surface where the wave function is zero as it changes sign (as at $r = 2a_0$, in the 2*s* orbital); this requires that $\Psi = 0$, and the probability of finding the electron at that point is also zero.

If the probability of finding an electron is zero ($\Psi^2 = 0$), Ψ must also be equal to zero. Because

$$\Psi(r, \theta, \phi) = R(r)Y(\theta, \phi)$$

in order for $\Psi = 0$, either $R(r) = 0$ or $Y(\theta, \phi) = 0$. We can therefore determine nodal surfaces by determining under what conditions $R = 0$ or $Y = 0$.

Table 2.5 summarizes the nodes for several orbitals. Note that the total number of nodes in any orbital is $n - 1$ if the conical nodes of some *d* and *f* orbitals count as two nodes.[21]

TABLE 2.5 Nodal Surfaces

Radial Nodes [$R(r) = 0$]					
Examples (number of radial nodes)					
1*s*	0	2*p*	0	3*d*	0
2*s*	1	3*p*	1	4*d*	1
3*s*	2	4*p*	2	5*d*	2

Angular Nodes [$Y(\theta, \phi) = 0$]	
Examples (number of angular nodes)	
s orbitals	0
p orbitals	1 plane for each orbital
d orbitals	2 planes for each orbital except d_{z^2}
	1 conical surface for d_{z^2}

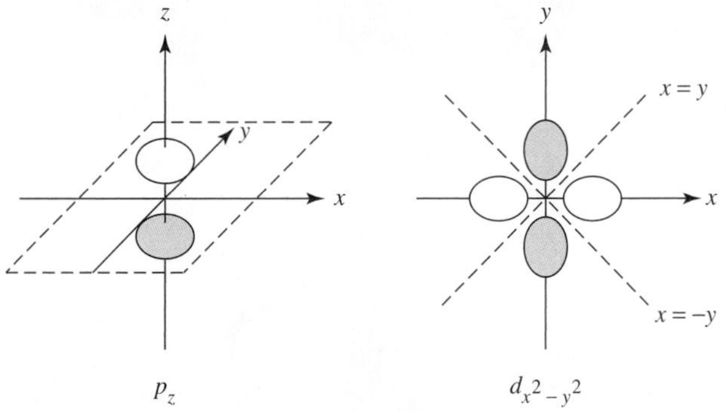

p_z $d_{x^2-y^2}$

[21] Mathematically, the nodal surface for the d_{z^2} orbital is one surface, but in this instance, it fits the pattern better if thought of as two nodes.

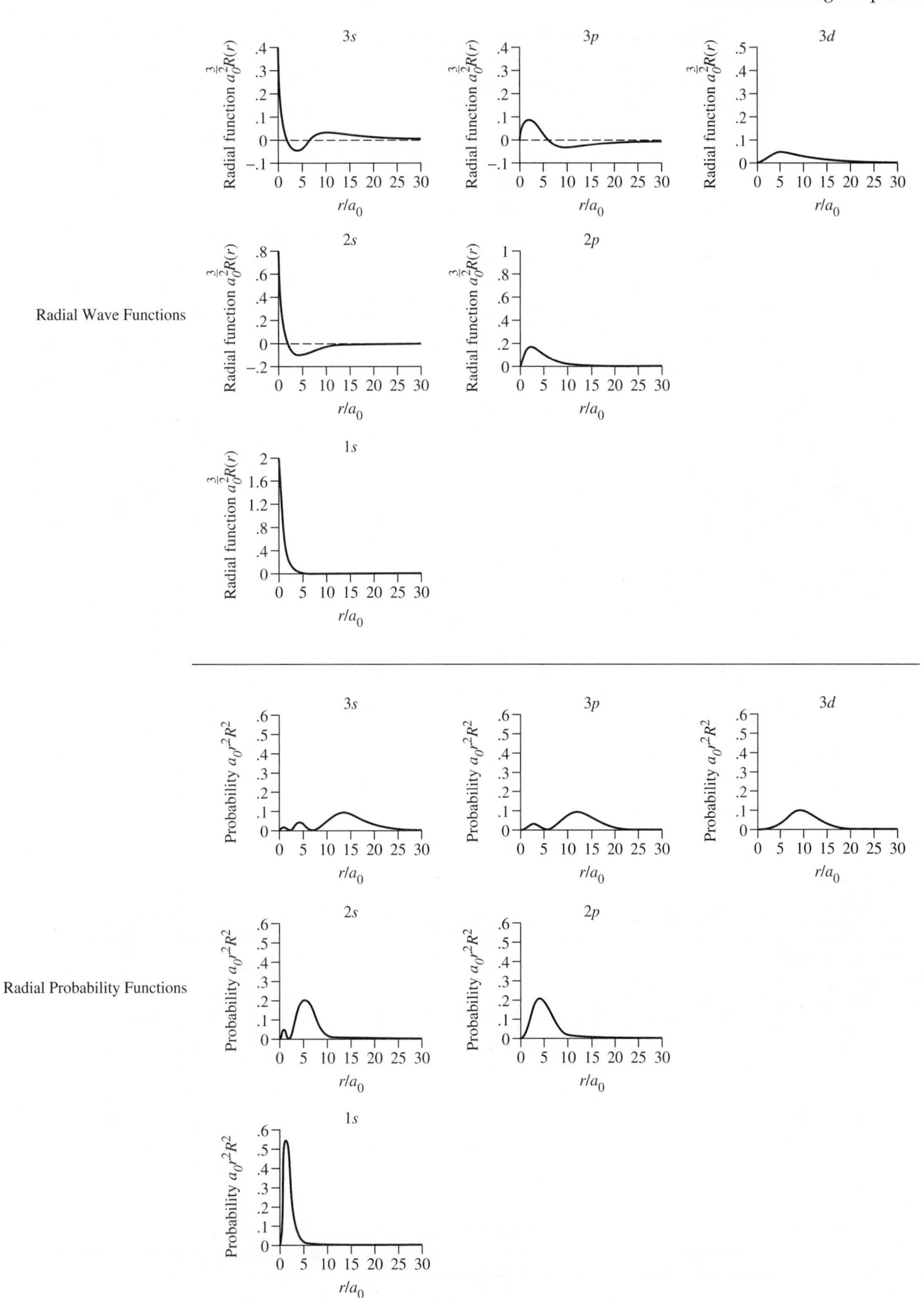

FIGURE 2.7 Radial Wave Functions and Radial Probability Functions.

Angular nodes result when $Y = 0$, and they are planar or conical. Angular nodes can be determined in terms of θ and ϕ but may be easier to visualize if Y is expressed in Cartesian (x, y, z) coordinates (see Table 2.3). In addition, the regions where the wave function is positive and where it is negative can be found. This information will be useful in working with molecular orbitals in later chapters. There are l angular nodes in any orbital, with the conical surface in the d_{z^2} orbitals—and other orbitals having conical nodes—counted as two nodes.

Radial nodes, or **spherical nodes**, result when $R = 0$. They give the atom a layered appearance, shown in Figure 2.8 for the 3s and $3p_z$ orbitals. These nodes occur when the radial function changes sign; they are depicted in the radial function graphs by $R(r) = 0$ and in the radial probability graphs by $4\pi r^2 R^2 = 0$. The lowest energy orbitals of each classification (1s, 2p, 3d, 4f, etc.) have no radial nodes. The number of radial nodes

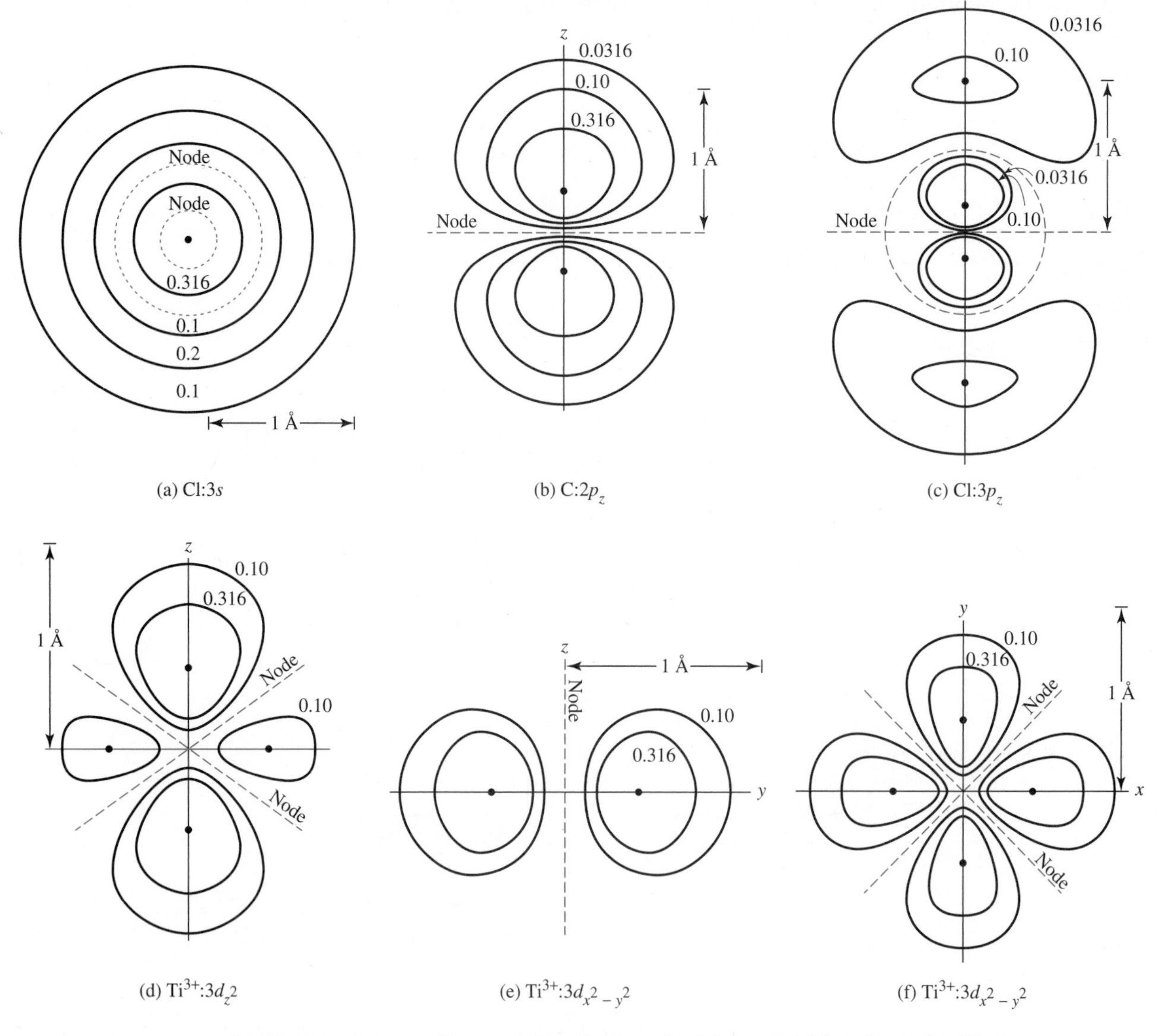

FIGURE 2.8 Constant Electron Density Surfaces for Selected Atomic Orbitals. (a)–(d) The cross-sectional plane is any plane containing the z axis. (e) The cross section is taken through the xz or yz plane. (f) The cross section is taken through the xy plane.

(Figures (b)–(f) reproduced with permission from E. A. Orgyzlo and G. B. Porter, *J. Chem. Educ.*, **1963**, *40*, 258.)

increases as n increases; the number of radial nodes for a given orbital is always equal to $n - l - 1$.[22]

Nodal surfaces can be puzzling. For example, a p orbital has a nodal plane through the nucleus. How can an electron be on both sides of a node at the same time without ever having been at the node, at which the probability is zero? One explanation is that the probability does not go quite to zero.[23]

Another explanation is that such a question really has no meaning for an electron behaving as a wave. Recall the particle-in-a-box example. Figure 2.4 shows nodes at $x/a = 0.5$ for $n = 2$ and at $x/a = 0.33$ and 0.67 for $n = 3$. The same diagrams could represent the amplitudes of the motion of vibrating strings at the fundamental frequency ($n = 1$) and multiples of 2 and 3. A plucked violin string vibrates at a specific frequency, and nodes at which the amplitude of vibration is zero are a natural result. Zero amplitude does not mean that the string does not exist at these points but simply that the magnitude of the vibration is zero. An electron wave exists at the node as well as on both sides of a nodal surface, just as a violin string exists at the nodes and on both sides of points having zero amplitude.

Still another explanation, in a lighter vein, was suggested by R. M. Fuoss to one of the authors in a class on bonding. Paraphrased from St. Thomas Aquinas, "Angels are not material beings. Therefore, they can be first in one place and later in another without ever having been in between." If the word "electrons" replaces the word "angels," a semitheological interpretation of nodes would result.

EXAMPLES

p_z The angular factor Y is given in Table 2.3 in terms of Cartesian coordinates:

$$Y = \frac{1}{2}\sqrt{\frac{3}{\pi}}\,\frac{z}{r}$$

This orbital is designated p_z because z appears in the Y expression. For an angular node, Y must equal zero, which is true only if $z = 0$. Therefore, $z = 0$ (the xy plane) is an angular nodal surface for the p_z orbital, as shown in Table 2.5 and Figure 2.8. The wave function is positive where $z > 0$ and negative where $z < 0$. In addition, a $2p_z$ orbital has no radial (spherical) nodes, a $3p_z$ orbital has one radial node, and so on.

$d_{x^2-y^2}$

$$Y = \frac{1}{4}\sqrt{\frac{15}{\pi}}\,\frac{(x^2 - y^2)}{r^2}$$

Here, the expression $x^2 - y^2$ appears in the equation, so the designation is $d_{x^2-y^2}$. Because there are two solutions to the equation $Y = 0$ (setting $x^2 - y^2 = 0$, the solutions are $x = y$ and $x = -y$), the planes defined by these equations are the angular nodal surfaces. They are planes containing the z axis and making 45° angles with the x and y axes (see Table 2.5). The function is positive where $x > y$ and negative where $x < y$. In addition, a $3d_{x^2-y^2}$ orbital has no radial nodes, a $4d_{x^2-y^2}$ has one radial node, and so on.

▶ **Exercise 2.2** Describe the angular nodal surfaces for a d_{z^2} orbital, whose angular wave function is

$$Y = \frac{1}{4}\sqrt{\frac{5}{\pi}}\,\frac{(2z^2 - x^2 - y^2)}{r^2}$$

[22] Again, counting a conical nodal surface, such as for a d_{z^2} orbital, as two nodes.

[23] A. Szabo, *J. Chem. Educ.*, **1969**, *46*, 678, uses relativistic arguments to explain that the electron probablity at a nodal surface has a very small but finite value.

▶ **Exercise 2.3** Describe the angular nodal surfaces for a d_{xz} orbital, whose angular wave function is

$$Y = \frac{1}{2}\sqrt{\frac{15}{\pi}}\,\frac{xz}{r^2}$$

The result of the calculations is the set of atomic orbitals familiar to all chemists. Figure 2.6 shows diagrams of *s*, *p*, and *d* orbitals, and Figure 2.8 shows lines of constant electron density in several orbitals. The different signs on the wave functions are shown by different shadings of the orbital lobes in Figure 2.6, and the outer surfaces shown enclose 90% of the total electron density of the orbitals. The orbitals we use are the common ones used by chemists; others that are also solutions of the Schrödinger equation can be chosen for special purposes.[24]

Angular functions for *f* orbitals are provided in Appendix B-8. The reader is encouraged to make use of Internet resources that display a wide range of atomic orbitals—including *f*, *g*, and higher orbitals—show radial and angular nodes, and provide a variety of additional information.[25]

2.2.3 The Aufbau Principle

Limitations on the values of the quantum numbers lead to the familiar **aufbau** (German, *Aufbau, building up*) **principle**, where the buildup of electrons in atoms results from continually increasing the quantum numbers. The energy level pattern in Figure 2.2 describes electron behavior in a hydrogen atom, where there is only one electron. However, interactions between electrons in polyelectronic atoms require that the order of filling orbitals be specified when more than one electron is in the same atom. In this process, we start with the lowest n, l, and m_l values (1, 0, and 0, respectively) and either of the m_s values (we will arbitrarily use $+\frac{1}{2}$ first). Three rules will then give us the proper order for the remaining electrons, as we increase the quantum numbers in the order m_l, m_s, l, and n.

1. Electrons are placed in orbitals to give the lowest total energy to the atom. This means that the lowest values of n and l are filled first. Because the orbitals within each set (p, d, etc.) have the same energy, the orders for values of m_l and m_s are indeterminate.
2. The **Pauli exclusion principle**[26] requires that each electron in an atom have a unique set of quantum numbers. At least one quantum number must be different from those of every other electron. This principle does not come from the Schrödinger equation, but from experimental determination of electronic structures.
3. **Hund's rule of maximum multiplicity**[27] requires that electrons be placed in orbitals so as to give the maximum total spin possible (or the maximum number of parallel spins). Two electrons in the same orbital have a higher energy than two electrons in different orbitals, caused by electrostatic repulsion; electrons in the same orbital repel each other more than electrons in separate orbitals. Therefore, this rule is a consequence of the lowest possible energy rule (Rule 1). When there are one to six electrons in p orbitals, the required arrangements are those given in Table 2.6. The **multiplicity** is the number of unpaired electrons plus 1, or $n + 1$. This is the number of possible energy levels that depend on the orientation of the net magnetic moment in a magnetic field. Any other arrangement of electrons results in fewer unpaired electrons.[28]

---

[24] R. E. Powell, *J. Chem. Educ.*, **1968**, *45*, 45.
[25] Two examples are http://www.orbitals.com and http://winter.group.shef.ac.uk/orbitron.
[26] W. Pauli, *Z. Physik*, **1925**, *31*, 765.
[27] F. Hund, *Z. Physik*, **1925**, *33*, 345.
[28] This is only one of Hund's rules; others are described in Chapter 11.

TABLE 2.6 Hund's Rule and Multiplicity

Number of Electrons	Arrangement	Unpaired e^-	Multiplicity
1	↑__ __ __	1	2
2	↑__ ↑__ __	2	3
3	↑__ ↑__ ↑__	3	4
4	↑↓__ ↑__ ↑__	2	3
5	↑↓__ ↑↓__ ↑__	1	2
6	↑↓__ ↑↓__ ↑↓__	0	1

This rule is a consequence of the energy required for pairing electrons in the same orbital. When two electrons occupy the same part of the space around an atom, they repel each other, because of their mutual negative charges, with a **Coulombic energy of repulsion,** Π_c, per pair of electrons. As a result, this repulsive force favors electrons in different orbitals (different regions of space) over electrons in the same orbitals.

In addition, there is an **exchange energy,** Π_e, which arises from purely quantum mechanical considerations. This energy depends on the number of possible exchanges between two electrons with the same energy and the same spin. For example, the electron configuration of a carbon atom is $1s^2\,2s^2\,2p^2$. Three arrangements of the $2p$ electrons can be considered:

(1) ↑↓__ __ __ (2) ↑__ ↓__ __ (3) ↑__ ↑__ __

The first arrangement involves Coulombic energy, Π_c, because it is the only one that pairs electrons in the same orbital. The energy of this arrangement is higher than that of the other two by Π_c as a result of electron–electron repulsion.

In the first two cases, there is only one possible way to arrange the electrons to give the same diagram, because there is only a single electron in each having + or – spin. However, in the third case, there are two possible ways in which the electrons can be arranged:

↑1__ ↑2__ __ ↑2__ ↑1__ __ (one exchange of electrons)

The exchange energy is Π_e per possible exchange of parallel electrons, and it is negative. The higher the number of possible exchanges, the lower the energy. Consequently, the third configuration is lower in energy than the second by Π_e. These results may be summarized in an energy diagram:

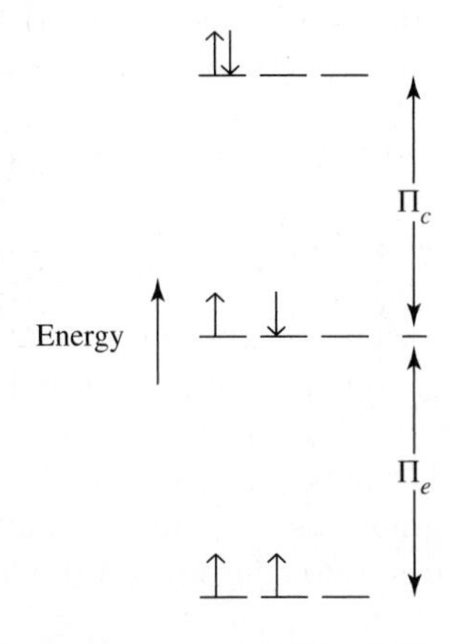

These two pairing terms add to produce the total pairing energy, Π:

$$\Pi = \Pi_c + \Pi_e$$

The Coulombic energy, Π_c, is positive and is nearly constant for each pair of electrons. The exchange energy, Π_e, is negative and is also nearly constant for each possible exchange of electrons with the same spin. When the orbitals are **degenerate** (have the same energy), both Coulombic and pairing energies favor the unpaired configuration over the paired configuration. If there is a difference in energy between the levels involved, this difference, in combination with the total pairing energy, determines the final configuration. For atoms, this usually means that one set of orbitals is filled before another has any electrons. However, this breaks down in some of the transition elements, because the 4*s* and 3*d* (or the higher corresponding levels) are so close in energy that the pairing energy is nearly the same as the difference between levels. Section 2.2.4 explains what happens in these cases.

EXAMPLE

Oxygen

With four *p* electrons, oxygen could have two unpaired electrons (↑↓ ↑ ↑), or it could have no unpaired electrons (↑↓ ↑↓ __). Find the number of electrons that could be exchanged in each case, and find the Coulombic and exchange energies for the atom.

↑↓ ↑ ↑ This configuration has one pair, energy contribution Π_c.

↑↓ ↑ ↑ One electron with ↓ spin and no possibility of exchange.

↑ ↑ ↑ Four possible arrangements; three exchange possibilities (1–2, 1–3, 2–3); energy contribution 3 Π_e.

↑1 ↑2 ↑3 ↑2 ↑1 ↑3 ↑3 ↑2 ↑1 ↑1 ↑3 ↑2

Overall, 3 Π_e + Π_c.

↑↓ ↑↓ __ has one exchange possibility for each spin pair and two pairs.

Overall, 2 Π_e + 2 Π_c.

Because Π_c is positive and Π_e is negative, the energy of the first arrangement is lower than the second; ↑↓ ↑ ↑ has the lower energy.

▶ **Exercise 2.4** A nitrogen atom with three *p* electrons could have three unpaired electrons (↑ ↑ ↑), or it could have one unpaired electron (↑↓ ↑ __). Find the number of electrons that could be exchanged in each case and the Coulombic and exchange energies for the atom. Which arrangement would be lower in energy?

Many schemes have been used to predict the order of filling of atomic orbitals. One, known as Klechkowsky's rule, states that the order of filling of the orbitals proceeds from the lowest available value for the sum $n + l$. When two combinations have the same value, the one with the smaller value of n is filled first. Combined with the other rules, this gives the order of filling of most of the orbitals.

One of the simplest methods that fits most atoms uses the periodic table blocked out as in Figure 2.9. The electron configurations of hydrogen and helium are clearly $1s^1$ and $1s^2$. After that, the elements in the first two columns on the left (Groups 1 and 2) are filling

Groups (IUPAC)																		
1	2	3		4	5	6	7	8	9	10	11	12	13	14	15	16	17	18
(US traditional)																		
IA	IIA	IIIB		IVB	VB	VIB	VIIB		VIIIB		IB	IIB	IIIA	IVA	VA	VIA	VIIA	VIIIA
1s																		1s
2s	2s												2p	2p	2p	2p	2p	2p
3s	3s												3p	3p	3p	3p	3p	3p
4s	4s	3d		3d	3d	3d	3d	3d	3d	3d	3d	3d	4p	4p	4p	4p	4p	4p
5s	5s	4d		4d	4d	4d	4d	4d	4d	4d	4d	4d	5p	5p	5p	5p	5p	5p
6s	6s	5d	*	5d	5d	5d	5d	5d	5d	5d	5d	5d	6p	6p	6p	6p	6p	6p
7s	7s	6d	**	6d	6d	6d	6d	6d	6d	6d	6d	6d						

*	4f	4f	4f	4f	4f	4f	4f	4f	4f	4f	4f	4f	4f	4f
**	5f	5f	5f	5f	5f	5f	5f	5f	5f	5f	5f	5f	5f	5f

s block   *p* block   *d* block   *f* block

FIGURE 2.9 Atomic Orbital Filling in the Periodic Table.

s orbitals, with $l = 0$; those in the six columns on the right (Groups 13 to 18) are filling *p* orbitals, with $l = 1$; and the ten in the middle (the transition elements, Groups 3 to 12) are filling *d* orbitals, with $l = 2$. The lanthanide and actinide series (numbers 58 to 71 and 90 to 103) are filling *f* orbitals, with $l = 3$. Either of these two methods is too simple, as shown in the following paragraphs, but they do fit most atoms and provide starting points for the others.

2.2.4 Shielding

In atoms with more than one electron, energies of specific levels are difficult to predict quantitatively. A useful approach to such predictions uses the concept of shielding: each electron acts as a shield for electrons farther from the nucleus, reducing the attraction between the nucleus and the more distant electrons.

Although the quantum number n is most important in determining the energy, quantum number l must also be included in calculating the energy in atoms having more than one electron. As the atomic number increases, electrons are drawn toward the nucleus, and the orbital energies become more negative. Although the energies decrease with increasing Z, the changes are somewhat irregular because of the shielding of outer electrons by inner electrons. The resulting order of orbital filling for the electrons is shown in Table 2.7.

As a result of shielding and other more subtle interactions between electrons, the simple order of energies of orbitals—higher energy with higher quantum number n—holds only for orbitals with lowest values of n (see Figure 2.10). For higher values of n, as the split in orbital energies with different values of quantum number l becomes comparable in magnitude to the differences in energy caused by n, the simplest order may not hold.

For example, consider the $n = 3$ and $n = 4$ sets in Figure 2.10. For many atoms the 4*s* orbital orbital is lower in energy than the 3*d* orbitals; consequently the order of filling is ...3*s*, 3*p*, 4*s*, 3*d*, 4*p*... rather than the "simplest" ...3*s*, 3*p*, 3*d*, 4*s*, 4*p*...

Similarly, 5*s* begins to fill before 4*d*, and 6*s* before 5*d*. Other examples can also be found in the figure.

TABLE 2.7 Electron Configurations of the Elements

Element	Z	Configuration	Element	Z	Configuration
H	1	$1s^1$	Cs	55	$[Xe]6s^1$
He	2	$1s^2$	Ba	56	$[Xe]6s^2$
Li	3	$[He]2s^1$	La	57	*$[Xe]6s^2 5d^1$
Be	4	$[He]2s^2$	Ce	58	*$[Xe]6s^2 4f^1 5d^1$
B	5	$[He]2s^2 2p^1$	Pr	59	$[Xe]6s^2 4f^3$
C	6	$[He]2s^2 2p^2$	Nd	60	$[Xe]6s^2 4f^4$
N	7	$[He]2s^2 2p^3$	Pm	61	$[Xe]6s^2 4f^5$
O	8	$[He]2s^2 2p^4$	Sm	62	$[Xe]6s^2 4f^6$
F	9	$[He]2s^2 2p^5$	Eu	63	$[Xe]6s^2 4f^7$
Ne	10	$[He]2s^2 2p^6$	Gd	64	*$[Xe]6s^2 4f^7 5d^1$
			Tb	65	$[Xe]6s^2 4f^9$
Na	11	$[Ne]3s^1$	Dy	66	$[Xe]6s^2 4f^{10}$
Mg	12	$[Ne]3s^2$	Ho	67	$[Xe]6s^2 4f^{11}$
Al	13	$[Ne]3s^2 3p^1$	Er	68	$[Xe]6s^2 4f^{12}$
Si	14	$[Ne]3s^2 3p^2$	Tm	69	$[Xe]6s^2 4f^{13}$
P	15	$[Ne]3s^2 3p^3$	Yb	70	$[Xe]6s^2 4f^{14}$
S	16	$[Ne]3s^2 3p^4$	Lu	71	$[Xe]6s^2 4f^{14} 5d^1$
Cl	17	$[Ne]3s^2 3p^5$	Hf	72	$[Xe]6s^2 4f^{14} 5d^2$
Ar	18	$[Ne]3s^2 3p^6$	Ta	73	$[Xe]6s^2 4f^{14} 5d^3$
			W	74	$[Xe]6s^2 4f^{14} 5d^4$
K	19	$[Ar]4s^1$	Re	75	$[Xe]6s^2 4f^{14} 5d^5$
Ca	20	$[Ar]4s^2$	Os	76	$[Xe]6s^2 4f^{14} 5d^6$
Sc	21	$[Ar]4s^2 3d^1$	Ir	77	$[Xe]6s^2 4f^{14} 5d^7$
Ti	22	$[Ar]4s^2 3d^2$	Pt	78	*$[Xe]6s^1 4f^{14} 5d^9$
V	23	$[Ar]4s^2 3d^3$	Au	79	*$[Xe]6s^1 4f^{14} 5d^{10}$
Cr	24	*$[Ar]4s^1 3d^5$	Hg	80	$[Xe]6s^2 4f^{14} 5d^{10}$
Mn	25	$[Ar]4s^2 3d^5$	Tl	81	$[Xe]6s^2 4f^{14} 5d^{10} 6p^1$
Fe	26	$[Ar]4s^2 3d^6$	Pb	82	$[Xe]6s^2 4f^{14} 5d^{10} 6p^2$
Co	27	$[Ar]4s^2 3d^7$	Bi	83	$[Xe]6s^2 4f^{14} 5d^{10} 6p^3$
Ni	28	$[Ar]4s^2 3d^8$	Po	84	$[Xe]6s^2 4f^{14} 5d^{10} 6p^4$
Cu	29	*$[Ar]4s^1 3d^{10}$	At	85	$[Xe]6s^2 4f^{14} 5d^{10} 6p^5$
Zn	30	$[Ar]4s^2 3d^{10}$	Rn	86	$[Xe]6s^2 4f^{14} 5d^{10} 6p^6$
Ga	31	$[Ar]4s^2 3d^{10} 4p^1$			
Ge	32	$[Ar]4s^2 3d^{10} 4p^2$	Fr	87	$[Rn]7s^1$
As	33	$[Ar]4s^2 3d^{10} 4p^3$	Ra	88	$[Rn]7s^2$
Se	34	$[Ar]4s^2 3d^{10} 4p^4$	Ac	89	*$[Rn]7s^2 \quad 6d^1$
Br	35	$[Ar]4s^2 3d^{10} 4p^5$	Th	90	*$[Rn]7s^2 \quad 6d^2$
Kr	36	$[Ar]4s^2 3d^{10} 4p^6$	Pa	91	*$[Rn]7s^2 5f^2 6d^1$
			U	92	*$[Rn]7s^2 5f^3 6d^1$
Rb	37	$[Kr]5s^1$	Np	93	*$[Rn]7s^2 5f^4 6d^1$
Sr	38	$[Kr]5s^2$	Pu	94	$[Rn]7s^2 5f^6$
			Am	95	$[Rn]7s^2 5f^7$
Y	39	$[Kr]5s^2 4d^1$	Cm	96	*$[Rn]7s^2 5f^7 6d^1$
Zr	40	$[Kr]5s^2 4d^2$	Bk	97	$[Rn]7s^2 5f^9$
Nb	41	*$[Kr]5s^1 4d^4$	Cf	98	*$[Rn]7s^2 5f^9 6d^1$
Mo	42	*$[Kr]5s^1 4d^5$	Es	99	$[Rn]7s^2 5f^{11}$
Tc	43	$[Kr]5s^2 4d^5$	Fm	100	$[Rn]7s^2 5f^{12}$
Ru	44	*$[Kr]5s^1 4d^7$	Md	101	$[Rn]7s^2 5f^{13}$
Rh	45	*$[Kr]5s^1 4d^8$	No	102	$[Rn]7s^2 5f^{14}$
Pd	46	*$[Kr]4d^{10}$	Lr	103	$[Rn]7s^2 5f^{14} 6d^1$
Ag	47	*$[Kr]5s^1 4d^{10}$	Rf	104	$[Rn]7s^2 5f^{14} 6d^2$
Cd	48	$[Kr]5s^2 4d^{10}$	Db	105	$[Rn]7s^2 5f^{14} 6d^3$
In	49	$[Kr]5s^2 4d^{10} 5p^1$	Sg	106	$[Rn]7s^2 5f^{14} 6d^4$
Sn	50	$[Kr]5s^2 4d^{10} 5p^2$	Bh	107	$[Rn]7s^2 5f^{14} 6d^5$
Sb	51	$[Kr]5s^2 4d^{10} 5p^3$	Hs	108	$[Rn]7s^2 5f^{14} 6d^6$
Te	52	$[Kr]5s^2 4d^{10} 5p^4$	Mt	109	$[Rn]7s^2 5f^{14} 6d^7$
I	53	$[Kr]5s^2 4d^{10} 5p^5$	Ds	110	*$[Rn]7s^1 5f^{14} 6d^9$
Xe	54	$[Kr]5s^2 4d^{10} 5p^6$	Rg	111	$[Rn]7s^1 5f^{14} 6d^{10}$
			Uub[a]	112	$[Rn]7s^2 5f^{14} 6d^{10}$

*Elements with configurations that do not follow the simple order of orbital filling.

[a] The name "copernicium" and symbol Cn for this element have been recommended provisionally by the IUPAC. In addition, evidence has been reported for elements having atomic numbers 113 through 116 and 118, but these have not been authenticated by the IUPAC.

Source: Actinide configurations are from J. J. Katz, G. T. Seaborg, and L. R. Morss, *The Chemistry of the Actinide Elements*, 2nd ed., Chapman and Hall, New York and London, 1986. Configurations for elements 100 to 112 are predicted, not experimental.

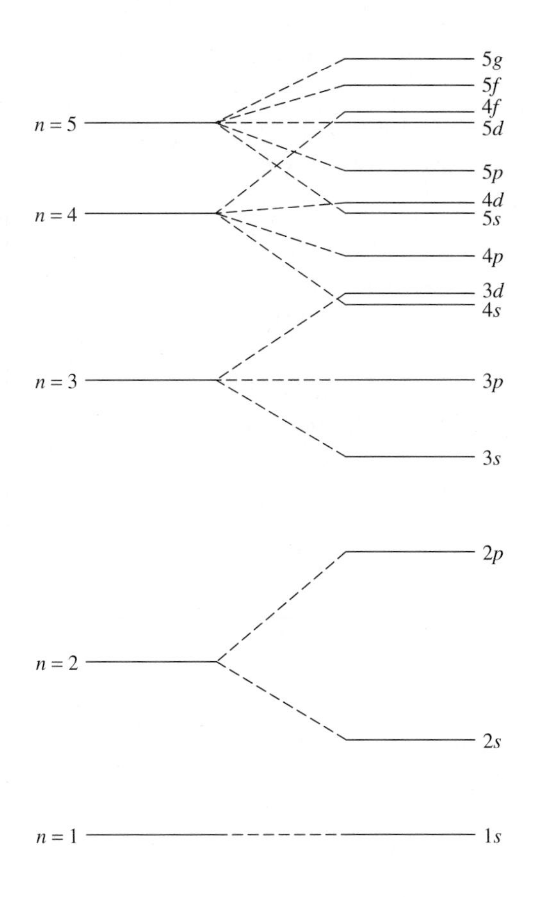

FIGURE 2.10 Energy Level Splitting and Overlap. The differences between the upper levels are exaggerated for easier visualization. This diagram provides unambiguous electron configurations for elements hydrogen to vanadium.

Slater[29] formulated a set of simple rules that serve as an approximate guide to this effect. These rules define the effective nuclear charge Z^* as a measure of the attraction of the nucleus for a particular electron:

Effective nuclear charge $Z^* = Z - S$, where Z = nuclear charge
S = shielding constant

Slater's rules for determining S for a specific electron:[30]

1. The atom's electronic structure is written in order of increasing quantum numbers n and l, grouped as follows:

   (1*s*) (2*s*, 2*p*) (3*s*, 3*p*) (3*d*) (4*s*, 4*p*) (4*d*) (4*f*) (5*s*, 5*p*) (5*d*) (and so on)

2. Electrons in groups to the right in this list do not shield electrons to their left.
3. The shielding constant S for electrons in these groups can now be determined. For *ns* and *np* valence electrons:
   a. Each electron in the same group contributes 0.35 to the value of S for each other electron in the group.

   Exception: A 1*s* electron contributes 0.30 to S for another 1*s* electron.

   Example: For a configuration $2s^2 2p^5$, a particular 2*p* electron has six other electrons in the (2*s*, 2*p*) group. Each of these contributes 0.35 to the value of S, for a total contribution to S of $6 \times 0.35 = 2.10$.

[29] J.C. Slater. *Phys. Rev.*, **1930**, *36*, 57.
[30] Slater's original numbering scheme has been changed for convenience.

b. Each electron in $n - 1$ groups contribute 0.85 to S.

> **Example:** For the 3*s* electron of sodium, there are eight electrons in the (2*s*, 2*p*) group. Each of these electrons contributes 0.85 to the value of S, a total contribution of $8 \times 0.85 = 6.80$.

c. Each electron in $n - 2$ or lower groups contributes 1.00 to S.

4. For *nd* and *nf* valence electrons:

a. Each electron in the same group contributes 0.35 to the value of S for each other electron in the group. (Same rule as 3.a.)

b. Each electron in groups to the left contributes 1.00 to S.

These rules are used to calculate the shielding constant S for valence electrons. Subtracting S from the total nuclear charge Z gives the effective nuclear charge Z^* on the selected electron:

$$Z^* = Z - S$$

Examples of calculations of S and Z^* follow.

EXAMPLES

Oxygen

Use Slater's rules to calculate the shielding constant and effective nuclear charge of a 2*p* electron.

Rule 1: The electron configuration is written using Slater's groupings, in order: $(1s^2)\ (2s^2, 2p^4)$

To calculate S for a valence 2*p* electron:

Rule 3a: Each other electron in the $(2s^2, 2p^4)$ group contributes 0.35 to S. Total contribution $= 5 \times 0.35 = 1.75$

Rule 3b: Each 1*s* electron contributes 0.85 to S. Total contribution $= 2 \times 0.85 = 1.70$

Total $S = 1.75 + 1.70 = 3.45$

Effective nuclear charge $Z^* = 8 - 3.45 = 4.55$

So rather than feeling the full +8 nuclear charge, a 2*p* electron is calculated to feel a charge of +4.55, or about 57% of the full nuclear charge.

Nickel

Use Slater's rules to calculate the shielding constant and effective nuclear charge of a 3*d* and 4*s* electron.

Rule 1: The electron configuration is written $(1s^2)\ (2s^2, 2p^6)\ (3s^2, 3p^6)\ (3d^8)\ (4s^2)$ For a 3*d* electron:

Rule 4a: Each other electron in the $(3d^8)$ group contributes 0.35 to S. Total contribution $= 7 \times 0.35 = 2.45$

Rule 4b: Each electron in groups to the left of $(3d^8)$ contributes 1.00 to S. Total contribution $= 18 \times 1.00 = 18.00$

Total $S = 2.45 + 18.00 = 20.45$

Effective nuclear charge $Z^* = 28 - 20.45 = 7.55$

For a 4*s* electron:

Rule 3a: The other electron in the $(4s^2)$ group contributes 0.35 to S.

Rule 3b: Each electron in the $(3s^2, 3p^6)(3d^8)$ groups $(n - 1)$ contributes 0.85. Total contribution $= 16 \times 0.85 = 13.60$

Rule 3c: Each other electron to the left contributes 1.00. Total contribution $= 10 \times 1.00 = 10.00$

Total $S = 0.35 + 13.60 + 10.00 = 23.95$

Effective nuclear charge $Z^* = 28 - 23.95 = 4.05$

The effective nuclear charge for the 4*s* electron is considerably smaller than the value for the 3*d* electron. This is equivalent to stating that the 4*s* electron is held less tightly than the 3*d* and should therefore be the first removed in ionization. This is consistent with experimental observations on nickel compounds. Ni^{2+}, the most common oxidation state of nickel, has a configuration of $[Ar]3d^8$, rather than $[Ar]3d^6 4s^2$, corresponding to loss of the 4*s* electrons from nickel atoms. All the transition metal atoms follow this same pattern of losing *ns* electrons more readily than $(n - 1)d$ electrons.

▶ **Exercise 2.5** Calculate the effective nuclear charge on a 5*s*, 5*p*, and 4*d* electron in a tin atom.

▶ **Exercise 2.6** Calculate the effective nuclear charge on a 7*s*, 5*f*, and 6*d* electron in a uranium atom.

Justification for Slater's rules, aside from the fact that they work, comes from the electron probability curves for the orbitals. The *s* and *p* orbitals have higher probabilities near the nucleus than do *d* orbitals of the same *n*, as shown earlier in Figure 2.7. Therefore, the shielding of 3*d* electrons by (3*s*, 3*p*) electrons is calculated as 100% effective, a contribution of 1.00. At the same time, shielding of 3*s* or 3*p* electrons by (2*s*, 2*p*) electrons is only 85% effective, a contribution of 0.85, because the 3*s* and 3*p* orbitals have regions of significant probability close to the nucleus. Therefore, electrons in these orbitals are not completely shielded by (2*s*, 2*p*) electrons.

A complication arises at Cr (Z = 24) and Cu (Z = 29) in the first transition series and in an increasing number of atoms under them in the second and third transition series. This effect places an extra electron in the 3*d* level and removes one electron from the 4*s* level. Cr, for example, has a configuration of $[Ar]4s^1 3d^5$ rather than $[Ar]4s^2 3d^4$. Traditionally, this phenomenon has often been explained as a consequence of the "special stability of half-filled subshells." Half-filled and filled *d* and *f* subshells are, in fact, fairly common, as shown in Figure 2.11. A more accurate explanation considers both the effects of increasing nuclear charge on the energies of the 4*s* and 3*d* levels and the interactions (repulsions) between the electrons sharing the same orbital.[31] This approach requires totaling the energies of all the electrons with their interactions, including consideration of Coulombic and exchange energies; results of the complete calculations match the experimental results.

Slater's rules have been refined to improve their match with experimental data. One relatively simple refinement is based on the ionization energies for the elements hydrogen through xenon, and it provides a calculation procedure similar to that proposed by Slater.[32] A more elaborate method incorporates exponential screening and provides energies that are in closer agreement with experimental values.[33]

Another explanation that is more pictorial and considers the electron–electron interactions was proposed by Rich.[34] He explained electronic structures of atoms by

---

[31] L. G. Vanquickenborne, K. Pierloot, and D. Devoghel, *J. Chem. Educ.*, **1994**, *71*, 469.
[32] J. L. Reed, *J. Chem. Educ.*, **1999**, *76*, 802.
[33] W. Eek, S. Nordholm, and G. B. Bacskay, *Chem. Educator*, **2006**, *11*, 235.
[34] R. L. Rich, *Periodic Correlations*, W. A. Benjamin, Menlo Park, CA, 1965, pp. 9–11.

FIGURE 2.11 Electron Configurations of Transition Metals, Including Lanthanides and Actinides. Solid lines surrounding elements designate filled (d^{10} or f^{14}) or half-filled (d^5 or f^7) subshells. Dashed lines surrounding elements designate irregularities in sequential orbital filling, which is also found within some of the solid lines.

					Half-filled *d*					Filled *d*							
Na	Mg											Al	Si	P	S	Cl	Ar
K	Ca	Sc $3d^1$	Ti $3d^2$	V $3d^3$	Cr $3d^5$ $4s^1$	Mn $3d^5$ $4s^2$	Fe $3d^6$	Co $3d^7$	Ni $3d^8$	Cu $3d^{10}$ $4s^1$	Zn $3d^{10}$ $4s^2$	Ga	Ge	As	Se	Br	Kr
Rb	Sr	Y $4d^1$	Zr $4d^2$	Nb $4d^4$ $5s^1$	Mo $4d^5$ $5s^1$	Tc $4d^5$ $5s^2$	Ru $4d^7$ $5s^1$	Rh $4d^8$ $5s^1$	Pd $4d^{10}$	Ag $4d^{10}$ $5s^1$	Cd $4d^{10}$ $5s^2$	In	Sn	Sb	Te	I	Xe
Cs	Ba	La $5d^1$	* Hf $4f^{14}$ $5d^2$	Ta $4f^{14}$ $5d^3$	W $4f^{14}$ $5d^4$	Re $5d^5$ $6s^2$	Os $5d^6$	Ir $5d^7$	Pt $5d^9$ $6s^1$	Au $5d^{10}$ $6s^1$	Hg $5d^{10}$ $6s^2$	Tl	Pb	Bi	Po	At	Rn
Fr	Ra	Ac $6d^1$	** Rf $5f^{14}$ $6d^2$	Db $5f^{14}$ $6d^3$	Sg $5f^{14}$ $6d^4$	Bh $5f^{14}$ $6d^5$	Hs $5f^{14}$ $6d^6$	Mt $5f^{14}$ $6d^7$	Ds $6d^9$	Rg $6d^{10}$ $7s^1$	Uub $6d^{10}$ $7s^2$		Uuq		Uuh		Uuo

						Half-filled *f*							Filled *f*	
*	Ce $4f^1$ $5d^1$	Pr $4f^3$	Nd $4f^4$	Pm $4f^5$	Sm $4f^6$	Eu $4f^7$	Gd $4f^7$ $5d^1$	Tb $4f^9$	Dy $4f^{10}$	Ho $4f^{11}$	Er $4f^{12}$	Tm $4f^{13}$	Yb $4f^{14}$	Lu $4f^{14}$ $5d^1$
**	Th $6d^2$	Pa $5f^2$ $6d^1$	U $5f^3$ $6d^1$	Np $5f^4$ $6d^1$	Pu $5f^6$	Am $5f^7$	Cm $5f^7$ $6d^1$	Bk $5f^9$	Cf $5f^9$ $6d^1$	Es $5f^{11}$	Fm $5f^{12}$	Md $5f^{13}$	No $5f^{14}$	Lr $5f^{14}$ $6d^1$

considering the difference in energy between the energy of one electron in an orbital and two electrons in the same orbital. Although the orbital itself is usually assumed to have only one energy, the electrostatic repulsion of the two electrons in one orbital adds the electron-pairing energy described in Section 2.2.3 as part of Hund's rule. We can visualize two parallel energy levels, each with electrons of only one spin, separated by the electron-pairing energy, as shown in Figure 2.12. As the nuclear charge increases, the electrons are more strongly attracted, and the energy levels decrease in energy—becoming more stable—with the *d* orbitals changing more rapidly than the *s* orbitals, because the *d* orbitals are not shielded as well from the nucleus. Electrons fill the lowest available orbitals in order up to their capacity, with the results shown in Figure 2.12 and in Table 2.7.

The schematic diagram in Figure 2.12(a) shows the order in which the levels fill, from bottom to top in energy. For example, Ti has two 4*s* electrons, one in each spin level, and two 3*d* electrons, both with the same spin. Fe has two 4*s* electrons, one in each spin level, five 3*d* electrons with spin $-\frac{1}{2}$, and one 3*d* electron with spin $+\frac{1}{2}$.

For vanadium, the first two electrons enter the 4*s*, $-\frac{1}{2}$ and 4*s*, $+\frac{1}{2}$ levels; the next three are all in the 3*d*, $-\frac{1}{2}$ level, and vanadium has the configuration $4s^2, 3d^3$. The 3*d*, $-\frac{1}{2}$ line crosses the 4*s*, $+\frac{1}{2}$ line between V and Cr. When the six electrons of chromium are filled in from the lowest level, chromium has the configuration $4s^1\,3d^5$. A similar crossing gives copper its $4s^1\,3d^{10}$ structure. This explanation does not depend on the stability of half-filled shells or other additional factors; those explanations break down for zirconium $\left(5s^2\,4d^2\right)$, niobium $\left(5s^1\,4d^4\right)$, and others in the lower periods.

Formation of a positive ion by removal of an electron reduces the overall electron repulsion and lowers the energy of the *d* orbitals more than that of the *s* orbitals, as shown in Figure 2.12(b). As a result, the remaining electrons occupy the *d* orbitals, and we can use the shorthand notion that the electrons with highest *n*—in this case, those in the *s* orbitals—are always removed first when ions are formed from the transition elements. This effect is even stronger for 2+ ions. *Transition metal cations have no s electrons, only d electrons in their outer levels.*

A similar, but more complex, crossing of levels appears in the lanthanide and actinide series. The simple explanation would have these elements start filling *f* orbitals

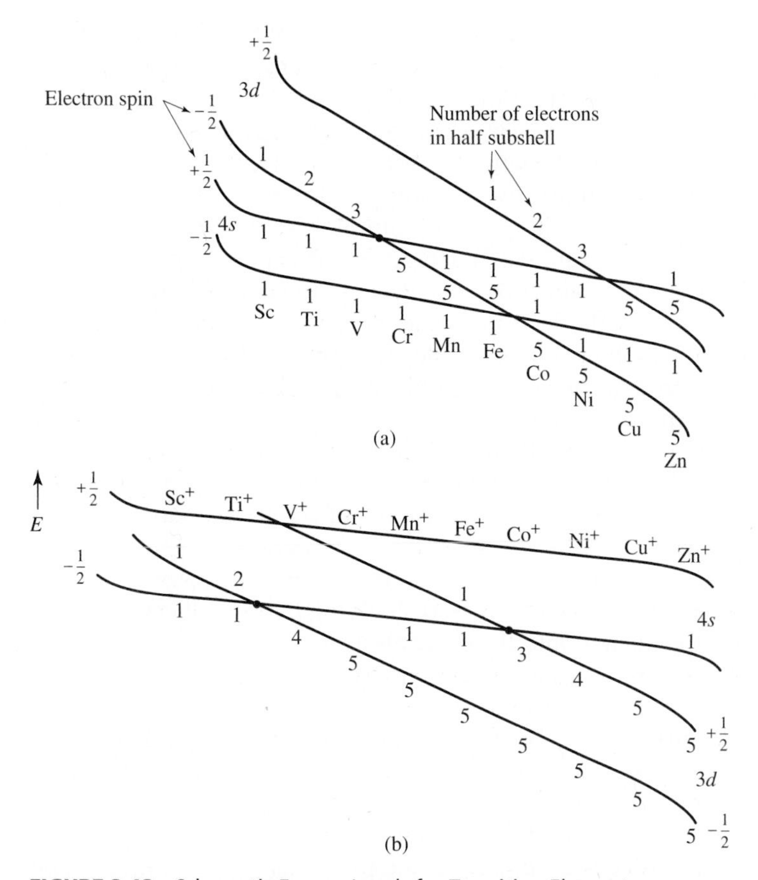

FIGURE 2.12 Schematic Energy Levels for Transition Elements. (a) Schematic interpretation of electron configurations for transition elements in terms of intraorbital repulsion and trends in subshell energies. (b) A similar diagram for ions, showing the shift in the crossover points on removal of an electron. The diagram shows that *s* electrons are removed before *d* electrons. The shift is even more pronounced for metal ions having 2+ or greater charges. As a consequence, transition-metal ions with 2+ or greater charges have no *s* electrons, only *d* electrons in their outer levels. Similar diagrams, although more complex, can be drawn for the heavier transition elements and the lanthanides.

(Reprinted with permission from R. L. Rich, *Periodic Correlations*, W. A. Benjamin, Menlo Park, CA, 1965, pp. 9–10.)

at lanthanum (57) and actinium (89), but these atoms have one *d* electron instead. Other elements in these series also show deviations from the "normal" sequence. Rich has shown how these may also be explained by similar diagrams, and the reader should refer to his book for further details.

2.3 PERIODIC PROPERTIES OF ATOMS

2.3.1 Ionization Energy

The ionization energy, also known as the *ionization potential*, is the energy required to remove an electron from a gaseous atom or ion:

$$A^{n+}(g) \longrightarrow A^{(n+1)+}(g) + e^- \qquad \text{ionization energy} = \Delta U$$

where $n = 0$ (first ionization energy), $n = 1$ (second ionization energy), and so on.

As would be expected from the effects of shielding, the ionization energy varies with different nuclei and different numbers of electrons. Trends for the first ionization energies of the early elements in the periodic table are shown in Figure 2.13. The general trend across a period is an increase in ionization energy as the nuclear charge increases. A plot of Z^*/r, the potential energy for attraction between an electron and the shielded nucleus, is nearly a straight line, with approximately the same slope as the shorter segments—boron through nitrogen, for example—shown in Figure 2.13 (a different representation is shown, in Figure 8.3). However, the experimental values show a break in the trend at boron and again at oxygen. Because the new electron in boron is in a new *p* orbital that has most of its electron density farther away from the nucleus than the other electrons, its ionization energy is smaller than that of the $2s^2$ electrons of beryllium. At the fourth *p* electron, at oxygen, a similar drop in ionization energy occurs. Here, the new electron shares an orbital with one of the previous $2p$ electrons, and the fourth *p* electron has a higher energy than the trend would indicate, because it must be paired with another in the same *p* orbital. The pairing energy, or repulsion between two electrons in the same region of space, makes loss of an electron easier, reducing the ionization energy. Similar patterns appear in the other periods. The transition elements have smaller differences in ionization energies, usually with a lower value for heavier atoms in the same group because of increased shielding by inner electrons and increased distance between the nucleus and the outer electrons.

Much larger decreases in ionization energy occur at the start of each new period, because the change to the next major quantum number requires that the new *s* electron have a much higher energy. The maxima at the noble gases decrease with increasing *Z*, because the outer electrons are farther from the nucleus in the heavier elements. Overall, the trends are toward higher ionization energy from left to right in the periodic table (the major change) and lower ionization energy from top to bottom (a minor change). The differences described in the previous paragraph are superimposed on these more general changes.

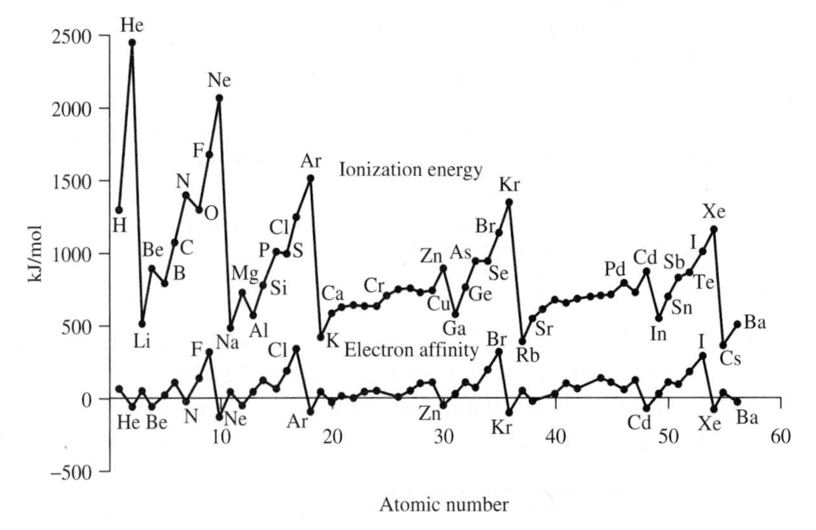

FIGURE 2.13 Ionization Energies and Electron Affinities. Ionization energy = ΔU for $M(g) \longrightarrow M^+(g) + e$

(Data from C.E. Moore, *Ionization Potentials and Ionization Limits, National Standards Reference Data Series*, U. S. National Bureau of Standards, Washington, DC, **1970**, NSRDS-NBS 34) Electron affinity = ΔU for $M^-(g) \longrightarrow M(g) + e^-$ (Data from H. Hotop and W. C. Lineberger, *J. Phys. Chem. Ref. Data*, **1985**, *14*, 731). Numerical values are in Appendices B-2 and B-3.

2.3.2 Electron Affinity

Electron affinity can be defined as the energy required to remove an electron from a negative ion:[35]

$$A^-(g) \longrightarrow A(g) + e^- \qquad \text{electron affinity} = \Delta U \text{ (or } EA\text{)}$$

Because of the similarity of this reaction to the ionization for an atom, electron affinity is sometimes described as the *zeroth ionization energy*. This reaction is endothermic (positive ΔU) except for the noble gases and the alkaline earth elements. The pattern of electron affinities with changing Z, shown in Figure 2.13, is similar to that of the ionization energies, but for one larger Z value (one more electron for each species) and with much smaller absolute numbers. For either of the reactions, removal of the first electron past a noble gas configuration is easy, so the noble gases have the lowest electron affinities. The electron affinities are all much smaller than the corresponding ionization energies, because electron removal from a negative ion is easier than removal from a neutral atom.

2.3.3 Covalent and Ionic Radii

The sizes of atoms and ions are also related to the ionization energies and electron affinities. As the nuclear charge increases, the electrons are pulled in toward the center of the atom, and the size of any particular orbital decreases. On the other hand, as the nuclear charge increases, more electrons are added to the atom, and their mutual repulsion keeps the outer orbitals large. The interaction of these two effects, increasing nuclear charge and increasing number of electrons, results in a gradual decrease in atomic size across each period. Table 2.8 gives nonpolar covalent radii, calculated for ideal molecules with no polarity. There are other measures of atomic size, such as the van der Waals radius, in which collisions with other atoms are used to define the size. It is difficult to obtain consistent data for any such measure, because the polarity, chemical structure, and physical

TABLE 2.8 Nonpolar Covalent Radii (pm)

1	2	3	4	5	6	7	8	9	10	11	12	13	14	15	16	17	18
H																	He
32																	31
Li	Be											B	C	N	O	F	Ne
123	89											82	77	75	73	71	69
Na	Mg											Al	Si	P	S	Cl	Ar
154	136											118	111	106	102	99	98
K	Ca	Sc	Ti	V	Cr	Mn	Fe	Co	Ni	Cu	Zn	Ga	Ge	As	Se	Br	Kr
203	174	144	132	122	118	117	117	116	115	117	125	126	122	120	117	114	111
Rb	Sr	Y	Zr	Nb	Mo	Tc	Ru	Rh	Pd	Ag	Cd	In	Sn	Sb	Te	I	Xe
216	191	162	145	134	130	127	125	125	128	134	148	144	140	140	136	133	126
Cs	Ba	La	Hf	Ta	W	Re	Os	Ir	Pt	Au	Hg	Tl	Pb	Bi	Po	At	Ra
235	198	169	144	134	130	128	126	127	130	134	149	148	147	146	(146)	(145)	

Source: R. T. Sanderson, *Inorganic Chemistry*, Reinhold, New York, 1967, p. 74; and E. C. M. Chen, J. G. Dojahn, and W. E. Wentworth, *J. Phys. Chem. A*, **1997**, *101*, 3088.

[35] Historically, the definition has been $-\Delta U$ for the reverse reaction, adding an electron to the neutral atom. The definition we use avoids the sign change.

state of molecules change drastically from one compound to another. The numbers shown here are sufficient for a general comparison of one element with another.

There are similar problems in determining the size of ions. Because the stable ions of the different elements have different charges and different numbers of electrons, as well as different crystal structures for their compounds, it is difficult to find a suitable set of numbers for comparison. Earlier data were based on Pauling's approach, in which the ratio of the radii of isoelectronic ions was assumed to be equal to the ratio of their effective nuclear charges. More recent calculations are based on a number of considerations, including electron density maps from X-ray data that show larger cations and smaller anions than those previously found. Those in Table 2.9 and Appendix B were called "crystal radii" by Shannon[36] and are generally different from the older values of "ionic radii" by +14 pm for cations and –14 pm for anions, as well as being

TABLE 2.9 Crystal Radii for Selected Ions

	Z	Element	Radius (pm)
Alkali metal ions	3	Li^{+}	90
	11	Na^{+}	116
	19	K^{+}	152
	37	Rb^{+}	166
	55	Cs^{+}	181
Alkaline earth ions	4	Be^{2+}	59
	12	Mg^{2+}	86
	20	Ca^{2+}	114
	38	Sr^{2+}	132
	56	Ba^{2+}	149
Other cations	13	Al^{3+}	68
	30	Zn^{2+}	88
Halide ions	9	F^{-}	119
	17	Cl^{-}	167
	35	Br^{-}	182
	53	I^{-}	206
Other anions	8	O^{2-}	126
	16	S^{2-}	170

Source: R. D. Shannon, *Acta Crystallogr*. **1976**, *A32*, 751. A longer list is given in Appendix B-1. All the values are for 6-coordinate ions.

[36] R. D. Shannon, *Acta Crystallogr*., **1976**, A32, 751.

revised to accommodate more recent measurements. The radii in Table 2.9 and Appendix B-1 can be used for rough estimation of the packing of ions in crystals and other calculations, as long as the "fuzzy" nature of atoms and ions is kept in mind.

Factors that influence ionic size include the coordination number of the ion, the covalent character of the bonding, distortions of regular crystal geometries, and delocalization of electrons (metallic or semiconducting character, described in Chapter 7). The radius of the anion is also influenced by the size and charge of the cation (the anion exerts a smaller influence on the radius of the cation).[37] The table in Appendix B-1 shows the effect of coordination number.

The values in Table 2.10 show that anions are generally larger than cations with similar numbers of electrons. The radius decreases as nuclear charge increases for ions with the same electronic structure, with the charge on cations having a strong effect, for example in the series Na^+, Mg^{2+}, Al^{3+}. Within a group, the ionic radius increases as Z increases because of the larger number of electrons in the ions and, for the same element, the radius decreases with increasing charge on the cation. Examples of these trends are shown in Tables 2.10, 2.11, and 2.12.

TABLE 2.10 Crystal Radius and Nuclear Charge

Ion	Protons	Electrons	Radius (pm)
O^{2-}	8	10	126
F^-	9	10	119
Na^+	11	10	116
Mg^{2+}	12	10	86
Al^{3+}	13	10	68

TABLE 2.11 Crystal Radius and Total Number of Electrons

Ion	Protons	Electrons	Radius (pm)
O^{2-}	8	10	126
S^{2-}	16	18	170
Se^{2-}	34	36	184
Te^{2-}	52	54	207

TABLE 2.12 Crystal Radius and Ionic Charge

Ion	Protons	Electrons	Radius (pm)
Ti^{2+}	22	20	100
Ti^{3+}	22	19	81
Ti^{4+}	22	18	75

[37] O. Johnson, *Inorg. Chem.*, **1973**, *12*, 780.

General References

Additional information on the history of atomic theory can be found in J. R. Partington, *A Short History of Chemistry*, 3rd ed., Macmillan, London, 1957, reprinted by Harper & Row, New York, 1960, and in the *Journal of Chemical Education*. For an introduction to atomic theory and orbitals, see V. M. S. Gil, *Orbitals in Chemistry: A Modern Guide for Students*, Cambridge University Press, Cambridge, UK, pp. 1–69. A more thorough treatment of the electronic structure of atoms is in M. Gerloch, *Orbitals, Terms, and States*, John Wiley & Sons, New York, 1986. Many Internet sites provide images of atomic orbitals, their wave equations, nodal behavior, and other characteristics. Two examples are http://www.orbitals.com and http://winter.group.shef.ac.uk/orbitron.

Problems

2.1 Determine the de Broglie wavelength of
- **a.** an electron moving at one-tenth the speed of light.
- **b.** a 400 g Frisbee moving at 10 km/h.
- **c.** an 8.0-pound bowling ball rolling down the lane with a velocity of 2.0 meters per second.
- **d.** a 13.7 g hummingbird flying at a speed of 30.0 miles per hour.

2.2 Using the equation $E = R_H\left(\frac{1}{2^2} - \frac{1}{n_h^2}\right)$, determine the energies and wavelengths of the visible emission bands in the atomic spectrum of hydrogen arising from $n_h = 4$, 5, and 6. (The red line, corresponding to $n_h = 3$, was calculated in Exercise 2.1.)

2.3 The transition from the $n = 7$ to the $n = 2$ level of the hydrogen atom is accompanied by the emission of radiation slightly beyond the range of human perception, in the ultraviolet region. Determine the energy and wavelength.

2.4 Emissions are observed at wavelengths of 383.65 and 379.90 nm for transitions from excited states of the hydrogen atom to the $n = 2$ state. Determine the quantum numbers n_h for these emissions.

2.5 What is the least amount of energy that can be emitted by an excited electron in a hydrogen atom falling from an excited state directly to the $n = 3$ state? What is the quantum number n for the excited state? Humans cannot visually observe the photons emitted in this process. Why not?

2.6 The details of several steps in the particle-in-a-box model in this chapter have been omitted. Work out the details of the following steps:
- **a.** Show that if $\Psi = A \sin rx + B \cos sx$ (A, B, r, and s are constants) is a solution to the wave equation for the one-dimensional box, then

$$r = s = \sqrt{2mE}\left(\frac{2\pi}{h}\right)$$

- **b.** Show that if $\Psi = A \sin rx$, the boundary conditions ($\Psi = 0$ when $x = 0$ and $x = a$) require that $r = \pm\frac{n\pi}{a}$, where n = any integer other than zero.
- **c.** Show that if $r = \pm\frac{n\pi}{a}$, the energy levels of the particle are given by

$$E = \frac{n^2 h^2}{8ma^2}$$

- **d.** Show that substituting the value of r given in into $\Psi = A \sin rx$ and applying the normalizing requirement gives $A = \sqrt{2/a}$.

2.7 For the $3p_z$ and $4d_{xz}$ hydrogen-like atomic orbitals, sketch the following:
- **a.** The radial function R
- **b.** The radial probability function $a_0 r^2 R^2$
- **c.** Contour maps of electron density.

2.8 Repeat the exercise in Problem 2.7 for the $4s$ and $5d_{x^2-y^2}$ orbitals.

2.9 Repeat the exercise in Problem 2.7 for the $5s$ and $4d_{z^2}$ orbitals.

2.10 The $4f_{z(x^2-y^2)}$ orbital has the angular function $Y = (\text{constant})\ z\,(x^2 - y^2)/r^3$.
- **a.** How many radial nodes does this orbital have?
- **b.** How many angular nodes does it have?
- **c.** Write equations to define the angular nodal surfaces. What shapes are these surfaces?
- **d.** Sketch the shape of the orbital, and show all radial and angular nodes.

2.11 Repeat the exercise in Problem 2.10 for the $5f_{xyz}$ orbital, which has $Y = (\text{constant})\ xyz/r^3$.

2.12 The label for an f_{z^3} orbital, like that for a d_{z^2} orbital, is an abbreviation. The actual angular function for this orbital is $Y = (\text{constant}) \times z(5z^2 - 3r^2)/r^3$. Repeat the exercise in Problem 2.10 for a $4f_{z^3}$ orbital. (*Note*: recall that $r^2 = x^2 + y^2 + z^2$).

2.13
- **a.** Determine the possible values for the l and m_l quantum numbers for a $5d$ electron, a $4f$ electron, and a $7g$ electron.
- **b.** Determine the possible values for all four quantum numbers for a $3d$ electron.
- **c.** What values of m_l are possible for f orbitals?
- **d.** At most, how many electrons can occupy a $4d$ orbital?

2.14 **a.** What are the values of quantum numbers l and n for a $5d$ electron?

b. At most, how many $4d$ electrons can an atom have? Of these electrons how many, at most, can have $m_s = -\frac{1}{2}$?

c. A $5f$ electron has what value of quantum number l? What values of m_l may it have?

d. What values of the quantum number m_l are possible for a subshell having $l = 4$?

2.15 **a.** At most, how many electrons in an atom can have both $n = 5$ and $l = 3$?

b. A $5d$ electron has what possible values of the quantum number m_l?

c. What value of quantum number l do p orbitals have? For what values of n do p orbitals occur?

d. What is the quantum number l for g orbitals? How many orbitals are in a g subshell?

2.16 Determine the Coulombic and exchange energies for the following configurations, and determine which configuration is favored (of lower energy):

a. ↑ ___ ↑ ___ and ↑↓ ___ ___

b. ↑ ___ ↑ ___ ↑ ___ and ↑↓ ___ ↑ ___ ___

2.17 Two excited states for a d^4 configuration are shown. Which is likely to have lower energy? Explain your choice in terms of Coulombic and exchange energies.

W: ↑ ↑ ↓ ↓ ___ X: ↑ ↑ ↑ ↓ ___

2.18 Two excited states for a d^5 configuration are shown. Which is likely to have lower energy? Why? Explain your choice in terms of Coulombic and exchange energies.

Y: ↑ ↑ ↑ ↓ ↓ Z: ↑ ↑ ↑ ↑ ↓

2.19 Provide explanations of the following phenomena:

a. The electron configuration of Cr is $[Ar]4s^1 3d^5$ rather than $[Ar]4s^2 3d^4$.

b. The electron configuration of Ti is $[Ar]4s^2 3d^2$, but that of Cr^{2+} is $[Ar]3d^4$.

2.20 Give electron configurations for the following:

a. V **b.** Br **c.** Ru^{3+}

d. Hg^{2+} **e.** Sb

2.21 Predict the electron configurations of the following metal anions:

a. Rb^-

b. Pt^{2-} (See: A. Karbov, J. Nuss, U. Weding, and M. Jansen, *Angew Chem. Int. Ed.*, **2003**, *42*, 4818.)

2.22 Briefly explain the following on the basis of electron configurations:

a. Fluorine forms an ion having a charge of 1–.

b. The most common ion formed by zinc has a 2+ charge.

c. The electron configuration of the molybdenum atom is [Kr] $5s^1 4d^5$ rather than [Kr] $5s^2 4d^4$.

2.23 Briefly explain the following on the basis of electron configurations:

a. The most common ion formed by silver has a 1+ charge.

b. Cm has the outer electron configuration $s^2d^1f^7$ rather than s^2f^8.

c. Sn often forms an ion having a charge of 2+ (the *stannous* ion).

2.24 **a.** Which 2+ ion has two $3d$ electrons? Which has eight $3d$ electrons?

b. Which is the more likely configuration for Mn^{2+}: $[Ar]4s^2 3d^3$ or $[Ar]3d^5$?

2.25 Using Slater's rules, determine Z^* for

a. a $3p$ electron in P, S, Cl, and Ar. Is the calculated value of Z^* consistent with the relative sizes of these atoms?

b. a $2p$ electron in O^{2-}, F^-, Na^+, and Mg^{2+}. Is the calculated value of Z^* consistent with the relative sizes of these ions?

c. a $4s$ and a $3d$ electron of Cu. Which type of electron is more likely to be lost when copper forms a positive ion?

d. a $4f$ electron in Ce, Pr, and Nd. There is a decrease in size, commonly known as the **lanthanide contraction**, with increasing atomic number in the lanthanides. Are your values of Z^* consistent with this trend?

2.26 A sample calculation in this chapter showed that, according to Slater's rules, a $3d$ electron of nickel has a higher effective nuclear charge than a $4s$ electron. Is the same true for early first-row transition metals? Using Slater's rules, calculate S and Z^* for $4s$ and $3d$ electrons of Sc and Ti, and comment on the similarities or differences with Ni.

2.27 Ionization energies should depend on the effective nuclear charge that holds the electrons in the atom. Calculate Z^* (Slater's rules) for N, P, and As. Do their ionization energies seem to match these effective nuclear charges? If not, what other factors influence the ionization energies?

2.28 Prepare a diagram such as the one in Figure 2.12(a) for the fifth period in the periodic table, elements Zr through Pd. The configurations in Table 2.7 can be used to determine the crossover points of the lines. Can a diagram be drawn that is completely consistent with the configurations in the table?

2.29 Why are the ionization energies of the alkali metals in the order Li > Na > K > Rb?

2.30 The second ionization of carbon ($C^+ \longrightarrow C^{2+} + e^-$) and the first ionization of boron ($B \longrightarrow B^+ + e^+$) both fit the reaction $1s^2 2s^2 2p^1 = 1s^2 2s^2 + e^-$. Compare the two ionization energies (24.383 eV and 8.298 eV, respectively) and the effective nuclear charge Z^*. Is this an adequate explanation of the difference in ionization energies? If not, suggest other factors.

2.31 The second ionization energy involves removing an electron from a positively charged ion in the gas phase (see preceding problem). How would a graph of second ionization energy vs. atomic number for the elements helium through neon compare with the graph of first ionization energy in Figure 2.13? Be specific in comparing the positions of peaks and valleys.

2.32 In each of the following pairs, pick the element with the higher ionization energy and explain your choice.
a. Fe, Ru **b.** P, S **c.** K, Br
d. C, N **e.** Cd, In **f.** Cl, F

2.33 On the basis of electron configurations, explain why
a. sulfur has a lower electron affinity than chlorine.
b. iodine has a lower electron affinity than bromine.
c. boron has a lower ionization energy than beryllium.
d. sulfur has a lower ionization energy than phosphorus.

2.34 **a.** The graph of ionization energy versus atomic number for the elements Na through Ar (Figure 2.13) shows maxima at Mg and P and minima at Al and S. Explain these maxima and minima.
b. The graph of electron affinity versus atomic number for the elements Na through Ar (Figure 2.13) also shows maxima and minima, but shifted by one element in comparison with the ionization energy graph. Why are the maxima and minima shifted in this way?

2.35 The second ionization energy of He is almost exactly four times the ionization energy of H, and the third ionization energy of Li is almost exactly nine times the ionization energy of H:

	IE (MJ mol^{-1})
$H(g) \longrightarrow H^+(g) + e^-$	1.3120
$He^+(g) \longrightarrow He^{2+}(g) + e^-$	5.2504
$Li^{2+}(g) \longrightarrow Li^{3+}(g) + e^-$	11.8149

Explain this trend on the basis of the Bohr equation for energy levels of single-electron systems.

2.36 The size of the transition-metal atoms decreases slightly from left to right in the periodic table. What factors must be considered in explaining this decrease? In particular, why does the size decrease at all, and why is the decrease so gradual?

2.37 Predict the largest and smallest radius in each series, and account for your choices:
a. Se^{2-} Br^- Rb^+ Sr^{2+}
b. Y^{3+} Zr^{4+} Nb^{5+}
c. Co^{4+} Co^{3+} Co^{2+} Co

2.38 Select the best choice, and briefly indicate the reason for each choice:
a. Largest radius: Na^+ Ne F^-
b. Greatest volume: S^{2-} Se^{2-} Te^{2-}
c. Highest ionization energy: Na Mg Al
d. Most energy necessary to remove an electron: Fe Fe^{2+} Fe^{3+}
e. Highest electron affinity: O F Ne

2.39 Select the best choice, and briefly indicate the reason for your choice:
a. Smallest radius: Sc Ti V
b. Greatest volume: S^{2-} Ar Ca^{2+}
c. Lowest ionization energy: K Rb Cs
d. Highest electron affinity: Cl Br I
e. Most energy necessary to remove an electron: Cu Cu^+ Cu^{2+}

2.40 There are a number of Web sites that display atomic orbitals. Use a search engine to find a complete set of the *f* orbitals.
a. How many orbitals are there in one set (for example, a set of 4*f* orbitals)?
b. Describe the angular nodes of the orbitals.
c. Observe what happens to the number of radial nodes as the principal quantum number is increased.

Include the URL for the site you used for each, along with sketches or printouts of the orbitals. (Two useful Web sites at this writing are http://www.orbitals.com and http://winter.group.shef.ac.uk/orbitron.)

2.41 Repeat the exercise in Problem 2.40, this time for a set of *g* orbitals.

3

Simple Bonding Theory

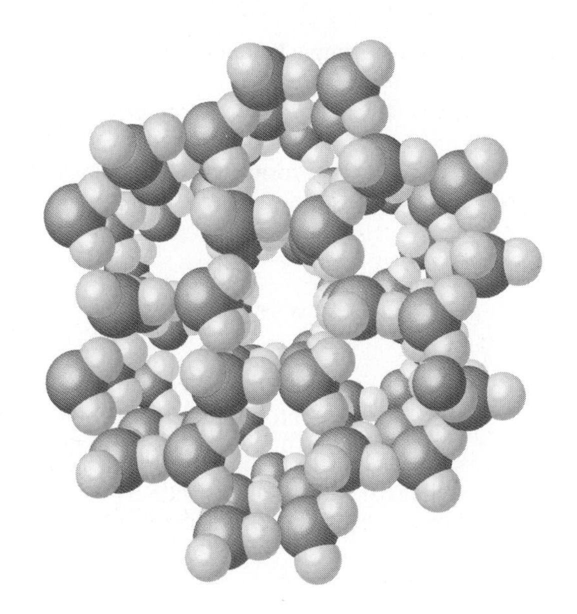

We now turn from the use of quantum mechanics and its description of the atom to an elementary description of molecules. Although most of the discussion of chemical bonding in this book uses the molecular orbital approach, simpler methods that provide approximate pictures of the shapes and polarities of molecules are also useful. This chapter provides an overview of Lewis dot structures, valence-shell electron-pair repulsion (VSEPR), and related topics. The molecular orbital descriptions of some of the same molecules are presented in Chapter 5 and later chapters, but the ideas of this chapter provide a starting point for that more modern treatment.

Ultimately, any description of bonding must be consistent with experimental data on bond lengths, bond angles, and bond strengths. Angles and distances are most frequently determined by diffraction (X-ray crystallography, electron diffraction, neutron diffraction) or spectroscopic (microwave, infrared) methods. For many molecules, there is general agreement on the nature of the bonding, although there are alternative ways to describe it. For others, there is considerable difference of opinion on the best way to describe the bonding. In this chapter and in Chapter 5, we describe some useful qualitative approaches, including some of the opposing views.

3.1 LEWIS ELECTRON-DOT DIAGRAMS

Lewis electron-dot diagrams, although very much oversimplified, provide a good starting point for analyzing the bonding in molecules. Credit for their initial use goes to G. N. Lewis,[1] an American chemist who contributed much to the understanding of thermodynamics and chemical bonding in the early years of the twentieth century. In Lewis diagrams, bonds between two atoms exist when they share one or more pairs of electrons. In addition, some molecules have nonbonding pairs, also called *lone pairs*, of electrons on atoms. These electrons contribute to the shape and reactivity of the molecule but do not directly bond the atoms together. Most Lewis structures are based on the concept that eight **valence electrons**, corresponding to *s* and *p* electrons outside the noble gas core, form a particularly stable arrangement, as in the noble gases with $s^2 p^6$ configurations. An exception is hydrogen, which is stable with two valence electrons. Also, some molecules require more than eight electrons around a given central atom.

[1] G. N. Lewis, *J. Am. Chem. Soc.*, **1916**, *38*, 762; *Valence and the Structure of Atoms and Molecules*, Chemical Catalogue Co., New York, 1923.

Simple molecules such as water follow the **octet rule**, in which eight electrons surround the central atom. The hydrogen atoms share two electrons each with the oxygen, forming the familiar picture with two bonds, each of which consists of an electron pair and two lone pairs:

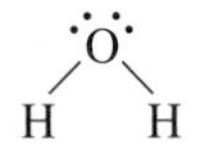

Shared electrons are considered to contribute to the electron requirements of both atoms involved; thus, the electron pairs shared by H and O in the water molecule are counted toward both the 8-electron requirement of oxygen and the 2-electron requirement of hydrogen.

Some bonds are double bonds, containing four electrons, or triple bonds, containing six electrons:

$$:\ddot{O}=C=\ddot{O}: \qquad H-C\equiv C-H$$

3.1.1 Resonance

In many molecules, the choice of which atoms are connected by multiple bonds is arbitrary. When several choices exist, all of them should be drawn. For example, as shown in Figure 3.1, three drawings (resonance structures) of CO_3^{2-} are needed to show the double bond in each of the three possible C—O positions. In fact, experimental evidence shows that all the C—O bonds are identical, with bond lengths (129 pm) between double-bond and single-bond distances (116 pm and 143 pm, respectively). None of the drawings alone is adequate to describe the structure, which is a combination of all three types, not an equilibrium between them. This is called **resonance**, to signify that there is more than one possible way in which the valence electrons can be placed in a Lewis structure. Note that in resonance structures, such as those shown for CO_3^{2-} in Figure 3.1, the electrons are drawn in different places, but the atomic nuclei remain in fixed positions.

The species CO_3^{2-}, NO_3^-, and SO_3 are **isoelectronic**; they have the same electronic structure. Their Lewis diagrams are identical except for the identity of the central atom.

When a molecule has several resonance structures, its overall electronic energy is lowered, making it more stable. Just as the energy levels of a particle in a box are lowered by making the box larger, the electronic energy levels of the bonding electrons are lowered when the electrons can occupy a larger space. The molecular orbital description of this effect is presented in Chapter 5.

3.1.2 Expanded Shells

When it is impossible to draw a structure consistent with the octet rule, it is necessary to increase the number of electrons around the central atom. An option limited to elements of the third and higher periods is to use *d* orbitals for this expansion, although theoretical work suggests that expansion beyond the *s* and *p* orbitals is unnecessary for most main group molecules.[2] In most cases, two or four added electrons will complete the

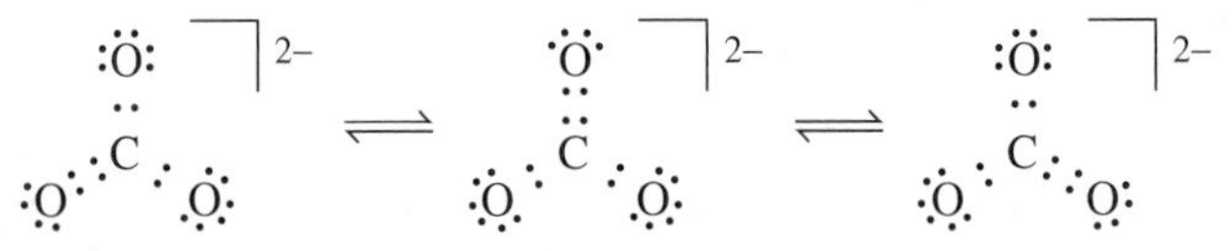

FIGURE 3.1 Lewis Diagrams for CO_3^{2-}.

[2] L. Suidan, J. K. Badenhoop, E. D. Glendening, and F. Weinhold, *J. Chem. Educ.*, **1995**, *72*, 583; J. Cioslowski and S. T. Mixon, *Inorg. Chem.*, **1993**, *32*, 3209; E. Magnusson, *J. Am. Chem. Soc.*, **1990**, *112*, 7940.

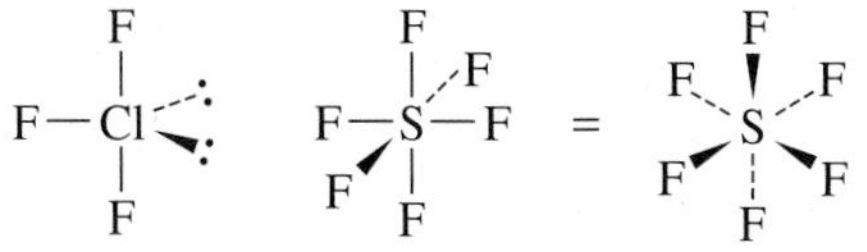

FIGURE 3.2 Structures of ClF_3 and SF_6.

bonding, but more can be added if necessary. For example, 10 electrons are required around chlorine in ClF_3 and 12 around sulfur in SF_6 (Figure 3.2). The increased number of electrons is described as an *expanded shell* or an *expanded electron count*. The term **hypervalent** is sometimes used to describe central atoms that have electron counts greater than the atom's usual requirement.

There are examples with even more electrons around the central atom, such as IF_7 (14 electrons), $[TaF_8]^{3-}$ (16 electrons), and $[XeF_8]^{2-}$ (18 electrons). There are rarely more than 18 electrons (2 for *s*, 6 for *p*, and 10 for *d* orbitals) around a single atom in the top half of the periodic table, and crowding of the outer atoms usually keeps the number below this, even for much heavier atoms that have *f* orbitals energetically available.

3.1.3 Formal Charge

Formal charge is the apparent electronic charge of each atom in a molecule, based on the electron-dot structure. Formal charges can be used to help assess resonance structures and molecular topology, and they are presented here as a simplified method of describing structures, just as the Bohr atom is a simple method of describing electronic configurations in atoms. Both of these methods are approximate, and newer approaches are more accurate, but they can be useful as long as their limitations are kept in mind.

Formal charges can help to understand bonding when there are several possibilities, for example, in eliminating the least likely resonance structures and, in some cases, suggesting multiple bonds beyond those required by the octet rule. It is essential, however, to remember that formal charge is only a tool for assessing Lewis structures, not a measure of any actual charge on the atoms. The number of valence electrons available in a free atom of an element minus the total for that atom in the molecule—determined by counting lone pairs as two electrons and bonding pairs as one electron assigned to each atom—is the formal charge on the atom:

$$\text{Formal charge} = \begin{pmatrix}\text{number of valence}\\ \text{electrons in a free}\\ \text{atom of the element}\end{pmatrix} - \begin{pmatrix}\text{number of unshared}\\ \text{electrons on the atom}\end{pmatrix} - \begin{pmatrix}\text{number of bonds}\\ \text{to the atom}\end{pmatrix}$$

In addition,

$$\text{Charge on molecule or ion} = \text{sum of formal charges}$$

Structures minimizing formal charges, placing negative formal charges on more electronegative elements (in the upper right-hand part of the periodic table), and those with smaller separation of charges tend to be favored. Three examples—SCN^-, OCN^-, and CNO^-—will illustrate the use of formal charges in describing electronic structures.

EXAMPLES

SCN^-

In the thiocyanate ion, SCN^-, three resonance structures are consistent with the electron-dot method, as shown in Figure 3.3. Structure A has only one negative formal charge on the nitrogen atom, the most electronegative atom in the ion, and it fits the rules well. Structure B has a single negative charge on the S, which is less electronegative

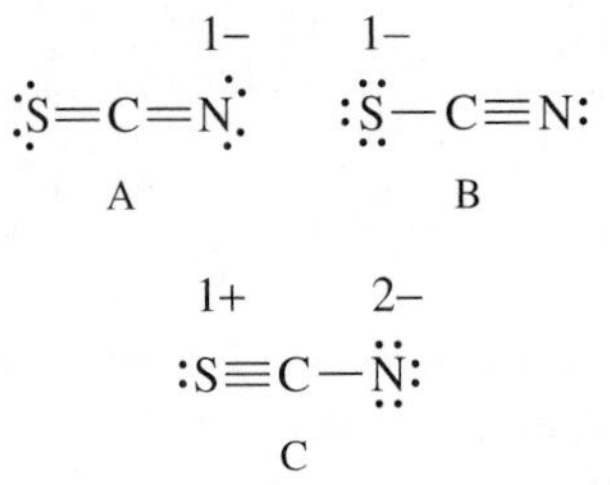

FIGURE 3.3
Resonance Structures of Thiocyanate, SCN^-.

A	B	C
$\ddot{S}=C=\overset{1-}{\ddot{N}}$	$\overset{1-}{\ddot{S}}-C\equiv N$:	$\overset{1+}{:S}\equiv C-\overset{2-}{\ddot{N}}$:

TABLE 3.1 Table of S—C and C—N Bond Lengths (pm)

	S—C	C—N
SCN^- (in NaSCN)	165	118
HNCS	156	122
Single bond	181	147
Double bond	155	128 (approximate)
Triple bond		116

Source: A. F. Wells, *Structural Inorganic Chemistry*, 5th ed., Oxford University Press, New York, 1984, pp. 807, 926, 934–936.

than N. Structure C has charges of 2− on N and 1+ on S, consistent with the relative electronegativities of these atoms but with a larger charge, and greater charge separation, than the others. Therefore these structures lead to the prediction that structure A is most important, structure B is next in importance, and any contribution from C is minor.

The bond lengths in Table 3.1 are somewhat consistent with this conclusion, with bond lengths between those of structures A and B. Protonation of the ion forms HNCS, consistent with a negative charge on N in SCN^-. The bond lengths in HNCS are close to those of double bonds, consistent with the structure H—N=C=S.

OCN^-

The isoelectronic cyanate ion, OCN^- (Figure 3.4), has the same possibilities, but the larger electronegativity of O is expected to make structure B more important than in thiocyanate. The protonated form contains two isomers, 97% HNCO and 3% HOCN, consistent with structure A and with a small contribution from B. The bond lengths in OCN^- and HNCO in Table 3.2 are consistent with this picture but do not agree perfectly. Although formal charge arguments provide a good starting point to assess Lewis structures, reactivity patterns can be more useful in gaining insight about electron distributions.

CNO^-

The isomeric fulminate ion, CNO^- (Figure 3.5), can be drawn with three similar structures, but the resulting formal charges are larger than in the case of OCN^-. Because the order of electronegativities is C < N < O, none of these are satisfactory structures, and

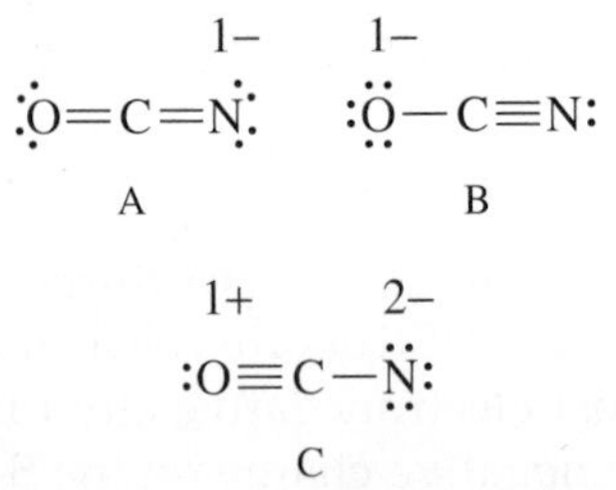

FIGURE 3.4
Resonance Structures of Cyanate, OCN^-.

A	B	C
$\ddot{O}=C=\overset{1-}{\ddot{N}}$	$\overset{1-}{\ddot{O}}-C\equiv N$:	$\overset{1+}{:O}\equiv C-\overset{2-}{\ddot{N}}$:

TABLE 3.2 Table of O—C and C—N Bond Lengths (pm)

	O—C	C—N
OCN^-	126	117
HNCO	118	120
Single bond	143	147
Double bond	116 (CO_2)	128 (approximate)
Triple bond	113 (CO)	116

Source: A. F. Wells, *Structural Inorganic Chemistry*, 5th ed., Oxford University Press, New York, 1984, pp. 807, 926, 933–934; S. E. Bradforth, E. H. Kim, E. W. Arnold, and D. M. Neumark, *J. Chem. Phys.*, **1993**, *98*, 800.

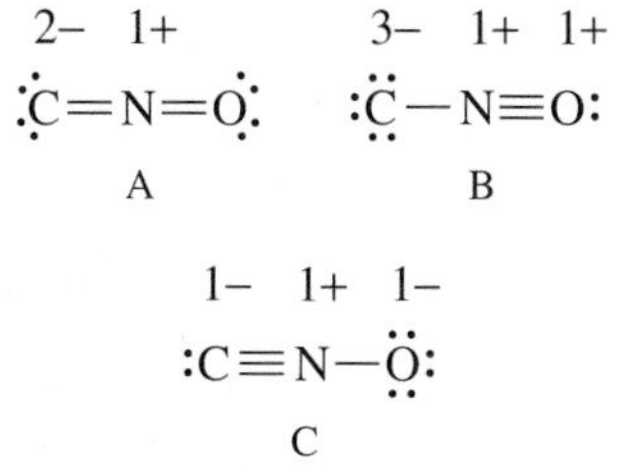

2– 1+
:C=N=O:
A

3– 1+ 1+
:C—N≡O:
B

1– 1+ 1–
:C≡N—O:
C

FIGURE 3.5 Resonance Structures of Fulminate, CNO^-.

the ion is predicted to be unstable. The only common fulminate salts are of mercury and silver; both are explosive. Fulminic acid is linear HCNO in the vapor phase, consistent with structure C; coordination complexes of CNO^- with many transition-metal ions are known with MCNO structures.[3]

▶ **Exercise 3.1** Use electron-dot diagrams and formal charges to find the bond order for each bond in POF_3, SOF_4, and SO_3F^-.

Some molecules have satisfactory electron-dot structures with octets but have better structures with expanded electron counts when formal charges are considered. In each of the cases in Figure 3.6, the observed structures are consistent with electron counts greater than 8 on the central atom and with the resonance structure that uses multiple bonds to minimize formal charges. The multiple bonds may also influence the shapes of the molecules.

3.1.4 Multiple Bonds in Be and B Compounds

A few molecules—such as BeF_2, $BeCl_2$, and BF_3—seem to require multiple bonds to satisfy the octet rule for Be and B, even though we do not usually expect multiple bonds for F and Cl. Structures minimizing formal charges for these molecules have only four electrons in the valence shell of Be and six electrons in the valence shell of B, in both cases fewer than the usual octet. The alternative, requiring eight electrons on the central

[3] A. G. Sharpe, "Cyanides and Fulminates," in G. Wilkinson, R. D. Gillard, and J. S. McClevert, eds., *Comprehensive Coordination Chemistry*, Vol. 2, Pergamon Press, New York, 1987, pp. 12–14.

	Octet			*Expanded*			
Molecule		*Atom*	*Formal Charge*		*Atom*	*Formal Charge*	*Expanded to:*
SNF_3	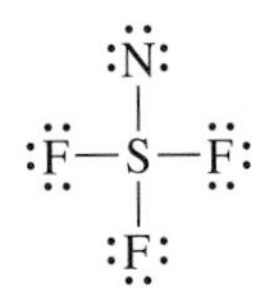	S N	2+ 2−	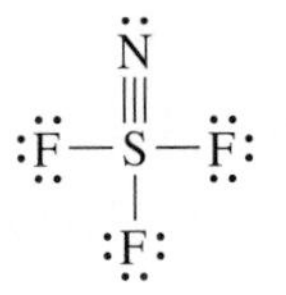	S N	0 0	12
SO_2Cl_2	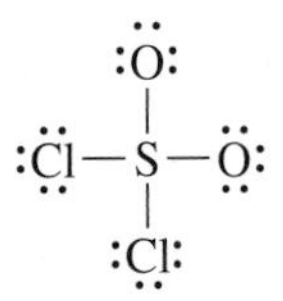	S O	2+ 1−	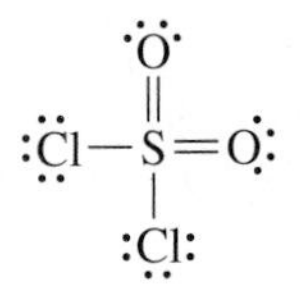	S O	0 0	12
XeO_3	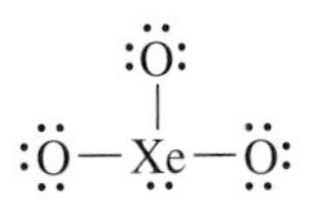	Xe O	3+ 1−	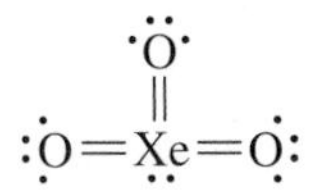	Xe O	0 0	14
SO_4^{2-}	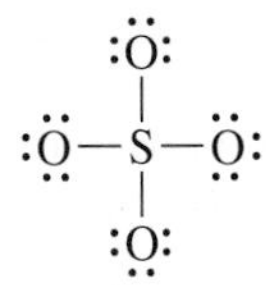	S O	2+ 1−	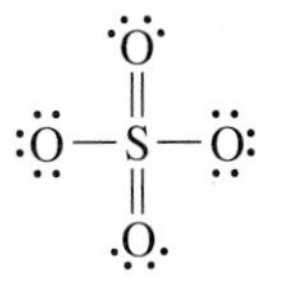	S O	0 0,1−	12
SO_3^{2-}	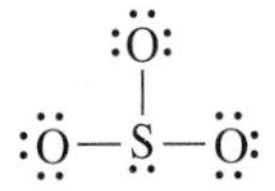	S O	1+ 1−	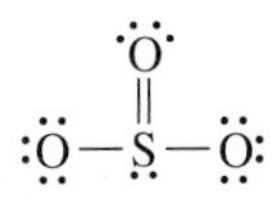	S O	0 0,1−	10

FIGURE 3.6 Formal Charge and Expanded Electron Counts on Central Atom.

atom, predicts multiple bonds, with BeF_2 analogous to CO_2 and BF_3 analogous to SO_3 (Figure 3.7). These structures, however, result in formal charges (2− on Be and 1+ on F in BeF_2, and 1− on B and 1+ on the double-bonded F in BF_3), which are unlikely by the usual rules.

In the solid, a complex network is formed with coordination number 4 for the Be atom (see Figure 3.7). $BeCl_2$ tends to dimerize to a 3-coordinate structure in the vapor phase, but the linear monomer is also known at high temperatures. The monomeric structure is unstable; in the dimer and polymer, the halogen atoms share lone pairs with the Be atom and bring it closer to the octet structure. The monomer is still frequently drawn as a singly bonded structure, with only four electrons around the beryllium and the ability to accept more from lone pairs of other molecules (Lewis acid behavior, discussed in Chapter 6).

Bond lengths in all the boron trihalides are shorter than expected for single bonds, so the partial double-bond character predicted seems reasonable in spite of the formal charges. While a small amount of double bonding is possible in these molecules, the strong polarity of the B–halogen bonds and the ligand close-packing (LCP) model (Section 3.2.4) have been used to account for the short bonds without the need to invoke multiple bonding. The boron trihalides combine readily with

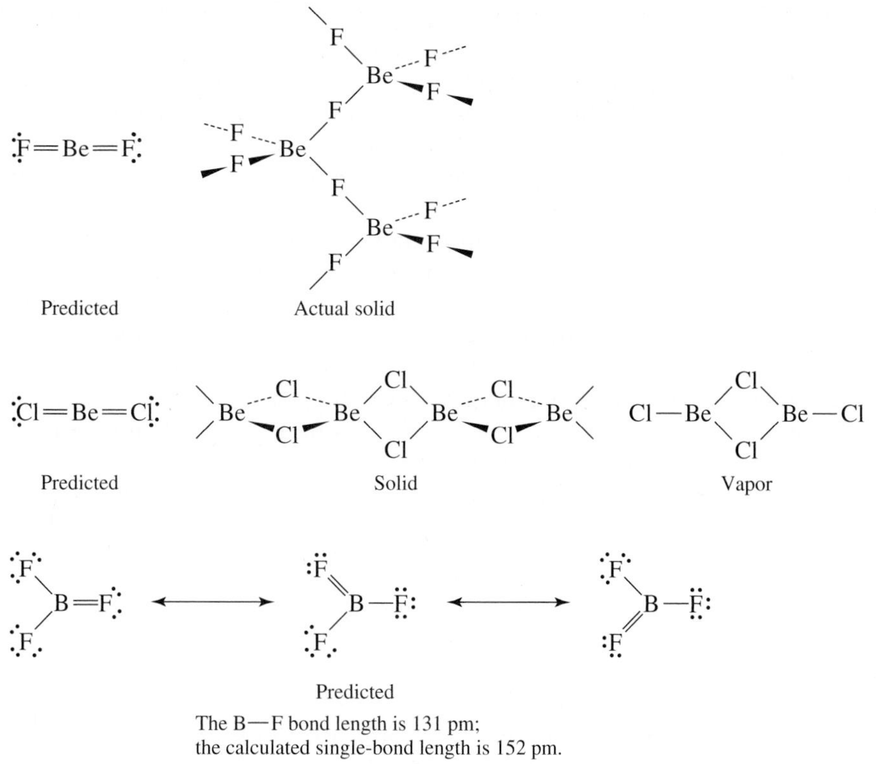

Predicted Actual solid

Predicted Solid Vapor

Predicted

The B—F bond length is 131 pm;
the calculated single-bond length is 152 pm.

FIGURE 3.7 Structures of BeF_2, $BeCl_2$, and BF_3. (A. F. Wells, *Structural Inorganic Chemistry*, 5th ed., Oxford University Press, Oxford, England, 1984, pp. 412, 1047.)

other molecules that can contribute a lone pair of electrons (Lewis bases), forming a roughly tetrahedral structure with four bonds:

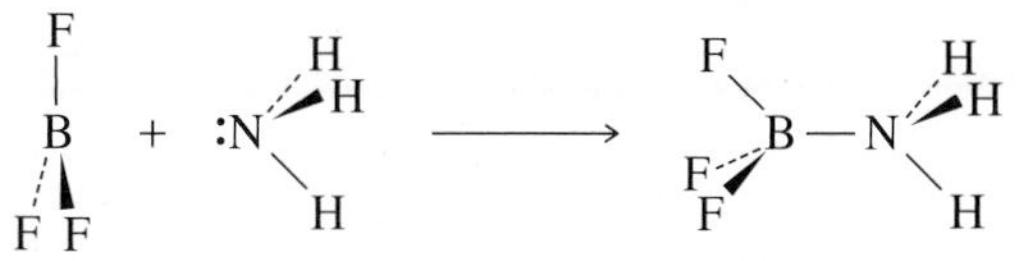

$$BF_3 + :NH_3 \longrightarrow F_3B{-}NH_3$$

Because of this tendency, the boron trihalides are frequently drawn with only six electrons around the boron.

Other boron compounds that do not fit simple electron-dot structures include the hydrides, such as B_2H_6, and a large array of more complex molecules. Their structures are discussed in Chapters 8 and 15.

3.2 VALENCE SHELL ELECTRON-PAIR REPULSION THEORY

Valence shell electron-pair repulsion theory (VSEPR) provides a method for predicting the shape of molecules based on the electron-pair electrostatic repulsion described by Sidgwick and Powell[4] in 1940 and further developed by Gillespie and Nyholm[5] in 1957 and in the succeeding decades. In spite of this method's very simple approach,

[4] N. V. Sidgwick and H. M. Powell, *Proc. R. Soc.*, **1940**, *A176*, 153.

[5] R. J. Gillespie and R. S. Nyholm, *Q. Rev. Chem. Soc.*, **1957**, *XI*, 339; R. J. Gillespie, *J. Chem. Educ.*, **1970**, *47*, 18; and R. J. Gillespie, *Coord. Chem. Rev.*, **2008**, *252*, 1315. The last reference provides a useful synopsis of 50 years of the VSEPR model.

based on Lewis electron-dot structures, the VSEPR method in most cases predicts shapes that compare favorably with those determined experimentally. However, this approach at best provides approximate shapes for molecules. The most common method of determining the actual structures is X-ray diffraction, although electron diffraction, neutron diffraction, and many types of spectroscopy are also used.[6] In Chapter 5, we will provide some of the molecular orbital approaches to describe bonding in simple molecules.

Electrons repel each other because they are negatively charged. Quantum mechanical rules dictate that some of them must be fairly close to each other in bonding pairs or lone pairs, but each pair repels all other pairs. According to the VSEPR model, therefore, molecules adopt geometries such that valence electron pairs position themselves as far from each other as possible to minimize electron–electron repulsions. A molecule can be described by the generic formula AX_mE_n, where *A* is the central atom, *X* stands for any atom or group of atoms surrounding the central atom, and *E* represents a lone pair of electrons. The **steric number**[7] (**SN** = $m + n$) is the number of positions occupied by atoms or lone pairs around a central atom; lone pairs and bonding pairs both influence the molecular shape.

O=C=O

Carbon dioxide is an example of a molecule with two bonding positions (SN = 2) on the central atom and double bonds in each direction. The electrons in each double bond must be between C and O, and the repulsion between the electrons in the double bonds forces a linear structure on the molecule. Sulfur trioxide has three bonding positions (SN = 3), with partial double-bond character in each. The best positions for the oxygens to minimize electron–electron repulsions in this molecule are at the corners of an equilateral triangle, with O—S—O bond angles of 120°. The multiple bonding does not affect the geometry, because it is shared equally among the three bonds.

The same pattern of finding the Lewis structure and then matching it to a geometry that minimizes the repulsive energy of bonding electrons is followed through steric numbers 4, 5, 6, 7, and 8, as shown in Figure 3.8.

The structures for two, three, four, and six electron pairs are completely regular, with all bond angles and distances the same. Neither 5- nor 7-coordinate structures can have uniform angles and distances, because there are no regular polyhedra with these numbers of vertices. The 5-coordinate molecules have a trigonal bipyramidal structure, with a central triangular plane of three positions plus two other positions above and below the center of the plane. The 7-coordinate molecules have a pentagonal bipyramidal structure, with a pentagonal plane of five positions and positions above and below the center of the plane. The regular square antiprism structure (SN = 8) is like a cube that has had the top face twisted 45° into the antiprism arrangement, as shown in Figure 3.9. It has three different bond angles for adjacent fluorines. $[TaF_8]^{3-}$ has square antiprism geometry but is distorted from this ideal in the solid.[8] (A simple cube has 109.5°, 99.6°, and 70.5° bond angles, measured between two corners and the center of the cube, and any square face can be taken as the bottom or top.)

[6] G. M. Barrow, *Physical Chemistry*, 6th ed., McGraw-Hill, New York, 1988, pp. 567–699; R. S. Drago, *Physical Methods for Chemists*, 2nd ed., Saunders College Publishing, Philadelphia, 1977, pp. 689–711.
[7] The *steric number* is also called the *number of electron pair domains*.
[8] J. L. Hoard, W. J. Martin, M. E. Smith, and J. F. Whitney, *J. Am. Chem. Soc.*, **1954**, *76*, 3820.

Steric Number	Geometry	Examples	Calculated Bond Angles	
2	Linear	CO_2	180°	O=C=O
3	Trigonal (triangular)	SO_3	120°	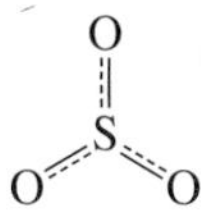
4	Tetrahedral	CH_4	109.5°	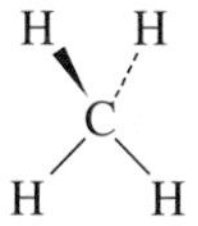
5	Trigonal bipyramidal	PCl_5	120°, 90°	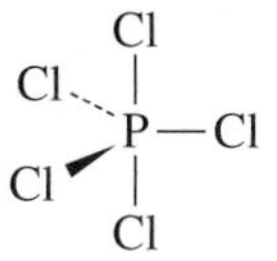
6	Octahedral	SF_6	90°	
7	Pentagonal bipyramidal	IF_7	72°, 90°	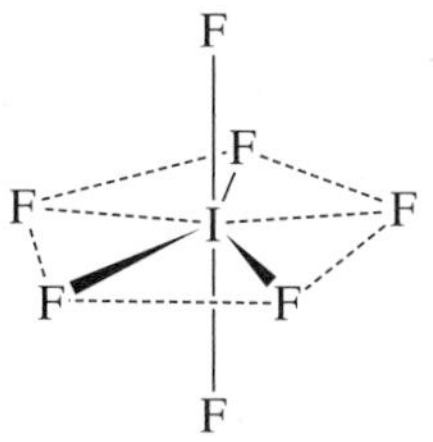
8	Square antiprismatic	TaF_8^{3-}	70.5°, 99.6°, 109.5°	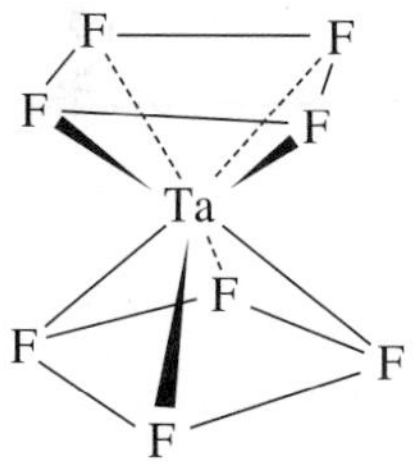

FIGURE 3.8 VSEPR Predictions.

3.2.1 Lone-Pair Repulsion

We must keep in mind the need to match our explanations to experimental data. New theories are continually being suggested and tested. Because we are working with such a wide variety of atoms and molecular structures, it is unlikely that a single, simple approach will work for all of them. Although the fundamental ideas of atomic and molecular structures are relatively simple, their application to complex molecules is not. To a first approximation, lone pairs, single bonds, double bonds, and triple bonds can all be treated similarly when predicting molecular shapes. However, better predictions of overall shapes can be made by considering some important differences between lone pairs and bonding pairs. These methods are sufficient to show the trends and explain the bonding, as in explaining why the H—N—H angle in ammonia is smaller than the tetrahedral angle in methane and larger than the H—O—H angle in water.

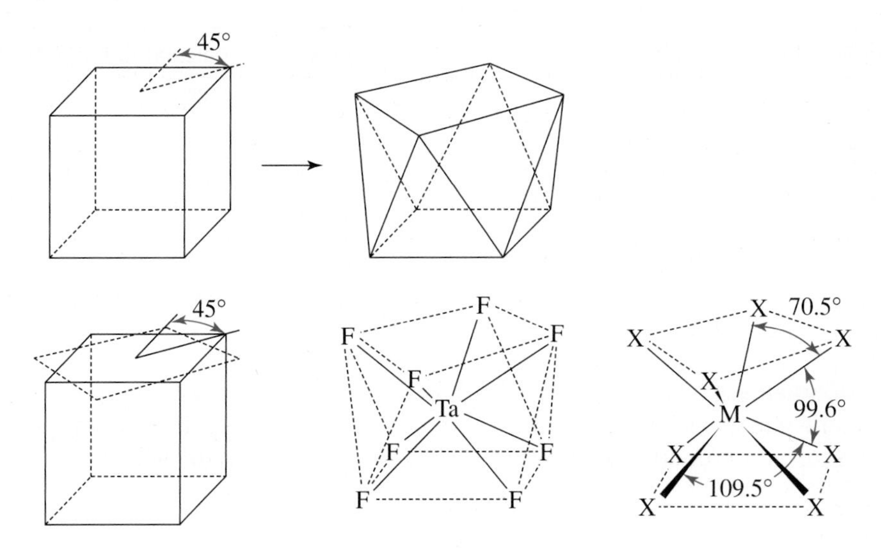

FIGURE 3.9 Conversion of a Cube into a Square Antiprism.

As a general guideline, the VSEPR model predicts that electron-pair repulsions involving lone pairs (*lp*) are stronger than those involving bonding pairs (*bp*) in the order

$$lp{-}lp \text{ repulsions} > lp{-}bp \text{ repulsions} > bp{-}bp \text{ repulsions}$$

STERIC NUMBER = 4 The isoelectronic molecules CH_4, NH_3, and H_2O (Figure 3.10) illustrate the effect of lone pairs on molecular shape. Methane has four identical bonds between carbon and each of the hydrogens. When the four pairs of electrons are arranged as far from each other as possible, the result is the familiar tetrahedral shape. The tetrahedron, with all H—C—H angles measuring 109.5°, has four identical bonds.

Ammonia also has four pairs of electrons around the central atom, but three are bonding pairs between N and H, and the fourth is a lone pair on the nitrogen. The nuclei form a trigonal pyramid with the three bonding pairs; with the lone pair, they make a nearly tetrahedral shape. Because each of the three bonding pairs is attracted by two positively charged nuclei (H and N), these pairs are largely confined to the regions between the H and N atoms. The lone pair, on the other hand, is concentrated near the nitrogen; it has no second nucleus to confine it to a small region of space. Consequently, the lone pair tends to spread out and to occupy more space around the nitrogen than the bonding pairs. As a result, the H—N—H angles are 106.6°, nearly 3° smaller than the angles in methane.

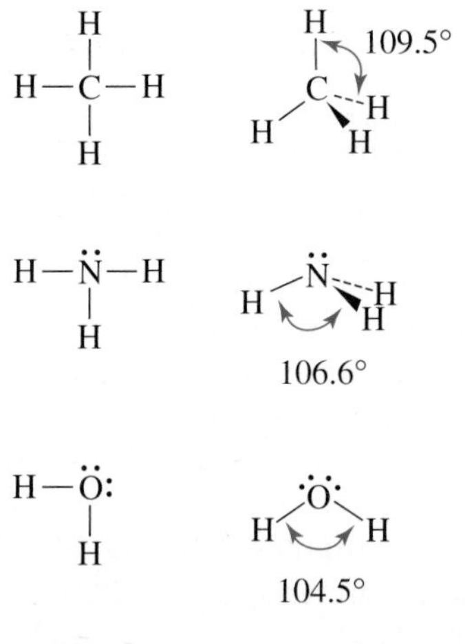

FIGURE 3.10 Shapes of Methane, Ammonia, and Water.

109.5°

106.6°

104.5°

The same principles apply to the water molecule, in which two lone pairs and two bonding pairs repel each other. Again, the electron pairs have a nearly tetrahedral arrangement, with the atoms arranged in a V shape. The angle of largest repulsion, between the two lone pairs, is not directly measurable. However, the lone pair–bonding pair (*lp–bp*) repulsion is greater than the bonding pair–bonding pair (*bp–bp*) repulsion; as a result, the H—O—H bond angle is only 104.5°, another 2.1° decrease from the ammonia angles. The net result is that we can predict approximate molecular shapes by assigning more space to lone electron pairs; being attracted to one nucleus rather than two, the lone pairs are able to spread out and occupy more space.

STERIC NUMBER = 5 For trigonal bipyramidal geometry, there are two possible locations of lone pairs, axial and equatorial. If there is a single lone pair, for example in SF_4, the lone pair occupies an equatorial position. This position provides the lone pair with the most space and minimizes the interactions between the lone pair and bonding pairs. If the lone pair were axial, it would have three 90° interactions with bonding pairs; in an equatorial position, it has only two such interactions, as shown in Figure 3.11. The actual structure is distorted by the lone pair as it spreads out in space and effectively squeezes the rest of the molecule together.

ClF_3 provides a second example of the influence of lone pairs in molecules having a steric number of 5. There are three possible structures for ClF_3, as shown in Figure 3.12.

In determining the structure of molecules, the lone pair–lone pair interactions are most important, with the lone pair–bonding pair interactions next in importance. In addition, interactions at angles of 90° or less are most important; larger angles generally have less influence. In ClF_3, structure B can be eliminated quickly because of the 90° *lp–lp* angle. The *lp–lp* angles are large for A and C, so the choice must come from the *lp–bp* and *bp–bp* angles. Because the *lp–bp* angles are more important, C, which has only four 90° *lp–bp* interactions, is favored over A, which has six such interactions. Experiments have confirmed that the structure is based on C, with slight distortions due to the lone pairs. The lone pair–bonding pair repulsion causes the *lp–bp* angles to be larger than 90° and the *bp–bp* angles to be less than 90° (actually,

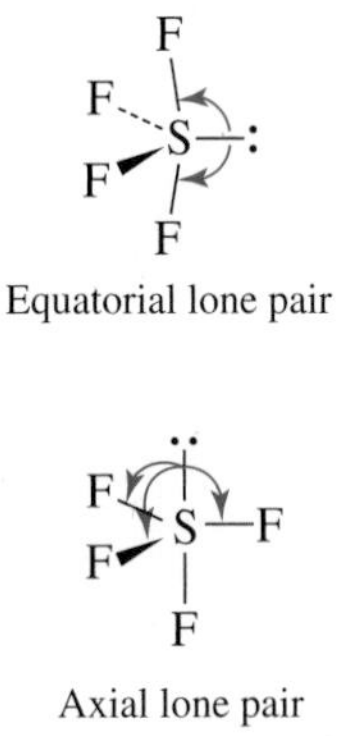

Equatorial lone pair

Axial lone pair

FIGURE 3.11 Structure of SF_4.

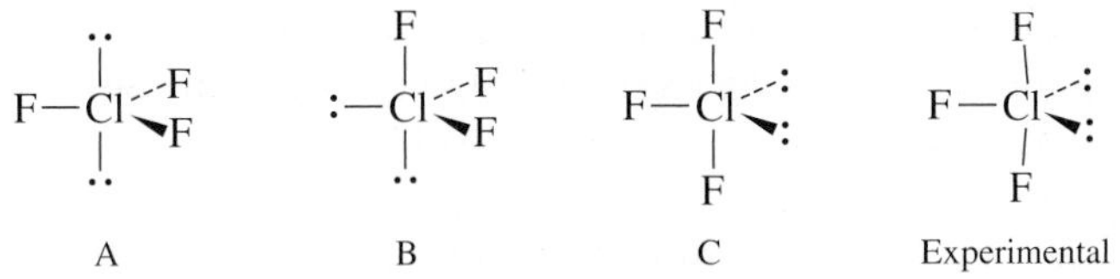

A B C Experimental

Interaction	Calculated A	Calculated B	Calculated C	Experimental
lp-lp	180°	90°	120°	Cannot be determined
lp-bp	6 at 90°	3 at 90°	4 at 90°	Cannot be determined
		2 at 120°	2 at 120°	
bp-bp	3 at 120°	2 at 90°	2 at 90°	2 at 87.5°
		1 at 120°		Axial Cl—F 169.8 pm
				Equatorial Cl—F 159.8 pm

FIGURE 3.12 Possible Structures of ClF_3.

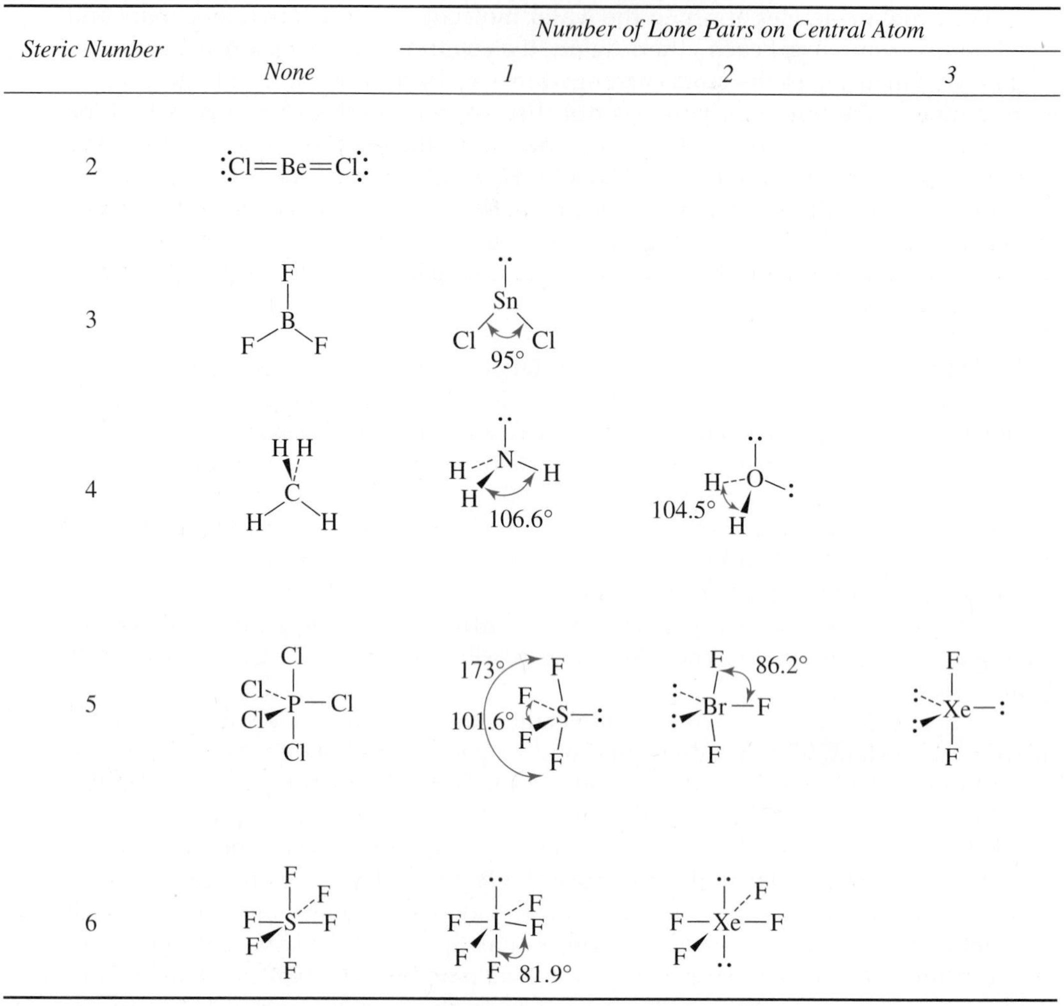

Steric Number	*None*	*Number of Lone Pairs on Central Atom* 1	2	3
2	:Cl=Be=Cl:			
3	F–B(F)–F	Cl–Sn(:)–Cl 95°		
4	H, H, C, H, H	H–N(:)–H, H 106.6°	H–O(:)–:, H 104.5°	
5	Cl, Cl, P–Cl, Cl, Cl	173°, F, F, S–:, F, F, 101.6°	F 86.2°, :Br–F, :, F	F, :Xe–:, :, F
6	F, F, F–S–F, F, F	F–I(:)–F, F, F, F 81.9°	F–Xe(:)–F, F, F, :	

FIGURE 3.13 Structures Containing Lone Pairs.

87.5°). The Cl—F bond distances show the repulsive effects as well, with the axial fluorines (approximately 90° *lp–bp* angles) at 169.8 pm and the equatorial fluorine (in the plane with two lone pairs) at 159.8 pm.[9] Angles involving lone pairs cannot be determined experimentally.

Additional examples of structures with lone pairs are illustrated in Figure 3.13. Notice that the structures based on a trigonal bipyramidal arrangement of electron pairs around a central atom always place any lone pairs in the equatorial plane, as in SF_4, BrF_3, and XeF_2. These are the shapes that minimize both lone pair–lone pair and lone pair–bonding pair repulsions. The shapes are called *seesaw* (SF_4), *distorted T* (BrF_3), and *linear* (XeF_2).

STERIC NUMBERS = 6 AND 7 In octahedral structures, all six positions are equivalent. When a single lone pair is present, it typically causes the opposite site of the molecule to become more compressed, reducing bond angles accordingly, as for IF_5 in Figure 3.13. In octahedron-based structures with two lone pairs, the lone pair–lone pair repulsion is

[9] A. F. Wells, *Structural Inorganic Chemistry*, 5th ed., Oxford University Press, New York, 1984, p. 390.

minimized if these pairs are *trans*, and this is the shape that is adopted. Square planar XeF_4, also shown in Figure 3.13, is an example. Apparently no structures having a steric number of 6 and three lone pairs have been reported.

The shape that minimizes electron-pair repulsions for a steric number of 7 is the pentagonal bipyramid, shown in Figure 3.8. IF_7 and TeF_7^{2-} are examples of this shape, with both axial and equatorial fluorines. If a single lone pair is present, in some cases the lone pair causes distortion (the nature of which is not always easy to ascertain; XeF_6 is a classic example[10]). In other cases the structure is octahedral (see problem 3.23). Two lone pairs minimize their repulsions by adopting axial (*trans*) positions, with the atoms all in the equatorial plane. The two known examples are XeF_5^- and IF_5^{2-}.

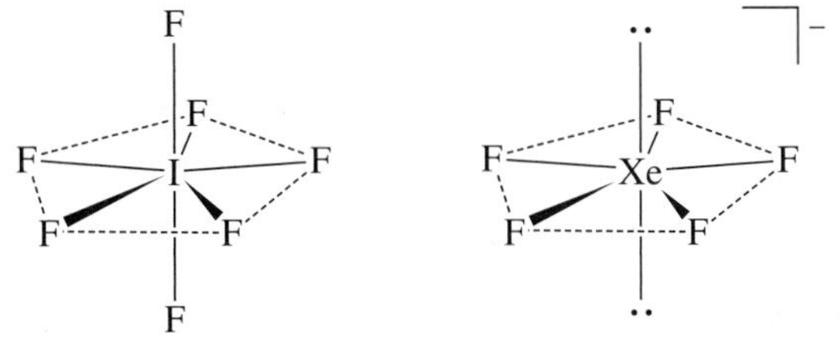

EXAMPLES

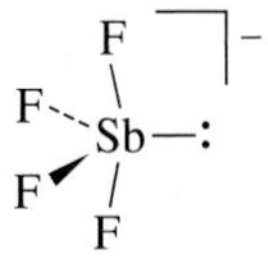

SbF_4^- has a single lone pair on Sb. Its structure is therefore similar to SF_4, with a lone pair occupying an equatorial position. This lone pair causes considerable distortion, giving an F—Sb—F (axial positions) angle of 155° and an F—Sb—F (equatorial) angle of 90°.

SF_5^- has a single lone pair. Its structure is based on an octahedron, with the ion distorted away from the lone pair, as in IF_5.

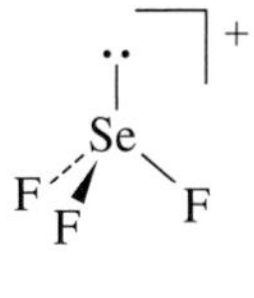

SeF_3^+ has a single lone pair. This lone pair reduces the F—Se—F bond angle significantly, to 94°.

▶ **Exercise 3.2** Predict the structures of the following ions. Include a description of distortions from the ideal angles (for example, less than 109.5° because . . .).

NH_2^- NH_4^+ I_3^- PCl_6^-

3.2.2 Multiple Bonds

The VSEPR model considers double and triple bonds to have slightly greater repulsive effects than single bonds because of the repulsive effect of π electrons. For example, the $H_3C—C—CH_3$ angle in $(CH_3)_2C{=}CH_2$ is smaller, and the $H_3C—C{=}CH_2$ angle is larger than the trigonal 120° (Figure 3.14).[11]

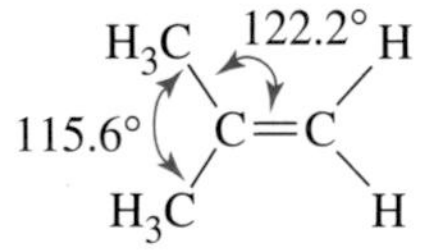

Additional examples of the effect of multiple bonds on molecular geometry are shown in Figure 3.15. Comparing Figures 3.14 and 3.15, we see that multiple bonds tend to occupy the same positions as lone pairs. For example, the double bonds to oxygen in SOF_4, ClO_2F_3, and XeO_3F_2 are all equatorial, as are the lone pairs in the

FIGURE 3.14 Bond Angles in $(CH_3)_2C{=}CH_2$.

[10] See, for example, S. Hoyer, T. Emmler, and K. Seppelt, *J. Fluorine Chem.*, **2006**, *127*, 1415, and references therein on the various structural modifications of XeF_6.

[11] R. J. Gillespie and I. Hargittai, *The VSEPR Model of Molecular Geometry*, Allyn & Bacon, Boston, 1991, p. 77.

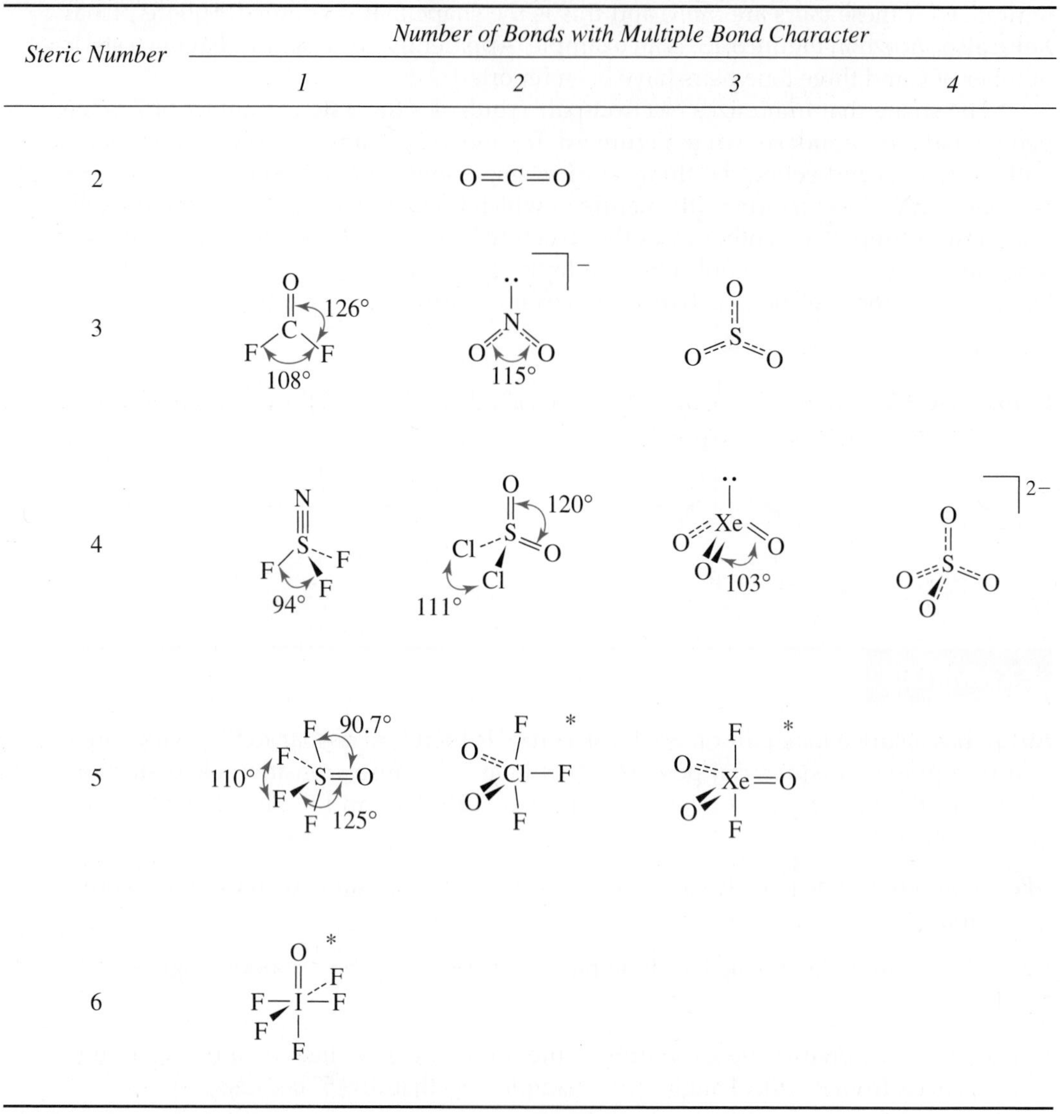

Steric Number	*Number of Bonds with Multiple Bond Character*			
	1	*2*	*3*	*4*
2		O=C=O		
3	O, C, F, F, 126°, 108°	N, O, O, 115° (anion, −)	O, S, O, O	
4	N, S, F, F, F, 94°	O, S, Cl, Cl, O, 120°, 111°	Xe, O, O, O, 103°	O, S, O, O, O (2−)
5	F, F, F, F, S=O, 90.7°, 110°, 125°	F, O, Cl—F, O, F *	F, O, Xe=O, O, F *	
6	O, F, F—I—F, F, F *			

* The bond angles of these molecules have not been determined accurately. However, spectroscopic measurements are consistent with the structures shown.

FIGURE 3.15 Structures Containing Multiple Bonds.

matching compounds of steric number 5, SF_4, BrF_3, and XeF_2. Also, multiple bonds, like lone pairs, tend to occupy more space than single bonds and to cause distortions that in effect squeeze the rest of the molecule together. In molecules that have both lone pairs and multiple bonds, these features may compete for space; examples are shown in Figure 3.16.

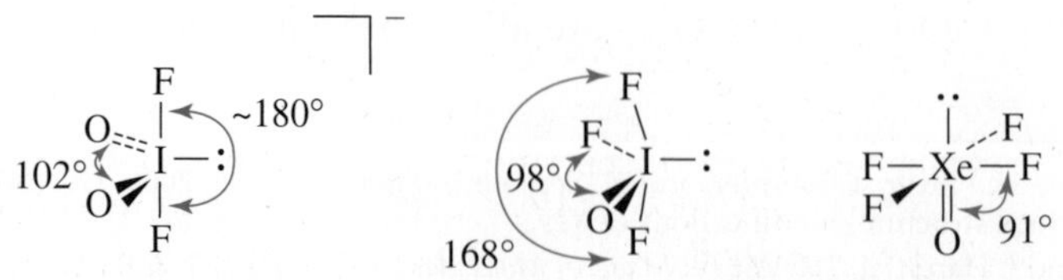

FIGURE 3.16 Structures Containing Both Lone Pairs and Multiple Bonds.

F, O, I, O, F, ~180°, 102° (anion, −)	F, F, I, O, F, 98°, 168°	F, F—Xe—F, F, O, 91°

EXAMPLES

HCP, like HCN, is linear, with a triple bond: H—C≡P.

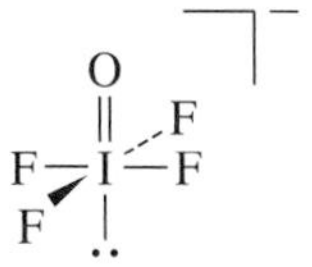

$\mathbf{IOF_4}^-$ has a single lone pair on the side opposite the oxygen. The lone pair has a slightly greater repulsive effect than the double bond to oxygen, as shown by the average O—I—F angle of 89°. (The extra repulsive character of the I═O bond places it opposite the lone pair.)

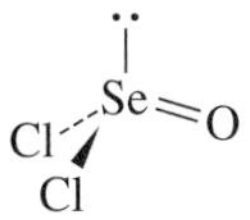

$\mathbf{SeOCl_2}$ has both a lone pair and double bonding to the oxygen. The lone pair has a greater effect than the double bond; the Cl—Se—Cl angle is reduced to 97° by this effect, and the Cl—Se—O angle is 106°.

▶ **Exercise 3.3** Predict the structures of the following. Indicate the direction of distortions from the regular structures.

$XeOF_2$ $ClOF_3$ $SOCl_2$

3.2.3 Electronegativity and Atomic Size Effects

Electronegativity was mentioned earlier as a guide in the use of formal charges. It also can play an important role in determining the arrangement of outer atoms around a central atom and in influencing bond angles. The effects of electronegativity and atomic size frequently parallel each other, but in some cases, the sizes of outer atoms and groups may play the more important role.

ELECTRONEGATIVITY SCALES The concept of electronegativity was first introduced by Linus Pauling in the 1930s as a means of describing bond energies. Bond energies of polar bonds (formed by atoms with different electronegativities) are larger than the average of the bond energies of the two homonuclear species. For example, HCl has a bond energy of 428 kJ/mol, compared to a calculated value of 336 kJ/mol, the average of the bond energies of H_2 (432 kJ/mol) and Cl_2 (240 kJ/mol). From data like these, Pauling calculated electronegativity values that could be used to predict other bond energies, using an arbitrary scale based on a value of 4.0 assigned as the electronegativity of fluorine. More recent values have been derived from other molecular properties and from atomic properties, such as ionization energy and electron affinity. Regardless of the method of calculation, the scale used is usually adjusted to give values near those of Pauling's to enable better comparison. Table 3.3 summarizes approaches used for determining different scales. We will use the values determined by Allen and colleagues; electronegativities of main group elements are in Table 3.4, and a more complete table of values is in Appendix B.4. A graphic representation of electronegativity is in Figure 8.2.

Pauling's calculation of electronegativities from bond energies requires averaging over a number of compounds to cancel out experimental uncertainties and other minor effects. Methods that use ionization energies and other atomic properties can be calculated more directly. The electronegativities reported here and in Appendix B.4 are suitable for most uses, but the actual values for atoms in specific molecules may differ from these values, depending on their electronic environment.

Many of those interested in electronegativity agree that it depends on the structure of the molecule as well as the atom. Jaffé used this idea to develop a theory of the electronegativity of *orbitals* rather than *atoms*. Such theories are useful in detailed calculations of properties that change with subtle changes in structure, but they are beyond

TABLE 3.3 Electronegativity Scales

Principal Authors	Method of Calculation or Description
Pauling[12]	Bond energies
Mulliken[13]	Average of electron affinity and ionization energy
Allred & Rochow[14]	Electrostatic attraction proportional to Z^*/r^2
Sanderson[15]	Electron densities of atoms
Pearson[16]	Average of electron affinity and ionization energy
Allen[17]	Average energy of valence shell electrons, configuration energies
Jaffé[18]	Orbital electronegativities

TABLE 3.4 Electronegativity (Pauling Units)

1	2	12	13	14	15	16	17	18
H 2.300								He 4.160
Li 0.912	Be 1.576		B 2.051	C 2.544	N 3.066	O 3.610	F 4.193	Ne 4.787
Na 0.869	Mg 1.293		Al 1.613	Si 1.916	P 2.253	S 2.589	Cl 2.869	Ar 3.242
K 0.734	Ca 1.034	Zn 1.588	Ga 1.756	Ge 1.994	As 2.211	Se 2.424	Br 2.685	Kr 2.966
Rb 0.706	Sr 0.963	Cd 1.521	In 1.656	Sn 1.824	Sb 1.984	Te 2.158	I 2.359	Xe 2.582
Cs 0.659	Ba 0.881	Hg 1.765	Tl 1.789	Pb 1.854	Bi (2.01)	Po (2.19)	At (2.39)	Rn (2.60)

Source: J. B. Mann, T. L. Meek, and L. C. Allen, *J. Am. Chem. Soc.*, **2000**, *122*, 2780, Table 2.

[12] L. Pauling, *The Nature of the Chemical Bond*, 3rd ed., 1960, Cornell University Press, Ithaca, NY; A. L. Allred, *J. Inorg. Nucl. Chem.*, **1961**, *17*, 215.
[13] R. S. Mulliken, *J. Chem. Phys.*, **1934**, *2*, 782; **1935**, *3*, 573; W. Moffitt, *Proc. R. Soc. (London)*, **1950**, *A202*, 548; R. G. Parr, R. A. Donnelly, M. Levy, and W. E. Palke, *J. Chem. Phys.*, **1978**, *68*, 3801; R. G. Pearson, *Inorg. Chem.*, **1988**, *27*, 734; S. G. Bratsch, *J. Chem. Educ.*, **1988**, *65*, 34, 223.
[14] A. L. Allred and E. G. Rochow, *J. Inorg. Nucl. Chem.*, **1958**, *5*, 264.
[15] R. T. Sanderson, *J. Chem. Educ.*, **1952**, *29*, 539; **1954**, *31*, 2, 238; *Inorganic Chemistry*, Van Nostrand-Reinhold, New York, 1967.
[16] R. G. Pearson, *Acc. Chem. Res.*, **1990**, *23*, 1.
[17] L. C. Allen, *J. Am. Chem. Soc.*, **1989**, *111*, 9003; J. B. Mann, T. L. Meek, and L. C. Allen, *J. Am. Chem. Soc.*, **2000**, *122*, 2780; J. B. Mann, T. L. Meek, E. T. Knight, J. F. Capitani, and L. C. Allen, *J. Am. Chem. Soc.*, **2000**, *122*, 5132.
[18] J. Hinze and H. H. Jaffé, *J. Am. Chem. Soc.*, **1962**, *84*, 540; *J. Phys. Chem.*, **1963**, *67*, 1501; J. E. Huheey, *Inorganic Chemistry*, 3rd ed., Harper & Row, New York, 1983, pp. 152–156.

the scope of this text. In most cases the different methods give similar values of electronegativity, except for those of the transition metals.[19]

It is important to emphasize that all electronegativities are measures of an atom's ability to attract electrons from a neighboring atom *to which it is bonded*. A critique of all electronegativity scales, and particularly Pauling's, describes conditions that all scales should meet and deficiencies of the various methods.[20]

With the exception of helium and neon, which have large calculated electronegativities and no known stable compounds, fluorine has the largest value, and electronegativity decreases toward the lower left corner of the periodic table. Although usually classified with Group 1 (IA), hydrogen is quite dissimilar from the alkali metals in its electronegativity, as well as in many other properties, both chemical and physical. Hydrogen's chemistry is distinctive from all the groups.

Electronegativities of the noble gases can be calculated more easily from ionization energies than from bond energies. Because the noble gases have higher ionization energies than the halogens, calculations have suggested that the electronegativities of the noble gases may exceed those of the halogens[21] (Table 3.4). The noble gas atoms are somewhat smaller than the neighboring halogen atoms—for example, Ne is smaller than F—as a consequence of a greater effective nuclear charge. This charge, which is able to attract noble gas electrons strongly toward the nucleus, is also likely to exert a strong attraction on electrons of neighboring atoms; hence, the high electronegativities predicted for the noble gases are reasonable.

ELECTRONEGATIVITY AND BOND ANGLES By the VSEPR approach, trends in many bond angles can be explained by electronegativity. Consider the bond angles in the following molecules:

Molecule	X–P–X Angle (°)	Molecule	X–S–X Angle (°)
PF_3	97.8	OSF_2	92.3
PCl_3	100.3	$OSCl_2$	96.2
PBr_3	101	$OSBr_2$	98.2

As the electronegativity of the halogen increases, the halogen exerts a stronger pull on electron pairs it shares with the central atom. This effect reduces the concentration of electrons near the central atom, allowing the lone pair to spread out and reducing the halogen–central atom–halogen angles. Consequently, the molecules with the most electronegative *outer* atoms, PF_3 and OSF_2, have the smallest angles.

If the central atom remains the same, molecules that have a larger difference in electronegativity values between their central and outer atoms have smaller bond angles. The atom with larger electronegativity draws the shared electrons toward itself and away from the central atom, reducing the repulsive effect of these electrons. The compounds of the halogens in Table 3.5 show this effect; the compounds containing fluorine have smaller angles than those containing chlorine, which in turn have smaller

[19] J. B. Mann, T. L. Meek, E. T. Knight, J. F. Capitani, and L. C. Allen, *J. Am. Chem. Soc.*, **2000**, *122*, 5132.
[20] L. R. Murphy, T. L. Meek, A. L. Allred, and L. C. Allen, *J. Am. Chem. Soc.*, **2000**, *122*, 5867.
[21] L. C. Allen and J. E. Huheey, *J. Inorg. Nucl. Chem.*, **1980**, *42*, 1523.

TABLE 3.5 Bond Angles and Lengths

Molecule	Bond Angle (°)	Bond Length (pm)	Molecule	Bond Angle (°)	Bond Length (pm)	Molecule	Bond Angle (°)	Bond Length (pm)	Molecule	Bond Angle (°)	Bond Length (pm)
H_2O	104.5	97	OF_2	103.3	96	OCl_2	110.9	170			
H_2S	92.1	135	SF_2	98.0	159	SCl_2	102.7	201			
H_2Se	90.6	146				$SeCl_2$	99.6	216			
H_2Te	90.2	169				$TeCl_2$	97.0	233			
NH_3	106.6	101.5	NF_3	102.2	137	NCl_3	106.8	175			
PH_3	93.2	142	PF_3	97.8	157	PCl_3	100.3	204	PBr_3	101.0	220
AsH_3	92.1	151.9	AsF_3	95.8	170.6	$AsCl_3$	98.9	217	$AsBr_3$	99.8	236
SbH_3	91.6	170.7	SbF_3	87.3	192	$SbCl_3$	97.2	233	$SbBr_3$	98.2	249

Source: N. N. Greenwood and A. Earnshaw, *Chemistry of the Elements*, 2nd ed., Butterworth-Heinemann, Oxford, 1997, pp. 557, 767; A. F. Wells, *Structural Inorganic Chemistry*, 5th ed., Oxford University Press, Oxford, 1987, pp. 705, 793, 846, and 879; R. J. Gillespie and I. Hargittai, *The VSEPR Model of Molecular Geometry*, Allyn and Bacon, Needham Heights, MA, 1991.

angles than those containing bromine or iodine. As a result, the lone pair effect is relatively larger and forces smaller bond angles.

Additional examples of halogen compounds showing the effects of electronegativity on bond angles are given in Table 3.5. In each case, the compounds with fluorine as an outer atom have smaller angles than those with chlorine, which in turn have smaller angles than those of bromine (and iodine, not shown). It is worth noting that this trend is also obtained if size is considered: as the size of the outer atom increases in the order $F < Cl < Br < I$, the bond angle increases.

Similar considerations can be made in situations where the outer atoms remain the same, but the central atom is changed, for example:

Molecule	Bond Angle (°)	Molecule	Bond Angle (°)
H_2O	104.5	NCl_3	106.8
H_2S	92.1	PCl_3	100.3
H_2Se	90.6	$AsCl_3$	98.9

In these cases, as the central atom becomes more electronegative, it pulls electrons in bonding pairs more strongly toward itself, increasing the concentration of electrons near the central atom. The net effect is that an increase in bonding pair–bonding pair repulsions near the central atom increases the bond angles. In these situations the molecule with the most electronegative *central* atom has the largest bond angles. Additional examples can be found in Table 3.5, where molecules having the same outer atoms, but different central atoms, are shown in the same column.

▶ **Exercise 3.4** Which molecule has the smallest bond angle in each series?

a. $OSeF_2$ $OSeCl_2$ $OSeBr_2$ (halogen–Se–halogen angle)
b. $SbCl_3$ $SbBr_3$ SbI_3
c. PI_3 AsI_3 SbI_3

EFFECTS OF SIZE In the examples considered so far, electronegativity and size have gone hand in hand; the most electronegative atoms have also been the smallest. For example, the smallest halogen, fluorine, is also the most electronegative. Consequently, we could have predicted the trends in bond angles on the basis of atomic size, with the smallest atoms capable of being crowded together most closely. In this connection, it is useful to consider situations in which size and electronegativity might have opposite effects, where a smaller outer group is *less* electronegative than a larger group attached to a central atom. For example:

Molecule	C–N–C Angle (°)
$N(CH_3)_3$	110.9
$N(CF_3)_3$	117.9

In this case VSEPR alone would predict that the more electronegative CF_3 groups would lead to a smaller bond angle because they would withdraw electrons more strongly than CH_3 groups. That the bond angle in $N(CF_3)_3$ is actually 7° larger than in $N(CH_3)_3$ suggests that in this case, size is the more important factor, with the larger CF_3 groups requiring more space. The point at which the size of outer atoms and groups becomes more important than electronegativity can be difficult to predict, but the potential of large outer atoms and groups to affect molecular shape should not be dismissed.

MOLECULES HAVING STERIC NUMBER = 5 For main group central atoms having a steric number of 5, it is instructive to consider the relative bond lengths for axial and equatorial positions. For example, in PCl_5, SF_4, and ClF_3, the central atom–axial distances are longer than the distances to equatorial atoms, as shown in Figure 3.17. This effect has been attributed to the greater repulsion of lone and bonding pairs with atoms in axial positions (three 90° interactions) than with atoms in equatorial positions (two 90° interactions).

In addition, there is a tendency for less electronegative groups to occupy equatorial positions, similar to lone pairs and multiply bonded atoms. For example, in phosphorus compounds having both fluorine and chlorine atoms, in each case the chlorines occupy equatorial positions (Figure 3.18). The same tendency is shown in compounds having formulas PF_4CH_3, $PF_3(CH_3)_2$, and $PF_2(CH_3)_3$, with the less electronegative CH_3 groups also equatorial (Figure 3.19). One can envision the P—less electronegative atom bond being concentrated closer to the phosphorus in such cases, leading to a preference for equatorial positions by similar reasoning applied to lone pairs and multiple bonds.

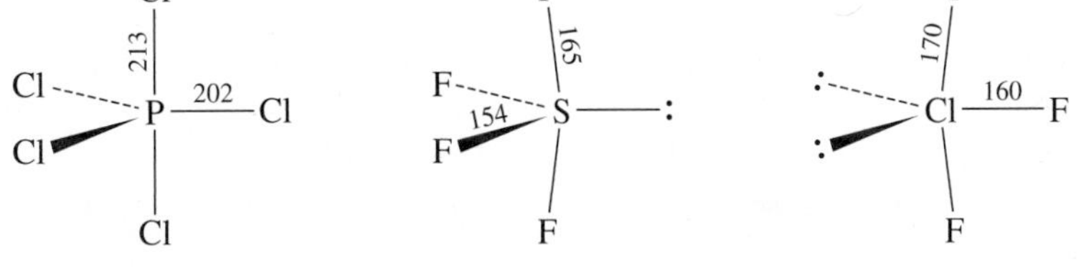

FIGURE 3.17 Bond Distances in PCl_5, SF_4, and ClF_3.

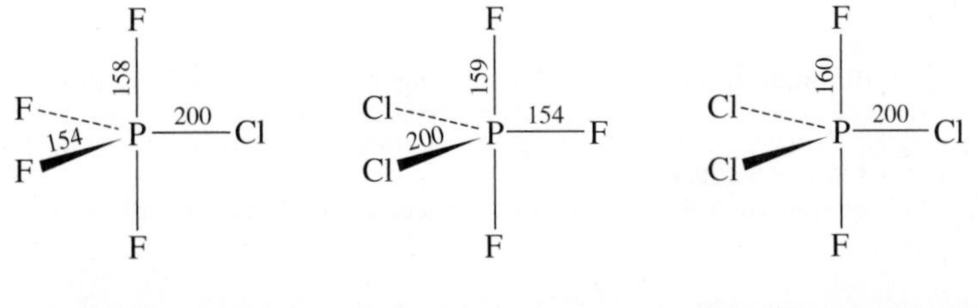

FIGURE 3.18 $PClF_4$, PCl_2F_3, and PCl_3F_2.

FIGURE 3.19 PF_4CH_3, $PF_3(CH_3)_2$, and $PF_2(CH_3)_3$.

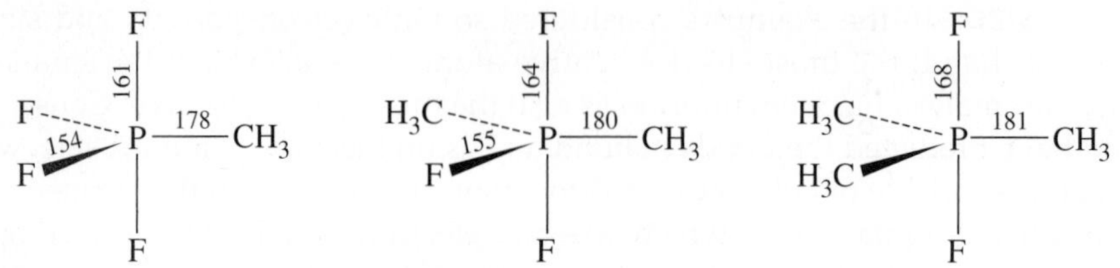

The effects on bond angles by less electronegative atoms are, however, typically less than for lone pairs and multiple bonds. For example, the bond angle to equatorial positions opposite the Cl atom in PF_4Cl is only slightly less than 120°, in contrast to the greater reduction in comparable angles in SF_4 and SOF_4 (Figure 3.20).

Predicting structures in some cases can prove challenging. Phosphorus compounds containing both fluorine atoms and CF_3 groups provide an intriguing example. CF_3 is an electron withdrawing group whose electronegativity has been calculated to be comparable to the more electronegative halogen atoms.[22] Does CF_3 favor equatorial positions more strongly than F? Forms of PF_4CF_3 having both axial and equatorial CF_3 groups are known (Figure 3.21). When two or three CF_3 groups are present, the orientations are truly a challenge to explain: these groups are axial in $PF_3(CF_3)_2$ but equatorial in $PF_2(CF_3)_3$! In both cases the more symmetrical structure, with identical equatorial groups, is preferred.[23]

▶ **Exercise 3.5** Briefly account for the following observations:

a. The bond angle in NCl_3 is nearly 5 degrees larger than in NF_3.
b. The S–F axial distance in SOF_4 is longer than the S–F equatorial distance.
c. In $Te(CH_3)_2I_2$ the methyl groups are in equatorial, rather than axial, positions.

3.2.4 Ligand Close Packing

The **ligand close-packing (LCP)** model developed by Gillespie[24] uses the distances between outer atoms in molecules as a guide to molecular shapes. For a series of molecules having the same central atom, the *non*bonded distances[25] between the outer atoms are consistent, but the bond angles and bond lengths change. The results of the LCP

FIGURE 3.20 Bond angles in PF_4Cl, SF_4, and SOF_4.

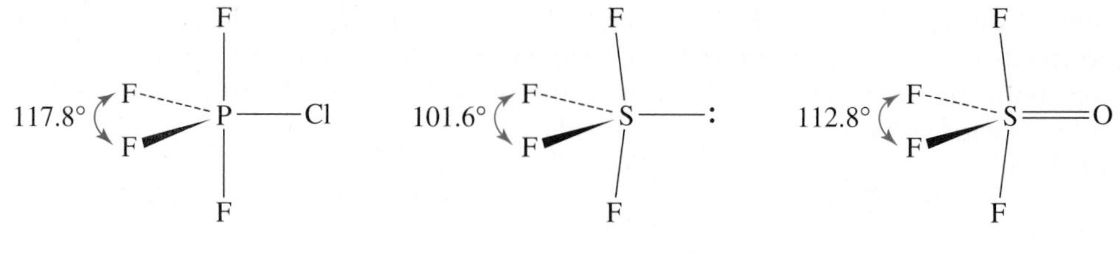

FIGURE 3.21 PF_4CF_3, $PF_3(CF_3)_2$, and $PF_2(CF_3)_3$.

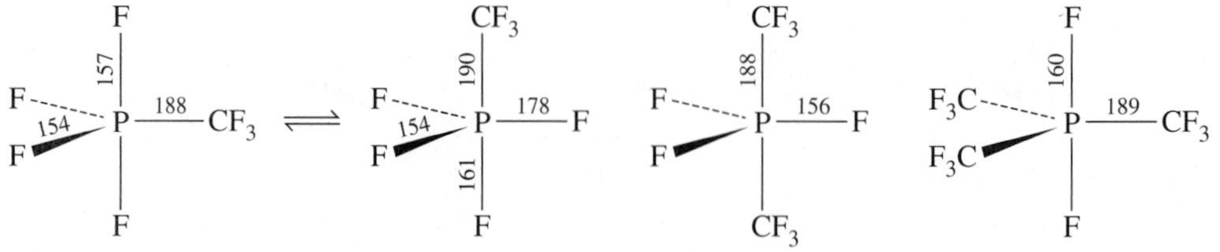

[22] For an analysis of different approaches to determining the electronegativity of CF_3, see J. E. True, T. D. Thomas, R. W. Winter, and G. L. Gard, *Inorg. Chem.*, **2003**, *42*, 4437.
[23] See H. Oberhammer, J. Grobe, and D. Le Van, *Inorg. Chem.*, **1982**, *21*, 275 for a discussion of these structures.
[24] R. J. Gillespie, *Coord. Chem. Rev.*, **2000**, *197*, 51.
[25] Three dots (···) will be used to designate distances between atoms that are not directly covalently bonded to each other.

TABLE 3.6 Ligand Close-Packing Data

Molecule	Coordination Number of B	B—F Distance (pm)	FBF Angle (°)	F···F Distance (pm)
BF_3	3	130.7	120.0	226
BF_2OH	3	132.3	118.0	227
BF_2NH_2	3	132.5	117.9	227
BF_2Cl	3	131.5	118.1	226
BF_2H	3	131.1	118.3	225
BF_2BF_2	3	131.7	117.2	225
BF_4^-	4	138.2	109.5	226
$BF_3CH_3^-$	4	142.4	105.4	227
$BF_3CF_3^-$	4	139.1	109.9	228
BF_3PH_3	4	137.2	112.1	228
BF_3NMe_3	4	137.2	111.5	229

Source: R. J. Gillespie and P. L. A. Popelier, *Chemical Bonding and Molecular Geometry*, Oxford University Press, New York, 2001, p. 119; Table 5.3, R. J. Gillespie, *Coord. Chem. Rev.*, **2000**, *197*, 51.

approach are in many ways consistent with those of the VSEPR model but focus primarily on the outer atoms rather than on the immediate environment of the central atom.

For example, it was found that in a series of boron compounds, BF_2X and BF_3X, the fluorine–fluorine distance remained nearly constant for a wide variety of X groups, even if the steric number changed from 3 to 4, as shown in Table 3.6. Similar results were obtained for a variety of other central atoms: chlorine–chlorine *non*bonded distances were similar in compounds in which the central atom was carbon, oxygen–oxygen nonbonded distances were similar when the central atom was beryllium, and so forth.[26]

In the LCP model, ligands (outer atoms) are viewed as exhibiting a specific radius when bonded to a certain central atom.[27] If the outer atoms pack tightly together, as assumed in this model, the distance between the nuclei of the atoms will then be the sum of these ligand radii. For example, a fluorine atom, when attached to a central boron, has a ligand radius of 113 pm. When two fluorines are attached to a central boron, as in the examples in Table 3.6, the distance between their nuclei will be the sum of the ligand radii, in this case 226 pm. This value matches the F···F distances of the examples in the table. Examples of how this approach can be used to describe molecular shapes are presented in the following discussion.

LIGAND CLOSE PACKING AND BOND DISTANCES The LCP model predicts that nonbonded atom–atom distances in molecules remain approximately the same, even if the bond angles around the central atom are changed. For example, the fluorine–fluorine distances in NF_4^+ and NF_3 are both 212 pm, even though the F—N—F bond angles are significantly different (Figure 3.22).

VSEPR predicts that NF_3 should have the smaller bond angle, and it does: 102.3° in comparison with the tetrahedral angle of 109.5° in NF_4^+. Because the F···F distance remains essentially unchanged, the N—F distance in NF_3 must be longer than the 130 pm

[26] R. J. Gillespie and P. L. A. Popelier, *Chemical Bonding and Molecular Geometry*, Oxford, New York, 2001, pp. 113–133 and references cited therein.

[27] Values of ligand radii can be found in R. J. Gillespie and E. A. Robinson, *Compt. Rend. Chimie*, **2005**, *8*, 1631.

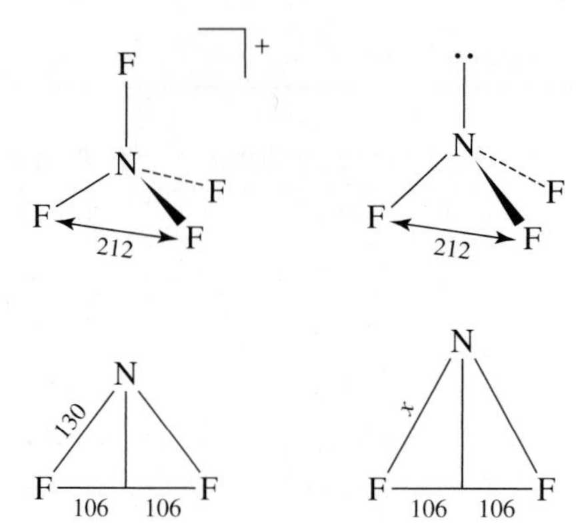

FIGURE 3.22 NF_4^+ and NF_3.

in NF_4^+. This can be illustrated using simple trigonometry, as also shown in the figure. In NF_3, because

$$\sin\left(\frac{102.3°}{2}\right) = \frac{\frac{1}{2}\left(\text{F} \cdots \text{F distance}\right)}{x},$$

$$x = \text{N—F bond distance} = \frac{106 \text{ pm}}{\sin 51.15°} = 136 \text{ pm (experimental: 136.5 pm)}$$

As expected, the smaller the F—N—F angle, the longer the N—F bond must be.

In short, the LCP model complements the VSEPR approach; whereas VSEPR predicts that a lone pair will cause a smaller bond angle on the opposite side, LCP predicts that the outer atom $\cdots$ outer atom distances should remain essentially unchanged, requiring longer central atom–outer atom distances. An atom that is multiply bonded to a central atom has a similar effect, as shown in the following example.

EXAMPLE

In PF_4^+ the F $\cdots$ F and P—F distances are 238 pm and 145.7 pm, respectively. Predict the P—F distance in POF_3, which has an F—P—F angle of 101.1°.

SOLUTION

The LCP model predicts that the F $\cdots$ F distances should be approximately the same in both structures. Sketches similar to those in Figure 3.22 can be drawn to illustrate the angles opposite the double bond:

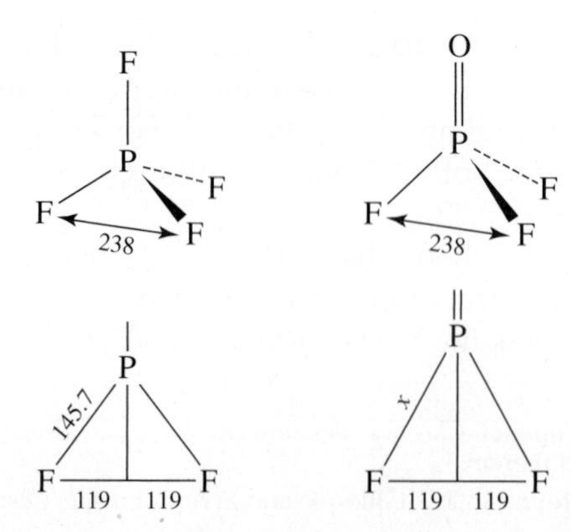

In this situation:

$$x = \text{P—F bond distance} = \frac{119 \text{ pm}}{\sin 50.55°} = 154 \text{ pm (experimental: 152 pm)}$$

(The actual F···F distance in POF_3 is 236 pm, slightly shorter than in PF_4^+. If this value is used in the calculation x = 153 pm, a better match for the experimental value.)

In the example, the two structures are predicted by the LCP model to have approximately the same size trigonal base of fluorines. A longer P—F distance and smaller F—P—F angle are consistent with the VSEPR approach, which would predict a smaller bond angle arising as a consequence of repulsion by the electrons in the double bond.

▶ **Exercise 3.6** Does this approach also work for different steric numbers? BCl_3 has B—Cl and Cl···Cl distances of 174 and 301 pm. Using the LCP model, predict the B—Cl distance in BCl_4^- and compare with the experimental value of 183 pm.

Gillespie and Popelier have also described several other approaches to molecular geometry, together with their advantages and disadvantages.[28]

3.3 MOLECULAR POLARITY

Whenever atoms with different electronegativities combine, the resulting molecule has polar bonds, with the electrons of the bond concentrated, perhaps very slightly, on the more electronegative atom; the greater the difference in electronegativity, the more polar the bond. As a result, the bonds are dipolar, with relatively positive and negative ends. This polarity can cause specific interactions between molecules, depending on the overall structure of the molecule.

Experimentally, the polarity of molecules is measured indirectly by measuring the dielectric constant, which is the ratio of the capacitance of a cell filled with the substance to be measured to the capacitance of the same cell with a vacuum between the electrodes. Orientation of polar molecules in the electric field partially cancels out the effect of the field and results in a larger dielectric constant. Measurements at different temperatures allow calculation of the **dipole moment** for the molecule, defined as

$$\mu = Qr,$$

where Q is the charge on each of two atoms separated by a distance r.[29] Dipole moments of diatomic molecules can be calculated directly. In more complex molecules, vector addition of the individual bond dipole moments gives the net molecular dipole moment. However, it is usually not possible to calculate molecular dipoles directly from bond dipoles. Table 3.7 shows experimental and calculated dipole moments of chloromethanes. The values calculated from vectors use C—H and C—Cl bond dipole moments of 1.3 and 4.9×10^{-30} Cm, respectively, and tetrahedral bond angles. Part of the discrepancy arises from bond angles that differ from the tetrahedral, but the column of data from PC Spartan,[30] a molecular modeling program, shows the difficulty of calculating dipoles. Clearly, calculating dipole moments is more complex than simply adding the vectors for individual bond moments. However, for most purposes, a qualitative approach is sufficient.

---

[28] R. J. Gillespie and P. L. A. Popelier, *Chemical Bonding and Molecular Geometry*, Oxford University Press, New York, 2001, pp. 113–133.

[29] The SI unit for dipole moment is a Coulomb meter (Cm), but a commonly used unit is the debye (D). One D = 3.33564×10^{-30} Cm.

[30] See http://www.wavefun.com.

TABLE 3.7 Dipole Moments of Chloromethanes

Molecule	Experimental (D)	Calculated (D)	
		Calculated from Vectors	Calculated by PC Spartan
CH_3Cl	1.87	1.77	1.51
CH_2Cl_2	1.60	2.08	1.50
$CHCl_3$	1.01	1.82	1.16

Source: Experimental data, *Handbook of Chemistry and Physics*, 66th ed., CRC Press, Cleveland, OH, 1985–1986, p. E-58 (from NBS table NSRDS-NBS 10); PC Spartan, see Footnote 30.

The dipole moments of NH_3, H_2O, and NF_3 (Figure 3.23) reveal the effect of lone pairs, which can be dramatic. In ammonia, the averaged N—H bond polarities and the lone pair all point in the same direction, resulting in a large dipole moment. Water has an even larger dipole moment, because the polarities of the O—H bonds and the two lone pairs result in polarities all reinforcing each other. On the other hand, NF_3 has a very small dipole moment, the result of the polarity of the three N—F bonds opposing polarity of the lone pair. The sum of the three N—F bond moments is larger than the lone pair effect, and the lone pair is the positive end of the molecule. In cases such as those of NF_3 and SO_2, the direction of the dipole is not easily predicted because of the opposing polarities. SO_2 has a large dipole moment (1.63 D), with the polarity of the lone pair prevailing over that of the S—O bonds.

Molecules with dipole moments interact electrostatically with each other and with other polar molecules. When the dipoles are large enough, the molecules orient themselves with the positive end of one molecule toward the negative end of another, and higher melting and boiling points result. Details of the most dramatic effects are given in the discussion of hydrogen bonding later in this chapter and in Chapter 6.

On the other hand, if the molecule has a highly symmetric structure, or if the polarities of different bonds cancel each other out, the molecule as a whole may be nonpolar, even though the individual bonds are quite polar. Tetrahedral molecules such as CH_4 and CCl_4, trigonal molecules and ions such as SO_3, NO_3^-, and CO_3^{2-}, and molecules having identical outer atoms for steric numbers 5 and 6 such as PCl_5 and SF_6, are all nonpolar. The C—H bond has very little polarity, but the bonds in the other molecules and ions are quite polar. In all these cases, the sum of all the polar bonds is zero because of the symmetry of the molecules, as shown in Figure 3.24.

Nonpolar molecules, whether they have polar bonds or not, still have intermolecular attractive forces acting on them. Small fluctuations in the electron density in such molecules create small temporary dipoles with extremely short lifetimes. These dipoles in turn attract or repel electrons in adjacent molecules, setting up dipoles in them as well. The result is an

FIGURE 3.23 Bond Dipoles and Molecular Dipoles.

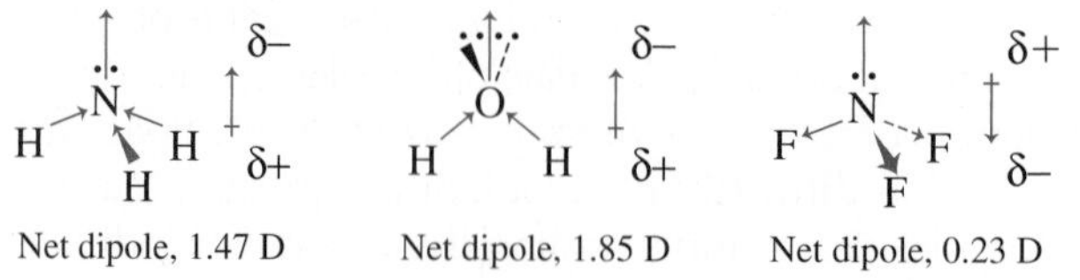

Net dipole, 1.47 D   Net dipole, 1.85 D   Net dipole, 0.23 D

FIGURE 3.24 Cancellation of Bond Dipoles Due to Molecular Symmetry.

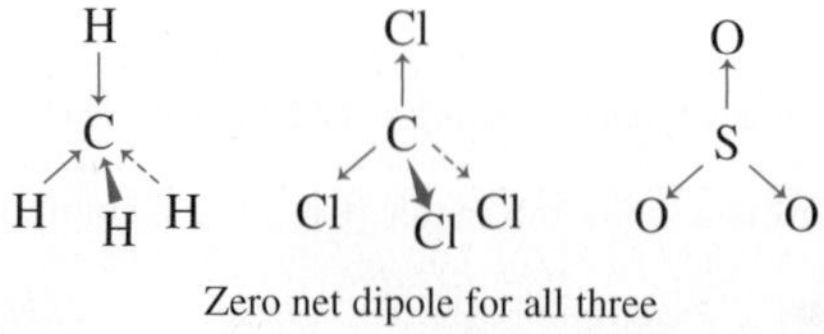

Zero net dipole for all three

overall attraction among molecules. These attractive forces are called **London forces** or **dispersion forces,** and they make liquefaction of the noble gases and nonpolar molecules—such as hydrogen, nitrogen, and carbon dioxide—possible. As a general rule, London forces are more important when there are more electrons in a molecule, because the attraction of the nuclei is shielded by inner electrons, and the electron cloud is more polarizable.

3.4 HYDROGEN BONDING

Ammonia, water, and hydrogen fluoride all have much higher boiling points than other similar molecules, as shown in Figure 3.25. In water and hydrogen fluorine, these high boiling points are caused by hydrogen bonds, in which hydrogen atoms bonded to oxygen or fluorine also form weaker bonds to a lone pair of electrons on another oxygen or fluorine. Bonds between hydrogen and these strongly electronegative atoms are very polar, with a partial positive charge on the hydrogen. This partially positive hydrogen is strongly attracted to the partially negative oxygen or fluorine of neighboring molecules. In the past, the attractions among these molecules were considered primarily electrostatic in nature, but an alternative molecular orbital approach (described in Chapters 5 and 6) gives a more complete description of this phenomenon. Regardless of the detailed explanation of the forces involved in hydrogen bonding, the strongly positive hydrogen and the strongly negative lone pairs tend to line up and hold the molecules together. Other atoms with high electronegativity, such as chlorine, can also enable formation of hydrogen bonds in strongly polar molecules such as chloroform, $CHCl_3$.

In general, boiling points rise with increasing molecular weight, both because the additional mass requires higher temperature for rapid movement of the molecules and because the larger number of electrons in the heavier molecules provides larger London forces. The difference in temperature between the actual boiling point of water and the extrapolation of the line connecting the boiling points of the heavier analogous compounds is almost 200° C. Ammonia and hydrogen fluoride have similar but smaller differences from the extrapolated values for their families. Water has a much larger effect, because each molecule can have as many as four hydrogen bonds (two through the lone

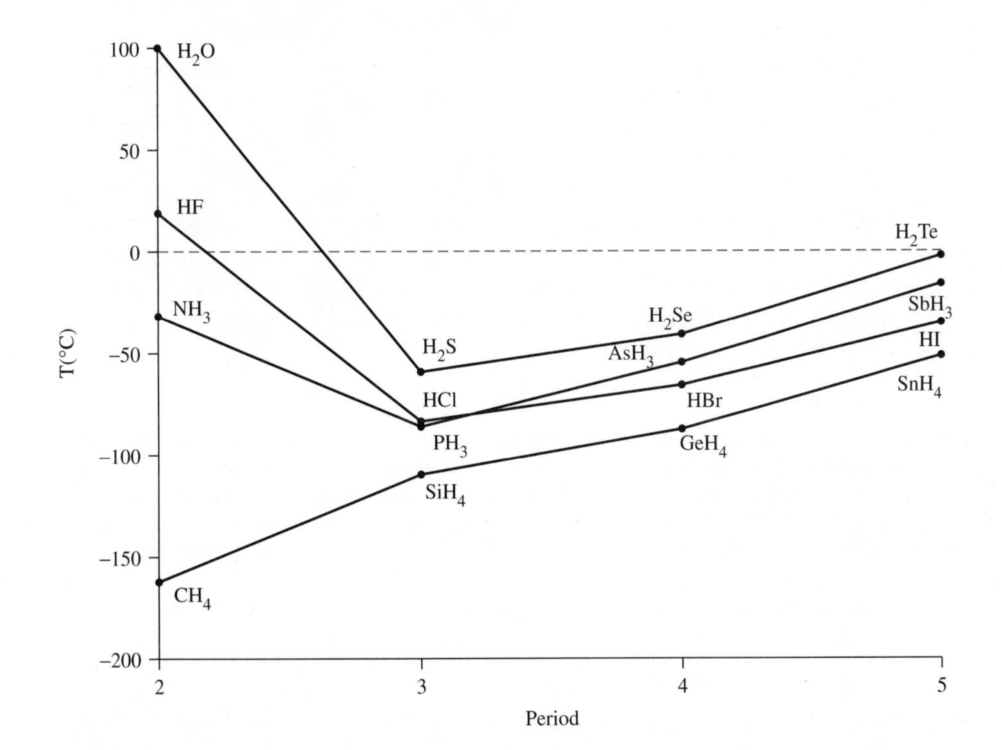

FIGURE 3.25 Boiling Points of Hydrogen Compounds.

pairs and two through the hydrogen atoms). Hydrogen fluoride can average no more than two hydrogen bonds, because hydrogen fluoride has only one hydrogen available.

Water has other unusual properties because of hydrogen bonding. For example, the freezing point of water is much higher than that of similar molecules. An even more striking feature is the decrease in density as water freezes. The tetrahedral structure around each oxygen atom, with two regular bonds to hydrogen and two hydrogen bonds to other molecules, requires a very open structure with large spaces between ice molecules (Figure 3.26). This makes the solid less dense than the more random liquid water surrounding it, so ice floats. Life on Earth would be very different if this were not so. Lakes, rivers, and oceans would freeze from the bottom up, ice cubes would sink, and ice fishing would be impossible. The results are difficult to imagine, but would certainly require a much different biology and geology. The same forces cause coiling of protein (Figure 3.27) and polynucleic acid molecules; a combination of hydrogen bonding with other dipolar forces imposes considerable secondary structure on these large molecules. In Figure 3.27(a), hydrogen bonds between carbonyl oxygen atoms and hydrogens attached to nitrogen atoms hold the molecule in a helical structure. In Figure 3.27(b), similar hydrogen bonds hold the parallel peptide chains together; the bond angles of the chains result in the pleated appearance of the sheet formed by the peptides. These are two of the many different structures that can be formed from peptides, depending on the side-chain groups R and the surrounding environment.

Another example is a theory of anesthesia by non–hydrogen bonding molecules such as cyclopropane, chloroform, and nitrous oxide, proposed by Pauling.[31] These molecules are of a size and shape that can fit neatly into a hydrogen-bonded water

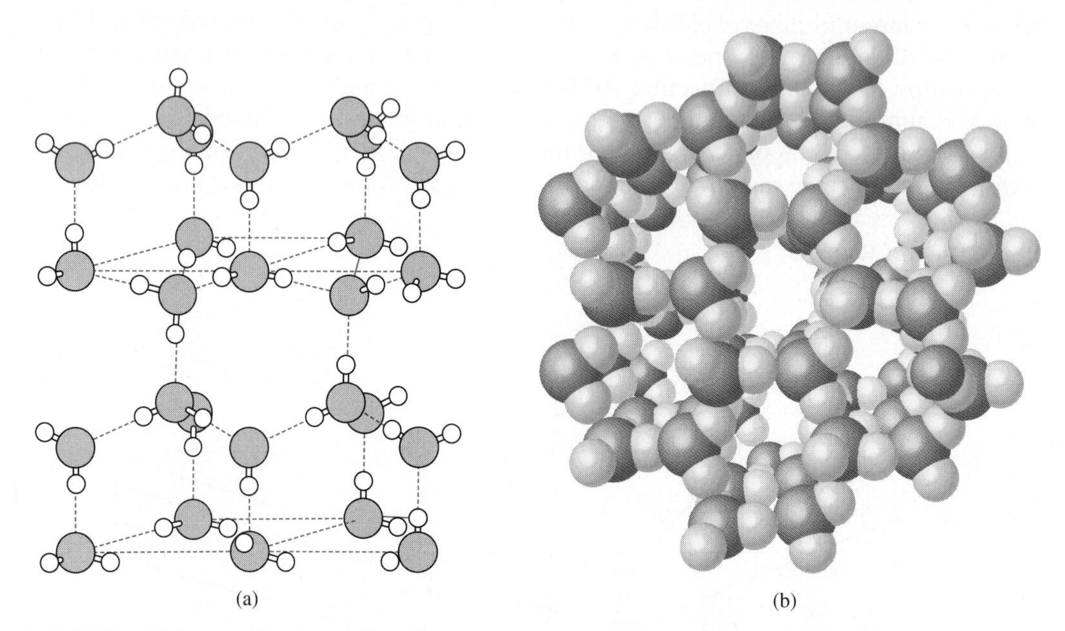

FIGURE 3.26 Two Drawings of Ice. (a) From T. L. Brown and H. E. LeMay, Jr., *Chemistry, The Central Science*, Prentice Hall, Englewood Cliffs, NJ, 1988, p. 628. Reproduced with permission. The rectangular lines are included to aid visualization; all bonding is between hydrogen and oxygen atoms. (b) Copyright © 1976 by W. G. Davies and J. W. Moore, used by permission; reprinted from J. W. Moore, W. G. Davis, and R. W. Collins, *Chemistry*, McGraw-Hill, New York, 1978. All rights reserved.

[31] L. Pauling, *Science*, **1961**, *134*, 15.

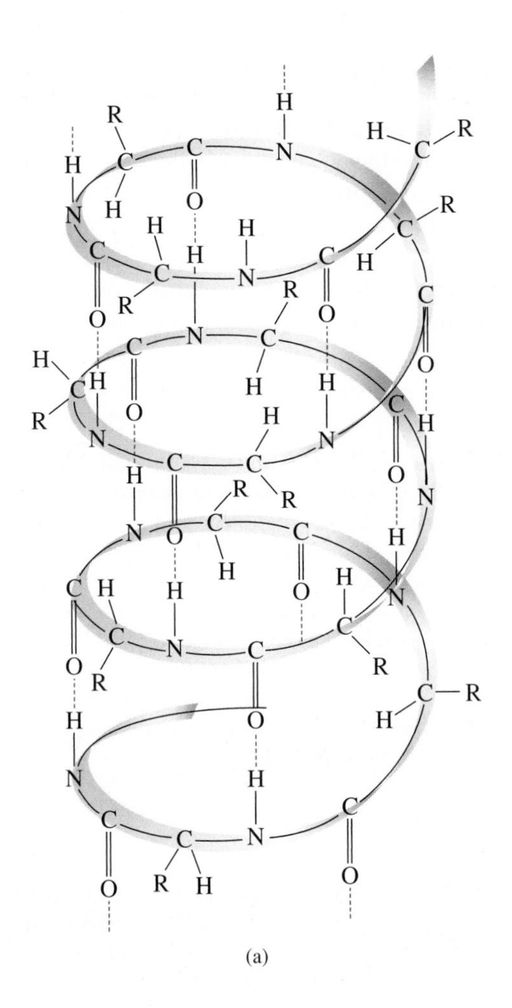

(a)

(b)

FIGURE 3.27 Hydrogen-Bonded Protein Structures. (a) A protein α helix. Peptide carbonyls and N—H hydrogens on adjacent turns of the helix are hydrogen bonded. (From T. L. Brown and H. E. LeMay, Jr., *Chemistry, the Central Science*, Prentice Hall, Englewood Cliffs, NJ, 1988, p. 946. Reproduced with permission.) The pleated sheet arrangement is shown in (b). Each peptide carbonyl group is hydrogen bonded to an N—H hydrogen on an adjacent peptide chain. (Reproduced with permission from L. G. Wade, Jr., *Organic Chemistry*, Prentice Hall, Englewood Cliffs, NJ, 1988, pp. 1255–1256.)

structure with even larger open spaces than ordinary ice. Such structures, with molecules trapped in holes in a solid, are called **clathrates**. Pauling proposed that similar hydrogen-bonded microcrystals form even more readily in nerve tissue because of the presence of other solutes in the tissue. These microcrystals could then interfere with the transmission of nerve impulses. Similar structures of methane and water are believed to hold large quantities of methane in the polar ice caps. The amount of methane in such crystals can be so great that they burn if ignited.[32]

More specific interactions involving the sharing of electron pairs between molecules are discussed in connection with acid–base chemistry in Chapter 6.

General References

Good sources for bond lengths and bond angles are the works of Wells, Greenwood and Earnshaw, and Cotton and Wilkinson cited in Chapter 1. Reviews of electron-dot diagrams and formal charges can be found in most general chemistry texts. One of the best VSEPR references is still the early paper by R. J. Gillespie and R. S. Nyholm, *Q. Rev. Chem. Soc.* **1957**, *XI*, 339–380. More recent expositions of the theory are in R. J. Gillespie and I. Hargittai, *The VSEPR Model of Molecular Geometry*, Allyn & Bacon, Boston, 1991, and R. J. Gillespie and P. L. A. Popelier, *Chemical Bonding and Molecular Geometry: From Lewis to Electron Densities*, Oxford University Press, New York, 2001. The latter also presents a useful introduction to the Ligand Close-Packing (LCP) model. Gillespie has also provided a retrospective on VSEPR in R. J. Gillespie, *Coord. Chem. Rev.*, **2008**, *252*, 1315, and he and Robinson have compared the VSEPR, LCP, and valence-bond descriptions of molecular geometry in R. J. Gillespie and E. A. Robinson, *Chem. Soc. Rev.*, **2005**, *34*, 396. Molecular orbital arguments for the shapes of many of the same molecules are presented in B. M. Gimarc, *Molecular Structure and Bonding*, Academic Press, New York, 1979, and J. K. Burdett, *Molecular Shapes*, John Wiley & Sons, New York, 1980.

Problems

3.1 The dimethyldithiocarbamate ion, $[S_2CN(CH_3)_2]^-$, has the following skeletal structure:

```
S\        /CH3
   C—N
S/        \CH3
```

a. Give the important resonance structures of this ion, including any formal charges where necessary. Select the resonance structure likely to provide the best description of this ion.

b. Repeat for the dithiocarbamate ion, $[OSCN(CH_3)_2]^-$.

3.2 Several resonance structures are possible for each of the following ions. For each, draw these resonance structures, assign formal charges, and select the resonance structure likely to provide the best description for the ion.

a. Selenocyanate ion, $SeCN^-$

b. Thioformate ion, H—C(O)(S) $^-$

c. Dithiocarbonate, $[S_2CO]^{2-}$ (C is central)

3.3 Draw the resonance structures for the isoelectronic ions NSO^- and SNO^-, and assign formal charges. Which ion is likely to be more stable?

3.4 Three isomers having the formula N_2CO are known: ONCN (nitrosyl cyanide), ONNC (nitrosyl isocyanide), and NOCN (isonitrosyl cyanide). Draw the most important resonance structures of these isomers, and determine the formal charges. Which isomer do you predict to be the most stable (lowest energy) form? (See G. Maier, H. P. Reinsenauer, J. Eckwert, M. Naumann, and M. De Marco, *Angew. Chem., Int. Ed.*, **1997**, *36*, 1707.)

3.5 Show the possible resonance structures for nitrous oxide, N_2O (the central atom is nitrogen). Indicate nonzero formal charges where they are present. Which resonance structure gives the best representation of this molecule?

3.6 Nitric acid, which exists as HNO_3 molecules in the absence of water, has the skeletal structure shown. Show the important resonance structures of HNO_3, and designate the formal charges on each atom.

```
        O
H  O  N
        O
```

[32] L. A. Stern, S. H. Kirby, W. B. Durham, *Science*, **1996**, *273*, 1765 (cover picture), 1843.

3.7 L. C. Allen has suggested that a more meaningful formal charge can be obtained by taking into account the electronegativities of the atoms involved. Allen's formula for this type of charge—referred to as the *Lewis-Langmuir (L–L) charge*—of an atom, A, bonded to another atom, B, is

$$\text{L–L charge} = \frac{\text{(US) group}}{\text{number of A}} - \frac{\text{number of unshared}}{\text{electrons on A}} - 2\sum_{B} \frac{\chi_A}{\chi_A + \chi_B}\begin{pmatrix}\text{number of bonds}\\ \text{between A and B}\end{pmatrix}$$

where χ_A and χ_B designate the electronegativities. Using this equation, calculate the L–L charges for CO, NO^-, and HF, and compare the results with the corresponding formal charges. Do you think the L–L charges are a better representation of electron distribution? (See L. C. Allen, *J. Am. Chem. Soc.*, **1989**, *111*, 9115; L. D. Garner, T. L. Meek, and B. G. Patrick, *THEOCHEM*, **2003**, *620*, 43.)

3.8 Give Lewis dot structures and sketch the shapes of the following:

a. $SeCl_4$ **b.** I_3^-
c. $PSCl_3$ (P is central) **d.** IF_4^-
e. PH_2^- **f.** TeF_4^{2-}
g. N_3^- **h.** $SeOCl_4$ (Se is central)
i. PH_4^+

3.9 Give Lewis dot structures and sketch the shapes of the following:

a. ICl_2^-
b. H_3PO_3 (one H is bonded to P)
c. BH_4^- **d.** $POCl_3$ **e.** IO_4^- **f.** $IO(OH)_5$
g. $SOCl_2$ **h.** $ClOF_4^-$ **i.** XeO_2F_2

3.10 Give Lewis dot structures and sketch the shapes of the following:

a. SOF_6 (one F is attached to O)
b. POF_3 **c.** ClO_2 **d.** NO_2
e. $S_2O_4^{2-}$ (symmetric, with an S—S bond)
f. N_2H_4 (symmetric, with an N—N bond)
g. $ClOF_2^+$ **h.** CS_2 **i.** $XeOF_5^-$

3.11 Explain the trends in bond angles and bond lengths of the following ions:

	X—O (pm)	O—X—O Angle
ClO_3^-	149	107°
BrO_3^-	165	104°
IO_3^-	181	100°

3.12 Select from each set the molecule or ion having the smallest bond angle, and briefly explain your choice:

a. NH_3, PH_3, or AsH_3

b. $SOCl_2$ SOF_2

(halogen—sulfur—halogen angle)

c. NO_2^- or O_3
d. ClO_3^- or BrO_3^-

3.13 **a.** Compare the structures of the azide ion, N_3^-, and the ozone molecule, O_3.
b. How would you expect the structure of the ozonide ion, O_3^-, to differ from that of ozone?

3.14 Consider the series OCl_2, $O(CH_3)_2$, and $O(SiH_3)_2$, which have bond angles at the oxygen atom of 110.9°, 111.8°, and 144.1° respectively. Account for this trend.

3.15 Two ions isoelectronic with carbon suboxide, C_3O_2, are N_5^+ and $OCNCO^+$. Whereas C_3O_2 is linear, both N_5^+ and $OCNCO^+$ are bent at the central nitrogen. Suggest an explanation. Also predict which has the smaller outer atom—N—outer atom angle and explain your reasoning. (See I. Bernhardi, T. Drews, and K. Seppelt, *Angew. Chem., Int. Ed.*, **1999**, *38*, 2232; K. O. Christe, W. W. Wilson, J. A. Sheehy, and J. A. Boatz, *Angew. Chem., Int. Ed.*, **1999**, *38*, 2004.)

3.16 Explain the following:

a. Ethylene, C_2H_4, is a planar molecule, but hydrazine, N_2H_4, is not.
b. ICl_2^- is linear, but NH_2^- is bent.
c. Of the compounds mercury(II) cyanate, $Hg(OCN)_2$, and mercury(II) fulminate, $Hg(CNO)_2$, one is highly explosive, and the other is not.

3.17 Explain the following:

a. PCl_5 is a stable molecule, but NCl_5 is not.
b. SF_4 and SF_6 are known, but OF_4 and OF_6 are not.

3.18 X-ray crystal structures of ClF_3O and BrF_3O have been determined.

a. Would you expect the lone pair on the central halogen to be axial or equatorial in these molecules? Why?
b. Which molecule would you predict to have the smaller $F_{equatorial}$–central atom–oxygen angle? Explain your reasoning. (See A. Ellern, J. A. Boatz, K. O. Christe, T. Drews, and K. Seppelt, *Z. Anorg. Allg. Chem.*, **2002**, *628*, 1991.)

3.19 Predict and sketch the structure of the (as yet) hypothetical ion IF_3^{2-}.

3.20 A solution containing the $IO_2F_2^-$ ion reacts slowly with excess fluoride ion to form $IO_2F_3^{2-}$.

a. Sketch the isomers that might be possible matching the formula $IO_2F_3^{2-}$.
b. Of these structures, which do you think is most likely? Why?
c. Propose a formula of a xenon compound or ion isoelectronic with $IO_2F_3^{2-}$.

(See J. P. Mack, J. A. Boatz, and M. Gerken, *Inorg. Chem.*, **2008**, *47*, 3243.)

3.21 Predict the structure of $I(CF_3)Cl_2$. Do you expect the CF_3 group to be in an axial or equatorial position? Why? (See R. Minkwitz and M. Merkei, *Inorg. Chem.*, **1999**, *38*, 5041.)

3.22 **a.** Which has the longer axial P—F distance, $PF_2(CH_3)_3$ or $PF_2(CF_3)_3$? Explain briefly.

b. has oxygen as central atom. Predict the approximate bond angle in this molecule and explain your answer.

c. Predict the structure of CAl_4. (See X. Li, L-S. Wang, A. I. Boldyrev, and J. Simons, *J. Am. Chem. Soc.*, **1999**, *121*, 6033.)

3.23 $SeCl_6^{2-}$, $TeCl_6^{2-}$, and ClF_6^- are all octahedral, but SeF_6^{2-} and IF_6^- are distorted, with a lone pair on the central atom apparently influencing the shape. Suggest a reason for the difference in shape of these two groups of ions. (See J. Pilmé, E. A. Robinson, and R. J. Gillespie, *Inorg. Chem.*, **2006**, *45*, 6198.)

3.24 When XeF_4 is reacted with a solution of water in CH_3CN solvent, the product $F_2OXeN{\equiv}CCH_3$ is formed. Applying a vacuum to crystals of this product resulted in slow removal of CH_3CN:

$$F_2OXeN{\equiv}CCH_3 \longrightarrow XeOF_2 + CH_3CN$$

Propose structures for $F_2OXeN{\equiv}CCH_3$ and $XeOF_2$. (See D. S. Brock, V. Bilir, H. P. A. Mercier, and G. J. Schrobilgen, *J. Am. Chem. Soc.*, **2007**, *129*, 3598.)

3.25 The thiazyl dichloride ion, $NSCl_2^-$, is isoelectronic with thionyl dichloride, $OSCl_2$.

a. Which of these species has the smaller Cl—S—Cl angle? Explain briefly.

b. Which do you predict to have the longer S—Cl bond? Why? (See E. Kessenich, F. Kopp, P. Mayer, and A. Schulz, *Angew. Chem., Int. Ed.*, **2001**, *40*, 1904.)

3.26 Sketch the most likely structure of PCl_3Br_2 and explain your reasoning.

3.27 a. Are the CF_3 groups in $PCl_3(CF_3)_2$ more likely axial or equatorial? Explain briefly.

b. Are the axial or equatorial bonds likely to be longer $SbCl_5$? Explain briefly.

3.28 Which has the smaller F—P—F angle, PF_4^+ or PF_3O? Which has the longer fluorine–fluorine distance? Explain briefly.

3.29 Account for the trend in P—F_{axial} distances in the compounds $PF_4(CH_3)$, $PF_3(CH_3)_2$, and $PF_2(CH_3)_3$. (See Figure 3.19.)

3.30 Although the C—F distances and the F—C—F bond angles differ considerably in $F_2C{=}CF_2$, F_2CO, CF_4, and F_3CO^- (C—F distances 131.9 to 139.2 pm; F—C—F bond angles 101.3° to 109.5°), the F···F distance in all four structures is very nearly the same (215 to 218 pm). Explain, using the LCP model of Gillespie. (See R. J. Gillespie, *Coord. Chem. Rev.*, **2000**, *197*, 51.)

3.31 The Cl···Cl distance in CCl_4 is 289 pm, and the C—Cl bond distance is 171.1 pm. Using the LCP model, calculate the C—Cl distance in Cl_2CO, which has a Cl—C—Cl angle of 111.8°.

3.32 Compounds in which hydrogen is the outer atom can provide challenges to theories of chemical bonding. Consider the following molecules. Using one or more of the approaches described in this chapter, provide a rationale for HOF having the smallest bond angle in this set.

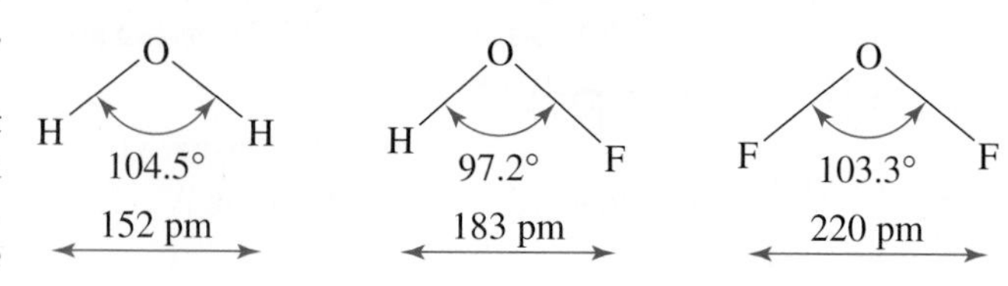

3.33 For each of the following bonds, indicate which atom is more negative, then rank the series in order of polarity.

a. C—N **b.** N—O **c.** C—I
d. O—Cl **e.** P—Br **f.** S—Cl

3.34 Give Lewis dot structures and shapes for the following:

a. $VOCl_3$ **b.** PCl_3 **c.** SOF_4
d. SO_3 **e.** ICl_3 **f.** SF_6
g. IF_7 **h.** XeO_2F_4 **i.** CF_2Cl_2
j. P_4O_6

(P_4O_6 is a closed structure with overall tetrahedral arrangement of phosphorus atoms; an oxygen atom bridges each pair of phosphorus atoms.)

3.35 Give Lewis dot structures and sketch the shapes for the following:

a. PH_3 **b.** H_2Se **c.** SeF_4
d. PF_5 **e.** IF_5 **f.** XeO_3
g. BF_2Cl **h.** $SnCl_2$ **i.** KrF_2
j. $IO_2F_5^{2-}$

3.36 Which of the molecules in Problem 3.34 are polar?

3.37 Which of the molecules in Problem 3.35 are polar?

3.38 Provide explanations for the following:

a. Methanol, CH_3OH, has a much higher boiling point than methyl mercaptan, CH_3SH.

b. Carbon monoxide has slightly higher melting and boiling points than N_2.

c. The *ortho* isomer of hydroxybenzoic acid [$C_6H_4(OH)(CO_2H)$] has a much lower melting point than the *meta* and *para* isomers.

d. The boiling points of the noble gases increase with atomic number.

e. Acetic acid in the gas phase has a significantly lower pressure (approaching a limit of one half) than predicted by the ideal gas law.

f. Mixtures of acetone and chloroform exhibit significant negative deviations from Raoult's law, which states that the vapor pressure of a volatile liquid is proportional to its mole fraction. For example, an equimolar mixture of acetone and chloroform has a lower vapor pressure than either of the pure liquids.

g. Carbon monoxide has a greater bond-dissociation energy (1072 kJ/mol) than molecular nitrogen (945 kJ/mol).

4

Symmetry and Group Theory

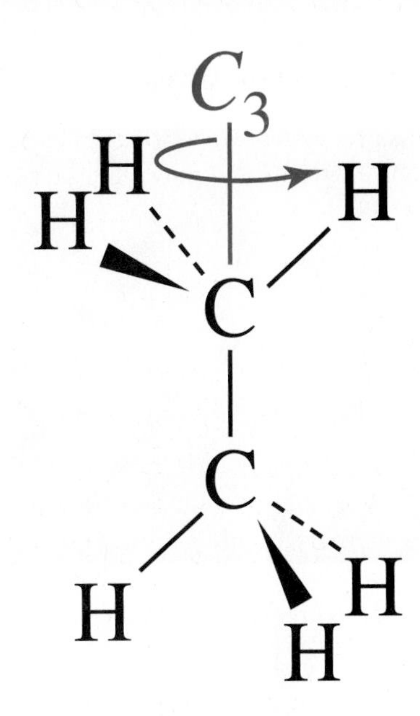

Symmetry is a phenomenon of the natural world, as well as the world of human invention (Figure 4.1). In nature, many types of flowers and plants, snowflakes, insects, certain fruits and vegetables, and a wide variety of microscopic plants and animals exhibit characteristic symmetry. Many engineering achievements have a degree of symmetry that contributes to their esthetic appeal. Examples include cloverleaf intersections, the pyramids of ancient Egypt, and the Eiffel Tower.

Symmetry concepts can be extremely useful in chemistry. By analyzing the symmetry of molecules, we can predict infrared spectra, describe the types of orbitals used in bonding, predict optical activity, interpret electronic spectra, and study a number of additional molecular properties. In this chapter, we first define symmetry very specifically in terms of five fundamental symmetry operations. We then describe how molecules can be classified on the basis of the types of symmetry they possess. We conclude with examples of how symmetry can be used to predict optical activity of molecules and to determine the number and types of infrared- and Raman-active molecular vibrations.

In later chapters, symmetry will be a valuable tool in the construction of molecular orbitals (Chapters 5 and 10) and in the interpretation of electronic spectra of coordination compounds (Chapter 11) and vibrational spectra of organometallic compounds (Chapter 13).

A molecular model kit is a very useful study aid for this chapter, even for those who can visualize three-dimensional objects easily. We strongly encourage the use of such a kit.

4.1 SYMMETRY ELEMENTS AND OPERATIONS

All molecules can be described in terms of their symmetry, even if it is only to say they have none. Molecules or any other objects may contain **symmetry elements** such as mirror planes, axes of rotation, and inversion centers. The actual reflection, rotation, or inversion is called the **symmetry operation**. To contain a given symmetry element, a molecule must have exactly the same appearance after the operation as before. In other words, photographs of the molecule (if such photographs were possible) taken from the same location before and after the symmetry operation would be indistinguishable. If a symmetry operation yields a molecule that can be distinguished from the original in any way, that operation is *not* a symmetry operation of the molecule. The examples in Figures 4.2 through 4.6 illustrate the possible types of molecular symmetry operations and elements.

The **identity operation** (E) causes no change in the molecule. It is included for mathematical completeness. An identity operation is characteristic of every molecule, even if it has no other symmetry.

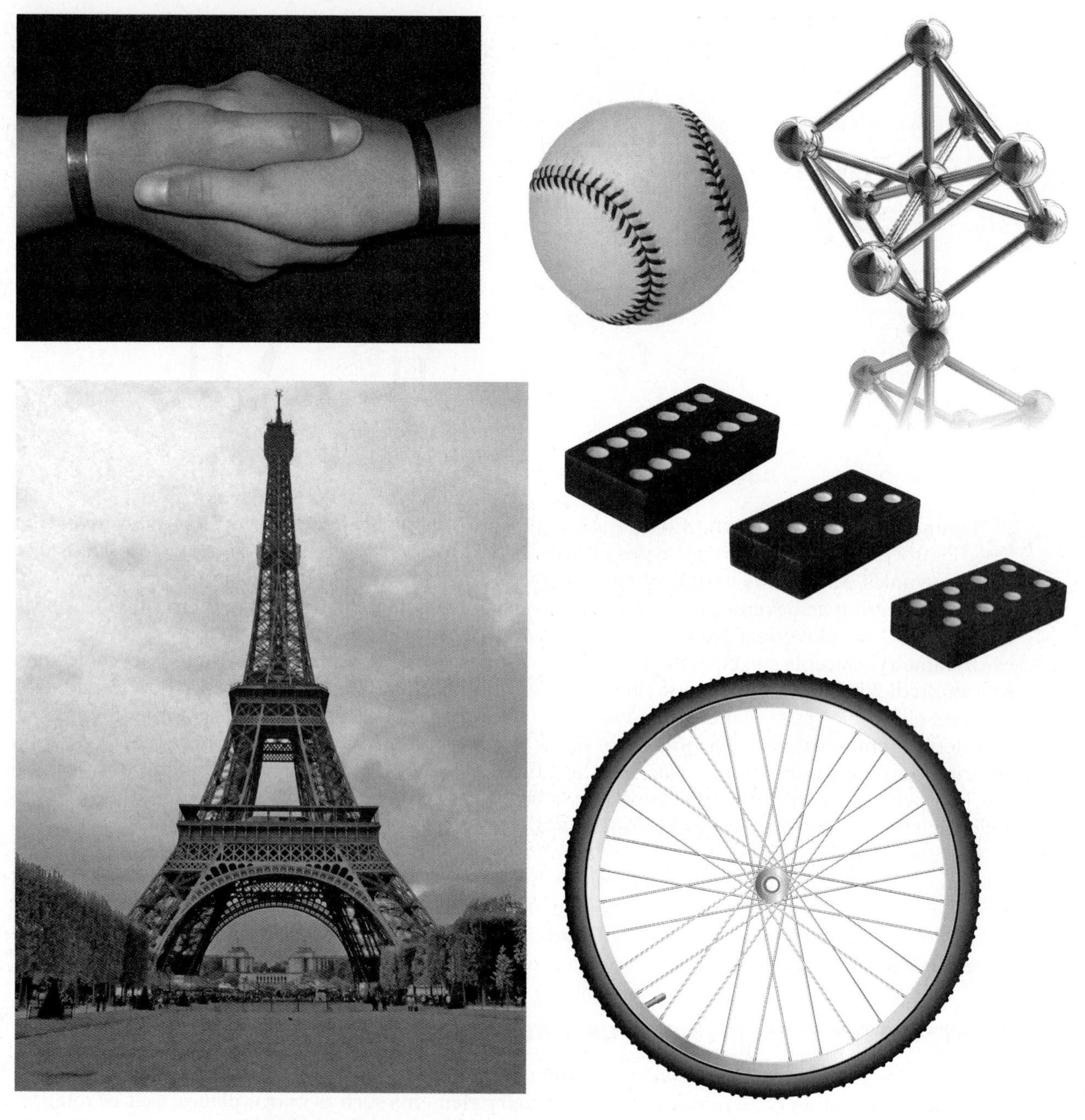

FIGURE 4.1 Examples of Symmetry.

The **rotation operation** (C_n), also called **proper rotation**, is rotation through $360°/n$ about a rotation axis. We use counterclockwise rotation as a positive rotation. An example of a molecule having a threefold (C_3) axis is $CHCl_3$. The rotation axis is coincident with the C—H bond axis, and the rotation angle is $360°/3 = 120°$. Two C_3 operations may be performed consecutively to give a new rotation of 240°. The resulting operation is designated $C_3{}^2$ and is also a symmetry operation of the molecule. Three successive C_3 operations are the same as the identity operation ($C_3{}^3 \equiv E$). The identity operation is included in all molecules.

Many molecules and other objects have multiple rotation axes. Snowflakes are a case in point, with complex shapes that are nearly always hexagonal and nearly planar. The line

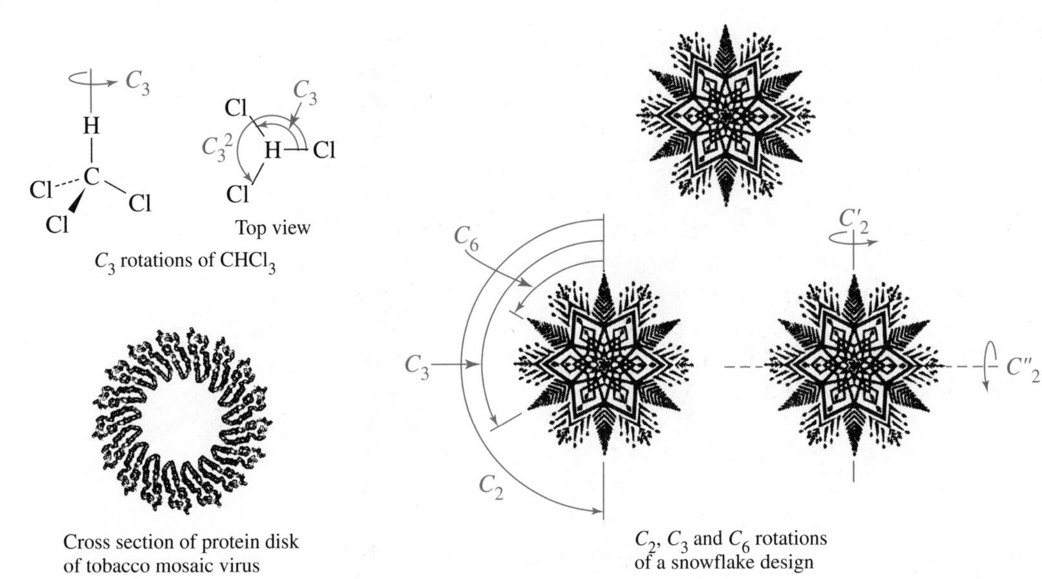

FIGURE 4.2 Rotations. The cross section of the tobacco mosaic virus is a cover diagram from *Nature*, 1976, *259*. © 1976, Macmillan Journals Ltd. Reproduced with permission of Aaron Klug.

through the center of the flake perpendicular to the plane of the flake contains a twofold (C_2) axis, a threefold (C_3) axis, and a sixfold (C_6) axis. Rotations by 240° (C_3^2) and 300° (C_6^5) are also symmetry operations of the snowflake.

Rotation Angle	Symmetry Operation
60°	C_6
120°	C_3 ($\equiv C_6^2$)
180°	C_2 ($\equiv C_6^3$)
240°	C_3^2 ($\equiv C_6^4$)
300°	C_6^5
360°	E ($\equiv C_6^6$)

There are also two sets of three C_2 axes in the plane of the snowflake, one set through opposite points and one through the cut-in regions between the points. One of each of these axes is shown in Figure 4.2. In molecules with more than one rotation axis, the C_n axis having the largest value of n is the **highest order rotation axis** or **principal axis**. The highest order rotation axis for a snowflake is the C_6 axis. (In assigning Cartesian coordinates, the highest order C_n axis is usually chosen as the z axis.) When necessary, the C_2 axes perpendicular to the principal axis are designated with primes; a single prime (C_2') indicates that the axis passes through several atoms of the molecule, whereas a double prime (C_2'') indicates that it passes between the outer atoms.

Finding rotation axes for some three-dimensional figures is more difficult but the same in principle. Remember that nature is not always simple when it comes to symmetry—the protein disk of the tobacco mosaic virus has a 17-fold rotation axis!

In the **reflection operation** (σ) the molecule contains a mirror plane. If details such as hair style and location of internal organs are ignored, the human body has a left–right mirror plane, as in Figure 4.3. Many molecules have mirror planes, although they may not be immediately obvious. The reflection operation exchanges left and right, as if each point had moved perpendicularly through the plane to a position exactly as far from the plane as when it started. Linear objects, such as a round wood

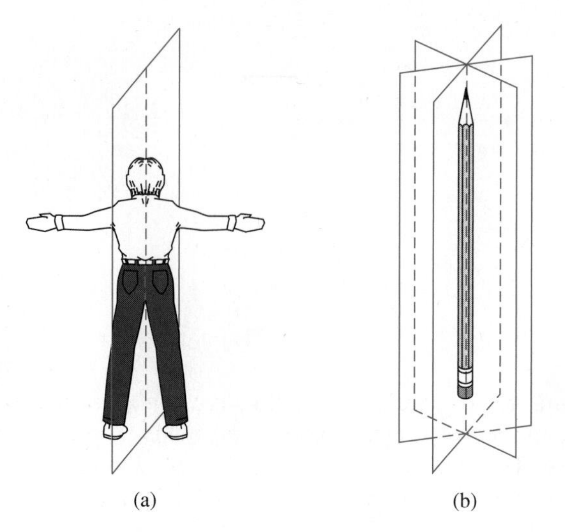

FIGURE 4.3 Reflections.

pencil, or molecules, such as acetylene and carbon dioxide, have an infinite number of mirror planes that include the center line of the object.

When the plane is perpendicular to the principal axis of rotation, it is called σ_h (horizontal). Other planes, which contain the principal axis of rotation, are labeled σ_v or σ_d.

Inversion (i) is a more complex operation. Each point moves through the center of the molecule to a position opposite the original position and as far from the central point as when it started.[1] An example of a molecule having a center of inversion is ethane in the staggered conformation, for which the inversion operation is shown in Figure 4.4.

Many molecules that seem at first glance to have an inversion center do not; for example, methane and other tetrahedral molecules lack inversion symmetry. To see this, hold a methane model with two hydrogen atoms in the vertical plane on the right and two hydrogen atoms in the horizontal plane on the left, as in Figure 4.4. Inversion results in two hydrogen atoms in the horizontal plane on the right and two hydrogen atoms in the vertical plane on the left. Inversion is therefore *not* a symmetry operation of methane, because the orientation of the molecule following the i operation differs from the original orientation.

Squares, rectangles, parallelograms, rectangular solids, octahedra, and snowflakes have inversion centers; tetrahedra, triangles, and pentagons do not (Figure 4.5).

A **rotation-reflection operation** (S_n), sometimes called **improper rotation**, requires rotation of $360°/n$, followed by reflection through a plane perpendicular to the axis of

FIGURE 4.4 Inversion.

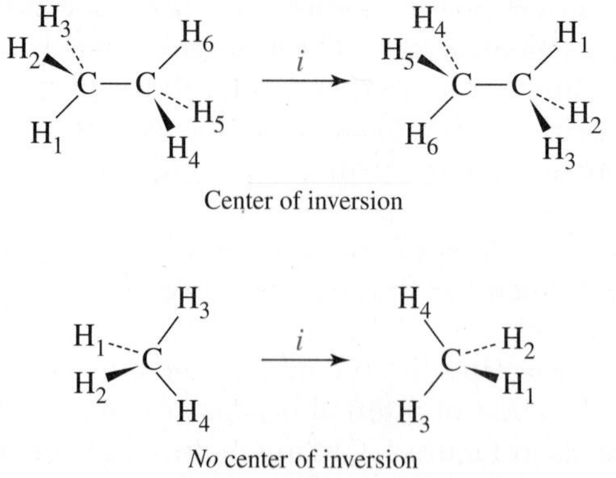

Center of inversion

No center of inversion

[1] This operation must be distinguished from the inversion of a tetrahedral carbon in a bimolecular reaction, which is more like that of an umbrella in a high wind.

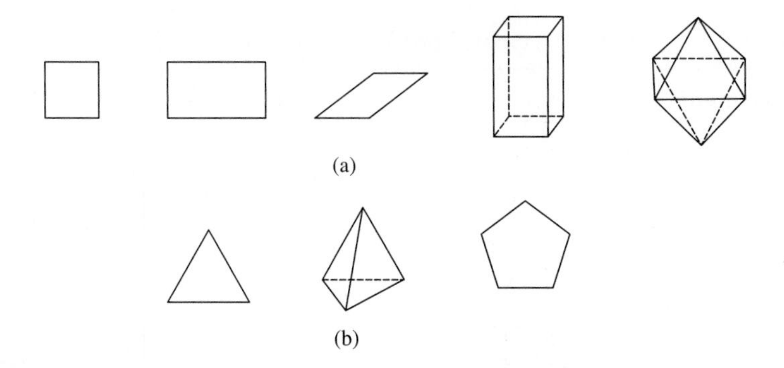

FIGURE 4.5 Figures (a) With and (b) Without Inversion Centers.

rotation. In methane, for example, a line through the carbon and bisecting the angle between two hydrogen atoms on each side is an S_4 axis. There are three such lines, for a total of three S_4 axes. The operation requires a 90° rotation of the molecule, followed by reflection through the plane perpendicular to the axis of rotation. Two S_n operations in succession generate a $C_{n/2}$ operation. In methane, two S_4 operations generate a C_2. These operations are shown in Figure 4.6, along with a table of C and S equivalences for methane.

Molecules sometimes have an S_n axis that is coincident with a C_n axis. For example, in addition to the rotation axes described previously, snowflakes have S_2 ($\equiv i$), S_3, and S_6 axes coincident with the C_6 axis. Molecules may also have S_{2n} axes coincident with C_n; methane is an example, with S_4 axes coincident with C_2 axes, as shown in Figure 4.6.

Note that an S_2 operation is the same as inversion, and an S_1 operation is the same as a reflection plane. The i and σ notations are preferred in these cases. Symmetry elements and operations are summarized in Table 4.1.

Rotation angle	*Symmetry operation*
90°	S_4
180°	C_2 $(= S_4^2)$
270°	S_4^3
360°	E $(= S_4^4)$

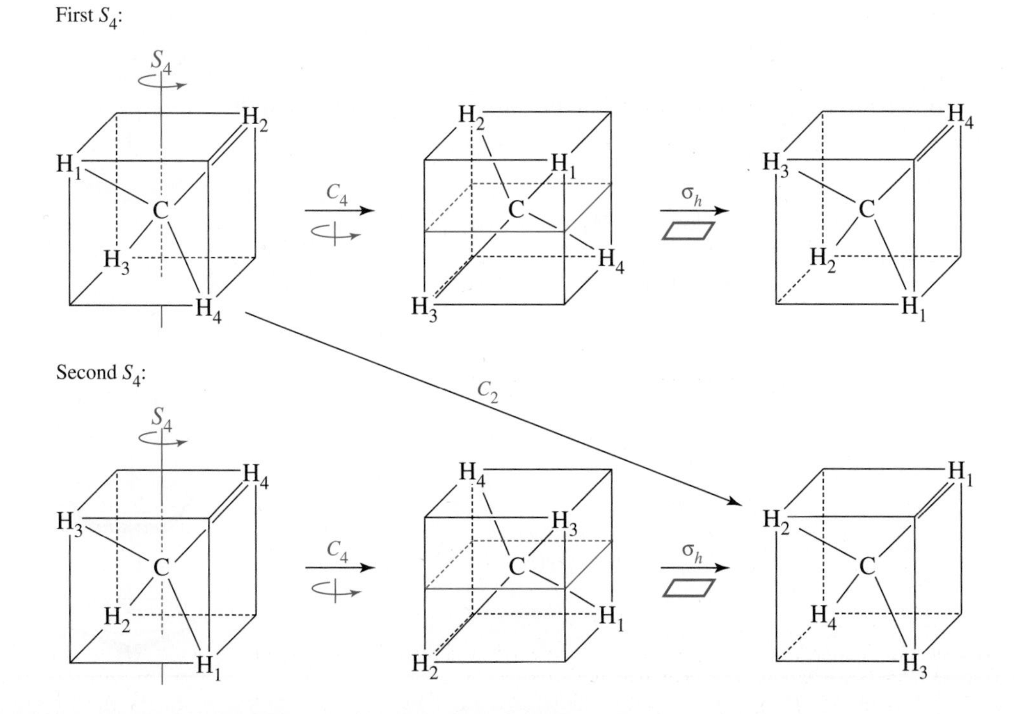

FIGURE 4.6 Improper Rotation or Rotation-Reflection.

TABLE 4.1 Summary Table of Symmetry Elements and Operations

Symmetry Operation	Symmetry Element	Operation	Examples
Identity, E	None	All atoms unshifted	CHFClBr 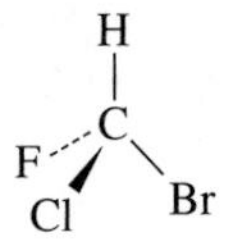
Rotation, C_2	Rotation axis	Rotation by $360°/n$	*p*-dichlorobenzene 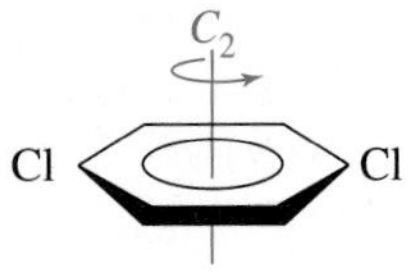
C_3			NH_3 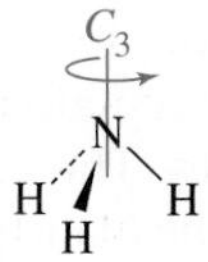
C_4			$[PtCl_4]^{2-}$ 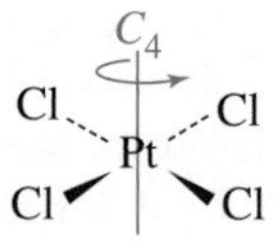
C_5			Cyclopentadienyl group 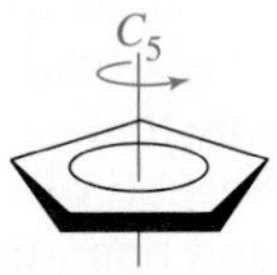
C_6			Benzene 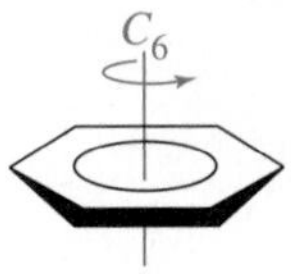
Reflection, σ	Mirror plane	Reflection through a mirror plane	H_2O 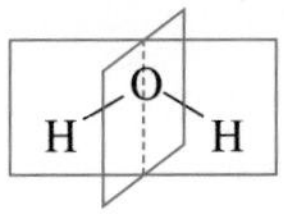
Inversion, i	Inversion center (point)	Inversion through the center	Ferrocene (staggered) 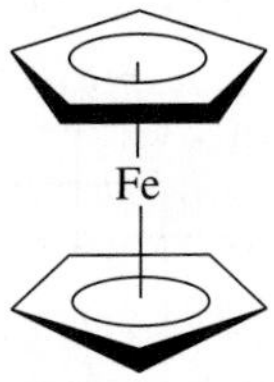
Rotation-reflection, S_4	Rotation-reflection axis (improper axis)	Rotation by $360°/n$, followed by reflection in the plane perpendicular to the rotation axis	CH_4 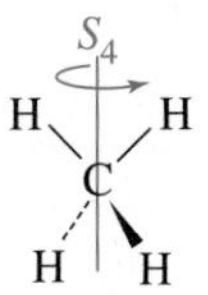
S_6			Ethane (staggered) 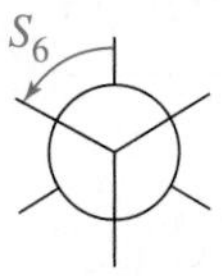
S_{10}			Ferrocene (staggered) 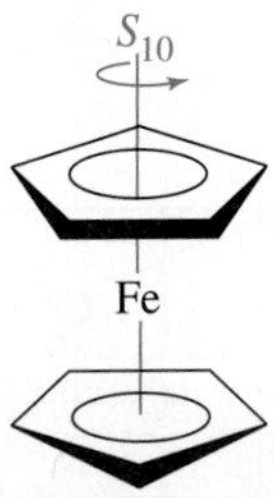

EXAMPLES

Find all the symmetry elements in the following molecules; consider only the atoms when assigning symmetry. Lone pairs influence shapes, but molecular symmetry is based on the geometry of the atoms.

H_2O

H_2O has two planes of symmetry, one in the plane of the molecule and one perpendicular to the molecular plane, as shown in Table 4.1. It also has a C_2 axis collinear with the intersection of the mirror planes. H_2O has no inversion center.

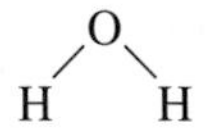

***p*-Dichlorobenzene**

This molecule has three mirror planes: the molecular plane; a plane perpendicular to the molecule, passing through both chlorines; and a plane perpendicular to the first two, bisecting the molecule between the chlorines. It also has three C_2 axes, one perpendicular to the molecular plane (see Table 4.1) and two within the plane, one passing through both chlorines and one perpendicular to the axis passing through the chlorines. Finally, *p*-dichlorobenzene has an inversion center.

Ethane (staggered conformation)

Ethane has three mirror planes, each containing the C—C bond axis and passing through two hydrogens on opposite ends of the molecule. It has a C_3 axis collinear with the carbon-carbon bond and three C_2 axes bisecting the angles between the mirror planes. (Use of a model is especially helpful for viewing the C_2 axes.) Ethane also has a center of inversion and an S_6 axis collinear with the C_3 axis (see Table 4.1).

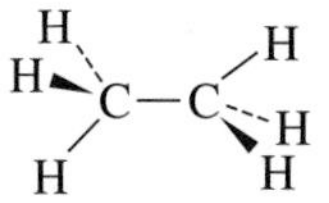

▶ **Exercise 4.1** Using diagrams as necessary, show that $S_2 \equiv i$ and $S_1 \equiv \sigma$.

▶ **Exercise 4.2** Find all the symmetry elements in the following molecules:

NH_3 Cyclohexane (boat conformation) Cyclohexane (chair conformation) XeF_2

4.2 POINT GROUPS

Each molecule has a set of symmetry operations that describes the molecule's overall symmetry. This set of symmetry operations is called the **point group** of the molecule. **Group theory**, the mathematical treatment of the properties of groups, can be used to determine the molecular orbitals, vibrations, and other properties of the molecule. With only a few exceptions, the rules for assigning a molecule to a point group are simple and straightforward. We need only to follow these steps in sequence until a final classification of the molecule is made. A diagram of these steps is shown in Figure 4.7.

1. Determine whether the molecule belongs to one of the cases of very low symmetry (C_1, C_s, C_i) or high symmetry (T_d, O_h, $C_{\infty v}$, $D_{\infty h}$, or I_h) described in Tables 4.2 and 4.3.
2. For all remaining molecules, find the rotation axis with the highest n, the highest order C_n axis for the molecule.

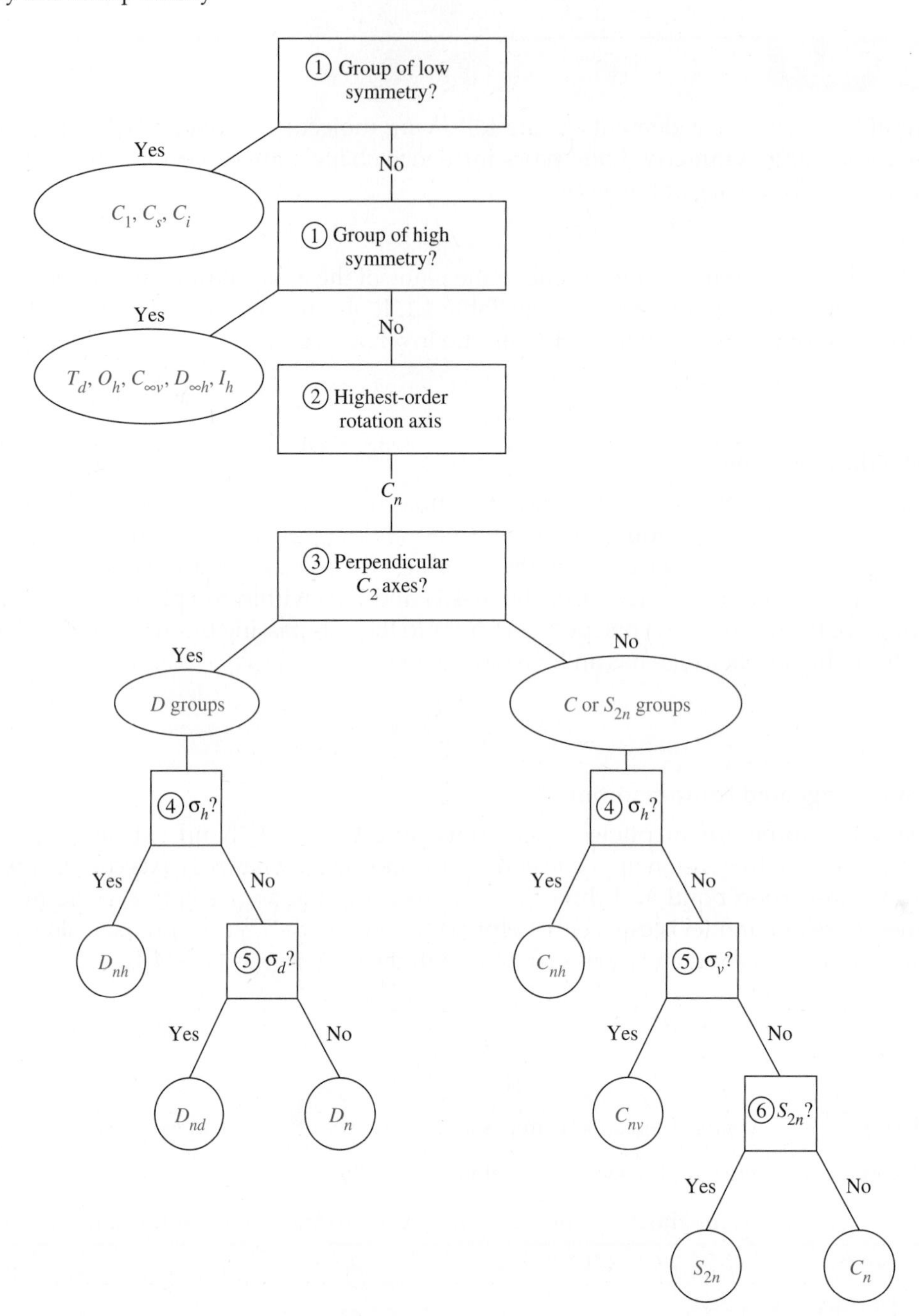

FIGURE 4.7 Diagram of the Point Group Assignment Method.

3. Does the molecule have any C_2 axes perpendicular to the C_n axis? If it does, there will be n of such C_2 axes, and the molecule is in the D set of groups. If not, it is in the C or S set.
4. Does the molecule have a mirror plane (σ_h) perpendicular to the C_n axis? If so, it is classified as C_{nh} or D_{nh}. If not, continue with Step 5.
5. Does the molecule have any mirror planes that contain the C_n axis (σ_v or σ_d)? If so, it is classified as C_{nv} or D_{nd}. If not, but it is in the D set, it is classified as D_n. If the molecule is in the C or S set, continue with Step 6.
6. Is there an S_{2n} axis collinear with the C_n axis? If so, it is classified as S_{2n}. If not, the molecule is classified as C_n.

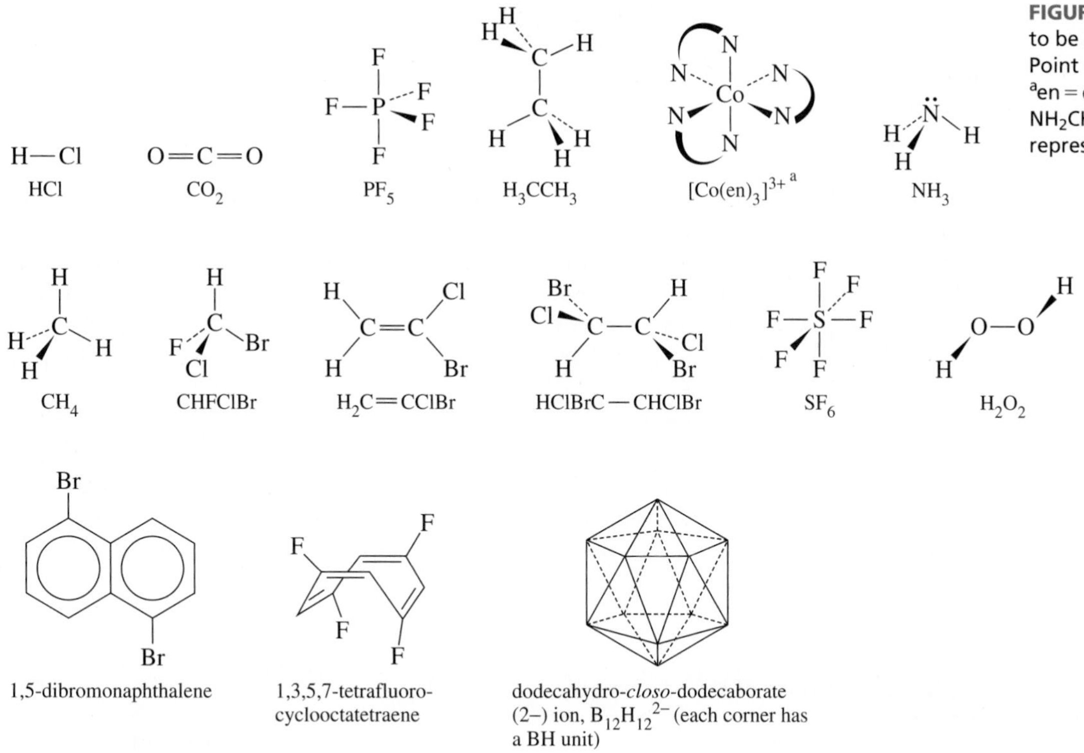

HCl CO_2 PF_5 H_3CCH_3 $[Co(en)_3]^{3+}$ [a] NH_3

CH_4 CHFClBr $H_2C{=}CClBr$ HClBrC—CHClBr SF_6 H_2O_2

1,5-dibromonaphthalene

1,3,5,7-tetrafluoro-cyclooctatetraene

dodecahydro-*closo*-dodecaborate (2−) ion, $B_{12}H_{12}^{2-}$ (each corner has a BH unit)

FIGURE 4.8 Molecules to be Assigned to Point Groups.
[a]en = ethylenediamine = $NH_2CH_2CH_2NH_2$, represented by N⌒N.

Each step is illustrated in the following text by assigning the molecules in Figure 4.8 to their point groups. The low- and high-symmetry cases are treated differently because of their special nature. Molecules that are not in one of these low- or high-symmetry point groups can be assigned to a point group by following steps 2 through 6.

4.2.1 Groups of Low and High Symmetry

> **1.** Determine whether the molecule belongs to one of the special cases of low or high symmetry.

First, inspection of the molecule will determine if it fits one of the low-symmetry cases. These groups have few or no symmetry operations and are described in Table 4.2.

TABLE 4.2 Groups of Low Symmetry

Group	Symmetry	Examples
C_1	No symmetry other than the identity operation	CHFClBr 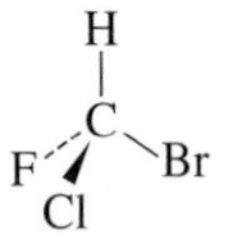
C_s	Only one mirror plane	$H_2C{=}CClBr$ 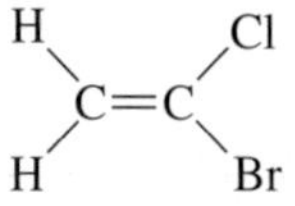
C_i	Only an inversion center; few molecular examples	HClBrC—CHClBr (staggered conformation) 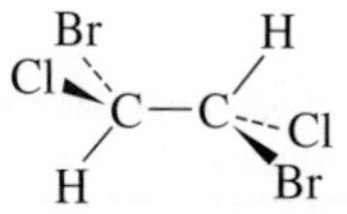

LOW SYMMETRY CHFClBr has no symmetry other than the identity operation and has C_1 symmetry, $H_2C{=}CClBr$ has only one mirror plane and C_s symmetry, and HClBrC—CHClBr in the conformation shown has only a center of inversion and C_i symmetry.

HIGH SYMMETRY Molecules with many symmetry operations may fit one of the high-symmetry cases of linear, tetrahedral, octahedral, or icosahedral symmetry with the characteristics described in Table 4.3. Molecules with very high symmetry are of two types, linear and polyhedral. Linear molecules having a center of inversion have $D_{\infty h}$ symmetry; those lacking an inversion center have $C_{\infty v}$ symmetry. The highly symmetric point groups T_d, O_h, and I_h are described in Table 4.3. It is helpful to note the C_n axes of these molecules. Molecules with T_d symmetry have only C_3 and C_2 axes; those with O_h symmetry have C_4 axes in addition to C_3 and C_2; and I_h molecules have C_5, C_3, and C_2 axes.

HCl has $C_{\infty v}$ symmetry, CO_2 has $D_{\infty h}$ symmetry, CH_4 has tetrahedral (T_d) symmetry, SF_6 has octahedral (O_h) symmetry, and $B_{12}H_{12}^{2-}$ has icosahedral (I_h) symmetry.

There are now seven molecules left to be assigned to point groups out of the original 15.

TABLE 4.3 Groups of High Symmetry

Group	Description	Examples
$C_{\infty v}$	These molecules are linear, with an infinite number of rotations and an infinite number of reflection planes containing the rotation axis. They do not have a center of inversion.	C_∞ H—Cl
$D_{\infty h}$	These molecules are linear, with an infinite number of rotations and an infinite number of reflection planes containing the rotation axis. They also have perpendicular C_2 axes, a perpendicular reflection plane, and an inversion center.	C_∞ O=C=O, C_2
T_d	Most (but not all) molecules in this point group have the familiar tetrahedral geometry. They have four C_3 axes, three C_2 axes, three S_4 axes, and six σ_d planes. They have no C_4 axes.	CH_4 (H, C, H, H, H)
O_h	These molecules include those of octahedral structure, although some other geometrical forms, such as the cube, share the same set of symmetry operations. Among their 48 symmetry operations are four C_3 rotations, three C_4 rotations, and an inversion.	SF_6 (F, F, F—S—F, F, F)
I_h	Icosahedral structures are best recognized by their six C_5 axes, as well as many other symmetry operations—120 in all.	$B_{12}H_{12}^{2-}$ with BH at each vertex of an icosahedron

In addition, there are four other groups, T, T_h, O, and I, which are rarely seen in nature. These groups are discussed at the end of this section.

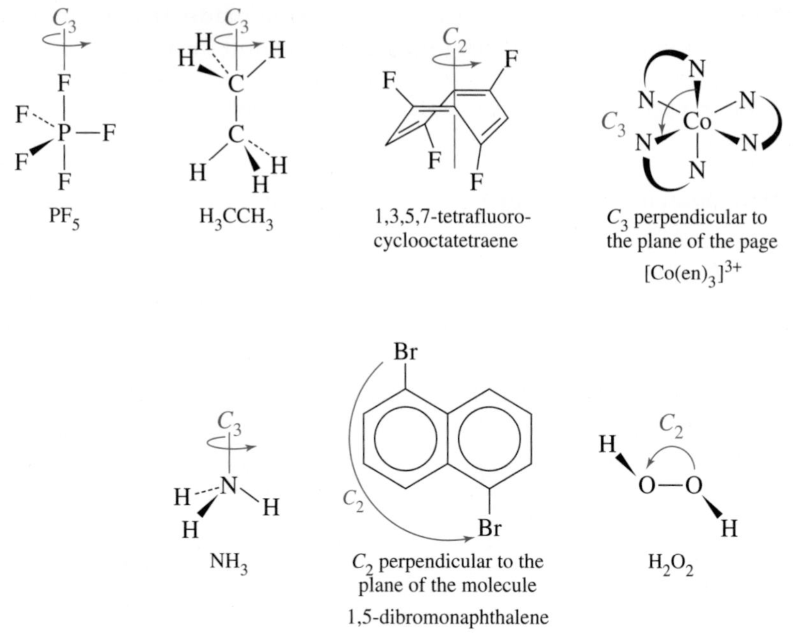

PF_5 — H_3CCH_3 — 1,3,5,7-tetrafluorocyclooctatetraene — C_3 perpendicular to the plane of the page $[Co(en)_3]^{3+}$

NH_3 — C_2 perpendicular to the plane of the molecule 1,5-dibromonaphthalene — H_2O_2

FIGURE 4.9 Rotation Axes.

4.2.2 Other Groups

> **2.** Find the rotation axis with the highest n, the highest order C_n axis for the molecule. This is the principal axis of the molecule.

The rotation axes for the examples are shown in Figure 4.9. If they are all equivalent, any one can be chosen as the principal axis.

> **3.** Does the molecule have any C_2 axes perpendicular to the C_n axis?

The C_2 axes are shown in Figure 4.10.

Yes *D Groups*

PF_5, H_3CCH_3, $[Co(en)_3]^{3+}$

Molecules with C_2 axes perpendicular to the principal axis are in one of the groups designated by the letter *D*; there are n C_2 axes.

No *C or S Groups*

NH_3, 1,5-dibromonaphthalene, H_2O_2, 1,3,5,7-tetrafluorocyclooctatetraene

Molecules with no perpendicular C_2 axes are in one of the groups designated by the letters *C* or *S*.

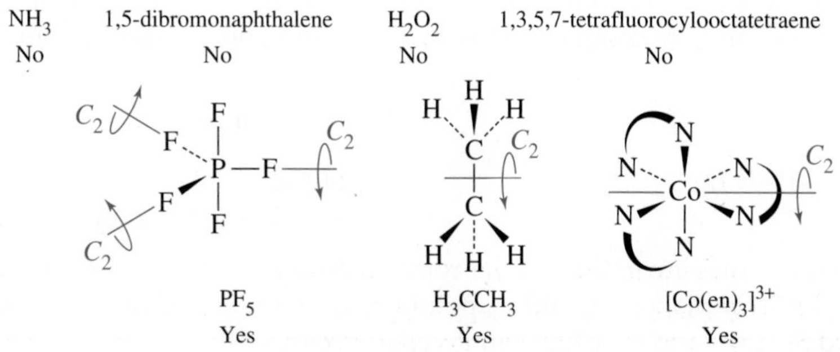

NH_3 No — 1,5-dibromonaphthalene No — H_2O_2 No — 1,3,5,7-tetrafluorocylooctatetraene No

PF_5 Yes — H_3CCH_3 Yes — $[Co(en)_3]^{3+}$ Yes

FIGURE 4.10 Perpendicular C_2 Axes.

No final assignments of point groups have been made, but the molecules have now been divided into two major categories, the *D* set and the *C* or *S* set.

> **4.** Does the molecule have a mirror plane (σ_h horizontal plane) perpendicular to the C_n axis?

The horizontal mirror planes are shown in Figure 4.11.

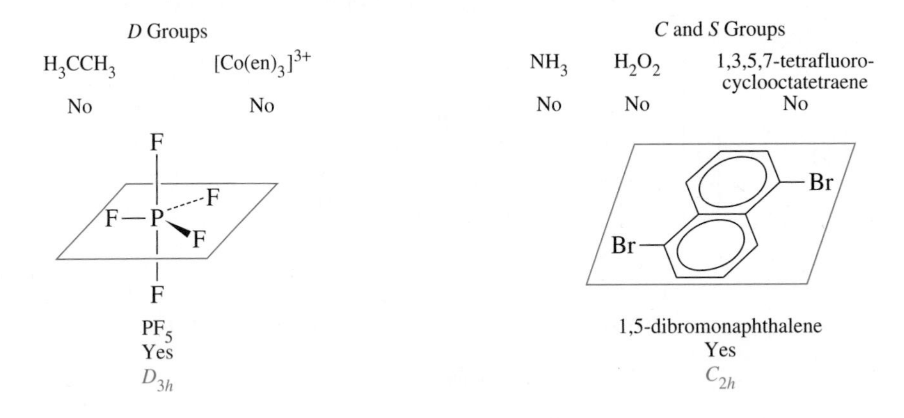

FIGURE 4.11 Horizontal Mirror Planes.

***D* Groups**

Yes D_{nh}

PF_5 is D_{3h}

***C* and *S* Groups**

Yes C_{nh}

1,5-dibromonaphthalene is C_{2h}

These molecules are now assigned to point groups and need not be considered further. Both have horizontal mirror planes.

No D_n or D_{nd}

H_3CCH_3, $[Co(en)_3]^{3+}$

No C_n, C_{nv}, or S_{2n}

NH_3, H_2O_2,
1,3,5,7-tetrafluorocyclooctatetraene

None of these have horizontal mirror planes; they must be carried further in the process.

> **5.** Does the molecule have any mirror planes that contain the C_n axis?

These mirror planes are shown in Figure 4.12.

***D* Groups**

Yes D_{nd}

H_3CCH_3 (staggered) is D_{3d}

***C* and *S* Groups**

Yes C_{nv}

NH_3 is C_{3v}

These molecules have mirror planes containing the major C_n axis, but no horizontal mirror planes, and are assigned to the corresponding point groups. There will be *n* of these planes.

No D_n

$[Co(en)_3]^{3+}$ is D_3

No C_n or S_{2n}

H_2O_2,
1,3,5,7-tetrafluorocyclooctatetraene

These molecules are in the simpler rotation groups D_n, C_n, and S_{2n} because they do not have any mirror planes. D_n and C_n point groups have *only* C_n axes. S_{2n} point groups have C_n and S_{2n} axes and may have an inversion center.

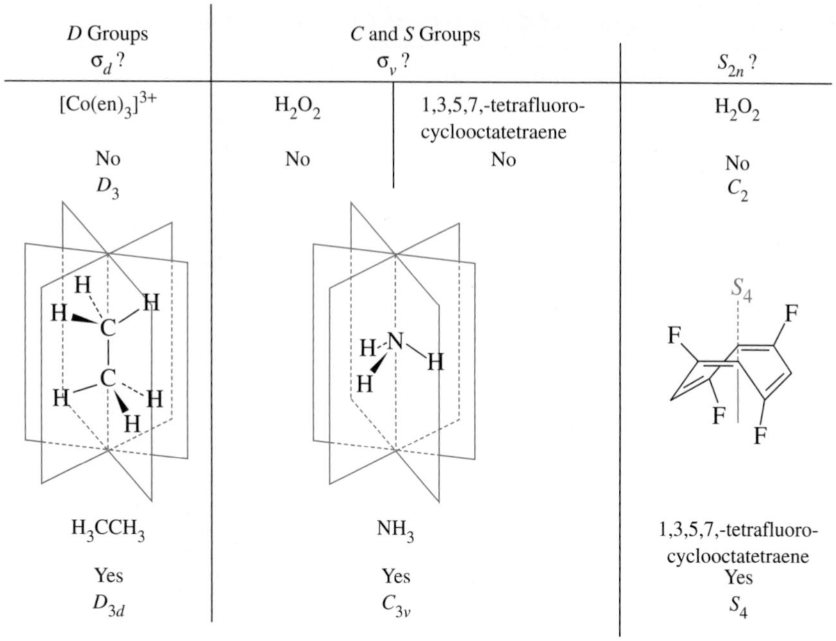

D Groups σ_d?	C and S Groups σ_v?		S_{2n}?
$[Co(en)_3]^{3+}$	H_2O_2	1,3,5,7,-tetrafluoro-cyclooctatetraene	H_2O_2
No D_3	No	No	No C_2
H_3CCH_3	NH_3		1,3,5,7,-tetrafluoro-cyclooctatetraene
Yes D_{3d}	Yes C_{3v}		Yes S_4

FIGURE 4.12 Vertical or Dihedral Mirror Planes or S_{2n} Axes.

6. Is there an S_{2n} axis collinear with the C_n axis?

***D* Groups**

Any molecules in this category that have S_{2n} axes have already been assigned to groups. There are no additional groups to be considered here.

***C* and *S* Groups**

Yes S_{2n}

1,3,5,7-tetrafluorocyclooctatetraene is S_4

No C_n

H_2O_2 is C_2

We have only one example in our list that falls into the S_{2n} groups, as seen in Figure 4.12.

A branching diagram that summarizes this method of assigning point groups was given in Figure 4.7, and more examples are given in Table 4.4.

TABLE 4.4 Further Examples of *C* and *D* Point Groups

General Label	Point Group and Example	
C_{nh}	C_{2h}	difluorodiazene 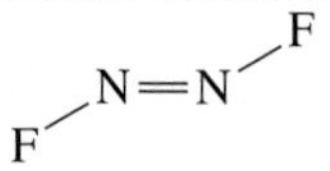
	C_{3h}	$B(OH)_3$, planar 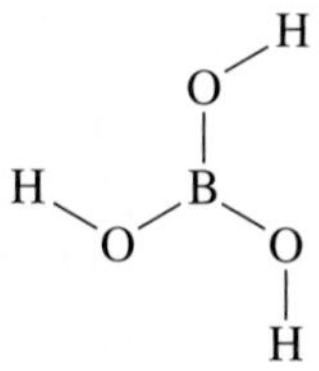
C_{nv}	C_{2v}	H_2O 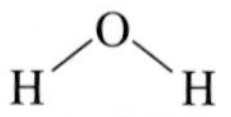

Continued

TABLE 4.4 Further Examples of *C* and *D* Point Groups—continued

General Label	Point Group and Example	
	C_{3v}	PCl_3 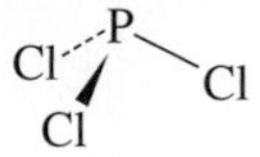
	C_{4v}	BrF_5 (square pyramid) 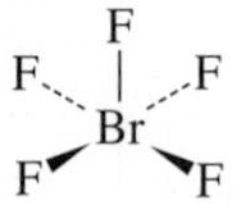
	$C_{\infty v}$	HF, CO, HCN
C_n	C_2	N_2H_4, which has a *gauche* conformation 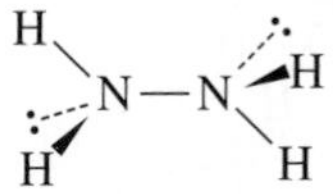
	C_3	$P(C_6H_5)_3$, which is like a three-bladed propeller distorted out of the planar shape by a lone pair on the P 
D_{nh}	D_{3h}	BF_3 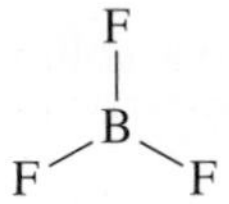
	D_{4h}	$PtCl_4^{2-}$ 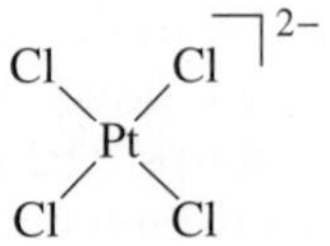
	D_{5h}	$Os(C_5H_5)_2$ (eclipsed) 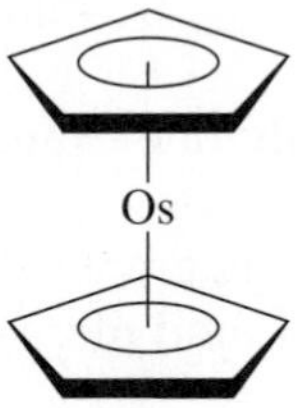
	D_{6h}	benzene 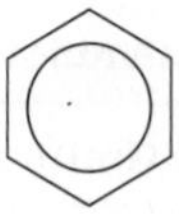
	$D_{\infty h}$	F_2, N_2
		acetylene (C_2H_2)
D_{nd}	D_{2d}	$H_2C=C=CH_2$, allene 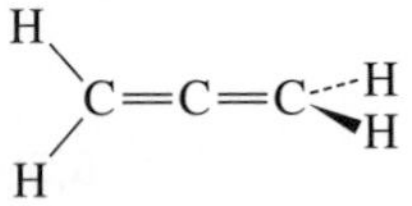

Continued

TABLE 4.4 Further Examples of *C* and *D* Point Groups—continued

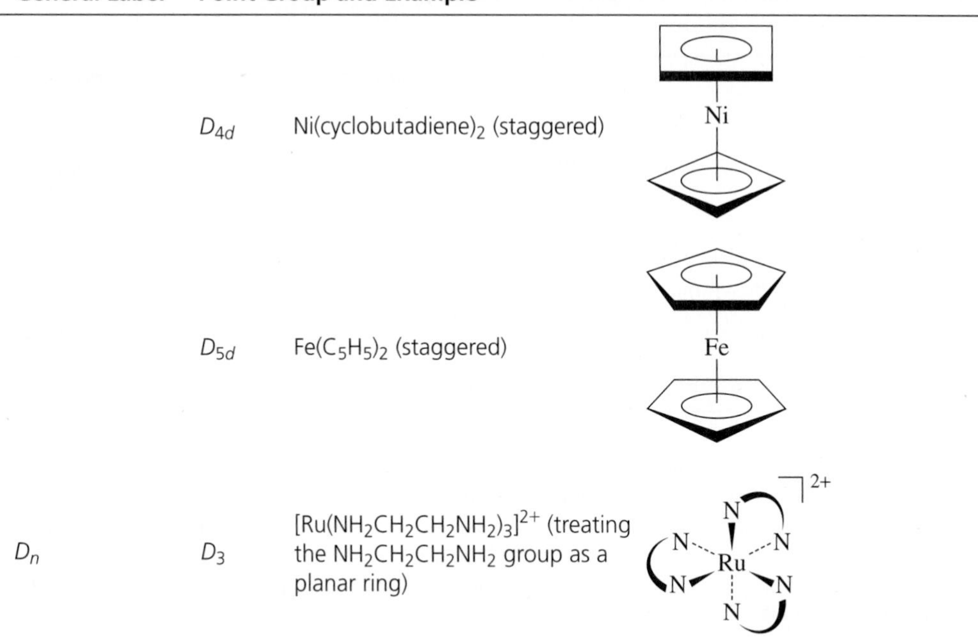

General Label	Point Group and Example		
	D_{4d}	Ni(cyclobutadiene)$_2$ (staggered)	Ni
	D_{5d}	$Fe(C_5H_5)_2$ (staggered)	Fe
D_n	D_3	$[Ru(NH_2CH_2CH_2NH_2)_3]^{2+}$ (treating the $NH_2CH_2CH_2NH_2$ group as a planar ring)	Ru, N, 2+

EXAMPLES

Determine the point groups of the following molecules and ions from Figures 3.13 and 3.16:

XeF_4

1. XeF_4 is not in the groups of low or high symmetry.
2. Its highest order rotation axis is C_4.
3. It has four C_2 axes perpendicular to the C_4 axis and is therefore in the *D* set of groups.
4. It has a horizontal plane perpendicular to the C_4 axis. Therefore its point group is D_{4h}.

SF_4

1. SF_4 is not in the groups of high or low symmetry.
2. Its highest order (and only) rotation axis is a C_2 axis passing through the lone pair.
3. The ion has no other C_2 axes and is therefore in the *C* or *S* set.
4. It has no mirror plane perpendicular to the C_2.
5. It has two mirror planes containing the C_2 axis. Therefore, the point group is C_{2v}.

IOF_3

1. The molecule has no symmetry (other than *E*). Its point group is C_1.

▶ **Exercise 4.3** Use the procedure described above to verify the point groups of the molecules in Table 4.4.

C VERSUS D POINT GROUP CLASSIFICATIONS All molecules having these classifications must have a C_n axis. If more than one C_n axis is found, the highest order axis (largest value of n) is used as the reference axis. In general, it is useful to orient this axis vertically.

	***D* Classifications**	***C* Classifications**
General Case:		
Look for C_2 axes perpendicular to the highest order C_n axis.	nC_2 axes $\perp$ C_n axis	No C_2 axes $\perp$ C_n axis
Subcategories:		
If a horizontal plane of symmetry exists:	$\mathbf{D_{nh}}$	$\mathbf{C_{nh}}$
If n vertical planes exist:	$\mathbf{D_{nd}}$	$\mathbf{C_{nv}}$
If no planes of symmetry exist:	$\mathbf{D_n}$	$\mathbf{C_n}$

NOTES:

1. Vertical planes contain the highest order C_n axis. In the D_{nd} case, the planes are designated *dihedral* because they are between the C_2 axes—thus, the subscript *d*.
2. The presence of a C_n axis does not guarantee that a molecule will be in a *D* or *C* category; the high-symmetry T_d, O_h, and I_h point groups and related groups have a large number of C_n axes.
3. When in doubt, you can always check the character tables (Appendix C) for a complete list of symmetry elements for any point group.

GROUPS RELATED TO I_h, O_h, AND T_d GROUPS The high-symmetry point groups I_h, O_h, and T_d are well known in chemistry and are represented by such classic molecules as C_{60}, SF_6, and CH_4. For each of these point groups, there is also a purely rotational subgroup (I, O, and T, respectively) in which the only symmetry operations other than the identity operation are proper axes of rotation. The symmetry operations for these point groups are in Table 4.5.

We are not yet finished with high-symmetry point groups. One more group, T_h, remains. The T_h point group is derived by adding a center of inversion to the T point group; adding i generates the additional symmetry operations S_6, S_6^5, and σ_h.

T_h symmetry is rare but is known for a few molecules. The compound shown in Figure 4.13 is an example. I, O, and T symmetry are rarely if ever encountered in chemistry.

That's all there is to it! It takes a fair amount of practice, preferably using molecular models, to learn the point groups well, but once you know them, they can be

TABLE 4.5 Symmetry Operations for High-Symmetry Point Groups and Their Rotational Subgroups

Point Group	Symmetry Operations									
I_h	E	$12C_5$	$12C_5^2$	$20C_3$	$15C_2$	i	$12S_{10}$	$12S_{10}^3$	$20S_6$	15σ
I	E	$12C_5$	$12C_5^2$	$20C_3$	$15C_2$					
O_h	E	$8C_3$	$6C_2$	$6C_4$	$3C_2$ $(\equiv C_4^2)$	i	$6S_4$	$8S_6$	$3\sigma_h$	$6\sigma_d$
O	E	$8C_3$	$6C_2$	$6C_4$	$3C_2$ $(\equiv C_4^2)$					
T_d	E	$8C_3$	$3C_2$				$6S_4$			$6\sigma_d$
T	E	$\overbrace{4C_3\ 4C_3^2}$	$3C_2$							
T_h	E	$4C_3\ 4C_3^2$	$3C_2$			i	$4S_6$	$4S_6^5$	$3\sigma_h$	

FIGURE 4.13 $W[N(CH_3)_2]_6$, a Molecule with T_h Symmetry.

extremely useful. Several practical applications of point groups appear later in this chapter, and additional applications are included in later chapters.

4.3 PROPERTIES AND REPRESENTATIONS OF GROUPS

All mathematical groups, of which point groups are special types, must have certain properties. These properties are listed and illustrated in Table 4.6, using the symmetry operations of NH_3 in Figure 4.14 as an example.

4.3.1 Matrices

Important information about the symmetry aspects of point groups is summarized in character tables, described later in this chapter. To understand the construction and use of character tables, we first need to consider the properties of matrices, which are the basis for the tables.[2]

By **matrix** we mean an ordered array of numbers, such as

$$\begin{bmatrix} 3 & 7 \\ 2 & 1 \end{bmatrix} \quad \text{or} \quad [2 \;\; 0 \;\; 1 \;\; 3 \;\; 5]$$

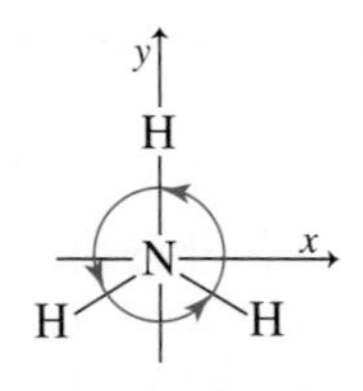

C_3 rotation about the z axis

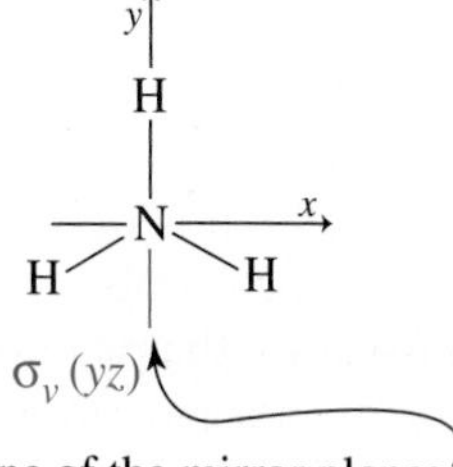

One of the mirror planes

NH_3 after E · NH_3 after C_3 · NH_3 after $\sigma_v(yz)$

FIGURE 4.14 Symmetry Operations for Ammonia. (Top view) NH_3 is of point group C_{3v}, with the symmetry operations E, C_3, C_3^2, σ_v, σ_v', and σ_v'' usually written as E, $2C_3$, and $3\sigma_v$ (note that $C_3^3 \equiv E$).

[2] More details on matrices and their manipulation are available in Appendix 1 of F. A. Cotton, *Chemical Applications of Group Theory*, 3rd ed., John Wiley & Sons, New York, 1990, and in linear algebra and finite mathematics textbooks.

TABLE 4.6 Properties of a Group

Property of Group	Examples from Point Group
1. Each group must contain an **identity** operation that commutes (in other words, $EA = AE$) with all other members of the group and leaves them unchanged ($EA = AE = A$).	C_{3v} molecules (and *all* molecules) contain the identity operation E.
2. Each operation must have an **inverse** that, when combined with the operation, yields the identity operation (sometimes a symmetry operation may be its own inverse). *Note:* By convention, we perform combined symmetry operations *from right to left* as written.	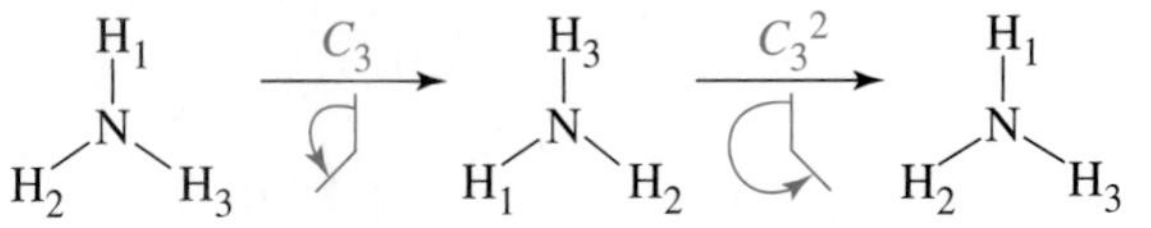 $C_3^2C_3 = E$ (C_3 and C_3^2 are inverses of each other) 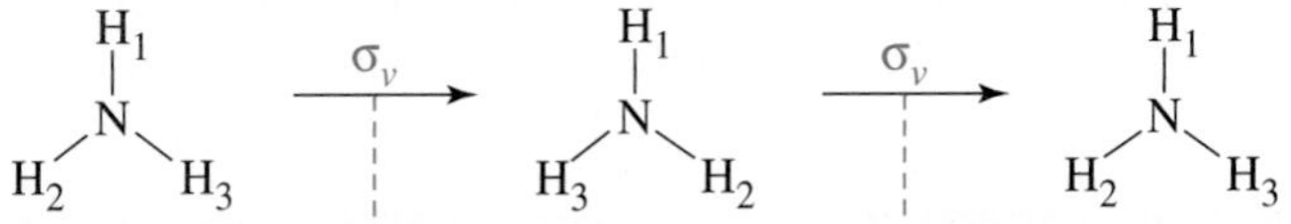 $\sigma_v\sigma_v = E$ (mirror planes are shown as dashed lines; σ_v is its own inverse)
3. The product of any two group operations must also be a member of the group. This includes the product of any operation with itself.	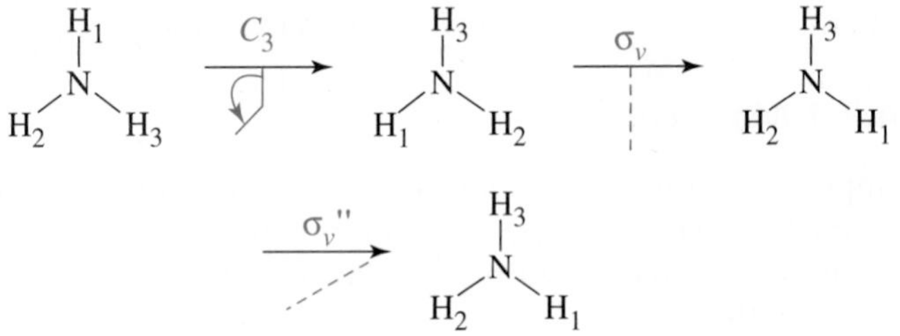 σ_vC_3 has the same overall effect as σ_v'', therefore we write $\sigma_vC_3 = \sigma_v''$. It can be shown that the products of any two operations in C_{3v} are also members of C_{3v}.
4. The associative property of combination must hold. In other words, $A(BC) = (AB)C$.	$C_3(\sigma_v\sigma_v') = (C_3\sigma_v)\sigma_v'$

To multiply matrices, it is first required that the number of vertical columns of the first matrix be equal to the number of horizontal rows of the second matrix. To find the product add, term by term, the products of each *row* of the first matrix by each *column* of the second (each term in a row must be multiplied by its corresponding term in the appropriate column of the second matrix). Place the resulting sum in the product matrix with the row determined by the row of the first matrix and the column determined by the column of the second matrix:

$$C_{ij} = \Sigma A_{ik} \times B_{kj}$$

Here C_{ij} = product matrix, with i rows and j columns

A_{ik} = initial matrix, with i rows and k columns

B_{kj} = initial matrix, with k rows and j columns

EXAMPLES

$$\overset{k}{i\begin{bmatrix} 1 & 5 \\ 2 & 6 \end{bmatrix}} \times \overset{j}{\begin{bmatrix} 7 & 3 \\ 4 & 8 \end{bmatrix}k} = \overset{j}{\begin{bmatrix} (1)(7)+(5)(4) & (1)(3)+(5)(8) \\ (2)(7)+(6)(4) & (2)(3)+(6)(8) \end{bmatrix}i} = \overset{j}{\begin{bmatrix} 27 & 43 \\ 38 & 54 \end{bmatrix}i}$$

This example has two rows and two columns in each initial matrix, so it has two rows and two columns in the product matrix; $i = j = k = 2$.

$$\overset{k}{i[1 \quad 2 \quad 3]}\overset{j}{\begin{bmatrix} 1 & 0 & 0 \\ 0 & -1 & 0 \\ 0 & 0 & 1 \end{bmatrix}k} =$$

$$\overset{j}{[(1)(1)+(2)(0)+(3)(0) \quad (1)(0)+(2)(-1)+(3)(0) \quad (1)(0)+(2)(0)+(3)(1)]i} = \overset{j}{[1 \quad -2 \quad 3]i}$$

Here, $i = 1, j = 3$, and $k = 3$, so the product matrix has 1 row (i) and 3 columns (j).

$$\overset{k}{i\begin{bmatrix} 1 & 0 & 0 \\ 0 & -1 & 0 \\ 0 & 0 & 1 \end{bmatrix}}\overset{j}{\begin{bmatrix} 1 \\ 2 \\ 3 \end{bmatrix}k} = \overset{j}{\begin{bmatrix} (1)(1)+(0)(2)+(1)(3) \\ (0)(1)+(-1)(2)+(0)(3) \\ (0)(1)+(0)(2)+(1)(3) \end{bmatrix}i} = \overset{j}{\begin{bmatrix} 1 \\ -2 \\ 3 \end{bmatrix}i}$$

Here $i = 3, j = 1$, and $k = 3$, so the product matrix has three rows (i) and one column (j).

▶ **Exercise 4.4** Do the following multiplications:

a. $\begin{bmatrix} 5 & 1 & 3 \\ 4 & 2 & 2 \\ 1 & 2 & 3 \end{bmatrix} \times \begin{bmatrix} 2 & 1 & 1 \\ 1 & 2 & 3 \\ 5 & 4 & 3 \end{bmatrix}$

b. $\begin{bmatrix} 1 & -1 & -2 \\ 0 & 1 & -1 \\ 1 & 0 & 0 \end{bmatrix} \times \begin{bmatrix} 2 \\ 1 \\ 3 \end{bmatrix}$

c. $[1 \quad 2 \quad 3] \times \begin{bmatrix} 1 & -1 & -2 \\ 2 & 1 & -1 \\ 3 & 2 & 1 \end{bmatrix}$

4.3.2 Representations of Point Groups

SYMMETRY OPERATIONS: MATRIX REPRESENTATIONS Consider the effects of the symmetry operations of the C_{2v} point group on the set of x, y, and z coordinates. The water molecule is an example of a molecule having C_{2v} symmetry. It has a C_2 axis through the oxygen and in the plane of the molecule, no perpendicular C_2 axes, and no horizontal mirror plane; but it does have two vertical mirror planes, as shown in Table 4.1 and

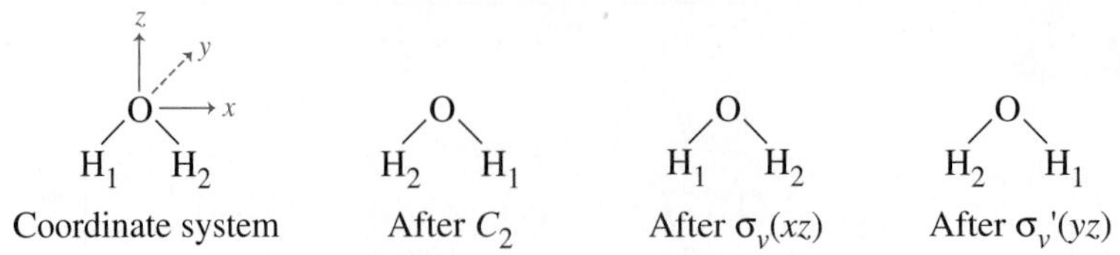

FIGURE 4.15
Symmetry Operations of the Water Molecule.

Figure 4.15. The z axis is usually chosen as the axis of highest rotational symmetry; for H_2O, this is the *only* rotational axis. The other axes are arbitrary. We will use the xz plane as the plane of the molecule.[3] This set of axes is chosen to obey the right-hand rule (the thumb and first two fingers of the right hand, held perpendicular to each other, are labeled x, y, and z, respectively).

Each symmetry operation may be expressed as a **transformation matrix** as follows:

$$[\text{New coordinates}] = [\text{transformation matrix}][\text{old coordinates}]$$

As examples, consider how transformation matrices can be used to represent the symmetry operations of the C_{2v} point group:

C_2: Rotate a point having coordinates (x, y, z) about the $C_2(z)$ axis. The new coordinates are given by

$$\begin{aligned} x' &= \text{new } x = -x \\ y' &= \text{new } y = -y \\ z' &= \text{new } z = z \end{aligned} \qquad \begin{bmatrix} -1 & 0 & 0 \\ 0 & -1 & 0 \\ 0 & 0 & 1 \end{bmatrix} \text{ Transformation matrix for } C_2$$

In matrix notation,

$$\begin{bmatrix} x' \\ y' \\ z' \end{bmatrix} = \begin{bmatrix} -1 & 0 & 0 \\ 0 & -1 & 0 \\ 0 & 0 & 1 \end{bmatrix} \begin{bmatrix} x \\ y \\ z \end{bmatrix} = \begin{bmatrix} -x \\ -y \\ z \end{bmatrix} \text{ or } \begin{bmatrix} x' \\ y' \\ z' \end{bmatrix} = \begin{bmatrix} -x \\ -y \\ z \end{bmatrix}$$

$$\begin{bmatrix} \text{New} \\ \text{coordinates} \end{bmatrix} = \begin{bmatrix} \text{transformation} \\ \text{matrix} \end{bmatrix} \begin{bmatrix} \text{old} \\ \text{coordinates} \end{bmatrix} = \begin{bmatrix} \text{new coordinates} \\ \text{in terms of old} \end{bmatrix}$$

$\sigma_v(xz)$: Reflect a point with coordinates (x, y, z) through the xz plane.

$$\begin{aligned} x' &= \text{new } x = x \\ y' &= \text{new } y = -y \\ z' &= \text{new } z = z \end{aligned} \qquad \begin{bmatrix} 1 & 0 & 0 \\ 0 & -1 & 0 \\ 0 & 0 & 1 \end{bmatrix} \text{ Transformation matrix for } \sigma_v(xz)$$

The matrix equation is

$$\begin{bmatrix} x' \\ y' \\ z' \end{bmatrix} = \begin{bmatrix} 1 & 0 & 0 \\ 0 & -1 & 0 \\ 0 & 0 & 1 \end{bmatrix} \begin{bmatrix} x \\ y \\ z \end{bmatrix} = \begin{bmatrix} x \\ -y \\ z \end{bmatrix} \text{ or } \begin{bmatrix} x' \\ y' \\ z' \end{bmatrix} = \begin{bmatrix} x \\ -y \\ z \end{bmatrix}$$

The transformation matrices for the four symmetry operations of the group follow:

$$E: \begin{bmatrix} 1 & 0 & 0 \\ 0 & 1 & 0 \\ 0 & 0 & 1 \end{bmatrix} \quad C_2: \begin{bmatrix} -1 & 0 & 0 \\ 0 & -1 & 0 \\ 0 & 0 & 1 \end{bmatrix} \quad \sigma_v(xz): \begin{bmatrix} 1 & 0 & 0 \\ 0 & -1 & 0 \\ 0 & 0 & 1 \end{bmatrix} \quad \sigma_v'(yz): \begin{bmatrix} -1 & 0 & 0 \\ 0 & 1 & 0 \\ 0 & 0 & 1 \end{bmatrix}$$

[3] Some sources use yz as the plane of the molecule. The assignment of B_1 and B_2 in Section 4.3.3 is reversed with this choice.

▶ **Exercise 4.5** Verify the transformation matrices for the E and $\sigma_v'(yz)$ operations of the C_{2v} point group.

This set of matrices satisfies the properties of a mathematical **group**. We call this a **matrix representation** of the C_{2v} point group. This representation is a set of matrices, each corresponding to an operation in the group; these matrices combine in the same way as the operations themselves. For example, multiplying two of the matrices is equivalent to carrying out the two corresponding operations and results in a matrix that corresponds to the resulting operation (the operations are carried out right to left, so $C_2 \times \sigma_v$ means σ_v followed by C_2):

$$C_2 \times \sigma_v(xz) = \begin{bmatrix} -1 & 0 & 0 \\ 0 & -1 & 0 \\ 0 & 0 & 1 \end{bmatrix} \begin{bmatrix} 1 & 0 & 0 \\ 0 & -1 & 0 \\ 0 & 0 & 1 \end{bmatrix} = \begin{bmatrix} -1 & 0 & 0 \\ 0 & 1 & 0 \\ 0 & 0 & 1 \end{bmatrix} = \sigma_v'(yz)$$

The matrices of the matrix representation of the C_{2v} group also describe the operations of the group shown in Figure 4.15. The C_2 and $\sigma_v'(yz)$ operations interchange H_1 and H_2, whereas E and $\sigma_v(xz)$ leave them unchanged.

CHARACTERS The **character**, defined only for a square matrix, is the trace of the matrix, or the sum of the numbers on the diagonal from upper left to lower right. For the C_{2v} point group, the following characters are obtained from the preceding matrices:

E	C_2	$\sigma_v(xz)$	$\sigma_v'(yz)$
3	−1	1	1

We can say that this set of characters also forms a **representation**. It is an alternate shorthand version of the matrix representation. Whether in matrix or character format, this representation is called a **reducible representation**, a combination of more fundamental **irreducible representations** as described in the next section. Reducible representations are frequently designated with a capital gamma (Γ).

REDUCIBLE AND IRREDUCIBLE REPRESENTATIONS Each transformation matrix in the C_{2v} set above is "block diagonalized"; that is, it can be broken down into smaller matrices along the diagonal, with all other matrix elements equal to zero:

$$E: \begin{bmatrix} [1] & 0 & 0 \\ 0 & [1] & 0 \\ 0 & 0 & [1] \end{bmatrix} \quad C_2: \begin{bmatrix} [-1] & 0 & 0 \\ 0 & [-1] & 0 \\ 0 & 0 & [1] \end{bmatrix} \quad \sigma_v(xz): \begin{bmatrix} [1] & 0 & 0 \\ 0 & [-1] & 0 \\ 0 & 0 & [1] \end{bmatrix} \quad \sigma_v'(yz): \begin{bmatrix} [-1] & 0 & 0 \\ 0 & [1] & 0 \\ 0 & 0 & [1] \end{bmatrix}$$

All the nonzero elements become 1×1 matrices along the principal diagonal.

When matrices are block diagonalized in this way, the x, y, and z coordinates are also block diagonalized. As a result, the x, y, and z coordinates are independent of each other. The matrix elements in the 1,1 positions (numbered as row, column) describe the results of the symmetry operations on the x coordinate, those in the 2,2 positions describe the results of the operations on the y coordinate, and those in the 3,3 positions describe the results of the operations on the z coordinate. The four matrix elements for x form a representation of the group, those for y form a second representation, and those for z form a third representation, all shown in the following table:

	E	C_2	$\sigma_v(xz)$	$\sigma_v'(yz)$	Coordinate Used
	1	−1	1	−1	x
	1	−1	−1	1	y
	1	1	1	1	z
Γ	3	−1	1	1	

Irreducible representations of the C_{2v} point group add to make up the reducible representation Γ.

Each row is an irreducible representation: it cannot be simplified further. The characters of these three irreducible representations added together under each operation (column) make up the characters of the reducible representation Γ, just as the combination of all the matrices for the x, y, and z coordinates makes up the matrices of the reducible representation. For example, the sum of the three characters for x, y, and z under the C_2 operation is -1, the character for Γ under this same operation.

The set of 3×3 matrices obtained for H_2O is called a reducible representation because it is the sum of irreducible representations (the block-diagonalized 1×1 matrices), which cannot be reduced to smaller component parts. The set of characters of these matrices also forms the reducible representation Γ, for the same reason.

4.3.3 Character Tables

Three of the representations for C_{2v}, labeled A_1, B_1, and B_2 below, have been determined so far. The fourth, called A_2, can be found by using the properties of a group described in Table 4.7. A complete set of irreducible representations for a point group is called the **character table** for that group. The character table for each point group is unique; character tables for the common point groups are included in Appendix C.

The complete character table for C_{2v} with the irreducible representations in the order commonly used, follows:

C_{2v}	E	C_2	$\sigma_v(xz)$	$\sigma_v'(yz)$		
A_1	1	1	1	1	z	x^2, y^2, z^2
A_2	1	1	−1	−1	R_z	xy
B_1	1	−1	1	−1	x, R_y	xz
B_2	1	−1	−1	1	y, R_x	yz

The labels used with character tables are as follows:

x, y, z	transformations of the x, y, z coordinates or combinations thereof
R_x, R_y, R_z	rotation about the x, y, and z axes
$\boldsymbol{R}$	any symmetry operation, such as C_2 or $\sigma_v(xz)$
χ	character of an operation
i and j	designation of different representations, such as A_1 or A_2
h	order of the group (the total number of symmetry operations in the group)

The labels in the left column, used to designate the representations, will be described later in this section. Other useful terms are defined in Table 4.7.

TABLE 4.7 Properties of Characters of Irreducible Representations in Point Groups

Property	Example: C_{2v}
1. The total number of symmetry operations in the group is called the **order (h)**. To determine the order of a group, simply total the number of symmetry operations listed in the top row of the character table.	Order = 4 four symmetry operations: E, C_2, $\sigma_v(xz)$, and $\sigma_v'(yz)$
2. Symmetry operations are arranged in **classes**. All operations in a class have identical characters for their transformation matrices and are grouped in the same column in character tables.	Each symmetry operation is in a separate class; therefore, there are four columns in the character table.
3. The number of irreducible representations equals the number of classes. This means that character tables have the same number of rows and columns (they are square).	Because there are four classes, there must also be four irreducible representations—and there are.
4. The sum of the squares of the **dimensions** (characters under E) of each of the irreducible representations equals the order of the group. $h = \sum_i [\chi_i(E)]^2$	$1^2 + 1^2 + 1^2 + 1^2 = 4 = h$, the order of the group.
5. For any irreducible representation, the sum of the squares of the characters multiplied by the number of operations in the class (see Table 4.8 for an example), equals the order of the group. $h = \sum_R [\chi_i(\mathbf{R})]^2$	For A_2, $1^2 + 1^2 + (-1)^2 + (-1)^2 = 4 = h$. Each operation is its own class in this group.
6. Irreducible representations are **orthogonal** to each other. The sum of the products of the characters, multiplied together for each class, for any pair of irreducible representations is 0. $\sum_R \chi_i(\mathbf{R})\chi_j(\mathbf{R}) = 0, \text{when } i \neq j$ Taking any pair of irreducible representations, multiplying together the characters for each class, multiplying by the number of operations in the class (see Table 4.8 for an example), and adding the products gives zero.	B_1 and B_2 are orthogonal: $\underset{E}{(1)(1)} + \underset{C_2}{(-1)(-1)} + \underset{\sigma_v(xz)}{(1)(-1)} + \underset{\sigma_v'(yz)}{(-1)(1)} = 0$ Each operation is its own class in this group.
7. A **totally symmetric representation**, with characters of 1 for all operations, is included in all groups.	C_{2v} has A_1, in which all characters = 1.

The A_2 representation of the C_{2v} group can now be explained. The character table has four columns; it has four classes of symmetry operations (Property 2 in Table 4.7). It must therefore have four irreducible representations (Property 3). The sum of the products of the characters of any two representations must equal zero (orthogonality, Property 6). Therefore, a product of A_1 and the unknown representation must have 1 for two of the characters and -1 for the other two. The character for the identity operation of this new representation must be 1 $[\chi(E) = 1]$ to have the sum of the squares of these characters equal 4 (required by Property 4). Because no two representations can be the same, A_2 must then have $\chi(E) = \chi(C_2) = 1$, and $\chi(\sigma_{xz}) = \chi(\sigma_{yz}) = -1$. This representation is also orthogonal to B_1 and B_2, as required.

The relationships between symmetry operations, matrix representations, reducible and irreducible representations, and character tables are conveniently illustrated in a flow chart, as shown for C_{2v} symmetry in Table 4.8.

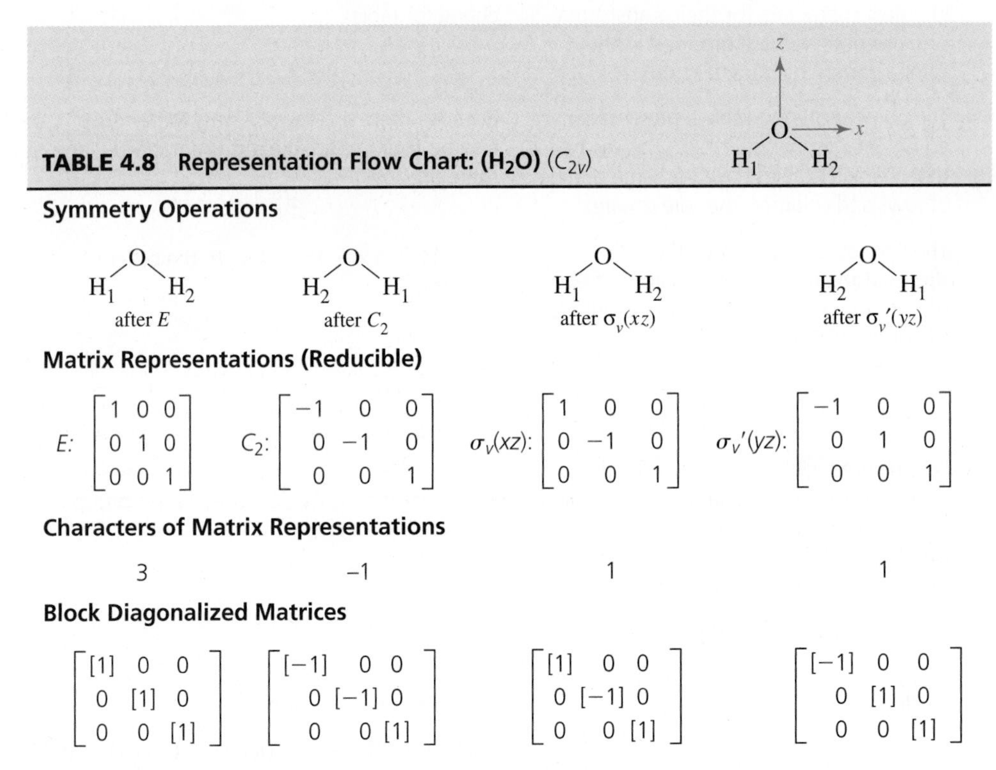

TABLE 4.8 Representation Flow Chart: (H_2O) (C_{2v})

Symmetry Operations

Matrix Representations (Reducible)

$$E: \begin{bmatrix} 1 & 0 & 0 \\ 0 & 1 & 0 \\ 0 & 0 & 1 \end{bmatrix} \quad C_2: \begin{bmatrix} -1 & 0 & 0 \\ 0 & -1 & 0 \\ 0 & 0 & 1 \end{bmatrix} \quad \sigma_v(xz): \begin{bmatrix} 1 & 0 & 0 \\ 0 & -1 & 0 \\ 0 & 0 & 1 \end{bmatrix} \quad \sigma_v'(yz): \begin{bmatrix} -1 & 0 & 0 \\ 0 & 1 & 0 \\ 0 & 0 & 1 \end{bmatrix}$$

Characters of Matrix Representations

3	–1	1	1

Block Diagonalized Matrices

$$\begin{bmatrix} [1] & 0 & 0 \\ 0 & [1] & 0 \\ 0 & 0 & [1] \end{bmatrix} \quad \begin{bmatrix} [-1] & 0 & 0 \\ 0 & [-1] & 0 \\ 0 & 0 & [1] \end{bmatrix} \quad \begin{bmatrix} [1] & 0 & 0 \\ 0 & [-1] & 0 \\ 0 & 0 & [1] \end{bmatrix} \quad \begin{bmatrix} [-1] & 0 & 0 \\ 0 & [1] & 0 \\ 0 & 0 & [1] \end{bmatrix}$$

Irreducible Representations

E	C_2	$\sigma_v(xz)$	$\sigma_v'(yz)$	Coordinate Used
1	–1	1	–1	x
1	–1	–1	1	y
1	1	1	1	z
3	–1	1	1	

Character Table

C_{2v}	E	C_2	$\sigma_v(xz)$	$\sigma_v'(yz)$	Matching Functions	
A_1	1	1	1	1	z	x^2, y^2, z^2
A_2	1	1	–1	–1	R_z	xy
B_1	1	–1	1	–1	x, R_y	xz
B_2	1	–1	–1	1	y, R_x	yz

▶ **Exercise 4.6** Prepare a representation flow chart according to the format of Table 4.8 for *trans*$-N_2F_2$, which has C_{2h} symmetry.

ANOTHER EXAMPLE: $C_{3v}(NH_3)$ Full descriptions of the matrices for the operations in this group will not be given, but the characters can be found by using the properties of a group. Consider the C_3 rotation shown in Figure 4.16. Rotation of 120° results in a new x' and y' as shown, which can be described in terms of the vector sums of x and y by using trigonometric functions:

$$x' = x\cos\frac{2\pi}{3} - y\sin\frac{2\pi}{3} = -\frac{1}{2}x - \frac{\sqrt{3}}{2}y$$

$$y' = x\sin\frac{2\pi}{3} + y\cos\frac{2\pi}{3} = \frac{\sqrt{3}}{2}x - \frac{1}{2}y$$

The transformation matrices for the symmetry operations shown are as follows:

$$E: \begin{bmatrix} 1 & 0 & 0 \\ 0 & 1 & 0 \\ 0 & 0 & 1 \end{bmatrix} \quad C_3: \begin{bmatrix} \cos\frac{2\pi}{3} & -\sin\frac{2\pi}{3} & 0 \\ \sin\frac{2\pi}{3} & \cos\frac{2\pi}{3} & 0 \\ 0 & 0 & 1 \end{bmatrix} = \begin{bmatrix} -\frac{1}{2} & -\frac{\sqrt{3}}{2} & 0 \\ \frac{\sqrt{3}}{2} & -\frac{1}{2} & 0 \\ 0 & 0 & 1 \end{bmatrix} \quad \sigma_{v(xz)}: \begin{bmatrix} 1 & 0 & 0 \\ 0 & -1 & 0 \\ 0 & 0 & 1 \end{bmatrix}$$

In the C_{3v} point group $\chi(C_3^2) = \chi(C_3)$, which means that they are in the same class and described as $2C_3$ in the character table. In addition, the three reflections have identical characters and are in the same class, described as $3\sigma_v$.

The transformation matrices for C_3 and C_3^2 cannot be block diagonalized into 1×1 matrices, because the C_3 matrix has off-diagonal entries; however, the matrices, can be block diagonalized into 2×2 and 1×1 matrices, with all other matrix elements equal to zero:

$$E: \begin{bmatrix} \begin{bmatrix} 1 & 0 \\ 0 & 1 \end{bmatrix} & \begin{matrix} 0 \\ 0 \end{matrix} \\ \begin{matrix} 0 & 0 \end{matrix} & [1] \end{bmatrix} \quad C_3: \begin{bmatrix} \begin{bmatrix} -\frac{1}{2} & -\frac{\sqrt{3}}{2} \\ \frac{\sqrt{3}}{2} & -\frac{1}{2} \end{bmatrix} & \begin{matrix} 0 \\ 0 \end{matrix} \\ \begin{matrix} 0 & 0 \end{matrix} & [1] \end{bmatrix} \quad \sigma_{v(xz)}: \begin{bmatrix} \begin{bmatrix} 1 & 0 \\ 0 & -1 \end{bmatrix} & \begin{matrix} 0 \\ 0 \end{matrix} \\ \begin{matrix} 0 & 0 \end{matrix} & [1] \end{bmatrix}$$

General case: $x' = x\cos\theta - y\sin\theta$
$y' = x\sin\theta + y\cos\theta$
For C_3: $\theta = 2\pi/3 = 120°$

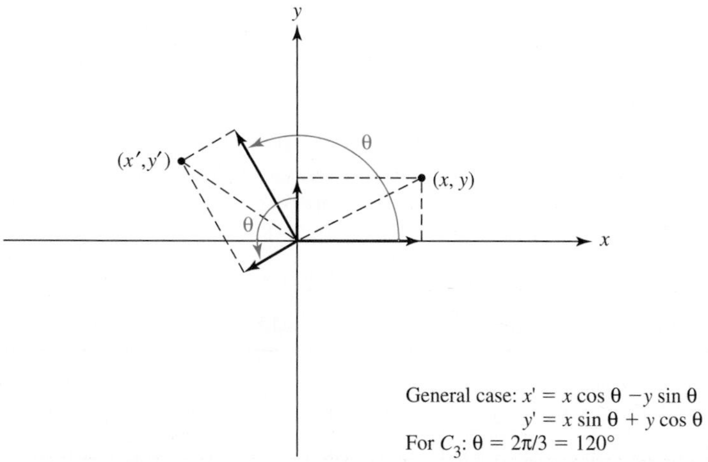

FIGURE 4.16 Effect of Rotation on Coordinates of a Point.

The C_3 matrix must be blocked this way because the (x, y) combination is needed for the new x' and y'; the other matrices must follow the same pattern for consistency across the representation. In this case, x and y are not independent of each other.

The characters of the matrices are the sums of the numbers on the principal diagonal (from upper left to lower right). The set of 2×2 matrices has the characters corresponding to the E representation in the following character table; the set of 1×1 matrices matches the A_1 representation. The third irreducible representation, A_2, can be found by using the defining properties of a mathematical group, as in the C_{2v} example above. Table 4.9 gives the properties of the characters for the C_{3v} point group.

C_{3v}	E	$2C_3$	$3\sigma_v$		
A_1	1	1	1	z	$x^2 + y^2, z^2$
A_2	1	1	-1	R_z	
E	2	-1	0	$(x, y), (R_x, R_y)$	$(x^2 - y^2, xy), (xz, yz)$

TABLE 4.9 Properties of the Characters for the C_{3v} Point Group

Property	C_{3v} Example
1. Order	6 (6 symmetry operations)
2. Classes	3 classes: E $2C_3\ (= C_3, C_3^2)$ $3\sigma_v\ (= \sigma_v, \sigma_v', \sigma_v'')$
3. Number of irreducible representations	3 (A_1, A_2, E)
4. Sum of squares of dimensions equals the order of the group	$1^2 + 1^2 + 2^2 = 6$
5. Sum of squares of characters multiplied by the number of operations in each class equals the order of the group	$E \quad 2C_3 \quad 3\sigma_v$ A_1: $1^2 + 2(1)^2 + 3(1)^2 = 6$ A_2: $1^2 + 2(1)^2 + 3(-1)^2 = 6$ E: $2^2 + 2(-1)^2 + 3(0)^2 = 6$ (Multiply the squares by the number of symmetry operations in each class.)
6. Orthogonal representations	The sum of the products of any two representations multiplied by the number of operations in each class equals 0. Example of $A_2 \times E$: $(1)(2) + 2(1)(-1) + 3(-1)(0) = 0$
7. Totally symmetric representation	A_1, with all characters = 1

ADDITIONAL FEATURES OF CHARACTER TABLES

1. When operations such as C_3 are in the same class, the listing in a character table is $2C_3$, indicating that the results are the same, whether rotation is in a clockwise or counterclockwise direction (or, alternately, that C_3 and C_3^2 give the same result). In either case, this is equivalent to two columns in the table being shown as one. Similar notation is used for multiple reflections.
2. When necessary, the C_2 axes perpendicular to the principal axis (in a *D* group) are designated with primes; a single prime indicates that the axis passes through several atoms of the molecule, whereas a double prime indicates that it passes between the atoms.
3. When the mirror plane is perpendicular to the principal axis, or horizontal, the reflection is called σ_h. Other planes are labeled σ_v or σ_d (see the character tables in Appendix C).
4. The expressions listed to the right of the characters indicate the symmetry of mathematical functions of the coordinates x, y, and z and of rotation about the axes (R_x, R_y, R_z). These can be used to find the orbitals that match the representation. For example, x with positive and negative directions matches the p_x orbital with positive and negative lobes in the quadrants in the xy plane, and the product xy with alternating signs on the quadrants matches lobes of the d_{xy} orbital, as in Figure 4.17. In all cases, the totally symmetric *s* orbital matches the first representation in the group, one of the *A* set. The rotational functions are used to describe the rotational motions of the molecule. Rotation and other motions of the water molecule are discussed in Section 4.4.2.

   In the C_{3v} example described previously, the x and y coordinates appeared together in the *E* irreducible representation. The notation for this is to group them as (x, y) in this section of the table. This means that x and y together have the same symmetry properties as the *E* irreducible representation. Consequently, the p_x and p_y orbitals together have the same symmetry as the *E* irreducible representation in this point group.
5. Matching the symmetry operations of a molecule with those listed in the top row of the character table will confirm any point group assignment.
6. Irreducible representations are assigned labels according to the following rules, in which *symmetric* means a character of 1 and *antisymmetric* a character of -1 (see the character tables in Appendix C for examples).
   a. Letters are assigned according to the dimension of the irreducible representation (the character for the identity operation).

Dimension	Symmetry Label	
1	*A*	If the representation is symmetric to the principal rotation operation ($\chi(C_n) = 1$).
	B	If it is antisymmetric ($\chi(C_n) = -1$).[4]
2	*E*	
3	*T*	

[4] In a few cases, such as D_{nd} (n = even) and S_{2n} point groups, the highest order axis is an S_{2n}. This axis takes priority, so the classification is *B* if the character is –1 for the S_{2n} operation, even if the character is +1 for the highest order C_n axis.

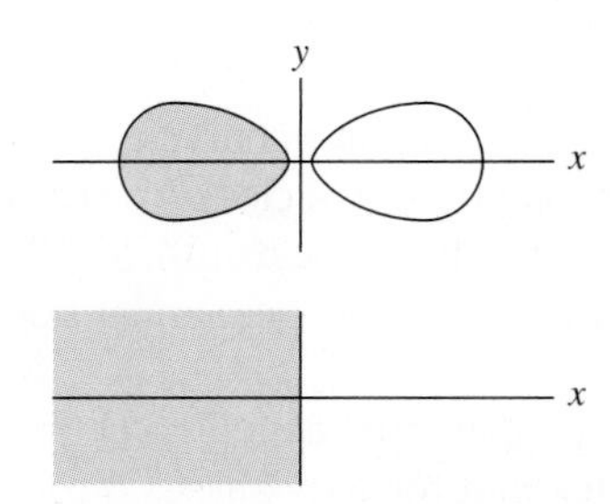

p_x orbitals have the same symmetry as x (positive in half the quadrants, negative in the other half).

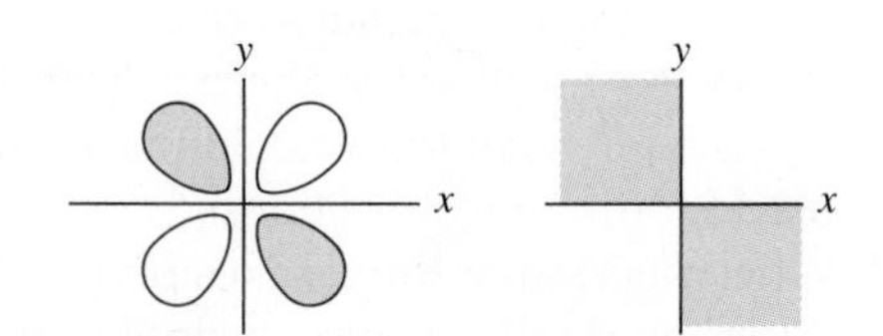

d_{xy} orbitals have the same symmetry as the function xy (sign of the function in the four quadrants).

FIGURE 4.17 Orbitals and Representations.

b. Subscript 1 designates a representation symmetric to a C_2 rotation perpendicular to the principal axis, and subscript 2 designates a representation antisymmetric to the C_2. If there are no perpendicular C_2 axes, 1 designates a representation symmetric to a vertical plane, and 2 designates a representation antisymmetric to a vertical plane.

c. Subscript g (*gerade*) designates representations symmetric to inversion, and subscript u (*ungerade*) designates representations antisymmetric to inversion.

d. Single primes are symmetric to σ_h and double primes are antisymmetric to σ_h when a distinction between representations is needed (C_{3h}, C_{5h}, D_{3h}, D_{5h}).

4.4 EXAMPLES AND APPLICATIONS OF SYMMETRY

4.4.1 Chirality

Many molecules are not superimposable on their mirror image. Such molecules, labeled **chiral** or **dissymmetric**, may have important chemical properties as a consequence of this nonsuperimposability. An example of a chiral organic molecule is CBrClFI, and many examples of chiral objects can also be found on the macroscopic scale, as in Figure 4.18.

Chiral objects are termed *dissymmetric*. This term does not imply that these objects necessarily have *no* symmetry. For example, the propellers shown in Figure 4.18 each have a C_3 axis, yet they are nonsuperimposable (if both were spun in a clockwise direction, they would move an airplane in opposite directions). In general, we can say that a molecule or some other object is chiral if it has no symmetry operations (other than E), or if it has *only proper rotation axes*.

▶ **Exercise 4.7** Which point groups are possible for chiral molecules? (Hint: Refer to the character tables in Appendix C.)

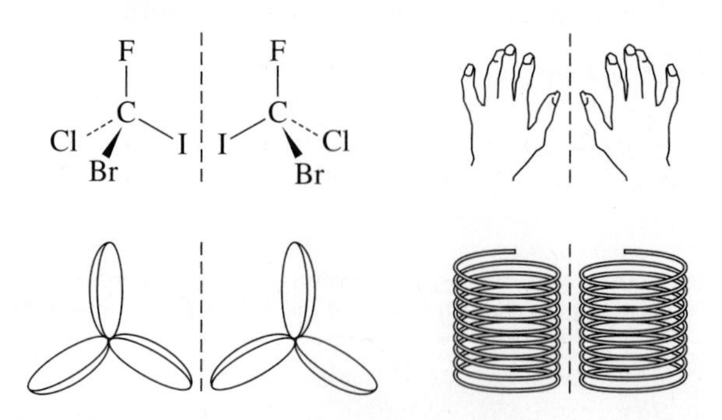

FIGURE 4.18 A Chiral Molecule and Other Chiral Objects.

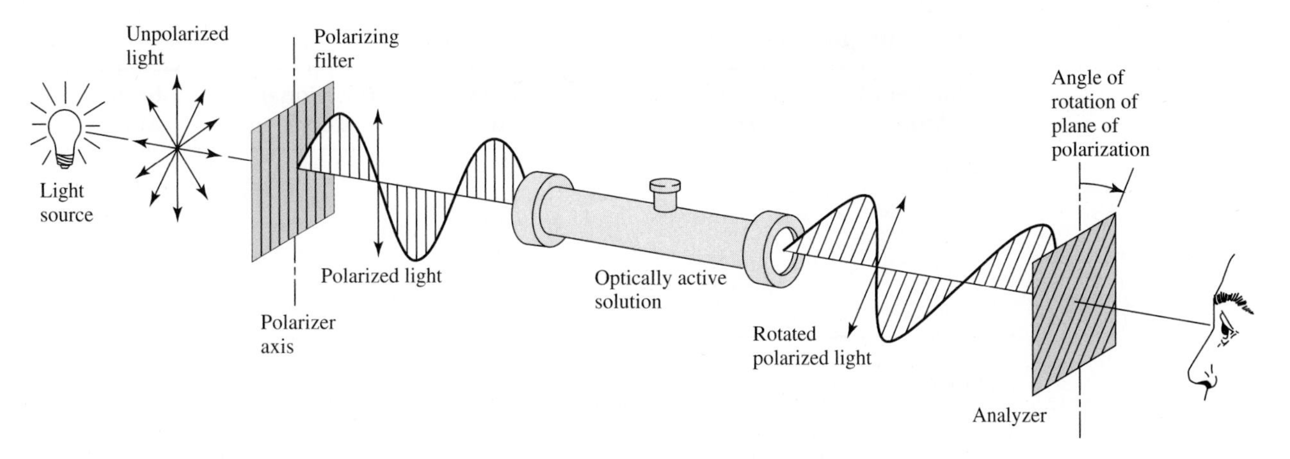

FIGURE 4.19 Rotation of Plane-Polarized Light.

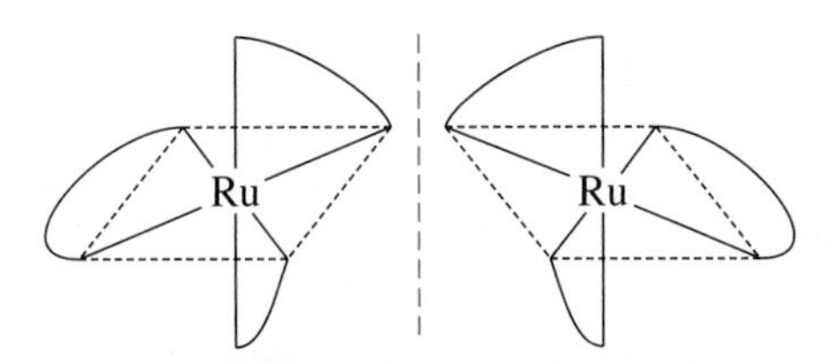

FIGURE 4.20 Chiral Isomers of $[Ru(NH_2CH_2CH_2NH_2)_3]^{2+}$.

Air blowing past the stationary propellers in Figure 4.18 will be rotated in either a clockwise or counterclockwise direction. By the same token, plane-polarized light will be rotated on passing through chiral molecules (Figure 4.19); clockwise rotation is designated **dextrorotatory**, and counterclockwise rotation is **levorotatory**. The ability of chiral molecules to rotate plane-polarized light is termed **optical activity**, which may be measured experimentally.

Many coordination compounds are chiral and thus exhibit optical activity if they can be resolved into the two isomers. One of these is $[Ru(NH_2CH_2CH_2NH_2)_3]^{2+}$, with D_3 symmetry (Figure 4.20). Mirror images of this molecule look much like left- and right-handed three-bladed propellers. Further examples will be discussed in Chapter 9.

4.4.2 Molecular Vibrations

Symmetry can be helpful in determining the modes of vibration of molecules. Vibrational modes of water and the stretching modes of CO in carbonyl complexes are examples that can be treated quite simply, as described in the following pages. Other molecules can be studied using the same methods.

WATER (C_{2v} SYMMETRY) Because the study of vibrations is the study of motion of the individual atoms in a molecule, we must first attach a set of x, y, and z coordinates to each atom. For convenience, we assign the z axes parallel to the C_2 axis of the molecule, the x axes in the plane of the molecule, and the y axes perpendicular to the plane (Figure 4.21). Each atom can move in all three directions, so a total of nine transformations (motion of each atom in the x, y, and z directions) must be considered. For N atoms in a molecule, there are $3N$ total motions, known as **degrees of freedom**. Degrees of freedom for different geometries are summarized in Table 4.10. Because water has three atoms, there must be nine different motions.

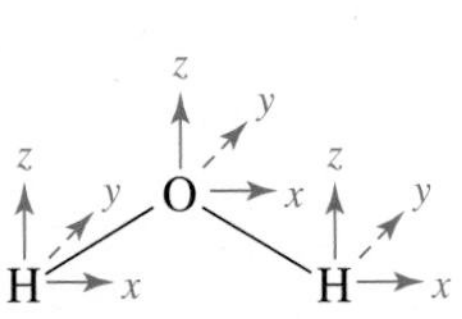

FIGURE 4.21 A Set of Axes for the Water Molecule.

We will use transformation matrices to determine the symmetry of all nine motions and then assign them to translation, rotation, and vibration. Fortunately, it is

TABLE 4.10 Degrees of Freedom

Number of Atoms	Total Degrees of Freedom	Translational Modes	Rotational Modes	Vibrational Modes
N (linear)	$3N$	3	2	$3N - 5$
3 (HCN)	9	3	2	4
N (nonlinear)	$3N$	3	3	$3N - 6$
3 (H_2O)	9	3	3	3

only necessary to determine the characters of the transformation matrices, not the individual matrix elements.

In this case, the initial axes make a column matrix with nine elements, and each transformation matrix is 9×9. A nonzero entry appears along the diagonal of the matrix only for an atom that does not change position. If the atom changes position during the symmetry operation, a zero is entered. If the atom remains in its original location and the vector direction is unchanged, a 1 is entered. If the atom remains, but the vector direction is reversed, a -1 is entered. (Because all the operations change vector directions by 0° or 180° in the C_{2v} point group, these are the only possibilities.) When all nine vectors are summed, the character of the reducible representation Γ is obtained. The full 9×9 matrix for C_2 is shown as an example; note that only the diagonal entries are used in finding the character.

$$\begin{matrix} \text{O} \\ \\ \\ \text{H}_a \\ \\ \\ \text{H}_b \\ \\ \end{matrix} \begin{bmatrix} x' \\ y' \\ z' \\ x' \\ y' \\ z' \\ x' \\ y' \\ z' \end{bmatrix} = \begin{bmatrix} -1 & 0 & 0 & 0 & 0 & 0 & 0 & 0 & 0 \\ 0 & -1 & 0 & 0 & 0 & 0 & 0 & 0 & 0 \\ 0 & 0 & 1 & 0 & 0 & 0 & 0 & 0 & 0 \\ 0 & 0 & 0 & 0 & 0 & 0 & -1 & 0 & 0 \\ 0 & 0 & 0 & 0 & 0 & 0 & 0 & -1 & 0 \\ 0 & 0 & 0 & 0 & 0 & 0 & 0 & 0 & 1 \\ 0 & 0 & 0 & -1 & 0 & 0 & 0 & 0 & 0 \\ 0 & 0 & 0 & 0 & -1 & 0 & 0 & 0 & 0 \\ 0 & 0 & 0 & 0 & 0 & 1 & 0 & 0 & 0 \end{bmatrix} \begin{bmatrix} x \\ y \\ z \\ x \\ y \\ z \\ x \\ y \\ z \end{bmatrix} \begin{matrix} \text{O} \\ \\ \\ \text{H}_a \\ \\ \\ \text{H}_b \\ \\ \end{matrix}$$

The H_a and H_b entries are not on the principal diagonal, because H_a and H_b exchange with each other in a C_2 rotation, and $x'(H_a) = -x(H_b)$, $y'(H_a) = -y(H_b)$, and $z'(H_a) = z(H_b)$. Only the oxygen atom contributes to the character for this operation, for a total of -1.

The other entries for Γ can also be found without writing out the matrices, as follows:

E: All nine vectors are unchanged in the identity operation, so the character is 9.

C_2: The hydrogen atoms change position in a C_2 rotation, so all their vectors have zero contribution to the character. The oxygen atom vectors in the x and y directions are reversed, each contributing -1, and the vector in the z direction remains the same, contributing 1 for a total of -1. The sum of the principal diagonal $= \chi(C_2) = (-1) + (-1) + (1) = -1$.

$\sigma_v(xz)$: Reflection in the plane of the molecule changes the direction of all the y vectors and leaves the x and z vectors unchanged, for a total of $3 - 3 + 3 = 3$.

$\sigma_v'(yz)$: Finally, reflection perpendicular to the plane of the molecule changes the position of the hydrogens so their contribution is zero; the x vector on the oxygen changes direction, and the y and z vectors are unchanged for a total of 1.

▶ **Exercise 4.8** Write transformation matrices for the $\sigma(xz)$ and $\sigma(yz)$ operations in C_{2v} symmetry.

Because all nine direction vectors are included in this representation, it represents all the motions of the molecule, three translations, three rotations, and (by difference) three vibrations. The characters of the reducible representation Γ are shown as the last row below the irreducible representations in the C_{2v} character table.

C_{2v}	E	C_2	$\sigma_v(xz)$	$\sigma_v'(yz)$		
A_1	1	1	1	1	z	x^2, y^2, z^2
A_2	1	1	−1	−1	R_z	xy
B_1	1	−1	1	−1	x, R_y	xz
B_2	1	−1	−1	1	y, R_x	yz
Γ	9	−1	3	1		

REDUCING REPRESENTATIONS TO IRREDUCIBLE REPRESENTATIONS The next step is to separate this representation into its component irreducible representations. This requires another property of groups. The number of times any irreducible representation appears in a reducible representation is equal to the sum of the products of the characters of the reducible and irreducible representations multiplied by the number of operations in the class, taken one operation at a time, divided by the order of the group. This may be expressed in equation form, with the sum taken over all symmetry operations of the group.[5]

$$\begin{pmatrix}\text{Number of irreducible}\\ \text{representations of}\\ \text{a given type}\end{pmatrix} = \frac{1}{\text{order}}\sum_R\left[\begin{pmatrix}\text{number}\\ \text{of operations}\\ \text{in the class}\end{pmatrix}\times\begin{pmatrix}\text{character of}\\ \text{reducible}\\ \text{representation}\end{pmatrix}\times\begin{pmatrix}\text{character of}\\ \text{irreducible}\\ \text{representation}\end{pmatrix}\right]$$

In the water example, the order of C_{2v} is 4, with one operation in each class $(E, C_2, \sigma_v, \sigma_v')$. The results are then as follows:

$$n_{A_1} = \frac{1}{4}[(9)(1) + (-1)(1) + (3)(1) + (1)(1)] = 3$$

$$n_{A_2} = \frac{1}{4}[(9)(1) + (-1)(1) + (3)(-1) + (1)(-1)] = 1$$

$$n_{B_1} = \frac{1}{4}[(9)(1) + (-1)(-1) + (3)(1) + (1)(-1)] = 3$$

$$n_{B_2} = \frac{1}{4}[(9)(1) + (-1)(-1) + (3)(-1) + (1)(1)] = 2$$

The reducible representation for all motions of the water molecule is therefore reduced to $3A_1 + A_2 + 3B_1 + 2B_2$.

Examination of the columns on the far right in the character table shows that translation along the x, y, and z directions is $A_1 + B_1 + B_2$ (translation is motion along the x, y, and z directions, so it transforms in the same way as the three axes). Rotation in the three directions (R_x, R_y, R_z) is $A_2 + B_1 + B_2$. Subtracting these from the total above leaves $2A_1 + B_1$, the three vibrational modes shown in Table 4.11. The number of vibrational modes equals $3N - 6$, as described earlier. Two of the modes are totally

[5] This procedure should yield an integer for the number of irreducible representations of each type; obtaining a fraction in this step indicates a calculation error.

TABLE 4.11 Symmetry of Molecular Motions of Water

All Motions	Translation (x, y, z)	Rotation (R_x, R_y, R_z)	Vibration (Remaining Modes)
$3A_1$	A_1		$2A_1$
A_2		A_2	
$3B_1$	B_1	B_1	B_1
$2B_2$	B_2	B_2	

TABLE 4.12 The Vibrational Modes of Water

Symmetry	Mode	Description
A_1	H–O–H	Symmetric stretch: change in dipole moment; more distance between positive hydrogens and negative oxygen *IR active*
B_1	H–O–H	Antisymmetric stretch: change in dipole moment; change in distances between positive hydrogens and negative oxygen *IR active*
A_1	H–O–H	Symmetric bend: change in dipole moment; angle between H—O vectors changes *IR active*

symmetric (A_1) and do not change the symmetry of the molecule, but one is antisymmetric to C_2 rotation and to reflection perpendicular to the plane of the molecule (B_1). These modes are illustrated as symmetric stretch, symmetric bend, and antisymmetric stretch in Table 4.12.

EXAMPLE

Using the x, y, and z coordinates for each atom in XeF_4, determine the reducible representation for all molecular motions; reduce this representation to its irreducible components; and classify these representations into translational, rotational, and vibrational modes.

First, it is useful to assign x, y, and z coordinate axes to each atom, as shown in Figure 4.22.

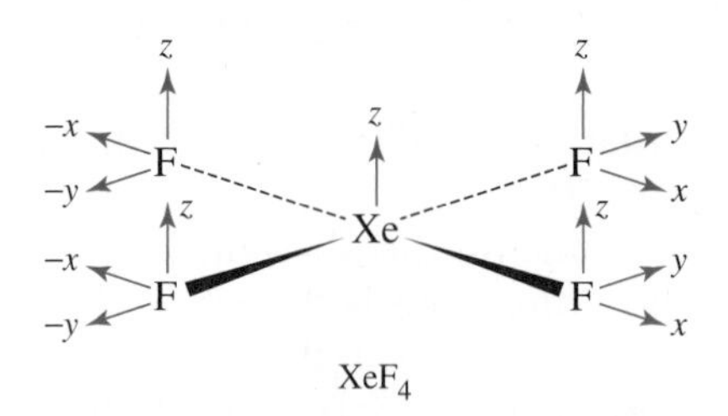

FIGURE 4.22 Coordinate Axes for XeF_4.

It is essential to recognize that *only the coordinates on atoms that do not move when symmetry operations are applied can give rise to nonzero elements along the diagonals of transformation matrices*. For example, if a symmetry operation applied to XeF_4 causes all F atoms to change position, these atoms cannot give rise to elements along the diagonal, so they can be ignored; only coordinates of Xe would need to be considered.

In addition: If a symmetry operation leaves the direction of a coordinate unchanged, it gives a character of 1 along the diagonal.

For example, the identity operation on XeF_4 leaves the coordinates x, y, and z unchanged; each of these has a diagonal element of 1 for each atom.

$$\underset{1}{x \longrightarrow x} \qquad \underset{1}{y \longrightarrow y} \qquad \underset{1}{z \longrightarrow z}$$

If a symmetry operation reverses the direction of a coordinate, this corresponds to a diagonal element of –1.

The σ_h operation on XeF_4 reverses the direction of the z axis for each atom.

$$\underset{-1}{z \longrightarrow -z}$$

If a symmetry operation transforms a coordinate into another coordinate, this gives a diagonal element of zero.

If XeF_4 is rotated about its C_4 axis, the x and y coordinates of Xe are interchanged; they contribute zero to the character.

Examining each of the D_{4h} symmetry operations in turn generates the following reducible representation for all the molecular motions of XeF_4:

D_{4h}	E	$2C_4$	C_2	$2C_2'$	$2C_2''$	i	$2S_4$	σ_h	$2\sigma_v$	$2\sigma_d$
Γ	15	1	–1	–3	–1	–3	–1	5	3	1

The character under E tells that there are 15 possible motions to be considered. By the procedure illustrated in the preceding example, this representation reduces to:

$$\Gamma = A_{1g} + A_{2g} + B_{1g} + B_{2g} + E_g + 2\,A_{2u} + B_{2u} + 3\,E_u$$

These can be classified as follows:

Translational Motion. This is motion through space with x, y, and z components. The irreducible representations matching these components have the labels x, y, and z on the right side of the D_{4h} character table: A_{2u} (matching x) and E_u (doubly degenerate, matching x and y together). These three motions can be represented as shown in Figure 4.23.

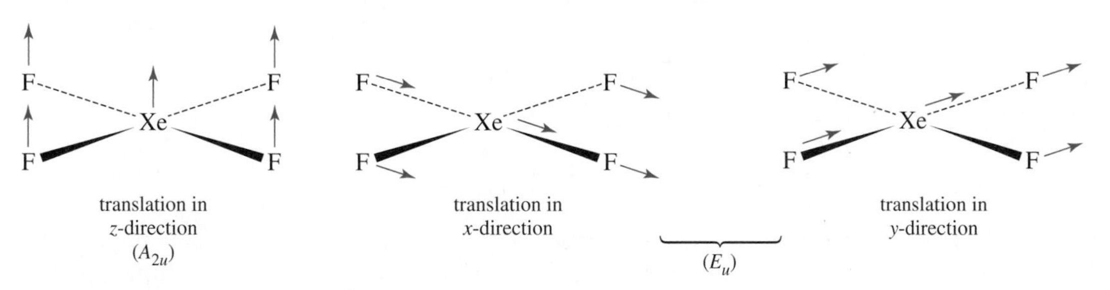

FIGURE 4.23 Translational Modes of XeF_4.

Rotational Motion. This type of motion can be factored into rotation about the mutually orthogonal x, y, and z axes. The matching expressions in the character table are R_x, R_y, and R_z, representing rotation about these three axes, respectively. The irreducible representations are A_{2g} (R_z, rotation about the z axis), and E_g (R_x, R_y), doubly degenerate rotations about the x and y axes) as shown in Figure 4.24.

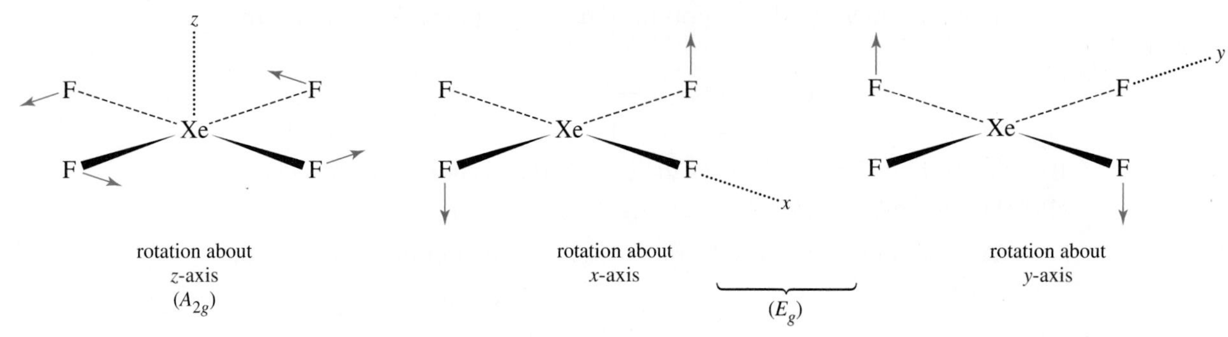

FIGURE 4.24 Rotational Modes of XeF_4.

Vibrational Motion. The remaining nine motions (15 total – 3 translations – 3 rotations) are vibrational. They involve changes in bond lengths and angles and motions both within and out of the molecular plane. For example, symmetrical stretching of all four Xe—F bonds matches the A_{1g} irreducible representation, symmetrical stretching of opposite bonds matches B_{1g}, and simultaneous opening of opposite bond angles matches B_{2g}, as shown in Figure 4.25:

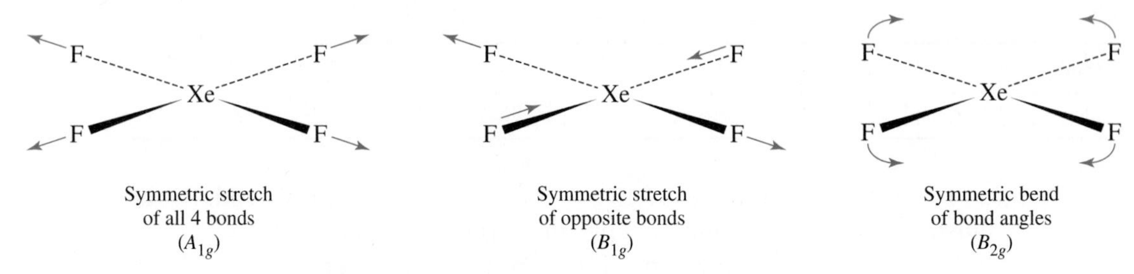

FIGURE 4.25 Selected Vibrational Modes of XeF_4.

Table 4.13 summarizes the classification of irreducible representations according to mode.

TABLE 4.13 Symmetry of Molecular Motions of XeF_4

	Γ (all modes)	Translation	Rotation	Vibration
	A_{1g}			A_{1g}
	A_{2g}		A_{2g}	
	B_{1g}			B_{1g}
	B_{2g}			B_{2g}
	E_g		E_g	
	2 A_{2u}	A_{2u}		A_{2u}
	B_{2u}			B_{2u}
	3 E_u	E_u		2 E_u
Total	15	3	3	9

▶ **Exercise 4.9** Using the x, y, and z coordinates for each atom in N_2O_4, which is planar and has a nitrogen–nitrogen bond, determine the reducible representation for all molecular motions. Reduce this representation to its irreducible components, and classify these representations into translational, rotational, and vibrational modes.

EXAMPLES

Reduce the following representations to their irreducible representations in the point group indicated (refer to the character tables in Appendix C):

C_{2h}	E	C_2	i	σ_h
Γ	4	0	2	2

SOLUTION

$$n_{A_g} = \frac{1}{4}[(4)(1) + (0)(1) + (2)(1) + (2)(1)] = 2$$

$$n_{B_g} = \frac{1}{4}[(4)(1) + (0)(-1) + (2)(1) + (2)(-1)] = 1$$

$$n_{A_u} = \frac{1}{4}[(4)(1) + (0)(1) + (2)(-1) + (2)(-1)] = 0$$

$$n_{B_u} = \frac{1}{4}[(4)(1) + (0)(-1) + (2)(-1) + (2)(1)] = 1$$

Therefore, $\Gamma = 2A_g + B_g + B_u$

C_{3v}	E	$2C_3$	$3\sigma_v$
Γ	6	3	−2

SOLUTION

$$n_{A_1} = \frac{1}{6}[(6)(1) + (2)(3)(1) + (3)(-2)(1)] = 1$$

$$n_{A_2} = \frac{1}{6}[(6)(1) + (2)(3)(1) + (3)(-2)(-1)] = 3$$

$$n_E = \frac{1}{6}[(6)(2) + (2)(3)(-1) + (3)(-2)(0)] = 1$$

Therefore, $\Gamma = A_1 + 3A_2 + E$.

Be sure to include the number of symmetry operations in a class (column) of the character table. This means that the second term in the C_{3v} calculation must be multiplied by 2 ($2C_3$; there are two operations in this class); the third term must be multiplied by 3, as shown.

▶ **Exercise 4.10** Reduce the following representations to their irreducible representations in the point groups indicated:

T_d	E	$8C_3$	$3C_2$	$6S_4$	$6\sigma_d$
Γ_1	4	1	0	0	2

D_{2d}	E	$2S_4$	C_2	$2C_2'$	$2\sigma_d$
Γ_2	4	0	0	2	0

C_{4v}	E	$2C_4$	C_2	$2\sigma_v$	$2\sigma_d$
Γ_3	7	−1	−1	−1	−1

INFRARED SPECTRA A molecular vibration is infrared active (i.e., it has an infrared absorption) only if it results in a change in the dipole moment of the molecule. The three vibrations of water (Table 4.12) can be analyzed in this way to determine their infrared behavior.

Group theory can give us the same information and can account for the more complicated cases as well; in fact, group theory in principle can account for *all* vibrational modes of a molecule. In group theory terms, a vibrational mode is active in the infrared *if it corresponds to an irreducible representation that has the same symmetry (or transforms) as the Cartesian coordinates x, y, or z,* because a vibrational motion that shifts the center of charge of the molecule in any of the x, y, or z directions results in a change in dipole moment. Otherwise, the vibrational mode is not infrared active.

▶ **Exercise 4.11** Which of the nine vibrational modes of XeF_4 (Table 4.13) are infrared active?

▶ **Exercise 4.12** Analysis of the x, y, and z coordinates of each atom in NH_3 gives the following representation:

C_{3v}	E	$2C_3$	$3\sigma_v$
Γ	12	0	2

a. Reduce Γ to its irreducible representations.
b. Classify the irreducible representations into translational, rotational, and vibrational modes.
c. Show that the total number of degrees of freedom $= 3N$.
d. Which vibrational modes are infrared active?

SELECTED VIBRATIONAL MODES It is often useful to consider a particular type of vibrational mode for a compound. For example, useful information often can be obtained from the C—O stretching bands in infrared spectra of metal complexes containing CO (carbonyl) ligands. The following example of *cis*- and *trans*-dicarbonyl square planar

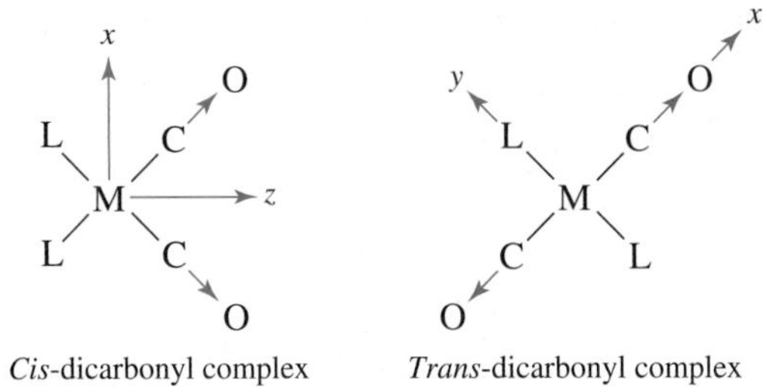

Cis-dicarbonyl complex

Trans-dicarbonyl complex

FIGURE 4.26 Carbonyl Stretching Vibrations of *cis*- and *trans*-Dicarbonyl Square Planar Complexes.

complexes shows the procedure. For these complexes,[6] a simple IR spectrum can distinguish whether a sample is *cis*- or *trans*-$ML_2(CO)_2$; the number of C—O stretching bands is determined by the geometry of the complex (Figure 4.26).

***cis*-$ML_2(CO)_2$, point group C_{2v}.** The principal axis (C_2) is the z axis, with the xz plane assigned as the plane of the molecule. Possible C—O stretching motions are shown by arrows in Figure 4.27; either an increase or decrease in the C—O distance is possible. These vectors are used to create the reducible representation below, using the symmetry operations of the C_{2v} point group. A C—O bond will transform with a character of 1 *if it remains unchanged* by the symmetry operations, and with a character of 0 *if it is changed*. These operations and their characters are shown in Figure 4.27. Both stretches are unchanged in the identity operation and in the reflection through the plane of the molecule, so each contributes 1 to the character, for a total of 2 for each operation. Both vectors move to new locations on rotation or reflection perpendicular to the plane of the molecule, so these two characters are 0.

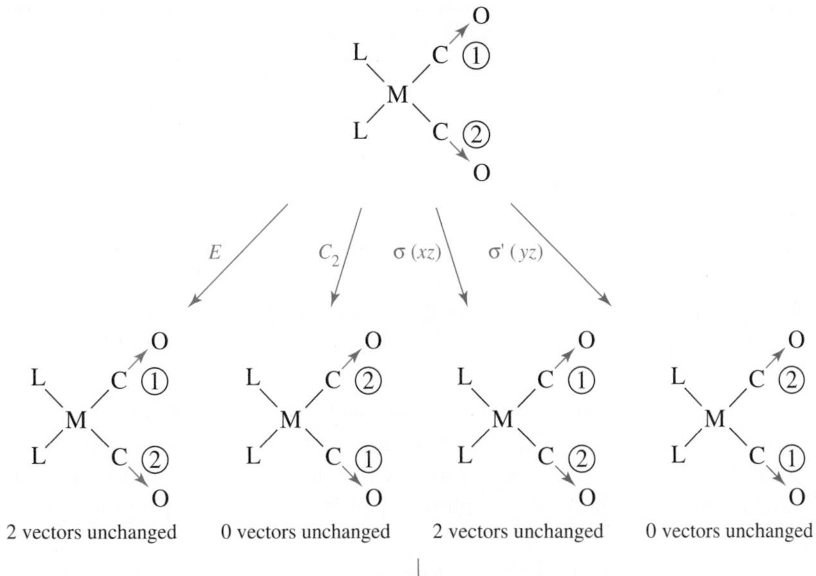

2 vectors unchanged | 0 vectors unchanged | 2 vectors unchanged | 0 vectors unchanged

	E	C_2	$\sigma(xz)$	$\sigma'(yz)$
Γ	2	0	2	0

FIGURE 4.27 Symmetry Operations and Characters for *cis* – $ML_2(CO)_2$.

[6] M represents any metal and L any ligand other than CO in these formulas.

The reducible representation Γ reduces to $A_1 + B_1$:

C_{2v}	E	C_2	$\sigma_v(xz)$	$\sigma_v'(yz)$		
Γ	2	0	2	0		
A_1	1	1	1	1	z	x^2, y^2, z^2
B_1	1	-1	1	-1	x, R_y	xz

A_1 is an appropriate irreducible representation for an IR-active band, because it transforms as (has the symmetry of) the Cartesian coordinate z. Furthermore, the vibrational mode corresponding to B_1 should be IR active, because it transforms as the Cartesian coordinate x.

In summary, there are two vibrational modes for C—O stretching, one having A_1 symmetry and one B_1 symmetry. Both modes are IR active, and we therefore expect to see two C—O stretches in the IR. This assumes that the C—O stretches are not sufficiently similar in energy to overlap in the infrared spectrum.

***trans*-$ML_2(CO)_2$, point group D_{2h}.** The principal axis, C_2, is again chosen as the z axis, which this time makes the plane of the molecule the xy plane. Using the symmetry operation of the D_{2h} point group, we obtain a reducible representation for C—O stretches that reduces to $A_g + B_{3u}$:

D_{2h}	E	$C_2(z)$	$C_2(y)$	$C_2(x)$	i	$\sigma(xy)$	$\sigma(xz)$	$\sigma(yz)$	
Γ	2	0	0	2	0	2	2	0	
A_g	1	1	1	1	1	1	1	1	x^2, y^2, z^2
B_{3u}	1	-1	-1	1	-1	1	1	-1	x

The vibrational mode of A_g symmetry is not IR active, because it does not have the same symmetry as a Cartesian coordinate x, y, or z (this is the IR-inactive symmetric stretch). The mode of symmetry B_{3u}, on the other hand, is IR active, because it has the same symmetry as x.

In summary, there are two vibrational modes for C—O stretching, one having the same symmetry as A_g, and one the same symmetry as B_{3u}. The A_g mode is IR inactive (it does not have the symmetry of x, y, or z); the B_{3u} mode is IR active (it has the symmetry of x). We therefore expect to see one C—O stretch in the IR.

It is therefore possible to distinguish *cis*- and *trans*-$ML_2(CO)_2$ by taking an IR spectrum. If one C—O stretching band appears, the molecule is *trans*; if two bands appear, the molecule is *cis*. A significant distinction can be made by a very simple measurement.

EXAMPLE

Determine the number of IR-active CO stretching modes for *fac*-$Mo(CO)_3(NCCH_3)_3$, as shown in the diagram.

This molecule has C_{3v} symmetry. The operations to be considered are E, C_3, and σ_v. E leaves the three bond vectors unchanged, giving a character of 3. C_3 moves all three vectors, giving a character of 0. Each σ_v plane passes through one of the CO groups, leaving it unchanged, while interchanging the other two. The resulting character is 1.

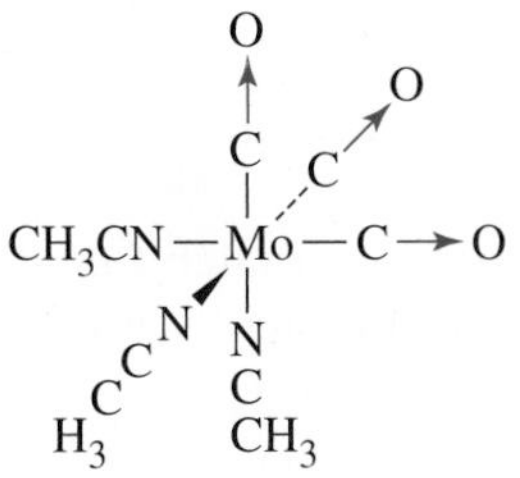

O
C C O
CH_3CN—Mo—C→O
N N
C C
H_3 CH_3

The representation to be reduced, therefore, is as follows:

E	$2C_3$	$3\sigma_v$
3	0	1

This reduces to $A_1 + E$. A_1 has the same symmetry as the Cartesian coordinate z and is therefore IR active. E has the same symmetry as the x and y coordinates together and is also IR active. It represents a degenerate pair of vibrations, which appear as one absorption band.

▶ **Exercise 4.13** Determine the number of IR-active C—O stretching modes for $Mn(CO)_5Cl$.

RAMAN SPECTROSCOPY This method of spectroscopy uses a different approach to observe molecular vibrations. Rather than observing absorption of infrared radiation directly as in IR spectroscopy, in Raman spectroscopy higher energy radiation, ordinarily from a laser, excites molecules to higher electronic states, envisioned as short-lived "virtual" states. Scattered radiation from decay of these excited states to the various vibrational states provides information about vibrational energy levels that is complementary to information gained from IR spectroscopy. In general, a vibration can give rise to a line in a Raman spectrum if it causes a change in polarizability.[7] From a symmetry standpoint, vibrational modes are Raman active if they correspond to any of the functions xy, xz, yz, x^2, y^2, or z^2 or a linear combination of any of these; if vibrations match these functions, they also occur with a change in polarizability. These functions are among those commonly listed in character tables.

For example, vibrational spectroscopy has played a rôle in supporting the tetrahedral structure of the highly explosive XeO_4. Raman spectroscopy has shown two bands in the region expected for Xe=O stretching vibrations, at 776 and 878 cm^{-1}.[8] Is this consistent with the proposed T_d structure?

To address this question, we need to once again create a representation, this time using the Xe=O stretches as a basis in the T_d point group. The resulting representation is as follows:

	E	$8C_3$	$3C_2$	$6S_4$	$6\sigma_d$
Γ	4	1	0	0	2

[7] For more details, see D. J. Willock, *Molecular Symmetry*, John Wiley & Sons, Chichester, UK, 2009, pp. 177–184.
[8] M. Gerken and G. J. Schrobilgen, *Inorg. Chem.*, **2002**, *41*, 198.

This reduces to $A_1 + T_2$:

A_1	1	1	1	1	1		$x^2 + y^2 + z^2$
T_2	3	0	−1	−1	1	(x, y, z)	(xy, xz, yz)

Both the A_1 and T_2 representations match functions necessary for Raman activity. The presence of these two bands is consistent with the proposed T_d symmetry.

$[IO_2F_5]^{2-}$

▶ **Exercise 4.14** Vibrational spectroscopy has played a role in supporting the pentagonal bipyramidal structure of the ion $IO_2F_5^{2-}$.[9] Raman spectroscopy of the tetramethylammonium salt of this ion shows a single absorption in the region expected for I=O stretching vibrations, at 789 cm^{-1}. Is a single Raman band consistent with the proposed *trans* orientation of the oxygen atoms?

General References

There are several helpful books on molecular symmetry and its applications. Good examples are D. J. Willock, *Molecular Symmetry*, John Wiley & Sons, Chichester, UK, 2009; F. A. Cotton, *Chemical Applications of Group Theory*, 3rd ed., John Wiley & Sons, New York, 1990; S. F. A. Kettle, *Symmetry and Structure: Readable Group Theory for Chemists*, 2nd ed., John Wiley & Sons, New York, 1995; and I. Hargittai and M. Hargittai, *Symmetry Through the Eyes of a Chemist*, 2nd ed., Plenum Press, New York, 1995. The last two also provide information on space groups used in solid state symmetry, and all give relatively gentle introductions to the mathematics of the subject.

Problems

4.1 Determine the point groups for
a. Ethane (staggered conformation)
b. Ethane (eclipsed conformation)
c. Chloroethane (staggered conformation)
d. 1,2-Dichloroethane (staggered *anti* conformation)

4.2 Determine the point groups for
a. Ethylene (H₂C=CH₂)
b. Chloroethylene
c. The possible isomers of dichloroethylene

4.3 Determine the point groups for
a. Acetylene
b. H—C≡C—F
c. H—C≡C—CH_3
d. H—C≡C—CH_2Cl
e. H—C≡C—Ph (Ph = phenyl)

4.4 Determine the point groups for
a. Naphthalene

[9] J. A. Boatz, K. O. Christe, D. A. Dixon, B. A. Fir, M. Gerken, R. Z. Gnann, H. P. A. Mercier, and G. J. Schrobilgen, *Inorg. Chem.*, **2003**, *42*, 5282.

b. 1,8-Dichloronaphthalene

c. 1,5-Dichloronaphthalene

d. 1,2-Dichloronaphthalene

4.5 Determine the point groups for

a. 1,1′ – Dichloroferrocene

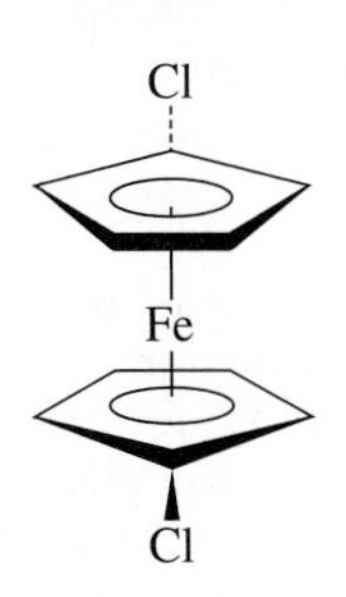

b. Dibenzenechromium (eclipsed conformation)

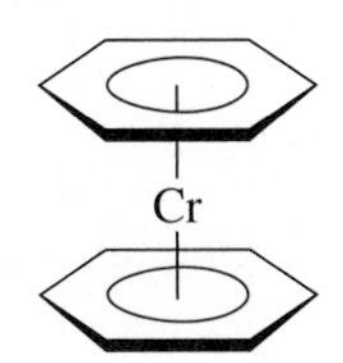

c.

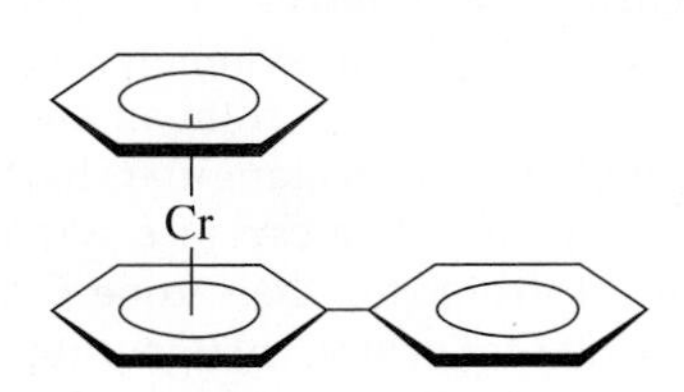

d. H_3O^+

e. O_2F_2

f. Formaldehyde, H_2CO

g. S_8 (puckered ring)

h. Borazine (planar)

i. $[Cr(C_2O_4)_3]^{3-}$

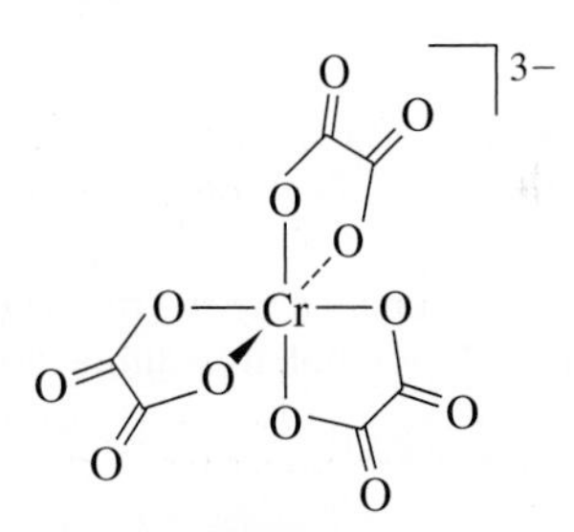

j. A tennis ball (ignoring the label, but including the pattern on the surface)

4.6 Determine the point groups for

a. Cyclohexane (chair conformation)

b. Tetrachloroallene $Cl_2C{=}C{=}CCl_2$

c. SO_4^{2-}

d. A snowflake

e. Diborane

f. The possible isomers of tribromobenzene

g. A tetrahedron inscribed in a cube in which alternate corners of the cube are also corners of the tetrahedron.

h. B_3H_8

4.7 Determine the points groups for

a. A sheet of typing paper

b. An Erlenmeyer flask (no label)

c. A screw

d. The number 96

e. Five examples of objects from everyday life; select items from five different point groups

f. A pair of eyeglasses, assuming lenses of equal strength
g. A five-pointed star
h. A fork with no decoration
i. Captain Hook, whose hand was devoured by a hungry crocodile
j. A metal washer

4.8 Determine the points groups for
a. A flat oval running track
b. A jack (child's toy)

c. A person's two hands, palm to palm
d. A rectangular towel, blue on one side, white on the other
e. A hexagonal pencil with a round eraser
f. The recycle symbol, in 3 dimensions

g. The meander motif

h. An open, eight-spoked umbrella with a straight handle
i. A round toothpick
j. A tetrahedron with one green face, the others red

4.9 Determine the point groups for
a. A triangular prism
b. A plus sign
c. A t-shirt with the letter T on the front
d. Set of three wind turbine blades
e. A spade design (as on a deck of playing cards)
f. A sand dollar

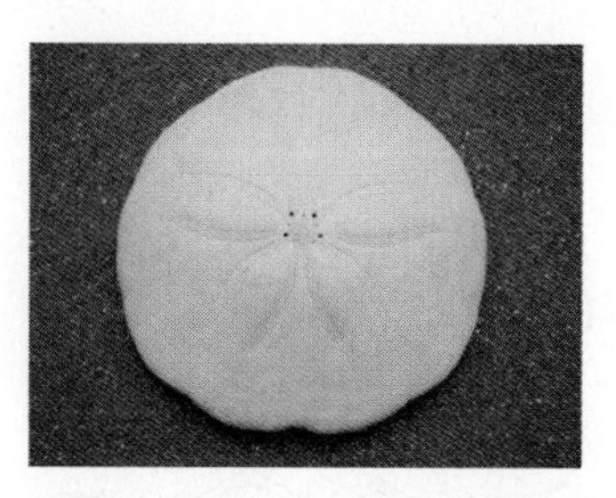

g. A pillar of the Parthenon (assume it is undamaged)
h. An octahedron with one blue face, the others yellow
i. A hula hoop
j. A coiled spring

4.10 Determine the point groups for the examples of symmetry in Figure 4.1.

4.11 Determine the point groups of the molecules in the following end-of-chapter problems from Chapter 3:
a. Problem 3-34 **b.** Problem 3-35

4.12 Determine the point groups of the molecules and ions in
a. Figure 3-8 **b.** Figure 3-15

4.13 Determine the point groups of the following atomic orbitals, including the signs on the orbital lobes:
a. p_x **b.** d_{xy} **c.** $d_{x^2-y^2}$
d. d_{z^2} **e.** f_{xyz}

4.14 **a.** Show that a cube has the same symmetry elements as an octahedron.
b. Suppose a cube has four dots arranged in a square on each face as shown. What is the point group?
c. Suppose that this set of dots is rotated as a set 10° clockwise on each face. Now what is the point group?

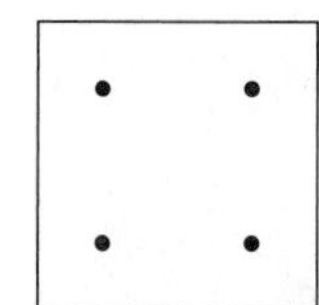

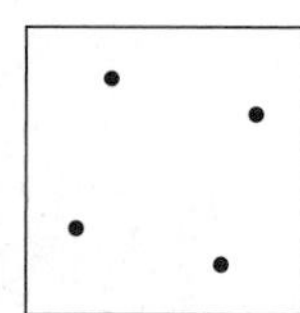

4.15 Baseball is a wonderful game, particularly for someone interested in symmetry. Where else can one watch a batter step from an on-deck circle of **a** symmetry to a rectangular batter's box of **b** symmetry, adjust a cap of **c** symmetry (it has OO on the front, for the Ozone City Oxygens), swing a bat of **d** symmetry (ignoring the label and grain of the wood) across a home plate of **e** symmetry at a baseball that has **f** symmetry (also ignoring the label) that has been thrown by a chiral pitcher having **g** symmetry, hit a towering fly ball that bounces off the fence, and race around the bases, only to be called out at home plate by an umpire who may have no appreciation for symmetry at all.

4.16 Determine the point groups for the following flags or parts of flags. You will need to look up images of flags not shown.

a. Botswana

b. Finland

c. Honduras

d. Field of stars in flag of Micronesia

e. Central design on the Ethiopian flag

f. Turkey
g. Japan
h. Switzerland
i. United Kingdom (be careful!)

4.17 Prepare a representation flow chart according to the format of Table 4.8 for SNF_3.

4.18 For *trans*-1,2-dichloroethylene, which has C_{2h} symmetry:

a. List all the symmetry operations for this molecule.
b. Write a set of transformation matrices that describe the effect of each symmetry operation in the C_{2h} group on a set of coordinates x, y, z for a point (your answer should consist of four 3×3 transformation matrices).
c. Using the terms along the diagonal, obtain as many irreducible representations as possible from the transformation matrices. You should be able to obtain three irreducible representations in this way, but two will be duplicates. You may check your results using the C_{2h} character table.
d. Using the C_{2h} character table, verify that the irreducible representations are mutually orthogonal.

4.19 Ethylene has D_{2h} symmetry.

a. List all the symmetry operations of ethylene.
b. Write a transformation matrix for each symmetry operation that describes the effect of that operation on the coordinates of a point x, y, z.
c. Using the characters of your transformation matrices, obtain a reducible representation.
d. Using the diagonal elements of your matrices, obtain three of the D_{2h} irreducible representations.
e. Show that your irreducible representations are mutually orthogonal.

4.20 Using the D_{2d} character table:

a. Determine the order of the group.
b. Verify that the E irreducible representation is orthogonal to each of the other irreducible representations.
c. For each of the irreducible representations, verify that the sum of the squares of the characters equals the order of the group.
d. Reduce the following representations to their component irreducible representations:

D_{2d}	E	$2S_4$	C_2	$2C_2'$	$2\sigma_d$
Γ_1	6	0	2	2	2
Γ_2	6	4	6	2	0

4.21 Reduce the following representations to irreducible representations:

C_{3v}	E	$2C_3$	$3\sigma_v$
Γ_1	6	3	2
Γ_2	5	−1	−1

O_h	E	$8C_3$	$6C_2$	$6C_4$	$3C_2$	i	$6S_4$	$8S_6$	$3\sigma_h$	$6\sigma_d$
Γ	6	0	0	2	2	0	0	0	4	2

4.22 For D_{4h} symmetry use sketches to show that d_{xy} orbitals have B_{2g} symmetry and that $d_{x^2-y^2}$ orbitals have B_{1g} symmetry. (Hint: you may find it useful to select a molecule that has D_{4h} symmetry as a reference for the operations of the D_{4h} point group. Observe how the signs on the orbital lobes change as the symmetry operations are applied.)

4.23 Which items in Problems 4.5 through 4.9 are chiral? List three items *not* from this chapter that are chiral.

4.24 $XeOF_4$ has one of the more interesting structures among noble gas compounds. On the basis of its symmetry:

a. Obtain a representation based on *all* the motions of the atoms in $XeOF_4$.

b. Reduce this representation to its component irreducible representations.

c. Classify these representations, indicating which are for translational, rotational, and vibrational motion.

d. Determine the irreducible representation matching the xenon–oxygen stretching vibration. Is this vibration IR active?

4.25 Repeat the procedure from the previous problem, parts a through c, for the SF_6 molecule and determine which vibrational modes are IR active.

4.26 For the following molecules, determine the number of IR-active C—O stretching vibrations:

a.

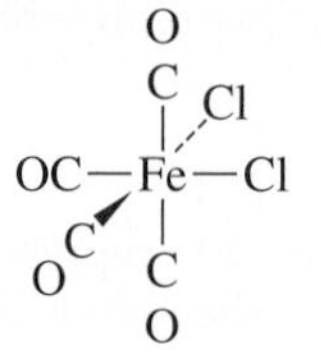

b.

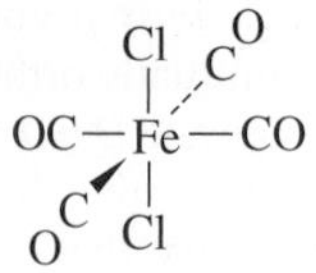

c. $Fe(CO)_5$

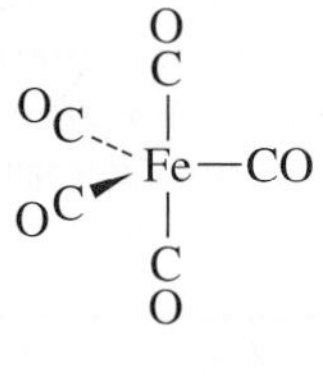

4.27 Repeat problem 4.24 to determine the number of Raman-active C—O stretching vibrations.

4.28 The structure of 1,1,2,2-tetraiododisilane is shown below. (Reference: T. H. Johansen, K. Hassler, G. Tekautz, and K. Hagen, J. *Mol. Struct.*, **2001**, *598*, 171).

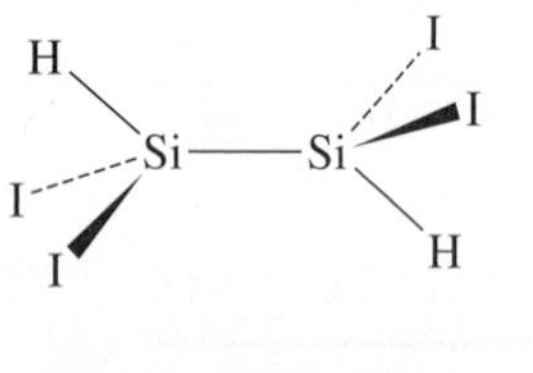

a. What is the point group of this molecule?

b. Predict the number of IR-active Si—I stretching vibrations.

c. Predict the number of Raman-active Si—I stretching vibrations.

4.29 Both *cis* and *trans* isomers of $IO_2F_4^-$ have been observed. Can IR spectra distinguish between these? Explain, supporting your answer on the basis of group theory. (Reference: K. O. Christe, R. D. Wilson, and C. J. Schack, *Inorg. Chem.*, **1981**, *20*, 2104.)

4.30 Three isomers of $W_2Cl_4(NHEt)_2(PMe_3)_2$ have been reported. These isomers have the core structures shown below. Determine the point group of each (Reference: F. A. Cotton, E. V. Dikarev, and W-Y. Wong, *Inorg. Chem.*, **1997**, *36*, 2670.)

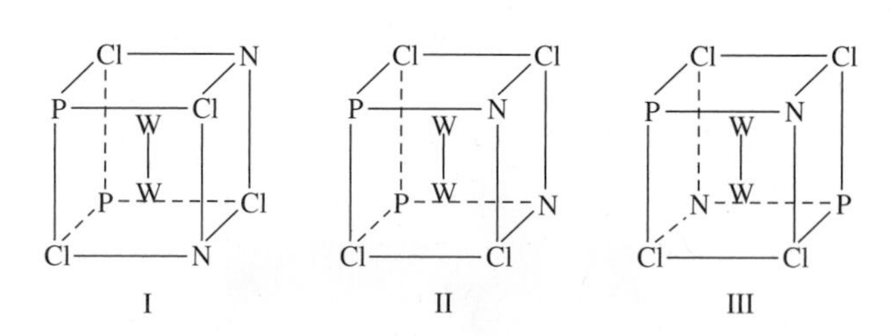

4.31 Derivatives of methane can be obtained by replacing one or more hydrogen atoms with other atoms, such as F, Cl, or Br. Suppose you had a supply of methane and the necessary chemicals and equipment to make derivatives of methane containing all possible combinations of the elements H, F, Cl, and Br. What would be the point groups of the molecules you could make? There are many possible molecules, and they can be arranged into five sets for assignment of point groups.

4.32 Determine the point groups of the following molecules:

a. F_3SCCF_3, with a triple S≡C bond

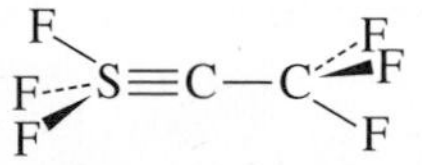

b. $C_6H_6F_2Cl_2Br_2$, a derivative of cyclohexane, in a chair conformation

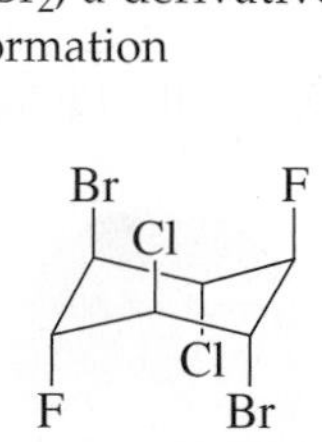

c. $M_2Cl_6Br_4$, where M is a metal atom

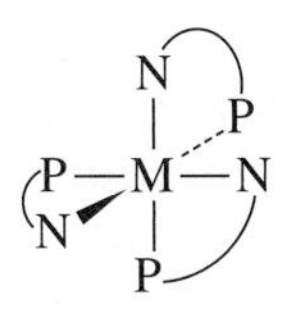

d. $M(NH_2C_2H_4PH_2)_3$, considering the $NH_2C_2H_4PH_2$ rings as planar

e. PCl_2F_3 (the most likely isomer)

4.33 Determine the point groups of the following:

a. The cluster anion $[Re_3(\mu_3{-}S)(\mu{-}S)_3Br_9]^{2-}$ [10]

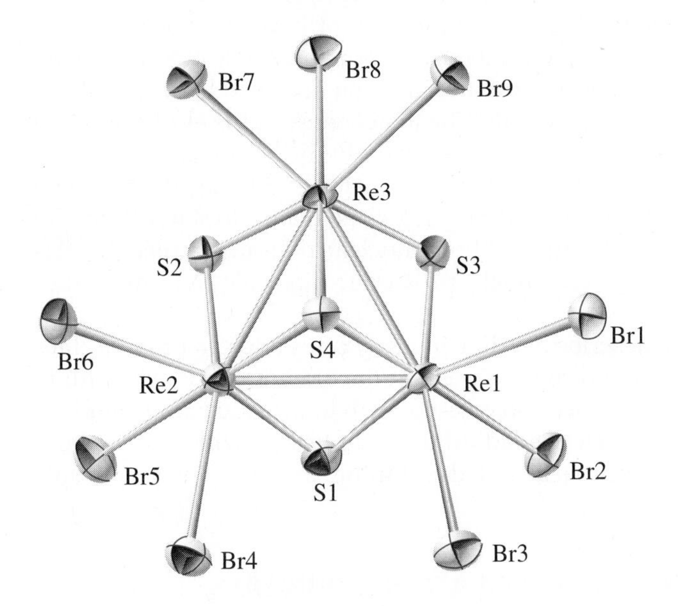

b. The cluster anion $[Co@Ge_{10}]^{3-}$ [11]

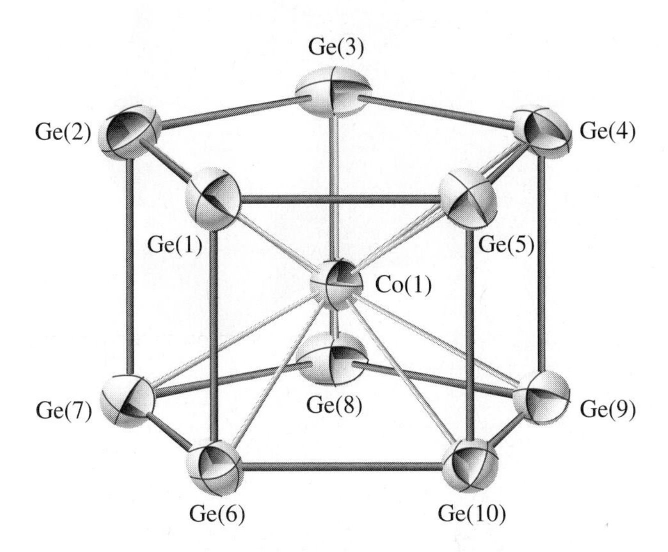

4.34 Use the Internet to search for molecules with the symmetry of

a. The S_6 point group

b. The T point group

c. The I_h point group

d. The T_h point group

Report the molecules, the URL of the Web site where you found them, and the search strategy you used.

[10] Reproduced with permission from H. Sakamoto, Y. Watanabe, and T. Sato, *Inorg. Chem.*, **2006**, *45*, 4578. ©2006, American Chemical Society.

[11] Reproduced with permission from J.-Q. Wang, S. Stegmaier, and T.F. Fassler, "$[Co@Ge_{10}]^{3-}$: An Intermetalloid Cluster with Archimedean Pentagonal Prismatic Structure," *Angew. Chem. Int. Ed.*, **2009**, *48*, 1998. ©Wiley-VCH Verlag GmbH & Co. KgaA.

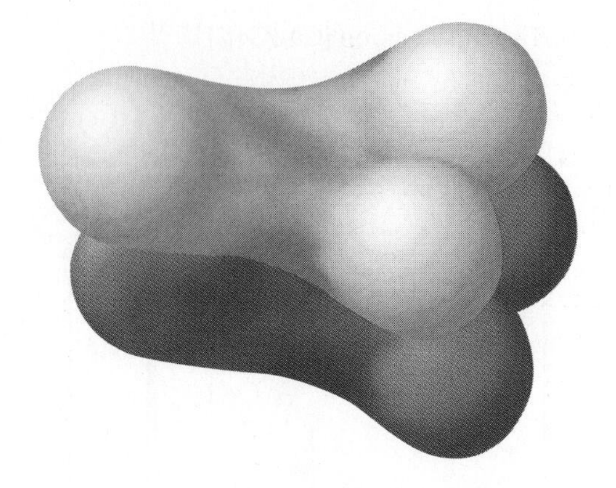

5 Molecular Orbitals

Molecular orbital theory uses the methods of group theory to describe the bonding in molecules; it complements and extends the simple pictures of bonding introduced in Chapter 3. The symmetry properties and relative energies of atomic orbitals determine how they interact to form molecular orbitals. These molecular orbitals are then occupied by the available electrons according to the same rules used for atomic orbitals. The total energy of the electrons in the molecular orbitals is compared with the initial total energy of electrons in the atomic orbitals. If the total energy of the electrons in the molecular orbitals is less than in the atomic orbitals, the molecule is stable compared with the separate atoms; if not, the molecule is unstable and does not form. We will first describe the bonding, or lack of it, in the first 10 homonuclear diatomic molecules (H_2 through Ne_2) and then expand the discussion to heteronuclear diatomic molecules and molecules having more than two atoms.

A simple pictorial approach is adequate to describe bonding in many cases and can provide clues to more complete descriptions of bonding in more difficult cases. On the other hand, it is helpful to know how a more elaborate group theoretical approach can be used, both to provide background for the simpler approach and to have it available when needed. In this chapter, we will describe both approaches, showing the simpler, pictorial approach and developing the symmetry concepts required for more complex cases.

5.1 FORMATION OF MOLECULAR ORBITALS FROM ATOMIC ORBITALS

As in the case of atomic orbitals, Schrödinger equations can be written for electrons in molecules. Approximate solutions to these molecular Schrödinger equations can be constructed from **linear combinations of atomic orbitals (LCAO)**, the sums and differences of the atomic wave functions. For diatomic molecules such as H_2, such wave functions have the form

$$\Psi = c_a\psi_a + c_b\psi_b$$

where Ψ is the molecular wave function, ψ_a and ψ_b are atomic wave functions for atoms a and b, and c_a and c_b are adjustable coefficients. The coefficients can be equal or unequal, positive or negative, depending on the individual orbitals and their energies. As the distance between two atoms is decreased, their orbitals overlap, with significant probability for electrons from both atoms in the region of overlap. As a result, **molecular orbitals** form. Electrons in bonding molecular orbitals occupy the space between the nuclei; the electrostatic forces between the electrons and the two positive nuclei hold the atoms together.

Three conditions are essential for overlap to lead to bonding. First, the symmetry of the orbitals must be such that regions with the same sign of ψ overlap. Second, the energies of the atomic orbitals must be similar. When the energies differ greatly, the change in the energy of electrons on formation of molecular orbitals is small, and the net reduction in energy of the electrons is too small for significant bonding. Third, the distance between the atoms must be short enough to provide good overlap of the orbitals, but not so short that repulsive forces of other electrons or the nuclei interfere. When these three conditions are met, the overall energy of the electrons in the occupied molecular orbitals is lower in energy than the overall energy of the electrons in the original atomic orbitals, and the resulting molecule has a lower total energy than the separated atoms.

5.1.1 Molecular Orbitals from *s* Orbitals

We will consider first the interactions between two s orbitals, as in H_2. For convenience, we label the atoms of a diatomic molecule a and b, so the atomic orbital wave functions are $\psi(1s_a)$ and $\psi(1s_b)$. We can visualize the two atoms approaching each other, until their electron clouds overlap and merge into larger molecular electron clouds. The resulting molecular orbitals are linear combinations of the atomic orbitals, the sum of the two orbitals and the difference between them:

In general terms | For H_2

$$\Psi(\sigma) = N[c_a\psi(1s_a) + c_b\psi(1s_b)] = \frac{1}{\sqrt{2}}[\psi(1s_a) + \psi(1s_b)] \quad (\mathrm{H}_a + \mathrm{H}_b)$$

$$\text{and } \Psi(\sigma^*) = N[c_a\psi(1s_a) - c_b\psi(1s_b)] = \frac{1}{\sqrt{2}}[\psi(1s_a) - \psi(1s_b)] \quad (\mathrm{H}_a - \mathrm{H}_b)$$

where N = normalizing factor, so $\int \Psi\Psi^*\, d\tau = 1$

c_a and c_b = adjustable coefficients

In this case, the two atomic orbitals are identical, and the coefficients are nearly identical as well.[1] These orbitals are depicted in Figure 5.1. In this diagram, as in all the orbital diagrams in this book (such as Table 2.3 and Figure 2.6), the signs of orbital lobes are indicated by shading. Light and dark lobes indicate opposite signs

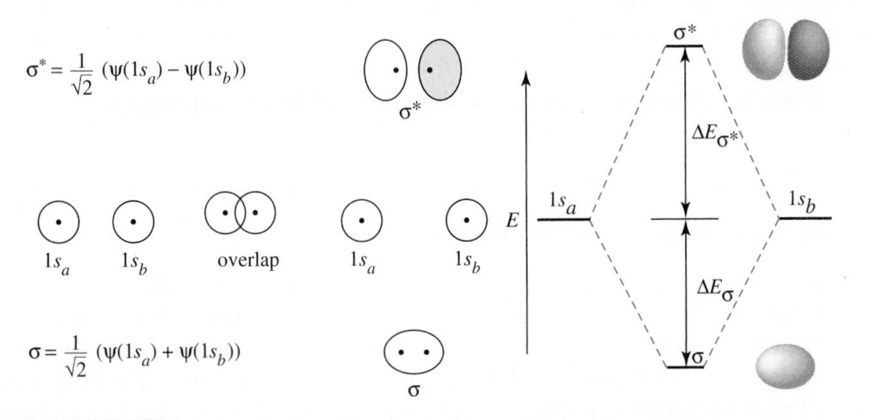

FIGURE 5.1 Molecular Orbitals from Hydrogen 1*s* Orbitals.

[1] More precise calculations show that the coefficients of the σ^* orbital are slightly larger than those for the σ orbital; but for the sake of simplicity, we will generally not focus on this. For identical atoms, we will use $c_a = c_b = 1$ and $N = 1/\sqrt{2}$. The difference in coefficients for the σ and σ^* orbitals also results in a larger change in energy (increase) from the atomic to the σ^* molecular orbitals than for the σ orbitals (decrease). In other words, $\Delta E_{\sigma^*} > \Delta E_{\sigma}$, as shown in Figure 5.1.

of Ψ. The choice of positive and negative for specific atomic orbitals is arbitrary; what is important is how they fit together to form molecular orbitals. In the diagrams on the right side in the figure, light and dark shading show opposite signs of the wave function.

Because the σ molecular orbital is the sum of the two atomic orbitals, $\frac{1}{\sqrt{2}}[\psi(1s_a) + \psi(1s_b)]$, and results in an increased concentration of electrons between the two nuclei, where both atomic wave functions contribute, it is a **bonding molecular orbital** and has a lower energy than the original atomic orbitals. The σ^* molecular orbital is the difference of the two atomic orbitals, $\frac{1}{\sqrt{2}}[\psi(1s_a) - \psi(1s_b)]$. It has a node with zero electron density between the nuclei—by cancellation of the two wave functions—and a higher energy; it is therefore called an **antibonding orbital**. Electrons in bonding orbitals are concentrated between the nuclei and attract the nuclei, holding them together. Antibonding orbitals have one or more nodes between the nuclei; electrons in these orbitals cause a mutual repulsion between the atoms. **Nonbonding orbitals** are also possible. The energy of a nonbonding orbital is essentially that of an atomic orbital, usually because the orbital on one atom has a symmetry that does not match any orbitals on the other atom. In some cases the energy of the molecular orbital matches that of the atomic orbital by coincidence, and such an orbital may be essentially nonbonding.

The σ (sigma) notation indicates orbitals that are symmetric to rotation about the line connecting the nuclei:

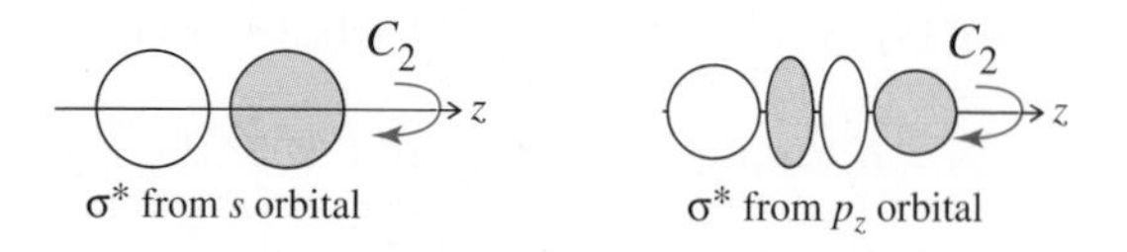

An asterisk is frequently used to indicate antibonding orbitals. Because the bonding, nonbonding, or antibonding nature of a molecular orbital is sometimes uncertain, we will use the asterisk notation only in the simpler cases in which the bonding and antibonding characters are clear.

The pattern described for H_2 is the usual model for combining two orbitals: two atomic orbitals combine to form two molecular orbitals, one bonding orbital with a lower energy and one antibonding orbital with a higher energy. Regardless of the number of orbitals, the unvarying rule is that the number of resulting molecular orbitals is the same as the initial number of atomic orbitals.

5.1.2 Molecular Orbitals from *p* Orbitals

Molecular orbitals formed from *p* orbitals are more complex. The algebraic sign of the wave function must be included when interactions between the orbitals are considered. When two orbitals overlap, and the overlapping regions have the same sign, the sum of the two orbitals has an increased electron probability in the overlap region. When two regions of opposite sign overlap, the combination has a decreased electron probability in the overlap region. Figure 5.1 shows this effect for the 1*s* orbitals of H_2; similar effects result from overlapping lobes of *p* orbitals with their alternating signs. The interactions of *p* orbitals are shown in Figure 5.2. For convenience, we will choose a common *z* axis connecting the nuclei and assign *x* and *y* axes as shown in the figure.

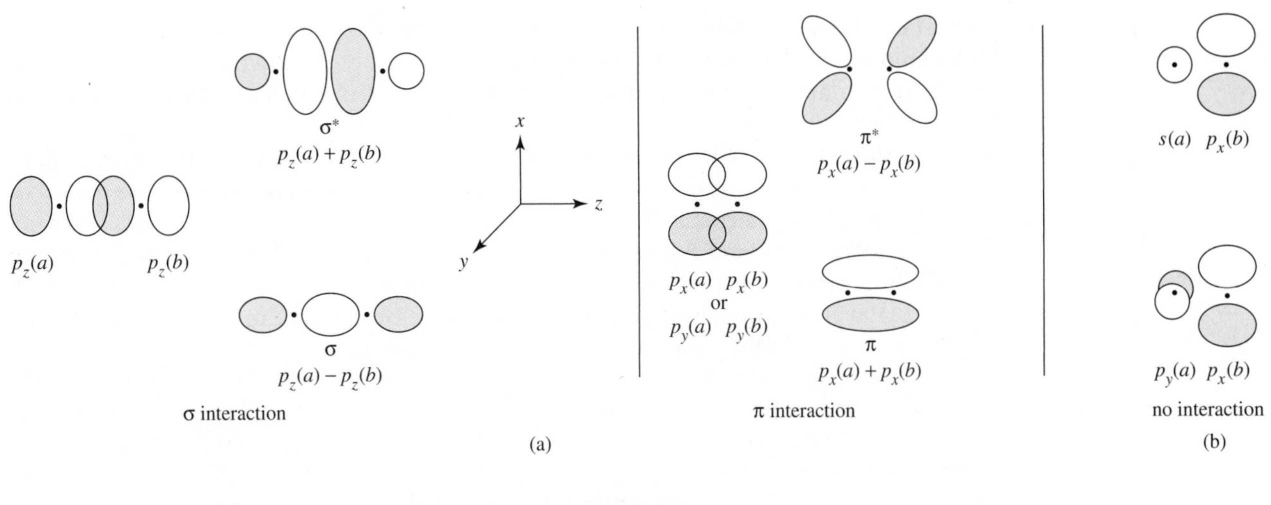

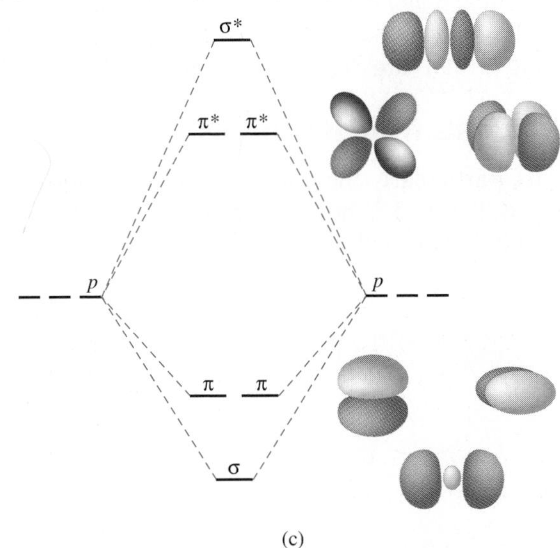

FIGURE 5.2 Interactions of *p* Orbitals. (a) Formation of molecular orbitals. (b) Orbitals that do not form molecular orbitals. (c) Energy level diagram.

When we draw the *z* axes for the two atoms pointing in the same direction,[2] the p_z orbitals subtract to form σ and add to form σ^* orbitals, both of which are symmetric to rotation about the *z* axis, with nodes perpendicular to the line that connects the nuclei. Interactions between p_x and p_y orbitals lead to π and π^* orbitals. The π (pi) notation indicates a change in sign of the wave function with C_2 rotation about the bond axis:

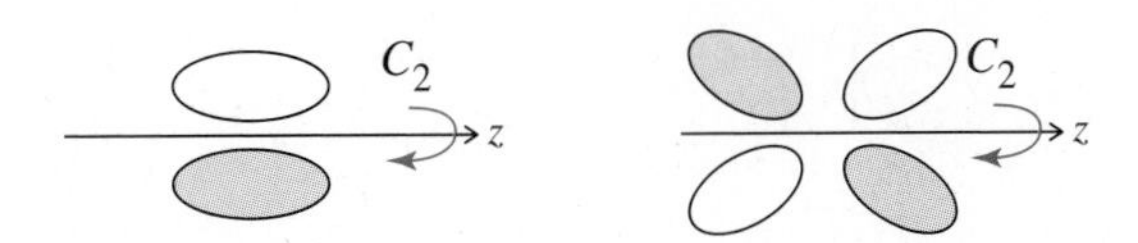

[2] The choice of direction of the *z* axes is arbitrary. When both are positive in the same direction, → →, **the difference between the p_z orbitals is the bonding combination**. When the positive *z* axes are chosen to point toward each other → ←, the sum of the p_z orbitals is the bonding combination. We have chosen to have them positive in the same direction for consistency with our treatment of triatomic and larger molecules.

As in the case of the *s* orbitals, the overlap of two regions with the same sign leads to an increased concentration of electrons, and the overlap of two regions of opposite sign leads to a node of zero electron density. In addition, the nodes of the atomic orbitals become the nodes of the resulting molecular orbitals. In the π^* antibonding case, four lobes result that are similar in appearance to an elongated *d* orbital, as in Figure 5.2(c).

The p_x, p_y, and p_z orbital pairs need to be considered separately. Because the *z* axis was chosen as the internuclear axis, the orbitals derived from the p_z orbitals are symmetric to rotation around the bond axis and are labeled σ and σ^* for the bonding and antibonding orbitals respectively. Similar combinations of the p_y orbitals form orbitals whose wave functions change sign with C_2 rotation about the bond axis; they are labeled π and π^*. In the same way, the p_x orbitals also form π and π^* orbitals.

When orbitals overlap equally with both the same and opposite signs, as in the $s + p_x$ example in Figure 5.2(b), the bonding and antibonding effects cancel, and no molecular orbital results. An equivalent description is that, because the symmetry properties of the orbitals do not match, no combination is possible. If the symmetry of an atomic orbital does not match *any* orbital of the other atom, it is called a nonbonding orbital. Homonuclear diatomic molecules have only bonding and antibonding molecular orbitals; nonbonding orbitals are described further in Sections 5.1.4, 5.2.2, and 5.4.3.

5.1.3 Molecular Orbitals from *d* Orbitals

In the heavier elements, particularly the transition metals, *d* orbitals can be involved in bonding in a similar way. Figure 5.3 shows the possible combinations. When the *z* axes are collinear, two d_{z^2} orbitals can combine end-on for σ bonding. The d_{xz} and d_{yz} orbitals

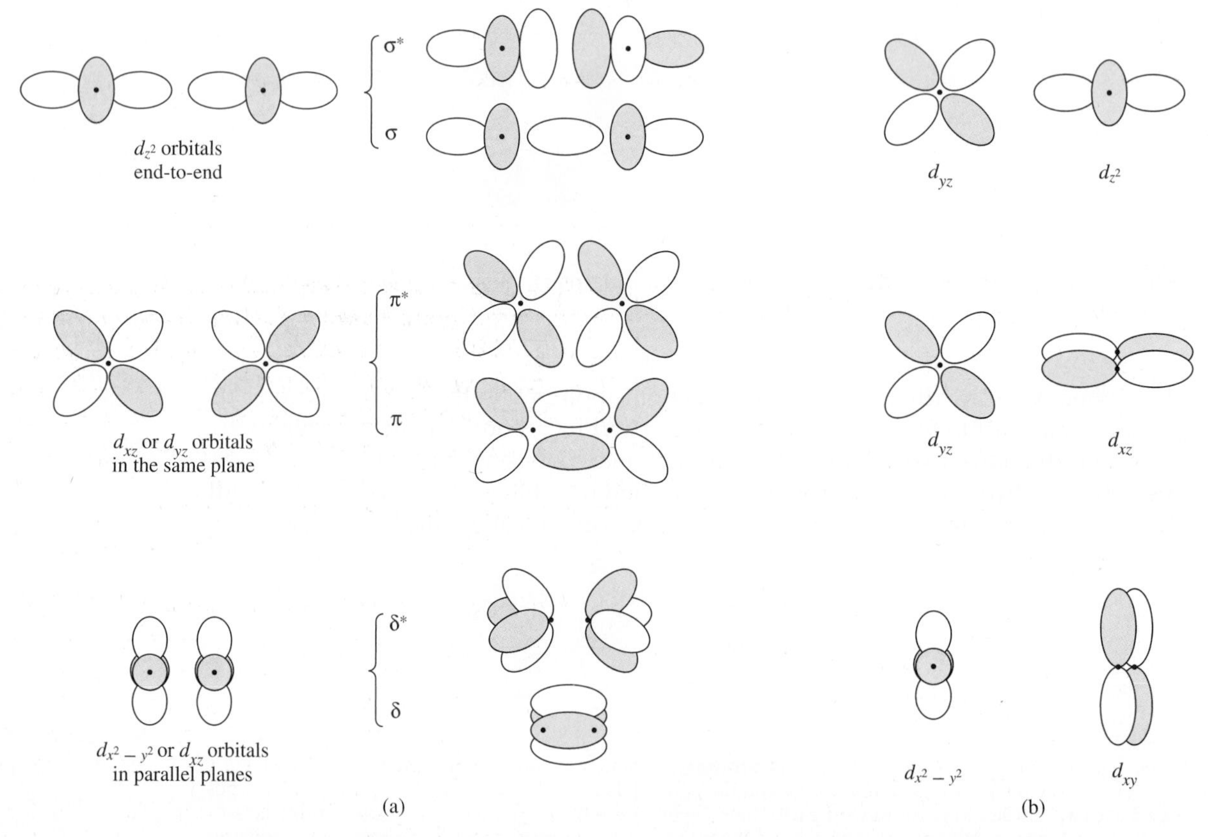

FIGURE 5.3 Interactions of *d* Orbitals. (a) Formation of molecular orbitals. (b) Orbitals that do not form molecular orbitals.

form π orbitals. When atomic orbitals meet from two parallel planes and combine side to side, as do the $d_{x^2-y^2}$ and d_{xy} orbitals with collinear z axes, they form (δ) delta orbitals. (The δ notation indicates sign changes on C_4 rotation about the bond axis.) Sigma orbitals have no nodes that include the line of centers of the atoms, pi orbitals have one node that includes the line of centers, and delta orbitals have two nodes that include the line of centers. Combinations of orbitals involving overlapping regions with opposite signs cannot form molecular orbitals; for example, p_z and d_{xz} have zero net overlap, one region with overlapping regions of the same sign and another with opposite signs.

EXAMPLE

Sketch the overlap regions of the following combination of orbitals, all with collinear z axes, and classify the interactions.

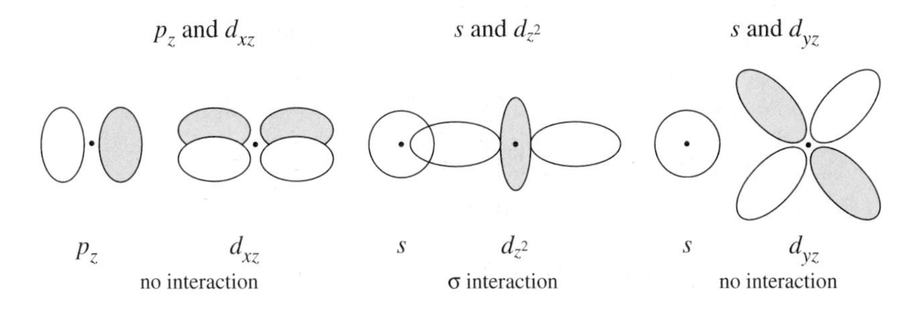

▶ **Exercise 5.1** Repeat the process in the preceding example for the following orbital combinations, again using collinear z axes.

p_x and d_{xz} $\quad$ p_z and d_{z^2} $\quad$ s and $d_{x^2-y^2}$

5.1.4 Nonbonding Orbitals and Other Factors

As mentioned previously, there can also be nonbonding molecular orbitals whose energy is essentially that of the original atomic orbitals. These can form in larger molecules, for example when there are three atomic orbitals of the same symmetry and similar energies, a situation that requires the formation of three molecular orbitals. One molecular orbital formed is a low-energy bonding orbital, one is a high-energy antibonding orbital, and one is of intermediate energy and is a nonbonding orbital. Examples will be considered in Section 5.4 and in later chapters.

In addition to symmetry, the second major factor that must be considered in forming molecular orbitals is the relative energy of the atomic orbitals. As shown in Figure 5.4, when the interacting atomic orbitals have the same energy, the resulting interaction is strong, and the resulting molecular orbitals have energies well below (bonding) and above (antibonding) that of the original atomic orbitals. When the two atomic orbitals have quite different energies, the interaction is weaker, and the resulting molecular orbitals have energies and shapes closer to the original atomic orbitals. For example, although they have the same symmetry, 1s orbitals do not combine significantly with 2s orbitals of the other atom in diatomic molecules such as N_2, because their energies are too far apart. The general rule is that the closer the energy match, the stronger the interaction.

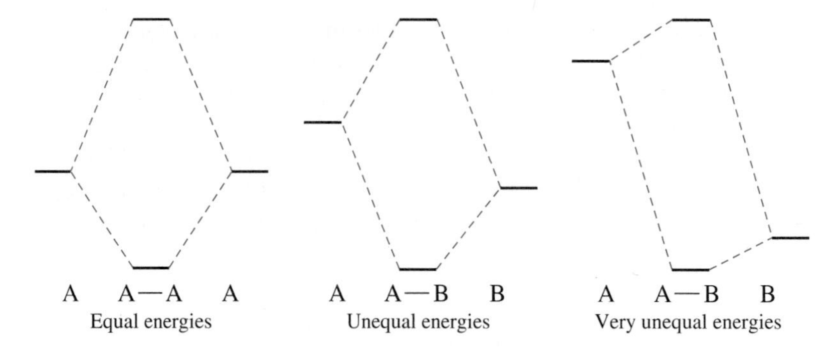

FIGURE 5.4 Energy Match and Molecular Orbital Formation.

5.2 HOMONUCLEAR DIATOMIC MOLECULES

5.2.1 Molecular Orbitals

Although apparently satisfactory Lewis electron-dot diagrams of N_2, O_2, and F_2 can be drawn, the same is not true of Li_2, Be_2, B_2, and C_2, which cannot show the usual octet structure. In addition, the Lewis diagram of O_2 shows a simple double-bonded molecule with all electrons paired (:Ö=Ö:), but experiment has shown it to have two unpaired electrons, making it paramagnetic (in fact, liquid oxygen poured between the poles of a large horseshoe magnet is attracted into the field and held there). As we will see, the molecular orbital description is more in agreement with experiment. Figure 5.5 shows the full set of molecular orbitals for the homonuclear diatomic molecules of the first 10 elements, based on the energies appropriate for O_2. The diagram shows the order of energy levels for the molecular orbitals, assuming interactions only between atomic orbitals of identical energy. The energies of the molecular orbitals change with increasing atomic number, but the general pattern remains similar (with some subtle changes, as will be described in several examples) even for heavier atoms lower in the periodic table. Electrons fill the molecular orbitals according to the same rules that govern the filling of atomic orbitals, filling from lowest to highest energy (aufbau principle), maximum spin multiplicity consistent with the lowest net energy (Hund's rules), and no two electrons with identical quantum numbers (Pauli exclusion principle).

The overall number of bonding and antibonding electrons determines the number of bonds (bond order):

$$\text{Bond order} = \frac{1}{2}\left[\begin{pmatrix}\text{number of electrons}\\ \text{in bonding orbitals}\end{pmatrix} - \begin{pmatrix}\text{number of electrons}\\ \text{in antibonding orbitals}\end{pmatrix}\right]$$

In most cases it is sufficient to consider only valence electrons. For example, O_2, with 10 electrons in bonding orbitals and 6 electrons in antibonding orbitals, has a bond order of 2, a double bond. Counting only valence electrons, 8 bonding and 4 antibonding, gives the same result. Because the molecular orbitals derived from the 1*s* orbitals have the same number of bonding and antibonding electrons, they have no net effect on the bonding.

Additional labels are helpful in describing the orbitals. The subscripts *g* for *gerade*, orbitals symmetric to inversion, and *u* for *ungerade*, orbitals antisymmetric to inversion (those whose signs change on inversion), are commonly used.[3] The *g* or *u* notation describes the symmetry of the orbitals without a judgment as to their relative energies.

[3] See end of Section 4.3.3 for more details on symmetry labels.

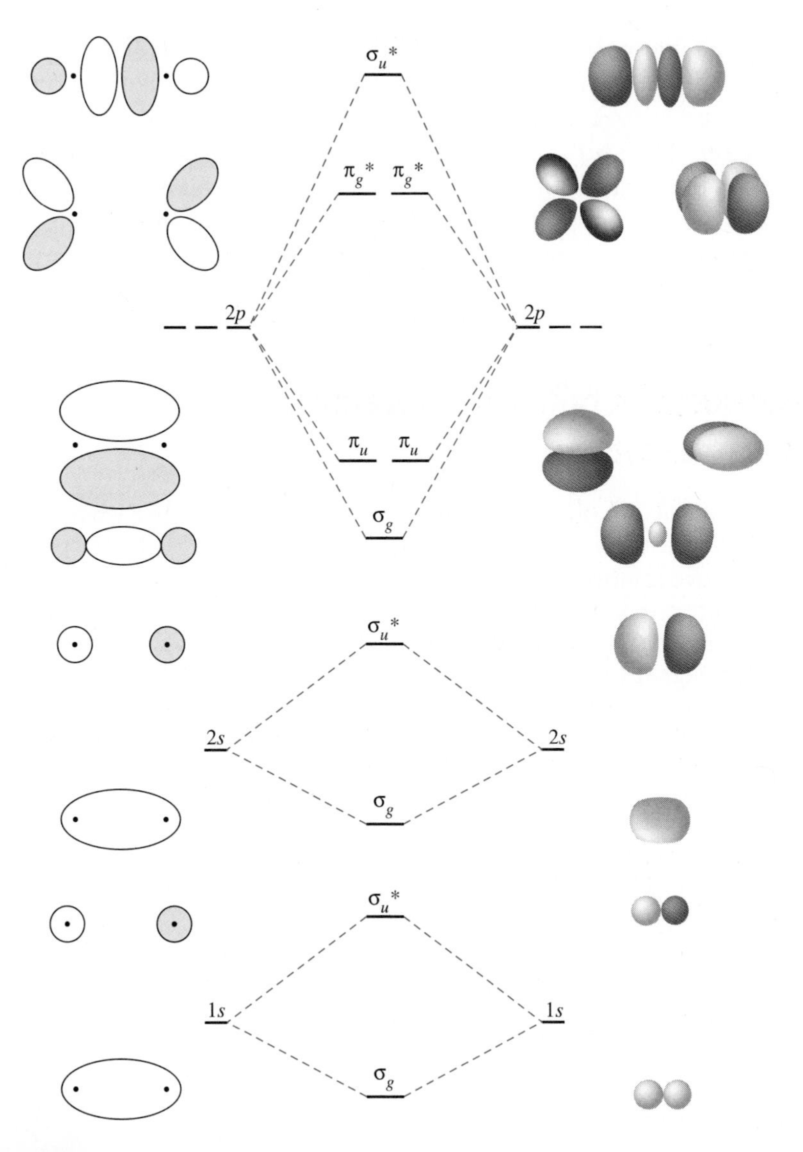

FIGURE 5.5 Molecular Orbitals for the First 10 Elements, Assuming Interactions Only Between Atomic Orbitals of Identical Energy.

EXAMPLE

Add a *g* or *u* label to each of the molecular orbitals in the energy-level diagram in Figure 5.2.

From top to bottom, the orbitals are $\sigma_u{}^*$, $\pi_g{}^*$, π_u, and σ_g.

▶ **Exercise 5.2** Add a *g* or *u* label to each of the molecular orbitals in Figure 5.3(a).

5.2.2 Orbital Mixing

So far, we have primarily considered interactions between orbitals of identical energy. However, orbitals with similar, but not equal, energies interact if they have appropriate symmetries. We will outline two approaches to analyzing this interaction, one in which the molecular orbitals interact and one in which the atomic orbitals interact directly.

When two molecular orbitals of the same symmetry have similar energies, they interact to lower the energy of the lower orbital and raise the energy of the higher orbital. For example, in the homonuclear diatomics, the $\sigma_g(2s)$ and $\sigma_g(2p)$ orbitals both have σ_g symmetry (symmetric to infinite rotation and inversion); these orbitals interact to lower the energy of the $\sigma_g(2s)$ and to raise the energy of the $\sigma_g(2p)$, as shown in Figure 5.6(b). Similarly, the $\sigma_u{}^*(2s)$ and $\sigma_u{}^*(2p)$ orbitals interact to lower the energy of the $\sigma_u{}^*(2s)$ and to raise the energy of the $\sigma_u{}^*(2p)$. This phenomenon is called **mixing**, which takes into account that molecular orbitals with similar energies interact if they have appropriate symmetry, a factor ignored in Figure 5.5. When two molecular orbitals of the same symmetry mix, the one with higher energy moves still higher in energy, and the one with lower energy moves lower.

Alternatively, we can consider that the four molecular orbitals (MOs) result from combining the four atomic orbitals (two 2*s* and two $2p_z$) that have similar energies. The resulting molecular orbitals have the following general form, where *a* and *b* identify the two atoms:

$$\Psi = c_1\psi(2s_a) \pm c_2\psi(2s_b) \pm c_3\psi(2p_a) \pm c_4\psi(2p_b)$$

For homonuclear diatomic molecules, $c_1 = c_2$ and $c_3 = c_4$ in each of the four MOs. The lowest energy MO has larger values of c_1 and c_2, the highest has larger values of c_3 and c_4, and the two intermediate MOs have intermediate values for all four coefficients. The symmetry of these four orbitals is the same as those without mixing, but their shapes are changed somewhat by having the mixture of *s* and *p* characters. In addition, the energies are shifted: higher for the two upper and lower for the two lower energy orbitals.

As we will see, *s-p* mixing can have an important influence on the energy of molecular orbitals. For example, in the early part of the second period diatomics (Li_2 to N_2), the σ_g orbital formed from $2p_z$ orbitals is higher in energy than the π_u orbitals formed from

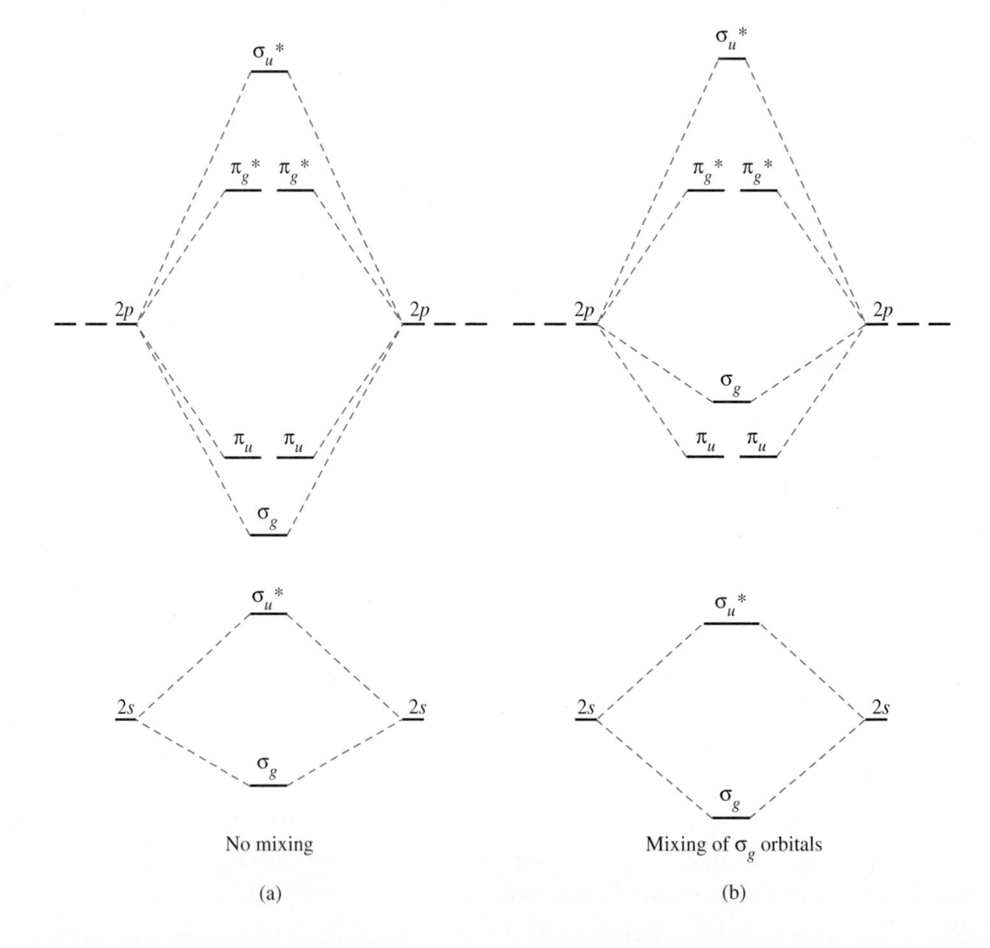

FIGURE 5.6 Interaction between Molecular Orbitals. Mixing molecular orbitals of the same symmetry results in a greater energy difference between the orbitals. The σ orbitals mix strongly; the σ^* orbitals differ more in energy and mix weakly.

the $2p_x$ and $2p_y$ orbitals. This is an inverted order from that expected without mixing (Figure 5.6). For B_2 and C_2, mixing affects the magnetic properties of the molecules. In addition, mixing changes the bonding–antibonding nature of some of the orbitals. The orbitals with intermediate energies may have either a slightly bonding or slightly antibonding character and contribute in minor ways to the bonding. Each orbital must be considered separately on the basis of its energy and electron distribution.

5.2.3 Diatomic Molecules of the First and Second Periods

Before proceeding with examples of homonuclear diatomic molecules, it is necessary to define two types of magnetic behavior, **paramagnetic** and **diamagnetic**. Paramagnetic compounds are attracted by an external magnetic field. This attraction is a consequence of one or more unpaired electrons behaving as tiny magnets. Diamagnetic compounds, on the other hand, have no unpaired electrons and are repelled slightly by magnetic fields. (An experimental measure of the magnetism of compounds is the **magnetic moment**, a term that will be described further in Chapter 10 in the discussion of the magnetic properties of coordination compounds.)

H_2, He_2, and the homonuclear diatomic species shown in Figure 5.7 will be discussed in the following pages. In the progression across the periodic table, the energy of all the orbitals decreases as the increased nuclear charge attracts the electrons more strongly. As shown in Figure 5.7, the decrease in energy is larger for σ orbitals than for π orbitals, resulting from the greater overlap of the atomic orbitals that participate in σ interactions.

$H_2[\sigma_g^2(1s)]$ This the simplest of the diatomic molecules. The MO description (Figure 5.1) shows a single σ orbital containing one electron pair; the bond order is 1, representing a single bond. The ionic species H_2^+, with a single electron in the a σ orbital and a bond order of $\frac{1}{2}$, has been detected in low-pressure gas-discharge systems. As expected, it is less stable than H_2 and has a considerably longer bond distance than H_2 (105.2 pm vs. 74.1 pm).

$He_2[\sigma_g^2\sigma_u^{*2}(1s)]$ The molecular orbital description of He_2 predicts two electrons in a bonding orbital and two in an antibonding orbital, with a bond order of zero—in other words, no bond. This is what is observed experimentally. The noble gas He has no significant tendency to form diatomic molecules and, like the other noble gases, exists in the form of free atoms. He_2 has been detected only in very low-pressure and low-temperature molecular beams. It has an extremely low binding energy,[4] approximately 0.01 J/mol; for comparison, H_2 has a bond energy of 436 kJ/mol.

$Li_2[\sigma_g^2(2s)]$ As shown in Figure 5.7, the MO model predicts a single Li—Li bond in Li_2, in agreement with gas–phase observations of the molecule.

$Be_2[\sigma_g^2\sigma_u^{*2}(2p)]$ Be_2 has the same number of antibonding and bonding electrons and consequently a bond order of zero. Hence, like He_2, Be_2 is not a stable chemical species.[5]

$B_2[\pi_u^1\pi_u^1(2p)]$ Here is an example in which the MO model has a distinct advantage over the Lewis dot picture. B_2 is found only in the gas phase; solid boron is found in several very hard forms with complex bonding, primarily involving B_{12} icosahedra.

---

[4] F. Luo, G. C. McBane, G. Kim, C. F. Giese, and W. R. Gentry, *J. Chem. Phys.*, **1993**, *98*, 3564; see also L. L. Lohr and S. M. Blinder, *J. Chem. Educ.*, **2007**, *84*, 860, and references cited therein.

[5] Be_2 is calculated to have a very weak bond when effects of higher energy, unoccupied orbitals are taken into account. See A. Krapp, F. M. Bickelhaupt, and G. Frenking, *Chem. Eur. J.*, **2006**, *12*, 9196.

FIGURE 5.7 Energy Levels of the Homonuclear Diatomics of the Second Period.

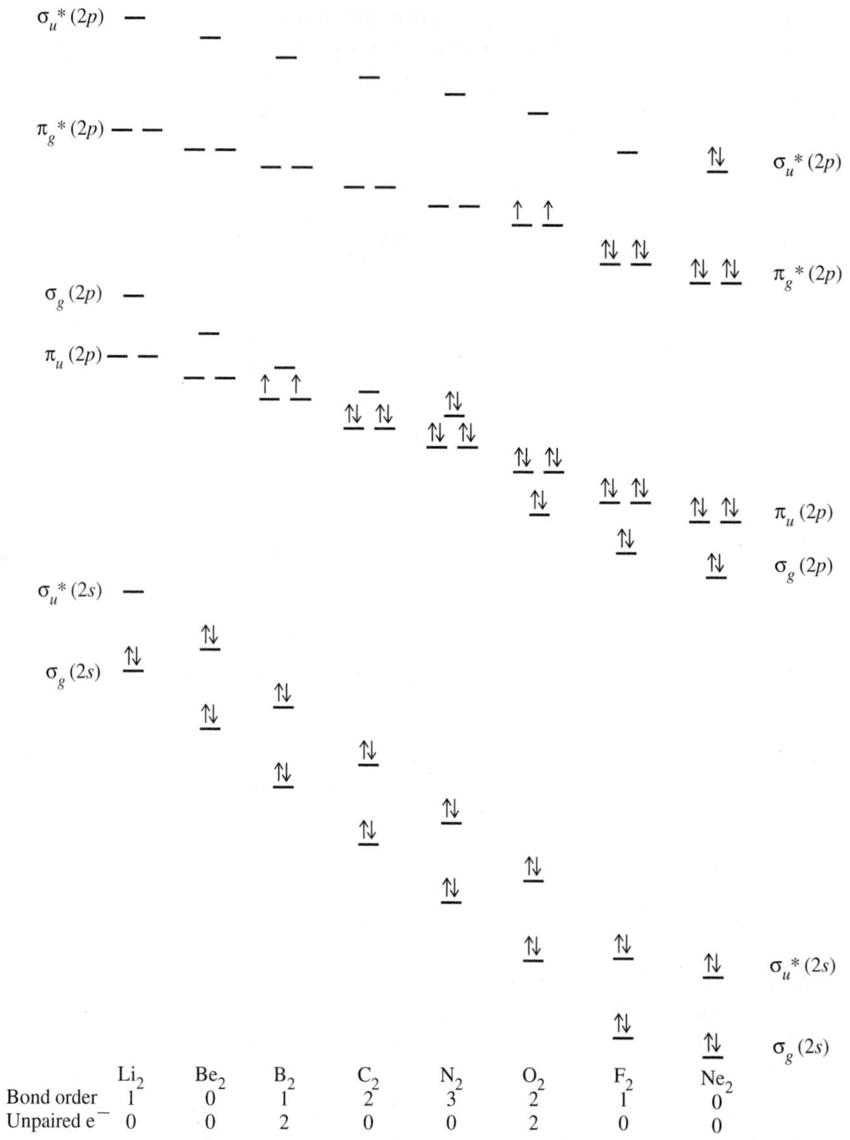

	Li_2	Be_2	B_2	C_2	N_2	O_2	F_2	Ne_2
Bond order	1	0	1	2	3	2	1	0
Unpaired e^-	0	0	2	0	0	2	0	0

B_2 is paramagnetic. This behavior can be explained if its two highest energy electrons occupy separate π orbitals, as shown. The Lewis dot model cannot account for the paramagnetic behavior of this molecule.

B_2 is also a good example of the energy-level shift caused by the mixing of s and p orbitals. In the absence of mixing, the $\sigma_g(2p)$ orbital would be expected to be lower in energy than the $\pi_u(2p)$ orbitals. Were this order followed, the molecule would be diamagnetic. However, mixing of the $\sigma_g(2s)$ orbital with the $\sigma_g(2p)$ orbital (Figure 5.6) lowers the energy of the $\sigma_g(2s)$ orbital and increases the energy of the $\sigma_g(2p)$ orbital to a higher level than the π orbitals, giving the order of energies shown in Figure 5.7. As a result, the last two electrons are unpaired in the **degenerate** (having the same energy) π orbitals, and the molecule is paramagnetic. Overall, the bond order is one, even though the two π electrons are in different orbitals.

$C_2[\pi_u^2\pi_u^2(2p)]$ The simple MO picture of C_2 predicts a doubly bonded molecule, with all electrons paired, but with both **highest occupied molecular orbitals (HOMOs)** having π symmetry. C_2 is unusual, because it has two π bonds and no σ bond.

Although C_2 is not a commonly encountered chemical species (carbon is more stable as diamond, graphite, and the fullerenes described in Chapter 8), the acetylide ion, C_2^{2-}, is well known, particularly in compounds with alkali metals, alkaline earths, and lanthanides. According to the molecular orbital model, C_2^{2-} should have a bond order of 3 (configuration $\pi_u^2\pi_u^2\sigma_g^2$). This is supported by the similar C—C distances in acetylene and calcium carbide (acetylide)[6, 7]:

	C—C Distance (pm)
C=C (gas phase)	124.2
H—C≡C—H	120.5
CaC_2	119.1

$N_2[\pi_u^2\pi_u^2\sigma_g^2(2p)]$ N_2 has a triple bond according to both the Lewis and the molecular orbital models. This is in agreement with its very short N—N distance (109.8 pm) and extremely high bond-dissociation energy (942 kJ/mol). Atomic orbitals decrease in energy with increasing nuclear charge Z as shown in Figure 5.7; as the effective nuclear charge increases, all orbitals are pulled to lower energies. The shielding effect and electron–electron interactions described in Section 2.2.4 cause an increase in the difference between the 2*s* and 2*p* energies as Z increases, from 5.7 eV for boron to 8.8 eV for carbon and 12.4 eV for nitrogen. (A table of these energies is given in Table 5.1 in Section 5.3.1.) As a result, the $\sigma_g(2s)$ and $\sigma_g(2p)$ levels of N_2 interact (mix) less than the B_2 and C_2 levels, and the $\sigma_g(2p)$ and $\pi_u(2p)$ are very close in energy. The order of energies of these orbitals has been a matter of controversy and will be discussed in more detail in Section 5.2.4 on photoelectron spectroscopy.

$O_2[\sigma_g^2\pi_u^2\pi_u^2\pi_g^{*1}\pi_g^{*1}(2p)]$ O_2 is paramagnetic. As for B_2, this property cannot be explained by the traditional Lewis dot structure :Ö=Ö:, but it is evident from the MO picture, which assigns two electrons to the degenerate π_g^* orbitals. The paramagnetism can be demonstrated by pouring liquid O_2 between the poles of a strong magnet; some of the O_2 will be held between the pole faces until it evaporates. Several ionic forms of diatomic oxygen are known, including O_2^+, O_2^-, and O_2^{2-}. The internuclear O—O distance can be conveniently correlated with the bond order predicted by the molecular orbital model, as shown in the following table.[8]

	Bond Order	Internuclear Distance (pm)
O_2^+ (dioxygenyl)	2.5	111.6
O_2 (dioxygen)	2.0	120.8
O_2^- (superoxide)	1.5	135
O_2^{2-} (peroxide)	1.0	149

NOTE: Oxygen–oxygen distances in O_2^- and O_2^{2-} are influenced by the cation. This influence is especially strong in the case of O_2^{2-} and is one factor in its unusually long bond distance, which should be considered approximate.

[6] M. Atoji, *J. Chem. Phys.*, **1961**, *35*, 1950.
[7] J. Overend and H. W. Thompson, *Proc. R. Soc. London*, **1954**, *A234*, 306.
[8] See Table 5.1 for references.

The extent of mixing is not sufficient in O_2 to push the $\sigma_g(2p)$ orbital to higher energy than the $\pi_g(2p)$ orbitals. The order of molecular orbitals shown is consistent with the photoelectron spectrum, discussed in Section 5.2.4.

$F_2[\sigma_g{}^2\pi_u{}^2\pi_u{}^2\pi_g{}^{*2}\pi_g{}^{*2}(2p)]$ The MO picture of F_2 shows a diamagnetic molecule having a single fluorine–fluorine bond, in agreement with experimental data on this highly reactive molecule.

The net bond order in N_2, O_2, and F_2 is the same whether or not mixing is taken into account, but the order of the $\sigma_g(2p)$ and $\pi_u(2p)$ orbitals is different. The switching of the order of these orbitals can occur because they are close in energy; minor changes in the $\sigma_g(2p)$ orbital can switch their order. The energy difference between the 2*s* and 2*p* orbitals of the atoms increases with increasing nuclear charge, from 5.7 eV in boron to 21.5 eV in fluorine (details are in Section 5.3.1). Because the difference becomes greater, the *s-p* interaction (mixing) decreases, and the "normal" order of molecular orbitals returns in O_2 and F_2. The higher σ_g orbital is seen again in CO, described in Section 5.3.1.

Ne_2 All the molecular orbitals are filled, there are equal numbers of bonding and antibonding electrons, and the bond order is therefore zero. The Ne_2 molecule is a transient species, if it exists at all.

One triumph of molecular orbital theory is its prediction of two unpaired electrons for O_2. Oxygen had long been known to be paramagnetic, but early explanations for this phenomenon were difficult. One idea used a special "three-electron bond"[9] to explain this phenomenon. On the other hand, the molecular orbital description provides for the unpaired electrons directly. In other cases, the experimental observations (paramagnetic B_2, diamagnetic C_2) require a shift of orbital energies, raising σ_g above π_u, but they do not require addition of any different type of orbitals or bonding.

BOND LENGTHS IN HOMONUCLEAR DIATOMIC MOLECULES Figure 5.8 shows the variation of bond distance with the number of valence electrons in second-period *p*-block homonuclear diatomic molecules having 6 to 14 valence electrons. Beginning at the

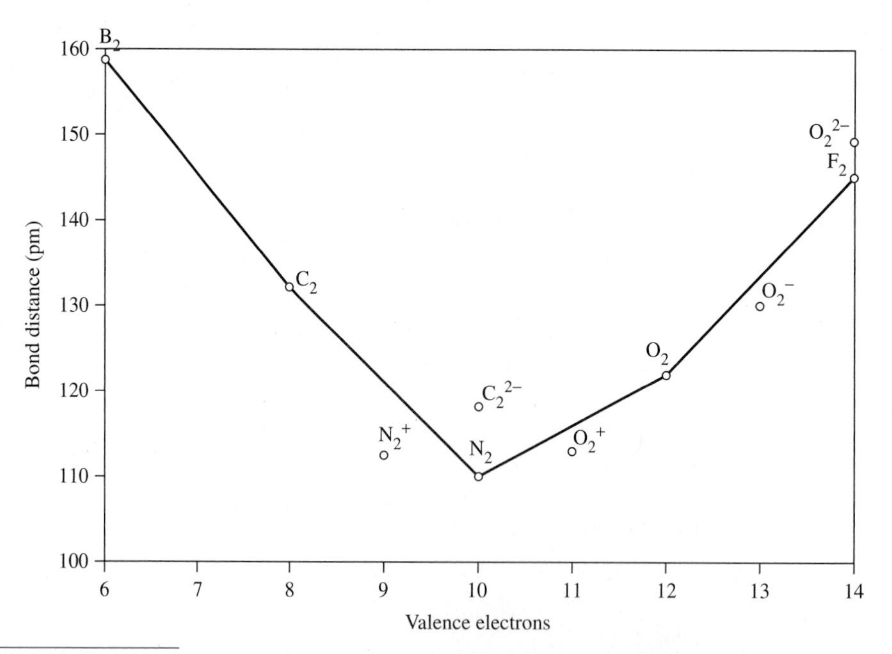

FIGURE 5.8 Bond Distances of Homonuclear Diatomic Molecules and Ions.

[9] L. Pauling, *The Nature of the Chemical Bond*, 3rd ed., Cornell University Press, Ithaca, NY, 1960, pp. 340–354.

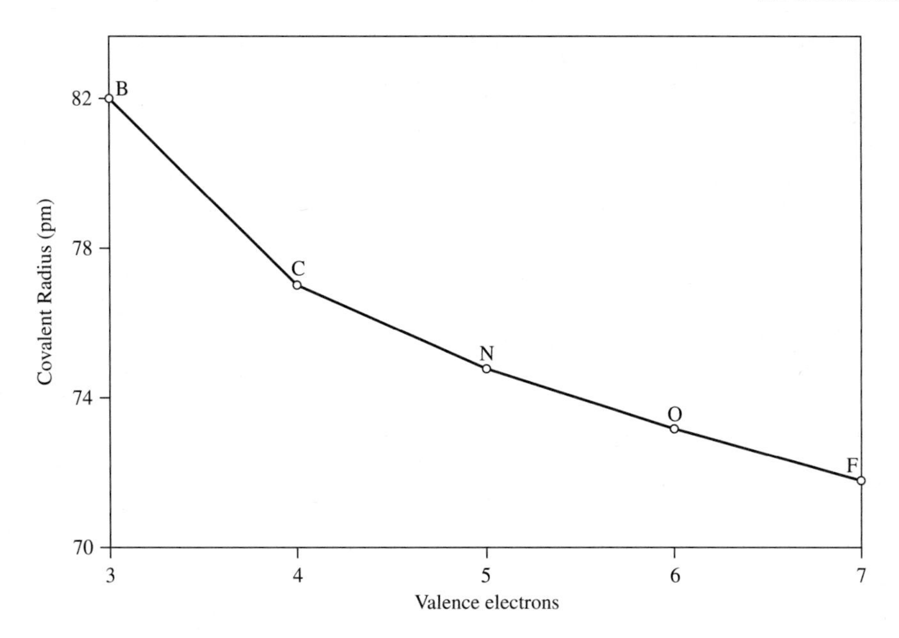

FIGURE 5.9 Covalent Radii of Second-Period Atoms.

left, as the number of electrons increases the number in bonding orbitals also increases; the bond strength becomes greater, and the bond length becomes shorter. This continues up to 10 valence electrons in N_2, where the trend reverses, because the additional electrons occupy antibonding orbitals. The ions N_2^+, O_2^+, O_2^-, and O_2^{2-} are also shown in the figure and follow a similar trend.

The minimum in Figure 5.9 occurs even though the radii of the free atoms decrease steadily from B to F. Figure 5.9 shows the change in covalent radius for the atoms, decreasing as the number of valence electrons increases primarily because the increasing nuclear charge pulls the electrons closer to the nucleus. For the elements boron through nitrogen, the trends shown in Figures 5.8 and 5.9 are similar: as the covalent radius of the atom decreases, the bond distance of the matching diatomic molecule also decreases. However, beyond nitrogen these trends diverge. Even though the covalent radii of the free atoms continue to decrease (N $>$ O $>$ F), the bond distances in their diatomic molecules increase ($N_2 < O_2 < F_2$) with the increasing population of antibonding orbitals. In general the bond order is the more important factor, overriding the covalent radii of the component atoms. Bond lengths of homo-nuclear and heteronuclear diatomic species are given in Table 5.1.

5.2.4 Photoelectron Spectroscopy

In addition to data on bond distances and energies, specific information about the energies of electrons in orbitals can be determined from photoelectron spectroscopy,[10] one of the more direct methods for determining orbital energies. In this technique, ultraviolet (UV) light or X-rays dislodge electrons from molecules:

$$O_2 + h\nu \text{ (photons)} \longrightarrow O_2^+ + e^-$$

The kinetic energy of the expelled electrons can be measured; the difference between the energy of the incident photons and this kinetic energy equals the ionization energy (binding energy) of the electron:

$$\text{Ionization energy} = h\nu \text{ (energy of photons)} - \text{kinetic energy of the expelled electron}$$

[10] E. A. V. Ebsworth, D. W. H. Rankin, and S. Cradock, *Structural Methods in Inorganic Chemistry*, 2nd ed., CRC Press, Boca Raton, FL, 1991, pp. 255–279. Pages 274 and 275 discuss the spectra of N_2 and O_2.

TABLE 5.1 Bond Distances in Diatomic Species[a]

Formula	Valence Electrons	Internuclear Distance (pm)
H_2^+	1	105.2
H_2	2	74.1
B_2	6	159.0
C_2	8	124.2
C_2^{2-}	10	119.1[b]
N_2^+	9	111.6
N_2	10	109.8
O_2^+	11	111.6
O_2	12	120.8
O_2^-	13	135
F_2	14	141.2
CN	9	117.2
CN^-	10	115.9[c]
CO	10	112.8
NO^+	10	106.3
NO	11	115.1
NO^-	12	126.7

[a] Except as noted data are from K. P. Huber and G. Herzberg, *Molecular Spectra and Molecular Structure. IV. Constants of Diatomic Molecules,* Van Nostrand Reinhold Company, New York, 1979. Additional data on diatomic species can be found in R. Janoscheck, *Pure Appl. Chem.*, **2001**, *73*, 1521.

[b] Distance in CaC_2 in M. J. Atoji, *J. Chem. Phys.*, **1961**, *35*, 1950.

[c] Distance in low-temperature orthorhombic phase of NaCN in T. Schräder, A. Loidl, and T. Vogt, *Phys. Rev. B*, **1989**, *39*, 6186.

UV light removes outer electrons; X-rays are more energetic and remove inner electrons as well. Figure 5.10 and 5.11 show photoelectron spectra for N_2 and O_2 and the relative energies of the highest occupied orbitals of the ions. The lower energy peaks (at the top in the figure) are for the higher energy orbitals (less energy required to remove electrons). If the energy levels of the ionized molecule are assumed to be essentially the same as those of the uncharged molecule, the observed energies can be directly correlated with the molecular orbital energies. The levels in the N_2 spectrum are much closer together than those in the O_2 spectrum, and some theoretical calculations have disagreed about the order of the highest occupied orbitals. A paper by Stowasser and Hoffmann[11] has compared different calculation methods and showed that the different order of energy levels was simply a consequence of the method of calculation used; the methods favored by the authors agree with the experimental results, with σ_g above π_u.

[11] R. Stowasser and R. Hoffmann, *J. Am. Chem. Soc.*, **1999**, *121*, 3414.

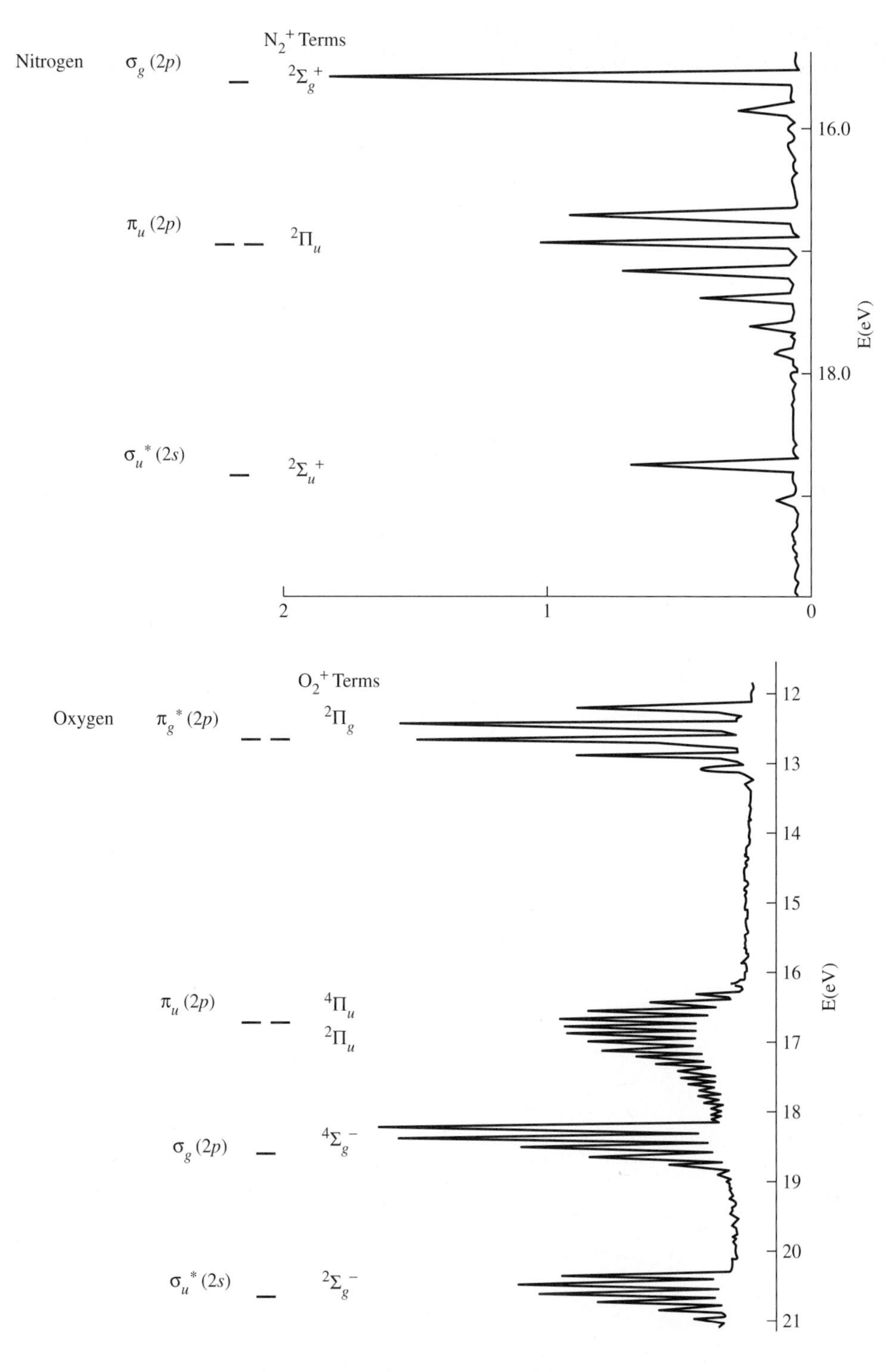

FIGURE 5.10 Photoelectron Spectrum and Molecular Orbital Energy Levels of N_2. (Photoelectron spectrum reproduced with permission from J. L. Gardner and J. A. R. Samson, *J. Chem. Phys.*, **1975**, *62*, 1447.)

FIGURE 5.11 Photoelectron Spectrum and Molecular Orbital Energy Levels of O_2. (Photoelectron spectrum reproduced with permission from J. H. D. Eland, *Photoelectron Spectroscopy*, Butterworths, London, 1974, p. 10.)

The photoelectron spectrum shows the π_u lower (Figure 5.10). In addition to the ionization energies of the orbitals, the spectrum shows the interaction of the electronic energy with the vibrational energy of the molecule. Because vibrational energy levels are much closer in energy than electronic levels, any collection of molecules has an energy distribution through many different vibrational levels. Because of this, transitions from electronic levels can originate from different vibrational levels, resulting in multiple peaks for a single electronic transition. Orbitals that are strongly involved in bonding have vibrational fine structure (multiple peaks); orbitals that are less involved in bonding have only a few individual peaks at each energy level.[12] The N_2 spectrum indicates that the π_u orbitals are more involved in the bonding than either of the σ orbitals. The CO photoelectron spectrum (Figure 5.14) has a similar pattern. The O_2 photoelectron spectrum (Figure 5.11) has much more vibrational fine structure for all the energy levels, with the π_u levels again more involved in bonding than the other orbitals. The photoelectron spectra of O_2 and of CO show the expected order of energy levels for these molecules, and the vibrational fine structure indicates that all the orbitals are important to bonding.

5.2.5 Correlation Diagrams

Mixing of orbitals of the same symmetry, as in the examples of Section 5.2.3, is seen in many other molecules. A **correlation diagram**[13] for this phenomenon is shown in Figure 5.12. This diagram shows the calculated effect of moving two atoms together, from a large interatomic distance on the right, with no interatomic interaction, to zero interatomic distance on the left, where the two nuclei become, in effect, a single

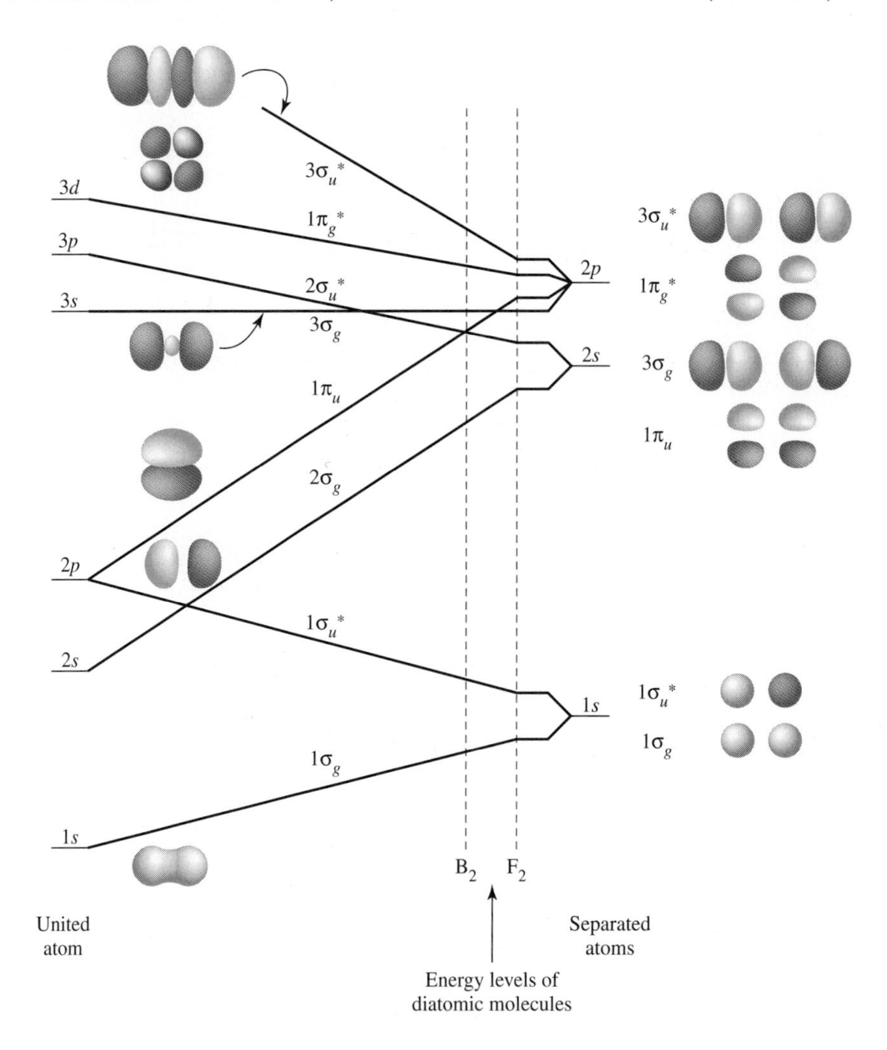

FIGURE 5.12 Correlation Diagram for Homonuclear Diatomic Molecular Orbitals.

[12] R. S. Drago, *Physical Methods in Chemistry*, 2nd ed., Saunders College Publishing, Philadelphia, 1992, pp. 671–677.

[13] R. McWeeny, *Coulson's Valence*, 3rd ed., Oxford University Press, Oxford, 1979, pp. 97–103.

nucleus. The simplest example has two hydrogen atoms on the right and a helium atom on the left. Naturally, such merging of two atoms into one never happens outside the realm of high-energy physics, but we consider the effects on orbitals as if it could. The diagram shows how the energies of the orbitals change with the internuclear distance and from the order of atomic orbitals on the left to the order of molecular orbitals of similar symmetry on the right. We will consider this correlation diagram beginning on the right side and moving toward the left.

On the right are the usual atomic orbitals—1*s*, 2*s*, and 2*p* for each of the two separated atoms. As the atoms approach each other, their atomic orbitals interact to form molecular orbitals.[14] The 1*s* orbitals form $1\sigma_g$ and $1\sigma_u^*$, 2*s* orbitals form $2\sigma_g$ and $2\sigma_u^*$, and 2*p* orbitals form $3\sigma_g$, $1\pi_u$, $1\pi_g^*$ and $3\sigma_u^*$. As the atoms move closer together (toward the left in the diagram), the bonding MOs decrease in energy, and the antibonding MOs increase in energy. At the far left, the MOs become the atomic orbitals of a united atom with twice the nuclear charge.

Symmetry is used to connect the molecular orbitals with the atomic orbitals of the united atom. Consider the $1\sigma_u^*$ orbital as an example. It is formed as the antibonding orbital from two 1*s* orbitals, as shown on the right side of the diagram. It has the same symmetry as a $2p_z$ atomic orbital (where *z* is the axis through both nuclei), which is the limit on the left side of the diagram. The degenerate $1\pi_u$ MOs are also connected to the 2*p* orbitals of the united atom, because they have the same symmetry as a $2p_x$ or $2p_y$ orbital (see Figure 5.2).

As another example, the degenerate pair of $1\pi_g^*$ MOs, formed by $2p_x$ or $2p_y$ orbitals of the separate atoms, is connected to the 3*d* orbitals on the left side, because the $1\pi_g^*$ orbitals have the same symmetry as the d_{xz} or d_{yz} orbitals (see Figures 5.2 and 5.3). The π orbitals formed from p_x and p_y orbitals are degenerate (have the same energy), as are the *p* orbitals of the merged atom; and the π^* orbitals from the same atomic orbitals are degenerate, as are the *d* orbitals of the merged atom.

Another consequence of this phenomenon is called the **noncrossing rule**, which states that orbitals of the same symmetry interact so that their energies never cross.[15] This rule helps in assigning correlations. If two sets of orbitals of the same symmetry seem to result in crossing in the correlation diagram, the matchups must be changed to prevent it.

The actual energies of molecular orbitals for diatomic molecules are intermediate between the extremes of this diagram, approximately in the region set off by the vertical lines. Toward the right within this region, closer to the separated atoms, the energy sequence is the "normal" one of O_2 and F_2; further to the left, the order of molecular orbitals is that of B_2, C_2, and N_2, with $\sigma_g(2p)$ above $\pi_u(2p)$.

5.3 HETERONUCLEAR DIATOMIC MOLECULES

5.3.1 Polar Bonds

Heteronuclear diatomic molecules follow the same general bonding pattern as homonuclear molecules, but a greater nuclear charge on one of the atoms lowers its atomic energy levels and shifts the resulting molecular orbital levels. In dealing with heteronuclear molecules, it is necessary to have a way to estimate the energies of the atomic orbitals that may interact. For this purpose, the orbital potential energies, given

---

[14] Molecular orbitals are labeled in many different ways. Most in this book are numbered within each set of the same symmetry ($1\sigma_g$, $2\sigma_g$ and $1\sigma_u^*$, $2\sigma_u^*$). In some figures, $1\sigma_g$ and $1\sigma_u^*$ MOs from 1*s* atomic orbitals are understood to be at lower energies than the MOs shown and are omitted.

[15] C. J. Ballhausen and H. B. Gray, *Molecular Orbital Theory*, W. A. Benjamin, New York, 1965, pp. 36–38.

in Table 5.2 and Figure 5.13, are useful.[16] These potential energies are negative, because they represent attraction between valence electrons and atomic nuclei. The values are the average energies for all electrons in the same level (for example, all $3p$ electrons), and they are weighted averages of all the energy states possible. These states are called *terms* and are explained in Chapter 11. For this reason, the values do not show the variations of the ionization energies seen in Figure 2.10 but steadily become more negative from left to right within a period, as the increasing nuclear charge attracts all the electrons more strongly.

TABLE 5.2 Orbital Potential Energies

		Orbital Potential Energy (eV)						
Atomic Number	**Element**	**1s**	**2s**	**2p**	**3s**	**3p**	**4s**	**4p**
1	H	−13.61						
2	He	−24.59						
3	Li		−5.39					
4	Be		−9.32					
5	B		−14.05	−8.30				
6	C		−19.43	−10.66				
7	N		−25.56	−13.18				
8	O		−32.38	−15.85				
9	F		−40.17	−18.65				
10	Ne		−48.47	−21.59				
11	Na				−5.14			
12	Mg				−7.65			
13	Al				−11.32	−5.98		
14	Si				−15.89	−7.78		
15	P				−18.84	−9.65		
16	S				−22.71	−11.62		
17	Cl				−25.23	−13.67		
18	Ar				−29.24	−15.82		
19	K						−4.34	
20	Ca						−6.11	
30	Zn						−9.39	
31	Ga						−12.61	−5.93
32	Ge						−16.05	−7.54
33	As						−18.94	−9.17
34	Se						−21.37	−10.82
35	Br						−24.37	−12.49
36	Kr						−27.51	−14.22

Source: J. B. Mann, T. L. Meek, and L. C. Allen, *J. Am. Chem. Soc.*, **2000**, *122*, 2780.

NOTE: All energies are negative, representing average attractive potentials between the electrons and the nucleus for all terms of the specified orbitals.

[16] A more complete listing of orbital potential energies is in Appendix B-9.

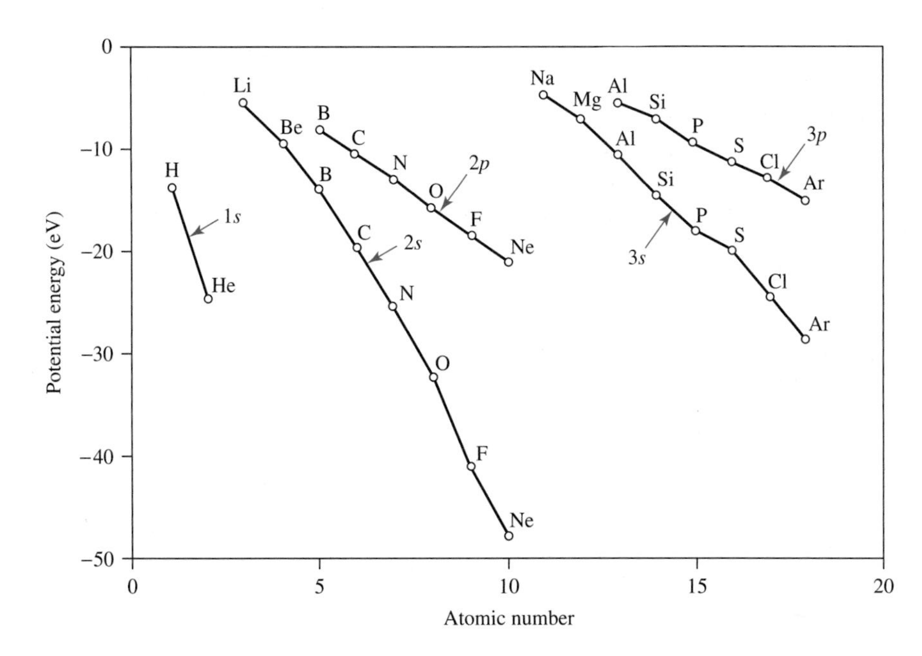

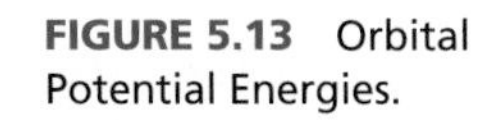
FIGURE 5.13 Orbital Potential Energies.

The atomic orbitals of homonuclear diatomic molecules have identical energies, and both atoms contribute equally to a given MO. Therefore, in the equations for the molecular orbitals, the coefficients for the two atomic orbitals are identical. In heteronuclear diatomic molecules, such as CO and HF, the atomic orbitals have different energies, and a given MO receives unequal contributions from the atomic orbitals; the equation for that MO has a different coefficient for each of the atomic orbitals that compose it. As the energies of the atomic orbitals get farther apart, the magnitude of the interaction decreases. The atomic orbital closer in energy to an MO contributes more to the MO, and its coefficient is larger in the wave equation.

CARBON MONOXIDE The simplest approach to bonding in heteronuclear diatomic molecules follows the pattern for homonuclear diatomics with one exception: the more electronegative element has atomic orbitals at lower potential energies than the less electronegative element. Carbon monoxide, shown in Figure 5.14, shows this effect, with oxygen having lower energies for its 2*s* and 2*p* orbitals than the matching orbitals of carbon. The result is that the orbital interaction diagram for CO looks like that for a homonuclear diatomic (Figure 5.5), with the right (more electronegative) side pulled down in comparison with the left. In CO, as in N_2, the $\pi_u(2p)$ orbitals are lower in energy than the σ_g (2*p*). This can be viewed as the consequence of significant interactions between the $2p_z$ orbital of oxygen and both the 2*s* and $2p_z$ orbitals of carbon. Oxygen's $2p_z$ orbital (−15.85 eV) is intermediate in energy between carbon's 2*s* (−19.43 eV) and $2p_z$ (−10.66 eV), so the energy match for both interactions is favorable.

The bonding orbital 2σ has more contribution from (and is closer in energy to) the lower energy oxygen 2*s* atomic orbital; the antibonding $2\sigma^*$ orbital has more contribution from (and is closer in energy to) the higher energy carbon 2*s* atomic orbital. In the simplest case, the bonding orbital is similar in energy and shape to the lower energy atomic orbital, and the antibonding orbital is similar in energy and shape to the higher energy atomic orbital. In more complicated cases, such as the $2\sigma^*$ orbital of CO, other orbitals (the oxygen $2p_z$ orbital) contribute, and the orbital shapes and energies are not as easily predicted. As a practical matter, atomic orbitals with energy differences greater than about 10 eV to 14 eV usually do not interact significantly.

Mixing of the two σ levels and the two σ^* levels, like that seen in the homonuclear σ_g and σ_u orbitals, causes a larger split in energy between them, and the 3σ is higher

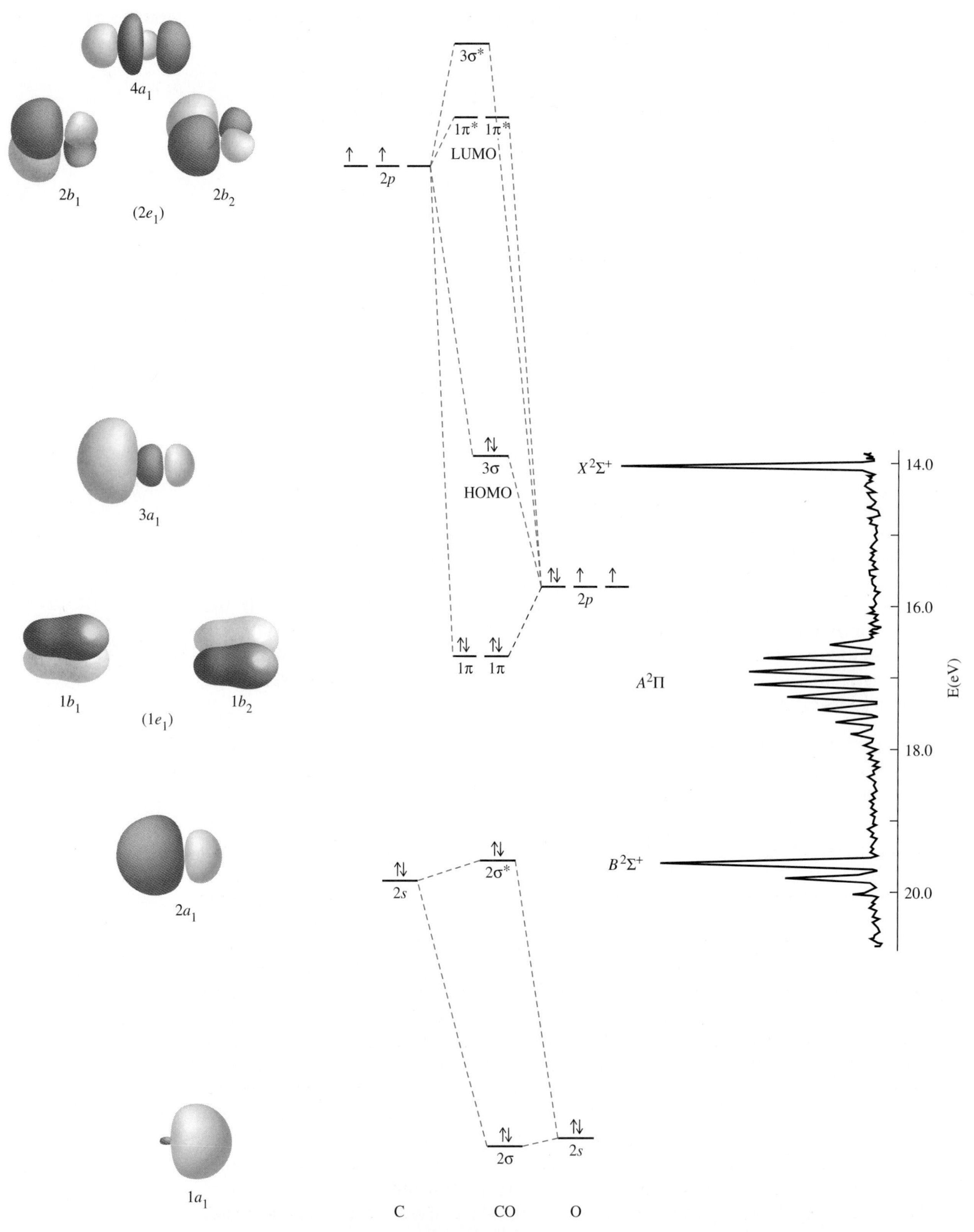

FIGURE 5.14 Molecular Orbitals and Photoelectron Spectrum of CO. Molecular orbitals 1σ and 1σ* are from the 1*s* orbitals and are not shown.
(Photoelectron spectrum reproduced with permission from J. L. Gardner and J. A. R. Samson, *J. Chem. Phys.*, **1975**, *62*, 1447.)

than the π levels. The p_x and p_y orbitals also form four molecular π orbitals, two bonding and two antibonding. When the electrons are filled in, as in Figure 5.14, the valence orbitals form four bonding pairs and one antibonding pair for a net bond order of 3.

EXAMPLE

Molecular orbitals for HF can be found by using the approach used for CO. The 2*s* orbital of the fluorine atom has an energy more than 26 eV lower than that of the hydrogen 1*s*, so there is very little interaction between them. The fluorine $2p_z$ orbital (–18.65 eV) and the hydrogen 1*s* (–13.61 eV), on the other hand, have similar energies, allowing them to combine into bonding σ and antibonding σ^* orbitals. The flourine $2p_x$ and $2p_y$ orbitals remain nonbonding, each with a pair of electrons. Overall, there is one bonding pair of electrons and three lone pairs.

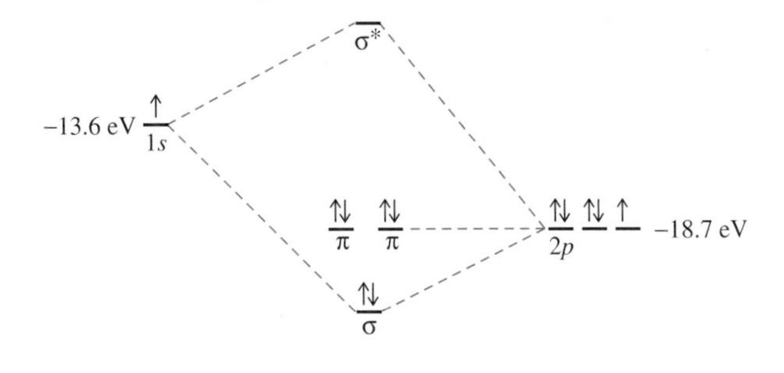

▶ **Exercise 5.3** Use a similar approach to the discussion of HF to explain the bonding in the hydroxide ion OH^-.

The molecular orbitals that are typically of greatest interest for reactions between molecules are the **highest occupied molecular orbital (HOMO)** and the **lowest unoccupied molecular orbital (LUMO)**, collectively known as **frontier orbitals** because they lie at the occupied–unoccupied frontier. The MO diagram of CO helps explain its reaction chemistry with transition metals, which is not that predicted by simple electronegativity considerations that place more electron density on the oxygen. If this were true, compounds in which CO is bonded to metals, called *carbonyl complexes*, should bond as M—O—C with the negative oxygen attached to the positive metal. The actual structure of metal carbonyl complexes, such as $Ni(CO)_4$ and $Mo(CO)_6$, has atoms in the order M—C—O. The HOMO of CO is 3σ, with a higher electron density and a larger lobe on the carbon. The lone pair in this orbital forms a bond with a vacant orbital on the metal. The interaction between CO and metal orbitals is enormously important in the field of organometallic chemistry and will be discussed in detail in Chapter 13.

In simple cases, bonding MOs have a greater contribution from the lower energy atomic orbital, and their electron density is concentrated on the atom with the lower energy levels or higher electronegativity (see Figure 5.14). If this is so, why does the HOMO of CO, a bonding MO, have greater electron density on carbon, which has the higher energy levels? The answer lies in the way the atomic orbital contributions are divided. The p_z of oxygen has an energy that enables it to contribute to the $2\sigma^*$, the 3σ (the HOMO), and the $3\sigma^*$ MOs. The higher energy carbon p_z, however, only contributes significantly to the latter two. Because the p_z of the oxygen atom is divided among three MOs, it has a relatively weaker contribution to each one, and the p_z of the carbon atom has a relatively stronger contribution to each of the two orbitals to which it contributes.

The LUMOs are the $1\pi^*$ orbitals and are concentrated on carbon, as expected. The frontier orbitals can contribute electrons (HOMO) or accept electrons (LUMO) in reactions. These effects will be discussed in Chapters 10 and 13.

5.3.2 Ionic Compounds and Molecular Orbitals

Ionic compounds can be considered the limiting form of polarity in heteronuclear diatomic molecules. As the atoms differ more in electronegativity, the difference in energy of the orbitals also increases, and the concentration of electrons shifts toward the more electronegative atom. At the limit, the electron is transferred completely to the more electronegative atom to form a negative ion, leaving a positive ion with a high-energy vacant orbital. When two elements with a large difference in their electronegativities (such as Li and F) combine, the result is often considered an ionic compound. However, in molecular orbital terms, we can also consider an ion pair as if it were a covalent compound. In Figure 5.15, the atomic orbitals and an approximate indication of molecular orbitals for such a diatomic molecule, LiF, are given. On formation of the compound LiF, the electron from the Li 2*s* orbital is transferred to the bonding orbital formed from interaction between the Li 2*s* orbital and the F $2p_z$ orbital. Both electrons, the one originating from Li and the one originating from the F $2p_z$ orbital, are stabilized.

In a more accurate picture of ionic crystals, the ions are held together in a three-dimensional lattice by a combination of electrostatic attraction and covalent bonding. Although there is a small amount of covalent character in even the most ionic compounds, there are no directional bonds, and each Li^+ ion is surrounded by six F^- ions, each of which in turn is surrounded by six Li^+ ions. The molecular orbitals in the crystal form energy bands, described in Chapter 7.

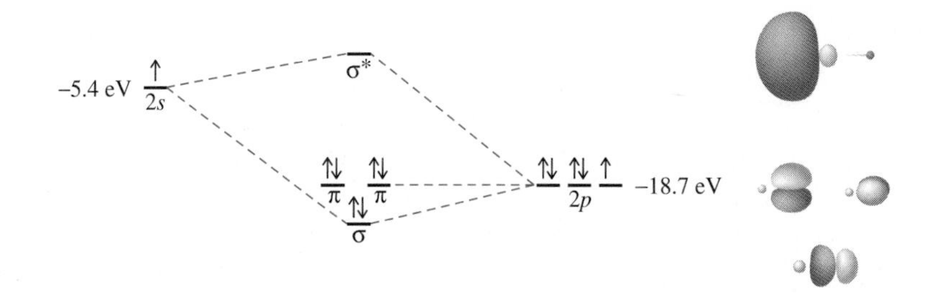

FIGURE 5.15 Approximate LiF Molecular Orbitals.

Formation of the ions can be described as a sequence of elementary steps, beginning with solid Li and gaseous F_2:

$Li(s) \longrightarrow Li(g)$	161 kJ/mol	(sublimation)
$Li(g) \longrightarrow Li^+(g) + e^-$	531 kJ/mol	(ionization, IE)
$\frac{1}{2} F_2(g) \longrightarrow F(g)$	79 kJ/mol	(dissociation)
$F(g) + e^- \longrightarrow F^-(g)$	−328 kJ/mol	(ionization, −EA)
$Li(s) + \frac{1}{2}F_2(g) \longrightarrow Li^+(g) + F^-(g)$	443 kJ/mol	

For a reaction to proceed spontaneously, the free energy change ($\Delta G = \Delta H - T\Delta S$) must be negative. Although the entropy change for the overall reaction above is positive, the very large positive ΔH results in a positive ΔG. If this were the final result, Li^+ and F^- would not form. However, the large attraction between the ions results in the release of 709 kJ/mol on formation of a single Li^+F^- ion pair and 1239 kJ/mol on formation of a crystal:

$Li^+(g) + F^-(g) \longrightarrow LiF(g)$	−709 kJ/mol (ion pairs)
$Li^+(g) + F^-(g) \longrightarrow LiF(s)$	−1239 kJ/mol (lattice enthalpy)

The **lattice enthalpy** for crystal formation is large enough to overcome all the endothermic processes and the negative entropy change, and to make formation of LiF from the elements a highly favorable reaction.

5.4 MOLECULAR ORBITALS FOR LARGER MOLECULES

The methods described for bonding in diatomic molecules can be extended to molecules consisting of three or more atoms, but this approach becomes more difficult as molecules become more complex. We will first consider several examples of linear molecules to illustrate the concept of group orbitals then proceed to molecules that can benefit from the use of more formal methods of group theory.

5.4.1 FHF^-

The linear ion FHF^-, an example of very strong hydrogen bonding,[17] provides a convenient introduction to the concept of **group orbitals**, collections of matching orbitals on outer atoms. To generate a set of group orbitals, we will use the valence orbitals of the fluorine atoms, as shown in Figure 5.16. We will then examine which central-atom orbitals have the proper symmetry to interact with the group orbitals.

The lowest energy group orbitals are those from the 2*s* orbitals of the fluorine atoms. These orbitals may either have matching signs of their wave functions (group orbital 1) or opposite signs (group orbital 2). These group orbitals should be viewed as collections of orbitals that potentially could interact with central atom orbitals. Group orbitals are the same combinations that formed bonding and antibonding orbitals in diatomic molecules (e.g., $p_{xa} + p_{xb}, p_{xa} - p_{xb}$), but they are now separated by the central hydrogen atom. Group orbitals 3 and 4 are derived from the $2p_z$ orbitals of the fluorines, again in one case having lobes with matching signs pointing toward the center

[17] J. H. Clark, J. Emsley, D. J. Jones, and R. E. Overill, *J. Chem. Soc.*, **1981**, 1219; J. Emsley, N. M. Reza, H. M. Dawes, and M. B. Hursthouse, *J. Chem. Soc. Dalton Trans.*, **1986**, 313; N. Elghobashi and L. González, *J. Chem. Phys.*, **2006**, *124*, 174308, and references cited therein.

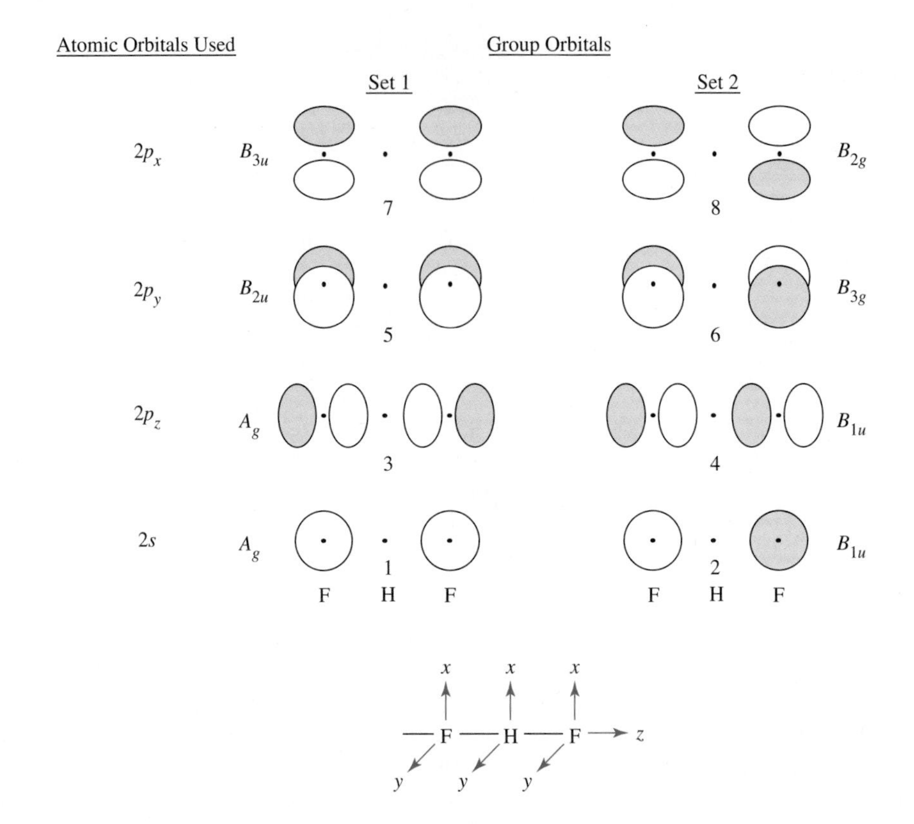

FIGURE 5.16 Group Orbitals.

(orbital 3), in the other case having opposite signs pointing toward the center (orbital 4). Group orbitals 5 through 8 are derived from the $2p_x$ and $2p_y$ orbitals of the fluorines, which are parallel to each other and can be paired according to matching (orbitals 5 and 7) or opposite (orbitals 6 and 8) signs of their wave functions.

The central hydrogen atom in FHF^-, with only its 1*s* orbital available, can in principle only interact with group orbitals 1 and 3; the 1*s* orbital is nonbonding with respect to the other group orbitals. The bonding and antibonding combinations are shown in Figure 5.17.

Which interaction is likely to be stronger? The potential energy of the 1*s* orbital of hydrogen (–13.61 eV) is a much better match for the fluorines' $2p_z$ orbitals (–18.65 eV) than their 2*s* orbitals (–40.17 eV). Consequently, we expect the interaction with the $2p_z$ orbitals (group orbital 3) to be much stronger than with the 2*s* (group orbital 1). The 1*s* orbital of hydrogen cannot interact with group orbitals 5 through 8, so these orbitals are nonbonding.

The overall set of molecular orbitals for FHF^- is shown in Figure 5.18. In sketching molecular orbital energy diagrams of polyatomic species, we will show the orbitals of the central atom on the far left, the group orbitals of the surrounding atoms on the far right, and the resulting molecular orbitals in the middle.

Five of the six group orbitals derived from the 2*p* orbitals of the fluorines do not interact with the central atom; these orbitals remain essentially nonbonding and

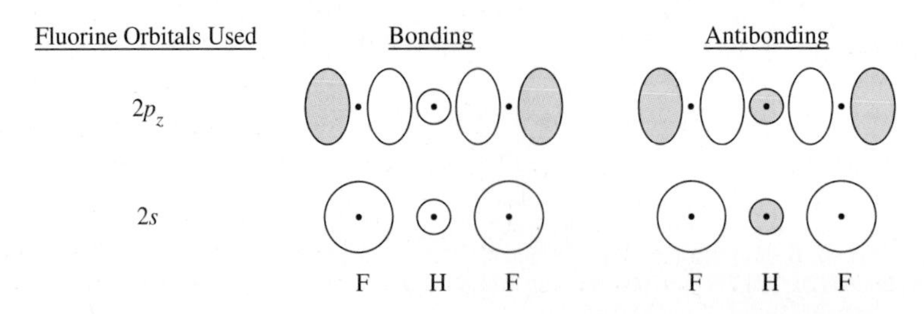

FIGURE 5.17 Interaction of Fluorine Group Orbitals with the Hydrogen 1*s* Orbital.

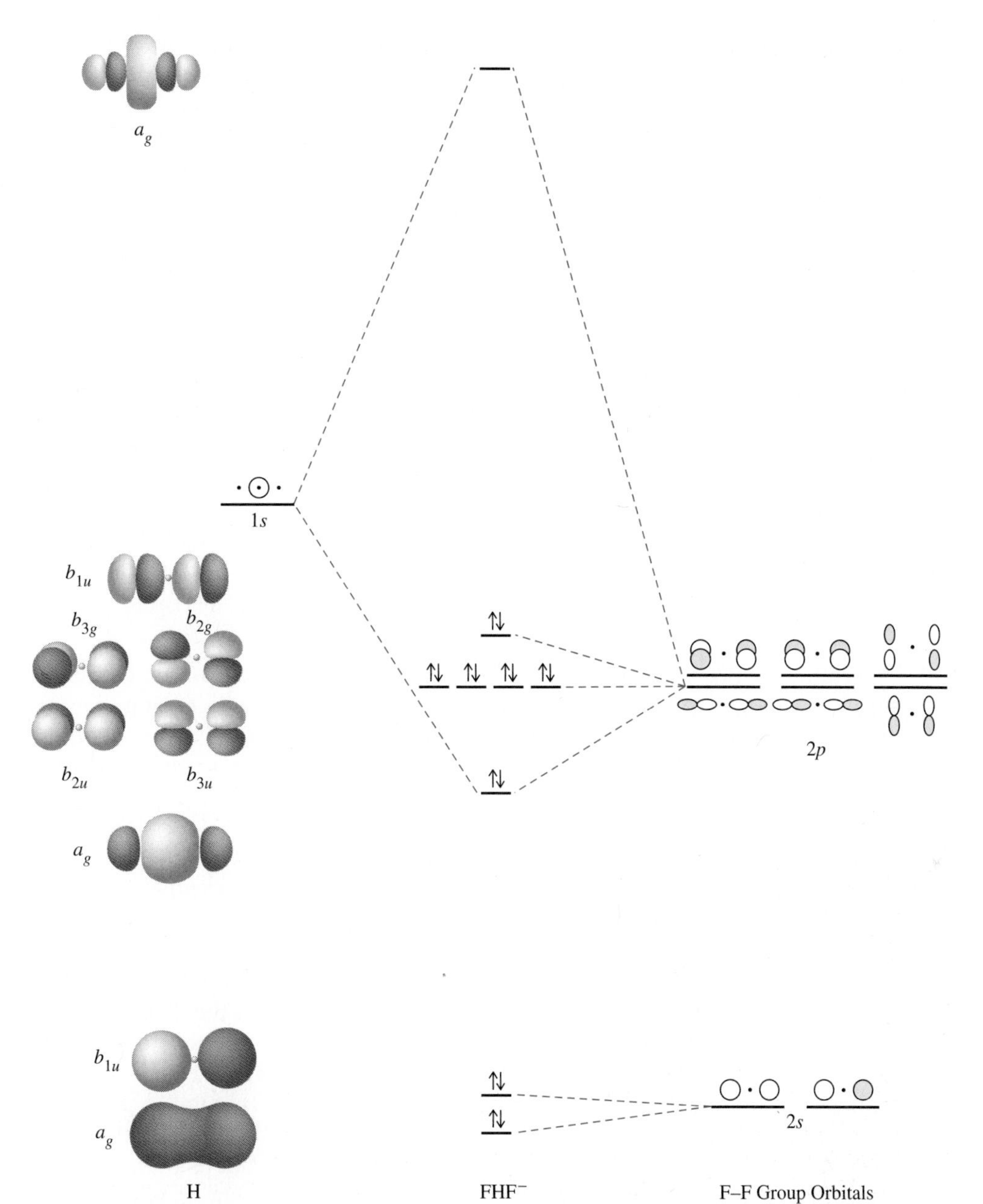

FIGURE 5.18 Molecular Orbital Diagram of FHF^-.

contain lone pairs of electrons. There is a slight interaction between orbitals on the non-neighboring fluorine atoms, but not enough to change their energies significantly. As already described, the sixth $2p$ group orbital, the remaining $2p_z$ (number 3), interacts with the $1s$ orbital of hydrogen to give two molecular orbitals, one bonding and one antibonding. An electron pair occupies the bonding orbital. The group orbitals from the $2s$ orbitals of the fluorine atoms are much lower in energy than the $1s$ orbital of the hydrogen atom and are essentially nonbonding.

The Lewis approach to bonding requires two electrons to represent a single bond between two atoms and would result in four electrons around the hydrogen atom of FHF^-. The molecular orbital picture, on the other hand, has a 2-electron bond delocalized over *three* atoms (a 3-center, 2-electron bond). The bonding MO in Figures 5.17 and 5.18 shows how the molecular orbital approach represents such a bond: two electrons occupy a low-energy orbital formed by the interaction of all three atoms (a central atom and a two-atom group orbital). The remaining electrons are in the group orbitals derived from the p_x and p_y orbitals of the fluorine at essentially the same energy as that of the atomic orbitals.

In general, bonding molecular orbitals derived from three or more atoms, like the one in Figure 5.18, usually have lower energies than those that include orbitals from only two atoms, but the total energy of a molecule is the sum of the energies of all of the electrons in all the orbitals. FHF^- has a bond energy of 212 kJ/mol and F—H distances of 114.5 pm. HF has a bond energy of 574 kJ/mol and an F—H bond distance of 91.7 pm.[18]

▶ **Exercise 5.4** Sketch the energy levels and the molecular orbitals for the H_3^+ ion using linear geometry.

5.4.2 CO_2

The approach used so far can also be used for other linear species—such as CO_2, N_3^-, and BeH_2—to consider how molecular orbitals can be constructed on the basis of interactions of group orbitals with central atom orbitals. However, we also need a way to deal with more complex molecules. We will first illustrate this approach using carbon dioxide, another linear molecule, that has a more complicated molecular orbital description than FHF^-, then move on to more complex, nonlinear examples.

We will use the following steps:

1. Determine the point group of the molecule. If it is a linear molecule, substituting a simpler point group that retains the symmetry of the orbitals (ignoring the signs) makes the process easier. It is useful to substitute D_{2h} for $D_{\infty h}$ and C_{2v} for $C_{\infty v}$. This substitution retains the symmetry of the orbitals without the need to use infinite-fold rotation axes.
2. Assign x, y, and z coordinates to the atoms, chosen for convenience. Experience is the best guide here. The general rule in all the examples in this book is that *the highest order rotation axis of the molecule is chosen as the z axis of the central atom.* In nonlinear molecules, the y axes of the outer atoms are chosen to point toward the central atom.
3. Construct a (reducible) representation for the combination of the valence s orbitals on the outer atoms. If the outer atom is not hydrogen, repeat the process, finding the representations for each of the other sets of outer atom orbitals (for example, p_x, p_y, and p_z). As in the case of the vectors described in Chapter 4, any orbital that changes position during a symmetry operation contributes zero to the character of the resulting representation; any orbital that remains in its original position—such as a p orbital that maintains its position and direction (signs of its orbital lobes)—contributes 1; and any orbital that remains in the original position, with the signs of its lobes reversed, contributes -1.
4. Reduce each representation from Step 3 to its irreducible representations. This is equivalent to finding the symmetry of the **group orbitals** or the **symmetry-adapted linear combinations (SALCs)** of the orbitals. The group orbitals are then the combinations of atomic orbitals that match the symmetry of the irreducible representations.
5. Identify the atomic orbitals of the central atom with the same symmetries (irreducible representations) as those found in Step 4.
6. Combine the atomic orbitals of the central atom and those of the group orbitals with matching symmetry and similar energy to form molecular orbitals. The total number of molecular orbitals formed must equal the number of atomic orbitals used from all the atoms.[19]

---

[18] M. Mautner, *J. Am. Chem. Soc.*, **1984**, *106*, 1257.

[19] We use lowercase labels on the molecular orbitals, with uppercase for the atomic orbitals and for representations in general. This practice is common but not universal.

In summary, the process used in creating molecular orbitals is to match the symmetries of the group orbitals, using their irreducible representations, with the symmetries of the central atom orbitals. If the symmetries match and the energies are similar, there is an interaction—both bonding and antibonding—if not, there is no interaction.

In CO_2 the group orbitals for the oxygen atoms are identical to the group orbitals for the fluorine atoms in FHF^- (Figure 5.16), but the central carbon atom in CO_2 has both *s* and *p* orbitals capable of interacting with the 2*p* group orbitals on the oxygen atoms. As in the discussion of FHF^-, the group orbital-central orbital interactions of CO_2 will be the focus.

1. **Point Group:** CO_2 has $D_{\infty h}$ symmetry. To simplify the analysis, the D_{2h} point group can be used.
2. **Coordinate System:** The *z* axis is chosen as the C_∞ axis, and the *y* and *z* coordinates are chosen similarly to the FHF^- example (Figure 5.19).
3. **Reducible Representations for Outer Atom Orbitals:** In CO_2 these are the 2*s* and 2*p* orbitals of the oxygens. These can be grouped into four sets according to their orientation. For example, the pair of 2*s* orbitals on the oxygen atoms has the following representation:

D_{2h}	E	$C_2(z)$	$C_2(y)$	$C_2(x)$	i	$\sigma(xy)$	$\sigma(xz)$	$\sigma(yz)$
Γ (2*s*)	2	2	0	0	0	0	2	2

The other outer atom orbitals have the following representations:

D_{2h}	E	$C_2(z)$	$C_2(y)$	$C_2(x)$	i	$\sigma(xy)$	$\sigma(xz)$	$\sigma(yz)$
Γ ($2p_z$)	2	2	0	0	0	0	2	2
Γ ($2p_x$)	2	−2	0	0	0	0	2	−2
Γ ($2p_y$)	2	−2	0	0	0	0	−2	2

4. **Group Orbitals from Reducible Representations:** Each of the representations from Step 3 can be reduced by the procedure described in Section 4.4.2. For example, the representation Γ (2*s*) reduces to $A_g + B_{1u}$:

D_{2h}	E	$C_2(z)$	$C_2(y)$	$C_2(x)$	i	$\sigma(xy)$	$\sigma(xz)$	$\sigma(yz)$
A_g	1	1	1	1	1	1	1	1
B_{1u}	1	1	−1	−1	−1	−1	1	1

When this procedure is conducted for each of the representations, the group orbitals of the oxygen atoms are of the same type as those for the fluorine atoms in FHF^-; these are shown for CO_2, with the appropriate labels for D_{2h}, in Figure 5.19.

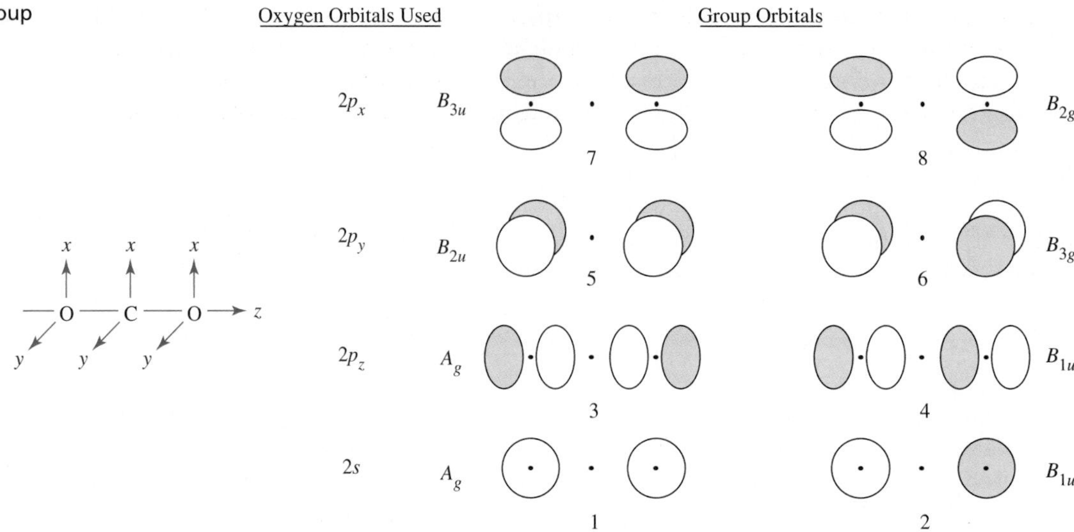

FIGURE 5.19 Group Orbital Symmetry in CO_2.

▶ **Exercise 5.5** Using the D_{2h} character table shown, verify that the group orbitals shown in Figure 5.19 match the labels of their irreducible representations.

D_{2h}	E	$C_2(z)$	$C_2(y)$	$C_2(x)$	i	$\sigma(xy)$	$\sigma(xz)$	$\sigma(yz)$		
A_g	1	1	1	1	1	1	1	1		x^2, y^2, z^2
B_{1g}	1	1	−1	−1	1	1	−1	−1	R_z	xy
B_{2g}	1	−1	1	−1	1	−1	1	−1	R_y	xz
B_{3g}	1	−1	−1	1	1	−1	−1	1	R_x	yz
A_u	1	1	1	1	−1	−1	−1	−1		
B_{1u}	1	1	−1	−1	−1	−1	1	1	z	
B_{2u}	1	−1	1	−1	−1	1	−1	1	y	
B_{3u}	1	−1	−1	1	−1	1	1	−1	x	

5. **Matching Orbitals on the Central Atom:** To determine which atomic orbitals of carbon are of correct symmetry to interact with the group orbitals, we will consider each of the group orbitals in turn. The carbon atomic orbitals are shown in Figure 5.20 with their symmetry labels for the D_{2h} point group.

   The character table of this group shows the symmetry of these orbitals. For example, B_{1u} has the symmetry of the z axis and of the p_z orbitals on the fluorines; they are unchanged by the E, $C_2(z)$, $\sigma(xz)$, and $\sigma(yz)$ operations, and the $C_2(y)$, $C_2(x)$, i, and $\sigma(xy)$ operations reverse their signs.

A_g B_{1u} B_{3u} B_{2u}
$2s$ $2p_z$ $2p_x$ $2p_y$

FIGURE 5.20 Symmetry of the Carbon Atomic Orbitals in the D_{2h} Point Group.

6. Formation of Molecular Orbitals: Group orbitals 1 and 2 in Figure 5.21, formed by adding and subtracting the oxygen 2*s* orbitals, have A_g and B_{1u} symmetry, respectively. Group orbital 1 is of appropriate symmetry to interact with the 2*s* orbital of carbon (both have A_g symmetry), and group orbital 2 is of appropriate symmetry to interact with the $2p_z$ orbital of carbon (both have B_{1u} symmetry).

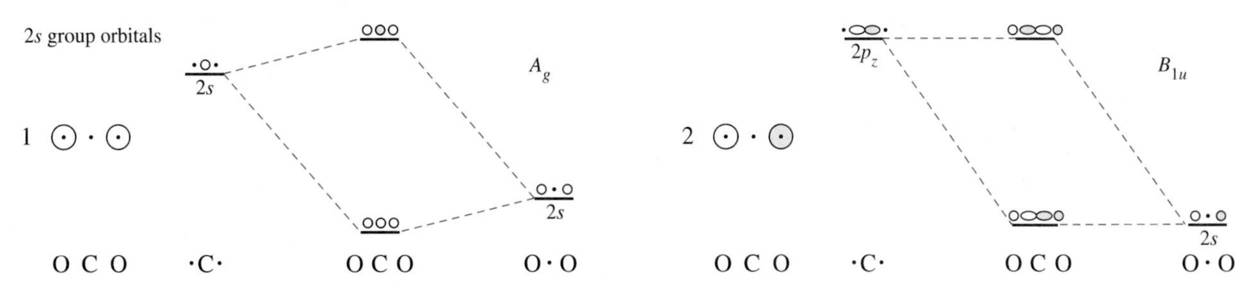

FIGURE 5.21 Group Orbitals 1 and 2 for CO_2.

Group orbitals 3 and 4 in Figure 5.22, formed by adding and subtracting the oxygen $2p_z$ orbitals, have the same A_g and B_{1u} symmetries. As in the first two, group orbital 3 can interact with the 2*s* of carbon, and group orbital 4 can interact with the carbon $2p_z$.

The 2*s* and $2p_z$ orbitals of carbon, therefore, have two possible sets of group orbitals with which they may interact. In other words, all four interactions in Figures 5.21 and 5.22 are symmetry allowed. It is then necessary to estimate which interactions can be expected to be the strongest from the potential energies of the 2*s* and 2*p* orbitals of carbon and oxygen given in Figure 5.23.

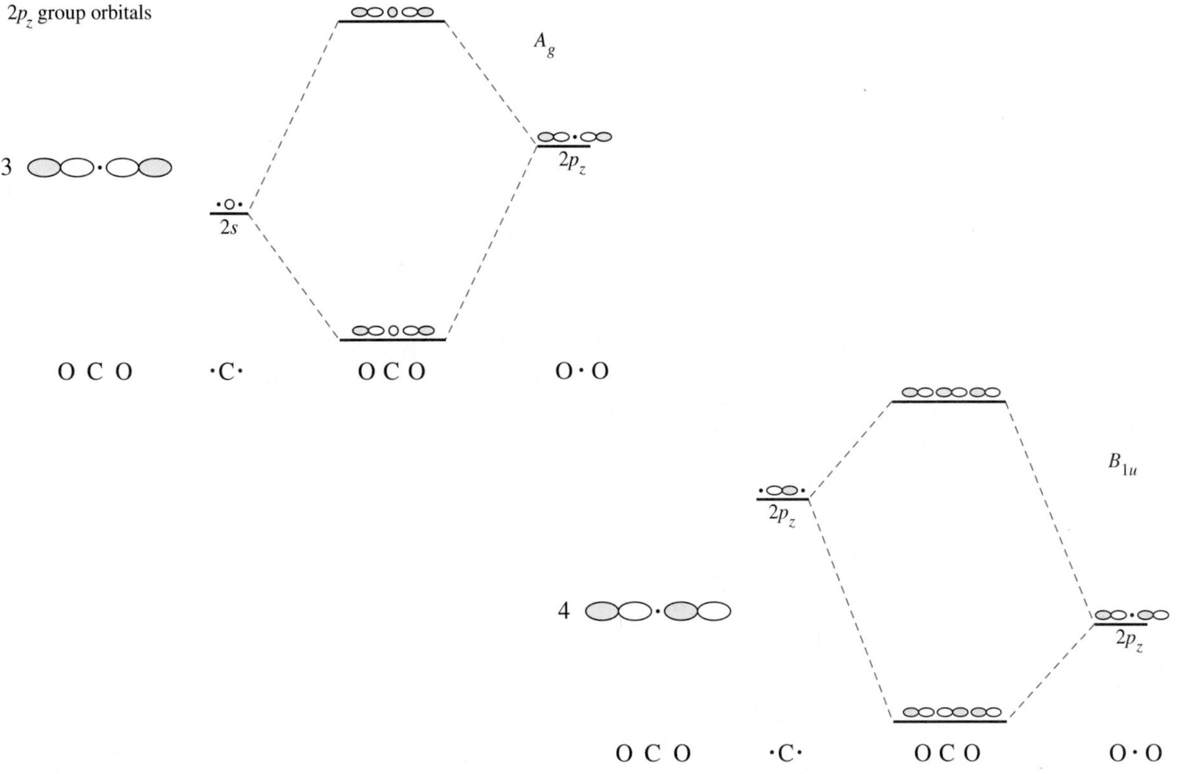

FIGURE 5.22 Group Orbitals 3 and 4 for CO_2.

FIGURE 5.23 Orbital Potential Energies of Carbon and Oxygen.

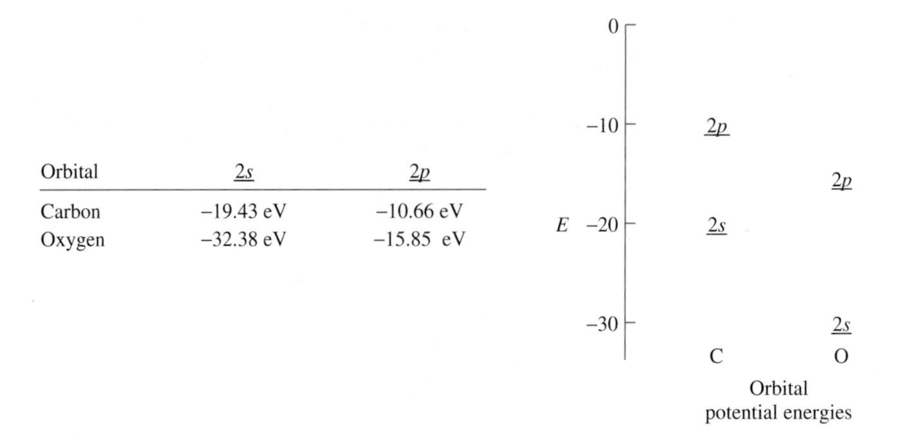

Orbital	$2s$	$2p$
Carbon	−19.43 eV	−10.66 eV
Oxygen	−32.38 eV	−15.85 eV

Interactions are strongest for orbitals having similar energies. Both group orbital 1, from the 2*s* orbitals of the oxygen, and group orbital 3, from the $2p_z$ orbitals, have the proper symmetry to interact with the 2*s* orbital of carbon. However, the energy match between group orbital 3 and the 2*s* orbital of carbon is much better (a difference of 3.58 eV) than the energy match between group orbital 1 and the 2*s* of carbon (a difference of 12.95 eV); therefore, the primary interaction is between the $2p_z$ orbitals of oxygen and the 2*s* orbital of carbon. Group orbital 2 also has energy too low for strong interaction with the carbon p_z (a difference of 21.72 eV), so the final molecular orbital diagram in Figure 5.26 shows no interaction with carbon orbitals for group orbitals 1 and 2.

▶ **Exercise 5.6** Using orbital potential energies, show that group orbital 4 is more likely than group orbital 2 to interact strongly with the $2p_z$ orbital of carbon.

The $2p_y$ orbital of carbon has B_{2u} symmetry and interacts with group orbital 5 (Figure 5.24). The result is the formation of two π molecular orbitals, one bonding and one antibonding. However, there is no orbital on carbon with B_{3g} symmetry to interact with group orbital 6, formed by combining $2p_y$ orbitals of oxygen. Therefore, group orbital 6 is nonbonding.

Interactions of the $2p_x$ orbitals are similar to those of the $2p_y$ orbitals. Group orbital 7, with B_{2u} symmetry, interacts with the $2p_x$ orbital of carbon to form π bonding and antibonding orbitals, whereas group orbital 8 is nonbonding (Figure 5.25).

The overall molecular orbital diagram of CO_2 is shown in Figure 5.26. The 16 valence electrons occupy, from the bottom, two essentially nonbonding σ orbitals, two bonding σ orbitals, two bonding π orbitals, and two nonbonding π orbitals. In other words, two of the bonding electron pairs are in σ orbitals, two are in π orbitals, and

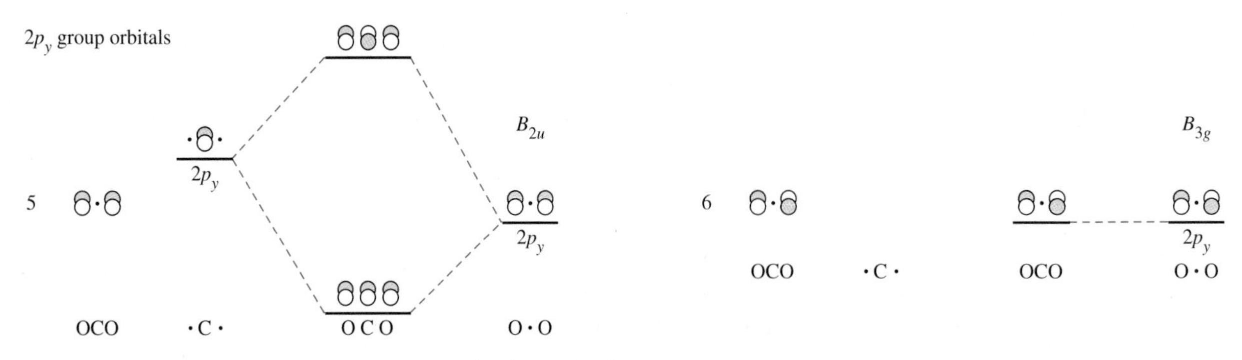

FIGURE 5.24 Group Orbitals 5 and 6 for CO_2.

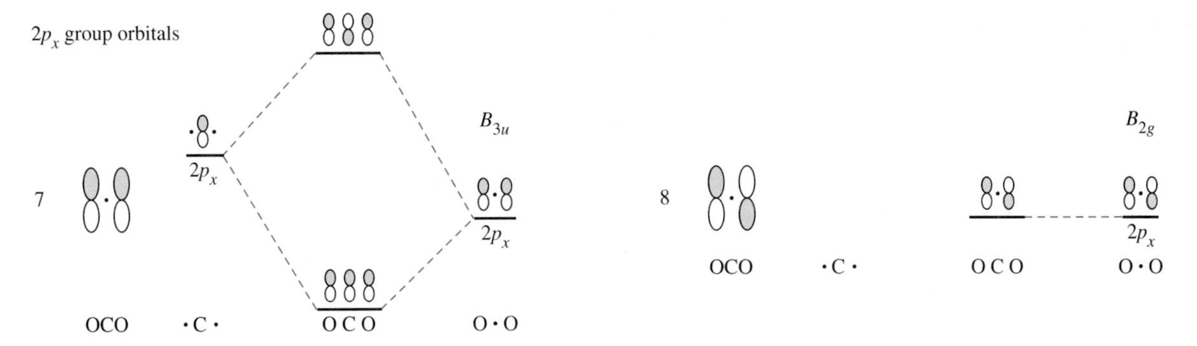

FIGURE 5.25 Group Orbitals 7 and 8 for CO_2.

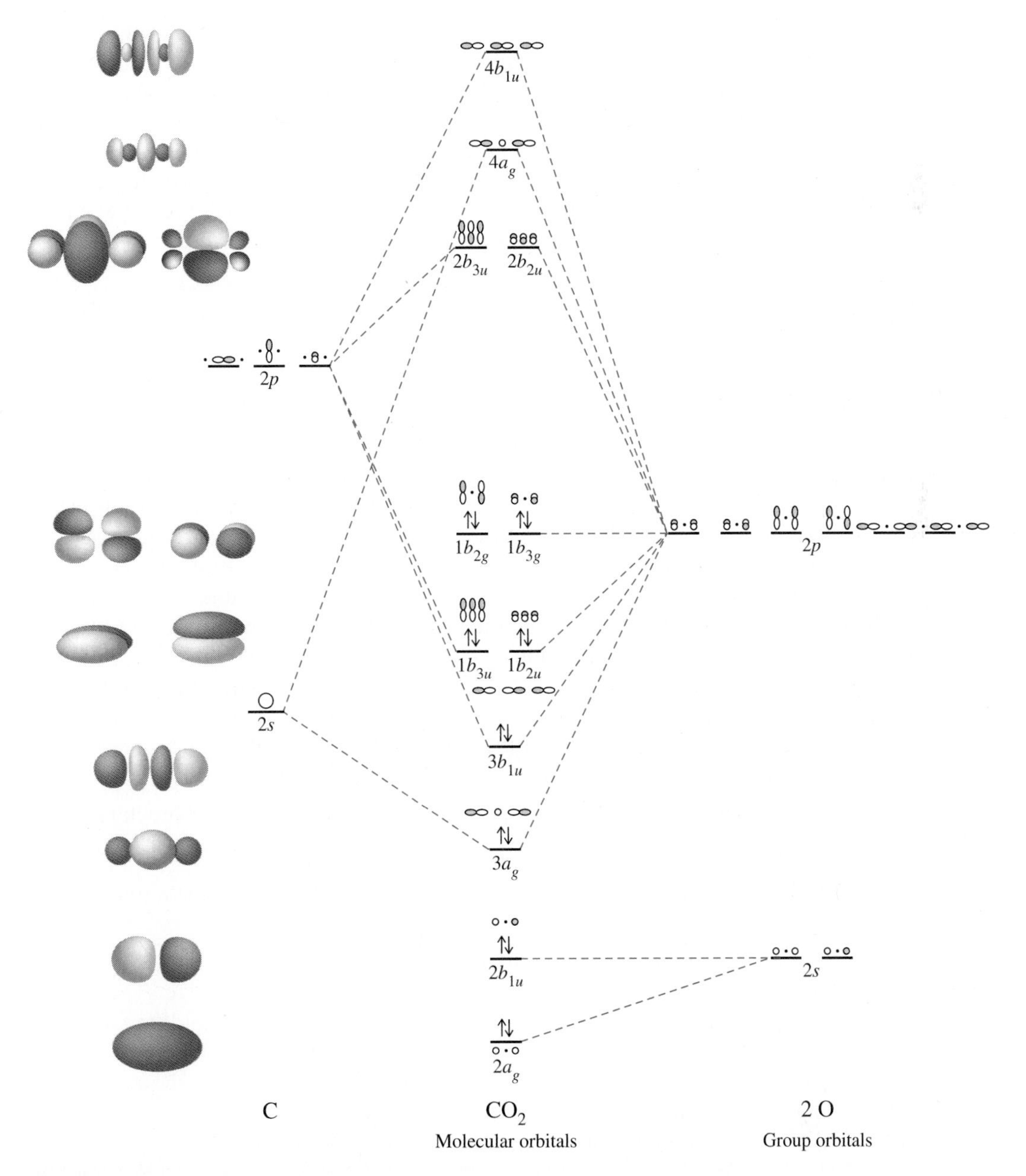

FIGURE 5.26 Molecular Orbitals of CO_2.

there are four bonds in the molecule, as expected. As in the FHF^- case, all the occupied molecular orbitals are 3-center, 2-electron orbitals and all are more stable (have lower energy) than 2-center orbitals.

The molecular orbital picture of other linear triatomic species—such as N_3^-, CS_2, and OCN^-—can be determined similarly. Likewise, the molecular orbitals of longer polyatomic species can be described by a similar method. Examples of bonding in linear π systems will be considered in Chapter 13.

▶ **Exercise 5.7** Prepare a molecular orbital diagram for the azide ion, N_3^-.

▶ **Exercise 5.8** Prepare a molecular orbital diagram for the BeH_2 molecule. (Assume an orbital potential energy of –6.0 eV for $2p$ orbitals of Be. This orbital set should be taken into account, even though it is unoccupied in a free Be atom.)

The process can be carried further to obtain numerical values of the coefficients of the atomic orbitals used in the molecular orbitals.[20] The coefficients may be small or large, positive or negative, similar or quite different, depending on the characteristics of the orbital under consideration. Computer software packages are available that will calculate these coefficients and generate the pictorial diagrams that describe the molecular orbitals. Examples of problems that use molecular modeling software to generate molecular orbitals of a variety of molecules are included in the problems at the end of this chapter and in later chapters. Discussion of the details of the calculations used by such software is beyond the scope of this text.

5.4.3 H_2O

Molecular orbitals of nonlinear molecules can be determined by similar procedures. Water is a useful example. Following the steps of the previous section:

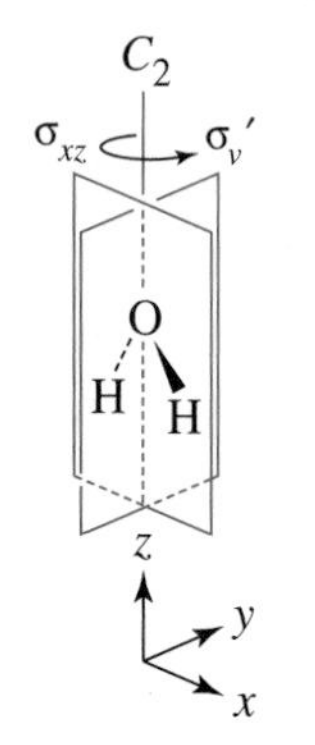

FIGURE 5.27 Symmetry of the Water Molecule.

1. Water is a bent molecule with a C_2 axis through the oxygen and two mirror planes that intersect along this axis, as shown in Figure 5.27. The point group is therefore C_{2v}.
2. The C_2 axis is chosen as the z axis and the xz plane as the plane of the molecule.[21] Because the hydrogen 1s orbitals have no directionality, it is not necessary to assign axes to the hydrogens.
3. Because the hydrogen atoms determine the symmetry of the molecule, we will use their orbitals as the basis of a reducible representation, Γ. The characters for each operation for the 1s orbitals of the hydrogen atoms can be obtained easily. The sum of the contributions to the character (1, 0, or -1, as described previously) for each symmetry operation is the character for that operation in the representation, and the complete list for all operations of the group is the reducible representation for the atomic orbitals:

   The E operation leaves both hydrogen orbitals unchanged, for a character of 2.

   C_2 rotation interchanges the orbitals, so each contributes 0, for a total character of 0.

   Reflection in the plane of the molecule (σ_v) leaves both hydrogens unchanged, for a character of 2.

   Reflection perpendicular to the plane (σ_v') switches the two orbitals, for a character of 0.

   Step 3 is summarized in Table 5.3.

[20] F. A. Cotton, *Chemical Applications of Group Theory*, 3rd ed., John Wiley & Sons, New York, 1990, pp. 133–188; D. J. Willock, *Molecular Symmetry*, John Wiley & Sons, Chichester, UK, 2009, pp. 195–212.
[21] One can also select the yz plane as the plane of the molecule. This results in $\Gamma = A_1 + B_2$ and switches the b_1 and b_2 labels of the molecular orbitals.

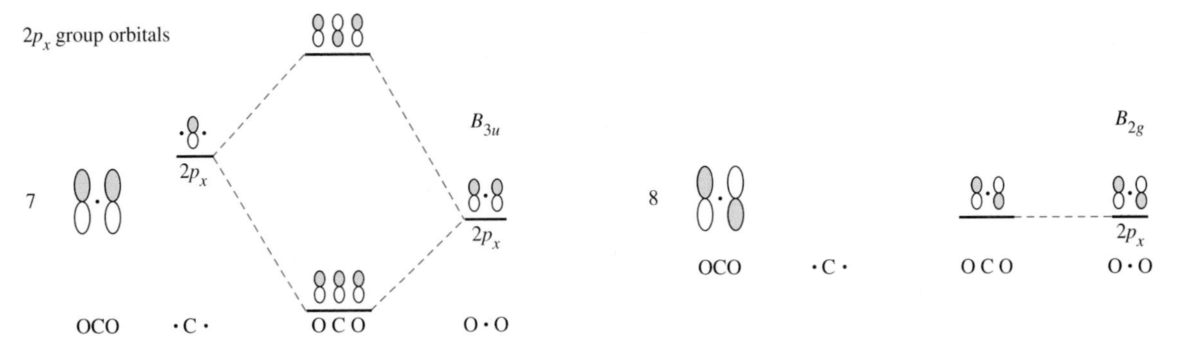

FIGURE 5.25 Group Orbitals 7 and 8 for CO_2.

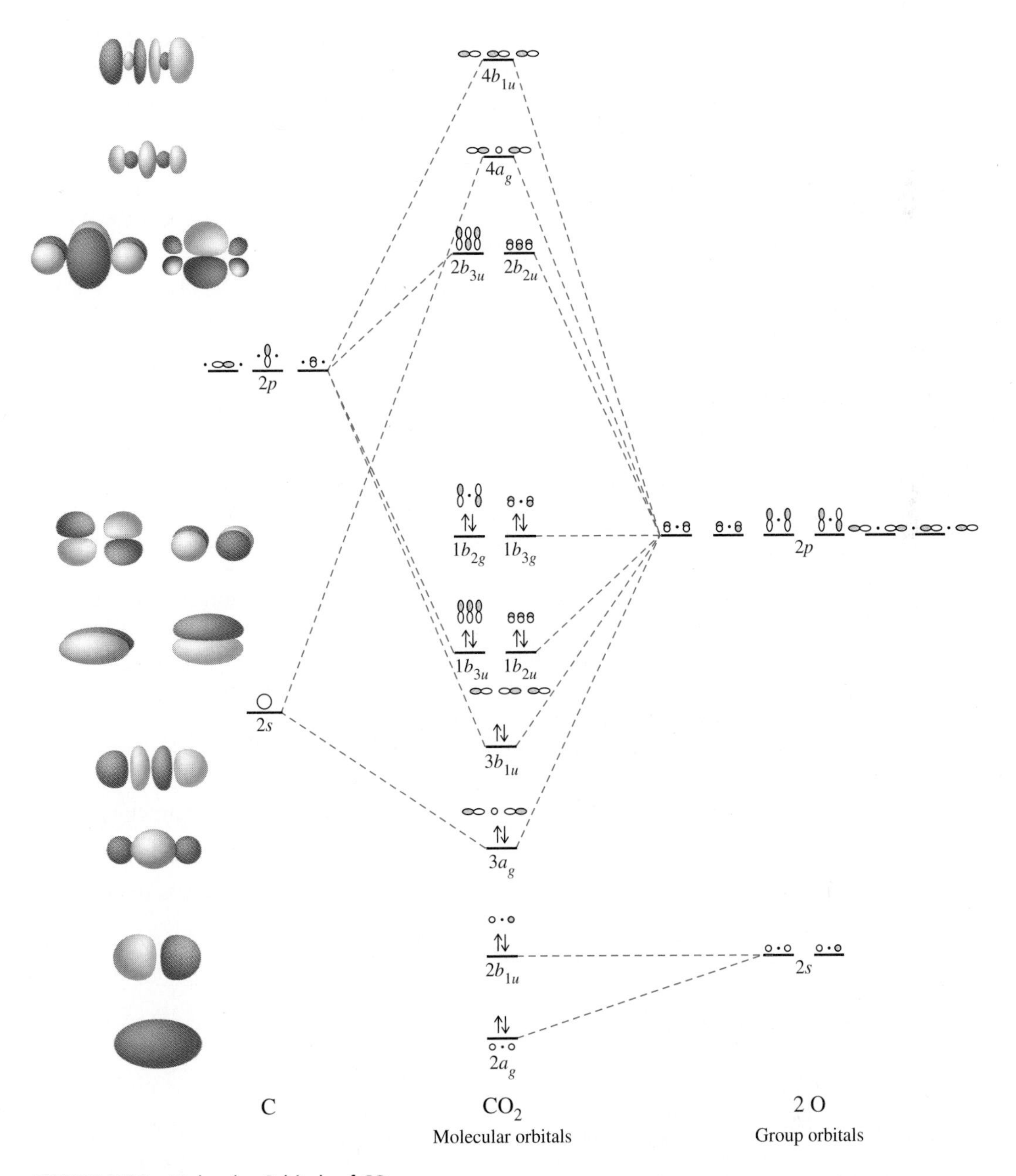

FIGURE 5.26 Molecular Orbitals of CO_2.

there are four bonds in the molecule, as expected. As in the FHF^- case, all the occupied molecular orbitals are 3-center, 2-electron orbitals and all are more stable (have lower energy) than 2-center orbitals.

The molecular orbital picture of other linear triatomic species—such as N_3^-, CS_2, and OCN^-—can be determined similarly. Likewise, the molecular orbitals of longer polyatomic species can be described by a similar method. Examples of bonding in linear π systems will be considered in Chapter 13.

▶ **Exercise 5.7** Prepare a molecular orbital diagram for the azide ion, N_3^-.

▶ **Exercise 5.8** Prepare a molecular orbital diagram for the BeH_2 molecule. (Assume an orbital potential energy of –6.0 eV for 2*p* orbitals of Be. This orbital set should be taken into account, even though it is unoccupied in a free Be atom.)

The process can be carried further to obtain numerical values of the coefficients of the atomic orbitals used in the molecular orbitals.[20] The coefficients may be small or large, positive or negative, similar or quite different, depending on the characteristics of the orbital under consideration. Computer software packages are available that will calculate these coefficients and generate the pictorial diagrams that describe the molecular orbitals. Examples of problems that use molecular modeling software to generate molecular orbitals of a variety of molecules are included in the problems at the end of this chapter and in later chapters. Discussion of the details of the calculations used by such software is beyond the scope of this text.

5.4.3 H_2O

Molecular orbitals of nonlinear molecules can be determined by similar procedures. Water is a useful example. Following the steps of the previous section:

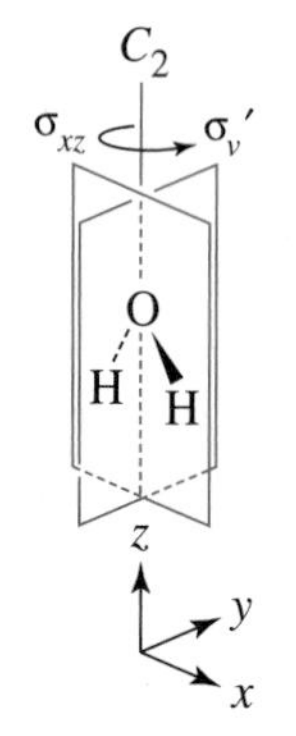

FIGURE 5.27
Symmetry of the Water Molecule.

1. Water is a bent molecule with a C_2 axis through the oxygen and two mirror planes that intersect along this axis, as shown in Figure 5.27. The point group is therefore C_{2v}.
2. The C_2 axis is chosen as the *z* axis and the *xz* plane as the plane of the molecule.[21] Because the hydrogen 1*s* orbitals have no directionality, it is not necessary to assign axes to the hydrogens.
3. Because the hydrogen atoms determine the symmetry of the molecule, we will use their orbitals as the basis of a reducible representation, Γ. The characters for each operation for the 1*s* orbitals of the hydrogen atoms can be obtained easily. The sum of the contributions to the character (1, 0, or -1, as described previously) for each symmetry operation is the character for that operation in the representation, and the complete list for all operations of the group is the reducible representation for the atomic orbitals:

   The *E* operation leaves both hydrogen orbitals unchanged, for a character of 2.

   C_2 rotation interchanges the orbitals, so each contributes 0, for a total character of 0.

   Reflection in the plane of the molecule (σ_v) leaves both hydrogens unchanged, for a character of 2.

   Reflection perpendicular to the plane (σ_v') switches the two orbitals, for a character of 0.

   Step 3 is summarized in Table 5.3.

[20] F. A. Cotton, *Chemical Applications of Group Theory*, 3rd ed., John Wiley & Sons, New York, 1990, pp. 133–188; D. J. Willock, *Molecular Symmetry*, John Wiley & Sons, Chichester, UK, 2009, pp. 195–212.
[21] One can also select the *yz* plane as the plane of the molecule. This results in $\Gamma = A_1 + B_2$ and switches the b_1 and b_2 labels of the molecular orbitals.

TABLE 5.3 Representations for C_{2v} Symmetry Operations for Hydrogen Atoms in Water

C_{2v} Character Table

C_{2v}	E	C_2	$\sigma_v(xz)$	$\sigma_v'(yz)$		
A_1	1	1	1	1	z	x^2, y^2, z^2
A_2	1	1	−1	−1	R_z	xy
B_1	1	−1	1	−1	x, R_y	xz
B_2	1	−1	−1	1	y, R_x	yz

$$\begin{bmatrix} H_a' \\ H_b' \end{bmatrix} = \begin{bmatrix} 1 & 0 \\ 0 & 1 \end{bmatrix} \begin{bmatrix} H_a \\ H_b \end{bmatrix} \text{ for the identity operation}$$

$$\begin{bmatrix} H_a' \\ H_b' \end{bmatrix} = \begin{bmatrix} 0 & 1 \\ 1 & 0 \end{bmatrix} \begin{bmatrix} H_a \\ H_b \end{bmatrix} \text{ for the } C_{2v} \text{ operation}$$

$$\begin{bmatrix} H_a' \\ H_b' \end{bmatrix} = \begin{bmatrix} 1 & 0 \\ 0 & 1 \end{bmatrix} \begin{bmatrix} H_a \\ H_b \end{bmatrix} \text{ for the } \sigma_v \text{ reflection } (xz \text{ plane})$$

$$\begin{bmatrix} H_a' \\ H_b' \end{bmatrix} = \begin{bmatrix} 0 & 1 \\ 1 & 0 \end{bmatrix} \begin{bmatrix} H_a \\ H_b \end{bmatrix} \text{ for the } \sigma_v' \text{ reflection } (yz \text{ plane})$$

The reducible representation $\Gamma = A_1 + B_1$:

C_{2v}	E	C_2	$\sigma_v(xz)$	$\sigma_v'(yz)$	
Γ	2	0	2	0	
A_1	1	1	1	1	z
B_1	1	−1	1	−1	x

4. The representation Γ can be reduced to the irreducible representations $A_1 + B_1$, representing the symmetries of the group orbitals. In Step 5 these group orbitals will be matched with orbitals of matching symmetries on oxygen.

5. In finding molecular orbitals, the first step is to combine the two hydrogen 1*s* orbitals. The sum of the two, $\frac{1}{\sqrt{2}}[\Psi(H_a) + \Psi(H_b)]$, has symmetry A_1 (this is the group orbital in which the 1*s* wave functions have matching signs); the difference, $\frac{1}{\sqrt{2}}[\Psi(H_a) - \Psi(H_b)]$, has symmetry B_1 (the group orbital in which the 1*s* wave functions have opposite signs) as can be seen by examining Figure 5.28. These group orbitals, or symmetry-adapted linear combinations, are each then treated as if they were atomic orbitals. In this case, the atomic orbitals are identical and have equal coefficients, so they contribute equally to the group orbitals. The normalizing factor is $\frac{1}{\sqrt{2}}$. In general, the normalizing factor for a group orbital is

$$N = \frac{1}{\sqrt{\sum c_i^2}}$$

where c_i = the coefficients on the atomic orbitals. Again, each group orbital is treated as a single orbital in combining with the oxygen orbitals. Both 2*s* and $2p_z$ orbitals have A_1 symmetry, and the $2p_x$ orbital has B_1 symmetry.

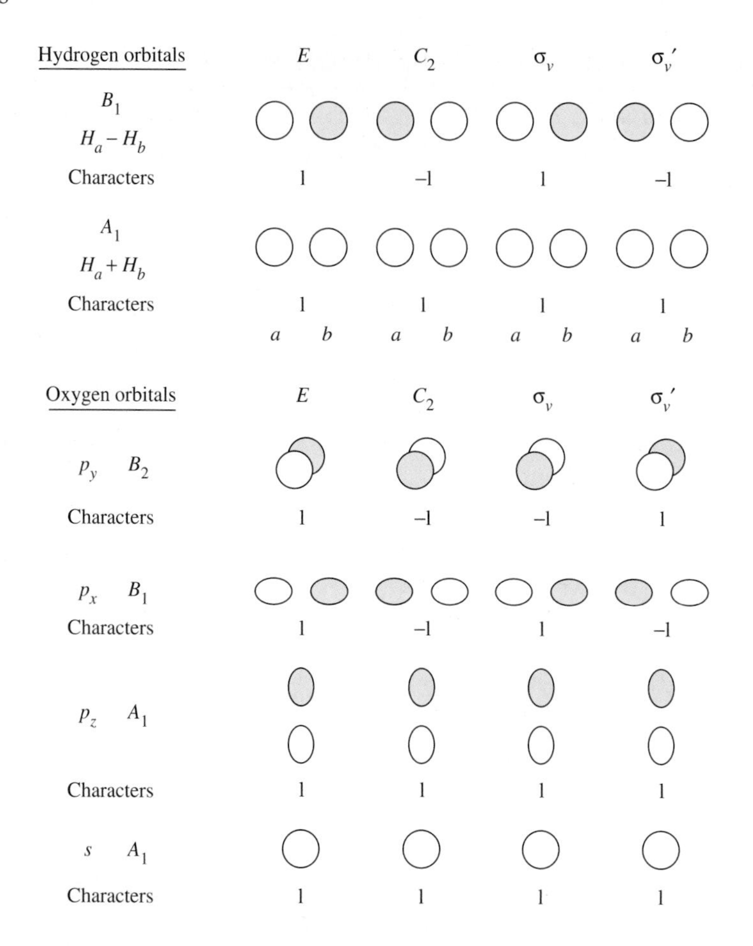

FIGURE 5.28 Symmetry of Atomic and Group Orbitals in the Water Molecule.

The same type of analysis is applied to the oxygen orbitals. Each orbital can be treated independently. The x, y, and z variables and the more complex functions in the character tables assist in assigning representations to the atomic orbitals.

The s orbital is unchanged by all the operations, so it has A_1 symmetry.

The p_x orbital has the B_1 symmetry of the x axis.

The p_y orbital has the B_2 symmetry of the y axis.

The p_z orbital has the A_1 symmetry of the z axis.

6. The atomic and group orbitals with the same symmetry are combined into molecular orbitals, as listed in Table 5.4 and shown in Figure 5.29. They are numbered Ψ_1 through Ψ_6 in order of their energy, with 1 the lowest and 6 the highest.

The A_1 group orbital combines with the s and p_z orbitals of the oxygen to form three molecular orbitals: one bonding, one nearly nonbonding (slightly bonding), and one antibonding (three atomic or group orbitals forming three molecular orbitals, Ψ_1, Ψ_3, and Ψ_5). The oxygen p_z has only minor contributions from the other orbitals in the weakly bonding Ψ_3 orbital, and the oxygen s and the hydrogen group orbitals combine weakly to form bonding and antibonding Ψ_1 and Ψ_5 orbitals that are changed only slightly from the atomic orbital energies.

The hydrogen B_1 group orbital combines with the oxygen p_x orbital to form two MOs, one bonding and one antibonding (Ψ_2 and Ψ_6). The oxygen p_y (Ψ_4, with B_2 symmetry) does not have the same symmetry as any of the hydrogen 1s group orbitals and is therefore nonbonding. Overall, there are two bonding orbitals, two nonbonding or nearly nonbonding orbitals, and two antibonding orbitals. The

TABLE 5.4 Molecular Orbitals for Water

Symmetry	Molecular Orbitals		Oxygen Atomic Orbitals		Group Orbitals from Hydrogen Atoms	Description
B_1	Ψ_6	=	$c_9\,\psi(p_x)$	+	$c_{10}\,[\psi(H_a) - \psi(H_b)]$	Antibonding (c_{10} is negative)
A_1	Ψ_5	=	$c_7\,\psi(s)$	+	$c_8\,[\psi(H_a) + \psi(H_b)]$	Antibonding (c_8 is negative)
B_2	Ψ_4	=	$\psi(p_y)$			Nonbonding
A_1	Ψ_3	=	$c_5\,\psi(p_z)$	+	$c_6\,[\psi(H_a) + \psi(H_b)]$	Nearly nonbonding (slightly bonding; c_6 is very small)
B_1	Ψ_2	=	$c_3\,\psi(p_x)$	+	$c_4\,[\psi(H_a) - \psi(H_b)]$	Bonding (c_4 is positive)
A_1	Ψ_1	=	$c_1\,\psi(s)$	+	$c_2\,[\psi(H_a) + \psi(H_b)]$	Bonding (c_2 is positive)

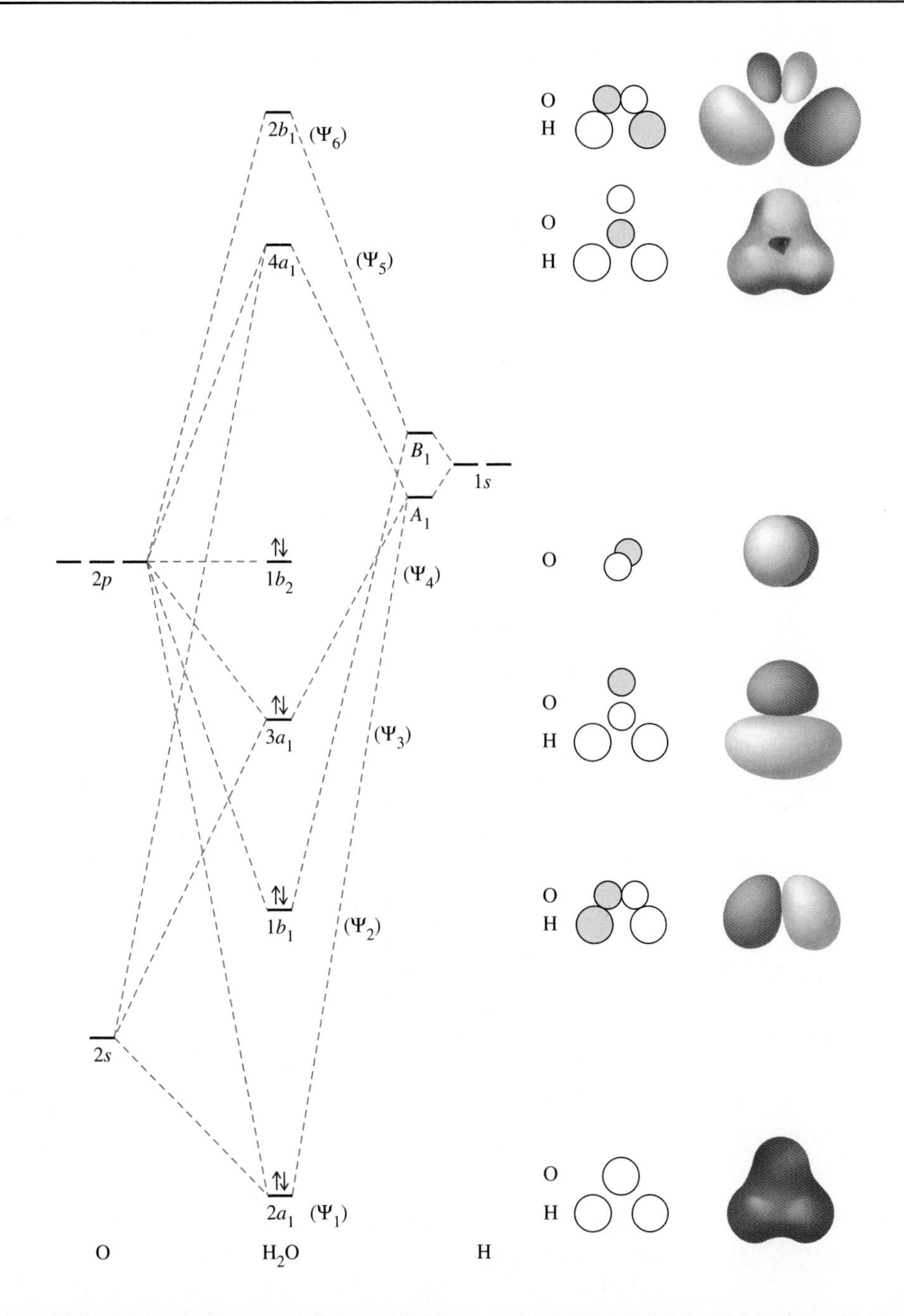

FIGURE 5.29 Molecular Orbitals of H_2O.

oxygen $2s$ orbital (–32.38 eV) is nearly 20 eV below the hydrogen orbitals in energy (–13.61 eV), so it has very little interaction with them. The oxygen $2p$ orbitals (–15.85 eV) are a good match for the hydrogen $1s$ energy, allowing formation of the bonding b_1 and a_1 molecular orbitals.

When the eight valence electrons available are added, two pairs occupy bonding orbitals, and two pairs occupy nonbonding orbitals; these may be compared to the two bonds and two lone pairs of the Lewis electron-dot structure. The resulting molecular orbital diagram is shown in Figure 5.29.

The molecular orbital picture differs from the common conception of the water molecule as having two equivalent lone electron pairs and two equivalent O—H bonds. In the MO picture, the highest energy electron pair, designated b_2, is truly nonbonding, occupying the oxygen $2p_y$ orbital with its axis perpendicular to the plane of the molecule. The two pairs next highest in energy are bonding pairs, resulting from overlap of the $2p_z$ and $2p_x$ orbital with the $1s$ orbitals of the hydrogens. The lowest energy pair is derived from the valence s orbitals and concentrated on the $2s$ orbital of oxygen. Here, all four occupied molecular orbitals are different.

5.4.4 NH_3

The VSEPR approach describes ammonia as a pyramidal molecule with a lone pair of electrons and C_{3v} symmetry. For the purpose of obtaining a molecular orbital picture of NH_3, it is convenient to view this molecule looking down on the lone pair (down the C_3, or z, axis) and with the yz plane passing through one of the hydrogens, as shown in Figure 5.30.

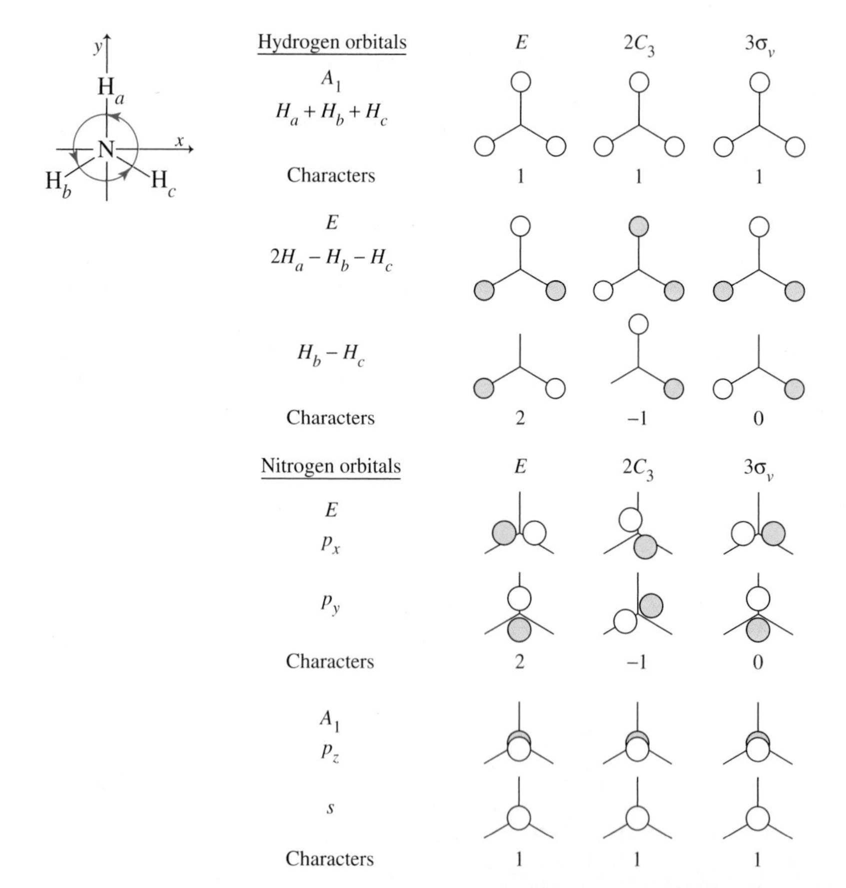

FIGURE 5.30 Coordinate System and Group Orbitals for NH_3.

TABLE 5.5 Representations for Atomic Orbitals in Ammonia

C_{3v} Character Table

C_{3v}	E	$2C_3$	$3\sigma_v$		
A_1	1	1	1	z	$x^2 + y^2, z^2$
A_2	1	1	−1		
E	2	−1	0	$(x, y), (R_x, R_y)$	$(x^2 - y^2, xy)\ (xz, yz)$

The reducible representation $\Gamma = A_1 + E$:

C_{3v}	E	$2C_3$	$3\sigma_v$		
Γ	3	0	1		
A_1	1	1	1	z	$x^2 + y^2, z^2$
E	2	−1	0	$(x, y,), (R_x, R_y)$	$(x^2 - y^2, xy)\ (xz, yz)$

The reducible representation for the three hydrogen atom 1*s* orbitals is given in Table 5.5. It can be reduced by the methods given in Chapter 4 to the A_1 and E irreducible representations, with the orbital combinations in Figure 5.30. Because three hydrogen 1*s* orbitals are to be considered, there must be three group orbitals formed from them, one with A_1 symmetry and two with E symmetry.

Up to this point, it has been a simple matter to obtain a description of the group orbitals. Each polyatomic example considered (FHF^-, CO_2, H_2O) has had two atoms attached to a central atom, and the group orbitals could be obtained by matching atomic orbitals on the terminal atoms in both a bonding and antibonding sense. In NH_3, this is no longer possible. The A_1 symmetry of the sum of the three hydrogen 1*s* orbitals is easily seen, but the two group orbitals of E symmetry are more difficult to visualize. (The matrix description of C_3 rotation for the x and y axes in Section 4.3.3 may also be helpful.)

One condition of the equations describing the molecular orbitals is that the sum of the squares of the coefficients of each of the atomic orbitals in the LCAOs equals 1 for each atomic orbital. A second condition is that the symmetry of the central atom orbitals matches the symmetry of the group orbitals with which they are combined. In this case, the E symmetry of the SALCs must match the E symmetry of the nitrogen p_x, p_y group orbitals that are being combined. This condition requires one node for each of the E group orbitals. With three atomic orbitals, the appropriate combinations are then

$$\frac{1}{\sqrt{6}}[2\Psi(H_a) - \Psi(H_b) - \Psi(H_c)] \quad \text{and} \quad \frac{1}{\sqrt{2}}[\Psi(H_b) - \Psi(H_c)]$$

The coefficients in these group orbitals result in equal contribution by each atomic orbital when each term is squared (as is done in calculating probabilities) and the terms for each orbital are summed.

For H_a, the contribution is $\left(\frac{2}{\sqrt{6}}\right)^2 = \frac{2}{3}$

For H_b and H_c, the contribution is $\left(\frac{1}{\sqrt{6}}\right)^2 + \left(\frac{1}{\sqrt{2}}\right)^2 = \frac{2}{3}$

H_a, H_b, and H_c each also have a contribution of $\frac{1}{3}$ in the A_1 group orbital,

$$\frac{1}{\sqrt{3}}[\Psi(H_a) + \Psi(H_b) + \Psi(H_c)], \quad \left(\frac{1}{\sqrt{3}}\right)^2 = \frac{1}{3}$$

giving a total contribution of 1 by each of the atomic orbitals.

The s and p_z orbitals of nitrogen both have A_1 symmetry, and the pair p_x, p_y has E symmetry, exactly the same as the representations of the hydrogen 1s orbitals. Therefore, all orbitals of nitrogen are capable of combining with the hydrogen orbitals; there are symmetry matches for both A_1 and E.

Again, each group orbital is treated as a single orbital, as shown in Figures 5.30 and 5.31, in combining with the nitrogen orbitals. The nitrogen s and p_z orbitals combine with the hydrogen A_1 group orbital to give three a_1 orbitals: one bonding, one nonbonding, and one antibonding. The nonbonding orbital is almost entirely nitrogen p_z, with the

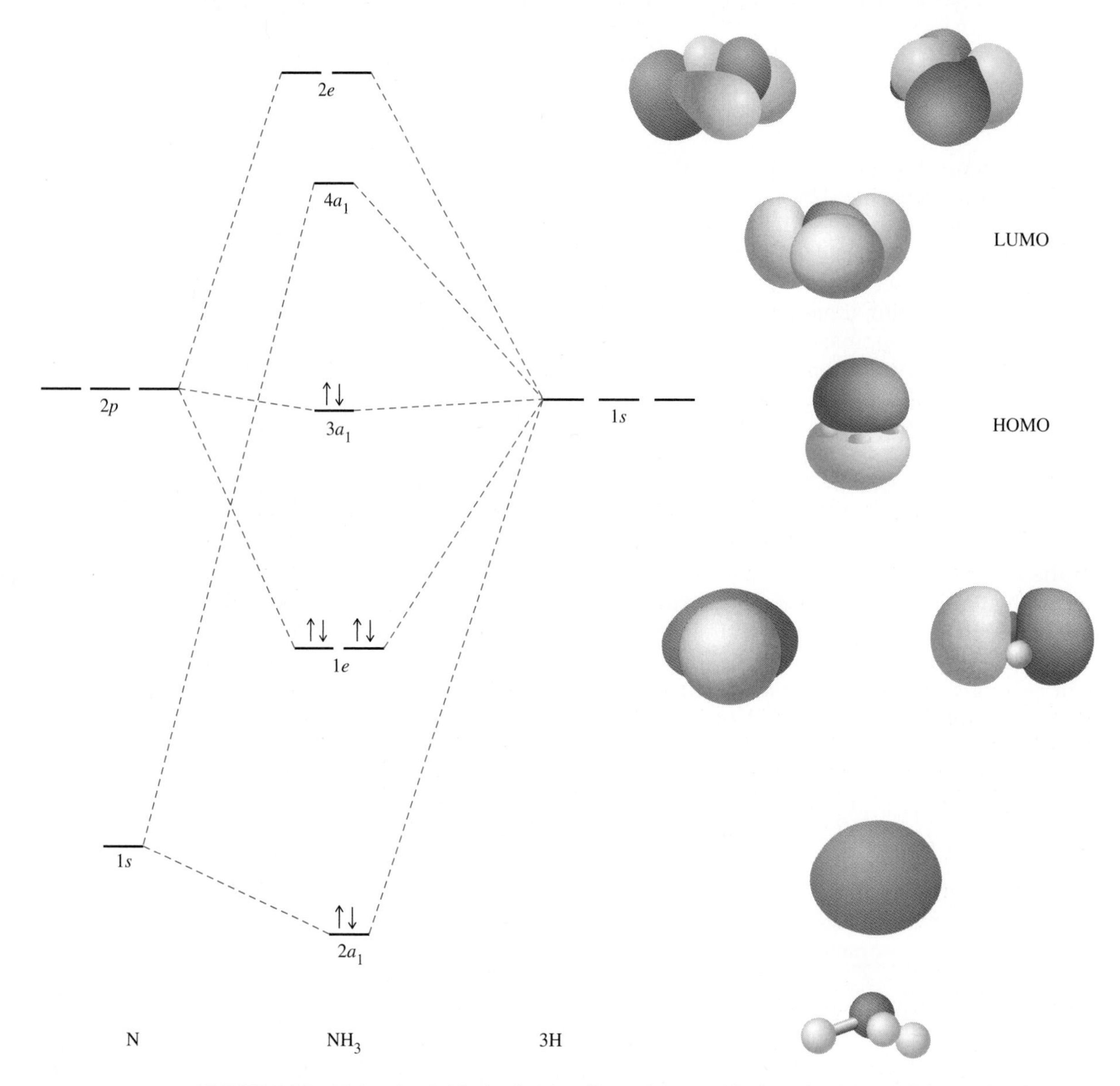

FIGURE 5.31 Molecular Orbitals of NH_3. All are shown with the orientation of the molecule at the bottom.

nitrogen s orbital combining effectively with the hydrogen group orbital for the bonding and antibonding orbitals.

The nitrogen p_x and p_y orbitals combine with the E group orbitals

$$\frac{1}{\sqrt{6}}[2\Psi(H_a) - \Psi(H_b) - \Psi(H_c)] \quad \text{and} \quad \frac{1}{\sqrt{2}}[\Psi(H_b) - \Psi(H_c)]$$

to form four e orbitals, two bonding and two antibonding (e has a dimension of 2, which requires a pair of degenerate orbitals).

When eight electrons are put into the lowest energy levels, three bonds and one essentially nonbonded lone pair are obtained, as suggested by the Lewis electron-dot structure. The $1s$ orbital energies (–13.61 eV) of the hydrogen atoms match well with the energies of the nitrogen $2p$ orbitals (–13.18 eV), resulting in large differences between the bonding and antibonding orbital energies. The nitrogen $2s$ has an energy low enough (–25.56 eV) that its interaction with the hydrogen orbitals is quite small, and the molecular orbital has nearly the same energy as the nitrogen $2s$ orbital.

The HOMO of NH_3 is slightly bonding, because it contains an electron pair in an orbital resulting from interaction of the $2p_z$ orbital of nitrogen with the $1s$ orbitals of the hydrogens (from the zero-node group orbital). This is the lone pair of the electron-dot and VSEPR models. It is also the pair donated by ammonia when it functions as a Lewis base (discussed in Chapter 6).[22]

5.4.5 BF_3

Boron trifluoride is a classic Lewis acid, an electron-pair acceptor. Therefore, an accurate molecular orbital picture of BF_3 should show, among other things, an orbital capable of acting as such an acceptor. The VSEPR shape is trigonal, consistent with experimental observations.

Although both molecules have threefold symmetry, the procedure for describing molecular orbitals of BF_3 differs from NH_3, because the fluorine atoms surrounding the central boron atom have $2p$ as well as $2s$ electrons to be considered. As is customary, we assign the highest order rotation axis, the C_3, to be in the z direction. The p_y axes of the fluorine atoms are chosen so that they are pointing toward the boron atom, and the p_x axes are in the plane of the molecule. The group orbitals and their symmetry in the D_{3h} point group are shown in Figure 5.32. The molecular orbitals are shown in Figure 5.33 (omitting sketches of the five nonbonding $2p$ group orbitals of the fluorine atoms for clarity).

As discussed in Chapter 3, resonance structures may be drawn for BF_3 showing this molecule to have some double-bond character in the B—F bonds. The molecular orbital view of BF_3 has an electron pair in a bonding π orbital with a_2'' symmetry delocalized over all four atoms; this is the orbital slightly below the five nonbonding electron pairs in energy. Overall, BF_3 has three bonding σ orbitals (a_1' and e') and one slightly bonding π orbital (a_2'') occupied by electron pairs, together with eight nonbonding pairs on the fluorine atoms. The greater than 10 eV difference between the boron and fluorine p orbital energies means that this π orbital is only slightly bonding.

The LUMO of BF_3 is worth particular note. It is an empty π orbital (a_2''), which has antibonding interactions between the $2p_z$ orbital on boron and the $2p_z$ orbitals of the surrounding fluorines. This orbital has very large, empty lobes on boron and can act as

[22] Lewis acids and bases will be discussed in Chapter 6.

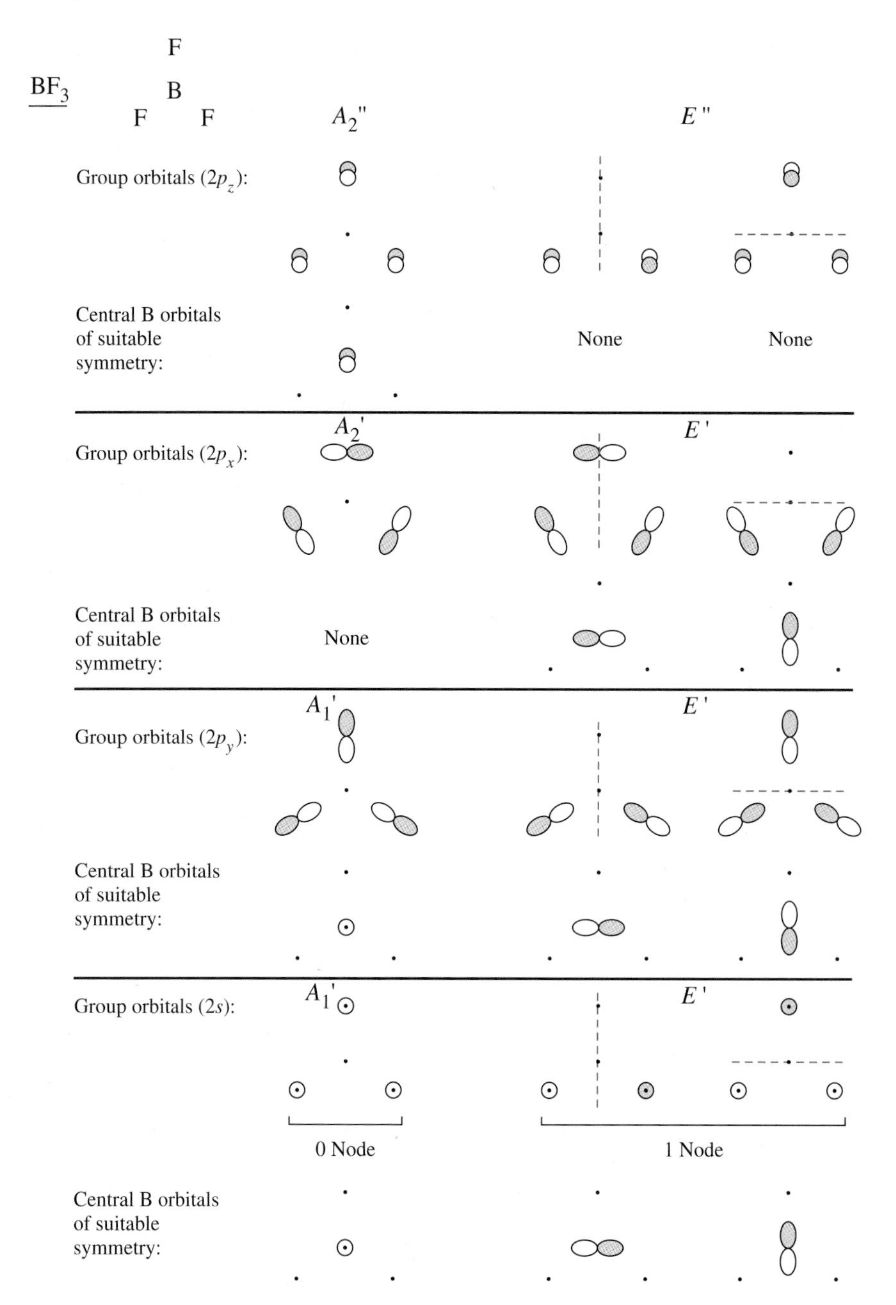

FIGURE 5.32 Group Orbitals for BF_3.

an electron-pair acceptor (for example, from the HOMO of NH_3) using these lobes in Lewis acid–base interactions.

The molecular orbitals of other trigonal species can be treated using similar procedures. The trigonal planar SO_3, NO_3^-, and CO_3^{2-} are isoelectronic with BF_3, with three electron pairs in σ-bonding orbitals and one pair in a π-bonding orbital. Group orbitals can also be used to derive molecular orbital descriptions of more complicated molecules. The simple approach described in this chapter, making minimal use of group theory, can lead conveniently to a qualitatively useful description of bonding in simple molecules. More advanced methods based on computer calculations are necessary to deal with more complex molecules and to obtain wave equations for the molecular orbitals. These more advanced methods often use molecular symmetry and group theory as well.

These qualitative methods do not allow us to determine the energies of the molecular orbitals, but we can place them in approximate order from their shapes

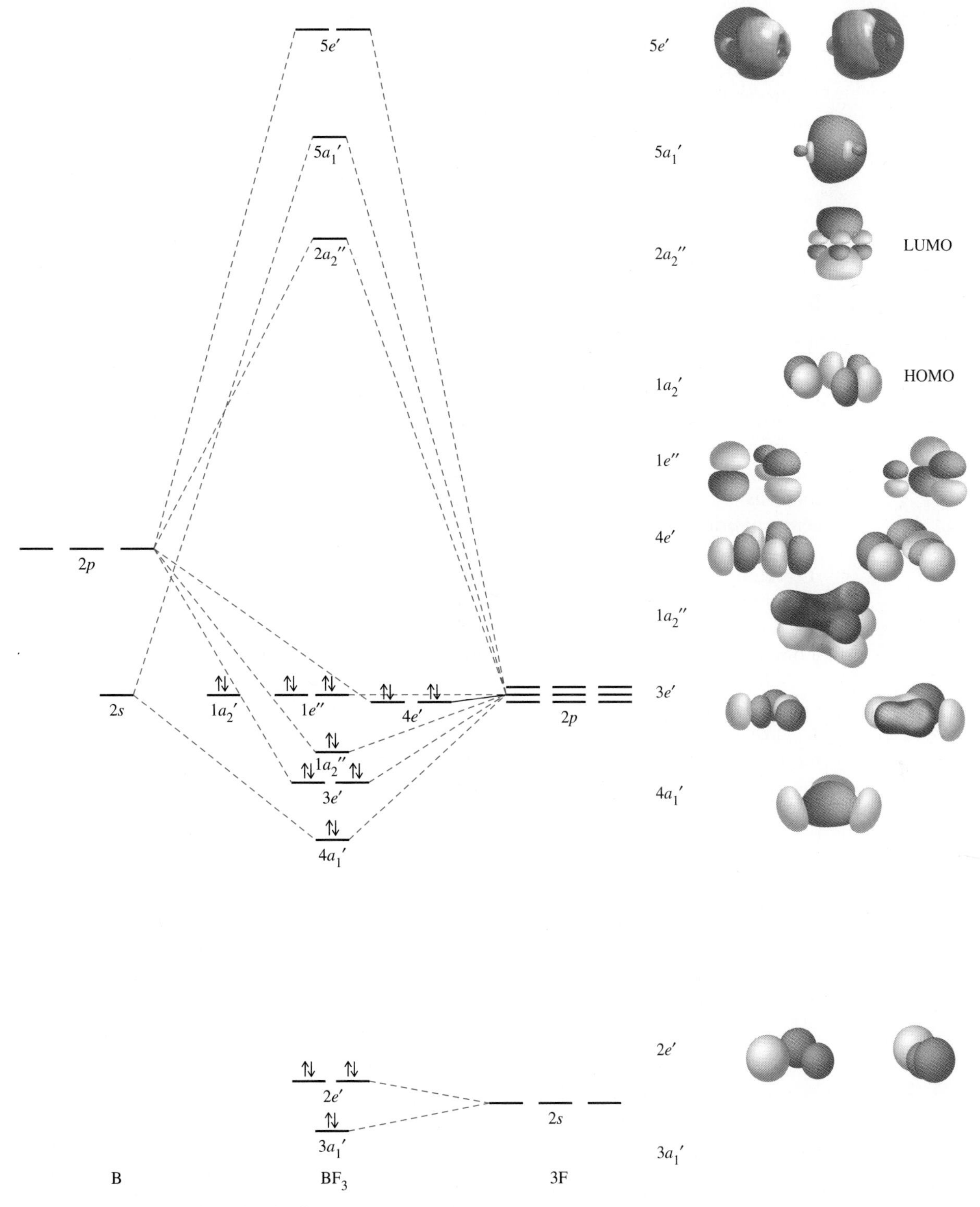

FIGURE 5.33 Molecular Orbitals of BF_3.

and expected orbital overlaps. The intermediate energy levels in particular are difficult to place in order. Whether an individual orbital is precisely nonbonding, slightly bonding, or slightly antibonding is likely to make little difference in the overall energy of the molecule. Such intermediate orbitals can be described as essentially nonbonding.

Differences in energy between two clearly bonding orbitals are likely to be more significant in the overall energy of a molecule. Because π interactions are generally weaker than σ interactions, a double bond made up of one σ orbital and one π orbital is not twice as strong as a single bond. In addition, single bonds between the same atoms may have widely different energies. For example, the C—C bond is usually described as having an energy near 345 kJ/mol, a value averaged from a large number of different molecules. These individual values may vary tremendously, with some as low as 63 and others as high as 628 kJ/mol.[23] The low value is for hexaphenyl ethane, $(C_6H_5)_3C—C(C_6H_5)_3$, and the high is for diacetylene, $H—C\equiv C—C\equiv C—H$, which are examples of extremes in steric crowding and bonding, respectively, on either side of the C—C bond.

5.4.6 Hybrid Orbitals

It is sometimes convenient to label the atomic orbitals that combine to form molecular orbitals as **hybrid orbitals**, or **hybrids**. In the hybrid concept, the orbitals of the central atom are combined into sets of equivalent hybrids. These hybrid orbitals are then used to form bonds with orbitals of other atoms. The hybrid model is especially useful in organic chemistry where, for example, it predicts four equivalent C—H bonds in methane, but it does not account well for energies of orbitals and gives results that are inconsistent with photoelectron spectroscopy. Like the Lewis-dot and resonance concepts, hybrids can be useful so long as their limits are recognized.

Hybrid orbitals are localized in space and are directional, pointing in a specific direction. In general, these hybrids point from a central atom toward surrounding atoms or lone pairs. Therefore, the symmetry properties of a set of hybrid orbitals will be identical to the properties of a set of vectors with origins at the nucleus of the central atom and pointing toward the surrounding atoms.

Methane provides a useful example. For methane, the vectors point at the corners of a tetrahedron or at alternate corners of a cube (Figure 5.34).

Using the T_d point group, we can use these four vectors as the basis of a reducible representation. As usual, the character for each vector is 1 if it remains unchanged by the symmetry operation, and 0 if it changes position in any other way (reversing direction is not an option for hybrids). The reducible representation for these four vectors is then $\Gamma = A_1 + T_2$:

T_d	E	$8C_3$	$3C_2$	$6S_4$	$6\sigma_d$		
Γ	4	1	0	0	2		
A_1	1	1	1	1	1		$x^2 + y^2 + z^2$
T_2	3	0	−1	−1	1	(x, y, z)	(xy, xz, yz)

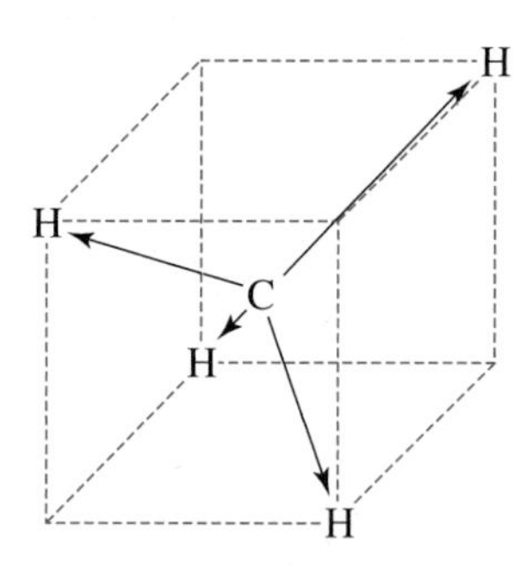

FIGURE 5.34 Bond Vectors in Methane.

[23] S. W. Benson, *J. Chem. Educ.*, **1965**, *42*, 502.

In terms of hybrids, this means that the atomic orbitals of carbon used in the hybrids *must* have symmetry matching $A_1 + T_2$; more specifically, one orbital must match A_1, and a set of three (degenerate) orbitals must match T_2.

A_1, the totally symmetric representation, has the same symmetry as the 2*s* orbital of carbon; T_2 has the same symmetry as the three 2*p* orbitals taken together (*x*, *y*, *z*) or the d_{xy}, d_{xz}, and d_{yz} orbitals taken together. Because the *d* orbitals of carbon are at much higher energy, and are therefore a poor match for the energies of the 1*s* orbitals of the hydrogens, the hybridization for methane must be sp^3, combining all four atomic orbitals—one *s* and three *p*—into four equivalent hybrid orbitals, one directed toward each hydrogen atom.

Ammonia fits the same pattern. Bonding in NH_3 uses all the nitrogen valence orbitals, so the hybrids are sp^3, incorporating one *s* orbital and all three *p* orbitals, with overall tetrahedral symmetry. The predicted HNH angle is 109.5°, narrowed to the actual 106.6° by repulsion from the lone pair, which is also viewed to occupy an sp^3 orbital.

There are two alternative approaches to hybridization for the water molecule. For example, the electron pairs around the oxygen atom in water can be considered as having nearly tetrahedral symmetry (counting the two lone pairs and the two bonds equally). All four valence orbitals of oxygen are used, and the hybrid orbitals are sp^3. The predicted bond angle is then the tetrahedral angle of 109.5°, compared with the experimental value of 104.5°. Repulsion by the lone pairs, as described in the VSEPR section of Chapter 3, is one explanation for this smaller angle.

In the other approach, which is closer to the molecular orbital description of Section 5.4.3, the bent planar shape indicates that the oxygen orbitals used in molecular orbital bonding in water are the 2*s*, $2p_x$, and $2p_z$ (in the plane of the molecule). As a result, the hybrids could be described as sp^2, a combination of one *s* orbital and two *p* orbitals. Three sp^2 orbitals have trigonal symmetry and a predicted H—O—H angle of 120°, considerably larger than the experimental value. Repulsion by the lone pairs on the oxygen—one in an sp^2 orbital, one in the remaining p_y orbital—forces the angle to be smaller. Note that the $1b_2$ orbital in the molecular orbital picture of H_2O (Figure 5.29) is a nonbonding $2p_y$ orbital occupied by a pair of electrons.

Similarly, CO_2 uses *sp* hybrids, and SO_3 uses sp^2 hybrids. Only the σ bonding is considered when determining the orbitals used in hybridization; *p* orbitals not used in the hybrids are available for π interactions. The number of atomic orbitals used in the hybrids is frequently the same as the steric number in the VSEPR method. The common hybrids are summarized in Figure 5.35. The group theory approach to hybridization is described in the following example.

EXAMPLE

Determine the types of hybrid orbitals for boron in BF_3.

For a trigonal planar molecule such as BF_3, the orbitals likely to be involved in bonding are the 2*s*, $2p_x$, and $2p_y$. This can be confirmed by finding the reducible representation in the D_{3h} point group of vectors pointing at the three fluorines and reducing it to its irreducible representations. The procedure for doing this is outlined as follows.

Step 1 Use VSEPR to determine the shape of the molecule, considering each sigma bond and lone pair on the central atom to be a vector pointing out from the center.

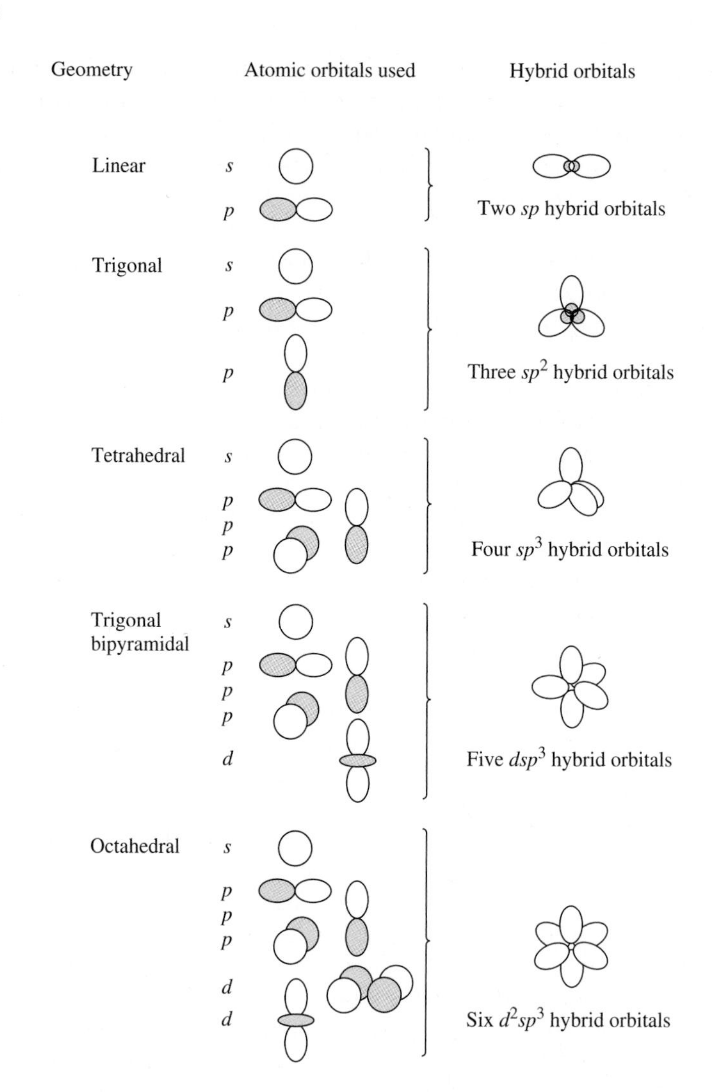

FIGURE 5.35 Hybrid Orbitals. Each single hybrid has the general shape. The figures here show all the resulting hybrids combined, omitting the smaller lobe in the sp^3 and higher orbitals.

Step 2 Determine the reducible representation for the vectors, using the appropriate point group and character table, and find the irreducible representations that combine to form the reducible representation.

Step 3 The atomic orbitals that match the irreducible representations are those used in the hybrid orbitals.

Using the symmetry operations of the D_{3h} group, we find that the reducible representation $\Gamma = A_1' + E'$.

D_{3h}	E	$2C_3$	$3C_2$	σ_h	$2S_3$	$3\sigma_v$		
Γ	3	0	1	3	0	1		
A_1'	1	1	1	1	1	1		$x^2 + y^2, z^2$
E'	2	−1	0	2	−1	0	(x, y)	$(x^2 - y^2, xy)$

This means that the atomic orbitals in the hybrids must have the same symmetry properties as A_1' and E'. More specifically, it means that one orbital must have the same symmetry as A_1' (which is one-dimensional) and two orbitals must have the same symmetry, collectively, as E' (which is two-dimensional). We must therefore select one orbital with A_1 symmetry and one *pair* of orbitals that collectively have E' symmetry. Examining the functions in the right-hand column of the character table, we see that the s orbital (not listed, but understood to be present for the totally symmetric representation) and the d_{z^2} orbital match the A_1' symmetry. However, the $3d$ orbitals, the lowest possible d orbitals, are too high in energy for bonding in BF_3 compared with the $2s$ and $2p$. Therefore, the $2s$ orbital is the contributor, with A_1' symmetry.

The functions listed for E' symmetry match the p_x, p_y set and the $d_{x^2-y^2}, d_{xy}$ set. The d orbitals are too high in energy, so the $2p_x$ and $2p_y$ orbitals are used by the central atom.[24]

Overall, the orbitals used in the hybridization are the $2s$, $2p_x$, and $2p_y$ orbitals of boron, comprising the familiar sp^2 hybrids. The difference between this approach and the molecular orbital approach is that these orbitals are combined to form the hybrids before considering their interactions with the fluorine orbitals. Because the overall symmetry is trigonal planar, the resulting hybrids must have that same symmetry, so the three sp^2 orbitals point at the three corners of a triangle, and each interacts with a fluorine p orbital to form the three σ bonds. The $2p_z$ orbital is not involved in the bonding and, according to the hybrid approach, is empty; this orbital serves as an acceptor in acid–base reactions.

▶ **Exercise 5.9** Determine the types of hybrid orbitals that are consistent with the symmetry of the central atom in

a. PF_5

b. $[PtCl_4]^{2-}$, a square planar ion

The procedure just described for determining hybrids is in some respects similar to that used in the molecular orbital approach. Hybridization uses vectors pointing toward the outlying atoms and usually deals only with σ bonding. Once the σ hybrids are known, π bonding is added, using orbitals that do not participate in the hybridization. It is also possible to use hybridization techniques for π bonding, but that approach will not be discussed here.[25] Hybridization may be quicker than the molecular orbital approach, because the molecular orbital approach uses all the atomic orbitals of the atoms and includes both σ and π bonding directly. Both methods are useful, and the choice of method depends on the particular problem and personal preference.

▶ **Exercise 5.10** Find the reducible representation for all the σ bonds, reduce it to its irreducible representations, and determine the sulfur orbitals used in bonding for $SOCl_2$.

General References

There are many books describing bonding and molecular orbitals, with levels ranging from those even more descriptive and qualitative than the treatment in this chapter to those designed for the theoretician interested in the latest methods. A classic that starts at the level of this chapter and includes many more details is R. McWeeny's revision of *Coulson's Valence*, 3rd ed., Oxford University Press, Oxford, 1979. A different approach that uses the concept of generator orbitals is that of J. G. Verkade in *A Pictorial Approach to Molecular Bonding and Vibrations*, 2nd ed., Springer-Verlag, New York, 1997. The group theory approach in this chapter is similar to that of F. A. Cotton in *Chemical Applications of Group Theory*, 3rd ed.,

[24] A combination of one p orbital and one d orbital cannot be chosen, because orbitals in parentheses must always be taken together.

[25] F. A. Cotton, *Chemical Applications of Group Theory*, 3rd ed., John Wiley & Sons, New York, 1990, pp. 227–230.

John Wiley & Sons, New York, 1990. A more recent book that extends the description is Y. Jean and F. Volatron, *An Introduction to Molecular Orbitals*, translated and edited by J. K. Burdett, Oxford University Press, Oxford, 1993. J. K. Burdett's *Molecular Shapes*, John Wiley & Sons, New York, 1980, and B. M. Gimarc's, *Molecular Structure and Bonding*, Academic Press, New York, 1979, are both good introductions to the qualitative molecular orbital description of bonding.

Problems

5.1 Expand the list of orbitals considered in Figures 5.2 and 5.3 by using all three *p* orbitals of atom A and all five *d* orbitals of atom B. Which of these have the necessary match of symmetry for bonding and antibonding orbitals? These combinations are rarely seen in simple molecules but can be important in transition metal complexes.

5.2 On the basis of molecular orbitals, predict the shortest bond, and provide a brief explanation.
a. Li_2^+ Li_2
b. F_2^+ F_2
c. He_2^+ HHe^+ H_2^+

5.3 On the basis of molecular orbitals, predict the weakest bond, and provide a brief explanation.
a. P_2 S_2 Cl_2
b. S_2^+ S_2 S_2^-
c. NO^- NO NO^+

5.4 Compare the bonding in O_2^{2-}, O_2^-, and O_2. Include Lewis structures, molecular orbital structures, bond lengths, and bond strengths in your discussion.

5.5 Although the peroxide ion, O_2^{2-}, and the acetylide ion, C_2^{2-}, have long been known, the diazenide ion N_2^{2-} has only been prepared much more recently. By comparison with the other diatomic species, predict the bond order, bond distance, and number of unpaired electrons for N_2^{2-}. (See G. Auffermann, Y. Prots, and R. Kniep, *Angew. Chem., Int. Ed.*, **2001**, *40*, 547.)

5.6 High-resolution photoelectron spectroscopy has provided information on the energy levels and bond distance in the ion Ar_2^+. Prepare a molecular orbital energy-level diagram for this ion. How would you expect the bond distance in Ar_2^+ to compare with 198.8 pm, the bond distance in Cl_2? (See A Wüst and F. Merkt, *J. Chem. Phys.*, **2004**, *120*, 638.)

5.7 **a.** Prepare a molecular orbital energy-level diagram for NO, showing clearly how the atomic orbitals interact to form MOs.
b. How does your diagram illustrate the difference in electronegativity between N and O?
c. Predict the bond order and the number of unpaired electrons.
d. NO^+ and NO^- are also known. Compare the bond orders of these ions with the bond order of NO. Which of the three would you predict to have the shortest bond? Why?

5.8 **a.** Prepare a molecular orbital energy-level diagram for the cyanide ion. Use sketches to show clearly how the atomic orbitals interact to form MOs.
b. What is the bond order for cyanide, and how many unpaired electrons does cyanide have?
c. Which molecular orbital of CN^- would you predict to interact most strongly with a hydrogen 1*s* orbital to form an H—C bond in the reaction $CN^- + H^+ \longrightarrow HCN$? Explain.

5.9 The hypofluorite ion, OF^-, can be observed only with difficulty.
a. Prepare a molecular orbital energy level diagram for this ion.
b. What is the bond order, and how many unpaired electrons are in this ion?
c. What is the most likely position for adding H^+ to the OF^- ion? Explain your choice.

5.10 Reaction of KrF_2 with AsF_5 at temperatures between –78 and –53° C yields $[KrF][AsF_6]$, a compound in which KrF^+ interacts strongly with AsF_6^- through a fluorine bridge, as shown. Would you predict the Kr–F bond to be shorter in KrF^+ or in KrF_2? Provide a brief explanation. (See J. F. Lehmann, D. A. Dixon, and G. J. Schrobilgen, *Inorg. Chem.*, **2001**, *40*, 3002.)

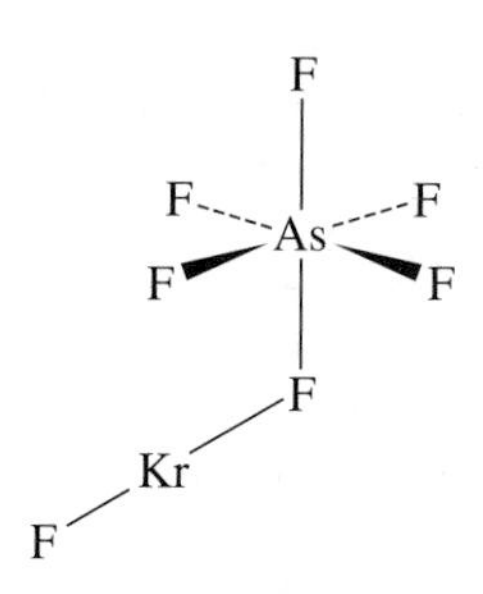

5.11 Although KrF^+ and XeF^+ have been studied, $KrBr^+$ has not yet been prepared. For $KrBr^+$:
a. Propose a molecular orbital diagram, showing the interactions of the valence shell *s* and *p* orbitals to form molecular orbitals.
b. Toward which atom would the HOMO be polarized? Why?
c. Predict the bond order.
d. Which is more electronegative, Kr or Br? Explain your reasoning.

5.12 Prepare a molecular orbital energy level diagram for SH^-, including sketches of the orbital shapes and the number of electrons in each of the orbitals. If a program for calculating molecular orbitals is available, use it to confirm your predictions or to explain why they differ.

5.13 Methylene, CH_2, plays an important role in many reactions. One possible structure of methylene is linear.

a. Construct a molecular orbital energy-level diagram for this species. Include sketches of the group orbitals, and indicate how they interact with the appropriate orbitals of carbon.

b. Would you expect linear methylene to be diamagnetic or paramagnetic?

5.14 Beryllium hydride, BeH_2, is linear in the gas phase.

a. Construct a molecular orbital energy level diagram for BeH_2. Include sketches of the group orbitals, and indicate how they interact with the appropriate orbitals of Be.

b. If you have worked Problem 5.13, compare the results of these two problems.

5.15 In the gas phase, BeF_2 forms linear monomeric molecules. Prepare a molecular orbital energy-level diagram for BeF_2, showing clearly which atomic orbitals are involved in bonding and which are nonbonding.

5.16 For the compound XeF_2:

a. Sketch the valence shell group orbitals for the fluorine atoms (with the *z* axes collinear with the molecular axis).

b. For each of the group orbitals, determine which outermost *s*, *p*, and *d* orbitals of xenon are of suitable symmetry for interaction and bonding.

5.17 TaH_5 has been predicted to have C_{4v} symmetry. Using the six step approach described in Section 5.4.2, describe the bonding in TaH_5 on the basis of matching group orbitals and central atom orbitals according to their symmetry. (See C. A. Bayse and M. B. Hall, *J. Am. Chem. Soc.*, **1999**, *121*, 1348.)

5.18 Describe the bonding in ozone, O_3, on the basis of matching group orbitals and central-atom orbitals according to their symmetry. Include both σ and π interactions, and try to put the resulting orbitals in approximate order of energy.

5.19 Describe the bonding in SO_3 by using group theory to find the molecular orbitals. Include both σ and π interactions, and try to put the resulting orbitals in approximate order of energy. (The actual results are more complex because of mixing of orbitals, but a simple description can be found by the methods given in this chapter.)

5.20 The ion H_3^+ has been observed, but its structure has been the subject of some controversy. Prepare a molecular orbital energy level diagram for H_3^+, assuming a cyclic structure. (The same problem for a linear structure is given in Exercise 5.4 in Section 5.4.2.)

5.21 Use molecular orbital arguments to explain the structures of SCN^-, OCN^-, and CNO^-, and compare the results with the electron-dot pictures of Chapter 3.

5.22 Thiocyanate and cyanate ions both bond to H^+ through the nitrogen atoms (HNCS and HNCO), whereas SCN^- forms bonds with metal ions through either nitrogen or sulfur, depending on the rest of the molecule. What does this suggest about the relative importance of S and N orbitals in the MOs of SCN^-? (Hint: See the discussion of CO_2 bonding in Section 5.4.2.)

5.23 The thiocyanate ion, SCN^-, can form bonds to metals through either S or N (See Problem 5.22). What is the likelihood of cyanide, CN^-, forming bonds to metals through N as well as C?

5.24 The isomeric ions NSO^- (thiazate) and SNO^- (thionitrite) ions have been reported. (S. P. So, *Inorg. Chem.*, **1989**, *28*, 2888).

a. On the basis of the resonance structures of these ions, predict which would be more stable.

b. Sketch the approximate shapes of the π and π^* orbitals of these ions.

c. Predict which ion would have the shorter N—S bond and which would have the higher energy N—S stretching vibration? (Stronger bonds have higher energy vibrations.)

5.25 SF_4 has C_{2v} symmetry. Predict the possible hybridization schemes for the sulfur atom in SF_4.

5.26 Consider a square pyramidal AB_5 molecule. Using the C_{4v} character table, determine the possible hybridization schemes for central atom A. Which of these would you expect to be most likely?

5.27 In coordination chemistry, many square-planar species are known (for example, $PtCl_4^{2-}$). For a square planar molecule, use the appropriate character table to determine the types of hybridization possible for a metal surrounded in a square-planar fashion by four ligands; consider hybrids used in σ bonding only.

5.28 For the molecule PCl_5:

a. Using the character table for the point group of PCl_5, determine the possible type(s) of hybrid orbitals that can be used by P in forming σ bonds to the five Cl atoms.

b. What type(s) of hybrids can be used in bonding to the axial chlorine atoms? To the equatorial chlorine atoms?

c. Considering your answer to part b, explain the experimental observation that the axial P—Cl bonds (219 pm) are longer than the equatorial bonds (204 pm).

5.29 Although the Cl_2^+ ion has not been isolated, it has been detected in the gas phase by UV spectroscopy. An attempt to prepare this ion by reaction of Cl_2 with IrF_6 yielded not Cl_2^+, but the rectangular ion Cl_4^+. (See S. Seidel and K. Seppelt, *Angew. Chem., Int. Ed.*, **2000**, *39*, 3923.)

a. Compare the bond distance and bond energy of Cl_2^+ with Cl_2.

b. Account for the bonding in Cl_4^+. This ion contains two short Cl—Cl bonds and two much longer ones. Would you expect the shorter Cl—Cl distances in Cl_4^+ to be longer or shorter than the Cl—Cl distance in Cl_2^+? Explain.

5.30 BF_3 is often described as a molecule in which boron is electron deficient, with an electron count of six. However, resonance structures can be drawn in which boron has an octet, with delocalized π electrons.

a. Draw these structures.

b. Find the molecular orbital in Figure 5.33 that shows this delocalization and explain your choice.

c. BF_3 is *the* classic Lewis acid, accepting a pair of electrons from molecules with lone pairs. Find the orbital in Figure 5.33 that is this acceptor; explain your choice, including why it looks like a good electron acceptor.

d. What is the relationship between the orbitals identified in Parts b and c?

The following problems require the use of molecular modeling software.

5.31 **a.** Identify the point group of the $1a_2''$, $2a_2''$, $1a_2'$, and $1e''$ molecular orbitals in Figure 5.33.

b. Use molecular modeling software to calculate and view the molecular orbitals of BF_3.

c. Do any of the molecular orbitals show π interactions between B and F ?

d. Print out the contributions of the atomic orbitals to the $3a_1'$, $4a_1'$, $1a_2''$, $1a_2'$, and $2a_2''$ molecular orbitals, confirming (if you can) the atomic orbital combinations shown in Figure 5.33.

5.32 The ions and molecules NO^+, CN^-, CO, and N_2 form an isoelectronic series. The changing nuclear charges will also change the molecular energy levels of the orbitals formed from the $2p$ atomic orbitals (1π, 3σ, and $1\pi^*$). Use molecular modeling software for the following:

a. Calculate and display the shapes of these three molecular orbitals for each species (CO and N_2 are included in this chapter).

b. Compare the shapes of each of the orbitals for each of the species (for example, the shapes of the 1π orbitals for each). What trends do you observe?

c. Compare the energies of each of the orbitals. For which do you see evidence of mixing?

5.33 Molecular modeling software is typically capable of calculations on molecules that are hypothetical, even seemingly bizarre, in their structures. Beginning with N_2, calculate and display molecular orbitals of the isoelectronic CO, BF, and BeNe (which is truly hypothetical!). Compare the shapes of the matching molecular orbitals in this series. What trends do you observe?

5.34 Calculate and display the orbitals for the linear molecule BeH_2. Describe how they illustrate the interaction of the outer group orbitals with the orbitals on the central atom. Compare your results with the answer to Problem 5.14.

5.35 Calculate and display the orbitals for the linear molecule BeF_2. Compare the orbitals and their interactions with those of BeH_2 from Problem 5.34. In particular, indicate the outer group orbitals that do not interact with orbitals on the central atom.

5.36 The azide ion, N_3^-, is another linear triatomic species. Calculate and display the orbitals for this ion, and compare the three highest energy occupied orbitals with those of BeF_2. How do the outer atom group orbitals differ in their interactions with the central atom orbitals? How do the orbitals compare with the CO_2 orbitals discussed in Section 5.4.2?

5.37 Calculate and display the molecular orbitals of the ozone molecule, O_3. Which orbitals show π interactions? Compare your results with your answer to Problem 5.18.

5.38 **a.** Calculate and display the molecular orbitals for linear and cyclic H_3^+.

b. Which species is more likely to exist (i.e., which is more stable)?

5.39 Diborane, B_2H_6, has the structure shown.

a. Using the point group of the molecule, create a representation using the $1s$ orbitals on the hydrogens as a basis. Reduce this representation, and sketch group orbitals matching each of the irreducible representations. (Suggestion: Treat the bridging and terminal hydrogens separately.)

b. Calculate and display the molecular orbitals. Compare the software-generated images with the group orbital sketches from part a, and explain how hydrogen can form "bridges" between two B atoms. (This type of bonding is discussed in Chapter 8.)

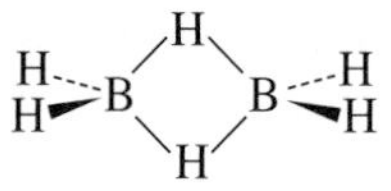

6

Acid–Base and Donor–Acceptor Chemistry

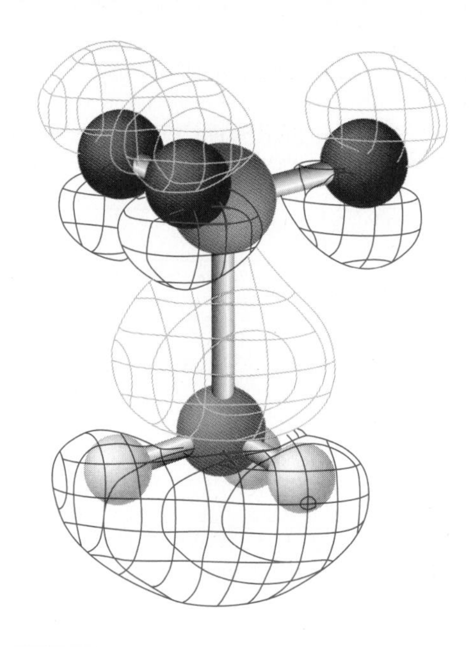

6.1 ACID–BASE CONCEPTS AS ORGANIZING CONCEPTS

The concept of acids and bases has been important since ancient times. It has been used to correlate large amounts of data and to predict trends. Jensen[1] has described a useful approach in the preface to his book on the Lewis acid–base concept:

> . . . acid–base concepts occupy a somewhat nebulous position in the logical structure of chemistry. They are, strictly speaking, neither facts nor theories and are, therefore, never really "right" or "wrong." Rather they are classificatory definitions or organizational analogies. They are useful or not useful . . . acid-base definitions are always a reflection of the facts and theories current in chemistry at the time of their formulation and . . . they must, necessarily, evolve and change as the facts and theories themselves evolve and change . . .

The changing definitions described in this chapter have generally led to a more inclusive and useful approach to acid–base concepts. Most of this chapter is concerned with the Lewis definition, its more recent explanation in terms of molecular orbitals, and its application to inorganic chemistry.

6.1.1 History

Practical acid–base chemistry was known in ancient times and developed gradually during the time of the alchemists. During the early development of acid–base theory, experimental observations included the sour taste of acids and bitter taste of bases, color changes in indicators caused by acids and bases, and the reaction of acids with bases to form salts. Partial explanations included the idea that all acids contained oxygen: oxides of nitrogen, phosphorus, sulfur, and the halogens all form acids in water. But by the early nineteenth century, many acids that do not contain oxygen were known. By 1838, Liebig defined acids as "compounds containing hydrogen, in which the hydrogen can be replaced by a metal,"[2] a definition that still works well in many instances.

Although many other acid–base definitions have been proposed and have been useful in particular types of reactions, only a few have been widely adopted for general use. Among these are the ones attributed to Arrhenius (based on hydrogen and hydroxide ion formation), Brønsted–Lowry (hydrogen ion donors and acceptors), and Lewis (electron-pair acceptors and donors). Others have received less attention

[1] W. B. Jensen, *The Lewis Acid–Base Concepts*, Wiley InterScience, New York, 1980, p. vii.
[2] R. P. Bell, *The Proton in Chemistry*, 2nd ed., Cornell University Press, Ithaca, NY, 1973, p. 9.

TABLE 6.1 Comparison of Acid–Base Definitions

		Definitions		Examples	
Description	**Date**	**Acid**	**Base**	**Acid**	**Base**
Lavoisier	~1776	Oxide of N, P, S	Reacts with acid	SO_3	NaOH
Liebig	1838	H replaceable by metal	Reacts with acid	HNO_3	NaOH
Arrhenius	1894	Forms hydronium ion	Forms hydroxide ion	H^+	OH^-
Brønsted–Lowry	1923	Hydrogen-ion donor	Hydrogen-ion acceptor	H_3O^+	H_2O
				H_2O	OH^-
				NH_4^+	NH_3
Lewis	1923	Electron-pair acceptor	Electron-pair donor	Ag^+	NH_3
Ingold–Robinson	1932	Electrophile (electron-pair acceptor)	Nucleophile (electron-pair donor)	BF_3	NH_3
Lux–Flood	1939	Oxide-ion acceptor	Oxide-ion donor	SiO_2	CaO
Usanovich	1939	Electron acceptor	Electron donor	Cl_2	Na
Solvent system	1950s	Solvent cation	Solvent anion	BrF_2^+	BrF_4^-
Frontier orbitals	1960s	LUMO of acceptor	HOMO of donor	BF_3	NH_3

or are useful only in a narrow range of situations. For example, the Lux-Flood definition[3] is based on oxide ion O^{2-} as the unit transferred between acids (oxide ion acceptors) and bases (oxide ion donors). The Usanovich definition[4] proposes that any reaction leading to a salt (including oxidation–reduction reactions) should be considered an acid–base reaction. This definition could include nearly all reactions and has been criticized for this all-inclusive approach. The Usanovich definition is rarely used today, but it fits the frontier orbital approach described in Section 6.2.5. The electrophile–nucleophile approach of Ingold[5] and Robinson,[6] widely used in organic chemistry, is essentially the Lewis theory with terminology related to reactivity: electrophilic reagents are acids, nucleophilic reagents are bases. Another approach that is described later in this chapter is an extension of the Lewis definition in terms of frontier orbitals. Table 6.1 summarizes these acid–base definitions.

6.2 MAJOR ACID–BASE CONCEPTS

6.2.1 Arrhenius Concept

Acid–base chemistry was first satisfactorily explained in molecular terms after Ostwald and Arrhenius established the existence of ions in aqueous solution in 1880–1890 (after much controversy and many professional difficulties, Arrhenius received the 1903 Nobel Prize in Chemistry for this theory). As defined at that time, **Arrhenius acids** *form hydrogen ions* (now frequently considered hydronium or oxonium[7] ions, H_3O^+)

[3] H. Lux, *Z. Electrochem.*, **1939**, *45*, 303; H. Flood and T. Förland, *Acta Chem. Scand.*, **1947**, *1*, 592, 718; W. B. Jensen, *The Lewis Acid–Base Concepts*, Wiley Interscience, New York, 1980, pp. 54–55.
[4] M. Usanovich, *Zh. Obshch. Khim.*, **1939**, *9*, 182; H. Gehlen, *Z. Phys. Chem.*, **1954**, *203*, 125; H. L. Finston and A. C. Rychtman, *A New View of Current Acid–Base Theories*, John Wiley & Sons, New York, 1982, pp. 140–146.
[5] C. K. Ingold, *J. Chem. Soc.*, **1933**, 1120; *Chem. Rev.*, **1934**, *15*, 225; *Structure and Mechanism in Organic Chemistry*, Cornell University Press, Ithaca, NY, 1953, Chapter V; W. B. Jensen, *The Lewis Acid–Base Concepts*, Wiley Interscience, New York, 1980, pp. 58–59.
[6] R. Robinson, *Outline of an Electrochemical (Electronic) Theory of the Course of Organic Reactions*, Institute of Chemistry, London, 1932, pp. 12–15; W. B. Jensen, *The Lewis Acid–Base Concepts*, Wiley Interscience, New York, 1980, pp. 58–59.
[7] In American practice, H_3O^+ is frequently called the *hydronium ion*. The International Union of Pure and Applied Chemistry (IUPAC) now recommends *oxonium* for this species. In many equations, the shorthand H^+ notation is used, for which the IUPAC recommends the terms *hydron* or *hydrogen ion*, rather than *proton*.

in aqueous solution; **Arrhenius bases** *form hydroxide ions in aqueous solution.* The reaction of hydrogen ions and hydroxide ions to form water is the universal aqueous acid–base reaction. The ions accompanying the hydrogen and hydroxide ions form a salt, so the overall Arrhenius acid–base reaction can be written

$$\text{acid} + \text{base} \longrightarrow \text{salt} + \text{water}$$

For example,

$$\text{hydrochloric acid} + \text{sodium hydroxide} \longrightarrow \text{sodium chloride} + \text{water.}$$

$$H^+ + Cl^- + Na^+ + OH^- \longrightarrow Na^+ + Cl^- + H_2O$$

This explanation works well in aqueous solution, but it is inadequate for nonaqueous solutions and for gas-phase and solid-phase reactions in which H^+ and OH^- may not exist. Definitions by Brønsted and Lowry and Lewis are more appropriate for general use.

6.2.2 Brønsted–Lowry Concept

In 1923, Brønsted[8] and Lowry[9] defined an **acid** as *a species with a tendency to lose a hydrogen ion* and a **base** as *a species with a tendency to gain a hydrogen ion.* This definition expanded the Arrhenius list of acids and bases to include the gases HCl and NH_3, along with many other compounds. It also introduced the concept of **conjugate acids and bases**, differing only in the presence or absence of a proton, and described all reactions as occurring between a stronger acid and base to form a weaker acid and base:

$$\underset{\text{acid 1}}{H_3O^+} + \underset{\text{base 2}}{NO_2^-} \longrightarrow \underset{\text{base 1}}{H_2O} + \underset{\text{acid 2}}{HNO_2}$$

Conjugate acid–base pairs:

Acid	Base
H_3O^+	H_2O
HNO_2	NO_2^-

In water, HCl and NaOH react as the acid H_3O^+ and the base OH^- to form water, which is the conjugate base of H_3O^+ *and* the conjugate acid of OH^-. Reactions in nonaqueous solvents having ionizable hydrogens parallel those in water. An example of such a solvent is liquid ammonia, in which NH_4Cl and $NaNH_2$ react as the acid NH_4^+ and the base NH_2^- to form NH_3, which is both a conjugate base and a conjugate acid:

$$NH_4^+ + Cl^- + Na^+ + NH_2^- \longrightarrow Na^+ + Cl^- + 2\,NH_3$$

with the net reaction

$$\underset{\text{acid}}{NH_4^+} + \underset{\text{base}}{NH_2^-} \longrightarrow \underset{\substack{\text{conjugate base}\\ \text{and conjugate acid}}}{2\,NH_3}$$

In any solvent, the direction of the reaction always favors the formation of acids or bases weaker than the reactants. In the two examples above, H_3O^+ is a stronger acid than HNO_2, NH_2^- is a stronger base than NH_3, and NH_4^+ is a stronger acid than NH_3, so the reactions favor formation of HNO_2 and ammonia.

[8] J. N. Brønsted, *Rec. Trav. Chem.*, **1923**, *42*, 718.
[9] T. M. Lowry, *Chem. Ind. (London)*, **1923**, *42*, 43.

6.2.3 Solvent-System Concept

Aprotic nonaqueous solutions require a similar approach, but with a different definition of acid and base. The **solvent-system** definition applies to any solvent that can dissociate into a *cation* and an *anion* (**autodissociation**). The resulting **cation** *is the acid* and the **anion** *is the base*. Solutes that *increase the concentration of the cation of the solvent are considered acids* and solutes that *increase the concentration of the anion are considered bases.*

The classic solvent system is water, which undergoes autodissociation:

$$2\ H_2O \rightleftharpoons H_3O^+ + OH^-$$

By the solvent-system definition, the cation, H_3O^+, is the acid, and the anion, OH^-, is the base. For example, in the reaction

$$H_2SO_4 + H_2O \longrightarrow H_3O^+ + HSO_4^-$$

sulfuric acid increases the concentration of the hydronium ion and is an acid by any of the three definitions given.

The solvent-system approach can also be used with solvents that do not contain hydrogen. For example, BrF_3 also undergoes autodissociation:

$$2\ BrF_3 \rightleftharpoons BrF_2^+ + BrF_4^-$$

Solutes that increase the concentration of the acid BrF_2^+ are considered acids. For example, SbF_5 is an acid in BrF_3:

$$SbF_5 + BrF_3 \longrightarrow BrF_2^+ + SbF_6^-$$

Solutes such as KF that increase the concentration of BrF_4^- are considered bases.

$$F^- + BrF_3 \longrightarrow BrF_4^-$$

Acid–base reactions in the solvent-system concept are the reverse of autodissociation.

$$H_3O^+ + OH^- \longrightarrow 2\ H_2O$$

$$BrF_2^+ + BrF_4^- \longrightarrow 2\ BrF_3$$

The Arrhenius, Brønsted–Lowry, and solvent-system neutralization reactions can be compared as follows:

Arrhenius:	acid + base	$\longrightarrow$	salt + water
Brønsted	acid 1 + base 2	$\longrightarrow$	base 1 + acid 2
Solvent system:	acid + base	$\longrightarrow$	solvent

▶ **Exercise 6.1** IF_5 undergoes autodissociation into $IF_4^+ + IF_6^-$. SbF_5 acts as an acid and KF acts as a base when dissolved in IF_5. Write balanced chemical equations for these reactions.

Table 6.2 gives some properties of common solvents. The pK_{ion} is the autodissociation constant for the pure solvent, indicating that, among these acids, sulfuric acid autodissociates by far to the greatest extent, and acetonitrile autodissociates the least. The boiling points are given to provide an estimate of the conditions under which each solvent might be used.

Caution is needed in interpreting these reactions. For example, $SOCl_2$ and SO_3^{2-} react as both acid and base in SO_2 solvent, with the reaction apparently occurring as

$$SOCl_2 + SO_3^{2-} \rightleftharpoons 2\ SO_2 + 2\ Cl^-$$

TABLE 6.2 Properties of Solvents

Protic Solvents				
Solvent	**Acid Cation**	**Base Anion**	**pK_{ion} (25° C)**	**Boiling Point (°C)**
Sulfuric acid, H_2SO_4	$H_3SO_4^+$	HSO_4^-	3.4 (10°)	330
Hydrogen fluoride, HF	H_2F^+	HF_2^-	~12 (0°)	19.5
Water, H_2O	H_3O^+	OH^-	14	100
Acetic acid, CH_3COOH	$CH_3COOH_2^+$	CH_3COO^-	14.45	118.2
Methanol, CH_3OH	$CH_3OH_2^+$	CH_3O^-	18.9	64.7
Ammonia, NH_3	NH_4^+	NH_2^-	27	−33.4
Acetonitrile, CH_3CN	CH_3CNH^+	CH_2CN^-	28.6	81

Aprotic Solvents	
Solvent	**Boiling Point (° C)**
Sulfur dioxide, SO_2	−10.2
Dinitrogen tetroxide, N_2O_4	21.2
Pyridine, C_5H_5N	115.5
Bromine trifluoride, BrF_3	127.6
Diglyme, $CH_3(OCH_2CH_2)_2OCH_3$	162

Source: Data from W. L. Jolly, *The Synthesis and Characterization of Inorganic Compounds*, Prentice Hall, Englewood Cliffs, NJ, 1970, pp. 99–101. Data for many other solvents are also given by Jolly.

It was at first believed that $SOCl_2$ dissociated and that the resulting SO^{2+} reacted with SO_3^{2-} as shown:

$$SOCl_2 \rightleftharpoons SO^{2+} + 2\,Cl^-$$

$$SO^{2+} + SO_3^{2-} \rightleftharpoons 2\,SO_2$$

However, the reverse reactions should lead to the exchange of oxygen atoms between SO_2 and $SOCl_2$, but no exchange is observed.[10] The details of the $SOCl_2 + SO_3^{2-}$ reaction are still uncertain but may involve dissociation of only one chloride, as in

$$SOCl_2 \rightleftharpoons SOCl^+ + Cl^-$$

▶ **Exercise 6.2** Show that the reverse of the following reactions should lead to oxygen atom exchange between SO_2 and $SOCl_2$, if one of them initially contains ^{18}O.

$$SOCl_2 \rightleftharpoons SO^{2+} + 2\,Cl^-$$

$$SO^{2+} + SO_3^{2-} \rightleftharpoons 2\,SO_2$$

[10] W. L. Jolly, *The Synthesis and Characterization of Inorganic Compounds*, Prentice Hall, Englewood Cliffs, NJ, 1970, pp. 108–109; R. E. Johnson, T. H. Norris, and J. L. Huston, *J. Am. Chem. Soc.*, **1951**, *73*, 3052.

6.2.4 Lewis Concept

Lewis[11] defined a base as an **electron-pair donor** and an acid as an **electron-pair acceptor.**[12] This definition further expands the list to include metal ions and other electron-pair acceptors as acids and provides a handy framework for nonaqueous reactions. Most of the acid–base descriptions in this book will use the Lewis definition, which encompasses the Brønsted–Lowry and solvent-system definitions. In addition to all the reactions discussed previously, the Lewis definition includes reactions such as

$$Ag^+ + 2\,{:}NH_3 \longrightarrow [H_3N{:}Ag{:}NH_3]^+$$

with the silver ion (or another cation) as an acid and ammonia (or another electron-pair donor) as a base. In reactions such as this one, the product is often called an **adduct**, a product of the reaction of a Lewis acid and base to form a new combination. Another example is the boron trifluoride–ammonia adduct, $BF_3 \cdot NH_3$. The BF_3 molecule described in Sections 3.1.4, 3.2.3, and 5.4.5 has a triangular structure. Because fluorine is the most electronegative element, the boron atom in BF_3 is quite positive, and the boron is frequently described as electron deficient. The lone pair in the HOMO of the ammonia molecule interacts with the empty LUMO of the BF_3—which has very large, empty orbital lobes on boron—to form the adduct. The molecular orbitals involved are depicted in Figure 6.1, and the energy levels of these orbitals are shown in Figure 6.2. The B—F bonds in the product are bent away from the ammonia into a nearly tetrahedral geometry around the boron. Similar interactions—in which electrons are donated or accepted completely (oxidation–reduction reactions) or shared, as in this reaction—are described in more detail in Sections 6.2.5 through 6.2.7.

Another common adduct, the boron trifluoride–diethyl ether adduct, $BF_3 \cdot O(C_2H_5)_2$, is frequently used in synthesis. Lone pairs on the oxygen of the diethyl ether are attracted to the boron; the result is that one of the lone pairs bonds to boron, changing the geometry around B from planar to nearly tetrahedral, as shown in Figure 6.3. As a result, BF_3, with a boiling point of –99.9° C, and diethyl ether, with a boiling point of 34.5° C, form an adduct with a boiling point of 125° to 126° C, at which temperature it decomposes into its two components. The formation of the adduct raises the boiling point enormously, a common result of such reactions.

Lewis acid–base adducts involving metal ions are called **coordination compounds**, and bonds formed with both electrons from one atom are called *coordinate bonds*; their chemistry will be discussed in Chapters 9 through 14.

FIGURE 6.1 Donor–Acceptor Bonding in $BF_3 \cdot NH_3$.

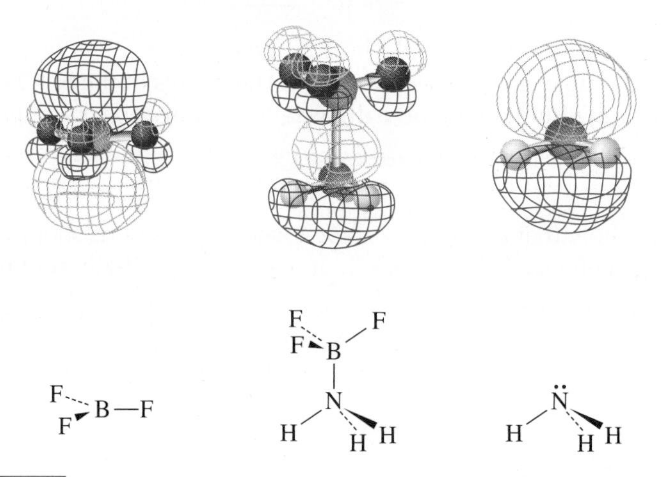

[11] G. N. Lewis, *Valence and the Structure of Atoms and Molecules*, Chemical Catalog, New York, 1923, pp. 141–142; *J. Franklin Inst.*, **1938**, *226*, 293.

[12] A Lewis base is also called a **nucleophile**, and a Lewis acid is also called an **electrophile**.

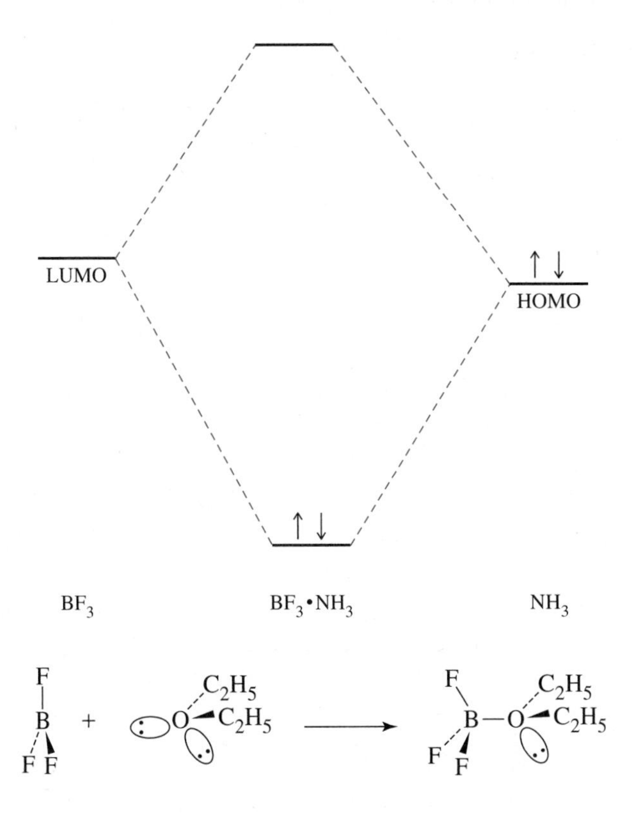

FIGURE 6.2 Energy Levels for the $BF_3 \cdot NH_3$ Adduct.

$BF_3 + O(C_2H_5)_2 \longrightarrow F_3B{-}O(C_2H_5)_2$

FIGURE 6.3 Boron Trifluoride–Ether Adduct.

6.2.5 Frontier Orbitals and Acid–Base Reactions[13]

The molecular orbital description of acid–base reactions mentioned in Section 6.2.4 uses **frontier molecular orbitals**, those at the occupied–unoccupied frontier, which can be illustrated by the simple reaction $NH_3 + H^+ \longrightarrow NH_4^+$. In this reaction, the a_1 orbital containing the lone-pair electrons of the ammonia molecule (see Figure 5.31) combines with the empty 1*s* orbital of the hydrogen ion to form bonding and antibonding orbitals. The lone pair in the a_1 orbital of NH_3 is stabilized by this interaction, as shown in Figure 6.4. The NH_4^+ ion has the same molecular orbital structure as methane, CH_4, with four bonding orbitals (a_1 and t_2) and four antibonding orbitals (also a_1 and t_2). Combining the seven NH_3 orbitals and the one H^+ orbital, accompanied by the change in symmetry from C_{3v} to T_d, gives the eight orbitals of NH_4^+. When the eight valence electrons are placed in these orbitals, one pair enters the bonding a_1 orbital, and three pairs enter bonding t_2 orbitals. The net result is a lowering of energy as the nonbonding a_1 becomes a bonding t_2, making the combined NH_4^+ more stable than the separated $NH_3 + H^+$. This is an example of the interaction of the HOMO of the base NH_3 and the LUMO of the acid H^+, accompanied by a change in symmetry to make a new set of orbitals, one bonding and one antibonding.

In most acid–base reactions, *a HOMO–LUMO combination forms new HOMO and LUMO orbitals of the product*. We can see that orbitals whose shapes allow significant overlap, and whose energies are similar, form useful bonding and antibonding orbitals. On the other hand, if the orbital combinations have no useful overlap, no net bonding is possible (as shown in Chapter 5), and they cannot form acid–base products.[14]

[13] W. B. Jensen, *The Lewis Acid–Base Concepts*, Wiley Interscience, New York, 1980, pp. 112–155.

[14] In a few cases, the orbitals with the required geometry and energy do not include the HOMO; this possibility should be kept in mind. When this happens, the HOMO is usually a lone pair that does not have the geometry needed for bonding with the acid.

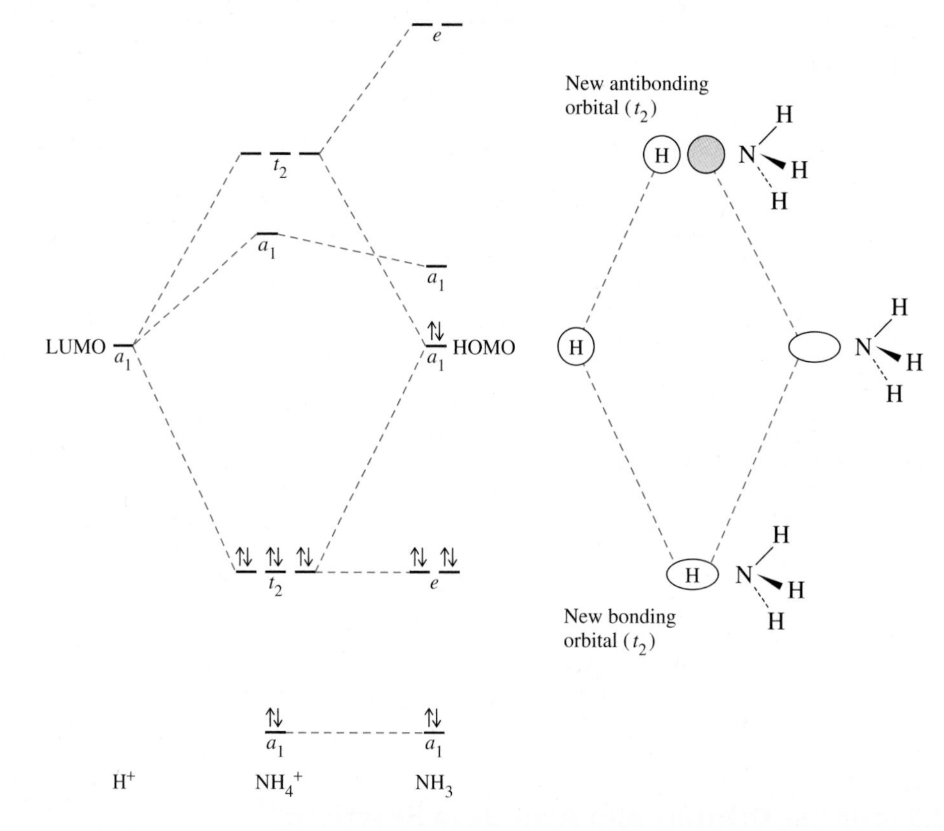

FIGURE 6.4 $NH_3 + H^+ \longrightarrow NH_4^+$ Molecular Energy Levels.

Even when the orbital shapes match, several reactions may be possible, depending on the relative energies. A single species can act as an oxidizing agent, an acid, a base, or a reducing agent, depending on the other reactant. These possibilities are shown in Figure 6.5. Although predictions using this approach may be difficult when the orbital energies are not known, they still provide useful context for these reactions.

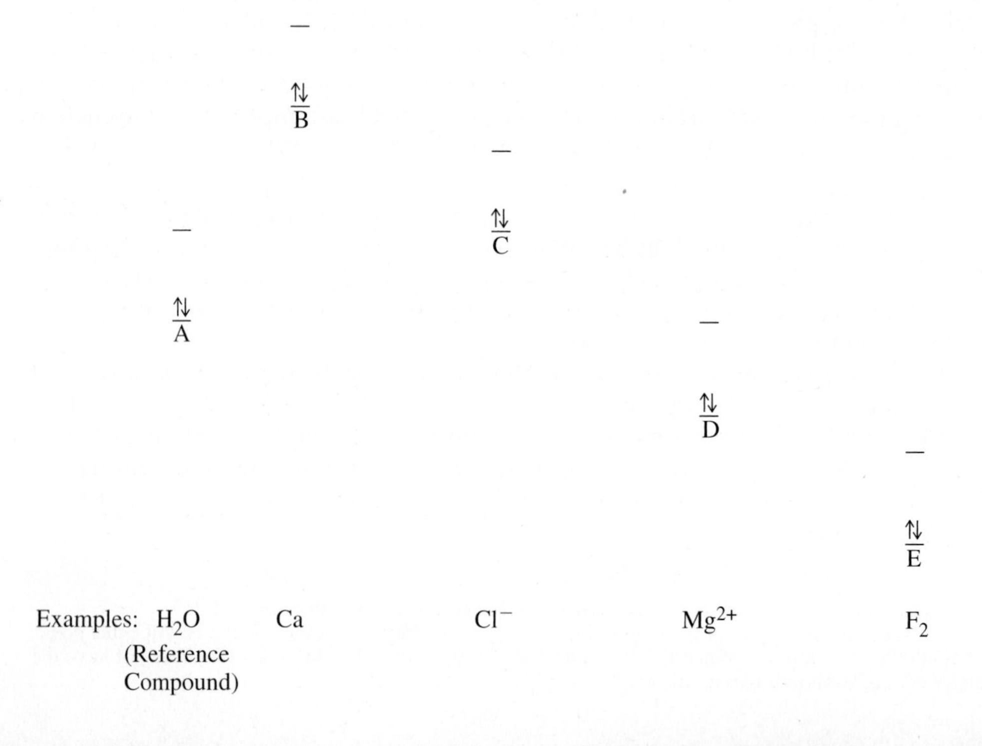

FIGURE 6.5 HOMO–LUMO Interactions.

(Adapted with permission from W. B. Jensen, *The Lewis Acid–Base Concepts*, Wiley Interscience, New York, 1980, p. 140, Figure 4-6.)

EXAMPLE

Consider water as an example of reactant A.

Reaction with B: Water as oxidizing agent. The first combination of reactants, A + B, has all the B orbitals at a much higher energy than those of water (calcium would be an example; the alkali metals react similarly but have only one electron in their highest *s* orbital). The energies are so different that no adduct can form, but a transfer of electrons can take place from B to A. From simple electron transfer, we might expect formation of H_2O^-, but reduction of water yields hydrogen gas instead. As a result, water is reduced to H_2 and OH^-, and Ca is oxidized to Ca^{2+}.

$$2\,H_2O + Ca \longrightarrow Ca^{2+} + 2\,OH^- + H_2 \text{ (water as oxidant)}$$

Reaction with C: Solvation of anion. If orbitals with matching shapes have similar energies, the resulting bonding orbitals will have lower energy than the reactant HOMOs, and a net decrease in energy (stabilization of electrons in the new HOMOs) results. An adduct is formed, with its stability dependent on the difference between the total energy of the product and the total energy of the reactants.

An example with water as acceptor (with lower energy orbitals) is the interaction with the chloride ion:

$$n\,H_2O + Cl^- \longrightarrow [Cl(H_2O)_n]^- \text{ (water as acid)}$$

The product is the solvated chloride ion. In this case, water is the acceptor, using as LUMO an antibonding orbital centered primarily on the hydrogen atoms (the chloride HOMO is a $3p$ orbital occupied by an electron pair).

Reaction with D: Solvation of cation. A reactant with orbitals lower in energy than those of water (for example, Mg^{2+}, D in Figure 6.5) allows water to act as a donor to solvate metal cations.

$$6\,H_2O + Mg^{2+} \longrightarrow [Mg(H_2O)_6]^{2+} \text{ (water as base)}$$

Here, water is the donor, contributing a lone pair primarily from the HOMO, which has a large contribution from the p_x orbital on the oxygen atom (the magnesium ion LUMO is the vacant $3s$ orbital). The molecular orbital levels that result from reactions with B or C are similar to those in Figures 6.7 and 6.8 for hydrogen bonding.

Reaction with E: Water as reducing agent. Finally, if the reactant has orbitals much lower than the water orbitals (F_2 for example), water can act as a reductant and transfer electrons to the other reactant. The product is not the simple result of electron transfer (H_2O^+) but the result of the breakup of the water molecule to molecular oxygen and hydrogen ions.

$$2\,H_2O + 2\,F_2 \longrightarrow 4\,F^- + 4\,H^+ + O_2 \text{ (water as reductant)}$$

Similar reactions can be described for other species, and the adducts formed in the acid–base reactions can be quite stable or very unstable, depending on the exact relationship between the orbital energies.

We are now in a position to reformulate the Lewis definition of acids and bases in terms of frontier orbitals: *A base has an electron pair in a HOMO of suitable symmetry to interact with the LUMO of the acid.* The better the energy match between the base's HOMO and the acid's LUMO, the stronger the interaction.

6.2.6 Hydrogen Bonding

The effects of hydrogen bonding were described in Section 3.4. In this section, the molecular orbital basis for hydrogen bonding is described as an introduction to the frontier molecular orbital approach to acid–base behavior.

The molecular orbitals for the symmetric FHF^- ion were described in Chapter 5 (Figure 5.18) as combinations of the atomic orbitals. They may also be generated by combining the molecular orbitals (MOs) of hydrogen fluoride (see example preceding Exercise 5.3) with F^-, as shown in Figure 6.6. The p_x and p_y lone-pair orbitals on the fluorines of both F^- and HF can be ignored, because there are no matching orbitals on the H atom. The shapes of the other orbitals are appropriate for bonding; overlap of the σ F^- orbital ($2p_z$) with the σ and σ^* HF orbitals forms the three product orbitals. These three orbitals are all symmetric about the central H nucleus. The lowest orbital is distinctly bonding, with all three component orbitals contributing and no nodes between the atoms. The middle (HOMO) orbital is essentially nonbonding, with nodes through each of the nuclei. The highest energy orbital (LUMO) is antibonding, with nodes between each pair of atoms. When three atomic orbitals are used—$2p$ orbitals from each F^- and the $1s$ orbital from H^+—the resulting pattern is one

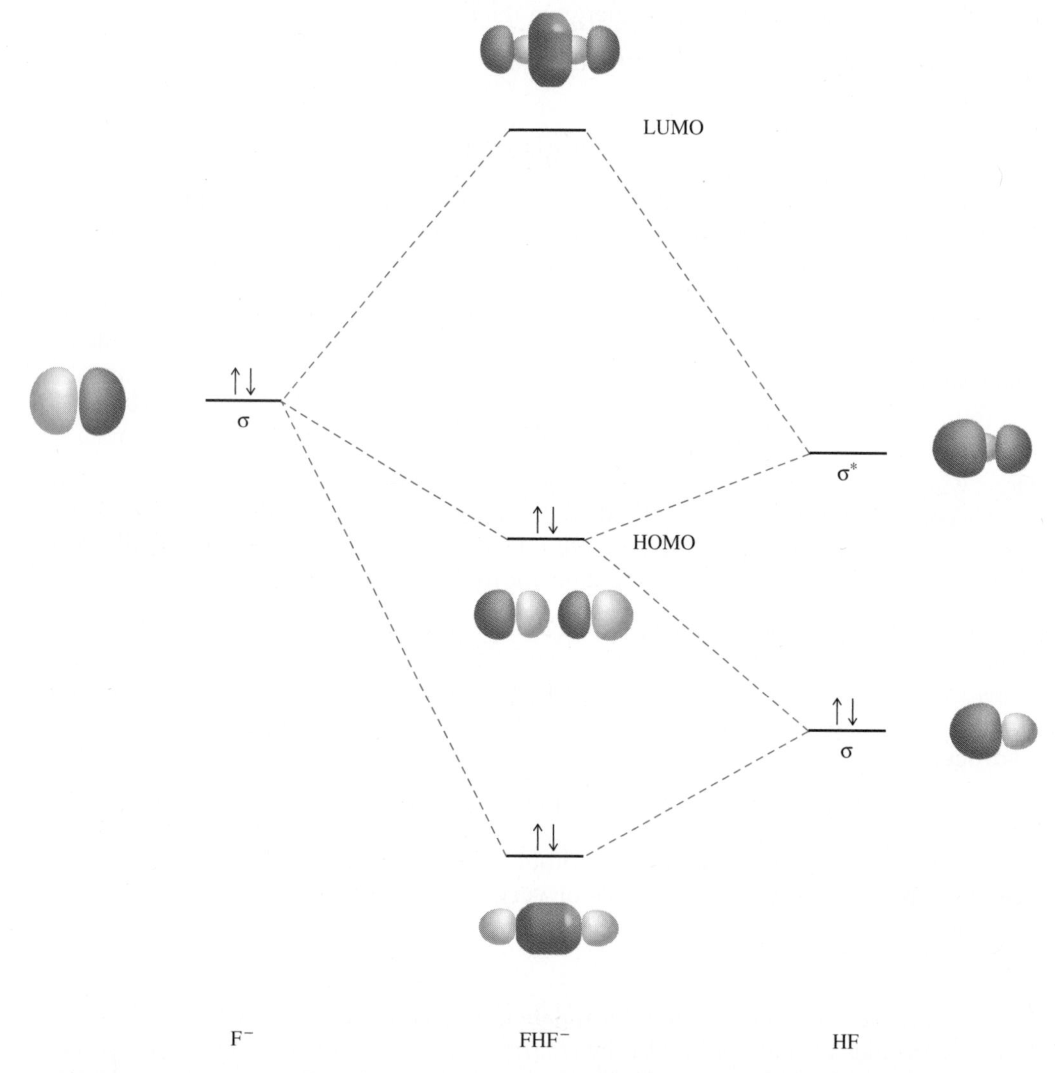

FIGURE 6.6 Molecular Orbitals for Hydrogen Bonding in FHF^-. The $1s$ and $2p$ nonbonding orbitals of F^- are omitted. Figure 5.18 shows the full set of molecular orbitals.

low-energy molecular orbital, one intermediate-energy orbital, and one high-energy orbital, with the number of nodes increasing with higher energy. The intermediate orbital may be slightly bonding, slightly antibonding, or nonbonding; we describe such orbitals as *essentially nonbonding*. (The two lowest energy orbitals of Figure 5.18 are essentially nonbonding; the fluorine 2*s* orbitals are the major contributors.)

For unsymmetrical hydrogen bonding, such as that of B + HA $\rightleftharpoons$ BHA shown in Figure 6.7, the pattern is similar. The two electron pairs in the lower orbitals have a lower total energy than the total for the electrons in the two reactants.

Regardless of the exact energies and location of the nodes, the general pattern is the same. The resulting FHF^- or BHA structure has a total energy lower than the sum of the energies of the reactants. For the general case of B + HA, three possibilities exist, as illustrated in Figure 6.8: First, for a poor match of energies when the occupied reactant orbitals are lower in total energy than those of the possible hydrogen-bonded product, no new product will be formed; there will be no hydrogen bonding (a). Second, for a good match of energies, the occupied product orbitals are lower in energy, and a hydrogen-bonded product forms (b); the greater the lowering of energies of these orbitals, the stronger the hydrogen bonding. Finally, for a very poor match of energies (c), occupied orbitals of the species BH + A are lower than those of B + HA; in this case, complete hydrogen ion transfer occurs.

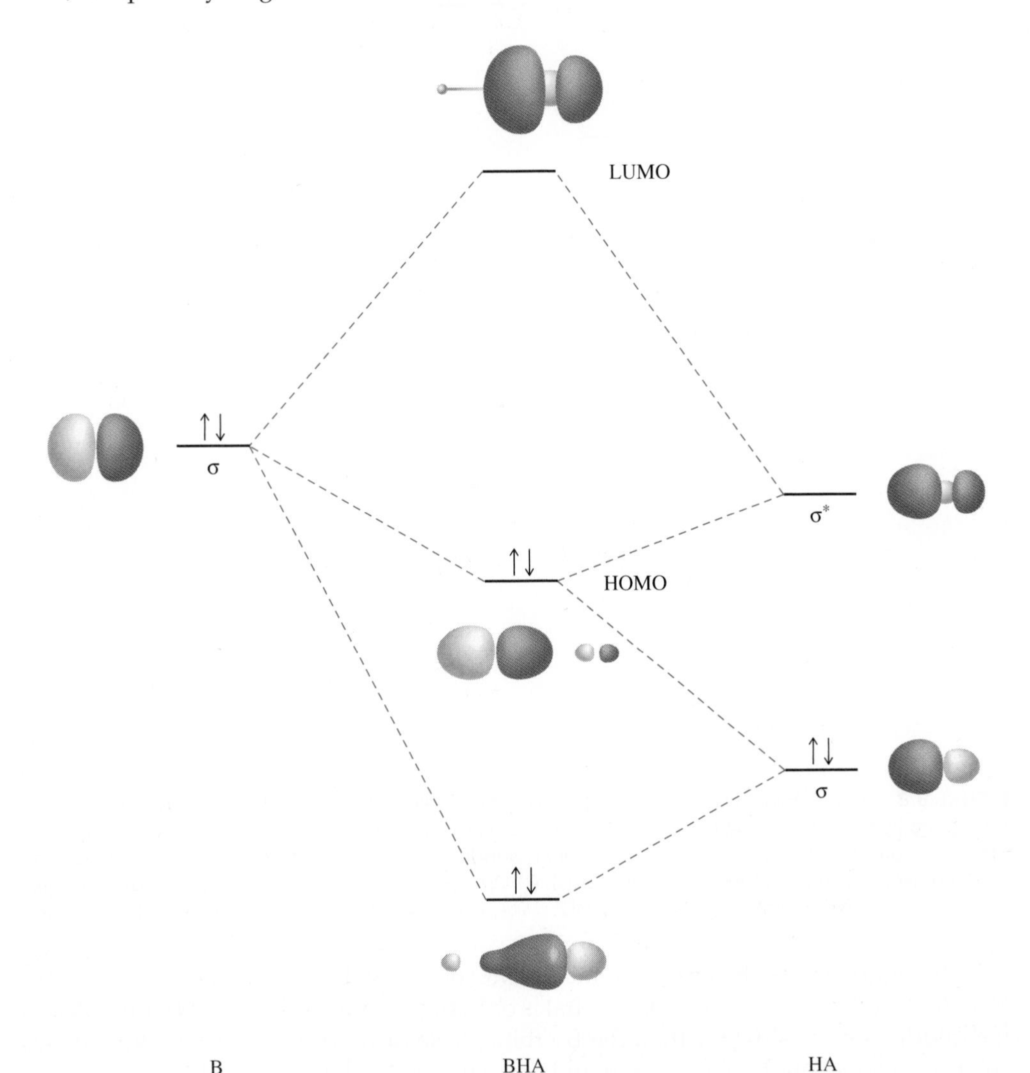

FIGURE 6.7 Molecular Orbitals for Asymmetric Hydrogen Bonding.

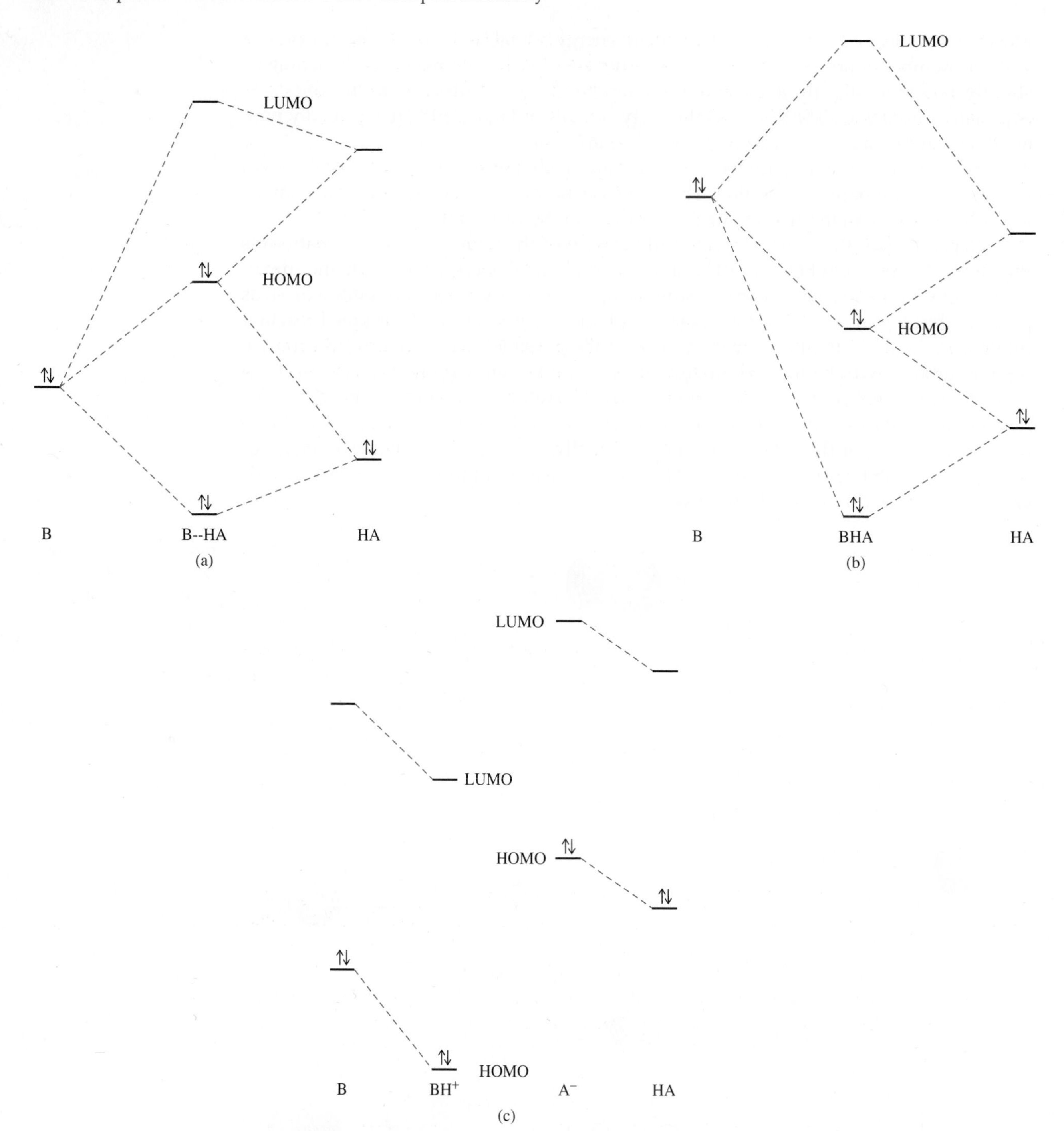

FIGURE 6.8 Orbital Possibilities for Hydrogen Bonding. (a) Poor match of HOMO–LUMO energies, little or no hydrogen bonding (HOMO of B well below LUMO of HA; reactants' energy below that of BHA). (b) Good match of energies, good hydrogen bonding (HOMO of B at nearly the same energy as LUMO of HA; BHA energy lower than reactants). (c) Very poor match of energies, transfer of hydrogen ion (HOMO of B below both LUMO and HOMO of HA; $BH^+ + A^-$ energy lower than B + HA or BHA).

In Figure 6.8 (a), the energy of the HOMO of B is well below that of the LUMO of HA. Because the lowest molecular orbital is only slightly lower than the HA orbital, and the middle orbital is higher than the B orbital, little or no reaction occurs. In aqueous solution, interactions between water and molecules with almost no acid–base character, such as CH_4, fit this group. Little or no interaction occurs between the hydrogens of the methane molecule and the lone pairs of surrounding water molecules.

In Figure 6.8 (b), the LUMO of HA and the HOMO of B have similar energies, both occupied product orbitals are lower than the respective reactant orbitals, and a hydrogen-bonded product forms with a lower total energy than the reactants. The node of the product HOMO is near the H atom, and the hydrogen-bonded product has a B—H bond similar in strength to the H—A bond. If the B HOMO is slightly higher than the HA LUMO, as in the figure, the H—A portion of the hydrogen bond is stronger. If the B HOMO is lower than the HA LUMO, the B—H portion is stronger (the product HOMO consists of more B than A orbital). Weak acids such as acetic acid, CH_3COOH, are examples of hydrogen-bonding solutes in water. Acetic acid hydrogen bonds strongly with water, and to some extent with other acetic acid molecules, with a small amount of hydrogen ion transfer to water to give hydronium and acetate ions.

In Figure 6.8 (c), the HOMO–LUMO energy match is so poor that no useful adduct orbitals can be formed. The product MOs here are those of A^- and BH^+, and the hydrogen ion is transferred from A to B. Strong acids such as HCl will donate their hydrogen ions completely to water, after which the H_3O^+ formed will hydrogen-bond strongly with other water molecules.

In all these diagrams, either HA or BH or both may have a positive charge, and either A or B or both may have a negative charge, depending on the circumstances.

When A is a highly electronegative element—such as F, O, or N—the highest occupied orbital of A has lower energy than the hydrogen 1*s* orbital, and the H—A bond is relatively weak, with most of the electron density near A and with H somewhat positively charged. This favors the hydrogen-bonding interaction by lowering the overall energy of the HA bonding orbital and improving overlap with the B orbital. In other words, when the reactant HA has a structure close to $H^+\cdots A^-$, hydrogen bonding is more likely. This explains the strong hydrogen bonding in cases with hydrogen bridging between F, O, and N atoms in molecules and the typically much weaker or nonexistent hydrogen bonding between H and other atoms. The description above can be viewed as a three-center, four-electron model,[15] with a bond angle at the hydrogen within 10° to 15° of a linear 180° angle.

6.2.7 Electronic Spectra

One reaction that shows the effect of adduct formation dramatically is the reaction of I_2 as an acid with different solvents and ions that act as bases. The changes in spectra and visible color caused by changes in electronic energy levels, shown in Figures 6.9 and 6.10, are striking. The upper molecular orbitals of I_2 are shown on the left in Figure 6.9, with a net single bond due to the filled $9\sigma_g$ orbital and lone pairs in the $4\pi_u$ and $4\pi_g^*$ orbitals. In the gas phase, I_2 is violet, absorbing light near 500 nm because of promotion of an electron from the $4\pi_g^*$ level to the $9\sigma_u^*$ level. This absorption removes the middle yellow, green, and blue parts of the visible spectrum, leaving red and violet at opposite ends of the spectrum to combine in the violet color that is seen.

In nondonor solvents such as hexane, the iodine color remains essentially the same violet; in benzene and other π-electron solvents, it becomes more reddish; and in good donors—such as ethers, alcohols, and amines—the color becomes distinctly brown. The solubility of I_2 also increases with increasing donor character of the solvent. Interaction of the donor orbital of the solvent with the $9\sigma_u^*$ orbital results in a lower occupied bonding orbital and a higher unoccupied antibonding orbital. As a result, the $\pi_g^* \longrightarrow \sigma_u^*$ transition for the I_2 + donor (Lewis base) has a higher energy, and an absorbance peak shifted toward the blue. The transmitted color shifts toward brown

---

[15] R. L. DeKock and W. B. Bosma, *J. Chem. Educ.*, **1988**, *65*, 194.

FIGURE 6.9 Electronic Transitions in I_2 Adducts.

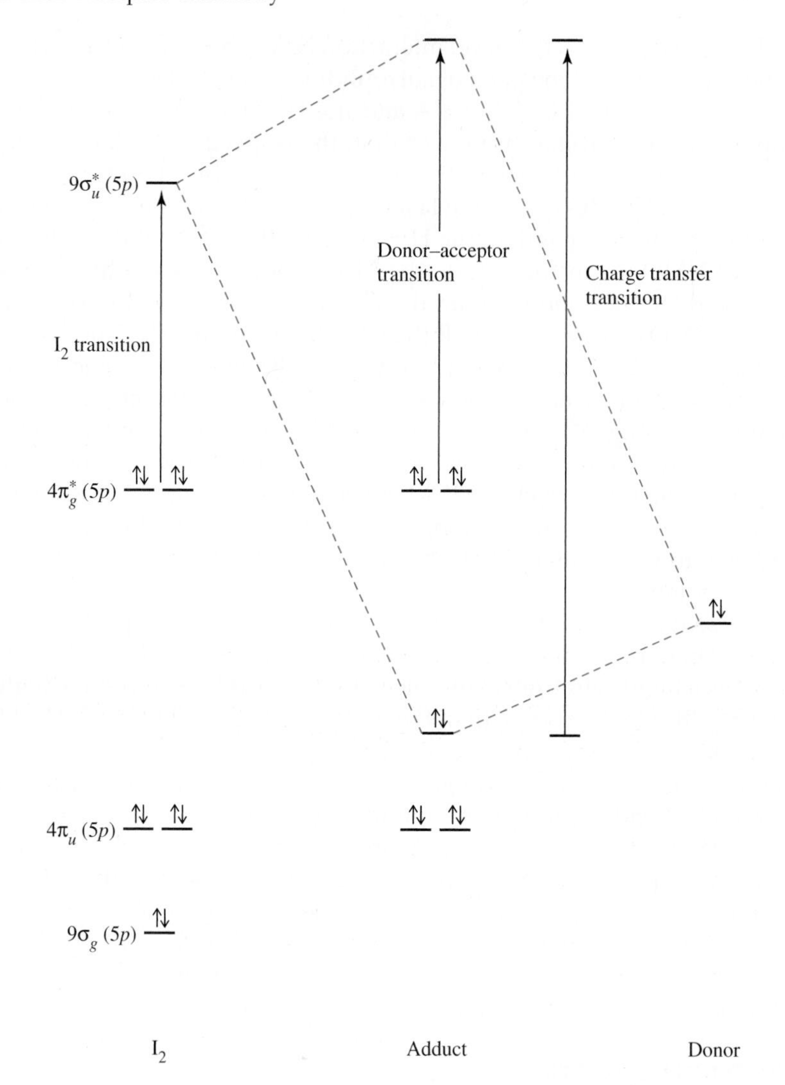

(combined red, yellow, and green), as more of the yellow and green light passes through. Water is also a donor, but not a very good one; I_2 is only slightly soluble in water, and the solution is yellow brown. I^-, a very good donor, reacts with I_2 to form I_3^-, which is brown and, being ionic, is very soluble in water. When the interaction between the donor and I_2 is strong, the LUMO of the adduct has a higher energy, and the energy of the donor–acceptor transition ($\pi_g{}^* \longrightarrow \sigma_u{}^*$) increases.

In addition to these shifts, a new band appears at the edge of the ultraviolet (230 to 400 nm, marked *CT* in Figure 6.10). This band is due to the transition $\sigma \longrightarrow \sigma^*$ between the two new orbitals formed by the interaction. Because the σ orbital has a larger proportion of the donor (solvent or I^-) orbital, and the σ^* orbital has a larger proportion of the I_2 orbital, the transition transfers an electron from an orbital that is primarily of donor composition to one that is primarily of acceptor composition; hence, the name **charge transfer (CT)** for this transition. The energy of this transition is less predictable, because it depends on the energy of the donor orbital. The transition may be shown schematically as

$$I_2 \cdot \text{Donor} \xrightarrow[\text{CT}]{h\nu} [I_2]^- \cdot [\text{Donor}]^+$$

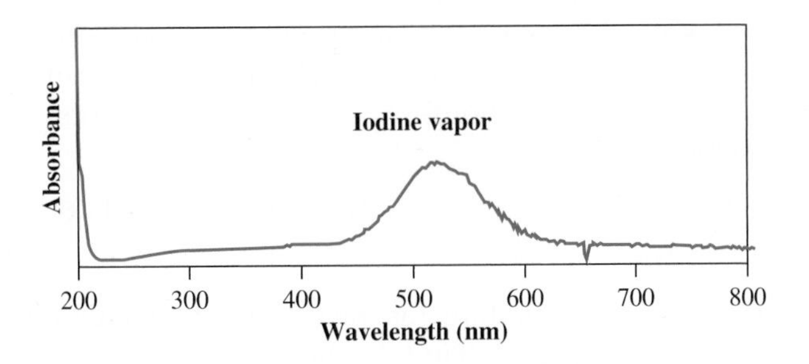

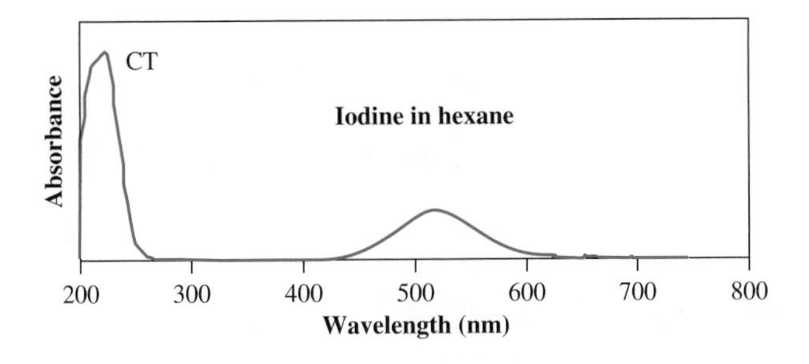

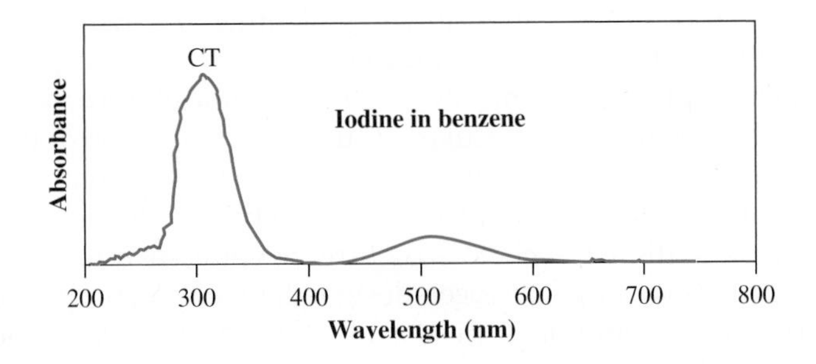

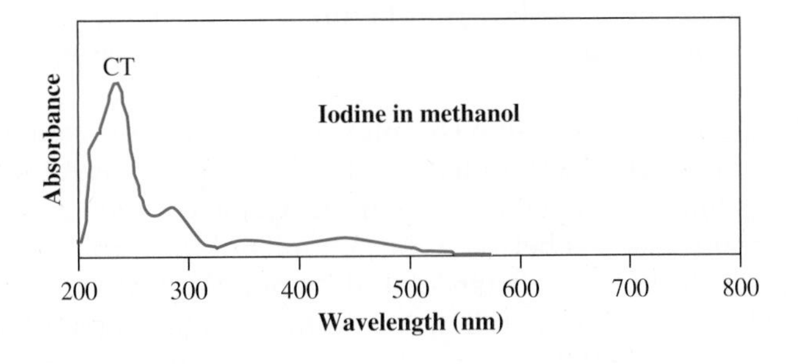

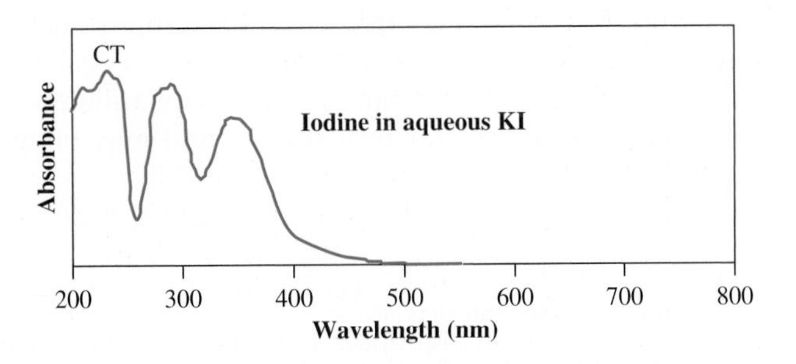

FIGURE 6.10
Spectra of I_2 with Different Bases.
I_2 vapor is purple or violet, absorbing near 520 nm, with no charge-transfer bands.
I_2 in hexane is purple or violet, absorbing near 520 nm, with a charge-transfer band at about 225 nm.
I_2 in benzene is red violet, absorbing near 500 nm, with a charge-transfer band at about 300 nm.
I_2 in methanol is yellow brown, absorbing near 450 nm, with a charge-transfer band near 240 nm and a shoulder at 290 nm.
I_2 in aqueous KI is brown, absorbing near 360 nm, with charge-transfer bands at higher energy.

The charge-transfer phenomenon also appears in many other adducts. If the charge-transfer transition actually transfers the electron permanently, the result is an oxidation–reduction reaction: the donor is oxidized, and the acceptor is reduced. The sequence of reactions

$$\underset{\text{Acid}}{[Fe(H_2O)_6]^{3+}} + \underset{\text{Base}}{X^-} \longrightarrow [Fe(H_2O)_5X]^{2+} + H_2O$$

illustrates the entire range of possibilities, as the energy of the HOMO of the halide ion increases from F^- to I^-. In concentrated iodide, there is complete transfer of the electron through the reaction

$$2\,[Fe(H_2O)_6]^{3+} + 2\,I^- \longrightarrow 2\,[Fe(H_2O)_6]^{2+} + I_2$$

Charge-transfer interactions of transition-metal complexes will be discussed in more detail in Chapter 11.

6.2.8 Receptor–Guest Interactions[16]

Another important type of interaction can occur between molecules having extended pi systems, when their pi systems interact with each other to hold molecules or portions of molecules together. Such interactions may be important on a large scale—for example, as a component of protein folding and other biochemical processes—and on a truly small scale, as in the function of molecular electronic devices. One of the most interesting recent areas of study in the realm of π–π interactions has been the quest to design concave receptors that can wrap around and attach to fullerenes such as C_{60} in what has been described as a "ball-and-socket" structure. Several such receptors, sometimes called *molecular tweezers* or *clips*, have been designed, involving porphyrin rings, corannulene and its derivatives, and other pi systems.[17] The first crystal structure of such a receptor–guest complex involving C_{60}, dubbed by the authors a "double-concave hydrocarbon buckycatcher," is shown in Figure 6.11; it was synthesized by mixing approximately equimolar quantities of corannulene derivative $C_{60}H_{24}$ and C_{60} in toluene solution.[18]

The product, called an **inclusion complex**, has two concave corannulene "hands" wrapped around the buckminsterfullerene; the distance between the carbons in the corannulene rings and the matching carbons on C_{60} is consistent with the distance expected for π–π interactions between the subunits of the structure (approximately 350 pm), with a shortest $C \cdots C$ distance of 312.8 pm. With no direct C—C covalent bonding, the binding between the fullerene and the corannulene units is attributed to pure π–π interactions. The electronic structure of the complex has been studied in connection with the potential use of such complexes as building units in molecular electronics.[19] Similar receptor–guest complexes involving other molecular tweezers and C_{60} have shown evidence of electron transfer from receptor to the fullerene on absorption of light (another example of charge-transfer transitions), and may be useful in the construction of photovoltaic devices.[20]

---

[16] Also called *receptor–substrate* or *host–guest* interactions.
[17] S. S. Gayathri, M. Wielopolski, E. M. Pérez, G. Fernández, L. Sánchez, R. Viruela, E. Ortí, D. M. Guldi, and N. Martín, *Angew. Chem. Int. Ed.*, **2009**, *48*, 815, and references therein.
[18] A. Sygula, F. R. Fronczek, R. Sygula, P. W. Rabideau, and M. M. Olmstead, *J. Am. Chem. Soc.*, **2007**, *129*, 3842.
[19] A. A. Voityuk and M. Duran, *J. Phys. Chem. C*, **2008**, *112*, 1672.
[20] A. Molina-Ontoria, G. Fernández, M. Wielopolski, C. Atienza, L. Sánchez, A. Gouloumis, T. Clark, N. Martín, and D. M. Guldi, *J. Am. Chem. Soc.*, **2009**, *131*, 12218.

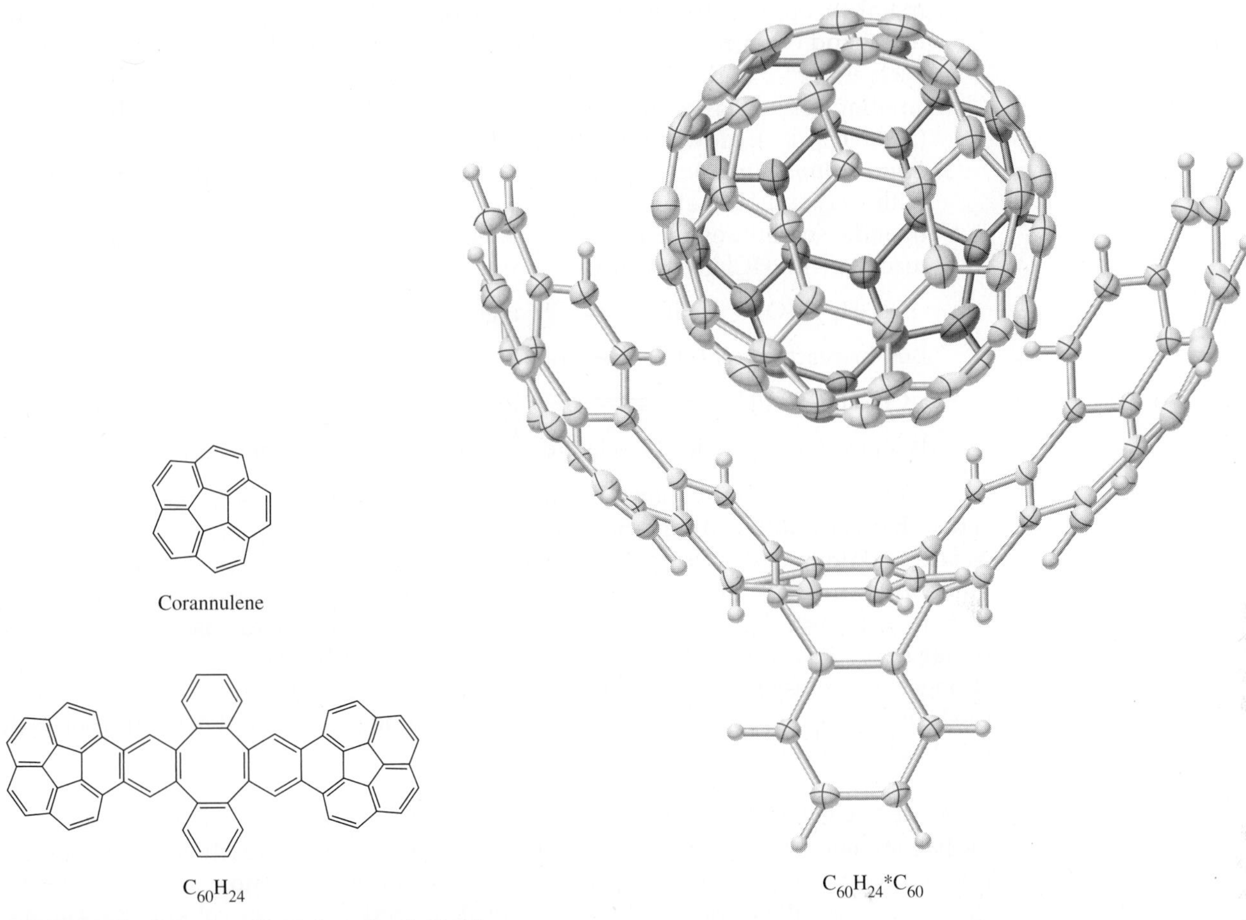

FIGURE 6.11 Receptor–Guest Complex $C_{60}H_{24}$*C_{60}.

Reproduced with permission from A. Sygula, F. R. Fronczek, R. Sygula, P. W. Rabideau, and M. M. Olmstead, *J. Am. Chem. Soc.*, **2007**, *129*, 3842. ©American Chemical Society.

6.3 HARD AND SOFT ACIDS AND BASES

To introduce this section, let us consider several experimental observations.

1. ***Relative solubilities of halides.*** The solubilities of silver halides in water decrease, going down the column of halogens in the periodic table:

$$AgF(s) + H_2O \longrightarrow Ag^+(aq) + F^-(aq) \qquad K_{sp} = 205$$

$$AgCl(s) + H_2O \longrightarrow Ag^+(aq) + Cl^-(aq) \qquad K_{sp} = 1.8 \times 10^{-10}$$

$$AgBr(s) + H_2O \longrightarrow Ag^+(aq) + Br^-(aq) \qquad K_{sp} = 5.2 \times 10^{-13}$$

$$AgI(s) + H_2O \longrightarrow Ag^+(aq) + I^-(aq) \qquad K_{sp} = 8.3 \times 10^{-17}$$

   Mercury(I) halides have a similar trend, with Hg_2F_2 the most soluble and Hg_2I_2 the least soluble. However, LiF is by far the *least* soluble of the lithium halides; its K_{sp} is only 1.8×10^{-3}, but the other lithium halides are highly soluble in water. Similarly, MgF_2 and AlF_3 are less soluble than the corresponding chlorides, bromides, and iodides. How can one account for these divergent trends?

2. ***Coordination of thiocyanate to metals.*** As we will consider in Chapter 9 and later chapters, numerous ions and other groups can act as **ligands**, forming bonds to metal ions. The thiocyanate ion, SCN^-, has the capacity to bond to a

metal through either its sulfur or nitrogen, depending on the nature of the metal and other factors. When it bonds to a large, highly polarizable metal ion such as Hg^{2+}, it attaches through sulfur ($[Hg(SCN)_4]^{2-}$); but when it bonds to smaller, less polarizable metals such as Zn^{2+}, it attaches through nitrogen ($[Zn(NCS)_4]^{2-}$). How can this be explained?

3. ***Equilibrium constants of exchange reactions.*** When the ion $[CH_3Hg(H_2O)]^+$—with CH_3 and H_2O attached as ligands to Hg^{2+}—is reacted with other potential ligands, sometimes the reaction is favorable, sometimes not. For example, the reaction with HCl goes nearly to completion.

$$[CH_3Hg(H_2O)]^+ + HCl \rightleftharpoons CH_3HgCl + H_3O^+ \qquad K = 1.8 \times 10^{12}$$

But the reaction with HF does not.

$$[CH_3Hg(H_2O)]^+ + HF \rightleftharpoons CH_3HgF + H_3O^+ \qquad K = 4.5 \times 10^{-2}$$

Is it possible to predict the relative magnitudes of such equilibrium constants?

To account for observations such as these, in 1963, R. G. Pearson presented the concept of **hard and soft acids and bases (HSABs)**, designating polarizable acids and bases as **soft** and nonpolarizable acids and bases as **hard**.[21] This concept has been a useful guide in a variety of aspects of acid–base chemistry and other chemical phenomena.[22] Stated most simply by Pearson, "Hard acids prefer to bind to hard bases, and soft acids prefer to bind to soft bases." Interactions between two hard or two soft species are stronger than those between one hard and one soft species. The three examples above can be interpreted in such terms, with reactions tending to favor hard–hard and soft–soft combinations. Let us examine each of these in turn.

RELATIVE SOLUBILITIES In the examples presented, the metal cation is the Lewis acid and the halide anion is the Lewis base. In the series of silver ion–halide reactions, the iodide ion is much softer (more polarizable) than the others and interacts more strongly with the silver ion, a soft cation. The result is a more covalent bond in AgI. The colors of the salts are also worth noting. AgI is yellow, AgBr is slightly yellow, and AgCl and AgF are white. Color depends on the difference in energy between occupied and unoccupied orbitals. A large difference results in absorption in the ultraviolet region of the spectrum; a smaller difference in energy levels moves the absorption into the visible region. Compounds absorbing violet are perceived as yellow, the complementary color to violet; as the absorption band moves toward lower energy, the color shifts and becomes more intense.

The lithium halides have solubilities roughly in the reverse order: LiBr > LiCl > LiI > LiF. The strong hard–hard interaction in LiF overcomes the tendency of LiF to be solvated by water. The weaker hard–soft interactions between Li^+ and the other halides are not strong enough to prevent solvation, and these halides are more soluble than LiF. LiI is out of order, probably because of the poor solvation of the very large iodide ion, but it is still much more soluble than LiF.

Ahrland, Chatt, and Davies[23] classified some of the same phenomena, as well as others by dividing metal ions into two classes:

Class (a) ions	Class (b) ions
Most metals	Cu^{2+}, Pd^{2+}, Ag^+, Pt^{2+}, Au^+, Hg_2^{2+}, Hg^{2+}, Tl^+, Tl^{3+}, Pb^{2+}, and heavier transition-metal ions

[21] R. G. Pearson, *J. Am. Chem. Soc.*, **1963**, *85*, 3533.
[22] For early discussions of the principles and theories of the HSAB concept, see R. G. Pearson, *J. Chem. Educ.*, **1968**, *45*, 581 and 643.
[23] S. Ahrland, J. Chatt, and N. R. Davies, *Q. Rev. Chem. Soc.*, **1958**, *12*, 265.

The members of class (b) are located primarily in a small region in the periodic table at the lower right-hand side of the transition metals. The periodic table in Figure 6.12 identifies the elements that are always in class (b) and those that are commonly in class (b) when they have low or zero oxidation states. In addition, the transition metals have class (b) character in compounds in which their oxidation state is zero (primarily organometallic compounds). The class (b) ions form halides whose solubility is generally in the order $F^- > Cl^- > Br^- > I^-$; the solubility of class (a) halides in water is typically in the reverse order. The class (b) metal ions also have a larger enthalpy of reaction with phosphorus donors than with nitrogen donors, again the opposite of the reactions of class (a) metal ions.

Ahrland, Chatt, and Davies explained the class (b) metals as having *d* electrons available for π bonding.[24] Elements farther left in the table have more class (b) character in low or zero oxidation states when more *d* electrons are present. Donor molecules or ions that have the most favorable enthalpies of reaction with class (b) metals are those that are readily polarizable and may have vacant *d* or π^* orbitals available for π bonding.

COORDINATION OF THIOCYANATE TO METALS We can now account for the two bonding modes of the SCN^- ion on the basis of the HSAB approach. The Hg^{2+} ion is much larger and more polarizable (softer) than the smaller, harder Zn^{2+}. The softer end of the thiocyanate ion is the sulfur. Consequently, the mercury ion $[Hg(SCN)_4]^{2-}$ is a soft–soft combination, and the zinc ion $[Zn(NCS)_4]^{2-}$ is a hard–hard combination, consistent with the HSAB prediction. In addition, other soft cations such as Pd^{2+} and Pt^{2+} form thiocyanate complexes attached through the softer sulfur; harder cations such as Ni^{2+} and Cu^{2+} form N-bonded thiocyanates. Intermediate transition-metal ions can in some cases bond to either end of thiocyanate. For example, both $[Co(NH_3)_5(SCN)]^{2+}$ and $[Co(NH_3)_5(NCS)]^{2+}$ are known, forming a classic example of "linkage" isomers (these and other types of isomers are discussed in Chapter 9).

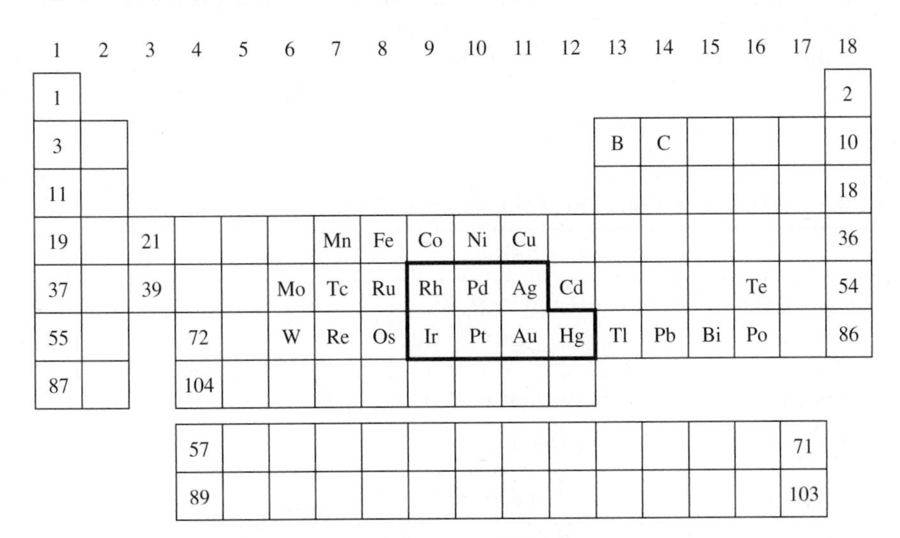

FIGURE 6.12 Location of Class (b) Metals in the Periodic Table. Those in the outlined region are always class (b) acceptors. Others indicated by their symbols are borderline elements, whose behavior depends on their oxidation state and the donor. The remainder (blank) are class (a) acceptors.

(Adapted with permission from S. Ahrland, J. Chatt, and N. R. Davies, *Q. Rev. Chem. Soc.*, **1958**, *12*, 265.)

[24] A discussion of metal–ligand bonding is included in Chapters 10 and 13.

EQUILIBRIUM CONSTANTS OF EXCHANGE REACTIONS Consider the data in Table 6.3 for reactions of aqueous methylmercury(I) cations of the form

$$[CH_3Hg(H_2O)]^+ + BH^+ \rightleftharpoons [CH_3HgB]^+ + H_3O^+.$$

Such reactions may be considered *exchange reactions* in that they involve an exchange of water and base B on the mercury. They may also be considered examples of competition between H_2O and B for a position of attachment on mercury—and also between H_2O and B for attachment to H^+.

In reactions 1 through 4 the trend is clear: as the halide becomes larger and more polarizable ($F^- \rightarrow I^-$), the tendency for attachment to mercury(II) grows stronger—a soft–soft interaction between mercury(II) and the increasingly soft halide ion. Among the other examples, the relatively soft sulfur atoms in reactions 6 and 7 also can be viewed as leading to a soft–soft interaction with Hg^{2+} and a large equilibrium constant. In reactions 1 and 5, on the other hand, the harder F and O atoms from HF and H_2O are less able to compete for attachment to the soft mercury(II), and the equilibrium constants are small.

6.3.1 Theory of Hard and Soft Acids and Bases

Pearson[26] designated the class (a) metal ions of Ahrland, Chatt, and Davies as *hard acids* and the class (b) ions as *soft acids*. Bases are also classified as hard or soft in terms of polarizability: the halide ions range from F^-, a very hard base, through less hard Cl^- and Br^- to I^-, a soft base. Reactions are more favorable for hard–hard and soft–soft interactions than for a mix of hard and soft in the reactants.

Much of the hard–soft distinction depends on polarizability, the degree to which a molecule or ion is easily distorted by interaction with other molecules or ions. Electrons in polarizable molecules can be attracted or repelled by charges on other molecules, forming slightly polar species that can then interact with the other molecules. Hard acids and bases are relatively small, compact, and nonpolarizable; soft acids and bases are larger and more polarizable. The hard acids include cations with a large positive charge (3+ or larger) or those whose *d* electrons are relatively unavailable for π bonding (e.g., alkaline earth ions, Al^{3+}). Other hard acid cations that do not fit this description are Cr^{3+}, Mn^{2+}, Fe^{3+}, and Co^{3+}. Soft acids are those whose *d* electrons or orbitals are readily available for

TABLE 6.3 Equilibrium Constants for Reactions of Mercury Complexes[25]

Reaction	*K*
1. $[CH_3Hg(H_2O)]^+ + HF \rightleftharpoons CH_3HgF + H_3O^+$	4.5×10^{-2}
2. $[CH_3Hg(H_2O)]^+ + HCl \rightleftharpoons CH_3HgCl + H_3O^+$	1.8×10^{12}
3. $[CH_3Hg(H_2O)]^+ + HBr \rightleftharpoons CH_3HgBr + H_3O^+$	4.2×10^{15}
4. $[CH_3Hg(H_2O)]^+ + HI \rightleftharpoons CH_3HgI + H_3O^+$	1×10^{18}
5. $[CH_3Hg(H_2O)]^+ + H_2O \rightleftharpoons CH_3HgOH + H_3O^+$	5×10^{-7}
6. $[CH_3Hg(H_2O)]^+ + SH^- \rightleftharpoons [CH_3HgS]^- + H_3O^+$	1×10^{7}
7. $[CH_3Hg(H_2O)]^+ + HSCN \rightleftharpoons CH_3HgSCN + H_3O^+$	5×10^{6}

[25] G. Schwarzenbach and M. Schellenberg, *Helv. Chim. Acta*, **1965**, *48*, 28.

[26] R. G. Pearson, *J. Am. Chem. Soc.*, **1963**, *85*, 3533; *Chem. Br.*, **1967**, *3*, 103; R. G. Pearson, ed., *Hard and Soft Acids and Bases*, Dowden, Hutchinson & Ross, Stroudsburg, PA, 1973. The terms *hard* and *soft* are attributed to D. H. Busch in the first paper of this footnote.

π bonding (neutral and 1+ cations, heavier 2+ cations). In addition, the larger and more massive the atom, the softer it is likely to be, because the large numbers of inner electrons shield the outer ones and make the atom more polarizable. This description fits the class (b) ions well, because they are primarily 1+ and 2+ ions with filled or nearly filled *d* orbitals, and most are in the second and third rows of the transition elements, with 45 or more electrons. Tables 6.4 and 6.5 list bases and acids in terms of their hardness or softness.

The trends in bases are easier to see: fluoride is hard and iodide is soft. Again, more electrons and larger sizes lead to softer behavior. S^{2-} is softer than O^{2-} because it has more electrons spread over a slightly larger volume, making S^{2-} more polarizable. Within a group, such comparisons are easy; as the electronic structure and size change, comparisons become more difficult but are still possible. Thus, S^{2-} is softer than Cl^-, which has the same electronic structure, because S^{2-} has a smaller nuclear charge and a slightly larger size. As a result, the negative charge is more available for polarization. Soft acids tend to react with soft bases, and hard acids with hard bases, so the reactions produce hard–hard and soft–soft combinations. Quantitative measures of hard–soft parameters are described in Section 6.3.2.

TABLE 6.4 Hard and Soft Bases

Hard Bases	Borderline Bases	Soft Bases
		H^-
F^-, Cl^-	Br^-	I^-
H_2O, OH^-, O^{2-}		H_2S, SH^-, S^{2-}
ROH, RO^-, R_2O, CH_3COO^-		RSH, RS^-, R_2S
NO_3^-, ClO_4^-	NO_2^-, N_3^-	SCN^-, CN^-, RNC, CO
CO_3^{2-}, SO_4^{2-}, PO_4^{3-}	SO_3^{2-}	$S_2O_3^{2-}$
NH_3, RNH_2, N_2H_4	$C_6H_5NH_2$, C_5H_5N, N_2	PR_3, $P(OR)_3$, AsR_3, C_2H_4, C_6H_6

Source: Adapted from R. G. Pearson, *J. Chem. Educ.*, **1968**, *45*, 581.

TABLE 6.5 Hard and Soft Acids

Hard Acids	Borderline Acids	Soft Acids
H^+, Li^+, Na^+, K^+		
Be^{2+}, Mg^{2+}, Ca^{2+}, Sr^2		
BF_3, BCl_3, $B(OR)_3$	$B(CH_3)_3$	BH_3, Tl^+, $Tl(CH_3)_3$
Al^{3+}, $Al(CH_3)_3$, $AlCl_3$, AlH_3		
Cr^{3+}, Mn^{2+}, Fe^{3+}, Co^{3+}	Fe^{2+}, Co^{2+}, Ni^{2+}, Cu^{2+}, Zn^{2+} Rh^{3+}, Ir^{3+}, Ru^{3+}, Os^{2+}	Cu^+, Ag^+, Au^+, Cd^{2+}, Hg_2^{2+}, Hg^{2+}, CH_3Hg^+, $[Co(CN)_5]^{3-}$, Pd^{2+}, Pt^{2+}, Pt^{4+},
Ions with formal oxidation states of 4 or higher		Br_2, I_2
HX (hydrogen-bonding molecules)		Metals with zero oxidation state
		π acceptors: e.g., trinitrobenzene, quinones, tetracyanoethylene

Source: Adapted from R. G. Pearson, *J. Chem. Educ.*, **1968**, *45*, 581.

EXAMPLE

Is OH^- or S^{2-} more likely to form insoluble salts with 3+ transition-metal ions? Which is more likely to form insoluble salts with 2+ transition-metal ions?

Because OH^- is hard and S^{2-} is soft, OH^- is more likely to form insoluble salts with 3+ transition-metal ions (hard), and S^{2-} is more likely to form insoluble salts with 2+ transition-metal ions (borderline or soft).

▶ **Exercise 6.3** Some of the products of the following reactions are insoluble, and some form soluble adducts. Consider only the HSAB characteristics in your answers.

a. Will Cu^{2+} react more strongly with OH^- or NH_3? With O^{2-} or S^{2-}?
b. Will Fe^{3+} react more strongly with OH^- or NH_3? With O^{2-} or S^{2-}?
c. Will Ag^+ react more strongly with NH_3 or PH_3?
d. Will Fe, Fe^{2+}, or Fe^{3+} react more strongly with CO?

More detailed comparisons are possible, but another factor, called the *inherent acid–base strength*, must also be kept in mind in these comparisons. An acid or a base may be either hard or soft and at the same time be either strong or weak. The strength of the acid or base may be more important than the hard–soft characteristics; both must be considered. If two soft bases are in competition for the same acid, the one with more inherent base strength may be favored, unless there is considerable difference in softness. As an example, consider the following reaction. Two hard–soft combinations react to give a hard–hard and a soft–soft combination, even though ZnO is composed of the strongest acid (Zn^{2+}) and the strongest base (O^{2-}):

$$\underset{\text{soft-hard}}{ZnO} + \underset{\text{hard-soft}}{2\,LiC_4H_9} \rightleftharpoons \underset{\text{soft-soft}}{Zn(C_4H_9)_2} + \underset{\text{hard-hard}}{Li_2O}$$

In this case, the HSAB parameters are more important than acid–base strength, because Zn^{2+} is considerably softer than Li^+. As a general rule, hard–hard combinations are more favorable energetically than soft–soft combinations.

EXAMPLE

Qualitative Analysis

The traditional qualitative analysis scheme can be used to show how the HSAB concept can be used to correlate solubility behavior; it also can show some of the difficulties with such correlations. In qualitative analysis for metal ions, the cations are successively separated into groups by precipitation for further analysis. The details differ with the specific reagents used but generally fall into the categories in Table 6.6.

In the usual analysis, the reagents are used in the order given from left to right. Ag^+, Pb^{2+}, and $Hg_2{}^{2+}$ (Group 1) are the only metal ions that precipitate with Cl^-, even though they are considered soft acids, and chloride is a marginally hard base. Apparently, the sizes of the ions permit strong bonding in the crystal lattice in spite of this mismatch, partly because their interaction with water—another hard base—is not strong enough to prevent precipitation. The reaction

$$M^{n+}(H_2O)_m + n\,Cl^-(H_2O)_p \longrightarrow MCl_n\downarrow + (m + p)\,H_2O$$

is favorable, although $PbCl_2$ is appreciably soluble in hot water.

TABLE 6.6 HSAB and Qualitative Analysis

Qualitative Analysis Separation

	Group 1	Group 2	Group 3	Group 4	Group 5
HSAB acids	Soft	Borderline and soft	Borderline	Hard	Hard
Reagent	HCl	H_2S (acidic)	H_2S (basic)	$(NH_4)_2CO_3$	Soluble
Precipitates	AgCl	HgS	MnS	$CaCO_3$	Na^+
	$PbCl_2$	CdS	FeS	$SrCO_3$	K^+
	Hg_2Cl_2	CuS	CoS	$BaCO_3$	NH_4^+
		SnS	NiS		
		As_2S_3	ZnS		
		Sb_2S_3	$Al(OH)_3$		
		Bi_2S_3	$Cr(OH)_3$		

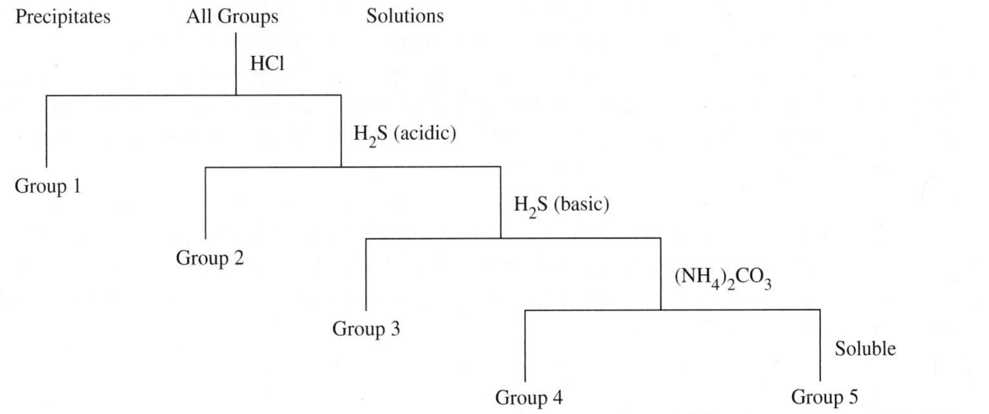

Precipitates
All Groups
Solutions
HCl
H_2S (acidic)
Group 1
H_2S (basic)
Group 2
$(NH_4)_2CO_3$
Group 3
Soluble
Group 4
Group 5

Group 2 is made up of borderline and soft acids that are readily precipitated in acidic H_2S solution, in which the S^{2-} concentration is very low because the equilibrium

$$H_2S \rightleftharpoons 2\,H^+ + S^{2-}$$

lies far to the left in acid solution. The metal ions in this group are soft enough that a low concentration of the soft sulfide is sufficient to precipitate them. Group 3 cleans up the remaining transition metals in the list, all of which are borderline acids. In the basic H_2S solution, the equilibrium above lies far to the right, and the high sulfide ion concentration precipitates even these cations. Al^{3+} and Cr^{3+} are hard enough that they prefer OH^- over S^{2-} and precipitate as hydroxides. Another hard acid could be Fe^{3+}, but it is reduced by S^{2-}, and iron precipitates as FeS. Group 4 is a clear-cut case of hard–hard interactions; both the alkaline earth ions and carbonate are hard. Finally, Group 5 cations are larger, with only a single electronic charge, and thus have small electrostatic attractions for anions. For this reason, they do not precipitate except with certain highly specific reagents, such as perchlorate, ClO_4^-, for potassium and tetraphenylborate, $[B(C_6H_5)_4]^-$, or zinc uranyl acetate, $[Zn(UO_2)_3(C_2H_3O_2)_9]^-$, for sodium.

This quick summary of the analysis scheme shows where hard–hard or soft–soft combinations lead to insoluble salts, but it also shows that the rules have limitations. Some cations considered hard will precipitate under the same conditions as others that are clearly soft. For this reason, any solubility predictions based on HSAB must be considered tentative, and solvent and other interactions must be considered carefully.

HSAB arguments such as these explain the formation of some metallic ores described in Chapter 1 (soft and borderline metal ions form sulfide ores; hard metal ions form oxide ores; some ores, such as bauxite, result from the leaching away of soluble salts) and some of the reactions of ligands with metals (borderline acid cations of Co, Ni, Cu, and Zn tend to form –NCS complexes, whereas softer acid cations of Rh, Ir, Pd, Pt, Au, and Hg tend to form –SCN complexes). Few of these cases involve only this one factor, but it is important in explaining trends in many reactions.

A somewhat oversimplified way to look at the hard–soft question considers the hard–hard interactions as simple electrostatic interactions, with the LUMO of the acid far above the HOMO of the base and relatively little change in orbital energies on adduct formation.[27] A soft–soft interaction involves HOMO and LUMO energies that are much closer and give a large change in orbital energies on adduct formation. Diagrams of such interactions are shown in Figure 6.13, but they need to be used with caution. The small drop in energy in the hard–hard case that seems to indicate only small interactions is not necessarily the entire story. The hard–hard interaction depends on a longer range electrostatic force, and this interaction can be quite strong. Many comparisons of hard–hard and soft–soft interactions indicate that the hard–hard combination is stronger and is the primary driving force. The contrast between the hard–hard product and the hard–soft reactants in such cases provides the net energy difference that leads to the products. One should also remember that many reactions to which the HSAB approach is applied involve competition between two different conjugate acid–base pairs; only in a limited number of cases is one interaction large enough to overwhelm the others and determine whether the reaction will proceed.

6.3.2 Quantitative Measures

There are two major approaches to quantitative measures of acid–base reactions. One, developed by Pearson,[28] uses the hard–soft terminology and defines the **absolute hardness**, η, as half the difference between the ionization energy and the electron affinity (both in eV):

$$\eta = \frac{I - A}{2}$$

This definition of hardness is related to Mulliken's definition of electronegativity, called *absolute electronegativity* by Pearson:

$$\chi = \frac{I + A}{2}$$

This approach describes a hard acid or base as a species that has a large difference between its ionization energy and its electron affinity. Ionization energy is assumed to

[27] Jensen, pp. 262–265; C. K. Jørgensen, *Struct. Bonding (Berlin)*, **1966**, *1*, 234.

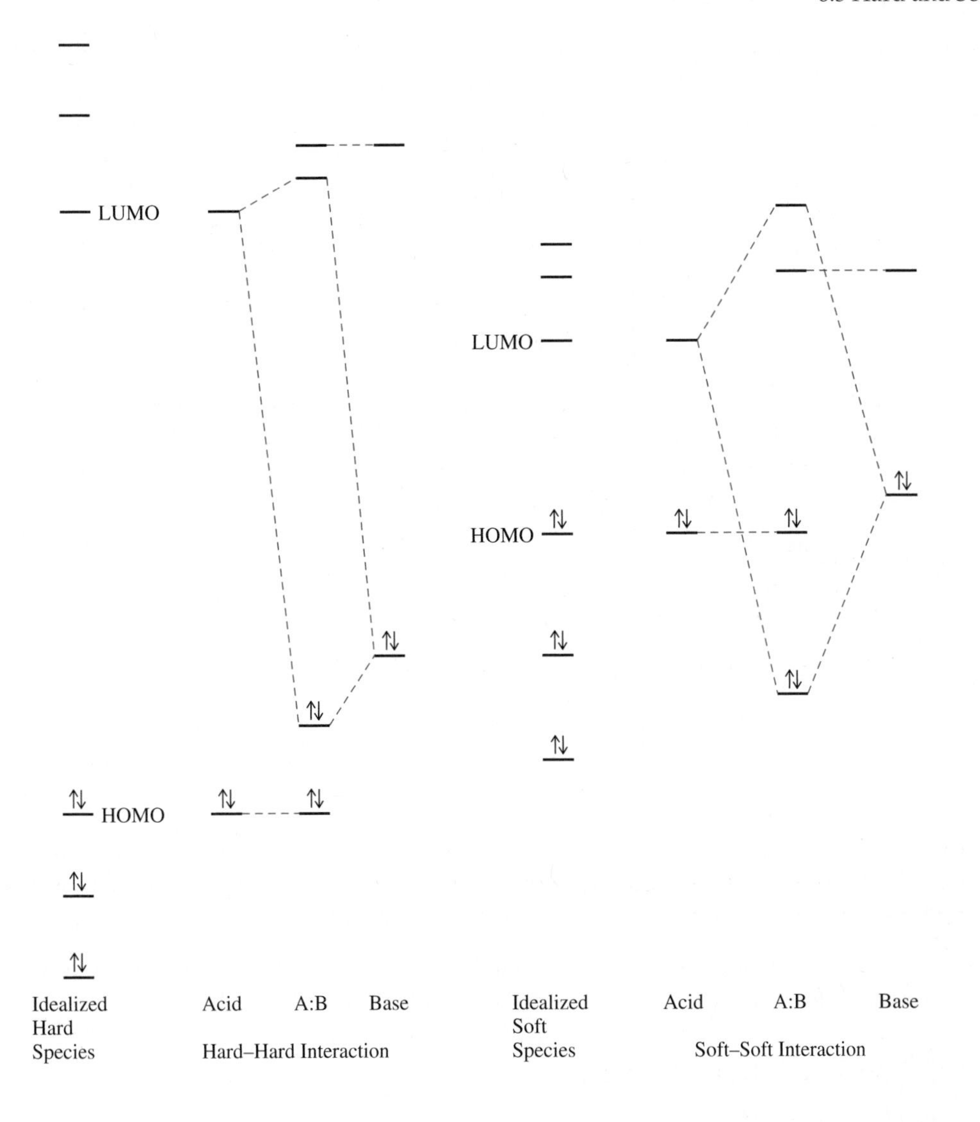

FIGURE 6.13 HOMO–LUMO Diagrams for Hard–Hard and Soft–Soft Interactions.

(Adapted with permission from W. B. Jensen, *The Lewis Acid-Base Concepts*, Wiley Interscience, New York, 1980, pp. 262–263. Copyright © 1980, John Wiley & Sons, Inc. Reprinted by permission of John Wiley & Sons, Inc.)

measure the energy of the HOMO, and electron affinity is assumed to measure the LUMO for a given molecule:

$$E_{HOMO} = -I$$
$$E_{LUMO} = -A.$$

Softness is defined as the inverse of hardness, $\sigma = \frac{1}{\eta}$. Because there are no electron affinities for anions, Pearson uses the values for the atoms as approximate equivalents.

The halogen molecules offer good examples of the use of these orbital arguments to illustrate HSAB. For the halogens, the trend in hardness (η) parallels the change in HOMO energies, because the LUMO energies are nearly the same, as shown in Figure 6.14. Fluorine is the most electronegative halogen. It is also the smallest and least polarizable halogen and is therefore the hardest. In orbital terms, the LUMOs of all the halogen molecules are nearly identical, and the HOMOs increase in energy from F_2 to I_2. The absolute electronegativities decrease in the order $F_2 > Cl_2 > Br_2 > I_2$ as

[28] R. G. Pearson, *Inorg. Chem.*, **1988**, *27*, 734.

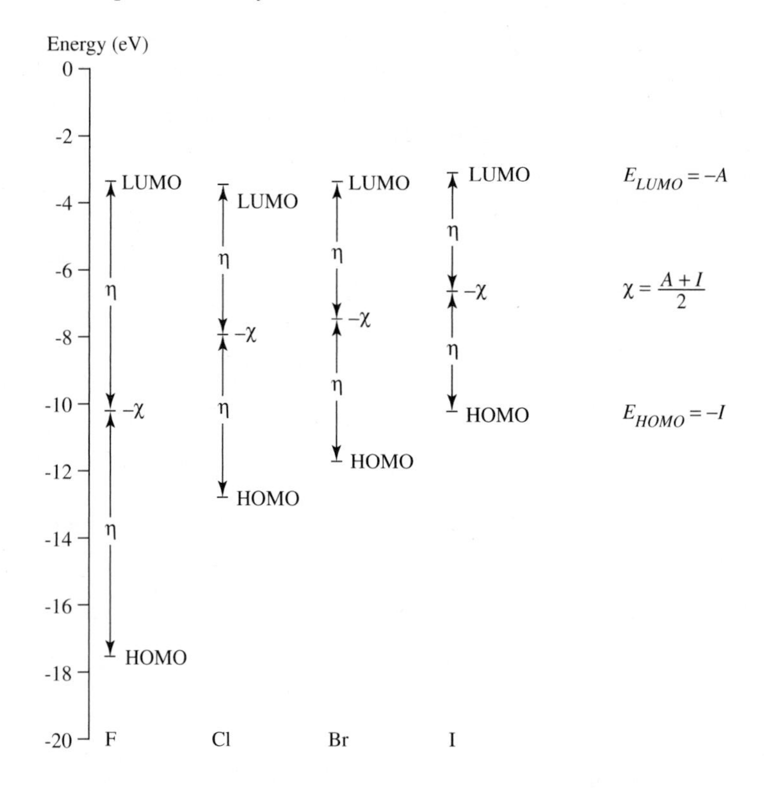

FIGURE 6.14 Energy Levels for Halogens. Relationships between absolute electronegativity (χ), absolute hardness (η), and HOMO and LUMO energies for the halogens.

the HOMO energies increase. The hardness also decreases in the same order in which the difference between the HOMO and LUMO decreases. Data for a number of other species are given in Table 6.7, and more are given in Appendix B-5.

▶ **Exercise 6.4** Confirm the absolute electronegativity and absolute hardness values for the following species, using data from Table 6.7 and Appendix B-5:

a. Al^{3+}, Fe^{3+}, Co^{3+}
b. OH^-, Cl^-, NO_2^-
c. H_2O, NH_3, PH_3

The absolute hardness is not enough to fully describe reactivity—for example, some hard acids are weak acids, and some are strong—and it deals only with gas-phase conditions. Drago and Wayland[29] have proposed a quantitative system of acid–base parameters to account more fully for reactivity by including electrostatic and covalent factors. This approach uses the equation

$$-\Delta H = E_A E_B + C_A C_B$$

where ΔH is the enthalpy of the reaction $A + B \longrightarrow AB$ in the gas phase or in an inert solvent, and E and C are parameters calculated from experimental data. E is a measure of the capacity for electrostatic (ionic) interactions, and C is a measure of the tendency to form covalent bonds. The subscripts refer to values assigned to the acid and base,

[29] R. S. Drago and B. B. Wayland, *J. Am. Chem. Soc.*, **1965**, *87*, 3571; R. S. Drago, G. C. Vogel, and T. E. Needham, *J. Am. Chem. Soc.*, **1971**, *93*, 6014; R. S. Drago, *Struct. Bonding (Berlin)*, **1973**, *15*, 73; R. S. Drago, L. B. Parr, and C. S. Chamberlain, *J. Am. Chem. Soc.*, **1977**, *99*, 3203.

TABLE 6.7 Hardness Parameters (eV)

Ion	I	A	χ	η
Al^{3+}	119.99	28.45	74.22	45.77
Li^{+}	75.64	5.39	40.52	35.12
Mg^{2+}	80.14	15.04	47.59	32.55
Na^{+}	47.29	5.14	26.21	21.08
Ca^{2+}	50.91	11.87	31.39	19.52
Sr^{2+}	43.6	11.03	27.3	16.3
K^{+}	31.63	4.34	17.99	13.64
Zn^{2+}	39.72	17.96	28.84	10.88
Hg^{2+}	34.2	18.76	26.5	7.7
Ag^{+}	21.49	7.58	14.53	6.96
Pd^{2+}	32.93	19.43	26.18	6.75
Rh^{2+}	31.06	18.08	24.57	6.49
Cu^{+}	20.29	7.73	14.01	6.28
Sc^{2+}	24.76	12.80	18.78	5.98
Ru^{2+}	28.47	16.76	22.62	5.86
Au^{+}	20.5	9.23	14.90	5.6
BF_3	15.81	−3.5	6.2	9.7
H_2O	12.6	−6.4	3.1	9.5
NH_3	10.7	−5.6	2.6	8.2
PF_3	12.3	−1.0	5.7	6.7
$(CH_3)_3N$	7.8	−4.8	1.5	6.3
PH_3	10.0	−1.9	4.1	6.0
$(CH_3)_3P$	8.6	−3.1	2.8	5.9
SO_2	12.3	1.1	6.7	5.6
C_6H_6	9.3	−1.2	4.1	5.3
C_5H_5N	9.3	−0.6	4.4	5.0
F^{-}	17.42	3.40	10.41	7.01
OH^{-}	13.17	1.83	7.50	5.67
CN^{-}	14.02	3.82	8.92	5.10
Cl^{-}	13.01	3.62	8.31	4.70
Br^{-}	11.84	3.36	7.60	4.24
NO_2^{-}	> 10.1	2.30	> 6.2	> 3.9
I^{-}	10.45	3.06	6.76	3.70

NOTE: The anion values are calculated from data for the radicals or atoms.

Source: Data from R. G. Pearson, *Inorg. Chem.*, **1988**, *27*, 734.

with I_2 chosen as the reference acid and N,N-dimethylacetamide and diethyl sulfide as reference bases. The defined values (in units of kcal/mol) are

	C_A	E_A	C_B	E_B
I_2	1.00	1.00		
N,N-dimethylacetamide				1.32
Diethyl sulfide			7.40	

Values of E_A and C_A for selected acids and E_B and C_B for selected bases are given in Table 6.8, and a longer list is in Appendix B-6. Combining the values of these parameters for acid-base pairs gives the enthalpy of reaction in kcal/mol;

TABLE 6.8 C_A, E_A, C_B, and E_B Values (kcal/mol)

Acid	C_A	E_A
Trimethylboron, $B(CH_3)_3$	1.70	6.14
Boron trifluoride (gas), BF_3	1.62	9.88
Trimethylaluminum, $Al(CH_3)_3$	1.43	16.9
Iodine (standard), I_2	1.00*	1.00*
Trimethylgallium, $Ga(CH_3)_3$	0.881	13.3
Iodine monochloride, ICl	0.830	5.10
Sulfur dioxide, SO_2	0.808	0.920
Phenol, C_6H_5OH	0.442	4.33
tert-butyl alcohol, C_4H_9OH	0.300	2.04
Pyrrole, C_4H_4NH	0.295	2.54
Chloroform, $CHCl_3$	0.159	3.02
Base	C_B	E_B
1-Azabicyclo[2.2.2] octane,		
Quinuclidine, $HC(C_2H_4)_3N$	13.2	0.704
Trimethylamine, $(CH_3)_3N$	11.54	0.808
Triethylamine, $(C_2H_5)_3N$	11.09	0.991
Dimethylamine, $(CH_3)_2NH$	8.73	1.09
Diethyl sulfide, $(C_2H_5)_2S$	7.40*	0.399
Pyridine, C_5H_5N	6.40	1.17
Methylamine, CH_3NH_2	5.88	1.30
Ammonia, NH_3	3.46	1.36
Diethyl ether, $(C_2H_5)_2O$	3.25	0.963
N,N-dimethylacetamide, $(CH_3)_2NCOCH_3$	2.58	1.32*
Benzene, C_6H_6	0.681	0.525

NOTE: *Reference values.

Source: Data from R. S. Drago, *J. Chem. Educ.*, **1974**, *51*, 300.

multiplying by 4.184 J/cal converts to joules (although we use joules in this book, these numbers were originally derived for calories, and we have chosen to leave them unchanged).

Examination of the data shows that most acids have lower C_A values and higher E_A values than I_2. Because I_2 has no permanent dipole, it has little electrostatic attraction for bases and has a low E_A. On the other hand, it has a strong tendency to bond with some other bases, indicated by a relatively large C_A. Because 1.00 was chosen as the reference value for both parameters for I_2, most C_A values are below 1, and most E_A values are above 1. For C_B and E_B, this relationship is reversed.

The example of iodine and benzene shows how these tables can be used.

$$\underset{\text{acid}}{I_2} + \underset{\text{base}}{C_6H_6} \longrightarrow I_2 \cdot C_6H_6$$

$$-\Delta H = E_A E_B + C_A C_B \quad \text{or} \quad \Delta H = -(E_A E_B + C_A C_B)$$

$$\Delta H = -([1.00 \times 0.681] + [1.00 \times 0.525]) = -1.206 \text{ kcal/mol, or } -5.046 \text{ kJ/mol}$$

The experimental value of ΔH is -1.3 kcal/mol, or -5.5 kJ/mol, 9% larger.[30] This is a weak adduct (other bases combining with I_2 have enthalpies 10 times as large), and the calculation does not agree with experiment as well as many. Because there can be only one set of numbers for each compound, Drago developed statistical methods for averaging experimental data from many different combinations. In many cases, the agreement between calculated and experimental enthalpies is within 5%.

One phenomenon not well accounted for by other approaches is seen in Table 6.9.[31] It shows a series of four acids and five bases in which both *E* and *C* increase. In most descriptions of bonding, as electrostatic (ionic) bonding increases, covalent bonding decreases, but these data show both increasing at the same time. Drago argued that this means that the *E* and *C* approach explains acid–base adduct formation better than the HSAB theory described earlier.

TABLE 6.9 Acids and Bases with Parallel Changes in *E* and *C*

Acids	C_A	E_A
$CHCl_3$	0.154	3.02
C_6H_5OH	0.442	4.33
m-$CF_3C_6H_4OH$	0.530	4.48
$B(CH_3)_3$	1.70	6.14
Bases	C_B	E_B
C_6H_6	0.681	0.525
CH_3CN	1.34	0.886
$(CH_3)_2CO$	2.33	0.987
$(CH_3)_2SO$	2.85	1.34
NH_3	3.46	1.36

[30] R. M. Keefer and L. J. Andrews, *J. Am. Chem. Soc.*, **1955**, *77*, 2164.
[31] R. S. Drago, *J. Chem. Educ.*, **1974**, *51*, 300.

EXAMPLE

Calculate the enthalpy of adduct formation predicted by Drago's *E*, *C* equation for the reactions of I_2 with diethyl ether and diethyl sulfide.

	E_A E_B C_A C_B ΔH (kcal/mol)	Experimental ΔH
Diethyl ether	$-([1.00 \times 0.963] + [1.00 \times 3.25]) = -4.21$	-4.2
Diethyl sulfide	$-([1.00 \times 0.339] + [1.00 \times 7.40]) = -7.74$	-7.8

Agreement is very good, with the product $C_A \times C_B$ by far the dominant factor. The softer sulfur reacts more strongly with the soft I_2.

▶ **Exercise 6.5** Calculate the enthalpy of adduct formation predicted by Drago's *E* and *C* equation for the following combinations, and explain the trends in terms of the electrostatic and covalent contributions:

a. BF_3 reacting with ammonia, methylamine, dimethylamine, and trimethylamine
b. Pyridine reacting with trimethylboron, trimethylaluminum, and trimethylgallium

Drago's system emphasized the two factors involved in acid–base strength (electrostatic and covalent) in the two terms of his equation for enthalpy of reaction. Pearson's system put more obvious emphasis on the covalent factor. Pearson[32] proposed the equation $\log K = S_A S_B + \sigma_A \sigma_B$, with the inherent strength *S* modified by a softness factor σ. Larger values of strength and softness then lead to larger equilibrium constants or rate constants. Although Pearson attached no numbers to this equation, it does show the need to consider more than just hardness or softness in working with acid–base reactions. However, his more recent development of absolute hardness based on orbital energies returns to a single parameter and considers only gas-phase reactions. Both Drago's *E* and *C* parameters and Pearson's HSAB are useful, but neither covers every case, and it is usually necessary to make judgments about reactions for which information is incomplete. With *E* and *C* numbers available, quantitative comparisons can be made. When they are not, the qualitative HSAB approach can provide a rough guide for predicting reactions. Examination of the tables also shows little overlap of the examples chosen by Drago and Pearson.

An additional factor that has been mentioned frequently in this chapter is solvation. Neither of the two quantitative theories takes this factor into account. Under most conditions, reactions will be influenced by solvent interactions, and they can promote or hinder reactions, depending on the details of these interactions.

6.4 ACID AND BASE STRENGTH

6.4.1 Measurement of Acid–Base Interactions

Interaction between acids and bases can be measured in many ways:

1. Changes in boiling or melting points can indicate the presence of adducts. Hydrogen-bonded solvents, such as water and methanol, and adducts such as $BF_3 \cdot O(C_2H_5)_2$ have higher boiling points or melting points than would otherwise be expected.

[32] R. G. Pearson, *J. Chem. Educ.* **1968**, *45*, 581.

2. Direct calorimetric methods or temperature dependence of equilibrium constants can be used to measure enthalpies and entropies of acid–base reactions. The following section gives more details on use of data from these measurements.
3. Gas-phase measurements of the formation of protonated species can provide similar thermodynamic data.
4. Infrared spectra can provide indirect measures of bonding in acid–base adducts by showing changes in bond force constants. For example, free CO has a C—O stretching band at 2143 cm^{-1}, and CO in $Ni(CO)_4$ has a C—O band at 2058 cm^{-1}.
5. NMR coupling constants provide a similar indirect measure of changes in bonding on adduct formation.
6. Ultraviolet or visible spectra can show changes in energy levels in the molecules as they combine.

Different methods of measuring acid–base strength yield different results, which is not surprising, when the physical properties being measured are considered. Some aspects of acid–base strength are discussed in the following section, with brief explanations of the experimental methods used.

6.4.2 Thermodynamic Measurements

The enthalpy change of some reactions can be measured directly, but for those that do not go to completion, as is common in acid–base reactions, thermodynamic data from reactions that do go to completion can be combined using Hess's law to obtain the needed data. For example, the enthalpy and entropy of ionization of a weak acid, HA, can be found by measuring (1) the enthalpy of reaction of HA with NaOH, (2) the enthalpy of reaction of a strong acid (such as HCl) with NaOH, and (3) the equilibrium constant for dissociation of the acid, usually determined from the titration curve.

	Reaction	Enthalpy Change
1.	$HA + OH^- \longrightarrow A^- + H_2O$	$\Delta H_1{}^\circ$
2.	$H_3O^+ + OH^- \longrightarrow 2\,H_2O$	$\Delta H_2{}^\circ$
3.	$HA + H_2O \overset{K_a}{\rightleftharpoons} H_3O^+ + A^-$	$\Delta H_3{}^\circ$

From the usual thermodynamic relationships,

4. $$\Delta H_3{}^\circ = \Delta H_1{}^\circ - \Delta H_2{}^\circ$$
[because Reaction 3 = Reaction 1 − Reaction 2]

5. $$\Delta S_3{}^\circ = \Delta S_1{}^\circ - \Delta S_2{}^\circ$$

6. $$\Delta G_3{}^\circ = -RT \ln K_a = \Delta H_3{}^\circ - T\Delta S_3{}^\circ$$

Rearranging:

7. $$\ln K_a = -\Delta H_3{}^\circ/RT + \Delta S_3{}^\circ/R$$

Naturally, the final calculation can be more complex than this, when HA is already partly dissociated in the first reaction, but the approach remains the same. It is also possible to measure the equilibrium constant at different temperatures and use Equation 6 to calculate ΔH° and ΔS°. On a plot of $\ln K_a$ versus $1/T$, the slope is $-\Delta H_3{}^\circ/R$ and the intercept is $\Delta S_3{}^\circ/R$. This method works as long as ΔH° and ΔS° do not change appreciably over the temperature range used. This is sometimes a difficult condition. Data for ΔH°, ΔS°, and K_a for acetic acid are given in Table 6.10.

▶ **Exercise 6.6** Use the data in Table 6.10 to calculate the enthalpy and entropy of reaction for dissociation of acetic acid using (a) Equations 4 and 5 and (b) the temperature dependence of K_a of Equation 7 by graphing $\ln K_a$ versus $1/T$.

TABLE 6.10 Thermodynamics of Acetic Acid Dissociation

	$\Delta H°$ (kJ mol^{-1})	$\Delta S°$ (JK^{-1} mol^{-1})
$H_3O^+ + OH^- \rightleftharpoons 2\,H_2O$	−55.9	−80.4
$HOAc + OH^- \rightleftharpoons H_2O + OAc^-$	−56.3	−12.0

$HOAc \rightleftharpoons H^+ + OAc^-$

T (K)	303	308	313	318	323
K_a ($\times 10^{-5}$)	1.750	1.728	1.703	1.670	1.633

NOTE: ΔH° and ΔS° for these reactions change rapidly with temperature. Calculations based on these data are valid only over the limited temperature range given above.

6.4.3 Proton Affinity

One of the purest measures of acid–base strength, but one difficult to relate to solution reactions, is gas-phase proton affinity:[33]

$$BH^+(g) \longrightarrow B(g) + H^+(g) \qquad \text{proton affinity} = \Delta H$$

A large proton affinity means it is difficult to remove the hydrogen ion; this means that B is a strong base, and BH^+ is a weak acid in the gas phase. In favorable cases, mass spectroscopy and ion cyclotron resonance spectroscopy[34] can be used to measure the reaction indirectly. The voltage of the ionizing electron beam in mixtures of B and H_2 is changed, until BH^+ appears in the output from the spectrometer. The enthalpy of formation for BH^+ can then be calculated from the voltage of the electron beam and combined with enthalpies of formation of B and H^+ to calculate the enthalpy change for the reaction.

In spite of the simple concept, the measured values of proton affinities have large uncertainties, because the molecules involved frequently are in excited states (with excess energy above their normal ground states), and some species do not yield BH^+ as a fragment. In addition, under common experimental conditions, the proton affinity must be combined with solvent or other environmental effects to fit the actual reactions. However, gas-phase proton affinities are useful in sorting out the different factors influencing acid–base behavior. For example, the alkali metal hydroxides, which are of equal basicity in aqueous solution, have gas-phase basicities in the order LiOH < NaOH < KOH < CsOH. This order matches the increase in the electron-releasing ability of the cation in these hydroxides. Proton affinity studies have also shown that pyridine and aniline, shown in Figure 6.15, are stronger bases than ammonia in the gas phase, but they are weaker than ammonia in aqueous solution,[35] presumably because the interaction of the ammonium ion with water is more favorable

FIGURE 6.15 Pyridine and Aniline Structures.

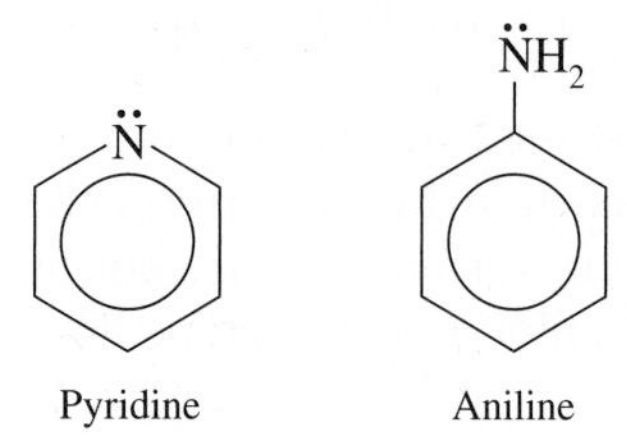

[33] H. L. Finston and A. C. Rychtman, *A New View of Current Acid–Base Theories*, John Wiley & Sons, New York, 1982, pp. 53–62.
[34] R. S. Drago, *Physical Methods in Chemistry*, W. B. Saunders, Philadelphia, 1977, pp. 552–565.
[35] H. L. Finston and A. C. Rychtman, *A New View of Current Acid–Base Theories*, John Wiley & Sons, New York, 1982, pp. 59–60.

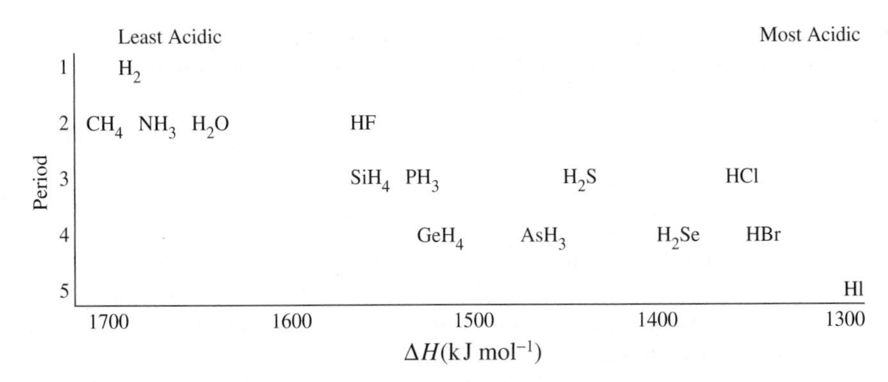

FIGURE 6.16 Acidity of Binary Hydrogen Compounds. Enthalpy of dissociation in kJ/mol for the reaction $AH(g) \longrightarrow A^-(g) + H^+(g)$, numerically the same as the proton affinity.

(Data from J. E. Bartmess, J. A. Scott, and R. T. McIver, Jr., *J. Am. Chem. Soc.*, **1979**, *101*, 6046; AsH_3 value from J. E. Bartmess and R. T. McIver, Jr., *Gas Phase Ion Chemistry*, M. T. Bowers, ed., Academic Press, New York, 1979, p. 87.)

than the interaction with the pyridinium or anilinium ions. Other comparisons of gas-phase data with solution data allow at least partial separation of the factors influencing reactions.

6.4.4 Acidity and Basicity of Binary Hydrogen Compounds

The binary hydrogen compounds (compounds containing only hydrogen and one other element) range from the strong acids HCl, HBr, and HI to the weak base NH_3. Others, such as CH_4, show almost no acid-base properties. Some of these molecules—in order of increasing gas phase acidities, from left to right—are shown in Figure 6.16.

Two apparently contradictory trends are seen in these data. Acidity increases with increasing numbers of electrons in the central atom, either going across the table or down; but the electronegativity effects are opposite for the two directions, as shown in Figure 6.17.

Within each column of the periodic table, acidity increases going down the series, as in $H_2Se > H_2S > H_2O$. The strongest acid is the largest, heaviest member, low in the periodic table, containing the nonmetal of lowest electronegativity of the group. An explanation of this is that the conjugate bases (SeH^-, SH^-, and OH^-) of the larger molecules have lower charge density and therefore a smaller attraction for hydrogen ions (the H—O bond is stronger than the H—S bond, which in turn is stronger than the H—Se bond). As a result, the larger molecules are stronger acids, and their conjugate bases are weaker.

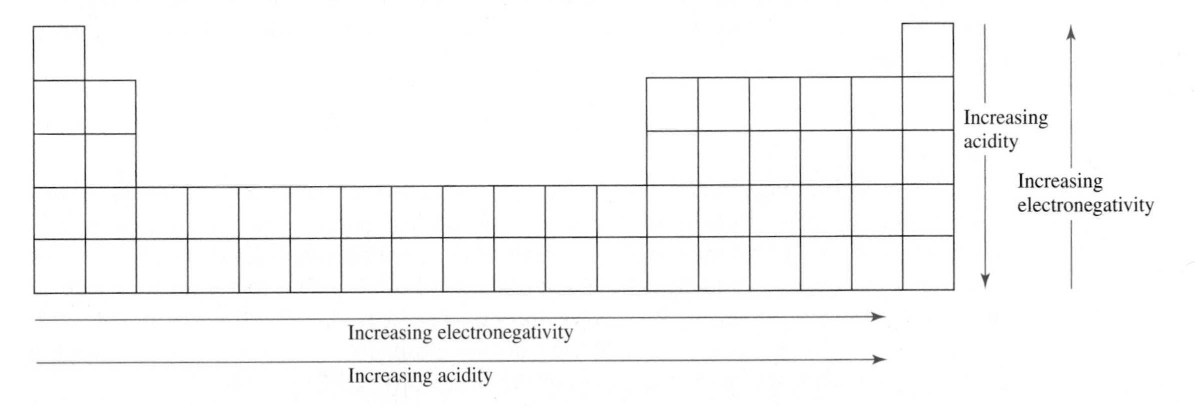

FIGURE 6.17 Trends in Acidity and Electronegativity of Binary Hydrides.

On the other hand, within a period, acidity is greatest for the compounds of elements toward the right, with greater electronegativity. The electronegativity argument cannot be used, because in this series, the more electronegative elements form the stronger acids. Although it may have no fundamental significance, one explanation that assists in remembering the trends divides the –1 charge of each conjugate base evenly among the lone pairs. Thus, NH_2^- has a charge of –1 spread over two lone pairs, or $-\frac{1}{2}$ on each, OH^- has a charge of –1 spread over three lone pairs, or $-\frac{1}{3}$ on each, and F^- has a charge of –1 spread over four lone pairs, or $-\frac{1}{4}$ on each. The amide ion, NH_2^-, has the strongest attraction for protons and is therefore the strongest of these three conjugate bases, and ammonia is the weakest acid of the three. The order of acid strength follows this trend: $NH_3 < H_2O < HF$.

The same general trends persist when the acidity of these compounds is measured in aqueous solution. The reactions are more complex, forming aquated ions, but the overall effects are similar. The three heaviest hydrohalic acids—HCl, HBr, and HI—are equally strong in water because of the leveling effect of the water. (The leveling effect and other solvent effects are considered in greater detail in Sections 6.4.9 and 6.4.10.) All the other binary hydrogen compounds are weaker acids, their strength decreasing toward the left in the periodic table. Methane and ammonia exhibit no acidic behavior in aqueous solution, nor do silane (SiH_4) and phosphine (PH_3).

6.4.5 Inductive Effects

Substitution of electronegative atoms or groups, such as fluorine or chlorine, in place of hydrogen on ammonia or phosphine results in weaker bases. The electronegative atom draws electrons toward itself, and as a result, the nitrogen or phosphorus atom has less negative charge, and its lone pair is less readily donated to an acid. For example, PF_3 is a much weaker base than PH_3.

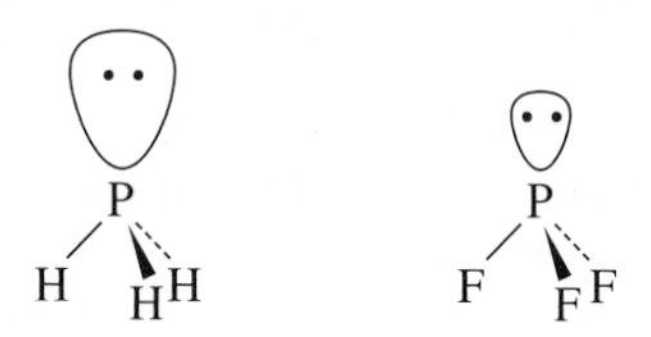

A similar effect in the reverse direction results from substitution of alkyl groups for hydrogen. For example, in amines, the alkyl groups contribute electrons to the nitrogen, increasing its negative character and making it a stronger base. Additional substitutions increase the effect, with the following resulting order of base strength in the gas phase:

$$NMe_3 > NHMe_2 > NH_2Me > NH_3$$

These **inductive effects** are similar to the effects seen in organic molecules containing electron-contributing or electron-withdrawing groups. Once again, caution is required in applying this idea to other compounds. The boron halides do not follow this argument, because BF_3 and BCl_3 have significant π bonding that increases the electron density on the boron atom. Inductive effects would make BF_3 the strongest acid, because the large electronegativity of the fluorine atoms draws electrons away from the boron atom. In fact, the acid strength is in the order $BF_3 < BCl_3 \leq BBr_3$.

6.4.6 Strength of Oxyacids

In the series of oxyacids of chlorine, the acid strength in aqueous solution is in the order

$$HClO_4 > HClO_3 > HClO_2 > HOCl$$

```
          O                  O                  O
          |                  |                  |
H—O—Cl—O       H—O—Cl—O       H—O—Cl        H—O—Cl
          |
          O
```

Pauling suggested a rule that predicts the strength of such acids semiquantitatively, based on n, the number of nonhydrogenated oxygen atoms per molecule. Pauling's equation described the acidity at 25°C as $pK_a \approx 9 - 7n$. Several other equations have been proposed; $pK_a \approx 8 - 5n$ fits some acids better. (Remember: the stronger the acid, the smaller the pK_a.) Estimated and experimental pK_a values of the acids above are as shown.

Acid	**Strongest** $HClO_4$	$HClO_3$	$HClO_2$	***Weakest*** $HOCl$
n	3	2	1	0
pK_a (calculated by $9 - 7n$)	–12	–5	2	9
pK_a (calculated by $8 - 5n$)	–7	–2	3	8
pK_a (experimental)	(–10)	–1	2	7.2

Neither equation is very accurate, but either provides approximate values. For oxyacids with more than one ionizable hydrogen, the pK_a values increase by about 5 units with each successive proton removal:

	H_3PO_4	$H_2PO_4^-$	HPO_4^{2-}	H_2SO_4	HSO_4^-
pK_a (by $9 - 7n$)	2	7	12	–5	0
pK_a (by $8 - 5n$)	3	8	13	–2	3
pK_a (experimental)	2.15	7.20	12.37	<0	2

The molecular explanation for these approximations hinges on electronegativity. Because each nonhydrogenated oxygen is highly electronegative, it draws electrons away from the central atom, increasing the positive charge on the central atom. This positive charge in turn draws the electrons of the hydrogenated oxygen toward itself. The net result is a weaker O—H bond (lower electron density in this bond), which makes it easier for the molecule to act as an acid by losing the H^+. As the number of highly electronegative oxygens increases, the acid strength of the molecule also increases.

The same argument can be seen from the point of view of the conjugate base. The negative charge of the conjugate base is spread over all the nonhydrogenated oxygens. The larger the number of these oxygens to share the negative charge, the more stable and weaker the conjugate base, and the stronger the hydrogenated acid. This explanation gives the same result as the first: the larger the number of nonhydrogenated oxygens, the stronger the acid.

▶ **Exercise 6.7**

a. Calculate approximate pK_a values for H_2SO_3, using both equations in this section.

b. H_3PO_3 has one hydrogen bonded directly to the phosphorus. Calculate approximate pK_a values for H_3PO_3 using both equations.

6.4.7 Acidity of Cations in Aqueous Solution

Many positive ions exhibit acidic behavior in solution. For example, Fe^{3+} in water forms an acidic solution, with yellow or brown iron species formed by reactions such as

$$[Fe(H_2O)_6]^{3+} + H_2O \rightleftharpoons [Fe(H_2O)_5(OH)]^{2+} + H_3O^+$$

$$[Fe(H_2O)_5(OH)]^{2+} + H_2O \rightleftharpoons [Fe(H_2O)_4(OH)_2]^{+} + H_3O^+$$

In less acidic (or more basic), solutions, hydroxide or oxide bridges between metal atoms form, the high positive charge promotes more hydrogen ion dissociation, and a large aggregate of hydrated metal hydroxide precipitates. A possible first step in this process is:

$$2\,[Fe(H_2O)_5(OH)]^{2+} \rightleftharpoons [(H_2O)_4Fe(\mu\text{-}OH)_2Fe(H_2O)_4]^{4+} + 2\,H_2O$$

In general, metal ions with larger charges and smaller radii are stronger acids. The alkali metals show essentially no acidity, the alkaline earth metals show it only slightly, 2+ transition-metal ions are weakly acidic, 3+ transition-metal ions are moderately acidic, and ions that would have charges of 4+ or higher as monatomic ions are such strong acids in aqueous solutions that they exist only as oxygenated ions. Some examples of acid-dissociation constants are given in Table 6.11.

Solubility of the metal hydroxide is also a measure of cation acidity. The stronger the cation acid, the less soluble the hydroxide. Generally, transition metal 3+ ions are acidic enough to form hydroxides that precipitate even in the slightly acidic solutions formed when their salts are dissolved in water. The yellow color of iron(III) solutions mentioned earlier is an example. A slight precipitate is also formed in concentrated solutions, unless acid is added. When acid is added, the precipitate dissolves, and the color disappears. Fe(III) is very faintly violet in concentrated solutions and colorless in dilute solutions. The 2+ *d*-block ions and Mg^{2+} precipitate as hydroxides in neutral or slightly basic solutions, and the alkali and remaining alkaline earth ions are so weakly acidic that no pH effects are measured. Some solubility products are given in Table 6.12.

At the highly charged extreme, the free metal cation is no longer a detectable species. Instead, ions such as permanganate (MnO_4^-), chromate (CrO_4^{2-}), uranyl

TABLE 6.11 Hydrated Metal Ion Acidities

Metal Ion	K_a	Metal Ion	K_a
Fe^{3+}	6.7×10^{-3}	Fe^{2+}	5×10^{-9}
Cr^{3+}	1.6×10^{-4}	Cu^{2+}	5×10^{-9}
Al^{3+}	1.1×10^{-5}	Ni^{2+}	5×10^{-10}
Sc^{3+}	1.1×10^{-5}	Zn^{2+}	2.5×10^{-10}

NOTE: These are equilibrium constants for $[M(H_2O)_m]^{n+} + H_2O \rightleftharpoons [M(H_2O)_{m-1}(OH)]^{(n-1)+} + H_3O^+$.

TABLE 6.12 Solubility Product Constants

Metal Hydroxide	K_{sp}	Metal Hydroxide	K_{sp}
$Fe(OH)_3$	6×10^{-38}	$Fe(OH)_2$	8×10^{-16}
$Cr(OH)_3$	7×10^{-31}	$Cu(OH)_2$	2.2×10^{-20}
$Al(OH)_3$	1.4×10^{-34}	$Ni(OH)_2$	2×10^{-15}
		$Zn(OH)_2$	7×10^{-18}
		$Mg(OH)_2$	1.1×10^{-11}

NOTE: These are equilibrium constants for the reaction $M(OH)_n(s) \rightleftharpoons M^{n+}(aq) + n\,OH^-$.

($UO_2^{\ +}$), dioxovanadium ($VO_2^{\ +}$), and vanadyl (VO^{2+}) are formed, with oxidation numbers of 7, 6, 5, 5, and 4 for the metals, respectively. Permanganate and chromate are strong oxidizing agents, particularly in acidic solutions. These ions are also very weak bases. For example,

$$CrO_4^{\ 2-} + H^+ \rightleftharpoons HCrO_4^{\ -} \quad K_b = 3.2 \times 10^{-8}$$

$$HCrO_4^{\ -} + H^+ \rightleftharpoons H_2CrO_4 \quad K_b = 5.6 \times 10^{-14}$$

In concentrated acid, the dichromate ion is formed by loss of water:

$$2\,HCrO_4^{\ -} \rightleftharpoons Cr_2O_7^{\ 2-} + H_2O$$

6.4.8 Steric Effects

There are also steric effects that influence acid–base behavior. When bulky groups are forced together by adduct formation, their mutual repulsion makes the reaction less favorable. Brown has contributed a great deal to these studies.[36] He described molecules as having front (F) strain or back (B) strain, depending on whether the bulky groups interfere directly with the approach of an acid and a base to each other, or whether the bulky groups interfere with each other when VSEPR effects force them to bend away from the other molecule forming the adduct. He also called effects from electronic differences within similar molecules internal (I) strain. Many reactions involving substituted amines and pyridines were used to sort out these effects.

EXAMPLE

Reactions of a series of substituted pyridines with hydrogen ions show the order of base strengths to be as shown.

2,6-dimethylpyridine > 2-methylpyridine > 2-*t*-butylpyridine > pyridine

H_3C N CH_3 H_3C N H_3C CH_3 C N H_3C N

[36] H.C. Brown, *J. Chem. Soc.*, **1956**, 1248.

This matches the expected order for electron donation (induction) by alkyl groups; the *t*-butyl group has counterbalancing inductive and steric effects. However, reaction with larger acids, such as BF_3 or BMe_3, shows the following order of basicity:

pyridine > 2-methylpyridine > 2,6-dimethylpyridine > 2-*t*-butylpyridine

Explain the difference between these two series.

The larger fluorine atoms or methyl groups attached to the boron and the groups on the *ortho* position of the substituted pyridines interfere with each other when the molecules approach each other, so reaction with the substituted pyridines is less favorable. Interference is greater with the 2,6-substituted pyridine and greater still for the *t*-butyl substituted pyridine. This is an example of F strain.

▶ **Exercise 6.8** Based on inductive arguments, would you expect boron trifluoride or trimethylboron to be the stronger acid in reaction with NH_3? Is this the same order expected for reaction with the bulky bases in the preceding example?

Gas-phase measurements of proton affinity show the sequence of basic strength $Me_3N > Me_2NH > MeNH_2 > NH_3$, as predicted on the basis of electron donation (induction) by the methyl groups, resulting in increased electron density and basicity of the nitrogen.[37] When larger acids are used, the order changes, as shown in Table 6.13. With both BF_3 and BMe_3, Me_3N is a much weaker base, very nearly the same as $MeNH_2$. With the even more bulky acid tri(*t*-butyl)boron, the order is nearly reversed from the proton affinity order, although ammonia is still weaker than methylamine. Brown has argued that these effects are from crowding of the methyl groups at the back of the nitrogen as the adduct is formed (B strain). It may also be argued that some direct interference is also present.

When triethylamine is used as the base, it does not form an adduct with trimethylboron, although the enthalpy change for such a reaction is slightly favorable. Initially,

TABLE 6.13 Amine Reactions

			ΔH of Adduct Formation[c]		
Amine	**ΔH of Hydrogen Ion Addition[a] (kJ/mol)**	**pK_b (Aqueous)[b]**	**BF_3 (Order)**	**BMe_3 (kJ/mol)**	**$B(t\text{–}Bu)_3$ (Order)**
NH_3	–846	4.75	4	–57.53	2
CH_3NH_2	–884	3.38	2	–73.81	1
$(CH_3)_2NH$	–912	3.23	1	–80.58	3
$(CH_3)_3N$	–929	4.20	3	–73.72	4
$(C_2H_5)_3N$	–958			~–42	
Quinuclidine	–967			–84	
Pyridine	–912			–74.9	

NOTE:[a] P. Kebarle, *Ann. Rev. Phys. Chem.*, **1977**, *28*, 445; [b]N. S. Isaacs, *Physical Organic Chemistry*, Longman/Wiley, New York, 1987, p. 213; [c]H. C. Brown, *J. Chem. Soc.*, **1956**, 1248.

[37] M. S. B. Munson, *J. Am. Chem. Soc.*, **1965**, *87*, 2332; J. I. Brauman and L. K. Blair, *J. Am. Chem. Soc.*, **1968**, *90*, 6561; J. I. Brauman, J. M. Riveros, and L. K. Blair, *J. Am. Chem. Soc.*, **1971**, *93*, 3914.

this seems to be another example of B strain, but examination of molecular models shows that one ethyl group is normally twisted out to the front of the molecule, where it interferes with adduct formation. When the alkyl chains are linked into rings, as in quinuclidine (1-azabicyclo[2.2.2]octane), adduct formation is more favorable, because the potentially interfering chains are pinned back and do not change on adduct formation. The proton affinities of quinuclidine and triethylamine are nearly identical, 967 and 958 kJ/mol. When mixed with trimethylboron, whose methyl groups are large enough to interfere with the ethyl groups of triethylamine, the quinuclidine reaction is twice as favorable as that of triethylamine (–84 versus –42 kJ/mol for adduct formation). Whether the triethylamine effect is due to interference at the front or the back of the amine is a subtle question, because the interference at the front is indirectly caused by other steric interference at the back between the ethyl groups.

6.4.9 Solvation and Acid–Base Strength

A further complication appears in the amine series. In aqueous solution, the methyl-substituted amines have basicities in the order $Me_2NH > MeNH_2 > Me_3N > NH_3$, as given in Table 6.13 (a smaller pK_b indicates a stronger base); ethyl-substituted amines are in the order $Et_2NH > EtNH_2 = Et_3N > NH_3$. In both series, the trisubstituted amines are weaker bases than expected, because of the reduced solvation of their protonated cations. Solvation energies (absolute values) for the reaction

$$R_nH_{4-n}N^+(g) + H_2O \longrightarrow R_nH_{4-n}N^+(aq)$$

are in the order $RNH_3^+ > R_2NH_2^+ > R_3NH^+$.[38] Solvation is dependent on the number of hydrogen atoms available for hydrogen bonding to water to form $H{-}O\cdots H{-}N$ hydrogen bonds. With fewer hydrogens available for such hydrogen bonding, the more highly substituted molecules are less basic. Competition between the two effects, induction and solvation, gives the scrambled order of solution basicity.

6.4.10 Nonaqueous Solvents and Acid–Base Strength

Reactions of acids or bases with water are only one aspect of solvent effects. Any acid will react with a basic solvent, and any base will react with an acidic solvent, with the extent of the reaction varying with their relative strengths. For example, acetic acid (a weak acid) will react with water to a very slight extent, but hydrochloric acid (a strong acid) reacts completely. Both form H_3O^+, together with the acetate ion and chloride ion, respectively.

$$HOAc + H_2O \rightleftharpoons H_3O^+ + OAc^- \quad \text{(about 1.3\% in 0.1 M solution)}$$

$$HCl + H_2O \rightleftharpoons H_3O^+ + Cl^- \quad \text{(100\% in 0.1 M solution)}$$

Similarly, water will react slightly with the weak base ammonia and completely with the strong base sodium oxide, forming hydroxide ion in both cases, together with the ammonium ion and the sodium ion:

$$NH_3 + H_2O \rightleftharpoons NH_4^+ + OH^- \quad \text{(about 1.3\% in 0.1 M solution)}$$

$$Na_2O + H_2O \rightleftharpoons 2\,Na^+ + 2\,OH^- \quad \text{(100\% in 0.1 M solution)}$$

These reactions show that water is **amphoteric**, with both acidic and basic properties.

The strongest acid possible in water is the hydronium (oxonium) ion, and the strongest base is the hydroxide ion, so they are formed in reactions with the stronger

[38] E. M. Arnett, *J. Chem. Educ.*, **1985**, 62, 385 reviews the effects of solvation, with many references.

acid HCl and the stronger base Na_2O, respectively. Weaker acids and bases react similarly, but only to a small extent. In glacial acetic acid solvent (100% acetic acid), only the strongest acids can force another hydrogen ion onto the acetic acid molecule, but acetic acid will react readily with any base, forming the conjugate acid of the base and the acetate ion:

$$H_2SO_4 + HOAc \rightleftharpoons H_2OAc^+ + HSO_4^-$$

$$NH_3 + HOAc \rightleftharpoons NH_4^+ + OAc^-$$

The strongest base possible in pure acetic acid is the acetate ion; any stronger base reacts with acetic acid solvent to form acetate ion, as in

$$OH^- + HOAc \longrightarrow H_2O + OAc^-$$

This is called the **leveling effect**, in which acids or bases are brought down to the limiting conjugate acid or base of the solvent. Because of this, nitric, sulfuric, perchloric, and hydrochloric acids are all equally strong acids in dilute aqueous solutions, reacting to form H_3O^+, the strongest acid possible in water. In acetic acid, their acid strength is in the order $HClO_4 > HCl > H_2SO_4 > HNO_3$, based on their ability to force a second hydrogen ion onto the carboxylic acid to form H_2OAc^+. Therefore, acidic solvents allow separation of strong acids in order of strength; basic solvents allow a similar separation of bases in order of strength. On the other hand, even weak bases appear strong in acidic solvents, and weak acids appear strong in basic solvents. This concept provides information that is frequently useful, both in choosing solvents for specific reactions and in describing the pH range that is possible for different solvents, as shown in Figure 6.18.

Inert solvents, with neither acidic nor basic properties, allow a wider range of acid–base behavior. For example, hydrocarbon solvents do not limit acid or base strength, because they do not form solvent acid or base species. In such solvents, the

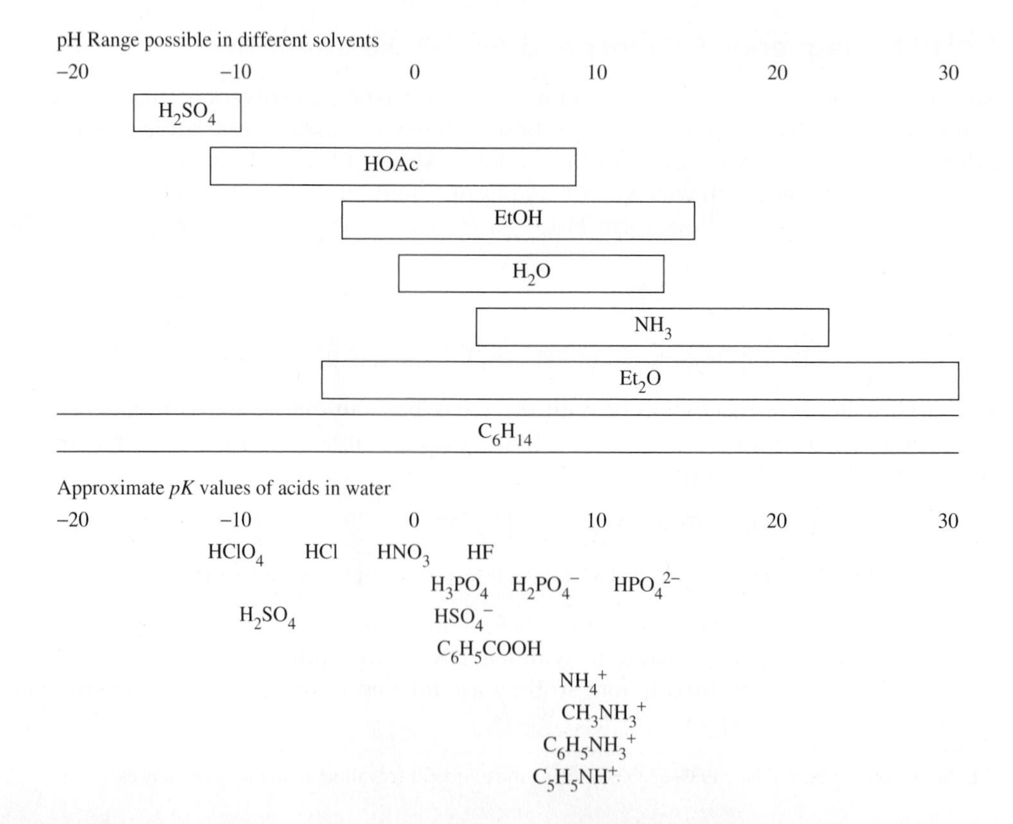

FIGURE 6.18 The Leveling Effect and Solvent Properties.

(Adapted from R. P. Bell, *The Proton in Chemistry*, 2nd ed., 1973, p. 50. Second edition Copyright © 1973 by R. P. Bell. Used by permission of Cornell University Press.)

acid or base strengths of the solutes determine the reactivity, and there is no leveling effect. Balancing the possible acid–base effects of a solvent with requirements for solubility, safety, and availability is one of the challenges for experimental chemists.

EXAMPLE

What are the reactions that take place, and what are the major species in solution at the beginning, midpoint, and end of the titration of a solution of ammonia in water by hydrochloric acid in water?

Beginning

NH_3 and a very small amount of NH_4^+ and OH^- are present. As a weak base, ammonia dissociates very little.

Midpoint

The reaction taking place during the titration is $H_3O^+ + NH_3 \longrightarrow NH_4^+ + H_2O$, because HCl is a strong acid and completely dissociated. At the midpoint, equal amounts of NH_3 and NH_4^+ are present, along with about 5.4×10^{-10} M H_3O^+ and 1.8×10^{-5} M OH^- (because pH = pK_a at the midpoint, pH = 9.3). Cl^- is the major anion present.

End point

All NH_3 has been converted to NH_4^+, so NH_4^+ and Cl^- are the major species in solution, along with about 2×10^{-6} M H_3O^+ (pH about 5.7).

After the end point

Excess HCl has been added, so the H_3O^+ concentration is now larger, and the pH is lower. NH_4^+ and Cl^- are still the major species.

▶ **Exercise 6.9** What are the reactions that take place and the major species in solution at the beginning, midpoint, and end of the following titrations? Include estimates of the extent of reaction (i.e., the acid dissociates completely, to a large extent, or very little).

a. Titration of a solution of acetic acid in water by sodium hydroxide in water
b. Titration of a solution of acetic acid in pyridine by tetramethylammonium hydroxide in pyridine

6.4.11 Superacids

Acid solutions more acidic than sulfuric acid are called **superacids**,[39] the discovery of which won George Olah the Nobel Prize in Chemistry in 1994. The acidity of such solutions is frequently measured by the *Hammett acidity function*:[40]

$$H_0 = pK_{BH^+} - \log \frac{[BH^+]}{[B]}$$

where B and BH^+ are a nitroaniline indicator and its conjugate acid. The stronger the acid, the more negative its H_0 value. On this scale, pure sulfuric acid has an H_0 of –11.9. Fuming sulfuric acid (oleum) is made by dissolving SO_3 in sulfuric acid. This solution contains $H_2S_2O_7$ and higher polysulfuric acids, all of them stronger acids than H_2SO_4. Other superacid solutions and their acidities are given in Table 6.14.

[39] G. Olah and G. K. S. Prakash, *Superacids*, John Wiley & Sons, New York, 1985; G. Olah, G. K. S. Prakash, and J. Sommer, Science, 1979, 206, 13; R. J. Gillespie, *Acc. Chem. Res.*, **1968**, *1*, 202.
[40] L. P. Hammett and A. J. Deyrup, *J. Am. Chem. Soc.*, **1932**, *54*, 2721.

TABLE 6.14 Superacids

Acid		H_0
Sulfuric acid	H_2SO_4	−11.9
Hydrofluoric acid	HF	−11.0
Perchloric acid	$HClO_4$	−13.0
Fluorosulfonic acid	HSO_3F	−15.6
Trifluoromethanesulfonic acid (triflic acid)	HSO_3CF_3	−14.6
Magic Acid*	$HSO_3F\text{-}SbF_5$	−21.0 to −25 (depending on concentration)
Fluoroantimonic acid	$HF\text{-}SbF_5$	−21 to −28 (depending on concentration)

NOTE: *Magic Acid is a registered trademark of Cationics, Inc., Columbia, SC.

The Lewis superacids formed by the fluorides are a result of transfer of anions to form complex fluoro anions.

$$\underset{\text{acid}}{2\,HF} + \underset{\text{base}}{2\,SbF_5} \rightleftharpoons \underset{\text{acid}}{H_2F^+} + \underset{\text{base}}{Sb_2F_{11}^-}$$

$$\underset{\text{acid}}{2\,HSO_3F} + \underset{\text{base}}{2\,SbF_5} \rightleftharpoons \underset{\text{acid}}{H_2SO_3F^+} + \underset{\text{base}}{Sb_2F_{10}(SO_3F)^-}$$

These acids are very strong Friedel–Crafts catalysts. For this purpose, the term *superacid* applies to any acid stronger than $AlCl_3$, the most common Friedel–Crafts catalyst. Other fluorides, such as those of arsenic, tantalum, niobium, and bismuth, also form superacids. Many other compounds exhibit similar behavior; additions to the list of superacids include $HSO_3F\text{-}Nb(SO_3F)_5$ and $HSO_3F\text{-}Ta(SO_3F)_5$, synthesized by oxidation of niobium and tantalum in HSO_3F by $S_2O_6F_2$.[41] Their acidity is explained by reactions similar to those for SbF_5 in fluorosulfonic acid. Crystal structures of a number of the oxonium salts of $Sb_2F_{11}^-$ and cesium salts of several fluorosulfato ions have more recently been determined,[42] and AsF_5 and SbF_5 in HF have been used to protonate H_2Se, H_3As, H_3Sb, H_2Se, H_4P_2, H_2O_2, and H_2S_2.[43]

General References

W. B. Jensen, *The Lewis Acid–Base Concepts: An Overview*, Wiley Interscience, New York, 1980, and H. L Finston and Allen C. Rychtman, *A New View of Current Acid-Base Theories*, John Wiley & Sons, New York, 1982, provide good overviews of the history of acid–base theories and critical discussions of the different theories. R. G. Pearson's *Hard and Soft Acids and Bases*, Dowden, Hutchinson, & Ross, Stroudsburg, PA, 1973, is a review by one of the leading exponents of HSAB. For other viewpoints, the references provided in this chapter should be consulted.

[41] W. V. Cicha and F. Aubke, *J. Am. Chem. Soc.*, **1989**, *111*, 4328.
[42] D. Zhang, S. J. Rettig, J. Trotter, and F. Aubke, *Inorg. Chem.*, **1996**, *35*, 6113.
[43] R. Minkwitz, A. Kormath, W. Sawodny, and J. Hahn, *Inorg. Chem.*, **1996**, *35*, 3622, and references therein.

Problems

Additional acid–base problems may be found at the end of Chapter 8.

6.1 For each of the following reactions, identify the acid and the base. Also indicate which acid–base definition (Lewis, solvent system, Brønsted–Lowry) applies. In some cases, more than one definition may apply.

a. $BF_3 + 2\,ClF \longrightarrow [Cl_2F]^+ + [BF_4]^-$
b. $HClO_4 + CH_3CN \longrightarrow CH_3CNH^+ + ClO_4^-$
c. $PCl_5 + ICl \longrightarrow [PCl_4]^+ + [ICl_2]^-$
d. $NOF + ClF_3 \longrightarrow [NO]^+ + [ClF_4]^-$
e. $2\,ClO_3^- + SO_2 \longrightarrow 2\,ClO_2 + SO_4^{2-}$
f. $Pt + XeF_4 \longrightarrow PtF_4 + Xe$

6.2 For each of the following reactions, identify the acid and the base. Also indicate which acid–base definition (Lewis, solvent system, Brønsted–Lowry) applies. In some cases, more than one definition may apply.

a. $XeO_3 + OH^- \longrightarrow [HXeO_4]^-$
b. $2\,HF + SbF_5 \longrightarrow [H_2F]^+ + [SbF_6]^-$
c. $2\,NOCl + Sn \longrightarrow SnCl_2 + 2\,NO$ (in N_2O_4 solvent)
d. $PtF_5 + ClF_3 \longrightarrow [ClF_2]^+ + [PtF_6]^-$
e. $(benzyl)_3N + CH_3COOH \longrightarrow (benzyl)_3NH^+ + CH_3COO^-$
f. $BH_4^- + 4H_2O \longrightarrow B(OH)_4^- + 4H_2$

6.3 Baking powder is a mixture of aluminum sulfate and sodium hydrogencarbonate, which generates a gas and makes bubbles in biscuit dough. Explain what the reactions are.

6.4 The conductivity of BrF_3 is increased by adding either AgF or SnF_4. Explain this increase, using the appropriate chemical equations.

6.5 The conductivity of ICl is increased by adding either NaCl or $AlCl_3$.

a. Suggest an equation to describe the autodissociation of ICl.
b. Account for the increase in conductivity by the two solutes.

6.6 Titration of NH_4Cl with $SnCl_4$ in ICl requires 2 moles of NH_4Cl for every mole of $SnCl_4$ to reach the end point. Explain, using chemical equations.

6.7 Dissolution of KF in IF_5 increases the conductivity of IF_5. Suggest an explanation.

6.8 The following reaction can be conducted as a titration in liquid BrF_5:

$$2\,Cs[\quad]^- + [\quad]^+[Sb_2F_{11}]^- \longrightarrow 3\,BrF_5 + 2\,CsSbF_6$$

a. The ions in brackets contain both bromine and fluorine. Fill in the most likely formulas of these ions.
b. What are the point groups of the cation and anion identified in part a?
c. Is the cation in part a serving as an acid or base?

6.9 Anhydrous H_2SO_4 and anhydrous H_3PO_4 both have high electrical conductivities. Explain.

6.10 The X-ray structure of $Br_3As \cdot C_6Et_6 \cdot AsBr_3$ (Et = ethyl) has been reported (H. Schmidbaur, W. Bublak, B. Huber, and G. Müller, *Angew. Chem., Int. Ed.*, **1987**, *26*, 234).

a. What is the point group of this structure?
b. Propose an explanation of how the frontier orbitals of $AsBr_3$ and C_6Et_6 can interact to form chemical bonds that stabilize this structure.

6.11 When $AlCl_3$ and $OPCl_3$ are mixed, the product, $Cl_3Al{-}O{-}PCl_3$, has a nearly linear $Al{-}O{-}P$ arrangement (bond angle 176°). Suggest an explanation for this unusually large angle. (See N. Burford, A. D. Phillips, R. W. Schurko, R. E. Wasilishen, and J. F. Richardson, *Chem. Commun. (Cambridge)*, **1997**, 2363.)

6.12 Of the donor–acceptor complexes $(CH_3)_3N{-}SO_3$ and $H_3N{-}SO_3$ in the gas phase,

a. Which has the longer $N{-}S$ bond?
b. Which has the larger $N{-}S{-}O$ angle?

Explain your answers briefly. (See D. L. Fiacco, A. Toro, and K. R. Leopold, *Inorg. Chem.*, **2000**, *39*, 37.)

6.13 Xenon difluoride, XeF_2, can act as a Lewis base toward metal cations such as Ag^+ and Cd^{2+}.

a. In these cases, do you expect the XeF_2 to exert its basicity through the lone pairs on Xe or those on F?
b. $[Ag(XeF_2)_2]AsF_6$ and $[Cd(XeF_2)_2](BF_4)_2$ have both been synthesized. In which case, AsF_6^- or BF_4^-, do you expect the fluorines to act as stronger Lewis bases? Explain briefly. (See G. Tavcar and B. Zemva, *Inorg. Chem.*, **2005**, *44*, 1525.)

6.14 The ion NO^- can react with H^+ to form a chemical bond. Which structure is more likely, HON or HNO? Explain your reasoning.

6.15 The absorption spectra of solutions containing Br_2 are solvent dependent. When elemental bromine is dissolved in nonpolar solvents such as hexane, a single absorption band in the visible spectrum is observed near 500 nm. When Br_2 is dissolved in methanol, however, this absorption band shifts and a new band is formed.

a. Account for the appearance of the new band.
b. Is the 500 nm band likely to shift to a longer or shorter wavelength in methanol? Why?

In your answers, you should show clearly how appropriate orbitals of Br_2 and methanol interact.

6.16 AlF_3 is insoluble in liquid HF but dissolves if NaF is present. When BF_3 is added to the solution, AlF_3 precipitates. Explain.

6.17 Why were most of the metals used in early prehistory class (b) (*soft*, in HSAB terminology) metals?

6.18 The most common source of mercury is cinnabar (HgS), whereas Zn and Cd in the same group occur as sulfide, carbonate, silicate, and oxide. Why?

6.19 The difference between melting point and boiling point (in °C) is given below for each of the Group IIB halides.

	F^-	Cl^-	Br^-	I^-
Zn^{2+}	630	405	355	285
Cd^{2+}	640	390	300	405
Hg^{2+}	5	25	80	100

What deductions can you draw?

6.20 **a.** Use Drago's E and C parameters to calculate ΔH for the reactions of pyridine and BF_3 and of pyridine and $B(CH_3)_3$. Compare your results with the reported experimental values of –71.1 and –64 kJ/mol for pyridine–$B(CH_3)_3$ and –105 kJ/mol for pyridine–BF_3.
b. Explain the differences found in part a in terms of the structures of BF_3 and $B(CH_3)_3$.
c. Explain the differences in terms of HSAB theory.

6.21 Repeat the calculations of the preceding problem using NH_3 as the base, and put the four reactions in order of the magnitudes of their ΔH values.

6.22 Compare the results of Problems 6.20 and 6.21 with the absolute hardness parameters of Appendix B-5 for BF_3, NH_3, and pyridine (C_5H_5N). What value of η would you predict for $B(CH_3)_3$? Compare NH_3 and $N(CH_3)_3$ as a guide.

6.23 CsI is much less soluble in water than CsF, and LiF is much less soluble than LiI. Why?

6.24 Rationalize the following data in HSAB terms:

	ΔH (kcal)
$CH_3CH_3 + H_2O \longrightarrow CH_3OH + CH_4$	12
$CH_3COCH_3 + H_2O \longrightarrow CH_3COOH + CH_4$	–13

6.25 Predict the order of solubility in water of each of the following series, and explain the factors involved.
a. $MgSO_4$ $CaSO_4$ $SrSO_4$ $BaSO_4$
b. $PbCl_2$ $PbBr_2$ PbI_2 PbS

6.26 In some cases CO can act as a bridging ligand between main-group and transition-metal atoms. When it forms a bridge between Al and W in the compound having the formula $(C_6H_5)_3Al$—[bridging CO]—$W(CO)_2(C_5H_5)$, is the order of atoms in the bridge Al—CO—W or Al—OC—W? Briefly explain your choice.

6.27 Choose and explain:
a. Strongest Brønsted acid: SnH_4 SbH_3 TeH_2
b. Strongest Brønsted base: NH_3 PH_3 SbH_3
c. Strongest base to H^+ (gas phase): NH_3 $CH_3NH_2(CH_3)_2$ $NH(CH_3)_3$ N
d. Strongest base to BMe_3: pyridine 2-methylpyridine 4-methylpyridine

6.28 B_2O_3 is acidic, Al_2O_3 is amphoteric, and Sc_2O_3 is basic. Why?

6.29 Predict the reactions of the following hydrogen compounds with water, and explain your reasoning.
a. CaH_2 **b.** HBr **c.** H_2S **d.** CH_4

6.30 List the following acids in order of their acid strength when reacting with NH_3.

BF_3 $B(CH_3)_3$ $B(C_2H_5)_3$ $B(C_6H_2(CH_3)_3)_3$
($C_6H_2(CH_3)_3$ is 2,4,6-trimethylphenyl)

6.31 Choose the stronger acid or base in the following pairs, and explain your choice.
a. CH_3NH_2 or NH_3 in reaction with H^+
b. Pyridine or 2-methylpyridine in reaction with trimethylboron
c. Triphenylboron or trimethylboron in reaction with ammonia

6.32 List the following acids in order of acid strength in aqueous solution:
a. $HMnO_4$ H_3AsO_4 H_2SO_3 H_2SO_4
b. HClO $HClO_4$ $HClO_2$ $HClO_3$

6.33 Solvents can change the acid–base behavior of solutes. Compare the acid–base properties of dimethylamine in water, acetic acid, and 2-butanone.

6.34 HF has $H_0 = -11.0$. Addition of 4% SbF_5 lowers H_0 to –21.0. Explain why this should be true, and why the resulting solution is so strongly acidic that it can protonate alkenes.

$$(CH_3)_2C{=}CH_2 + H^+ \longrightarrow (CH_3)_3C^+$$

6.35 The reasons behind the relative Lewis acidities of the boron halides BF_3, BCl_3, and BBr_3 with respect to NH_3 have been controversial. Although BF_3 might be expected to be the strongest Lewis acid on the basis of electronegativity, the Lewis acidity order is $BBr_3 > BCl_3 > BF_3$. Consult the references listed below to address the following questions. (See also J. A. Plumley and J. D. Evanseck, *J. Phys. Chem. A*, **2009**, *113*, 5985.)
a. How does the LCP approach account for a Lewis acidity order of $BBr_3 > BCl_3 > BF_3$? (See B. D. Rowsell, R. J. Gillespie, and G. L. Heard, *Inorg. Chem.*, **1999**, *38*, 4659.)
b. What explanation has been offered on the basis of the calculations presented in F. Bessac and G. Frenking, *Inorg. Chem.*, **2003**, *42*, 7990?

6.36 A Lewis base can be "frustrated" if, because of crowding, it cannot get its donor pair close enough to a Lewis acid for a reaction to occur. For example, $P(t\text{-}C_4H_9)_3$ and $B(C_6F_5)_3$ do not react directly because of crowding between the lone pair of phosphorus and the boron. However, if an equimolar mixture of $P(t\text{-}C_4H_9)_3$ and $B(C_6F_5)_3$ is mixed with 1 bar of the gas N_2O in bromobenzene solution, a white product is formed in good yield. A variety of NMR evidence has been gathered on the product: there is a single ^{31}P NMR resonance; ^{11}B and ^{19}F NMR are consistent with a 4-coordinate boron atom; and ^{15}N NMR indicates two nonequivalent nitrogen atoms. In addition, no gas is released in the reaction.

a. Suggest the role of N_2O in this reaction.

b. Propose a structure of the product. (See E. Otten, R. C. Neu, and D. W. Stephan, *J. Am. Chem. Soc.*, **2009**, *131*, 9918.)

6.37 What is a "frustrated Lewis pair" (FLP)? What are some potential ways to exploit such pairs? (See D. W. Stephan, *Dalton Trans.*, **2009**, *3129*.)

The following problems use molecular modeling software.

6.38 **a.** Calculate and display the molecular orbitals of NO^-. Show how the reaction of NO^- and H^+ can be described as a HOMO–LUMO interaction.

b. Calculate and display the molecular orbitals of HNO and HON. On the basis of your calculations, and your answer to part a, which structure is favored?

6.39 Calculate and display the frontier orbitals of Br_2, methanol, and the Br_2–methanol adduct to show how the orbitals of the reactants interact.

6.40 **a.** Calculate and display the molecular orbitals of BF_3, NH_3, and the $F_3B{-}NH_3$ Lewis acid–base adduct.

b. Examine the bonding and antibonding orbitals involved in the $B{-}N$ bond in $F_3B{-}NH_3$. Is the bonding orbital polarized toward the B or the N? The antibonding orbital? Explain briefly.

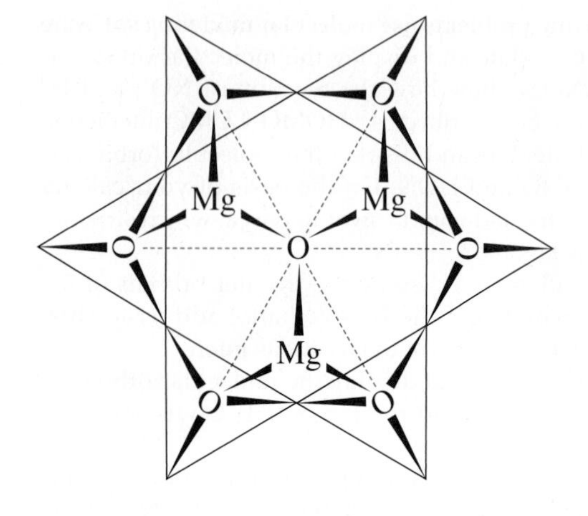

7 The Crystalline Solid State

Solid-state chemistry uses the same principles for bonding as those for molecules; the differences arise from the magnitude of the "molecules" in the solid state. In many cases, a macroscopic crystal can reasonably be described as a single molecule with molecular orbitals extending throughout. This description leads to significant differences in the molecular orbitals and behavior of solids in comparison with small molecules. There are two major classifications of solid materials: crystals and amorphous materials. Our attention in this chapter is on crystalline solids composed of atoms or ions.

We will first describe the common structures of crystals and then consider their bonding according to the molecular orbital approach. Finally, we will describe some of the thermodynamic and electronic properties of these materials and their uses.

7.1 FORMULAS AND STRUCTURES

Crystalline solids have atoms, ions, or molecules packed in regular geometric arrays, with the structural unit called the **unit cell**. Some of the common crystal geometries are described in this section. In addition, we will consider the role of the relative sizes of the components in determining structures. Use of a model kit, such as the one available from ICE,[1] makes the study of these structures much easier.

7.1.1 Simple Structures

The crystal structures of metals are relatively simple. Those of some minerals can be complex, but minerals usually have simpler structures that can be recognized within the more complex structure. The unit cell is a structural component that, when repeated in all directions, results in a macroscopic crystal. Structures of the 14 possible crystal structures (Bravais lattices) are shown in Figure 7.1. Several different unit cells are possible for some structures; the one used may be chosen for convenience, depending on the particular application. The atoms (or ions) on the corners, edges, or faces of the unit cell are shared with other unit cells as follows:

> Atoms at the corners of rectangular unit cells are shared equally by eight unit cells and contribute $\frac{1}{8}$ to each ($\frac{1}{8}$ of the atom is counted as part of each cell). The total for a single unit cell is $8 \times \frac{1}{8} = 1$ atom for all of the corners.

[1] Institute for Chemical Education, Department of Chemistry, University of Wisconsin–Madison, 1101 University Ave., Madison, WI 53706. Sources for other model kits are given in A. B. Ellis, M. J. Geselbracht, B. J. Johnson, G. C. Lisensky, and W. R. Robinson, *Teaching General Chemistry: A Materials Science Companion*, American Chemical Society, Washington, DC, 1993.

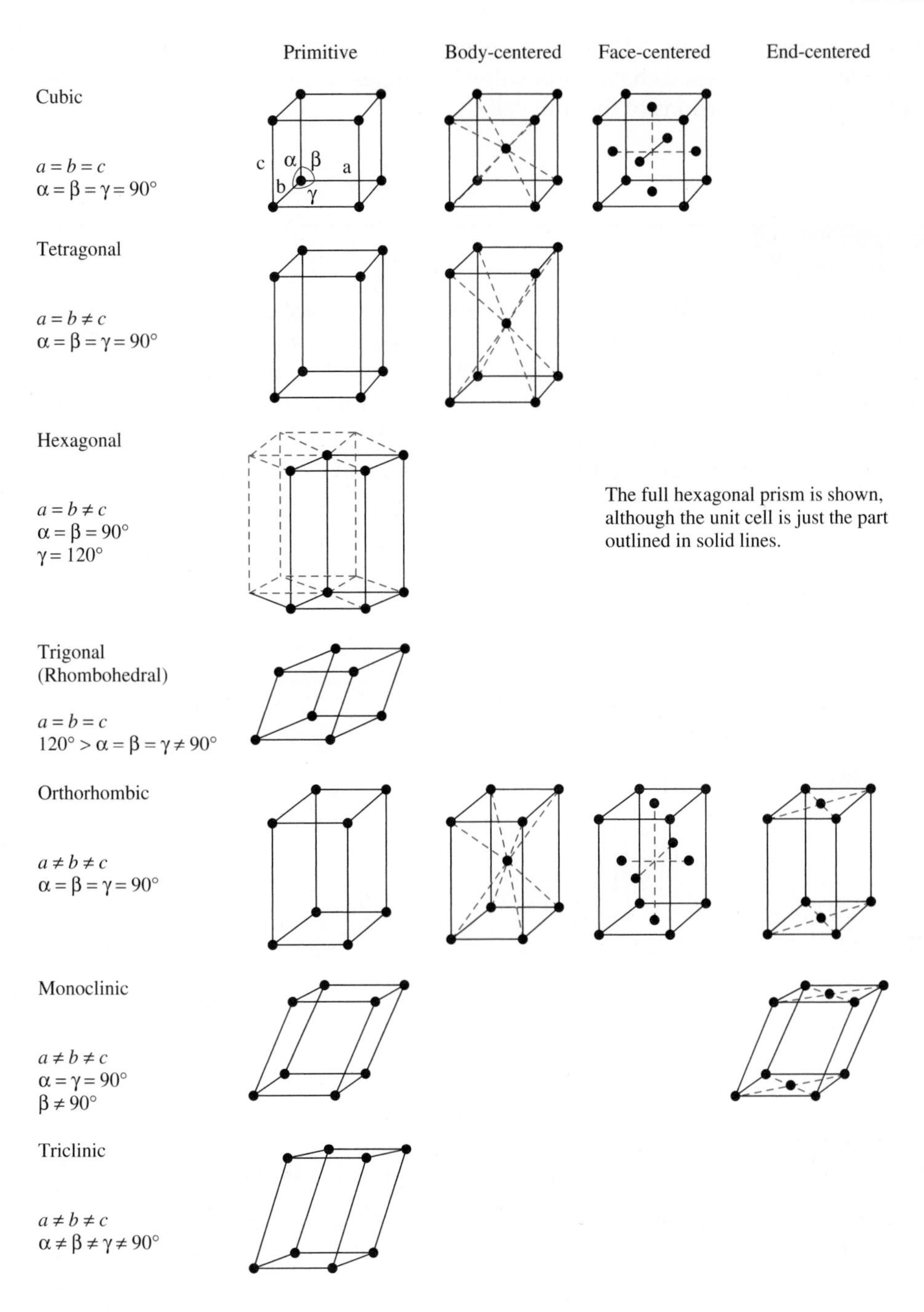

FIGURE 7.1 The Seven Crystal Classes and Fourteen Bravais Lattices. The points shown are not necessarily individual atoms but are included to show the necessary symmetry.

Atoms at the corners of nonrectangular unit cells also contribute one atom total to the unit cell; small fractions on one corner are matched by larger fractions on another.

Atoms on edges of unit cells are shared by four unit cells—two in one layer, two in the adjacent layer—and contribute $\frac{1}{4}$ to each.

Atoms on faces of unit cells are shared between two unit cells and contribute $\frac{1}{2}$ to each.

As can be seen in Figure 7.1, unit cells need not have equal dimensions or angles. For example, triclinic crystals have three different angles and may have three different distances for the dimensions of the unit cell.

EXAMPLE

The diagram below shows a space-filling diagram of a face-centered cubic unit cell cut to show only the part of each atom that is inside the unit-cell boundaries. The corner atoms are each shared among eight unit cells, so $\frac{1}{8}$ of the atom is in the unit cell shown. The face-centered atoms are shared between two unit cells, so $\frac{1}{2}$ of the atom is in the unit cell shown. The eight corners of the unit cell then total $8 \times \frac{1}{8} = 1$ atom, and the six faces total $6 \times \frac{1}{2} = 3$ atoms; a total of four atoms are in the unit cell.

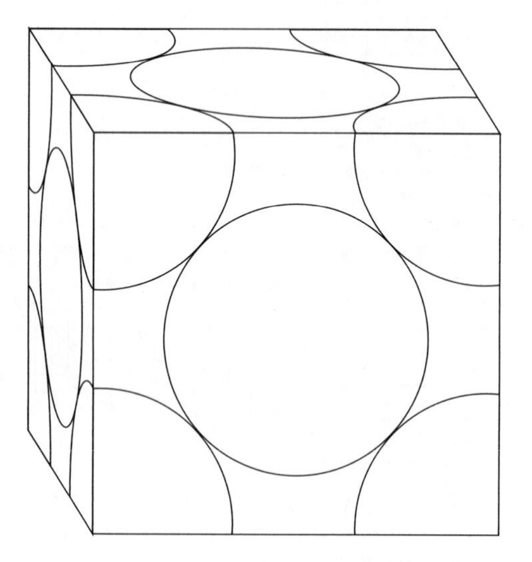

▶ **Exercise 7.1** Calculate the number of atoms in each unit cell of

a. A body-centered cubic structure
b. A hexagonal structure

The structures are shown in Figure 7.1.

The positions of atoms are frequently described in **lattice points**, expressed as fractions of the unit cell dimensions. For example, the body-centered cube has atoms at the origin [the corner at which $x = 0, y = 0, z = 0$, or (0, 0, 0)] and at the center of the cube $\left[x = \frac{1}{2}, y = \frac{1}{2}, z = \frac{1}{2}, \left(\frac{1}{2}, \frac{1}{2}, \frac{1}{2}\right)\right]$. The other atoms can be generated by moving these two atoms in each direction in increments of one cell length.

CUBIC The most basic crystal structure is the simple cube, called the **primitive cubic** structure, with atoms at the eight corners. It can be described by specifying the length of one side, the angle 90°, and the single lattice point (0, 0, 0). Because each of the atoms is shared between eight cubes, four in one layer and four in the layer above or below, the total number of atoms in the unit cell is $8 \times \frac{1}{8} = 1$, the number of lattice points required. Each atom is surrounded by six others, for a **coordination number** (CN) of 6. This structure is not efficiently packed, because the spheres occupy only 52.4% of the total volume. In the center of the cube is a vacant space that has eight nearest neighbors, or a coordination number of 8. Calculation shows that a sphere with a radius $0.73r$, where r is the radius of the corner spheres, would fit in the center of this cube if the corner spheres are in contact with each other.

BODY-CENTERED CUBIC If another sphere is added in the center of the simple cubic structure, the result is called **body-centered cubic (bcc)**. If the added sphere has the same radius as the others, the size of the unit cell expands so that the diagonal distance through the cube is $4r$, where r is the radius of the spheres. The corner atoms are no longer in contact with each other. The new unit cell is $2.31r$ on each side and contains two atoms, because the body-centered atom is completely within the unit cell. This cell has two lattice points, at the origin (0, 0, 0) and at the center of the cell $\left(\frac{1}{2}, \frac{1}{2}, \frac{1}{2}\right)$.

▶ Exercise 7.2 Show that the side of the unit cell for a body-centered cubic crystal is 2.31 times the radius of the atoms in the crystal.

CLOSE-PACKED STRUCTURES When marbles or ball bearings are poured into a flat box, they tend to form a close-packed layer, in which each sphere is surrounded by six others in the same plane. This arrangement provides the most efficient packing possible for a single layer. When three or more close-packed layers are placed on top of each other systematically, two structures are possible. When the third layer is placed with all atoms directly above those of the first layer, the result is an ABA structure called **hexagonal close packing (hcp)**. When the third layer is displaced, so each atom is above a hole in the first layer, the resulting ABC structure is called **cubic close packing (ccp)** or **face-centered cubic (fcc)**. In both **hcp** and **ccp/fcc** structures, the coordination number for each atom is 12, six in its own layer, three in the layer above, and three in the layer below. These are shown in Figure 7.2. In both these structures, there are two tetrahedral holes per atom (coordination number 4, formed by three atoms in one layer and one in the layer above or below) and one octahedral hole per atom (three atoms in each layer, a total coordination number of 6).

Hexagonal close packing is relatively easy to see, with hexagonal prisms sharing vertical faces in the larger crystal (Figure 7.3). The minimal unit cell is smaller than the hexagonal prism; taking any four atoms that all touch each other in one layer and extending lines up to the third layer will generate a unit cell with a parallelogram as the base. As shown in Figure 7.3, this cell contains half an atom in the first layer (four atoms averaging $\frac{1}{8}$ each), four similar atoms in the third layer, and one atom from the second layer—whose center is within the unit cell—for a total of two atoms in the unit cell. The unit cell has dimensions of $2r$, $2r$, and $2.83r$, an angle of 120° between the first two axes in the basal plane, and 90° between each of these axes and the third, vertical axis. The atoms are at the lattice points (0, 0, 0) and $\left(\frac{1}{3}, \frac{2}{3}, \frac{1}{2}\right)$.

The cube in cubic close packing is more difficult to see when each of the layers is close-packed. The unit-cell cube rests on one corner, with four close-packed layers

FIGURE 7.2 Close-Packed Structures.

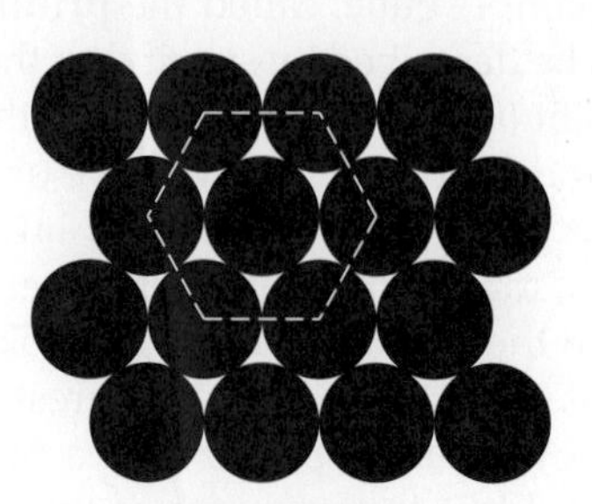

A single close-packed layer, A, with the hexagonal packing outlined.

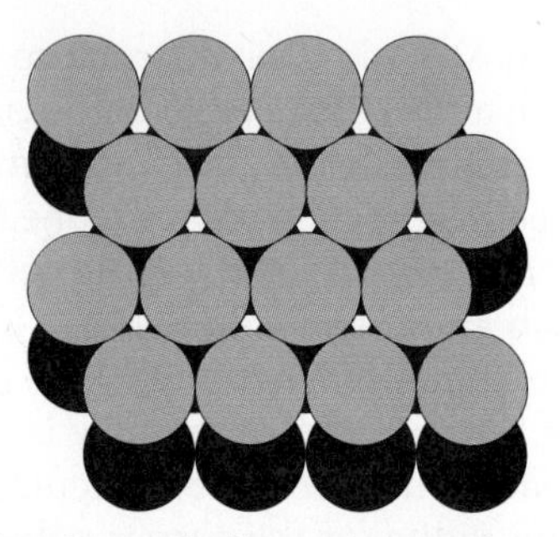

Two close-packed layers, A and B. Octahedral holes can be seen extending through both layers surrounded by three atoms in each layer. Tetrahedral holes are under each atom of the second layer and over each atom of the bottom layer. Each is made up of three atoms from one layer and one from the other.

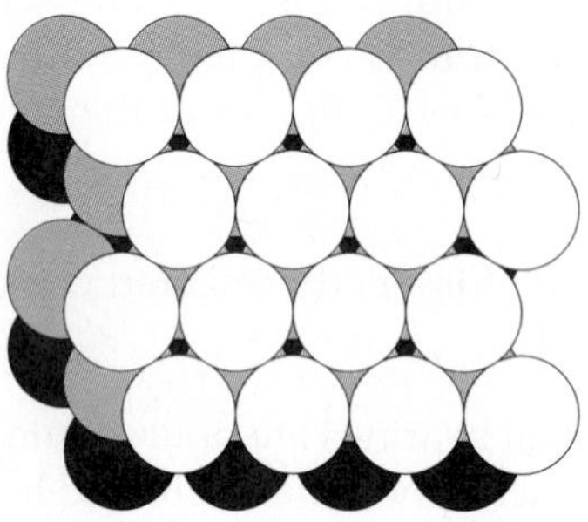

Cubic close-packed layers, in an ABC pattern. Octahedral holes are offset, so no hole extends through all three layers.

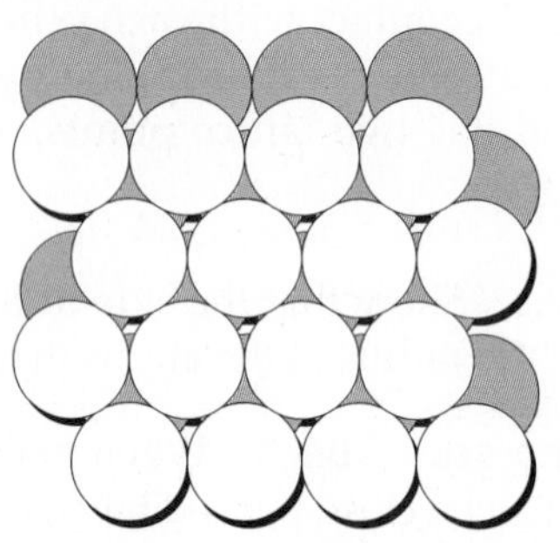

Hexagonal close-packed layers. The third layer is exactly over the first layer in this ABA pattern. Octahedral holes are aligned exactly over each other, one set between the first two layers A and B, the other between the second and third layers, B and A.

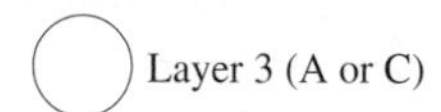

required to complete the cube. The first layer has only one sphere, and the second has six in a triangle, as shown in Figure 7.4(a). The third layer has another six-membered triangle, with the vertices rotated 60° from the one in the second layer, and the fourth layer again has one sphere. The cubic shape of the cell is easier to see if the faces are placed in the conventional horizontal and vertical directions, as in Figure 7.4(b).

The unit cell of the cubic close-packed structure is a face-centered cube, with spheres at the corners and in the middle of each of the six faces. The lattice points are at $(0, 0, 0)$, $\left(\frac{1}{2}, \frac{1}{2}, 0\right)$, $\left(\frac{1}{2}, 0, \frac{1}{2}\right)$, and $\left(0, \frac{1}{2}, \frac{1}{2}\right)$, for a total of four atoms in the unit cell, $\frac{1}{8} \times 8$ at the corners and $\frac{1}{2} \times 6$ in the faces. In both close-packed structures, the spheres occupy 74.0% of the total volume.

Ionic crystals can also be described in terms of the interstices, or holes, in the structures. Figure 7.5 shows the location of tetrahedral and octahedral holes in close-packed structures. Whenever an atom is placed in a new layer over a close-packed layer, it creates a tetrahedral hole surrounded by three atoms in the first layer and one in the second (CN = 4). When additional atoms are added to the second layer, they also create tetrahedral holes surrounded by one atom in the one

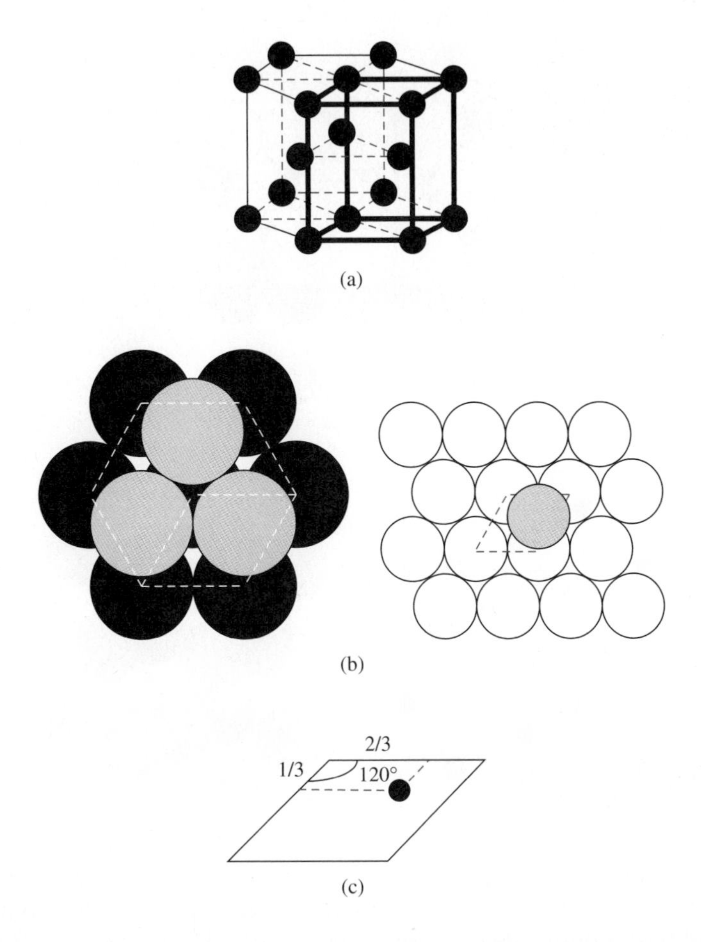

FIGURE 7.3 Hexagonal Close Packing. (a) The hexagonal prism with the unit cell outlined in bold. (b) Two layers of an hcp unit cell. The parallelogram is the base of the unit cell. The third layer is identical to the first. (c) Location of the atom in the second layer.

layer and three in the other. In addition, there are octahedral holes (CN = 6) surrounded by three atoms in each layer. Overall, close-packed structures have two tetrahedral holes and one octahedral hole per atom. These holes can be filled by smaller ions: the tetrahedral holes by ions with radius $0.225r$ for ions in contact, where r is the radius of the larger ions, and the octahedral holes by ions with radius $0.414r$. In more complex crystals, even if the ions are not in contact with each other, the geometry is described in the same terminology. For example, NaCl has chloride ions in a cubic close-packed array, with sodium ions, also in a ccp array, in the octahedral holes. The sodium ions have a radius 0.695 times the radius of the chloride ion ($r_+ = 0.695r_-$), large enough to force the chloride ions apart, but not large enough to allow a coordination number larger than 6.

METALLIC CRYSTALS Except for the actinides, most metals crystallize in body-centered cubic, cubic close-packed, and hexagonal close-packed structures, with approximately equal numbers of each type. In addition, changes in pressure or temperature can change many metallic crystals from one structure to another. This variability shows that we should not think of these metal atoms as hard spheres that pack together in crystals independent of electronic structure. Instead, the sizes and packing of atoms are somewhat variable. Atoms attract each other at moderate distances and repel each other when they are close enough that their electron clouds overlap too much. The balance between these forces, modified by the specific electronic configuration of the atoms, determines the net forces between them and the structure that is most stable. Simple geometric calculations alone are not sufficient.

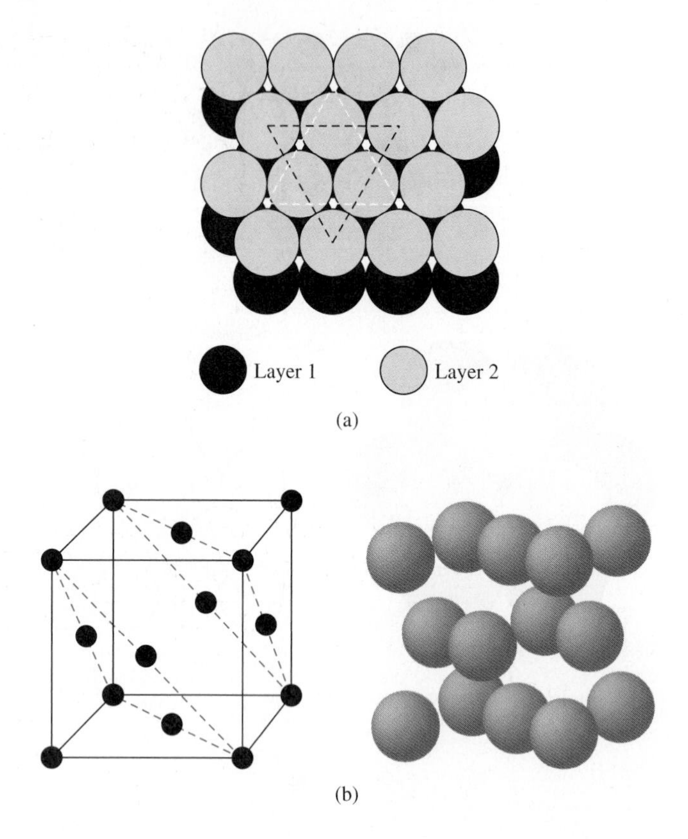

FIGURE 7.4 Cubic Close Packing. (a) Two layers of a ccp (or fcc) cell. The atom in the center of the triangle in the first layer and the six atoms connected by the triangle form half the unit cell. The other half, in the third and fourth layers, is identical, but with the direction of the triangle reversed. (b) Two views of the unit cell, with the close-packed layers marked in the first.

PROPERTIES OF METALS The single property that best distinguishes metals from nonmetals is conductivity. Metals have high conductivity (low resistance) to the passage of electricity and heat; nonmetals have low conductivity (high resistance). One exception is diamond, which has low electrical conductivity and high heat conductivity. Conductivity is discussed further in Section 7.3 on the electronic structure of metals and semiconductors.

Aside from conductivity, metals have quite varied properties. Some are soft and easily deformed by pressure or impact, or malleable (Cu, an fcc structure), whereas others are hard and brittle, more likely to break rather than bend (Zn, an hcp structure). However, most can be shaped by hammering or bending. This is possible because the bonding in metals is nondirectional; each atom is bonded to all neighboring atoms, rather than to individual atoms, as is the case in discrete molecules. When force is applied, the atoms can slide over each other and realign into new structures with nearly

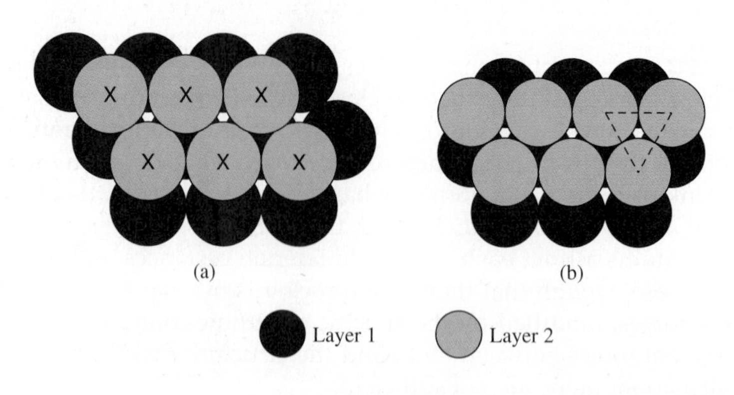

FIGURE 7.5 Tetrahedral and Octahedral Holes in Close-Packed Layers. (a) Tetrahedral holes are under each x and at each point where an atom of the first layer appears in the triangle between three atoms in the second layer. (b) An octahedral hole is outlined, surrounded by three atoms in each layer.

the same overall energy. This effect is facilitated by **dislocations**, where the crystal is imperfect and atoms are out of place; they occupy those positions because of the rigidity of the rest of the crystal. The effects of these discontinuities are increased by impurity atoms, especially those with a size different from that of the host. These atoms tend to accumulate at discontinuities in the crystal, making it even less uniform. Such imperfections allow gradual slippage of layers, rather than movement of an entire layer at the same time. Some metals can be work-hardened by repeated deformation. When the metal is hammered, the defects tend to group together, eventually resisting deformation. Heating can restore flexibility by redistributing the dislocations and reducing their numbers. For different metals or alloys (mixtures of metals) heat treatment and slow or fast cooling can lead to much different results. Some metals can be tempered to be harder and hold a sharp edge better, others can be heat-treated to be more resilient and flex without being permanently bent. Still others can be treated to have "shape memory." These alloys can be bent but return to their initial shape on moderate heating.

DIAMOND The final simple structure we will consider is that of diamond (Figure 7.6), which has the same overall geometry as zinc blende (described later) but with all atoms identical. If a face-centered cubic crystal is divided into eight smaller cubes by planes cutting through the middle, and additional atoms are added in the centers of four of the smaller cubes, none of them adjacent, the diamond structure is the result. Each carbon atom is bonded tetrahedrally to its four nearest neighbors, and the bonding between them is similar to ordinary carbon–carbon single bonds. The strength of the crystal is a consequence of the covalent nature of the bonding; each carbon has its complete set of four bonds. Although there are cleavage planes in diamond, the structure has essentially the same strength in all directions. In addition to carbon, three other elements in the same group—silicon, germanium, and α-tin—have the same structure. Ice also has the same crystal symmetry (see Figure 3.25), with O—H—O hydrogen bonds between all the oxygens. The ice structure is more open because of the greater distance between oxygen atoms.

7.1.2 Structures of Binary Compounds

Binary compounds, those consisting of two elements, may have very simple crystal structures and can be described in several different ways. Two simple structures are shown in Figure 7.7. As described in Section 7.1.1, in close-packed structures, there are two tetrahedral holes and one octahedral hole per atom. If the larger ions (usually the anions) are in close-packed structures, ions of the opposite charge occupy these holes, depending primarily on two factors:

1. *The relative sizes of the atoms or ions.* The **radius ratio** (usually r_+/r_- but sometimes r_-/r_+, where r_+ is the radius of the cation and r_- is the radius of the anion) is generally used to measure this. Small cations can fit in the tetrahedral or octahedral

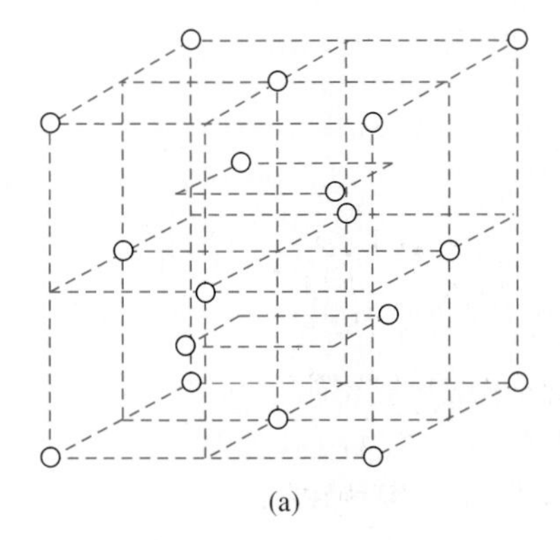

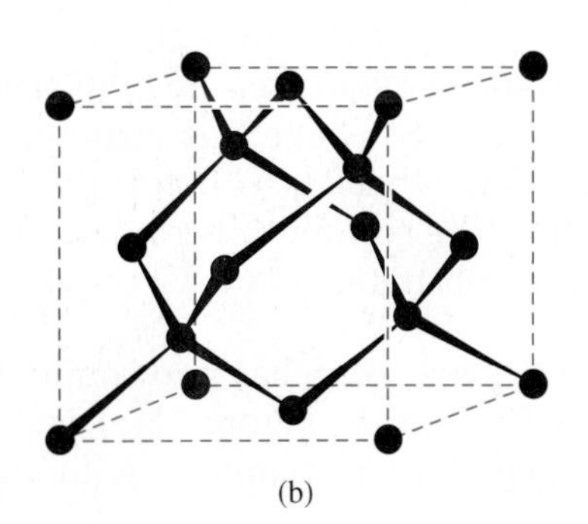

FIGURE 7.6 The Structure of Diamond. (a) Subdivision of the unit cell, with atoms in alternating smaller cubes. (b) The tetrahedral coordination of carbon is shown for the four interior atoms.

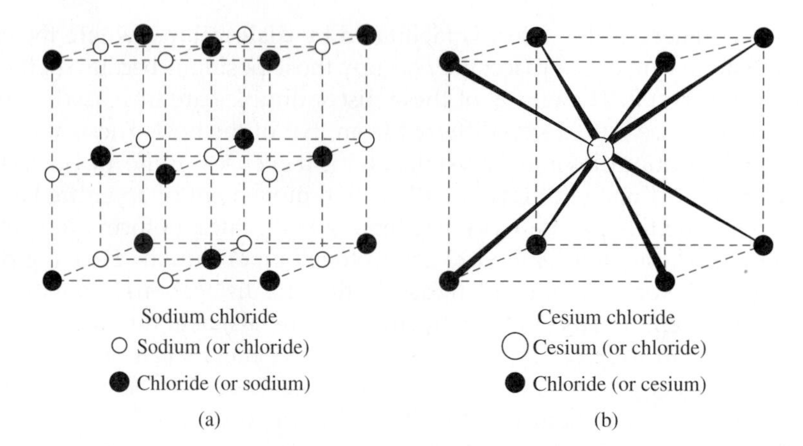

FIGURE 7.7 Sodium Chloride and Cesium Chloride Unit Cells.

holes of a close-packed anion lattice. Somewhat larger cations can fit in the octahedral holes, but not in tetrahedral holes, of the same lattice. Still larger cations force a change in structure. This will be explained more fully in Section 7.1.4.

2. *The relative numbers of cations and anions.* For example, a formula of M_2X will not allow a close-packed anion lattice and occupancy of all of the octahedral holes by the cations, because there are too many cations. The structure must either have the cations in tetrahedral holes, have many vacancies in the anion lattice, or have an anion lattice that is not close-packed.

The structures described in this section are generic, named for the most common compound with the structure. Although some structures are also influenced by the electronic structure of the ions, particularly when there is a high degree of covalency, this effect will not be considered here.

SODIUM CHLORIDE, NaCl NaCl, Figure 7.7(a), is made up of face-centered cubes of sodium ions and face-centered cubes of chloride ions, offset by half a unit-cell length in one direction, so that the sodium ions are centered in the edges of the chloride lattice and vice versa. If all the ions were identical, the NaCl unit cell would be made up of eight simple, cubic unit cells. Many alkali halides and other simple compounds share this same structure. For these crystals, the ions tend to have quite different sizes, usually with the anions larger than the cations. Each sodium ion is surrounded by six nearest-neighbor chloride ions, and each chloride ion is surrounded by six nearest-neighbor sodium ions.

CESIUM CHLORIDE, CsCl As mentioned previously, a sphere of radius $0.73r$ will fit exactly in the center of a cubic structure. Although the fit is not perfect, this is what happens in CsCl, Figure 7.7(b), where the chloride ions form simple cubes with cesium ions in the centers. In the same way, the cesium ions form simple cubes with chloride ions in the centers. The average chloride ion radius is 0.83 times as large as the cesium ion (167 pm and 202 pm, respectively), but the interionic distance in CsCl is 356 pm, about 3.5% smaller than the sum of the average ionic radii. Only CsCl, CsBr, CsI, TlCl, TlBr, TlI, and CsSH have this structure at ordinary temperatures and pressures, although some other alkali halides have this structure at high pressure and high temperature. The cesium salts can also be made to crystallize in the NaCl lattice on NaCl or KBr substrates, and CsCl converts to the NaCl lattice at about 469° C.

ZINC BLENDE, ZnS ZnS has two common crystalline forms, both with coordination number 4. Zinc blende, Figure 7.8(a), is the most common zinc ore and has essentially the same geometry as diamond, with alternating layers of Zn and S. It can also be

FIGURE 7.8
ZnS Crystal Structures.
(a) Zinc blende.
(b, c) Wurtzite.

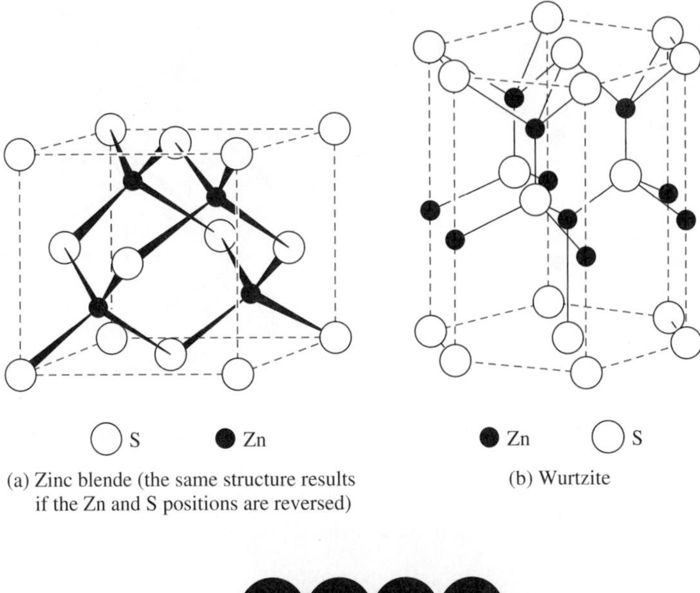

(a) Zinc blende (the same structure results if the Zn and S positions are reversed)

(b) Wurtzite

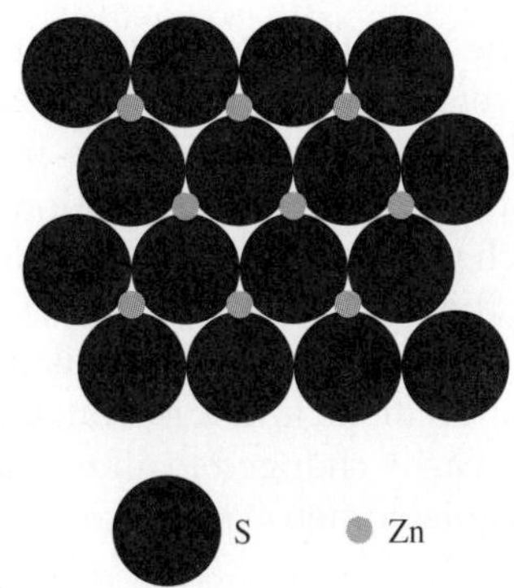

(c) One sulfide layer and one zinc layer of wurtzite. The third layer contains sulfide ions, directly above the zinc ions. The fourth layer is zinc ions, directly above the sulfides of the first layer.

described as having zinc ions and sulfide ions, each in face-centered lattices, so that each ion is in a tetrahedral hole of the other lattice. The stoichiometry requires half of these tetrahedral holes to be occupied, with alternating occupied and vacant sites.

WURTZITE, ZnS The wurtzite form of ZnS, Figure 7.8(b, c), is much rarer than zinc blende and is formed at higher temperatures. It also has zinc and sulfide, each in a tetrahedral hole of the other lattice, but each type of ion forms a hexagonal close-packed lattice. As in zinc blende, half of the tetrahedral holes in each lattice are occupied.

FLUORITE, CaF_2 The fluorite structure in Figure 7.9 can be described as having the calcium ions in a cubic close-packed lattice, with eight fluoride ions surrounding each one and occupying all of the tetrahedral holes. An alternative description of the same structure, shown in Figure 7.9(b), has the fluoride ions in a simple cubic array, with calcium ions in alternate body centers. The ionic radii are nearly perfect fits for this geometry. There is also an *antifluorite* structure, in which the cation–anion stoichiometry is reversed. This structure is found in all the oxides and sulfides of Li, Na, K, and Rb and in Li_2Te and Be_2C. In the antifluorite structure, every tetrahedral hole in the anion

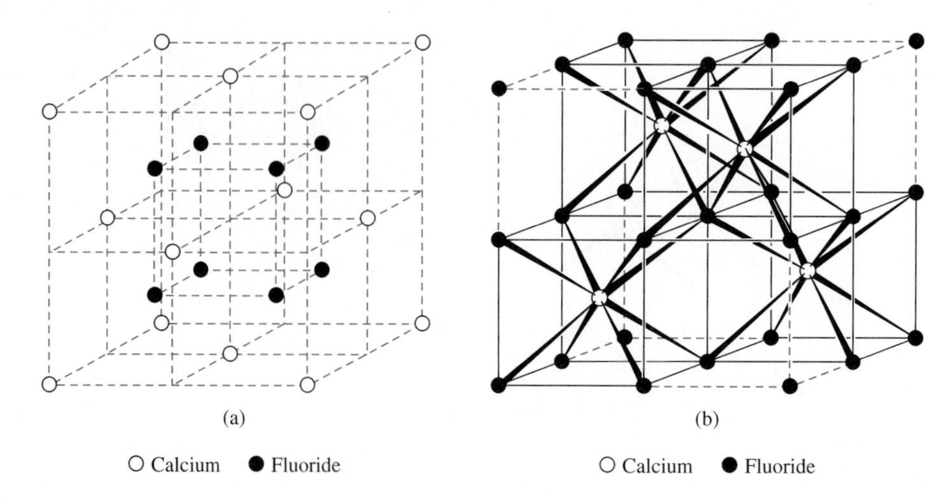

FIGURE 7.9 Fluorite and Antifluorite Crystal Structures. (a) Fluorite shown as Ca^{2+} in a cubic close-packed lattice, each surrounded by eight F^- in the tetrahedral holes. (b) Fluorite shown as F^- in a simple cubic array, with Ca^{2+} in alternate body centers. Solid lines enclose the cubes containing Ca^{2+} ions. If the positive and negative ion positions are reversed, as in Li_2O, the structure is known as antifluorite.

lattice is occupied by a cation, in contrast to the ZnS structures, in which half the tetrahedral holes of the sulfide ion lattice are occupied by zinc ions.

NICKEL ARSENIDE, NiAs The nickel arsenide structure (Figure 7.10) has arsenic atoms in identical close-packed layers stacked directly over each other with nickel atoms in all the octahedral holes. The larger arsenic atoms are in the center of trigonal prisms of nickel atoms. Both types of atoms have coordination number 6, with layers of nickel atoms close enough that each nickel can also be considered as bonded to two others. An alternate description is that the nickel atoms occupy all the octahedral holes of a hexagonal close-packed arsenic lattice. This structure is also adopted by many MX compounds, where M is a transition metal and X is from Groups 14, 15, or 16 (Sn, As, Sb, Bi, S, Se, or Te). This structure is easily changed to allow for larger amounts of the nonmetal to be incorporated into nonstoichiometric materials.

RUTILE, TiO_2 TiO_2 in the rutile structure (Figure 7.11) has distorted TiO_6 octahedra that form columns by sharing edges, resulting in coordination numbers of 6 and 3 for titanium and oxygen, respectively. Adjacent columns are connected by sharing corners of

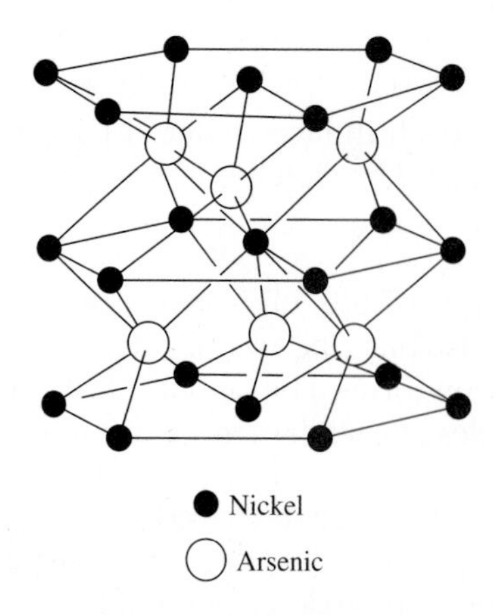

FIGURE 7.10 NiAs Crystal Structure.

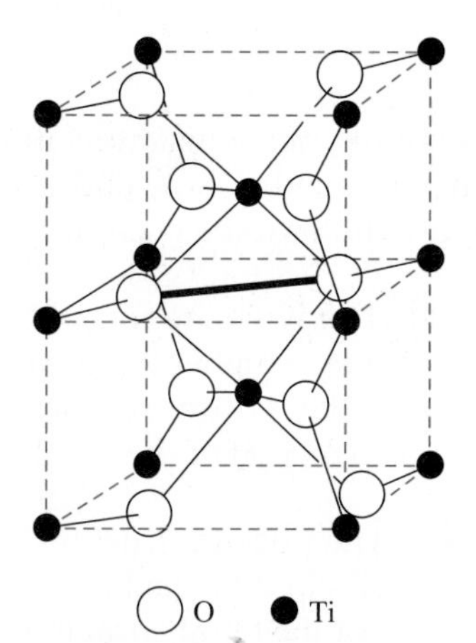

FIGURE 7.11 Rutile (TiO_2) Crystal Structure. The figure shows two unit cells of rutile. The heavy line across the middle shows the edge shared between two TiO_6 octahedra.

the octahedra. The oxide ions have three nearest-neighbor titanium ions in a planar configuration, one at a slightly greater distance than the other two. The unit cell has titanium ions at the corners and in the body center, two oxygens in opposite quadrants of the bottom face, two oxygens directly above the first two in the top face, and two oxygens in the plane with the body-centered titanium forming the final two positions of the oxide octahedron. The same geometry is found for MgF_2, ZnF_2, and some transition-metal fluorides. Compounds that contain larger metal ions adopt the fluorite structure with coordination numbers of 8 and 4.

7.1.3 More Complex Compounds

It is possible to form many compounds by substitution of one ion for another in part of the locations in a lattice. If the charges and ionic sizes are the same, there may be a wide range of possibilities. If the charges or sizes differ, the structure must change, sometimes balancing the charge by leaving vacancies and frequently adjusting the lattice to accommodate larger or smaller ions. When the anions are complex and nonspherical, the crystal structure must accommodate the shape by distortions, and large cations may require increased coordination numbers. A large number of salts ($LiNO_3$, $NaNO_3$, $MgCO_3$, $CaCO_3$, $FeCO_3$, $InBO_3$, YBO_3) adopt the calcite structure, Figure 7.12(a), named for a hexagonal form of calcium carbonate, in which the metal has six nearest-neighbor oxygens. A smaller number (KNO_3, $SrCO_3$, $LaBO_3$), with larger cations, adopt the aragonite structure shown in Figure 7.12(b), an orthorhombic form of $CaCO_3$ that has 9-coordinate metal ions.

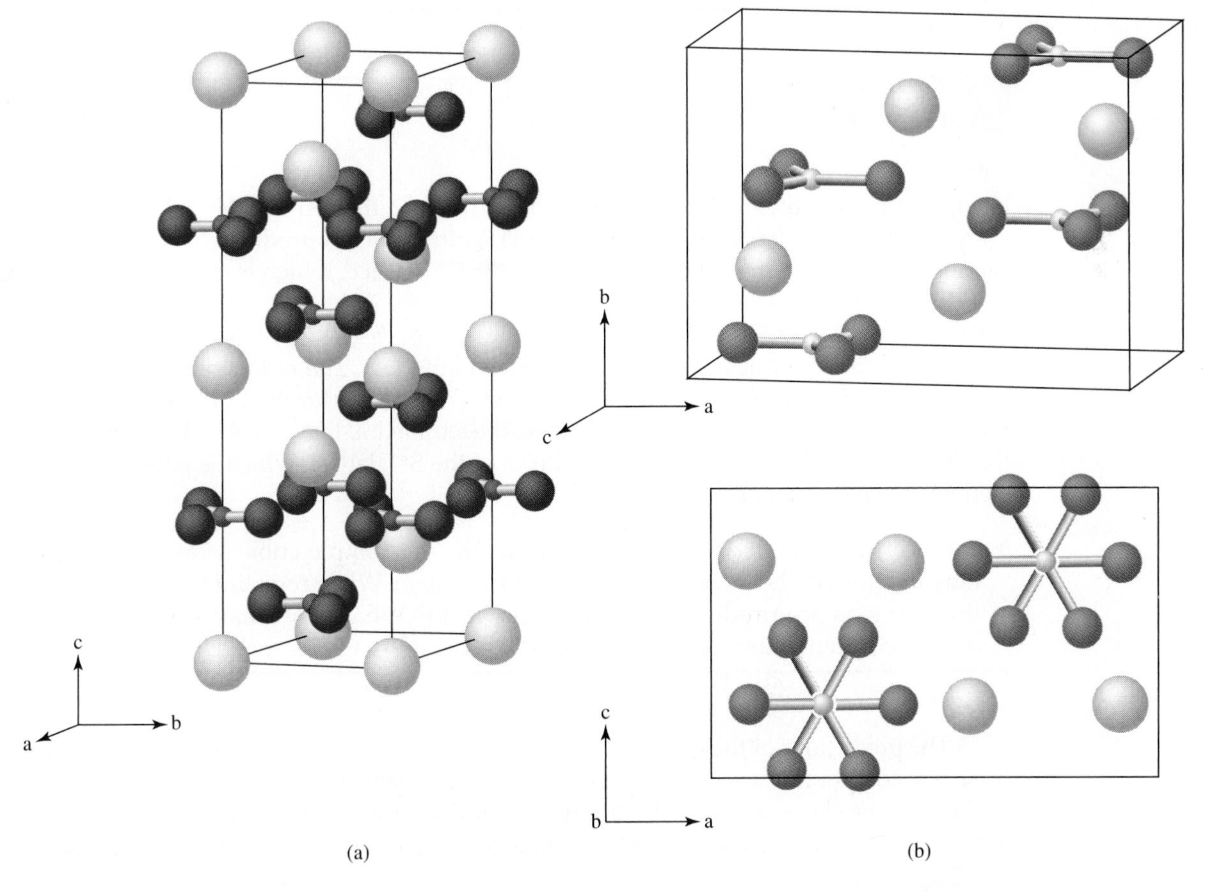

FIGURE 7.12 Structures of Calcium Carbonate, $CaCO_3$. (a) Calcite. (b) Two views of aragonite.

7.1.4 Radius Ratio

Coordination numbers in different crystals depend on the sizes and shapes of the ions or atoms, their electronic structures, and, in some cases, on the temperature and pressure under which they were formed. A simple, but at best approximate, approach to predicting coordination numbers uses the radius ratio, r_+/r_-. Simple calculation from tables of ionic radii allows prediction of possible structures, treating the ions as if they were hard spheres. For hard spheres, the ideal size for a smaller cation in an octahedral hole of an anion lattice is a radius of $0.414r_-$. Similar calculations for other geometries result in the radius ratios (r_+/r_-) shown in Table 7.1.

TABLE 7.1 Radius Ratios and Coordination Numbers

Radius Ratio Limiting Values	Coordination Number	Geometry	Ionic Compounds
	4	Tetrahedral	ZnS
0.414			
	4	Square planar	None
	6	Octahedral	NaCl, TiO_2 (rutile)
0.732			
	8	Cubic	CsCl, CaF_2 (fluorite)
1.00			
	12	Cubooctahedron	No ionic examples, but many metals are 12-coordinate.

EXAMPLE

NaCl

Using the radius of the Na^+ cation (Appendix B-1) for either CN = 4 or CN = 6, $r_+/r_- = 113/167 = 0.677$ or $116/167 = 0.695$, both of which predict CN = 6. The Na^+ cation fits easily into the octahedral holes of the Cl^- lattice, which is ccp.

ZnS

The zinc ion radius varies more with coordination number. The radius ratios are $r_+/r_- = 74/170 = 0.435$ for the CN = 4 and $r_+/r_- = 88/170 = 0.518$ for the CN = 6 radius. Both predict CN = 6, but the smaller one is close to the tetrahedral limit of 0.414. Experimentally, the Zn^{2+} cation fits into the tetrahedral holes of the S^{2-} lattice, which is either ccp (zinc blende) or hcp (wurtzite).

▶ **Exercise 7.3** Fluorite (CaF_2) has fluoride ions in a simple cubic array and calcium ions in alternate body centers, with $r_+ / r_- = 0.97$. What coordination numbers of the two ions are predicted by the radius ratio? What coordination numbers are observed? Predict the coordination number of Ca^{2+} in $CaCl_2$ and $CaBr_2$.

The predictions of the example and exercise match reasonably well with the facts for these two compounds, even though ZnS is largely covalent rather than ionic. However, all radius ratio predictions should be used with caution, because ions are not hard spheres, and there are many cases in which the radius ratio predictions are not correct. One study[2]

[2] L. C. Nathan, *J. Chem. Educ.*, **1985**, *62*, 215.

reported that the actual structure matches the predicted structure in about two thirds of the cases, with a higher fraction correct at CN = 8 and a lower fraction correct at CN = 4.

There are also compounds in which the cations are larger than the anions; in these cases, the appropriate radius ratio is r_- / r_+, which determines the CN of the anions in the holes of a cation lattice. Cesium fluoride is an example, with $r_- / r_+ = 119 / 181 = 0.657$, which places it in the six-coordinate range, consistent with the NaCl structure observed for this compound.

When the ions are nearly equal in size, a cubic arrangement of anions with the cation in the body center results, as in cesium chloride with CN = 8. Although a close-packed structure (ignoring the difference between cations and anions) would seem to give greater attractive forces, the CsCl structure separates ions of the same charge, reducing the repulsive forces between them.

Compounds whose stoichiometry is not 1:1, such as CaF_2 and Na_2S, may either have different coordination numbers for the cations and anions or structures in which only a fraction of the possible sites are occupied. Details of such structures are available in Wells[3] and other references.

7.2 THERMODYNAMICS OF IONIC CRYSTAL FORMATION

Formation of ionic compounds from the elements appears to be one of the simpler chemical reactions, but can also be written as a series of steps adding up to the overall reaction. The Born–Haber cycle is the process of considering the series of component reactions that can be imagined as the individual steps in compound formation. In the example of lithium fluoride, the first five reactions added together result in the sixth overall reaction.

$Li(s) \longrightarrow Li(g)$	$\Delta H_{sub} = 161$ kJ/mol	Sublimation	(1)
$\frac{1}{2} F_2(g) \longrightarrow F(g)$	$\Delta H_{dis} = 79$ kJ/mol	Dissociation	(2)
$Li(g) \longrightarrow Li^+(g) + e^-$	$\Delta H_{ion} = 531$ kJ/mol	Ionization energy	(3)
$F(g) + e^- \longrightarrow F^-(g)$	$\Delta H_{ion} = -328$ kJ/mol	−Electron affinity	(4)
$Li^+(g) + F^-(g) \longrightarrow LiF(s)$	$\Delta H_{xtal} = -1239$ kJ/mol	Lattice enthalpy	(5)
$Li(s) + \frac{1}{2} F_2(g) \longrightarrow LiF(s)$	$\Delta H_{form} = -796$ kJ/mol	Formation	(6)

Historically, such calculations were used to determine electron affinities when the enthalpies for all the other reactions could either be measured or calculated. Calculated lattice enthalpies were combined with experimental values for the other reactions and for the overall reaction of $Li(s) + \frac{1}{2}F_2(g) \longrightarrow LiF(s)$. Now that it is easier to measure electron affinities, the complete cycle can be used to determine more accurate lattice enthalpies. Although this is a simple calculation, it can be powerful in calculating thermodynamic properties for reactions that are difficult to measure directly.

7.2.1 Lattice Energy and the Madelung Constant

At first glance, calculation of the lattice energy of a crystal may seem simple: just take every pair of ions and calculate the sum of the electrostatic energy between each pair, using the following equation.

$$\Delta U = \frac{Z_i Z_j}{r_0}\left(\frac{e^2}{4\pi\,\varepsilon_0}\right)$$

[3] A. F. Wells, *Structural Inorganic Chemistry*, 5th ed., Oxford University Press, New York, 1988.

where Z_i, Z_j = ionic charges in electron units

r_0 = distance between ion centers

e = electronic charge = 1.602×10^{-19} C

$4\pi\varepsilon_0$ = permittivity of a vacuum = $1.11 \times 10^{-10}\ C^2\ J^{-1}\ m^{-1}$

$$\frac{e^2}{4\pi\varepsilon_0} = 2.307 \times 10^{-28}\ J\ m$$

Summing the energy for interactions between nearest neighbors is insufficient, because significant energy is involved in longer-range interactions between the ions. For a crystal as simple as NaCl, the closest neighbors to a sodium ion are six chloride ions at half the unit-cell distance, but the set of next-nearest neighbors is a set of 12 sodium ions at 0.707 times the unit-cell distance, and the numbers rise rapidly from there. The sum of all these geometric factors carried out until the interactions become infinitesimal is called the **Madelung constant**. It is used in the similar equation for the molar energy

$$\Delta U = \frac{NMZ_+Z_-}{r_0}\left(\frac{e^2}{4\pi\varepsilon_0}\right)$$

where N is Avogadro's number and M is the Madelung constant. Repulsion between close neighbors is a more complex function, frequently involving an inverse sixth- to twelfth-power dependence on the distance. The Born–Mayer equation, a simple and usually satisfactory equation, corrects for this using only the distance and a constant, ρ

$$\Delta U = \frac{NMZ_+Z_-}{r_0}\left(\frac{e^2}{4\pi\varepsilon_0}\right)\left(1 - \frac{\rho}{r_0}\right)$$

For simple compounds, $\rho = 30$ pm works well when r_0 is also in pm. Lattice enthalpies are twice as large when charges of 2 and 1 are present, and four times as large when both ions are doubly charged. Madelung constants for some crystal structures are given in Table 7.2.

The lattice enthalpy is $\Delta H_{xtal} = \Delta U + \Delta(PV) = \Delta U + \Delta nRT$, where Δn is the change in the number of gas-phase particles on formation of the crystal (e.g., –2 for

TABLE 7.2 Madelung Constants

Crystal Structure	Madelung Constant, M
NaCl	1.74756
CsCl	1.76267
ZnS (zinc blende)	1.63805
ZnS (wurtzite)	1.64132
CaF_2	2.51939
TiO_2 (rutile)	2.3850
Al_2O_3 (corundum)	4.040

Source: D. Quane, *J. Chem. Educ.*, **1970**, *47*, 396, has described this definition and several others, which include all or part of the charge (Z) in the constant. Caution is needed when using M because of the different possible definitions.

AB compounds, –3 for AB_2 compounds). The value of ΔnRT is small (–4.95 kJ/mol for AB, –7.43 kJ/mol for AB_2); for approximate calculations, $\Delta H_{xtal} \approx \Delta U$.

▶ **Exercise 7.4** Calculate the lattice energy for NaCl, using the ionic radii from Appendix B-1.

7.2.2 Solubility, Ion Size, and HSAB

Thermodynamic calculations can also be used to show the effects of solvation and solubility. For the overall reaction $AgCl(s) + H_2O \longrightarrow Ag^+(aq) + Cl^-(aq)$, the following reactions can be used:

$AgCl(s) \longrightarrow Ag^+(g) + Cl^-(g)$	$\Delta H =$ 917 kJ/mol	–Lattice enthalpy
$Ag^+(g) + H_2O \longrightarrow Ag^+(aq)$	$\Delta H =$ −475 kJ/mol	Solvation
$Cl^-(g) + H_2O \longrightarrow Cl^-(aq)$	$\Delta H =$ −369 kJ/mol	Solvation
$AgCl(s) + H_2O \longrightarrow Ag^+(aq) + Cl^-(aq)$	$\Delta H =$ 73 kJ/mol	Dissolution

If any three of the four reactions can be measured or calculated, the fourth can be found by completing the cycle. It has been possible to estimate the solvation effects of many ions by comparing similar measurements on a number of different compounds. Naturally, the entropy of solvation also needs to be included to calculate the free energy change on dissolving.

Many factors are involved in the thermodynamics of solubility, including ionic size and charge, the hardness or softness of the ions (HSAB), the crystal structure of the solid, and the electronic structure of each ion. Small ions have a strong electrostatic attraction for each other and for water molecules, whereas large ions have a weaker attraction for each other and for water molecules, but can accommodate more water molecules around each ion. These factors work together to make compounds formed of two large ions (soft) or of two small ions (hard) less soluble than compounds containing one large ion and one small ion, particularly when they have the same charge magnitude. In the examples given by Basolo[4], LiF, with two small ions, and CsI, with two large ions, are less soluble than LiI and CsF, which have one large and one small ion. For the small ions, the larger lattice energy overcomes the larger hydration enthalpies, and for the large ions, the smaller hydration enthalpies allow the lattice energy to dominate. An example of the significance of entropy is that a saturated CsI solution is about 15 times as concentrated as a solution of LiF (in molarity) in spite of the less favorable enthalpy change for the former.

Cation	Hydration Enthalpy (kJ/mol)	Anion	Hydration Enthalpy (kJ/mol)	Lattice Energy (kJ/mol)	Net Enthalpy of Solution (kJ/mol)
Li^+	–519	F^-	–506	–1025	0
Li^+	–519	I^-	–293	–745	–67
Cs^+	–276	F^-	–506	–724	–58
Cs^+	–276	I^-	–293	–590	+21

[4] F. Basolo, *Coord. Chem. Rev.*, **1968**, 3, 213.

In this same set of four compounds, the reaction LiI(*s*) + CsF(*s*) ⟶ CsI(*s*) + LiF(*s*) is exothermic ($\Delta H = -146$ kJ/mol) because of the large lattice enthalpy of LiF. This is contrary to the simple electronegativity notion that the most electropositive and the most electronegative elements form the most stable compounds. However, these same compounds fit the hard–soft model, with LiF, the hard–hard combination, and CsI, the soft–soft combination, the least soluble salts (Section 6.3). Sometimes these factors are also modified by particular interactions because of the electronic structures of the ions.

7.3 MOLECULAR ORBITALS AND BAND STRUCTURE

When molecular orbitals are formed from two atoms, each type of atomic orbital (such as 2*s*) gives rise to two molecular orbitals (σ_{2s} and σ_{2s}^*). When n atoms are used, the same approach results in n molecular orbitals. In the case of solids, n is very large—Avogadro's number for a mole of atoms. If the atoms were all in a one-dimensional row, the lowest energy orbital would have no nodes, and the highest would have $n - 1$ nodes; in a three-dimensional solid, the nodal structure is more complex but still just an extension of this linear model. Because the number of atoms is large, the number of orbitals and energy levels with closely spaced energies is also large. The result is a **band** of orbitals of similar energy, rather than the discrete energy levels of small molecules.[5] These bands then contain the electrons from the atoms. The highest energy band containing electrons is called the **valence band**; the next higher, empty band is called the **conduction band**.

In elements with filled valence bands, and a large energy difference between the highest valence band and the lowest conduction band, this **band gap** prevents motion of the electrons, and the material is an **insulator**, with the electrons restricted in their motion. In cases with partly filled orbitals, the distinction between the valence and conduction bands is blurred, and very little energy is required to move some electrons to higher energy levels within the band. As a result, they are then free to move throughout the crystal, as are the **holes** (electron vacancies) left behind in the occupied portion of the band. These materials are **conductors** of electricity, because the electrons and holes are both free to move. They are also usually good conductors of heat, because the electrons are free to move within the crystal and transmit energy. As required by the usual rules of electrons occupying the lowest energy levels, the holes tend to be in the upper levels within a band. The band structures of insulators and conductors are shown in Figure 7.13.

The concentration of energy levels within bands is described as the **density of states**, *N(E)*, actually determined for a small increment of energy, d*E*. Figure 7.14 shows three examples, two with distinctly separate bands and one with overlapping bands.

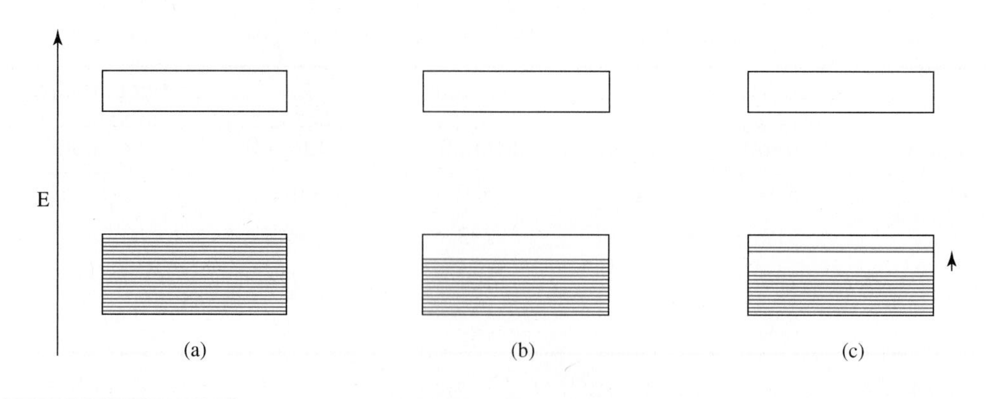

FIGURE 7.13 Band Structure of Insulators and Conductors. (a) Insulator. (b) Metal with no voltage applied. (c) Metal with electrons excited by applied voltage.

[5] R. Hoffmann, *Solids and Surfaces: A Chemist's View of Bonding in Extended Structures*, VCH Publishers, New York, 1988, pp. 1–7.

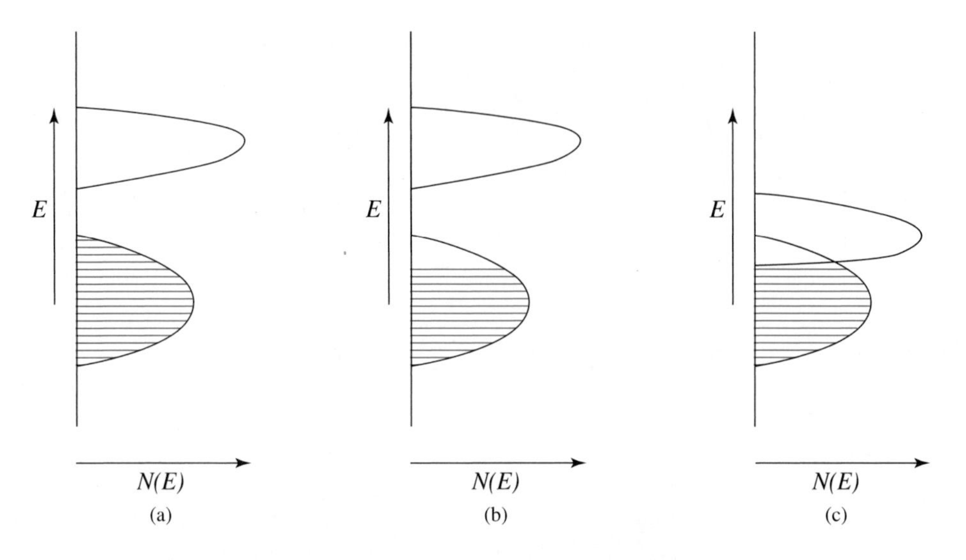

FIGURE 7.14 Energy Bands and Density of States. (a) An insulator with a filled valence band. (b) A metal with a partly filled valence band and a separate empty band. (c) A metal with overlapping bands caused by similar energies of the initial atomic orbitals.

The shaded portions of the bands are occupied, and the unshaded portions are empty. The figure shows an insulator with a filled valence band and a metal in which the valence band is partly filled. When an electric potential is applied, some of the electrons can move to slightly higher energies, leaving vacancies or holes in the lower part of the band. The electrons at the top of the filled portion can then move in one direction, and the holes can move in the other, conducting electricity. In fact, the holes appear to move because an electron moving to fill one hole creates another in its former location.

▶ **Exercise 7.5** Hoffmann uses a linear chain of hydrogen atoms as a starting model for his explanation of band theory. Using a chain of eight hydrogen atoms, sketch the phase relationships (positive and negative signs) of all the molecular orbitals that can be formed. These orbitals, bonding at the bottom and antibonding at the top, form a band.

The conductance of metals decreases with increasing temperature, because the increasing vibrational motion of the atoms interferes with the motion of the electrons and increases the resistance to electron flow. High conductance (low resistance) in general, and decreasing conductance with increasing temperature, are characteristics of metals. Some elements have bands that are either completely filled or completely empty, but they differ from insulators by having the bands very close in energy (approximately 2 eV or less). Silicon and germanium are examples: their diamond structure crystals have bonds that are more nearly like ordinary covalent bonds, with four bonds to each atom. At very low temperatures, they are insulators, but the conduction band is very near the valence band in energy. At higher temperatures, when a potential is placed across the crystal, a few electrons can jump into the higher (vacant) conduction band, as in Figure 7.15(a). These electrons are then free to move through the crystal. The vacancies, or holes, left in the lower energy band can also appear to move as electrons move into them. In this way, a small amount of current can flow. When the temperature is raised, more electrons are excited into the upper band, more holes are created in the lower band, and conductance *increases* (resistance decreases). This is the distinguishing characteristic of **semiconductors**. They have much higher conductivity than insulators and much lower conductivity than conductors.

It is possible to change the properties of semiconductors within very close limits. As a result, the flow of electrons can be controlled by applying the proper voltages to some of these modified semiconductors. The entire field of solid-state electronics (transistors and integrated circuits) depends on these phenomena. Silicon and germanium are **intrinsic semiconductors**, meaning that the pure materials have semiconductive

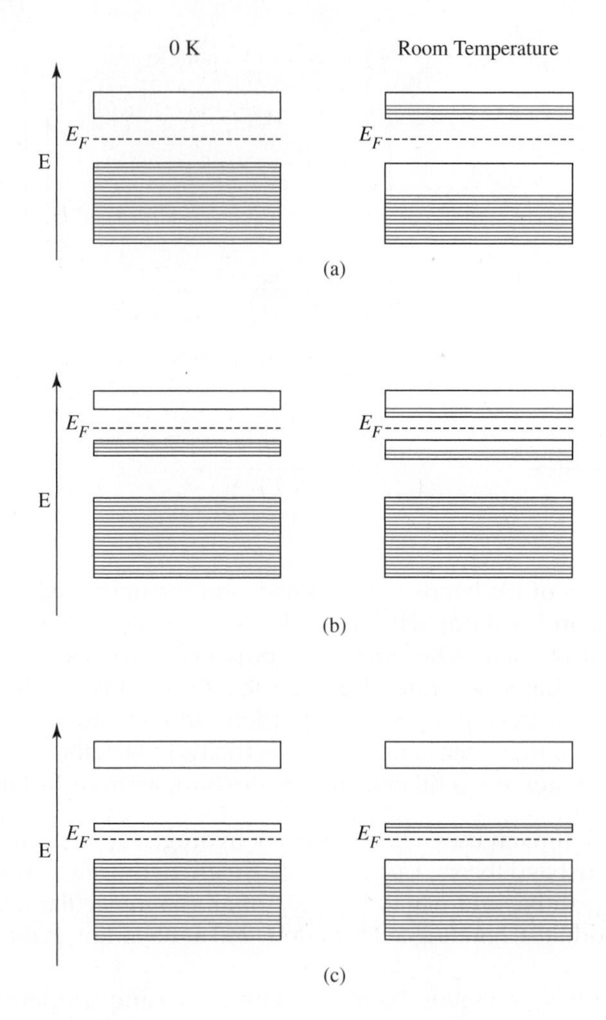

FIGURE 7.15 Semiconductor Bands at 0 K and at Room Temperature. (a) Intrinsic semiconductor, (b) an *n*-type semiconductor, and (c) a *p*-type semiconductor.

properties. Both molecular and nonmolecular compounds can also be semiconductors. A short list of some of the nonmolecular compounds and their band gaps is given in Table 7.3. Other elements that are not semiconductors in the pure state can be modified by adding a small amount of another element with energy levels close to those of the host, to make **doped semiconductors**. *Doping* can be thought of as replacing a few atoms of the original element with atoms having either more or fewer electrons. If the added material has more electrons in the valence shell than the host material, the result is an ***n*-type semiconductor** (*n* for *negative,* adding electrons), shown in Figure 7.15(b). Phosphorus is an example in a silicon host, with five valence electrons, compared with four in silicon. These electrons have energies just slightly lower in energy than the conduction band of silicon. With the addition of a small amount of energy, electrons from this added energy level can jump up into the empty band of the host material, resulting in higher conductance.

If the added material has fewer valence electrons than the host, it adds positive holes and the result is a ***p*-type semiconductor**, shown in Figure 7.15(c). Aluminum is a *p*-type dopant in a silicon host, with three electrons instead of four in a band very close in energy to that of the silicon valence band. Addition of a small amount of energy boosts electrons from the host valence band into this new level and generates more holes in the valence band of the host, thus increasing the conductance. With careful doping, the conductance can be carefully tailored to the need. Layers of intrinsic, *n*-type, and *p*-type semiconductors, together with insulating materials, are used to create the integrated circuits that are so essential to the electronics industry. Controlling the voltage applied to the junctions between the different layers controls conductance through the device.

TABLE 7.3 Semiconductors

Material	Band Gap (eV)
Elemental	
Si	1.11
Ge	2.2
Group 13–15 Compounds	
GaP	2.25
GaAs	1.42
InSb	0.17
Group 12–16 Compounds	
CdS	2.40
ZnTe	2.26

The number of electrons that are able to make the jump between the valence and the conduction band depends on the temperature and on the energy gap between the two bands. In an intrinsic semiconductor, the **Fermi level** (E_F, Figure 7.15), the energy at which an electron is equally likely to be in each of the two levels, is near the middle of the band gap. Addition of an *n*-type dopant raises the Fermi level to an energy near the middle of the band gap between the new band and the conduction band of the host. Addition of a *p*-type dopant lowers the Fermi level to a point near the middle of the band gap between the new conduction band and the valence band of the host.

7.3.1 Diodes, the Photovoltaic Effect, and Light-Emitting Diodes

Putting layers of *p*-type and *n*-type semiconductors together creates a *p–n* junction. A few of the electrons in the conduction band of the *n*-type material can migrate to the valence band of the *p*-type material, leaving the *n* type positively charged and the *p* type negatively charged. An equilibrium is quickly established, because the electrostatic forces are too large to allow much charge to accumulate. The separation of charges then prevents transfer of any more electrons. At this point, the Fermi levels are at the same energy, as shown in Figure 7.16. The band gap remains the same in both layers, with the energy levels of the *n*-type layer lowered by the buildup of positive

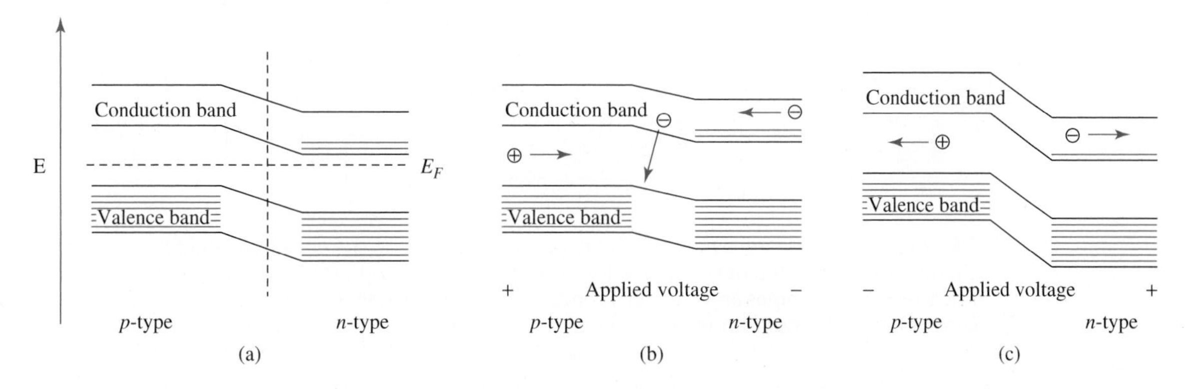

FIGURE 7.16 Band-Energy Diagram of a *p–n* Junction. (a) At equilibrium, the two Fermi levels are at the same energy, changing from the pure *n*-type or *p*-type Fermi levels, because a few electrons can move across the boundary (vertical dashed line). (b) With forward bias, current flows readily. (c) With reverse bias, very little current flows.

charge. If a negative potential is applied to the n-type side of the junction and a positive potential is applied to the p-type side, it is called a *forward bias*. The excess electrons raise the level of the n-type conduction band and then have enough energy to move into the p-type side. Holes move toward the junction from the left, and electrons move toward the junction from the right, canceling each other at the junction, and current can flow readily. If the potential is reversed (reverse bias), the energy of the n-type levels is lowered compared with the p-type levels, the holes and electrons both move away from the junction, and very little current flows. This is the description of a **diode**, which allows current to flow readily in one direction but has a high resistance to current flow in the opposite direction, as in Figure 7.17.

A junction of this sort can be used as a light-sensitive switch. With a reverse bias applied (extra electrons supplied to the p side), no current would flow, as described for diodes. However, if the difference in energy between the valence band and the conduction band of a semiconductor is small enough, light of visible wavelengths is energetic enough to lift electrons from the valence band into the conduction band, as shown in Figure 7.18. Light falling on the junction increases the number of electrons in the conduction band and the number of holes in the valence band, allowing current to flow in spite of the reverse bias. Such a junction then acts as a photoelectric switch, passing current when light strikes it.

If no external voltage is applied, and if the gap has the appropriate energy, light falling on the junction can increase the transfer of electrons from the p-type material into the conduction band of the n-type material. If external connections are made to the two layers, current can flow through this external circuit. **Photovoltaic cells** of this sort are commonly used in calculators and other "solar" devices in remote locations—for example, emergency telephones and low-energy lighting—and are increasingly being used to generate electricity for home and commercial use.

A forward-biased junction can reverse this process and emit light as a **light-emitting diode (LED)**. The current is carried by holes on the p-type side and by electrons on the n-type side. When electrons move from the n-type layer to the p-type layer, they recombine with the holes. If the resulting energy change is of the right magnitude, it can be released as visible light (luminescence), and an LED results. In practice, $GaP_x As_{1-x}$ with $x = 0.40$ to 1.00 can be used for LEDs that emit red light (band gap of 1.88 eV) to green light (2.23 eV). The energy of the light emitted can be changed by adjusting the composition of the material. GaAs has a band gap of about 1.4 eV; GaP has a band gap of about 2.25 eV. The band gap increases steadily as the fraction of phosphorus is increased, with an abrupt change in slope at $x = 0.45$, where there is a change from a direct band gap to an indirect band gap.[6] In arsenic-rich materials, the electrons drop directly across the energy gap into holes in the lower level (a direct band gap), and the light is emitted with high efficiency. In

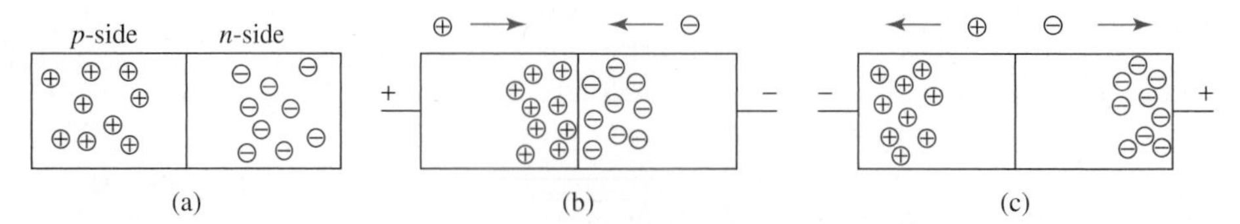

FIGURE 7.17 Diode Behavior. (a) With no applied voltage, no current flows, and few charges are neutralized near the junction by transfer of electrons. (b) Forward bias: current flows readily, with holes and electrons combining at the junction. (c) Reverse bias: very little current can flow, because the holes and electrons move away from each other.

[6] A. G. Thompson, M. Cardona, K. L. Shaklee, and J. C. Wooley, *Phys. Rev.*, **1966**, *146*, 601; H. Mathieu, P. Merle, and E. L. Ameziane, *Phys. Rev.*, **1977**, *B15*, 2048; M. E. Staumanis, J. P. Krumme, and M. Rubenstein, *J. Electrochem. Soc.*, **1977**, *146*, 640.

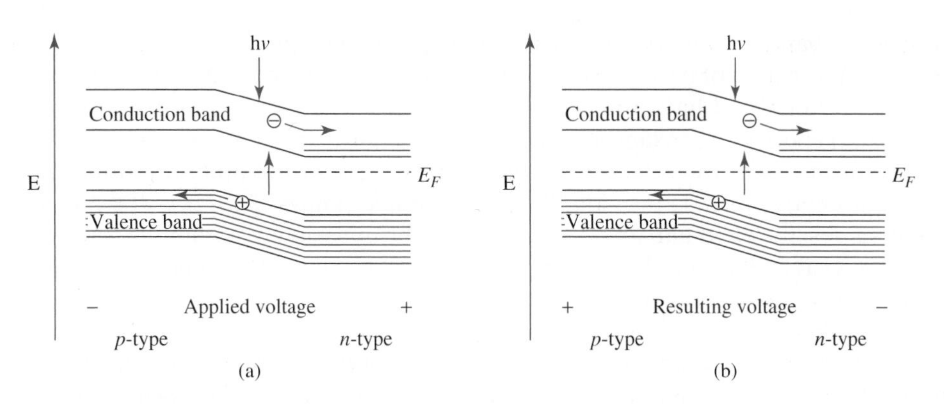

FIGURE 7.18 The Photovoltaic Effect. (a) As a light-activated switch. (b) Generating electricity. Light promotes electrons into the conduction band in the junction.

phosphorus-rich materials, this process must be accompanied by a change in vibrational energy of the crystal (an indirect band gap). This indirect process is less efficient and requires addition of a dopant to give efficient emission by relaxing these rules. These materials also have more complex emission and absorption spectra because of the addition of the dopant, in contrast to the arsenic-rich materials, which have spectra with one simple band. The efficiency of emission is also improved for both types at lower temperatures, at which the intensity of vibrations is reduced. Similar behavior is observed in $Al_x Ga_{1-x} As$ LEDs; emission bands (from 840 nm for $x = 0.05$ to 625 nm for $x = 0.35$) shift to shorter wavelengths, and much greater intensity, on cooling to the temperature of liquid nitrogen (77 K).

Adding a third layer with a larger band gap and making the device exactly the right dimensions changes the behavior of an LED into a solid-state laser. Gallium arsenide doped to provide an *n*-type layer, a *p*-type layer, and then a larger band gap in a *p*-type layer, with Al added to the GaAs, is a commonly used combination. The general behavior is the same as in the LED, with a forward bias on the junction creating luminescence. The larger band gap added to the *p*-type layer prevents the electrons from moving out of the middle *p*-type layer. If the length of the device is exactly a half-integral number of wavelengths of the light emitted, photons released by recombination of electrons and holes are reflected by the edges and stimulate the release of more photons in phase with the first ones. The net result is a large increase in the number of photons and a specific direction for their release in a laser beam. The commonly seen red laser light used for pointers and in supermarket scanners uses this phenomenon.

7.3.2 Quantum Dots

If samples of semiconductors are prepared in smaller and smaller sizes, at some point the bulk properties of the sample will no longer show a continuum of states, as described in previous sections, but will begin to exhibit quantized energy states; the limiting case would be a single molecule with its molecular orbitals. This is called the *quantum confinement effect* and results in very small particles showing a discrete energy level structure rather than a continuum. Nanoparticles showing this effect that have diameters smaller than approximately 10 nm are often called **quantum dots**; because of their size, they behave differently than bulk solids.

The energy-level spacings of quantum dots are related to their size; experiments have shown that the difference in energy between the valence band and conduction band increases as the particle size gets smaller, and the bulk semiconductor becomes more like a single molecule. Consequently, for smaller particles, more energy is needed for excitation and, similarly, more energy is emitted as electrons return to the valence band. More specifically, when an electron is excited, it leaves a hole in the valence band. This excited electron–hole combination is called an **exciton**, and it has an energy slightly lower than the lowest energy of the conduction band. Decay of the exciton to the valence band causes photoemission of a specific energy. Because the energy level spacing is a function of nanoparticle size, it is possible to take advantage of this effect to prepare particles that have the capacity to emit electromagnetic radiation of specific (quantized) energies, for example, where light of a particular color is needed.

An example of the relationship between the size of the quantum dots and electronic emission spectra is provided by zinc selenide (Figure 7.19).[7] As the nanocrystal size increases, the band gap grows smaller, and the emission maximum moves from the UV to the visible (left to right in the figure), consistent with closer energy-level spacing as the particle size increases.

Considerable effort has been devoted to developing processes for preparing quantum dots of consistent and reproducible sizes and shapes and using different materials for optimum optical properties.[8] For example, ZnSe quantum dots emit in the violet and UV, PbS in the near-IR and visible, and CdSe throughout the range of visible light. Numerous uses have been proposed, for example the conversion of solar energy to electricity, in data processing and recording, and in a variety of uses as biosensors. Among medical applications has been the tracking of the uptake of nanoparticles of different sizes by tumors, to investigate whether there is an optimum size for drug delivery[9], and labeling cell surface proteins to track their motion within cell membranes.[10] Because many of the nanoparticles are potentially toxic, effort has also been directed to coating them to reduce potential medical and environmental effects.[11] One of the broadest potential impacts of quantum dot technology

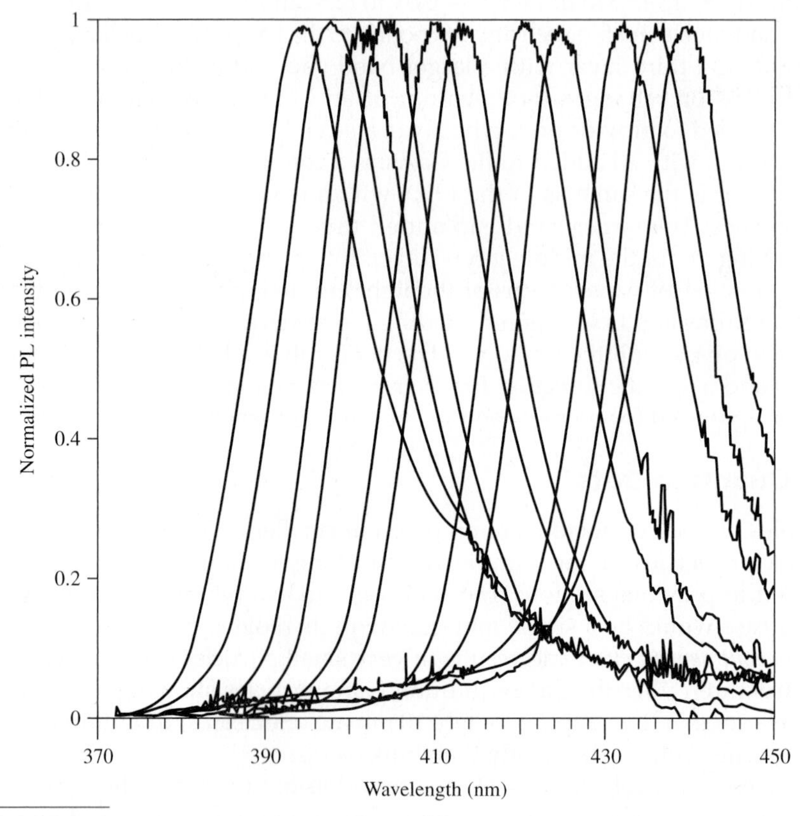

FIGURE 7.19 Photoluminescence Emission Spectrum of ZnSe Quantum Dots. (Reproduced with permission from P. Reiss, G. Quemard, S. Carayon, J. Bleuse, F. Chandezon, and A. Pron, *Mater. Chem. Phys.*, **2004**, *84*, 10.)

[7] V. V. Nikesh, A. D. Lad, S. Kimura, S. Nozaki, and S. Mahamuni, *J. Appl. Phys.*, **2006**, *100*, 113520; P. Reiss, *New. J. Chem.*, **2007**, *31*, 1843 and references cited therein.
[8] G. D. Scholes, *Adv. Funct. Mater.*, **2008**, *18*, 1157.
[9] M. Stroh, J. P. Zimmer, D. G. Duda, T. S. Levchenko, K. S. Cohen, E. B. Brown, D. T. Scadden, V. P. Torchilin, M. G. Bawendi, D. Fukumora, and R. K. Jain, *Nature Medicine*, **2005**, *11*, 678.
[10] M. Howarth, K. Takao, Y. Hayashi, and A. Y. Ting, *Proc. Natl. Acad. Sci. U.S.A.*, **2005**, *102*, 7583.
[11] J. Drbohlavova, V. Adam, R. Kizek, and J. Hubalek, *Int. J. Mol. Sci.*, **2009**, *10*, 656.

may be in the realm of highly efficient lighting. Currently LEDs have a wide variety of uses—automobiles, video displays, sensors, traffic signals—but they emit light of too narrow a spectrum to be attractive as high-efficiency general replacements for incandescent and fluorescent lighting; in addition, LEDs remain expensive. Coupling LEDs with quantum dots, perhaps in coatings that use a variety of particle sizes for a range of emission colors, may provide a pathway to more efficient solid-state lighting for general use.

7.4 SUPERCONDUCTIVITY

The conductivity of some metals changes abruptly near liquid helium temperatures (frequently below 10 K), as in Figure 7.20, and they become **superconductors**, an effect discovered by Kammerling Onnes in 1911[12] while studying mercury at liquid helium temperature. In this state, these metals offer no resistance to the flow of electrons, and currents started in a loop will continue to flow indefinitely—several decades at least—without significant change. For chemists, one of the most common uses of this effect is in superconducting magnets used in nuclear magnetic resonance instruments, in which it allows generation of much larger magnetic fields than can be obtained with ordinary electromagnets.

7.4.1 Low-Temperature Superconducting Alloys

Some of the most common superconducting materials are alloys of niobium, particularly Nb–Ti alloys, which can be formed into wire and handled with relative ease. These Type I superconductors have the additional property of expelling all magnetic flux when cooled below the **critical temperature**, T_c. This is called the *Meissner effect*. It prevails until the magnetic field reaches a critical value, H_c, at which point the applied field destroys the superconductivity. As in the temperature dependence, the change between superconducting and normal conduction is abrupt rather than gradual. The highest T_c found for niobium alloys is 23.3 K for Nb_3Ge.[13]

Type II superconductors have a more complicated field dependence. Below a given critical temperature they exclude the magnetic field completely. Between this first critical temperature and a second critical temperature, they allow partial penetration by the field; above this second critical temperature, they lose their superconductivity and

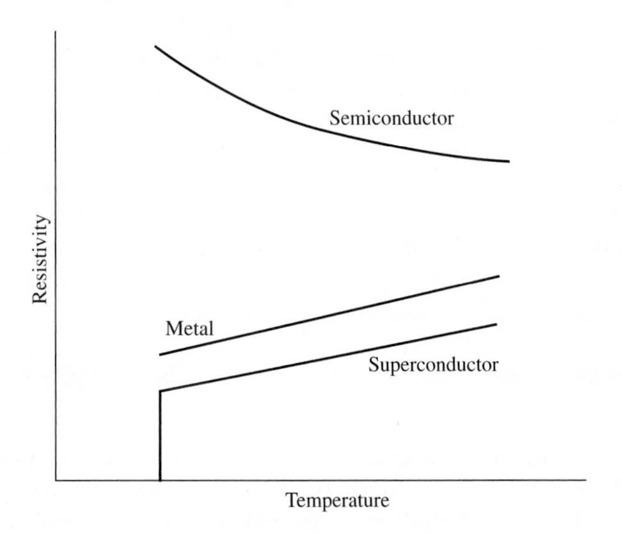

FIGURE 7.20 Temperature Dependence of Resistivity in Semiconductors, Metals, and Superconductors.

[12] H. Kammerlingh Onnes, *Akad. Van Wetenschappen (Amsterdam)*, **1911**, *14*, 113, and *Leiden Comm.*, **1911**, *122b*, 124c.

[13] C. P. Poole, Jr., H. A. Farach, and R. J. Creswick, *Superconductivity*, Academic Press, San Diego, 1995, p. 22.

display normal conductance behavior. In the intermediate temperature region, these materials seem to have a mixture of superconducting and normal regions.

The Meissner effect is being explored for practical use in many areas, including magnetic levitation of trains, although other electromagnetic effects are presently being used for this. A common demonstration is to cool a small piece of superconducting material below its critical temperature and then place a small magnet above it. The magnet is suspended above the superconductor, because the superconductor repels the magnetic flux of the magnet. As long as the superconductor remains below its critical temperature, it expels the magnetic flux from its interior and holds the magnet at a distance.

The levitation demonstration works only with Type II superconductors, because the magnetic field lines that do enter the superconductor resist sideways motion and allow the balance of magnetic repulsion and gravitation to "float" the magnet above the superconductor. With Type I superconductors, the magnetic field lines cannot enter the superconductor at all and, because there is no resistance to sideways motion, the magnet will not remain stationary over the superconductor.

The materials used in the coils of superconducting magnets are frequently Nb–Ti–Cu or Nb_3Sn–Cu mixtures, providing a balance between T_c, which is about 10 K for these materials, and ductility for easier formation into wire.

Superconducting magnets allow very high currents to flow with no change indefinitely as long as the magnet is kept cold enough. In practice, an outer Dewar flask containing liquid nitrogen (boiling point 77.3 K) reduces boil-off of liquid helium (boiling point 4.23 K) from the inner Dewar flask surrounding the magnet coils. A power supply is attached to the magnet, and electrical current is supplied to bring it to the proper field. When the power supply is removed, current flows continuously, maintaining the magnetic field.

A major goal of superconductor research is finding a material that is superconducting at higher temperatures, to remove the need for liquid helium and liquid nitrogen for cooling.

7.4.2 The Theory of Superconductivity (Cooper Pairs)

In the late 1950s, more than 40 years after its discovery, Bardeen, Cooper, and Schrieffer[14] (BCS) provided a theory to explain superconductivity. Their BCS theory postulated that electrons travel through a material in pairs, in spite of their mutual electrostatic repulsion, as long as the two have opposite spins. The formation of these *Cooper pairs* is assisted by small vibrations of the atoms in the lattice; as one electron moves past, the nearest positively charged atoms are drawn very slightly toward it. This increases the positive charge density, which attracts the second electron. This effect then continues through the crystal, in a manner somewhat analogous to a sports crowd doing the wave. The attraction between the two electrons is small, and they change partners frequently, but the overall effect is that the lattice helps them on their way rather than interfering, as is the case with metallic conductivity. If the temperature rises above T_c, the thermal motion of the atoms is sufficient to overcome the slight attraction between the electrons, and the superconductivity ceases.

7.4.3 High-Temperature Superconductors: $YBa_2Cu_3O_7$ and Related Compounds

In 1986, Bednorz and Müller discovered that the ceramic oxide La_2CuO_4 was superconducting above 30 K when doped with Ba, Sr, or Ca to form compounds such as $(La_{2-x}Sr_x)CuO_4$.[15] This opened many more possibilities for the use of superconductivity.

---

[14] J. Bardeen, L. Cooper, and J. R. Schrieffer, *Phys. Rev.*, **1957**, *108*, 1175; J. R. Schrieffer, *Theory of Superconductivity*, W. A. Benjamin, New York, 1964; A. Simon, *Angew. Chem., Int. Ed.*, **1997**, *36*, 1788.

[15] J. G. Bednorz and K. A. Müller, *Z. Phys. B*, **1986**, *64*, 189.

Then, in 1987, $YBa_2Cu_3O_7$ was discovered to have an even higher T_c, 93 K.[16] This material, called *1-2-3* for the stoichiometry of the metals in it, is a Type II superconductor, which expels magnetic flux at low fields, but allows some magnetic field lines to enter at higher fields, and consequently ceases to be superconducting at high fields. A number of other similar compounds have since been prepared and found to be superconducting at these or even higher temperatures. These high-temperature superconductors are of great practical interest, because they would allow cooling with liquid nitrogen, rather than liquid helium, a much more expensive coolant. However, the ceramic nature of these materials makes them more difficult to work with than metals. They are brittle and cannot be drawn into wire, making fabrication a problem. Researchers are working to overcome these problems by modifying the formulas or by depositing the materials on a flexible substrate. The present record is a critical temperature of 164 K for $HgBa_2Ca_2Cu_3O_{8-\delta}$ under pressure.[17]

The structures of all the high-temperature superconductors are related, most with copper oxide planes and chains, as shown in Figure 7.21. In $YBa_2Cu_3O_7$, these are stacked with square-pyramidal, square-planar, and inverted square-pyramidal units. The copper atoms in the top and bottom layers of Figure 7.21(a) are those in the square-planar units of Figure 7.21(b); two units are shown. In a related tetragonal structure, the oxygen atoms of the top and bottom planes in 7.21(a) are randomly dispersed in the four equivalent edges of the plane; the resulting material is not superconducting. Oxygen-deficient structures are also possible and are superconducting until about $\delta = 0.65$; materials closer to the formula $YBa_2Cu_3O_6$ are not superconducting.

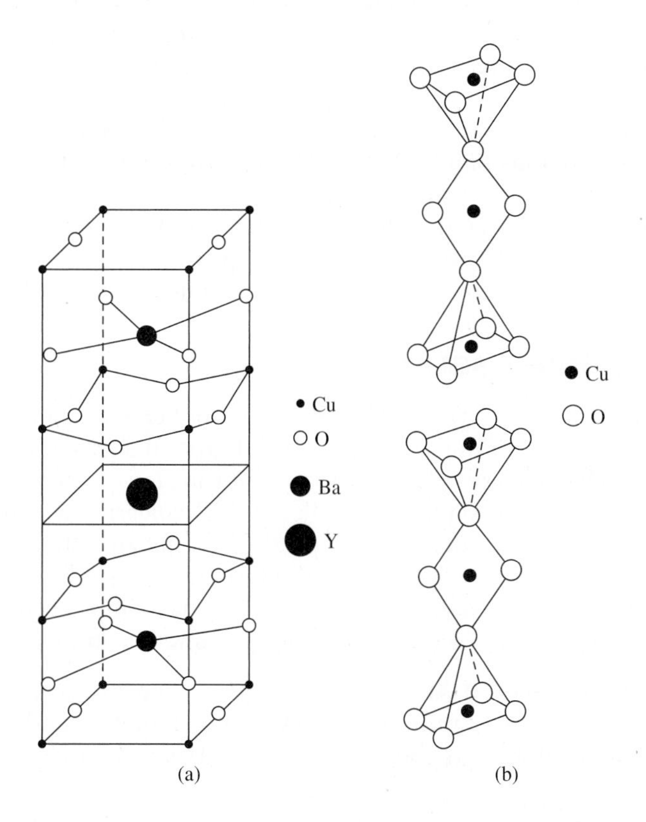

FIGURE 7.21 Two Views of Orthorhombic $YBa_2Cu_3O_7$. (a) The unit cell. The Y atom in the middle is in a reflection plane. (Adapted from C. P. Poole, Jr., H. A. Farach, and R. J. Creswick, *Superconductivity*, Academic Press, San Diego, 1995, p. 181, with permission.) (b) Stacking of copper oxide units. (Adapted from C. P. Poole, Jr., T. Datta, and H. A. Farach, *Copper Oxide Superconductors*, John Wiley & Sons, New York, 1988, p. 100. © John Wiley & Sons, used by permission.)

[16] M. K. Wu, J. R. Ashburn, C. J. Torng, P. H. Hor, R. L. Meng, L. Gao, Z. J. Huang, Y. Q. Wang, and C. W. Chu, *Phys. Rev. Lett.*, **1987**, *58*, 908.

[17] L. Gao, Y. Y. Xue, F. Chen, Q. Ziong, R. L. Meng, D. Ramirez, C. W. Chu, J. H. Eggert, and H. K. Mao, *Phys. Rev. B*, **1994**, *50*, 4260.

The understanding of superconductivity in high-temperature superconductors is incomplete, but at this point, an extension of the BCS theory seems to fit much of the known data. The mechanism of electron pairing and the details of the behavior of the electron pairs are less clear.

7.5 BONDING IN IONIC CRYSTALS

The simplest picture of bonding in ionic crystals is of hard-sphere ions held together by purely electrostatic forces. This picture is far too simple, even for compounds such as NaCl that are expected to be strongly ionic in character. It is the deviation from this simple model that makes questions about ion sizes so difficult. For example, the Pauling radius of Li^+ is 60 pm. The crystal radius given by Shannon (Appendix B-1) for a six-coordinate structure is 90 pm, a value that is much closer to the position of minimum electron density between ions, determined by X-ray crystallography. The four-coordinate Li^+ has a radius of 73 pm, and estimates by Goldschmidt and Ladd are between 73 and 90 pm.[18] The sharing of electrons or the transfer of charge back from the anion to the cation varies from a few percent in NaCl to as much as 0.33 electrons per atom in LiI. Each set of radii is self-consistent, but mixing some radii from one set with some from another does not work.

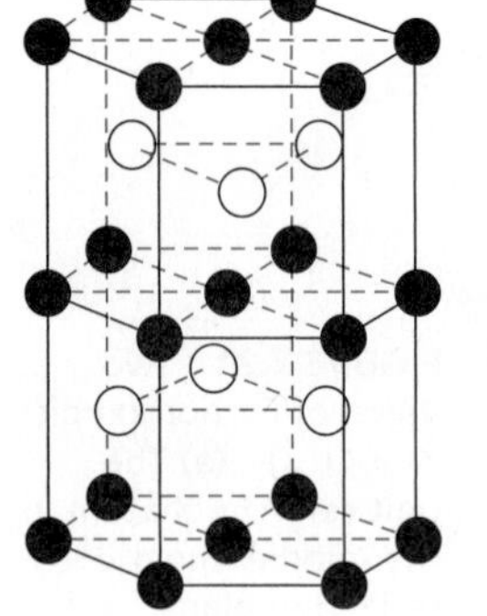

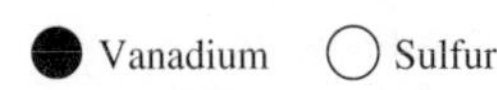

FIGURE 7.22 Structure of Vanadium Sulfide.

Some of the structures shown earlier in this chapter (Figures 7.7 through 7.11) are given as if the components were simple ions, even though the bonding is strongly covalent. In any of these structures, this ambiguity must be kept in mind. The band structures described previously are much more complete in their descriptions of the bonding. Hoffmann[19] has described the bands in vanadium sulfide, an example of the NiAs structure. The crystal has layers that could be described as ABACA in the hexagonal unit cell, with the identical A layers made up of a hexagonal array of V atoms, and the B and C layers made up of S atoms in the alternate trigonal prisms formed by the metal (Figure 7.22). In this structure, both atoms are six-coordinate, with V atoms octahedrally coordinated to S atoms and S atoms in a trigonal prism of V atoms. The very complex band structure derived from this structure has been analyzed by Hoffmann in terms of smaller components of the crystal.

Hoffmann has also shown that the contributions to the density of states of specific orbitals can be calculated.[20] In rutile, TiO_2, a clear separation of the *d* orbital contribution into t_{2g} and e_g parts can be seen, as predicted by ligand field theory (Chapter 10).

7.6 IMPERFECTIONS IN SOLIDS

In practice, all crystals have imperfections. If a substance crystallizes rapidly, it is likely to have many more imperfections, because crystal growth starts at many sites almost simultaneously. Each small crystallite grows until it runs into its neighbors; the boundaries between these small crystallites are called *grain boundaries*, which can be seen on microscopic examination of a polished surface. Slow crystal growth reduces the number of grain boundaries, because crystal growth starts from a smaller number of sites. However, even if a crystal appears to be perfect, it will have imperfections on an atomic level caused by impurities in the material or by dislocations within the lattice.

VACANCIES AND SELF-INTERSTITIALS Vacancies are missing atoms and are the simplest defects. Because higher temperatures increase vibrational motion and expand a crystal, more vacancies are formed at higher temperatures. However, even near the melting

---

[18] N. N. Greenwood and A. Earnshaw, *Chemistry of the Elements*, 2nd ed., Butterworth–Heinemann, Oxford, 1997, p. 81.
[19] R. Hoffmann, *Solids and Surfaces: A Chemist's View of Bonding in Extended Structures*, VCH Publishers, New York, 1988, pp. 102–107.
[20] R. Hoffmann, *Solids and Surfaces*, p. 34.

point, the number of vacancies is small relative to the total number of atoms—on the order of 1 in 10,000. The effect of a vacancy on the rest of the lattice is small, because it is a localized defect, and the rest of the lattice remains unaffected. Self-interstitials are atoms displaced from their normal location that appear in one of the interstices in the lattice. Here, the distortion spreads at least a few layers in the crystal, because the atoms are much larger than the available space. In most cases, the number of these defects is much smaller than the number of vacancies.

SUBSTITUTIONS Substitution of one atom for another is a common phenomenon. Such mixtures are also called *solid solutions*. For example, nickel and copper atoms have similar sizes and electronegativities and the same FCC crystal structures. Mixtures of the two are stable in any proportion, with random arrangement of the atoms in the alloys. Other combinations that can work well have a very small atom in a lattice of larger atoms. In this case, the small atom occupies one of the interstices in the larger lattice, with small effects on the rest of the lattice but potentially large effects on behavior of the mixture. If the impurity atoms are larger than the holes, lattice strains result, and a new solid phase may be formed.

DISLOCATIONS Edge dislocations result when atoms in one layer do not match up precisely with those of the next. As a result, the distances between the dislocated atoms and atoms in adjacent rows are larger than usual, and the angles between atoms are distorted for a number of rows on either side of the dislocation. A screw dislocation is one that has part of one layer shifted a fraction of a cell dimension. This kind of dislocation frequently causes a rapidly growing site during crystal growth, and it forms a helical path, hence the name. Because they provide sites that allow atoms from the solution or melt to fit into a corner where attractions from three directions can hold them in place, screw dislocations are frequently growth sites for crystals.

In general, dislocations are undesirable in crystals. Mechanically, they can lead to weakness that can cause fracture. Electrically, they interfere with conduction of electrons and reduce reliability, reproducibility, and efficiency in semiconductor devices. For example, one of the challenges of photocell manufacture is to raise the efficiency of cells made of polycrystalline silicon to levels that are reached by single crystals.

7.7 SILICATES

Oxygen, silicon, and aluminum are the most abundant elements in the surface of the earth, and more than 80% of the atoms in the solid crust are oxygen or silicon, mostly in the form of silicates. The number of compounds and minerals that contain these elements is very large, and their importance in industrial uses matches their number. We can give only a very brief description of some of these compounds so we will focus on a few of the silicates.

Silica, SiO_2, has three crystalline forms: quartz at temperatures below 870° C, tridymite from 870° to 1470° C, and cristobalite from 1470° to 1710° C, at which temperature it melts. The high viscosity of molten silica makes crystallization slow; instead of crystallizing, it frequently forms a glass, which softens near 1500° C. Conversion from one crystalline form to another is difficult and slow, even at high temperatures, because it requires breaking Si–O bonds. All forms contain SiO_4 tetrahedra sharing oxygen atoms, with Si—O—Si angles of 143.6°.

Quartz is the most common form of silica and contains helical chains of SiO_4 tetrahedra, which are chiral with clockwise or counterclockwise twists. Each full turn of the helix contains three Si atoms and three O atoms, and six of these helices combine to form the overall hexagonal shape (Figure 7.23).[21]

---

[21] This figure was prepared with the assistance of Robert M. Hanson's Origami program and the Chime plug-in (MDL) to Netscape.

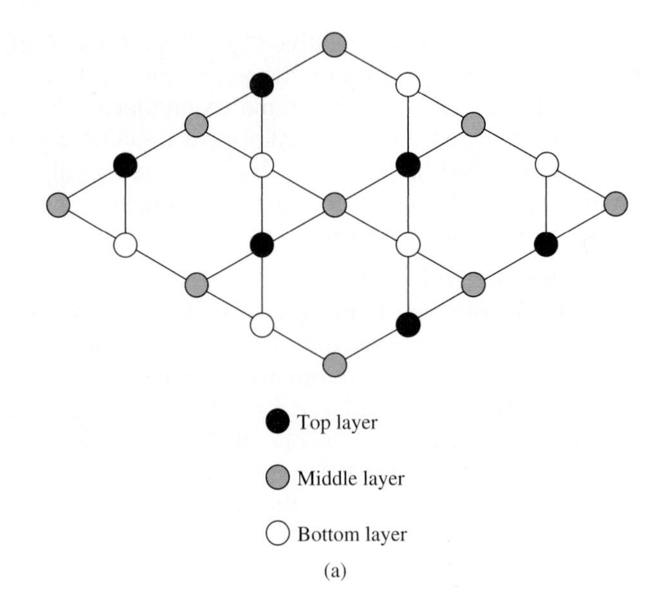

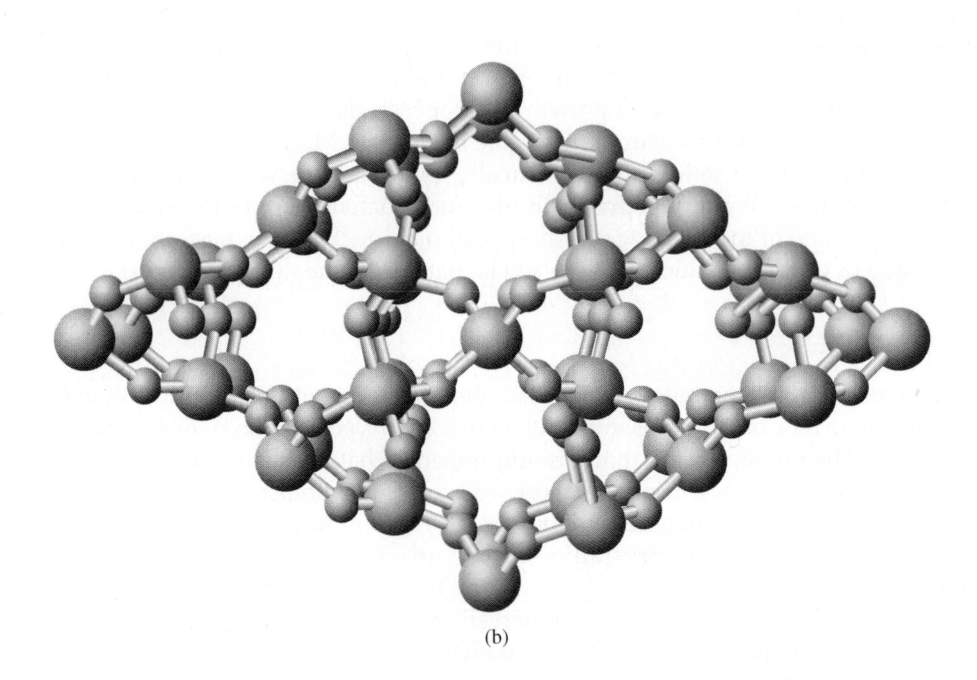

FIGURE 7.23 Crystal Structure of β-Quartz. (a) Overall structure, showing silicon atoms only. (b) Three-dimensional representation with both silicon (larger) and oxygen atoms. There are six triangular units surrounding and forming each hexagonal unit. Each triangular unit is helical, with a counterclockwise twist, and three silicon atoms and three oxygen atoms per turn; α-Quartz has a similar, but less regular, structure.

The four-coordination of silicon is also present in the silicates, forming chains, double chains, rings, sheets, and three-dimensional arrays. Al^{3+} can substitute for Si^{4+} but requires the addition of another cation to maintain charge balance. Aluminum, magnesium, iron, and titanium are common cations that occupy octahedral holes in the aluminosilicate structure, although any metal cation can be present. Some of the simpler examples of silicate structures are shown in Figure 7.24. These subunits pack together to form octahedral holes to accommodate the cations required to balance the charge. As mentioned previously, aluminum can substitute for silicon. A series of minerals is known with similar structures but different ratios of silicon to aluminum.

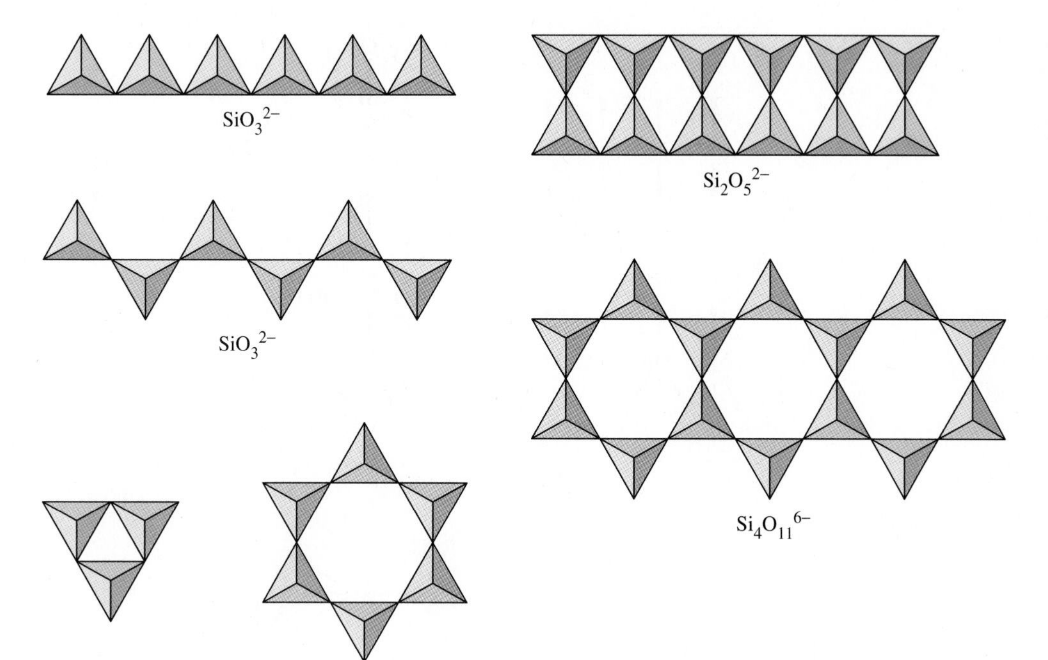

FIGURE 7.24 Common Silicate Structures. (Reproduced with permission from N. N. Greenwood and A. Earnshaw, *Chemistry of the Elements*, Pergamon Press, Elmsford, NY, 1984, pp. 403, 405; and from A. F. Wells, *Structural Inorganic Chemistry*, 5th ed., Oxford University Press, New York, 1984, pp. 1006, 1024.)

EXAMPLE

Relate the formulas of SiO_3^{2-} and $Si_2O_5^{2-}$ to the number of corners shared in the structures shown in Figure 7.24.

Consider the first tetrahedron in the chains of SiO_3^{2-} to have four oxygen atoms, or SiO_4. Extending the chain by adding SiO_3 units, with the fourth position sharing an oxygen atom of the previous tetrahedron, results in an infinite chain with the formula SiO_3. The charge can be calculated based on Si^{4+} and O^{2-}.

$Si_2O_5^{2-}$ can be described similarly. Beginning with one Si_2O_7 unit can start the chain. Adding Si_2O_5 units—two tetrahedra sharing one corner, and each with a vacant corner for sharing with the previous unit—can continue the chain indefinitely. Again, the charge can be calculated from the formula, Si_2O_5 based on Si^{4+} and O^{2-}.

▶ **Exercise 7.6** Describe the structure of $Si_3O_9^{6-}$ in a similar fashion.

One common family has units of two layers of silicates in the $Si_4O_{11}^{6-}$ geometry bound together by Mg^{2+}, Al^{3+}, or other metal ions, and hydroxide ions to form $Mg_3(OH)_4Si_2O_5$ or $Al_4(OH)_8Si_4O_{10}$ (kaolinite). Kaolinite is a china-clay mineral that forms very small hexagonal plates. If three magnesium ions substitute for two aluminum ions (for charge balance), the result is talc, $Mg_3(OH)_2Si_4O_{10}$. In either case, the oxygen atoms of the silicate units that are not shared between silicon atoms are in a hexagonal array that fits with the positions of hydroxide ions around the cation. The result is hydroxide ion bridging between Al or Mg and Si, as shown in Figure 7.25(a). The layers in talc are (1) all oxygen, the three shared by silicate tetrahedra; (2) all silicon; (3) oxygen and hydroxide in a 2:1 ratio, shared by silicon and magnesium; (4) magnesium; and (5) hydroxide shared between magnesium ions. If another silicate layer—made up of individual layers 3, 2, and 1—is on top of these layers, as in kaolinite, the mineral is called *pyrophyllite*. In both pyrophyllite and talc, the outer surfaces of these

FIGURE 7.25 Layer Structure of $Mg(OH)_2$-Si_2O_5 Minerals. (a) Side view of the separate layers. (b) Separate views of the layers. (c) The two layers superimposed, showing the sharing of O and OH between them.

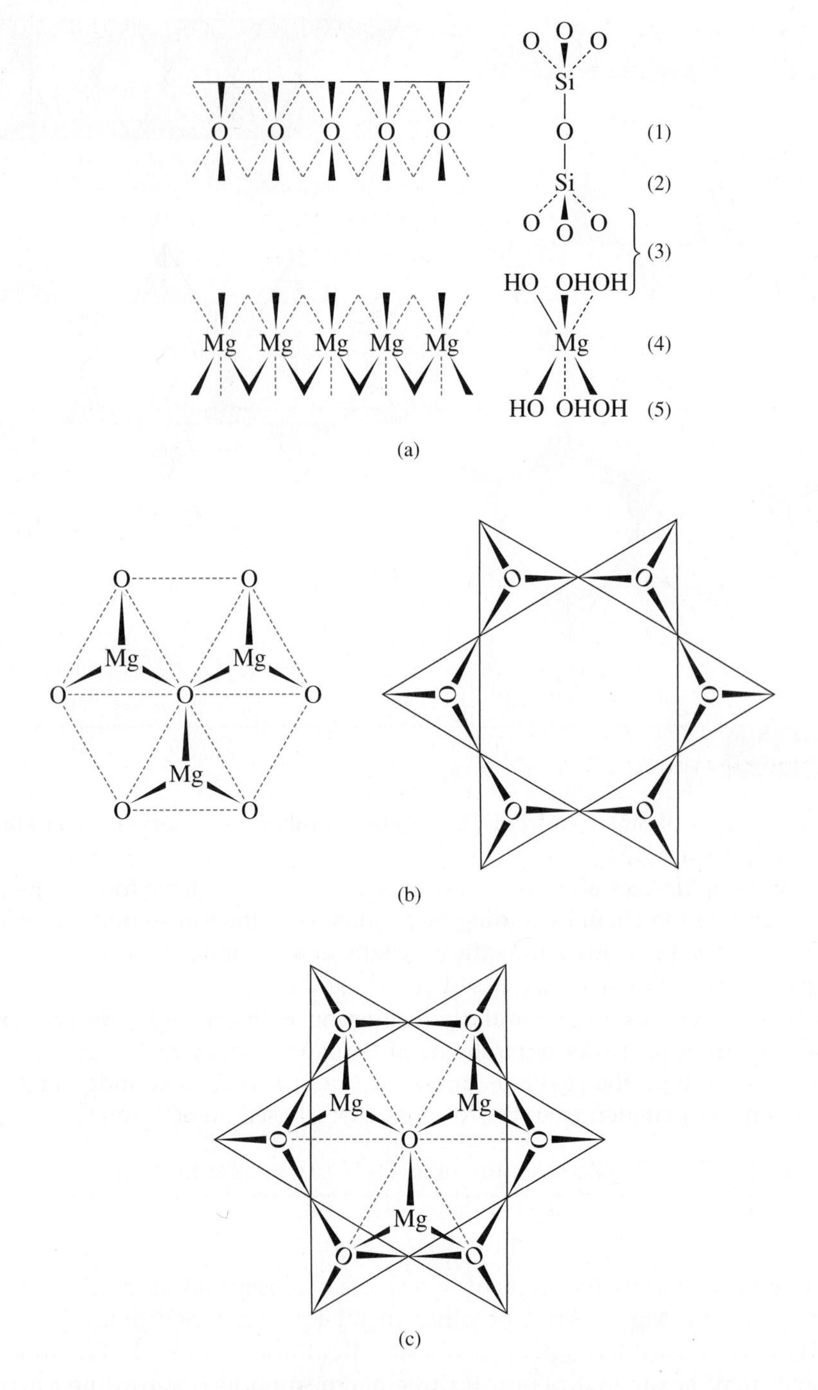

layered structures are the oxygen atoms of silicate tetrahedra, resulting in weak attractive forces and very soft materials. Soapstone and the talc used in cosmetics, paints, and ceramics are commercial products with these structures.

Hydrated montmorillonite has water between the silicate-aluminate-silicate layers. The micas (e.g., muscovite) have potassium ions in comparable positions and also have aluminum substituting for silicon in about 25% of the silicate sites. Changes in the proportions of aluminum and silicon in either of these allow the introduction of other cations and the formation of a large number of minerals. The layered structures of some micas are pronounced, allowing them to be cleaved into sheets used for high-temperature applications in which a transparent window is needed. They also have valuable insulating properties and are used in electrical devices.

If the *octahedral* Al^{3+} is partially replaced by Mg^{2+}, additional cations with charges of 1+ or 2+ are also added to the structures, and montmorillonites are the result. These clays swell on the absorption of water, act as cation exchangers, and have **thixotropic** properties: They are gels when undisturbed but become liquid when stirred, making them useful as oil field "muds" and in paints. Their formulas are variable, with $Na_{0.33}[Mg_{0.33}Al_{1.67}(OH)_2(Si_4O_{10})] \cdot n\ H_2O$ as an example. The cations can include Mg, Al, and Fe in the framework and H, Na, K, Mg, or Ca in the exchangeable positions.

The term *asbestos* is usually applied to a fibrous group of minerals that includes the amphiboles, such as tremolite, $Ca_2(OH)_2Mg_5(Si_4O_{11})_2$, with double-chain structures, and chrysotile, $Mg_3(OH)_4Si_2O_5$. In chrysotile, the dimensions of the silicate and magnesium layers are different, resulting in a curling that forms the characteristic cylindrical fibers.

The final group we will consider are the *zeolites*, mixed aluminosilicates containing $(Si, Al)_nO_{2n}$ frameworks with cations added to maintain charge balance. These minerals contain cavities that are large enough for other molecules to enter. Synthetic zeolites can be made with cavities tailored for specific purposes. The holes that provide entrances to the cavities can have from 4 to 12 silicon atoms around them. A common feature of many of these is a cubo-octahedral cavity formed from 24 silicate tetrahedra, each sharing oxygens on three corners. These units can then be linked by sharing of the external oxygen atoms to form cubic or tetrahedral units with still larger cavities. These minerals exhibit ion-exchange properties in which alkali and alkaline earth metal cations can exchange, depending on concentration. They were used in water softeners to remove excess Ca^{2+} and Mg^{2+} before the development of polystyrene ion-exchange resins. They can also be used to absorb water, oil, and other molecules and are known in the laboratory as "molecular sieves." A larger commercial market is as cat litter and oil absorbent, and they are also used in the petroleum industry as catalysts and as supports for other surface catalysts. A large number of zeolites have been described and illustrated in the *Atlas of Zeolite Structure Types*.[22] The references by Wells, Greenwood and Earnshaw cited previously also provide more information about these essential materials.

Figure 7.26 shows an example of the type of structure possible in the zeolites. Others have larger or smaller pores and larger or smaller entries into the pores.

The extreme range of sizes for the pores (260 to 1120 pm) makes it possible to control entry to and escape from the pores based on the size and branching geometry of the added material. In addition, the surfaces of the pores can be prepared with reactive

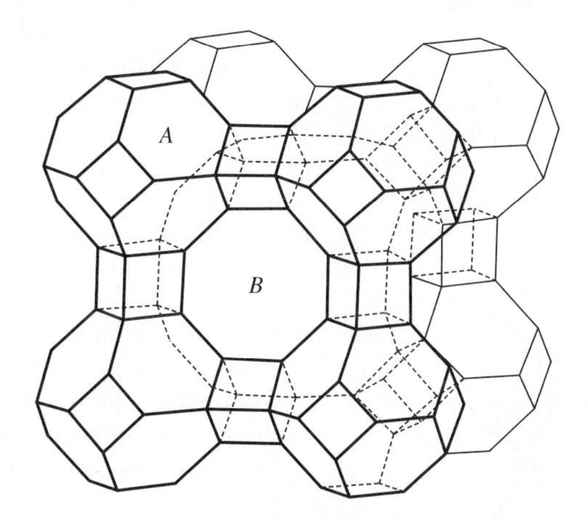

FIGURE 7.26 An Example of an Aluminosilicate Structure. Illustrated is the space-filling arrangement of truncated octahedra, cubes, and truncated cubooctahedra. (Reproduced with permission from A. F. Wells, *Structural Inorganic Chemistry*, 5th ed., Oxford University Press, Oxford, 1975, p. 1039.)

[22] W. M. Meier and D. H. Olson, *Atlas of Zeolite Structure Types*, 2nd ed., Structure Commission of the International Zeolite Commission, Butterworths, London, 1988.

metal atoms, providing opportunities for surface-catalyzed reactions. Although much of the design of these catalytic zeolites is of the "try it and see what happens" variety, patterns are emerging from the extensive base of data, and planned synthesis of catalysts is possible in some cases.

General References

Good introductions to most of the topics in this chapter are in A. B. Ellis et al., *Teaching General Chemistry: A Materials Science Companion*, American Chemical Society, Washington, DC, 1993; P. A. Cox, *Electronic Structure and Chemistry of Solids*, Oxford University Press, Oxford, 1987; and L. Smart and E. Moore, *Solid State Chemistry*, Chapman & Hall, London, 1992. Cox presents more of the theory, and Smart and Moore present more description of structures and their properties. Superconductivity is described by C. P. Poole, Jr., H. A. Farach, and R. J. Creswick in *Superconductivity*, Academic Press, San Diego, 1995; and G. Burns' *High-Temperature Superconductivity*, Academic Press, San Diego, 1992. A. F. Wells' *Structural Inorganic Chemistry*, 5th ed., Clarendon Press, Oxford, 1984; and N. N. Greenwood and A. Earnshaw's *Chemistry of the Elements*, 2nd ed., Butterworth–Heinemann, Oxford, 1997; describe the structures of a very large number of solids and discuss the bonding in them. A very good Web site on superconductors is http://www.superconductors.org.

Problems

7.1 Determine the point groups of the following unit cells:
  a. Face-centered cubic
  b. Body-centered tetragonal
  c. CsCl (Figure 7.7)
  d. Diamond (Figure 7.6)
  e. Nickel arsenide (Figure 7.10)

7.2 Show that atoms occupy only 52.4% of the total volume in a primitive cubic structure in which all the atoms are identical.

7.3 Show that a sphere of radius $0.73r$, where r is the radius of the corner atoms, will fit in the center of a primitive cubic structure.

7.4 **a.** Show that spheres occupy 74.0% of the total volume in a face-centered cubic structure in which all atoms are identical.
  b. What percent of the total volume is occupied by spheres in a body-centered cube in which all atoms are identical?

7.5 Using the diagrams of unit cells shown below, count the number of atoms at each type of position (corner, edge, face, internal) and each atom's fraction in the unit cell to determine the formulas ($M_m X_n$) of the compounds represented. Open circles represent cations, and closed circles represent anions.

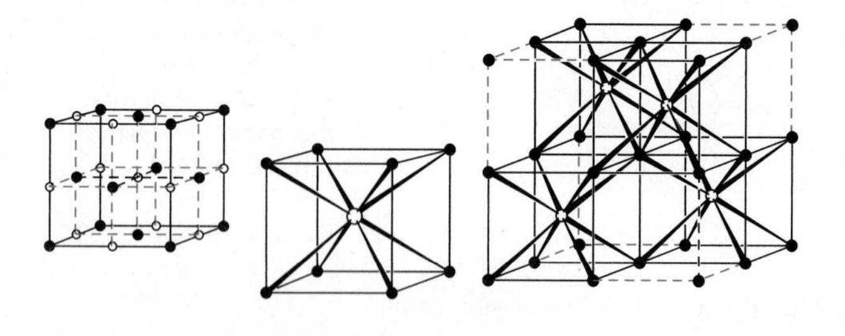

7.6 LiBr has a density of 3.464 g/cm^3 and the NaCl crystal structure. Calculate the interionic distance, and compare your answer with the value from the sum of the ionic radii found in Appendix B-1.

7.7 Compare the CsCl and CaF_2 lattices, particularly their coordination numbers.

7.8 Show that the zinc blende structure can be described as having zinc and sulfide ions each in face-centered lattices, merged so that each ion is in a tetrahedral hole of the other lattice.

7.9 Graphite has a layered structure, with each layer made up of six-membered rings of carbon fused with other similar rings on all sides. The Lewis structure shows alternating single and double bonds. Diamond is an insulator, and graphite is a moderately good conductor. Explain these facts in terms of the bonding in each. (Conductance of graphite is significantly lower than metals but is higher than most nonmetals.) What behavior would you predict for carbon nanotubes, the cylindrical form of fullerenes?

7.10 What experimental evidence is there for the model of alkali halides as consisting of positive and negative ions?

7.11 Mercury(I) chloride and all other Hg(I) salts are diamagnetic. Explain how this can be true. You may want to check the molecular formulas of these compounds.

7.12 **a.** Formation of anions from neutral atoms results in an increase in size, but formation of cations from neutral atoms results in a decrease in size. What causes these changes?
  b. The oxide ion and the fluoride ion both have the same electronic structure, but the oxide ion is larger. Why?

7.13 Calculate the radius ratios for the alkali halides. Which fit the radius ratio rules, and which violate them? (See L. C. Nathan, *J. Chem. Educ.*, **1985**, *62*, 215.)

7.14 Comment on the trends in the following values for interionic distances (pm):

LiF	201	NaF	231	AgF	246
LiCl	257	NaCl	281	AgCl	277
LiBr	275	NaBr	298	AgBr	288

7.15 Calculate the electron affinity of Cl from the following data for NaCl, and compare your result with the value in Appendix B-1: Cl_2 bond energy = 239 kJ/mol; ΔH_f (NaCl) = −413 kJ/mol; ΔH_{sub}(Na) = 109 kJ/mol; *IE*(Na) = 5.14 eV; and $r_+ + r_- = 281$ pm.

7.16 CaO is harder and has a higher melting point than KF, and MgO is harder and has a higher melting point than CaF_2. CaO, KF, and MgO have the NaCl structure. Explain these differences.

7.17 Calculate the lattice energies of the hypothetical compounds $NaCl_2$ and MgCl, assuming that the Mg^+ and Na^+ ions and the Na^{2+} and Mg^{2+} ions have the same radii. How do these results explain the compounds that are found experimentally? Use the following data in the calculation: Second ionization energies ($M^+ \longrightarrow M^{2+} + e^-$): Na, 4562 kJ/mol; Mg, 1451 kJ/mol; enthalpy of formation: NaCl, –411 kJ/mol; $MgCl_2$, –642 kJ/mol.

7.18 Use the Born–Haber cycle to calculate the enthalpy of formation of KBr, which crystallizes in the NaCl lattice. Use these data in the calculation: ΔH_{vap} (Br_2) = 29.8 kJ/mol; Br_2 bond energy = 190.2 kJ/mol; and ΔH_{sub} (K) = 79 kJ/mol.

7.19 Use the Born–Haber cycle to calculate the enthalpy of formation of MgO, which crystallizes in the rutile lattice. Use these data in the calculation: O_2 bond energy = 247 kJ/mol; ΔH_{sub} (Mg) = 37 kJ/mol. Second ionization energy of Mg = 1451 kJ/mol; second electron affinity of O = −744 kJ/mol.

7.20 Using crystal radii from Appendix B-1, calculate the lattice energy for PbS, which crystallizes in the NaCl structure. Compare the results with the Born–Haber cycle values obtained using the ionization energies and the following data for enthalpies of formation. Remember that enthalpies of formation are calculated beginning with the stable form of the elements. Use these data: ΔH_f: S^{2-} (*g*), 535 kJ/mol; Pb(g), 196 kJ/mol; PbS, –98 kJ/mol. The second ionization energy of Pb = 15.03 eV.

7.21 In addition to the doping described in this chapter, *n*-type semiconductors can be formed by increasing the amount of metal in ZnO or TiO_2, and *p*-type semiconductors can be formed by increasing the amount of nonmetal in Cu_2S, CuI, or ZnO. Explain how this is possible.

7.22 Explain how Cooper pairs can exist in superconducting materials, even though electrons repel each other.

7.23 Referring to other references if necessary, explain how zeolites that contain sodium ions can be used to soften water.

7.24 CaC_2 is an insulating ionic crystal. However, $Y_2Br_2C_2$, which can be described as containing C_2^{4-} ions, is metallic in two dimensions and becomes superconducting at 5 K. Describe the possible electronic structure of C_2^{4-}. In the crystal, monoclinic crystal symmetry leads to distortion of the Y_6 surrounding structure. How might this change the electronic structure of the ion? (See A. Simon, *Angew. Chem., Int. Ed.*, **1997**, *36*, 1788.)

7.25 Gallium arsenide is used in LEDs that emit red light. Would gallium nitride be expected to emit higher or lower energy light than gallium arsenide? How might such a process (emission of light from gallium nitride) be useful?

7.26 A series of ZnSe quantum dots was prepared of a range of sizes, with diameters from approximately 1.5 to 4.5 nm, and the photoluminescence emission spectra were recorded. Were the lowest energy emission bands produced by the largest or smallest quantum dots? Explain. (See V. V. Nikesh, A. D. Lad, S. Kimura, S. Nozaki, and S. Mahamuni, *J. Appl. Phys.*, **2006**, *100*, 113520.)

7.27 Medical studies on applications of quantum dots are progressing rapidly. Using appropriate search tools, such as Web of Science and SciFinder, find and briefly describe medical applications of quantum dots other than those mentioned in this chapter. Be sure to cite the references you consult.

7.28 Determine the formulas of the following silicates (*c* is a chain, extending vertically).

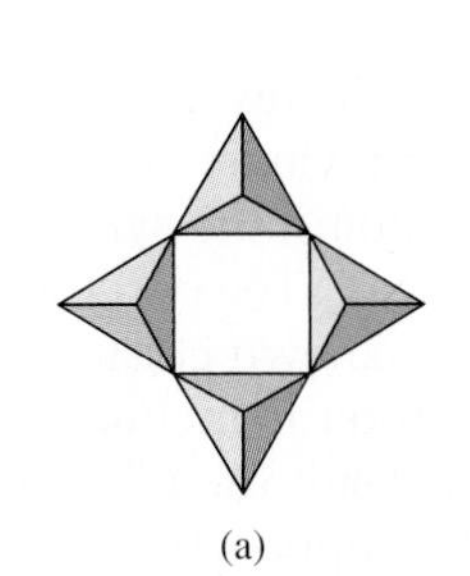
(a)

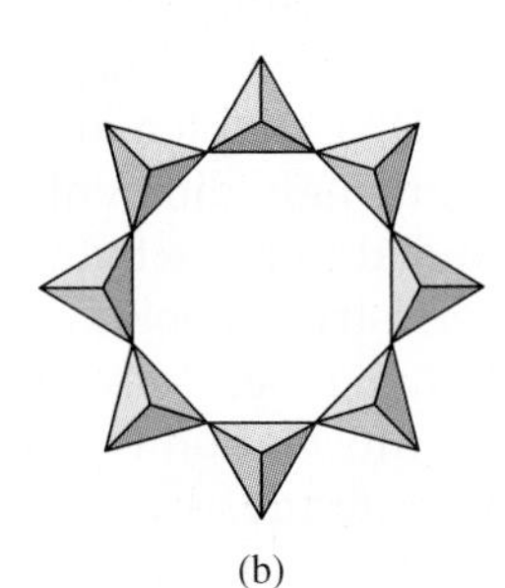
(b)

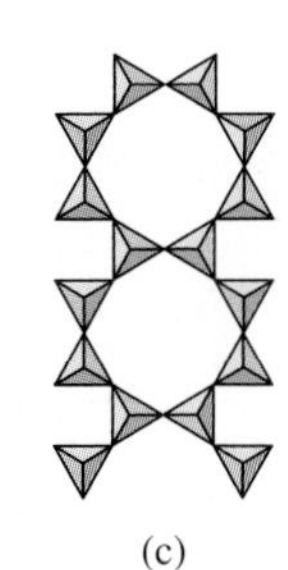
(c)

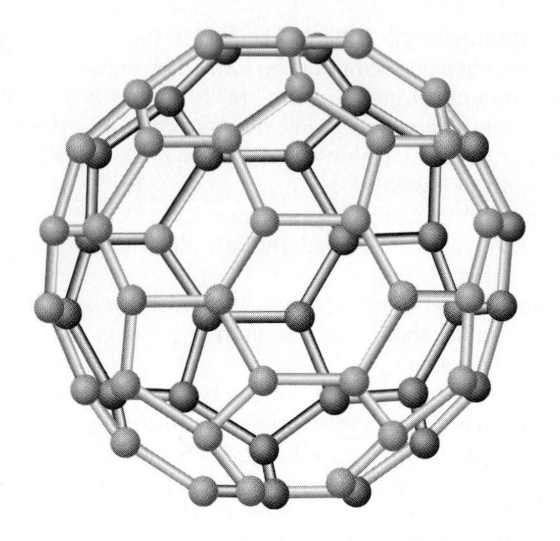

8 Chemistry of the Main Group Elements

This chapter presents some of the most significant physical and chemical data on each of the main groups of elements, also known as the *representative elements*, treating hydrogen first and continuing in sequence from Groups 1, 2, and 13 through 18 (Groups IA through VIIIA in common American notation).

The 20 industrial chemicals and minerals produced in the greatest amounts in the United States are main group elements or compounds (Table 8.1), and eight of the top ten may be classified as "inorganic"; numerous other compounds of these elements are of great commercial importance.

A discussion of main group chemistry provides a useful context in which to introduce a variety of topics not covered previously in this text. These topics may be particularly characteristic of main group chemistry but may be applicable to the chemistry of other elements as well. For example, many examples are known of atoms that form bridges between other atoms. Main group examples include the following:

```
  H        Cl        F
 / \      /  \      / \
B   B    Al   Al   Be  Be
```

In this chapter, we will discuss in some detail one important type of bridge: the hydrogens that form bridges between boron atoms in boranes. A similar approach can be used to describe bridges formed by other atoms and by groups such as CO (CO bridges between transition-metal atoms will be discussed in Chapter 13).

This chapter also provides examples in which modern chemistry has developed in ways surprisingly different from previously held ideas. Examples include compounds in which carbon is bonded to more than four atoms, the synthesis of alkali metal anions, and the now fairly extensive chemistry of noble gas elements. The past two decades have also seen the remarkable development of the fullerenes, previously unknown clusters of carbon atoms, and related forms of carbon. Much of the information in this chapter is included for the sake of handy reference; for more details, the interested reader should consult the references listed at the end of this chapter. The bonding and structures of main group compounds (Chapters 3 and 5) and acid–base reactions involving these compounds (Chapter 6) have already been discussed in this text.

TABLE 8.1 Top 20 Industrial Chemicals Produced in the United States, 2008

Rank	Chemical	Production ($\times 10^9$ kg)
1	Sodium chloride, $NaCl$	46.0[a]
2	Sulfuric acid, H_2SO_4	32.4
3	Phosphate rock, MPO_4	29.7
4	Ethylene, $H_2C{=}CH_2$	22.6
5	Lime, CaO	19.8[a]
6	Propylene, $H_2C{=}CH{-}CH_3$	14.8
7	Sodium carbonate, Na_2CO_3	11.2[a]
8	Chlorine, Cl_2	9.6
9	Ammonia, NH_3	9.5
10	Phosphoric acid, H_3PO_4	9.2
11	Sulfur, S_8	9.2[a]
12	Dichloroethane, $H_2ClC{-}CH_2Cl$	9.0
13	Nitric acid, HNO_3	7.5
14	Ammonium nitrate, NH_4NO_3	7.3
15	Sodium hydroxide, $NaOH$	7.3
16	Benzene, C_6H_6	5.6
17	Urea, $(NH_2)_2C{=}O$	5.3
18	Ethylbenzene, $C_2H_5C_6H_5$	4.1
19	Styrene, $C_6H_5CH{=}CH_2$	4.1
20	Hydrochloric acid, HCl	3.8

Sources: Chem. Eng. News, July 6, 2009, pp. 53, 56; estimated values from U. S. Department of the Interior, U. S. Geological Survey, *Mineral Commodity Summaries 2009.*

NOTE: [a] Estimated value.

8.1 GENERAL TRENDS IN MAIN GROUP CHEMISTRY

8.1.1 Physical Properties

The main group elements complete their electron configurations using *s* and *p* electrons. The total number of such electrons in the outermost shell is conveniently given by the traditional American group numbers in the periodic table. It is also the last digit in the group numbers recommended by the IUPAC (Groups 1, 2, and 13 through 18).[1] These elements range from the most metallic to the most nonmetallic, with elements of intermediate

[1] G. J. Leigh, ed., *Nomenclature of Inorganic Chemistry, Recommendations 1990,* International Union of Pure and Applied Chemistry, Blackwell Scientific Publications, Oxford UK, pp. 41–43.

properties, the semimetals—also known as metalloids—in between. On the far left, the alkali metals and alkaline earths exhibit the expected metallic characteristics of luster, high ability to conduct heat and electricity, and malleability. The distinction between metals and nonmetals is best illustrated by their difference in conductance. In Figure 8.1, electrical resistivities (inversely proportional to conductivities) of the solid main group elements are plotted.[2] At the far left are the alkali metals, having low resistivities (high conductances); at the far right are the nonmetals. Metals contain loosely bound valence electrons that are relatively free to move and thereby conduct current. In most cases, nonmetals contain much more localized lone electron pairs and covalently bonded pairs that are less mobile. An exception, as we will see, is graphite, a form of carbon that has a much greater ability to conduct than most nonmetals because of delocalized electron pairs.

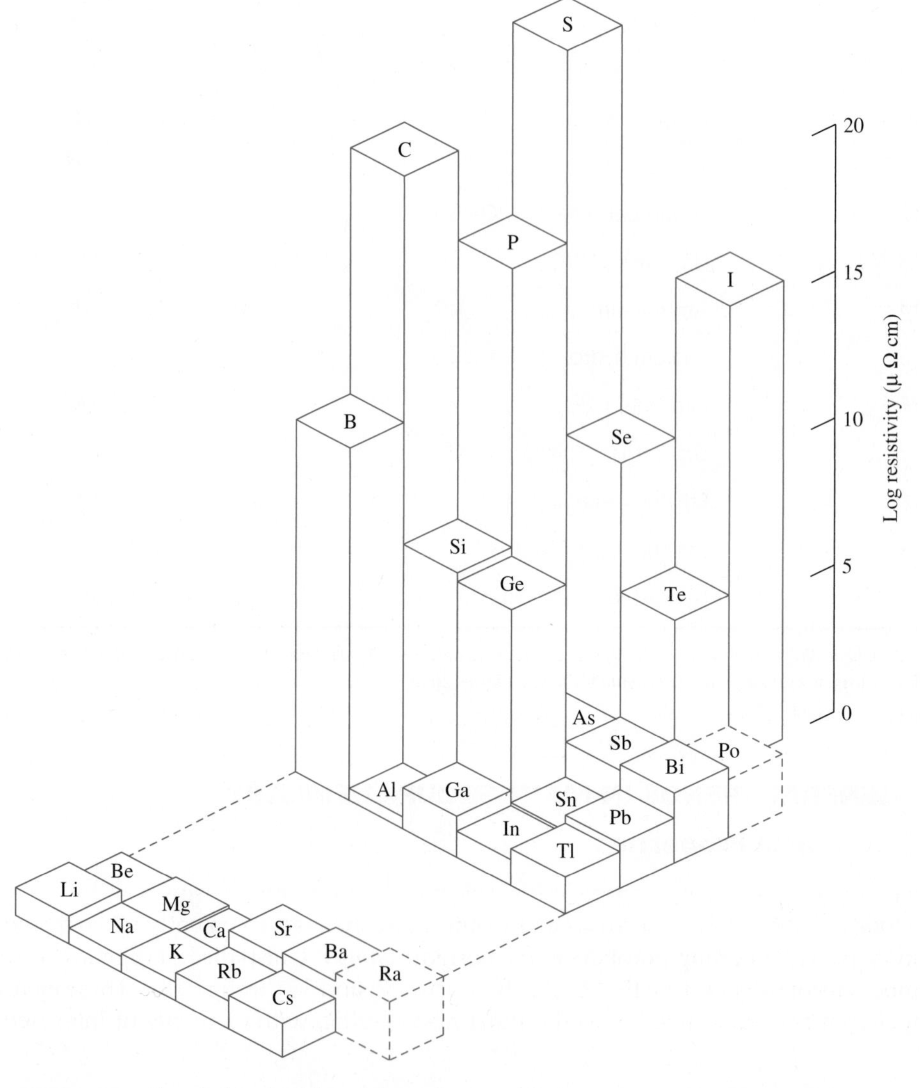

FIGURE 8.1 Electrical Resistivities of the Main Group Elements. Dashed lines indicate estimated values. (Data from J. Emsley, *The Elements*, Oxford University Press, New York, 1989.)

[2] The electrical resistivity shown for carbon is for the diamond allotrope. Graphite, another allotrope of carbon, has a resistivity between that of metals and semiconductors.

Elements along a rough diagonal from boron to polonium are intermediate in behavior, in some cases having both metallic and nonmetallic allotropes (elemental forms); these elements are designated as **metalloids** or **semimetals**. As described in Chapter 7, some elements, such as silicon and germanium, are capable of having their conductivity finely tuned by the addition of small amounts of impurities and are consequently of enormous importance in the manufacture of semiconductors in the electronics industry.

Some of the columns of main group elements have long been designated by common names (e.g., the halogens); names for others have been suggested, and some have been used more frequently in recent years:

Group	Common Name	Group	Common Name
1(I)	Alkali metals	15(V)	Pnicogens, pnictogens
2(II)	Alkaline earths	16(VI)	Chalcogens
13(III)	Triel elements	17(VII)	Halogens
14(IV)	Tetrel elements	18(VIII)	Noble gases

8.1.2 Electronegativity

Electronegativity, shown in Figure 8.2, also provides a guide to the chemical behavior of the main group elements. The extremely high electronegativity of the nonmetal fluorine and the noble gases helium and neon are evident, with a steady decline in electronegativity toward the left and the bottom of the periodic table. The semimetals form a diagonal of intermediate electronegativity. Definitions of electronegativity have been given in Chapter 3 (Section 3.2.3) and tabulated values for the elements are given in Table 3.3 and Appendix B.4.

Although usually classified with Group 1 (IA), hydrogen is quite dissimilar from the alkali metals in its electronegativity and in many other properties, both chemical and physical. Hydrogen's chemistry is distinctive from all the groups, so this element will be discussed separately in this chapter.

The noble gases have higher ionization energies than the halogens, and calculations have suggested that the electronegativities of the noble gases may match or even exceed those of the halogens.[3] The noble gas atoms are somewhat smaller than the neighboring halogen atoms (e.g., Ne is smaller than F) as a consequence of a greater effective nuclear charge. This charge, which attracts noble gas electrons strongly toward the nucleus, is also likely to exert a strong attraction on electrons of neighboring atoms; hence, high electronegativities predicted for the noble gases are reasonable. Estimated values of these electronegativities are included in Figure 8.2 and Appendix B.4.

8.1.3 Ionization Energy

Ionization energies of the main group elements exhibit trends similar to those of electronegativity, as shown in Figure 8.3. There are some subtle differences, however.

As discussed in Section 2.3.1, although a general increase in ionization energy occurs toward the upper right-hand corner of the periodic table, three of the Group 13 (IIIA) elements have lower ionization energies than the preceding Group 2 (IIA) elements, and three Group 16 (VIA) elements have lower ionization energies than the preceding Group 15 (VA) elements. For example, the ionization energy of boron is lower

[3] L. C. Allen and J. E. Huheey, *J. Inorg. Nucl. Chem.*, **1980**, *42*, 1523; T. L. Meek, *J. Chem. Educ.*, **1995**, *72*, 17.

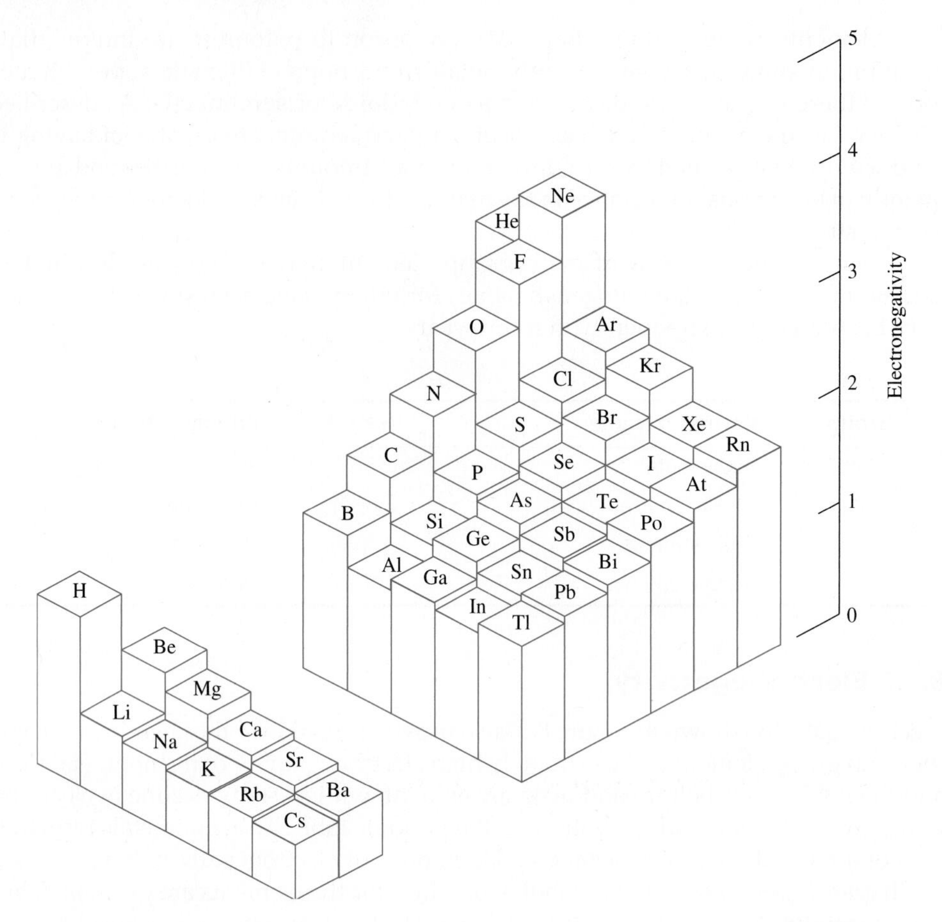

FIGURE 8.2 Electronegativities of the Main Group Elements. (Data from J. B. Mann, T. L. Meek, and L. C. Allen, *J. Am. Chem. Soc.*, **2000**, *122*, 2780.)

than that of beryllium, and the ionization energy of oxygen is lower than that of nitrogen (see also Figure 2.13). Be and N have electron subshells that are completely filled ($2s^2$ for Be) or half-filled ($2p^3$ for N). The next atoms (B and O) have an additional electron that is lost with comparative ease. In boron, the outermost electron, a $2p$, has significantly higher energy (higher quantum number l) than the filled $1s$ and $2s$ orbitals and is thus more easily lost than a $2s$ electron of Be. In oxygen, the fourth $2p$ electron must pair with another $2p$ electron; occupation of this orbital by two electrons is accompanied by an increase in electron–electron repulsions that facilitates loss of an electron. Additional examples of this phenomenon can be seen in Figures 8.3 and 2.13 (tabulated values of ionization energies are included in Appendix B.2).

8.1.4 Chemical Properties

Efforts to find similarities in the chemistry of the main group elements began well before the formulation of the modern periodic table. The strongest parallels are within each group: the alkali metals most strongly resemble other alkali metals, halogens resemble other halogens, and so on. In addition, certain similarities have been recognized between some elements along diagonals (upper left to lower right) in the periodic table. One example is that of electronegativities. As can be seen from Figure 8.2, electronegativities along diagonals are similar; for example, values along the diagonal from B to Te are in the range 1.9 to 2.2. Other "diagonal" similarities include the unusually low solubilities of LiF and MgF_2, (a consequence of the small sizes of Li^+ and Mg^{2+}, which lead to high

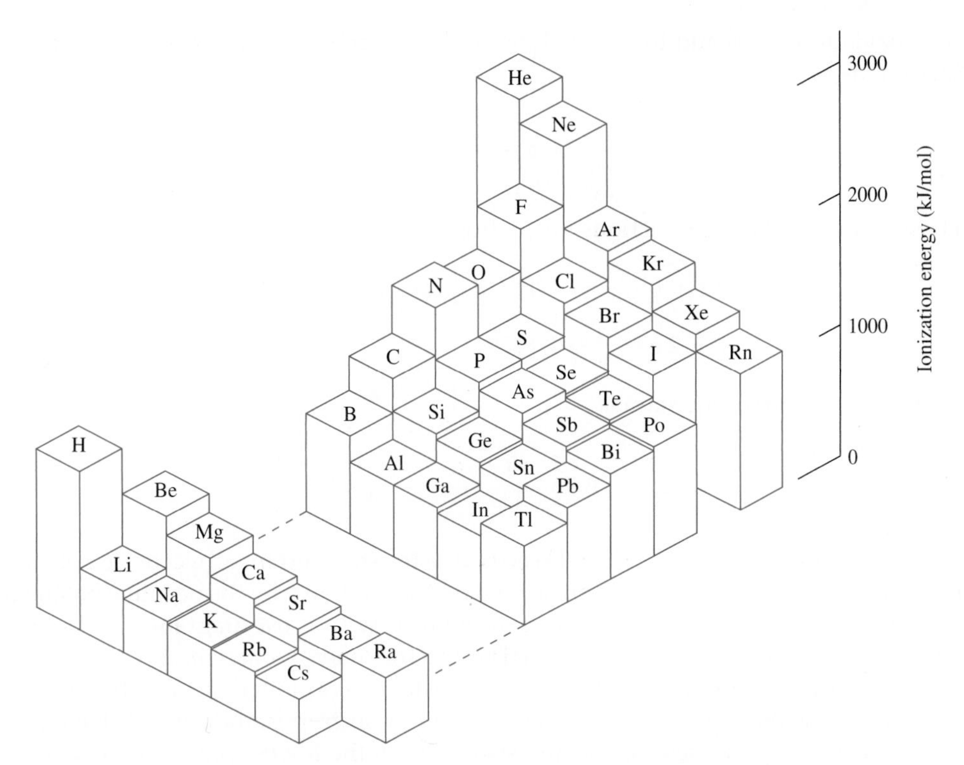

FIGURE 8.3 Ionization Energies of the Main Group Elements. (Data from C. E. Moore, *Ionization Potentials and Ionization Limits Derived from the Analyses of Optical Spectra*, National Standard Reference Data Series, U. S. National Bureau of Standards, NSRDS-NBS 34, Washington, DC, 1970.)

lattice energies in these ionic compounds), similarities in solubilities of carbonates and hydroxides of Be and Al, and the formation of complex three-dimensional structures based on SiO_4 and BO_4 tetrahedra. These parallels are interesting but somewhat limited in scope; they can often be explained on the basis of similarities in sizes and electronic structures of the compounds in question.

The main group elements show the "first-row anomaly" (counting the elements Li through Ne as the first row). Properties of elements in this row are often significantly different from properties of other elements in the same group. For example, consider the following: F_2 has a much lower bond energy than expected by extrapolation of the bond energies of Cl_2, Br_2, and I_2; HF is a weak acid in aqueous solution, whereas HCl, HBr, and HI are all strong acids; multiple bonds between carbon atoms are much more common than multiple bonds between other elements in Group 14 (IVA); and hydrogen bonding is much stronger for compounds of F, O, and N than for compounds of other elements in their groups. No single explanation accounts for all the differences between elements in this row and other elements. However, in many cases the distinctive chemistry of the first-row elements is related to the small atomic sizes and the related high electronegativities of these elements.

OXIDATION–REDUCTION REACTIONS Oxidation–reduction reactions of inorganic species can be described in many different ways. For example, hydrogen exhibits oxidation states of −1, 0, and +1. In acidic aqueous solution, these oxidation states occur in the half-reactions

$$2\,H^+ + 2\,e^- \longrightarrow H_2 \quad \mathscr{E}^\circ = 0\text{ V}$$

$$H_2 + 2\,e^- \longrightarrow 2\,H^- \quad \mathscr{E}^\circ = -2.25\text{ V}$$

These oxidation states and their matching reduction potentials are shown in a **Latimer diagram**[4] as

$$\begin{array}{ccccc} +1 & & 0 & & -1 \\ H^+ & \xrightarrow{0} & H_2 & \xrightarrow{-2.25} & H^- \end{array} \longleftarrow \text{Oxidation states}$$

In basic solution, the half-reactions for hydrogen are

$$H_2O + e^- \longrightarrow OH^- + \tfrac{1}{2}H_2 \qquad \mathscr{E}^\circ = -0.828 \text{ V}$$

$$H_2 + 2\,e^- \longrightarrow 2\,H^- \qquad \mathscr{E}^\circ = -2.25 \text{ V}$$

The matching Latimer diagram is

$$\begin{array}{ccccc} +1 & & 0 & & -1 \\ H_2O & \xrightarrow{-0.828} & H_2 & \xrightarrow{-2.25} & H^- \end{array}$$

The half-reaction $2\,H^+ + 2\,e^- \longrightarrow H_2$ is used as the standard for all electrode potentials in acid solutions; the others shown are less favorable, as shown by the negative potentials.

Another way to describe the same reactions is with **Frost diagrams**, as shown in Figure 8.4, in which $-n\mathscr{E}^\circ$ (proportional to the free energy change, $\Delta G^\circ = -n\mathscr{F}\mathscr{E}^\circ$, where n is the number of moles of electrons transferred) is plotted against the oxidation state in the same order as in Latimer diagrams. In these graphs, the neutral element has a value of zero on both scales, and the species with the lowest potential (lowest free energy) is the most stable. Similar diagrams for oxygen show that the most stable form in either acid or base is water or hydroxide ion.

The Latimer diagrams for oxygen are

$$\begin{array}{cccccl} 0 & & -1 & & -2 & \\ O_2 & \xrightarrow{-0.0695} & H_2O_2 & \xrightarrow{1.763} & H_2O & \text{in acid} \\ 0 & & -1 & & -2 & \\ O_2 & \xrightarrow{-0.0649} & HO_2^- & \xrightarrow{0.867} & OH^- & \text{in base} \end{array}$$

and the Frost diagrams are given in Figure 8.5. Species such as HO_2, OH, and O are omitted from these diagrams for simplicity. These species are germane to gas phase reactions but are not ordinarily encountered in aqueous solution chemistry.

Latimer diagrams for many elements are in Appendix B.7.

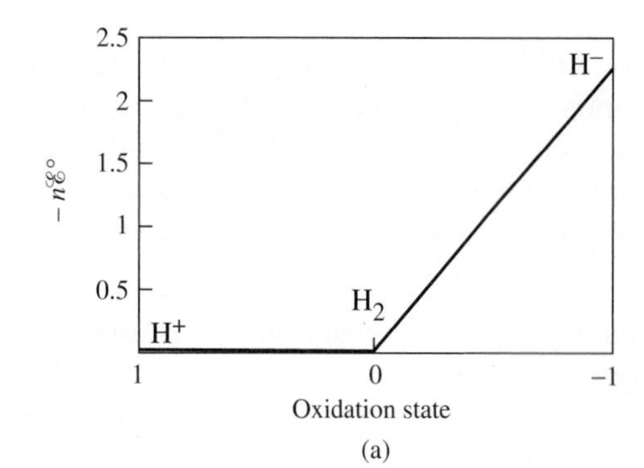

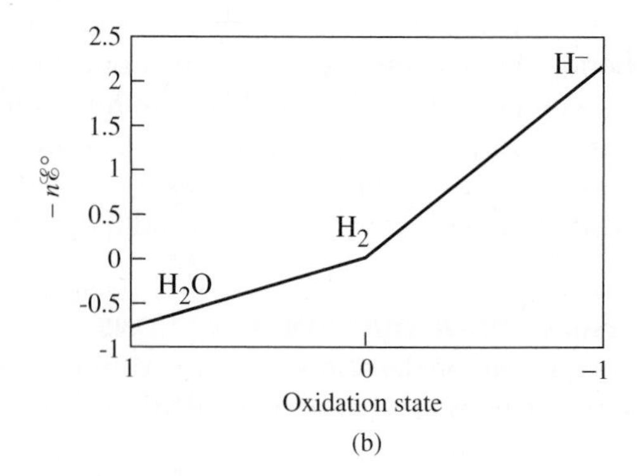

FIGURE 8.4 Frost Diagrams for Hydrogen. (a) Acidic solution. (b) Basic solution.

[4] W. M. Latimer, *Oxidation Potentials*, Prentice Hall, Englewood Cliffs, NJ, 1952.

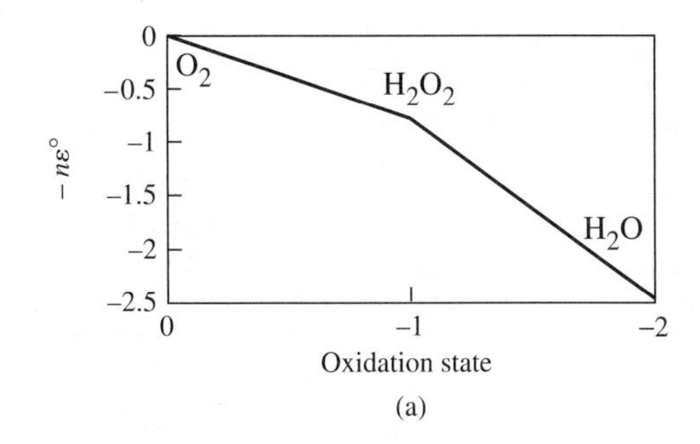

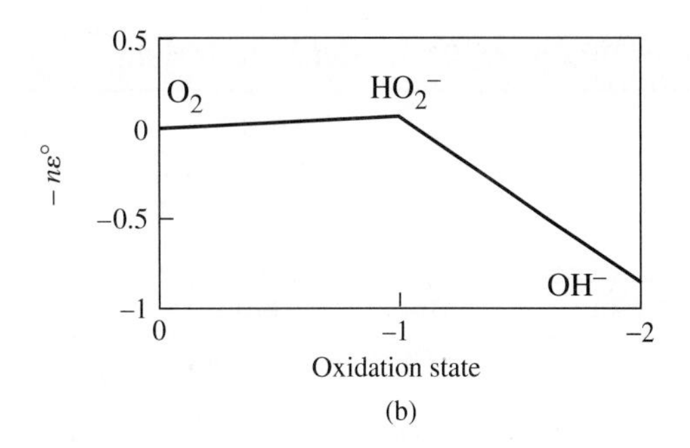

FIGURE 8.5 Frost Diagrams for Oxygen. (a) Acidic solution. (b) Basic solution.

8.2 HYDROGEN

The most appropriate position of hydrogen in the periodic table has been a matter of some dispute among chemists. Its electron configuration, $1s^1$, is similar to the valence electron configurations of the alkali metals (ns^1); hence, hydrogen is most commonly listed in the periodic table at the top of Group 1 (IA). However, it has little chemical similarity to the alkali metals. Hydrogen is also one electron short of a noble gas configuration and could conceivably be classified with the halogens. Although hydrogen has some similarities with the halogens—for example, in forming a diatomic molecule and an ion of 1− charge—these similarities are limited. A third possibility is to place hydrogen in Group 14 (IVA) above carbon: both elements have half-filled valence electron shells, are of similar electronegativity, and usually form covalent rather than ionic bonds. We prefer not to attempt to fit hydrogen into any particular group in the periodic table, because it is a unique element in many ways and deserves separate consideration.

Hydrogen is by far the most abundant element in the universe (and the sun) and, primarily in its compounds, is the third most abundant element in the Earth's crust. The element occurs as three isotopes: ordinary hydrogen, or *protium*, ^1H; *deuterium*, ^2H or D; and *tritium*, ^3H or T. Both ^1H and ^2H have stable nuclei; ^3H undergoes β decay

$$^{3}_{1}\mathrm{H} \longrightarrow {}^{3}_{2}\mathrm{He} + {}^{0}_{-1}\mathrm{e}$$

and has a half-life of 12.35 years. Naturally occurring hydrogen is 99.985% ^1H and essentially all the remainder is ^2H; only traces of the radioactive ^3H are found on earth. Deuterium compounds are used extensively as solvents for nuclear magnetic resonance (NMR) spectroscopy and in kinetic studies on reactions involving bonds to hydrogen (deuterium isotope effects). Tritium is produced in nuclear reactors by bombardment of ^6Li nuclei with neutrons:

$$^{6}_{3}\mathrm{Li} + {}^{1}_{0}\mathrm{n} \longrightarrow {}^{4}_{2}\mathrm{He} + {}^{3}_{1}\mathrm{H}$$

It has many applications as a tracer—for example, to study the movement of ground water—and to study the *ab*sorption of hydrogen by metals and the *ad*sorption of hydrogen on metal surfaces. Many deuterated and tritiated compounds have been synthesized and studied. Some of the important physical properties of the isotopes of hydrogen are listed in Table 8.2.

8.2.1 Chemical Properties

Hydrogen can gain an electron to achieve a noble gas configuration in forming the hydride ion, H^-. Many metals, such as the alkali metals and alkaline earths, form hydrides that are essentially ionic and contain discrete H^- ions. The hydride ion is a

TABLE 8.2 Properties of Hydrogen, Deuterium, and Tritium

			Properties of Molecules, X_2			
Isotope	**Abundance (%)**	**Atomic Mass**	**Melting Point (K)**	**Boiling Point (K)**	**Critical Temperature (K)[a]**	**Enthalpy of Dissociation ($kJ\ mol^{-1}$ at 25 °C)**
Protium (^1H), H	99.985	1.007825	13.957	20.30	33.19	435.88
Deuterium (^2H), D	0.015	2.014102	18.73	23.67	38.35	443.35
Tritium (^3H), T	$\sim 10^{-16}$	3.016049	20.62	25.04	40.6 (calc)	446.9

Sources: Abundance and atomic mass data from I. Mills, T. Cuitoš, K. Homann, N. Kallay, and K. Kuchitsu, eds., *Quantities, Units, and Symbols in Physical Chemistry*, International Union of Pure and Applied Chemistry, Blackwell Scientific Publications, Oxford UK, 1988. Other data are from N. N. Greenwood and A. Earnshaw, *Chemistry of the Elements*, Pergamon Press, Elmsford, NY, 1984.

NOTE: [a]The highest temperature at which a gas can be condensed to a liquid

powerful reducing agent (for); it reacts, for example, with water and other protic solvents to generate H_2:

$$2\,H^- + H_2O \longrightarrow H_2 + 2\,OH^-$$

In many other cases, bonding to hydrogen atoms is essentially covalent—for example, in compounds with carbon and other nonmetals. Hydride ions may also act as ligands in bonding to metals, with as many as nine hydrogens on a single metal, as in $ReH_9{}^{2-}$. Many complex hydrides, such as $BH_4{}^-$ and $AlH_4{}^-$, serve as important reagents in organic and inorganic synthesis. Although such complexes may be described formally as hydrides, their bonding is essentially covalent.

Lithium aluminum hydride can be prepared by the treatment of lithium hydride with a solution of aluminum chloride in ether:[5]

$$4\,LiH + AlCl_3 \longrightarrow LiAlH_4 + 3\,LiCl$$

$LiAlH_4$ is a versatile reducing agent for many organic compounds, such as ketones, aldehydes, nitriles, and nitro compounds. This ion also has many applications in inorganic synthesis. Examples of the inorganic conversions effected by $LiAlH_4$ include

$$SiCl_4 \longrightarrow SiH_4 \qquad BCl_3 \longrightarrow B_2H_6 \qquad NO \longrightarrow HON{=}NOH$$

Reference to the "hydrogen ion," H^+, is also common. However, in the presence of solvent, the extremely small size of the proton (radius approximately 1.5×10^{-3} pm) requires that it be associated with solvent molecules or other dissolved species. In aqueous solution, a more correct description is $H_3O^+(aq)$, although larger species such as $H_9O_4{}^+$ have also been identified. Another important characteristic of H^+ that is a consequence of its small size is its ability to form hydrogen bonds.

The ready combustibility of hydrogen, together with the lack of potentially polluting by-products, has led to the proposal to use hydrogen as a fuel. For example, as a potential fuel for automobiles, H_2 can provide a greater amount of energy per unit mass than gasoline without producing such environmentally damaging by-products as carbon monoxide, sulfur dioxide, and unburned hydrocarbons. A challenge for

[5] A. E. Finholt, A. C. Bond, Jr., and H. I. Schlesinger, *J. Am. Chem. Soc.*, **1947**, *69*, 1199.

chemists is to develop practical thermal or photochemical processes for generating hydrogen from its most abundant source—water.

Small amounts of H_2 can be generated in the laboratory by reacting "active" metals such as zinc, magnesium, or aluminum with acid:

$$Zn(s) + 2\,H^+(aq) \longrightarrow H_2(g) + Zn^{2+}(aq)$$

Commercially, H_2 is frequently prepared by "cracking" petroleum hydrocarbons with solid catalysts, also forming alkenes

$$C_2H_6 \longrightarrow C_2H_4 + H_2$$

or by steam reforming of natural gas, typically using a nickel catalyst

$$CH_4 + H_2O \longrightarrow CO + 3H_2$$

Molecular hydrogen is also an important reagent, especially in the industrial hydrogenation of unsaturated organic molecules. Examples of such processes involving transition–metal catalysts are discussed in Chapter 14.

8.3 GROUP 1 (IA): THE ALKALI METALS

Alkali metal salts, in particular sodium chloride, have been known and used since antiquity. In early times, long before the chemistry of these compounds was understood, salt was used in the preservation and flavoring of food and even as a medium of exchange. However, because of the difficulty of reducing the alkali metal ions, the elements were not isolated until comparatively recently, well after many other elements. Two of the alkali metals, sodium and potassium, are essential for human life; their careful regulation is often important in treating a variety of medical conditions.

8.3.1 The Elements

Potassium and sodium were first isolated within a few days of each other in 1807 by Humphry Davy as products of the electrolysis of molten KOH and NaOH. In 1817, J. A. Arfvedson, a young chemist working with J. J. Berzelius, recognized similarities between the solubilities of compounds of lithium and those of sodium and potassium. The following year, Davy also became the first to isolate lithium, this time by electrolysis of molten Li_2O. Cesium and rubidium were discovered with the help of the spectroscope in 1860 and 1861, respectively; they were named after the colors of the most prominent emission lines (Latin, *caesius*, "sky blue," *rubidus*, "deep red"). Francium was not identified until 1939 as a short-lived radioactive isotope from the nuclear decay of actinium.

The alkali metals are silvery—except for cesium, which has a golden appearance—highly reactive solids having low melting points. They are ordinarily stored under nonreactive oil to prevent air oxidation and are soft enough to be easily cut with a knife or spatula. Their melting points decrease with increasing atomic number, because metallic bonding between the atoms becomes weaker with increasing atomic size. Physical properties of the alkali metals are summarized in Table 8.3.

8.3.2 Chemical Properties

The alkali metals are very similar in their chemical properties, which are governed in large part by the ease with which they can lose one electron (the alkali metals have the lowest ionization energies of all the elements) and thereby achieve a noble gas configuration. All

TABLE 8.3 Properties of the Group 1(IA) Elements: The Alkali Metals

Element	Ionization Energy ($kJ\ mol^{-1}$)	Electron Affinity ($kJ\ mol^{-1}$)	Melting Point (°C)	Boiling Point (°C)	Electro-negativity	$\mathscr{E}°$ ($M^+ \longrightarrow M$) (V)[a]
Li	520	60	180.5	1347	0.912	−3.04
Na	496	53	97.8	881	0.869	−2.71
K	419	48	63.2	766	0.734	−2.92
Rb	403	47	39.0	688	0.706	−2.92
Cs	376	46	28.5	705	0.659	−2.92
Fr	400[b, c]	60[b, d]	27		0.7[b]	−2.9[d]

Sources: Ionization energies cited in this chapter are from C. E. Moore, *Ionization Potentials and Ionization Limits Derived from the Analyses of Optical Spectra*, National Standard Reference Data Series, U.S. National Bureau of Standards, NSRDS-NBS 34, Washington, DC, 1970, unless noted otherwise. Electron affinity values listed in this chapter are from H. Hotop and W. C. Lineberger, *J. Phys. Chem. Ref. Data*, **1985**, *14*, 731. Standard electrode potentials listed in this chapter are from A. J. Bard, R. Parsons, and J. Jordan, eds., *Standard Potentials in Aqueous Solutions*, Marcel Dekker (for IUPAC), New York, 1985. Electronegativities cited in this chapter are from J. B. Mann, T. L. Meek, and L. C. Allen, *J. Am. Chem. Soc.*, **2000**, *122*, 2780, Table 2. Other data are from N. N. Greenwood and A. Earnshaw, *Chemistry of the Elements*, Pergamon Press, Elmsford, NY, 1984, except where noted.

NOTES: [a]Aqueous solution, 25 °C

[b]Approximate value

[c]J. Emsley, *The Elements*, Oxford University Press, New York, 1989.

[d]S. G. Bratsch, *J. Chem. Educ.*, **1988**, *65*, 34.

are highly reactive metals and are excellent reducing agents. The metals react vigorously with water to form hydrogen; for example,

$$2\,Na + 2\,H_2O \longrightarrow 2\,NaOH + H_2$$

This reaction is highly exothermic, and the hydrogen formed may ignite in air, sometimes explosively, if a large quantity of sodium is used. Consequently, special precautions must be taken to prevent these metals from coming into contact with water when they are stored.

Alkali metals react with oxygen to form oxides, peroxides, and superoxides, depending on the metal. Combustion in air yields the following products:[6]

Principal Combustion Product (Minor Product)			
Alkali Metal	Oxide	Peroxide	Superoxide
Li	Li_2O	(Li_2O_2)	
Na	(Na_2O)	Na_2O_2	
K			KO_2
Rb			RbO_2
Cs			CsO_2

Alkali metals dissolve in liquid ammonia and other donor solvents, such as aliphatic amines (NR_3, in which R = alkyl) and $OP(NMe_2)_3$, hexamethylphosphoramide, to give blue solutions believed to contain solvated electrons:

$$Na + x\,NH_3 \longrightarrow Na^+ + e(NH_3)_x^-$$

[6] Additional information on peroxide, superoxide, and other oxygen-containing ions is provided in Table 8.12.

Because of these solvated electrons, dilute solutions of alkali metals in ammonia conduct electricity far better than completely dissociated ionic compounds in aqueous solutions. As the concentration of the alkali metals is increased, the conductivity first declines, then increases. At sufficiently high concentration, the solution acquires a bronze metallic luster and a conductivity comparable to a molten metal. Dilute solutions are paramagnetic, with approximately one unpaired electron per metal atom, corresponding to one solvated electron per metal atom; this paramagnetism decreases at higher concentrations. One interesting aspect of these solutions is that they are less dense than liquid ammonia itself. The solvated electrons may be viewed as creating cavities for themselves (estimated radius of approximately 300 pm) in the solvent, thus increasing the volume significantly. The blue color, corresponding to a broad absorption band near 1500 nm that extends into the visible range, is attributed to the solvated electron (alkali metal ions are colorless). At higher concentrations these solutions have a coppery color and contain alkali metal anions, M^-.

Not surprisingly, solutions of alkali metals in liquid ammonia are excellent reducing agents. Following are examples of reductions that can be effected by these solutions:

$$RC\equiv CH + e^- \longrightarrow RC\equiv C^- + \tfrac{1}{2}H_2$$

$$NH_4^+ + e^- \longrightarrow NH_3 + \tfrac{1}{2}H_2$$

$$S_8 + 2\,e^- \longrightarrow S_8^{2-}$$

$$Fe(CO)_5 + 2\,e^- \longrightarrow [Fe(CO)_4]^{2-} + CO$$

The solutions of alkali metals are unstable and undergo slow decomposition to form amides:

$$M + NH_3 \longrightarrow MNH_2 + \tfrac{1}{2}H_2$$

Other metals—especially the alkaline earths Ca, Sr, and Ba and the lanthanides Eu and Yb (both of which can form 2+ ions)—can also dissolve in liquid ammonia to give the solvated electron; however, the alkali metals undergo this reaction more efficiently and have been used far more extensively for synthetic purposes.

Alkali metal atoms have very low ionization energies and readily lose their outermost (ns^1) electron to form their common ions of 1+ charge. These ions can form complexes with a variety of Lewis bases (ligands, to be discussed more fully in Chapters 9 through 14). Of particular interest are cyclic Lewis bases that have several donor atoms that can surround, or trap, cations. Examples of such molecules are shown in Figure 8.6. The first of these is one of a large group of cyclic ethers, commonly known as "crown" ethers, which donate electron density to metals through their oxygen atoms. The second, one of a family of cryptands (or cryptates), can be

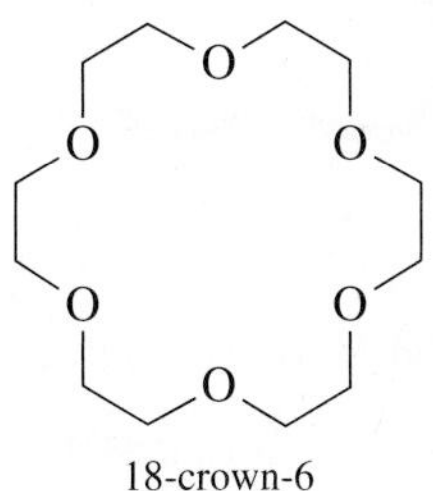

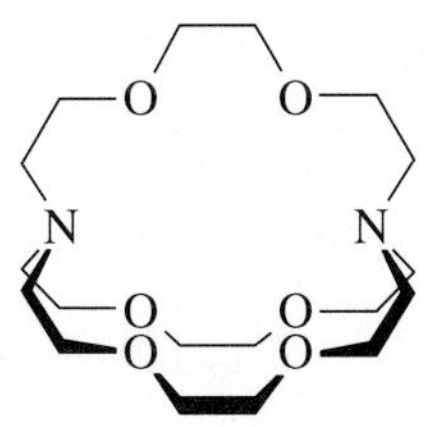

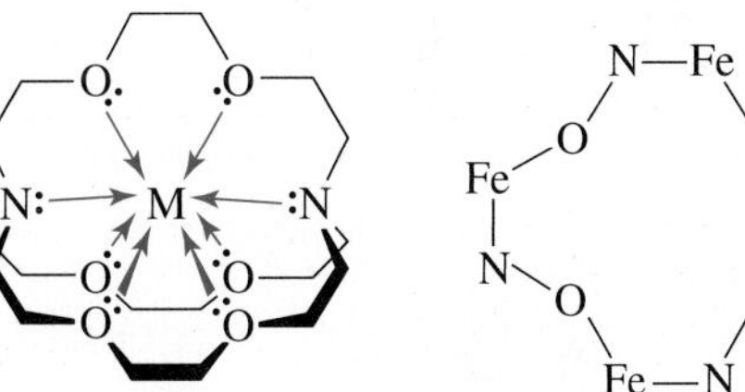

FIGURE 8.6 A Crown Ether, a Cryptand, a Metal Ion Encased in a Cryptand, and a Metallacrown.

even more effective as a cage with eight donor atoms surrounding a central metal. Metallacrowns, which incorporate metals into the crown structure, have also been developed.[7] An example of the framework structure of an iron-containing metallacrown is also shown in Figure 8.6. The importance of these structures was recognized when D. J. Cram, C. J. Pedersen, and J.-M. Lehn won the Nobel Prize in Chemistry in 1987 for work with these compounds.[8]

The ability of a cryptand to trap an alkali metal cation depends on the sizes of both the cage and the metal ion: the better the match between these sizes, the more effectively the ion can be trapped. This effect is shown graphically for the alkali metal ions in Figure 8.7.

The largest of the alkali metal cations, Cs^+, is trapped most effectively by the largest cryptand ([3.2.2]), and the smallest, Li^+, by the smallest cryptand ([2.1.1]).[9] Other correlations can easily be seen in Figure 8.6. Cryptands have played an important role in the study of a characteristic of the alkali metals that was not recognized until rather recently: a capacity to form negatively charged ions.

Although the alkali metals are known primarily for their formation of unipositive ions, numerous examples of alkali metal anions (alkalides) have been reported since 1974. The first of these was the sodide ion, Na^-, formed from the reaction of sodium with the cryptand $N\{(C_2H_4O)_2C_2H_4\}_3N$ in the presence of ethylamine:

$$2\,Na + \underset{\text{cryptand[2.2.2]}}{N\{(C_2H_4O)_2C_2H_4\}_3N} \longrightarrow \underset{[Na(\text{cryptand[2.2.2]})]^+}{[Na\,N\{(C_2H_4O)_2C_2H_4\}_3N]^+} + Na^-$$

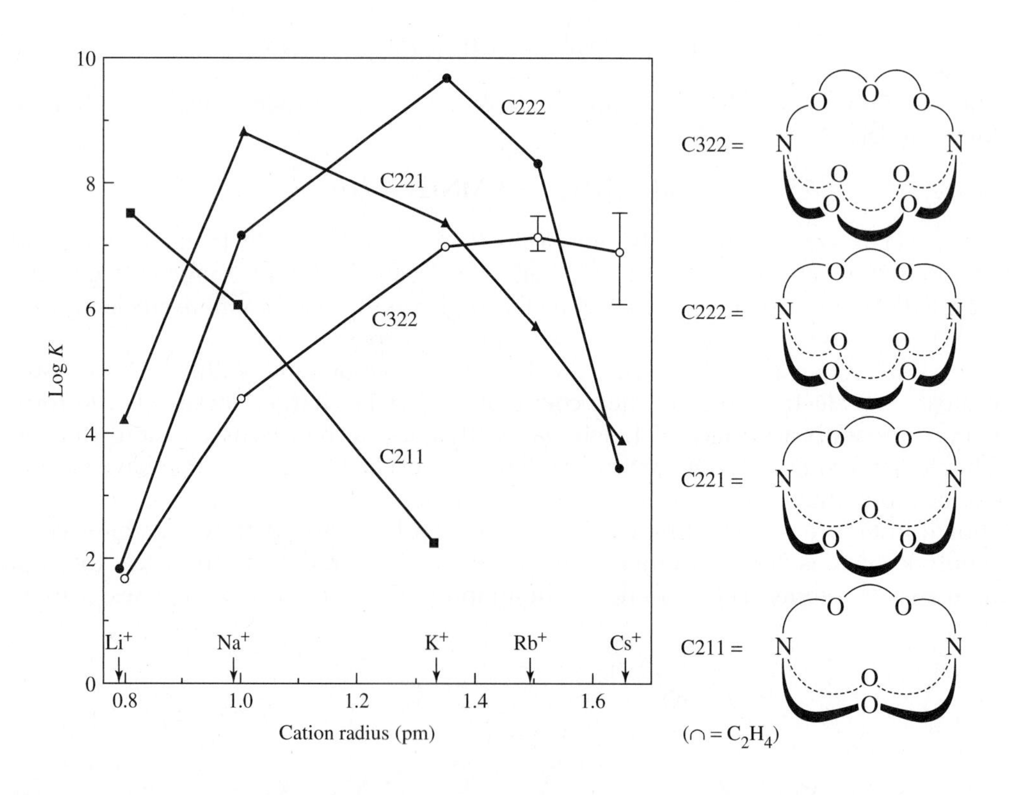

FIGURE 8.7 Formation Constants of Alkali Metal Cryptands. (From J. L. Dye, *Progr. Inorg. Chem.*, **1984** *32*, 337. © 1984, John Wiley & Sons. Reproduced by permission of John Wiley & Sons, Inc.)

---

[7] V. L. Pecoraro, A. J. Stemmler, B. R. Gibney, J. J. Bodwin, H. Wang, J. W. Kampf, and A. Barwinski, *Progr. Inorg. Chem.*, **1997**, *45*, 83.

[8] Their Nobel Prize lectures: D. J. Cram, *Angew. Chem.*, **1988**, *100*, 1041; C. J. Pedersen, *Angew. Chem.*, **1988**, *100*, 1053; J.-M. Lehn, *Angew. Chem.*, **1988**, *100*, 91.

[9] The numbers indicate the number of oxygen atoms in each bridge between the nitrogens. Thus, cryptand [3.2.2] has one bridge with three oxygens and two bridges with two oxygens, as shown in Figure 8.7.

In this complex, the Na^- occupies a site sufficiently remote from the coordinating N and O atoms of the cryptand that it can be viewed as a separate entity; it is formed as the result of disproportionation of Na into Na^+ (surrounded by the cryptand) plus Na^-. Alkalide ions are also known for the other members of Group 1 (IA) and for other metals, especially those for which a 1– charge gives rise to an s^2d^{10} electron configuration. As might be expected, alkalide ions are powerful reducing agents. This means that the cryptand or other cyclic group must be highly resistant to reduction to avoid being reduced by the alkalide ion. Even if such groups are carefully chosen, most alkalides are rather unstable and are subject to irreversible decomposition.

The crystal structure of the crown ether sandwich electride $Cs^+(15C5)_2\,e^-$ in Figure 8.8(a) shows both the coordination of two 15C5 rings to each Cs^+ ion and the cavity occupied by the electron e^-.[10]

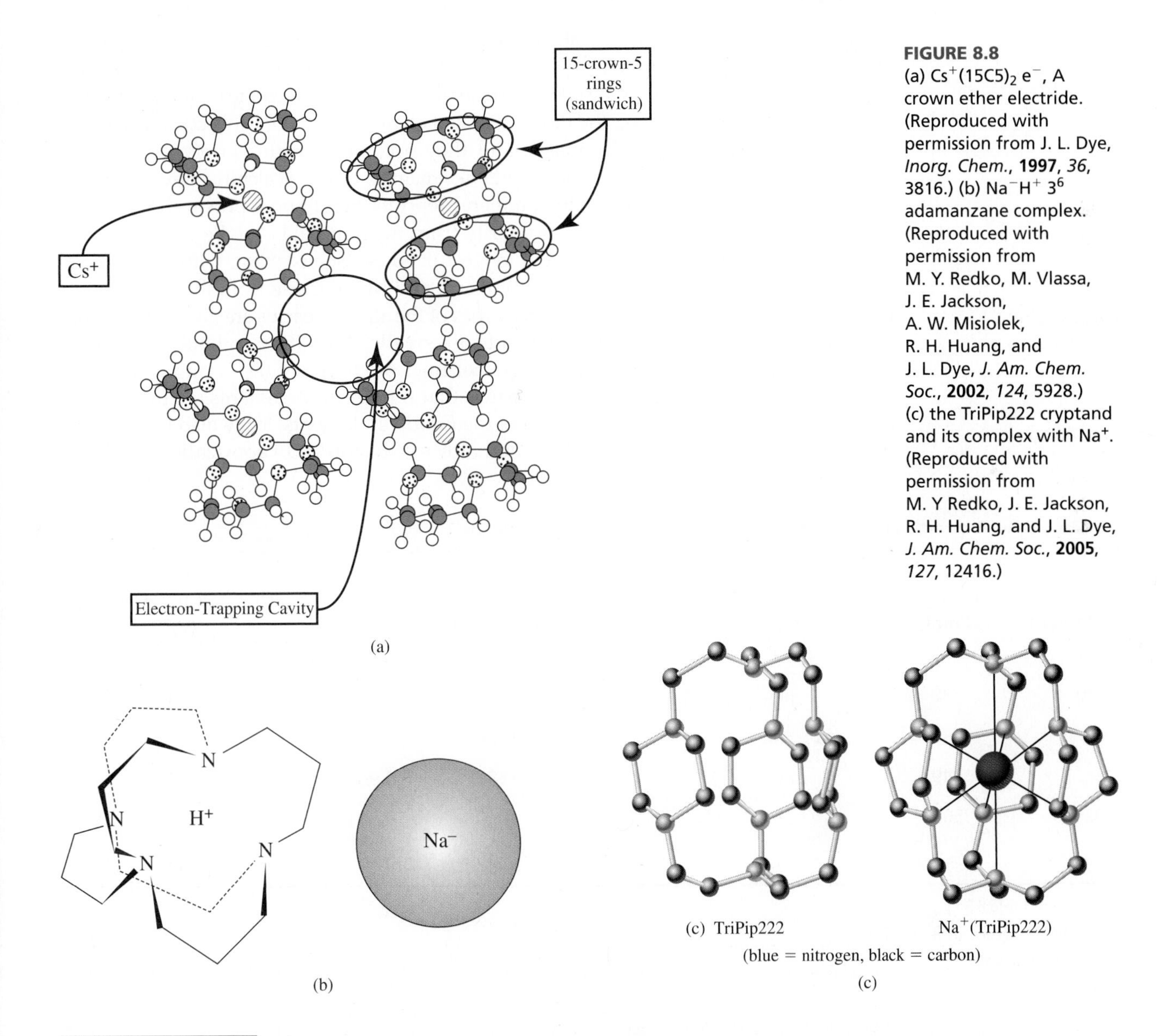

FIGURE 8.8
(a) $Cs^+(15C5)_2\,e^-$, A crown ether electride. (Reproduced with permission from J. L. Dye, *Inorg. Chem.*, **1997**, *36*, 3816.) (b) $Na^-H^+\,3^6$ adamanzane complex. (Reproduced with permission from M. Y. Redko, M. Vlassa, J. E. Jackson, A. W. Misiolek, R. H. Huang, and J. L. Dye, *J. Am. Chem. Soc.*, **2002**, *124*, 5928.) (c) the TriPip222 cryptand and its complex with Na^+. (Reproduced with permission from M. Y Redko, J. E. Jackson, R. H. Huang, and J. L. Dye, *J. Am. Chem. Soc.*, **2005**, *127*, 12416.)

[10] J. L. Dye, *Inorg. Chem.*, **1997**, *36*, 3816.

Among the intriguing developments in alkalide chemistry has been the synthesis of "inverse sodium hydride," which contains a sodide ion, Na^-, and an H^+ ion encapsulated in 3^6 adamanzane.[11] The H^+ in this structure is strongly coordinated by four nitrogen atoms in the adamanzane ligand, shown in Figure 8.8(b).

In 2005 the first electride to be thermally stable at room temperature was reported, together with its X-ray crystal structure. This structure used the per-aza cryptand TriPip222, shown in Figure 8.8(c), which is able to strongly complex Na^+ ions and enables electrons to be trapped in the cavities. The isostructural sodide, with Na^- trapped in the cavities, was also prepared.[12]

8.4 GROUP 2 (IIA): THE ALKALINE EARTHS

8.4.1 The Elements

Compounds of magnesium and calcium have been used since antiquity. For example, the ancient Romans used mortars containing lime (CaO) mixed with sand, and the ancient Egyptians used gypsum ($CaSO_4 \cdot 2\,H_2O$) in the plasters used to decorate their tombs. These two alkaline earths are among the most abundant elements in the Earth's crust (calcium is fifth and magnesium sixth, by mass), and they occur in a wide variety of minerals. Strontium and barium are less abundant; but like magnesium and calcium, they commonly occur as sulfates and carbonates in their mineral deposits. Beryllium is fifth in abundance of the alkaline earths and is obtained primarily from the mineral beryl, $Be_3Al_2(SiO_3)_6$. All isotopes of radium are radioactive (the longest lived isotope is ^{226}Ra, with a half-life of 1600 years). Radium was first isolated by Pierre and Marie Curie from the uranium ore pitchblende in 1898. Selected physical properties of the alkaline earths are given in Table 8.4.

Atoms of the Group 2 (IIA) elements are smaller than the neighboring Group 1 (IA) elements as a consequence of the greater nuclear charge of the Group 2 elements. The observed result of this decrease in size is that the Group 2 elements are more dense and have higher ionization energies than the Group 1 elements. They also have higher melting and boiling points and higher enthalpies of fusion and vaporization, as can be

TABLE 8.4 Properties of the Group 2 (IIA) Elements: The Alkaline Earths

Element	Ionization Energy ($kJ\ mol^{-1}$)	Electron Affinity ($kJ\ mol^{-1}$)^b	Melting Point (°C)	Boiling Point (°C)	Electronegativity	$\mathscr{E}°$ ($M^{2+} + 2\,e^- \longrightarrow M$) (V)^a
Be	899	−50	1287	2500^b	1.576	−1.97
Mg	738	−40	649	1105	1.293	−2.36
Ca	590	−30	839	1494	1.034	−2.84
Sr	549	−30	768	1381	0.963	−2.89
Ba	503	−30	727	1850^b	0.881	−2.92
Ra	509	−30	700^b	1700^b	0.9^b	−2.92

Source: See Table 8.3

NOTES: [a]Aqueous solution, 25 °C

[b]Approximate values

[11] M. Y. Redko, M. Vlassa, J. E. Jackson, A. W. Misiolek, R. H. Huang, and J. L. Dye, *J. Am. Chem. Soc.*, **2002**, *124*, 5928.

[12] M. Y. Redko, J. E. Jackson, R. H. Huang, and J. L. Dye, *J. Am. Chem. Soc.*, **2005**, *127*, 12416.

seen from Tables 8.3 and 8.4. Beryllium, the lightest of the alkaline earth metals, is widely used in alloys with copper, nickel, and other metals. When added in small amounts to copper, for example, beryllium increases the strength of the metal dramatically and improves the corrosion resistance, while preserving high conductivity and other desirable properties. Emeralds and aquamarine are obtained from two types of beryl, the mineral source of beryllium; the vivid green and blue colors of these stones are the result of small amounts of chromium and other impurities. Magnesium, with its alloys, is used widely as a strong, but very light, construction material; its density is less than one fourth that of steel. The other alkaline earth metals are used occasionally, but in much smaller amounts, in alloys. Radium has been used in the treatment of cancerous tumors, but its use has largely been superseded by other radioisotopes.

8.4.2 Chemical Properties

The elements in Group 2 (IIA), with the exception of beryllium, have very similar chemical properties, with much of their chemistry governed by their tendency to lose two electrons to achieve a noble gas electron configuration. In general, therefore, elements in this group are good reducing agents. Although not as violently reactive toward water as the alkali metals, the alkaline earths react readily with acids to generate hydrogen:

$$Mg + 2\,H^+ \longrightarrow Mg^{2+} + H_2$$

The reducing ability of these elements increases with atomic number. As a consequence, calcium and the heavier alkaline earths react directly with water in a reaction that can conveniently generate small quantities of hydrogen:

$$Ca + 2\,H_2O \longrightarrow Ca(OH)_2 + H_2$$

Beryllium is distinctly different from the other alkaline earths in its chemical properties. The smallest of the alkaline earths, it participates primarily in covalent, rather than ionic, bonding. Although the ion $[Be(H_2O)_4]^{2+}$ is known, free Be^{2+} ions are rarely if ever encountered. Beryllium and its compounds are extremely toxic, and special precautions are required in their handling. As discussed in Section 3.1.4, although beryllium halides of formula BeX_2 may be monomeric and linear in the gas phase at high temperature, in the solid phase, the molecules polymerize to form halogen-bridged chains with tetrahedral coordination around beryllium, as shown in Figure 8.9. Beryllium hydride, BeH_2, is also polymeric in the solid with bridging hydrogens. The three-center bonding involved in bridging by halogens, hydrogen, and other atoms and groups is also commonly encountered in the chemistry of the Group 13 (IIIA) elements and will be discussed more fully with those elements in Section 8.5.

Among the most chemically useful magnesium compounds are the Grignard reagents, of general formula RMgX (X = alkyl or aryl). These reagents are complex in their structure and function, consisting of a variety of species in solution linked by equilibria such as those shown in Figure 8.10. The relative positions of these equilibria, and hence the concentrations of the various species, are affected by the nature of the R group and the halogen, solvent, and temperature. Grignard reagents are versatile and can be used to synthesize a vast range of organic compounds, including alcohols, aldehydes,

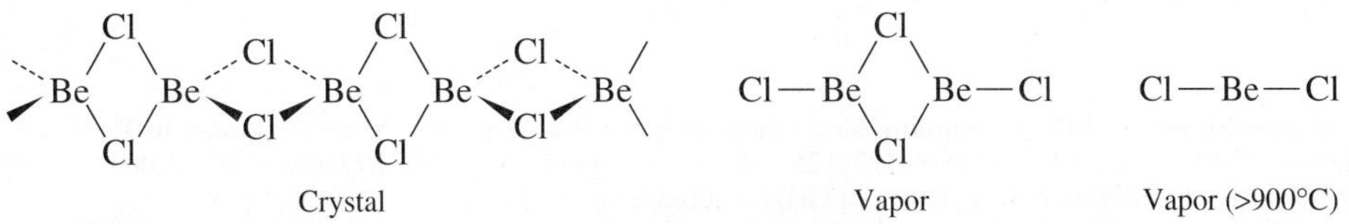

Crystal | Vapor | Vapor (>900°C): Cl—Be—Cl

FIGURE 8.9 Structure of $BeCl_2$.

FIGURE 8.10 Grignard Reagent Equilibria.

$$2\,RMg^+ + 2\,X^-$$

$$\rightleftharpoons$$

$$R{-}Mg(\mu{-}X)_2Mg{-}R \rightleftharpoons 2\,RMgX \rightleftharpoons MgR_2 + MgX_2$$

$$\rightleftharpoons$$

$$RMg^+ + RMgX_2^- \rightleftharpoons XMg(\mu{-}X)(\mu{-}R)MgR$$

ketones, carboxylic acids, esters, thiols, and amines. Details of these syntheses are presented in many organic chemistry texts.[13]

Chlorophylls contain magnesium coordinated by chlorin groups. These compounds, essential in photosynthesis, will be discussed in Chapter 16.

Portland cement—a complex mixture of calcium silicates, aluminates, and ferrates—is one of the world's most important construction materials, with annual worldwide production in excess of 10^{12} kg. When mixed with water and sand, it changes by slow hydration to concrete. Water and hydroxide link the other components into larger, very strong crystals.

8.5 GROUP 13 (IIIA)

8.5.1 The Elements

Elements in this group include one nonmetal, boron, and four elements that are primarily metallic in their properties. Physical properties of these elements are shown in Table 8.5.

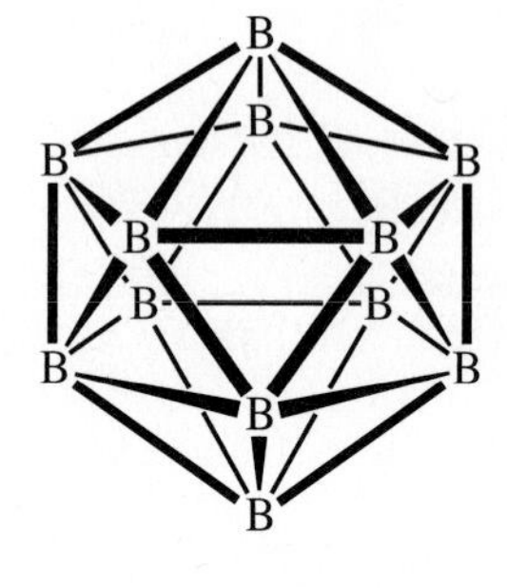

BORON Boron's chemistry is so different from that of the other elements in this group that it deserves separate discussion. Chemically, boron is a nonmetal; in its tendency to form covalent bonds, it shares more similarities with carbon and silicon than with aluminum and the other Group 13 elements. Like carbon, boron forms many hydrides; like

TABLE 8.5 Properties of the Group 13 (IIIA) Elements

Element	Ionization Energy ($kJ\ mol^{-1}$)	Electron Affinity ($kJ\ mol^{-1}$)	Melting Point (°C)	Boiling Point (°C)	Electronegativity
B	801	27	2180	3650[a]	2.051
Al	578	43	660	2467	1.613
Ga	579	30[a]	29.8	2403	1.756
In	558	30[a]	157	2080	1.656
Tl	589	20[a]	304	1457	1.789

Source: See Table 8.3

NOTE: [a]Approximate values

[13] The development of these reagents since their original discovery by Victor Grignard in 1900 has been reviewed. See *Bull. Soc. Chim. France*, **1972**, 2127–2186. See also D. Seyferth, *Organometallics*, **2009**, *28*, 1598 for a discussion of highlights of the history of Grignard reagents.

silicon, it forms oxygen-containing minerals with complex structures (borates). Compounds of boron have been used since ancient times in the preparation of glazes and borosilicate glasses, but the element itself has proven extremely difficult to purify. The pure element has a wide diversity of allotropes, different forms of the pure element, many of which are based on the icosahedral B_{12} unit.

In the boron hydrides, called *boranes*, hydrogen often serves as a bridge between boron atoms, a function rarely performed by hydrogen in carbon chemistry. How is it possible for hydrogen to serve as a bridge? One way to address this question is to consider the bonding in diborane, B_2H_6:

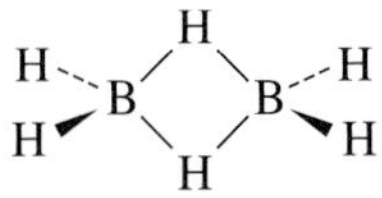

H H H
B B
H H H

Diborane

Diborane has 12 valence electrons. By the Lewis approach to bonding, eight of these electrons are involved in bonding to the terminal hydrogens. Thus, four electrons remain to account for bonding in the bridges. This type of bonding, involving three atoms and two bonding electrons per bridge, is described as *three-center, two-electron bonding*.[14] To understand how this type of bonding is possible, we need to consider the orbital interactions in this molecule.

Diborane has D_{2h} symmetry. Focusing on the boron atoms and the bridging hydrogens, we can use the approach of Chapter 5 to sketch the group orbitals and determine their matching irreducible representations, as shown in Figure 8.11. The possible interactions between the boron group orbitals and the group orbitals of the bridging hydrogens can be determined by matching the labels of the irreducible representations. For example, one group orbital in each set has B_{3u} symmetry. This involves the hydrogen group orbital with lobes of opposite sign and one of the boron group orbitals derived from p_x atomic orbitals. The results, shown in Figure 8.12, are two molecular orbitals of b_{3u} symmetry, one bonding and one antibonding. The bonding orbital, with lobes on the top and bottom spanning the B—H—B bridges, is one of the orbitals chiefly responsible for the stability of the bridges.

The other hydrogen group orbital has A_g symmetry. Two boron group orbitals have A_g symmetry: one is derived from p_z orbitals, and one is derived from *s* orbitals. All three group orbitals have similar energy. The result of the A_g interactions is the formation of three molecular orbitals: one strongly bonding, one weakly bonding, and one antibonding.[15] (The other boron group orbitals, with B_{1u} and B_{2g} symmetry, do not participate in interactions with the bridging hydrogens.) These interactions are summarized in Figure 8.13.[16] In contrast to the simple model (two three-center, two-electron bonds), three bonding orbitals play a significant role in joining the boron atoms through the hydride bridges, two of a_g symmetry and one of b_{3u} symmetry. The shapes of these orbitals are shown in Figure 8.14.

Similar bridging hydrogen atoms occur in many other boranes, as well as in carboranes, which contain both boron and carbon atoms arranged in clusters. In addition,

[14] W. N. Lipscomb, *Boron Hydrides*, W. A. Benjamin, New York, 1963.

[15] One of the group orbitals on the terminal hydrogens also has A_g symmetry. The interaction of this group orbital with the other orbitals of A_g symmetry influences the energy and shape of the a_g molecular orbitals, shown in Figure 8.12, and generates a fourth, antibonding a_g molecular orbital (not shown in the figure).

[16] This figure does not show interactions with terminal hydrogens. One terminal atom group orbital has B_{1u} symmetry and therefore interacts with the B_{1u} group orbitals of boron, resulting in molecular orbitals that are no longer nonbonding.

FIGURE 8.11 Group Orbitals of Diborane.

Reducible representation for p orbitals involved in bonding with bridging hydrogens:

p_z: B B (H above and below); p_x: B B (H above and below); axes x, z

	E	$C_2(z)$	$C_2(y)$	$C_2(x)$	i	$\sigma(xy)$	$\sigma(xz)$	$\sigma(yz)$	
$\Gamma(p_z)$	2	2	0	0	0	0	2	2	$= A_g + B_{1u}$
$\Gamma(p_x)$	2	−2	0	0	0	0	2	−2	$= B_{2g} + B_{3u}$

The irreducible representations have the following symmetries:

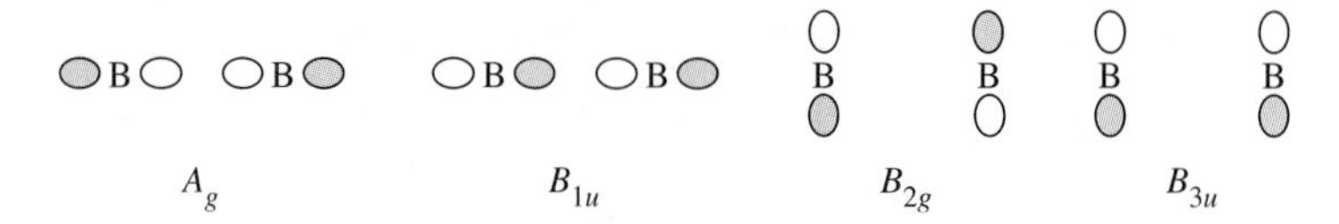

Reducible representation for 1s orbitals of bridging hydrogens:

H
B B
H

	E	$C_2(z)$	$C_2(y)$	$C_2(x)$	i	$\sigma(xy)$	$\sigma(xz)$	$\sigma(yz)$
$\Gamma(1s)$	2	0	0	2	0	2	2	0

This reduces to $A_g + B_{3u}$, which have the following symmetries:

B B B B

A_g B_{3u}

bridging hydrogens and alkyl groups are frequently encountered in aluminum chemistry. A few examples of these compounds are shown in Figure 8.15.

The boranes, carboranes, and related compounds are also of interest in the field of cluster chemistry, the chemistry of compounds containing metal–metal bonds. The bonding in these compounds will be discussed and compared with the bonding in transition-metal cluster compounds in Chapter 15.

Boron has two stable isotopes, ^{11}B (80.4% abundance) and ^{10}B (19.6%). ^{10}B has a very high neutron absorption cross section; it is a good absorber of neutrons. This property has been developed for use in the treatment of cancerous tumors in a process called *boron neutron capture therapy* (BNCT).[17] Boron-containing compounds having a strong preference for attraction to tumor sites, rather than healthy sites, can be irradiated with beams of neutrons. The subsequent nuclear decay emits high-energy particles, $^{7}_{3}Li$ and $^{4}_{2}He$ (alpha particles), which can kill the adjacent cancerous tissue:

$$^{10}_{5}B + ^{1}_{0}n \longrightarrow ^{11}_{5}B$$

$$^{11}_{5}B \longrightarrow ^{7}_{3}Li + ^{4}_{2}He$$

[17] M. F. Hawthorne, *Angew. Chem., Int. Ed.*, **1993**, *32*, 950.

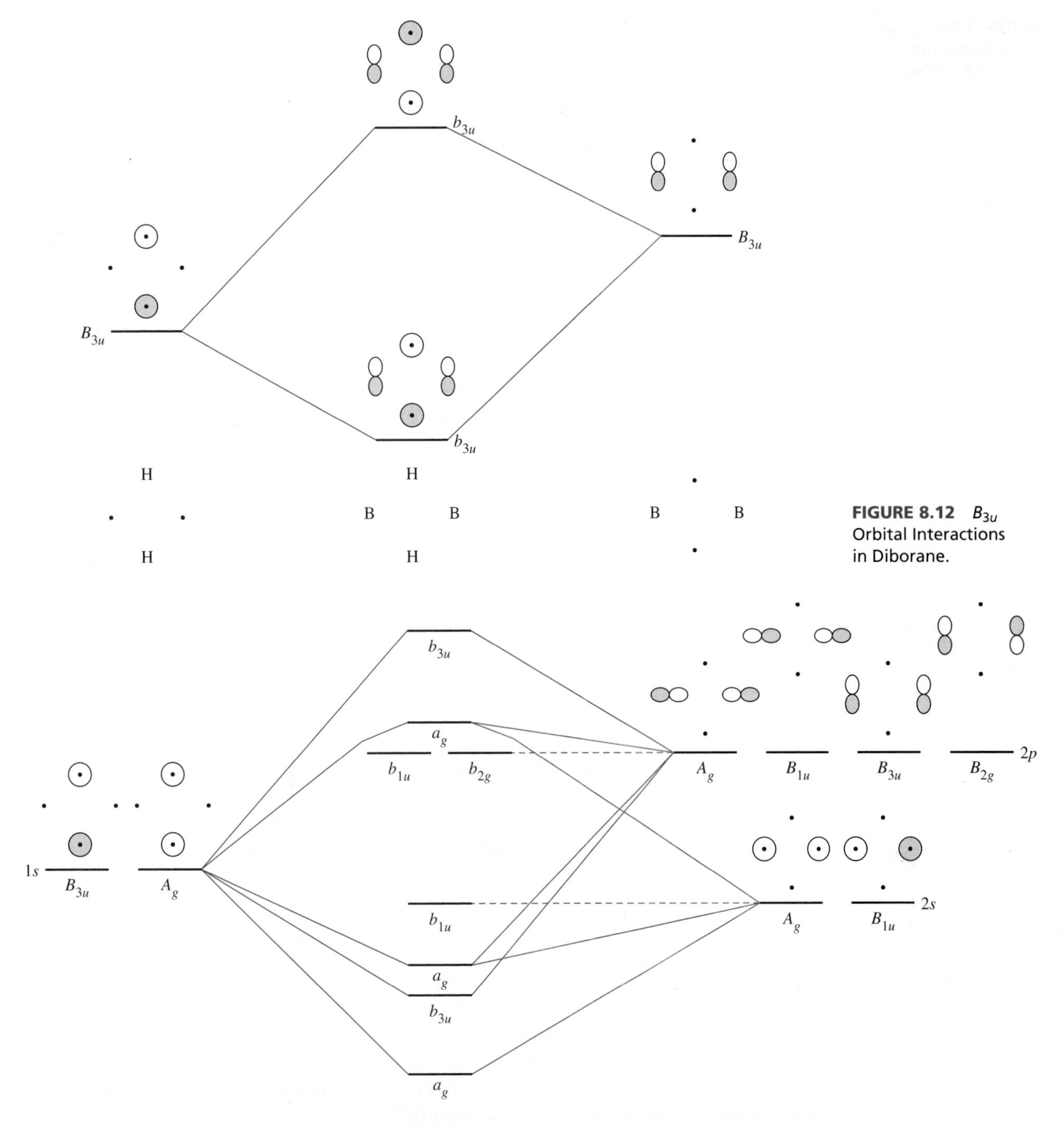

FIGURE 8.12 B_{3u} Orbital Interactions in Diborane.

FIGURE 8.13 Bridging Orbital Interactions in Diborane.

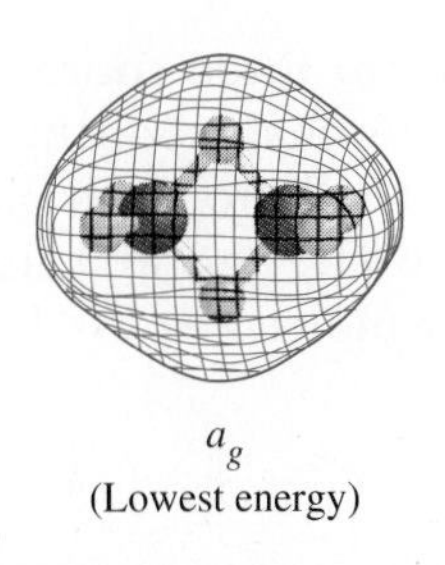

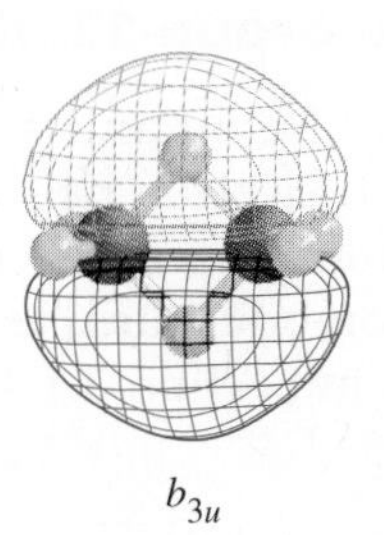

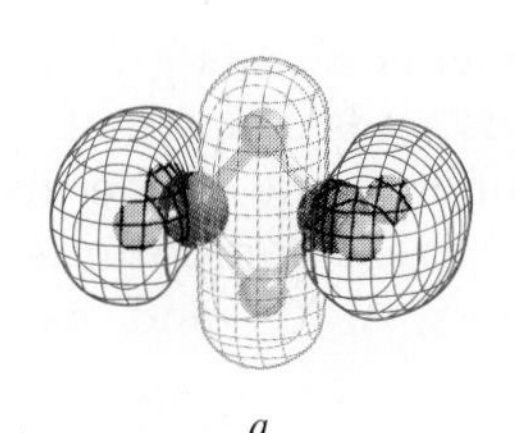

FIGURE 8.14 Bonding Orbitals Involved in Hydrogen Bridges in Diborane.

FIGURE 8.15 Boranes, Carboranes, and Bridged Aluminum Compounds.

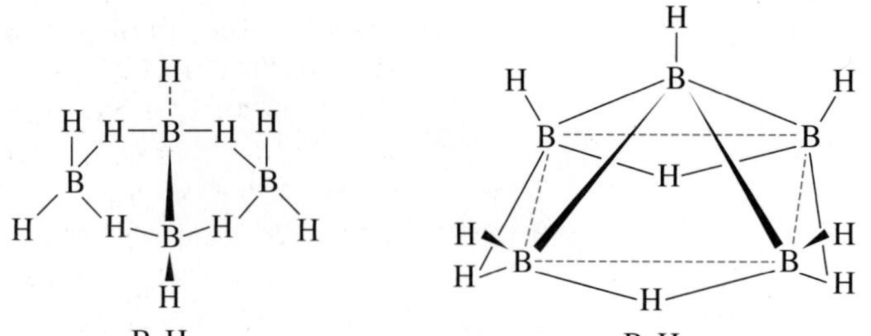

B_4H_{10} B_5H_9

Boranes

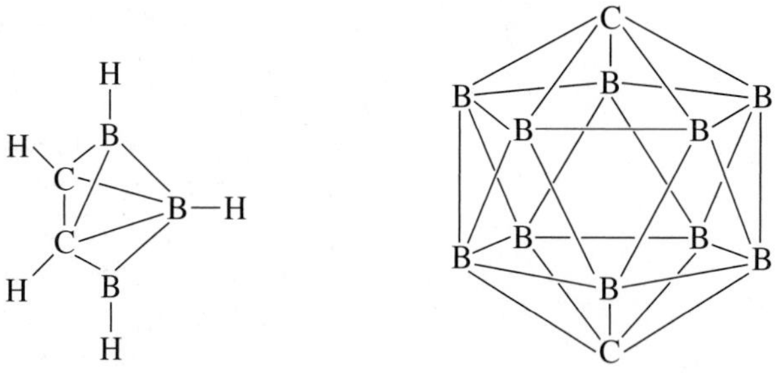

$C_2B_3H_5$ *para*-$C_2B_{10}H_{12}$ (one H on each C and B)

Carboranes

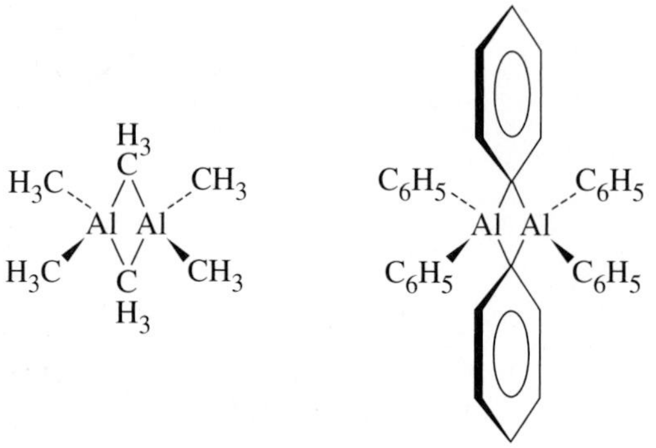

Bridged aluminum compounds

The challenge to chemists has been to develop boron-containing reagents that can be selectively concentrated in cancerous tissue while avoiding healthy tissue. Various approaches to this task have been attempted.[18]

8.5.2 Other Chemistry of the Group 13 (IIIA) Elements

Elements in this group, especially boron and aluminum, form three-coordinate Lewis acids capable of accepting an electron pair and increasing their coordination number. Some of the most commonly used Lewis acids are the boron trihalides, BX_3. These compounds are monomeric—unlike diborane, B_2H_6, and aluminum halides, Al_2X_6—and, as discussed in Section 3.1.4, are planar molecules. As Lewis

[18] See S. B. Kahl and J. Li, *Inorg. Chem.*, **1996**, *35*, 3878, and references therein.

acids, they can accept an electron pair from a halide ion to form tetrahaloborate ions, BX_4^-. The Lewis acid behavior of these compounds has been discussed in Chapter 6.

Boron halides can also act as halide ion acceptors when they serve as catalysts; for example, in the Friedel–Crafts alkylation of aromatic hydrocarbons:

$$\begin{array}{lrcl} & BF_3 + RX & \longrightarrow & R^+ + BF_3X^- \\ & R^+ + PhH & \longrightarrow & H^+ + RPh \\ & H^+ + BF_3X^- & \longrightarrow & HX + BF_3 \\ \hline \text{Net:} & RX + PhH & \longrightarrow & RPh + HX \end{array}$$

The metallic nature of the elements in Group 13 (IIIA) increases going down the group. Aluminum, gallium, indium, and thallium commonly form 3+ ions by loss of their valence *p* electron and both valence *s* electrons. Thallium also forms a 1+ ion by losing its *p* electron and retaining its two *s* electrons. This is the first case we have encountered of the **inert pair effect**, in which a metal has an oxidation state that is two less than the traditional American group number. For example, Pb is in Group IVA according to the traditional numbering system (Group 14 in IUPAC), and it has a 2+ ion as well as a 4+ ion. The effect is commonly ascribed to the stability of an electron configuration with entirely filled subshells: in the inert pair effect, a metal loses all the *p* electrons in its outermost subshell, leaving a filled s^2 subshell; the pair of *s* electrons seems relatively "inert" and is less easily removed. The actual reasons for this effect are considerably more complex than described here.[19]

Parallels between main group and organic chemistry can be instructive. One of the best known of these parallels is between the organic molecule benzene and the isoelectronic borazine (alias "inorganic benzene"), $B_3N_3H_6$. Some of the similarities in physical properties between these two are striking, as shown in Table 8.6.

Despite these parallels, the chemistry of these two compounds is quite different. In borazine, the difference in electronegativity between boron (2.051) and nitrogen (3.066) adds considerable polarity to the B—N bonds and makes the molecule much more susceptible to attack by nucleophiles (at the more positive boron) and electrophiles (at the more negative nitrogen) than benzene.

Parallels between benzene and isoelectronic inorganic rings remain of interest. Some examples include reports on boraphosphabenzenes (containing B_3P_3 rings)[20] and $[(CH_3)AlN(2,6\text{-diisopropylphenyl})]_3$ containing an Al_3N_3 ring.[21] Calculations on borazine, $B_3P_3H_6$, and a variety of other candidate "inorganic benzenes" have indicated that borazine is not aromatic but $B_3P_3H_6$, Si_6H_6, N_6, and P_6 have some aromatic character.[22]

Another interesting parallel between boron-nitrogen chemistry and carbon chemistry is offered by boron nitride, BN. Like carbon (Section 8.6), boron nitride exists in a diamond-like form and in a form similar to graphite. In the diamond-like (cubic) form, each nitrogen is coordinated tetrahedrally by four borons and each boron by four nitrogens. As in diamond, such coordination gives high rigidity to the structure and makes BN comparable to diamond in hardness. In the graphite-like hexagonal form, BN also occurs in extended fused-ring systems. However, there is much less delocalization of pi electrons in this form and, unlike graphite, hexagonal BN is a poor conductor. As in the case of diamond, the harder, more dense form (cubic) can be formed from the less dense form (hexagonal) under high pressure.

---

[19] See, for example, N. N. Greenwood and A. Earnshaw, *Chemistry of the Elements*, Pergamon Press, Elmsford, NY, 1984, pp. 255–256.

[20] H. V. R. Dias and P. P. Power, *Angew. Chem., Int. Ed.*, **1987**, *26*, 1270; *J. Am. Chem. Soc.*, **1989**, *111*, 144.

[21] K. M. Waggoner, H. Hope, and P. P. Power, *Angew. Chem., Int. Ed.*, **1988**, *27*, 1699.

[22] J. J. Engelberts, R. W. A. Havenith, J. H. van Lenthe, L. W. Jenneskens, and P. W. Fowler, *Inorg. Chem.*, **2005**, *44*, 5266.

TABLE 8.6 Benzene and Borazine

Property	Benzene	Borazine
Melting point (°C)	6	−57
Boiling point (°C)	80	55
Density (g cm^{-3})[a]	0.81	0.81
Surface tension (N m^{-1})[a]	0.0310	0.0311
Dipole moment	0	0
Internuclear distance in ring (pm)	142	144
Internuclear distance, bonds to H (pm)	C—H: 108	B—H: 120 N—H: 102

Source: Data from N. N. Greenwood and A. Earnshaw, *Chemistry of the Elements*, Pergamon Press, Elmsford, NY, 1984, p. 238.
NOTE: [a]At the melting point

8.6 GROUP 14 (IVA)

8.6.1 The Elements

Elements in this group range from a nonmetal, carbon, to the metals tin and lead, with the intervening elements showing semimetallic behavior. Carbon has been known from prehistory as the charcoal resulting from partial combustion of organic matter. In recorded history, diamonds have been prized as precious gems for thousands of years. Neither form of carbon, however, was recognized as a chemical element until late in the eighteenth century. Tools made of flint, primarily SiO_2, were used throughout the Stone Age. However, free silicon was not isolated until 1823, when J. J. Berzelius obtained it by reducing K_2SiF_6 with potassium. Tin and lead have also been known since ancient times. A major early use of tin was in combination with copper in the alloy bronze; weapons and tools containing bronze date back more than 5000 years. Lead was used by the ancient Egyptians in pottery glazes and by the Romans for plumbing and other purposes. In recent decades, the toxic effects of lead and lead compounds in the environment have gained increasing attention and have led to restrictions on the use of lead compounds; for example, in paint pigments and in gasoline additives, primarily tetraethyllead, $(C_2H_5)_4Pb$. Germanium was a "missing" element for a number of years. Mendeleev accurately predicted the properties of this then-unknown element in 1871 ("eka-silicon"), but it was not discovered until 1886, by C. A. Winkler. Properties of the Group 14 (IVA) elements are summarized in Table 8.7.

Although carbon occurs primarily as the isotope ^{12}C, whose atomic mass serves as the basis of the modern system of atomic mass, two other isotopes, ^{13}C and ^{14}C, are important as well. ^{13}C, which has a natural abundance of 1.11 percent, has a nuclear spin of $\frac{1}{2}$ in contrast to ^{12}C, which has zero nuclear spin. This means that even though

TABLE 8.7 Properties of the Group 14 (IVA) Elements

Element	Ionization Energy ($kJ\ mol^{-1}$)	Electron Affinity ($kJ\ mol^{-1}$)	Melting Point (°C)	Boiling Point (°C)	Electro-negativity
C	1086	122	4100	[a]	2.544
Si	786	134	1420	3280[b]	1.916
Ge	762	120	945	2850	1.994
Sn	709	120	232	2623	1.824
Pb	716	35	327	1751	1.854

Source: See Table 8.3
NOTES: [a]Sublimes
[b]Approximate value

^{13}C comprises only about 1 part in 90 of naturally occurring carbon, it can be used as the basis of NMR observations for the characterization of carbon-containing compounds. With the advent of Fourier transform technology, ^{13}C NMR spectrometry has become a valuable tool in both organic and inorganic chemistry. Uses of ^{13}C NMR in organometallic chemistry are described in Chapter 13.

^{14}C is formed in the atmosphere from nitrogen by thermal neutrons from the action of cosmic rays:

$$^{14}_{7}N + ^{1}_{0}n \longrightarrow ^{14}_{6}C + ^{1}_{1}H$$

^{14}C is formed by this reaction in comparatively small amounts—approximately 1.2×10^{-10} percent of atmospheric carbon; it is incorporated into plant and animal tissues by biological processes. When a plant or animal dies, the process of exchange of its carbon with the environment by respiration and other biological processes ceases, and the ^{14}C in its system is effectively trapped. However, ^{14}C decays by beta emission, with a half-life of 5730 years:

$$^{14}_{6}C \longrightarrow ^{14}_{7}N + ^{0}_{-1}e$$

Therefore, by measuring the remaining amount of ^{14}C one can determine to what extent this isotope has decayed and, in turn, the time elapsed since death. Often called simply *radiocarbon dating*, this procedure has been used to estimate the ages of many archeological samples, including Egyptian remains, charcoal from early campfires, and the Shroud of Turin.

EXAMPLE

What fraction of ^{14}C remains in a sample that is 50,000 years old?

This is 50,000/5,730 = 8.73 half-lives. For first-order reactions, such as radioactive decay, the initial amount decreases by $\frac{1}{2}$ during each half-life, so the fraction remaining is $\left(\frac{1}{2}\right)^{8.73} = 2.36 \times 10^{-3}$.

▶ **Exercise 8.1** A sample of charcoal from an archeological site has a remaining fraction of ^{14}C of 3.5×10^{-2}. What is its age?

DIAMOND AND GRAPHITE Until 1985, carbon was encountered primarily in two allotropes, diamond and graphite. The diamond structure is rigid, with each atom surrounded tetrahedrally by four other atoms in a structure that has a cubic unit cell; as a result, diamond is extremely hard, the hardest of all naturally occurring substances. Graphite, on the other hand, consists of layers of fused, six-membered rings of carbon atoms. The carbon atoms in these layers may be viewed as being sp^2 hybridized. The remaining, unhybridized p orbitals are perpendicular to the layers and participate in extensive π bonding, with pi electron density delocalized over the layers. Because of the relatively weak interactions between the layers, they are free to slip with respect to each other, and pi electrons are free to move within each layer, making graphite a good lubricant and electrical conductor. The structures of diamond and graphite are shown in Figure 8.16, and their important physical properties are given in Table 8.8.

At room temperature, graphite is thermodynamically the more stable form. However, the density of diamond is much greater than that of graphite, and graphite can be converted to diamond at very high pressure (high temperature and molten metal catalysts are also used to facilitate this conversion). Since the first successful synthesis of diamonds from graphite in the mid-1950s, the manufacture of diamonds for industrial (rather than gemstone) use has developed rapidly, and the majority of industrial diamonds are now produced synthetically.

A thin layer of hydrogen bonded to a diamond surface significantly reduces the coefficient of friction of the surface compared to a clean diamond surface, presumably because the clean surface provides sites for attachment of molecules—bonds that must be broken for surfaces to be able to slip with respect to each other.[23]

GRAPHENE Graphite consists of multiple layers of carbon atoms, as shown in Figure 8.16. A single such layer, free of other layers, is called **graphene**, also shown in the figure. First prepared in 2004,[24] graphene has been the center of considerable research, both to study its properties and to develop efficient ways to prepare graphene sheets, still a technical challenge. Graphene is remarkably resistant to fracture and deformation, has a high thermal conductivity, and has a conduction band that touches its valence band; its electrical properties can be influenced by adding various functional groups. Graphene has been prepared both by mechanically peeling away layers of graphite and by a variety of specialized techniques, including chemical vapor deposition on metal substrates, the chemical reduction of graphite oxide (an oxidation product of graphite that contains OH groups and bridging

TABLE 8.8 Physical Properties of Diamond and Graphite

Property	Diamond	Graphite
Density (g cm^{-3})	3.513	2.260
Electrical resistivity (Ωm)	10^{11}	1.375×10^{-5}
Standard molar entropy (J mol^{-1} K^{-1})	2.377	5.740
C_p at 25° C (J mol^{-1} K^{-1})	6.113	8.527
C—C distance (pm)	154.4	141.5 (within layer) 335.4 (between layers)

Source: J. Elmsley, *The Elements*, Oxford University Press, New York, 1989, p. 44.

[23] R. J. A. van den Oetelaar and C. F. J. Flipse, *Surf. Sci.*, **1997**, *384*, L828.
[24] A. K. Geim, et al., *Science*, **2004**, *306*, 666.

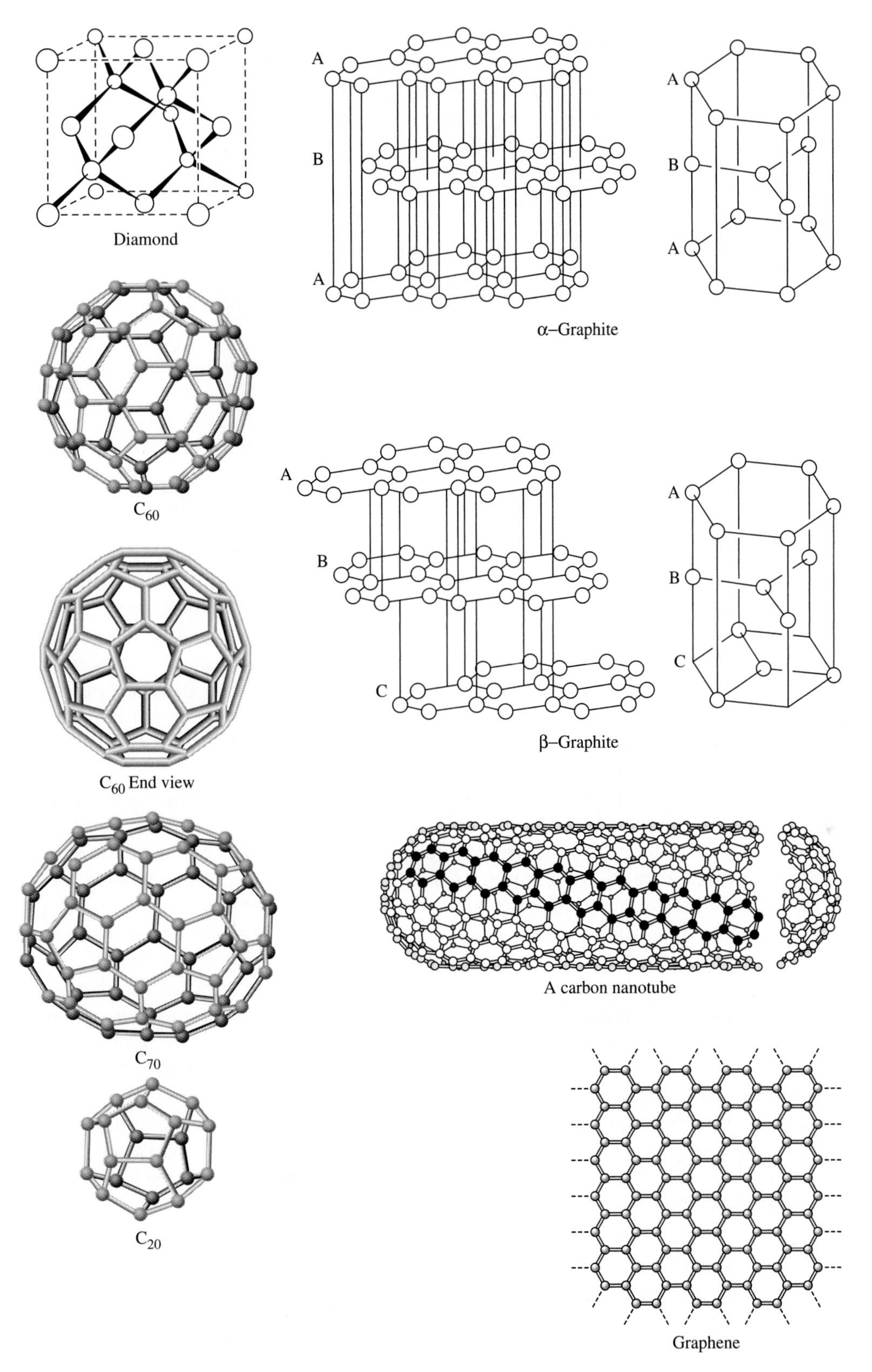

FIGURE 8.16 Diamond, Graphite, Fullerenes, and Graphene.

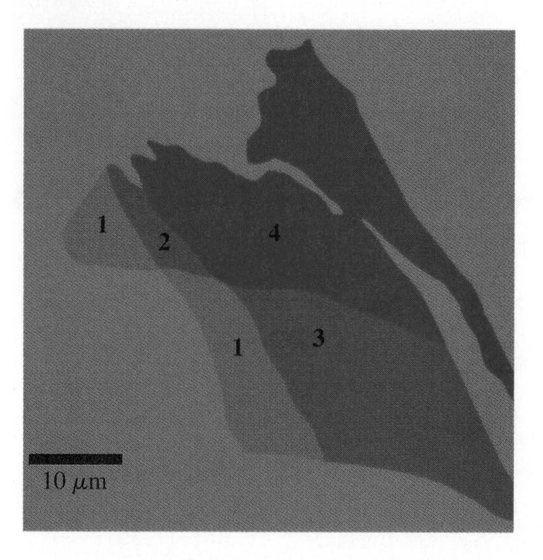

FIGURE 8.17 Optical Image of Graphene Showing Numbers of Layers. Reprinted with permission from Z. H. Ni et al., *Nano Letters*, **2007**, *7*, 2759. Copyright 2007 American Chemical Society.

oxygens), and sonication of colloidal suspensions of graphite oxide.[25,26] In addition to single-layer graphene, samples with two (bi-layer graphene) or more layers have also been studied.

Graphene represents an essentially two-dimensional structure with a thickness of approximately 340 pm. Its honeycomb-like surface has been imaged using scanning transmission microscopy (STM), and it can be viewed optically as well, with the level of contrast indicating how many layers are present, as shown in Figure 8.17.[27] Numerous potential applications of graphene have been proposed, including their use in energy-storage materials, microsensor devices, liquid crystal displays, polymer composites, and a wide range of electronic devices.

NANORIBBONS AND NANOTUBES Graphene can be cut into thin strips, dubbed **nanoribbons**, by lithographic techniques.[28] These ribbons are described by their edges, either *zigzag* or *armchair*, as shown in Figure 8.18. If the nanoribbons are sufficiently narrow, they have a band gap between their conduction and valence bands; as in the case of quantum dots (Section 7.3.2), the quantum confinement effect becomes more pronounced as the dimension—in this case, the width of the ribbon—becomes smaller, and the band gap in these semiconductors becomes wider.[29] In orbital terms, the smaller the dimension, the more the energy levels show distinct, quantized energies, rather than a continuum of energy levels, as in a larger structure. Interest in nanoribbons is strong because of their potential applications in nanoelectronics and composite materials.

One can envision joining the edges of nanoribbons together to form tubes. Such **nanotubes** can be formed in several ways: If the edges that are joined are zigzag, the ends of the tube have carbons in a chair arrangement; if armchair edges are joined, the ends of the tube have a zigzag arrangement, as shown in Figure 8.19. Furthermore, if the edges that are brought together are from different rows, the nanotube that is formed will be helical, and therefore chiral, also shown in the figure.[30]

---

[25] See C. N. R. Rao, K. Biswas, K. S. Subrahmanyam, and A. Govindaraj, *J. Mater. Chem.*, **2009**, *19*, 2457 for a useful review of graphene's synthesis and properties, including electronic properties that are beyond the scope of this text.
[26] S. Park and R. S. Ruoff, *Nature Nanotech.*, **2009**, *4*, 217.
[27] Z. H. Ni, H. M. Wang, J. Kasim, H. M. Fan, T. Yu, Y. H. Wu, Y. P. Feng, and Z. X. Shen, *Nano Lett.*, **2007**, *7*, 2758.
[28] W. A. de Heer, *Science*, **2006**, *312*, 1191.
[29] Y.-W. Son, M. L. Cohen, and S. G. Louie, *Phys. Rev. Lett.*, **2006**, *97*, 216803.
[30] M. S. Dresselhaus, G. Dresselhaus, and R. Saito, *Carbon*, **1995**, *33*, 883.

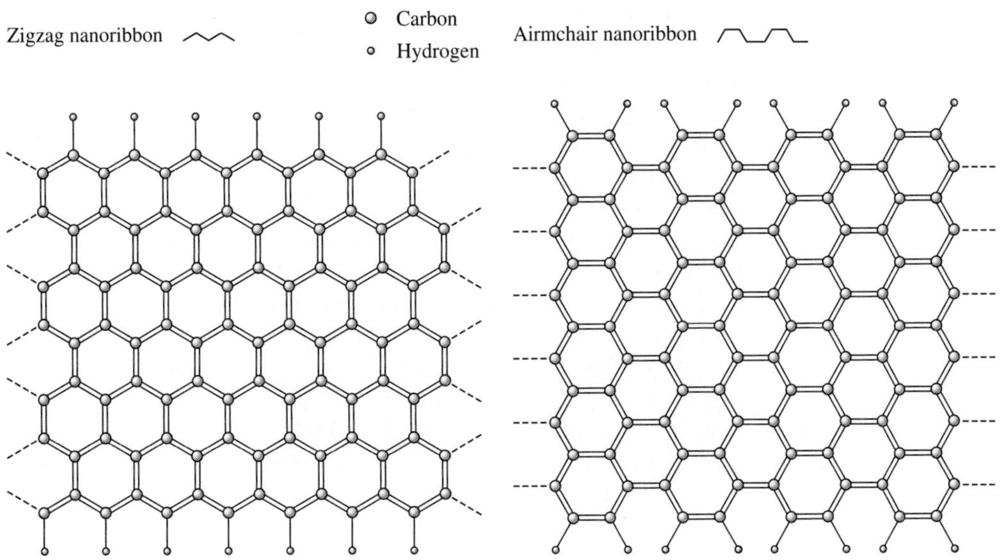

Zigzag nanoribbon

Carbon

Hydrogen

Airmchair nanoribbon

FIGURE 8.18 Zigzag and Armchair Nanoribbons.

Both single-walled nanotubes, as shown in Figure 8.16, and multiple-walled nanotubes with multiple, concentric layers of carbons have been prepared and studied extensively. Commercially they are prepared by a variety of methods, including applications of electric arcs to graphite electrodes, laser ablation of graphite, and catalytic pyrolysis of gaseous hydrocarbons. In some processes, they must be separated from fullerenes, which may be produced simultaneously; and individual nanotubes may

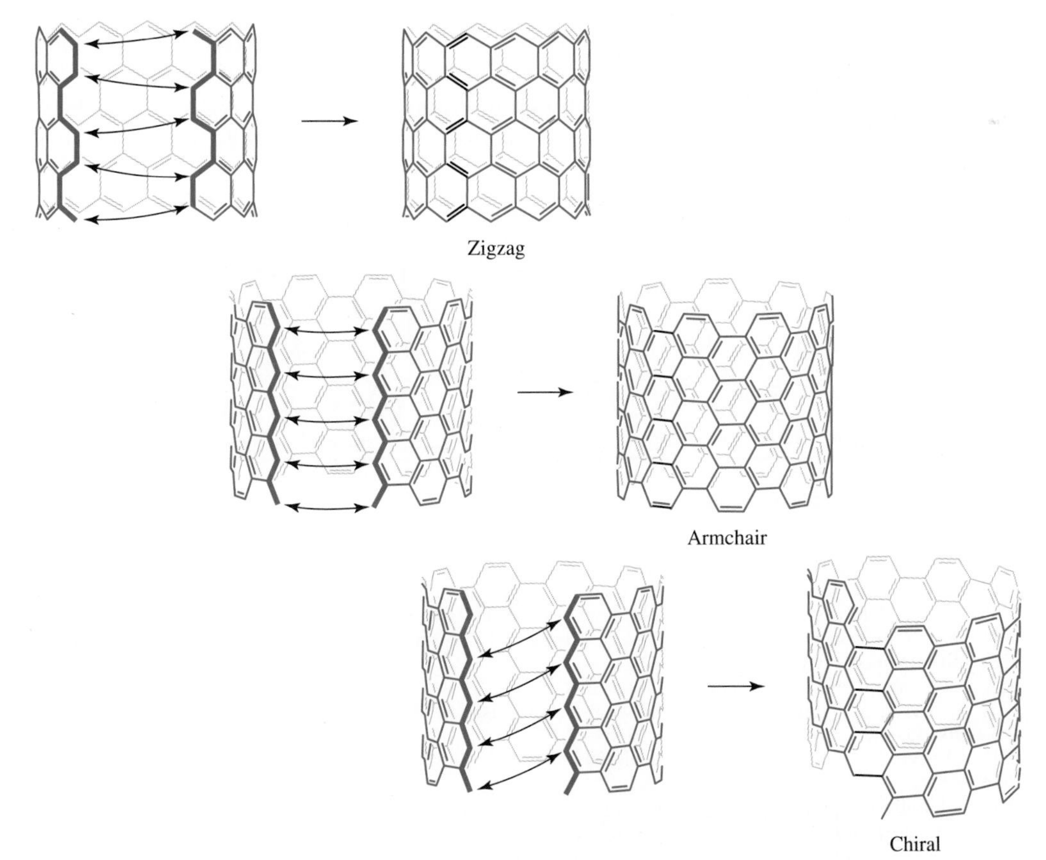

FIGURE 8.19 Conformations of Nanotubes.

need to be unbundled from others, for example by sonication. The conductivity of carbon nanotubes varies with the diameter of the tubes and their chirality, spanning the range from semiconductors to metallic conductors, up to approximately a thousand times the conductivity of copper. They are also highly elastic and extremely strong. They can be small in diameter, on the order of the diameter of a fullerene molecule (see the following section) or much larger.[31]

Many potential applications of nanotubes have been proposed, and some have been implemented. Some of the most promising are in the electronics industry, where, for example, nanotubes have been cited as the leading candidate to replace silicon, when the size limit on miniaturization of silicon chips has been reached.[32] A particularly interesting application is the development of field-effect transistors that contain fullerenes inside the tubes, commonly called *carbon peapods* (Figure 8.20).[33] They have also been used as scaffolding for light-harvesting assemblies, including quantum dots,[34] and for delivery of the anticancer agent cisplatin, $PtCl_2(NH_3)_2$, to cancer cells.[35] Because of their strength, carbon nanotubes have also been proposed for use in body armor, safety shields, and other protective devices.

A particularly clever application of nanotubes has been their use in the synthesis of nanoribbons by chemically "unzipping" the nanotubes into ribbons. This process involves the strong oxidizing agent combination $KMnO_4/H_2SO_4$ to form rows of parallel carbonyls on both sides of a slit, as it is being made along the length of the nanotube, with the mechanism proposed in Figure 8.21.[36] The result is a nanoribbon with a width having the same number of carbons as the circumference of the original nanotube.

FULLERENES One of the most fascinating developments in modern chemistry has been the synthesis of buckminsterfullerene[37], C_{60}, and the related *fullerenes*, molecules having near-spherical shapes resembling geodesic domes. First reported by Kroto, Curl, Smalley, and colleagues[38] in 1985, C_{60}, C_{70}, C_{80}, and a variety of related species were soon synthesized; examples of their structures are shown in Figure 8.16. Subsequent work has been extensive, and many compounds of fullerenes containing groups attached to the outside of

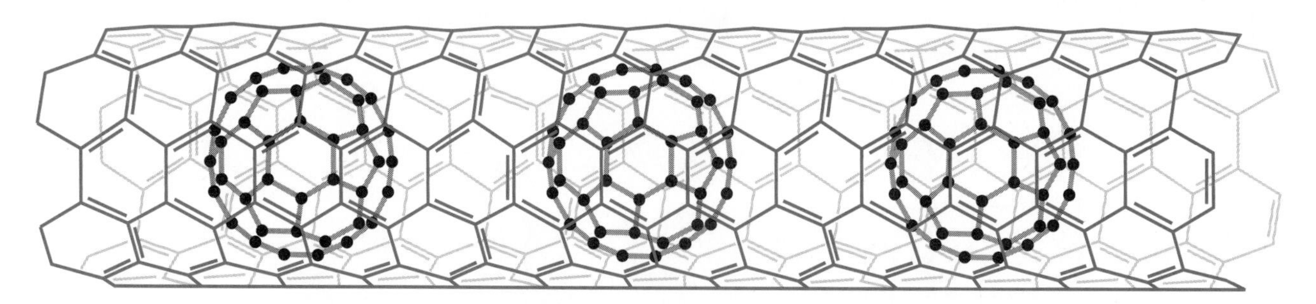

FIGURE 8.20 A Carbon Peapod.

[31] See D. R. Mitchell, R. M. Brown, Jr., T. L. Spires, D. K. Romanovicz, and R. J. Lagow, *Inorg. Chem.*, **2001**, *40*, 2751 for a variety of electronic microscopic images of larger diameter nanotubes.
[32] V. Derycke, R. Martel, J. Appenzeller, and P. Avouris, *Nano Lett.*, **2001**, *1*, 453.
[33] L. Ge, J. H. Jefferson, B. Montanari, N. M. Harrison, D. G. Pettifor, and G. A. D. Briggs, *ACS Nano*, **2009**, *3*, 1069 and references cited therein.
[34] A. Kongkanand and P. V. Kamat, *ACS Nano*, **2007**, *1*, 13.
[35] A. A. Bhirde, V. Patel, J. Gavard, G. Zhang, A. A. Sousa, A. Masedunskas, R. D. Leapman, R. Weigert, J. S. Gutkind, and J. F. Rusling, *ACS Nano*, **2009**, *3*, 307.
[36] D. V. Kosynkin, A. L. Higginbotham, A. Sinitskii, J. R. Lomeda, A. Dimiev, B. K. Price, and J. M. Tour, *Nature*, **2009**, *458*, 872.
[37] More familiarly known as a "buckyball."
[38] H. W. Kroto, J. R. Heath, S. C. O'Brien, R. F. Curl, and R. E. Smalley, *Nature (London)*, **1985**, *318*, 162.

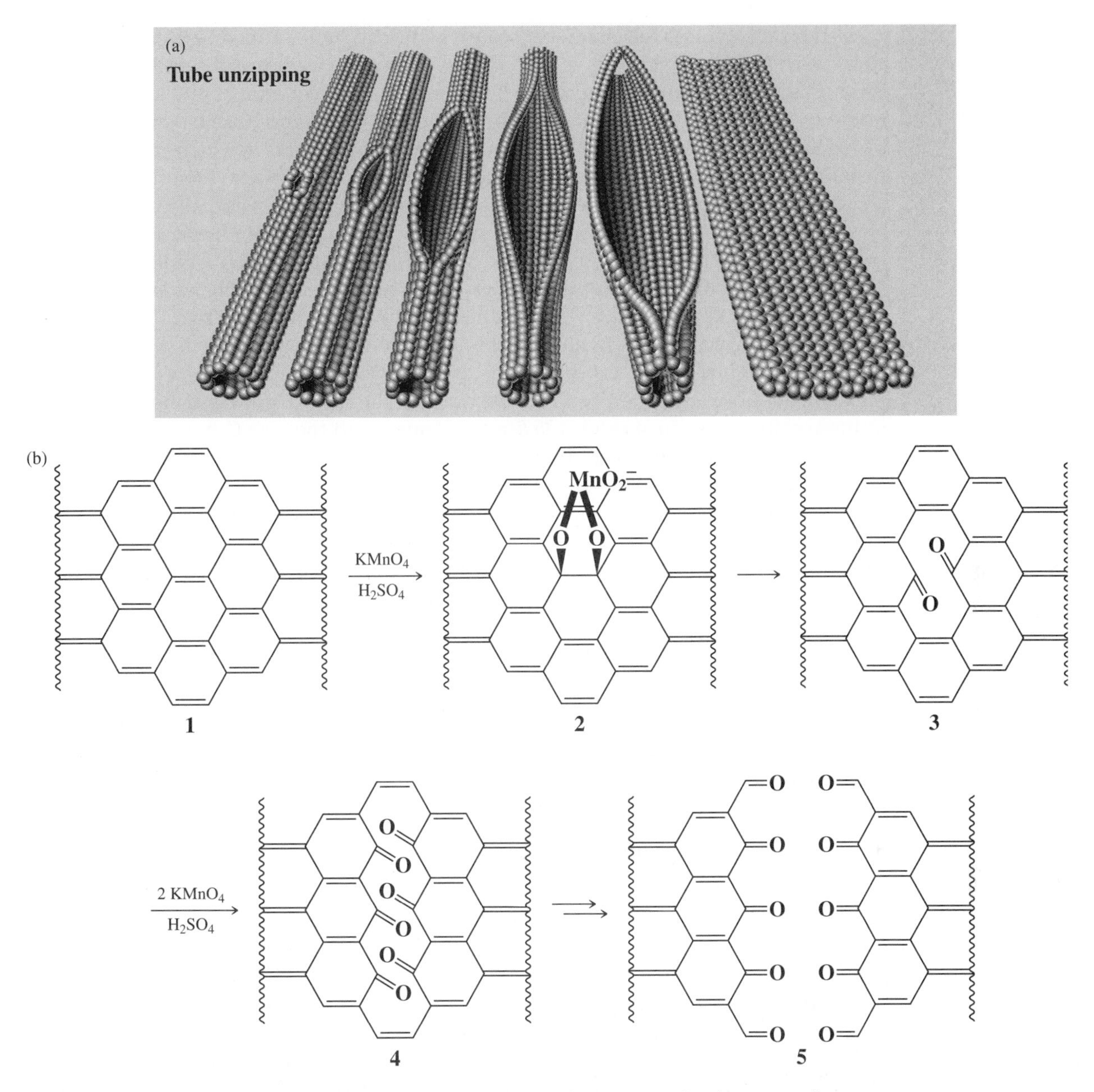

FIGURE 8.21 Nanotube Unzipping. (a) Proposed mechanism of unzipping of carbon nanotube; (b) representation of nanotube unzipping. Reprinted by permission from Macmillan Publishers Ltd: D. V. Kosynkin, A. L. Higginbotham, A. Sinitskii, J. R. Lomeda, A.Dimiev, B. K. Price, and J. M. Tour, *Nature*, **2009**, *458*, 872. Copyright 2009.

these large clusters have been synthesized. In addition, small atoms and molecules have been trapped inside fullerene cages. Remarkably, roughly 9 years after the first synthesis of fullerenes, natural deposits of these molecules were discovered at the impact sites of ancient meteorites.[39] The development of large-scale synthetic procedures for fullerenes has been a challenging undertaking, with most methods to date involving condensation of carbon in an inert atmosphere, from laser or other high-energy vaporization of graphite, or from controlled pyrolysis of aromatic hydrocarbons.[40]

[39] L. Becker, J. L. Bada, R. E. Winans, J. E. Hunt, T. E. Bunch, and B. M. French, *Science*, **1994**, *265*, 642; D. Heymann, L. P. F. Chibante, R. R. Brooks, W. S. Wolbach, and R. E. Smalley, *Science*, **1994**, *265*, 645.
[40] J. R. Bowser, *Adv. Inorg. Chem.*, **1994**, *36*, 61–62 and references therein.

The prototypical fullerene, C_{60}, consists of fused five- and six-membered carbon rings. Each six-membered ring is surrounded alternately by hexagons and pentagons of carbons; each pentagon is fused to five hexagons. The consequence of this structural motif is that each hexagon is like the base of a bowl: the three pentagons fused to this ring, linked by hexagons, force the structure to curve (in contrast to graphite, in which each hexagon is fused to six surrounding hexagons in the same plane). This phenomenon, best seen by assembling a model of C_{60}, results in a domelike structure that eventually curves around on itself to give a structure resembling a sphere.[41] The shape resembles a soccer ball (the most common soccer ball has an identical arrangement of pentagons and hexagons on its surface); all 60 atoms are equivalent and give rise to a single ^{13}C NMR resonance.

Although all atoms in C_{60} are equivalent, the bonds are not. Two types of bonds occur (best viewed using a model), at the fusion of two six-membered rings and at the fusion of five- and six-membered rings. X-ray crystallographic studies on C_{60} complexes have shown that the C—C bond lengths at the fusion of two six-membered rings in these complexes are shorter, 135.5 pm, compared to the C—C bond lengths at the fusion of five- and six-membered rings, 146.7 pm.[42] This indicates a greater degree of pi bonding at the fusion of the six-membered rings.

Surrounding each six-membered ring with two pentagons (on opposite sides) and four hexagons (with each pentagon, as in C_{60}, fused to five hexagons) gives a slightly larger, somewhat prolate structure with 70 carbon atoms. C_{70} is often obtained as a by-product of the synthesis of C_{60} and is among the most stable of the fullerenes. Unlike C_{60}, five different types of carbon are present in C_{70}, giving rise to five ^{13}C NMR resonances.[43]

Structural variations on fullerenes have evolved well beyond the individual clusters themselves. The following are a few examples.

Polymers. The rhombohedral polymer of C_{60} shown in Figure 8.22(a) has been reported to act as a ferromagnet at room temperature and above.[44] Linear chain polymers have also been reported.[45]

Nano "onions." These are spherical particles based on multiple carbon layers surrounding a C_{60} or other fullerene core. One proposed use is in lubricants.[46]

Other linked structures. These include fullerene rings,[47] linked "ball-and-chain" dimers,[48] and an increasing variety of other forms. Examples are shown in Figure 8.22.

The smallest known fullerene is C_{20} (Figure 8.16), synthesized by replacing the hydrogen atoms of dodecahedrane, $C_{20}H_{20}$, with bromines, followed by debromination.[49]

During the past quarter of a century, fullerene chemistry has developed from its infancy into a broadly studied realm of science that is increasingly focused on practical applications, although challenges remain for many of these. Fullerenes also form a variety of chemical compounds via reactions on their surfaces and can trap atoms and small molecules inside; these features of fullerene chemistry will be considered in Chapter 13.

---

[41] The structure of C_{60} has the same symmetry as an icosahedron.

[42] These distances were obtained for a twinned crystal of C_{60} at 110 K. (S. Liu, Y. Lu, M. M. Kappes, and J. A. Ibers, *Science*, **1991**, *254*, 408.) Neutron diffraction data at 5 K give slightly different results: 139.1 pm at the fusion of the 6-membered rings and 145.5 pm at the fusion of 5- and 6-membered rings (W. I. F. David, R. M. Ibberson, J. C. Matthew, K. Pressides, T. J. Dannis, J. P. Hare, H. W. Kroto, R. Taylor, and D. C. M. Walton, *Nature*, **1991**, *353*, 147).

[43] R. Taylor, J. P. Hare, A. K. Abdul-Sada, and H. W. Kroto, *Chem. Commun. (Cambridge)*, **1990**, 1423.

[44] T. L. Makarova, B. Sundqvist, R. Höhne, P. Esquinazi, Y. Kopelevich, P. Scharff, V. A. Davydov, L. S. Kashevarova, and A. V. Rakhmanina, *Nature (London)*, **2001**, *413*, 716; *Chem. Eng. News*, **2001**, *79*, 10.

[45] H. Brumm, E. Peters, and M. Jansen, *Angew. Chem., Int. Ed.*, **2001**, *40*, 2069.

[46] N. Sano, H. Wang, M. Chhowalla, I. Alexandrou, and G. A. J. Amaratunga, *Nature (London)*, **2001**, *414*, 506.

[47] Y. Li, Y. Huang, S. Du, and R. Liu, *Chem. Phys. Lett.*, **2001**, *335*, 524.

[48] A. A. Shvartsburg, R. R. Hudgins, R. Gutierrez, G. Jungnickel, T. Frauenheim, K. A. Jackson, and M. F. Jarrold, *J. Phys. Chem. A*, **1999**, *103*, 5275.

[49] H. Prinzbach, A. Weller, P. Landenberger, F. Wahl, J. Wörth, L. T. Scott, M. Gelmont, D. Olevano, and B. Issendorff, *Nature (London)*, **2000**, *407*, 60.

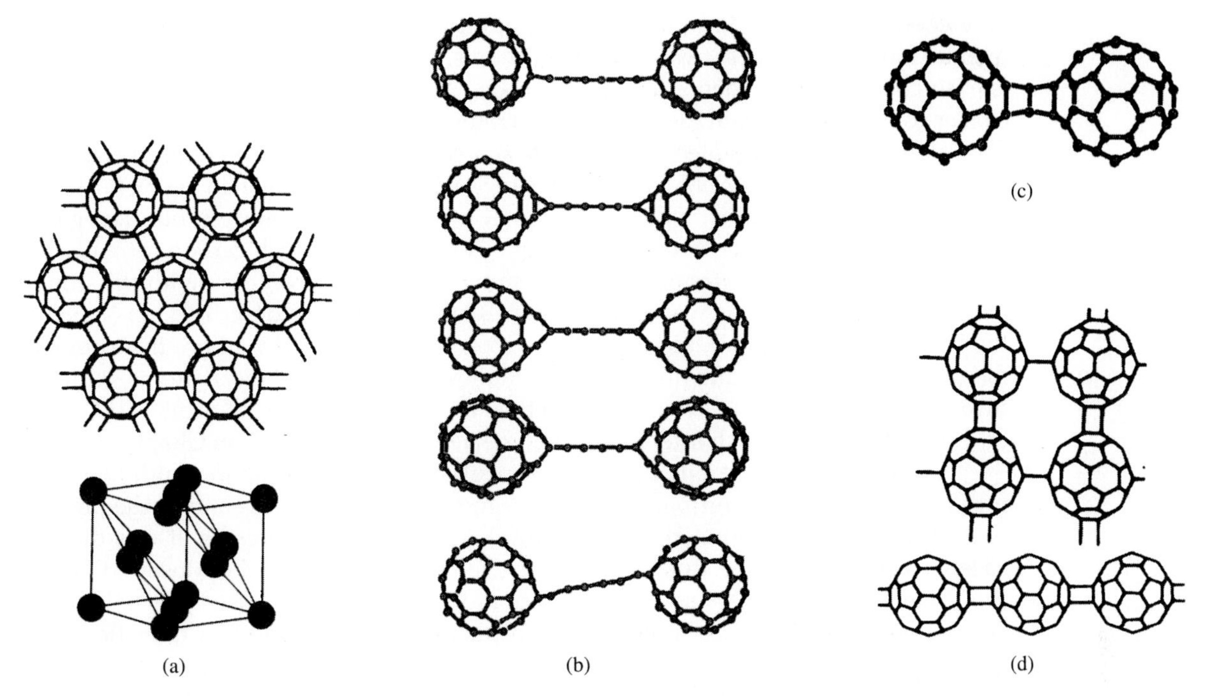

FIGURE 8.22 Polymers of C_{60}. (a) Rhombohedral polymer. (b) Ball-and-chain dimers. *Top to bottom, sp,* closed-66 *sp*^{2}, open-66 *sp*^{2}, open-56 *sp*^{2}, open-56 *sp*^{2} in a distorted configuration. (c) Double [2 + 2] closed-66 isomer for C_{122}. (d) Other linked and linear chains. The *56* and *66* labels indicate that the bonds are to carbons common to five- and six-membered rings and to two six-membered rings, respectively. (Figures (a) and (d) from M. Núñez-Tegueiro, L. Marques, J. L. Hodeau, O. Béthoux, and M. Perroux, *Phys. Rev. Lett.*, **1995**, *74*, 278; (b) and (c) from A. A. Shvartsburg, R. R. Hudgins, R. Gutierrez, G. Jungnickel, T. Frauenheim, K. A. Jackson, and M. F. Jarrold, *J. Phys. Chem. A*, **1999**, *103*, 5275. Reproduced with permission.)

Perhaps one of the most promising applications of fullerene chemistry was introduced in 2006, when it was found that multiple copies of a water-soluble C_{60} derivative could be attached to a melanoma antibody; loading this derivative with anticancer drug molecules can deliver the drug directly into the melanoma.[50] The use of fullerenes, nanotubes, quantum dots, and other nanoparticles for drug delivery has recently been reviewed.[51]

Silicon and germanium crystallize in the diamond structure. However, they have somewhat weaker covalent bonds than carbon as a consequence of less efficient orbital overlap. These weaker bonds result in lower melting points for silicon—1420 °C for Si and 945 °C for Ge, compared with 4100 °C for diamond—and greater chemical reactivity. Both silicon and germanium are semiconductors, as described in Chapter 7.

On the other hand, tin has two allotropes: a diamond (α) form more stable below 13.2 °C and a metallic (β) form more stable at higher temperatures.[52] Lead is entirely metallic and is among the most dense, and most poisonous, of the metals.

8.6.2 Compounds

A common misconception is that carbon can, at most, be four-coordinate. Although carbon is bonded to four or fewer atoms in the vast majority of its compounds, many examples are now known in which carbon has coordination numbers of 5, 6, or higher.

[50] J. M. Ashcroft, D. A. Tsyboulski, K. B. Hartman, T. Y. Zakharian, J. W. Marks, R. B. Weisman, M. G. Rosenblum, and L. J. Wilson, *Chem. Commun.*, **2006**, 3004.
[51] R. Singh and J. W. Lillard, Jr., *Exp. Mol. Pathology*, **2009**, *86*, 215.
[52] These forms are *not* similar to the α and β forms of graphite (Figure 8.16).

Five-coordinate carbon is actually rather common, with methyl and other groups frequently forming bridges between two metal atoms, as in $Al_2(CH_3)_6$ (see Figure 8.15). There is even considerable evidence for the five-coordinate ion CH_5^+.[53] Many organometallic cluster compounds contain carbon atoms surrounded by polyhedra of metal atoms. Such compounds, often designated *carbide clusters*, are discussed in Chapter 15. Examples of carbon atoms having coordination numbers of 5, 6, 7, and 8 are shown in Figure 8.23.

The two most familiar oxides of carbon, CO and CO_2, are colorless, odorless gases. Carbon monoxide is a rarity of sorts, a stable compound in which carbon formally has only three bonds. It is extremely toxic, forming a bright red complex with the iron in hemoglobin, which has a greater affinity for CO than for O_2. As described in Chapter 5, the highest occupied molecular orbital of CO is concentrated on carbon; this provides the molecule an opportunity to interact strongly with a variety of metal atoms, which in turn can donate electron density through their *d* orbitals to empty π^* orbitals (LUMOs) on CO. The details of such interactions will be described more fully in Chapter 13.

Carbon dioxide is familiar as a component of Earth's atmosphere—although only fifth in abundance, after nitrogen, oxygen, argon, and water vapor—and as the product of respiration, combustion, and other natural and industrial processes. It was the first gaseous component to be isolated from air, the "fixed air" isolated by Joseph Black in 1752. More recently, CO_2 has gained international attention because of its role in the "greenhouse" effect and the potential atmospheric warming and other climatic consequences of an increase in CO_2 abundance. Because of the energies of carbon dioxide's vibrational levels, it absorbs a significant amount of thermal energy and, hence, acts as a sort of atmospheric blanket. Since the beginning of the Industrial Revolution, the carbon dioxide concentration in the atmosphere has increased substantially, an increase that will continue indefinitely unless major policy changes are made by the industrialized nations. A start was made on policies for greenhouse gas reduction at an international conference in Kyoto, Japan, in 1997. The consequences of a continuing increase in atmospheric CO_2 are difficult to forecast; the dynamics of the atmosphere are extremely complex, and the interplay between atmospheric composition, human activity, the oceans, solar cycles, and other factors is not yet well understood.

FIGURE 8.23 High Coordination Numbers of Carbon.

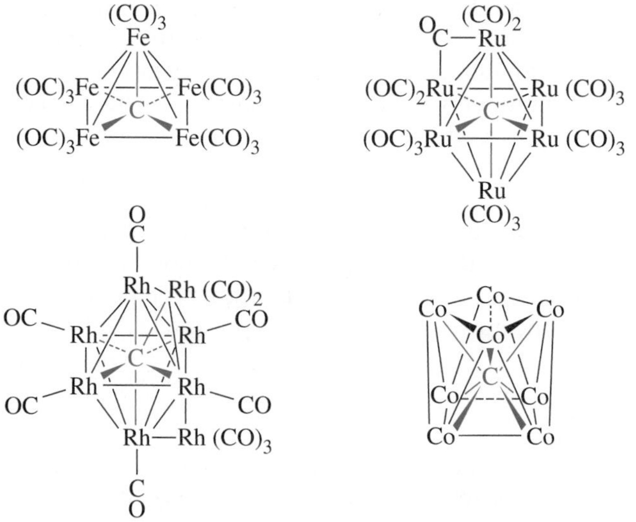

$Rh_8C(CO)_{19}$
Not shown: Eight COs bridging edges of polyhedron

$[Co_8C(CO)_{18}]^{2-}$
COs not shown: One on each cobalt; ten bridging edges of polyhedron

[53] G. A. Olah and G. Rasul, *Acc. Chem. Res.*, **1997**, *30*, 245.

Although only two forms of elemental carbon are common, carbon forms several anions, especially in combination with the most electropositive metals. In these compounds, called collectively the *carbides*, there is considerable covalent as well as ionic bonding, with the proportion of each depending on the metal. The best characterized carbide ions are shown here.

Ion	Common Name	Systematic Name	Example	Major Hydrolysis Product
C^{4-}	Carbide or methanide	Carbide	Al_4C_3	CH_4
C_2^{2-}	Acetylide	Dicarbide (2–)	CaC_2	$H{-}C{\equiv}C{-}H$
C_3^{4-}		Tricarbide (4–)	Mg_2C_3[a]	$H_3C{-}C{\equiv}C{-}H$

NOTE: [a]This is the only known compound containing the C_3^{4-} ion.

These carbides, as indicated, liberate organic molecules on reaction with water. For example:

$$Al_4C_3 + 12\,H_2O \longrightarrow 4\,Al(OH)_3 + 3\,CH_4$$

$$CaC_2 + 2\,H_2O \longrightarrow Ca(OH)_2 + HC{\equiv}CH$$

Calcium carbide, CaC_2, is the most important of the metal carbides. Its crystal structure resembles that of NaCl, with parallel C_2^{2-} units, as shown in Figure 8.24. Before compressed gases were readily available, calcium carbide was commonly used as a source of acetylene for lighting and welding; early automobiles had carbide headlights.

It may seem surprising that carbon, with its vast range of literally millions of compounds, is not the most abundant element in this group. By far the most abundant Group 14 (IVA) element on Earth is silicon, which comprises 27% of Earth's crust by mass and is second in abundance after oxygen; carbon is only seventeenth in abundance. Silicon, with its semimetallic properties, is of enormous importance in the semiconductor industry, with wide applications in such fields as computers and solar energy collection.

In nature, silicon occurs almost exclusively in combination with oxygen, with many minerals containing tetrahedral SiO_4 structural units. Silicon dioxide, SiO_2,

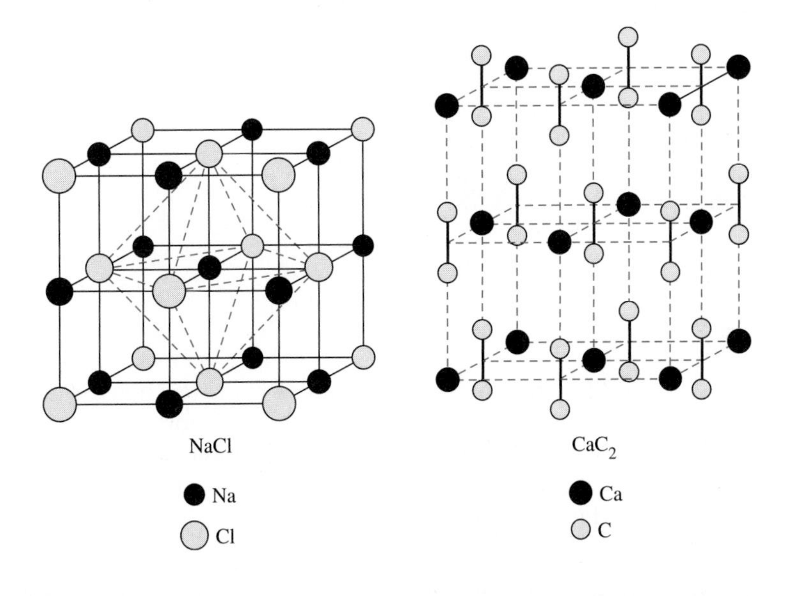

FIGURE 8.24 Crystal Structures of NaCl and CaC_2.

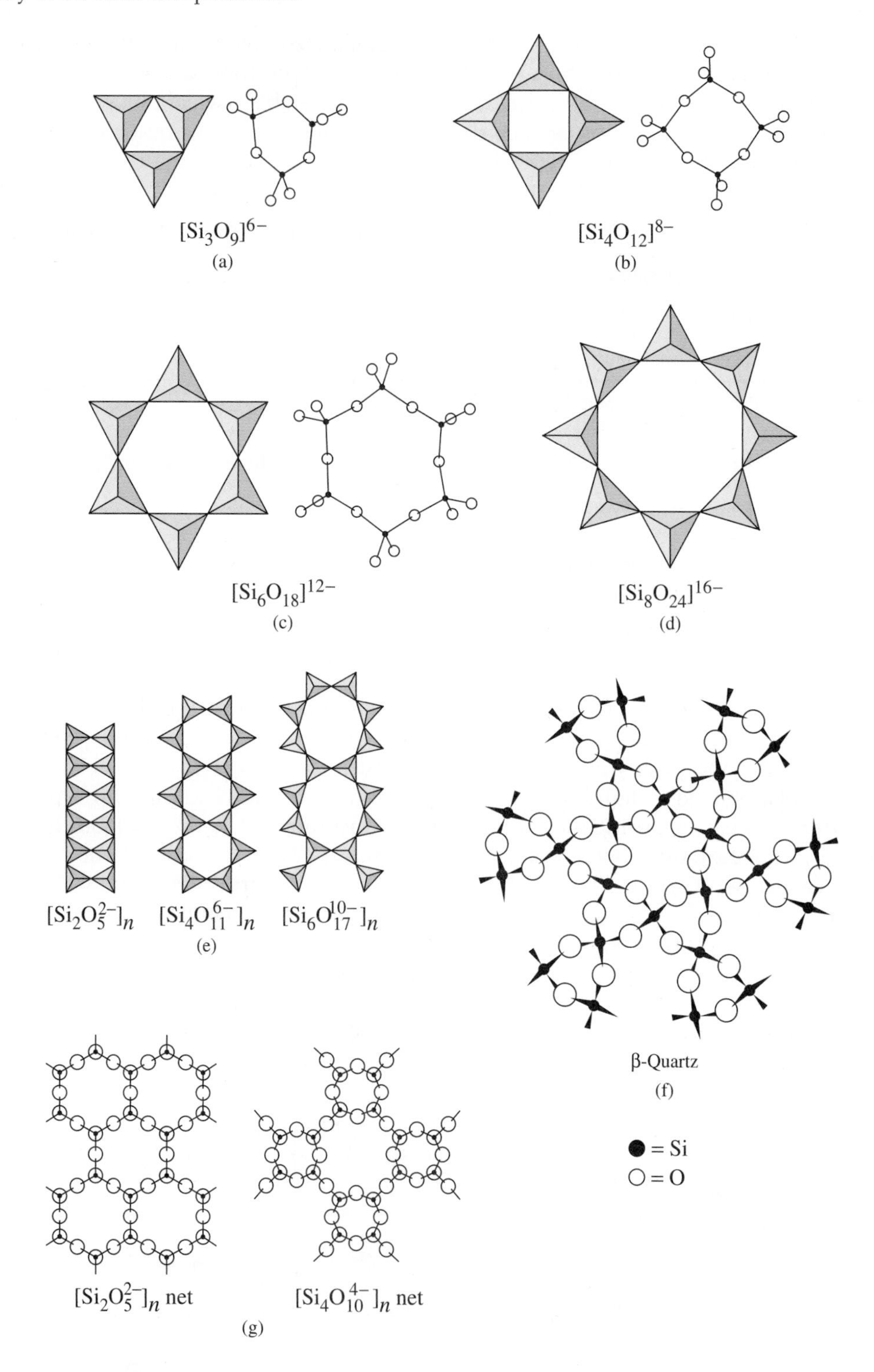

FIGURE 8.25 Examples of Silicate Structures. (Structures shown in **a**, **b**, **c**, **d**, and **e** were reproduced with permission from N. N. Greenwood and A. Earnshaw, *Chemistry of the Elements*, Pergamon Press, Elmsford, NY, 1984, pp. 403, 405, © 1984; structures **f** and **g** are from A. F. Wells, *Structural Inorganic Chemistry*, 5th ed., Oxford University Press, New York, 1984, pp. 1006, 1024.)

occurs in a variety of forms in nature, the most common of which is α-quartz, a major constituent of sandstone and granite. SiO_2 is industrially important as the major component of glass, in finely divided form as a chromatographic support (silica gel) and catalyst substrate, as a filtration aid (as diatomaceous earth, the remains of diatoms, tiny unicellular algae), and in many other applications.

The SiO_4 structural units occur in nature in silicates, compounds in which these units may be fused by sharing corners, edges, or faces in diverse ways. Examples of silicate structures are shown in Figure 8.25. The interested reader can find extensive discussions of these structures in the chemical literature.[54]

With carbon forming the basis for the colossal number of organic compounds, it is interesting to consider whether silicon or other members of this group can form the foundation for an equally vast array of compounds. Unfortunately, such does not seem the case; the ability to catenate (form bonds with other atoms of the same element) is much lower for the other Group 14 (IVA) elements than for carbon, and the hydrides of these elements are also much less stable.

Silane, SiH_4, is stable and, like methane, tetrahedral. However, although silanes (of formula Si_nH_{n+2}) up to eight silicon atoms in length have been synthesized, their stability decreases markedly with chain length; Si_2H_6, disilane, undergoes only very slow decomposition, but Si_8H_{18} decomposes rapidly. In recent years, a few compounds containing $Si{=}Si$ bonds have been synthesized, but there is no promise of a chemistry of multiply bonded Si species comparable at all in diversity with the chemistry of unsaturated organic compounds. Germanes of formulas GeH_4 to Ge_5H_{12} have been made, as have SnH_4 (stannane), Sn_2H_6, and possibly PbH_4 (plumbane), but the chemistry in these cases is even more limited than of the silanes.

Why are the silanes and other analogous compounds less stable (more reactive) than the corresponding hydrocarbons? First, the $Si{-}Si$ bond is slightly weaker than the $C{-}C$ bond (approximate bond energies, 340 and 368 kJ mol^{-1}, respectively), and $Si{-}H$ bonds are weaker than $C{-}H$ bonds (393 versus 435 kJ mol^{-1}). Silicon is less electronegative (1.92) than hydrogen (2.30) and is, therefore, more susceptible to nucleophilic attack; this is in contrast to carbon, which is more electronegative (2.54) than hydrogen. Silicon atoms are also larger and therefore provide greater surface area for attack by nucleophiles. In addition, silicon atoms have low-lying *d* orbitals that can act as acceptors of electron pairs from nucleophiles. Similar arguments can be used to describe the high reactivity of germanes, stannanes, and plumbanes. Silanes are believed to decompose by elimination of $:SiH_2$ by way of a transition state having a bridging hydrogen, as shown in Figure 8.26. This reaction, incidentally, can be used to prepare silicon of extremely high purity.

As mentioned, elemental silicon has the diamond structure. Silicon carbide, SiC, occurs in many crystalline forms, some based on the diamond structure and some on the wurtzite structure (see Figures 7.6 and 7.8b). It can be made from the elements at high temperature. Carborundum, one form of silicon carbide, is widely used as an abrasive, with a hardness nearly as great as diamond and a low chemical reactivity. SiC has now garnered interest as a high-temperature semiconductor.

The elements germanium, tin, and lead show increasing importance of the 2+ oxidation state, an example of the inert pair effect. For example, all three show two sets of

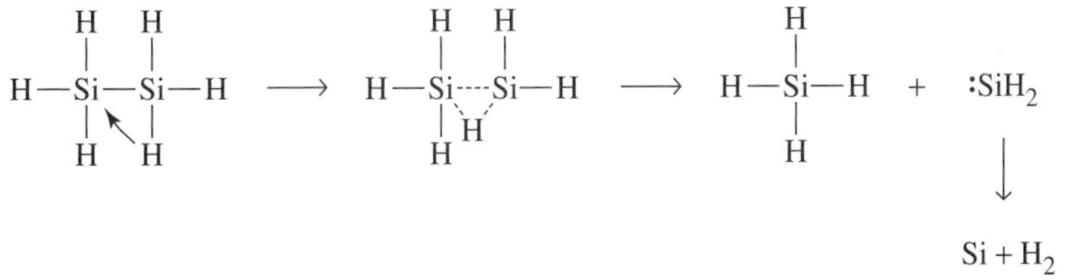

$$\mathrm{H_3Si{-}SiH_3} \longrightarrow \mathrm{H_2Si{\cdots}(H){\cdots}SiH_3} \longrightarrow \mathrm{H{-}SiH_2{-}H} + \mathrm{:SiH_2}$$

$$\mathrm{:SiH_2} \longrightarrow \mathrm{Si + H_2}$$

FIGURE 8.26 Decomposition of Silanes.

[54] A. F. Wells, *Structural Inorganic Chemistry*, 5th ed., Clarendon Press, Oxford, 1984, pp. 1009–1043.

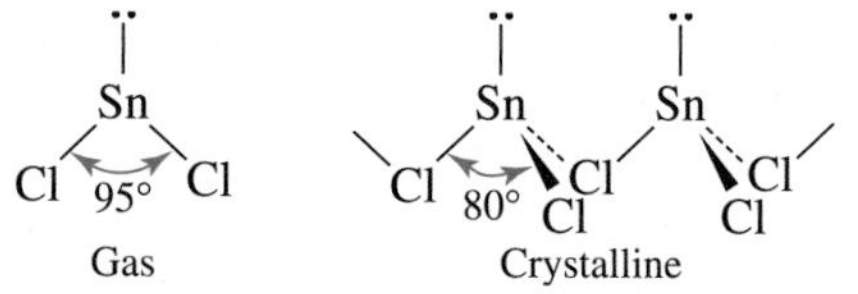

FIGURE 8.27 Structure of $SnCl_2$ in Gas and Crystalline Phases.

halides, of formula MX_4 and MX_2. For germanium, the most stable halides have the formula GeX_4; for lead, it is PbX_2. For the dihalides, the metal exhibits a stereochemically active lone pair. This leads to bent geometry for the free molecules and to crystalline structures in which the lone pair is evident, as shown for $SnCl_2$ in Figure 8.27.

8.7 GROUP 15 (VA)

Nitrogen is the most abundant component of Earth's atmosphere (78.1% by volume). However, the element was not successfully isolated from air until 1772, when Rutherford, Cavendish, and Scheele achieved the isolation nearly simultaneously by successively removing oxygen and carbon dioxide from air. Phosphorus was first isolated from urine by H. Brandt in 1669. Because the element glowed in the dark on exposure to air, it was named after the Greek *phos*, "light," and *phoros*, "bringing." Interestingly, the last three elements in Group 15 (VA)[55] had long been isolated by the time nitrogen and phosphorus were discovered. Their dates of discovery are lost in history, but all had been studied extensively, especially by alchemists, by the fifteenth century.

These elements again span the range from nonmetallic (nitrogen and phosphorus) to metallic (bismuth) behavior, with the elements in between (arsenic and antimony) having intermediate properties. Selected physical properties are given in Table 8.9.

8.7.1 The Elements

Nitrogen is a colorless diatomic gas. As discussed in Chapter 5, the dinitrogen molecule has a nitrogen–nitrogen triple bond of unusual stability. In large part, the stability of this bond is responsible for the low reactivity of this molecule, although it is by no means totally inert. Nitrogen is therefore suitable as an inert environment for many chemical studies of reactions that are either oxygen or moisture sensitive. Liquid nitrogen, at 77 K, is frequently used as a convenient, rather inexpensive coolant for studying

TABLE 8.9 Properties of the Group 15 (VA) Elements

Element	Ionization Energy ($kJ\ mol^{-1}$)	Electron Affinity ($kJ\ mol^{-1}$)	Melting Point (°C)	Boiling Point (°C)	Electro-negativity
N	1402	−7	−210	−195.8	3.066
P	1012	72	44[a]	280.5	2.053
As	947	78	b	b	2.211
Sb	834	103	631	1587	1.984
Bi	703	91	271	1564	2.01[c]

Source: See Table 8.3

NOTES: [a] α-P_4

[b] Sublimes at 615 °C

[c] Approximate value

[55] Elements in this group are sometimes called the *pnicogens* or *pnictogens*.

low-temperature reactions, trapping of solvent vapors, and cooling superconducting magnets (actually, for preserving the liquid helium coolant, which boils at 4 K).

Phosphorus has many allotropes. The most common of these is white phosphorus, which exists in two modifications, α-P_4 (cubic) and β-P_4 (hexagonal). Condensation of phosphorus from the gas or liquid phases, both of which contain tetrahedral P_4 molecules, gives primarily the α form, which slowly converts to the β form at temperatures above –76.9 °C. During slow air oxidation, α-P_4 emits a yellow-green light, an example of phosphorescence that has been known since antiquity (and is the source of the name of this element); to slow such oxidation, white phosphorus is commonly stored underwater. White phosphorus was once used in matches; however, its extremely high toxicity has led to its replacement by other materials, especially P_4S_3 and red phosphorus, which are much less toxic.

Heating of white phosphorus in the absence of air gives red phosphorus, an amorphous material that exists in a variety of polymeric modifications. Still another allotrope, black phosphorus, is the most thermodynamically stable form; it can be obtained from white phosphorus by heating at very high pressures. Black phosphorus converts to other forms at still higher pressures. Examples of these structures are shown in Figure 8.28. The interested reader can find more detailed information on allotropes of phosphorus in other sources.[56]

As mentioned, phosphorus exists as tetrahedral P_4 molecules in the liquid and gas phases. At very high temperatures, P_4 can dissociate into P_2:

$$P_4 \rightleftharpoons 2\,P{\equiv}P$$

At approximately 1800 °C, this dissociation reaches 50 percent.

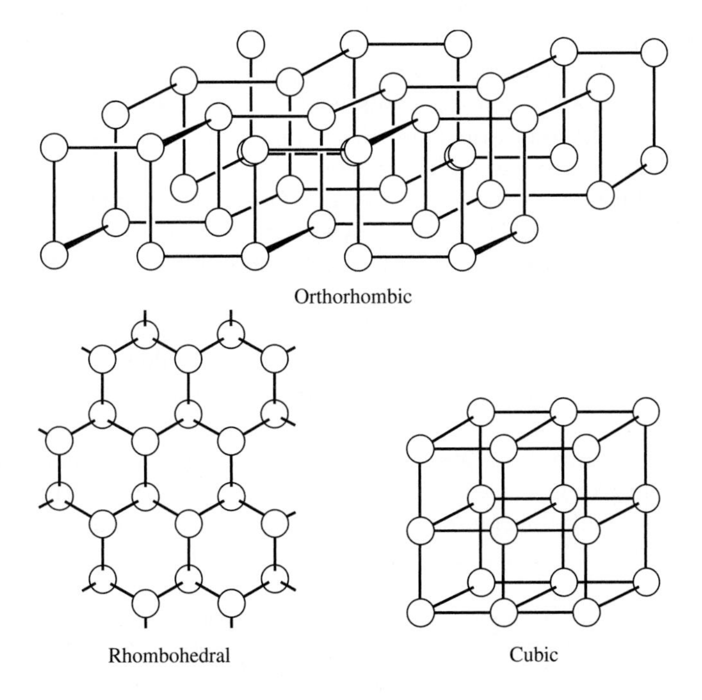

FIGURE 8.28 Allotropes of Phosphorus. (Reproduced with permission from N. N. Greenwood and A. Earnshaw, *Chemistry of the Elements*, Pergamon Press, Elmsford, NY, 1984, p. 558. © 1984, Pergamon Press PLC.)

[56] A. F. Wells, *Structural Inorganic Chemistry*, 5th ed., Clarendon Press, Oxford, 1984, pp. 838–840 and references therein.

Arsenic, antimony, and bismuth also exhibit a variety of allotropes. The most stable allotrope of arsenic is the gray (α) form, which is similar to the rhombohedral form of phosphorus. In the vapor phase, arsenic, like phosphorus, exists as tetrahedral As_4. Antimony and bismuth also have similar α forms. These three elements have a somewhat metallic appearance but are brittle and are only moderately good conductors. Arsenic, for example, is the best conductor in this group but has an electrical resistivity nearly 20 times as great as copper.

Bismuth is the heaviest element to have a stable, nonradioactive nucleus; polonium and all heavier elements are radioactive.

IONS During the past decade, the field of nitrogen anion chemistry has been an active one. For more than a century, the only isolable chemical species containing nitrogen and no other elements were N_2 and the nitride and azide ions, N^{3-} and N_3^-. Nitrides of primarily ionic character are formed by lithium and the Group 2 (IIA) elements; many other nitrides having a greater degree of covalence are also known. In addition, N^{3-} is a strong pi-donor ligand toward transition metals (metal–ligand interactions will be described in Chapter 10). Stable compounds containing the linear N_3^- ion include those of the Groups 1 and 2 (IA and IIA) metals. However, some other azides are explosive: $Pb(N_3)_2$, for example, is shock sensitive and is used as a primer for explosives.

Remarkably, in 1999, a new species, N_5^+, was reported, a product of the following reaction:

$$N_2F^+[AsF_6]^- + HN_3 \longrightarrow N_5^+[AsF_6]^- + HF$$

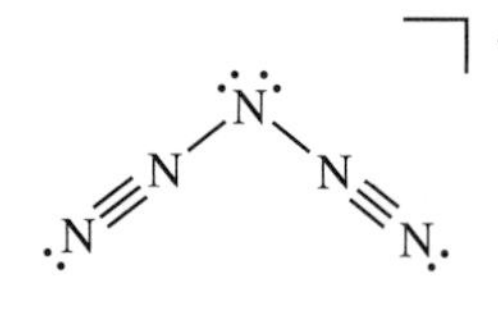

$N_5^+[AsF_6]^-$ is not stable at room temperature but can be preserved for weeks at –78 °C. The N_5^+ ion has a V-shaped structure, bent at the central nitrogen and linear at the neighboring atoms.[57] Furthermore, even though the ions C_2^{2-} and O_2^{2-} have been known for many years, the N_2^{2-} ion was not characterized until 2001.[58] In SrN_2, the bond distance in this ion is 122.4 pm, comparable to 120.8 pm in the isoelectronic O_2 molecule and much longer than the 109.8 pm in neutral N_2.

A longstanding challenge to chemists was the synthesis of what might seem a simple compound—sodium nitride, Na_3N. This explosive compound was not successfully prepared until 2002, from atomic beams of sodium and nitrogen at liquid nitrogen temperatures.[59] A synthesis 5 years later using sodium and plasma-activated nitrogen yielded a sample that was characterized by X-ray crystallography, giving the structure shown in Figure 8.29.[60] In this structure, each nitrogen is at the corner of a cube and surrounded by an octahedron of sodiums.

In recent years, considerable interest has been shown in preparing the pentazolate ion, *cyclo*-N_5^-, which would be electronically equivalent to the cyclopentadienide ion, $C_5H_5^-$, a key ligand in organometallic chemistry (which will be discussed in Chapter 13). To date, although N_5^- has been detected in the gas phase, attempts to isolate compounds containing this ion have not been successful.[61] However, the isoelectronic *cyclo*-P_5^- has been prepared, most notably as a ligand in the first carbon-free metallocene (see Section 15.2.2).[62] In addition, both P_4^{2-} and

[57] K. O. Christe, W. W. Wilson, J. A. Sheehy, and J. A. Boatz, *Angew. Chem., Int. Ed.*, **1999**, *38*, 2004; for more information on nitrogen-containing species, see T. M. Klapötke, *Angew. Chem., Int. Ed.*, **1999**, *38*, 2536.
[58] G. Auffermann, Y. Prots, and R. Kniep, *Angew. Chem., Int. Ed.*, **2001**, *40*, 547.
[59] D. Fischer and M. Jansen, *Angew. Chem. Int. Ed.*, **2002**, *41*, 1755.
[60] G. V. Vajenine, *Inorg. Chem.*, **2007**, *46*, 5146.
[61] I. Kobrsi, W. Zheng, J. E. Knox, M. J. Heeg, H. B. Schlegel, and C. H. Winter, *Inorg. Chem.*, **2006**, *45*, 8700.
[62] E. Urnezius, W. W. Brennessel, C. J. Cramer, J. E. Ellis, and P. v. R. Schleyer, *Science*, **2002**, *295*, 832.

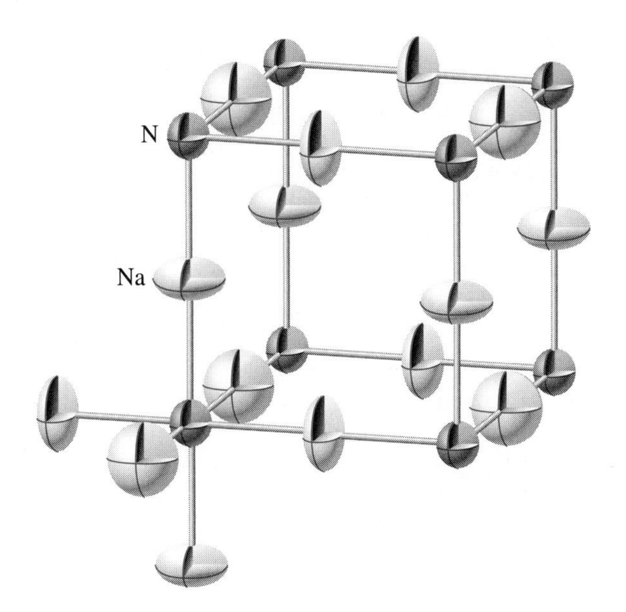

FIGURE 8.29 Na_3N. Reprinted with permission from Grigori V. Vajenine, *Inorganic Chemistry*, **2007**, *46*, 5147. Copyright 2007 American Chemical Society.

As_4^{2-} have been reported, both with square geometries,[63] and N_4^{4-} has been reported as a bridging ligand.[64]

Although phosphides, arsenides, and other Group 15 compounds are known with formulas that may suggest that they are ionic (e.g., Na_3P, Ca_3As_2), such compounds are generally lustrous and have good thermal and electrical conductivity, properties more consistent with metallic than with ionic bonding.

8.7.2 Compounds

HYDRIDES In addition to ammonia, nitrogen forms the hydrides N_2H_4 (hydrazine), N_2H_2 (diazene or diimide), and HN_3 (hydrazoic acid). Structures of these compounds are shown in Figure 8.30.

The chemistry of ammonia and the ammonium ion is vast; ammonia is of immense industrial importance and is synthesized in larger molar quantities than any other chemical. More than 80 percent of the ammonia produced is used in fertilizers, with additional uses that include the synthesis of explosives, the manufacture of synthetic fibers—such as rayon, nylon, and polyurethanes—and the synthesis of a wide variety of organic and inorganic compounds. As described in Chapter 6, liquid ammonia is used extensively as a nonaqueous ionizing solvent.

In nature, ammonia is produced by the action of nitrogen-fixing bacteria on atmospheric N_2 under very mild conditions (room temperature and 0.8 atm N_2 pressure). These bacteria contain nitrogenases, iron- and molybdenum-containing enzymes that catalyze the formation of NH_3. Industrially, NH_3 is synthesized from its elements by the Haber–Bosch process, which typically uses finely divided iron as catalyst:

$$N_2 + 3\,H_2 \longrightarrow 2\,NH_3$$

---

[63] F. Kraus, T. Hanauer, and N. Korber, *Inorg. Chem.*, **2006**, *45*, 1117.

[64] W. Massa, R. Kujanek, G. Baum, and K. Dehnicke, *Angew. Chem., Int. Ed.*, **1984**, *23*, 149.

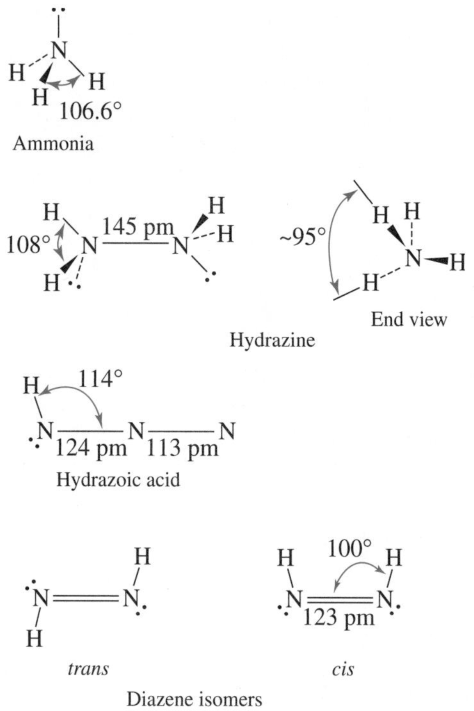

FIGURE 8.30
Nitrogen Hydrides. Some multiple bonds are not shown. (Bond angles and distances are from A. F. Wells, *Structural Inorganic Chemistry*, 5th ed., Oxford University Press, New York, 1984.)

Even with a catalyst, this process is far more difficult than the nitrogenase-catalyzed route in bacteria; typically, temperatures above 380 °C and pressures of approximately 200 atm are necessary. Fritz Haber won the 1918 Nobel Prize for this discovery and is credited both with making commercial fertilizers possible and for helping Germany in World War I to replace imported nitrates used in explosives.

The nitrogen for this process is obtained from fractional distillation of liquid air. Although originally obtained from electrolyis of water, H_2 is now obtained more economically from hydrocarbons (see Section 8.2.1).

Oxidation of hydrazine is highly exothermic:

$$N_2H_4 + O_2 \longrightarrow N_2 + 2\,H_2O \qquad \Delta H^\circ = -622 \text{ kJ mol}^{-1}$$

Advantage has been taken of this reaction in the major use of hydrazine and its methyl derivatives—in rocket fuels. Hydrazine is also a convenient and versatile reducing agent, capable of being oxidized by a wide variety of oxidizing agents, in acidic (as the protonated hydrazonium ion, $N_2H_5^+$) and basic solutions. It may be oxidized by one, two, or four electrons, depending on the oxidizing agent.

Oxidation Reaction	**$\mathscr{E}^\circ$ (Oxidation; V)**	**Examples of Oxidizing Agents**
$N_2H_5^+ \rightleftharpoons NH_4^+ + \frac{1}{2}N_2 + H^+ + e^-$	1.74	MnO_4^-, Ce^{4+}
$N_2H_5^+ \rightleftharpoons \frac{1}{2}NH_4^+ + \frac{1}{2}HN_3 + \frac{5}{2}H^+ + 2\,e^-$	–0.11	H_2O_2
$N_2H_5^+ \rightleftharpoons N_2 + 5\,H^+ + 4\,e^-$	0.23	I_2

Both the *cis* and *trans* isomers of diazene are known; they are unstable except at very low temperatures. The fluoro derivatives, N_2F_2, are more stable and have been characterized structurally. Both isomers of N_2F_2 show N—N distances consistent with double bonds (*cis*, 120.9 pm; *trans*, 122.4 pm).

Phosphine, PH_3, is a highly poisonous gas. It has significantly weaker intermolecular attractions than NH_3 in the solid state; consequently, its melting and boiling points are much lower than those of ammonia (–133.5 °C and –87.5 °C for PH_3 versus –77.8 °C and –34.5 °C for NH_3). Phosphine derivatives of formula PR_3 (phosphines; R = H, alkyl, or aryl) and $P(OR)_3$ (phosphites) are important ligands that form numerous coordination compounds. Examples of phosphine compounds will be discussed in Chapters 13 and 14. Arsines, AsR_3, and stibines, SbR_3, are also important ligands in coordination chemistry.

NITROGEN OXIDES AND OXYIONS Nitrogen oxides and ions containing nitrogen and oxygen are among the most frequently encountered species in inorganic chemistry. The most common of these are summarized in Table 8.10.

Nitrous oxide, N_2O, is commonly used as a mild dental anesthetic and propellant for aerosols; on atmospheric decomposition, it yields its innocuous parent gases and is therefore an environmentally acceptable substitute for chlorofluorocarbons. On the other hand, N_2O contributes to the greenhouse effect and is increasing in the atmosphere. Nitric oxide, NO, is an effective coordinating ligand; its function in this context is discussed in Chapter 13. It also has many biological functions, which are discussed in Chapter 16.

The gases N_2O_4 and NO_2 form an interesting pair. At ordinary temperatures and pressures, both exist in significant amounts in equilibrium:

$$N_2O_4(g) \rightleftharpoons 2\,NO_2(g) \qquad \Delta H^\circ = 57.20\ \text{kJ mol}^{-1}$$

Colorless, diamagnetic N_2O_4 has a weak N—N bond that can readily dissociate to give the brown, paramagnetic NO_2.

Nitric oxide is formed in the combustion of fossil fuels and is present in the exhausts of automobiles and power plants; it can also be formed from the action of lightning on atmospheric N_2 and O_2. In the atmosphere, NO is oxidized to NO_2. These gases, often collectively designated NO_x, contribute to the problem of acid rain, primarily because NO_2 reacts with atmospheric water to form nitric acid:

$$3\,NO_2 + H_2O \longrightarrow 2\,HNO_3 + NO$$

Nitrogen oxides are also believed to be instrumental in the destruction of the Earth's ozone layer, as will be discussed in the following section.

Nitric acid is of immense industrial importance, especially in the synthesis of ammonium nitrate and other chemicals. Ammonium nitrate is used primarily as a fertilizer. In addition, it is thermally unstable and undergoes violently exothermic decomposition at elevated temperature:

$$2\,NH_4NO_3 \longrightarrow 2\,N_2 + O_2 + 4\,H_2O$$

Because this reaction generates a large amount of gas in addition to being strongly exothermic, ammonium nitrate was recognized early as a potentially useful explosive. Its use in commercial explosives is now second in importance to its use as a fertilizer.

Nitric acid is also of interest as a nonaqueous solvent and undergoes the following autoionization:

$$2\,HNO_3 \rightleftharpoons H_2NO_3^{\,+} + NO_3^{\,-} \rightleftharpoons H_2O + NO_2^{\,+} + NO_3^{\,-}$$

TABLE 8.10 Compounds and Ions Containing Nitrogen and Oxygen

Formula	Name	Structure[a]	Notes
N_2O	Nitrous oxide	N=N=O	mp = −90.9 °C; bp = −88.5 °C
NO	Nitric oxide	115 N≡O	mp = −163.6 °C; bp = −151.8 °C; bond order approximately 2.5; paramagnetic
NO_2	Nitrogen dioxide	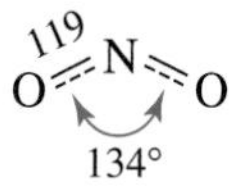 O=N=O (119; 134°)	Brown, paramagnetic gas; exists in equilibrium with N_2O_4; $2\ NO_2 \rightleftharpoons N_2O_4$
N_2O_3	Dinitrogen trioxide	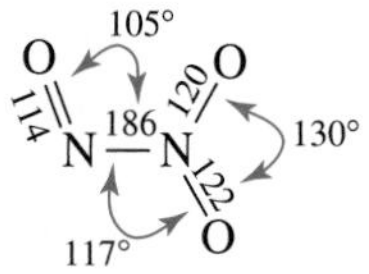 O=N—N(=O)O (105°; 114; 186; 120; 122; 130°; 117°)	mp = −100.1 °C; dissociates above melting point: $N_2O_3 \rightleftharpoons NO + NO_2$
N_2O_4	Dinitrogen tetroxide	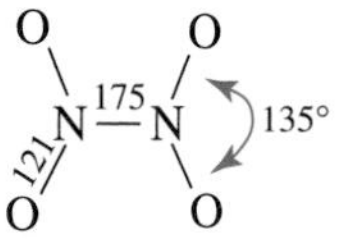 O_2N—NO_2 (175; 121; 135°)	mp = −11.2 °C; bp = −21.15 °C; dissociates into 2 NO_2 [ΔH(dissociation) = 57 kJ/mol]
N_2O_5	Dinitrogen pentoxide	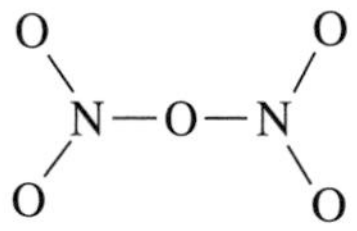 O_2N—O—NO_2	N—O—N bond may be bent; consists of $NO_2^+NO_3^-$ in the solid
NO^+	Nitrosonium or nitrosyl	106 N≡O	Isoelectronic with CO
NO_2^+	Nitronium or nitryl	115 O=N=O	Isoelectronic with CO_2
NO_2^-	Nitrite	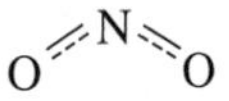 O=N=O (bent)	N—O distance varies from 113 to 123 pm, and bond angle varies from 116° to 132° depending on cation; versatile ligand (see Chapter 9)
NO_3^-	Nitrate	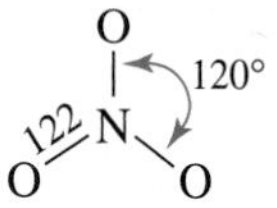 NO_3 (122; 120°)	Forms compounds with nearly all metals; as ligand, has a variety of coordination modes
$N_2O_2^{2-}$	Hyponitrite	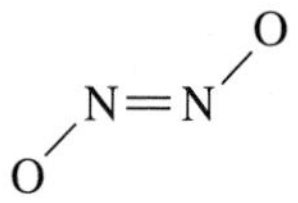 O—N=N—O	Useful reducing agent
NO_4^{3-}	Orthonitrate	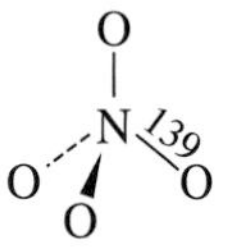 NO_4 (139)	Na and K salts known; decomposes in presence of H_2O and CO_2
HNO_2	Nitrous acid	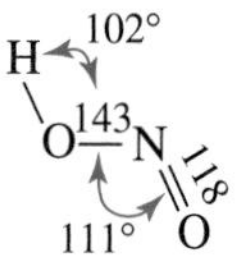 H—O—N=O (102°; 143; 118; 111°)	Weak acid (pK_a = 3.3 at 25 °C); disproportionates: $3\ HNO_2 \rightleftharpoons H_3O^+ + 2\ NO + NO_3^-$ in aqueous solution

Continued

TABLE 8.10 Compounds and Ions Containing Nitrogen and Oxygen—continued

Formula	Name	Structure[a]	Notes
HNO_3	Nitric acid	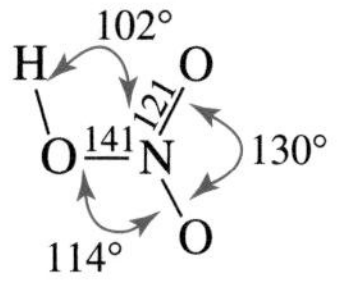 H–O–N(O)O; 102°, 121, 141, 130°, 114°	Strong acid in aqueous solution; concentrated aqueous solutions are strong oxidizing agents

Source: N. N. Greenwood and A. Earnshaw, *Chemistry of the Elements*, Pergamon Press, Elmsford, NY, 1984, pp. 508–545.
NOTES: [a]Distances in pm

Nitric acid is synthesized commercially via two nitrogen oxides. First, ammonia is reacted with oxygen using a platinum-rhodium gauze catalyst to form nitric oxide, NO:

$$4\,NH_3 + 5\,O_2 \longrightarrow 4\,NO + 6\,H_2O$$

The nitric oxide is then oxidized by air and water:

$$2\,NO + O_2 \longrightarrow 2\,NO_2$$

$$3\,NO_2 + H_2O \longrightarrow 2\,HNO_3 + NO$$

The first step, oxidation of NH_3, requires a catalyst that is specific for NO generation; otherwise, oxidation to form N_2 can occur:

$$4\,NH_3 + 3\,O_2 \longrightarrow 2\,N_2 + 6\,H_2O$$

An additional nitrogen oxyanion is peroxynitrite, $ONOO^-$, whose structure has different conformations.[65] The structure of one conformation of $ONOO^-$ is shown in Figure 8.31; a twisted form with different bond angles and a different N—O distance is also found in the crystal. Peroxynitrite may play important roles in cellular defense against infection and in environmental water chemistry.[66]

Nitrogen has a rich redox chemistry in aqueous solution, as shown in the Latimer and Frost diagrams in the following example and in Figure 8.32. The Frost diagram in Figure 8.32 shows that the ammonium ion and elemental N_2 are the most stable nitrogen species in acidic solution, and hydroxylammonium ion, NH_3OH^+, and nitrate ion, NO_3^-, are the least stable.

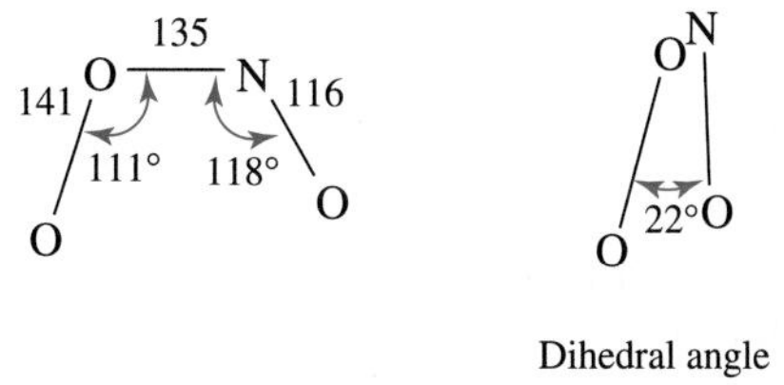

135
141
O
N
116
111°
118°
O
O
N
O
22°
O
O
Dihedral angle

FIGURE 8.31 Peroxynitrite Structure.

[65] M. Wörle, P. Latal, R. Kissner, R. Nesper, and W. H. Koppenol, *Chem. Res. Toxicol.*, **1999**, *12*, 305.
[66] O. V. Gerasimov and S. V. Lymar, *Inorg. Chem.*, **1999**, *38*, 4317; *Chem. Res. Toxicol.*, **1998**, *11*, 709.

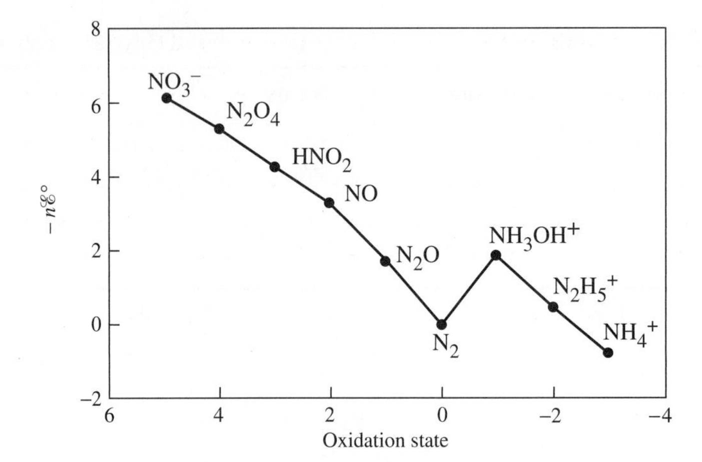

FIGURE 8.32 Frost Diagram for Nitrogen Compounds in Acid.

EXAMPLE

Combining half reactions to find the potentials for other reactions depends on the fact that free energies are additive, but potentials may not be. If an oxidation reaction and a reduction reaction add to give a balanced reaction (no electrons in the final reaction), the potentials are additive. If the result is another half reaction, the potentials are not additive, and the $n\mathscr{E}°$ values—proportional to free energy, which is additive—must be used. Part of the Latimer diagram for nitrogen in acidic solution is

$$\overset{+5}{NO_3^-} \xrightarrow{0.80} \overset{+4}{NO_2} \xrightarrow{1.07} \overset{+3}{HNO_2}$$

$$NO_3^- \xrightarrow{?} HNO_2$$

To calculate $\mathscr{E}°$ for the conversion of NO_3^- to HNO_2, it is necessary to find the change in free energy for each step:

	$\mathscr{E}°$	$\Delta G° = -n\mathscr{F}\mathscr{E}°$
$NO_3^- + e^- \longrightarrow NO_2$	0.80 V	$-(1)(\mathscr{F})(0.80\ V) = -0.80\ V\mathscr{F}$
$NO_2 + e^- \longrightarrow HNO_2$	1.07 V	$-(1)(\mathscr{F})(1.07\ V) = -1.07\ V\mathscr{F}$
$NO_3^- + 2\,e^- \longrightarrow HNO_2$		$-(2)\mathscr{F}\mathscr{E}° = -1.87\ V\mathscr{F}$

$$\text{Two electrons are transferred overall: } \mathscr{E}° = \frac{-1.87\ V\mathscr{F}}{-2\mathscr{F}} = -0.94\ V$$

▶ **Exercise 8.2** Use the same approach to find the potential for the $NO \longrightarrow N_2O$ reaction, given the following:

$$NO \longrightarrow N_2 \quad \mathscr{E}° = 1.68\ V$$

$$N_2O \longrightarrow N_2 \quad \mathscr{E}° = 1.77\ V$$

▶ **Exercise 8.3** Show whether the decomposition of NH_4NO_3 can be a spontaneous reaction, based on the potentials given in Appendix B.7.

Among all the acids, phosphoric acid, H_3PO_4, is second only to sulfuric acid in industrial production. Two methods are commonly used. The first of these involves the combustion of molten phosphorus, sprayed into a mixture of air and steam in a stainless steel chamber. The P_4O_{10} formed initially is converted into H_3PO_4:

$$P_4 + 5\,O_2 \longrightarrow P_4O_{10}$$

$$P_4O_{10} + 6\,H_2O \longrightarrow 4\,H_3PO_4$$

Alternatively, phosphoric acid is made by treating phosphate minerals with sulfuric acid. For example,

$$Ca_3(PO_4)_2 + 3\,H_2SO_4 \longrightarrow 2\,H_3PO_4 + 3\,CaSO_4$$

8.8 GROUP 16 (VIA)

8.8.1 The Elements

The first two elements of this group, occasionally designated the *chalcogen* group, are familiar as O_2, the colorless gas that makes up about 21 percent of Earth's atmosphere, and sulfur, a yellow solid of typical nonmetallic properties. The third element in this group, selenium, is perhaps not as well known but is important in the xerography process. A brilliant red formed by a combination of CdS and CdSe is used in colored glasses. Although elemental selenium is highly poisonous, trace amounts of the element are essential for life. Tellurium is of less commercial interest but is used in small amounts in metal alloys, tinting of glass, and catalysts in the rubber industry. All isotopes of polonium, a metal, are radioactive. The highly exothermic radioactive decay of this element has made it a useful power source for satellites.

Sulfur, which occurs as the free element in numerous natural deposits, has been known since prehistoric times; it is the "brimstone" of the Bible. Sulfur was of considerable interest to alchemists and, following the development of gunpowder (a mixture of sulfur, KNO_3, and powdered charcoal) in the thirteenth century, to military leaders as well. Although oxygen is widespread in Earth's atmosphere and, combined with other elements, in Earth's crust (which contains 46 percent oxygen by mass) and in bodies of water, the pure element was not isolated and, characterized until the 1770s by C. W. Scheele and J. Priestley. Priestley's classic synthesis of oxygen by heating HgO with sunlight focused by a magnifying glass was a landmark in the history of experimental chemistry. Selenium (1817) and tellurium (1782) were soon discovered and, because of their chemical similarities, were named after the moon (Greek, *selene*) and Earth (Latin, *tellus*). Polonium was discovered by Marie Curie in 1898; like radium, it was isolated in trace amounts from tons of uranium ore. Some important physical properties of these elements are summarized in Table 8.11.

OXYGEN Oxygen exists primarily in the diatomic form O_2, but traces of ozone, O_3, are found in the upper atmosphere and in the vicinity of electrical discharges. O_2 is paramagnetic and O_3 is diamagnetic. As discussed in Chapter 5, the paramagnetism of O_2 is the

TABLE 8.11 Properties of the Group 16 (VIA) Elements

Element	Ionization Energy ($kJ\ mol^{-1}$)	Electron Affinity ($kJ\ mol^{-1}$)	Melting Point (°C)	Boiling Point (°C)	Electro-negativity
O	1314	141	−218.8	−183.0	3.610
S	1000	200	112.8	444.7	2.589
Se	941	195	217	685	2.424
Te	869	190	452	990	2.158
Po	812	180[a]	250[a]	962	2.19[a]

Source: See Table 8.3
NOTE: [a]Approximate value

consequence of two electrons with parallel spin occupying $\pi^*(2p)$ orbitals. In addition, the two known excited states of O_2 have π^* electrons of opposite spin and are higher in energy as a consequence of the effects of pairing energy and exchange energy (see Section 2.2.3):

		Relative Energy ($kJ\ mol^{-1}$)
Excited states:	↑↓ ___	157.85
	↑ ___ ↓	94.72
Ground state:	↑ ___ ↑	0

The excited states of O_2 can be achieved when photons are absorbed in the liquid phase during molecular collisions; under these conditions, a single photon can simultaneously excite two colliding molecules. This absorption occurs in the visible region of the spectrum, at 631 and 474 nm, and gives rise to the blue color of the liquid.[67] The excited states are also important in many oxidation processes. Of course, O_2 is essential for respiration. The mechanism for oxygen transport to the cells via hemoglobin has received much attention and will be discussed briefly in Chapter 16.

Molecules still reveal surprises. Although the phases of molecular oxygen have been studied for many years, not until 2006 was it found that the molecular structure of one of the high-pressure phases, ε-O_2, consists of tetrameric $(O_2)_4$, or O_8, molecules.[68] A dark red allotrope observed at pressures between 10 and 96 GPa, O_8 consists of collections of parallel O_2 units, arranged as shown in Figure 8.33. The O_2 units have bond distances of 120(3) pm, comparable to the oxygen–oxygen distance in free O_2, 120.8 pm; the oxygen–oxygen distances between O_2 units are nearly an Ångstrom greater.

[67] E. A. Ogryzlo, *J. Chem. Educ.*, **1965**, *42*, 647.
[68] L. F. Lundegaard, G. Weck, M. I. McMahon, S. Desgreniers, and P. Loubeyre, *Nature*, **2006**, *443*, 201.

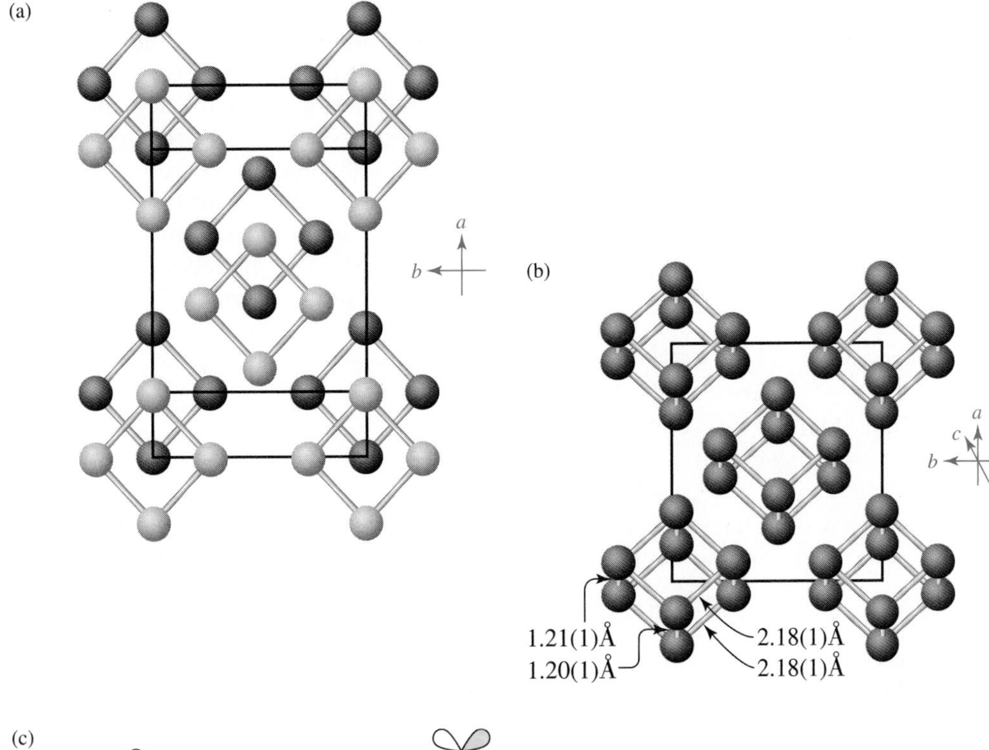

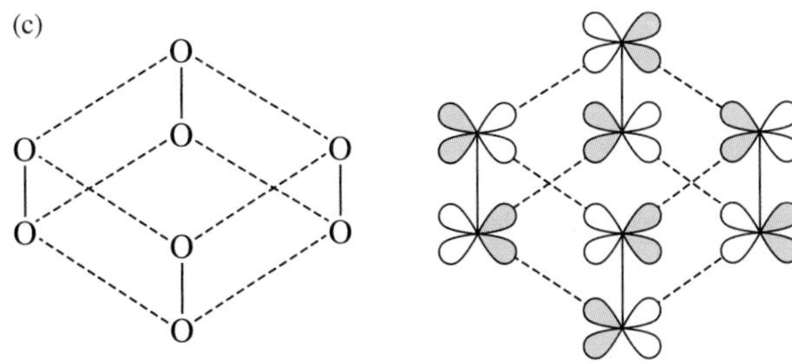

FIGURE 8.33 O_8. Parts (a) and (b) show two perspectives of the crystalline arrangement. An example of orbital interactions involving *p* orbitals is shown in (c). Reprinted by permission from Macmillan Publishers Ltd: Lars F. Lundegaard, Gunnar Weck, Malcolm I. McMahon, Serge Desgreniers, and Paul Loubeyre, *Nature*, **2006**, *443*, 203. Copyright 2006.

Bonding between O_2 units can be described as involving the π^* orbitals of O_2. Recall that these orbitals are singly occupied (Figure 5.7), accounting for O_2's paramagnetism. Overlap of the π^* orbitals in O_8, involving two such orbitals per O_2, generates eight new orbitals, four bonding and four antibonding. One example of such interactions is shown in Figure 8.33. The eight available electrons can then occupy the bonding orbitals, pairing all the electrons. This analysis is consistent with the diamagnetism observed for O_8.[69]

Ozone absorbs ultraviolet radiation below 320 nm. It thus forms an indispensable shield in the upper atmosphere, protecting Earth's surface from most of the potentially hazardous effects of such high-energy electromagnetic radiation. There is now increasing concern because atmospheric pollutants are depleting the ozone layer worldwide, with the most serious depletion over Antarctica as a result of seasonal variations in high-altitude air circulation. In the upper atmosphere, ozone is formed from O_2:

$$O_2 \xrightarrow{h\nu} 2\,O \qquad \lambda \leq 242 \text{ nm}$$

$$O + O_2 \longrightarrow O_3$$

[69] R. Steudel and M. W. Wong, *Angew. Chem. Int. Ed.*, **2007**, *46*, 1768.

Absorption of ultraviolet radiation by O_3 causes it to decompose to O_2. In the upper atmosphere, therefore, a steady-state concentration of ozone is achieved, a concentration ordinarily sufficient to provide significant ultraviolet protection of Earth's surface. However, some pollutants in the upper atmosphere catalyze the decomposition of ozone. Examples include nitrogen oxides from high-flying aircraft (nitrogen oxides are also produced in trace amounts naturally) and chlorine atoms from photolytic decomposition of chlorofluorocarbons from aerosols, refrigerants, and other sources. The overall processes governing the concentration of ozone in the atmosphere are extremely complex. The following reactions can be studied in the laboratory and are examples of the processes believed to be involved in the atmosphere:

$$\begin{aligned} NO_2 + O_3 &\longrightarrow NO_3 + O_2 \\ NO_3 &\longrightarrow NO + O_2 \\ NO + O_3 &\longrightarrow NO_2 + O_2 \\ \hline \text{Net:} \quad 2\,O_3 &\longrightarrow 3\,O_2 \end{aligned}$$

$$\begin{aligned} Cl + O_3 &\longrightarrow ClO + O_2 \\ ClO + O &\longrightarrow Cl + O_2 \\ \hline \text{Net:} \quad O_3 + O &\longrightarrow 2\,O_2 \end{aligned}$$

Cl formed from photodecomposition of chlorofluorocarbons

Ozone is a more potent oxidizing agent than O_2; in acidic solution, ozone is exceeded only by fluorine among the elements as an oxidizing agent.

Several diatomic and triatomic oxygen ions are known and are summarized in Table 8.12.

SULFUR More allotropes are known for sulfur than for any other element, with the most stable form at room temperature (orthorhombic, α-S_8) having eight sulfur atoms arranged in a puckered ring. Two of the most common sulfur allotropes are shown in Figure 8.34.[70]

TABLE 8.12 Neutral and Ionic O_2 and O_3 Species

Formula	Name	O—O Distance (pm)	Notes
O_2^+	Dioxygenyl	111.6	Bond order 2.5
O_2	Dioxygen	120.8	Coordinates to transition metals; singlet O_2 (excited state) important in photochemical reactions; oxidizing agent
O_2^-	Superoxide	135	Moderate oxidizing agent; most stable compounds are KO_2, RbO_2, CsO_2
O_2^{2-}	Peroxide	149	Forms ionic compounds with alkali metals, Ca, Sr, Ba; strong oxidizing agent
O_3	Ozone	127.8	Bond angle 116.8°; strong oxidizing agent; absorbs in UV (below 320 nm)
O_3^-	Ozonide	134	Formed from reaction of O_3 with dry alkali metal hydroxides, decomposes to O_2^-

Source: N. N. Greenwood and A. Earnshaw, *Chemistry of the Elements*, Pergamon Press, Elmsford, NY, 1984; K. P. Huber and G. Herzberg, *Molecular Spectra and Molecular Structure. IV. Constants of Diatomic Molecules*, Nostrand Reinhold Company, New York, 1979.

[70] B. Meyer, *Chem. Rev.*, **1976**, *76*, 367.

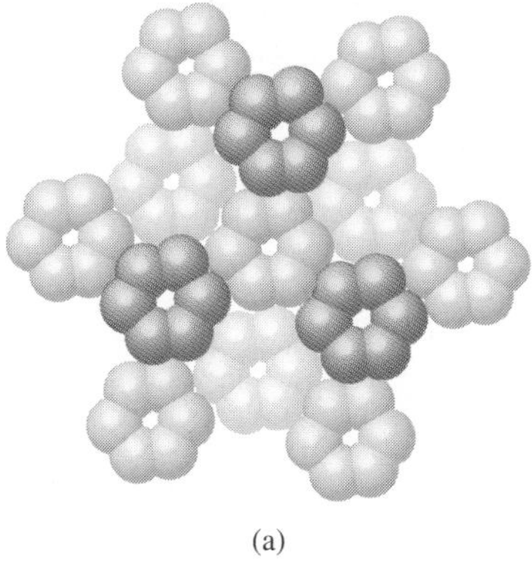

(a)

(b)

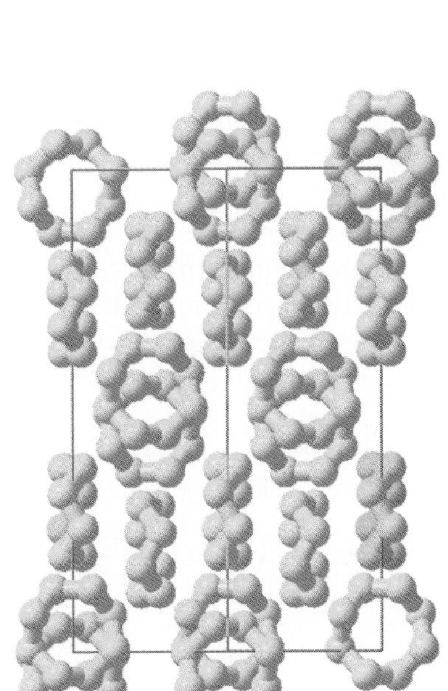

(c)

FIGURE 8.34 Allotropes of Sulfur. (a) S_6; (b) and (c) α-S_8, two different views. Figures in (a) and (b) reproduced with permission from M. Schmidt and W. Siebert, "Sulphur," in J. C. Bailar, Jr., H. C. Emeléus, R. Nyholm, and A. F. Trotman-Dickinson, eds., *Comprehensive Inorganic Chemistry*, vol. 2, Pergamon Press, Elmsford, NY, 1973, pp. 804, 806. © 1973, Pergamon Press PLC.)

Heating sulfur results in interesting changes in viscosity. At approximately 119 °C, sulfur melts to give a yellow liquid, whose viscosity gradually decreases because of greater thermal motion until approximately 155 °C (Figure 8.35). Further heating causes the viscosity to increase, dramatically so above 159 °C, until the liquid pours very sluggishly. Above about 200 °C, the viscosity again decreases, with the liquid eventually acquiring a reddish hue at higher temperatures.[71]

The explanation of these changes in viscosity involves the tendency of S—S bonds to break and to reform at high temperatures. Above 159 °C, the S_8 rings begin to open; the resulting S_8 chains can react with other S_8 rings to open them and form S_{16} chains, S_{24} chains, and so on:

$$S_8 \text{ (ring)} \longrightarrow \text{S–S–S–S–S–S–S–S (chain)} \longrightarrow S_{16} \longrightarrow S_{24} \longrightarrow \cdots$$

The longer the chains, the greater the viscosity (the more the chains can intertwine with each other). Large rings can also form by the linking of ends of chains. Chains exceeding 200,000 sulfur atoms are formed at the temperature of maximum viscosity, near 180 °C. At higher temperatures, thermal breaking of sulfur chains occurs more rapidly than

[71] W. N. Tuller, ed., *The Sulphur Data Book*, McGraw-Hill, New York, 1954.

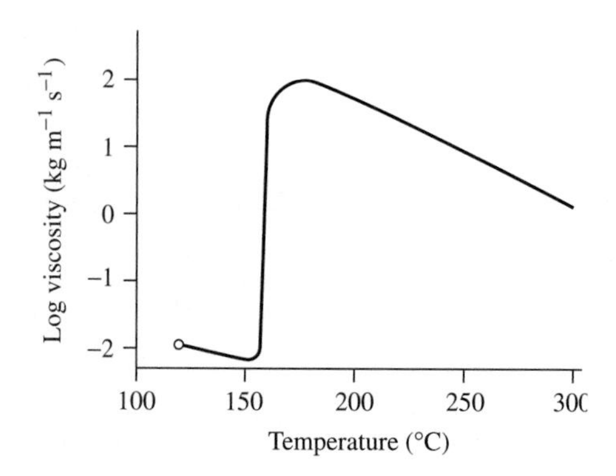

FIGURE 8.35 The Viscosity of Sulfur.

propagation of chains, and the average chain length decreases, accompanied by a decrease in viscosity. At very high temperatures, brightly colored species such as S_3 increase in abundance, and the liquid takes on a reddish coloration. When molten sulfur is poured into cold water, it forms a rubbery solid that can be molded readily. However, this form eventually converts to the yellow crystalline α form, the most thermodynamically stable allotrope, which consists again of the S_8 rings.

Sulfuric acid, produced in greater amounts than any other chemical, has been manufactured commercially for approximately 400 years. The modern process for producing H_2SO_4 begins with the synthesis of SO_2, either by combustion of sulfur or by roasting (heating in the presence of oxygen) of sulfide minerals:

$$S + O_2 \longrightarrow SO_2 \qquad \text{Combustion of sulfur}$$
$$M_xS_y + O_2 \longrightarrow y\,SO_2 + M_xO_y \qquad \text{Roasting of sulfide ore}$$

SO_2 is then converted to SO_3 by the exothermic reaction

$$2\,SO_2 + O_2 \longrightarrow 2\,SO_3$$

using V_2O_5 or another suitable catalyst in a multiple-stage catalytic converter (multiple stages are necessary to achieve high yields of SO_3). The SO_3 then reacts with water to form sulfuric acid:

$$SO_3 + H_2O \longrightarrow H_2SO_4$$

If SO_3 is passed directly into water, a fine aerosol of H_2SO_4 droplets is formed. To avoid this, the SO_3 is absorbed into 98 percent H_2SO_4 solution to form disulfuric acid, $H_2S_2O_7$ (oleum):

$$SO_3 + H_2SO_4 \longrightarrow H_2S_2O_7$$

The $H_2S_2O_7$ is then mixed with water to form sulfuric acid:

$$H_2S_2O_7 + H_2O \longrightarrow 2\,H_2SO_4$$

Sulfuric acid is a dense (1.83 g cm^{-3}) viscous liquid that reacts very exothermically with water. When concentrated sulfuric acid is diluted with water, it is therefore essential to add the acid carefully to the water; adding water to the acid is likely to lead to spattering, because the solution at the top may boil. Sulfuric acid also has a high affinity for water. For example, it causes sugar to char by removing water, leaving carbon behind, and it can cause rapid and serious burns to human tissue.

Anhydrous H_2SO_4 undergoes significant autoionization:

$$2\,H_2SO_4 \rightleftharpoons H_3SO_4^+ + HSO_4^- \qquad K = 2.7 \times 10^{-4} \text{ at } 25\,°\text{C}$$

Many compounds and ions containing sulfur and oxygen are known; many of these are important acids or conjugate bases. Some useful information about these compounds and ions is summarized in Table 8.13.

OTHER ELEMENTS Selenium, a highly poisonous element, and tellurium also exist in a variety of allotropic forms, whereas polonium, a radioactive element, exists in two metallic allotropes. Selenium is a photoconductor—a poor conductor ordinarily, but a

TABLE 8.13 Molecules and Ions Containing Sulfur and Oxygen

Formula	Name	Structure[a]	Notes
SO_2	Sulfur dioxide	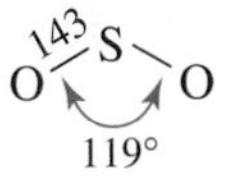 S–O 143; O–S–O 119°	mp = −75.5 °C; bp = −10.0 °C; colorless, choking gas; product of combustion of elemental sulfur
SO_3	Sulfur trioxide	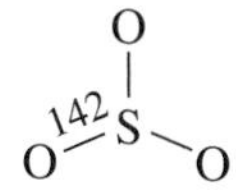 S–O 142	mp = −16.9 °C; bp = 44.6 °C; formed from oxidation of SO_2: $SO_2 + \frac{1}{2} O_2 \longrightarrow SO_3$; in equilibrium with trimer S_3O_9 in liquid and gas phases; reacts with water to form sulfuric acid
	Trimer	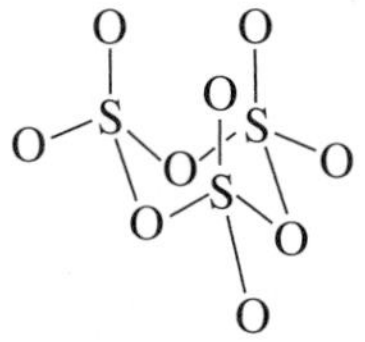	
SO_3^{2-}	Sulfite	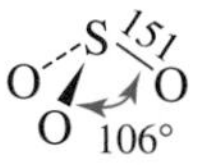 S–O 151; O–S–O 106°	Conjugate base of HSO_3^-, formed when SO_2 dissolves in water
SO_4^{2-}	Sulfate	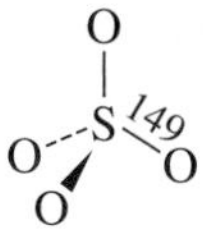 S–O 149	T_d symmetry, extremely common ion, used in gravimetric analysis
$S_2O_3^{2-}$	Thiosulfate	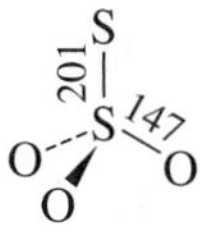 S–S 201; S–O 147	Moderate reducing agent, used in analytical determination of I_2: $I_2 + 2\, S_2O_3^{2-} \longrightarrow 2\, I^- + S_4O_6^{2-}$
$S_2O_4^{2-}$	Dithionite	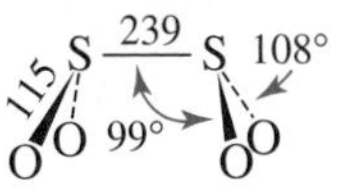 S–S 239; S–O 115; 108°; 99°	Very long S—S bond; dissociates into SO_2^-: $S_2O_4^{2-} \rightleftharpoons 2\, SO_2^-$; Zn and Na salts used as reducing agents
$S_2O_8^{2-}$	Peroxodisulfate	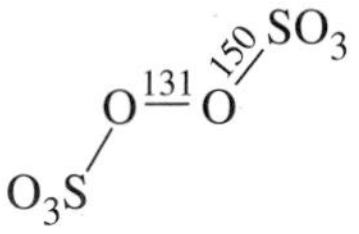 O_3S–O–O–SO_3; O–O 131; O–S 150	Useful oxidizing agent, readily reduced to sulfate: $S_2O_8^{2-} + 2\, e^- \rightleftharpoons 2\, SO_4^{2-}$, $\mathscr{E}° = 2.01$ V
H_2SO_4	Sulfuric acid	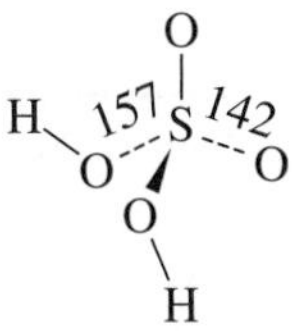 S–OH 157; S–O 142	C_2 symmetry; mp = 10.4 °C; bp = ~300 °C (dec); strong acid in aqueous solution; undergoes autoionization: $2\, H_2SO_4 \rightleftharpoons H_3SO_4^+ + HSO_4^-$, $pK = 3.57$ at 25 °C

Source: N. N. Greenwood and A. Earnshaw, *Chemistry of the Elements*, Pergamon Press, Elmsford, NY, 1984, pp. 821–854.

NOTE: [a]Distances in pm

good conductor in the presence of light. It is used extensively in xerography, photoelectric cells, and semiconductor devices.

8.9 GROUP 17 (VIIA): THE HALOGENS

8.9.1 The Elements

Compounds containing the halogens (Greek, *halos* + *gen*, "salt former") have been used since antiquity, with the first use probably that of rock or sea salt (primarily NaCl) as a food preservative. Isolation and characterization of the neutral elements, however, has occurred comparatively recently.[72] Chlorine was first recognized as a gas by J. B. van Helmont in approximately 1630, first studied carefully by C. W. Scheele in the 1770s (hydrochloric acid, which was used in these early syntheses, had been prepared by the alchemists around the year 900 C.E.). Iodine was next, obtained by Courtois in 1811 by subliming the product of the reaction of sulfuric acid with seaweed ash. A. J. Balard obtained bromine in 1826 by reacting chlorine with $MgBr_2$, which was present in saltwater marshes. Although hydrofluoric acid had been used to etch glass since the latter part of the seventeenth century, elemental fluorine was not isolated until 1886, when H. Moissan obtained a small amount of the very reactive gas by the electrolysis of KHF_2 in anhydrous HF. Astatine, one of the last of the nontransuranium elements to be produced, was first synthesized in 1940 by D. R. Corson, K. R. Mackenzie, and E. Segre by bombardment of ^{209}Bi with alpha particles. All isotopes of astatine are radioactive (the longest lived isotope has a half-life of 8.1 hours) and, consequently, the chemistry of this element has been studied only with the greatest difficulty.

All neutral halogens are diatomic and readily reduced to halide ions. All combine with hydrogen to form gases that, except for HF, are strong acids in aqueous solution. Some physical properties of the halogens are summarized in Table 8.14.

The chemistry of the halogens is governed in large part by their tendency to acquire an electron to attain a noble gas electron configuration. Consequently, the halogens are excellent oxidizing agents, with F_2 being the strongest oxidizing agent of all the

TABLE 8.14 Properties of the Group 17 (VIIA) Elements: The Halogens[a]

				Halogen Molecules, X_2			
Element	Ionization Energy (kJ mol^{-1})	Electron Affinity (kJ mol^{-1})	Electronegativity	Melting Point (°C)	Boiling Point (°C)	X—X Distance (pm)	ΔH of Dissociation (kJ mol^{-1})
F	1681	328	4.193	−218.6	−188.1	143	158.8
Cl	1251	349	2.869	−101.0	−34.0	199	242.6
Br	1140	325	2.685	−7.25	59.5	228	192.8
I	1008	295	2.359	113.6[a]	185.2	266	151.1
At	930[b]	270[b]	2.39[b]	302[b]			

Source: See Table 8.3. Ionization energy for At is from J. Emsley, *The Elements*, Oxford University Press, New York, 1989, p. 23.

NOTES: [a]Sublimes readily

[b]Approximate value

[72] M. E. Weeks, "The Halogen Family," in *Discovery of the Elements*, 7th ed, revised by H. M. Leicester, Journal of Chemical Education, Easton, PA, 1968, pp. 701–749.

elements. The tendency of the halogen atoms to attract electrons is also shown in their high electron affinities and electronegativities.

F_2 is extremely reactive and cannot be handled except by special techniques; it is ordinarily prepared by electrolysis of molten fluorides such as KF. Cl_2 is a yellow gas and has an odor that is recognizable as the characteristic scent of "chlorine" bleach (an alkaline solution of the hypochlorite ion, ClO^-, which exists in equilibrium with small amounts of Cl_2). Br_2 is a dark red liquid that evaporates easily and is also a strong oxidizing agent. I_2 is a black, lustrous solid, readily sublimable at room temperature to produce a purple vapor, and, like the other halogens, highly soluble in nonpolar solvents. The color of iodine solutions varies significantly with the donor ability of the solvent, typically giving vivid colors as a consequence of charge-transfer interactions, as described in Chapter 6. Iodine is also a moderately good oxidizing agent but the weakest of the halogens. Solutions of iodine in alcohol, commonly labeled "tincture of iodine," are commonly used as a household antiseptic. Because of its radioactivity, astatine has not been studied extensively; it would be interesting to be able to compare its properties and reactions with those of the other halogens.

Several trends in physical properties of the halogens are immediately apparent, as can be seen in Table 8.14. As the atomic number increases, the ability of the nucleus to attract outermost electrons decreases; consequently, fluorine is the most electronegative and has the highest ionization energy, and astatine is lowest in both properties. With increasing size and number of electrons of the diatomic molecules in going down the periodic table, the London interactions between the molecules increase: F_2 and Cl_2 are gases, Br_2 is a liquid, and I_2 is a solid as a consequence of these interactions. The trends are not entirely predictable, because fluorine and its compounds exhibit some behavior that is substantially different than would be predicted by extrapolation of the characteristics of the other members of the group.

One of the most striking properties of F_2 is its remarkably low bond dissociation enthalpy, an important factor in the high reactivity of this molecule. Extrapolation from the bond dissociation enthalpies of the other halogens would yield a value of approximately 290 kJ mol^{-1}, nearly double the actual value. Several suggestions have been made to account for this low value. It is likely that the weakness of the F—F bond is largely a consequence of repulsions between the nonbonding electron pairs.[73] The small size of the fluorine atom brings these pairs into close proximity when F—F bonds are formed. Electrostatic repulsions between these pairs on neighboring atoms result in weaker bonding and an equilibrium bond distance significantly greater than would be expected in the absence of such repulsions.

For example, the covalent radius obtained for other compounds of fluorine is 64 pm; an F—F distance of 128 pm would therefore be expected in F_2. However, the actual distance is 143 pm. In this connection, it is significant that oxygen and nitrogen share similar anomalies with fluorine; the O—O bonds in peroxides and the N—N bonds in hydrazines are longer than the sums of their covalent radii, and these bonds are weaker than the corresponding S—S and P—P bonds in the respective groups of these elements. In the case of oxygen and nitrogen, it is likely that the repulsion of electron pairs on neighboring atoms also plays a major role in the weakness of these bonds.[74] The weakness of the fluorine–fluorine bond, in combination with the small size and high electronegativity of fluorine, account in large part for the very high reactivity of F_2.

---

[73] J. Berkowitz and A. C. Wahl, *Adv. Fluorine Chem.*, **1973**, *7*, 147. See also R. Ponec and D. L. Cooper, *J. Phys. Chem. A*, **2007**, *111*, 11294 and references cited therein.

[74] Anomalous properties of fluorine, oxygen, and nitrogen have been discussed by P. Politzer in *J. Am. Chem. Soc.*, **1969**, *91*, 6235, and *Inorg. Chem.*, **1977**, *16*, 3350.

Of the hydrohalic acids, HF is by far the weakest in aqueous solution ($pK_a = 3.2$ at 25 °C); HCl, HBr, and HI are all strong acids. Although HF reacts with water, strong hydrogen bonding occurs between F^- and the hydronium ion ($F^- — H^+ — OH_2$) to form the ion pair $H_3O^+F^-$, reducing the activity coefficient of H_3O^+. As the concentration of HF increases, however, its tendency to form H_3O^+ increases as a result of further reaction of this ion pair with HF:

$$H_3O^+F^- + HF \rightleftharpoons H_3O^+ + HF_2^-$$

This view is supported by X-ray crystallographic studies of the ion pairs $H_3O^+F^-$ and $H_3O^+F_2^-$.[75]

Chlorine and chlorine compounds are used as bleaching and disinfecting agents in many industries. Perhaps the most commonly known of these compounds is hypochlorite, OCl^-, a common household bleach prepared by dissolving chlorine gas in sodium or calcium hydroxide:

$$Cl_2 + 2\,OH^- \longrightarrow Cl^- + ClO^- + H_2O$$

The redox potentials supporting this reaction and others are shown in the Frost diagram in Figure 8.36. The disproportionation of Cl_2 to Cl^- and OCl^- in basic solution can be seen in Figure 8.36, because Cl_2 is above the line between Cl^- and OCl^-. The free-energy change from Cl_2 to OCl^- is positive (higher on the $-n\mathscr{E}°$ scale), but the free-energy change from Cl_2 to Cl^- is negative and larger in magnitude, resulting in a net negative free-energy change and a spontaneous reaction. The oxidizing power of the higher oxidation number species in acid is also evident. Perchlorate is an extremely strong oxidizing agent, and ammonium perchlorate is used as a rocket fuel. In the fall of 2001, chlorine dioxide, ClO_2, was used to disinfect U.S. mail and at least one congressional office that may have been infected with anthrax. This gas is also used as an alternative to Cl_2 for purifying drinking water and as a bleaching agent in the paper industry.

POLYATOMIC IONS In addition to the common monatomic halide ions, numerous polyatomic species, both cationic and anionic, have been prepared. Many readers will be familiar with the brown triiodide ion, I_3^-, formed from I_2 and I^-:

$$I_2 + I^- \rightleftharpoons I_3^- \qquad K \approx 698 \text{ at } 25\text{ °C in aqueous solution}$$

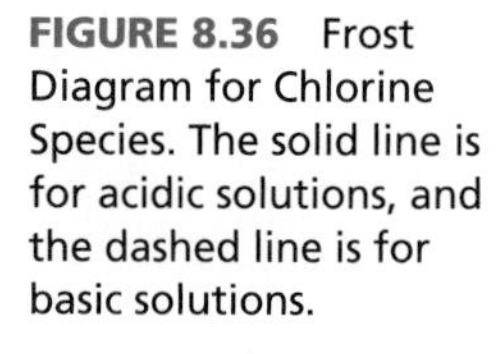

FIGURE 8.36 Frost Diagram for Chlorine Species. The solid line is for acidic solutions, and the dashed line is for basic solutions.

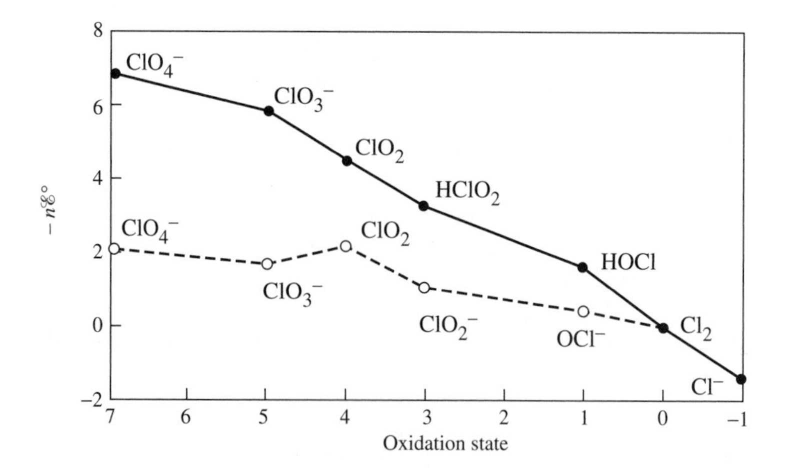

[75] D. Mootz, *Angew. Chem., Int. Ed.*, **1981**, *20*, 791.

Many other polyiodide ions have been characterized; in general, these may be viewed as aggregates of I_2 and I^- (sometimes I_3^-). Examples are shown in Figure 8.37.

The halogens Cl_2, Br_2, and I_2 can also be oxidized to cationic species. Examples include the diatomic ions Br_2^+ and I_2^+ (Cl_2^+ has been characterized in low-pressure discharge tubes but is much less stable), I_3^+ and I_5^+. I_2^+ dimerizes into I_4^{2+}:

$$2\,I_2^+ \rightleftharpoons I_4^{2+}$$

INTERHALOGENS Halogens form many compounds containing two or more different halogens. Like the halogens themselves, these may be diatomic, such as ClF, or polyatomic, such as ClF_3, BrF_5, or IF_7. In addition, polyatomic ions containing two or more halogens have been synthesized for many of the possible combinations. Selected neutral (boxed) and ionic interhalogen species are listed in Table 8.15. The effect of the size of the central atom can readily be seen, with iodine the only element able to have up to seven fluorine atoms in a neutral molecule, whereas chlorine and bromine have a maximum of five fluorines. The effect of size is also evident in the ions, with iodine being the only halogen large enough to exhibit ions of formula XF_6^+ and XF_8^-.

Neutral interhalogens can be prepared in a variety of ways, including by direct reaction of the elements (the favored product often depending on the ratio of halogens used) and reaction of halogens with metal halides or other halogenating agents. Examples include the following:

$$Cl_2 + F_2 \longrightarrow 2\,ClF \qquad T = 225\ °C$$

$$I_2 + 5\,F_2 \longrightarrow 2IF_5 \qquad \text{Room temperature}$$

$$I_2 + 3\,XeF_2 \longrightarrow 2\,IF_3 + 3\,Xe \qquad T < -30\ °C$$

$$I_2 + AgF \longrightarrow IF + AgI \qquad 0\ °C$$

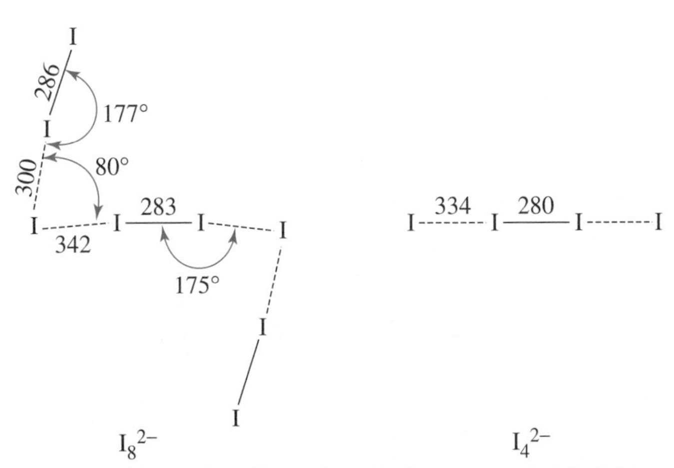

FIGURE 8.37 Polyiodide Ions. (Bond angles and distances, in pm, are from A. F. Wells, *Structural Inorganic Chemistry*, 5th ed., Oxford University Press, New York, 1984, pp. 396–399.)

*Distances in triiodide vary depending on the cation. In some cases both I—I distances are identical, but in the majority of cases they are different. Differences in I—I distances as great as 33 pm have been reported.

TABLE 8.15 Interhalogen Species

Formal Oxidation State of Central Atom	Number of Lone Pairs on Central Atom	Compounds and Ions						
+7	0			IF_7				
				IF_6^+				
				IF_8^-				
+5	1	ClF_5	BrF_5	IF_5				
		ClF_4^+	BrF_4^+	IF_4^+				
			BrF_6^-	IF_6^-				
+3	2	ClF_3	BrF_3	IF_3		I_2Cl_6		
		ClF_2^+	BrF_2^+	IF_2^+		ICl_2^+	IBr_2^+	$IBrCl^+$
		ClF_4^-	BrF_4^-	IF_4^-		ICl_4^-		
+1	3	ClF	BrF	IF	$BrCl$	ICl	IBr	
		ClF_2^-	BrF_2^-	IF_2^-	$BrCl_2^-$	ICl_2^-	IBr_2^-	
					Br_2Cl^-	I_2Cl^-	I_2Br^-	$IBrCl^-$

Interhalogens can also serve as intermediates in the synthesis of other interhalogens:

$$ClF + F_2 \longrightarrow ClF_3 \qquad T = 200\text{ °C to }300\text{ °C}$$
$$ClF_3 + F_2 \longrightarrow ClF_5 \qquad h\nu\text{, room temperature}$$

Several interhalogens undergo autoionization in the liquid phase and have been studied as nonaqueous solvents. Examples of these are

$$3\ IX \rightleftharpoons I_2X^+ + IX_2^- \quad (X = Cl, Br)$$
$$2\ BrF_3 \rightleftharpoons BrF_2^+ + BrF_4^-$$
$$I_2Cl_6 \rightleftharpoons ICl_2^+ + ICl_4^-$$
$$IF_5 \rightleftharpoons IF_4^+ + IF_6^-$$

Examples of acid–base reactions in the autoionizing solvents BrF_3 and IF_5 have been discussed in Chapter 6.

PSEUDOHALOGENS Parallels have been observed between the chemistry of the halogens and a number of other dimeric species. Dimeric molecules showing considerable similarity to the halogens are often called **pseudohalogens**. Some of the most important parallels in chemistry between the halogens and pseudohalogens include those illustrated for chlorine in Table 8.16.

For example, there are many similarities between the halogens and cyanogen, NCCN. The monoanion, CN^-, is, of course, well known; it combines with hydrogen to form the weak acid HCN and, with Ag^+ and Pb^{2+}, to form precipitates of low solubility in water. Interhalogen compounds such as FCN, ClCN, BrCN, and ICN are all known. Cyanogen, like the halogens, can add across double or triple carbon–carbon bonds. The pseudohalogen idea is a useful classification tool, although not many cases are known in which all six characteristics are satisfied. Some examples of pseudohalogens are given in Table 8.16.[76]

[76] For additional examples of pseudohalogens, see J. Ellis, *J. Chem. Educ.*, **1976**, *53*, 2.

TABLE 8.16 Pseudohalogens

Characteristics	Examples[a]		
Neutral diatomic species	Cl_2	$(CN)_2$	$[Co(CO)_4]_2$
Ion of 1− charge	Cl^-	CN^-	$[Co(CO)_4]^-$
Formation of hydrohalic acids	HCl	HCN	$HCo(CO)_4$ (strong)[b]
Formation of interhalogen compounds	ICl, BrCl, ClF	$Cl_2 + (CN)_2 \longrightarrow 2\ ClCN$	$[Co(CO)_4]_2 + I_2 \longrightarrow 2\ ICo(CO)_4$
Formation of heavy metal salts of low solubility	AgCl, $PbCl_2$	AgCN	$AgCo(CO)_4$
Addition to unsaturated species	$Cl_2 + \text{>C=C<} \longrightarrow -\text{C(Cl)}-\text{C(Cl)}-$		$[Co(CO)_4]_2 + F_2C{=}CF_2 \longrightarrow (OC)_4Co-CF_2-CF_2-Co(CO)_4$

NOTES: [a]Metal carbonyl (CO) compounds will be discussed in Chapters 13 to 15.
[b]However, $HCo(CO)_4$ is only slightly soluble in water.

8.10 GROUP 18 (VIIIA): THE NOBLE GASES

The elements in Group 18 (VIIIA), long designated the "inert" or "rare" gases, no longer satisfy these early labels. They are now known to have an interesting, although somewhat limited, chemistry, and they are rather abundant. Helium, for example, is the second most abundant element in the universe, and argon is the third most abundant component of dry air, approximately 24 times as abundant by volume as carbon dioxide.

8.10.1 The Elements

The first experimental evidence for the noble gases was obtained by Henry Cavendish in 1766. In a series of experiments on air, he was able to sequentially remove nitrogen (then known as "phlogisticated air"), oxygen ("dephlogisticated air"), and carbon dioxide ("fixed air") from air by chemical means; but a small residue, no more than one part in 120, resisted all attempts at reaction.[77] The nature of Cavendish's unreactive fraction of air remained a mystery for more than a century. This fraction was, of course, eventually shown to be a mixture of argon and other noble gases.[78]

During a solar eclipse in 1868, a new emission line, matching no known element, was found in the spectrum of the solar corona. J. N. Locklear and E. Frankland proposed the existence of a new element named, appropriately, helium (Greek, *helios*, sun). The same spectral line was subsequently observed in the gases of Mount Vesuvius.

In the early 1890s, Lord Rayleigh and William Ramsay observed a discrepancy in the apparent density of nitrogen isolated from air and from ammonia. The two

[77] H. Cavendish, *Philos. Trans.*, **1785**, *75*, 372.
[78] Cavendish's experiments and other early developments in noble gas chemistry have been described in E. N. Hiebert, "Historical Remarks on the Discovery of Argon: The First Noble Gas", in H. H. Hyman, ed., *Noble Gas Compounds*, University of Chicago Press, Chicago, 1963, pp. 3–20.

researchers independently performed painstaking experiments to isolate and characterize what seemed to be either a new form of nitrogen (the formula N_3 was one suggestion) or a new element. Eventually the two worked cooperatively, with Ramsay apparently the first to suggest that the unknown gas might fit into the periodic table after the element chlorine. In 1895, they reported the details of their experiments and evidence for the element they had isolated: argon (Greek, *argos*, no work, lazy).[79]

Within 3 years, Ramsay and M. W. Travers had isolated three additional elements by low-temperature distillation of liquid air: neon (Greek, *neos*, new), krypton (Greek, *kryptos*, concealed), and xenon (Greek, *xenos*, strange). The last of the noble gases, radon, was isolated as a nuclear decay product in 1902.

Helium is fairly rare on Earth, but it is the second most abundant element in the universe (76% H, 23% He) and is a major component of stars. Commercially, helium is obtained from natural gas. The other noble gases, with the exception of radon, are present in small amounts in air (see Table 8.17) and are commonly obtained by fractional distillation of liquid air. Helium is used as an inert atmosphere for arc welding, in weather and other balloons, and in gas mixtures used in deep-sea diving, but it is less soluble in blood than nitrogen. Recently, liquid helium (with a boiling point of 4.2 K) has increasingly been used as a coolant for superconducting magnets in NMR instruments. Argon, the least expensive noble gas, is commonly used as an inert atmosphere for studying chemical reactions, for high-temperature metallurgical processes, and for filling incandescent bulbs. One useful property of the noble gases is that they emit light of vivid colors when an electrical discharge is passed through them; neon's emission spectrum, for example, is responsible for the bright orange red of neon signs. Other noble gases are also used in discharge tubes, in which the color depends on the gases used. All isotopes of radon are radioactive; the longest lived isotope, ^{222}Rn, has a half-life of only 3.825 days. There has been concern regarding the level of radon in many homes. A potential cause of lung cancer, radon is formed from the decay of trace amounts of uranium in certain rock formations and itself undergoes alpha decay, leaving radioactive daughter isotopes in the lungs. Radon commonly enters homes through basement walls and floors.

Important properties of the noble gases are summarized in Table 8.17.

TABLE 8.17 Properties of the Group 18 (VIIIA) Elements: The Noble Gases

Element	Ionization Energy ($kJ\ mol^{-1}$)	Melting Point (°C)	Boiling Point (°C)	Enthalpy of Vaporization ($kJ\ mol^{-1}$)	Electronegativity	Abundance in Dry Air (% by Volume)
He[a]	2372	–	–268.93	0.08	4.160	0.000524
Ne	2081	–248.61	–246.06	1.74	4.787	0.001818
Ar	1521	–189.37	–185.86	6.52	3.242	0.934
Kr	1351	–157.20	–153.35	9.05	2.966	0.000114
Xe	1170	–111.80	–108.13	12.65	2.582	0.0000087
Rn	1037	–71	–62	18.1	2.60[b]	Trace

Source: See Table 8.3

NOTES: [a]Helium cannot be frozen at 1 atm pressure

[b]Approximate value

[79] Lord Rayleigh and W. Ramsay, *Philos. Trans. A*, **1895**, *186*, 187.

8.10.2 Chemistry

For many years, the group 18 elements were known as the "inert" gases, because they were believed to be totally unreactive as a consequence of the very stable "octet" valence electron configurations of their atoms. Their chemistry was simple: they had none!

The first chemical compounds containing noble gases were known as **clathrates**, "cage" compounds in which noble gas atoms could be trapped. Experiments begun in the late 1940s showed that when water or solutions containing hydroquinone (*p*-dihydroxybenzene, $HO—C_6H_4—OH$) were crystallized under high pressures of certain gases, hydrogen-bonded lattices having rather large cavities could be formed, with gas molecules of suitable size trapped in the cavities. Clathrates containing the noble gases argon, krypton, and xenon—as well as those containing small molecules such as SO_2, CH_4, and O_2—have been prepared. No clathrates have been found for helium and neon; these atoms are simply too small to be trapped.

Even though clathrates of three of the noble gases had been prepared by the beginning of the 1960s, no compounds containing covalently bonded noble gas atoms had been synthesized. Attempts had been made to react xenon with elemental fluorine, the most reactive of the elements, but without apparent success. However, in 1962, this situation changed dramatically. Neil Bartlett had observed that the compound PtF_6 changed color on exposure to air. With D. H. Lohmann, he demonstrated that PtF_6 was serving as a very strong oxidizing agent in this reaction and that the color change was due to the formation of $O_2{}^+[PtF_6]^-$.[80] Bartlett noted the similarity of the ionization energies of xenon (1169 kJ mol^{-1}) and O_2 (1175 kJ mol^{-1}) and repeated the experiment, reacting Xe with PtF_6. He observed a color change from the deep red of PtF_6 to orange yellow and reported the product as $Xe^+[PtF_6]^-$.[81] Although the product of this reaction later proved to be a complex mixture of several xenon compounds, these were the first covalently bonded noble gas compounds to be synthesized, and their discovery stimulated study of the chemistry of the noble gases in earnest. In a matter of months, the compounds XeF_2 and XeF_4 had been characterized, and other noble gas compounds soon followed.[82]

Scores of compounds of noble gas elements are now known, although the number remains modest in comparison with the other groups. The known noble gas compounds of xenon are by far the most diverse, and most of the other chemistry of this group is of compounds of krypton. There is evidence for the formation of such radon compounds as RnF_2, but the study of radon chemistry is hampered by the element's high radioactivity. In the year 2000, the first "stable" compound of argon, HArF, was reported.[83] This compound was synthesized by condensing a mixture of argon and an HF-pyridine polymer onto a CsI substrate at 7.5 K. Although stable at a low temperature, HArF decomposes at room temperature and above. Transient species containing helium and neon have been observed using mass spectrometry. However, most of the stable noble gas compounds are those of xenon with the highly electronegative elements F, O, and Cl; a few compounds have also been reported with Xe—N, Xe—C, and even Xe—transition metal bonds. Some of the compounds and ions of the noble gases are shown in Table 8.18.

Several of these compounds and ions have interesting structures which have provided tests for models of bonding. For example, structures of the xenon fluorides have been interpreted on the basis of the VSEPR model (Figure 8.38). XeF_2 and XeF_4 have structures entirely in accord with their VSEPR descriptions: XeF_2 is linear, with three lone pairs on Xe, and XeF_4 is planar, with two lone pairs.

---

[80] N. Bartlett and D. H. Lohmann, *Proc. Chem. Soc.*, **1962**, 115.
[81] N. Bartlett, *Proc. Chem. Soc.*, **1962**, 218.
[82] For a discussion of the development of the chemistry of xenon compounds, see P. Laszlo and G L Schrobilgen, *Angew. Chem., Int. Ed.*, **1988**, *27*, 479.
[83] L. Khriachtchev, M. Pettersson, N. Runeberg, J. Lundell, and M. Räsänen, *Nature (London)*, **2000**, *406*, 874.

TABLE 8.18 Example of Noble Gas Compounds and Ions

Formal Oxidation State of Noble Gas	Number of Lone Pairs on Central Atom	Compounds and Ions			
+2	3	KrF^+	XeF^+		
		KrF_2	XeF_2		
+4	2		$XeF_3{}^+$		
			XeF_4	$XeOF_2$	
			$XeF_5{}^-$		
+6	1		$XeF_5{}^+$	$XeOF_4$	XeO_3
			XeF_6	XeO_2F_2	
			$XeF_7{}^-$	XeO_2F^+	
			$XeF_8{}^{2-}$	XeO_3F^-	
				$XeOF_5{}^-$	
+8	0			XeO_3F_2	XeO_4
					$XeO_6{}^{4-}$

XeF_6 and $[XeF_8]^{2-}$, on the other hand, are more difficult to interpret by VSEPR. Each has a single lone pair on the central xenon. The VSEPR model would predict this lone pair to occupy a definite position on the xenon, as do single lone pairs in such molecules as NH_3, SF_4, and IF_5. However, no definite location is found for the central lone pair of XeF_6 or $XeF_8{}^{2-}$. One explanation is based on the degree of crowding around xenon. With a large number of fluorines attached to the central atom, repulsions between the electrons in the xenon–fluorine bonds are strong—too strong to enable a lone pair to occupy a well-defined position by itself. The central lone pair does play a role, however. In XeF_6, the structure is not octahedral but is somewhat distorted as a consequence of the lone pair on xenon. Although the structure of XeF_6 in the gas phase has been very difficult to determine, spectroscopic evidence indicates that the lowest energy form has C_{3v} symmetry, as shown in Figure 8.39. This is not a rigid structure, however; the molecule apparently undergoes rapid rearrangement from one C_{3v} structure to another—the lone pair appears to move from the center of one face to another—by way of intermediates having other symmetry.[84] Solid XeF_6 contains at

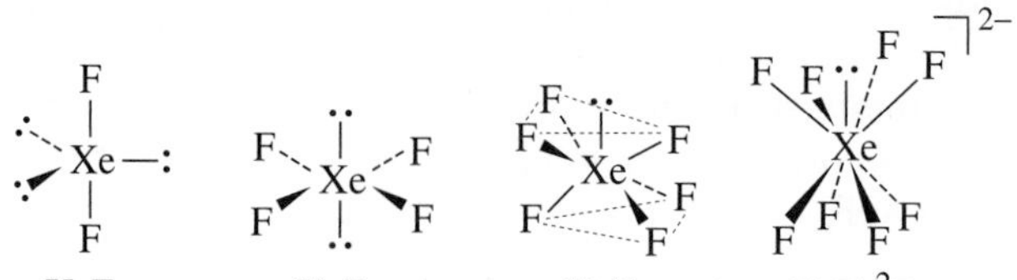

FIGURE 8.38 Structures of Xenon Fluorides.

XeF_2 XeF_4 XeF_6 $XeF_8{}^{2-}$

[84] K. Seppelt and D. Lentz, *Progr. Inorg. Chem.*, **1982**, *29*, 172–180; E. A. V. Ebsworth, D. W. H. Rankin, and S. Craddock, *Structural Methods in Inorganic Chemistry*, Blackwell Scientific Publications, Oxford, 1987, pp. 397–398.

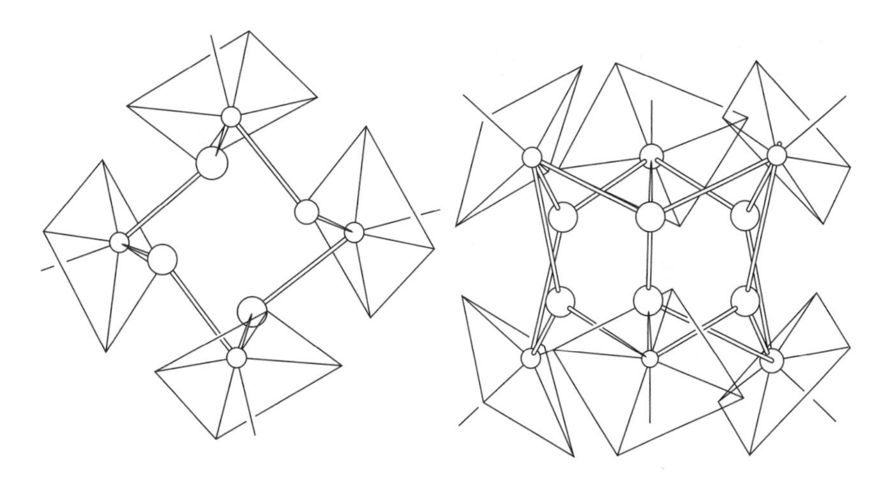

FIGURE 8.39 Xenon Hexafluoride (Crystalline Forms). (Reproduced with permission from R. D. Burbank and G. R. Jones, *J. Am. Chem. Soc.*, **1984**, *96*, 43. © 1974 American Chemical Society.)

least four phases, consisting of square-pyramidal XeF_5^+ ions bridged by fluoride ions, as shown for one of the phases in Figure 8.39.[85]

The structure of XeF_8^{2-} is also distorted, but very slightly. As shown in Figure 8.38, XeF_8^{2-} is nearly a square antiprism (D_{4d} symmetry), but one face is slightly larger than the opposite face, resulting in approximate C_{4v} symmetry.[86] Although this distortion may be a consequence of the way in which these ions pack in the crystal, it is also possible that the distortion is caused by a lone pair exerting some influence on the size of the larger face.[87]

Positive ions containing xenon are also known. For example, Bartlett's original reaction of xenon with PtF_6 is now believed to proceed as follows:

$$Xe + 2\,PtF_6 \longrightarrow [XeF]^+[PtF_6]^- + PtF_5 \longrightarrow [XeF]^+[Pt_2F_{11}]^-$$

The ion XeF^+ does not ordinarily occur as a discrete ion but rather is attached covalently to a fluorine on the anion; an example, $[XeF]^+[RuF_6]^-$, is shown in Figure 8.40.[88]

A remarkable aspect of the chemistry of xenon is its ability to act as a ligand toward Au^{2+}. Figure 8.41 shows square-planar $AuXe_4^{2-}$; other ions are *cis*-$[AuXe_2]^{2+}([Sb_2F_{11}]^-)_2$ and *trans*-$[AuXe_2]^{2+}([Sb_2F_6]^-)_2$.[89] Synthesis of $[AuXe_4]\,[Sb_2F_{11}]_2$ occurs in the very strong acid HF/SbF_5, in which Xe is a stronger base than HF and can displace HF from $[Au(HF)_n]^{2+}$ complexes. Xe also serves as a weak reducing agent, reducing Au^{3+} to Au^{2+}

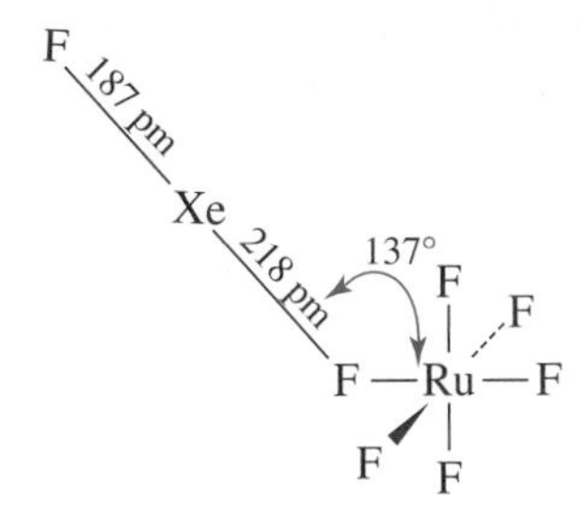

FIGURE 8.40 $[XeF]^+[RuF_6]^-$. (Data from N. Bartlett, M. Gennis, D. D. Gibler, B. K. Morrell, and A. Zalkin, *Inorg. Chem.*, **1973**, *12*, 1717.)

[85] R. D. Burbank and G. R. Jones, *J. Am. Chem. Soc.*, **1974**, *96*, 43.
[86] S. W. Peterson, J. H. Holloway, B. A. Coyle, and J. M. Williams, *Science*, **1971**, *173*, 1238.
[87] The effect of lone pairs can be difficult to predict. For examples of sterically active and inactive lone pairs in ions of formula AX_6^{n-}, see K. O. Christe and W. Wilson, *Inorg. Chem.*, **1989**, *28*, 3275 and references therein.
[88] N. Bartlett, *Inorg. Chem.*, **1973**, *12*, 1717.
[89] S. Seidel and K. Seppelt, *Science*, **2000**, *290*, 117; T. Drews, S. Seidel, and K. Seppelt, *Angew. Chem., Int. Ed.*, **2002**, *41*, 454.

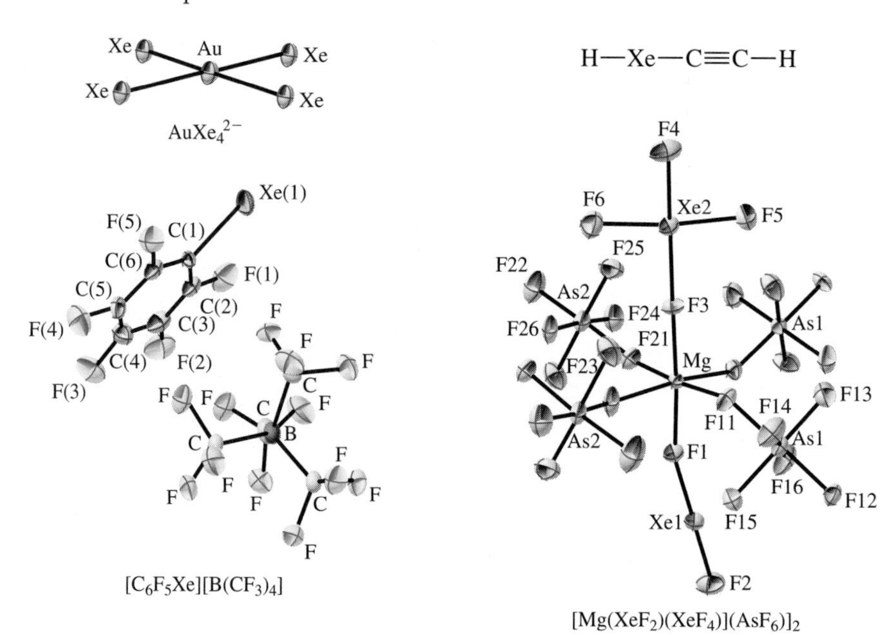

FIGURE 8.41 Examples of Structures of Xenon Compounds. T. Drews, S. Seidel, and K. Seppelt: Gold -Xenon Complexes. *Angew. Chem. Int. Ed.*, **2002**, *41*, 454. Copyright Wiley-VCH Verlag GmbH & Co. KGaA. Reproduced with permission. Reprinted with permission from K. Koppe, H.-J. Frohn, H. P. A. Mercier, and G. J. Schrobilgen, *Inorganic Chemistry*, **2008**, *47*, 3208. Copyright 2008 American Chemical Society. G. Tavčar and B. Žemva: XeF_4 as a Ligand for a Metal Ion. *Angew. Chem. Int. Ed.*, **2009**, *48*, 1432. Copyright Wiley-VCH Verlag GmbH & Co. KGaA. Reproduced with permission.

rather than to Au^+ as expected. Examples of notable xenon compounds prepared in recent years include a variety of salts of the perfluorophenylxenon ion $[C_6F_5Xe]^+$;[90] the compound HXeCCH, formed by insertion of xenon into a C—H bond of acetylene;[91] and both XeF_4 and XeF_2 serving as ligands in $[Mg(XeF_2)(XeF_4)](AsF_6)_2$.[92] These are also shown in Figure 8.41.

The chemistry of krypton is much more limited, although progress is being made. Krypton forms several species with fluorine, including the ions KrF^+ and $Kr_2F_3^+$ as well as the neutral KrF_2, the only known neutral halide. KrF_2 exists in two forms in the solid. In the alpha form, shown in Figure 8.42, all molecules are parallel to each other, with eight molecules centered at the corners of the unit cell and a ninth centered in the cell.[93] Examples of bonding to elements other than fluorine include $[F—Kr—N{\equiv}CH^+]\,AsF_6^-$,[94] $Kr(OTeF_5)_2$,[95] and HKrCCH.[96]

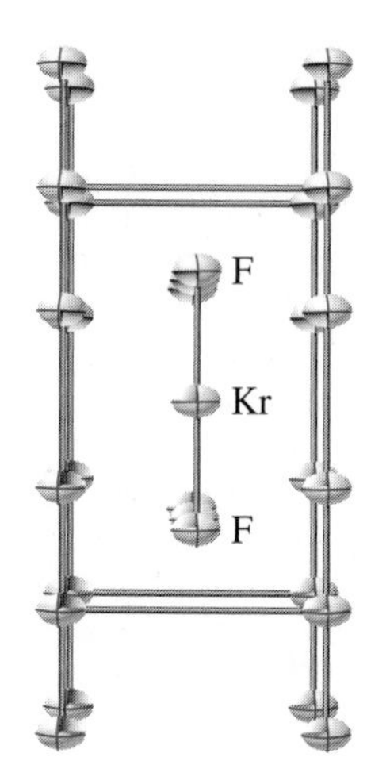

FIGURE 8.42 Krypton Fluoride Crystal Structure.

During the past 15 years, significant progress has been made in the development of noble gas hydrides, with 23 neutral species reported by the beginning of 2009.[97] Typically prepared by UV photolysis of precursor molecules in frozen noble gas matrices, hydrides are known for the elements argon (HArF, mentioned previously), krypton, and xenon and include both an example of a dihydride (HXeH) and compounds with bonds between noble gases and F, Cl, Br, I, C, N, O, and S.

The radioactivity of radon has made the study of its chemistry difficult; RnF_2 and a few other compounds have been observed through tracer studies.

Several reactions of the noble gas compounds are worth noting. Interest in using noble gas compounds as reagents in organic and inorganic synthesis has been stimulated

---

[90] K. Koppe, V. Bilir, H.-J. Frohn, H. P. A. Mercier, and G. J. Schrobilgen, *Inorg. Chem.*, **2007**, *46*, 9425.
[91] L. Khriachtchev, H. Tanskanen, J. Lundell, M. Pettersson, H. Kiljunen, and M.Räsänen, *J. Am. Chem. Soc.*, **2003**, *125*, 4696.
[92] G. Tavcar and B. Zemva, *Angew. Chem. Int. Ed.*, **2009**, *48*, 1432.
[93] J. F. Lehmann, D. A. Dixon, and G. J. Schrobilgen, *Inorg. Chem.*, **2001**, *40*, 3002. This reference also has structural data on compounds containing KrF^+ and $Kr_2F_3^+$.
[94] P. J. MacDougall, G. J. Schrobilgen, and R. F. W. Bader, *Inorg. Chem.*, **1989**, *28*, 763.
[95] J. C. P. Saunders and G. J. Schrobilgen, *Chem. Commun. (Cambridge)*, **1989**, 1576.
[96] L. Khriachtchev, H. Tanskanen, A. Cohen, R. B. Gerber, J. Lundell, M. Pettersson, H. Kilijunen, and M. Räsänen, *J. Am. Chem. Soc.*, **2003**, *125*, 6876.
[97] L. Khriachtchev, M. Räsänen, and R. B. Gerber, *Acc. Chem. Res.*, **2009**, *42*, 183. This article also provides a useful review of noble gas hydride chemistry.

in part because the byproduct of such reactions is often the noble gas itself. The xenon fluorides XeF_2, XeF_4, and XeF_6 have been used as fluorinating agents for both organic and inorganic compounds. For example,

$$2\,SF_4 + XeF_4 \longrightarrow 2\,SF_6 + Xe \qquad C_6H_5I + XeF_2 \longrightarrow C_6H_5IF_2 + Xe$$

XeF_4 can also selectively fluorinate aromatic positions in arenes such as toluene.

The oxides XeO_3 and XeO_4 are extremely explosive and must be handled using special precautions. XeO_3 is a powerful oxidizing agent in aqueous solution. The electrode potential of the half-reaction

$$XeO_3 + 6\,H^+ + 6\,e^- \longrightarrow Xe + 3\,H_2O$$

is 2.10 V. In basic solution, XeO_3 forms $HXeO_4^-$:

$$XeO_3 + OH^- \rightleftharpoons HXeO_4^- \quad K = 1.5 \times 10^{-3}$$

The $HXeO_4^-$ ion subsequently disproportionates to form the perxenate ion, XeO_6^{4-}:

$$2\,HXeO_4^- + 2\,OH^- \longrightarrow XeO_6^{4-} + Xe + O_2 + 2\,H_2O$$

The perxenate ion is an even more powerful oxidizing agent than XeO_3 and is capable of oxidizing Mn^{2+} to permanganate, MnO_4^- in acidic solution.

General References

More detailed descriptions of the chemistry of the main group elements can be found in N. N. Greenwood and A. Earnshaw, *Chemistry of the Elements,* 2nd ed., Butterworth-Heinemann, London, 1997, and in F. A. Cotton, G. Wilkinson, C. A. Murillo, and M. Bochman, *Advanced Inorganic Chemistry,* 6th ed., Wiley InterScience, New York, 1999. A handy reference on the properties of the elements themselves, including many physical properties, is J. Emsley's, *The Element*, 3rd ed., Oxford University Press, 1998. For extensive structural information on inorganic compounds, see A. F. Wells, *Structural Inorganic Chemistry,* 5th ed., Clarendon Press, Oxford, 1984. Three useful references on the chemistry of nonmetals are R. B. King's, *Inorganic Chemistry of Main Group Element*, VCH Publishers, New York, 1995; P. Powell and P. Timms, *The Chemistry of the Nonmetals*, Chapman and Hall, London, 1974; and R. Steudel, *Chemistry of the Non-Metals*, Walter de Gruyter, Berlin, 1976, English edition by F. C. Nachod and J. J. Zuckerman. The most complete reference on chemistry of the main group compounds through the early 1970s is the five-volume set by J. C. Bailar, Jr., H. C. Emeléus, R. Nyholm, and A. F. Trotman-Dickinson, editors, *Comprehensive Inorganic Chemistry*, Pergamon Press, Oxford, 1973. Two recent references on fullerene and related chemistry are A. Hirsch and M. Brettreich, *Fullerenes*, Wiley-VCH, Weinheim, Germany, 2005, and F. Langa and J.-F. Nierengarten, editors, *Fullerenes, Principles, and Applications*, RSC Publishing, Cambridge, UK, 2007.

Problems

8.1 The ions H_2^+ and H_3^+ have been observed in gas discharges.

a. H_2^+ has been reported to have a bond distance of 106 pm and a bond dissociation enthalpy of 255 kJ mol^{-1}. Comparable values for the neutral molecule are 74.2 pm and 436 kJ mol^{-1}. Are these values for H_2^+ in agreement with the molecular orbital picture of this ion? Explain.

b. Assuming H_3^+ to be triangular (the probable geometry), describe the molecular orbitals of this ion and determine the expected H—H bond order.

8.2 The species He_2^+ and HeH^+ have been observed spectroscopically. Prepare molecular orbital diagrams for these two ions. What would you predict for the bond order of each?

8.3 The chemical species IF_4^- and XeF_4 are known. Is the isoelectronic ion CsF_4^+ plausible? If it could be made, predict and sketch its most likely shape. Comment on whether such an ion might or might not be possible.

8.4 The equilibrium constant for the formation of the cryptand [Sr{cryptand(2.2.1)}]$^{2+}$ is larger than the equilibrium constants for the analogous calcium and barium cryptands. Suggest an explanation. (See E. Kauffmann, J-M. Lehn, and J-P. Sauvage, *Helv. Chim. Acta*, **1976**, *59*, 1099.)

8.5 Gas phase BeF_2 is monomeric and linear. Prepare a molecular orbital description of the bonding in BeF_2.

8.6 In the gas phase, $BeCl_2$ forms a dimer of the following structure:

Cl—Be(μ-Cl)₂Be—Cl

Describe the bonding of the chlorine bridges in this dimer in molecular orbital terms.

8.7 BF can be obtained by reaction of BF_3 with boron at 1850 °C and low pressure; BF is highly reactive but can be preserved at liquid nitrogen temperature (77 K). Prepare a molecular orbital diagram of BF. How would the molecular orbitals of BF differ from CO, with which BF is isoelectronic?

8.8 N-heterocyclic carbenes, such as the example shown here, have become increasingly important in both main group and transition-metal chemistry. For example, the first report of a stable neutral molecule having a boron–boron double bond used this N-heterocyclic carbene to bond to each of the borons in a product of the reaction of **R**BBr_3 with the powerful reducing agent KC_8 (potassium graphite) in diethylether solvent. (See Y. Wang, B. Quillian, P. Wei, C. S. Wannere, Y. Xie, R. B. King, H. F. Schaefer, III, P. v. R. Schleyer, and G H. Robinson, *J. Am. Chem. Soc.*, **2007**, *129*, 12412.)

a. What was the structure of this first molecule with a B=B bond? What evidence was cited for a double bond?

b. If **R**$SiCl_4$ was used in place of **R**BBr_3, an equally notable silicon compound was formed. What was this compound, and what parallels were noted with the reaction in Part a? (See Y. Wang, Y. Xie, P. Wei, R. B. King, H. F. Schaefer III, P. v. R. Schleyer, and G. H. Robinson, *Science*, **2008**, *321*, 1069.)

R, an N-heterocyclic carbene

8.9 $Al_2(CH_3)_6$ is isostructural with diborane, B_2H_6. On the basis of the orbitals involved, describe the Al—C—Al bonding for the bridging methyl groups in $Al_2(CH_3)_6$.

8.10 Referring to the description of bonding in diborane in Figure 8.12:

a. Show that the representation $\Gamma(p_z)$ reduces to $A_g + B_{1u}$.

b. Show that the representation $\Gamma(p_x)$ reduces to $B_{2g} + B_{3u}$.

c. Show that the representation $\Gamma(1s)$ reduces to $A_g + B_{3u}$.

d. Using the D_{2h} character table, verify that the sketches for the group orbitals match their respective symmetry designations (A_g, B_{2g}, B_{1u}, B_{3u}).

8.11 The compound $C(PPh_3)_2$ is bent at carbon; the P—C—P angle in one form of this compound has been reported as 130.1°. Account for the nonlinearity at carbon.

8.12 The C—C distances in carbides of formula MC_2 are in the range 119 to 124 pm if M is a Group 2 (IIA) metal or other metal commonly forming a 2+ ion, but in the approximate range 128 to 130 pm for Group 3 (IIIB) metals, including the lanthanides. Why is the C—C distance greater for the carbides of the Group 3 metals?

8.13 The half-life of ^{14}C is 5730 years. A sample taken for radiocarbon dating was found to contain 56 percent of its original ^{14}C. What was the age of the sample? (Radioactive decay of ^{14}C follows first-order kinetics.)

8.14 Prepare a model of buckminsterfullerene, C_{60}. By referring to the character table, verify that this molecule has I_h symmetry.

8.15 Determine the point groups of the following:

a. The unit cell of diamond

b. C_{20}

c. C_{70}

d. The nanoribbons shown in Figure 8.18

8.16 What is graph*ane*? How has it been synthesized, and what might be some of its potential uses? (See D. C. Elias, *et al.*, *Science*, **2009**, *323*, 610.)

8.17 Prepare a sheet showing an extended graphene structure, approximately 12 by 15 fused carbon rings or larger. Use this sheet to show how the graphene structure could be rolled up to form (**a**) a zigzag nanotube, (**b**) an armchair nanotube, and (**c**) a chiral nanotube. Is more than one chiral structure possible? (See M. S. Dresselhaus, G. Dresselhaus, and R. Saito, *Carbon*, **1995**, *33*, 883.)

8.18 To what extent has using carbon nanotubes for delivery of the anticancer agent cisplatin shown promise in the killing of cancer cells? (See J. F. Rusling, J. S. Gutkind, A. A. Bhirde, et al., *ACS Nano*, **2009**, *3*, 307). The reader is encouraged to search the recent literature to find more up to date references in this ongoing area of research.

8.19 Explain the increasing stability of the 2+ oxidation state for the Group 14 (IVA) elements with increasing atomic number.

8.20 1,2-diiododisilane has been observed in both *anti* and *gauche* conformations. (See K. Hassler, W. Koell, and K. Schenzel, *J. Mol. Struct.*, **1995**, *348*, 353.) For the *anti* conformation, shown here,

I H H Si—Si H H I

a. What is the point group?

b. Predict the number of infrared-active silicon–hydrogen stretching vibrations.

8.21 The reaction $P_4(g) \rightleftharpoons 2\,P_2(g)$ has $\Delta H =$ 217 kJ mol^{-1}. If the bond energy of a single phosphorus–phosphorus bond is 200 kJ mol^{-1}, calculate the bond energy of the P≡P bond. Compare the value you obtain with the bond energy in N_2 (946 kJ mol^{-1}), and suggest an explanation for the difference in bond energies in P_2 and N_2.

8.22 The azide ion, N_3^-, is linear with equal N—N bond distances.

a. Describe the pi molecular orbitals of azide.

b. Describe in HOMO-LUMO terms the reaction between azide and H^+ to form hydrazoic acid, HN_3.

c. The N—N bond distances in HN_3 are given in Figure 8.30. Explain why the terminal N—N distance is shorter than the central N—N distance in this molecule.

8.23 In aqueous solution, hydrazine is a weaker base than ammonia. Why? (pK_b values at 25 °C: NH_3, 4.74; N_2H_4, 6.07)

8.24 The bond angles for the hydrides of the Group 15 (VA) elements are as follows: NH_3, 107.8°; PH_3, 93.6°; AsH_3, 91.8°; and SbH_3, 91.3°. Account for this trend.

8.25 Gas-phase measurements show that the nitric acid molecule is planar. Account for the planarity of this molecule.

8.26 With the exception of NO_4^{3-}, all the molecules and ions in Table 8.10 are planar. Assign their point groups.

8.27 What type of interaction holds the O_2 units together in the O_8 structure? How does this interaction stabilize the larger molecule? (Hint: consider the molecular orbitals of O_2. See R. Steudel and M. W. Wong, *Angew. Chem. Int. Ed.*, **2007**, *46*, 1768.)

8.28 The sulfur–sulfur distance in S_2, the major component of sulfur vapor above ~720 °C, is 189 pm, significantly shorter than the sulfur–sulfur distance of 206 pm in S_8. Suggest an explanation for the shorter distance in S_2. (See C. L. Liao and C. Y. Ng, *J. Chem. Phys.*, **1986**, *84*, 778.)

8.29 Because of its high reactivity with most chemical reagents, F_2 is ordinarily synthesized electrochemically. However, the chemical synthesis of F_2 has been recorded via the reaction

$$2\,K_2MnF_6 + 4\,SbF_5 \longrightarrow 4\,KSbF_6 + 2\,MnF_3 + F_2$$

This reaction can be viewed as a Lewis acid–base reaction. Explain. (See K. O. Christe, *Inorg. Chem.*, **1986**, *25*, 3722.)

8.30 **a.** Chlorine forms a variety of oxides, among them Cl_2O and Cl_2O_7. Cl_2O has a central oxygen; Cl_2O_7 also has a central oxygen atom, bridging two ClO_3 groups. Of these two compounds, which would you predict to have the smaller Cl—O—Cl angle? Explain briefly.

b. The dichromate ion, $Cr_2O_7^{2-}$, has the same structure as Cl_2O_7, with oxygen now bridging two CrO_3 groups. Which of these two oxygen-bridged species do you expect to have the smaller outer atom–O–outer atom angle? Explain briefly.

8.31 The triiodide ion I_3^- is linear, but I_3^+ is bent. Explain.

8.32 Although B_2H_6 has D_{2h} symmetry, I_2Cl_6 is planar. Account for the difference in the structures of these two molecules.

8.33 BrF_3 undergoes autodissociation according to the equilibrium

$$2\,BrF_3 \rightleftharpoons BrF_2^+ + BrF_4^-$$

Ionic fluorides such as KF behave as bases in BrF_3, whereas some covalent fluorides such as SbF_5 behave as acids. On the basis of the solvent system concept, write balanced chemical equations for these acid–base reactions of fluorides with BrF_3.

8.34 The diatomic cations Br_2^+ and I_2^+ are both known.

a. On the basis of the molecular orbital model, what would you predict for the bond orders of these ions? Would you predict these cations to have longer or shorter bonds than the corresponding neutral diatomic molecules?

b. Br_2^+ is red, and I_2^+ is bright blue. What electronic transition is most likely responsible for absorption in these ions? Which ion has the more closely spaced HOMO and LUMO?

c. I_2 is violet, and I_2^+ is blue. On the basis of frontier orbitals (identify them), account for the difference in their colors.

8.35 I_2^+ exists in equilibrium with its dimer I_4^{2+} in solution. I_2^+ is paramagnetic and the dimer is diamagnetic. Crystal structures of compounds containing I_4^{2+} have shown this ion to be planar and rectangular, with two short I—I distances (258 pm) and two longer distances (326 pm).

a. Using molecular orbitals, propose an explanation for the interaction between two I_2^+ units to form I_4^{2+}.

b. Which form is favored at high temperature, I_2^+ or I_4^{2+}? Why?

8.36 How many possible isomers are there of the ion $IO_2F_3^{2-}$? Sketch these, and indicate the point group for each.
The observed ion has IR-active iodine–fluorine stretches at 802 and 834 cm^{-1}. What does this indicate about the most likely structure? (See J. P. Mack, J. A. Boatz, and M. Gerken, *Inorg. Chem.*, **2008**, *47*, 3243.

8.37 Bartlett's original reaction of xenon with PtF_6 apparently yielded products other than the expected $Xe^+PtF_6^-$. However, when xenon and PtF_6 are reacted in the presence of a large excess of sulfur hexafluoride, $Xe^+PtF_6^-$ is apparently formed. Suggest the function of SF_6 in this reaction. (See: K. Seppelt and D. Lentz, *Progr. Inorg. Chem.*, **1982**, *29*, 170–171.)

8.38 On the basis of VSEPR, predict the structures of $XeOF_2$, $XeOF_4$, XeO_2F_2, and XeO_3F_2. Assign the point group of each.

8.39 The sigma bonding in the linear molecule XeF_2 may be described as a three-center, four-electron bond. If the z axis is assigned as the internuclear axis, use the p_z orbitals on each of the atoms to prepare a molecular orbital description of the sigma bonding in XeF_2.

8.40 The $OTeF_5$ group can stabilize compounds of xenon in formal oxidation states IV and VI. On the basis of VSEPR, predict the structures of $Xe(OTeF_5)_4$ and $O{=}Xe(OTeF_5)_4$.

8.41 Write a balanced equation for the oxidation of Mn^{2+} to MnO_4^- by the perxenate ion in acidic solution; assume that neutral Xe is formed.

8.42 The ion XeF_5^- is a rare example of pentagonal planar geometry. On the basis of the symmetry of this ion, predict the number of IR-active Xe–F stretching bands.

8.43 $XeOF_4$ has one of the more interesting structures among noble gas compounds. On the basis of its symmetry,

a. Obtain a representation based on *all* the motions of the atoms in $XeOF_4$.

b. Reduce this representation to its component irreducible representations.

c. Classify these representations, indicating which are for translational, rotational, and vibrational motion.

8.44 As of this edition, the ion XeF_2^{2-} has not been reported. Would you expect this ion to be bent or linear? Suggest why it remains elusive.

8.45 Liquid NSF_3 reacts with $[XeF][AsF_6]$ to form $[F_3SNXeF][AsF_6]$, **1**. On mild warming in the solid state, compound **1** rearranges to $[F_4SNXe][AsF_6]$, **2**. Reaction of compound **2** with HF yields $[F_4SNH_2][AsF_6]$, **3**, $[F_5SN(H)Xe][AsF_6]$, **4**, and XeF_2. (See G. L. Smith, H. P. A. Mercier, and G. J. Schrobilgen, *Inorg. Chem.*, **2007**, *46*, 1369; *Inorg. Chem.*, **2008**, *47*, 4173; and *J. Am. Chem. Soc.*, **2009**, *131*, 4173.)

a. What is the bond order in $[XeF]^+$?

b. Of compounds **1** through **4**, which has the longest S–N distance? The shortest?

c. Which of the compounds **1** through **4** is most likely to have linear bonding around the nitrogen atom?

d. By the VSEPR approach, is the bonding around Xe expected to be linear or bent in compound **1**?

e. In compounds **2** and **3**, are the NXe and NH_2 groups likely to occupy axial or equatiorial sites on sulfur?

f. In compounds **2** and **3**, which bonds are more likely to be longer, the $S{-}F_{axial}$ bonds or $S{-}F_{equatorial}$ bonds?

The following problems require the use of molecular modeling software.

8.46 It has been proposed that salts containing the cation $[FBeNg]^+$, where Ng = He, Ne, or Ar, may be stable. Use molecular modeling software to calculate and display the molecular orbitals of $[FBeNe]^+$. Which molecular orbitals would be the primary ones engaged in bonding in this ion? (See M. Aschi and F. Grandinetti, *Angew. Chem., Int. Ed.*, **2000**, *39*, 1690.)

8.47 The dixenon cation Xe_2^+ has been characterized structurally. Using molecular modeling software, calculate and observe the energies and shapes of the molecular orbitals of this ion. Classify each of the seven highest-energy occupied orbitals as σ, π, or δ, and as bonding or antibonding. The bond in this compound was reported as the longest main group–main group bond observed to date. Account for this very long bond. (See T. Drews and K. Seppelt, *Angew. Chem., Int. Ed.*, **1997**, *36*, 273.)

8.48 To further investigate the bonding in the O_8 structure, construct an O_8 unit, and calculate and display its molecular orbitals. Which orbitals are involved in holding the O_2 units together? In each of these orbitals, identify the atomic orbitals that are primarily involved. (See Figure 8.33; additional structural details can be found in L. F. Lundegaard, G. Weck, M. I. McMahon, S. I. Desgreniers, and P. Loubeyre, *Science*, **2006**, *443*, 201 and R. Steudel and M. W. Wong, *Angew. Chem., Int. Ed.*, **2007**, *46*, 1768.)

9

Coordination Chemistry I: Structures and Isomers

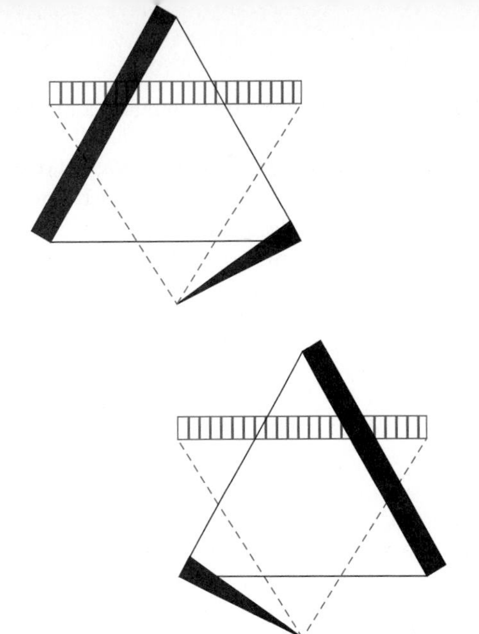

Coordination compounds, as the term is usually used in inorganic chemistry, include compounds composed of a metal atom or ion and one or more **ligands** (atoms, ions, or molecules) that formally donate electrons to the metal. This definition includes compounds with metal–carbon bonds, or **organometallic compounds**, which are described in Chapters 13 to 15.

The name *coordination compound* comes from the coordinate covalent bond, which historically was considered to form by donation of a pair of electrons from one atom to another. In coordination compounds the donors are usually the ligands, and the acceptors are the metals. Coordinate covalent bonds are identical to covalent bonds formally formed by combining one electron from each atom; only the formal electron counting distinguishes them. Coordination compounds are also acid–base adducts (described in Chapter 6), frequently called **complexes** or, if charged, **complex ions**.

9.1 HISTORY

Although the history of interpretation of structures and reactions of coordination compounds really begins with Alfred Werner (1866–1919), coordination compounds were known much earlier. Many have been used as pigments and dyes since antiquity. Examples still in use include Prussian blue ($KFe[Fe(CN)_6]$), aureolin $K_3[Co(NO_2)_6] \cdot 6H_2O$, yellow), and alizarin red dye (the calcium aluminum salt of 1,2-dihydroxy-9,10-anthraquinone). The striking colors of compounds such as these, and their color changes on reaction, were described in very early documents and provided the impetus for further studies. The ion known today as tetraamminecopper(II)—(actually $[Cu(NH_3)_4(H_2O)_2]^{2+}$ in solution), which has a striking royal blue color—was certainly known in prehistoric times. With the gradual development of analytical methods, the formulas of many of these compounds became known late in the nineteenth century, and theories of structure and bonding became possible.

Inorganic chemists tried to use the advances in organic bonding theory and the simple ideas of ionic charges to explain bonding in coordination compounds, but they found the existing theories inadequate. In a compound such as hexaamminecobalt(III) chloride, $[Co(NH_3)_6]Cl_3$, the early bonding theories allowed only three other atoms to be attached to the cobalt because of its "valence" of 3. By analogy with ordinary salts, such as $FeCl_3$, the chlorides were assigned this role. This left the six ammonia molecules with no means of participating in bonding, and it was necessary to develop new ideas. One theory, proposed first by C. W. Blomstrand[1] (1826–1894) and developed further by S. M. Jørgensen[2]

[1] C. W. Blomstrand, *Berichte*, **1871**, *4*, 40; translated by G. B. Kauffman, *Classics in Coordination Chemistry, Part 2*, Dover, New York, 1976, pp. 75–93.
[2] S. M. Jørgensen, *Z. Anorg. Chem.*, **1899**, *19*, 109; translated by G. B. Kauffman, *Classics in Coordination Chemistry, Part 2*, pp. 94–164.

(1837–1914), was that the nitrogens could form chains much like those of carbon, as shown in Table 9.1, and thus could have a valence of 5. According to this theory, chloride ions attached directly to cobalt were bonded more strongly than those bonded to nitrogen. Alfred Werner[3] (1866–1919) proposed instead that all six ammonias could bond directly to the cobalt ion. Werner allowed for a looser bonding of the chloride ions, and we now consider them independent ions.

The series of compounds in Table 9.1 illustrates how both the chain theory and Werner's coordination theory predict the number of ions to be formed by a series of cobalt complexes. Blomstrand's theory allowed dissociation of chlorides attached to ammonia but not of chlorides attached directly to cobalt. Werner's theory also included two kinds of chlorides. The number of chlorides attached to the cobalt, and therefore unavailable as ions, plus the number of ammonia molecules totaled six. The other chlorides were considered less firmly bound and could therefore form ions in solution. We now consider them to be ions in the solid state as well.

Except for the last compound in the table, the predictions of the two theories match, and the ionic behavior does not distinguish between them. Even with the last compound, experimental problems with purity and conductance measurements left some ambiguity. The argument between Jørgensen and Werner continued for many years, with each presenting data and explanations favoring his own position. This case illustrates some of the good features of such controversy. Werner was forced to develop his theory further, and synthesize new compounds to test his ideas, because Jørgensen defended the earlier theory so vigorously. Werner proposed an octahedral structure for compounds such as those in Table 9.1. He prepared and characterized many isomers, including both green and violet forms of $[Co(H_2NC_2H_4NH_2)_2Cl_2]^+$. He claimed that these compounds had the chlorides arranged ***trans*** (opposite each other) and ***cis*** (adjacent to each other) respectively, in an overall octahedral geometry, as in Figure 9.1.

TABLE 9.1 Comparison of Blomstrand's Chain Theory and Werner's Coordination Theory

Werner Formula (Modern Form)	Number of Ions Predicted	Blomstrand Chain Formula	Number of Ions Predicted
$[Co(NH_3)_6]\mathit{Cl}_3$	4	$NH_3—\mathit{Cl}$ $Co—NH_3—NH_3—NH_3—NH_3—\mathit{Cl}$ $NH_3—\mathit{Cl}$	4
$[Co(NH_3)_5Cl]\mathit{Cl}_2$	3	$NH_3—\mathit{Cl}$ $Co—NH_3—NH_3—NH_3—NH_3—\mathit{Cl}$ Cl	3
$[Co(NH_3)_4Cl_2]\mathit{Cl}$	2	Cl $Co—NH_3—NH_3—NH_3—NH_3—\mathit{Cl}$ Cl	2
$[Co(NH_3)_3Cl_3]$	0	Cl $Co—NH_3—NH_3—NH_3—\mathit{Cl}$ Cl	2

NOTE: The italicized chlorides dissociate in solution, according to the two theories.

[3] A. Werner, *Z. Anorg. Chem.*, **1893**, *3*, 267; *Berichte*, **1907**, *40*, 4817; **1911**, *44*, 1887; **1914**, *47*, 3087; A. Werner and A. Miolati, *Z. Phys. Chem.*, **1893**, *12*, 35; **1894**, *14*, 506; all translated by G. B. Kauffman, *Classics in Coordination Chemistry, Part 1*, New York, 1968.

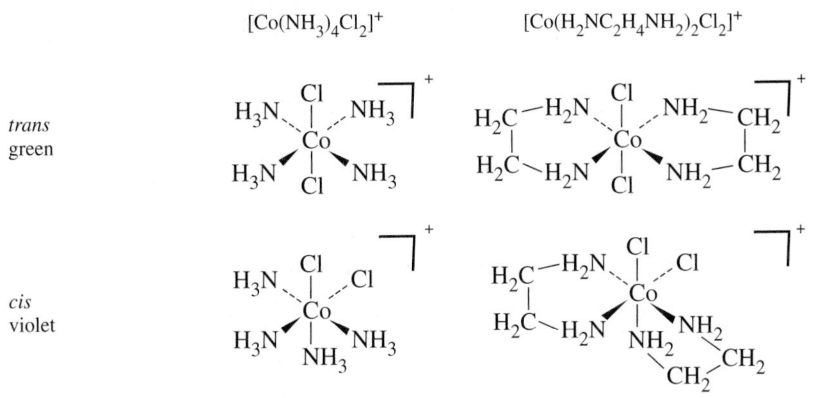

$[Co(NH_3)_4Cl_2]^+$ | $[Co(H_2NC_2H_4NH_2)_2Cl_2]^+$

trans green

cis violet

FIGURE 9.1 *cis* and *trans* Isomers.

Jørgensen offered alternative isomeric structures but finally conceded defeat in 1907, when Werner succeeded in synthesizing the green *trans* and the violet *cis* isomers of $[Co(NH_3)_4Cl_2]^+$, for which there were no counterparts in the chain theory.

However, even synthesis of this compound and the later discovery of optically active coordination compounds did not completely convince all chemists, although such compounds could not be explained directly by the chain theory. It was argued that Werner's optically active compounds still contained carbon and that their chirality could be due to the carbon atoms. Finally, Werner resolved the compound $[Co\{Co(NH_3)_4(OH)_2\}_3]Br_6$ (Figure 9.2), initially prepared by Jørgensen, into its two optically active forms, using *d*- and *l*-α-bromocamphor-π-sulfonate as the resolving agents. With this final proof of optical activity without carbon, the validity of Werner's theory was finally accepted. Pauling[4] extended the theory in terms of hybrid orbitals, and later theories[5] have adapted arguments first used for electronic structures of ions in crystals to coordination compounds.

The Werner theory of coordination compounds was based on a group of compounds that is relatively slow to react in solution and thus easier to study. For this reason, many of his examples were compounds of Co(III), Rh(III), Cr(III), Pt(II), and Pt(IV), which are kinetically inert or slow to react. Examination of more reactive compounds over the years has confirmed their similarity to those originally studied, so we will include examples of both types of compounds in the descriptions that follow.

Werner's theory required two kinds of bonding in the compound: a primary one, in which the positive charge of the central metal ion is balanced by negative ions in the compound, and a secondary one, in which molecules or ions—known collectively as

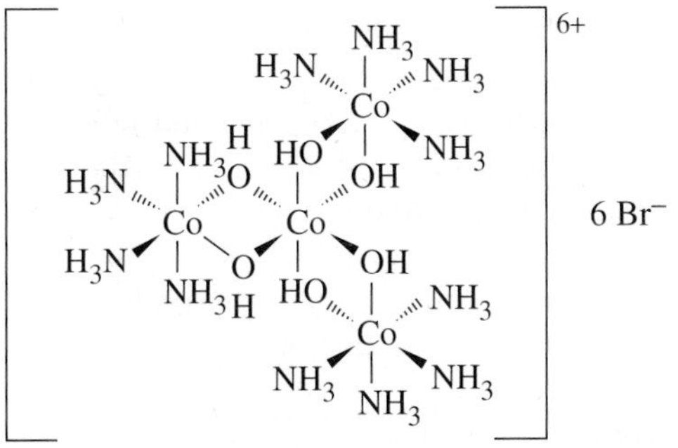

$6\ Br^-$

FIGURE 9.2 Werner's Totally Inorganic Optically Active Compound, $[Co\{Co(NH_3)_4(OH)_2\}_3]Br_6$

[4] L. Pauling, *J. Chem. Soc.*, **1948**, 1461; *The Nature of the Chemical Bond*, 3rd ed., Cornell University Press, Ithaca, NY, 1960, pp. 145–182.

[5] J. S. Griffith and L. E. Orgel, *Q. Rev. Chem. Soc.*, **1957**, *XI*, 381.

ligands—are attached directly to the transition-metal ion. The secondary bonded unit has been given many different names, such as the **complex ion** or the **coordination sphere**, and the formula is written with this part in brackets. Current practice considers this coordination sphere the more important, so the words *primary* and *secondary* no longer bear the same significance. In the examples in Table 9.1, the coordination sphere acts as a unit; the ions outside the brackets balance the charge and are free ions in solution. Depending on the nature of the metal and the ligands, the metal can have from one up to at least 16 atoms attached to it, with four and six the most common numbers.[6] Additional water molecules may be added to the coordination sphere when the compound is dissolved in water. We should include the water molecules specifically in the description of the compound, but in some cases they are omitted to concentrate on the other ligands. The discussion that follows concentrates on the coordination sphere; the other ions associated with it can frequently vary without changing the bonding between ligands and the central metal.

Werner used compounds with four or six ligands to develop his theories, with the shapes of the coordination compounds established by the synthesis of isomers. For example, he was able to synthesize only two isomers of the $[Co(NH_3)_4Cl_2]^+$ ion. The possible structures with six ligands are hexagonal, hexagonal pyramidal, trigonal prismatic, trigonal antiprismatic, and octahedral. Because there are two possible isomers for the octahedral shape and three for each of the others, as shown in Figure 9.3, Werner claimed that the structure was octahedral. Such an argument cannot be conclusive, because a missing isomer may simply be difficult to synthesize or isolate. However, later experiments confirmed the octahedral shape, with *cis* and *trans* isomers as shown.

Werner's synthesis and separation of optical isomers (Figure 9.2) proved the octahedral shape conclusively, because none of the other six-coordinate geometries could have similar optical activity.

In a similar way, other experiments were consistent with square-planar Pt(II) compounds, with the four ligands at the corners of a square. Only two isomers are found for $[Pt(NH_3)_2Cl_2]$. Although the two could have had different shapes (tetrahedral and square-planar, for example), Werner assumed that they had the same overall shape and, because only one tetrahedral structure is possible for this compound, he argued that they must have square-planar shapes with *cis* and *trans* geometries. Again, his arguments were correct, although the evidence he presented could not be conclusive. The possible structures are shown in Figure 9.4.

After Werner's evidence for the octahedral and square-planar natures of many complexes, it was clear that any acceptable theory needed to account for bonds between ligands and metals and that the number of bonds required was more than that commonly accepted at that time. Transition-metal compounds with six ligands, for example, cannot fit the simple Lewis theory with eight electrons around each atom, and even expanding the shell to 10 or 12 electrons does not work in cases such as $[Fe(CN)_6]^{4-}$, with a total of 18 electrons to accommodate. In fact, the **18-electron rule** is sometimes useful in accounting for the bonding in many coordination compounds in a simple way; the total number of valence electrons around the central atom is counted, with 18 as a common result. (This approach is more often used in organometallic compounds and is discussed in Chapter 13.)

Pauling[7] used his **valence bond** approach to explain differences in magnetic behavior among coordination compounds by use of either 3*d* or 4*d* orbitals of the metal ion.

---

[6] N. N. Greenwood and A. Earnshaw, *Chemistry of the Elements*, 2nd ed., Butterworth–Heinemann, Oxford, UK, 1997, p. 912. The larger numbers depend on how the number of donors in organometallic compounds are counted; some would assign smaller coordination numbers because of the special nature of the organic ligands.
[7] Pauling, *The Nature of the Chemical Bond*, pp. 145–182.

FIGURE 9.3 Possible Isomers for Hexacoordinate Complexes.

cis - and *trans* - Tetramminedichlorocobalt (III), $[Co(NH_3)_4Cl_2]^+$

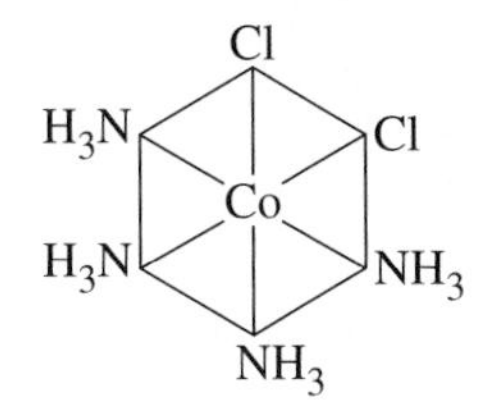

Hexagonal (three isomers)

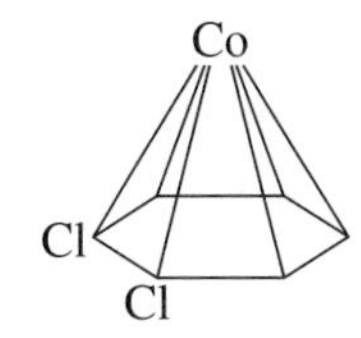

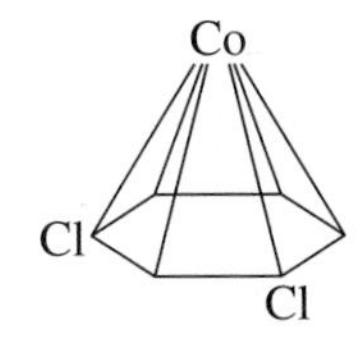

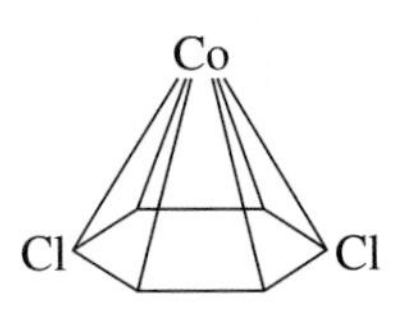

Hexagonal pyramidal (three isomers)

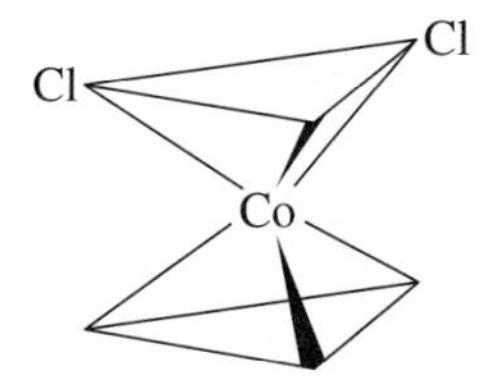

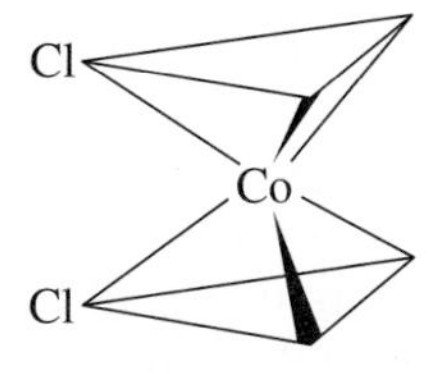

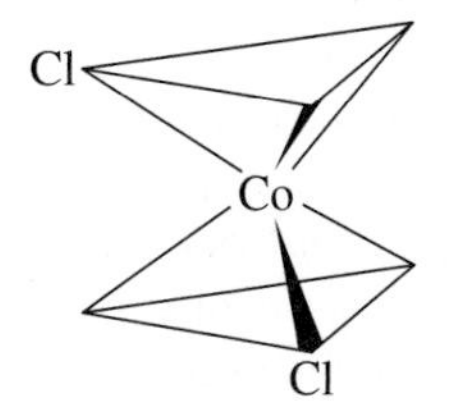

Trigonal prismatic (three isomers)

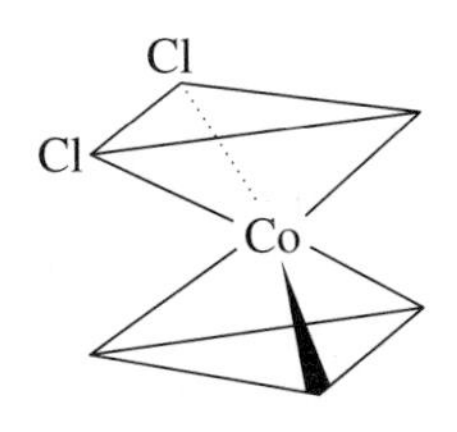

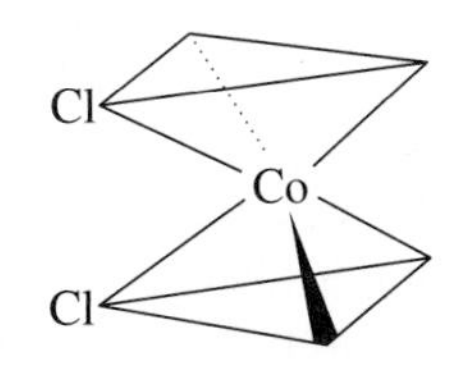

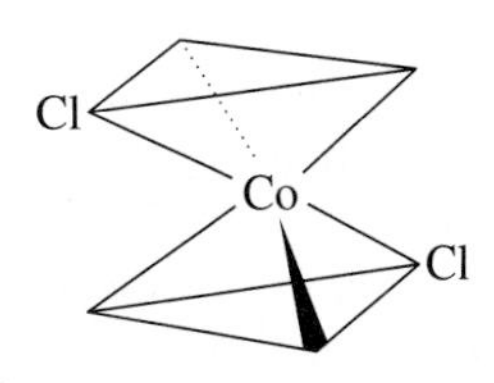

Trigonal antiprismatic (three isomers)

Octahedral (two isomers)

FIGURE 9.4 Possible Structures for Tetracoordinate Complexes.

cis- and *trans*- Diamminedichloroplatinum(II), $[PtCl_2(NH_3)_2]$

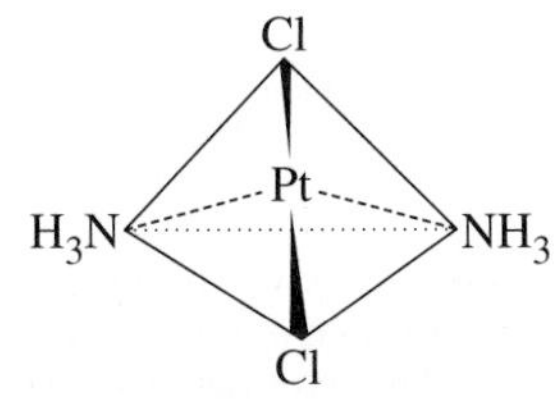

Tetrahedral (one isomer)

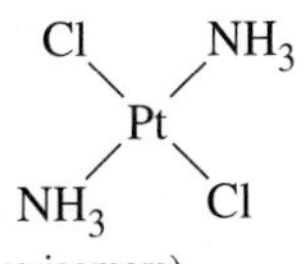

Square planar (two isomers)

Griffith and Orgel[8] developed and popularized the use of **ligand field theory**, derived from the **crystal field theory** of Bethe[9] and Van Vleck[10] on the behavior of metal ions in crystals and from the molecular orbital treatment of Van Vleck.[11] Several of these approaches are described in Chapter 10, with emphasis on the ligand field theory.

This chapter describes a sampling of the different shapes of coordination compounds. Because of the complex factors involved in determining shapes of coordination compounds, it can be difficult to predict shapes with confidence, except when compounds of similar composition are already known. It is possible, however, to relate some structures to the individual factors that interact to produce them. This chapter also describes some of the isomers possible for coordination compounds and some of the experimental methods used to study them. Structures of some organometallic compounds are even more difficult to predict, as will be seen in Chapters 13 through 15.

9.2 NOMENCLATURE

As in any field of study, careful attention to nomenclature is important. Some of the rules for names and formulas of coordination compounds are given here, with examples to show their use. The nomenclature of coordination chemistry has changed over time. In many cases, the notation used by those who first prepared a compound has been retained and expanded; in other cases, conflicting rules for names have been proposed by different people, and only after some time has a standard been established. The literature naturally includes papers using all the possible names, and sometimes careful research is necessary to interpret those names that had relatively short lifetimes.

Following are the major rules required to name coordination compounds in this text and those found in the general literature. Reference to more complete sources may be needed to determine the names of other compounds.[12]

Organic (and some inorganic) ligands are frequently named using older trivial names rather than International Union of Pure and Applied Chemistry (IUPAC) names. The IUPAC names are more generally accepted, but trivial names and abbreviations are still commonly used. Tables 9.2, 9.3, and 9.4 list some of the most common ligands. Those with two or more points of attachment to metal atoms are called **chelating ligands**, and their compounds are called **chelates** (pronounced key′-lates), a name derived from the Greek *khele*, the claw of a crab. Ligands such as ammonia are **monodentate**, with one point of attachment (literally, "one tooth"). Ligands are described as **bidentate** if they have two points of attachment, as in ethylenediamine ($NH_2CH_2CH_2NH_2$), which can bond to metals through the two nitrogens. The prefixes tri-, tetra-, penta-, and hexa- are used for three through six bonding positions, as shown in Table 9.3. **Chelate rings** may have any number of atoms; the most common contain five or six atoms, including the metal. Smaller rings have angles and distances that lead to strain; larger rings frequently result in crowding, both within the ring and between adjoining ligands. Some ligands can form more than one ring; ethylenediaminetetraacetate (EDTA) can form five by using the four carboxylate groups and the two amine nitrogens.

---

[8] Griffith and Orgel, *op. cit.*; L. E. Orgel, *An Introduction to Transition-Metal Chemistry*, Methuen, London, 1960.
[9] H. Bethe, *Ann. Phys.*, **1929**, *3*, 133.
[10] J. H. Van Vleck, *Phys. Rev.*, **1932**, *41*, 208.
[11] J. H. Van Vleck, *J. Chem. Phys.*, **1935**, *3*, 807.
[12] T. E. Sloan, "Nomenclature of Coordination Compounds," in G. Wilkinson, R. D. Gillard, and J. A. McCleverty, eds., *Comprehensive Coordination Chemistry*, Pergamon Press, Oxford, 1987, Vol. 1, pp. 109–134; G. J. Leigh, ed., International Union of Pure and Applied Chemistry, *Nomenclature of Inorganic Chemistry: Recommendations 1990*, Blackwell Scientific Publications, Cambridge, MA, 1990; J. A. McCleverty and N. G. Connelly, eds., International Union of Pure and Applied Chemistry, *Nomenclature of Inorganic Chemistry II: Recommendations 2000*, Royal Society of Chemistry, Cambridge, UK, 2001.

TABLE 9.2 Common Monodentate Ligands

Common Name	IUPAC Name	Formula
hydrido	hydrido	H^-
fluoro	fluoro	F^-
chloro	chloro	Cl^-
bromo	bromo	Br^-
iodo	iodo	I^-
nitrido	nitrido	N^{3-}
azido	azido	N_3^-
oxo	oxido	O^{2-}
cyano	cyano	CN^-
thiocyano	thiocyanato-*S* (S-bonded)	SCN^-
isothiocyano	thiocyanato-*N* (N-bonded)	NCS^-
hydroxo	hydroxo	OH^-
aqua	aqua	H_2O
carbonyl	carbonyl	CO
thiocarbonyl	thiocarbonyl	CS
nitrosyl	nitrosyl	NO^+
nitro	nitrito-*N* (N-bonded)	NO_2^-
nitrito	nitrito-*O* (O-bonded)	ONO^-
methyl isocyanide	methylisocyanide	CH_3NC
phosphine	phosphane	PR_3
pyridine	pyridine (abbrev. py)	C_5H_5N
ammine	ammine	NH_3
methylamine	methylamine	$MeNH_2$
amido	azanido	NH_2^-
imido	azanediido	NH^{2-}

TABLE 9.3 Common Chelating Amines

Chelating Points	Common Name	IUPAC Name	Abbreviation	Formula
monodentate	ammine, methylamine	ammine, methylamine		NH_3, CH_3NH_2
bidentate	ethylenediamine	1,2–ethanediamine	en	$NH_2CH_2CH_2NH_2$
tridentate	diethylenetriamine	2,2′-diaminodiethylamine or 1,4,7-triazaheptane	dien	$NH_2CH_2CH_2NHCH_2CH_2NH_2$

Continued

TABLE 9.3 Common Chelating Amines—continued

Chelating Points	Common Name	IUPAC Name	Abbreviation	Formula
tetradentate	triethylenetetraamine	1,4,7,10-tetraazadecane	trien	$NH_2CH_2CH_2NHCH_2CH_2NHCH_2CH_2NH_2$
	β, β', β''-triaminotriethylamine	β, β', β''-tris(2-aminoethyl)amine	tren	$NH_2CH_2CH_2N(CH_2CH_2NH_2)CH_2CH_2NH_2$
pentadentate	tetraethylenepentamine	1,4,7,10,13-pentaazatridecane		$NH_2CH_2CH_2NHCH_2CH_2NHCH_2CH_2NHCH_2CH_2NH_2$
hexadentate	ethylenediaminetetraacetate	1,2-ethanediyl (dinitrilo) tetraacetate	EDTA	$(^-OOCCH_2)_2NCH_2CH_2N(CH_2COO^-)_2$

TABLE 9.4 Common Multidentate (Chelating) Ligands

Common Name	IUPAC Name	Abbreviation	Formula and Structure
acetylacetonato	2,4-pentanediono	acac	$CH_3COCHCOCH_3^-$ 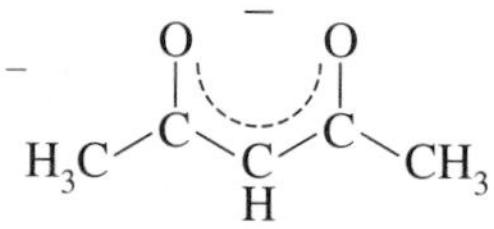
2,2′-bipyridine	2,2′-bipyridyl	bipy	$C_{10}H_8N_2$ 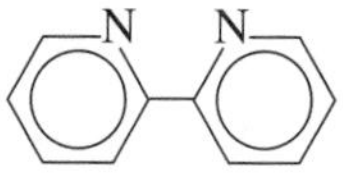
1,10-phenanthroline, *o*-phenanthroline	1,10-diaminophenanthrene	phen, *o*-phen	$C_{12}H_8N_2$ 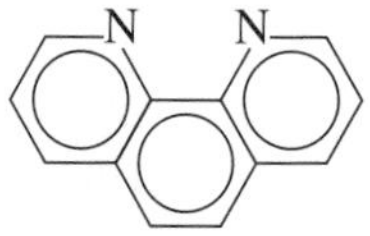
oxalato	oxalato	ox	$C_2O_4^{2-}$ 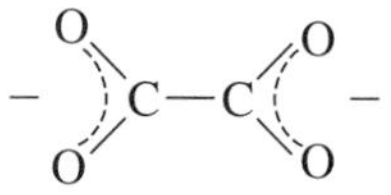
dialkyldithiocarbamato	dialkylcarbamodithioato	dtc	$S_2CNR_2^-$ 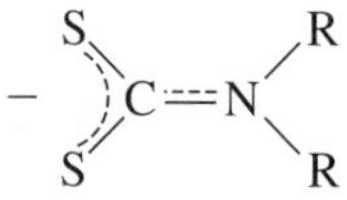
1,2-bis (diphenylphosphino)ethane	1,2-ethanediylbis (diphenylphosphane)	dppe	$Ph_2PC_2H_4PPh_2$ 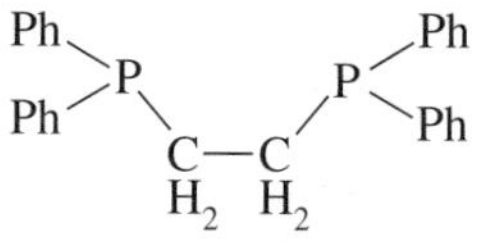

Continued

TABLE 9.4 Common Multidentate (Chelating) Ligands—continued

Common Name	IUPAC Name	Abbreviation	Formula and Structure
o-phenylenebis (dimethylarsine)	1,2-phenylenebis (dimethylarsane)	diars	$C_6H_4(As(CH_3)_2)_2$ 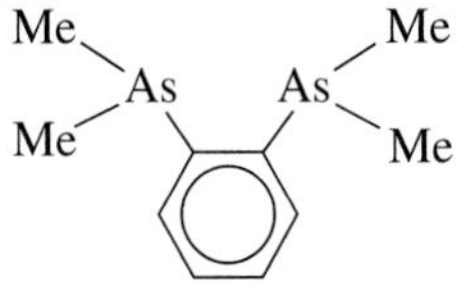
dimethylglyoximato	butanediene dioxime	DMG	$HONCC(CH_3)C(CH_3)NO^-$ 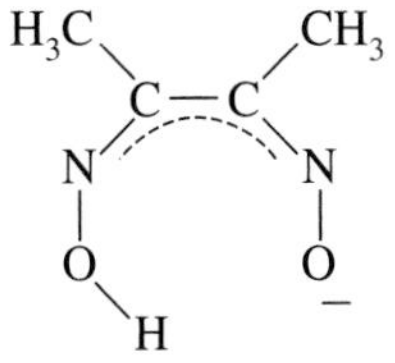
pyrazolylborato (scorpionate)	hydrotris-(pyrazo-1-yl)borato	Tp	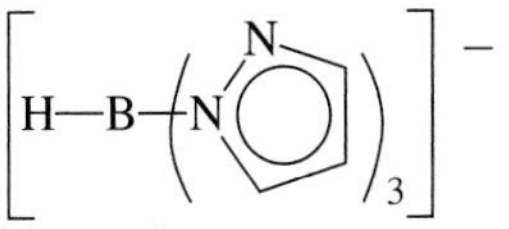

NOMENCLATURE RULES

1. The positive ion (cation) comes first, followed by the negative ion (anion). This is also the common order for simple salts.

   Examples: diamminesilver(I) chloride, $[Ag(NH_3)_2]Cl$

   potassium hexacyanoferrate(III), $K_3[Fe(CN)_6]$

2. The inner coordination sphere is enclosed in square brackets in the formula. Within the coordination sphere, the ligands are named before the metal; but in formulas, the metal ion is written first.

   Examples: tetraamminecopper(II) sulfate, $[Cu(NH_3)_4]SO_4$

   hexaamminecobalt(III) chloride, $[Co(NH_3)_6]Cl_3$

3. The number of ligands of one kind is given by the following prefixes. If the ligand name includes these prefixes or is complicated, it is set off in parentheses, and the second set of prefixes (all ending in *–is*) is used.

2	di	bis	7	hepta	heptakis
3	tri	tris	8	octa	octakis
4	tetra	tetrakis	9	nona	nonakis
5	penta	pentakis	10	deca	decakis
6	hexa	hexakis			

   Examples: dichlorobis(ethylenediamine)cobalt(III), $[Co(NH_2CH_2CH_2NH_2)_2Cl_2]^+$

   tris(bipyridine)iron(II), $[Fe(NH_4C_5\text{-}C_5H_4N)_3]^{2+}$

4. Ligands are named in alphabetical order—according to the name of the ligand, not the prefix—although exceptions to this rule are common. An earlier rule gave anionic ligands first, then neutral ligands, each listed alphabetically.

Examples: tetraamminedichlorocobalt(III), $[Co(NH_3)_4Cl_2]^+$ (tetraammine is alphabetized by *a* and dichloro by *c*, not by the prefixes)
amminebromochloromethylamineplatinum(II), $Pt(NH_3)BrCl(CH_3NH_2)$

5. Anionic ligands are given an *o* suffix. Neutral ligands retain their usual name. Coordinated water is called *aqua* and coordinated ammonia is called *ammine.*

Examples:
chloro, Cl^-
bromo, Br^-
methylamine, CH_3NH_2
ammine, NH_3 (the double *m* distinguishes NH_3 from alkyl amines)
sulfato, SO_4^{2-}
aqua, H_2O

6. Two systems exist for designating charge or oxidation number:

a. The *Stock system* puts the calculated oxidation number of the metal ion as a Roman numeral in parentheses after the name of the metal. This is the more common convention, although there are cases in which it is difficult to assign oxidation numbers.

b. The *Ewing-Bassett system* puts the charge on the coordination sphere in parentheses after the name of the metal. This convention is used by *Chemical Abstracts* and offers an unambiguous identification of the species.

In either case, if the charge is negative, the suffix *-ate* is added to the name of the coordination sphere.

Examples: tetraammineplatinum(II) or tetraammineplatinum (2+), $[Pt(NH_3)_4]^{2+}$
tetrachloroplatinate(II) or tetrachloroplatinate (2–), $[PtCl_4]^{2-}$
hexachloroplatinate(IV) or hexachloroplatinate (2–), $[PtCl_6]^{2-}$

7. The prefixes *cis-* and *trans-* designate adjacent and opposite geometric locations. Examples are in Figures 9.1 and 9.5. Other prefixes are used as well and will be introduced as needed.

Examples: *cis-* and *trans*-diamminedichloroplatinum(II), $[PtCl_2(NH_3)_2]$
cis- and *trans*-tetraamminedichlorocobalt(III), $[CoCl_2(NH_3)_4]^+$

cis *trans*

FIGURE 9.5 *cis* and *trans* Isomers of Diamminedichloroplat-inum(II), $[PtCl_2(NH_3)_2]$. The *cis* isomer, also known as *cisplatin*, is used in cancer treatment.

8. Bridging ligands between two metal ions, as in Figures 9.2 and 9.6, have the prefix μ-.

Examples: tris(tetraammine-μ-dihydroxocobalt)cobalt(6+), $[Co(Co(NH_3)_4(OH)_2)_3]^{6+}$

μ-amido-μ-hydroxobis(tetrammінecobalt)(4+), $[(NH_3)_4Co(OH)(NH_2)Co(NH_3)_4]^{4+}$

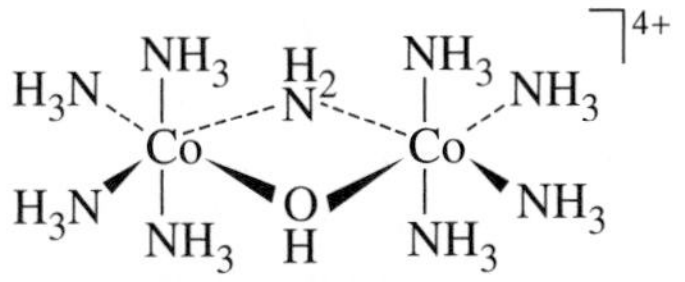

FIGURE 9.6 Bridging Amide and Hydroxide Ligands. In μ-amido-μ-hydroxobis (tetraamminecobalt) (4+), $[(NH_3)_4Co(OH)(NH_2)Co(NH_3)_4]^{4+}$.

9. When the complex is negatively charged, the names for the following metals are derived from the sources of their symbols, rather than from their English names:

iron (Fe)	ferrate	lead (Pb)	plumbate
silver (Ag)	argentate	tin(Sn)	stannate
		gold (Au)	aurate

Examples: tetrachloroferrate(III) or tetrachloroferrate(1–), $[FeCl_4]^-$

dicyanoaurate(I) or dicyanoaurate(1–), $[Au(CN)_2]^-$

▶ **Exercise 9.1** Name the following coordination complexes:

a. $Cr(NH_3)_3Cl_3$
b. $Pt(en)Cl_2$
c. $[Pt(ox)_2]^{2-}$
d. $[Cr(H_2O)_5Br]^{2+}$
e. $[Cu(NH_2CH_2CH_2NH_2)Cl_4]^{2-}$
f. $[Fe(OH)_4]^-$

▶ **Exercise 9.2** Give the structures of the following coordination complexes:

a. Tris(acetylacetonato)iron(III)
b. Hexabromoplatinate(2–)
c. Potassium diamminetetrabromocobaltate(III)
d. Tris(ethylenediamine)copper(II) sulfate
e. Hexacarbonylmanganese(I) perchlorate
f. Ammonium tetrachlororuthenate(1–)

9.3 ISOMERISM

The variety of coordination numbers in these compounds, as compared with organic compounds, provides a large number of **isomers**; the number of possible isomers increases with coordination number. We will focus our discussion on the more common coordination numbers, primarily 4 and 6. We will also limit our discussion to isomers with the same ligands arranged in different geometries and not consider cases where the ligands *themselves* are isomers. For example, coordination compounds of the ligands 1-aminopropane and 2-aminopropane are isomers, but we will not include them in our discussion.

Isomers in coordination chemistry include many types. **Hydrate** or **solvent isomers**, **ionization isomers**, and **coordination isomers** have the same overall formula but have different ligands attached to the central atom or ion. The terms **linkage isomerism** or **ambidentate isomerism** are used for cases of bonding through different atoms of the same ligand. **Stereoisomers** have the same ligands, but differ in the geometric arrangement of the ligands. The diagram and examples that follow may help make the distinctions clearer.

9.3.1 Stereoisomers

Stereoisomers include *cis* and *trans* isomers, chiral isomers, compounds with different conformations of chelate rings, and other isomers that differ only in the geometry of attachment to the metal ion. As mentioned at the beginning of this chapter, study of

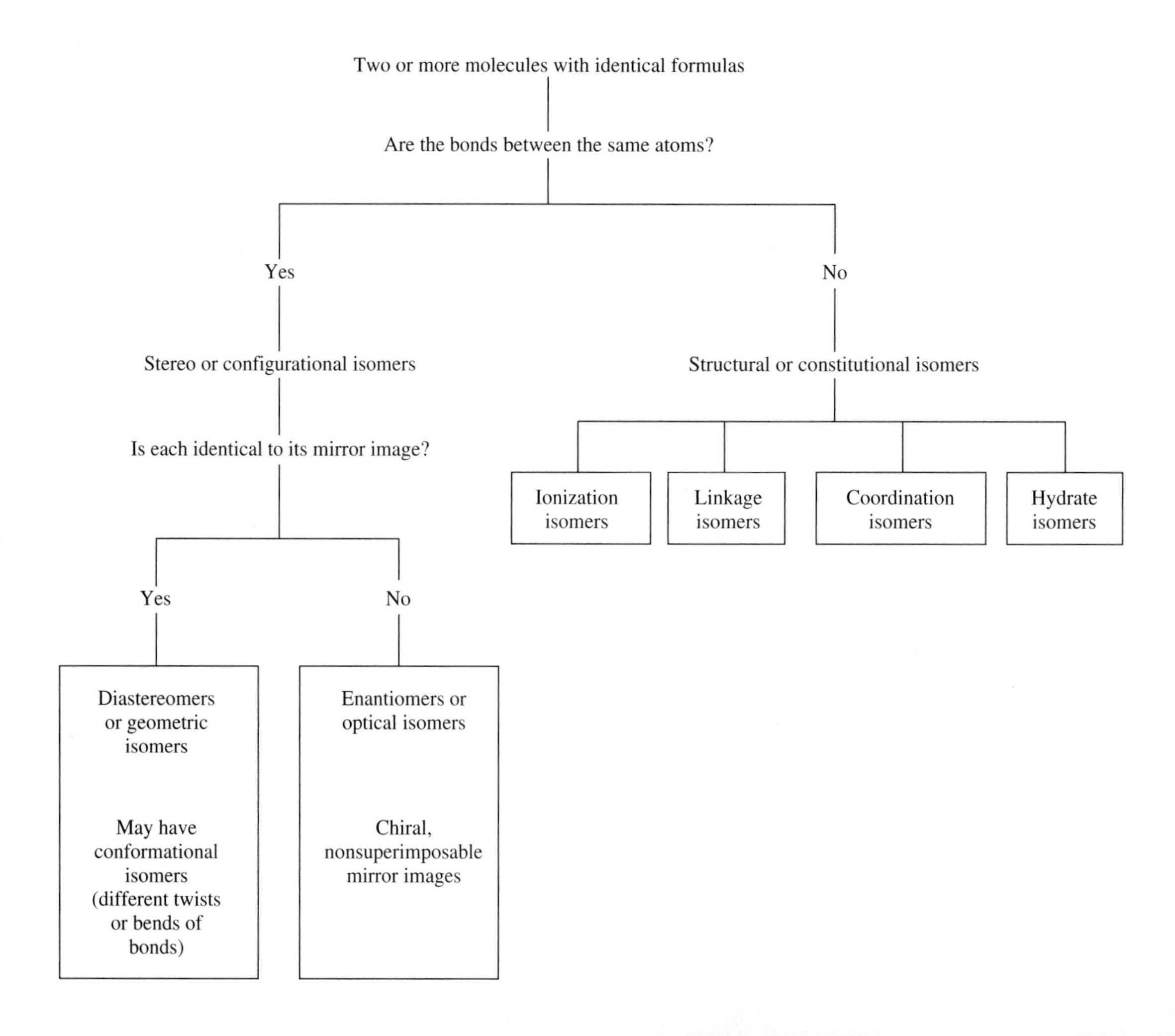

stereoisomers provided much of the experimental evidence used by Werner to develop and defend his coordination theory. Similar study of new compounds is useful in establishing structures and reactions, even though development of experimental methods, such as automated X-ray crystallography, can shorten the process considerably.

9.3.2 Four-Coordinate Complexes

Square-planar complexes may have *cis* and *trans* isomers as shown in Figure 9.4, but no chiral isomers are possible when the molecule has a mirror plane, as do many square-planar molecules. In making decisions about whether a molecule has a mirror plane, we usually ignore minor changes in the ligand, such as rotation of substituent groups, conformational changes in ligand rings, and bending of bonds. Examples of chiral square-planar complexes are the platinum(II) and palladium(II) isomers in Figure 9.7. In this case, the geometry of the ligands rules out mirror planes. If the complexes were tetrahedral, only one structure would be possible, with a mirror plane splitting the molecule between the two phenyl groups and between the two methyl groups.

Cis and *trans* isomers of square-planar complexes are common, with platinum(II) being one of the most common metal ions studied. The isomers of $[Pt(NH_3)_2Cl_2]$ are shown in Figure 9.4. The *cis* isomer is used in medicine as the antitumor agent cisplatin (see Chapter 16). Chelate rings can require the *cis* structure, because the chelating ligand is too small to span the *trans* positions. The distance across the two *trans* positions is too large for all but very large ligands.

9.3.3 Chirality

Chiral molecules (Greek, *kheir*, "hand") have a degree of asymmetry that makes their mirror images nonsuperimposable. This condition can also be expressed in terms of symmetry elements. A molecule can be chiral only if it has no rotation-reflection (S_n) axes. This means that chiral molecules either have no symmetry elements or have only axes of proper rotation (C_n); see Section 4.4.1. Tetrahedral molecules with four different ligands or with unsymmetrical chelating ligands can be chiral, as can octahedral molecules with bidentate or higher chelating ligands, or with $[Ma_2b_2c_2]$, $[Mabc_2d_2]$, $[Mabcd_3]$, $[Mabcde_2]$, or $[Mabcdef]$ structures, where M = metal and a, b, c, d, e, f are monodentate ligands. Not all isomers of such molecules are chiral, but the possibility must be considered for each.

Because of the inherent symmetry of the tetrahedron, the only isomers possible for tetrahedral complexes are chiral.

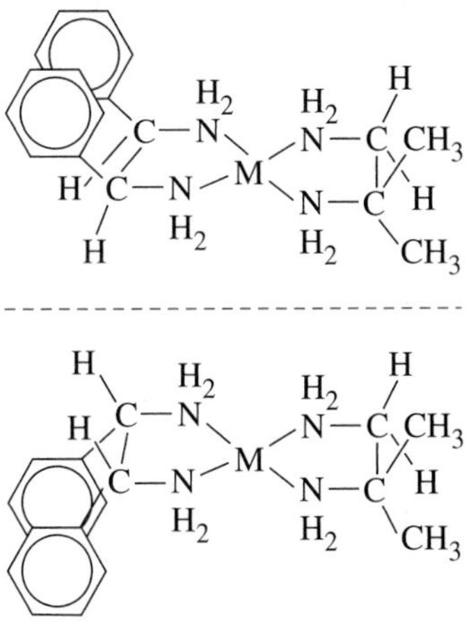

FIGURE 9.7 Chiral Isomers of Square-Planar Complexes. (*Meso*-stilbenediamine)(*iso*-butylenediamine) platinum(II) and palladium(II). (From W. H. Mills and T. H. H. Quibell, *J. Chem. Soc.*, **1935**, 839; A. G. Lidstone and W. H. Mills, *J. Chem. Soc.*, **1939**, 1754.)

9.3.4 Six-Coordinate Complexes

Complexes of the formula $ML_3L'_3$, where L and L′ are monodentate ligands, may have two isomeric forms called *fac*- (facial) and *mer*- (meridional). *Fac* isomers have three identical ligands on one triangular face; *mer* isomers have three identical ligands in a plane bisecting the molecule. Similar isomers are possible with some chelating ligands. Examples with monodentate and tridentate ligands are shown in Figure 9.8.

Special nomenclature has been proposed for other isomers of a similar type. For example, triethylenetetramine compounds have three forms: α, with all three chelate rings in different planes; β, with two of the rings coplanar; and *trans*, with all three rings coplanar, as in Figure 9.9. Additional isomeric forms are possible, some of which will be discussed later in this chapter (both α and β have chiral isomers, and all three have additional isomers that depend on the conformations of the individual rings). Even when one multidentate ligand has a single geometry, other ligands may result in isomers. For example, the β, β', β''-triaminotriethylamine (tren) ligand bonds to four adjacent sites, but an asymmetric ligand such as salicylate can then bond in the two ways shown in Figure 9.10, with the carboxylate either *cis* or *trans* to the tertiary nitrogen.

Other isomers are possible when the number of different ligands is increased. There have been several schemes for calculating the maximum number of isomers for each case,[13] although omissions were difficult to avoid until computer programs were used to assist in the process. One such program[14] begins with a single structure, generates all the others by switching ligands from one position to another, and then rotates the new form to all possible positions for comparison with the earlier structures. It is also possible to calculate the number of isomers using group theory, in a procedure developed by mathematician George Pólya.[15]

One approach to tabulating isomers is shown in Figure 9.11 and Table 9.5. The notation <ab> indicates that a and b are *trans* to each other; M is the metal; and a, b, c, d, e, and f are monodentate ligands. The $[M < ab > < cd > < ef >]$ isomers of $Pt(py)(NH_3)(NO_2)(Cl)(Br)(I)$ are examples,[16] shown in Figure 9.11. The six octahedral

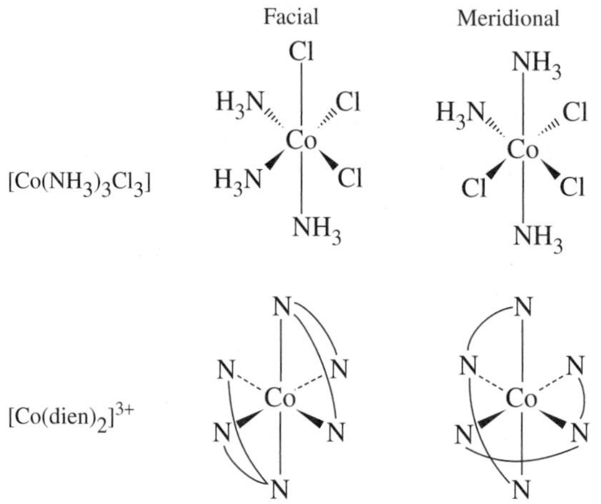

FIGURE 9.8 Facial and Meridional Isomers of $[Co(NH_3)_3Cl_3]$ and $[[Co(dien)_2]^{3+}$.

[13] J. C. Bailar, Jr., *J. Chem. Educ.*, **1957**, *34*, 334; S. A. Meyper, *J. Chem. Educ.*, **1957**, *34*, 623.
[14] W. E. Bennett, *Inorg. Chem.*, **1969**, *8*, 1325.
[15] S. Pevac and G. Crundwell, *J. Chem. Educ.*, **2000**, *77*, 1358; I. Baraldi and D. Vanossi, *J. Chem. Inf. Comput. Sci.*, **1999**, *40*, 386.
[16] L. N. Essen and A. D. Gel'man, *Zh. Neorg. Khim.*, **1956**, *1*, 2475.

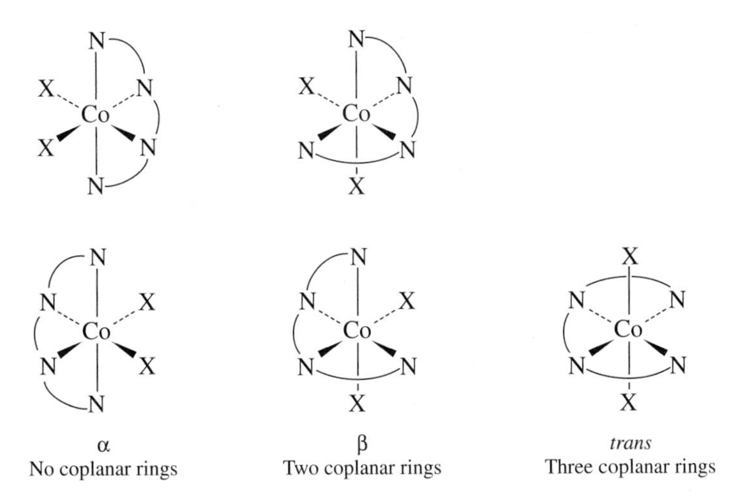

FIGURE 9.9 Isomers of Triethylenetetramine Complexes.

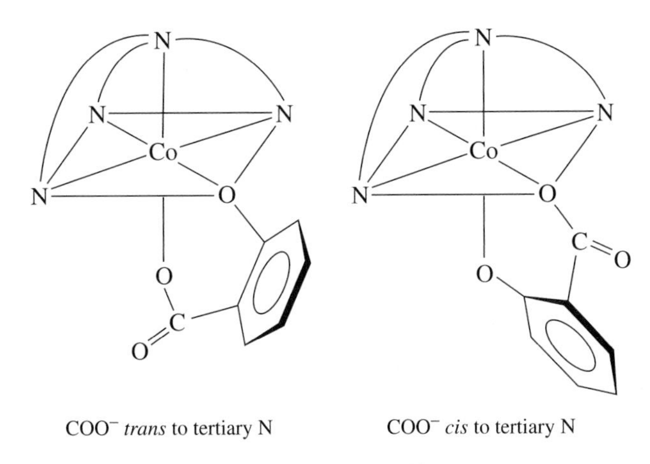

FIGURE 9.10 Isomers of $[Co(tren)(sal)]^+$.

positions are commonly numbered as in the figure, with positions 1 and 6 in axial positions and 2 through 5 in counterclockwise order as viewed from the 1 position.

If the ligands are completely scrambled, rather than limited to the *trans* pairs shown in Figure 9.11, there are 15 different diastereoisomers—structures that are not mirror images of each other—each of which has an enantiomer, or mirror image. This means that a complex with six different ligands in an octahedral shape can have 30 different isomers! The isomers of [Mabcdef] are given in Table 9.5. Each of the 15 entries represents an isomer and its enantiomer, for a total of 30 isomers.

Finding the number and identity of the isomers of a complex is primarily a matter of systematically listing the possible structures, then checking for duplicates and chirality. The method suggested by Bailar uses a list of isomers. One *trans* pair, such as <ab>, is held constant; the second pair has one component constant, and the other is systematically changed; and the third pair is whatever is left over. Then, the second component of the first pair is changed, and the process is continued.

This procedure generates the results in Table 9.5. Isomers A1 (a pair of enantiomers), are shown in Figure 9.11. The same approach can be used for chelating ligands, with limits on the location of the ring. For example, a normal bidentate chelate ring cannot connect *trans* positions.

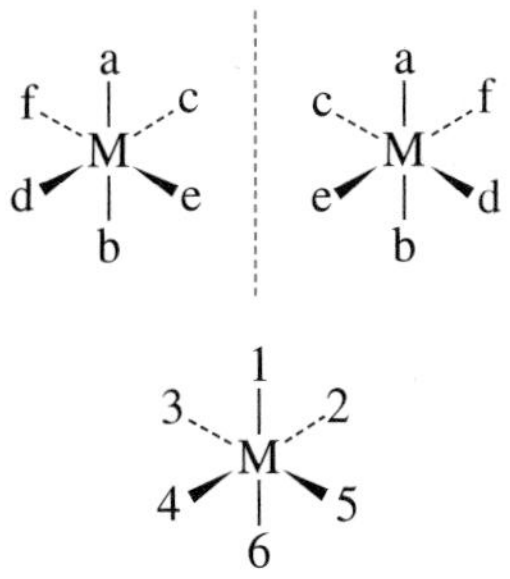

FIGURE 9.11 [M<ab><cd><ef>] Isomers and the Octahedral Numbering System.

TABLE 9.5 [Mabcdef] Isomers[a]

	A	B	C
1	ab cd ef	ab ce df	ab cf de
2	ac bd ef	ac be df	ac bf de
3	ad bc ef	ad be cf	ad bf ce
4	ae bc df	ae bf cd	ae bd cf
5	af bc de	af bd ce	af be cd

NOTE: [a]Each 1×3 box is a set of three *trans* pairs of ligands. For example, box C3 represents the two enantiomers of [M < ad > < bf > < ce >].

EXAMPLE

The isomers of $Ma_2b_2c_2$ can be found by this method. In each row below, the first pair of ligands is held constant: <aa>, <ab>, and <ac> in rows 1, 2, and 3, respectively. In column B, one component of the second pair is traded for a component of the third pair (for example, in row 2, <ab> and <cc> become <ac> and <bc>).

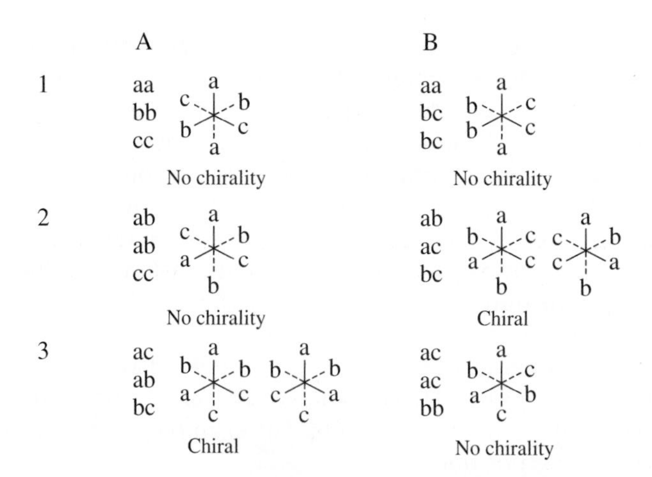

Once all the *trans* arrangements are listed, drawn, and checked for chirality, we can check for duplicates; in this case, A3 and B2 are identical. Overall, there are four nonchiral isomers and one chiral pair, for a total of six isomers.

▶ **Exercise 9.3** Find the number and identity of all the isomers of [Ma_2b_2cd].

After listing all the isomers without this restriction, those that are sterically impossible can be quickly eliminated and the others checked for duplicates and then for enantiomers. Table 9.6 lists the number of isomers and enantiomers for many general formulas, all calculated using a computer program similar to Bennett's.[17]

TABLE 9.6 Number of Possible Isomers for Specific Complexes

Formula	Number of Stereoisomers	Pairs of Enantiomers
Ma_6	1	0
Ma_5b	1	0
Ma_4b_2	2	0
Ma_3b_3	2	0
Ma_4bc	2	0
Ma_3bcd	5	1
Ma_2bcde	15	6
Mabcdef	30	15
$Ma_2b_2c_2$	6	1
Ma_2b_2cd	8	2
Ma_3b_2c	3	0
M(AA)(BC)de	10	5
M(AB)(AB)cd	11	5
M(AB)(CD)ef	20	10
$M(AB)_3$	4	2
M(ABA)cde	9	3
$M(ABC)_2$	11	5
M(ABBA)cd	7	3
M(ABCBA)d	7	3

NOTE: Uppercase letters represent chelating ligands, and lowercase letters represent monodentate ligands.

[17] W. E. Bennett, *Inorg. Chem.*, **1969**, *8*, 1325; B. A. Kennedy, D. A. MacQuarrie, and C. H. Brubaker, Jr., *Inorg. Chem.*, **1964**, *3*, 265.

EXAMPLE

A methodical approach is important in finding isomers. AA and BB must be in *cis* positions, because they are linked in the chelate ring. For M(AA)(BB)cd, we first try c and d in *cis* positions. One A and one B must be *trans* to each other:

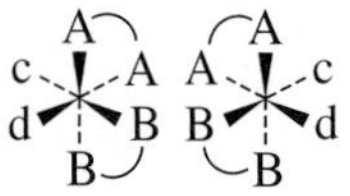

c opposite B
d opposite A

The mirror image is different, so there is a chiral pair.

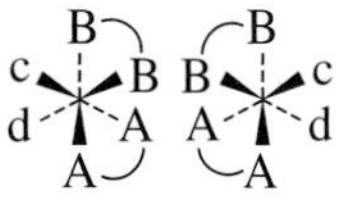

c opposite A
d opposite B

The mirror image is different, so there is a chiral pair.

Then, trying c and d in *trans* positions, where AA and BB are in the horizontal plane:

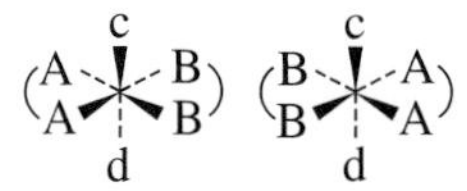

The mirror images are identical, so there is only one isomer. There are two chiral pairs and one individual isomer, for a total of five isomers.

▶ **Exercise 9.4** Find the number and identity of all isomers of [M(AA)bcde], where AA is a bidentate ligand with identical coordinating groups.

9.3.5 Combinations of Chelate Rings

Before discussing nomenclature rules for ring geometry, we need to establish clearly the idea of the handedness of propellers and helices. Consider the propellers shown in Figure 9.12. The first is a left-handed propeller, which means that rotating it *counterclockwise* in air or water would move it away from the observer. The second, a right-handed propeller, moves away on *clockwise* rotation. The tips of the propeller blades describe left- and right-handed helices, respectively. With rare exceptions, the threads on screws and bolts are right-handed helices; a clockwise twist with a screwdriver or wrench drives them into a nut or piece of wood. The same clockwise motion drives a nut onto a stationary bolt. Another example of a helix is a coil spring, which can usually have either handedness without affecting its operation.

Complexes with three rings, such as $[Co(en)_3]^{3+}$, can be treated like three-bladed propellers by looking at the molecule down a threefold axis. Figure 9.13 shows a number of different ways to draw these structures, all equivalent. The counterclockwise (Λ) or clockwise (Δ) character can also be found by the procedure in the next paragraph.

Complexes with two or more nonadjacent chelate rings may have chiral character. Any two non-coplanar and nonadjacent chelate rings (not sharing a common atom bonded to the metal) can be used to determine the handedness. Figure 9.14 illustrates the process, which can be summarized as follows:

1. Rotate the figure to place one ring horizontally across the back, at the top of one of the triangular faces.

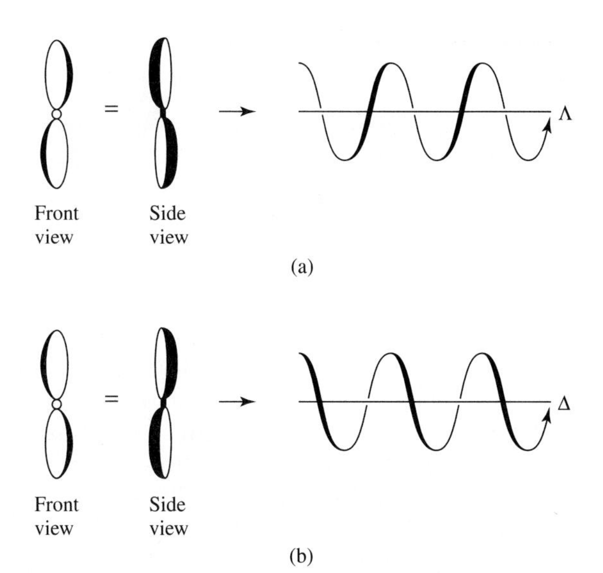

FIGURE 9.12 Right- and Left-Handed Propellers. (a) Left-handed propeller and helix traced by the tips of the blades. (b) Right-handed propeller and helix traced by the tips of the blades.

2. Imagine the ring in the front triangular face as having originally been parallel to the ring at the back. Determine what rotation of the front face is required to obtain the actual configuration.
3. If the rotation from Step 2 is counterclockwise, the structure is designated lambda (Λ). If the rotation is clockwise, the designation is delta (Δ).

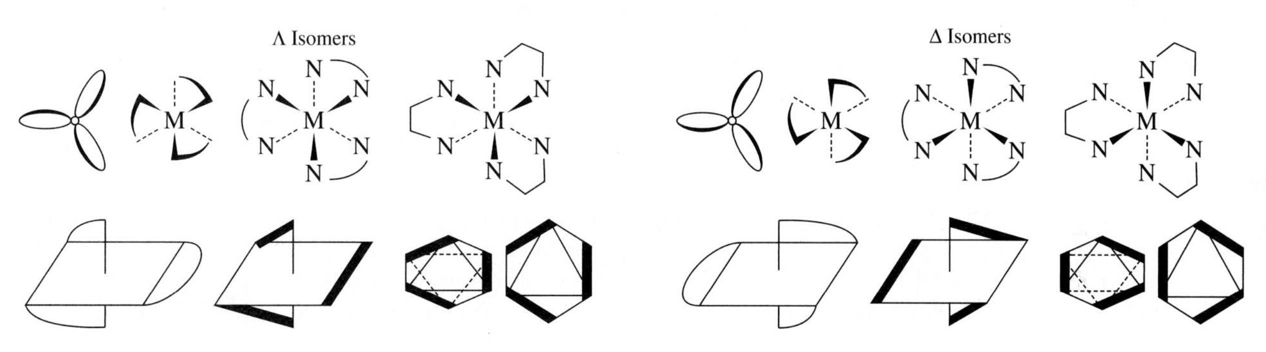

FIGURE 9.13 Left- and Right-Handed Chelates.

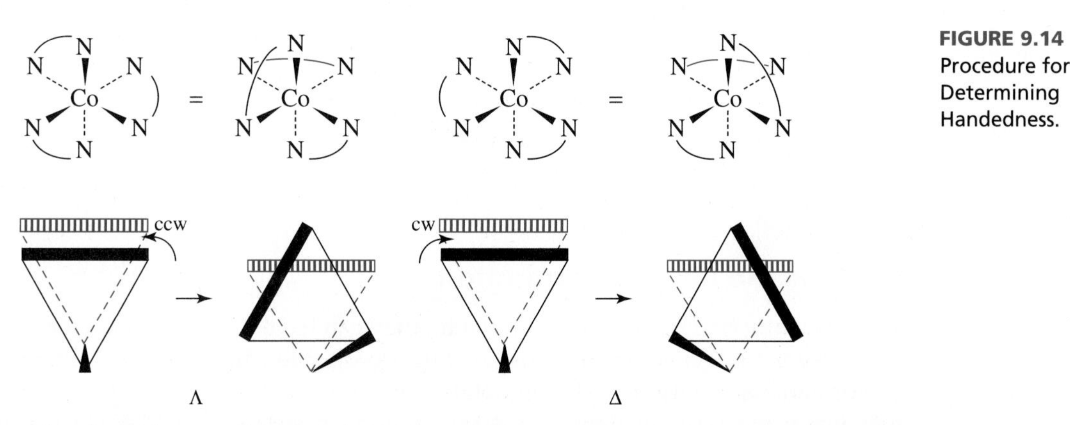

FIGURE 9.14 Procedure for Determining Handedness.

A molecule with more than one pair of rings may require more than one label, but it is treated similarly. The handedness of each pair of skew rings is determined, and the final description then includes all the designations. For example, an EDTA complex has six points of attachment and five rings. One isomer is shown in Figure 9.15, where the

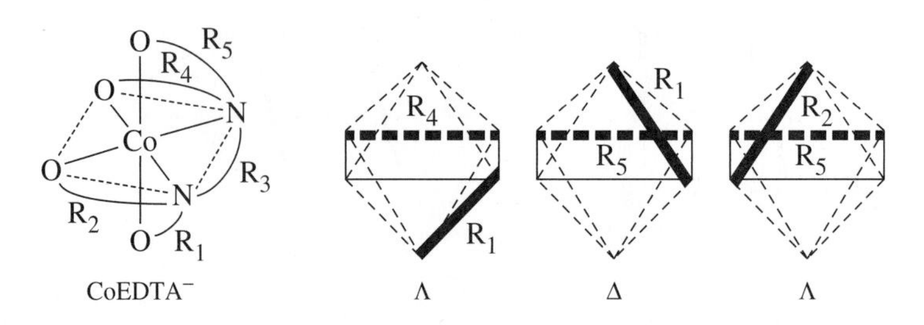

FIGURE 9.15 Labeling of Chiral Rings. The rings are numbered arbitrarily R_1 through R_5. The combination R_1-R_4 is Λ, R_1-R_5 is Δ, and R_2-R_5 is Λ. The notation for this structure is then $\Lambda\Delta\Lambda$-(ethylenediaminetetraacetato)cobaltate(III).

rings are numbered arbitrarily R_1 through R_5. All ring pairs that are not coplanar and are not connected at the same atom are used in the description. The N—N ring (R_3) is omitted, because it is connected at the same atom with each of the other rings. Considering only the four O—N rings, there are three useful pairs, R_1-R_4, R_1-R_5, and R_2-R_5. The fourth pair, R_2-R_4, is not used because the two rings are coplanar. The method described above gives Λ for R_1-R_4, Δ for R_1-R_5, and Λ for R_2-R_5. The notation for the compound given is then $\Lambda\Delta\Lambda$-(ethylenediaminetetraacetato)cobaltate(III). The order of the designations is arbitrary and could also be $\Lambda\Lambda\Delta$ or $\Delta\Lambda\Lambda$.

EXAMPLE

Determine the chirality label(s) for the complex shown:

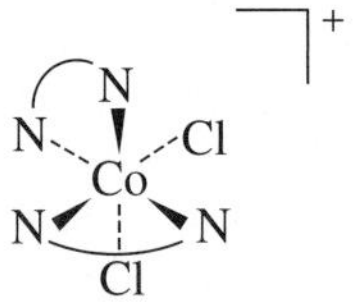

Rotating the figure 180° about the vertical axis puts one ring across the back and the other connecting the top and the front right positions. If this front ring were originally parallel to the back one, a clockwise rotation would put it into the correct position. Therefore, the structure is Δ-*cis*-dichlorobis(ethylenediamine)cobalt(III).

▶ **Exercise 9.5** Determine the chirality label(s) for the complex shown:

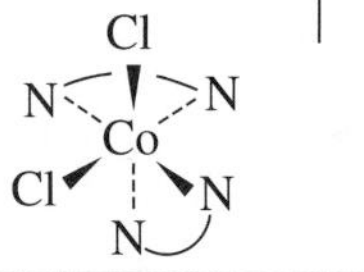

9.3.6 Ligand Ring Conformation

Because many chelate rings are not planar, they can have different conformations in different molecules, even in otherwise identical molecules. In some cases, these different conformations are also chiral. The notation used in these situations also requires using two lines to establish the handedness and the lowercase labels λ and δ. The first line connects the atoms bonded to the metal. In the case of ethylenediamine, this line connects the two nitrogen atoms. The second line connects the two carbon atoms of the ethylenediamine, and the handedness of the two rings is found by the method described in

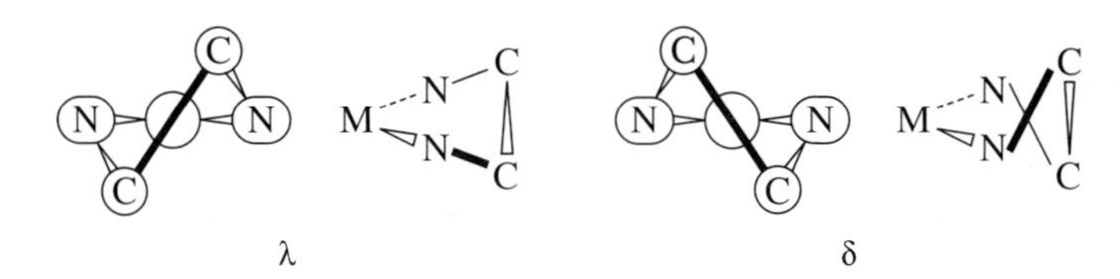

FIGURE 9.16 Chelate Ring Conformations.

Section 9.3.5 for separate rings. A counterclockwise rotation of the second line is called λ and a clockwise rotation is called δ, as shown in Figure 9.16. Complete description of a complex then requires identification of the overall chirality and the chirality of each ring.

Corey and Bailar[18] examined some examples and observed the same steric interactions found in cyclohexane and other ring structures. For example, the $\Delta\lambda\lambda\lambda$ form of $[Co(en)_3]^{3+}$ was calculated to be 7.5 kJ/mol more stable than the $\Delta\delta\delta\delta$ form because of interactions between protons on the nitrogens. For the Λ form, the $\delta\delta\delta$ ring conformations are more stable. Although there are examples in which this preference is not followed, in general, the experimental results have confirmed these calculations. In solution, the small difference in energy allows rapid interconversion of conformation between λ and δ, and the most abundant configuration for the Λ isomer is $\delta\delta\lambda$.[19]

An additional isomeric possibility arises because the symmetry of ligands can be changed by coordination. An example is a secondary amine in a ligand such as diethylenetriamine (dien) or triethylenetetraamine (trien). As a free ligand, inversion at the nitrogen is easy, and only one isomer is possible. After coordination there may be additional chiral isomers. If there are chiral centers on the ligands, either inherent in their structure or created by coordination (as in some secondary amines), their structure must be described by the R and S notation familiar from organic chemistry.[20] The trien structures are illustrated in Figures 9.17 and 9.18 and are described in the following example. The α, β, and *trans* structures appear in Figure 9.9 without the ring conformations.

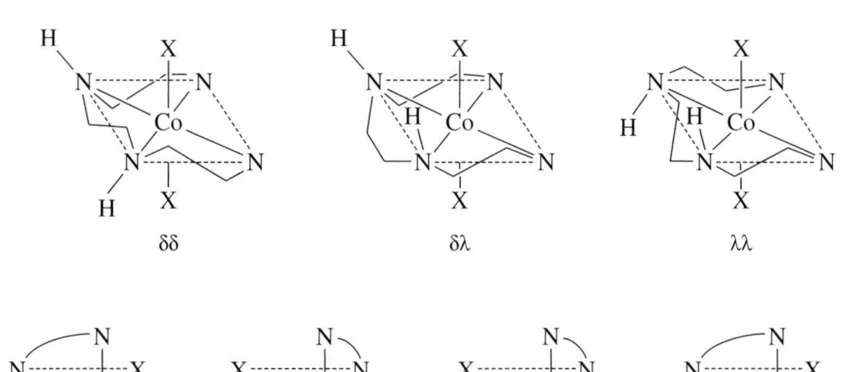

FIGURE 9.17 Chiral Structures of *trans*-$[CoX_2(trien)]^+$.

FIGURE 9.18 The α and β Forms of $[CoX_2(trien)]^+$.

[18] E. J. Corey and J. C. Bailar, Jr., *J. Am. Chem. Soc.*, **1959**, *81*, 2620.
[19] J. K. Beattie, *Acc. Chem. Res.*, **1971**, *4*, 253.
[20] R. S. Cahn and C. K. Ingold, *J. Chem. Soc.*, **1951**, 612; Cahn, Ingold, and V. Prelog, *Experientia*, **1956**, *12*, 81.

EXAMPLE

Confirm the chirality of the rings in the *trans*-$[CoX_2\,trien]^+$ structures in Figure 9.17.

Take the ring on the front edge of the first structure, with the line between the two nitrogens as the reference. If the line connecting the two carbons was originally parallel to the N—N line, a clockwise rotation is required to reach the actual conformation, so it is δ. The ring on the back of the molecule is the same, so it is also δ. The tetrahedral nature of the ligand N forces the hydrogens on the two secondary nitrogens into the positions shown, so the middle ring must be λ. However, it need not be labeled as such, because there is no other possibility. The label is then $\delta\delta$.

The same procedure on the other two structures results in labels of $\delta\lambda$ and $\lambda\lambda$, respectively. Again, the middle ring has only one possible structure, so it need not be labeled.

► **Exercise 9.6** $[Co(dien)_2]^{3+}$ can have several forms, two of which are shown below. Identify the Δ or Λ chirality of the rings, using all unconnected pairs. Each complex may have three labels.

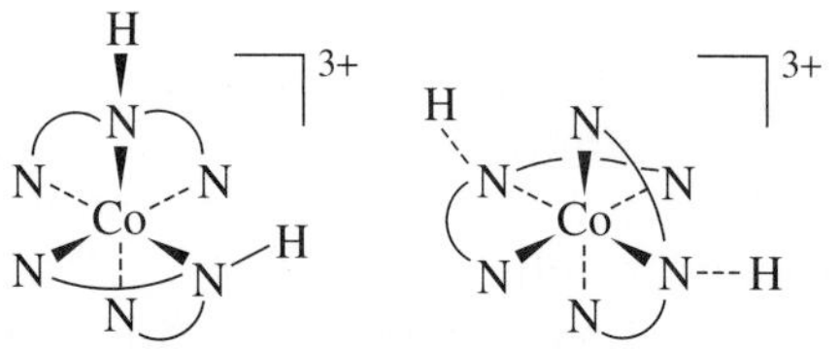

9.3.7 Constitutional Isomers

HYDRATE ISOMERISM Hydrate isomerism is not common but deserves mention, because it contributed to some of the confusion in describing coordination compounds before the Werner theory was generally accepted. Hydrate isomerism differs from other types of isomerism in that it has water as either a ligand or as an added part of the crystal structure, as in the hydrates of sodium sulfate (Na_2SO_4, $Na_2SO_4 \cdot 7\,H_2O$, and $Na_2SO_4 \cdot 10\,H_2O$ are known). More strictly, it should be called *solvent isomerism* to allow for the possibility of ammonia or other ligands also used as solvents to participate in the structure.

The standard example is $CrCl_3 \cdot 6\,H_2O$, which can have three distinctly different crystalline compounds, now known as $[Cr(H_2O)_6]Cl_3$ (violet), $[CrCl(H_2O)_5]Cl_2 \cdot H_2O$ (blue-green), and $[CrCl_2(H_2O)_4]Cl \cdot 2\,H_2O$ (dark green). A fourth isomer, $[CrCl_3(H_2O)_3]$ (yellow-green) also occurs at high concentrations of HCl.[21] The three cationic isomers can be separated by cation ion exchange from commercial $CrCl_3 \cdot 6\,H_2O$, in which the major component is $[CrCl_2(H_2O)_4]Cl \cdot 2\,H_2O$ in the *trans* configuration. Other examples are also known, and a few are listed below.

$[Co(NH_3)_4(H_2O)Cl]Cl_2$	and	$[Co(NH_3)_4Cl_2]Cl \cdot H_2O$
$[Co(NH_3)_5(H_2O)](NO_3)_3$	and	$[Co(NH_3)_5(NO_3)](NO_3)_2 \cdot H_2O$

IONIZATION ISOMERISM Compounds with the same formula, but which give different ions in solution, exhibit ionization isomerization. The difference is in which ion is included as a ligand and which is present to balance the overall charge. Some examples are also hydrate isomers, such as the first pair listed.

[21] S. Diaz-Moreno, A. Muñoz-Paez, J. M. Martinez, R. R. Pappalardo, and E. S. Marcos, *J. Am. Chem. Soc.*, **1996**, *118*, 12654.

$[Co(NH_3)_4(H_2O)Cl]Br_2$	and	$[Co(NH_3)_4Br_2]Cl \cdot H_2O$
$[Co(NH_3)_5SO_4]NO_3$	and	$[Co(NH_3)_5NO_3]SO_4$
$[Co(NH_3)_4(NO_2)Cl]Cl$	and	$[Co(NH_3)_4Cl_2]NO_2$

Many other examples exist. Enthusiasm for preparing and characterizing such compounds is not great at this time, and new examples are more likely to be discovered only as part of other studies.

COORDINATION ISOMERISM Examples of a complete series of coordination isomers require at least two metal ions and sometimes more. The total ratio of ligand to metal remains the same, but the ligands attached to a specific metal ion change. This is best described by example.

For the empirical formula $Pt(NH_3)_2Cl_2$, there are three possibilities:

$[Pt(NH_3)_2Cl_2]$	
$[Pt(NH_3)_3Cl][Pt(NH_3)Cl_3]$	(This compound apparently has not been reported, but the individual ions are known.)
$[Pt(NH_3)_4][PtCl_4]$	(Magnus's green salt, the first platinum ammine, was discovered in 1828.)

Other examples are possible with different metal ions and with different oxidation states:

$[Co(en)_3][Cr(CN)_6]$	and	$[Cr(en)_3][Co(CN)_6]$
$[Pt(NH_3)_4][PtCl_6]$	and	$[Pt(NH_3)_4Cl_2][PtCl_4]$
Pt(II) Pt(IV)		Pt(IV) Pt(II)

LINKAGE (AMBIDENTATE) ISOMERISM Some ligands can bond to the metal through different atoms. The most common early examples were thiocyanate, SCN^-, and nitrite, NO_2^-. In thiocyanate, class (a) metal ions (hard acids) tend to bond to the nitrogen and class (b) metal ions (soft acids) bond through the sulfur. Solvent can also influence the point of attachment. Compounds of rhodium and iridium with the general formula $[M(PPh_3)_2(CO)(NCS)_2]$ form M—S bonds in solvents of high dipole moment (for example, acetone and acetonitrile) and M—N bonds in solvents of low dipole moment (for example, benzene and CCl_4),[22] as shown in Figure 9.19(a). There are also compounds[23] that contain both M–SCN (thiocyanato) and M–NCS (isothiocyanato); for example, isothiocyanatothiocyanato(1-diphenylphosphino-3-dimethylaminopropane)palladium(II), Figure 9.19(b). M–NCS combinations are linear, and M–SCN combinations are bent at the S atom in all thiocyanate complexes. This bend means that the M–SCN isomer has a larger steric effect, because the CN part of the ligand can rotate about the M—S bond.

▶ **Exercise 9.7** Use the HSAB concept to account for the tendency of M-SCN complexes to be favored in solvents having high dipole moments and M–NCS complexes to be favored in solvents having low dipole moments.

---

[22] J. L. Burmeister, R. L. Hassel, and R. J. Phelen, *Inorg. Chem.*, **1971**, *10*, 2032; J. E. Huheey and S. O. Grim, *Inorg. Nucl. Chem. Lett.*, **1974**, *10*, 973.

[23] D. W. Meek, P. E. Nicpon, and V. I. Meek, *J. Am. Chem. Soc.*, **1970**, *92*, 5351; G. R. Clark and G. J. Palenik, *Inorg. Chem.*, **1970**, *9*, 2754.

FIGURE 9.19 Linkage (Ambidentate) Isomers.

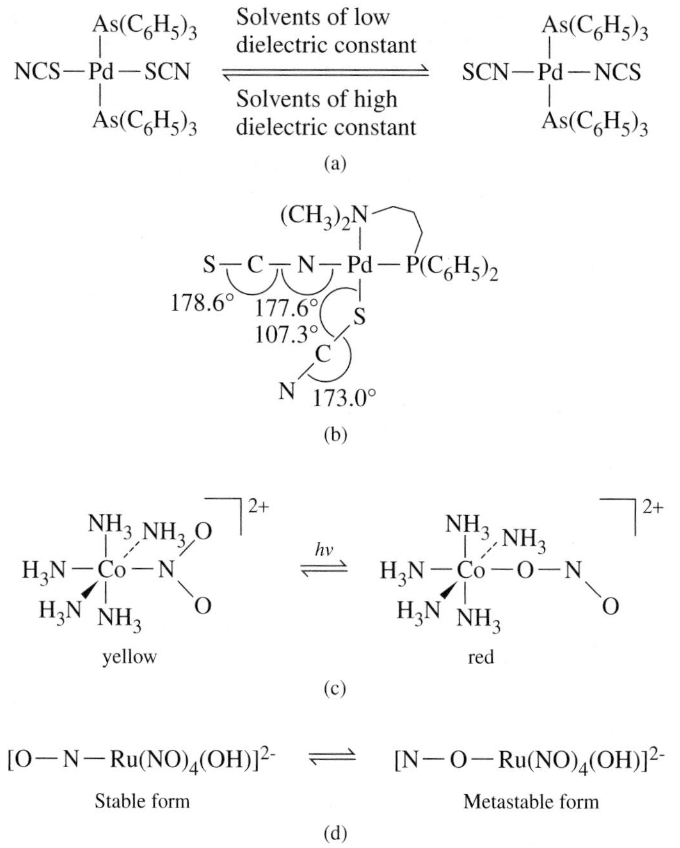

The nitrite isomers of $[Co(NH_3)_5NO_2]^{2+}$ were studied by Jørgensen and Werner, who observed that there were two compounds of the same chemical formula but of different colors [Figure 9.19(c)]. A red form of low stability converted readily to a yellow form. The red form was thought to be the M—ONO nitrito isomer and the yellow form the $M—NO_2$ nitro isomer, based on comparison with compounds of similar color. This conclusion was later confirmed, and kinetic[24] and ^{18}O labeling[25] experiments showed that conversion of one form to the other is strictly intramolecular, not a result of dissociation of the NO_2^- ion followed by reattachment. In a more recent example, the stable O—N—Ru form of $[Ru(NO)_5(OH)]^{2-}$ is in equilibrium with the metastable N—O—Ru form[26] [Figure 9.19(d)].

9.3.8 Experimental Separation and Identification of Isomers

Separation of geometric isomers frequently requires fractional crystallization with different counterions. Because different isomers have slightly different shapes, the packing in crystals depends on the fit of the ions and their overall solubility. One helpful idea, systematized by Basolo,[27] is that ionic compounds are least soluble when the positive and negative ions have the same size and magnitude of charge. For example, large cations of charge 2+ are best crystallized with large anions of charge 2−. Although not a surefire method to separate isomers, this method helps to decide what combinations to try.

---

[24] B. Adell, *Z. Anorg. Chem.*, **1944**, *252*, 277.
[25] R. K. Murmann and H. Taube, *J. Am. Chem. Soc.*, **1956**, *78*, 4886.
[26] D. V. Fornitchev and P. Coppens, *Inorg. Chem.*, **1996**, *35*, 7021.
[27] F. Basolo, *Coord. Chem. Rev.*, **1968**, *3*, 213.

Separating chiral isomers requires chiral counterions. Cations are frequently resolved by using anions *d*-tartrate, antimony *d*-tartrate, and α-bromocamphor-π-sulfonate; anionic complexes are resolved by the bases brucine or strychnine or by resolved cationic complexes such as $[Rh(en)_3]^{3+}$.[28] For compounds that racemize at appreciable rates, adding a chiral counterion may shift the equilibrium, even if it does not precipitate one form; interactions between the ions in solution may be sufficient to stabilize one form over the other.[29]

The best method of identifying isomers, when crystallization allows it, is X-ray crystallography. Current methods allow for the rapid determination of the absolute configuration at costs that compare favorably with more indirect methods.

Measurement of optical activity is a natural method for assigning absolute configuration to chiral isomers, but it usually requires more than simply determining molar rotation at a single wavelength. Optical rotation changes markedly with the wavelength of the light used in the measurement, and it changes sign near absorption peaks. Many organic compounds have their largest rotation in the ultraviolet, and the old standard of molar rotation at the sodium D wavelength (589.29 nm)[30] is a measurement of the tail of the much larger peak. Coordination compounds frequently have their major absorption (and therefore rotation) bands in the visible part of the spectrum, and it then becomes necessary to examine the rotation as a function of wavelength to determine the isomer present. Before the development of the X-ray methods now used to determine structures, debates over assignments of configuration were common, since comparison of similar compounds could lead to contradictory assignments, depending on which measurements and compounds were compared.

Polarized light can be either circularly polarized or plane polarized. When circularly polarized, the electric or magnetic vector rotates (right-handed if clockwise rotation when viewed facing the source, left-handed if counterclockwise) with a frequency related to the frequency of the light. Plane-polarized light is made up of both right- and left-handed components; when combined, the vectors reinforce each other at 0° and 180° and cancel at 90° and 270°, leaving a planar motion of the vector. When plane-polarized light passes through a chiral substance, the plane of polarization is rotated. This **optical rotatory dispersion (ORD)**, or optical rotation, is caused by a difference in the refractive indices of the right and left circularly polarized light, according to the equation

$$\alpha = \frac{\eta_l - \eta_r}{\lambda}$$

where η_l and η_r are the refractive indices for left and right circularly polarized light, and λ is the wavelength of the light. ORD is measured by passing light through a polarizing medium, then through the substance to be measured, and then through an analyzing polarizer. The polarizer is rotated until the angle at which the maximum amount of light passing through the substance is found, and the measurement is repeated at different wavelengths. ORD frequently shows a positive value on one side of an absorption maximum and a negative value on the other, passing through zero at or near the absorption maximum; it also frequently shows a long tail extending far from the absorption wavelength. When optical rotation of colorless compounds is measured using visible light, it is this tail that is measured, far from the ultraviolet absorption band. The variance with wavelength is known as the **Cotton effect**, positive when the rotation is positive (right-handed) at low energy and negative when it is positive at high energy.

---

[28] R. D. Gillard, D. J. Shepherd, and D. A. Tarr, *J. Chem. Soc., Dalton Trans.*, **1976**, 594.

[29] J. C. Bailar, ed., *Chemistry of the Coordination Compounds*, Reinhold Publishing, New York, **1956**, pp. 334–335, cites several instances, specifically $[Fe(phen)_3]^{2+}$ (Dwyer), $[Cr(C_2O_4)_3]^{3-}$ (King), and $[Co(en)_3]^{3+}$ (Jonassen, Bailar, and Huffmann).

[30] Actually a doublet at 588.99 and 589.59 nm.

Another measurement, **circular dichroism (CD)**, is caused by a difference in the absorption of right- and left-circularly polarized light, defined by the equation

$$\text{Circular dichroism} = \varepsilon_l - \varepsilon_r$$

where ε_l and ε_r are the molar absorption coefficients for left- and right-circularly polarized light. CD spectrometers have an optical system much like UV-visible spectrophotometers with the addition of a crystal of ammonium phosphate mounted to allow imposition of a large electrostatic field on it. When the field is imposed, the crystal allows only circularly polarized light to pass through; changing the direction of the field rapidly provides alternating left- and right-circularly polarized light. The light received by the detector is compared electronically and presented as the difference between the absorbances.

Circular dichroism is usually observed only in the vicinity of an absorption band: a positive Cotton effect shows a positive peak at the absorption maximum and a negative effect shows a negative peak. This simple spectrum makes CD more selective and easier to interpret than ORD. With improvements in instrumentation, it has become the method of choice for studying chiral complexes. Both ORD and CD spectra are shown in Figure 9.20.

Even with CD, spectra are not always easily interpreted, because there may be overlapping bands of different signs.[31] Interpretation requires determination of the overall symmetry around the metal ion and assignment of absorption spectra to specific transitions between energy levels (discussed in Chapter 11) in order to assign specific CD peaks to the appropriate transitions. Even then, there are cases in which the CD peaks do not match the absorption peaks, and interpretation is more difficult.

9.4 COORDINATION NUMBERS AND STRUCTURES

The isomers described to this point have had octahedral or square-planar geometry. In this section, we describe some other common geometries. Explanations for some of the shapes are easy and follow the VSEPR approach presented in Chapter 3, usually ignoring the *d* electrons of the metal. In these cases, three-coordinate complexes have a trigonal-planar shape, four-coordinate complexes are tetrahedral, and so forth, assuming that each ligand–metal bond results from a two-electron donor atom interacting with the metal. Some complexes do not follow these rules and either require more elaborate explanations or have no ready explanation.

The overall shape of a coordination compound is the product of several interacting factors. One factor may be dominant in one compound, with another factor dominant in another. Factors involved in determining the structures of coordination complexes include:

1. **The number of bonds.** Because bond formation is usually considered exothermic, more bonds should make for a more stable structure.
2. **VSEPR considerations,** as used in the simpler cases of the main group elements.
3. **Occupancy of *d* orbitals.** Examples of how the number of *d* electrons may affect the geometry (e.g., square-planar versus tetrahedral) will be discussed in Chapter 10.
4. **Steric interference,** by large ligands crowding each other around the central metal.
5. **Crystal packing effects.** These include the effects resulting from the sizes of ions and the overall shape of coordination complexes. The regular shape may be distorted when it is packed into a crystalline lattice, and it may be difficult to determine whether deviations from regular geometry are caused by effects within a given unit or by packing into a crystal.

[31] R. D. Gillard, "Optical Rotatory Dispersion and Circular Dichroism," in H. A. O. Hill and P. Day, eds., *Physical Methods in Advanced Inorganic Chemistry*, Wiley InterScience, New York, 1968, pp. 183–185; C. J. Hawkins, *Absolute Configuration of Metal Complexes*, Wiley InterScience, New York, 1971, p. 156.

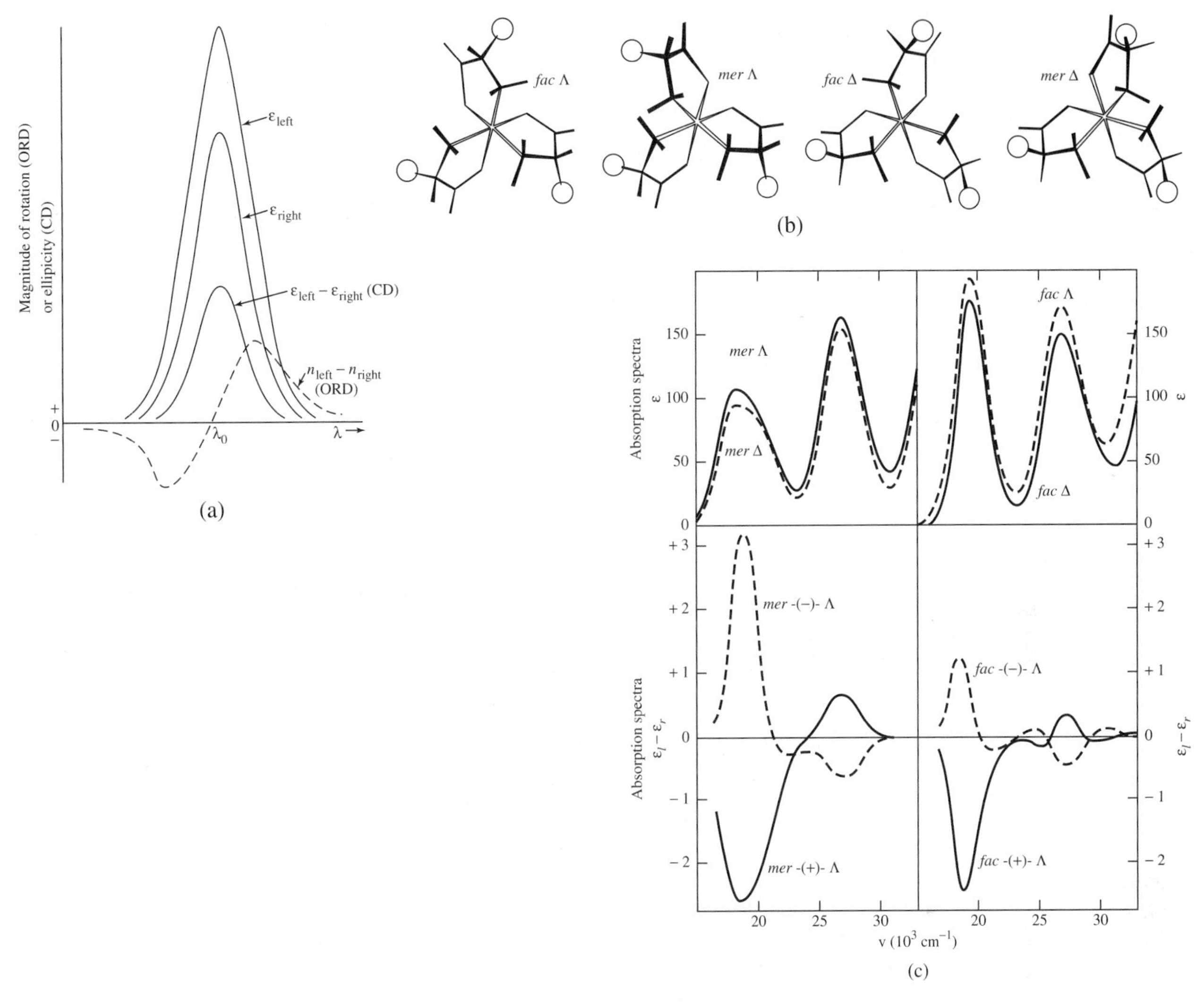

FIGURE 9.20 The Cotton Effect in ORD and CD. (a) Idealized optical rotatory dispersion (ORD) and circular dichroism (CD) curves at an absorption peak, with a positive Cotton effect. (b) Structures of tris-(S-alaninato)cobalt(III) complexes. (c) Absorption and circular dichroism spectra of the compounds in (b). (Data and structures in (b) adapted with permission from R. G. Denning and T. S. Piper, *Inorg. Chem.*, **1966**, *5*, 1056. © 1966 American Chemical Society. Curves in (c) adapted with permission from J. Fujita and Y. Shimura, "Optical Rotatory Dispersion and Circular Dichroism," in K. Nakamoto and P. J. McCarthy, eds., *Spectroscopy and Structure of Metal Chelate Compounds*, John Wiley & Sons Inc., New York, 1968, p. 193. © 1968 John Wiley & Sons, Inc. Reprinted by permission of John Wiley & Sons, Inc.)

The angles in a crystal lattice may fit none of the ideal cases. It is frequently difficult to predict shapes, and all predictions should be addressed skeptically unless backed by experimental evidence.

9.4.1 Coordination Numbers 1, 2, and 3

Coordination number 1 is rare, except in ion pairs in the gas phase. Even species in aqueous solution that seem to be singly coordinated usually have water attached as well and have an overall coordination number higher than 1. Two organometallic compounds

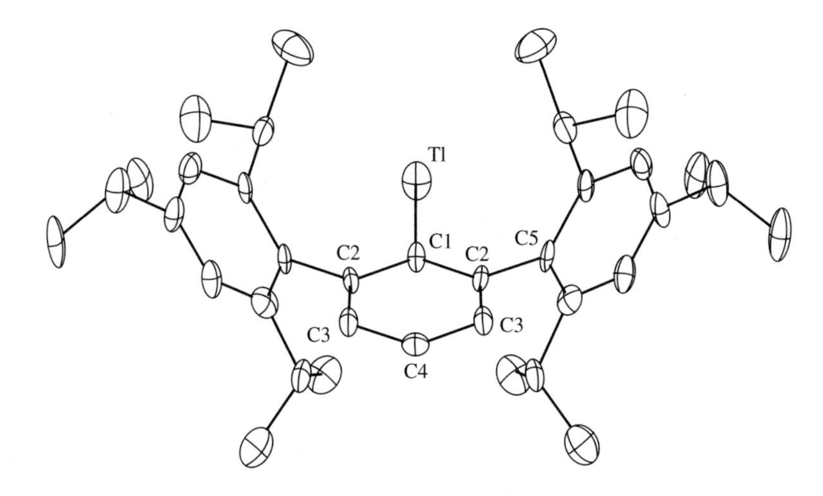

FIGURE 9.21 Coordination Number 1. Shown is 2,6- $Trip_2C_6H_3Tl$ (Trip = 2,4,6-$iPr_3C_6H_2$). (Reproduced with permission from M. Niemeyer and P. P. Power, *Angew. Chem., Int. Ed.*, **1998**, *37*, 1277.)

with coordination number 1 are the Tl(I) and In(I) complexes of 2,6-$Trip_2C_6H_3$.[32] The thallium compound is shown in Figure 9.21. In spite of the very bulky ligand that prevents any bridging between metals, the indium complex can also form a complex with $Mn(\eta^5\text{-}C_5H_5)(CO)_2$, in which indium is 2-coordinate, with an In—Mn bond. $Ga[C(SiMe_3)_3]$ is an example of a monomeric, singly coordinated organometallic in the gas phase.[33] A transient species that seems to be singly coordinated is VO^{2+}.

Coordination number 2 is also rare. The best known example is $[Ag(NH_3)_2]^+$, the diamminesilver(I) ion. The silver 1+ ion is d^{10} (a filled, spherical subshell), so the only electrons to be considered in the VSEPR treatment are those forming the bonds with the ammonia ligands, and the structure is linear as expected for two bonding positions. Other well-known examples are also d^{10} and linear, for example $[CuCl_2]^-$, $Hg(CN)_2$, and $[Au(CN)_2]^-$. A d^5 example is $Mn[N(SiMePh_2)_2]_2$, shown in Figure 9.22. Examples of 2-coordinate d^6 and d^7 complexes also exist.[34,35] Large, bulky ligands such as $N(SiMePh_2)_2$ help force a linear or near-linear arrangement.

Coordination number 3 also is more likely with d^{10} ions, with a trigonal-planar structure most common. Three-coordinate Au(I) and Cu(I) complexes that are known include $[Au(PPh_3)_3]^+$, $[Au(PPh_3)_2Cl]$, and $[Cu(SPPh_3)_3]^+$.[36,37] Most three-coordinate complexes seem to have a low coordination number because of ligand

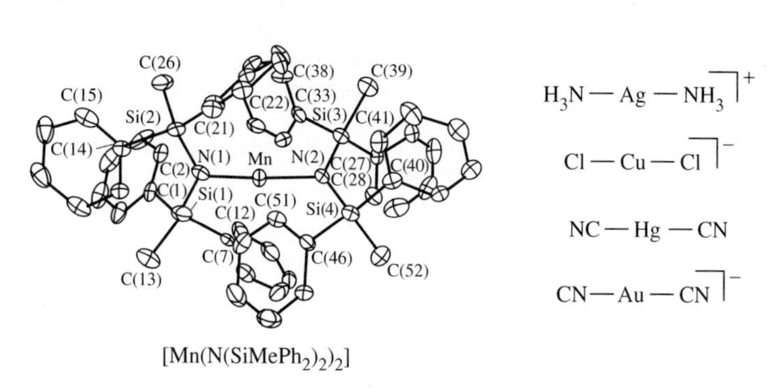

FIGURE 9.22 Complexes with Coordination Number 2. ($[Mn(N[SiMePh_2]_2)_2]$ reproduced with permission from H. Chen, R. A. Bartlett, H. V. R. Dias, M. M. Olmstead, and P. P. Power, *J. Am. Chem. Soc.*, **1989**, *111*, 4338. © 1989 American Chemical Society.)

[32] M. Niemeyer and P. P. Power, *Angew. Chem., Int. Ed.*, **1998**, *37*, 1277; S. T. Haubrich and P. P. Power, *J. Am. Chem. Soc.*, **1998**, *120*, 2202.

[33] A. Haaland, K.-G. Martinsen, H. V. Volden, W. Kaim, E. Waldhör, W. Uhl, and U. Schütz, *Organometallics*, **1996**, *15*, 1146.

[34] D. C. Bradley and K. J. Fisher, *J. Am. Chem. Soc.*, **1971**, *93*, 2058.

[35] H. Chen, R. A. Bartlett, H. V. R. Dias, M. M. Olmstead, and P. P. Power, *J. Am. Chem. Soc.*, **1989**, *111*, 4338.

[36] F. Klanberg, E. L. Muetterties, and L. J. Guggenberger, *Inorg. Chem.*, **1968**, *7*, 2273.

[37] N. C. Baenziger, K. M. Dittemore, and J. R. Doyle, *Inorg. Chem.*, **1974**, *13*, 805.

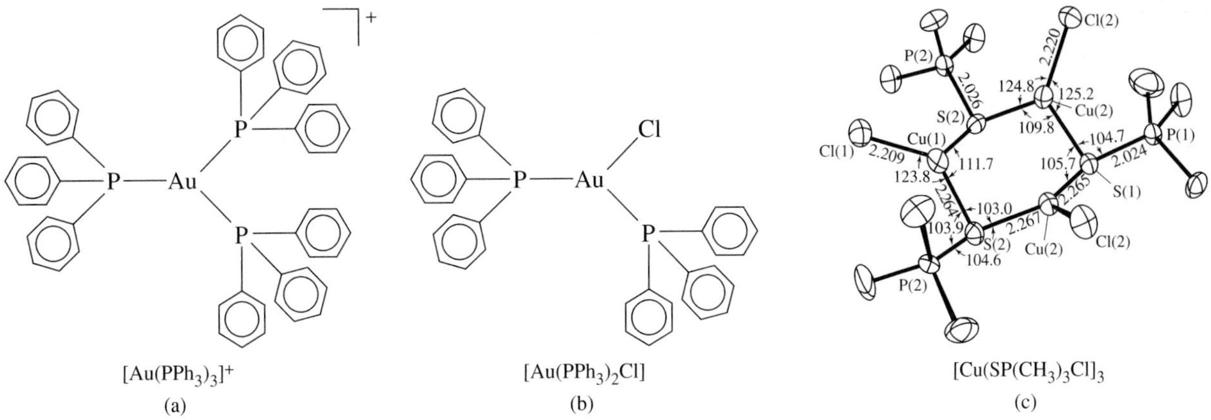

$[Au(PPh_3)_3]^+$ (a) $[Au(PPh_3)_2Cl]$ (b) $[Cu(SP(CH_3)_3Cl]_3$ (c)

FIGURE 9.23 Complexes with Coordination Number 3. [(c) Reproduced with permission from J. A. Tiethof, J. K. Stalick, and D. W. Meek, *Inorg. Chem.*, **1973**, *12*, 1170. © 1973 American Chemical Society.]

crowding. Ligands such as triphenylphosphine, PPh_3, and di(trimethylsilyl)amide, $N(SiMe_3)_2^-$, are bulky enough to prevent larger coordination numbers, even when the electronic structure favors them. All the first-row transition metals form three-coordinate complexes, either with three identical ligands or two of one ligand and one of the other. These complexes have a geometry close to trigonal planar around the metal. Other complexes with three ligands are MnO_3^+, HgI_3^-, and the cyclic compound $[Cu(SPMe_3)Cl]_3$. Examples of three-coordinate complexes are shown in Figure 9.23.

A final example, Figure 9.24 shows gold with three different geometries: linear Au(I), trigonal Au(I), and square-planar Au(III).[38]

9.4.2 Coordination Number 4[39]

Tetrahedral and square-planar structures are two common structures with four ligands. Another structure, with four bonds and one lone pair, appears in main group compounds—such as SF_4 and $TeCl_4$—giving a "seesaw" structure, as described in Chapter 3 (Figure 3.11). Crowding around small ions of high positive charge prevents higher coordination numbers for ions such as Mn(VII) and Cr(VI), and large ligands can prevent higher coordination for other ions. Many d^0 or d^{10} complexes have tetrahedral structures—such as MnO_4^-, CrO_4^{2-}, $Ni(CO)_4$, and $[Cu(py)_4]^+$. A few d^5 complexes, including $MnCl_4^{2-}$, are also tetrahedral. In such cases, the shape can be explained on the basis of VSEPR, because the *d* orbital occupancy is spherically symmetrical with zero, one, or two electrons in each *d* orbital. In addition, tetahedral structures also occur in some Co(II) d^7 species, such as $CoCl_4^{2-}$, and other transition metal complexes such as $Co(PF_3)_4$, $TiCl_4$, $[NiCl_4]^{2-}$, and $[NiCl_2(PPh_3)_2]$. Tetrahedral structures are also found in the tetrahalide complexes of Cu(II). $Cs_2[CuCl_4]$ and $(NMe_4)_2[CuCl_4]$ contain $CuCl_4^{2-}$ ions that are close to tetrahedral, as are the same ions in solution. The Jahn-Teller effect described in Chapter 10 causes distortion of the tetrahedron, with two of the Cl—Cu—Cl bond angles near 102° and two near 125°. The bromide complexes have similar structures. Examples of tetrahedral species are given in Figure 9.25.

Square-planar geometry is also possible for four-coordinate species, with the same geometric requirements imposed by octahedral geometry; both require 90° angles

[38] K. Chordroudis, T. J. McCarthy, and M. G. Kanatzidis, *Inorg. Chem.*, **1996**, *35*, 3451.
[39] M. C. Favas and D. L. Kepert, *Prog. Inorg. Chem.*, **1980**, *27*, 325.

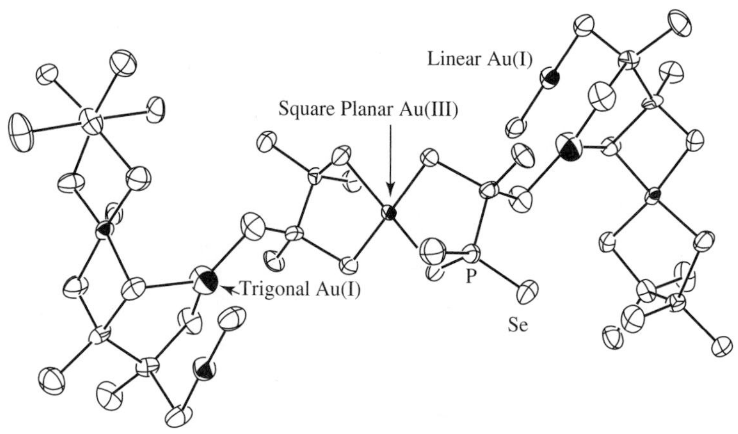

FIGURE 9.24 $K_2Au_2P_2Se_6$, a Gold Complex with Gold in Three Different Geometries. Dark ellipsoids, Au; large open ellipsoids, Se; small open ellipsoids, P. $[P_2Se_6]^{4-}$ ions bridge Au(I) in linear and trigonal geometries and Au(III) in square-planar geometry. The structure is a long chain, stacking to form open channels containing the K^+ ions. (Reproduced with permission from K. Chordroudis, T. J. McCarthy, and M. G. Kanatzidis, *Inorg. Chem.*, **1996**, *35*, 3451. © 1996 American Chemical Society.)

between ligands. The only common square-planar complexes whose structures are not imposed by a planar ligand contain d^8 ions—Ni(II), Pd(II), Pt(II), for example. Ni(II) and Cu(II) can have tetrahedral, square-planar, or intermediate shapes, depending on both the ligand and the counterion in the crystal. Cases such as these indicate that the energy difference between the two structures is small, and crystal packing can have a large influence on the choice. Pd(II) and Pt(II) complexes are square-planar, as are the d^8 complexes $[AgF_4]^-$, $[RhCl(PPh_3)_3]$, $[Ni(CN)_4]^{2-}$, and $[NiCl_2(PMe_3)_2]$. At least one compound, $[NiBr_2(P(C_6H_5)_2(CH_2C_6H_5)_2]$, has both square-planar and tetrahedral isomers in the same crystal.[40] Some square-planar complexes are shown in Figure 9.26.

FIGURE 9.25 Complexes with Tetrahedral Geometry.

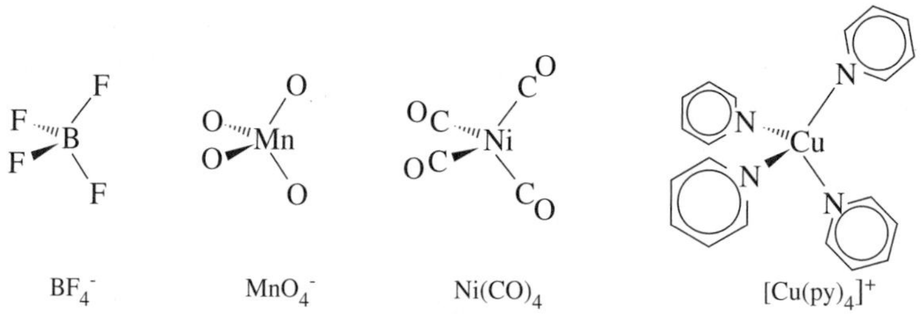

9.4.3 Coordination Number 5

Possible structures for coordination number 5 are the trigonal bipyramid, square pyramid, and pentagonal plane, which is unknown except for $[XeF_5]^-$ and $[IF_5]^{2-}$, probably because of the crowding that would be required of the ligands.[41] The energy difference between the trigonal bipyramid and the square pyramid is small. In fact, many molecules with five ligands either have structures between these two or can switch easily from one to the other in fluxional behavior. For example, $Fe(CO)_5$ and PF_5 have NMR spectra—using ^{13}C and ^{19}F, respectively—that show only one peak, indicating that the atoms are identical on the NMR time scale. Because both the trigonal bipyramid and the square pyramid have ligands in two different environments, the experiment shows that the compounds switch from one structure to another rapidly or that they have a solution structure intermediate between the two. In the solid state, both are trigonal bipyramids. $VO(acac)_2$ is a square pyramid, with the doubly bonded oxygen in the apical site. There is also evidence that $[Cu(NH_3)_5]^{2+}$ exists

[40] B. T. Kilbourn, H. M. Powell, and J. A. C. Darbyshire, *Proc. Chem. Soc.*, **1963**, 207.
[41] R. R. Holmes, *Prog. Inorg. Chem.*, **1984**, *32*, 119; T. P. E. Auf der Heyde and H.-B. Bürgi, *Inorg. Chem.*, **1989**, *28*, 3960.

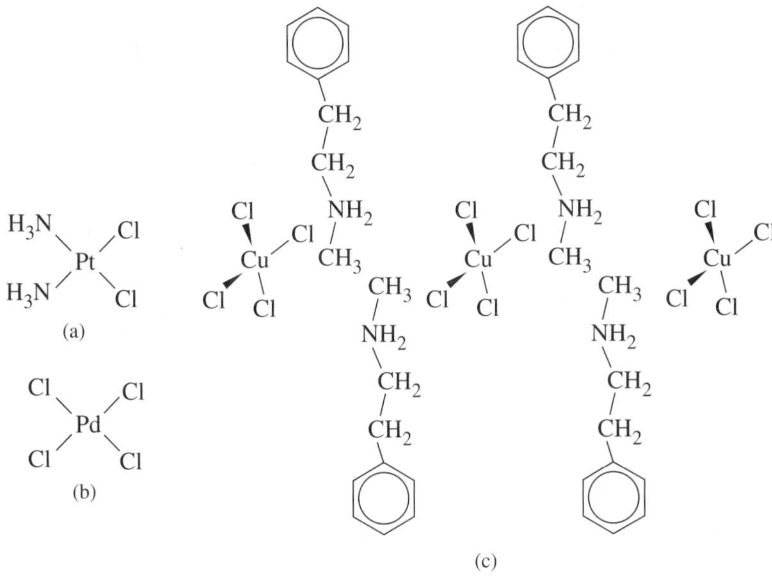

FIGURE 9.26 Complexes with Square-Planar Geometry. (a) $PtCl_2(NH_3)_2$ (b) $[PdCl_4]^{2-}$ (c) N-Methylphenethylammonium tetrachlorocuprate(II), green modification at 25 °C. At 80 °C, the color changes to yellow; the yellow modification has $CuCl_4{}^{2-}$ ions that are nearly tetrahedral. (Adapted with permission from R. L. Harlow, W. J. Wells, III, G. W. Watt, and S. H. Simonsen, *Inorg. Chem.*, **1974**, *13*, 2106. © 1974 American Chemical Society. The original reference has stereo images of both the green and yellow forms.)

as a square-pyramidal structure in liquid ammonia.[42] Other five-coordinate complexes are known for the full range of transition metals, including $[CuCl_5]^{3-}$ and $[FeCl(S_2C_2H_2)_2]$. Examples of five-coordinate complexes are shown in Figure 9.27.

9.4.4 Coordination Number 6

Six is the most common coordination number. The most common structure is octahedral, but some trigonal prismatic structures are also known. Octahedral compounds exist for all the transition metals with d^0 to d^{10} configurations.

Octahedral compounds have been used in many of the earlier illustrations in this chapter and in others. Other octahedral complexes include tris(ethylenediamine)-cobalt(III), $[Co(en)_3]^{3+}$, and hexanitritocobaltate(III), $[Co(NO_2)_6]^{3-}$, shown in Figure 9.28.

For complexes that are not regular octahedra, several types of distortion are possible. The first is elongation, leaving four short bonds in a square-planar arrangement together with two longer bonds above and below the plane. Second is the reverse, a compression with two short bonds at the top and bottom and four longer bonds in the plane. Either results in a tetragonal shape, as shown in Figure 9.29. Chromium dihalides exhibit tetragonal elongation; crystalline CrF_2 has a distorted rutile structure, with four Cr–F distances of 200 pm and two of 243 pm, and other chromium(II) halides have similar bond distances but different crystal structures.[43]

A trigonal elongation or compression results in a trigonal antiprism, when the angle between the top and bottom triangular faces is 60°, and a trigonal prism when the two triangular faces are eclipsed, as shown in Figure 9.30. Most trigonal prismatic complexes have three bidentate ligands—for example, a variety of dithiolates, $S_2C_2R_2$, and oxalates—linking the top and bottom triangular faces. Although similar in other ways, β-diketone complexes usually have skew conformations and near-octahedral symmetry around the metal. Trigonal prismatic dithiolate complexes are shown in Figure 9.30. The trigonal structures of complexes such as these may be

[42] M. Valli, S. Matsuo, H. Wakita, Y. Yamaguchi, and M. Nomura, *Inorg. Chem.*, **1996**, *35*, 5642.
[43] A. F. Wells, *Structural Inorganic Chemistry*, 5th ed., Oxford University Press, Oxford, 1984, p. 413.

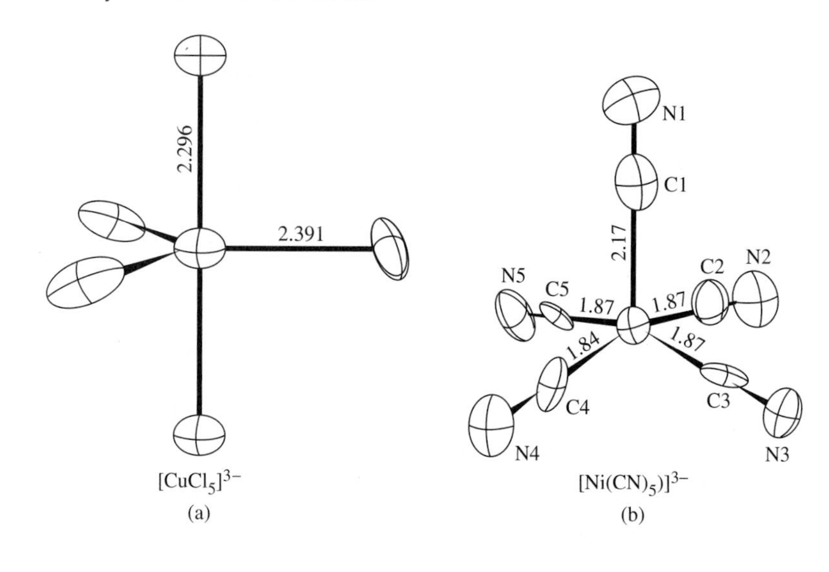

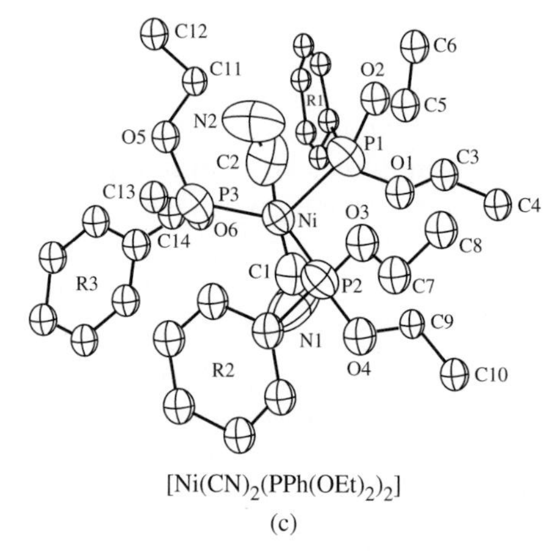

FIGURE 9.27 Complexes with Coordination Number 5. (a) $[CuCl_5]^{3-}$ (From $[Cr(NH_3)_6]$ $[CuCl_5]$, K. N. Raymond, D. W. Meek, and J. A. Ibers, *Inorg. Chem.*, **1968**, *7*, 1111.) (b) $[Ni(CN)_5]^{3-}$ (From $[Cr(en)_3][Ni(CN)_5]$, K. N. Raymond, P. W. R. Corfield, and J. A. Ibers, *Inorg. Chem.*, **1968**, *7*, 1362.) (c) $Ni(CN)_2[PPh(OEt)_2]_3$ (From J. K. Stalick and J. A. Ibers, *Inorg. Chem.*, **1969**, *8*, 1084.) (Reproduced with permission of the American Chemical Society. © 1968 and 1969.)

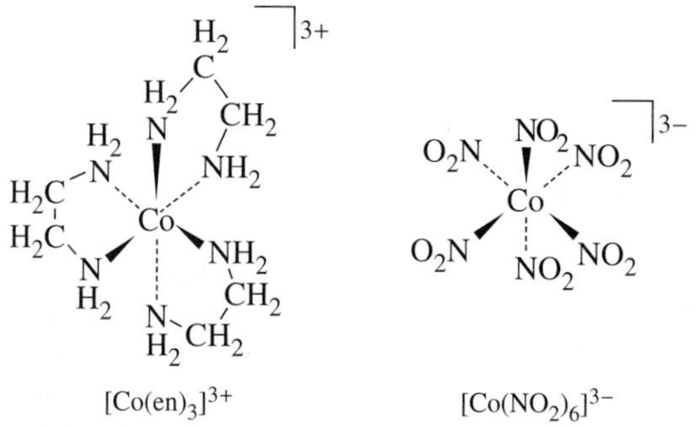

$[Co(en)_3]^{3+}$ $[Co(NO_2)_6]^{3-}$

FIGURE 9.28 Complexes with Octahedral Geometry.

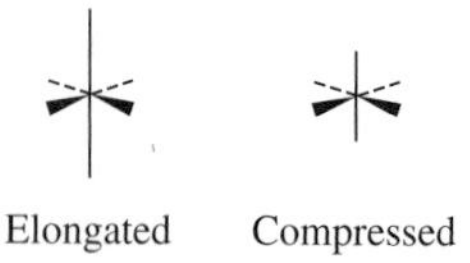

Elongated Compressed

FIGURE 9.29 Tetragonal Distortions of the Octahedron.

due to π interactions between adjacent sulfur atoms in the trigonal faces. Campbell and Harris[44] summarize the arguments for stability of the trigonal prismatic structure relative to the octahedral structure.

A number of complexes that appear to be 4-coordinate are more accurately described as 6-coordinate. Although $(NH_4)_2[CuCl_4]$ is frequently cited as having a square-planar $[CuCl_4]^{2-}$ ion, the ions in the crystal are packed so that two more chlorides are above and below the plane at considerably larger distances in a distorted octahedral structure. The

[44] S. Campbell and S. Harris, *Inorg. Chem.*, **1996**, *35*, 3285.

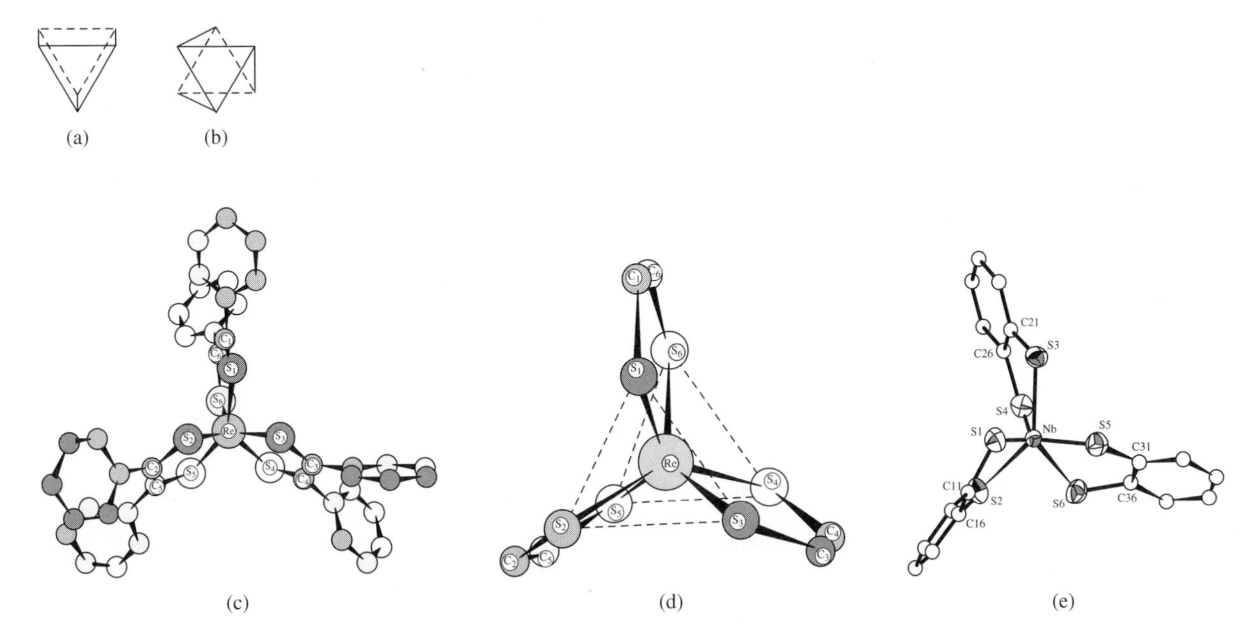

FIGURE 9.30 Complexes with Trigonal Prismatic Geometry. (a) A trigonal prism. (b) A trigonal antiprism. (c), (d) $Re[S_2C_2(C_6H_5)_2]_3$. Part (d) is a perspective drawing of the coordination geometry of (c) excluding the phenyl rings. (Reproduced with permission from R. Eisenberg and J. A. Ibers, *Inorg. Chem.*, **1966**, *5*, 411. © 1966 American Chemical Society.) (e) Tris(benzene-1,2-dithiolato)niobate(V), $[Nb(S_2C_6H_4)_3]^-$, omitting the hydrogens. (Reproduced with permission from M. Cowie and M. J. Bennett, *Inorg. Chem.*, **1976**, *15*, 1589. © 1976 American Chemical Society.)

Jahn-Teller effect described in Chapter 10 is the cause of this distortion. Similarly, $[Cu(NH_3)_4]SO_4 \cdot H_2O$ has the ammonias in a square-planar arrangement, but each copper is also connected to more distant bridging water molecules above and below the plane.

Another nonoctahedral six-coordinate ion is $[CuCl_6]^{4-}$ in the compound [*tris*(2-aminoethyl)amine$H_4]_2[CuCl_6]Cl_4 \cdot 2H_2O$.[45] In the $[CuCl_6]^{4-}$ ion there are three different Cu—Cl bond distances, in *trans* pairs at 225.1, 236.1, and 310.5 pm, resulting in approximately D_{2h} symmetry. Many hydrogen-bond interactions occur between the chlorides and the water molecules in this crystal; if the hydrogen bonds are strong, the Cu—Cl bonds are longer.

9.4.5 Coordination Number 7

Three structures are possible for seven-coordinate complexes, the pentagonal bipyramid, capped trigonal prism, and capped octahedron.[46] In the capped shapes, the seventh ligand is simply added to a face of the core structure, with related adjustments in the other angles to allow it to fit. Although seven-coordination is not common, all three shapes are found experimentally, with preference for a particular structure apparently resulting from different counterions and the steric requirements of the ligands, especially chelating ligands.

The pentagonal bipyramidal geometry found in IF_7 and a few other main group examples also occurs in transition-metal complexes such as $[UO_2F_5]^{3-}$, $[NbOF_6]^{3-}$, and the iron complex in Figure 9.31(a). Examples of capped trigonal prismatic structures include $[NiF_7]^{2-}$ and $[NbF_7]^{2-}$, in which the seventh fluoride caps a rectangular face of the prism (Figure 9.31(b)). Two examples of monocapped octahedral geometry are

[45] M. Wei, R. D. Willett, and K. W. Hipps, *Inorg. Chem.*, **1996**, *35*, 5300.
[46] D. L. Kepert, *Prog. Inorg. Chem.*, **1979**, *25*, 41.

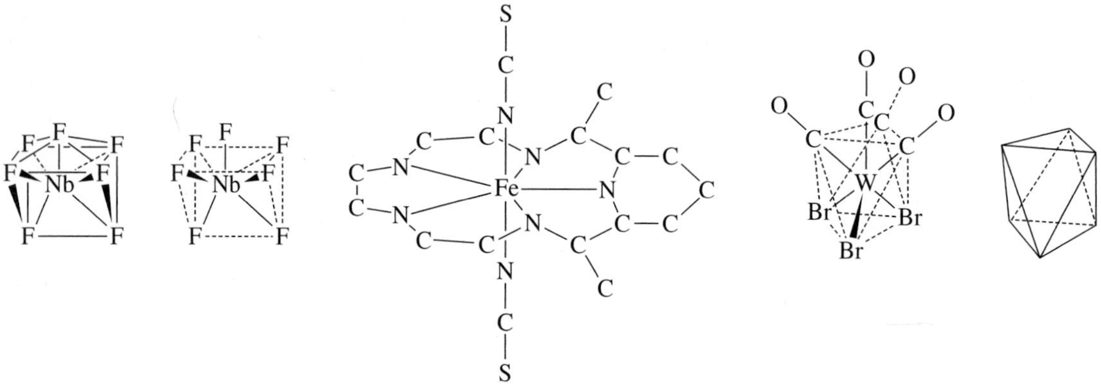

(a) (b) (c)

FIGURE 9.31 Complexes with Coordination Number 7. (a) Heptafluoroniobate(V), $[NbF_7]^{2-}$, a capped trigonal prism. The capping F is at the top. (b) 2,13-dimethyl-3,6,9,12, 18-pentaazabicyclo[12.3.1]-octadeca-1(18),2,12,14,16-pentaene complex of Fe(II) with two axial thiocyanates, a pentagonal bipyramid. (From E. Fleischer and S. Hawkinson, *J. Am. Chem. Soc.*, **1967**, 89, 720.) (c) Tribromotetracarbonyltungstate(II) anion, $[W(CO)_4Br_3]^-$, a capped octahedron, and an octahedron in the same orientation. The capping CO is at the top. (From M. G. B. Drew and A. P. Wolters, *Chem. Commun. (Cambridge UK)*, **1972**, 457.)

$[W(CO)_4Br_3]^-$ (Figure 9.31(c)) and the set of complexes $[M(trenpy)]^{2+}$, where M is any of the metals from Mn to Zn, and trenpy = $(C_5H_4NCH{=}NCH_2CH_2)_3N$, in which the central nitrogen of the ligand caps a trigonal face of an octahedron. An analysis of different geometries and many references has been presented by Lin and Bytheway.[47]

9.4.6 Coordination Number 8

Although the cube has eight-coordinate geometry, it exists only in simple ionic lattices such as CsCl. The square antiprism and dodecahedron are common in transition-metal complexes.[48] Because the central ion must be large in order to accommodate eight ligands, eight coordination is rare among the first-row transition metals (although it is likely in $[Fe(edta)(H_2O)_2]^+$ in solution). Solid-state examples include $Na_7Zr_6F_{31}$, which has square antiprisms of ZrF_8 units, and $[Zr(acac)_4]$, a regular dodecahedron. $[AmCl_2(H_2O)_6]^+$ has still another geometry, a trigonal prism of water ligands with chloride caps on the trigonal faces (bicapped trigonal prismatic). Three of these complexes are shown in Figure 9.32. $[Yb(NH_3)_8]^{3+}$ also has a square-antiprism structure.[49] In addition, coordination number 8 is observed in $[Mo(CN)_8]^{4-}$ ions[50] in a compressed square-antiprism structure, and when As_8 rings bond to transition metals in a "crown-like" configuration, as in $MoAs_8{}^{2-}$ (also shown in Figure 9.32) and similar complexes.[51]

9.4.7 Larger Coordination Numbers

Coordination numbers are known up to 16, but most over 8 are special cases.[52] Two examples are shown in Figure 9.33. $[La(NH_3)_9]^{3+}$ has a capped square-antiprism structure.[53]

---

[47] Z. Lin and I. Bytheway, *Inorg. Chem.*, **1996**, *35*, 594.
[48] D. L. Kepert, *Prog. Inorg. Chem.*, **1978**, *24*, 179.
[49] D. M. Young, G. L. Schimek, and J. W. Kolis, *Inorg. Chem.*, **1996**, *35*, 7620.
[50] W. Meske and D. Babel, *Z. Naturforsch., B: Chem. Sci.*, **1999**, *54*, 117.
[51] B. W. Eichhorn, S. P. Mattamana, D. R. Gardner, and J. C. Fettinger, *J. Am. Chem. Soc.*, **1998**, *120*, 9708; J. Li and K. Wu, *Inorg. Chem.*, **2000**, *39*, 1538.
[52] M. C. Favas and D. L. Kepert, *Prog. Inorg. Chem.*, **1981**, *28*, 309.
[53] D. M. Young, G. L. Schimek, and J. W. Kolis, *Inorg. Chem.*, **1996**, *35*, 7620.

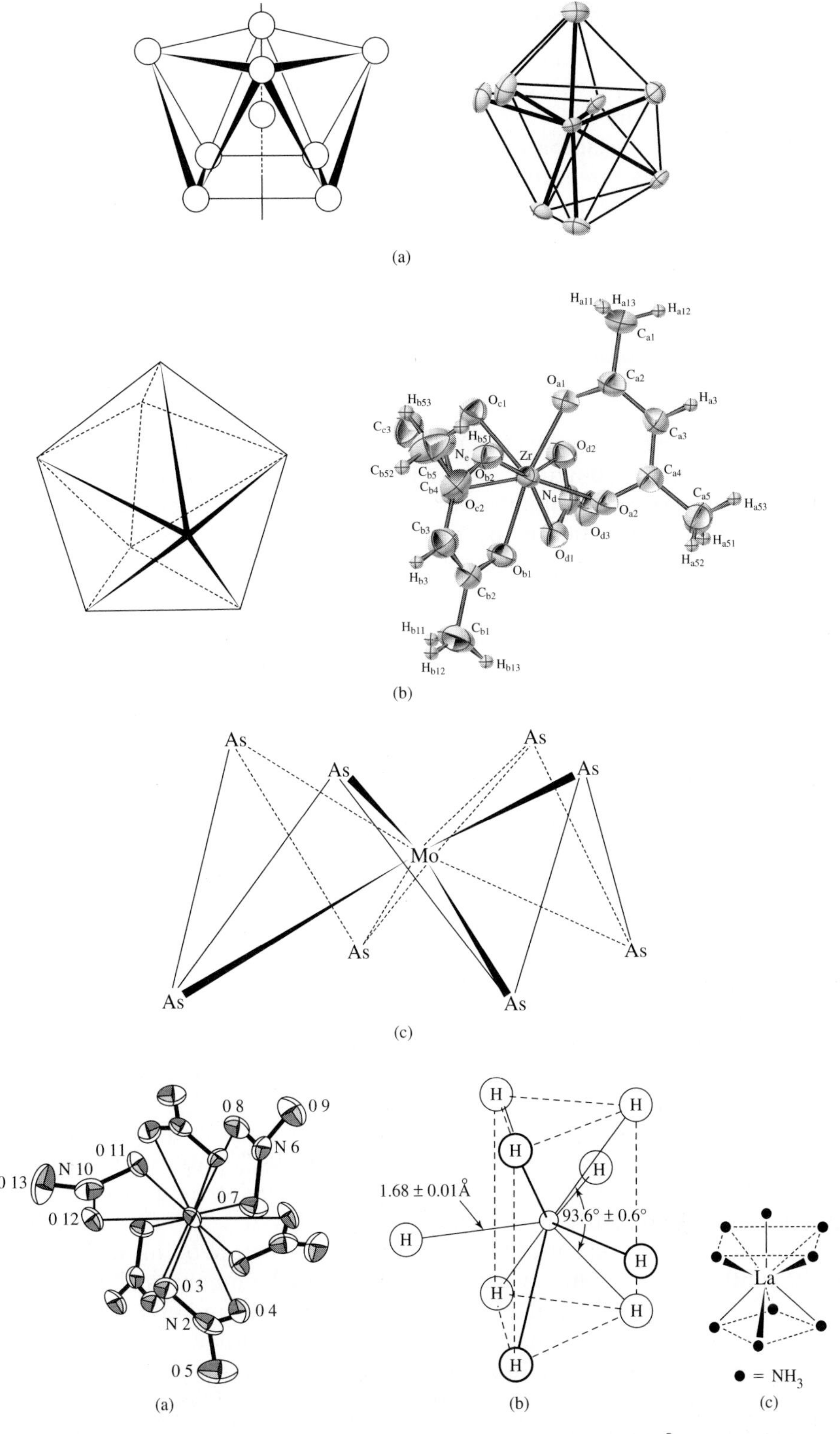

FIGURE 9.32 Complexes with Coordination Number 8. (a) $Na_7Zr_6F_{31}$, square antiprisms of ZrF_8. (Reproduced with permission from J. H. Burns, R. D. Ellison, and H. A. Levy, *Acta Crystallogr.*, **1968**, *B24*, 230.) (b) $[Zr(acac)_2(NO_3)_2]$, regular dodecahedron. (Reproduced with permission from V. W. Day and R. C. Fay, J. Am. Chem. Soc., **1975**, *97*, 5136. © 1975 American Chemical Society.) (c) $MoAs_8{}^{2-}$. (Redrawn from B. W. Eichhorn, S. P. Mattamana, D. R. Gardner, and J. C. Fettinger, *J. Am. Chem. Soc.*, **1998**, *120*, 9708.)

FIGURE 9.33 Complexes with Larger Coordination Numbers. (a) $[Ce(NO_3)_6]^{3-}$, with bidentate nitrates and coordination number 12. (Reproduced with permission from T. A. Beinecke and J. Delgaudio, *Inorg. Chem.*, **1968**, 7, 715. © 1968 American Chemical Society.) (b) $[ReH_9]^{2-}$, tricapped trigonal prism. (Reproduced with permission from S. C. Abrahams, A. P. Ginsberg, and K. Knox, *Inorg. Chem.*, **1964**, *3*, 558. © 1964 American Chemical Society.) (c) $[La(NH_3)_9]^{3+}$, capped square antiprism.

9.5 COORDINATION FRAMEWORKS

To this point, discussion has centered on coordination complexes that have a single metal. Numerous examples are also known in which ligands can act as bridges to create more extended structures. Many types of such structures are known; this section will briefly introduce a few examples.

Zeolites, porous three-dimensional aluminosilicate structures that have been used in ion exchange, catalysis, and other applications, were mentioned in Chapter 7. A more recent development of inorganic porous materials has been the construction of **coordination polymers**, in which coordination complexes are linked through ligands in infinite arrays. These polymers may be "one-dimensional" chains, with linear or zigzag linkages, or they may be two- or three-dimensional; the range of possibilities is broad and growing rapidly. We will confine our focus to structures in which coordination complexes are linked through organic molecules and ions or through donor groups on ligands.

Metal-organic frameworks (**MOFs**) are extended structures in which metal ions or clusters are linked through organic molecules that have two or more sites through which links can be formed. Common functional groups that can act as links include carboxylate, cyanide, phosphate, and pyridyl; these are molecules or ions that have two or more Lewis base sites through which they can link metals or metal clusters. Examples are shown in Figure 9.34

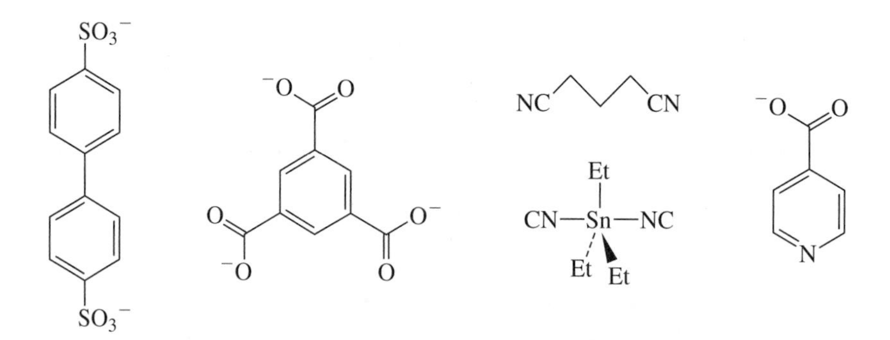

FIGURE 9.34 Examples of Linking Groups in Metal-Organic Frameworks.

Construction of frameworks of desired structures and properties can be viewed as judicious assembly of such "building blocks."[54] An example of an interesting three-dimensional network is provided by the structural unit shown in Figure 9.35. This structure was formed using *m*-benzenedicarboxylate (*m*-BDC, shown in the margin) and copper(II). Although the structure of this network is complex, its synthesis was relatively simple, from equimolar amounts of *m*-dicarboxylic acid and $Cu(NO_3)_2 \cdot 2.5\ H_2O$, in a solvent mixture of DMF and ethanol. The product contains pairs of copper(II) ions bridged by *m*-BDC "paddlewheel" units, as shown in the figure. The crystal structure (Figure 9.36) has twelve such units and an overall structure of a truncated cuboctahedron with a spherical void having a diameter of 1.5 nm.[55]

[54] O. M. Yaghi, M. O'Keeffe, N. W. Ockwig, H. K. Chae, M. Eddaoudi, and J. Kim, *Nature*, **2003**, *423*, 705.
[55] M. Eddaoudi, J. Kim, J. B. Wachter, H. K. Chae, M. O'Keefe, and O. M. Yaghi, *J. Am. Chem. Soc.*, **2001**, *123*, 4368.

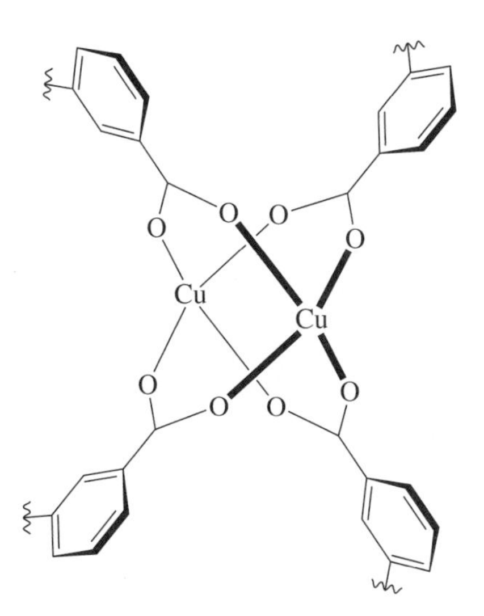

FIGURE 9.35 Pairs of Cu^{2+} Ions Bridged by *m*-BDC in "Paddlewheel" Structure.

O^- O^- O O

m-BDC

A detailed classification system for networks has been devised using three-letter codes to describe topologies, for example **dia** for networks based on the diamond structure, **pcu** for primitive cubic, **bcu** for body-centered cubic, and so forth.[56] The details of this system are beyond the scope of this text.

Coordination complexes with attached donor groups on their ligands, designated **metalloligands**, can also be used as building blocks in MOFs.[57] These complexes in essence are acting as ligands themselves. For example, the coordination complex $[Cr(ox)_3]^{3-}$ (ox = oxalate, $C_2O_4^{\ 2-}$) has electron pairs in its outer six oxygen atoms that can form bonds via donor–acceptor interactions to metal ions and can serve as a metalloligand in the formation of large, two-dimensional frameworks (nets) with cations such as Na^+, Zn^{2+}, and Cr^{3+}, as shown in Figure 9.37.

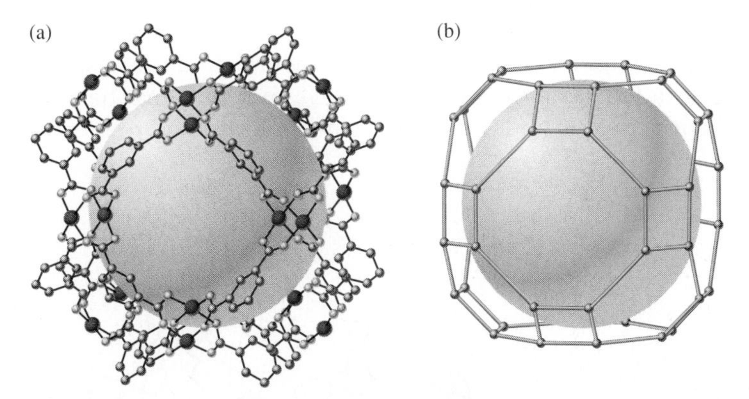

FIGURE 9.36 Crystal Structure of Metal-Organic Polyhedron. (a) Unit cell of Cu^{2+}–*m*-BDC network. (b) Cuboctahedron. Reproduced with permission from M. Eddaoudi, J. Kim, J. B. Wachter, H. K. Chae, M. O'Keefe, and O. M. Yaghi, *Journal of the American Chemical Society*, **2001**, *123*, 4369. Copyright 2001 American Chemical Society.

[56] N. W. Ockwig, O. Delgado-Friedrichs, M. O'Keeffe, and O. M. Yaghi, *Acc. Chem. Res.*, **2005**, *38*, 176. Supplemental information in this reference has illustrations of nets based on coordination numbers 3 through 6; M. O'Keeffe, M. A. Peskov, S. J. Ramsden, and O. M. Yaghi. *Acc. Chem. Res.*, **2008**, *41*, 1782.

[57] S. J. Garibay, J. R. Stork, and S. M. Cohen, *Progr. Inorg. Chem.*, **2009**, *56* 335.

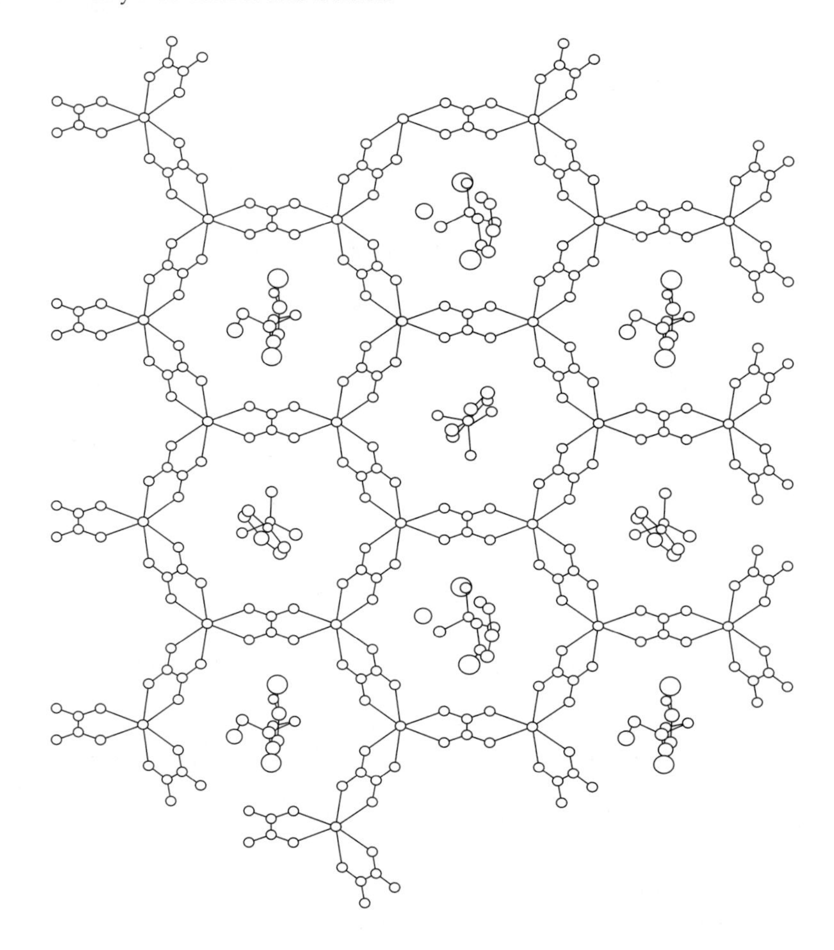

FIGURE 9.37 $[Cr(ox)_3]^{3-}$ and Net Formed with Cations. Reproduced with permission, R. P. Farrell, T. W. Hambley, and P. A. Lay, *Inorg. Chem.*, **1995**, 34, 757.

Depending on the number and orientation of available bonding sites on metalloligands, one-dimensional coordination polymers and three-dimensional networks with extended channels through them, can also be formed. Examples of these arrangements are shown in Figures 9.38 and 9.39.

Procedures to engineer coordination frameworks to desired specifications have progressed rapidly toward formation of robust frameworks that have very large pore volumes and surface areas and structures that can be crafted to meet specific needs.

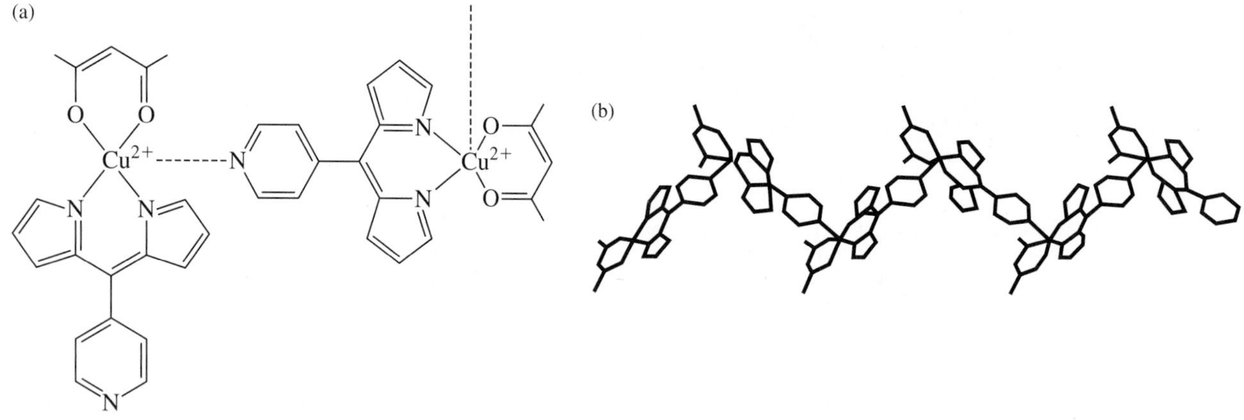

(a)

(b)

FIGURE 9.38 One-Dimensional Chain. Reproduced with permission, S. R. Halper, M. R. Malachowski, H. M. Delaney, and S. M. Cohen, *Inorg. Chem.*, **2004**, *43*, 1242.

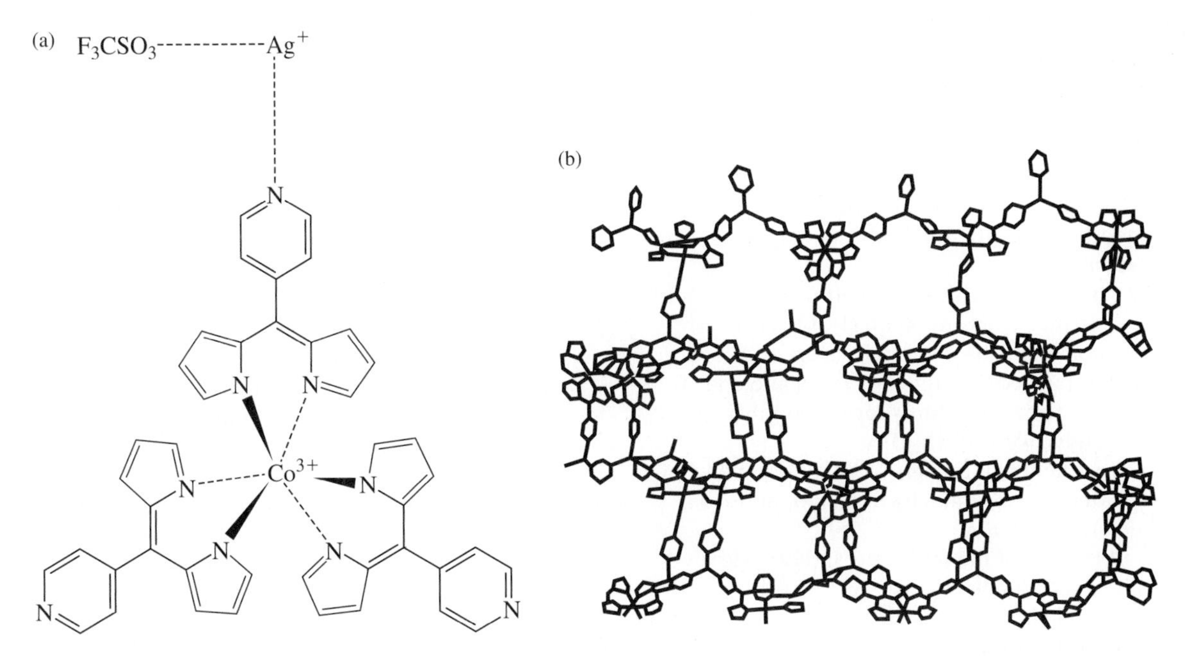

FIGURE 9.39 Three-Dimensional Network. Reproduced with permission, S. R. Halper and S. M. Cohen, *Inorg. Chem.*, **2005**, *44*, 486.

Areas of interest and potential applications include ion exchange, hydrogen storage, molecular sensing, drug delivery, medical imaging, and the design of chiral porous frameworks to perform chiral separations and to act as chiral catalysts.

General References

The official documents on IUPAC nomenclature are G. J. Leigh, editor, *Nomenclature of Inorganic Chemistry*, Blackwell Scientific Publications, Oxford, England, 1990 and J. A. McCleverty and N. G. Connelly, editors, *IUPAC, Nomenclature of Inorganic Chemistry II: Recommendations 2000*, Royal Society of Chemistry, Cambridge, UK, 2001. The best single reference for isomers and geometric structures is G. Wilkinson, R. D. Gillard, and J. A. McCleverty, editors, *Comprehensive Coordination Chemistry*, Pergamon Press, Oxford, 1987. The reviews cited in the individual sections are also very comprehensive. A useful recent reference on coordination polymers, metal-organic frameworks, and related topics is S. R. Batten, S. M. Neville, and D. R. Turner, *Coordination Polymers*, RSC Publishing, Cambridge, UK, 2009.

Problems

9.1 By determining the point groups, verify that none of the first four proposed structures for hexacoordinate complexes in Figure 9.3 would show optical activity.

9.2 Give chemical names for the following:
a. $[Fe(CN)_2(CH_3NC)_4]$ **b.** $Rb[AgF_4]$
c. $[Ir(CO)Cl(PPh_3)_2]$ (two isomers)
d. $[Co(N_3)(NH_3)_5]SO_4$ **e.** $[Ag(NH_3)_2][BF_4]$

9.3 Give chemical names for the following:
a. $[Co(N_3)(NH_3)_5]SO_4$ **b.** $Na[AlCl_4]$
c. $[Co(en)_2(CO_3)]Cl$ **d.** $[Ni(bipy)_3](NO_3)_2$
e. $[Co(en)_2CO_3]Br$

9.4 Give chemical names for the following:
a. $[Cu(NH_3)_4]^{2+}$ **b.** $[PtCl_4]^{2-}$
c. $Fe(S_2CNMe_2)_3$ **d.** $[Mn(CN)_6]^{4-}$
e. $[ReH_9]^{2-}$

9.5 Name all the complexes in Problem 9.12, omitting isomer designations.

9.6 Name all the complexes in Problem 9.17, omitting isomer designations.

9.7 Give structures for the following:
a. Bis(en)Co(III)-μ-amido-μ-hydroxobis(en)Co(III) ion
b. DiaquadiiododinitritoPd(IV), all isomers
c. $Fe(dtc)_3$, all isomers

dtc = $[S_2C{-}N(CH_3)(H)]^-$

9.8 Show structures for the following:
a. Triammineaquadichlorocobalt(III) chloride, all isomers
b. μ-oxo-bis[pentaamminechromium(III)] ion
c. Potassium diaquabis(oxalato)manganate(III)

9.9 Show structures for the following:
a. *cis*-Diamminebromochloroplatinum(II)
b. Diaquadiiododinitritopalladium(IV), all ligands *trans*
c. Tri-μ-carbonylbis(tricarbonyliron(0))

9.10 Glycine has the structure NH_2CH_2COOH. It can lose a proton from the carboxyl group and form chelate rings bonded through both the N and one of the O atoms. Draw structures for all possible isomers of tris(glycinato)cobalt(III).

9.11 Sketch structures of all isomers of $M(AB)_3$, in which AB is a bidentate unsymmetrical ligand, and label the structures *fac* or *mer*.

9.12 Sketch all isomers of the following. Indicate clearly each pair of enantiomers.
a. $[Pt(NH_3)_3Cl_3]^+$
b. $[Co(NH_3)_2(H_2O)_2Cl_2]^+$
c. $[Co(NH_3)_2(H_2O)_2BrCl]^+$
d. $[Cr(H_2O)_3BrClI]$
e. $[Pt(en)_2Cl_2]^{2+}$
f. $[Cr(o\text{-phen})(NH_3)_2Cl_2]^+$
g. $[Pt(bipy)_2BrCl]^{2+}$
h. $Re(arphos)_2Br_2$

arphos = (benzene ring with $As(CH_3)_2$ and $P(CH_3)_2$ substituents)

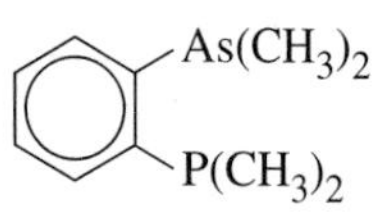

i. $Re(dien)Br_2Cl$

9.13 The (2-aminoethyl)phosphine ligand has the structure shown below; it often acts as a bidentate ligand toward transition metals. (See N. Komine, S. Tsutsuminai, M. Hirano, and S. Komiya, *J. Organomet. Chem.*, **2007**, *692*, 4486.)

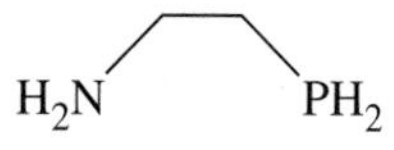

H_2N ⁄‾‾\ PH_2

a. When this ligand forms monodentate complexes with palladium, it bonds through its phosphorus atom rather than its nitrogen. Suggest an explanation.
b. How many possible isomers of dichlorobis[(2-aminoethyl)phosphine]nickel(II), an octahedral coordination complex in which (2-aminoethyl)phosphine is bidentate, are there? Sketch each isomer, and identify any pairs of enantiomers.
c. Classify the configuration of chiral isomers as Λ or Δ.

9.14 An octahedrally coordinated transition metal M has the following ligands:
Two chloro ligands
One (2-aminoethyl)phosphine ligand (see Problem 9.13)
One $[O{-}CH_2{-}CH_2{-}S]^{2-}$ ligand
a. Sketch all isomers, clearly indicating pairs of enantiomers.
b. Classify the configuration of chiral isomers as Λ or Δ.

9.15 Suppose a complex of formula $[Co(CO)_2(CN)_2Br_2]^-$ has been synthesized. In the infrared spectrum, it shows two bands attributable to C—O stretching but only one band attributable to C—N stretching. What is the most likely structure of this molecule? (See Section 4.4.2.)

9.16 How many possible isomers are there of an octahedral complex having the formula $M(ABC)(NH_3)(H_2O)Br$, where ABC is the tridentate ligand $H_2N{-}C_2H_4{-}PH{-}C_2H_4{-}AsH_2$? How many of these consist of pairs of enantiomers? Sketch all isomers, showing clearly any pairs of enantiomers. The tridentate ligand may be abbreviated as N—P—As for simplicity.

9.17 Assign absolute configurations (Λ or Δ) to the following:

a. $[Ru(CO)_2(S\frown S)_2]^+$ (structure: OC—Ru—S, with S, CO, S, S)
S⌒S = dimethyldithiocarbamate

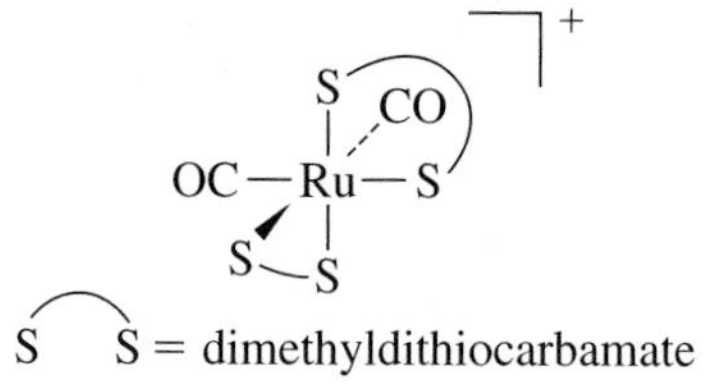

b. $[Co(O\frown O)_3]^{3-}$ (structure)
O⌒O = oxalate

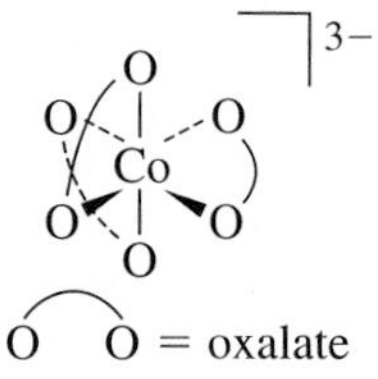

c. $[Ru(N\frown N)_3]^{2+}$ (structure)
N⌒N = ethylenediamine

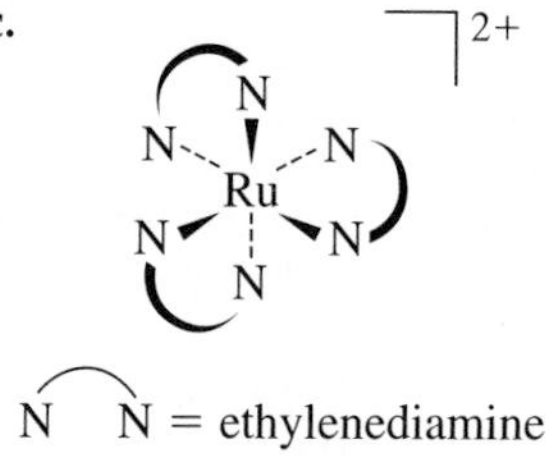

d. $Ni(N\frown N)_2Cl_2$ (structure: Cl, Cl, N, N, N, N around Ni)
N⌒N = 2, 2'-bipyridine

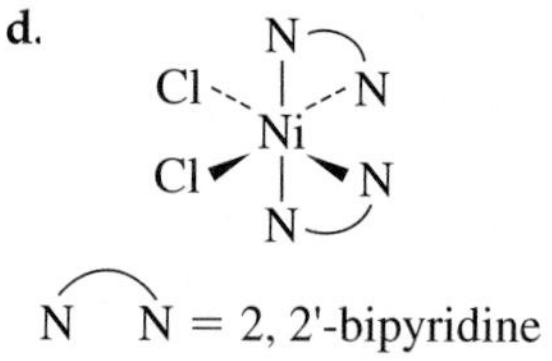

9.18 Which of the following molecules are chiral?

a. **b.**

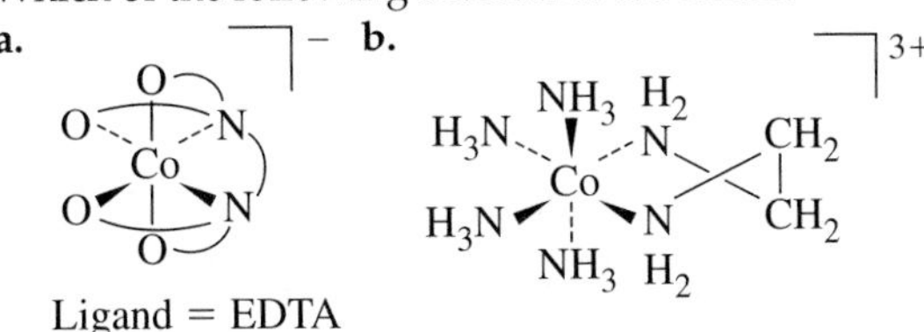

Ligand = EDTA

c.

Hydrogens omitted for clarity

9.19 Give the symmetry designations (λ or δ) for the chelate rings in Problem 9-18b and c.

9.20 When *cis*-OsO_2F_4 is dissolved in SbF_5, the cation $OsO_2F_3^+$ is formed. The ^{19}F NMR spectrum of this cation shows two resonances, a doublet and a triplet having relative intensities of 2:1. What is the most likely structure of this ion? What is its point group? (See W. J. Casteel, Jr., D. A. Dixon, H. P. A. Mercier, and G. J. Schrobilgen, *Inorg. Chem.*, **1996**, *35*, 4310.)

9.21 When solid $Cu(CN)_2$ was ablated with 1064 nm laser pulses, various ions containing two-coordinate Cu^{2+} bridged by cyanide ions were formed. These ions collectively have been dubbed a metal cyanide "abacus." What are the likely structures of such ions, including the most likely geometry around the copper ion? (See I. G. Dance, P. A. W. Dean, and K. J. Fisher, *Inorg. Chem.*, **1994**, *33*, 6261.)

9.22 Complexes with the formula $[Au(PR_3)_2]^+$, where R is mesityl, exhibit "propeller" isomerism at low temperature as a consequence of crowding around the phosphorus. How many such isomers are possible? (See A. Bayler, G. A. Bowmaker, and H. Schmidbaur, *Inorg. Chem.*, **1996**, *35*, 5959.)

9.23 One of the more striking hydride complexes is enneahydridorhenate, $[ReH_9]^{2-}$, which has tricapped trigonal prismatic geometry, as shown in Figure 9.33. Construct a representation using the hydrogen orbitals as a basis. Reduce this to its component irreducible representations, and indicate which orbitals of Re are of suitable symmetry to interact with the hydrogen orbitals.

9.24 The chromium(III) complex $[Cr(bipy)(ox)_2]^-$ can act as a metalloligand to form a coordination polymer chain with Mn(II) ions, in which each manganese ion is eight-coordinate (in flattened square-antiprism geometry) and bridges four $[Cr(bipy)(ox)_2]^-$ units; the ratio of Mn to Cr is 1:2. Sketch two units of this chain. (See F. D. Rochon, R. Melanson, and M. Andruh, *Inorg. Chem.*, **1996**, *35*, 6086.)

9.25 The metalloligand $Cu(acacCN)_2$ forms a two-dimensional "honeycomb" sheet with 2′, 4′, 6′-tri(pyridyl)triazine (tpt); each honeycomb "cell" has sixfold symmetry. Show how six metalloligands and six tpt molecules can form such a structure. (See J. Yoshida, S.-I. Nishikiori, and R. Kuroda, *Chem. Lett.*, **2007**, *36*, 678.)

$Cu(acacCN)_2$

tpt

9.26 Determine the point groups:

a. $Cu(acacCN)_2$ and tpt in Problem 9.25. (Assume delocalization of electrons in the O...O part of the acacCN ligands and in the aromatic rings of tpt.)

b. The molecular cartwheel shown here (note orientation of rings). (See H. P. Dijkstra, P. Steenwinkel, D. M. Grove, M. Lutz, A. L. Spek, and G. van Koten, *Angew. Chem. Int. Ed.*, **1999**, *38*, 2186.)

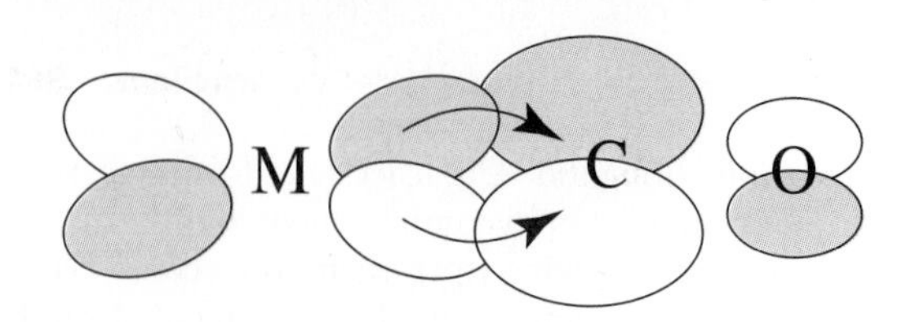

Coordination Chemistry II: Bonding

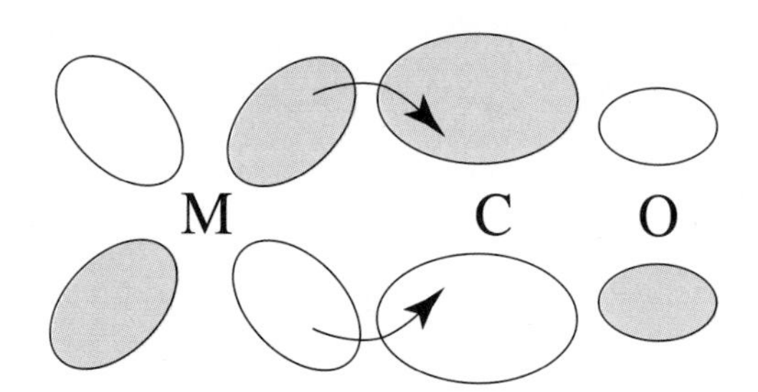

10.1 EXPERIMENTAL EVIDENCE FOR ELECTRONIC STRUCTURES

Any successful theory of bonding in coordination complexes must be consistent with experimental data regarding their behavior. This chapter provides a review of some of the types of experimental observations that have been made on coordination complexes, then describes theories of electronic structure and bonding that have been used to account for their properties.

10.1.1 Thermodynamic Data

One of the primary goals of a theory of bonding must be to explain the energies of chemical compounds. Inorganic chemists, and coordination chemists in particular, frequently use **stability constants**, sometimes called **formation constants**, as indicators of bonding strengths (Table 10.1). These are equilibrium constants for the formation of coordination complexes, usually measured in aqueous solution. Here are two examples of these reactions and their corresponding stability constant expressions:[1]

$$[Fe(H_2O)_6]^{3+} + SCN^-(aq) \rightleftharpoons [Fe(SCN)(H_2O)_5]^{2+} + H_2O \qquad K_1 = \frac{[FeSCN^{2+}]}{[Fe^{3+}][SCN^-]} = 9 \times 10^2$$

$$[Cu(H_2O)_6]^{2+} + 4\,NH_3(aq) \rightleftharpoons [Cu(NH_3)_4(H_2O)_2]^{2+} + 4\,H_2O \qquad K_4 = \frac{[Cu(NH_3)_4{}^{2+}]}{[Cu^{2+}][NH_3]^4} = 1 \times 10^{13}$$

The large stability constants indicate that bonding of the metal ions with the incoming ligand is much more favorable than bonding with water. In other words, the incoming ligands, SCN^- and NH_3, win the competition with H_2O to form bonds to the metal ions.

As described in Section 6.4.2, enthalpies of reaction can be measured directly, or the temperature dependence of equilibrium constants can be used to calculate enthalpies and entropies of reaction by plotting ln K versus $1/T$. Complicating factors such as solvent interaction with both reactants and products must be considered in any reactions in solution.

In practice, thermodynamic values alone are rarely sufficient to predict other properties of coordination complexes or their structures or formulas; they are more valuable in comparing similar complexes, such as a series of different metal ions all reacting with the same ligand or a series of different

[1] Water molecules have been omitted from the equilibrium constant expressions for simplicity.

ligands reacting with the same metal ion. In such cases, useful correlations between thermodynamic properties and electronic structure can often be made.

Table 10.1 provides equilibrium constants for reactions of Ag^+ and Cu^{2+} with four ligands. The metal ion–ammonia constants are similar (K for Cu^{2+} is about 8.5 times larger), as are the metal ion–fluoride constants (nearly a factor of 12), but the metal ion–chloride and metal ion–bromide constants are very different (factors of 1,000 and more than 22,000 in the opposite direction). Chloride and bromide compete much more effectively with water in bonding to silver ion than does fluoride, whereas fluoride interacts more strongly with Cu^{2+}. This can be viewed as another example of the HSAB approach: silver ion is a soft cation, and copper(II) is borderline. Neither bonds strongly to the hard fluoride ion, but Ag^+ bonds much more strongly with the softer bromide ion than does Cu^{2+}. Such qualitative descriptions are useful, but it is difficult to completely rationalize data such as these without extensive theoretical calculations.

TABLE 10.1 Formation Constants (*K*) at 25 °C

Cation	NH_3	F^-	Cl^-	Br^-
Ag^+	2,000	0.68	1,200	20,000
Cu^{2+}	17,000	8	1.2	0.9

Source: R. M. Smith and A. E. Martell, *Critical Stability Constants, Vol. 4, Inorganic Complexes*, Plenum Press, New York, 1976, pp. 40–42, 96–119. Not all ionic strengths were identical for these determinations, but the trends in K values shown here are consistent with determinations at a variety of ionic strengths.

An additional factor appears when there are two donor sites in the same ligand, such as in ethylenediamine (en), $NH_2CH_2CH_2NH_2$. After one amine nitrogen bonds with a metal ion, the proximity of the second nitrogen makes its interaction with the metal much easier. In one example of the difference between chelating and monodentate ligands, $[Ni(en)_3]^{2+}$ is stable in solution at high dilution; but under similar conditions, the complex with the mondenate ligand methylamine, $[Ni(CH_3NH_2)_6]^{2+}$, dissociates, and nickel hydroxide precipitates:

$$[Ni(CH_3NH_2)_6]^{2+} + 6\,H_2O \longrightarrow Ni(OH)_2(s) + 6\,CH_3NH_3{}^+ + 4\,OH^-$$

As another example, complexation reactions of Cd^{2+} with methylamine and ethylenediamine are compared in Table 10.2 for the reactions:

$$[Cd(H_2O)_6]^{2+} + 4\,CH_3NH_2 \longrightarrow [Cd(CH_3NH_2)_4(H_2O)_2]^{2+} + 4\,H_2O \text{ (no change in number of molecules)}$$

$$[Cd(H_2O)_6]^{2+} + 2\,\text{en} \longrightarrow [Cd(en)_2(H_2O)_2]^{2+} + 4\,H_2O \quad \text{(increase of two molecules)}$$

Because the enthalpy changes for the two reactions are similar, the large difference in equilibrium constants is a consequence of the large difference in entropy change: the second reaction has a positive ΔS accompanying a net increase of two moles in the reaction, in contrast to the first reaction, in which the number of moles is unchanged. This **chelate effect** is most prominent when the total ring size, including ligand atoms and the metal, is five or six atoms; smaller rings are strained, and for larger rings, the second complexing atom is farther away, and formation of the second bond may require awkward bends. In some cases, an enthalpy effect may also be present.[2]

[2] J. J. R. Fausto da Silva, *J. Chem. Educ.*, **1983**, *60*, 390; R. D. Hancock, *J. Chem. Educ.*, **1992**, *69*, 615.

TABLE 10.2 Formation Constants for Cd^{2+} Complexes $[Cd(CH_3NH_2)_4]^{2+}$ and $[Cd(en)_2]^{2+}$

Ligand	$\Delta H°$ (kJ/mol)	$\Delta S°$ (J/mol K)	$-T\Delta S°$ (kJ/mol)	$\Delta G°$ (kJ/mol)	log β
4 CH_3NH_2	−57.3	−67.3	+20.1	−37.2	6.52
2 en	−56.5	+14.1	−4.2	−60.7	10.6

Source: F. A. Cotton and G. Wilkinson, *Advanced Inorganic Chemistry*, 6th ed., 1999, Wiley InterScience, New York, p. 28.

NOTE: β is the overall equilibrium constant for formation of the complexes in solution.

10.1.2 Magnetic Susceptibility

Just as in the diatomic examples in Chapter 5, the magnetic properties of a coordination compound can provide indirect evidence of the orbital energy levels. Hund's rule requires the maximum number of unpaired electrons in energy levels with equal, or nearly equal, energies. Diamagnetic compounds, with all electrons paired, are slightly repelled by a magnetic field. When there are unpaired electrons, a compound is paramagnetic and is attracted into a magnetic field. The measure of this magnetism is called the **magnetic susceptibility**, χ.[3]

We will describe a modification of the Gouy method[4] for determining magnetic susceptibility that requires only an analytical balance and a small magnet (Figure 10.1).[5] In this approach, the solid sample is placed in a small glass sample tube. A small high-field U-shaped magnet is weighed four times: (1) alone, (2) with the sample suspended between the poles of the magnet, (3) with a reference compound of known magnetic susceptibility suspended in the gap, and finally (4) with the empty sample tube suspended in the gap (to correct for any magnetic effect in the sample tube). With a diamagnetic sample, the tube and magnet repel each other, and the magnet appears slightly heavier. With a paramagnetic sample, the tube and magnet attract each other, and the magnet appears lighter. The measurement of the known compound provides a standard from which the mass susceptibility (susceptibility per gram) of the sample can be calculated and converted to the molar susceptibility. More precise measurements require temperature control and measurement at different magnetic field strengths to correct for possible impurities.

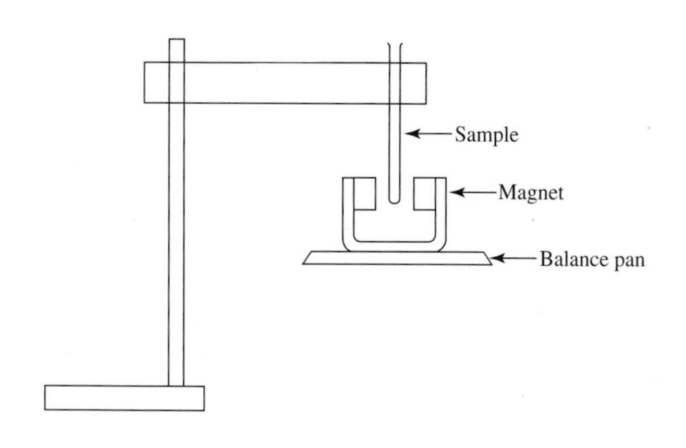

FIGURE 10.1 Modified Gouy Magnetic Susceptibility Apparatus. (Adapted with permission from S. S. Eaton and G. R. Eaton, *J. Chem. Educ.*, **1979**, *56*, 170.)

[3] D. P. Shoemaker, C. W. Garland, and J. W. Nibler, Experiments in Physical Chemistry, 5th ed., McGraw-Hill, New York, 1989, pp. 418–439.

[4] B. Figgis and J. Lewis, in H. Jonassen and A. Weissberger, eds., *Techniques of Inorganic Chemistry*, Vol. IV, Interscience, New York, 1965, p. 137.

[5] S. S. Eaton and G. R. Eaton, *J. Chem. Educ.*, **1979**, *56*, 170.

Magnetic susceptibility is related to the **magnetic moment, μ**, according to the relationship

$$\mu = 2.828(\chi T)^{\frac{1}{2}}$$

where χ = magnetic susceptibility (cm^3/mol)

T = temperature (Kelvin)

The unit of magnetic moment is the Bohr magneton, μ_B

$1\ \mu_B = 9.27 \times 10^{-24}\ \text{J T}^{-1}$ (joules/tesla).

Paramagnetism arises because electrons behave as tiny magnets. Although there is no direct evidence for spinning movement by electrons, a charged particle spinning rapidly would generate a **spin magnetic moment**, and the popular term has therefore become **electron spin**. As discussed in Section 2.2.2, electrons with $m_s = -\frac{1}{2}$ are said to have a negative spin, those with $m_s = +\frac{1}{2}$ a positive spin. The total spin magnetic moment is characterized by the spin quantum number S, which is equal to the maximum total spin, the sum of the m_s values.

For example, an isolated oxygen atom with electron configuration $1s^2 2s^2 2p^4$ in its ground state has one electron in each of two $2p$ orbitals and a pair in the third. The maximum total spin is $S = +\frac{1}{2} + \frac{1}{2} + \frac{1}{2} - \frac{1}{2} = 1$. The orbital angular momentum, characterized by the quantum number L, where L is equal to the maximum possible sum of the m_l values, results in an additional orbital magnetic moment. For the oxygen atom, the maximum possible sum of the m_l values for the p^4 electrons occurs when two electrons have $m_l = +1$ and one each have $m_l = 0$ and $m_l = -1$. In this case, $L = +1+0-1+1 = 1$. The combination of these two contributions to the magnetic moment, added as vectors, is the total magnetic moment of the atom or molecule. Additional details of quantum numbers S and L are provided in Chapter 11.

▶ **Exercise 10.1** Calculate L and S for the nitrogen atom.

The equation for the magnetic moment in terms of S and L is

$$\mu_{S+L} = g\sqrt{\left[S(S+1)\right]+\left[\tfrac{1}{4}L(L+1)\right]}$$

where μ = magnetic moment

g = gyromagnetic ratio (conversion to magnetic moment)

S = spin quantum number

L = orbital quantum number

Although detailed determination of the electronic structure requires consideration of the orbital moment, for most complexes of the first transition series, the spin-only moment is sufficient, because orbital contribution is small. The **spin-only magnetic moment, μ_S**, is therefore

$$\mu_S = g\sqrt{S(S+1)}$$

External fields from other atoms and ions may effectively quench the orbital moment in these complexes. For the heavier transition metals and the lanthanides, the orbital contribution is larger and must be taken into account. Because we are usually concerned primarily with the number of unpaired electrons in a compound, and the possible values of μ differ significantly for different numbers of unpaired electrons, the errors introduced by considering only the spin moment are usually not large enough to cause difficulty.

In Bohr magnetons, the gyromagnetic ratio, g, is 2.00023, frequently rounded to 2. The equation for μ_S then becomes

$$\mu_S = 2\sqrt{S(S + 1)} = \sqrt{4S(S + 1)}$$

Because $S = \frac{1}{2}, 1, \frac{3}{2}, \ldots$ for 1, 2, 3,. . . unpaired electrons, this equation can also be written

$$\mu_S = \sqrt{n(n + 2)}$$

where n = number of unpaired electrons. This is the equation that is used most frequently. Table 10.3 shows the change in μ_S and μ_{S+L} with n, along with some experimental moments.

▶ **Exercise 10.2** Show that $\sqrt{4S(S + 1)}$ and $\sqrt{n(n + 2)}$ are equivalent expressions.

▶ **Exercise 10.3** Calculate the spin-only magnetic moment for the following atoms and ions. (Remember the order of loss of electrons from transition metals described near the end of Section 2.2.4.)

$$\text{Fe} \quad \text{Fe}^{2+} \quad \text{Cr} \quad \text{Cr}^{3+} \quad \text{Cu} \quad \text{Cu}^{2+}$$

Several other techniques may be used to measure magnetic susceptibility, including nuclear magnetic resonance[6] and the Faraday method, using an unsymmetrical magnetic field.[7]

10.1.3 Electronic Spectra

Direct evidence of orbital energy levels can be obtained from electronic spectra. The energy of the light absorbed as electrons are raised to higher levels is the difference in energy between the states, which depends on the orbital energy levels and their occupancy. The observed spectra are frequently more complex than the simple energy diagrams used in this chapter seem to indicate; Chapter 11 gives a more complete picture of electronic spectra of coordination compounds. Much information about bonding and electronic structures in complexes has come from the study of electronic spectra.

10.1.4 Coordination Numbers and Molecular Shapes

Although a number of factors influence the number of ligands bonded to a metal and the shapes of the resulting species, in some cases we can predict which structure is favored from the electronic structure of a complex. For example, two four-coordinate structures are possible, tetrahedral and square planar. Some metals, such as Pt(II), almost exclusively form square-planar complexes. Others, such as Ni(II) and Cu(II), exhibit both structures, and sometimes intermediate structures, depending on the ligands. Subtle differences in electronic structure, described later in this chapter, help to explain these differences.

---

[6] D. F. Evans, *J. Chem. Soc.*, **1959**, 2003.
[7] L. N. Mulay and I. L. Mulay, *Anal. Chem.*, **1972**, *44*, 324R.

TABLE 10.3 Calculated and Experimental Magnetic Moments

Ion	n	S	L	μ_S	μ_{S+L}	Observed
V^{4+}	1	$\frac{1}{2}$	2	1.73	3.00	1.7–1.8
Cu^{2+}	1	$\frac{1}{2}$	2	1.73	3.00	1.7–2.2
V^{3+}	2	1	3	2.83	4.47	2.6–2.8
Ni^{2+}	2	1	3	2.83	4.47	2.8–4.0
Cr^{3+}	3	$\frac{3}{2}$	3	3.87	5.20	~ 3.8
Co^{2+}	3	$\frac{3}{2}$	3	3.87	5.20	4.1–5.2
Fe^{2+}	4	2	2	4.90	5.48	5.1–5.5
Co^{3+}	4	2	2	4.90	5.48	~ 5.4
Mn^{2+}	5	$\frac{5}{2}$	0	5.92	5.92	~ 5.9
Fe^{3+}	5	$\frac{5}{2}$	0	5.92	5.92	~ 5.9

Source: F. A. Cotton and G. Wilkinson, *Advanced Inorganic Chemistry*, 4th ed., Wiley, New York, 1980, pp. 627–628.

NOTE: All moments are given in Bohr magnetons.

10.2 THEORIES OF ELECTRONIC STRUCTURE

10.2.1 Terminology

Different names have been used for the theoretical approaches to the electronic structure of coordination complexes, and sometimes these names have been used inconsistently. We will discuss the following labels, in order of their historical development:

Valence Bond Theory. This method describes bonding using hybrid orbitals and electron pairs, as an extension of the electron-dot and hybrid orbital methods used for simpler molecules. Although the theory as originally proposed is less commonly used today, the hybrid notation is still common.

Crystal Field Theory. This is an electrostatic approach, used to describe the split in metal *d*-orbital energies. It provides an approximate description of the electronic energy levels that determine the ultraviolet and visible spectra, but it does not describe the bonding.

Ligand Field Theory. This is a more complete description of bonding in terms of the interactions between metal and ligand frontier orbitals to form molecular orbitals. It may use some of the terminology of crystal field theory but focuses on orbital interactions rather than attractions between ions.

Angular Overlap Method. This is a method of estimating the relative magnitudes of orbital energies in a molecular orbital calculation. It explicitly takes into account the bonding energy as well as the relative orientation of the frontier orbitals.

The advent of sophisticated computer software has made it possible to conduct a wide range of calculations to predict geometries, orbital shapes and energies, and a variety of other properties of molecules, including coordination complexes. Molecular orbital calculations are typically based on the Born–Oppenheimer approximation, which considers nuclei to be in fixed positions in comparison with rapidly moving electrons. Because such calculations are "many-body" problems that cannot be solved exactly, various approximate methods have been developed to simplify the calculations, and current journal articles routinely discuss the most recent such methods in comparison with other methods. The simplest of these approaches, using Extended Hückel Theory, in many cases generates useful three-dimensional images of molecular orbitals and is relatively inexpensive. Details of molecular orbital calculations are beyond the scope of this text; however, the reader is encouraged to make use of molecular modeling software where available to supplement the topics and images—some of which were generated using molecular modeling software—in this text. Suggested references on this topic are included in the General References at the end of this chapter.[8]

In the following pages, the valence bond theory and the crystal field theory are described very briefly to set more recent developments in their historical context. The rest of the chapter describes the ligand field theory and the method of angular overlap, which can be used to estimate the orbital energy levels. These two supply the basic approach to bonding in coordination compounds for the remainder of this book.

10.2.2 Historical Background

VALENCE BOND THEORY The valence bond theory, originally proposed by Pauling in the 1930s, uses the hybridization ideas presented in Chapter 5.[9] For octahedral complexes, d^2sp^3 hybrids of the metal orbitals are required. However, the *d* orbitals used by the first-row transition metals could be either 3*d* or 4*d*. Pauling originally described the structures resulting from these as covalent and ionic, respectively. He later changed the terms to "hyperligated" and "hypoligated," and they are also known as inner orbital (using 3*d*) and outer orbital (using 4*d*) complexes. The number of unpaired electrons, measured by the magnetic behavior of the compounds, determines which *d* orbitals are used.

For example, Fe(III) has five unpaired electrons as an isolated ion, one in each of the 3*d* orbitals. In octahedral coordination complexes, it may have either one or five unpaired electrons. In complexes with one unpaired electron, the ligand electrons force the metal *d* electrons to pair up, leaving two 3*d* orbitals available for hybridization and bonding. In complexes with five unpaired electrons, the ligands do not bond strongly enough to force pairing of the 3*d* electrons. Pauling proposed that the 4*d* orbitals could be used for bonding in such cases, with the arrangement of electrons shown in Figure 10.2.

When seven electrons must be provided for, as in Co(II), there are either one or three unpaired electrons. In the low-spin case with one unpaired electron, the seventh electron must go into a higher orbital (unspecified by Pauling, but presumed to be 5*s*).[10] In the high-spin case, with three unpaired electrons, the 4*d* or outer orbital hybrid must be used for bonding, leaving the metal electrons in the 3*d* levels.[11]

The valence bond theory was of great importance in the development of bonding theory for coordination compounds, but it is rarely used today except when discussing hybrid orbitals. Although it provided a set of orbitals for bonding, using the very high energy 4*d* orbitals seems unlikely, and the results do not lend themselves to a good explanation of the electronic spectra of complexes. Because much of our experimental data are derived from electronic spectra, this is a serious shortcoming.

[8] A brief introduction and comparison of various computational methods is in G. O. Spessard and G. L. Miessler, *Organometallic Chemistry*, Oxford University Press, New York, 2010, pp. 42–49.
[9] L. Pauling, *The Nature of the Chemical Bond*, 3rd ed., Cornell University Press, Ithaca, NY, 1960, Chapter 5.
[10] B. N. Figgis and R. S. Nyholm, *J. Chem. Soc.*, **1959**, 338; J. S. Griffith and L. E. Orgel, *Q. Rev. Chem. Soc.*, **1957**, XI, 381.
[11] The terms *high spin* and *low spin* are discussed in section 10.3-2.

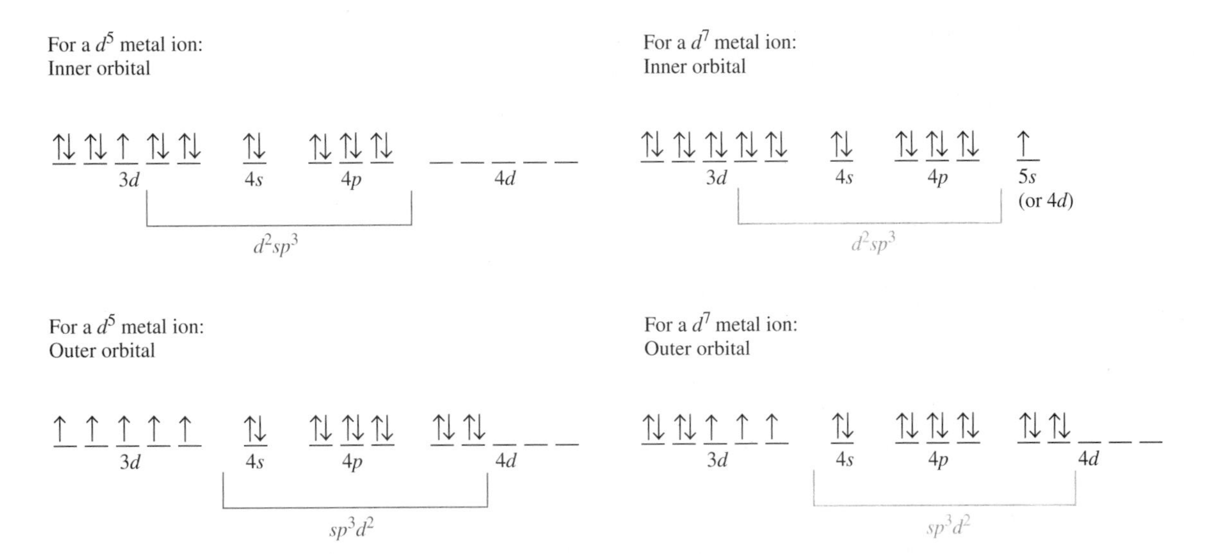

FIGURE 10.2 Inner and Outer Orbital Complexes. In each case, ligand electrons fill the d^2sp^3 bonding orbitals. The remaining orbitals contain the electrons from the metal.

CRYSTAL FIELD THEORY As originally developed, crystal field theory[12] was used to describe the electronic structure of metal ions in crystals, where they are surrounded by anions that create an electrostatic field with symmetry dependent on the crystal structure. The energies of the *d* orbitals of the metal ions are split by the electrostatic field, and approximate values for these energies can be calculated. No attempt was made to deal with covalent bonding, because the ionic crystals did not require it. Crystal field theory was developed in the 1930s. Shortly afterward, it was recognized that the same arrangement of electron-pair donor species around a metal ion existed in coordination complexes as well as in crystals, and a more complete molecular orbital theory was developed.[13] However, neither was widely used until the 1950s, when interest in coordination chemistry increased.

When the *d* orbitals of a metal ion are placed in an octahedral field of ligand electron pairs, any electrons in these orbitals are repelled by the field. As a result, the $d_{x^2-y^2}$ and d_{z^2} orbitals, which have e_g symmetry, are directed at the surrounding ligands and are raised in energy. The d_{xy}, d_{xz}, and d_{yz} orbitals (t_{2g} symmetry), directed between the ligands, are relatively unaffected by the field. The resulting energy difference is identified as Δ_o (*o* for octahedral; some older references use the term $10Dq$ instead of Δ_o). This approach provides a simple means of identifying the *d*-orbital splitting found in coordination complexes and can be extended to include more quantitative calculations. It requires extension to the more complete ligand field theory to include pi interactions.

The average energy of the five *d* orbitals is above that of the free ion orbitals, because the electrostatic field of the ligands raises their energy. The t_{2g} orbitals are $0.4\Delta_o$ below, and the e_g orbitals are $0.6\Delta_o$ above, this average energy, as shown in Figure 10.3. The three t_{2g} orbitals then have a total energy of $-0.4\Delta_o \times 3 = -1.2\Delta_o$ and the two e_g orbitals have a total energy of $+0.6\Delta_o \times 2 = +1.2\Delta_o$ compared with the average. The energy difference between the actual distribution of electrons and that for all electrons in the uniform field levels is called the **crystal field stabilization energy (CFSE)**. It is equal in magnitude to the ligand field stabilization energy (LFSE) described later in this chapter.

The chief drawbacks to the crystal field approach are in its concept of the repulsion of orbitals by the ligands and its lack of any explanation for bonding in coordination

[12] H. Bethe, *Ann. Phys.*, **1929**, *3*, 133.
[13] J. H. Van Vleck, *J. Chem. Phys.*, **1935**, *3*, 807.

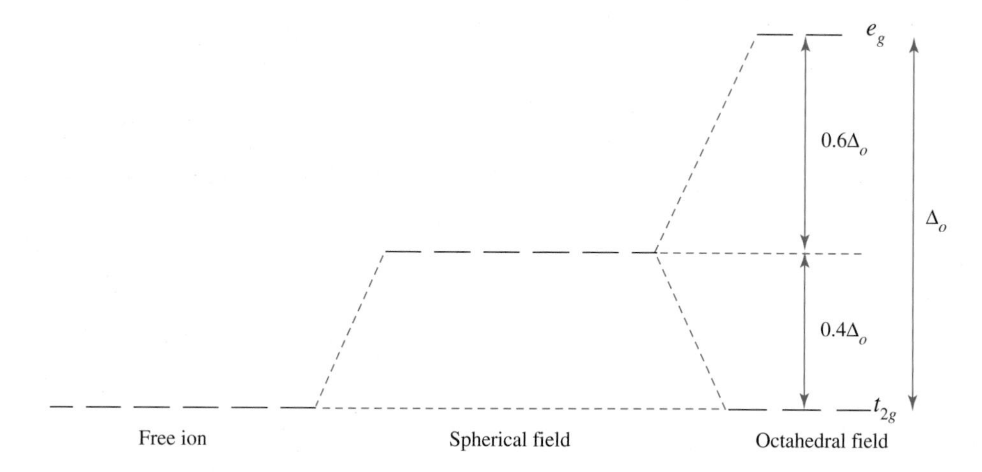

FIGURE 10.3 Crystal Field Splitting.

complexes. As we have seen in all our discussions of molecular orbitals, any interaction between orbitals leads to formation of both higher and lower energy molecular orbitals. The purely electrostatic approach does not allow for the lower (bonding) molecular orbitals, and thus fails to provide a complete picture of the electronic structure.

10.3 LIGAND FIELD THEORY

The electrostatic crystal field theory and the molecular orbital theory were combined into a more complete theory called **ligand field theory**, described qualitatively by Griffith and Orgel.[14] Many of the details presented here come from their work.

10.3.1 Molecular Orbitals for Octahedral Complexes

For octahedral complexes, ligands can interact with metals in a sigma fashion, donating electrons directly to metal orbitals, or in a pi fashion, with ligand–metal orbital interactions occurring in two regions off to the side. Examples of such interactions are shown in Figure 10.4.

As in Chapter 5 we will proceed to first consider group orbitals on ligands based on O_h symmetry, then we will consider how these group orbitals can interact with

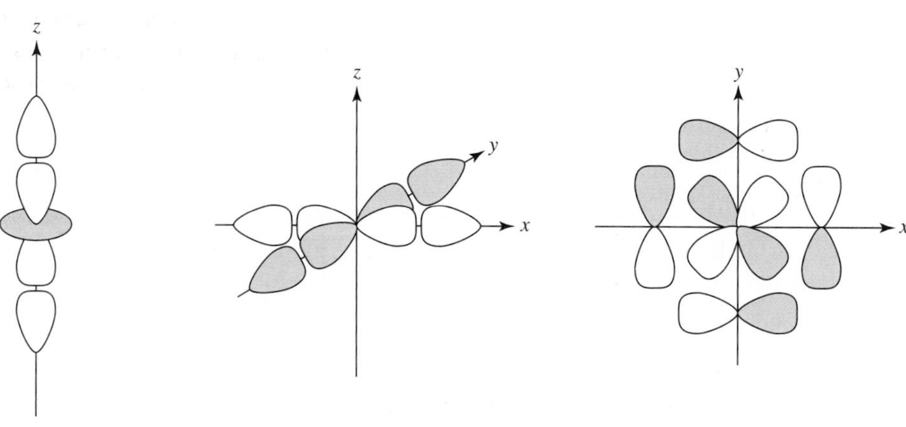

FIGURE 10.4 Orbital Interactions in Octahedral Complexes.

[14] J. S. Griffith and L. E. Orgel, *Q. Rev. Chem. Soc.*, **1957**, *XI*, 381.

TABLE 10.4 Character Table for O_h

O_h	E	$8C_3$	$6C_2$	$6C_4$	$3C_2(=C_4^2)$	i	$6S_4$	$8S_6$	$3\sigma_h$	$6\sigma_d$		
A_{1g}	1	1	1	1	1	1	1	1	1	1		
A_{2g}	1	1	−1	−1	1	1	−1	1	1	−1		
E_g	2	−1	0	0	2	2	0	−1	2	0		$(2z^2 - x^2 - y^2, x^2 - y^2)$
T_{1g}	3	0	−1	1	−1	3	1	0	−1	−1	(R_x, R_y, R_z)	
T_{2g}	3	0	1	−1	−1	3	−1	0	−1	1		(xy, xz, yz)
A_{1u}	1	1	1	1	1	−1	−1	−1	−1	−1		
A_{2u}	1	1	−1	−1	1	−1	1	−1	−1	1		
E_u	2	−1	0	0	2	−2	0	1	−2	0		
T_{1u}	3	0	−1	1	−1	−3	−1	0	1	1	(x, y, z)	
T_{2u}	3	0	1	−1	−1	−3	1	0	1	−1		

orbitals of matching symmetry on the central atom, in this case a transition metal. We will first consider sigma interactions then pi interactions. For reference, the character table for O_h symmetry is provided in Table 10.4.

SIGMA INTERACTIONS The basis for a reducible representation is a set of six donor orbitals on the ligands as, for example, σ donor orbitals on six NH_3 ligands. Using this set as a basis—or equivalently in terms of symmetry, a set of six vectors pointing toward the metal, as shown at right—the following representation can be obtained:

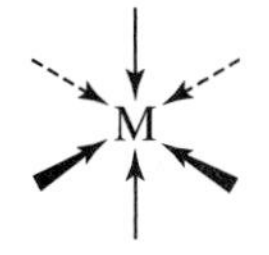

O_h	E	$8C_3$	$6C_2$	$6C_4$	$3C_2(=C_4^2)$	i	$6S_4$	$8S_6$	$3\sigma_h$	$6\sigma_d$
Γ_σ	6	0	0	2	2	0	0	0	4	2

This representation reduces to $A_{1g} + T_{1u} + E_g$:

O_h	E	$8C_3$	$6C_2$	$6C_4$	$3C_2(=C_4^2)$	i	$6S_4$	$8S_6$	$3\sigma_h$	$6\sigma_d$		
A_{1g}	1	1	1	1	1	1	1	1	1	1		$x^2 + y^2 + z^2$
T_{1u}	3	0	−1	1	−1	−3	−1	0	1	1	(x, y, z)	
E_g	2	−1	0	0	2	2	0	−1	2	0		$(2z^2 - x^2 - y^2, x^2 - y^2)$

▶ **Exercise 10.4** Verify the characters of this reducible representation Γ_σ and that it reduces to $A_{1g} + T_{1u} + E_g$.

THE *d* ORBITALS The *d* orbitals play key roles in transition-metal coordination chemistry, so it is useful to examine them first. According to the O_h character table, the *d* orbitals match the irreducible representations E_g and T_{2g}. For the E_g (d_{x2-y2} and d_{z2}) orbitals there is a match with the E_g ligand orbitals. Because the symmetries match, there is an interaction between the two sets of E_g orbitals to form a pair of bonding orbitals (e_g) and the counterpart pair of antibonding orbitals (e_g*). It is not surprising that significant interaction occurs between the d_{x2-y2} and d_{z2} orbitals and the sigma donor ligands—after all, the lobes of these *d* orbitals and the σ donor orbitals of the

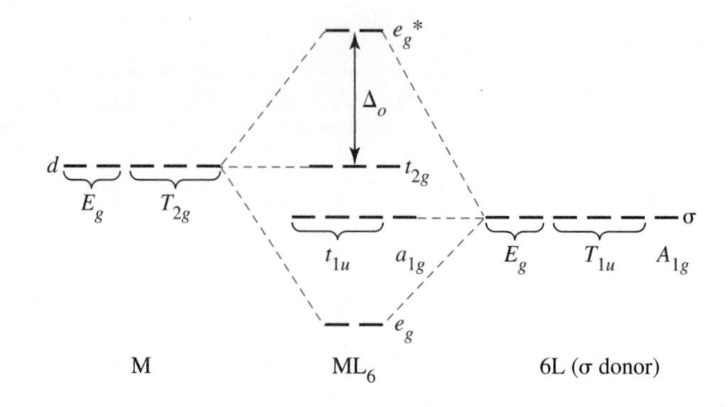

FIGURE 10.5 Sigma–Donor Interactions with Metal *d* Orbitals.

ligands point toward each other. On the other hand, there are no ligand orbitals matching the T_{2g} symmetry of the d_{xy}, d_{xz}, and d_{yz} orbitals—whose lobes point between the ligands—so these metal orbitals are nonbonding. The overall *d* interactions are shown in Figure 10.5.

THE *s* AND *p* ORBITALS The valence *s* and *p* orbitals of the metal have symmetry that matches the two remaining irreducible representations: *s* matches A_{1g} and the set of *p* orbitals matches T_{1u}. Because of the symmetry match, the A_{1g} interactions lead to the formation of bonding and antibonding orbitals (a_{1g} and a_{1g}*), and the T_{1u} interactions lead to formation of a set of three bonding orbitals (t_{1u}) and the matching three antibonding orbitals (t_{1u}*). These interactions, in addition to those already described for *d* orbitals, are shown in Figure 10.6. The molecular orbital energy-level diagram in this figure summarizes the interactions for octahedral complexes containing ligands that are exclusively sigma donors. As a result of interactions between the donor orbitals on the ligands and the *s*, *p*, and d_{x2-y2} and d_{z2} orbitals of the metal, six bonding orbitals are formed, occupied by the electrons donated by the ligands. These six electron pairs are stabilized in energy; they represent the sigma bonds stabilizing the complex.

The d_{xy}, d_{xz}, and d_{yz} orbitals are nonbonding, so their energies are unaffected by the σ donor orbitals; they are shown in the molecular orbital diagam with the symmetry label t_{2g}. At higher energy, above the t_{2g}, are the antibonding partners to the six bonding molecular orbitals.

An example of a complex that can be described by the energy level diagram in Figure 10.6 is the green ion $[Ni(H_2O)_6]^{2+}$. The six bonding orbitals (a_{1g}, e_g, t_{1u}), are occupied by the six electron pairs donated by the aqua ligands. In addition, the Ni^{2+} ion has eight *d* electrons.[15] In the complex, six of these electrons fill the t_{2g} orbitals, and the final two electrons occupy the e_g* (separately, with parallel spin). By fortunate circumstance the energy difference between the t_{2g} and e_g* orbitals in many transition-metal complexes is within the energy range of visible light. In $[Ni(H_2O)_6]^{2+}$ the difference in energy between the t_{2g} and e_g* is an approximate match for red light. Consequently, when white (full spectrum) light passes through a solution of $[Ni(H_2O)_6]^{2+}$, red light is absorbed and excites electrons from the t_{2g} to the e_g* orbitals; the light that passes through, now with some of its red light removed, is perceived as green, the complementary color to red. This phenomenon, which is more complicated than its brief description here, will be discussed in more detail in Chapter 11.

In Figure 10.6 we again see Δ_o, frequently used as a measure of the magnitude of metal–ligand interactions.

[15] Recall that in transition metal ions the valence electrons are all *d* electrons.

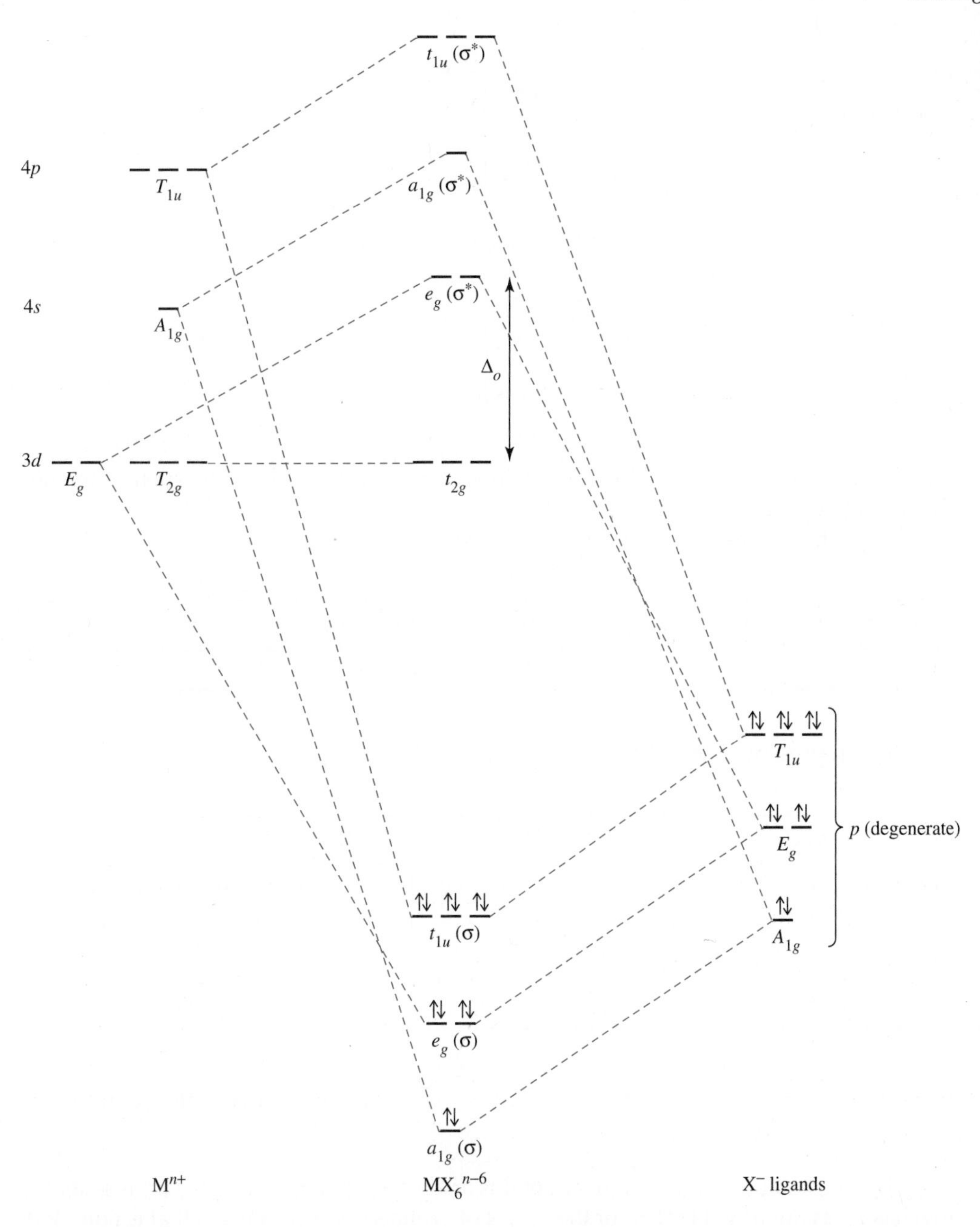

FIGURE 10.6 Sigma Donor Interactions with Metal *s*, *p*, and *d* Orbitals. As in Chapter 5, the symmetry labels of the atomic orbitals are capitalized, and the labels of the molecular orbitals are in lowercase. (Adapted from F. A. Cotton, *Chemical Applications of Group Theory*, 3rd ed., Wiley InterScience, New York, 1990, p. 232, omitting π orbitals. © 1990, John Wiley & Sons, Inc. Reprinted by permission of John Wiley & Sons, Inc.)

PI INTERACTIONS Although Figure 10.6 can be used as a guide to describe energy levels in octahedral transition-metal complexes, it needs to be modified when ligands that can engage in pi interactions with metals are involved. As we will see, pi interactions can have important effects on the t_{2g} orbitals, which to this point have been considered nonbonding.

$Cr(CO)_6$ is a good example of a molecule that has ligands that can engage in both sigma and pi interactions with a central metal. As has been shown in Figure 5.14, the CO ligand has large lobes in both its HOMO (3σ) and LUMO ($1\pi^*$) orbitals; it is both an effective σ donor and π acceptor. As a π acceptor, it has two orthogonal π^* orbitals, both of which can accept electron density from metal orbitals of matching symmetry.

Once again it is necessary to create a representation, this time using as basis the set of 12 π^* orbitals, two from each ligand, from the set of six CO ligands. In constructing

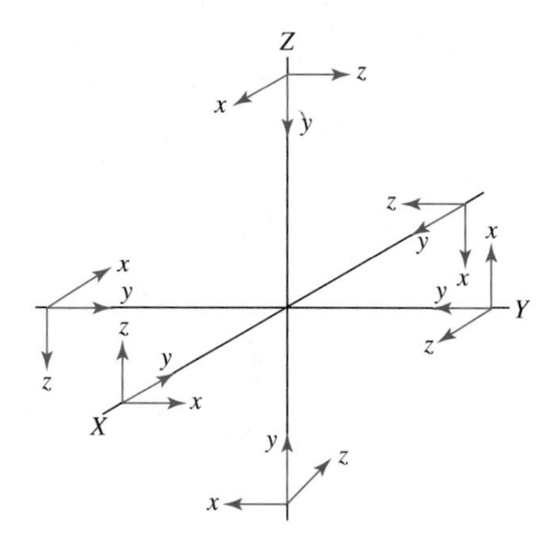

FIGURE 10.7 Coordinate System for Octahedral π Orbitals.

this representation, it is useful to have a consistent coordinate scheme, such as the one shown in Figure 10.7.

Using this set as a basis, the representation Γ_π can be obtained:

O_h	E	$8C_3$	$6C_2$	$6C_4$	$3C_2(=C_4^2)$	i	$6S_4$	$8S_6$	$3\sigma_h$	$6\sigma_d$
Γ_π	12	0	0	0	−4	0	0	0	0	0

This representation reduces to $T_{1g} + T_{2g} + T_{1u} + T_{2u}$:

O_h	E	$8C_3$	$6C_2$	$6C_4$	$3C_2(=C_4^2)$	i	$6S_4$	$8S_6$	$3\sigma_h$	$6\sigma_d$	
T_{1g}	3	0	−1	1	−1	3	1	0	−1	−1	
T_{2g}	3	0	1	−1	−1	3	−1	0	−1	1	(xy, xz, yz)
T_{1u}	3	0	−1	1	−1	−3	−1	0	1	1	(x, y, z)
T_{2u}	3	0	1	−1	−1	−3	1	0	1	−1	

▶ **Exercise 10.5** Verify the characters of Γ_π and that it reduces to $T_{1g} + T_{2g} + T_{1u} + T_{2u}$.

The most important consequence of this analysis is that it generates a representation that has T_{2g} symmetry, a match for the T_{2g} set of orbitals (d_{xy}, d_{xz}, and d_{yz}) that is nonbonding for ligands that are σ donors only. If the ligand is a π acceptor such as CO, the effect is to lower the energy of the t_{2g} orbitals, in forming bonding molecular orbitals, and to raise the energy of the (empty) t_{2g}* from the ligands, in forming antibonding orbitals. A less important interaction is between the T_{1u} of the ligands and the set of p orbitals on the metal, and there is also a T_{1u} sigma interaction. The T_{1g} and T_{2u} orbitals have no matching metal orbitals and are nonbonding. The overall result is shown in Figure 10.8.

Strong π acceptor ligands therefore have the capacity to increase the magnitude of Δ_o by lowering the energy of the t_{2g} orbitals. In the example of $Cr(CO)_6$ there are 12 electrons in the bonding a_{1g}, e_g, and t_{1u} orbitals at the bottom of the diagram. The next six electrons, from Cr, fill the three t_{2g} orbitals, which are also bonding. Because the energy difference between the t_{2g} and e_g* is much greater than it would have been in the absence of the π acceptor ligands, it takes much more energy to excite an electron from the lower to the higher levels than, for example, in $[Ni(H_2O)_6]^{2+}$—energy in the ultraviolet part of

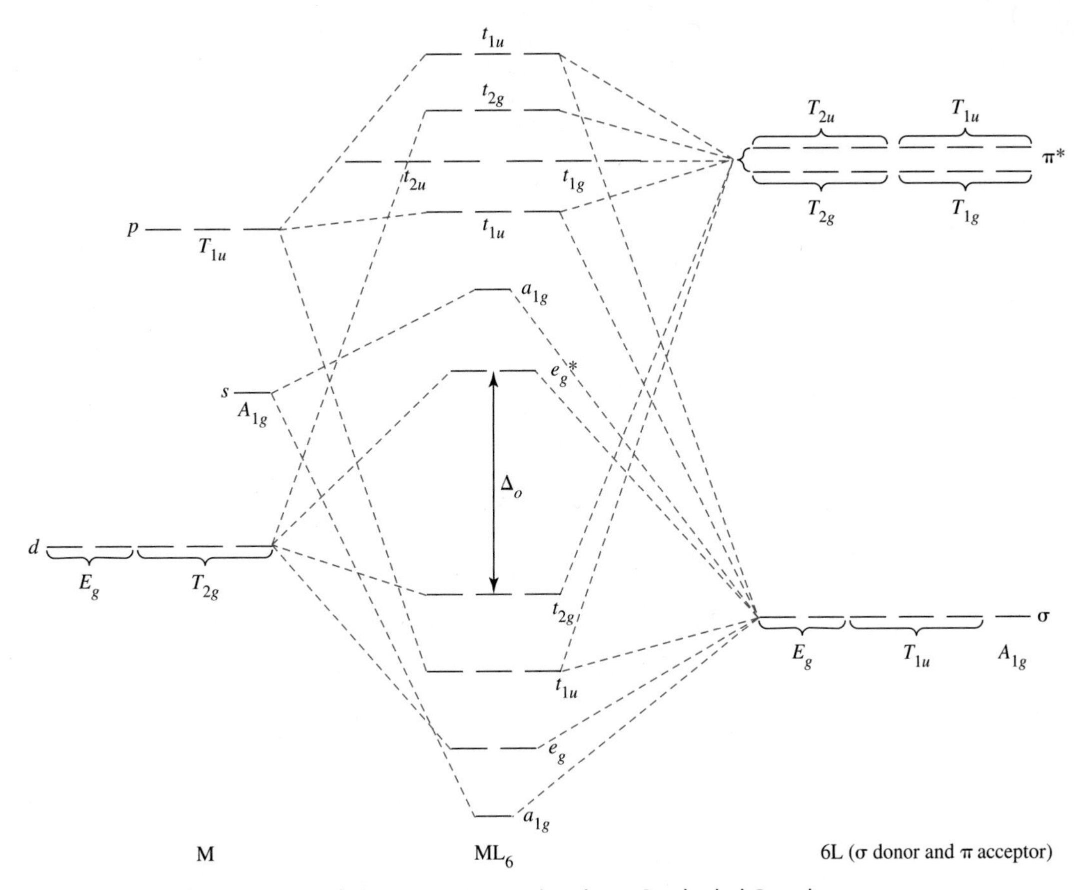

FIGURE 10.8 Sigma Donor and Pi Acceptor Interactions in an Octahedral Complex.

the spectrum; $Cr(CO)_6$ is colorless, because its energy levels are too far apart to absorb visible light.

Electrons in bonding orbitals provide the potential energy that holds molecules together. Electrons in the higher levels, affected by ligand field effects, help determine structural details, magnetic properties, electronic spectrum absorptions, and reactivity.

The cyanide ion (Figure 10.9) provides another example of a ligand that can engage in both sigma and pi interactions. The molecular orbital picture of CN^- is intermediate between those of N_2 and CO given in Chapter 5, because the energy differences between C and N orbitals are significant but less than those between C and O orbitals. The HOMO for CN^- is a σ orbital with considerable bonding character and a concentration of electron density on the carbon. This is the donor orbital used by CN^- in forming σ orbitals in the complex. Above the HOMO, the LUMO orbitals of CN^- are two empty π^* orbitals that can be used for π bonding with the metal. Overlap of ligand orbitals with metal *d* orbitals is shown in Figure 10.10.

The ligand π^* orbitals have energies higher than those of the metal t_{2g} (d_{xy}, d_{xz}, d_{yz}) orbitals, with which they overlap. As a result, they form molecular orbitals, with the bonding orbitals lower in energy than the initial metal t_{2g} orbitals. The corresponding antibonding orbitals are higher in energy than the $e_g{}^*$ orbitals. Metal *d* electrons occupy the bonding orbitals (now the HOMO), resulting in a larger value for Δ_o and increased bonding strength, as shown in Figure 10.11(a). Significant energy stabilization can result from this added π bonding. This **metal-to-ligand (M $\longrightarrow$ L) π bonding** is also called **π back-bonding,** with electrons from *d* orbitals of the metal donated back to the ligands. The ligand in this case is called a **π acceptor**.

FIGURE 10.9 Cyanide Molecular Orbitals.

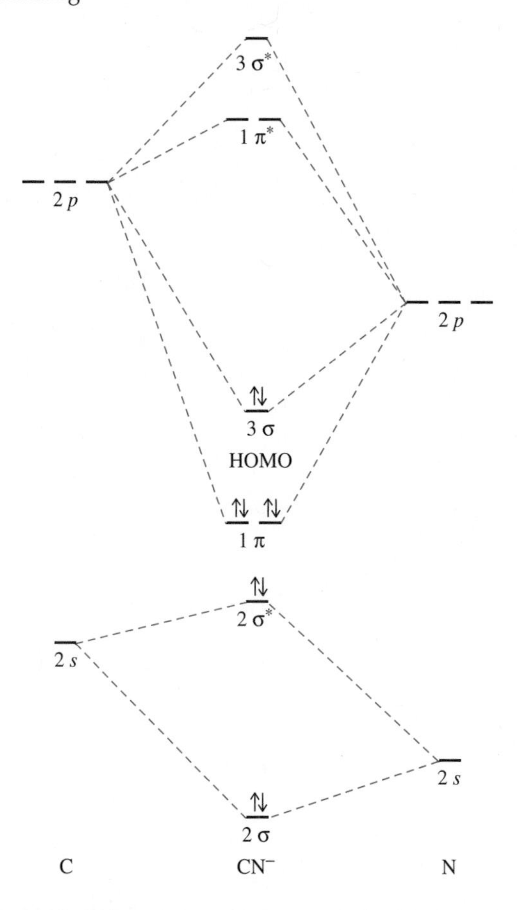

When the ligand has electrons in its p orbitals, as in F^- or Cl^-, the bonding molecular π orbitals are occupied by these electrons, and there are two net results: the t_{2g} bonding orbitals, derived primarily from ligand orbitals, strengthen the ligand–metal linkage slightly; and the corresponding $t_{2g}{}^*$ levels, derived primarily from metal d orbitals, are raised in energy and become antibonding. This reduces Δ_o, as in Figure 10.11(b). The metal ion d electrons are pushed into the higher $t_{2g}{}^*$ orbital by the ligand electrons. This is described as **ligand-to-metal (L $\longrightarrow$ M) π bonding**, with the π electrons from the ligands being donated to the metal ion. Ligands participating in such interactions are called **π-donor** ligands. The decrease in the energy of the bonding orbitals is partly counterbalanced by the increase in the energy of the $t_{2g}{}^*$ orbitals. In addition, the combined σ and π donations from the ligands give the metal more negative charge, which decreases attraction between the metal and the ligands and makes this type of bonding less favorable.

FIGURE 10.10 Overlap of d, π^*, and p Orbitals with Metal d Orbitals. Overlap is good with ligand d and π^* orbitals but poorer with ligand p orbitals.

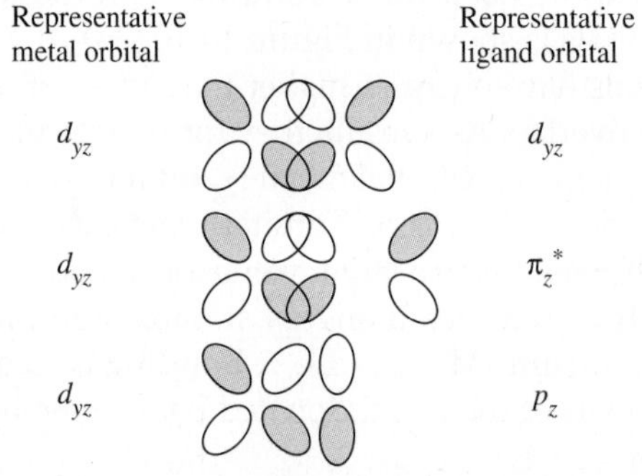

Representative metal orbital	Representative ligand orbital
d_{yz}	d_{yz}
d_{yz}	$\pi_z{}^*$
d_{yz}	p_z

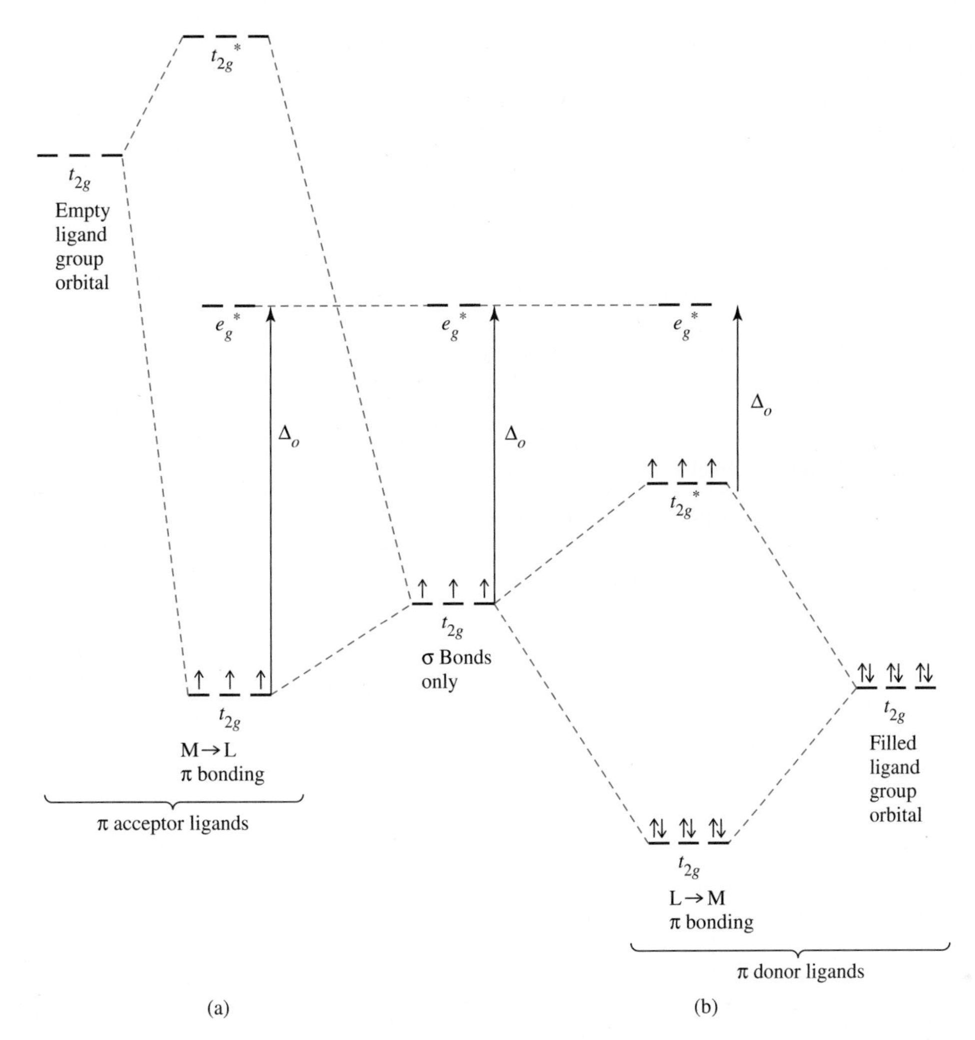

FIGURE 10.11 Effects of π Bonding on Δ_o using a d^3 Ion.

Overall, filled π^* or p orbitals on ligands, frequently with relatively low energy, result in L $\longrightarrow$ M π bonding and a smaller Δ_o for the complex. Empty higher-energy π or d orbitals on the ligands result in M $\longrightarrow$ L π bonding and a larger Δ_o for the complex. Ligand-to-metal π bonding usually gives decreased stability for the complex, favoring high-spin configurations; metal-to-ligand π bonding usually gives increased stability and favors low-spin configurations.

Part of the stabilizing effect of π back-bonding is a result of transfer of negative charge away from the metal ion. The positive ion accepts electrons from the ligands to form σ bonds. The metal is then left with a surplus of negative charge. When the π orbitals can be used to transfer part of this charge back to the ligands, the overall stability is improved. The π-acceptor ligands that can participate in π back-bonding are extremely important in organometallic chemistry and will be discussed further in Chapter 13.

10.3.2 Orbital Splitting and Electron Spin

In octahedral coordination complexes, electrons from the ligands fill all six bonding molecular orbitals, and any electrons from the metal ion occupy the t_{2g} and e_g^* orbitals. Ligands whose orbitals interact strongly with the metal orbitals are called **strong-field ligands**; with these, the split between the t_{2g} and e_g^* orbitals is large, so Δ_o is large.

Ligands with weak interactions are called **weak-field ligands**; the split between the t_{2g} and e_g orbitals is smaller and Δ_o is small. For d^0 through d^3 and d^8 through d^{10} ions, only one electron configuration is possible. On the other hand, the d^4 through d^7 ions exhibit **high-spin** and **low-spin** states, as shown in Table 10.5. Strong ligand fields lead to low-spin complexes, and weak ligand fields lead to high-spin complexes.

Terminology for these configurations is summarized as follows:

$$\text{Strong ligand field} \rightarrow \text{large } \Delta_o \rightarrow \text{low spin}$$
$$\text{Weak ligand field} \rightarrow \text{small } \Delta_o \rightarrow \text{high spin}$$

As explained in Section 2.2.3, the energy of pairing two electrons depends on the Coulombic energy of repulsion between two electrons in the same region of space, Π_c,

TABLE 10.5 Spin States and Ligand Field Strength

Complex with Weak-Field Ligands (High Spin)

Δ_o	d^1	d^2	d^3	d^4	d^5	d^6	d^7	d^8	d^9	d^{10}
upper (e_g)	— —	— —	— —	↑ —	↑ ↑	↑ ↑	↑ ↑	↑ ↑	↑↓ ↑	↑↓ ↑↓
lower (t_{2g})	↑ — —	↑ ↑ —	↑ ↑ ↑	↑ ↑ ↑	↑ ↑ ↑	↑↓ ↑ ↑	↑↓ ↑↓ ↑	↑↓ ↑↓ ↑↓	↑↓ ↑↓ ↑↓	↑↓ ↑↓ ↑↓

Complex with Strong Field Ligands (Low Spin)

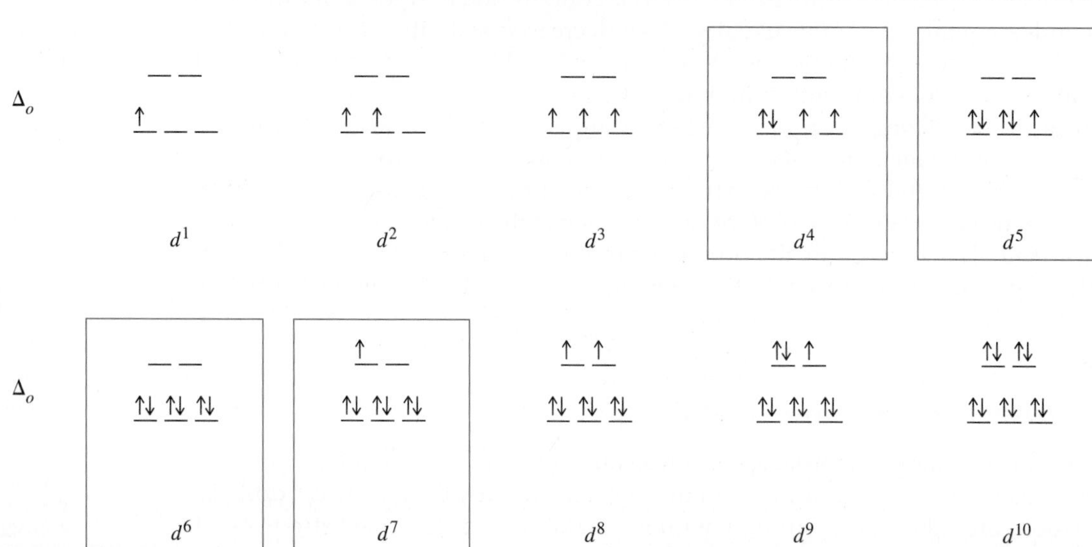

and the purely quantum mechanical exchange energy, Π_e. The relationship between the difference between the t_{2g} and e_g energy levels, the Coulombic energy, and the exchange energy— Δ_o, Π_c, and Π_e respectively—determines the orbital configuration of the electrons. The configuration with the lower total energy is the ground state for the complex. Π_c is a *positive* energy, indicating less stability, and Π_e is a *negative* energy, indicating more stability.

For example, a d^5 ion could have five unpaired electrons, three in t_{2g} and two in e_g orbitals, as a **high-spin** case; or it could have only one unpaired electron, with all five electrons in the t_{2g} levels, as a **low-spin** case. The possibilities for all cases, d^1 through d^{10}, are given in Table 10.5.

EXAMPLE

Determine the exchange energies for high-spin and low-spin d^6 ions in an octahedral complex.

In the high-spin complex, the electron spins are as shown on the right. The five ↑ electrons have exchangeable pairs 1-2, 1-3, 2-3, and 4-5, for a total of four. The exchange energy is therefore $4\Pi_e$. Only electrons at the same energy can exchange.

↑4 ↑5

↑1↓1 ↑2 ↑3

In the low-spin complex, as shown on the right, each set of three electrons with the same spin has exchangeable pairs 1-2, 1-3, and 2-3, for a total of six, and the exchange energy is $6\Pi_e$.

↑1↓1 ↑2↓2 ↑3↓3

The difference between the high-spin and low-spin complexes is two exchangeable pairs.

▶ **Exercise 10.6** Determine the exchange energy for a d^5 ion, both as a high-spin and as a low-spin complex.

Unlike the total pairing energy Π, Δ_o is strongly dependent on the ligands and on the metal. Table 10.6 presents values of Δ_o for aqueous ions, in which water is a relatively weak-field ligand (small Δ_o). The number of unpaired electrons in the complex depends on the balance between Δ_o and Π:

When $\Delta_o > \Pi$, there is a net loss in energy (an increase in stability) on pairing electrons in the lower levels; the low-spin configuration is more stable;

When $\Delta_o < \Pi$, the total energy is lower with more unpaired electrons; the high-spin configuration is more stable.

In Table 10.6, only Co^{3+} has Δ_o near the size of Π, and $[Co(H_2O)_6]^{3+}$ is the only low-spin aqua complex. All the other first-row transition metal ions require a stronger field ligand than water for a low-spin configuration. The tabulated Δ_o and Π energies for $[Co(H_2O)_6]^{3+}$ indicate that the relative magnitudes of these values provide a useful conceptual framework to rationalize high and low spin states but that experimental measurements, such as the determination of magnetic susceptibility, provide the most reliable data for assessing electronic configurations.

In general, the strength of the ligand–metal interaction is greater for metals having higher charges. This can be seen in the table: Δ_o for 3+ ions is larger than for 2+ ions. Also, values for d^5 ions are smaller than for d^4 and d^6 ions.

Another factor that influences electron configurations is the position of the metal in the periodic table. Metals from the second and third transition series form low-spin complexes more readily than metals from the first transition series. This is a consequence of

TABLE 10.6 Orbital Splitting (Δ_o) and Mean Pairing Energy (Π) for Aqueous Ions[a]

	Ion	Δ_o	Π	Ion	Δ_o	Π
d^1				Ti^{3+}	18,800	
d^2				V^{3+}	18,400	
d^3	V^{2+}	12,300		Cr^{3+}	17,400	
d^4	Cr^{2+}	9,250	23,500	Mn^{3+}	15,800	28,000
d^5	Mn^{2+}	7,850[b]	25,500	Fe^{3+}	14,000	30,000
d^6	Fe^{2+}	9,350	17,600	Co^{3+}	16,750	21,000
d^7	Co^{2+}	8,400	22,500	Ni^{3+}		27,000
d^8	Ni^{2+}	8,600				
d^9	Cu^{2+}	7,850				
d^{10}	Zn^{2+}	0				

Sources: For Δ_o: M^{2+} data from D. A. Johnson and P. G. Nelson, *Inorg. Chem.*, **1995**, *34*, 5666; M^{3+} data from D. A. Johnson and P. G. Nelson, *Inorg. Chem.*, **1999**, *38*, 4949. For Π: Data from D. S. McClure, The Effects of Inner-orbitals on Thermodynamic Properties, in T. M. Dunn, D. S. McClure, and R. G. Pearson, *Some Aspects of Crystal Field Theory*, Harper & Row, New York, 1965, p. 82.

NOTE: [a]Values given are in cm^{-1}.

[b] Estimated value

two cooperating effects: one is the greater overlap between the larger 4*d* and 5*d* orbitals and the ligand orbitals, and the other is a decreased pairing energy due to the larger volume available for electrons in the 4*d* and 5*d* orbitals as compared with 3*d* orbitals.

10.3.3 Ligand Field Stabilization Energy

The difference between (1) the total energy of a coordination complex with the electron configuration resulting from ligand field splitting of the orbitals and (2) the total energy for the same complex with all the *d* orbitals if they were equally populated is called the **ligand field stabilization energy (LFSE)**. The LFSE represents the stabilization of the *d* electrons because of the metal–ligand environment. A common way to calculate LFSE is shown for d^4 in Figure 10.12.

The interaction of the *d* orbitals of the metal with the ligand orbitals results in lower energy for the t_{2g} set of orbitals ($-\frac{2}{5}\Delta_o$ relative to the average energy of all t_{2g} and e_g orbitals) and increased energy for the e_g set ($\frac{3}{5}\Delta_o$). The total energy of a one-electron system would then be $-\frac{2}{5}\Delta_o$, and the total energy of a high-spin four-electron system would be $\frac{3}{5}\Delta_o + 3(-\frac{2}{5}\Delta_o) = -\frac{3}{5}\Delta_o$. An alternative method of arriving at these energies is given by Cotton.[16]

▶ **Exercise 10.7** Determine the LFSE for a d^6 ion for both high-spin and low-spin cases.

Table 10.7 has the LFSE values for σ-bonded octahedral complexes with 1-10 electrons in both high- and low-spin arrangements. The final columns show the pairing energies and

[16] F. A. Cotton, *J. Chem. Educ.*, **1964**, *41*, 466.

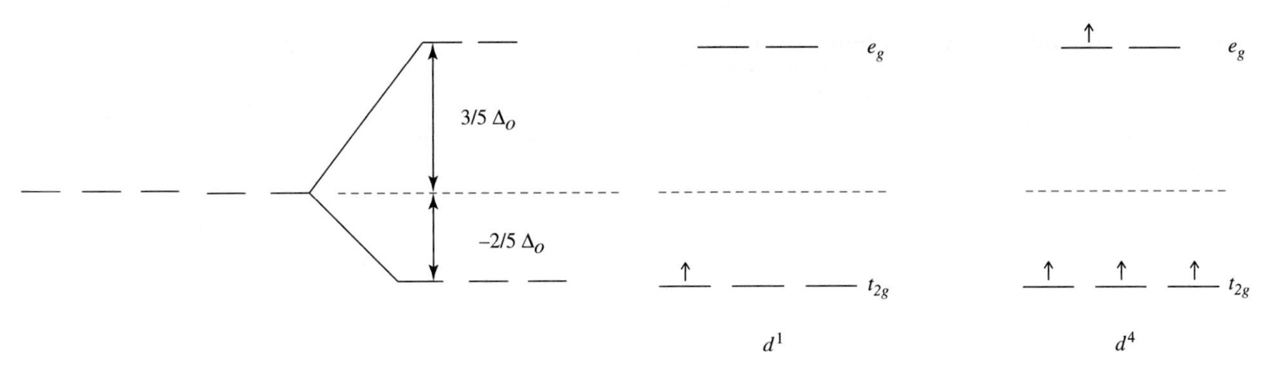

FIGURE 10.12 Splitting of Orbital Energies in a Ligand Field.

the difference in LFSE between low-spin and high-spin complexes with the same total number of *d* electrons. For one to three and eight to ten electrons, there is no difference in the number of unpaired electrons or the LFSE. For four to seven electrons, there is a significant difference in both, and high- and low-spin arrangements are possible.

The most commonly cited example of LFSE in thermodynamic data appears in the exothermic enthalpy of hydration of bivalent ions of the first transition series, usually assumed to have six waters of hydration:

$$M^{2+}(g) + 6\,H_2O(l) \longrightarrow [M(H_2O)_6]^{2+}(aq)$$

Ions with spherical symmetry should have ΔH becoming increasingly exothermic (more negative) continuously across the transition series, because of the decreasing radius of the ions with increasing nuclear charge and corresponding increase in electrostatic attraction for the ligands. Instead, the enthalpies show the characteristic double-loop shape shown in Figure 10.13. The almost linear curve of the "corrected" enthalpies is expected for ions with decreasing radius. The differences between this curve and the double-humped experimental values are approximately equal to the LFSE values in Table 10.7 for high-spin complexes,[17] with additional smaller corrections for spin-orbit splittings (0 to 16 kJ/mol), a relaxation effect caused by contraction of the metal–ligand distance (0 to 24 kJ/mol), and an interelectronic repulsion energy that depends on the exchange interactions between electrons with the same spins (0 to −19 kJ/mol for M^{2+}, 0 to −156 kJ/mol for M^{3+}).[18] The latter three effects are relatively minimal, but they improve the shape of the curve for the corrected values significantly. In the case of the hexaaqua and hexafluoro complexes of the 3+ transition-metal ions, the interelectronic repulsion energy, sometimes called the *nephelauxetic effect*, is larger and is required to remove the deviation from a smooth curve through the d^0, d^5, and d^{10} values.

Why do we care about LFSE? There are two principal reasons. First, it provides a more quantitative approach to the high-spin–low-spin electron configurations, helping predict which configuration will be more likely. Second, it is the basis for our discussion of the spectra of these complexes in Chapter 11. Measurements of Δ_o are commonly provided in studies of these complexes, with a goal of eventually allowing a better and more quantitative understanding of metal-ligand interactions. At this point, the relative sizes of Δ_o, Π_c, and Π_e are the important features.

[17] L. E. Orgel, *J. Chem. Soc.*, **1952**, 4756; P. George and D. S. McClure, *Prog. Inorg. Chem.*, **1959**, *1*, 381.
[18] D. A. Johnson and P. G. Nelson, *Inorg. Chem.*, **1995**, *34*, 3253; **1995**, *34*, 5666; **1999**, *38*, 4949.

TABLE 10.7 Ligand Field Stabilization Energies

Number of *d* Electrons	Weak-Field Arrangement t_{2g}	Weak-Field Arrangement e_g	LFSE (Δ_o)	Coulombic Energy	Exchange Energy
1	↑		$-\frac{2}{5}$		
2	↑ ↑		$-\frac{4}{5}$		Π_e
3	↑ ↑ ↑		$-\frac{6}{5}$		$3\Pi_e$
4	↑ ↑ ↑	↑	$-\frac{3}{5}$		$3\Pi_e$
5	↑ ↑ ↑	↑ ↑	0		$4\Pi_e$
6	↑↓ ↑ ↑	↑ ↑	$-\frac{2}{5}$	Π_c	$4\Pi_e$
7	↑↓ ↑↓ ↑	↑ ↑	$-\frac{4}{5}$	$2\Pi_c$	$5\Pi_e$
8	↑↓ ↑↓ ↑↓	↑ ↑	$-\frac{6}{5}$	$3\Pi_c$	$7\Pi_e$
9	↑↓ ↑↓ ↑↓	↑↓ ↑	$-\frac{3}{5}$	$4\Pi_c$	$7\Pi_e$
10	↑↓ ↑↓ ↑↓	↑↓ ↑↓	0	$5\Pi c$	$8\Pi_e$

Number of *d* Electrons	Strong-Field Arrangement t_{2g}	Strong-Field Arrangement e_g	LFSE (Δ_o)	Coulombic Energy	Exchange Energy	Strong Field – Weak Field
1	↑		$-\frac{2}{5}$			0
2	↑ ↑		$-\frac{4}{5}$		Π_e	0
3	↑ ↑ ↑		$-\frac{6}{5}$		$3\Pi_e$	0
4	↑↓ ↑ ↑		$-\frac{8}{5}$	Π_c	$3\Pi_e$	$-\Delta_o + \Pi_c$
5	↑↓ ↑↓ ↑		$-\frac{10}{5}$	$2\Pi_c$	$4\Pi_e$	$-2\Delta_o + 2\Pi_c$
6	↑↓ ↑↓ ↑↓		$-\frac{12}{5}$	$3\Pi_c$	$6\Pi_e$	$-2\Delta_o + 2\Pi_c + 2\Pi_e$
7	↑↓ ↑↓ ↑↓	↑	$-\frac{9}{5}$	$3\Pi_c$	$6\Pi_e$	$-\Delta_o + \Pi_c + \Pi_e$
8	↑↓ ↑↓ ↑↓	↑ ↑	$-\frac{6}{5}$	$3\Pi_c$	$7\Pi_e$	0
9	↑↓ ↑↓ ↑↓	↑↓ ↑	$-\frac{3}{5}$	$4\Pi_c$	$7\Pi_e$	0
10	↑↓ ↑↓ ↑↓	↑↓ ↑↓	0	$5\Pi_c$	$8\Pi_e$	0

NOTE: In addition to the LFSE, each pair formed has a positive Coulombic energy, Π_c, and each set of two electrons with the same spin has a negative exchange energy, Π_e. When $\Delta_o > \Pi_c$ for d^4 or d^5, or when $\Delta_o > \Pi_c + \Pi_e$ for d^6 or d^7, the strong-field arrangement (low spin) is favored.

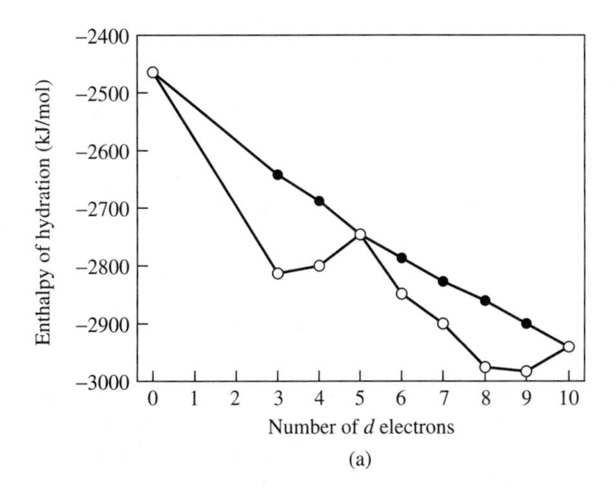

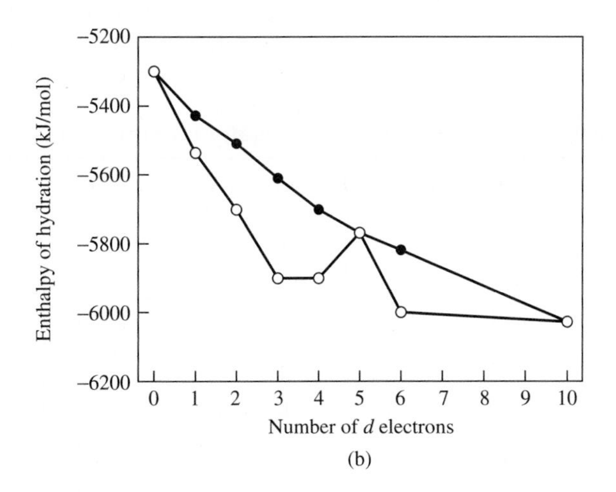

FIGURE 10.13 Enthalpies of Hydration of Transition-Metal Ions. The lower curves show experimental values; the upper curves result when contributions from spin-orbit splitting, a relaxation effect from contraction of the metal–ligand distance, and interelectronic repulsion energy are subtracted. (a) 2+ ions. (b) 3+ ions. (Reproduced with permission from D. A. Johnson and P. G. Nelson, *Inorg. Chem.*, **1995**, *34*, 5666 (M^{2+} data); and D. A. Johnson and P. G. Nelson, *Inorg. Chem.*, **1999**, 4949 (M^{3+} data). © 1995, 1999, American Chemical Society.)

10.3.5 Square-Planar Complexes

SIGMA BONDING The same general approach works for any geometry, although some are more complicated than others. Square-planar complexes such as $[Ni(CN)_4]^{2-}$, with D_{4h} symmetry, provide an example. As before, the axes for the ligand atoms are chosen for convenience. The y axis of each ligand is directed toward the central atom, the x axis is in the plane of the molecule, and the z axis is perpendicular to the plane of the molecule, as shown in Figure 10.14. The p_y set of ligand orbitals is used in σ bonding. Unlike the octahedral case, there are two distinctly different sets of potential π-bonding orbitals, the parallel set ($\pi_{\|}$ or p_x, in the molecular plane) and the perpendicular set ($\pi_{\perp}$ or p_z, perpendicular to the plane). By taking each set in turn, we can use the techniques of Chapter 4 to find the representations that fit the different symmetries. Table 10.8 gives the results.

SIGMA INTERACTIONS The matching metal orbitals for σ bonding in the first transition series are those with lobes in the x and y directions, $3d_{x^2-y^2}$, $4p_x$, and $4p_y$, with some contribution from the less directed $3d_{z^2}$ and $4s$. Ignoring the other orbitals for the moment, we can construct the energy-level diagram for the σ bonds, as in Figure 10.15.

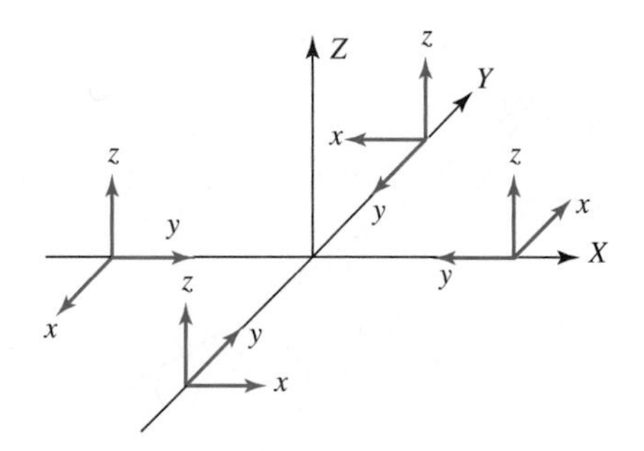

FIGURE 10.14 Coordinate System for Square-Planar Orbitals.

Comparing Figures 10.6 and 10.15, we see that the square-planar diagram looks more complex, because the lower symmetry results in sets with less degeneracy than in the octahedral case. D_{4h} symmetry splits the d orbitals into three single representations (a_{1g}, b_{1g}, and b_{2g}, for d_{z^2}, $d_{x^2-y^2}$, and d_{xy} respectively) and the degenerate e_g for the d_{xz}, d_{yz} pair. The b_{2g} and e_g levels are nonbonding (no ligand orbital matches their symmetry) and the difference between them and the antibonding a_{1g} level corresponds to Δ.

▶ **Exercise 10.8** Derive the reducible representations for square-planar bonding, and show that their component irreducible representations are those in Table 10.8.

TABLE 10.8 Representations and Orbital Symmetry for Square-Planar Complexes

D_{4h}	E	$2C_4$	C_2	$2C_2'$	$2C_2''$	i	$2S_4$	σ_h	$2\sigma_v$	$2\sigma_d$		
A_{1g}	1	1	1	1	1	1	1	1	1	1		x^2+y^2, z^2
A_{2g}	1	1	1	−1	−1	1	1	1	−1	−1	R_z	
B_{1g}	1	−1	1	1	−1	1	−1	1	1	−1		x^2-y^2
B_{2g}	1	−1	1	−1	1	1	−1	1	−1	1		xy
E_g	2	0	−2	0	0	2	0	−2	0	0	(R_x, R_y)	(xz, yz)
A_{1u}	1	1	1	1	1	−1	−1	−1	−1	−1		
A_{2u}	1	1	1	−1	−1	−1	−1	−1	1	1	z	
B_{1u}	1	−1	1	1	−1	−1	1	−1	−1	1		
B_{2u}	1	−1	1	−1	1	−1	1	−1	1	−1		
E_u	2	0	−2	0	0	−2	0	2	0	0	(x, y)	

D_{4h}	E	$2C_4$	C_2	$2C_2'$	$2C_2''$	i	$2S_4$	σ_h	$2\sigma_v$	$2\sigma_d$
Γ_σ (y)	4	0	0	2	0	0	0	4	2	0
$\Gamma_\parallel$ (x)	4	0	0	−2	0	0	0	4	−2	0
$\Gamma_\perp$ (z)	4	0	0	−2	0	0	0	−4	2	0

$\Gamma_\sigma = A_{1g} + B_{1g} + E_u$ — (σ) Matching orbitals on the central atom: s, d_{z^2}, $d_{x^2-y^2}$, (p_x, p_y)

$\Gamma_\parallel = A_{2g} + B_{2g} + E_u$ — ($\parallel$) Matching orbitals on the central atom: d_{xy}, (p_x, p_y)

$\Gamma_\perp = A_{2u} + B_{2u} + E_g$ — ($\perp$) Matching orbitals on the central atom: p_z, (d_{xz}, d_{yz})

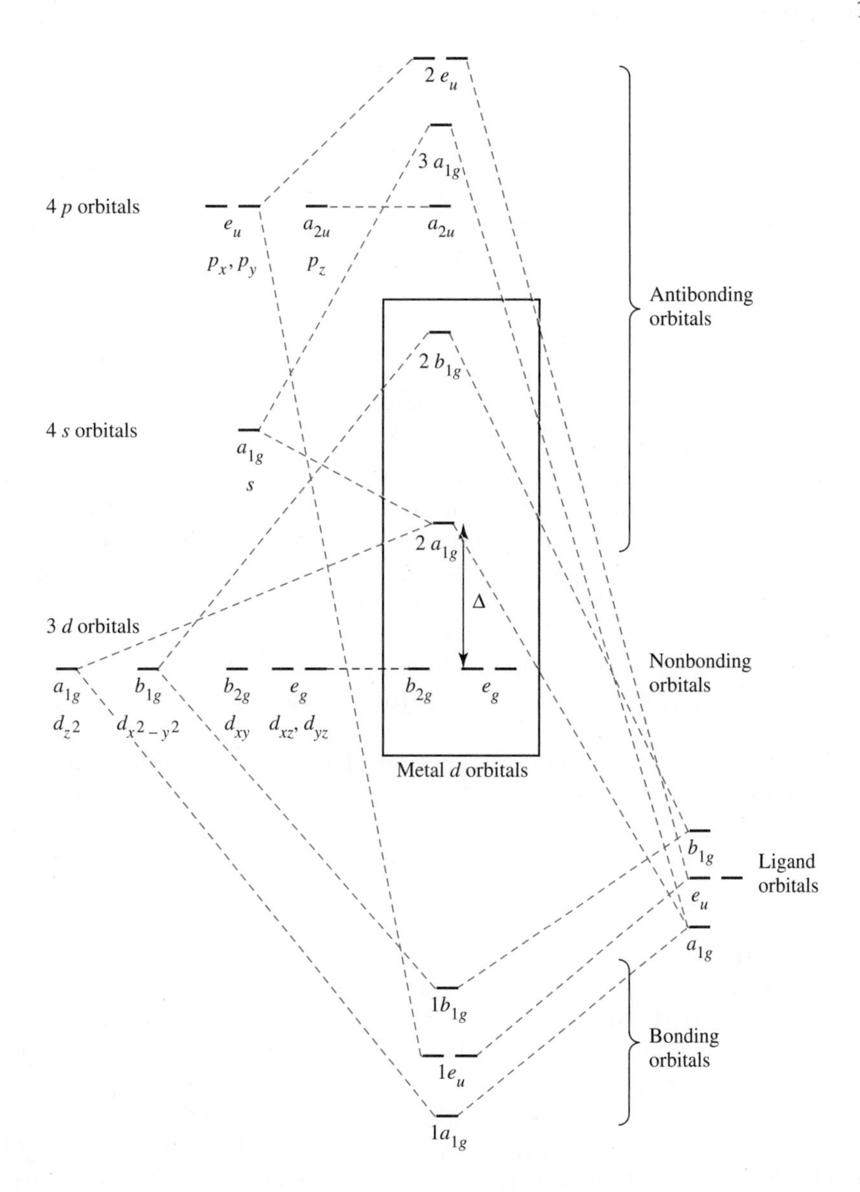

FIGURE 10.15 D_{4h} Molecular Orbitals, σ Orbitals Only. (Adapted from T. A. Albright, J. K. Burdett, and M.-Y. Whangbo, *Orbital Interactions in Chemistry*, Wiley InterScience, New York, 1985, p. 296. © 1985, John Wiley & Sons, Inc. Reprinted by permission of John Wiley & Sons, Inc.)

PI INTERACTIONS The π-bonding orbitals are also shown in Table 10.8. The d_{xy} (b_{2g}) orbital interacts with the p_x ($\pi_{||}$) ligand orbitals, and the d_{xz} and d_{yz} (e_g) orbitals interact with the p_z ($\pi_{\perp}$) ligand orbitals, as shown in Figure 10.16. The b_{2g} orbital is in the plane of the molecule, and the two e_g orbitals have lobes above and below the plane. The results of these interactions are shown in Figure 10.17, as calculated for $[Pt(CN)_4]^{2-}$.

This diagram should emphasize how complex molecular orbitals can be! Despite the complexity of this figure, key aspects of the orbitals can be seen by examining the sets of orbitals set off by boxes:

The lowest energy set contains the bonding orbitals, as in Figure 10.15. Eight electrons from the ligand orbitals fill them.

The next higher set consists of the eight π-donor orbitals, essentially lone pairs on a simple halide ion or π orbitals on CN^-. Their interaction with the metal orbitals is small and has the effect of decreasing the energy difference between the orbitals of the next higher set.

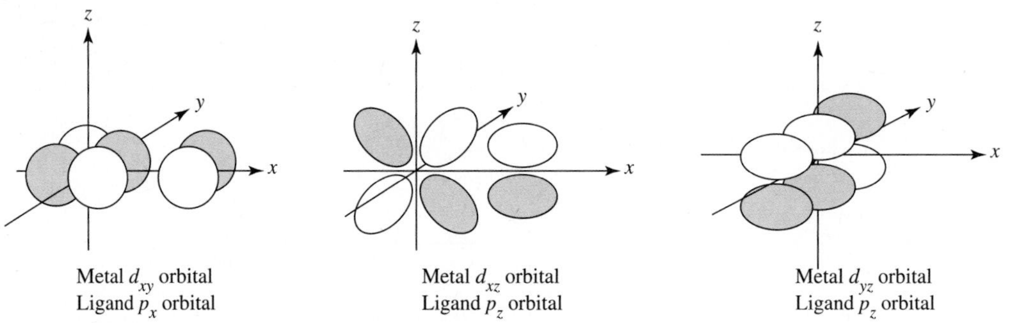

Metal d_{xy} orbital
Ligand p_x orbital

Metal d_{xz} orbital
Ligand p_z orbital

Metal d_{yz} orbital
Ligand p_z orbital

FIGURE 10.16 Pi Bonding Interactions in D_{4h} Molecules.

The third set of molecular orbitals consists primarily of metal d orbitals, modified by interaction with the ligand orbitals. The order of these orbitals has been described in several ways, depending on the detailed method used in the calculations.[19] The order shown is that found by using relativistic corrections in the calculations. In all cases, there is, however, agreement that the b_{2g}, e_g, and a_{1g} orbitals are all low and have small differences in energy, and the b_{1g} orbital has a much higher energy than all the others. In the $[Pt(CN)_4]^{2-}$ ion, it is described as being higher in energy than the a_{2u} (mostly from the metal p_z).

The relative energies of molecular orbitals derived from d orbital interactions may vary with different metals and ligands. For example, the order in $[Ni(CN)_4]^{2-}$ matches that for d orbitals in Figure 10.17 ($x^2 - y^2 \gg z^2 > xz, yz > xy$), but the a_{2u}, involving a p_z interaction, in the nickel complex is calculated to be higher in energy than the $d_{x^2-y^2}$ (b_{1g}).[20]

The remaining high-energy orbitals are important only in excited states and will not be considered further.

The important parts of Figure 10.17 are these major sets. Two electrons from each ligand form the σ bonds, the next four electrons from each ligand can either π bond slightly or remain essentially nonbonding, and the remaining electrons from the metal occupy the third set. In the case of Ni^{2+} and Pt^{2+}, there are eight d electrons, and there is a large gap in energy between their orbitals and the LUMO ($2a_{2u}$), leading to diamagnetic complexes. The effect of the π^* orbitals of the ligand is to increase the difference in energy between these orbitals. For example, in $[PtCl_4]^{2-}$, with no π^* orbitals, the difference between the $2e_g$ and $2a_{1g}$ orbitals is about 6000 cm^{-1}, and the difference between the $2a_{1g}$ and $2a_{2u}$ orbitals is about 23,500 cm^{-1}. The corresponding differences for $[Pt(CN)_4]^{2-}$ are 12,600 and more than 30,000 cm^{-1}.[21]

The energy differences between the orbitals in this set are labeled Δ_1, Δ_2, and Δ_3 from top to bottom. Because b_{2g} and e_g are π orbitals, their energies change significantly if the ligands are changed. We should also note that Δ_1 is related to Δ_o, is usually much larger than Δ_2 and Δ_3, and is almost always larger than Π, the pairing energy. This means that the b_{1g} or a_{2u} level, whichever is lower, is usually empty for metal ions with fewer than nine electrons.

[19] T. Ziegler, J. K. Nagle, J. G. Snijders, and E. J. Baerends, *J. Am. Chem. Soc.*, **1989**, *111*, 5631, and the references cited therein.
[20] P. Hummel, N. W. Halpern-Manners, and H. B. Gray, *Inorg. Chem.*, **2006**, *45*, 7397.
[21] H. B. Gray and C. J. Ballhausen, *J. Am. Chem. Soc.*, **1963**, *85*, 260.

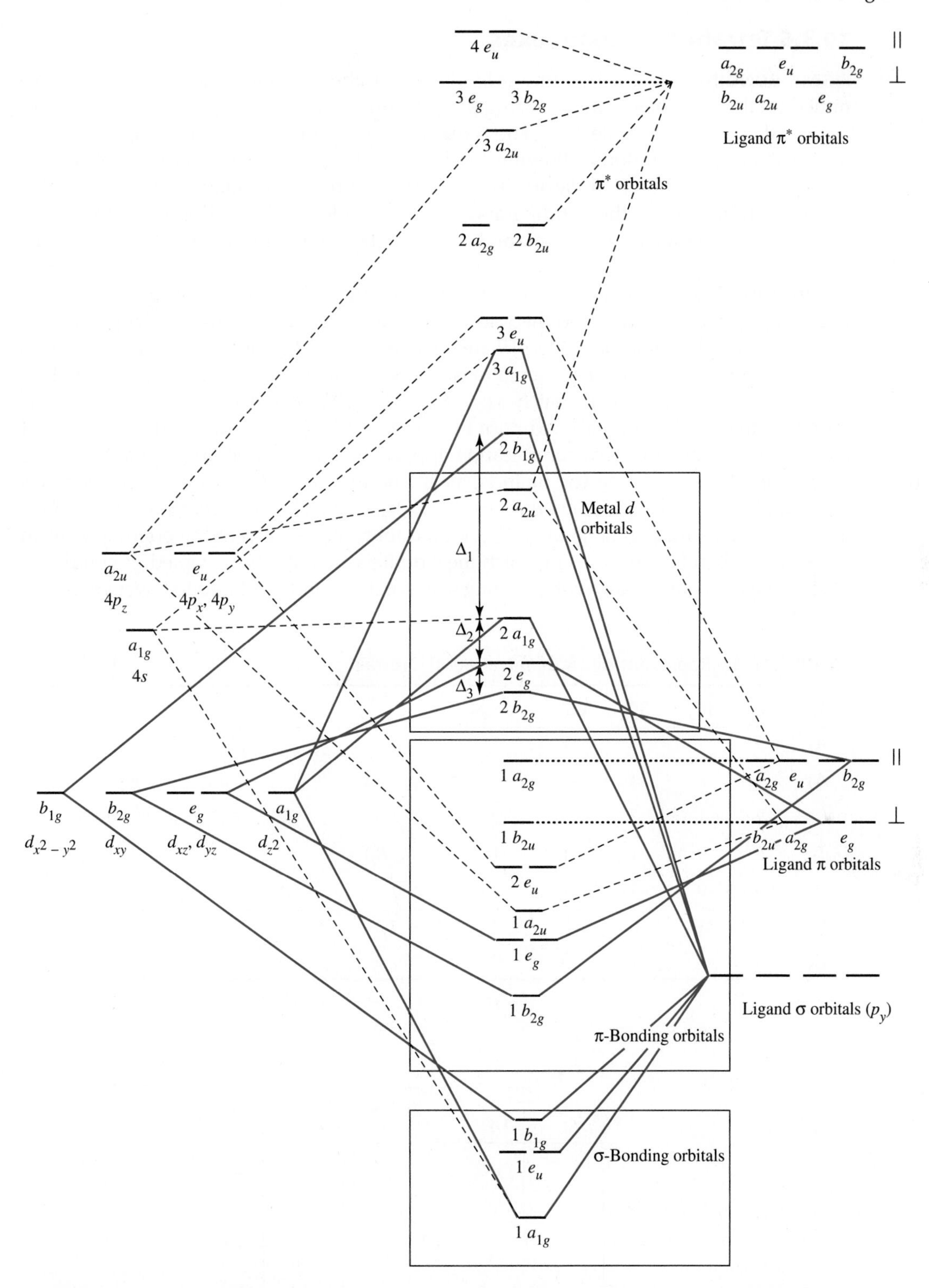

FIGURE 10.17 D_{4h} Molecular Orbitals, Including π Orbitals. Interactions with metal d orbitals are indicated by solid lines, interactions with metal s and p orbitals by dashed lines, and nonbonding orbitals by dotted lines.

10.3.6 Tetrahedral Complexes

SIGMA BONDING The σ-bonding orbitals for tetrahedral complexes are easily determined on the basis of symmetry, using the coordinate system illustrated in Figure 10.18 to give the results in Table 10.9. The reducible representation includes the A_1 and T_2 irreducible representations, allowing for four bonding MOs. The energy level picture for the d orbitals, shown in Figure 10.19, is inverted from the octahedral levels, with e the nonbonding and t_2 the bonding and antibonding levels. In addition, the split, now called Δ_t, is smaller than for octahedral geometry; the general result is $\Delta_t = \frac{4}{9}\Delta_o$.

PI BONDING The π orbitals are more difficult to see, but if the y axis of the ligand orbitals is chosen along the bond axis, and the x and z axes are arranged to allow the C_2 operation to work properly, the results in Table 10.9 are obtained. The reducible representation includes the E, T_1, and T_2 irreducible representations. The T_1 has no matching metal atom orbitals, E matches d_{z^2} and $d_{x^2-y^2}$, and T_2 matches d_{xy}, d_{xz}, and d_{yz}. The E and T_2 interactions lower the energy of the bonding orbitals and raise the corresponding antibonding orbitals, for a net increase in Δ_t. An additional complication appears when both bonding and antibonding π orbitals are available on the ligands, as is true for CO or CN^-. Figure 10.20 shows the orbitals and their relative energies for $Ni(CO)_4$, in which the interactions of the CO σ and π orbitals with the metal orbitals are probably small. Much of the bonding is from $M \longrightarrow L$ π bonding. In cases in which the d orbitals are not fully occupied, σ bonding is likely to be more important, with resulting shifts of the a_1 and t_2 orbitals to lower energies.

TABLE 10.9 Representations of Tetrahedral Orbitals

T_d	E	$8C_3$	$3C_2$	$6S_4$	$6\sigma_d$		
A_1	1	1	1	1	1		$x^2 + y^2 + z^2$
A_2	1	−1	1	−1	−1		
E	2	−1	2	0	0		$(2x^2 - x^2 - y^2, x^2 - y^2)$
T_1	3	0	−1	1	−1	(R_x, R_y, R_z)	
T_2	3	0	−1	−1	1	(x, y, z)	(xy, yz, xz)
Γ_σ	4	1	0	0	2	$A_1 + T_2$	
Γ_π	8	−1	0	0	0	$E + T_1 + T_2$	

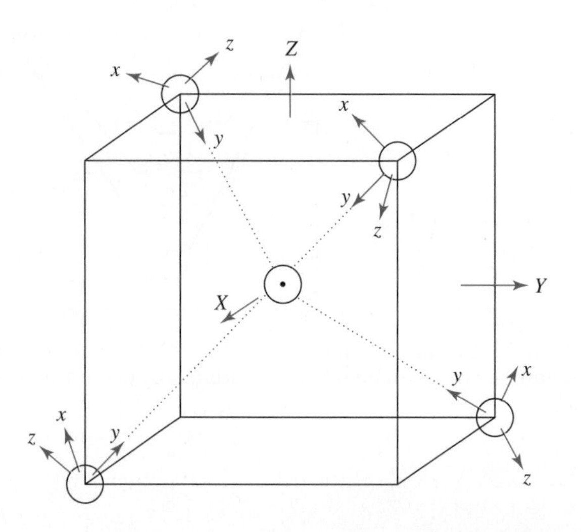

FIGURE 10.18 Coordinate System for Tetrahedral Orbitals.

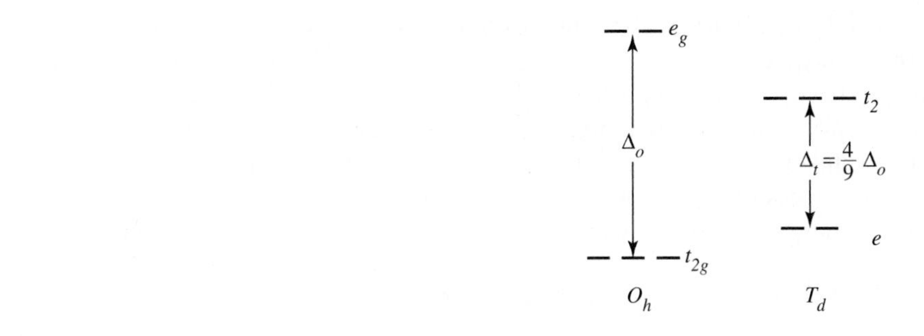

FIGURE 10.19 Orbital Splitting in Octahedral and Tetrahedral Geometries.

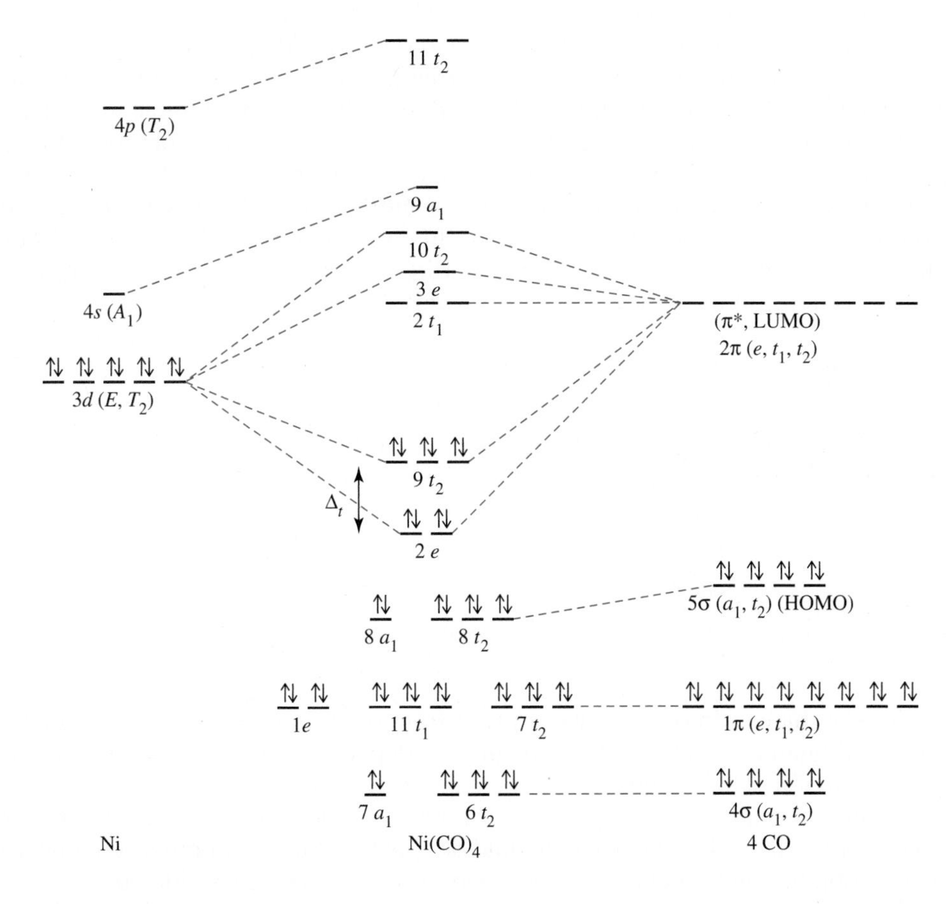

FIGURE 10.20 Molecular Orbitals for Tetrahedral Ni(CO)$_4$. C. W. Bauschlicher, Jr., and P. S. Bagus, *J. Chem. Phys.*, **1984**, *81*, 5889, argue that there is almost no σ bonding from the 4*s* and 4*p* orbitals of Ni, and that the d^{10} configuration is the best starting place for the calculations, as shown here. G. Cooper, K. H. Sze, and C. E. Brion, *J. Am. Chem. Soc.*, **1989**, *111*, 5051, include the metal 4*s* as a significant part of σ bonding but with essentially the same net result in molecular orbitals.

10.4 ANGULAR OVERLAP

The **angular overlap** model is an approach to bonding that is useful for making estimates of energies of orbitals in coordination complexes, and it has the flexibility to deal with a variety of possible geometries.[22, 23] This approach estimates the strength of interaction between individual ligand orbitals and metal *d* orbitals based on the overlap between them. Both sigma and pi interactions are considered, and different coordination numbers and geometries can be treated. The term angular overlap is used, because the amount of overlap depends strongly on the angular arrangement of the metal orbitals and the angles at which the ligands interact with metal orbitals.

---

[22] E. Larsen and G. N. La Mar, *J. Chem. Educ.*, **1974**, *51*, 633. (Note: There are misprints on pp. 635 and 636.)
[23] J. K. Burdett, *Molecular Shapes*, Wiley InterScience, New York, 1980.

In the angular overlap approach, the energy of a metal d orbital in a coordination complex is determined by summing the effects of each ligand on that orbital. Some ligands have a strong effect on a particular d orbital, some have a weaker effect, and some have no effect at all, because of their angular dependence. Similarly, both sigma and pi interactions must be taken into account to determine the final energy of a particular orbital. By systematically considering each of the five d orbitals, we can use this approach to determine the overall energy pattern corresponding to a particular coordination geometry.

10.4.1 Sigma-Donor Interactions

In the angular overlap model the strongest sigma interaction is considered to be between a metal d_{z^2} orbital and a ligand p orbital (or a hybrid ligand orbital of the same symmetry), as shown in Figure 10.21. The strength of all other sigma interactions is determined relative to the strength of this reference interaction. Interaction between these two orbitals results in a bonding orbital, which has a larger component of the ligand orbital, and an antibonding orbital, which is largely metal orbital in composition. Although the observed increase in energy of the antibonding orbital is greater than the decrease in energy of the bonding orbital, this model approximates the molecular orbital energies by an increase in the antibonding (mostly metal d) orbital of e_σ and a decrease in energy of the bonding (mostly ligand) orbital of e_σ.

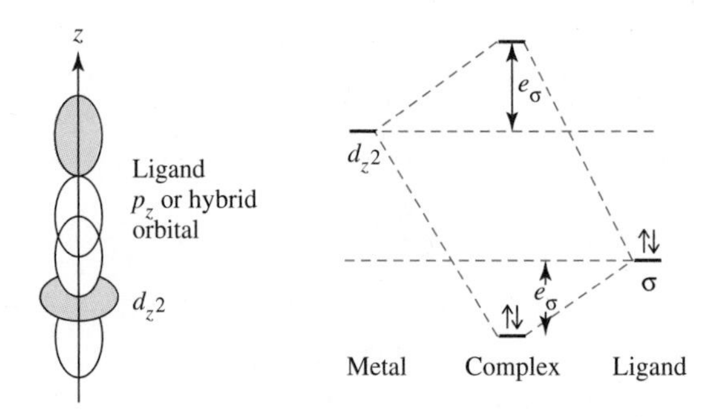

FIGURE 10.21
Sigma Interaction for Angular Overlap.

Similar changes in orbital energy result from other interactions between metal d orbitals and ligand orbitals, with the magnitude dependent on the ligand location and the specific d orbital being considered. Table 10.10 gives values of these energy changes for a variety of shapes. Calculation of the numbers in the table (in e_σ units) is beyond the scope of this book, but the reader should be able to justify the numbers qualitatively by comparing the amount of overlap between the orbitals being considered.

The angular overlap approach is best described by example. We will consider first the most common geometry for coordination complexes, octahedral.

EXAMPLE

$[M(NH_3)_6]^{n+}$

These ions are examples of octahedral complexes with only sigma interactions. The ammonia ligands have no π orbitals available for bonding with the metal ion. The donor orbital of NH_3 is mostly nitrogen p_z orbital in composition, and the other p orbitals are used in bonding to the hydrogens (see Figure 5.31).

In calculating the orbital energies in a complex, the value for a given d orbital is the sum of the numbers for the appropriate ligands in the vertical column for that orbital in Table 10.10. The change in energy for a specific ligand orbital is the sum of the numbers for all d orbitals in the horizontal row for the required ligand position.

Metal d Orbitals

d_{z^2} orbital: The interaction is strongest with ligands in positions 1 and 6, along the z axis. Each interacts with the orbital to raise its energy by e_σ. The ligands in positions 2, 3, 4, and 5 interact more weakly with the d_{z^2} orbital, each raising the energy of the orbital by $\frac{1}{4}e_\sigma$. Overall, the energy of the d_{z^2} orbital is increased by the sum of all these interactions, for a total of $3e_\sigma$.

$d_{x^2-y^2}$ orbital: The ligands in positions 1 and 6 do not interact with this metal orbital, but the ligands in positions 2, 3, 4, and 5 each interact to raise the energy of the metal orbital by $\frac{3}{4}e_\sigma$, for a total increase of $3e_\sigma$.

d_{xy}, d_{xz}, and d_{yz} orbitals: None of these orbitals interact in a sigma fashion with any of the ligand orbitals, so the energy of these metal orbitals remains unchanged.

Ligand Orbitals

The energy changes for the ligand orbitals are the same as those above for each interaction. The totals, however, are taken *across a row* of the Table 10.10, including each of the d orbitals.

Ligands in positions 1 and 6 interact strongly with d_{z^2} and are lowered by e_σ. They do not interact with the other d orbitals.

Ligands in positions 2, 3, 4, and 5 are lowered by $\frac{1}{4}e_\sigma$ by interaction with d_{z^2} and by $\frac{3}{4}e_\sigma$ by interaction with $d_{x^2-y^2}$, for a total stabilization of e_σ for each donor orbital.

Overall, each ligand orbital is lowered by e_σ.

The resulting energy pattern is shown in Figure 10.22. This is the same pattern obtained from the ligand field approach. Both describe how the metal complex is stabilized: as two of the d orbitals of the metal increase in energy, and three remain unchanged, the six ligand orbitals fall in energy, and electron pairs in those orbitals are stabilized in the formation of ligand–metal bonds. The net stabilization is $12e_\sigma$ for the bonding pairs; any d electrons in the upper (e_g*) level are destabilized by $3e_\sigma$ each.

The more complete molecular orbital picture that includes the metal s and p orbitals in the formation of molecular orbitals was shown in Figure 10.6. There are no examples of complexes with electrons in the antibonding orbitals from s and p orbitals, and these high-energy antibonding orbitals are not significant in describing the spectra of complexes, so we will not consider them further.

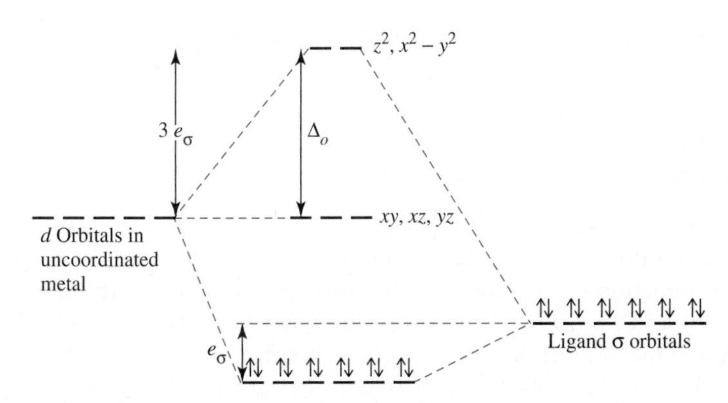

FIGURE 10.22 Energies of d Orbitals in Octahedral Complexes: Sigma-Donor Ligands. $\Delta_o = 3e_\sigma$. Metal s and p orbitals also contribute to the bonding molecular orbitals.

TABLE 10.10 Angular Overlap Parameters: Sigma Interactions

Octahedral Positions | Tetrahedral positions | Trigonal-Bipyramidal Positions

				Sigma Interactions (all in units of e_σ) Metal *d* Orbital				
CN	**Shape**	**Positions**	**Ligand Position**	z^2	$x^2 - y^2$	xy	xz	yz
2	Linear	1, 6	1	1	0	0	0	0
3	Trigonal	2, 11, 12	2	$\frac{1}{4}$	$\frac{3}{4}$	0	0	0
3	T shape	1, 3, 5	3	$\frac{1}{4}$	$\frac{3}{4}$	0	0	0
4	Tetrahedral	7, 8, 9, 10	4	$\frac{1}{4}$	$\frac{3}{4}$	0	0	0
4	Square planar	2, 3, 4, 5	5	$\frac{1}{4}$	$\frac{3}{4}$	0	0	0
5	Trigonal bipyramidal	1, 2, 6, 11, 12	6	1	0	0	0	0
5	Square pyramidal	1, 2, 3, 4, 5	7	0	0	$\frac{1}{3}$	$\frac{1}{3}$	$\frac{1}{3}$
6	Octahedral	1, 2, 3, 4, 5, 6	8	0	0	$\frac{1}{3}$	$\frac{1}{3}$	$\frac{1}{3}$
			9	0	0	$\frac{1}{3}$	$\frac{1}{3}$	$\frac{1}{3}$
			10	0	0	$\frac{1}{3}$	$\frac{1}{3}$	$\frac{1}{3}$
			11	$\frac{1}{4}$	$\frac{3}{16}$	$\frac{9}{16}$	0	0
			12	$\frac{1}{4}$	$\frac{3}{16}$	$\frac{9}{16}$	0	0

▶ **Exercise 10.9** Using the angular overlap model, determine the relative energies of *d* orbitals in a metal complex of formula ML_4 having tetrahedral geometry. Assume that the ligands are capable of sigma interactions only.

How does this result for Δ_t compare with the value for Δ_o?

10.4.2 Pi-Acceptor Interactions

Ligands such as CO, CN^-, and phosphines (PR_3) are π acceptors, with empty orbitals that can interact with metal *d* orbitals in a pi fashion. In the angular overlap model, the strongest pi interaction is considered to be between a metal d_{xz} orbital and a ligand π^* orbital, as shown in Figure 10.23. Because the ligand π^* orbitals are higher in energy than the original metal *d* orbitals, the resulting bonding molecular orbitals are lower in energy than the metal *d* orbitals (a difference of e_π) and the antibonding molecular orbitals are higher in energy.

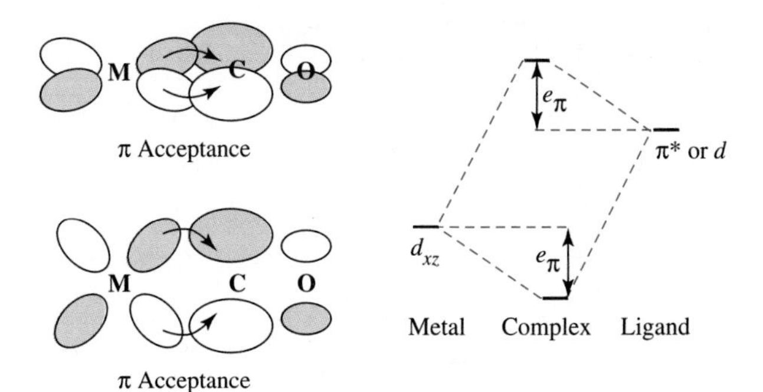

FIGURE 10.23 Pi-Acceptor Interactions.

Because the overlap for these orbitals is smaller than the sigma overlap described in the previous section, $e_\pi < e_\sigma$. The other pi interactions are weaker than this reference interaction, with the magnitudes depending on the degree of overlap between the orbitals. Table 10.11 gives values for ligands at the same angles as in Table 10.10.

The d_{z^2} and $d_{x^2-y^2}$ orbitals do not engage in pi interactions with the ligands in positions 1 through 6 (their parameters in the table are all zero). However, the d_{xy}, d_{xz}, and d_{yz} orbitals all have total interactions of $4e_\pi$; in the formation of molecular orbitals, these three

TABLE 10.11 Angular Overlap Parameters: Pi Interactions

Octahedral Positions — Tetrahedral Positions — Trigonal Bipyramidal Positions

			Pi Interactions (all in units of e_π) Metal *d* Orbital					
CN	**Shape**	**Positions**	**Ligand Position**	z^2	x^2-y^2	xy	xz	yz
2	Linear	1, 6	1	0	0	0	1	1
3	Trigonal	2, 11, 12	2	0	0	1	1	0
3	T shape	1, 3, 5	3	0	0	1	0	1
4	Tetrahedral	7, 8, 9, 10	4	0	0	1	1	0
4	Square planar	2, 3, 4, 5	5	0	0	1	0	1
5	Trigonal bipyramidal	1, 2, 6, 11, 12	6	0	0	0	1	1
5	Square pyramidal	1, 2, 3, 4, 5	7	$\frac{2}{3}$	$\frac{2}{3}$	$\frac{2}{9}$	$\frac{2}{9}$	$\frac{2}{9}$
6	Octahedral	1, 2, 3, 4, 5, 6	8	$\frac{2}{3}$	$\frac{2}{3}$	$\frac{2}{9}$	$\frac{2}{9}$	$\frac{2}{9}$
			9	$\frac{2}{3}$	$\frac{2}{3}$	$\frac{2}{9}$	$\frac{2}{9}$	$\frac{2}{9}$
			10	$\frac{2}{3}$	$\frac{2}{3}$	$\frac{2}{9}$	$\frac{2}{9}$	$\frac{2}{9}$
			11	0	$\frac{3}{4}$	$\frac{1}{4}$	$\frac{1}{4}$	$\frac{1}{4}$
			12	0	$\frac{3}{4}$	$\frac{1}{4}$	$\frac{1}{4}$	$\frac{1}{4}$

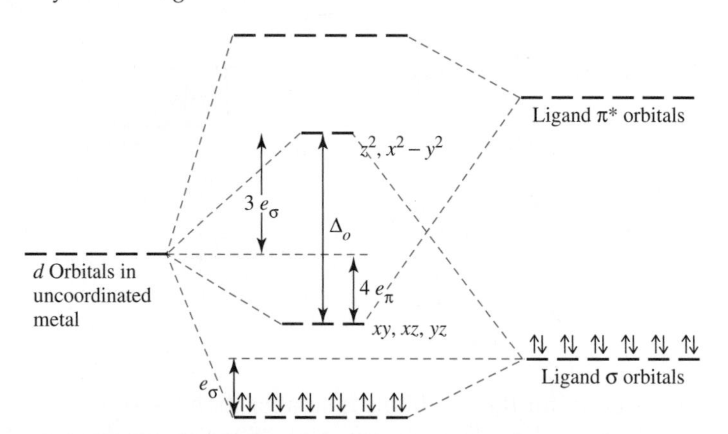

FIGURE 10.24 Energies of *d* Orbitals in Octahedral Complexes: Sigma-Donor and Pi-Acceptor Ligands. $\Delta_o = 3e_\sigma + 4e_\pi$. Metal *s* and *p* orbitals also contribute to the bonding molecular orbitals.

d orbitals undergo stabilization by this quantity, a change of energy of $-4e_\pi$, while the ligand orbitals involved in pi interactions are raised in energy. The *d* electrons then occupy the bonding MOs, with a net energy change of $-4e_\pi$ for each electron, as in Figure 10.24.

EXAMPLE

$[M(CN)_6]^{n-}$

The result of these interactions for $[M(CN)_6]^{n-}$ complexes is shown in Figure 10.24. The d_{xy}, d_{xz}, and d_{yz} orbitals are lowered by $4e_\pi$ each, and the six ligand positions have an average increase in orbital energy of $2e_\pi$ (from summing the rows for positions 1 through 6 in Table 10.11). The resulting ligand π^* orbitals have high energies and are not involved directly in the bonding. The net value of the $t_{2g} - e_g$ split is $\Delta_o = 3e_\sigma + 4e_\pi$.

10.4.3 Pi-Donor Interactions

The interactions between occupied ligand *p*, *d*, or π orbitals and metal *d* orbitals are similar to those in the π-acceptor case. In other words, the angular overlap model treats π-donor ligands similarly to π-acceptor ligands except that *for π-donor ligands, the signs of the changes in energy are reversed*, as shown in Figure 10.25. The metal *d* orbitals are raised in energy, whereas the ligand π orbitals are lowered in energy. The overall effect is shown in Figure 10.26.

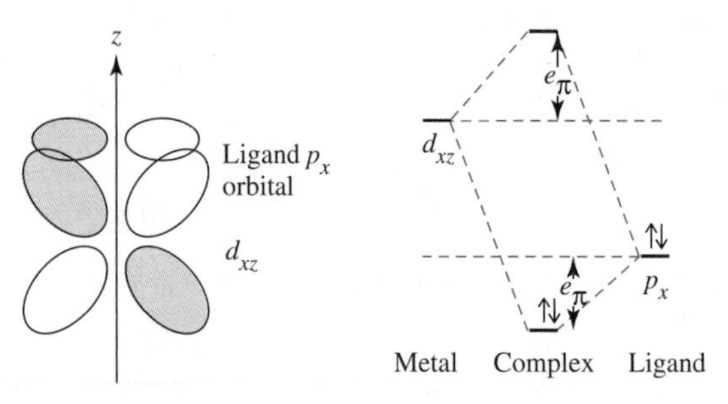

FIGURE 10.25 Pi-Donor Interactions.

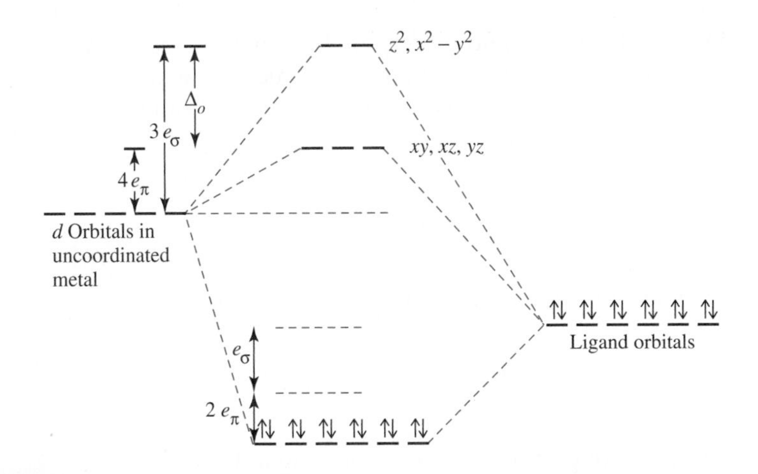

FIGURE 10.26 Energies of *d* Orbitals in Octahedral Complexes: Sigma-Donor and Pi-Donor Ligands. $\Delta_o = 3e_\sigma - 4e_\pi$. Metal *s* and *p* orbitals also contribute to the bonding molecular orbitals.

EXAMPLE

$[MX_6]^{n-}$

Halide ions donate electron density to a metal via p_y orbitals, a sigma interaction; the ions also have p_x and p_z orbitals that can interact with metal orbitals and donate additional electron density via pi interactions. We will use $[MX_6]^{n-}$ as our example, where X is a halide ion or other ligand that is both a σ and a π donor.

d_{z^2} and $d_{x^2-y^2}$ orbitals:

Neither of these orbitals has the correct orientation for pi interactions; therefore, the π orbitals have no effect on the energies of these *d* orbitals.

d_{xy}, d_{xz}, and d_{yz} orbitals:

Each of these orbitals interacts in a pi fashion with four of the ligands. For example, the d_{xy} orbital interacts with ligands in positions 2, 3, 4, and 5 with a strength of $1e_\pi$, resulting in a total increase of the energy of the d_{xy} orbital of $4e_\pi$ (the interaction with ligands at positions 1 and 6 is zero). The reader should verify by using Table 10.11 that the d_{xz} and d_{yz} orbitals are also raised in energy by $4e_\pi$.

▶ **Exercise 10.10** Using the angular overlap model, determine the splitting pattern of *d* orbitals for a tetrahedral complex of formula MX_4, where X is a ligand that can act as σ donor and π donor.

In many cases, in situations involving ligands that can behave as both π acceptors and π donors (such as CO and CN^-), the π-acceptor nature predominates. Although π-donor ligands cause the value of Δ_o to decrease, the larger effect of the π-acceptor ligands causes Δ_o to increase. Pi-acceptor ligands are better at splitting the *d* orbitals, causing larger values of Δ_o.

▶ **Exercise 10.11** Determine the energies of the *d* orbitals predicted by the angular overlap model for a square-planar complex:

a. Considering σ interactions only.
b. Considering both σ-donor and π-acceptor interactions.

10.4.4 Types of Ligands and the Spectrochemical Series

Ligands are frequently classified by their donor and acceptor capabilities. Some, like ammonia, are σ donors only, with no orbitals of appropriate symmetry for π interactions. Bonding by these ligands to metals is relatively simple, using only the σ orbitals identified in Figure 10.4. The ligand field split, Δ, then depends on the relative energies of the metal ion and ligand orbitals and on the degree of overlap. Ethylenediamine has a stronger effect than ammonia among these ligands, generating a larger Δ. This is also the order of their proton basicity:

$$\text{en} > NH_3$$

The halide ions have ligand field strengths in the order

$$F^- > Cl^- > Br^- > I^-$$

which is also the order of proton basicity of these ligands.

Ligands that have occupied p orbitals are potentially π donors. They tend to donate these electrons to the metal along with the σ-bonding electrons. As shown in Section 10.4.3, this π-donor interaction decreases Δ. As a result, most halide complexes have high-spin configurations. Other primarily σ-donor ligands that can also act as π donors include H_2O, OH^-, and RCO_2^-. They fit into the series in the order

$$H_2O > F^- > RCO_2^- > OH^- > Cl^- > Br^- > I^-$$

with OH^- below H_2O in the series because it has more π-donating tendency.

When ligands have vacant π^* or d orbitals of suitable energy, there is the possibility of π back-bonding, and the ligands may be π acceptors. This addition to the bonding scheme increases Δ. Ligands that do this very effectively include CN^-, CO, and many others. A selected list of these ligands in order is

$$CO, CN^- > \text{phenanthroline (phen)} > NO_2^- > NCS^-$$

When the lists of ligands are combined, the result is called the **spectrochemical series** and runs roughly in order from strong π-acceptor effect to strong π-donor effect:

$$CO, CN^- > \text{phen} > NO_2^- > \text{en} > NH_3 > NCS^- > H_2O > F^- > RCO_2^- > OH^- > Cl^- > Br^- > I^-$$

Low spin		High spin
Strong field		Weak field
Large Δ		Small Δ
π acceptors	σ donor only	π donors

Ligands high in the series tend to interact strongly with the orbitals of transition metals to cause large splitting of d-orbital energies (large values of Δ) and to favor low-spin complexes; ligands low in the series are not as effective at causing d-orbital splitting and yield lower values of Δ.

10.4.5 Magnitudes of e_σ, e_π, and Δ

CHARGE ON METAL Because changing the ligand or the metal affects the magnitudes of e_σ and e_π, the value of Δ also changes. One consequence may be a change in the number of unpaired electrons. For example, water is a relatively weak-field ligand.

When combined with Co^{2+} in an octahedral geometry, the result is high-spin $[Co(H_2O)_6]^{2+}$ with three unpaired electrons. Combined with Co^{3+}, water forms a low-spin complex with no unpaired electrons. The increase in charge on the metal changes Δ_o sufficiently to favor low spin, as shown in Figure 10.27.

DIFFERENT LIGANDS Similar effects appear when different ligands are used. For example, $[Fe(H_2O)_6]^{3+}$ is a high-spin species, and $[Fe(CN)_6]^{3-}$ is low-spin. Replacing H_2O with CN^- is enough to favor low spin; in this case, the change in Δ_o is caused solely by the ligand. As described in Section 10.3.2, the balance among Δ, Π_c, and Π_e determines whether a specific complex will be high or low spin. Because Δ_t is small, low-spin tetrahedral complexes are unlikely; ligands with strong enough fields to give low-spin complexes are likely to form low-spin octahedral complexes instead.

Tables 10.12 and 10.13 show values for some angular overlap parameters derived from electronic spectra. Several trends can be seen. First, e_σ is always larger than e_π, in some cases by a factor as large as 9, in others less than 2. This is as expected, because σ interactions are more direct, with orbital overlaps directly between nuclei, in contrast to π interactions, which have smaller overlap because the interacting orbitals are not directed toward each other. In addition, the magnitudes of both the σ and π parameters decrease with increasing size and decreasing electronegativity of the halide ions. Increasing the size of the ligand and the corresponding bond length leads to a smaller overlap with the metal d orbitals. In addition, decreasing the electronegativity decreases the pull that a ligand exerts on the metal d electrons, so the two effects reinforce each other.

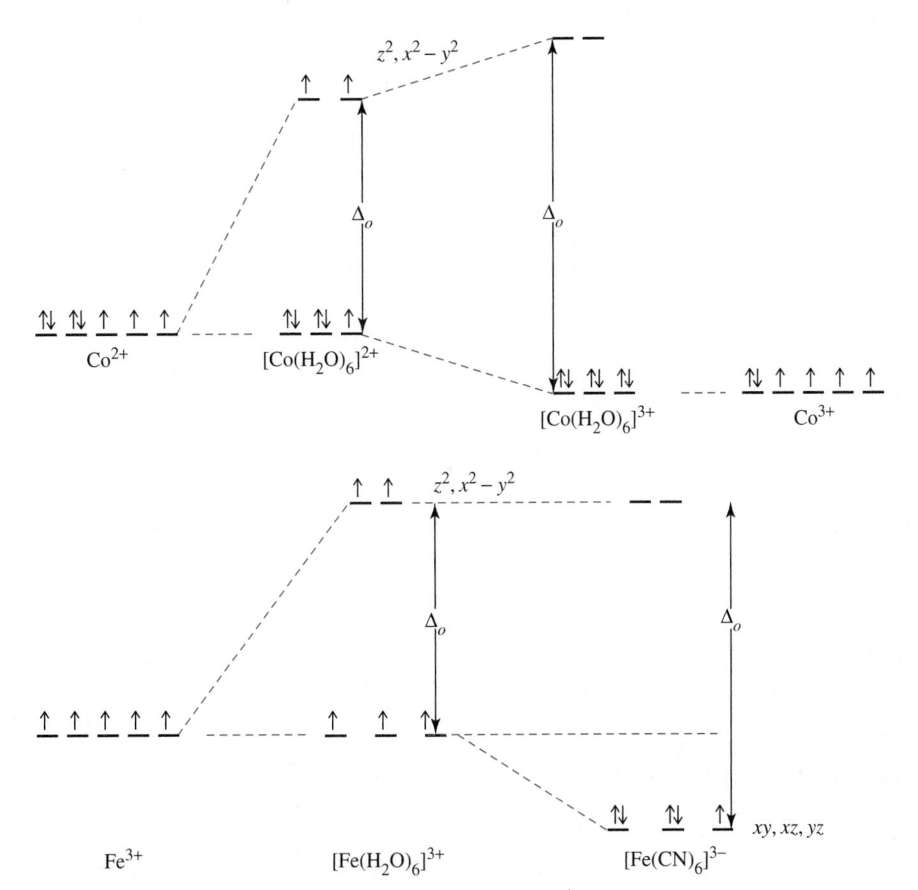

FIGURE 10.27 $[Co(H_2O)_6]^{2+}$, $[Co(H_2O)_6]^{3+}$, $[Fe(H_2O)_6]^{3+}$, $[Fe(CN)_6]^{3-}$, and Unpaired Electrons.

TABLE 10.12 Angular Overlap Parameters

Metal	X	e_σ (cm^{-1})	e_π (cm^{-1})	$\Delta_o = 3e_\sigma - 4e_\pi$
Octahedral MX_6 Complexes				
Cr^{3+}	CN^-	7530	–930	26,310
	en	7260		21,780
	NH_3	7180		21,540
	H_2O	7550	1850	15,250
	F^-	8200	2000	16,600
	Cl^-	5700	980	13,180
	Br^-	5380	950	12,340
	I^-	4100	670	9,620
Ni^{2+}	en	4000		12,000
	NH_3	3600		10,800

Source: Data from B. N. Figgis and M. A. Hitchman, *Ligand Field Theory and Its Applications*, Wiley-VCH, New York, 2000, p. 71, and references therein.

In Table 10.12, ligands in each group are listed in their order in the spectrochemical series. For example, for octahedral complexes of Cr^{3+}, CN^- is listed first; it is the highest in the spectrochemical series and is a π acceptor (e_π is negative). Ethylenediamine and NH_3 are next, listed in order of their e_σ values (which measure σ-donor ability). The halide ions are π donors as well as σ donors and as a group are at the bottom of the series.

Some measures of orbital interaction show different results. For example, these parameters derived from spectra show the order of interaction as $F^- > Cl^- > Br^-$, whereas the reverse is predicted on the basis of donor ability. This can be rationalized as resulting from measurements influenced by different orbitals. The spectral data are derived from transitions to antibonding orbitals; other measures may be derived from bonding molecular orbitals. In addition, the detailed calculation of the energies of the molecular orbitals shows that the antibonding orbital energy is more strongly influenced by the ligand orbitals, but the bonding orbital energy is more strongly influenced by the metal orbitals.[24] The magnitude of the antibonding effect is larger.

SPECIAL CASES The angular overlap model can describe the electronic energy of complexes with different shapes or with combinations of different ligands. It is possible to estimate approximately the magnitudes of e_σ and e_π with different ligands and to predict the effects on the electronic structure of complexes such as $[Co(NH_3)_4Cl_2]^+$. This complex, like nearly all Co(III) complexes except $[CoF_6]^{3-}$ and $[Co(H_2O)_3F_3]$, is low spin, so the magnetic properties do not depend on Δ_o. However, the magnitude of Δ_o does have a significant effect on the visible spectrum, as discussed in Chapter 11. Angular overlap can be used to help compare the energies of different geometries—for example, to predict whether a four-coordinate complex is likely to be tetrahedral or square planar, as described in Section 10.6. It is also possible to use the angular overlap model to estimate the energy change for reactions in which the transition state results in either a higher or lower coordination number, as described in Chapter 12.

[24] J. K. Burdett, *Molecular Shapes*, Wiley InterScience, New York, 1980, p. 157.

TABLE 10.13 Angular Overlap Parameters for MA_4B_2 Complexes

	Equatorial Ligands (A)			Axial Ligands (B)			Reference
		e_σ (cm^{-1})	e_π (cm^{-1})		e_σ (cm^{-1})	e_π (cm^{-1})	
Cr^{3+}, D_{4h}	en	7233	0	F^-	7811	2016	a
		7233		F^-	8033	2000	c
		7333		Cl^-	5558	900	a
		7500		Cl^-	5857	1040	c
		7567		Br^-	5341	1000	a
		7500		Br^-	5120	750	c
		6987		I^-	4292	594	b
		6840		OH^-	8633	2151	a
		7490		H_2O	7459	1370	a
		7833		H_2O	7497	1410	c
		7534		dmso	6769	1653	b
	H_2O	7626	1370 (assumed)	F^-	8510	2539	a
	NH_3	6967	0	F^-	7453	1751	a
Ni^{2+}, D_{4h}							
	py	4670	570	Cl^-	2980	540	c
		4500	500	Br^-	2540	340	c
	pyrazole	5480	1370	Cl^-	2540	380	c
		5440	1350	Br^-	1980	240	c
$[CuX_4]^{2-}$, D_{2d}							
	Cl^-	6764	1831				c
	Br^-	4616	821				c

Source: [a] M. Keeton, B. Fa-chun Chou, and A. B. P. Lever, *Can. J. Chem.* **1971**, *49*, 192; erratum, *ibid.*, **1973**, *51*, 3690.
[b] T. J. Barton and R. C. Slade, *J. Chem. Soc. Dalton Trans.*, **1975**, 650.
[c] M. Gerloch and R. C. Slade, *Ligand Field Parameters*, Cambridge University Press, London, 1973, p. 186.

10.5 THE JAHN–TELLER EFFECT

The Jahn–Teller theorem[25] states that there cannot be unequal occupation of orbitals with identical energies. To avoid such unequal occupation, the molecule distorts so that these orbitals are no longer degenerate. For example, octahedral Cu(II), a d^9 ion, would have three electrons in the two e_g levels without the Jahn–Teller effect, as in the center of Figure 10.28; but the effect requires that the shape of the complex change slightly, resulting in changes in the energies of the orbitals. The resulting distortion is most often an elongation along one axis, but compression along one axis is also possible. In octahedral complexes, where the e_g* orbitals are directed toward the ligands, distortion of the complex has a larger effect on these energy levels and a smaller effect when the t_{2g} orbitals

[25] H. A. Jahn and E. Teller, *Proc. R. Soc. London*, **1937**, *A161*, 220.

are involved. Strong Jahn–Teller effects occur when e_g* orbitals, pointed toward the ligands, are unequally occupied; weak effects, which may be difficult to observe experimentally, occur on uneven occupation of t_{2g} orbitals. The effect of both elongation and compression on *d*-orbital energies is shown in Figure 10.28, and the expected Jahn–Teller effects are summarized in the following table:

Number of electrons	1	2	3	4	5	6	7	8	9	10
High-spin Jahn–Teller	w	w		s		w	w		s	
Low-spin Jahn–Teller	w	w		w	w		s		s	

w = weak Jahn–Teller effect expected (t_{2g} orbitals unevenly occupied); s = strong Jahn–Teller effect expected (e_g orbitals unevenly occupied); No entry = no Jahn–Teller effect expected.

▶ **Exercise 10.12** Using the *d*-orbital splitting diagram in Table 10.5, show that the Jahn–Teller effects in the table match the guidelines in the preceding paragraph.

Examples of significant Jahn–Teller effects are found in complexes of Cr(II) (d^4), high-spin Mn(III) (d^4), and Cu(II) (d^9), Ni(III) (d^7), and low-spin Co(II) (d^7).

Low-spin Cr(II) complexes are octahedral with tetragonal distortion (distorted from O_h to D_{4h} symmetry). They show two absorption bands, one in the visible and one in the near-infrared region, because of this distortion. In a pure octahedral field, there should be only one *d–d* transition (see Chapter 11 for more details). Cr(II) also forms dimeric complexes with Cr—Cr bonds in many complexes. The acetate $Cr_2(OAc)_4$ is an example in which the acetate ions bridge between the two chromiums, with significant Cr—Cr bonding that results in a nearly diamagnetic complex.

Curiously, the $[Mn(H_2O)_6]^{3+}$ ion appears to form an undistorted octahedron in $CsMn(SO_4)_2 \cdot 12\,H_2O$, although other Mn(III) complexes show the expected distortion.[26, 27]

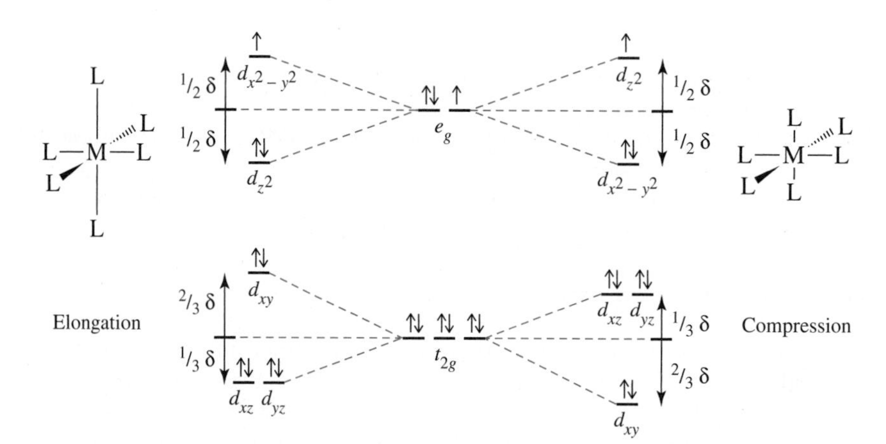

FIGURE 10.28 Jahn–Teller Effect on a d^9 Complex. Elongation along the *z* axis is coupled with a slight decrease in bond length for the other four bonding directions. Similar changes in energy result when the axial ligands have shorter bond distances. The resulting splits are larger for the e_g orbitals than for the t_{2g}. The energy differences are exaggerated in this figure.

[26] A. Avdeef, J. A. Costamagna, and J. P. Fackler, Jr., *Inorg. Chem.*, **1974**, *13*, 1854.
[27] J. P. Fackler, Jr., and A. Avdeef, *Inorg. Chem.*, **1974**, *13*, 1864.

Cu(II) forms the most common complexes with significant Jahn–Teller effects. In most cases, the distortion is an elongation of two bonds, but K_2CuF_4 forms a crystal with two shortened bonds in the octahedron. Elongation also affects equilibrium constants for complex formation. For example, $[Cu(NH_3)_4]^{2+}$ is readily formed in aqueous solution as a distorted octahedron with two water molecules at greater distances than the ammonias; liquid ammonia is required for formation of the hexammine complex. The formation constants for these reactions show the difficulty of putting the fifth and sixth ammonias on the metal.[28] Which factor is the cause and which the result is uncertain, but the bond distances for the two axial positions are longer than those of the four equatorial positions, and the equilibrium constants for placing NH_3 ligands in these positions are much smaller.

$$[Cu(H_2O)_6]^{2+} + NH_3 \rightleftharpoons [Cu(NH_3)(H_2O)_5]^{2+} + H_2O \qquad K_1 = 20{,}000$$

$$[Cu(NH_3)(H_2O)_5]^{2+} + NH_3 \rightleftharpoons [Cu(NH_3)_2(H_2O)_4]^{2+} + H_2O \qquad K_2 = 4{,}000$$

$$[Cu(NH_3)_2(H_2O)_4]^{2+} + NH_3 \rightleftharpoons [Cu(NH_3)_3(H_2O)_3]^{2+} + H_2O \qquad K_3 = 1{,}000$$

$$[Cu(NH_3)_3(H_2O)_3]^{2+} + NH_3 \rightleftharpoons [Cu(NH_3)_4(H_2O)_2]^{2+} + H_2O \qquad K_4 = 200$$

$$[Cu(NH_3)_4(H_2O)_2]^{2+} + NH_3 \rightleftharpoons [Cu(NH_3)_5(H_2O)]^{2+} + H_2O \qquad K_5 = 0.3$$

$$[Cu(NH_3)_5(H_2O)]^{2+} + NH_3 \rightleftharpoons [Cu(NH_3)_6]^{2+} + H_2O \qquad K_6 = \text{very small}$$

In many cases, Cu(II) complexes have square-planar or nearly square-planar geometry, with nearly tetrahedral shapes also possible. $[CuCl_4]^{2-}$, in particular shows structures ranging from tetrahedral through square planar to distorted octahedral depending on the cation present.[29]

10.6 FOUR- AND SIX-COORDINATE PREFERENCES

Angular overlap calculations of the energies expected for different numbers of *d* electrons and different geometries can give us some indication of relative stabilities. Here, we will consider the three major geometries, octahedral, square planar, and tetrahedral. In Chapter 12, similar calculations will be used to help describe reactions at the coordination sites.

Figure 10.29 shows the results of angular overlap calculations for d^0 through d^{10} electron configurations considering sigma interactions only. Figure 10.29(a) compares octahedral and square-planar geometries. Because of the greater number of bonds formed in the octahedral complexes, they are more stable (lower energy) for all configurations except d^8, d^9, and d^{10}. A low-spin square-planar geometry has the same net energy as either a high- or low-spin octahedral geometry for all three of these configurations. This indicates that these configurations are the most likely to have square-planar structures, although octahedral is equally probable from this approach.

[28] R. M. Smith and A. E. Martell, *Critical Stability Constants, Vol. 4, Inorganic Complexes*, Plenum Press, New York, 1976, p. 41.

[29] N. N. Greenwood and A. Earnshaw, *Chemistry of the Elements*, Pergamon Press, Elmsford, NY, 1984, pp. 1385–1386.

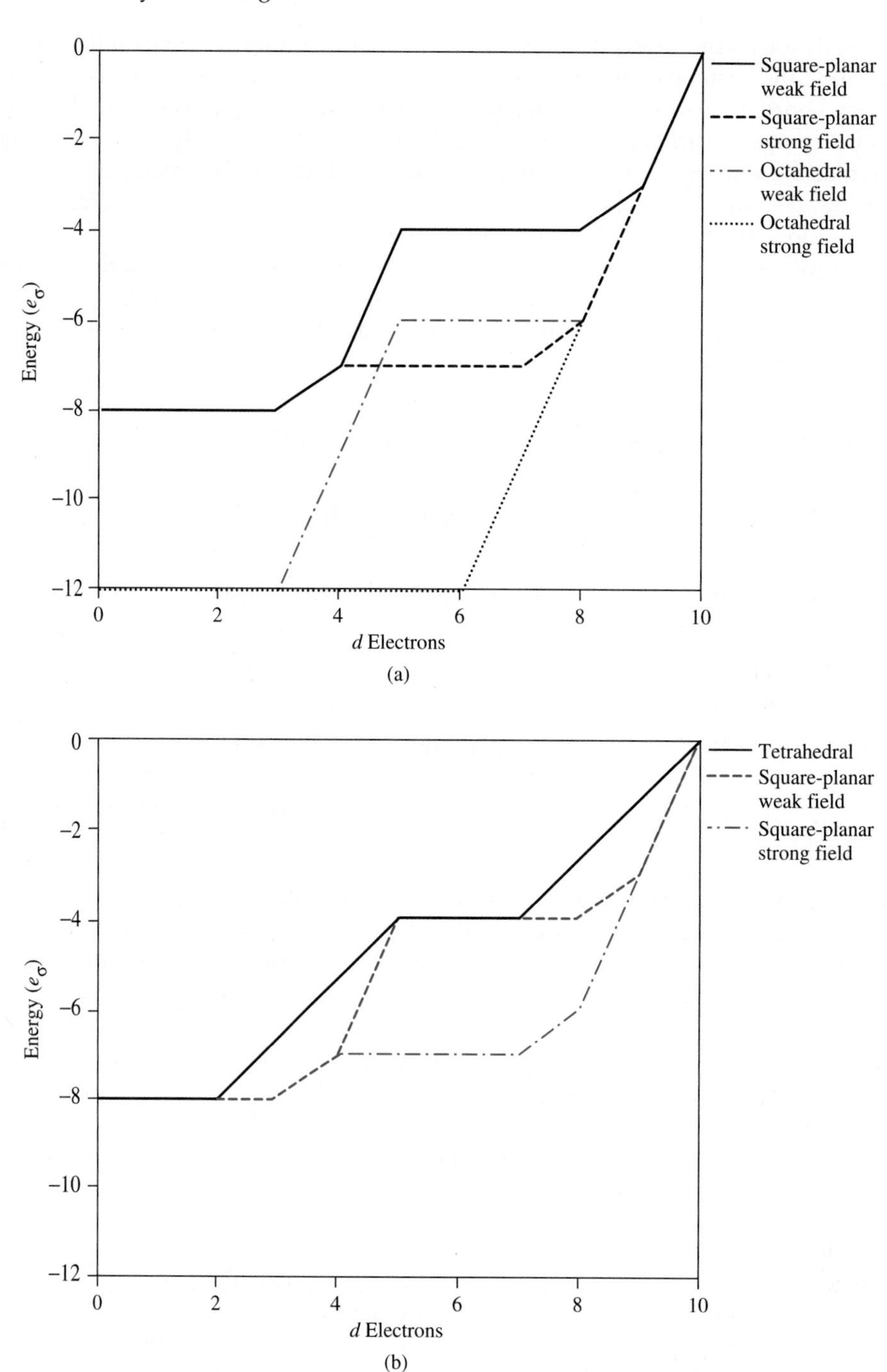

FIGURE 10.29 Angular Overlap Energies of Four- and Six-Coordinate Complexes. Only sigma bonding is considered. (a) Octahedral and square-planar geometries, both strong- and weak-field cases. (b) Tetrahedral and square-planar geometries, both strong- and weak-field cases. There are no known low-spin tetrahedral complexes.

Figure 10.29(b) compares square-planar and tetrahedral structures. For strong-field ligands, square planar is preferred in all cases except d^0, d^1, d^2, and d^{10}. In those cases, the angular overlap approach predicts that the two structures are equally probable. For weak-field ligands, tetrahedral and square-planar structures also have equal energies in the d^5, d^6, and d^7 cases.

How accurate are these predictions? Their success is variable, because other factors are also important. In addition, bond lengths for the same ligand-metal pair depend on the geometry of the complex. One factor that must be included in addition to the d electron energies is the interaction of the s and p orbitals of the metal with the ligand orbitals. The bonding orbitals from these interactions are at a lower energy than those from *d* orbital interactions and are therefore completely filled. Their overall energy is, then, a combination of the energy of the metal atomic orbitals (approximated by their orbital potential energies) and the ligand orbitals. Orbital potential energies for transition metals become more negative with increasing atomic number. As a result, the formation enthalpy for complexes also becomes more negative with increasing atomic number and increasing ionization energy. This trend provides a downward slope to the baseline under the contributions of the *d* orbital–ligand interactions shown in Figure 10.29(a). Burdett[30] has shown that the calculated values of enthalpy of hydration can reproduce the experimental values for enthalpy of hydration very well by using this technique. Figure 10.30 shows a simplified version of this, simply adding $-0.3e_\sigma$ (an arbitrary choice) to the total enthalpy for each increase in Z (which equals the number of *d* electrons). The parallel lines show this slope running through the d^0, d^5, and d^{10} points. Addition of a *d* electron beyond a completed spin set increases the hydration enthalpy until the next set is complete. Comparison with Figure 10.13, in which the experimental values are given, shows that the approach is at least approximately valid. Certainly other factors need to be included for complete agreement with experiment, but their influence seems small.

As expected from the values shown in Figure 10.29, Cu(II) (d^9) complexes show great variability in geometry. Complicating the simple picture used in this section is the change in bond distance that accompanies change in geometry. Overall, the two regular structures most commonly seen are tetragonal—four ligands in a square-planar geometry, with two axial ligands at greater distances—and tetrahedral, sometimes flattened to approximately square planar. There are also trigonal-bipyramidal $[CuCl_5]^{3-}$ ions in $[Co(NH_3)_6][CuCl_5]$. By careful selection of ligands, many of the transition-metal ions can form complexes with geometries other than octahedral. For d^8 ions, some of the simpler possibilities are the square-planar Au(III), Pt(II), and Pd(II) complexes. Ni(II) forms tetrahedral $[NiCl_4]^{2-}$, octahedral $[Ni(en)_3]^{2+}$, and square-planar $[Ni(CN)_4]^{2-}$ complexes, as well as other special cases such as the square-pyramidal $[Ni(CN)_5]^{3-}$. The d^7 Co(II) ion forms tetrahedral blue and octahedral pink complexes—$[CoCl_4]^{2-}$ and $[Co(H_2O)_6]^{2+}$ are simple examples—as well as square-planar complexes when the ligands have strong planar tendencies, such as [Co(salen)], where salen = bis (salicylaldehydeethylenediimine); and a few trigonal-bipyramidal structures such as $[Co(CN)_5]^{3-}$. Many other examples can be found; descriptive works such as that by Greenwood and Earnshaw[31] should be consulted for these.

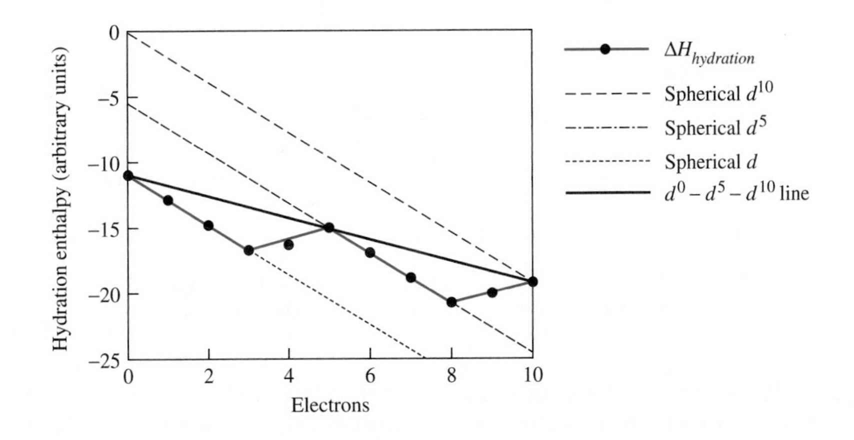

FIGURE 10.30 Simulated Hydration Enthalpies of M^{2+} Transition-Metal Ions.

[30] J. K. Burdett, *J. Chem. Soc. Dalton Trans.*, **1976**, 1725.
[31] N. N. Greenwood and A. Earnshaw, *Chemistry of the Elements*, 2nd ed., Butterworth-Heinemann, Oxford, 1997.

10.7 OTHER SHAPES

Group theory and angular overlap can also be used to determine which *d* orbitals interact with ligand σ orbitals and to obtain a rough idea of the energies of the resulting molecular orbitals for a wide variety of other geometries. As usual, the reducible representation for the ligand σ orbitals is determined and reduced to its irreducible representations. The character table is then used to determine which *d* orbitals match the representations. A qualitative estimate of the energies can usually be determined by examination of the shapes of the orbitals and their apparent overlap, confirmed by using the angular overlap tables.

As an example, consider a trigonal-bipyramidal complex ML_5, in which L is a σ donor only. The point group is D_{3h}, and the reducible and irreducible representations are:

D_{3h}	E	$2C_3$	$3C_2$	σ_h	$2S_3$	$3\sigma_v$	Orbitals
Γ	5	2	1	3	0	3	
A_1'	1	1	1	1	1	1	s
A_1'	1	1	1	1	1	1	d_{z^2}
A_2''	1	1	−1	−1	−1	1	p_z
E'	2	−1	0	2	−1	0	$(p_x, p_y), (d_{x^2-y^2}, d_{xy})$

The d_{z^2} orbital has two ligand orbitals overlapping with it and forms the highest energy molecular orbital. The $d_{x^2-y^2}$ and d_{xy} are in the plane of the three equatorial ligands, but overlap is small because of the angles. They form molecular orbitals relatively high in energy, but not as high as the d_{z^2}. The remaining two orbitals, d_{xz} and d_{yz}, do not have symmetry matching that of the ligand orbitals. These observations are enough to allow us to draw the diagram in Figure 10.31. The angular overlap method is consistent with these more qualitative results, with strong sigma interaction with d_{z^2}, somewhat weaker interaction with $d_{x^2-y^2}$ and d_{xy}, and no interaction with the d_{xz} and d_{yz} orbitals.

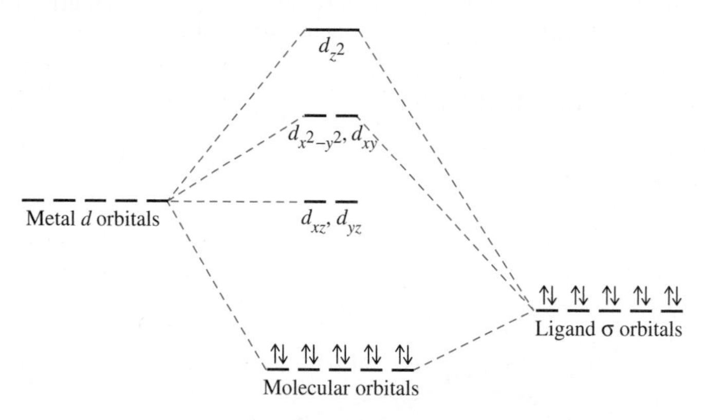

FIGURE 10.31 Trigonal-Bipyramidal Energy Levels. Metal *s* and *p* orbitals also contribute to the bonding molecular orbitals.

▶ **Exercise 10.13** On the basis of the angular overlap model, predict the energies of the *d* orbitals in a trigonal bipyramidal complex in which the ligands are sigma donors only, and compare the results with Figure 10.31. Also, using the D_{3h} character table, assign symmetry labels for the irreducible representations of the *d* orbitals in this figure.

General References

One of the best sources is G. Wilkinson, R. D. Gillard, and J. A. McCleverty, editors, *Comprehensive Coordination Chemistry*, Pergamon Press, Elmsford, NY, 1987; Vol. 1, *Theory and Background*, and Vol. 2, *Ligands*, are particularly useful. Others include the books cited in Chapter 4, which include chapters on coordination compounds. Some older, but still useful, sources are C. J. Ballhausen, *Introduction to Ligand Field Theory*, McGraw-Hill, New York, 1962; T. M. Dunn, D. S. McClure, and R. G. Pearson, *Crystal Field Theory*, Harper & Row, New York, 1965; and C. J. Ballhausen and H. B. Gray, *Molecular Orbital Theory*, W. A. Benjamin, New York, 1965.

More recent volumes include T. A. Albright, J. K. Burdett, and M. Y. Whangbo, *Orbital Interactions in Chemistry*, Wiley InterScience, New York, 1985; and the related text by T. A. Albright and J. K. Burdett, *Problems in Molecular Orbital Theory*, Oxford University Press, Oxford, 1992, which offers examples of many problems and their solutions. Discussions of angular overlap and a variety of other aspects of orbital interactions in coordination complexes are provided in J. K. Burdett, *Molecular Shapes*, John Wiley & Sons, New York, 1980, and B. N. Figgis and M. A. Hitchman, *Ligand Field Theory and Its Applications*, Wiley-VCH, New York, 2000. Discussions of computational method are provided in A. Leach, *Molecular Modeling: Principles and Applications*, 2nd ed., Prentice Hall, Upper Saddle River, NJ, 2001, and C. J. Cramer, *Essentials of Computational Chemistry: Theory and Models*, Wiley, Chichester, UK, 2002.

Problems

10.1 Predict the number of unpaired electrons for each of the following:

a. A tetrahedral d^6 ion
b. $[Co(H_2O)_6]^{2+}$
c. $[Cr(H_2O)_6]^{3+}$
d. A square-planar d^7 ion
e. A coordination compound with a magnetic moment of 5.1 Bohr magnetons

10.2 Identify the *first-row* transition metal **M** that satisfies the requirements given (more than one answer may be possible):

a. $[\mathbf{M}(H_2O)_6]^{3+}$ having one unpaired electron
b. $[\mathbf{M}Br_4]^-$ having the most unpaired electrons
c. Diamagnetic $[\mathbf{M}(CN_6)]^{3-}$
d. $[\mathbf{M}(H_2O)_6]^{2+}$ having LFSE $= -\frac{3}{5}\,\Delta_O$

10.3 Identify the most likely transition metal **M**:

a. $K_3[\mathbf{M}(CN)_6]$, in which **M** is a first-row transition metal and the complex has 3 unpaired electrons.
b. $[\mathbf{M}(H_2O)_6]^{3+}$, in which **M** is a second row transition metal and LFSE $= -2.4\,\Delta_o$.
c. Tetrahedral $[\mathbf{M}Cl_4]^-$, which has 5 unpaired electrons and first-row transition metal **M**.
d. The third row d^8 transition metal in the square-planar complex $\mathbf{M}Cl_2(NH_3)_2$, which has two M–Cl stretching bands in the IR.

10.4 The stepwise stability constants in aqueous solution at 25 °C for the formation of the ions $[M(en)(H_2O)_4]^{2+}$, $[M(en)_2(H_2O)_2]^{2+}$, and $[M(en)_3]^{2+}$ for copper and nickel are given in the table. Why is there such a difference in the third values? (Hint: Consider the special nature of d^9 complexes.)

	$[M(en)(H_2O)_4]^{2+}$	$[M(en)_2(H_2O)_2]^{2+}$	$[M(en)_3]^{2+}$
Cu	3×10^{10}	1×10^9	0.1 (estimated)
Ni	2×10^7	1×10^6	1×10^4

10.5 A first-row transition-metal complex of formula $[\mathbf{M}(H_2O)_6]^{2+}$ has a magnetic moment of 3.9 Bohr magnetons. Determine the most likely number of unpaired electrons and the identity of the metal.

10.6 Predict the magnetic moments (spin-only) of the following species:

a. $[Cr(H_2O)_6]^{2+}$
b. $[Cr(CN)_6]^{4-}$
c. $[FeCl_4]^-$
d. $[Fe(CN)_6]^{3-}$
e. $[Ni(H_2O)_6]^{2+}$
f. $[Cu(en)_2(H_2O)_2]^{2+}$

10.7 A compound with the empirical formula $Fe(H_2O)_4(CN)_2$ has a magnetic moment corresponding to $2\frac{2}{3}$ unpaired electrons per iron. How is this possible? (Hint: Two octahedral Fe(II) species are involved, each containing a single type of ligand.)

10.8 What are the possible magnetic moments of Co(II) in tetrahedral, octahedral, and square-planar complexes?

10.9 Monothiocarbamate complexes of Fe(III) have been prepared. (See K.R. Kunze, D.L. Perry, and L.J. Wilson, *Inorg. Chem.*, **1977**, *16*, 594.) For the methyl and ethyl complexes, the magnetic moment, μ, is 5.7 to 5.8 μ_B at 300 K; it changes to 4.70 to 5 μ_B at 150 K, and drops still further to 3.6 to 4 μ_B at 78 K. The color changes from red to orange as the temperature is lowered. With larger R groups (propyl, piperidyl, pyrrolidyl), $\mu > 5.3\,\mu_B$ at all temperatures and greater than 6 μ_B in some. Explain these changes. Monothiocarbamate complexes have the following structure.

$R_2N{-}C(S)(O){-}M$ (S and O both bound to M)

10.10 Show graphically how you would expect ΔH for the reaction

$$[M(H_2O)_6]^{2+} + 6\,NH_3 \longrightarrow [M(NH_3)_6]^{2+} + 6\,H_2O$$

to vary for the first transition series (M = Sc through Zn).

10.11 Using the coordinate system in Figure 10.18, verify the characters of Γ_σ and Γ_π in Table 10.9 and show that these representations reduce to $A_1 + T_2$ and $E + T_1 + T_2$, respectively.

10.12 Using the angular overlap model, determine the energies of the *d* orbitals of the metal for each of the following geometries, first for ligands that act as σ donors only and second for ligands that act as both σ donors and π acceptors.

a. Linear ML_2
b. Trigonal-planar ML_3
c. Square-pyramidal ML_5
d. Trigonal-bipyramidal ML_5
e. Cubic ML_8 (Hint: A cube is two superimposed tetrahedra.)

10.13 Use the angular overlap method to calculate the energies of both ligand and metal orbitals for *trans*-$[Cr(NH_3)_4Cl_2]^+$, taking into account that ammonia is a stronger σ donor ligand than chloride, but chloride is a stronger π donor. Use the 1 and 6 positions for the chloride ions.

10.14 Consider a transition-metal complex of formula ML_4L'. Using the angular overlap model and assuming trigonal-bipyramidal geometry, determine the energies of the *d* orbitals:

a. Considering sigma interactions only (assume L and L′ are similar in donor ability).
b. Considering L′ as a π acceptor as well. Consider L′ in both (1) axial and (2) equatorial positions.
c. Based on the preceding answers, would you expect π-acceptor ligands to preferentially occupy axial or equatorial positions in five-coordinate complexes? What other factors should be considered in addition to angular overlap?

10.15 On the basis of your answers to Problems 10.13 and 10.14, which geometry, square-pyramidal or trigonal-bipyramidal, is predicted to be more likely for five-coordinate complexes by the angular overlap model? Consider both σ-donor and combined σ-donor and π-acceptor ligands.

10.16 The common structures having CN = 4 for transition-metal complexes are tetrahedral and square planar. However, these are not the only conceivable structures. Examples of main group compounds having *seesaw* structures are known, and *trigonal pyramidal* structures may also be possible in some cases.

Seesaw **Trigonal Pyramidal**

a. For these structures, determine the relative energies of the *d* orbitals of a transition-metal complex of formula ML_4 in which L is a σ donor only.
b. Considering both high-spin and low-spin possibilities, calculate the energy of each configuration d^1 to d^{10} in terms of e_σ.
c. For which configurations is the seesaw structure favored? The trigonal pyramidal structure? Neither?

10.17 A possible geometry for an eight-coordinate complex ML_8 might be a hexagonal bipyramid:

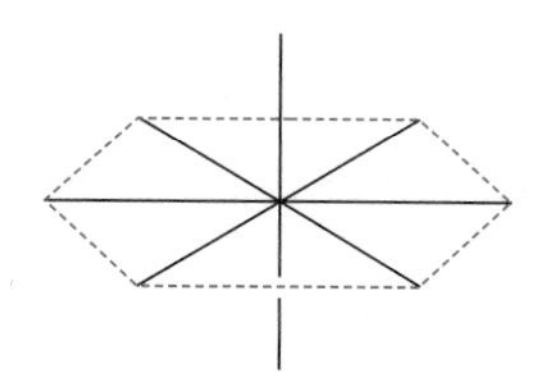

a. Predict the effect of the eight ligands on the energies of the *d* orbitals of a metal M, using the angular overlap model and assuming that the ligands are sigma donors only. [Note: To determine the values of e_σ, you will need to add two more positions to Table 10.11.]
b. Assign the symmetry labels of the *d* orbitals (labels of the irreducible representations).
c. Repeat the calculations in part a for a ligand that can act both as a σ donor and a π acceptor.
d. For this geometry, and assuming low spin, which d^n configurations would be expected to give rise to Jahn–Teller distortions?

10.18 $[Co(H_2O)_6]^{3+}$ is a strong oxidizing agent that will oxidize water, but $[Co(NH_3)_6]^{3+}$ is stable in aqueous solution. Explain this difference. Table 10.6 gives data on the aqueous complex; Δ_o for $[Co(NH_3)_6]^{2+}$ is 10,200 cm^{-1}, and Δ_o for $[Co(NH_3)_6]^{3+}$ is about 24,000 cm^{-1}. Both are low-spin complexes.

10.19 Explain the order of the magnitudes of the following Δ_o values for Cr(III) complexes in terms of the σ and π donor and acceptor properties of the ligands.

Ligand	F^-	Cl^-	H_2O	NH_3	en	CN^-
Δ_o (cm^{-1})	15,200	13,200	17,400	21,600	21,900	33,500

10.20 Oxygen is more electronegative than nitrogen; fluorine is more electronegative than the other halogens. Fluoride is a stronger field ligand than the other halides, but ammonia is a stronger field ligand than water. Why?

10.21 **a.** Explain the effect on the *d*-orbital energies when an octahedral complex is compressed along the *z* axis.

b. Explain the effect on the *d*-orbital energies when an octahedral complex is stretched along the *z* axis. In the limit, this results in a square-planar complex.

10.22 Solid CrF_3 contains a Cr(III) ion surrounded by six F^- ions in an octahedral geometry, all at distances of 190 pm. However, MnF_3 is in a distorted geometry, with Mn–F distances of 179, 191, and 209 pm (two of each). Suggest an explanation.

10.23 **a.** Determine the number of unpaired electrons, magnetic moment, and ligand field stabilization energy for each of the following complexes:

$[Co(CO)_4]^-$ $[Cr(CN)_6]^{4-}$ $[Fe(H_2O)_6]^{3+}$ $[Co(NO_2)_6]^{4-}$
$[Co(NH_3)_6]^{3+}$ MnO_4^- $[Cu(H_2O)_6]^{2+}$

b. Why are two of these complexes tetrahedral and the rest octahedral?

c. Why is tetrahedral geometry more stable for Co(II) than for Ni(II)?

10.24 The 2+ ions in the first transition series generally show a preference for octahedral geometry over tetrahedral geometry. Nevertheless, the number of tetrahedral complexes formed is in the order Co > Fe > Ni.

a. Calculate the ligand field stabilization energies for tetrahedral and octahedral symmetries for these ions. Do these numbers explain this order?

b. Does the angular overlap model offer any advantage in explaining this order?

10.25 Except in cases in which ligand geometry requires it, square-planar geometry occurs most commonly with d^7, d^8, and d^9 ions and with strong-field, π-acceptor ligands. Suggest why these conditions support square-planar geometry.

10.26 Use the group theory approach of Section 10.7 to prepare an energy level diagram for

a. a square-pyramidal complex.

b. a pentagonal-bipyramidal complex.

10.27 Nitrogen monofluoride, NF, can serve as a ligand in transition-metal complexes.

a. Prepare a molecular orbital energy-level diagram of the NF molecule, showing clearly how the atomic orbitals interact.

b. If NF can interact with a transition-metal ion to form a chemical bond, what type(s) of ligand–metal interactions would be most important? Would you expect NF to be high or low in the spectrochemical series? Explain.

10.28 Calculations have been reported on the changes that occur when the following compounds are oxidized by one electron. (See T. Leyssens, D. Peeters, A. G. Orpen, and J. N. Harvey, *New J. Chem.*, **2005**, *29*, 1424–1430.)

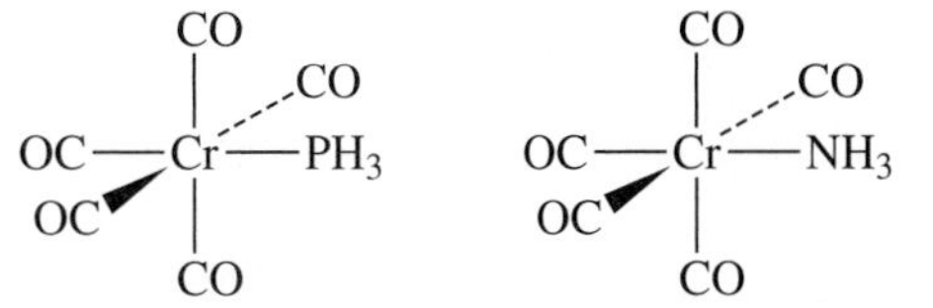

a. When these compounds are oxidized, what is the effect on the C–O distances? Explain.

b. When these compounds are oxidized, what is the effect on the Cr–P distance? On the Cr–N distance? Explain.

10.29 The linear molecule FeH_2 has been observed in the gas phase. (See H. Korsgen, W. Urban, and J. M. Brown, *J. Chem. Phys.* **1999**, *110*, 3861.) Assume that the iron atom can potentially use *s*, *p*, and *d* orbitals to interact with the hydrogens. If the *z* axis is collinear with the molecular axis:

a. Sketch the group orbitals of the hydrogen atoms that potentially could interact with the iron.

b. Show how the group orbitals and the central atom would interact.

c. Which interaction would you expect to be the strongest? The weakest? Explain briefly.

10.30 The ion $[Pt_2D_9]^{5-}$, shown below, has eclipsed geometry.

a. What is the point group of this ion?

b. Assume that the platinums can potentially use *s*, *p*, and *d* orbitals to interact with the central deuterium. If the *z* axis is chosen to be collinear with the principal axis of rotation:

1. Sketch the group orbitals of the platinum atoms that potentially could interact with the central D. Be sure to label all orbitals.
2. Show how the group orbitals and the central atom would interact.
3. Which interaction would you expect to be the strongest, and why?

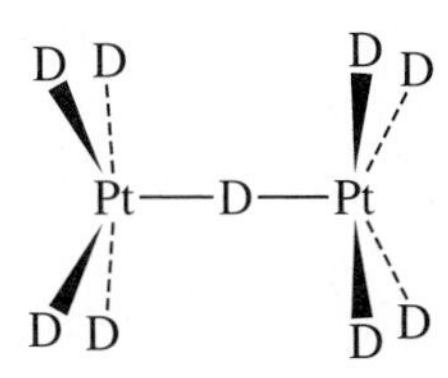

10.31 On the basis of molecular orbitals, explain why the Mn–O distance in $[MnO_4]^{2-}$ is longer (by 3.9 pm) than in $[MnO_4]^-$. (See G. J. Palenik, *Inorg. Chem.*, **1967**, *6*, 503, 507.)

The following problems require the use of molecular modeling software.

10.32 The ion $[TiH_6]^{2-}$ has been found to have O_h symmetry. (See I. B. Bersuker, N. B. Balabanov, D. Pekker, and J. E. Boggs, *J. Chem. Phys.* **2002**, *117*, 10478.)

a. Using the H orbitals of the ligands as a basis, construct a reducible representation (the symmetry equivalent of a collection of group orbitals) for this ion.

b. Reduce this representation to its irreducible components.

c. Which orbitals of Ti are suitable for interaction with each of the results from part **b**?

d. Show the interactions of *d* orbitals of Ti with the appropriate group orbital(s), labeled to show the matching irreducible representations, in an energy-level diagram. Identify Δ_o on this diagram.

e. Now use molecular modeling software to calculate and display the molecular orbitals of TiH_6^{2-}. Compare the results with your work in part **d** and with Figure 10.6, and comment on the similarities and differences.

10.33 Calculate and view the molecular orbitals of the octahedral ion $[TiF_6]^{3-}$.

a. Identify the t_{2g} and e_g bonding and antibonding orbitals, and indicate which *d* orbitals of Ti are involved in each.

b. Compare your results with Figures 10.6 and 10.8. Do they indicate that fluoride is acting as a π-donor as well as σ-donor?

10.34 Reaction of many iron(III) compounds with hydrochloric acid yield the tetrahedral $[FeCl_4]^-$ ion. Calculate and view the molecular orbitals of this ion.

a. Identify the *e* and t_2 orbitals involved in Fe–Cl bonding (see Figures 10.19 and 10.20), and indicate which *d* orbitals of Fe are involved in each.

b. Compare your results with Figure 10.20. Comment on the similarities and differences.

10.35 Table 10.12 provides values of Δ_o for eight octahedral complexes of chromium(III). Select three of the ligands listed, draw the structures of their octahedral complexes of Cr(III), and calculate and view the molecular orbitals. Identify the t_{2g} and e_g orbitals, record the energy of each, and determine the Δ_o values. Is your trend consistent with the values in the table? (Note: the results are likely to vary significantly with the level of sophistication of the software used. If you have several molecular modeling programs available, you may want to try different ones to compare their results.)

11

Coordination Chemistry III: Electronic Spectra

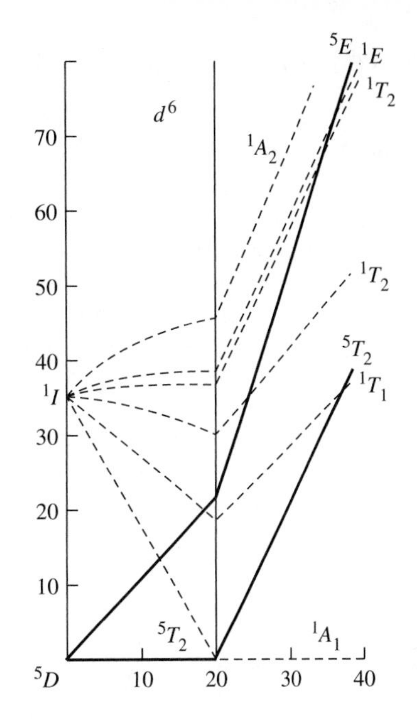

Perhaps the most striking aspect of many coordination compounds of transition metals is their vivid colors. The dye Prussian blue, for example, has been used as a pigment for more than two centuries and is still used in blueprints; it is a complicated coordination compound involving iron(II) and iron(III) coordinated octahedrally by cyanide. Many precious gems exhibit colors resulting from transition-metal ions incorporated into their crystalline lattices. For example, emeralds are green as a consequence of the incorporation of small amounts of chromium(III) into crystalline $Be_3Al_2Si_6O_{18}$; amethysts are violet as a result of the presence of small amounts of iron(II), iron(III), and titanium(IV) in an Al_2O_3 lattice; and rubies are red because of chromium(III), also in a lattice of Al_2O_3. The color of blood is caused by the red heme group, a coordination compound of iron present in hemoglobin. Most readers are probably familiar with blue $CuSO_4 \cdot 5\,H_2O$, a compound often used to demonstrate the growing of large, highly symmetric crystals.

It is desirable to understand why so many coordination compounds are colored, in contrast to most organic compounds, which are transparent, or nearly so, in the visible spectrum. We will first review the concept of light absorption and how it is measured. The ultraviolet and visible spectra of coordination compounds of transition metals involve transitions between the *d* orbitals of the metals. Therefore, we will need to look closely at the energies of these orbitals (as discussed in Chapter 10) and at the possible ways that electrons can be raised from lower to higher energy levels. The energy levels of *d* electron configurations—as opposed to the energies of *individual* electrons—are somewhat more complicated than might be expected, and we need to consider how electrons in atomic orbitals can interact with each other.

For many coordination compounds, the electronic absorption spectrum provides a convenient method for determining the magnitude of the effect of ligands on the *d* orbitals of the metal. Although in principle we can study this effect for coordination compounds of any geometry, we will concentrate on the most common geometry, octahedral, and will examine how the absorption spectrum can be used to determine the magnitude of the octahedral ligand field parameter Δ_o for a variety of complexes.

11.1 ABSORPTION OF LIGHT

In explaining the colors of coordination compounds, we are dealing with the phenomenon of *complementary colors*: if a compound absorbs light of one color, we see the complement of that color. For example, when white light (containing a broad spectrum of all visible wavelengths) passes through a substance that absorbs red light, the color observed is green. Green is the complement of red, so green

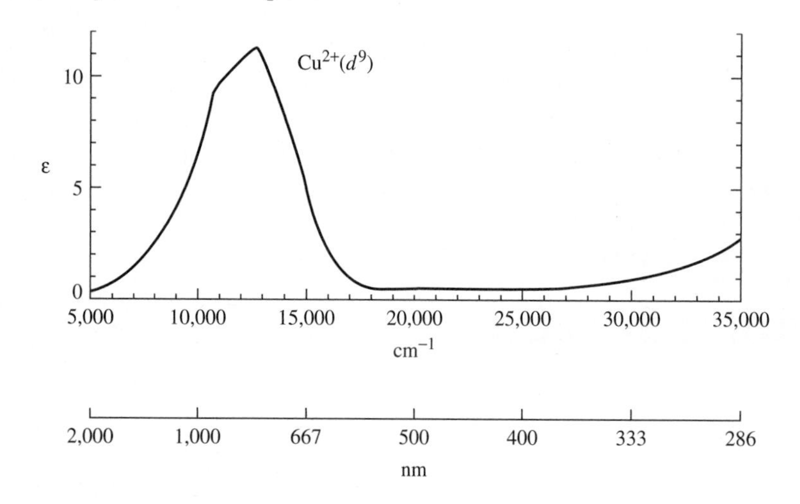

FIGURE 11.1 Absorption Spectrum of $[Cu(H_2O)_6]^{2+}$.
(Reproduced with permission from B. N. Figgis, *Introduction to Ligand Fields*, Wiley-InterScience, New York, 1966, p. 221.)

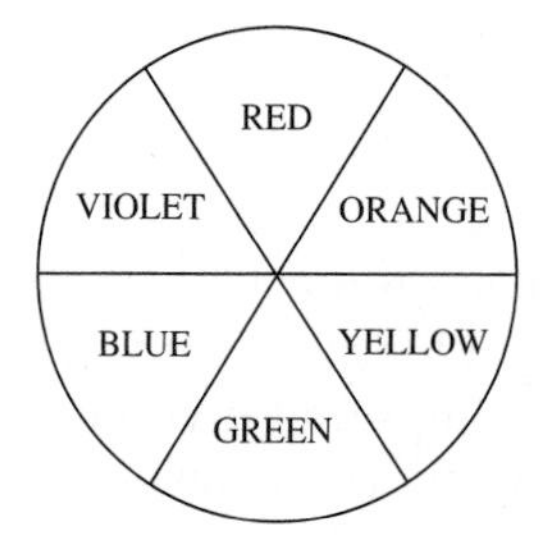

predominates visually when red light is subtracted from white. Complementary colors can conveniently be remembered as the color pairs on opposite sides of the color wheel shown in the margin.

An example from coordination chemistry is the deep blue color of aqueous solutions of copper(II) compounds, containing the ion $[Cu(H_2O)_6]^{2+}$. The blue color is a consequence of the absorption of light between approximately 600 and 1000 nm (maximum near 800 nm; Figure 11.1), in the yellow to infrared region of the spectrum. The color observed, blue, is the average complementary color of the light absorbed.

It is not always possible to make a simple prediction of color directly from the absorption spectrum, in large part because many coordination compounds contain two or more absorption bands of different energies and intensities. The net color observed is the color predominating after the various absorptions are removed from white light.

For reference, the approximate wavelengths and complementary colors to the principal colors of the visible spectrum are given in Table 11.1.

11.1.1 Beer–Lambert Absorption Law

If light of intensity I_o at a given wavelength passes through a solution containing a species that absorbs light, the light emerges with intensity I, which may be measured by a suitable detector (Figure 11.2).

TABLE 11.1 Visible Light and Complementary Colors

Wavelength Range (nm)	Wave Numbers (cm^{-1})	Color	Complementary Color
< 400	> 25,000	Ultraviolet	
400–450	22,000–25,000	Violet	Yellow
450–490	20,000–22,000	Blue	Orange
490–550	18,000–20,000	Green	Red
550–580	17,000–18,000	Yellow	Violet
580–650	15,000–17,000	Orange	Blue
650–700	14,000–15,000	Red	Green
> 700	< 14,000	Infrared	

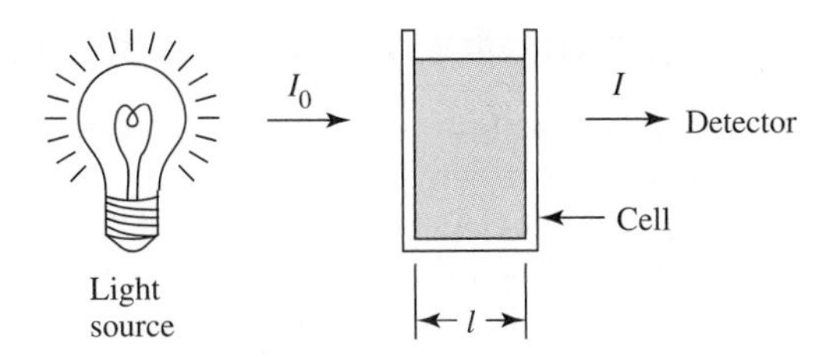

FIGURE 11.2 Absorption of Light by Solution.

The Beer–Lambert law may be used to describe the absorption of light (ignoring scattering and reflection of light from cell surfaces) at a given wavelength by an absorbing species in solution:

$$\log \frac{I_o}{I} = A = \varepsilon l c$$

where A = absorbance

ε = molar absorptivity ($L\ mol^{-1}\ cm^{-1}$) (also known as molar extinction coefficient)

l = length through solution (cm)

c = concentration of absorbing species ($mol\ L^{-1}$)

Absorbance is a dimensionless quantity. An absorbance of 1.0 corresponds to 90% absorption at a given wavelength,[1] an absorbance of 2.0 corresponds to 99% absorption, and so on. The most common units of the other quantities in the Beer–Lambert law are shown in parentheses above.

Spectrophotometers commonly obtain spectra as plots of absorbance versus wavelength. The molar absorptivity is a characteristic of the species that is absorbing the light and is highly dependent on wavelength. A plot of molar absorptivity versus wavelength gives a spectrum characteristic of the molecule or ion in question, as in Figure 11.1. As we will see, this spectrum is a consequence of transitions between states of different energies and can provide valuable information about those states and, in turn, about the structure and bonding of the molecule or ion.

Although the quantity most commonly used to describe absorbed light is the wavelength, energy and frequency are also used. In addition, the wavenumber, the number of waves per centimeter (a quantity proportional to the energy), is frequently used, especially in reference to infrared light. For reference, the relations between these quantities are given by the equations

$$E = h\nu = \frac{hc}{\lambda} = hc\left(\frac{1}{\lambda}\right) = hc\overline{\nu}$$

where E = energy

h = Planck constant = 6.626×10^{-34} J s

c = speed of light = $2.998 \times 10^{8}\ m\ s^{-1}$

ν = frequency (s^{-1})

λ = wavelength (often reported in nm)

$\frac{1}{\lambda} = \overline{\nu}$ = wavenumber (cm^{-1})

[1] For absorbance = 1.0, $\log (I_o / I) = 1.0$. Therefore, $I_o / I = 10$, and $I = 0.10\ I_o = 10\% \times I_o$; 10% of the light is transmitted, and 90% is absorbed.

11.2 QUANTUM NUMBERS OF MULTIELECTRON ATOMS

Absorption of light results in the excitation of electrons from lower to higher energy states; because such states are quantized, we observe absorption in "bands" (as in Figure 11.1), with the energy of each band corresponding to the difference in energy between the initial and final states. To gain insight into these states and the energy transitions between them, we first need to consider how electrons in atoms can interact with each other.

Although the quantum numbers and energies of individual electrons can be described in fairly simple terms, interactions between electrons complicate this picture. Some of these interactions were discussed in Section 2.2.3: as a result of repulsions between electrons (characterized by energy Π_c), electrons tend to occupy separate orbitals; as a result of exchange energy (Π_e), electrons in separate orbitals tend to have parallel spins.

Consider again the example of the energy levels of a carbon atom. Carbon has the electron configuration $1s^2\,2s^2\,2p^2$. At first glance, we might expect the p electrons to have the same energy. However, there are three major energy levels for the p^2 electrons, differing in energy by pairing and exchange energies (Π_c and Π_e). In addition, the lowest major energy level is split into three slightly different energies, for a total of five energy levels. As an alternative to the discussion presented in Section 2.2.3, each energy level can be described as a combination of the m_l and m_s values of the $2p$ electrons.

Independently, each of the $2p$ electrons could have any of six possible m_l, m_s combinations:

$n = 2, l = 1$	(quantum numbers defining $2p$ orbitals)
$m_l = +1, 0, \text{ or } -1$	(three possible values)
$m_s = +\frac{1}{2} \text{ or } -\frac{1}{2}$	(two possible values)

The $2p$ electrons are not independent of each other, however; the orbital angular momenta (characterized by m_l values) and the spin angular momenta (characterized by m_s values) of the $2p$ electrons interact in a manner called **Russell–Saunders coupling** or **LS coupling**.[2] The interactions produce atomic states called **microstates** that can be described by new quantum numbers:

$M_L = \Sigma m_l$	Total orbital angular momentum
$M_S = \Sigma m_s$	Total spin angular momentum

We need to determine how many possible combinations of m_l and m_s values there are for a p^2 configuration.[3] Once these combinations are known, we can determine the corresponding values of M_L and M_S. For shorthand, we will designate the m_s value of each electron by a superscript +, representing $m_s = +\frac{1}{2}$, or −, representing $m_s = -\frac{1}{2}$. For example, an electron having $m_l = +1$ and $m_s = +\frac{1}{2}$ will be written as 1^+.

One possible set of values for the two electrons in the p^2 configuration would be

$$\left.\begin{array}{lll} \text{First electron:} & m_l = +1 \text{ and } & m_s = +\frac{1}{2} \\ \text{Second electron:} & m_l = 0 \text{ and } & m_s = -\frac{1}{2} \end{array}\right\} \text{Notation: } 1^+0^-$$

Each set of possible quantum numbers, such as 1^+0^-, is called a microstate.

The next step is to tabulate the possible microstates. In doing this, we need to take two precautions: (1) to be sure that no two electrons in the same microstate have identical

[2] For a more advanced discussion of coupling and its underlying theory, see M. Gerloch, *Orbitals, Terms, and States*, Wiley InterScience, New York, 1986.

[3] Electrons in filled orbitals can be ignored, because their net spin and angular momenta are both zero.

quantum numbers (the Pauli exclusion principle applies); and (2) to count only the *unique* microstates. For example, the microstates 1^+0^- and 0^-1^+, 0^+0^- and 0^-0^+ in a p^2 configuration are duplicates and only one of each pair will be listed.

If we determine all possible microstates and tabulate them according to their M_L and M_S values, we obtain a total of 15 microstates.[4] These microstates can be arranged according to their M_L and M_S values and listed conveniently in a microstate table, as shown in Table 11.2.

TABLE 11.2 Microstate Table for p^2

M_L \ M_S	−1	0	+1
+2		$1^+\ 1^-$	
+1	$1^-\ 0^-$	$1^+\ 0^-$ $1^-\ 0^+$	$1^+\ 0^+$
0	$-1^-\ 1^-$	$-1^+\ 1^-$ $0^+\ 0^-$ $-1^-\ 1^+$	$-1^+\ 1^+$
−1	$-1^-\ 0^-$	$-1^+\ 0^-$ $-1^-\ 0^+$	$-1^+\ 0^+$
−2		$-1^+\ -1^-$	

EXAMPLE

Determine the possible microstates for an $s^1 p^1$ configuration, and use them to prepare a microstate table.

The *s* electron can have $m_l = 0$ and $m_s = \pm\frac{1}{2}$.

The *p* electron can have $m_l = +1, 0, -1$ and $m_s = \pm\frac{1}{2}$.

The resulting microstate table is then

M_L \ M_S	−1	0	+1
+1	$0^-\ 1^-$	$0^-\ 1^+$ $0^+\ 1^-$	$0^+\ 1^+$
0	$0^-\ 0^-$	$0^+\ 0^-$ $0^-\ 0^+$	$0^+\ 0^+$
−1	$0^-\ -1^-$	$0^-\ -1^+$ $0^+\ -1^-$	$0^+\ -1^+$

In this case, 0^+0^- and 0^-0^+ are different microstates, because the first electron is an *s* and the second electron is a *p*; both must be counted.

[4] The number of microstates $= i!/[j!(i-j)!]$, where i = number of m_l, m_s combinations (six here, because m_l can have values of 1, 0, and −1, and m_s can have values of $+\frac{1}{2}$ and $-\frac{1}{2}$) and j = number of electrons.

▶ **Exercise 11.1** Determine the possible microstates for a d^2 configuration and use them to prepare a microstate table. (Your table should contain 45 microstates!)

We have now seen how electronic quantum numbers m_l and m_s may be combined into atomic quantum numbers M_L and M_S, which describe atomic microstates. M_L and M_S in turn give atomic quantum numbers L, S, and J. These quantum numbers collectively describe the energy and symmetry of an atom or ion and determine the possible transitions between states of different energies. These transitions account for the colors observed for many coordination complexes, as will be discussed later in this chapter.

The quantum numbers that describe states of multielectron atoms are defined as follows:

L = total orbital angular momentum quantum number

S = total spin angular momentum quantum number

J = total angular momentum quantum number

These total angular momentum quantum numbers are determined by vector sums of the individual quantum numbers; determination of their values is described in this section and the next.

Quantum numbers L and S describe collections of microstates, whereas M_L and M_S describe the microstates themselves. L and S are the largest possible values of M_L and M_S. M_L is related to L much as m_l is related to l, and the values of M_S and m_s are similarly related:

Atomic States	Individual Electrons
$M_L = 0, \pm1, \pm2, \pm L$	$m_l = 0, \pm1, \pm2, \ldots, \pm l$
$M_S = S, S-1, S-2, \ldots, -S$	$m_s = +\frac{1}{2}, -\frac{1}{2}$

Just as the quantum number m_l describes the component of the quantum number l in the direction of a magnetic field for an electron, the quantum number M_L describes the component of L in the direction of a magnetic field for an atomic state. Similarly, m_s describes the component of an electron's spin in a reference direction, and M_S describes the component of S in a reference direction for an atomic state.

$L = 0$	S state
$L = 1$	P state
$L = 2$	D state
$L = 3$	F state

The values of L correspond to atomic states described as S, P, D, F, and higher states in a manner similar to the designation of atomic orbitals as s, p, d, and f. The values of S are used to calculate the **spin multiplicity**, defined as $2S + 1$. For example, states having spin multiplicities of 1, 2, 3, and 4 are described as *singlet*, *doublet*, *triplet*, and *quartet* states. The spin multiplicity is designated as a left superscript. Examples of atomic states are given in Table 11.3 and in the examples that follow.[5]

Atomic states characterized by S and L are often called **free-ion terms** (sometimes Russell–Saunders terms) because they describe individual atoms or ions, free of ligands. Their labels are often called **term symbols**.[6] Term symbols are composed of a letter

[5] Unfortunately, S is used in two ways: to designate the atomic spin quantum number and to designate a state having $L = 0$. Chemists are not always wise in choosing their symbols!

[6] Although *term* and *state* are often used interchangeably, *term* is suggested as the preferred label for the results of Russell–Saunders coupling just described, and *state* for the results of spin-orbit coupling, described in the following section, including the quantum number J. In most cases, the meaning of *term* and *state* can be deduced from the context. (See B. N. Figgis, "Ligand Field Theory," in G. Wilkinson, R. D. Gillard, and J. A. McCleverty, eds. *Comprehensive Coordination Chemistry*, Vol. 1, Pergamon Press, Elmsford, NY, 1987, p. 231.)

TABLE 11.3 Examples of Atomic States (Free-Ion Terms) and Quantum Numbers

Term	L	S
^1S	0	0
^2S	0	$\frac{1}{2}$
^3P	1	1
^4D	2	$\frac{3}{2}$
^5F	3	2

relating to the value of L and a left superscript for the spin multiplicity. For example, the term symbol ^3D corresponds to a state in which $L = 2$ and the spin multiplicity $(2S + 1)$ is 3; ^5F marks a state in which $L = 3$ and $2S + 1 = 5$.

Free-ion terms are very important in the interpretation of the spectra of coordination compounds. The following examples show how to determine the values of L, M_L, S, and M_S for a given term and how to prepare microstate tables from them.

EXAMPLES

^1S (singlet S)

An S term has $L = 0$ and must therefore have $M_L = 0$. The spin multiplicity (the superscript) is $2S + 1$. Because $2S + 1 = 1$, S must equal 0 (and $M_S = 0$). There can be only one microstate having $M_L = 0$ and $M_S = 0$ for a ^1S term. For the minimum configuration of two electrons we have the following:

		M_S
		0
M_L	0	0^+0^-

or

		M_S
		0
M_L	0	x

Each microstate is designated by x in the second form of the table.

^2P (doublet P)

A P term has $L = 1$; therefore, M_L can have three values: +1, 0, and –1. The spin multiplicity is $2 = 2S + 1$. Therefore, $S = \frac{1}{2}$, and M_S can have two values: $+\frac{1}{2}$ and $-\frac{1}{2}$. There are six microstates in a ^2P term (3 rows × 2 columns). For the minimum case of one electron we have the following:

		M_S	
		$-\frac{1}{2}$	$+\frac{1}{2}$
	1	1^-	1^+
M_L	0	0^-	0^+
	–1	-1^-	-1^+

or

		M_S	
		$-\frac{1}{2}$	$+\frac{1}{2}$
	1	x	x
M_L	0	x	x
	–1	x	x

The spin multiplicity is equal to the number of possible values of M_S; therefore, the spin multiplicity is simply the number of columns in the microstate table.

▶ **Exercise 11.2** For each of the following free-ion terms, determine the values of L, M_L, S, and M_S. Diagram the microstate table as in the preceding examples: ^2D, ^1P, and ^2S.

At last, we are in a position to return to the p^2 microstate table and reduce it to its constituent atomic states (terms). To do this, it is sufficient to designate each microstate simply by x; it is important to tabulate the number of microstates, but it is not necessary to write out each microstate in full.

To reduce this microstate table into its component free-ion terms, note that each of the terms described in the examples and in Exercise 11.2 consists of a rectangular array of microstates. To reduce the p^2 microstate table into its terms, all that is necessary is to find the rectangular arrays. This process is illustrated in Table 11.4. For each term, the spin multiplicity is the same as the number of columns of microstates: a singlet term (such as ^1D) has a single column, a doublet term has two columns, a triplet term (such as ^3P) has three columns, and so forth.

Therefore, the p^2 electron configuration gives rise to three free-ion terms, designated ^3P, ^1D, and ^1S. These terms have different energies; they represent three states with different degrees of electron–electron interactions. For our example of a p^2 configuration for a carbon atom, the ^3P, ^1D, and ^1S terms have three distinct energies—the three major energy levels observed experimentally.

TABLE 11.4 The Microstate Table for p^2 and Its Reduction to Free-Ion Terms

M_L \ M_S	−1	0	+1
+2		x	
+1	x	x x	x
0	x	x x (x)	x
−1	x	x x	x
−2		x	

M_L \ M_S	−1	0	+1
+2		x	
+1		x	
0		x	
−1		x	
−2		x	

^1D

M_L \ M_S	−1	0	+1
+2			
+1	x	x	x
0	x	x	x
−1	x	x	x
−2			

^3P

M_L \ M_S	−1	0	+1
+2			
+1			
0		x	
−1			
−2			

^1S

NOTE: The ^1S and ^1D terms have higher energy than the ^3P terms but cannot be identified with a single electron configuration. The relative energies of higher-energy terms like these also cannot be determined by simple rules.

The final step in this procedure is to determine which term has the lowest energy. This can be done by using two of **Hund's rules**:

1. The ground term (term of lowest energy) has the highest spin multiplicity. In our example of p^2, therefore, the ground term is the ^3P. This term can be identified as having the following configuration:

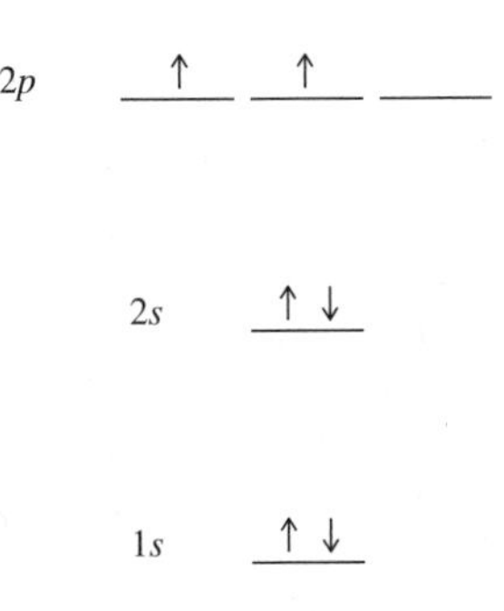

   This is sometimes called *Hund's rule of maximum multiplicity*, introduced in Section 2.2.3.
2. If two or more terms share the maximum spin multiplicity, the ground term is the one having the highest value of L. For example, if ^4P and ^4F terms are both found for an electron configuration, the ^4F has lower energy: ^4F has $L = 3$, and ^4P has $L = 1$.

EXAMPLE

Reduce the microstate table for the $s^1\,p^1$ configuration to its component free-ion terms, and identify the ground-state term.

The microstate table (prepared in the example preceding Exercise 11.1) is the sum of the microstate tables for the ^3P and ^1P terms:

^3P:

M_L \ M_S	−1	0	+1
+1	x	x	x
0	x	x	x
−1	x	x	x

^1P:

M_L \ M_S	−1	0	+1
+1		x	
0		x	
−1		x	

Hund's rule of maximum multiplicity requires ^3P as the ground state.

▶ **Exercise 11.3** In Exercise 11.1, you obtained a microstate table for the d^2 configuration. Reduce this to its component free-ion terms, and identify the ground-state term.

11.2.1 Spin-Orbit Coupling

Up to this point in the discussion of multielectron atoms, the spin and orbital angular momenta have been treated separately. An additional factor is important: the spin and orbital angular momenta couple with each other, a phenomenon known as *spin-orbit coupling*. In multielectron atoms, the S and L quantum numbers combine into the total angular momentum quantum number J. The quantum number J may have the following values:

$$J = L + S, L + S - 1, L + S - 2, \ldots, |L - S|$$

The value of J is given as a subscript.

EXAMPLE

Determine the possible values of J for the carbon terms.

For the term symbols just described for carbon, the ^1D and ^1S terms each have only one J value, whereas the ^3P term has three slightly different energies, each described by a different J. J can have only the value 0 for the ^1S term $(0 + 0)$ and only the value 2 for the ^1D term $(2 + 0)$. For the ^3P term, J can have the three values 2, 1, and 0 $(1 + 1$, $1 + 1 - 1$, and $1 + 1 - 2)$.

▶ **Exercise 11.4** Determine the possible values of J for the terms obtained from a d^2 configuration in Exercise 11.3.

Spin-orbit coupling acts to split free-ion terms into states of different energies. The ^3P term therefore splits into states of three different energies, and the total energy-level diagram for the carbon atom can be shown as follows:

LS coupling only	Spin–orbit coupling (exaggerated scale for ^3P)	Energy (cm^{-1})
^1S	$^1S_0$	21648.8
^1D	$^1D_2$	10193.7
^3P	$^3P_2$	43.5
	$^3P_1$	16.4
	$^3P_0$	0

These are the five energy states for the carbon atom referred to at the beginning of this section. The state of lowest energy (spin-orbit coupling included) can be predicted from **Hund's third rule**:

3. For subshells (such as p^2) that are less than half filled, the state having the lowest J value has the lowest energy ($^3P_0$ above); for subshells that are more than half filled, the state having the highest J value has the lowest energy. Half-filled subshells have only one possible J value.

Spin-orbit coupling can have significant effects on the electronic spectra of coordinations compounds, especially those involving fairly heavy metals (atomic number > 40). For example, Pb, the most metallic of the elements in carbon's group, has the same terms and energy level pattern as shown for carbon. However, the consequences of spin-orbit coupling for Pb are much larger: the $^3P_2$ and $^3P_1$ levels are 10,650.5 and 7,819.4 cm^{-1} respectively above the $^3P_0$ level (the $^1S_0$ is 29,466.8 cm^{-1} and the $^1D_2$ level is 21,457.9 cm^{-1} above the $^3P_0$).

11.3 ELECTRONIC SPECTRA OF COORDINATION COMPOUNDS

We can now make the connection between electron–electron interactions and the absorption spectra of coordination compounds. In Section 11.2, we considered a method for determining the microstates and free-ion terms for electron configurations. For example, a d^2 configuration gives rise to five free-ion terms—^3F, ^3P, ^1G, ^1D, and ^1S—with the ^3F term of lowest energy (Exercises 11.1 and 11.3). Absorption spectra of coordination compounds in most cases involve the d orbitals of the metal, and it is consequently important to know the free-ion terms for the possible d configurations. Determining the microstates and free-ion terms for configurations of three or more electrons can be a tedious process. For reference, therefore, these are listed for the possible d electron configurations in Table 11.5.

In the interpretation of spectra of coordination compounds, it is often important to identify the lowest-energy term. A quick and fairly simple way to do this is given here, using as an example a d^3 configuration in octahedral symmetry.

1. Sketch the energy levels, showing the d electrons.
   — —
   ↑ ↑ ↑
2. Spin multiplicity of lowest-energy state = number of unpaired electrons +1.[7]
   Spin multiplicity $= 3 + 1 = 4$
3. Determine the maximum possible value of M_L (sum of m_l values) for the configuration as shown. This determines the type of free-ion term (e.g., S, P, D).
   Maximum possible value of m_l for three electrons as shown:
   $2 + 1 + 0 = 3$
   therefore, F term
4. Combine results of Steps 2 and 3 to get the ground term.
   ^4F

TABLE 11.5 Free-Ion Terms for d^n Configurations

Configuration	Free-Ion Terms						
d^1	^2D						
d^2		$^1S\ ^1D\ ^1G$	$^3P\ ^3F$				
d^3	^2D		$^4P\ ^4F$	$^2P\ ^2D\ ^2F\ ^2G\ ^2H$			
d^4	^5D	$^1S\ ^1D\ ^1G$	$^3P\ ^3F$	$^3P\ ^3D\ ^3F\ ^3G\ ^3H$	$^1S\ ^1D\ ^1F\ ^1G\ ^1I$		
d^5	^2D		$^4P\ ^4F$	$^2P\ ^2D\ ^2F\ ^2G\ ^2H$	$^2S\ ^2D\ ^2F\ ^2G\ ^2I$	$^4D\ ^4G$	^6S
d^6	Same as d^4						
d^7	Same as d^3						
d^8	Same as d^2						
d^9	Same as d^1						
d^{10}	^1S						

NOTE: For any configuration, the free-ion terms are the sum of those listed; for example, for the d^2 configuration, the free-ion terms are $^1S + ^1D + ^1G + ^3P + ^3F$.

[7] This is equivalent to the spin multiplicity $= 2S + 1$, as shown previously.

Step 3 deserves elaboration. The maximum value of m_l for the first electron would be 2, the highest value possible for a d electron. Because the electron spins are parallel, the second electron cannot also have $m_l = 2$ (it would violate the exclusion principle); the highest value it can have is $m_l = 1$. Finally, the third electron cannot have $m_l = 2$ or 1, because it would then have the same quantum numbers as one of the first two electrons; the highest m_l value this electron could have would therefore be 0. Consequently, the maximum value of $M_L = 2 + 1 + 0 = 3$.

EXAMPLE

d^4 (low spin):

1.
   ___ ___

   ↑↓ ↑ ↑
2. Spin mulitiplicity $= 2 + 1 = 3$
3. Highest possible value of $M_L = 2 + 2 + 1 + 0 = 5$; therefore, H term.
   Note that here, $m_l = 2$ for the first two electrons does not violate the exclusion principle because the electrons have opposite spins.
4. Therefore, the ground term is ^3H.

▶ **Exercise 11.5** Determine the ground terms for high-spin and low-spin d^6 configurations in O_h symmetry.

With this review of atomic states, we may now consider the electronic states of coordination compounds and how transitions between these states can give rise to the observed spectra. Before considering specific examples of spectra, however, we must also consider which types of transitions are most probable and, therefore, give rise to the most intense absorptions.

11.3.1 Selection Rules

The relative intensities of absorption bands are governed by a series of selection rules. On the basis of the symmetry and spin multiplicity of ground and excited electronic states, two of these rules may be stated as follows:[8, 9]

1. Transitions between states of the same parity (symmetry with respect to a center of inversion) are forbidden. For example, transitions between d orbitals are forbidden ($g \longrightarrow g$ transitions; d orbitals are symmetric to inversion), but transitions between d and p orbitals are allowed ($g \longrightarrow u$ transitions; p orbitals are antisymmetric to inversion). This is known as the **Laporte selection rule**.
2. Transitions between states of different spin multiplicities are forbidden. For example, transitions between $^4A_2$ and $^4T_1$ states are "spin-allowed," but between $^4A_2$ and $^2A_2$ are "spin-forbidden." This is called the **spin selection rule**.

These rules would seem to rule out most electronic transitions for transition-metal complexes. However, many such complexes are vividly colored, a consequence

[8] B. N. Figgis and M. A. Hitchman, *Ligand Field Theory and its Applications*, Wiley-VCH, New York, 2000, pp. 181–183.
[9] B. N. Figgis, "Ligand Field Theory," in G. Wilkinson, R. D. Gillard, and J. A. McCleverty, eds., *Comprehensive Coordination Chemistry*, Vol. 1, Pergamon Press, Elmsford, NY, 1987, pp. 243–246.

of various mechanisms by which these rules can be relaxed. Some of the most important of these mechanisms are as follows:

1. The bonds in transition-metal complexes are not rigid but undergo vibrations that may temporarily change the symmetry. Octahedral complexes, for example, vibrate in ways in which the center of symmetry is temporarily lost; this phenomenon, called *vibronic coupling*, provides a way to relax the first selection rule. As a consequence, *d–d* transitions having molar absorptivities in the range of approximately 5–50 L mol^{-1} cm^{-1} commonly occur, and they are often responsible for the bright colors of many of these complexes.
2. Tetrahedral complexes often absorb more strongly than octahedral complexes of the same metal in the same oxidation state. Metal–ligand σ bonding in transition-metal complexes of T_d symmetry can be described as involving a combination of sp^3 and sd^3 hybridization of the metal orbitals; both types of hybridization are consistent with the symmetry. The mixing of *p*-orbital character (of *u* symmetry) with *d*-orbital character provides a second way of relaxing the first selection rule.
3. Spin-orbit coupling in some cases provides a mechanism of relaxing the second selection rule, with the result that transitions may be observed from a ground state of one spin multiplicity to an excited state of different spin multiplicity. Such absorption bands for first-row transition-metal complexes are usually very weak, with typical molar absorptivities less than 1 L mol^{-1} cm^{-1}. For complexes of second- and third-row transition metals, spin-orbit coupling can be more important.

Examples of spectra illustrating the selection rules and the ways in which they may be relaxed are given in the following sections of this chapter. Our first example will be a metal complex having a d^2 configuration and octahedral geometry, $[V(H_2O)_6]^{3+}$.

In discussing spectra, it will be particularly useful to be able to relate the electronic spectra of transition-metal complexes to the ligand field splitting, Δ_o for octahedral complexes. To do this it will be necessary to introduce two special types of diagrams: **correlation diagrams** and **Tanabe–Sugano diagrams**.

11.3.2 Correlation Diagrams

Figure 11.3 is an example of a correlation diagram for the configuration d^2. These diagrams make use of two extremes:

1. **Free ions (no ligand field).** In Exercise 11.4, the terms ^3F, ^3P, ^1G, ^1D, and ^1S were obtained for a d^2 configuration, with the ^3F term having the lowest energy. These terms describe the energy levels of a "free" d^2 ion—in our example, a V^{3+} ion—in the absence of any interactions with ligands. In correlation diagrams, we will show these free-ion terms on the far left.
2. **Strong ligand field.** There are three possible configurations for two *d* electrons in an octahedral ligand field:

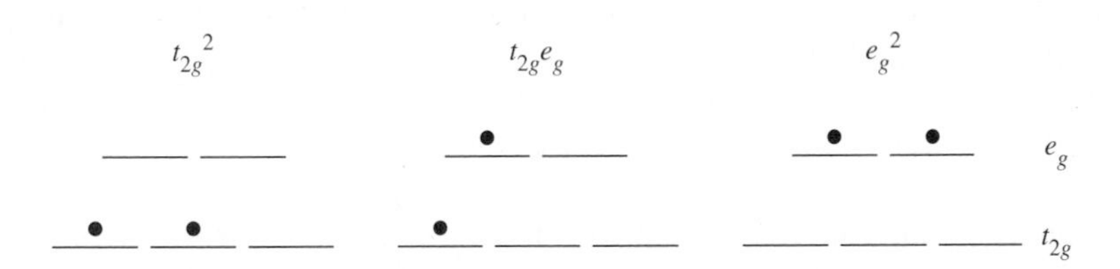

In our example, these would be the possible electron configurations of V^{3+} in an extremely strong ligand field (t_{2g}^2 would be the ground state; the others would be excited states). In correlation diagrams, we will show these states on the far right as the

FIGURE 11.3 Correlation Diagram for d^2 in Octahedral Ligand Field.

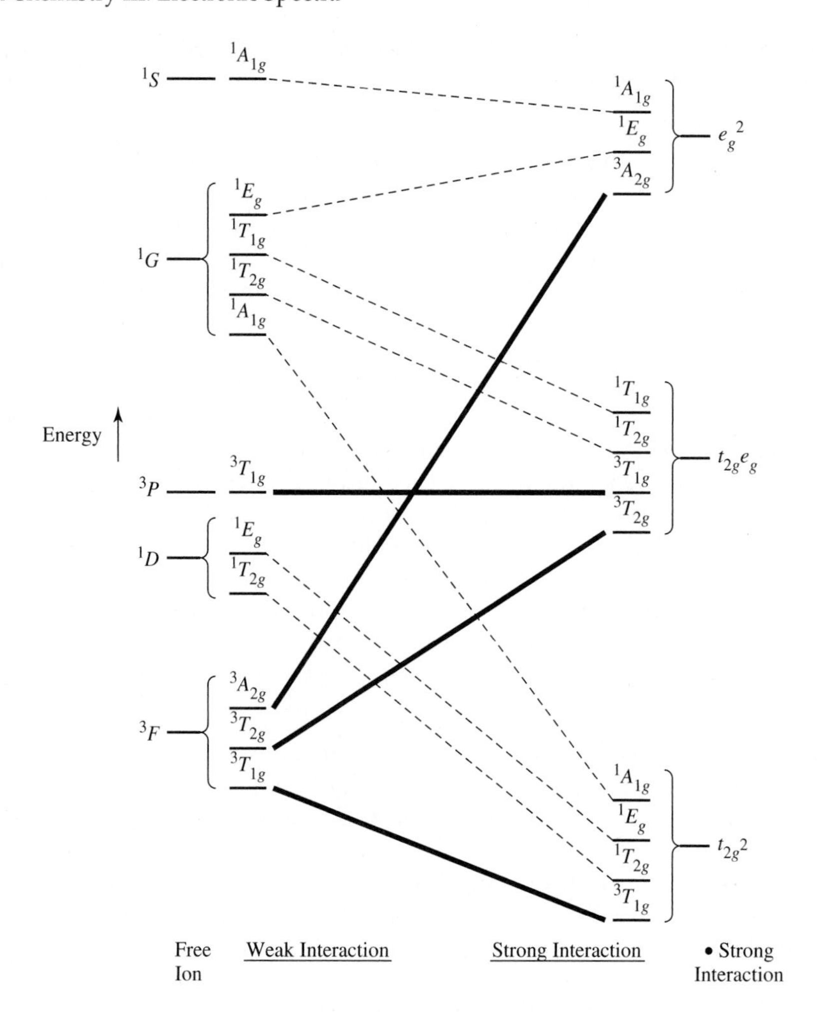

strong-field limit. Here, the effect of the ligands is so strong that it completely overrides the effects of *LS* coupling.

In actual coordination compounds, the situation is intermediate between these extremes. At zero field, the m_l and m_s values of the individual electrons couple to form, for d^2, the five terms ^3F, ^3P, ^1G, ^1D, and ^1S, representing five atomic states with different energies. At a very high ligand field strength, the $t_{2g}{}^2$, $t_{2g}\,e_g$, and $e_g{}^2$ configurations predominate. The correlation diagram shows the full range of in-between cases in which both factors are important.

Some details of the method for achieving this are beyond the scope of this text; the interested reader should consult the literature[10] for details omitted here. The aspect of this problem that is important to us is that free-ion terms, shown on the far left in the correlation diagrams, have symmetry characteristics that enable them to be reduced to their constituent irreducible representations; in our example, these will be irreducible representations in the O_h point group. In an octahedral ligand field, the free-ion terms will be split into states corresponding to the irreducible representations, as shown in Table 11.6.

Similarly, irreducible representations may be obtained for the strong-field limit configurations (in our example, $t_{2g}{}^2$, $t_{2g}\,e_g$, and $e_g{}^2$). The irreducible representations for the two limiting situations must match; each irreducible representation for the free ion

[10] F. A. Cotton, *Chemical Applications of Group Theory*, 3rd ed., Wiley InterScience, New York, 1990, Chapter 9, pp. 253–303.

TABLE 11.6 Splitting of Free-Ion Terms in Octahedral Symmetry

Term	Irreducible Representations
S	A_{1g}
P	T_{1g}
D	$E_g + T_{2g}$
F	$A_{2g} + T_{1g} + T_{2g}$
G	$A_{1g} + E_g + T_{1g} + T_{2g}$
H	$E_g + 2T_{1g} + T_{2g}$
I	$A_{1g} + A_{2g} + E_g + T_{1g} + 2T_{2g}$

NOTE: Although representations based on atomic orbitals may have either g or u symmetry, the terms given here are for d orbitals and as a result have only g symmetry. See F. A. Cotton, *Chemical Applications of Group Theory* (3rd ed., Wiley InterScience, New York, 1990, pp. 263–264) for a discussion of these labels.

must match, or correlate with, a representation for the strong-field limit. This is shown in the correlation diagram for d^2 in Figure 11.3.

Note especially the following characteristics of this correlation diagram:

1. The free-ion states (terms arising from LS coupling) are shown on the far left.
2. The extremely strong field states are shown on the far right.
3. Both the free-ion and strong-field states can be reduced to irreducible representations, as shown. Each free-ion irreducible representation is matched with (correlates with) a strong-field irreducible representation having the same symmetry (same label). As mentioned in Section 11.3.1, transitions to excited states having the same spin multiplicity as the ground state are more likely than transitions to states of different spin multiplicity. To emphasize this, the ground state and states of the same spin multiplicity as the ground state are shown as heavy lines, and states having other spin multiplicities are shown as dashed lines.

In the correlation diagram, the states are shown in order of energy. A noncrossing rule is observed: lines connecting states of the same symmetry designation do not cross. Correlation diagrams are available for other d-electron configurations.[11]

11.3.3 Tanabe–Sugano Diagrams

Tanabe–Sugano diagrams are special correlation diagrams that are particularly useful in the interpretation of electronic spectra of coordination compounds.[12] In Tanabe–Sugano diagrams, the lowest-energy state is plotted along the horizontal axis; consequently, the vertical distance above this axis is a measure of the energy of the excited state above the ground state. For example, for the d^2 configuration, the lowest-energy state is described by the line in the correlation diagram (Figure 11.3) joining the $^3T_{1g}$ state arising from the ^3F free-ion term with the $^3T_{1g}$ state arising from the strong-field term, $t_{2g}{}^2$. In the Tanabe–Sugano diagram (Figure 11.4), this line is made horizontal; it is labeled $^3T_{1g}$ (F) and is shown to arise from the ^3F term in the free-ion limit (left side of diagram).[13]

[11] B. N. Figgis and M. A. Hitchman, *Ligand Field Theory and Its Applications*, Wiley-VCH, New York, 2000, pp. 128–134.

[12] Y. Tanabe and S. Sugano, *J. Phys. Soc. Japan*, **1954**, *9*, 766.

[13] The F in parentheses distinguishes this $^3T_{1g}$ term from the higher energy $^3T_{1g}$ term arising from the ^3P term in the free-ion limit.

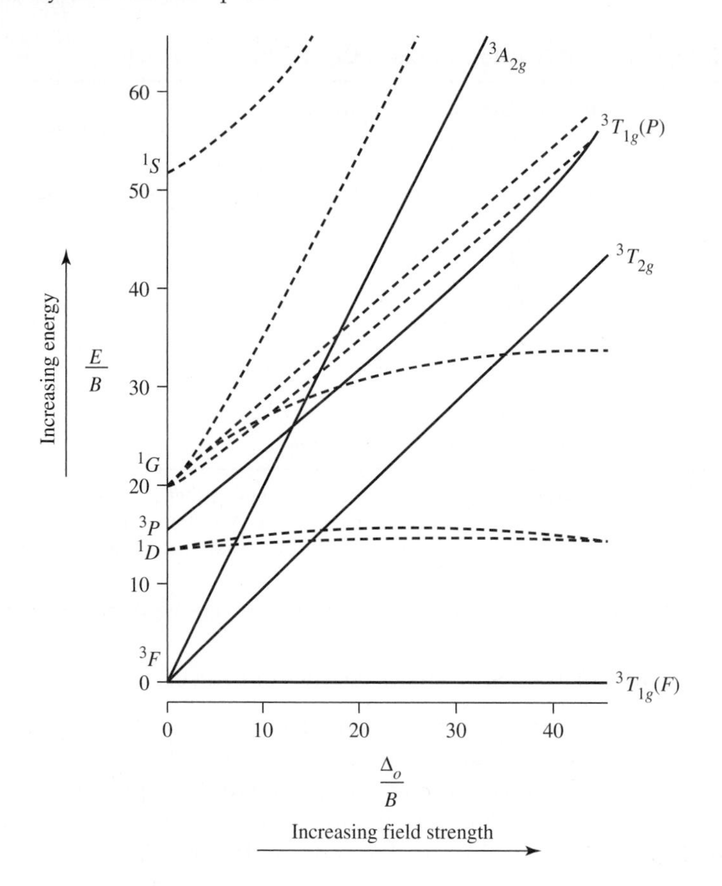

FIGURE 11.4 Tanabe–Sugano Diagram for d^2 in Octahedral Ligand Field.

The Tanabe–Sugano diagram also shows excited states. In the d^2 diagram, the excited states of the same spin multiplicity as the ground state are the ${}^3T_{2g}$, ${}^3T_{1g}$ (P), and the ${}^3A_{2g}$. The reader should verify that these are the same triplet excited states shown in the d^2 correlation diagram. Excited states of other spin multiplicities are also shown; but, as we will see, they are generally not as important in the interpretation of spectra.

The quantities plotted in a Tanabe–Sugano diagram are as follows:

Horizontal axis: $\frac{\Delta_o}{B}$ where Δ_o is the octahedral ligand field splitting, described in Chapter 10.

$B =$ Racah parameter, a measure of the repulsion between terms of the same multiplicity. For d^2, for example, the energy difference between ${}^3F$ and ${}^3P$ is 15B.

(${}^3P$ — 15 B — ${}^3F$)

Vertical axis: $\frac{E}{B}$ where E is the energy (of excited states) above the ground state.

As mentioned, one of the most useful characteristics of Tanabe–Sugano diagrams is that *the ground electronic state is always plotted along the horizontal axis*; this makes it easy to determine values of E/B above the ground state.[14]

[14] For a discussion of Racah parameters, see B.N. Figgis, "Ligand Field Theory," in *Comprehensive Coordination Chemistry*, Pergamon Press, Elmsford, NY, 1987 Vol. 1, p. 232.

EXAMPLE

$[V(H_2O)_6]^{3+}$ (d^2)

A good example of the utility of Tanabe–Sugano diagrams in explaining electronic spectra is provided by the d^2 complex $[V(H_2O)_6]^{3+}$. The ground state is $^3T_{1g}$ (F); under ordinary conditions, this is the only electronic state that is appreciably occupied. Absorption of light should occur primarily to excited states also having a spin multiplicity of 3. There are three of these: $^3T_{2g}$, $^3T_{1g}$ (P), and $^3A_{2g}$. Therefore, three allowed transitions are expected, as shown in Figure 11.5. Consequently, we expect three absorption bands for $[V(H_2O)_6]^{3+}$, one corresponding to each allowed transition. Is this actually observed for $[V(H_2O)_6]^{3+}$? Two bands are readily observed at 17,800 and 25,700 cm^{-1}, as can be seen in Figure 11.6.[15] A third band, at approximately 38,000 cm^{-1}, is apparently obscured in aqueous solution by charge-transfer bands nearby (charge-transfer bands of coordination compounds will be discussed later in this chapter). In the solid state, however, a band attributed to the $^3T_{1g} \longrightarrow {}^3A_{2g}$ transition is observed at 38,000 cm^{-1}. These bands match the transitions ν_1, ν_2, and ν_3 indicated on the Tanabe–Sugano diagram (Figure 11.5).

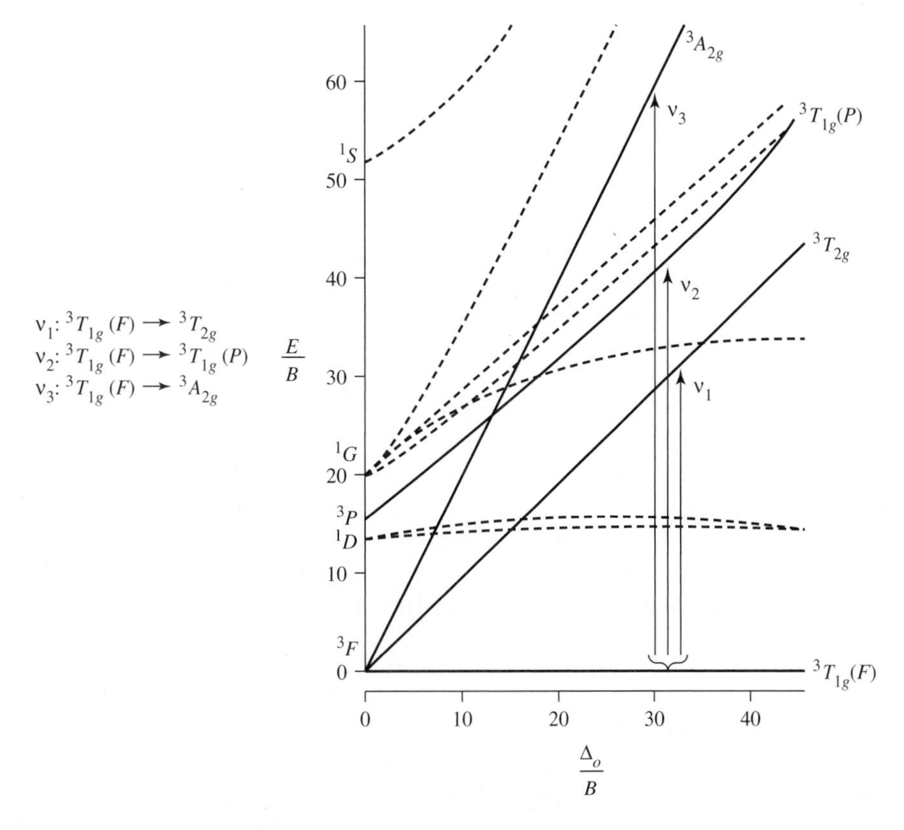

FIGURE 11.5 Spin-Allowed Transitions for d^2 Configuration.

OTHER ELECTRON CONFIGURATIONS Tanabe–Sugano diagrams for d^2 through d^8 are shown in Figure 11.7. The cases of d^1 and d^9 configurations will be discussed in Section 11.3.4. The diagrams for d^4, d^5, d^6, and d^7 have apparent discontinuities,

[15] The third band is in the ultraviolet and is off-scale to the right in the spectrum shown; see B. N. Figgis, *Introduction to Ligand Fields*, Wiley InterScience, New York, 1966, p. 219.

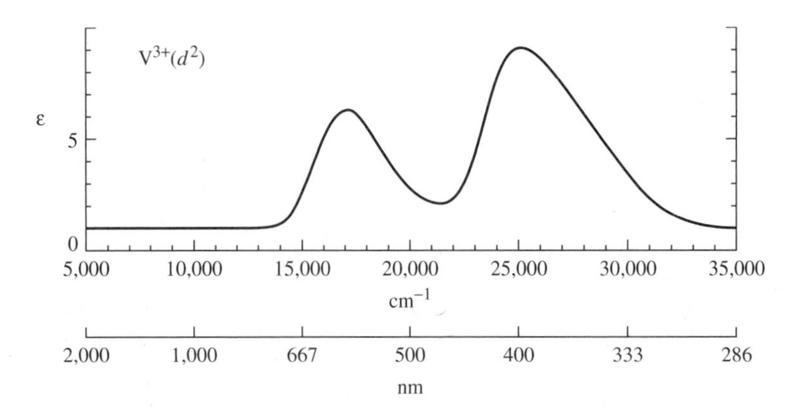

FIGURE 11.6 Absorption Spectrum of $[V(H_2O)_6]^{3+}$. (Reproduced with permission from B. N. Figgis, *Introduction to Ligand Fields*, Wiley InterScience, New York, 1966, p. 221.)

marked by vertical lines near the center. These are configurations for which low spin and high spin are both possible. For example, consider the configuration d^4:

High-spin (weak-field) d^4 has four unpaired electrons of parallel spin; such a configuration has a spin multiplicity of 5.

$S = 4(\frac{1}{2}) = 2$;
spin multiplicity $= 2S + 1 = 2(2) + 1 = 5$

Low-spin (strong-field) d^4, on the other hand, has only two unpaired electrons and a spin multiplicity of 3.

$S = 2(\frac{1}{2}) = 1$;
spin multiplicity $= 2S + 1 = 2(1) + 1 = 3$

In the weak-field part of the Tanabe–Sugano diagram (left of $\Delta_o / B = 27$), the ground state is ^5E_g, with the expected spin multiplicity of 5. On the right (strong-field) side of the diagram, the ground state is $^3T_{1g}$ (correlating with the ^3H term in the free-ion limit), with the required spin multiplicity of 3. The vertical line is thus a dividing line between weak- and strong-field cases: high-spin (weak-field) complexes are to the left of this line, and low-spin (strong-field) complexes are to the right. At the dividing line, the ground state changes from ^5E_g to $^3T_{1g}$. The spin multiplicity changes from 5 to 3 to reflect the change in the number of unpaired electrons.

Figure 11.8 shows absorption spectra of first-row transition-metal complexes of the formula $[M(H_2O)_6]^{n+}$. Because water is a rather weak-field ligand, these are all high-spin complexes, represented by the left side of the Tanabe–Sugano diagrams. It is an interesting exercise to compare the number of bands in these spectra with the number of bands expected from the respective Tanabe–Sugano diagrams. Note that in some cases absorption bands are off-scale, farther into the ultraviolet than the spectral region shown.

In Figure 11.8, molar absorptivities (extinction coefficients) are shown on the vertical scale. The absorptivities for most bands are similar (1 to 20 L mol^{-1} cm^{-1}) except for the spectrum of $[Mn(H_2O)_6]^{2+}$, which has much weaker bands. Solutions of $[Mn(H_2O)_6]^{2+}$ are an extremely pale pink, much more weakly colored than solutions of the other ions shown. Why is absorption by $[Mn(H_2O)_6]^{2+}$ so weak? To answer this question, it is useful to examine the corresponding Tanabe–Sugano diagram, in this case for a d^5 configuration. We expect $[Mn(H_2O)_6]^{2+}$ to be a high-spin complex, because H_2O is a relatively weak-field ligand. The ground state for weak-field d^5 is the $^6A_{1g}$. There are no excited states of the same spin multiplicity (6), and consequently there can be no spin-allowed absorptions. That $[Mn(H_2O)_6]^{2+}$ is colored at all is a consequence of very weak forbidden transitions to excited states of spin multiplicity other than 6 (there are many such excited states, hence the rather complicated spectrum).

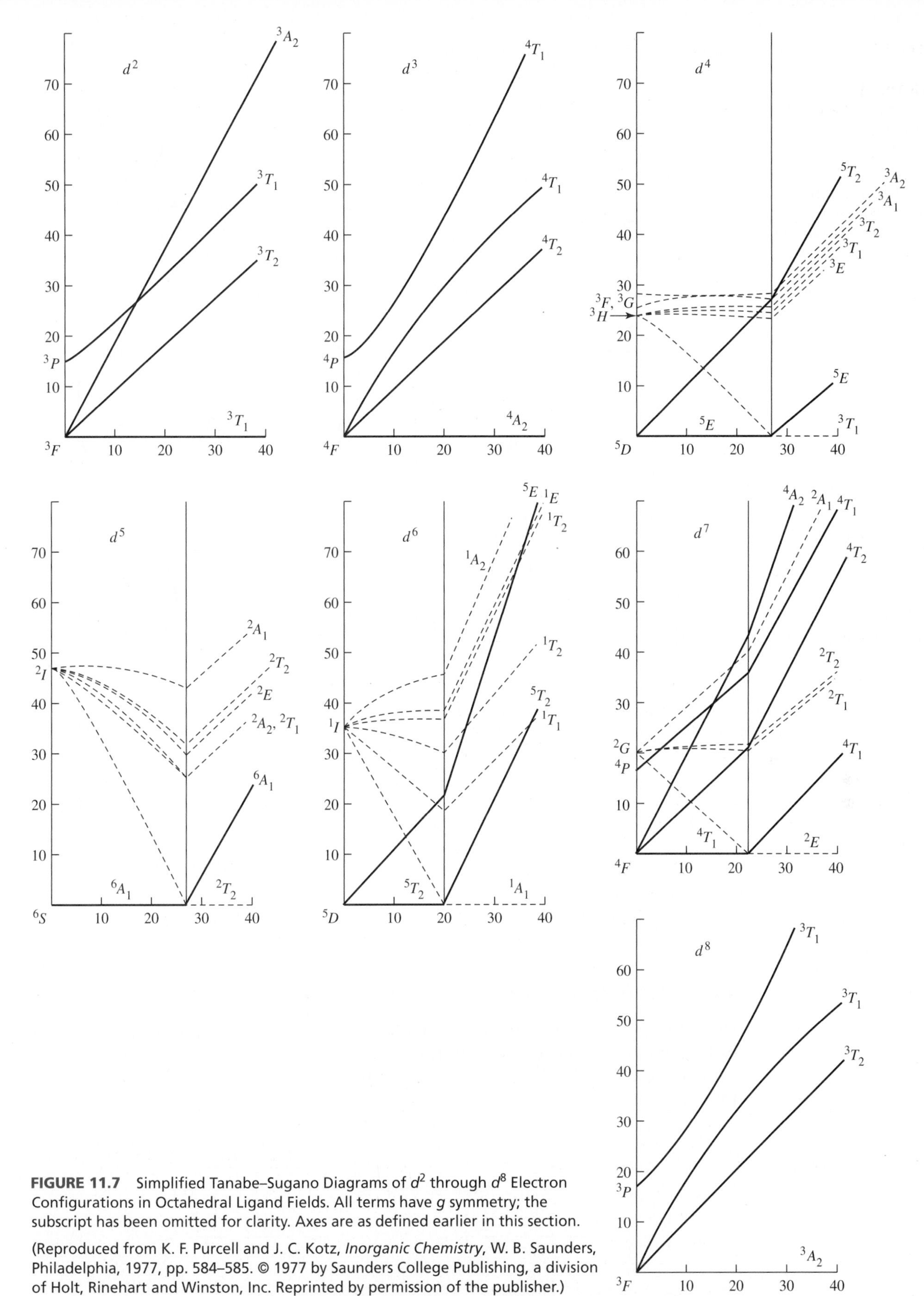

FIGURE 11.7 Simplified Tanabe–Sugano Diagrams of d^2 through d^8 Electron Configurations in Octahedral Ligand Fields. All terms have *g* symmetry; the subscript has been omitted for clarity. Axes are as defined earlier in this section.

(Reproduced from K. F. Purcell and J. C. Kotz, *Inorganic Chemistry*, W. B. Saunders, Philadelphia, 1977, pp. 584–585. © 1977 by Saunders College Publishing, a division of Holt, Rinehart and Winston, Inc. Reprinted by permission of the publisher.)

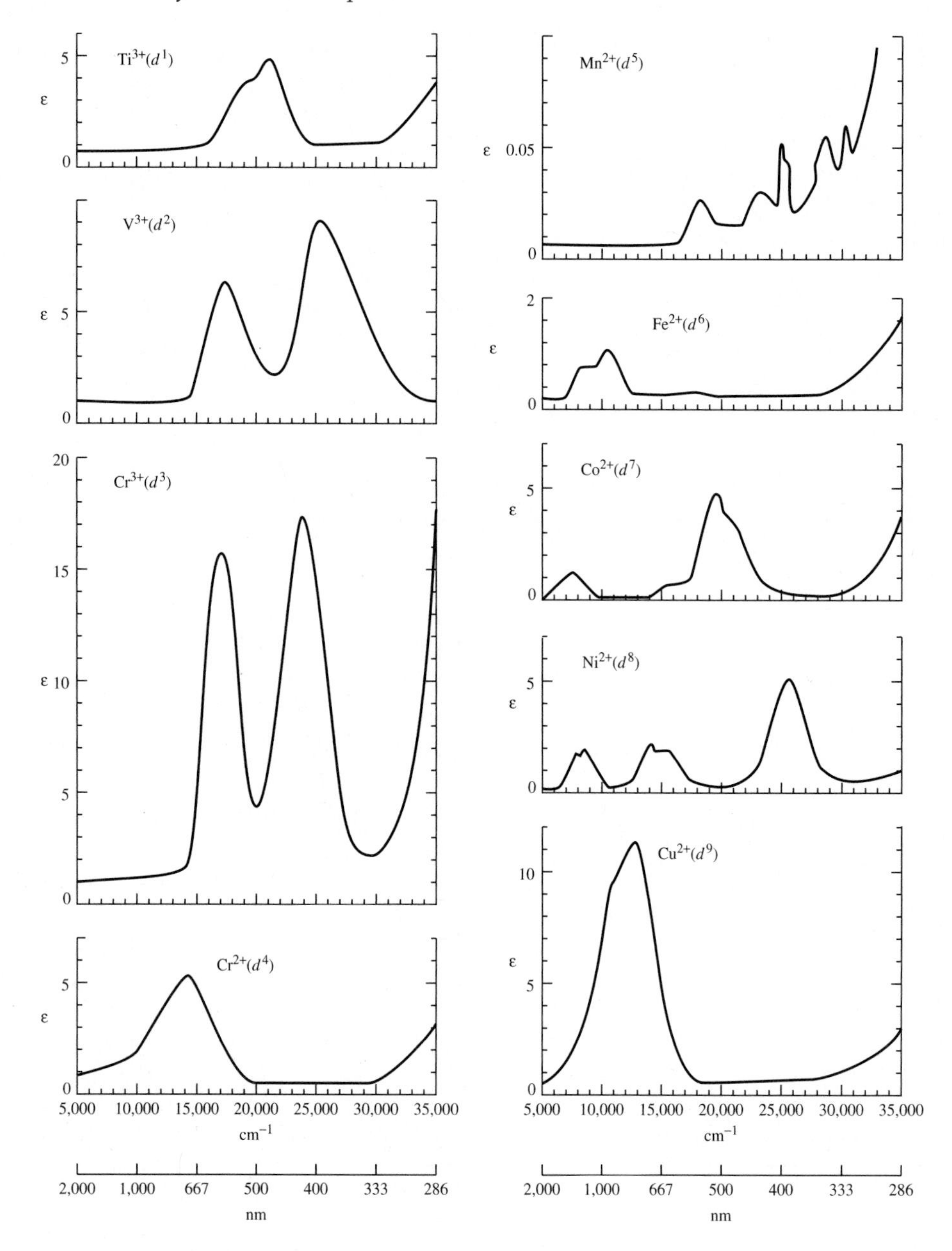

FIGURE 11.8 Electronic Spectra of First-Row Transition-Metal Complexes of Formula $[M(H_2O)_6]^{n+}$. (Reproduced with permission from B. N. Figgis, *Introduction to Ligand Fields*, Wiley InterScience, New York, 1966, pp. 221, 224.)

11.3.4 Jahn–Teller Distortions and Spectra

Up to this point, we have not discussed the spectra of d^1 and d^9 complexes. By virtue of the simple *d*-electron configurations for these cases, we might expect each to exhibit one absorption band corresponding to excitation of an electron from the t_{2g} to the e_g levels:

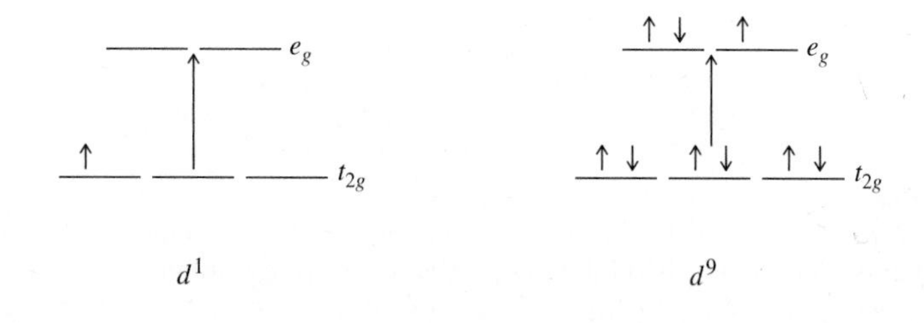

However, this view must be at least a modest oversimplification, because examination of the spectra of $[Ti(H_2O)_6]^{3+}$ (d^1) and $[Cu(H_2O)_6]^{2+}$ (d^9), shown in Figure 11.8, shows these coordination compounds to exhibit two closely overlapping absorption bands rather than a single band.

To account for the apparent splitting of bands in these examples, it is necessary to recall that, as described in Section 10.5, some configurations can cause complexes to be distorted. In 1937, Jahn and Teller showed that nonlinear molecules having a degenerate electronic state should distort to lower the symmetry of the molecule and to reduce the degeneracy; this is commonly called the **Jahn–Teller theorem.**[16] For example, a d^9 metal in an octahedral complex has the electron configuration $t_{2g}{}^6 e_g{}^3$; according to the Jahn–Teller theorem, such a complex should distort. If the distortion takes the form of an elongation along the z axis, the most common distortion observed experimentally, the t_{2g} and e_g orbitals are affected as shown in Figure 11.9. Distortion from O_h to D_{4h} symmetry results in stabilization of the molecule: the e_g pair of orbitals is split into a lower a_{1g} level and a higher b_{1g} level.

When degenerate orbitals are asymmetrically occupied, Jahn–Teller distortions are likely. For example, the first two configurations below should give distortions, but the third and fourth should not:

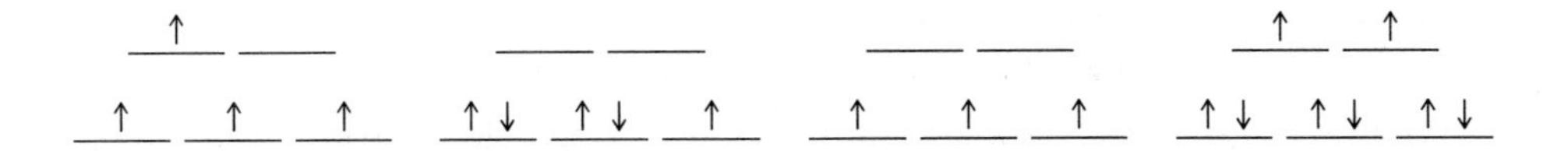

In practice, the only electron configurations for O_h symmetry that give rise to measurable Jahn–Teller distortions are those that have asymmetrically occupied e_g orbitals, such as the high-spin d^4 configuration. The Jahn–Teller theorem does not predict what the distortion will be; by far, the most common distortion observed is elongation along the z axis. Although the Jahn–Teller theorem predicts that configurations having asymmetrically occupied t_{2g} orbitals, such as the low-spin d^5 configuration, should also be distorted, such distortions are too small to be measured in most cases.

The Jahn–Teller effect on spectra can easily be seen from the example of $[Cu(H_2O)_6]^{2+}$, a d^9 complex. From Figure 11.9, which shows the effect on d orbitals of distortion from O_h to D_{4h} geometry, we can see the additional splitting of orbitals accompanying the reduction of symmetry.

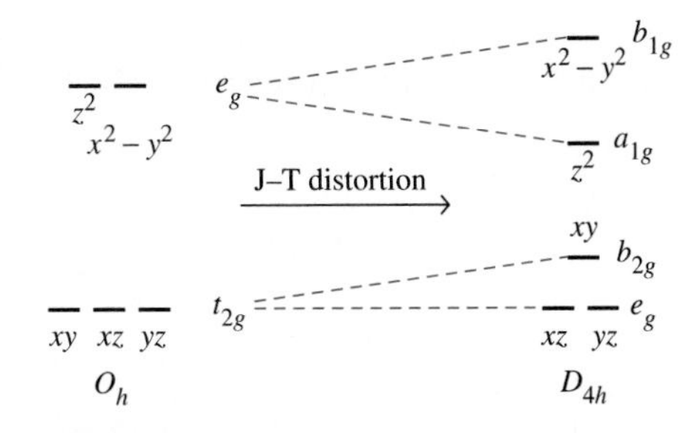

FIGURE 11.9 Effect of Jahn–Teller Distortion on d Orbitals of an Octahedral Complex.

[16] B. Bersucker, *Coord. Chem. Rev.*, **1975**, *14*, 357.

SYMMETRY LABELS FOR CONFIGURATIONS Electron *configurations* have symmetry labels that match their degeneracies, as follows:

		Examples	
T	Designates a triply degenerate asymmetrically occupied state.	__ __ • __ __	• • •• •• •
E	Designates a doubly degenerate asymmetrically occupied state.	• __ • • •	•• • •• •• ••
A or *B*	Designate a nondegenerate state. Each set of levels in an *A* or *B* state is symmetrically occupied.	__ __ • • •	• • • • •

▶ **Exercise 11.6** Identify the following configurations as *T*, *A*, or *E* states in octahedral complexes:

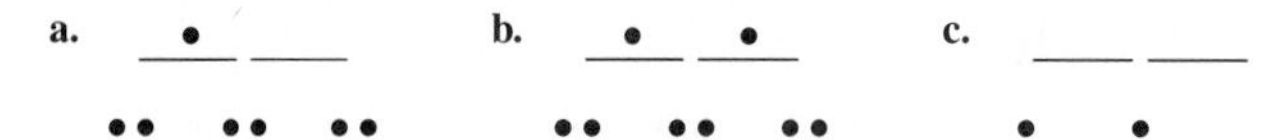

When a ^2D term for d^9 is split by an octahedral ligand field, two configurations result:

e_g •• •   •• ••

t_{2g} •• •• ••   •• •• •

Lower energy   Higher energy

The lower energy configuration is doubly degenerate in the e_g orbitals (occupation of the e_g orbitals could be •• • or • ••) and has the designation ^2E_g; the higher energy configuration is triply degenerate in the t_{2g} levels (three arrangements are possible in these levels: •• • ••, •• •• •, or • •• ••) and has the designation $^2T_{2g}$. Thus, the lower-energy configuration is the ^2E_g, and the higher-energy configuration is the $^2T_{2g}$, as in Figure 11.10. This is opposite to the order of energies of the orbitals (t_{2g} lower than e_g), shown in Figure 11.9.

Similarly, for distortion to D_{4h}, the order of labels of the orbitals in Figure 11.9 is the reverse of the order of labels of the energy configurations in Figure 11.10.

In summary, the ^2D free-ion term is split into ^2E_g and $^2T_{2g}$ by a field of O_h symmetry, and it is further split on distortion to D_{4h} symmetry. The labels of the states resulting

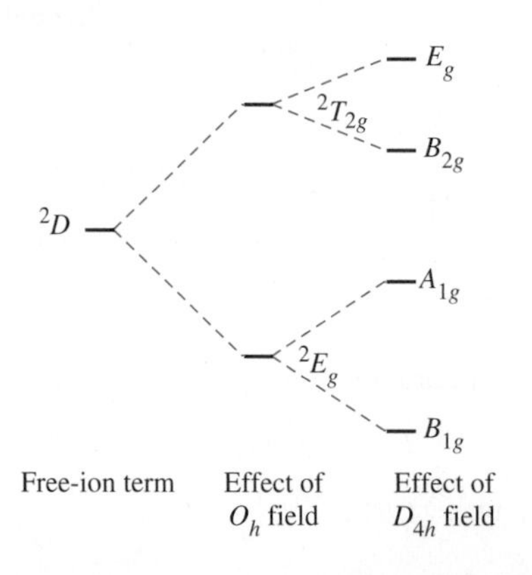

FIGURE 11.10 Splitting of Octahedral Free-Ion Terms on Jahn–Teller Distortion for d^9 Configuration.

from the free-ion term (Figure 11.10) are in reverse order to the labels on the orbitals; for example, the b_{1g} atomic orbital is of highest energy, whereas the B_{1g} state originating from the ^2D free-ion term is of lowest energy.[17]

For a d^9 configuration, the ground state in octahedral symmetry is a ^2E_g term, and the excited state is a $^2T_{2g}$ term. On distortion to D_{4h} geometry, these terms split, as shown in Figure 11.10. In an octahedral d^9 complex, we would expect excitation from the ^2E_g state to the $^2T_{2g}$ state and a single absorption band. Distortion of the complex to D_{4h} geometry splits the $^2T_{2g}$ level into two levels, the E_g and the B_{2g}. Excitation can now occur from the ground state (now the B_{1g} state) to the A_{1g}, the E_g, or the B_{2g} states (the splitting is exaggerated in Figure 11.10). The $B_{1g} \longrightarrow A_{1g}$ transition is too low in energy to be observed in the visible spectrum. If the distortion is strong enough, therefore, two separate absorption bands may be observed in the visible region, to the E_g or the B_{2g} levels (or a broadened or narrowly split peak is found, as in $[Cu(H_2O)_6]^{2+}$).

For a d^1 complex, a single absorption band, corresponding to excitation of a t_{2g} electron to an e_g orbital, might be expected:

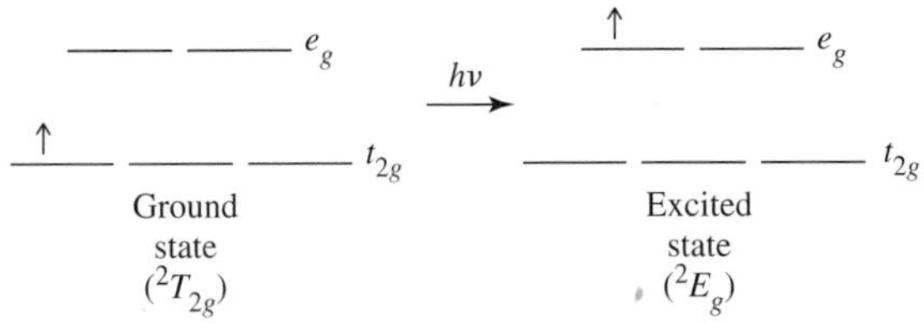

e_g $\quad$ $h\nu \longrightarrow$ $\quad$ e_g

t_{2g} $\quad\quad$ t_{2g}

Ground state ($^2T_{2g}$) $\quad$ Excited state (^2E_g)

However, the spectrum of $[Ti(H_2O)_6]^{3+}$, an example of a d^1 complex, shows two apparently overlapping bands rather than a single band. How is this possible?

One explanation commonly used is that the excited state can undergo Jahn–Teller distortion,[18] as in Figure 11.10. As in the examples considered previously, asymmetric occupation of the e_g orbitals can split these orbitals into two of slightly different energy (of A_{1g} and B_{1g} symmetry). Excitation can now occur from the t_{2g} level to either of these orbitals. Therefore, as in the case of the d^9 configuration, there are now two excited states of slightly different energy. The consequence may be a broadening of a spectrum into a two-humped peak, as in $[Ti(H_2O)_6]^{3+}$ or, in some cases, into two more clearly defined separate peaks.[19]

One additional point needs to be made in regard to Tanabe–Sugano diagrams. These diagrams, as shown in Figure 11.7, assume O_h symmetry in excited states as well as in ground states. The consequence is that the diagrams are useful in predicting the general properties of spectra; in fact, many complexes do have sharply defined bands that fit the Tanabe–Sugano description well (see the d^2, d^3, and d^4 examples in Figure 11.7). However, distortions from pure octahedral symmetry are rather common, and the consequence can be the splitting of bands—or, in some cases of severe distortion, situations in which the bands are difficult to interpret. Additional examples of spectra showing the splitting of absorption bands can be seen in Figure 11.8.

▶ **Exercise 11.7** $[Fe(H_2O)_6]^{2+}$ has a two-humped absorption peak near 1000 nm. By using the appropriate Tanabe–Sugano diagram, account for the most likely origin of this absorption. Also account for the splitting of the absorption band.

---

[17] B.N. Figgis, "Ligand Field Theory," in *Comprehensive Coordination Chemistry*, Vol. 1, pp. 252–253.

[18] C. J. Ballhausen, *Introduction to Ligand Field Theory*, McGraw-Hill, New York, 1962, p. 227, and references therein.

[19] F. A. Cotton and G. Wilkinson, *Advanced Inorganic Chemistry*, 4th ed., Wiley InterScience, New York, 1980, pp. 680–681.

11.3.5 Examples of Applications of Tanabe–Sugano Diagrams: Determining Δ_o from Spectra

Absorption spectra of coordination compounds can be used to determine the magnitude of the ligand field splitting, which is Δ_o for octahedral complexes. It should be made clear from the outset that the accuracy with which Δ_o can be determined is to some extent limited by the mathematical tools used to solve the problem. Absorption spectra often have overlapping bands; to determine the positions of the bands accurately, therefore, requires an appropriate mathematical technique for reducing overlapping bands into their individual components. Such analysis is beyond the scope of this text. However, we can often obtain Δ_o values (and sometimes values of the Racah parameter, B) of reasonable accuracy simply by using the positions of the absorption maxima taken directly from the spectra.

The ease with which Δ_o can be determined depends on the d-electron configuration of the metal; in some cases, Δ_o can be read easily from a spectrum, but in other cases a more complicated analysis is necessary. The following discussion will proceed from the simplest cases to the most complicated.

d^1, d^4 (HIGH SPIN), d^6 (HIGH SPIN), d^9 Each of these cases, as shown in Figure 11.11, corresponds to a simple excitation of an electron from a t_{2g} to an e_g orbital, with the final (excited) electron configuration having the same spin multiplicity as the initial configuration. In each case, there is a single excited state of the same spin multiplicity as the ground state. Consequently, there is a single spin-allowed absorption, with the energy of the absorbed light equal to Δ_o. Examples of such complexes include $[Ti(H_2O)_6]^{3+}$, $[Cr(H_2O)_6]^{2+}$, $[Fe(H_2O)_6]^{2+}$, and $[Cu(H_2O)_6]^{2+}$; note from Figure 11.8 that each of these complexes exhibits essentially a single absorption band. In some cases, splitting of bands due to Jahn–Teller distortion is observed, as discussed in Section 11.3.4.

d^3, d^8 These electron configurations have a ground-state F term. In an octahedral ligand field, an F term splits into three terms: an A_{2g}, a T_{2g}, and a T_{1g}. As shown in Figure 11.12, the A_{2g} is of lowest energy for d^3 or d^8. For these configurations, the difference in energy between the two lowest-energy terms, the A_{2g} and the T_{2g}, is equal to Δ_o. Therefore, to find Δ_o, we simply find the energy of the lowest-energy transition in the absorption spectrum. Examples include $[Cr(H_2O)_6]^{3+}$ and $[Ni(H_2O)_6]^{2+}$. In each case,

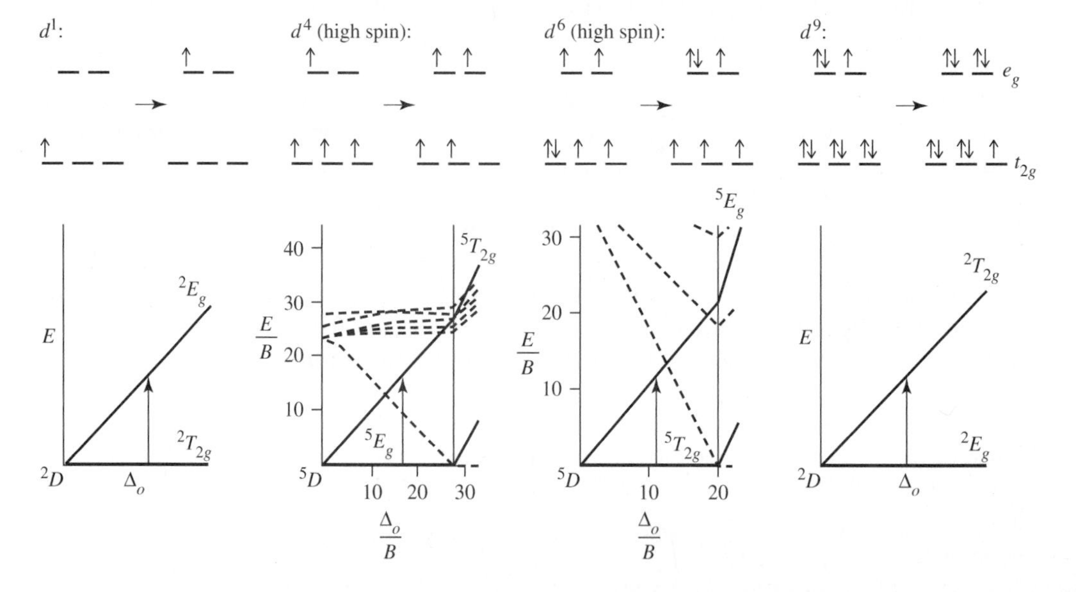

FIGURE 11.11 Determining Δ_o for d^1, d^4 (High Spin), d^6 (High Spin), and d^9 Configurations.

the lowest-energy band in the spectra of these complexes (Figure 11.8) is for the transition from the ${}^4A_{2g}$ ground state to the ${}^4T_{2g}$ excited state. The energies of these bands, approximately 17,500 and 8,500 cm^{-1}, respectively, are the corresponding values of Δ_o.

d^2, d^7 (HIGH SPIN) As in the case of d^3 and d^8, the ground free-ion terms for these two configurations are *F* terms. However, the determination of Δ_o is not as simple for d^2 and d^7. To explain this, it is necessary to take a close look at the Tanabe–Sugano diagrams. We will compare the d^3 and d^2 Tanabe–Sugano diagrams; the d^8 and d^7 (high-spin) cases can be compared in a similar fashion (note the similarity of the d^3 and d^8 Tanabe–Sugano diagrams and of the d^2 and d^7 [high-spin region] diagrams).

In the d^3 case, the ground state is a ${}^4A_{2g}$ state. There are three excited quartet states: ${}^4T_{2g}$, ${}^4T_{1g}$ (from ${}^4F$ term), and ${}^4T_{1g}$ (from ${}^4P$ term). Note the two states of the same symmetry (${}^4T_{1g}$). An important property of such states is that states of the same symmetry may mix. The consequence of such mixing is that, as the ligand field is increased, the states appear to repel each other; the lines in the Tanabe–Sugano diagram curve away from each other. This effect can easily be seen in the diagram for d^3 (see Figure 11.7). However, this causes no difficulty in obtaining Δ_o for a d^3 complex, because the lowest-energy transition (${}^4A_{2g} \longrightarrow {}^4T_{2g}$) is not affected by such curvature. (The Tanabe–Sugano diagram shows that the energy of the ${}^4T_{2g}$ state varies linearly with the strength of the ligand field.)

The situation in the d^2 case is not quite as simple. For d^2, the free-ion ${}^3F$ term is also split into ${}^3T_{1g} + {}^3T_{2g} + {}^3A_{2g}$; these are the same states obtained from d^3, but in reverse order (Figure 11.12). For d^2, the ground state is ${}^3T_{1g}$. It is tempting to simply determine the energy of the ${}^3T_{1g}\,(F) \longrightarrow {}^3T_{2g}$ band and assign this as the value of Δ_o. After all, the ${}^3T_{1g}\,(F)$ can be identified with the configuration $t_{2g}{}^2$ (see correlation diagram, Figure 11.3), and ${}^3T_{2g}$ with the configuration $t_{2g}\,e_g$; the difference between these states should give Δ_o. However, the ${}^3T_{1g}\,(F)$ state can mix with the ${}^3T_{1g}$ state arising from the ${}^3P$ free-ion term, causing a slight curvature of both in the Tanabe–Sugano diagram. This curvature can lead to some error in using the ground state to obtain values of Δ_o.

Therefore, we must resort to an alternative: to determine the difference in energy between the $t_{2g}\,e_g$ and $e_g{}^2$ configurations, which should also be equal to Δ_o (because the energy necessary to excite a single electron from a t_{2g} to an e_g orbital is equal to Δ_o). This means that we can use the difference between ${}^3T_{2g}$ (for the $t_{2g}\,e_g$ configuration) and ${}^3A_{2g}$ (for $e_g{}^2$; see Figure 11.3) to calculate Δ_o:

$$\begin{array}{rl} \text{energy of transition } {}^3T_{1g} \longrightarrow & {}^3A_{2g} \\ -\ \text{energy of transition } {}^3T_{1g} \longrightarrow & {}^3T_{2g} \\ \hline \Delta_o = \text{energy difference between } {}^3A_{2g} \text{ and } & {}^3T_{2g} \end{array}$$

(see Figure 11.13)

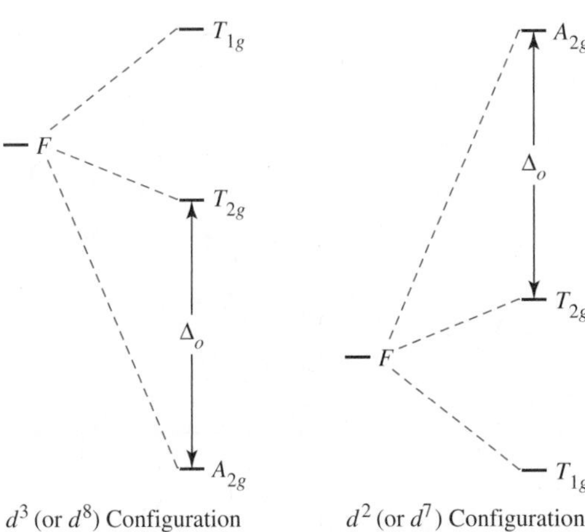

FIGURE 11.12 Splitting of *F* Terms in Octahedral Symmetry.

The difficulty with this approach is that two lines cross in the Tanabe–Sugano diagram. Therefore, the assignment of the absorption bands may be in question. From the diagram for d^2, we can see that although the lowest energy absorption band (to $^3T_{2g}$) is easily assigned, there are two possibilities for the next band: to $^3A_{2g}$ for very weak field ligands, or to $^3T_{1g}$ (*P*) for stronger-field ligands. In addition, the second and third absorption bands may overlap, making it difficult to determine the exact positions of the bands (the apparent positions of absorption maxima may be shifted if the bands overlap). In such cases a more complicated analysis—involving a calculation of the Racah parameter, *B*—may be necessary. This procedure is best illustrated by the following example.

EXAMPLE

$[V(H_2O)_6]^{3+}$ has absorption bands at 17,800 and 25,700 cm^{-1}. Using the Tanabe–Sugano diagram for d^2, estimate values of Δ_o and *B* for this complex.

From the Tanabe–Sugano diagram there are three possible spin-allowed transitions (Figure 11.13):

$^3T_{1g}(F) \longrightarrow {}^3T_{2g}(F)$ $\quad \nu_1$ $\quad$ (lowest energy)

$^3T_{1g}(F) \longrightarrow {}^3T_{1g}(P)$ $\quad \nu_2$
$^3T_{1g}(F) \longrightarrow {}^3A_{2g}(F)$ $\quad \nu_3$ $\quad$ (one of these must be the higher-energy band)

When working with spectra, it is often useful to determine the ratio of energies of the absorption bands. In this example,

$$\frac{25{,}700 \text{ cm}^{-1}}{17{,}800 \text{ cm}^{-1}} = 1.44$$

The ratio of energy of the higher-energy transition (ν_2 or ν_3) to the lowest-energy transition (ν_1) must therefore be approximately 1.44. From the Tanabe–Sugano diagram, we can see that the ratio of ν_3 to ν_1 is approximately 2, regardless of the strength of the

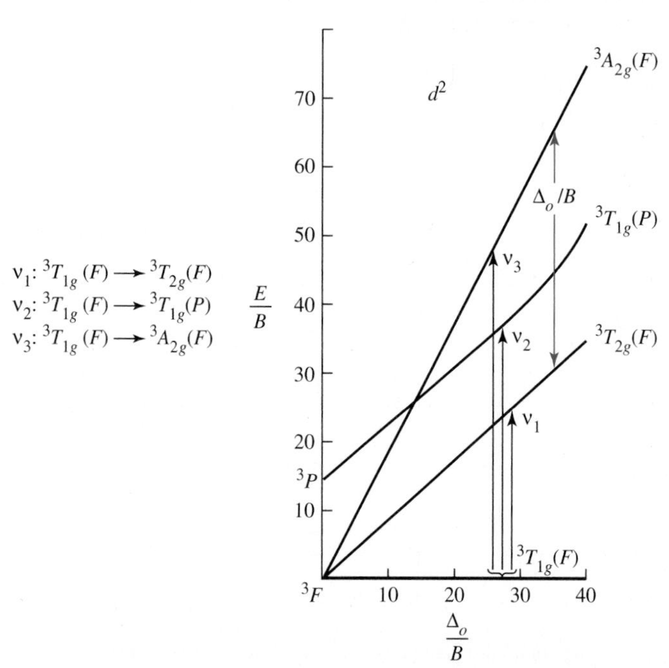

FIGURE 11.13 Spin-Allowed Transitions for d^2 Configuration.

ligand field; the slope of line associated with the ${}^3A_{2g}(F)$ state is approximately twice that of the line associated with the ${}^3T_{2g}(F)$ state. We can therefore eliminate ν_3 as the possible transition occurring at 25,700 cm^{-1}. This means that the 25,700 cm^{-1} band must be ν_2, corresponding to ${}^3T_{1g}(F) \longrightarrow {}^3T_{1g}(P)$, and $1.44 = \frac{\nu_2}{\nu_1}$.

The ratio ν_2 / ν_1 varies as a function of the strength of the ligand field. By plotting the ratio ν_2 / ν_1 versus Δ_o / B (Figure 11.14), we find that $\nu_2 / \nu_1 = 1.44$ at approximately $\Delta_o / B = 31$.[20, 21]

At $\frac{\Delta_o}{B} = 31$:

$$\nu_2: \quad \frac{E}{B} = 42 \text{ (approximately)}; \quad B = \frac{E}{42} = \frac{25{,}700 \text{ cm}^{-1}}{42} = 610 \text{ cm}^{-1}$$

$$\nu_1: \quad \frac{E}{B} = 29 \text{ (approximately)}; \quad B = \frac{E}{29} = \frac{17{,}800 \text{ cm}^{-1}}{29} = 610 \text{ cm}^{-1}$$

Because $\frac{\Delta_o}{B} = 31$:

$$\Delta_o = 31 \times B = 31 \times 610 \text{ cm}^{-1} = 19{,}000 \text{ cm}^{-1}$$

This procedure can be followed for d^2 and d^7 complexes of octahedral geometry to estimate values for Δ_o (and B).

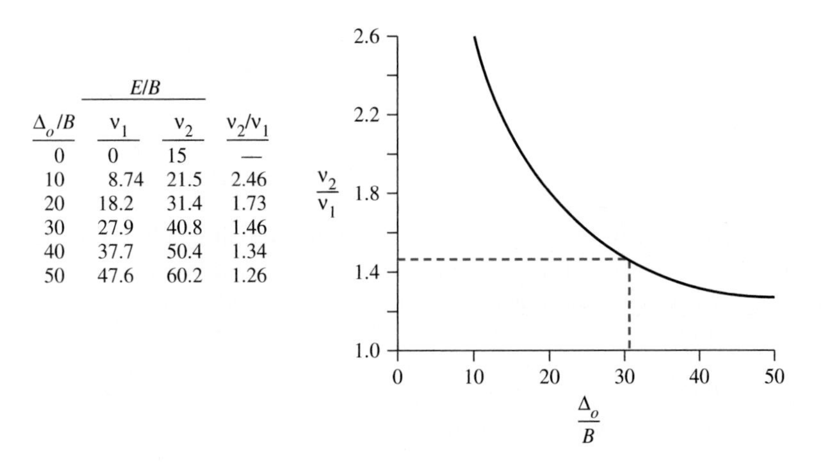

Δ_o/B	E/B: ν_1	E/B: ν_2	ν_2/ν_1
0	0	15	—
10	8.74	21.5	2.46
20	18.2	31.4	1.73
30	27.9	40.8	1.46
40	37.7	50.4	1.34
50	47.6	60.2	1.26

FIGURE 11.14 Value of ν_2 / ν_1 Ratio for d^2 Configuration.

▶ **Exercise 11.8** Use the Co(II) spectrum in Figure 11.8 and the Tanabe–Sugano diagrams of Figure 11.7 to find Δ_o and B. The broad band near 20,000 cm^{-1} can be considered to have the ${}^4T_{1g} \longrightarrow {}^4A_{2g}$ transition in the small shoulder near 16,000 cm^{-1} and the ${}^4T_{1g}(F) \longrightarrow {}^4T_{1g}(P)$ transition at the peak.[22]

OTHER CONFIGURATIONS: d^5 (HIGH SPIN), d^4 TO d^7 (LOW SPIN) As has been mentioned previously, high-spin d^5 complexes have no excited states of the same spin multiplicity (6) as the ground state. The bands that are observed are therefore the consequence of spin-forbidden transitions and are typically very weak as, for example, in $[Mn(H_2O)_6]^{2+}$.

[20] N. N. Greenwood and A. Earnshaw, *Chemistry of the Elements*, Pergamon Press, Elmsford, NY, 1984, p. 1161; B. N. Figgis and M. A. Hitchman, *Ligand Field Theory and Its Applications*, Wiley-VCH, New York, 2000, pp. 189–193.

[21] Different references report slightly different positions for the absorption bands of $[V(H_2O)_6]^{3+}$ and hence slightly different values of B and Δ_o.

[22] The ${}^4T_{1g} \longrightarrow {}^4A_{2g}$ transition is generally weak in octahedral complexes of Co^{2+}, because such a transition corresponds to simultaneous excitation of two electrons and is less probable than the other spin-allowed transitions, which are for excitations of single electrons.

The interested reader is referred to the literature[23] for an analysis of such spectra. In the case of low-spin d^4 to d^7 octahedral complexes, the analysis can be difficult, since there are many excited states of the same spin multiplicity as the ground state (see right side of Tanabe–Sugano diagrams for d^4 to d^7, Figure 11.7). Again, the chemical literature provides examples and analyses of the spectra of such compounds.[24]

11.3.6 Tetrahedral Complexes

In general, tetrahedral complexes have more intense absorptions than octahedral complexes. This is a consequence of the first (Laporte) selection rule (Section 11.3.1): transitions between *d* orbitals in a complex having a center of symmetry are forbidden. As a result, absorption bands for octahedral complexes are weak (small molar absorptivities); that they absorb at all is partly the result of vibrational motions that act continually to distort molecules slightly from pure O_h symmetry.

In tetrahedral complexes, the situation is different. The lack of a center of symmetry means that the Laporte selection rule does not apply, and transitions between *d* orbitals are more allowed; the consequence is that tetrahedral complexes often have much more intense absorption bands than octahedral complexes.[25]

As we have seen, the *d* orbitals for tetrahedral complexes are split in the opposite fashion to octahedral complexes:

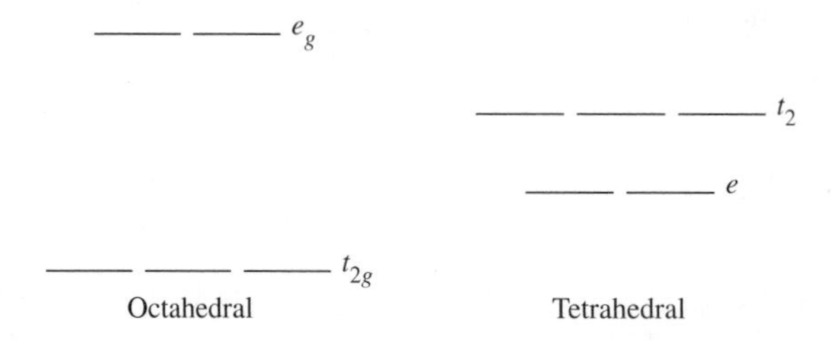

A useful comparison can be drawn between these by using what is called the **hole formalism**. This can best be illustrated by example. Consider a d^1 configuration in an octahedral complex. The one electron occupies an orbital in a triply degenerate set (t_{2g}). Now, consider a d^9 configuration in a tetrahedral complex. This configuration has a "hole" in a triply degenerate set of orbitals (t_2). It can be shown that, in terms of symmetry, the d^1 O_h configuration is analogous to the d^9 T_d configuration; the "hole" in d^9 results in the same symmetry as the single electron in d^1.

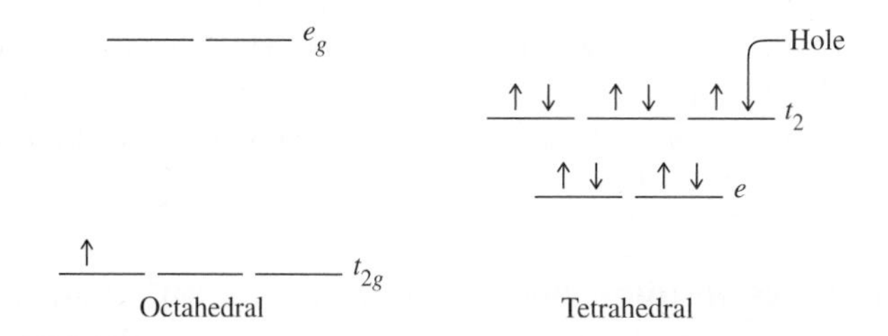

---

[23] B. N. Figgis and M. A. Hitchman, *Ligand Field Theory and Its Applications*, Wiley-VCH, New York, 2000, pp. 208–209.

[24] Figgis and Hitchman, *Ligand Field Theory and its Applications*, pp. 204–207; B. N. Figgis, in G. Wilkinson, R. D. Gillard, and J. A. McCleverty, eds., Comprehensive Coordination Chemistry, Vol. 1, Pergamon, Elmsord, NY, 1987, pp. 243–246.

[25] Two types of hybrid orbitals are possible for a central atom of T_d symmetry: sd^3 and sp^3 (see Chapter 5). These types of hybrids may be viewed as mixing, to yield hybrid orbitals that contain some *p* character (note that *p* orbitals are not symmetric to inversion), as well as *d* character. The mixing in of *p* character can be viewed as making transitions between these orbitals more allowed. For a more thorough discussion of this phenomenon, see F. A. Cotton, *Chemical Applications of Group Theory*, 3rd ed., Wiley InterScience, New York, 1990, pp. 295–296. Pages 289–297 of this reference also give a more detailed discussion of other selection rules.

In practical terms, this means that, for tetrahedral geometry, we can use the correlation diagram for the d^{10-n} configuration in octahedral geometry to describe the d^n configuration in tetrahedral geometry. Thus, for a d^2 tetrahedral case, we can use the d^8 octahedral correlation diagram; for the d^3 tetrahedral case, we can use the d^7 octahedral diagram, and so on. We can then identify the appropriate spin-allowed bands as in octahedral geometry, with allowed transitions occurring between the ground state and excited states of the same spin multiplicity.

Other geometries can also be considered according to the same principles as for octahedral and tetrahedral complexes. The interested reader is referred to the literature for a discussion of different geometries.[26]

11.3.7 Charge-Transfer Spectra

Examples of charge-transfer absorptions in solutions of halogens have been described in Chapter 6. In these cases, a strong interaction between a donor solvent and a halogen molecule, X_2, leads to the formation of a complex in which an excited state (primarily of X_2 character) can accept electrons from a HOMO (primarily of solvent character) on absorption of light of suitable energy:

$$X_2 \cdot \text{donor} \longrightarrow [\text{donor}^+][X_2^-]$$

The absorption band, known as a **charge-transfer band**, can be very intense; it is responsible for the vivid colors of some of the halogens in donor solvents.

It is common for coordination compounds also to exhibit strong charge-transfer absorptions, typically in the ultraviolet and/or visible portions of the spectrum. These absorptions may be much more intense than *d–d* transitions (which for octahedral complexes commonly have ε values of 20 L mol^{-1} cm^{-1} or less); molar absorptivities of 50,000 L mole^{-1} cm^{-1} or greater are not uncommon for these bands. Such absorption bands involve the transfer of electrons from molecular orbitals that are primarily ligand in character to orbitals that are primarily metal in character, or vice versa. For example, consider an octahedral d^6 complex with σ-donor ligands. The ligand electron pairs are stabilized, as shown in Figure 11.15.

The possibility exists that electrons can be excited, not only from the t_{2g} level to the e_g, but also from the σ orbitals originating from the ligands to the e_g. The latter excitation results in a charge-transfer transition; it may be designated as **charge transfer to metal (CTTM)** or **ligand to metal charge transfer (LMCT)**. This type of transition results in formal reduction of the metal. A CTTM excitation involving a cobalt (III) complex, for example, would exhibit an excited state having cobalt(II).

Examples of charge-transfer absorptions are numerous. For example, the octahedral complexes $IrBr_6^{2-}$ (d^5) and $IrBr_6^{3-}$ (d^6) both show charge-transfer bands. For $IrBr_6^{2-}$,

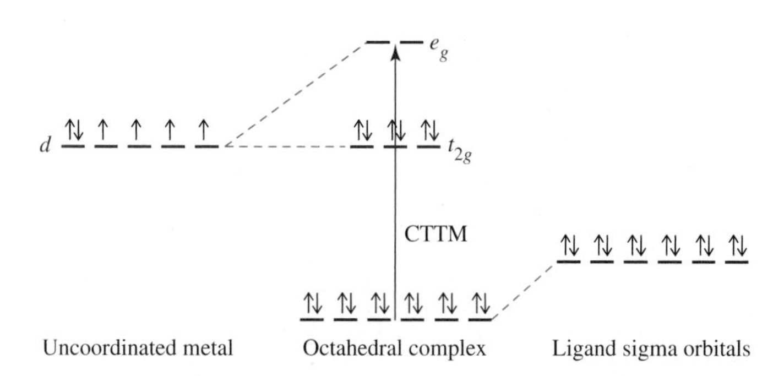

FIGURE 11.15 Charge Transfer to Metal.

[26] Figgis and Hitchman, *Ligand Field Theory and Its Applications*, pp. 211–214; Cotton, *Chemical Applications of Group Theory*, 3rd ed., pp. 295–303.

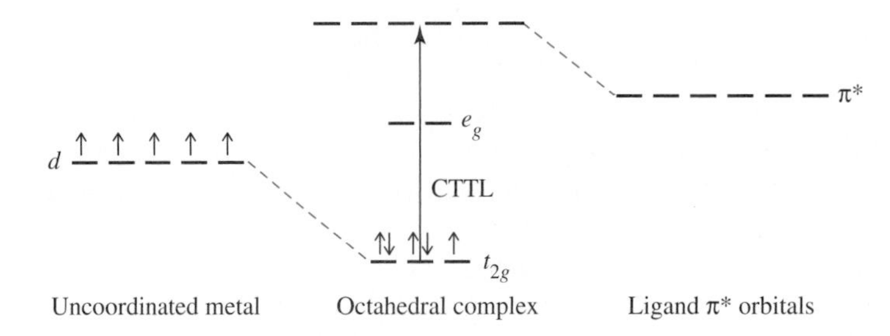

FIGURE 11.16 Charge Transfer to Ligand.

two bands appear, near 600 nm and near 270 nm; the former is attributed to transitions to the t_{2g} levels and the latter to the e_g. In $IrBr_6^{3-}$, the t_{2g} levels are filled, and the only possible CTTM absorption is therefore to the e_g. Consequently, no low-energy absorptions in the 600-nm range are observed, but strong absorption is seen near 250 nm, corresponding to charge transfer to e_g. A common example of tetrahedral geometry is the permanganate ion, MnO_4^-, which is intensely purple because of a strong absorption involving charge transfer from orbitals derived primarily from the filled oxygen *p* orbitals to empty orbitals derived primarily from the manganese(VII).

Similarly, it is possible for there to be **charge transfer to ligand (CTTL)** transitions, also known as **metal to ligand charge transfer (MLCT)**, in coordination compounds having π-acceptor ligands. In these cases, empty π^* orbitals on the ligands become the acceptor orbitals on absorption of light. Figure 11.16 illustrates this phenomenon for a d^5 complex.

CTTL results in oxidation of the metal; a CTTL excitation of an iron(III) complex would give an iron(IV) excited state. CTTL most commonly occurs with ligands having empty π^* orbitals, such as CO, CN^-, SCN^-, bipyridine, and dithiocarbamate ($S_2CNR_2^-$).

In complexes such as $Cr(CO)_6$ which have both σ-donor and π-acceptor orbitals, both types of charge transfer are possible. It is not always easy to determine the type of charge transfer in a given coordination compound. Many ligands give highly colored complexes that have a series of overlapping absorption bands in the ultraviolet part of the spectrum as well as the visible. In such cases, the *d–d* transitions may be completely overwhelmed and essentially impossible to observe.

Finally, the ligand itself may have a chromophore and still another type of absorption band, an **intraligand band**, may be observed. These bands may sometimes be identified by comparing the spectra of complexes with the spectra of free ligands. However, coordination of a ligand to a metal may significantly alter the energies of the ligand orbitals, and such comparisons may be difficult, especially if charge-transfer bands overlap the intraligand bands. Also, it should be noted that not all ligands exist in the free state; some owe their existence to the ability of metal atoms to stabilize molecules that are otherwise highly unstable. Examples of several such ligands will be discussed in later chapters.

▶ **Exercise 11.9** The isoelectronic ions VO_4^{3-}, CrO_4^{2-}, and MnO_4^- all have intense charge-transfer transitions. The wavelengths of these transitions increase in this series, with MnO_4^- having its charge-transfer absorption at the longest wavelength. Suggest a reason for this trend.

General References

B. N. Figgis and M. A. Hitchman, *Ligand Field Theory and Its Applications*, Wiley-VCH, New York, 2000; and B. N. Figgis, "Ligand Field Theory," in G. Wilkinson, R. D. Gillard, and J. A. McCleverty, eds., *Comprehensive Coordination Chemistry*, Vol. 1, Pergamon Press, Elmsford, NY, 1987, pp. 213–280, provide extensive background in the theory of electronic spectra, with numerous examples. Also useful is C. J. Ballhausen, *Introduction to Ligand Field Theory*, McGraw-Hill, New York, 1962. Important aspects of symmetry applied to this topic can be found in F. A. Cotton, *Chemical Applications of Group Theory*, 3rd ed., Wiley InterScience, New York, 1990.

Problems

11.1 For each of the following configurations, construct a microstate table and reduce the table to its constituent free-ion terms. Identify the lowest-energy term for each.

a. p^3

b. p^1d^1 (as in a $4p^1\ 3d^1$ configuration)

11.2 For each of the lowest-energy (ground state) terms in Problem 11.1, determine the possible values of J. Which J value describes the state with the lowest energy?

11.3 An excited state of calcium has the configuration $[Ar]4s^13d^1$. For an s^1d^1 configuration do the following:

a. Prepare a microstate table, showing each microstate.

b. Reduce the table to its free ion terms.

c. Determine the lowest energy term.

11.4 The outer electron configuration of the element cerium is d^1f^1. For this configuration do the following:

a. Construct a microstate table.

b. Reduce this table to its constituent free-ion terms (with labels).

c. Identify the lowest energy term (including J value).

11.5 The nitrogen atom is an example of a valence p^3 configuration. There are five energy levels associated with this configuration, with the energies shown below.

Energies (cm^{-1})
28839.31
28838.92
19233.18
19224.46
0

a. Account for these five energy levels.

b. Using information from Section 2.2.3, calculate Π_c and Π_e.

11.6 For each of the following free-ion terms, determine the values of L, M_L, S, and M_S:

a. ^2D (d^3)

b. ^3G (d^4)

c. ^4F (d^7)

11.7 For each of the free-ion terms in Problem 11.6, determine the possible values of J, and decide which is the lowest in energy.

11.8 The most intense absorption band in the visible spectrum of $[Mn(H_2O)_6]^{2+}$ is at 24,900 cm^{-1} and has a molar absorptivity of 0.038 L mol^{-1} cm^{-1}. What concentration of $[Mn(H_2O)_6]^{2+}$ would be necessary to give an absorbance of 0.10 in a cell of path length 1.00 cm?

11.9 **a.** Determine the wavelength and frequency of 24,900 cm^{-1} light.

b. Determine the energy and frequency of 366 nm light.

11.10 Determine the ground terms for the following configurations:

a. d^8 (O_h symmetry)

b. High-spin and low-spin d^5 (O_h symmetry)

c. d^4 (T_d symmetry)

d. d^9 (D_{4h} symmetry, square-planar)

11.11 The spectrum of $[Ni(H_2O)_6]^{2+}$ (Figure 11.8) shows three principal absorption bands, with two of the bands showing signs of further splitting. Referring to the Tanabe–Sugano diagram, estimate the value of Δ_o. Give a likely explanation for the further splitting of the spectrum.

11.12 From the following spectral data, and using Tanabe–Sugano diagrams (Figure 11.7), calculate Δ_o for the following:

a. $[Cr(C_2O_4)_3]^{3-}$, which has absorption bands at 23,600 and 17,400 cm^{-1}. A third band occurs well into the ultraviolet.

b. $[Ti(NCS)_6]^{3-}$, which has an asymmetric, slightly split band at 18,400 cm^{-1}. (Also, suggest a reason for the splitting of this band.)

c. $[Ni(en)_3]^{2+}$, which has three absorption bands: 11,200, 18,350, and 29,000 cm^{-1}.

d. $[VF_6]^{3-}$, which has absorption bands at 14,800 and 23,250 cm^{-1}, plus a third band in the ultraviolet. Also calculate B for this ion.

e. The complex $VCl_3(CH_3CN)_3$, which has absorption bands at 694 and 467 nm. Calculate Δ_o and B for this complex.

11.13 $[Co(NH_3)_6]^{2+}$ has absorption bands at 9,000 and 21,100 cm^{-1}. Calculate Δ_o and B for this ion. (Hints: The $^4T_{1g} \longrightarrow {}^4A_{2g}$ transition in this complex is too weak to be observed. The graph in Figure 11.13 may be used for d^7 as well as d^2 complexes.)

11.14 Classify the following configurations as A, E, or T in complexes having O_h symmetry. Some of these configurations represent excited states.

a. ${t_{2g}}^4{e_g}^2$ **b.** ${t_{2g}}^6$ **c.** ${t_{2g}}^3{e_g}^3$

d. ${t_{2g}}^5$ **e.** e_g

11.15 Of the first-row transition metal complexes of formula $[M(NH_3)_6]^{3+}$, which metals are predicted by the Jahn–Teller theorem to have distorted complexes?

11.16 ${MnO_4}^-$ is a stronger oxidizing agent than ${ReO_4}^-$. Both ions have charge-transfer bands; however, the charge-transfer band for ${ReO_4}^-$ is in the ultraviolet, whereas the corresponding band for ${MnO_4}^-$ is responsible for its intensely purple color. Are the relative positions of the charge-transfer absorptions consistent with the oxidizing abilities of these ions? Explain.

11.17 The complexes $[Co(NH_3)_5X]^{2+}$ (X = Cl, Br, I) have charge transfer to metal bands. Which of these complexes would you expect to have the lowest-energy charge-transfer band? Why?

11.18 $[Fe(CN)_6]^{3-}$ exhibits two sets of charge-transfer absorptions, one of lower intensity in the visible region of the spectrum, and one of higher intensity in the ultraviolet. $[Fe(CN)_6]^{4-}$, however, shows only the high-intensity charge transfer in the ultraviolet. Explain.

11.19 The complexes $[Cr(O)Cl_5]^{2-}$ and $[Mo(O)Cl_5]^{2-}$ have C_{4v} symmetry.

a. Use the angular overlap approach (Chapter 10) to estimate the relative energies of the *d* orbitals in these complexes.

b. Using the C_{4v} character table, determine the symmetry labels (labels of irreducible representations) of these orbitals.

c. The $^2B_2 \longrightarrow {}^2E$ transition occurs at 12,900 cm^{-1} for $[Cr(O)Cl_5]^{2-}$ and at 14,400 cm^{-1} for $[Mo(O)Cl_5]^{2-}$. Account for the higher energy for this transition in the molybdenum complex. (See W. A. Nugent and J. M. Mayer, *Metal-Ligand Multiple Bonds*, John Wiley & Sons, New York, 1988, pp. 33–35.)

11.20 For the isoelectronic series $[V(CO)_6]^-$, $Cr(CO)_6$, and $[Mn(CO)_6]^+$, would you expect the energy of metal to ligand charge-transfer bands to increase or decrease with increasing charge on the complex? Why? (See K. Pierloot, J. Verhulst, P. Verbeke, and L. G. Vanquickenborne, *Inorg. Chem.*, **1989**, *28*, 3059.)

11.21 The compound *trans*-Fe(*o*-phen)$_2$(NCS)$_2$ has a magnetic moment of 0.65 Bohr magneton at 80 K, increasing with temperature to 5.2 Bohr magnetons at 300 K.

a. Assuming a spin-only magnetic moment, calculate the number of unpaired electrons at these two temperatures.

b. How can the increase in magnetic moment with temperature be explained? (Hint: There is also a significant change in the UV-visible spectrum with temperature.)

11.22 The absorption spectrum of the linear ion $NiO_2{}^{2-}$ has bands attributed to *d–d* transitions at approximately 9,000 and 16,000 cm^{-1}.

a. Using the angular overlap model (Chapter 10), predict the expected splitting pattern of the *d* orbitals of nickel in this ion.

b. Account for the two absorption bands.

c. Calculate the approximate value of e_σ and e_π. (See M. A. Hitchman, H. Stratemeier, and R. Hoppe, *Inorg. Chem.*, **1988**, *27*, 2506.)

11.23 The electronic absorption spectra of a series of complexes of formula Re(CO)$_3$(L)(DBSQ) [DBSQ = 3,5-di*tert*-butyl-1,2-benzosemiquinone] show a single maximum in the visible spectrum. The absorption maxima for three of these complexes in benzene solution are shown below; typical molar absorptivities are in the range of 5,000 to 6,000 L mol^{-1} cm^{-1}.

L	ν_{max}, cm^{-1}
P(OPh$_3$)$_3$	18,250
PPh$_3$	17,300
NEt$_3$	16,670

Are these bands more likely due to charge transfer to metal or charge transfer to ligand? Explain briefly. (See F. Hartl and A. Vlcek, Jr., *Inorg. Chem.*, **1996**, *35*, 1257.)

11.24 The d^2 ions $CrO_4{}^{4-}$, $MnO_4{}^{3-}$, $FeO_4{}^{2-}$, and $RuO_4{}^{2-}$ have been reported.

a. Which of these has the largest value of Δ_t? Which has the smallest? Explain briefly.

b. Of the first three, which ion has the shortest metal–oxygen bond distance? Explain briefly.

c. The charge-transfer transitions for the first three complexes occur at 43,000, 33,000, and 21,000 cm^{-1}, respectively. Are these more likely to be ligand-to-metal or metal-to-ligand charge-transfer transitions? Explain briefly.

(See T. C. Brunhold and U. Güdel, *Inorg. Chem.*, **1997**, *36*, 2084.)

11.25 An aqueous solution of Ni(NO$_3$)$_2$ is green. Addition of aqueous NH_3 causes the color of the solution to change to blue. If ethylenediamine is added to the green solution, the color changes to violet. Account for the colors of these complexes. Are they consistent with the expected positions of these ligands in the spectrochemical series?

11.26 The pertechnetate ion, $TcO_4{}^-$, is often used to introduce the radioactive Tc into compounds, some of which are used as medical tracers. Unlike the isoelectronic, vividly purple permanganate ion, pertechnetate is very pale red.

a. Describe the most likely absorption that gives rise to colors in these ions. In addition to a written description, your answer should include an energy-level sketch showing how *d* orbitals on the metals interact with oxide orbitals to form molecular orbitals.

b. Suggest why $TcO_4{}^-$ is red but $MnO_4{}^-$ is purple.

c. The manganate ion, $MnO_4{}^{2-}$, is green. On the basis of your answers to a and b, provide an explanation for this color.

11.27 A 2.00×10^{-4} M solution of Fe(S$_2$CNEt$_2$)$_3$ (Et = C_2H_5) in CHCl$_3$ at 25 °C has absorption bands at 350 nm (A = 2.34), 514 nm (A = 0.532), 590 nm (A = 0.370), and 1540 nm (A = 0.0016).

a. Calculate the molar absorptivity for this compound at each wavelength.

b. Are these bands more likely due to *d–d* transitions or charge-transfer transitions? Explain.

11.28 Use the spectral data below to find the term symbols for the ground and excited states of each species, and calculate Δ_o and the Racah parameter, B, for each.

Species	Absorbance Bands (in cm^{-1})		
$[Ni(H_2O)_6]^{2+}$	8,500	15,400	26,000
$[Ni(NH_3)_6]^{2+}$	10,750	17,500	28,200
$[Ni(OS(CH_3)_2)_6]^{2+}$	7,728	12,970	24,038
$[Ni(dma)_6]^{2+}$	7,576	12,738	23,809

11.29 For each of the compounds $[Co(bipy)_3]^{2+}$ and $[Co(NH_3)_6]^{2+}$, do the following:

a. Find the ground-state term symbol.

b. Use the Tanabe–Sugano diagram to identify the predicted spectral bands.

c. Calculate the ligand field stabilization energy.

d. Do you expect broad or narrow absorption bands in the visible and UV regions?

e. Sketch an MO energy-level diagram for each.

	ν_1 (cm^{-1})	ν_3 (cm^{-1})
$[Co(bipy)_3]^{2+}$	11,300	22,000
$[Co(NH_3)_6]^{3-}$	9,000	21,100

11.30 In the complexes $FeL(SC_6H_5)$ and $NiL(SC_6H_5)$, where L = hydrotris-(pyrazoylborate) $HB(3,5\text{-}i\text{-}Pr_2pz)_3^{\ -}$, strong charge-transfer bands were observed in the regions 28,000 to 32,500 and 20,100 to 30,000 cm^{-1}, respectively. Were these more likely LMCT or MLCT bands? Explain, taking into account the relative energies of the metal orbitals in these complexes. (See S. I. Gorelsky, L. Basumallick, J. Vura-Weis, R. Sarangi, K. O. Hodgson, B. Hedman, K. Fujisawa, and E. I. Solomon, *Inorg. Chem.*, **2005**, *44*, 4947.)

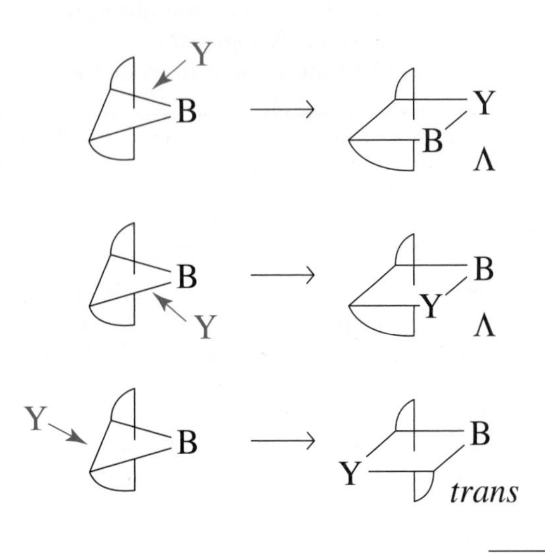

12

Coordination Chemistry IV: Reactions and Mechanisms

Reactions of coordination compounds share some characteristics with reactions of other molecules, both organic and inorganic, so an understanding of coordination compound reactions can draw on some familiar concepts. However, the chemistry of coordination compounds has some additional features, because the molecules have more complex geometries and more possibilities for rearrangement, the metal atoms exhibit more variability in their reactions, and different factors influence the course of reactions.

Reactions of coordination complexes can be conveniently divided into substitution reactions at the metal center, oxidation–reduction reactions, and reactions of the ligands that do not change the attachments to the metal center. Reactions that include more elaborate rearrangements of ligand structures are more often observed in organometallic compounds; a description of these reactions is given in Chapter 14.

12.1 BACKGROUND

Synthesis of coordination compounds has always been a major part of chemistry. Although early chemists did not know the structures of the compounds they worked with, they did learn how to make many of them and described them according to the colorful style of the time. The synthetic work done by Werner, Jørgensen, and others who established the foundations of coordination geometry began the systematic development of reactions for specific purposes. Many years of experimentation and consideration of possible reaction pathways have led to more modern ideas, and even now these ideas must be considered tentative and provisional in many cases. The unification of reaction theory is still a goal of chemists, whether they work with organic, inorganic, coordination, organometallic, polymeric, solid-state, liquid, or gaseous compounds; but reaching this goal is still in the future. Although the ability to predict products and choose appropriate reaction conditions to obtain desired products is still a matter of art as well as science, the list of known reactions is now long enough to provide considerable guidance.

The goals of those studying reaction kinetics and mechanisms vary, but a major underlying reason for such studies is to understand electronic structures of coordination compounds and their interactions. The information from these studies also allows more control of reactions and the design of reaction steps that may be useful for synthesis. A by-product of synthetic and kinetic studies is the esthetic pleasure of seeing the colors characteristic of many coordination compounds, and how they change with different ligands and metal ions.

We will first review some of the background needed to understand reaction mechanisms, then consider the major categories of such mechanisms, and finally describe some of the results of these mechanistic studies.

In general, chemical reactions move from one energy minimum (the reactants) through a higher energy structure (the transition state) to another energy minimum (the products). In simple cases, the energies and bond distances can be shown as a three-dimensional surface, with different bond distances along the base-plane axes (x and y) and free energy as the vertical dimension (z). The reaction $MX + Y \longrightarrow MY + X$ begins at a point representing the short M—X distance of the bond to be broken and the longer distance between the two reactants, MX and Y. As the M—X bond breaks and the M—Y bond forms, the reaction point moves to represent the short M—Y bond distance and the longer distance between the two products, MY and X. The free-energy surface usually has a saddle shape, much like a mountain pass between two valleys. For more complex reactions, such a visual representation is more difficult; but the path between the reactants and the products is always the lowest energy pathway and must be the same regardless of the direction of the reaction. This is the **principle of microscopic reversibility**, frequently described by the mountain-pass analogy; the lowest pass going in one direction must also be the lowest pass going in the opposite direction.

If the reaction is such that the conversion from reactants to products takes place with no hesitation at the transition point, as in Figure 12.1(a), the structure at that state is called the **transition state**. If there is a structure that lasts a bit longer, as in Figure 12.1(b), and particularly if it is detectable by some experimental means, it is called an

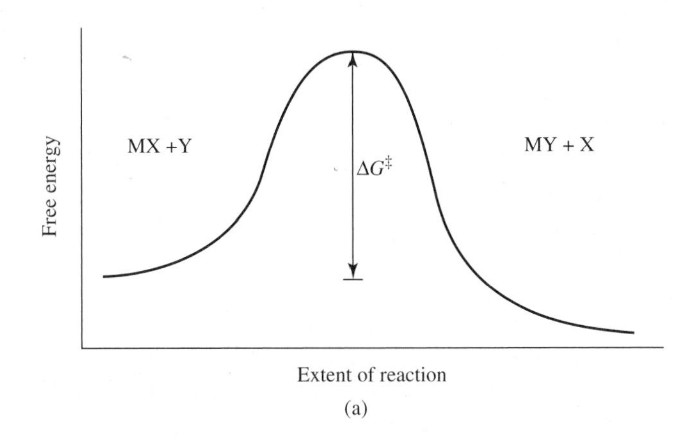

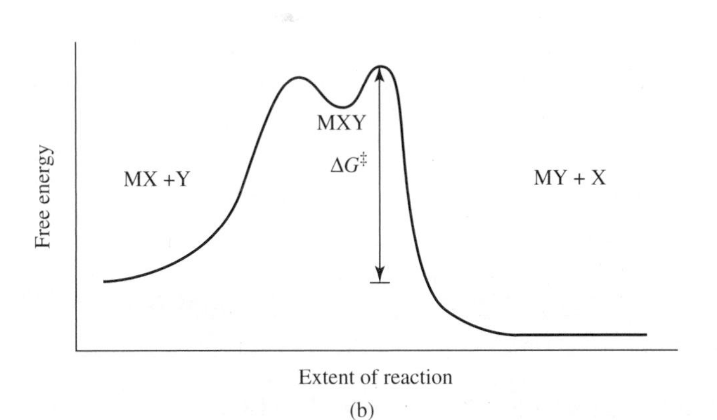

FIGURE 12.1 Energy Profiles and Intermediate Formation. (a) No intermediate. The activation energy is the energy difference between the reactants and the transition state. (b) An intermediate is present at the small minimum at the top of the curve. The activation energy is measured at the maximum point of the curve.

intermediate. Frequently, the kinetic equations include intermediates, even if they remain undetected. Their presence allows treatment by a **steady-state approximation**, in which the concentration of the intermediate is assumed to be small and essentially unchanging during much of the reaction. Details of this approach are described later.

A number of different parameters can be obtained from kinetics experiments. First, the **order** of the reaction, indicated by the power of the reactant concentration in the differential equation that describes it, can be determined, together with the **rate constant** that describes the speed of the reaction. By studying a reaction at different temperatures, the **free energy of activation** and the **enthalpy** (or **heat**) and **entropy of activation** can be found. These allow further interpretation of the mechanism and the energy surface. Including pressure dependence provides the **volume of activation**, which offers insight into whether the transition state is larger or smaller than the reactants.

12.2 SUBSTITUTION REACTIONS

12.2.1 Inert and Labile Compounds

Many synthetic reactions require substitution, replacing one ligand by another. This is particularly true when the starting material is in aqueous solution, where the metal ion is likely to be in the form $[M(H_2O)_m]^{n+}$. Some simpler reactions of this type produce colored products that can be used to identify metal ions:

$$\underset{\text{green}}{[Ni(H_2O)_6]^{2+}} + 6\,NH_3 \rightleftharpoons + \underset{\text{blue}}{[Ni(NH_3)_6]^{2+}} + 6\,H_2O$$

$$\underset{\text{very pale violet}}{[Fe(H_2O)_6]^{3+}} + SCN^- \rightleftharpoons + \underset{\text{red}}{[Fe(H_2O)_5(SCN)]^{2+}} + H_2O$$

These reactions, and others like them, are rapid and form species that can undergo a variety of reactions that are also very fast. Addition of $HNO_3(H^+)$, $NaCl(Cl^-)$, $H_3PO_4(PO_4{}^{3-})$, $KSCN(SCN^-)$, and $NaF(F^-)$ successively to a solution of $Fe(NO_3)_3 \cdot 9\,H_2O$ shows this clearly. The initial solution is yellow due to the presence of $[Fe(H_2O)_5(OH)]^{2+}$ and other "hydrolyzed" species containing both water and hydroxide ion. Although the exact species formed in this series depend on solution concentrations, the products in the reactions given here are representative:

$$\underset{\text{yellow}}{[Fe(H_2O)_5(OH)]^{2+}} + H^+ \longrightarrow \underset{\text{colorless (very pale violet)}}{[Fe(H_2O)_6]^{3+}}$$

$$[Fe(H_2O)_6]^{3+} + Cl^- \longrightarrow \underset{\text{yellow}}{[Fe(H_2O)_5(Cl)]^{2+}} + H_2O$$

$$[Fe(H_2O)_5(Cl)]^{2+} + PO_4^{3-} \longrightarrow \underset{\text{colorless}}{Fe(H_2O)_5(PO_4)} + Cl^-$$

$$Fe(H_2O)_5(PO_4) + SCN^- \longrightarrow \underset{\text{red}}{[Fe(H_2O)_5(SCN)]^{2+}} + PO_4{}^{3-}$$

$$[Fe(H_2O)_5(SCN)]^{2+} + F^- \longrightarrow \underset{\text{colorless}}{[Fe(H_2O)_5(F)]^{2+}} + SCN^-$$

Complexes such as these, which react rapidly, are called **labile**. In many cases, exchange of one ligand for another can take place in the time of mixing the solutions. Taube[1] has suggested a reaction half-life (the time it takes for the amount of the initial compound to decrease by one half) of one minute or less as the criterion for lability. Compounds that react more slowly are called **inert** or **robust** (a term used less often). An inert compound is not inert in the sense that no reaction can take place; it is simply slower to react. These kinetic terms must also be distinguished from the thermodynamic terms **stable** and **unstable**. A species such as $[Fe(H_2O)_5(F)]^{2+}$ is very stable—it has a large equilibrium constant for formation—but it is also labile. On the other hand, hexaaminecobalt (3+) is thermodynamically unstable in acid and can decompose to the equilibrium mixture shown.

$$[Co(NH_3)_6]^{3+} + 6\,H_3O^+ \rightleftharpoons [Co(H_2O)_6]^{3+} + 6\,NH_4{}^+ (\Delta G^\circ < 0)$$

But it reacts very slowly (has a very high activation energy) and is therefore called *inert* or *robust*. The possible confusion of terms is unfortunate. For clarity the complexes can be described as **substitutionally** or **kinetically labile** or **inert**.

Werner studied cobalt(III), chromium(III), platinum(II), and platinum(IV) compounds because they are inert and can be more readily characterized than labile compounds. This tendency has continued, and much of the discussion in this chapter is based on inert compounds because they can be more easily crystallized from solution and their structures determined. Labile compounds have also been studied extensively, but their study requires techniques capable of dealing with very short times.

Although there are exceptions, general rules can be given for inert and labile electronic structures. Inert octahedral complexes are generally those with high ligand field stabilization energies (Section 10.3.3), specifically those with d^3 or low-spin d^4 through d^6 electronic structures. Complexes with d^8 configurations generally react somewhat faster, but still slower than the d^7, d^9, or d^{10} complexes. With strong-field ligands, d^8 atoms often form square-planar complexes, many of which are inert. Compounds with other d configurations tend to be labile.

Slow Reactions (Inert)	Intermediate	Fast Reactions (Labile)
d^3, low-spin d^4, d^5, and d^6		d^1, d^2, high-spin d^4, d^5, and d^6
Strong-field d^8 (square planar)	Weak-field d^8	d^7, d^9, d^{10}

12.2.2 Mechanisms of Substitution

Langford and Gray[2] have described the range of possibilities for substitution reactions, listed in Table 12.1. At one extreme, the departing ligand leaves, and a discernible intermediate with a lower coordination number is formed, a mechanism labeled ***D*** for **dissociation**. At the other extreme, the incoming ligand adds to the complex, and an intermediate with an increased coordination number is formed in a mechanism labeled ***A*** for **association**. Between the two extremes is **interchange, *I***, in which the incoming ligand is presumed to assist in the reaction, but no detectable intermediates appear. When the degree of assistance is small and the reaction is primarily dissociative, it is called **dissociative interchange, I_d**. When the incoming ligand begins forming a bond

[1] H. Taube, *Chem. Rev.*, **1952**, *50*, 69.
[2] C. H. Langford and H. B. Gray, *Ligand Substitution Processes*, W. A. Benjamin, New York, 1966.

TABLE 12.1 Classification of Substitution Mechanisms

	Stoichiometric Mechanism		
Intimate Mechanism	**Dissociative 5-Coordinate Transition State for Octahedral Reactant**		**Associative 7-Coordinate Transition State for Octahedral Reactant**
Dissociative activation	D	I_d	
Associative activation		I_a	A
	Alternative Labels		
S_N1 lim (limiting first-order nucleophilic substitution)			S_N2 lim (limiting second-order nucleophilic substitution)

to the central atom before the departing ligand bond is weakened appreciably, it is called **associative interchange, I_a**. Many reactions are described by I_a or I_d mechanisms, rather than by *A* or *D*, when the kinetic evidence points to association or dissociation, but detection of intermediates is not possible. Langford and Gray call these categories the **stoichiometric mechanisms**; the distinction between activation processes that are associative and dissociative is called the **intimate mechanism**. The energy profiles for associative and dissociative reactions are shown in Figure 12.2. The clear separation of these two mechanisms in the figure should not be taken as an indication that the distinction is easily made. In many cases, there is no clear-cut evidence to distinguish them, and inferences must be made using the available evidence.

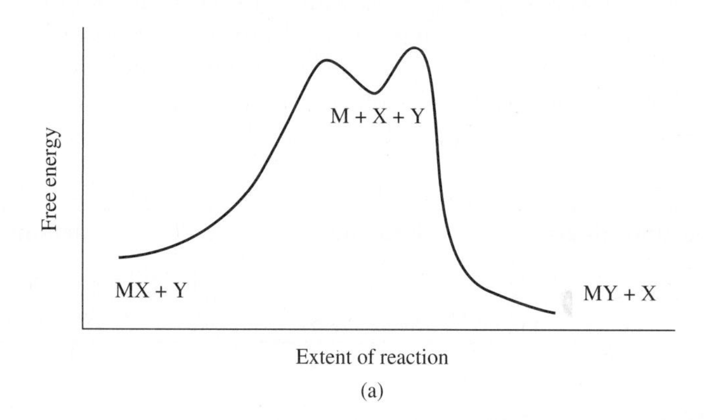

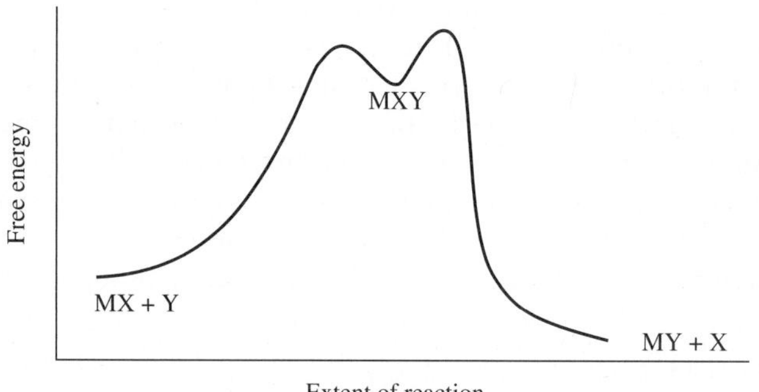

FIGURE 12.2 Energy Profiles for Dissociative and Associative Reactions. (a) Dissociative mechanism. The intermediate has a lower coordination number than the starting material. (b) Associative mechanism. The intermediate has a higher coordination number than the reactant.

Kinetic experiments are frequently carried out with great excess of the incoming reagent, Y. This simplifies the analysis of the progress of the reaction for each kinetic run, but it requires a number of runs at different concentrations of Y to determine the order of the reaction with respect to Y.

12.3 KINETIC CONSEQUENCES OF REACTION PATHWAYS

Although the kinetic rate law is helpful in determining the mechanism of a reaction, it does not always provide sufficient information. In cases of ambiguity, other evidence must be used. This chapter will describe a number of examples in which the rate law and other experimental evidence have been used to find the mechanism of a reaction. Our goal is to provide two related types of information: (1) the type of information used to determine mechanisms and (2) a selection of specific reactions for which the mechanisms seem to be fairly completely determined. The first is more important, because it enables chemists to examine data for other reactions critically and to evaluate the proposed mechanisms. The second is also helpful, because it provides part of the collection of knowledge required for designing new syntheses. Each of the substitution mechanisms will be described with its required rate law.[3]

12.3.1 Dissociation (*D*)

In a dissociative (*D*) reaction, loss of a ligand to form an intermediate with a lower coordination number is followed by addition of a new ligand to the intermediate:

$$ML_5X \underset{k_{-1}}{\overset{k_1}{\rightleftharpoons}} ML_5 + X$$

$$ML_5 + Y \xrightarrow{k_2} ML_5Y$$

The *stationary-state*, or *steady-state*, hypothesis assumes a very small concentration of the intermediate, ML_5, and requires that the rates of formation and reaction of the intermediate must be equal. This in turn requires that the rate of change of $[ML_5]$ be zero during much of the reaction. Expressed as a rate equation,

$$\frac{d[ML_5]}{dt} = k_1 [ML_5X] - k_{-1} [ML_5][X] - k_2 [ML_5][Y] = 0$$

Solving for $[ML_5]$,

$$[ML_5] = \frac{k_1 [ML_5X]}{k_{-1} [X] + k_2[Y]}$$

and substituting into the rate law for formation of the product,

$$\frac{d[ML_5Y]}{dt} = k_2[ML_5][Y]$$

[3] In the reactions of this chapter, X will indicate the ligand that is leaving a complex, Y the ligand that is entering, and L any ligands that are unchanged during the reaction. In cases of solvent exchange, all the X, Y, and L may be chemically the same species, but in the more general case, they may all be different. Charges will be omitted in the general case, but any of the species may be ions. The general examples will usually be six-coordinate, but other coordination numbers could be chosen, and the discussion would be similar.

leads to the rate law:

$$\frac{d[ML_5Y]}{dt} = \frac{k_2 k_1 [ML_5X][Y]}{k_{-1}[X] + k_2 [Y]}$$

One criterion for this mechanism is that the intermediate, ML_5, be detectable during the reaction. Direct detection at the low concentrations expected is a very difficult experimental challenge, and there are very few clear-cut dissociative reactions. More often, the evidence is indirect, but no intermediate has been found. Such reactions are usually classified as following an interchange mechanism.

12.3.2 Interchange (*I*)

In an interchange (*I*) reaction, a rapid equilibrium between the incoming ligand and the 6-coordinate reactant forms an ion pair or loosely bonded molecular combination. This species, which is not described as having an increased coordination number and is not directly detectable, then reacts to form product and release the initial ligand.

$$ML_5X + Y \underset{k_{-1}}{\overset{k_1}{\rightleftharpoons}} ML_5X \cdot Y$$

$$ML_5X \cdot Y \xrightarrow{k_2} ML_5Y + X$$

When $k_2 \ll k_{-1}$, the reverse reaction of the first step is fast enough that this step is independent of the second step, and the first step is an equilibrium with $K_1 = k_1 / k_{-1}$.

Applying the stationary-state hypothesis:

$$\frac{d[ML_5X \cdot Y]}{dt} = k_1 [ML_5X][Y] - k_{-1}[ML_5X \cdot Y] - k_2 [ML_5X \cdot Y] = 0$$

If [Y] is large compared to $[ML_5X]$, a common experimental condition, the concentration of the unstable transition species may be large enough to significantly change the concentration of the ML_5X but not that of Y. For this reason, we must solve for this species in terms of the total initial reactant concentrations of ML_5X and Y, which we will call $[M]_0$ and $[Y]_0$:

$$[M]_0 = [ML_5X] + [ML_5X \cdot Y]$$

Assuming that the concentration of the final product, $[ML_5Y]$, is too small to change the concentration of Y significantly, then

$$[Y]_0 \cong [Y]$$

From the stationary-state equation,

$$k_1 \left([M]_0 - [ML_5X \cdot Y]\right)[Y]_0 - k_{-1}[ML_5X \cdot Y] - k_2[ML_5X \cdot Y] = 0$$

The final rate equation then becomes

$$\frac{d[ML_5Y]}{dt} = k_2 [ML_5X \cdot Y] = \frac{k_2 K_1 [M]_0 [Y]_0}{1 + K_1 [Y]_0 + \left(k_2 / k_{-1}\right)} \cong \frac{k_2 K_1 [M]_0 [Y]_0}{1 + K_1 [Y]_0}$$

where k_2 / k_{-1} is very small and can be omitted, because $k_2 \ll k_{-1}$ is required for the first step to be an equilibrium.

K_1 can be measured experimentally in some cases and estimated theoretically in others from calculation of the electrostatic energy of the interaction. Such calculations are beyond the scope of this text.

Two variations on the interchange mechanism are I_d, *dissociative interchange*, and I_a, *associative interchange*. The difference between them is in the degree of bond formation in the first step of the mechanism. If bonding between the incoming ligand and the metal is more important, it is an I_a mechanism. If breaking the bond between the departing ligand and the metal is more important, it is an I_d mechanism. The distinction between them is subtle, and careful experimental design is necessary to determine which description fits a given reaction.

As can be seen from these equations, both *D* and *I* mechanisms have the same mathematical form for their rate laws. (If both the numerator and the denominator of the *D* rate law are divided by k_{-1}, the equations have the similar forms shown here.)

$$\text{Rate} = \frac{k[\text{M}][\text{Y}]}{[\text{X}] + k'[\text{Y}]} \qquad \text{Rate} = \frac{k[\text{M}]_0\,[\text{Y}]_0}{1 + k'[\text{Y}]_0}$$

At low [Y], the denominator simplifies to [X] for the dissociative and to 1 for the interchange equation. Both are then first order in M and Y (rate $= k[\text{M}]_0\,[\text{Y}]_0$ or $k[\text{M}]_0\,[\text{Y}]_0 / [\text{X}]$), with the rate of the dissociative reaction slowing as [X] increases.

At high [Y], a common condition in kinetic experiments, the second term in the denominator is larger, $[\text{X}] + k'[\text{Y}] = k'[\text{Y}]$ and $1 + k'[\text{Y}]_0 = k'[\text{Y}]_0$, and [Y] cancels, making the reaction first order in complex and zero order in Y (rate $= (k / k')[\text{M}]_0$).

The change from one rate law to the other depends on the specific values of the rate constants. The similarity of the rate laws limits their usefulness in determining the mechanism and requires other means of distinguishing between different mechanisms.

12.3.3 Association (*A*)

In an associative reaction, the first step, forming an intermediate with an increased coordination number, is the rate-determining step. It is followed by a faster reaction in which the exiting ligand is lost:

$$ML_5X + Y \underset{k_{-1}}{\overset{k_1}{\rightleftharpoons}} ML_5XY$$

$$ML_5XY \xrightarrow{k_2} ML_5Y + X$$

The same stationary-state approach used in the other rate laws results in the rate law

$$\frac{d[ML_5Y]}{dt} = \frac{k_1 k_2\,[ML_5X][Y]}{k_{-1} + k_2} = k[ML_5X][Y]$$

This is a second-order equation regardless of the concentration of Y.

▶ **Exercise 12.1** Show that the preceding equation is the result of the stationary-state approach for an associative reaction.

As with the dissociative mechanism, there are few clear examples of associative mechanisms in which the intermediate is detectable. Most reactions fit better between the two extremes, following associative or dissociative interchange mechanisms. The next section summarizes the evidence for the different mechanisms.

12.4 EXPERIMENTAL EVIDENCE IN OCTAHEDRAL SUBSTITUTION

12.4.1 Dissociation

Most substitution reactions of octahedral complexes are believed to be dissociative, with the complex losing one ligand to become a 5-coordinate square pyramid in the transition state and the incoming ligand filling the vacant site to form the new octahedral product. Theoretical justification for the inert and labile classifications of

Section 12.2.1 comes from ligand field theory, based on calculation of the change in LFSE between the octahedral reactant and the presumed 5-coordinate transition state, either square-pyramidal or trigonal-bipyramidal in shape. Table 12.2 gives the **ligand field activation energy** (LFAE), defined as the difference between the LFSE of the square-pyramidal transition state and the LFSE of the octahedral reactant. LFAEs calculated for trigonal-bipyramidal transition states are generally the same or larger than those for square-pyramidal transition states. These calculations provide estimates of the energy necessary to form the transition state. When combined with the general change in enthalpies of formation described in Section 10.6, and particularly Figure 10.27, the activation energies of the square-pyramidal transition state match the experimental observations (d^3 and d^8 complexes are inert in both the strong- and weak-field cases, and d^6 strong-field complexes are inert). Therefore, the calculation of LFAE supports a square-pyramidal geometry, and a dissociative mechanism, for the transition state. However, all these numbers assume an idealized geometry not likely to be found in practice, and the LFAE is only one factor that must be considered in any reaction.

Even for thermodynamically favorable reactions, a large activation energy means that the reaction will be slow. For thermodynamically unfavorable reactions, even a fast reaction (with small activation energy) would be unlikely to occur. The rate of reaction depends on the activation energy, as in the Arrhenius equation:

$$k = Ae^{-\frac{E_a}{RT}} \quad \text{or} \quad \ln k = \ln A - \frac{E_a}{RT}$$

Some of the possible energy relationships for reactions are shown in Figure 12.3. In (a) and (b), the reaction is exothermic ($\Delta H < 0$), and the equilibrium constant is large (if entropy effects are insignificant). In (a), the reaction is spontaneous ($\Delta H < 0$), but E_a is large, so few molecules have enough energy to get over the barrier, and the reaction is slow. In (b) the reaction is spontaneous, with an intermediate at the dip near the top of the activation energy curve. Intermediates of this sort are frequently described but can

TABLE 12.2 Ligand Field Activation Energies Calculated by Angular Overlap

	Strong Fields (units of e_σ)			Weak Fields (units of e_σ)		
System	**Octahedral LFSE**	**Square-Pyramidal LFSE**	**LFAE**	**Octahedral LFSE**	**Square-Pyramidal LFSE**	**LFAE**
d^0	−12	−10	2	−12	−10	2
d^1	−12	−10	2	−12	−10	2
d^2	−12	−10	2	−12	−10	2
d^3	−12	−10	2	−12	−10	2
d^4	−12	−10	2	−9	−8	1
d^5	−12	−10	2	−6	−5	1
d^6	−12	−10	2	−6	−5	1
d^7	−9	−8	1	−6	−5	1
d^8	−6	−5	1	−6	−5	1
d^9	−3	−3	0	−3	−3	0
d^{10}	0	0	0	0	0	0

NOTE: For a square-pyramidal transition state, LFAE = square pyramid LFSE – Octahedral LFSE, for σ donor only.

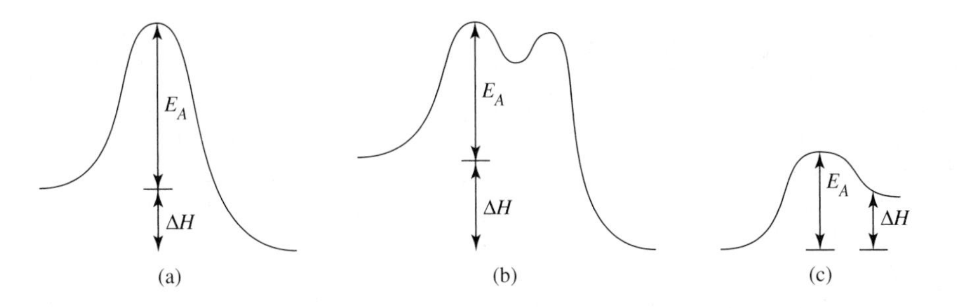

FIGURE 12.3 Activation Energies and Reaction Enthalpies. (a), (b), Large E_a, slow reaction; (c) Small E_a, fast reaction. (a), (b), $\Delta H < 0$, large equilibrium constant; (c) $\Delta H < 0$, small equilibrium constant. In (b), the intermediate is potentially detectable.

be detected and identified in only a few cases. In (c), the reaction can occur quickly because of the low activation energy, but it has a small equilibrium constant, because the overall enthalpy change is positive.

When *s*- and *p*-orbital influences are added, the results are similar to those of the thermodynamic case of enthalpy of hydration, shown in Figure 10.7, with long half-lives for relatively inert d^3 and d^8 configurations and short half-lives for d^0, d^4, d^9, and as shown in Figure 12.4.

Other metal-ion factors that affect reaction rates of octahedral complexes include the following (relative rates for ligand exchange are indicated by the inequalities):

1. ***Oxidation state of the central ion.*** Central atoms with higher oxidation states have slower ligand exchange rates.

$$\underset{3+}{[AlF_6]^{3-}} > \underset{4+}{[SiF_6]^{2-}} > \underset{5+}{[PF_6]^{-}} > \underset{6+}{SF_6}$$

$$\underset{1+}{[Na(H_2O)_n]^{+}} > \underset{2+}{[Mg(H_2O)_n]^{2+}} > \underset{3+}{[Al(H_2O)_6]^{3+}}$$

2. ***Ionic radius.*** Smaller ions have slower exchange rates.

$$\underset{112\text{ pm}}{[Sr(H_2O)_6]^{2+}} > \underset{99\text{ pm}}{[Ca(H_2O)_6]^{2+}} > \underset{66\text{ pm}}{[Mg(H_2O)_6]^{2+}}$$

Both effects can be attributed to a higher electrostatic attraction between the central atom and the attached ligands. A strong attraction between the two will slow a reaction, because reaction is presumed to require dissociation of a ligand from the complex.

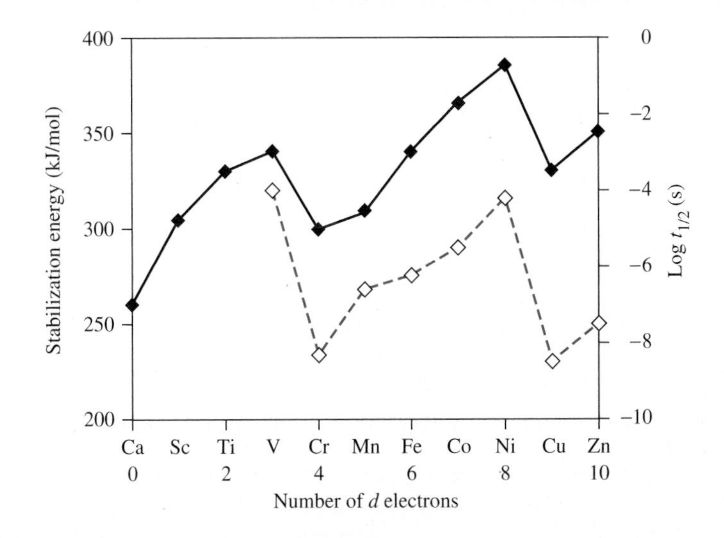

FIGURE 12.4 Stabilization Energy and Experimental Half-Lives for Water Exchange. (Angular overlap data [*solid line*] from J. K. Burdett, *J. Chem. Soc. Dalton*, **1976**, 1725. Half-lives for water exchange [*dashed line*] from F. Basolo and R. G. Pearson, *Mechanisms of Inorganic Reactions*, 2nd ed., John Wiley & Sons, New York, 1967, p. 155.)

Figure 12.4 shows the half-lives for exchange of water molecules on aquated 2+ transition-metal ions. All the ions in the figure are labile, with half-lives for the aqua complexes shorter than 1 second; measurement of such fast reactions is done by indirect methods, particularly relaxation methods.[4] The monovalent alkali metal cations have very short half-lives (10^{-9} second or less); of the common 2+ metal ions, only Be^{2+} and V^{2+} have half-lives as long as 0.01 second. For comparison, Al^{3+} has a half-life approaching 1 second, and Cr^{3+} has a half-life of 40 hours—the only inert aquated transition-metal ion.

The evidence for dissociative mechanisms can be grouped as follows:[5,6,7,8]

1. The rate of reaction changes only slightly with changes in the incoming ligand. In many cases, **aquation** (substitution by water) and **anation** (substitution by an anion) rates are comparable. If dissociation is the rate-determining reaction, the entering group should have no effect at all on the reaction rate. Although there is no specific criterion for this, changes in rate constants of less than a factor of 10 are generally considered to be insignificant for this purpose.
2. Decreasing negative charge or increasing positive charge on the reactant complex decreases the rate of substitution. Larger electrostatic attraction between the positive metal ion and the negative ligand should slow the dissociation.
3. Steric crowding on the reactant complex increases the rate of ligand dissociation. When ligands on the reactant are crowded, loss of one of the ligands is made easier. On the other hand, if the reaction has an A or I_a mechanism, steric crowding interferes with the incoming ligand and slows the reaction. (This phenomenon will be considered further in Chapter 14).
4. The rate of reaction correlates with the metal–ligand bond strength of the leaving group, in a linear free-energy relationship (LFER, explained in the next section).
5. Activation energies and entropies are consistent with dissociation, although interpretation of these parameters is difficult. Another activation parameter now being measured by experiments at increased pressure is the volume of activation, ΔV_{act}, the change in volume on forming the activated complex. Dissociative mechanisms generally result in positive values for ΔV_{act} because one species splits into two, and associative mechanisms result in negative values because two species combine into one, with a presumed volume smaller than the total for the reactants. However, caution is needed in interpreting volume effects, because solvation effects, particularly for highly charged ions, may be significant.

12.4.2 Linear Free-Energy Relationships

Many kinetic effects can be related to thermodynamic effects by a **linear free-energy relationship (LFER)**.[9] Such effects are seen when, for example, the bond strength of a metal-ligand bond (a thermodynamic function) plays a major role in determining the dissociation rate of a ligand (a kinetic function). When this is true, a plot of the logarithm of the rate constants (kinetic) for different leaving ligands versus the logarithm of the equilibrium constants (thermodynamic) for the same ligands in similar compounds is linear. The justification for this correlation is found in the Arrhenius

---

[4] F. Wilkinson, *Chemical Kinetics and Reactions Mechanisms*, Van Nostrand-Reinhold, New York, 1980, pp. 83–91.
[5] F. Basolo and R. G. Pearson, *Mechanisms of Inorganic Reactions*, 2nd ed., John Wiley & Sons, New York, 1967, pp. 158–170.
[6] R. G. Wilkins, *The Study of Kinetics and Mechanism of Reactions of Transition Metal Complexes*, Allyn and Bacon, Boston, 1974, pp. 193–196.
[7] J. D. Atwood, *Inorganic and Organometallic Reaction Mechanisms*, Brooks/Cole, Monterey, CA, 1985, pp. 82–83.
[8] C. H. Langford and T. R. Stengle, *Ann. Rev. Phys. Chem.*, **1968**, *19*, 193.
[9] J. W. Moore and R. G. Pearson, *Kinetics and Mechanism*, 3rd ed., John Wiley & Sons, New York, 1981, pp. 357–363.

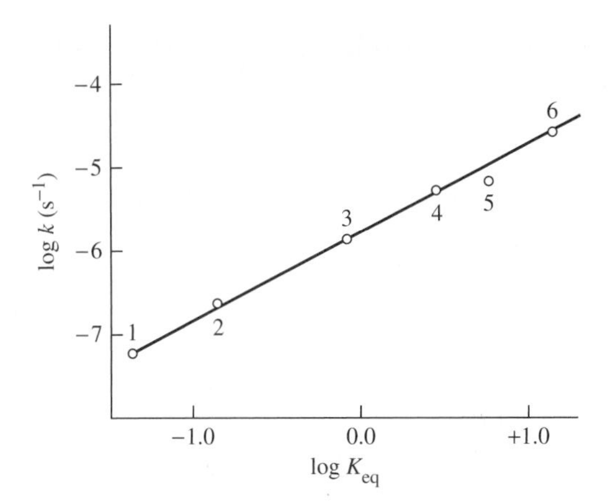

FIGURE 12.5 Linear Free Energy and $[Co(NH_3)_5X]^{2+}$ Hydrolysis. The log of the rate constant is plotted against the log of the equilibrium constant for the acid hydrolysis reaction of $[Co(NH_3)_5X]^{2+}$ ions. Measurements were made at 25.0 °C. Points are designated as follows: 1, $X^- = F^-$; 2, $X^- = H_2PO_4^-$; 3, $X^- = Cl^-$; 4, $X^- = Br^-$; 5, $X^- = I^-$; 6, $X^- = NO_3^-$.

(Reproduced with permission from C. H. Langford, *Inorg. Chem.*, **1965**, *4*, 265. Data for F^- from S. C. Chan, *J. Chem. Soc.*, **1964**, 2375, and for I^- from R. G. Yalman, *Inorg. Chem.*, **1962**, *1*, 16. All other data from A. Haim and H. Taube, *Inorg. Chem.*, **1964**, *3*, 1199.)

equation for temperature dependence of rate constants and the equation for temperature dependence of equilibrium constants. In logarithmic form, they are

$$\ln k = \ln A - \frac{E_a}{RT} \quad \text{and} \quad \ln K = \frac{-\Delta H°}{RT} + \frac{\Delta S°}{R}$$

kinetic thermodynamic

If the preexponential factor, A, and the entropy change, $\Delta S°$, are nearly constant, and the activation energy, E_a, depends on the enthalpy of reaction, $\Delta H°$, there will be a linear correlation between $\ln k$ and $\ln K$. A straight line on such a log–log plot is indirect evidence for a strong influence of the thermodynamic parameter, $\Delta H°$, on the activation energy of the reaction. In molecular bonding terms, a stronger bond between a metal and a leaving group results in a larger activation energy, a logical connection for a dissociative mechanism. Figure 12.5 shows an example from the hydrolysis of $[Co(NH_3)_5X]^{2+}$:

$$[Co(NH_3)_5X]^{2+} + H_2O \longrightarrow [Co(NH_3)_5(H_2O)]^{3+} + X^-$$

From this evidence, Langford[10] argued that the X^- group is essentially completely dissociated and acts as a solvated anion in the transition state of acid hydrolysis, and that water is, at most, weakly bound in the transition state. Another example from reactions of square-planar platinum complexes is given in Section 12.6.2.

Examples of the effect, or lack of effect, of incoming ligand are given in Tables 12.3 and 12.4. In Table 12.3, the data are for the first-order region (large [Y]). The k_1 column gives the rate constants for anion exchange; the $k_1/k_1(H_2O)$ shows the ratio of k_1 to the rate for water exchange. The rate constants are all relatively close to that for water exchange, as would be expected for a dissociative mechanism. Table 12.4 gives data for the second-order region for anation of $[Ni(H_2O)_6]^{2+}$. The second-order rate constant, k_0K_0, is the product of the ion-pair equilibrium constant, K_0, and the rate constant, k_0:

$$Ni{-}OH_2 + L \rightleftharpoons Ni{-}OH_2 \cdot L \qquad K_0$$

$$Ni{-}OH_2 \cdot L \longrightarrow Ni{-}L + H_2O \qquad k_0$$

K_0 is calculated from an electrostatic model that provides good agreement with the few cases in which experimental evidence is also available. The rate constant, k_0, varies by a factor of 5 or less and is close to the rate constant for the exchange of water. The close agreement for the wide variety of different ligands shows that the effect of the incoming ligand on the second step is minor, although the difference in ion-pair formation is significant. Both these reactions are consistent with D or I_d mechanisms, with ion-pair formation likely as the first step in the nickel reactions.

[10] C. H. Langford, *Inorg. Chem.*, **1965**, *4*, 265.

TABLE 12.3 Limiting Rate Constants for Anation or Water Exchange of $[Co(NH_3)_5H_2O]^{3+}$ at 45 °C.

$$[Co(NH_3)_5H_2O]^{3+} + Y^{m-} \longrightarrow [Co(NH_3)_5Y]^{(3-m)+} + H_2O$$

Y^{m-}	k_1 (10^{-6} s^{-1})	$k_1/k_1(H_2O)$	Reference
H_2O	100	1.0	a
N_3^-	100	1.0	b
SO_4^{2-}	24	0.24	c
Cl^-	21	0.21	d
NCS^-	16	0.16	d

Sources: [a]W. Schmidt and H. Taube, *Inorg. Chem.*, **1963**, 2, 698.
[b]H. R. Hunt and H. Taube, *J. Am. Chem. Soc.*, **1958**, *75*, 1463.
[c]T. W. Swaddle and G. Guastalla, *Inorg. Chem.*, **1969**, *8*, 1604.
[d]C. H. Langford and W. R. Muir, *J. Am. Chem. Soc.*, **1967**, *89*, 3141.

TABLE 12.4 Rate Constants for Substitution on $[Ni(H_2O)_6]^{2+}$

Y	k_0K_0(10^3 M^{-1} s^{-1})	K_0 (M^{-1})	k_0 (10^4 s^{-1})
$CH_3PO_4^{2-}$	290	40	0.7
CH_3COO^-	100	3	3
NCS^-	6	1	0.6
F^-	8	1	0.8
HF	3	0.15	2
H_2O			3
NH_3	5	0.15	3
C_5H_5N, pyridine	~4	0.15	~3
$C_4H_4N_2$, pyrazine	2.8	0.15	2
$NH_2(CH_2)_2NMe_3^+$	0.4	0.02	2

Source: Adapted with permission from R. G. Wilkins, *Acc. Chem. Res.*, **1970**, 3, 408; $C_4H_4N_2$ data are from J. M. Malin and R. E. Shepherd, *J. Inorg. Nucl. Chem.*, **1972**, *34*, 3203.

12.4.3 Associative Mechanisms

Associative reactions are also possible in octahedral substitution but are much less common.[11] Table 12.5 gives data for both dissociative and associative interchanges for similar reactants. In the case of water substitution by several different anions in $[Cr(NH_3)_5(H_2O)]^{3+}$, the rate constants are similar (within a factor of 6), indicative of an I_d mechanism. On the other hand, the same ligands reacting with $[Cr(H_2O)_6]^{3+}$ show a large variation in rates (more than a 2000-fold difference), indicative of an I_a mechanism. Data for similar Co(III) complexes are not conclusive, but their reactions generally seem to have I_d mechanisms.

[11] Atwood, *Inorganic and Organometallic Reaction Mechanisms*, p. 85.

TABLE 12.5 Effects of Entering Group on Rates

	Rate Constants for Anation	
Entering Ligand	$[Cr(H_2O)_6]^{3+}$ k (10^{-8} M^{-1} s^{-1})	$[Cr(NH_3)_5H_2O]^{3+}$ k (10^{-4} M^{-1} s^{-1})
NCS^-	180	4.2
NO_3^-	73	—
Cl^-	2.9	0.7
Br^-	1.0	3.7
I^-	0.08	—
CF_3COO^-	—	1.4

Source: Reproduced with permission from J. D. Atwood, *Inorganic and Organometallic Reaction Mechanisms*, Books/Cole, Monterey, CA, 1985, p. 85; data from D. Thusius, *Inorg. Chem.*, **1971**, *10*, 1106; T. Ramasami and A. G. Sykes, *Chem. Commun. (Cambridge)*, **1978**, 378.

Reactions of Ru(III) compounds frequently have associative mechanisms, and those of Ru(II) compounds generally have dissociative mechanisms. The entropies of activation for substitution reactions of $[Ru(III)(EDTA)(H_2O)]^-$ are negative, indicating association as part of the transition state. They also show a large range of rate constants depending on the incoming ligand (Table 12.6), as required for an I_a mechanism; but those of Ru(II) (Table 12.7) are nearly the same for different ligands, as required for an I_d mechanism. The reasons for this difference are unclear. Both complexes have a free carboxylate (the EDTA is pentadentate, with the sixth position occupied by a water molecule). Hydrogen bonding between this free carboxylate and the bound water may distort the shape sufficiently in the Ru(III) complex to open a place for entry by the incoming ligand. Although similar hydrogen bonding may be possible for the

TABLE 12.6 Rate Constants for $[Ru(III)(EDTA)(H_2O)]^-$ Substitution

Ligand	k_1(M^{-1} s^{-1})	$\Delta H^\ddagger$ (kJ mol^{-1})	$\Delta S^\ddagger$ (J mol^{-1} K^{-1})
Pyrazine	20,000 ± 1,000	5.7 ± 0.5	−20 ± 3
Isonicotinamide	8,300 ± 600	6.6 ± 0.5	−19 ± 3
Pyridine	6,300 ± 500		
Imidazole	1,860 ± 100		
SCN^-	270 ± 20	8.9 ± 0.5	−18 ± 3
CH_3CN	30 ± 7	8.3 ± 0.5	−24 ± 4

Source: T. Matsubara and C. Creutz, *Inorg. Chem.*, **1979**, *18*, 1956.

TABLE 12.7 Rate Constants for $[Ru(II)(EDTA)(H_2O)]^{2-}$ Substitution

Ligand	k_1 (M^{-1} s^{-1})
Isonicotinamide	30 ± 15
CH_3CN	13 ± 1
SCN^-	2.7 ± 0.2

Source: T. Matsubara and C. Creutz, *Inorg. Chem.*, **1979**, *18*, 1956.

Ru(II) complex, the increased negative charge may reduce the Ru—H_2O bond strength enough to promote dissociation.

12.4.4 The Conjugate Base Mechanism

Other cases in which second-order kinetics seemed to require an associative mechanism have subsequently been found to have a **conjugate base mechanism,**[12] called S_N1CB for substitution, nucleophilic, unimolecular, conjugate bases in Ingold's notation.[13] These reactions depend on amine, ammine, or aqua ligands that can lose protons to form amido or hydroxo species that are then more likely to lose one of the other ligands. If the structure allows it, the ligand *trans* to the amido or hydroxo group is frequently the one lost.

$$[Co(NH_3)_5X]^{2+} + OH^- \rightleftharpoons [Co(NH_3)_4(NH_2)X]^+ + H_2O \quad \text{(equilibrium)} \quad (1)$$

$$[Co(NH_3)_4(NH_2)X]^+ \longrightarrow [Co(NH_3)_4(NH_2)]^{2+} + X^- \quad \text{(slow)} \quad (2)$$

$$[Co(NH_3)_4(NH_2)]^{2+} + H_2O \longrightarrow [Co(NH_3)_5(OH)]^{2+} \quad \text{(fast)} \quad (3)$$

Overall,

$$[Co(NH_3)_5X]^{2+} + OH^- \longrightarrow [Co(NH_3)_5(OH)]^{2+} + X^-$$

In the third step, addition of a ligand other than water is also possible; in basic solution, the rate constant is k_{OH}, and the equilibrium constant for the overall reaction is K_{OH}.

Additional evidence for the conjugate base mechanism has been provided by several related studies:

1. Base-catalyzed exchange of hydrogen from the amine groups takes place under the same conditions as these reactions.
2. The isotope ratio ($^{18}O/^{16}O$) in the product in ^{18}O-enriched water is the same as that in the water regardless of the leaving group ($X^- = Cl^-, Br^-, NO_3^-$). If an incoming water molecule had a large influence (an associative mechanism), a higher concentration of ^{18}O should be in the product, because the equilibrium constant $K = 1.040$ for the reaction

$$H_2{}^{16}O + {}^{18}OH^- \rightleftharpoons H_2{}^{18}O + {}^{16}OH^-$$

3. RNH_2 complexes (R = alkyl) react faster than NH_3 complexes, possibly because steric crowding favors the 5-coordinate intermediate formed in Step 2.
4. The rate constants and dissociation constants for these complexes form a linear free-energy relationship (LFER), in which a plot of ln k_{OH} versus ln K_{OH} is linear.
5. When substituted amines are used, and there are no protons on the nitrogens available for ionization, the reaction is very slow or nonexistent.

Reactions with $[Co(tren)(NH_3)Cl]^{2+}$ isomers show that the position *trans* to the leaving group is the most likely deprotonation site for a conjugate base mechanism.[14] The reaction in Figure 12.6(a) is 10^4 times faster than that in Figure 12.6(b). In addition, most of the product in both reactions is best explained by a trigonal-bipyramidal

[12] Wilkins, *The Study of Kinetics and Mechanism of Reactions of Transition Metal Complexes*, pp. 207–210; Basolo and Pearson, Mechanisms of Inorganic Reactions, pp. 177–193.
[13] C. K. Ingold, *Structure and Mechanism in Organic Chemistry*, Cornell University Press, Ithaca, NY, 1953, Chapters 5 and 7.
[14] D. A. Buckingham, P. J. Cressell, and A. M. Sargeson, *Inorg. Chem.*, **1975**, *14*, 1485.

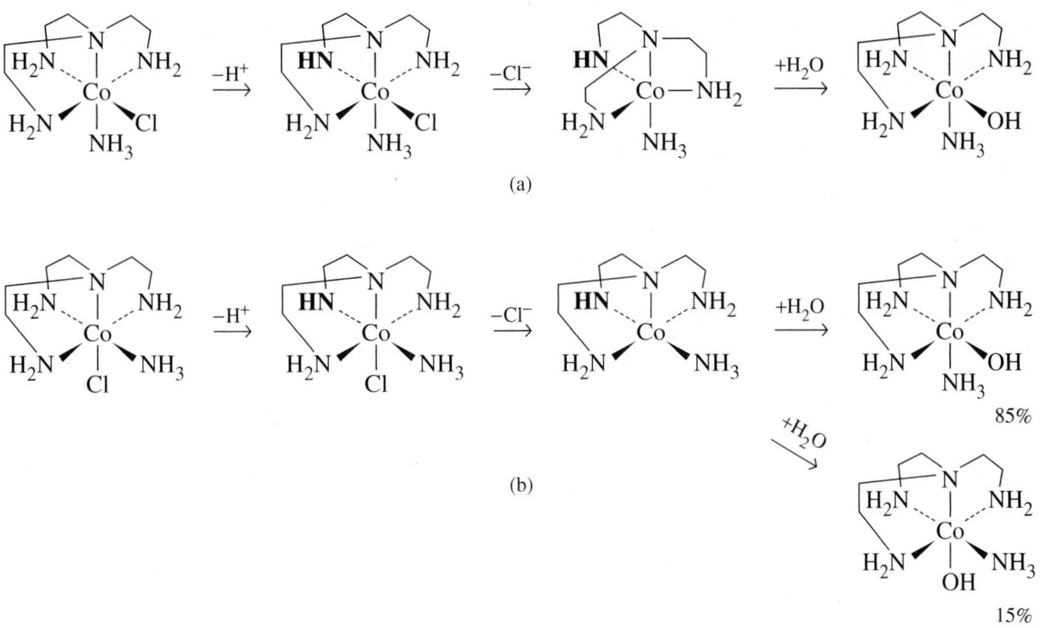

FIGURE 12.6 Base Hydrolysis of $[Co(tren)(NH_3)Cl]^{2+}$ Isomers. (a) Leaving group (Cl^-) *trans* to deprotonated nitrogen. (b) Leaving group (Cl^-) *cis* to deprotonated nitrogen. (a) Is 10^4 faster than (b), indicating that *trans* substitution is strongly favored. (Data from D. A. Buckingham, P. J. Creswell, and A. M. Sargeson, *Inorg. Chem.*, **1975**, *14*, 1485.)

intermediate or transition state with the deprotonated amine in the trigonal plane. The reaction in Figure 12.6(a) can form this state immediately; the reaction in Figure 12.6(b) requires rearrangement of an initial square-pyramidal structure.

Explanations of the promotional effect of the amido group center on its basic strength, either as a σ donor or because of ligand-to-metal π interaction. The π interaction is most effective when the amido group is part of the trigonal plane in a trigonal-bipyramidal geometry, but there is at least one case in which this geometry is not necessarily achieved.[15]

12.4.5 The Kinetic Chelate Effect

The thermodynamic chelate effect, which causes polydentate complexes to be thermodynamically more stable than their monodentate counterparts,[16] was described in Section 10.1.1. The difference in rate of attachment and dissociation of the second (and third, or higher numbered) point of attachment for the ligand is also observed kinetically.

Substitution for a chelated ligand is generally a slower reaction than that for a similar monodentate ligand. Explanations for this effect center on two factors, the increased energy needed to remove the first bound atom and the probability of a reversal of this first step.[17]

[15] D. A. Buckingham, P. A. Marzilli, and A. M. Sargeson, *Inorg. Chem.*, **1969**, *8*, 1595.
[16] Basolo and Pearson, *Mechanisms of Inorganic Reactions*, pp. 27, 223; G. Schwarzenbach, *Helv. Chim. Acta*, **1952**, *35*, 2344.
[17] D. W. Margerum, G. R. Cayley, D. C. Weatherburn, and G. K. Pagenkopf, "Kinetics of Complex Formation and Ligand Exchange," in A. E. Martell, ed., *Coordination Chemistry*, Vol. 2, American Chemical Society Monograph 174, Washington, DC, 1978, pp. 1–220.

The reaction must have two dissociation steps for a bidentate ligand, one for each bound atom (the addition of water in Steps 2 and 4 is likely to be fast, because of its high concentration):

$$\mathrm{M}\begin{matrix}\diagup \mathrm{NH_2{-}CH_2} \\ \diagdown \mathrm{NH_2{-}CH_2}\end{matrix}\!\Big| \;\xrightleftharpoons{\text{slow}}\; \mathrm{M}\begin{matrix}\diagdown \\ \mathrm{NH_2{-}CH_2}\diagup\end{matrix}\!\!\mathrm{CH_2{-}NH_2} \tag{1}$$

$$\mathrm{M{-}NH_2{-}CH_2{-}CH_2{-}NH_2} + \mathrm{H_2O} \;\xrightleftharpoons{\text{fast}}\; \mathrm{(H_2O)M{-}NH_2{-}CH_2{-}CH_2{-}NH_2} \tag{2}$$

$$\mathrm{(H_2O)M{-}NH_2{-}CH_2{-}CH_2{-}NH_2} \;\xrightleftharpoons{\text{slow}}\; \mathrm{M{-}OH_2} + \mathrm{NH_2CH_2CH_2NH_2} \tag{3}$$

$$\mathrm{M{-}OH_2} + \mathrm{H_2O} \;\xrightleftharpoons{\text{fast}}\; \mathrm{M(OH_2)_2} \tag{4}$$

The first dissociation (1) is expected to be slower than a similar dissociation of ammonia, because the ligand must bend and rotate to move the free amine group away from the metal. The second dissociation (3) is likely to be slow, because the concentration of the intermediate is low, and because the first dissociation can readily reverse. The uncoordinated nitrogen is held near the metal by the rest of the ligand, making reattachment more likely. Overall, this kinetic chelate effect reduces the rates of aquation reactions by factors from 20 to 10^5.

12.5 STEREOCHEMISTRY OF REACTIONS

A common assumption is that reactions with dissociative mechanisms are more likely to result in random isomerization or racemization, and associative mechanisms are more likely to result in single-product reactions; however, the evidence is much less clear-cut. Dissociative mechanisms can lead to single-product reactions with either retention of configuration or a change of configuration, depending on the circumstances. For example, base hydrolysis of Λ-*cis*-$[Co(en)_2Cl_2]^+$ in dilute (< 0.01 M) hydroxide yields Λ-*cis*-$[Co(en)_2(OH)_2]^+$, but in more concentrated (> 0.25 M) hydroxide, it gives Δ-*cis*-$[Co(en)_2(OH)_2]^+$ (Tables 12.8 and 12.9 and Figure 12.7).[18] A conjugate base mechanism is expected in both cases, with the hydroxide removing a proton from an ethylenediamine nitrogen, followed by loss of the chloride *trans* to the deprotonated nitrogen. In more concentrated base, the higher concentration of ion pairs ($[Co(en)_2Cl_2]^+ \cdot OH^-$) is assumed to result in a water molecule (from the OH^- and the H^+ removed from ethylenediamine), positioned for easy addition with inversion of the chiral center.

A similar change in product, this time dependent on temperature, takes place in the substitution of ammonia for both chlorides[19] in $[Co(en)_2Cl_2]^+$. At low temperatures (–33 °C or below, in liquid ammonia), there is inversion of configuration; at higher temperatures (above 25 °C in liquid ammonia, alcohol solution, or solid exposed to gaseous ammonia), there is retention. In both cases, a small fraction of the *trans* isomer also forms.

---

[18] L. J. Boucher, E. Kyuno, and J. C. Bailar, Jr., *J. Am. Chem. Soc.*, **1964**, *86*, 3656.

[19] J. C. Bailar, Jr., J. H. Haslam, and E. M. Jones, *J. Am. Chem. Soc.*, **1936**, *58*, 2226; E. Kyuno and J. C. Bailar, Jr., *J. Am. Chem. Soc.*, **1966**, *88*, 1125.

TABLE 12.8 Stereochemistry of Acid Aquation

$$[Co(en)_2LX]^{n+} + H_2O \longrightarrow [Co(en)_2LH_2O]^{(1+n)+} + X^-$$

cis-L	X	% *cis* Product	*trans*-L	X	% *cis* Product
OH^-	Cl^-	100	OH^-	Cl^-	75
OH^-	Br^-	100	OH^-	Br^-	73
Br^-	Cl^-	100	Br^-	Cl^-	50
Cl^-	Cl^-	100	Br^-	Br^-	30
Cl^-	Br^-	100	Cl^-	Cl^-	35
N_3^-	Cl^-	100	Cl^-	Br^-	20
NCS^-	Cl^-	100	NCS^-	Cl^-	50–70
NCS^-	Br^-	100	NH_3	Cl^-	0
NO_2^-	Cl^-	100	NO_2^-	Cl^-	0

Source: Data from F. Basolo and R. G. Pearson, *Mechanisms of Inorganic Reactions*, 2nd ed., J. Wiley & Sons, New York, 1967, p. 257.

TABLE 12.9 Stereochemistry of Base Substitution

$$[Co(en)_2LX]^{n+} + OH^- \longrightarrow [Co(en)_2LOH]^{n+} + X^-$$

		% *cis* Product					
cis-L	X	Δ	Racemic[a]	Λ	*trans*-L	X	% *cis* Product
OH^-	Cl^-	61		36	OH^-	Cl^-	94
OH^-	Br^-		96		OH^-	Br^-	90
Cl^-	Cl^-	21		16	Cl^-	Cl^-	5
Cl^-	Br^-		30		Cl^-	Br^-	5
Br^-	Cl^-		40		Br^-	Cl^-	0
N_3^-	Cl^-		51		N_3^-	Cl^-	13
NCS^-	Cl^-	56		24	NCS^-	Cl^-	76
NH_3	Br^-	59		26	NCS^-	Br^-	81
NH_3	Cl^-	60		24	NH_3	Cl^-	76
NO_2^-	Cl^-	46		20	NO_2^-	Cl^-	6

Source: Data from F. Basolo and R. G. Pearson, *Mechanisms of Inorganic Reactions*, 2nd ed., J. Wiley & Sons, New York, 1967, p. 262.

NOTE: The total % *cis* product is the sum of Δ and Λ obtained from the Δ-*cis* starting material. The optically inactive *trans* isomer will yield racemic *cis*; % *trans* = 100% – % *cis*.

[a] Racemic reactant, so the product is also racemic.

Although not a complete explanation of these reactions, all the reported inversion reactions occur under conditions in which a conjugate base mechanism is possible.[20] The orientation of the ligand entering the proposed trigonal-bipyramidal intermediate then dictates the configuration of the product. In some cases, a preferred orientation of the other ligands may dictate the product. For example, the β form of trien complexes is more stable than the α form; both are shown in Figure 9.18.

[20] Basolo and Pearson, *Mechanisms of Inorganic Reactions*, p. 272.

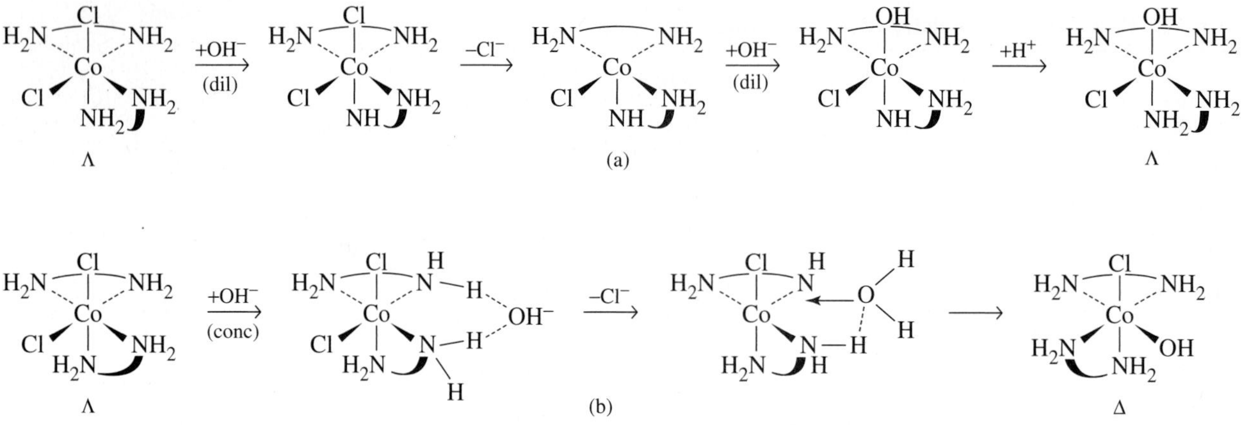

FIGURE 12.7 Mechanisms of Base Hydrolysis of Λ-*cis*-$[Co(en)_2Cl_2]^+$. (a) Retention of configuration in dilute hydroxide. (b) Inversion of configuration in concentrated hydroxide.

12.5.1 Substitution in *trans* Complexes

Substitution of Y for X in *trans*-$[M(LL)_2BX]$ (LL = a bidentate ligand such as en) can proceed by three different pathways. If dissociation of X from the reactant leaves a square-pyramidal intermediate that then adds the new ligand directly into the vacant site, the result is retention of configuration and the product, like the reactant, is *trans*, shown in Figure 12.8(a). A trigonal-bipyramidal intermediate with B in the trigonal plane leads to a mixture of *trans* and *cis*, as shown in Figure 12.8(b). The incoming ligand can enter along any of the three sides of the triangle, resulting in two *cis* possibilities and one *trans* possibility. Dissociation to form a trigonal pyramid with B in an axial position, Figure 12.8(c), allows two positions for attack by Y, both of which give *cis* products (the third side of the triangle is blocked by an LL ring). An intermediate with an axial B is less likely than one with an equatorial B, because an axial B requires more rearrangement of the ligands (a 90° change by one nitrogen and 30° changes by two others, in contrast to two 30° changes for the equatorial B) as well as a larger stretch for the LL ring in the equatorial plane. As a result, the statistical probability of a change from *trans* to *cis* is two thirds for a trigonal-bipyramidal intermediate.

▶ **Exercise 12.2** Starting with the structure shown here, follow the example of Figure 12.8(b) and show that the first two products would be Δ rather than Λ.

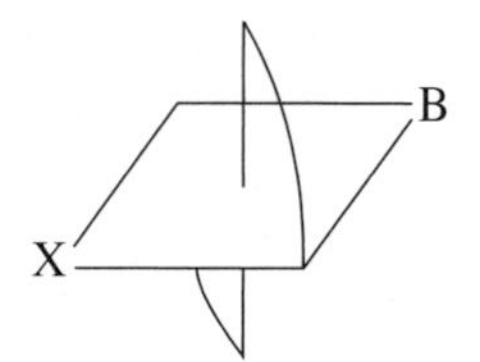

(Experimentally, the two chiral forms are equally likely, because how we draw the structures has no effect on the experimental result.)

In fact, experimental results indicate that the statistical distribution is seldom followed. With *trans* reactants, both acid aquation and base substitution reactions result in a mixture of isomers; the fractions of *cis* and *trans* depend on the retained ligand and range from 100% *trans* to 94% *cis*, as shown in Tables 12.8 and 12.9. Any *cis* isomer produced from the optically inactive *trans* reactants will be a racemic mixture of Λ and Δ. For both

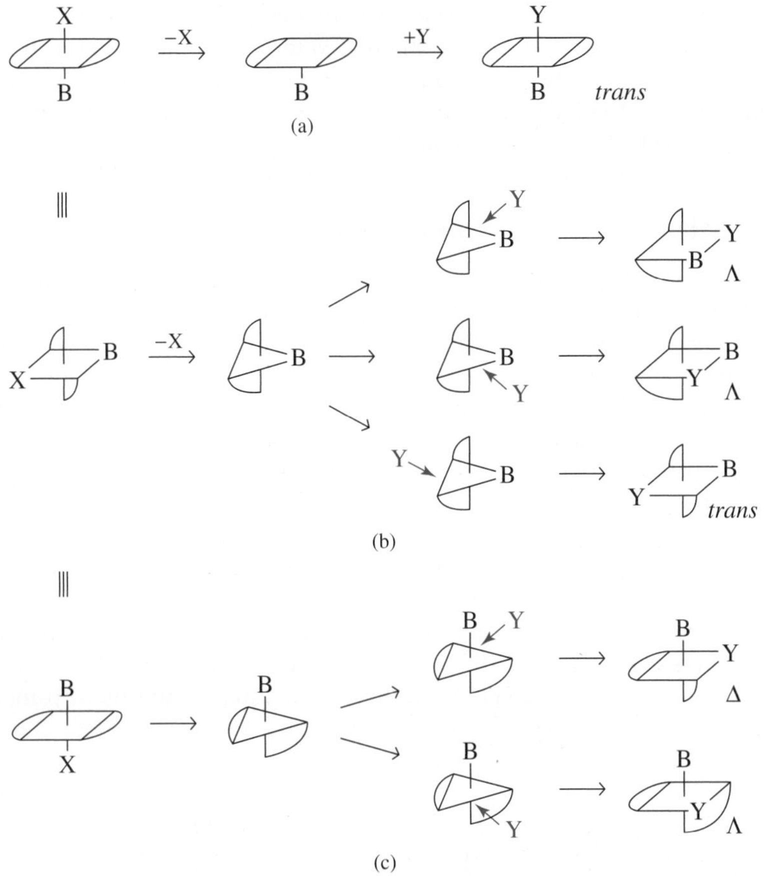

FIGURE 12.8 Dissociation Mechanism and Stereochemical Changes for *trans*-$[M(LL)_2BX]$. (a) Square-pyramidal intermediate (retention of configuration). (b) Trigonal-bipyramidal intermediate (three possible products). (c) Less likely trigonal-bipyramidal intermediate (two possible products).

the axial and equatorial types of trigonal-bipyramidal intermediates, the chiral form of the product is determined by which square plane becomes trigonal in the intermediate; Δ and Λ products are equally likely.

Other factors, such as the leaving ligand X, can strongly influence both mechanism and outcome, making the products deviate further from statistical probability. Data on several reactions for $[Co(en)_2(H_2O)X]^{n+}$ are shown in Table 12.10. For X = Cl^-, SCN^-,

TABLE 12.10 Rate Constants for Reactions of $[Co(en)_2(H_2O)X]^{n+}$ at 25 °C, $k(10^{-5}\ s^{-1})$

X	*cis* ⟶ *trans*	*trans* ⟶ *cis*	Racemization	H_2O Exchange
OH^-	200	300	—	160
Br^-	5.4	16.1	—	—
Cl^-	2.4	7.2	2.4	—
N_3^-	2.5	7.4	—	—
NCS^-	0.0014	0.071	0.022	0.13
H_2O	0.012	0.68	~ 0.015	1.0
NH_3	< 0.0001	0.002	0.003	0.10
NO_2^-	0.012	0.005	—	—

Source: Adapted with permission from R. G. Wilkins, *The Study of Kinetics and Mechanism of Reactions of Transition Metal Complexes*, Allyn and Bacon, Boston, 1974, p. 344. Data from M. L. Tobe, in J. H. Ridd, ed., *Studies in Structure and Reactivity*, Methuen, London, 1966, and M. N. Hughes, *J. Chem. Soc., A*, **1967**, 1284.

and H_2O, racemization and *cis* $\longrightarrow$ *trans* conversion are nearly equal in rate, making it likely that they have identical intermediates. Water exchange is faster than the other reactions for all but the hydroxide complex; simple exchange with the solvent requires no rearrangement of the ligands, and the high concentration of water makes it more likely. For X = NH_3, racemization and *trans* $\longrightarrow$ *cis* conversion are faster than *cis* $\longrightarrow$ *trans* conversion. The reasons for this difference are not clear.

12.5.2 Substitution in *CIS* Complexes

Substitution in *cis* complexes can proceed by the same three intermediates as in the *trans* complexes. Again, a square-pyramidal intermediate results in retention of configuration, providing a *cis* product in this case. If dissociation of X forms a trigonal bipyramid with B in the trigonal plane, there are three possible locations for the addition of Y, all in the same trigonal plane. Two of these result in *cis* products and one in a *trans* product. The less likely trigonal bipyramid with an axial B, whether derived from a *cis* or a *trans* reactant, produces two *cis* products that are half Δ and half Λ. These possibilities are all shown in Figure 12.9.

An optically active *cis* complex can yield products that retain the same configuration, convert to *trans* geometry, or create a racemic mixture. Statistically, the product of substitution of a *cis*-$[M(LL)_2BX]$ complex through a trigonal-bipyramidal intermediate should be one sixth *trans* if both intermediates were equally likely and one third *trans* if the axial B form is not formed at all. Experimentally, aquation of *cis*-$[M(LL)_2BX]$ in acid results in 100% *cis* isomer (Table 12.9), indicating a square-pyramidal transition state.

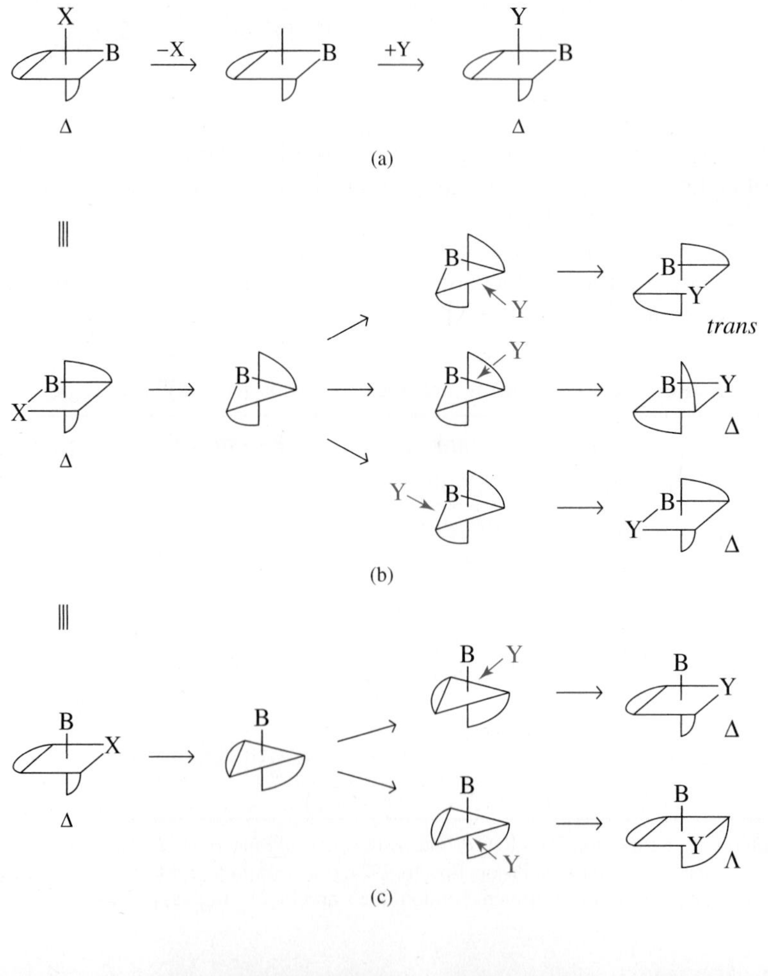

FIGURE 12.9 Dissociation Mechanism and Stereochemical Changes for *cis*-$[M(LL)_2BX]$. (a) Square-pyramidal intermediate (retention of configuration). (b) Trigonal-bipyramidal intermediate (three possible products). (c) Unlikely trigonal bipyramidal intermediate (two possible products).

Substitution of optically active *cis* complexes in base gives products ranging from 95% to 30% *cis*, with about 2:1 retention of chiral configuration (Table 12.9); four of the reactants listed in Table 12.9 are racemic, so the product is also a racemic *cis* mixture). Among the compounds that retain their optical activity and geometry on hydrolysis are $[M(en)_2Cl_2]^+$ with M = Co, Rh, and Ru.[21] As a general rule, *cis* reactants retain their *cis* configuration, but *trans* reactants are more likely to give a mixture of *cis* and *trans* products.

12.5.3 Isomerization of Chelate Rings

Isomerization has been described previously for a number of complexes with monodentate ligands or with two bidentate ligands. Similar reactions with three bidentate ligands or with more complex ligands can follow two different mechanisms. In some cases, one end of a chelate ring dissociates, and the resulting 5-coordinate intermediate rearranges before reattachment of the loose end. This mechanism does not differ appreciably from the substitution reactions described in Sections 12.5.1 and 12.5.2; the ligand that dissociates in the first step is the same one that adds in the final step, after rearrangement.

PSEUDOROTATION Other isomerization mechanisms involving compounds containing chelating ligands are different types of twists. A number of twist mechanisms have been described, with different movements of the rings; those most commonly considered are shown in Figure 12.10.

The trigonal, or Bailar, twist, Figure 12.10(a), requires twisting the two opposite trigonal faces through a trigonal prismatic transition state to the new structure. In the tetragonal twists, one chelate ring is held stationary, while the other two are twisted to the new structure. The first one illustrated, Figure 12.10(b), has a transition state with the stationary ring perpendicular to those being twisted. The second tetragonal twist, Figure 12.10(c), requires twisting the two rings through a transition state with all three rings parallel. Attempts have been made to determine which of these mechanisms is applicable, but the complexity of the reactions and the indirect means of measurement leave them subject to various interpretations. NMR study of tris(trifluoroacetylacetonato) metal(III) chelates shows that a trigonal twist mechanism is not possible

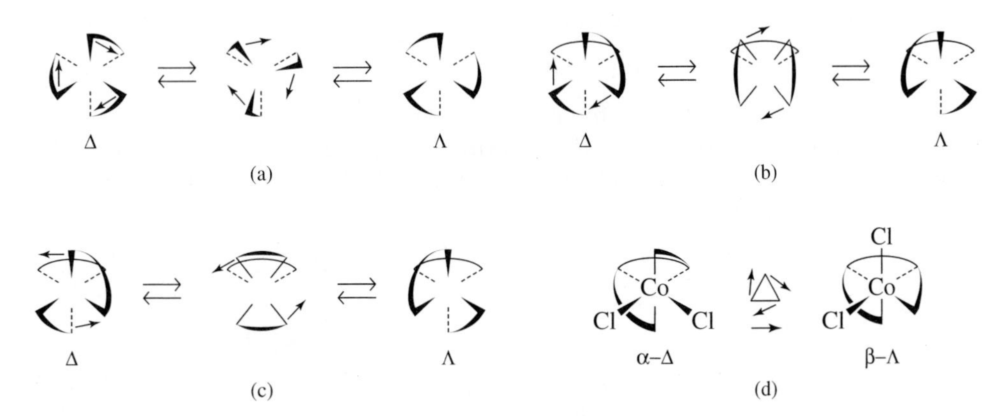

FIGURE 12.10 Twist Mechanisms for Isomerization of $M(LL)_3$ and $[Co(trien)Cl_2]^+$ Complexes. (a) Trigonal twist: the front triangular face rotates with respect to the back triangular face. (b) Tetragonal twist with perpendicular rings: the back ring remains stationary, as the front two rings rotate clockwise. (c) Tetragonal twist with parallel rings: the back ring remains stationary, as the front two rings rotate counterclockwise. (d) $[Co(trien)Cl_2]^+$ α-β isomerization: the connected rings limit this isomerization to a clockwise trigonal twist of the front triangular face.

[21] S. A. Johnson, F. Basolo, and R. G. Pearson, *J. Am. Chem. Soc.*, **1963**, *85*, 1741; J. A. Broomhead and L. Kane-Maguire, *Inorg. Chem.*, **1969**, *8*, 2124.

for M = Al, Ga, In, and *fac*-Cr but leaves it a possibility for *fac*-Co.[22] The multiple-ring structure of *cis*-α-$[Co(trien)Cl_2]^+$ allows only a trigonal twist in its conversion to the β isomer, as shown in Figure 12.10(d).

12.6 SUBSTITUTION REACTIONS OF SQUARE-PLANAR COMPLEXES

The products of substitution reactions of square-planar complexes—platinum(II) complexes are the primary examples—have the same configuration as the reactants, with direct replacement of the departing ligand by the new ligand. The rates vary enormously, and different compounds can be formed, depending both on the entering and the departing ligands. This section and Section 12.7 describe some of these effects.

12.6.1 Kinetics and Stereochemistry of Square-Planar Substitutions

Because many of the reactions studied have been with platinum compounds, we will use as our initial example the simplified reaction

$$\text{T—Pt—X} + \text{Y} \longrightarrow \text{T—Pt—Y} + \text{X}$$

where T is the ligand *trans* to the departing ligand X, and Y is the incoming ligand. We will also designate the plane of the molecule the xy plane and the Pt axis through T—Pt—X the x axis, as shown in Figure 12.11. The other two ligands, L, are of lesser importance and will be ignored for the moment.

It is generally accepted that reactions of square-planar complexes are associative, although there may be various degrees of association, and they are classified as I_a. Two mechanisms, both associative, are shown in Figure 12.11. In mechanism (a) the incoming ligand approaches along the z axis. As it bonds to the Pt, the complex rearranges to approximate a trigonal bipyramid with Pt, T, X, and Y in the trigonal plane. As X leaves, Y moves down into the plane of T, Pt, and the two L ligands. This same general description will fit whether the incoming ligand bonds strongly to Pt before the departing ligand bond is weakened appreciably (I_a), or the departing ligand bond is weakened considerably before the incoming ligand forms its bond (I_d). The solvent-assisted mechanism (b) follows the same pattern but requires two associative steps for completion.

Square-planar substitution reactions frequently show two-term rate laws, of the form

$$\text{Rate} = k_1\,[\text{Cplx}] + k_2\,[\text{Cplx}][\text{Y}]$$

where [Cplx] = concentration of the reactant complex and [Y] = concentration of the incoming ligand. Both pathways (both terms in the rate law) are considered to be associative, in spite of the difference in order. The k_2 term easily fits an associative

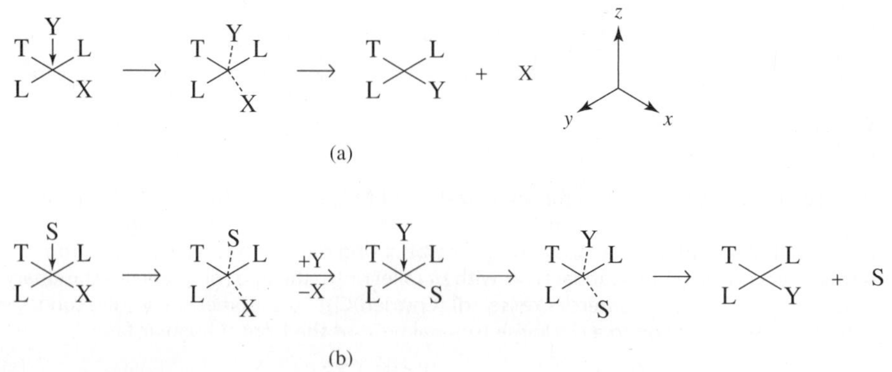

(a)

(b)

FIGURE 12.11 The Interchange Mechanism in Square-Planar Reactions. (a) Direct substitution by Y. (b) Solvent-assisted substitution.

[22] R. C. Fay and T. S. Piper, *Inorg. Chem.*, **1964**, *3*, 348.

mechanism (a) in which the incoming ligand Y and the reacting complex form a 5-coordinate transition state. The accepted explanation for the k_1 term is a solvent-assisted reaction (b), with solvent replacing X on the complex through a similar 5-coordinate transition state, and then itself being replaced by Y. The second step of this mechanism is presumed to be faster than the first, and the concentration of solvent is large and unchanging (leading to pseudo first order conditions), so the overall rate law for this path is approximated as first order in complex.

12.6.2 Evidence for Associative Reactions

The evidence for a 5-coordinate intermediate is very strong, including isolation of several 5-coordinate complexes with trigonal-bipyramidal geometry ($[Ni(CN)_5]^{3-}$, $[Pt(SnCl_3)_5]^{3-}$, and similar complexes), although Basolo and Pearson argue that the transition state may well be 6-coordinate, with assistance from solvent.[23] The highest energy transition state may be either during the formation of the intermediate or as the exiting ligand dissociates from the intermediate.

This mechanism explains naturally the effect of the incoming ligand. A strong Lewis base is likely to react readily, but the hard-soft nature of the base has an even greater effect. Pt(II) is generally a soft acid, so soft ligands react more readily with it. The order of ligand reactivity depends somewhat on the other ligands on the Pt, but the order for the reaction

$$trans\text{-}PtL_2Cl_2 + Y \longrightarrow trans\text{-}PtL_2ClY + Cl^-$$

for different Y in methanol was found to be as follows (see examples in Table 12.11):

$$PR_3 > CN^- > SCN^- > I^- > Br^- > N_3^- > NO_2^- > py > NH_3 \sim Cl^- > CH_3OH$$

A similar order, with some shuffling of the center of the list, is found for reactants with ligands other than chloride as T. The ratio of the rate constants for the extremes in the list is very large, with $k(PPh_3)\ /\ k(CH_3OH) = 9 \times 10^8$. Because T and Y have similar positions in the transition state, it is reasonable for them to have similar effects on the rate, and they do. Discussion of this ***trans* effect** is in the next section.

TABLE 12.11 Rate Constants and LFER Parameters for Entering Groups

$trans\text{-}PtL_2Cl_2 + Y \longrightarrow trans\text{-}PtL_2ClY + Cl^-$

	k ($10^{-3}\ M^{-1}\ s^{-1}$)		
Y	$L = py$ ($s = 1$)	$L = PEt_3$ ($s = 1.43$)	η_{Pt}
PPh_3	249,000		8.93
SCN^-	180	371	5.75
I^-	107	236	5.46
Br^-	3.7	0.93	4.18
N_3^-	1.55	0.2	3.58
NO_2^-	0.68	0.027	3.22
NH_3	0.47		3.07
Cl^-	0.45	0.029	3.04

Source: Rate constants from U. Belluco, L. Cattalini, F. Basolo, R. G. Pearson, and A. Turco, *J. Am. Chem. Soc.*, **1965**, *87*, 241; PPh_3 and η_{Pt} data from R. G. Pearson, H. Sobel, and J. Songstad, *J. Am. Chem. Soc.*, **1968**, *90*, 319.
NOTE: s and η are nucleophilic reaction parameters explained in the text.

[23] Basolo and Pearson, *Mechanisms of Inorganic Reactions*, pp. 377–379, 395.

By the same argument, the leaving group X should also have a significant influence on the rate, and it does (Table 12.12).[24] The order of ligands is nearly the reverse of that given above, with hard ligands such as Cl^-, NH_3, and NO_3^- leaving quickly. Soft ligands with considerable π bonding, such as CN^- and NO_2^-, leave reluctantly; in the reaction

$$[Pt(dien)X^+] + py \longrightarrow [Pt(dien)(py)]^{2+} + X^-$$

the rate increases by a factor of 10^5 with H_2O as compared with $X^- = CN^-$ or NO_2^- as the leaving group. The bond-strengthening effect of the metal-to-ligand π bonding reduces the reactivity of these ligands significantly. In addition, π bonding to the leaving group uses the same orbitals as those bonding to the entering group in the trigonal plane. These two effects result in the slow displacement of metal-to-ligand π-bonding ligands when compared with ligands with only σ bonding or ligand-to-metal π bonding.

Good leaving groups, those that leave easily, show little discrimination between entering groups. Apparently, the ease of breaking the Pt—X bond takes precedence over the formation of the Pt—Y bond. On the other hand, for complexes with less reactive leaving groups, the other ligands have a significant role; the softer PEt_3 and $AsEt_3$ ligands show a large selective effect when compared with the harder dien or en ligands. The LFER equation[25] for this comparison is

$$\log k_Y = s\,\eta_{Pt} + \log k_S$$

where

k_Y = rate constant for reaction with Y

k_S = rate constant for reaction with solvent

s = **nucleophilic discrimination factor** (for the complex)

η_{Pt} = **nucleophilic reactivity constant** (for the entering ligand)

The parameter s is defined as 1 for *trans*-$[Pt(py)_2Cl_2]$ and has values from 0.44 for the hard $[Pt(dien)H_2O]^{2+}$ to 1.43 for the soft *trans*-$[Pt(PEt_3)_2Cl_2]$. Values of η_{Pt} are found by the equation $\eta_{Pt} = \log(k_Y/k_S)$, where k_S refers to reaction with *trans*-$[Pt(py)_2Cl_2]$ in methanol at 30 °C. Table 12.11 shows both factors. For L = PEt_3, the change in rate constant is greater than for L = py because of the larger s value, and the increase in rate constants

TABLE 12.12 Rate Constants for Leaving Groups

$$[Pt(dien)X]^+ + py \longrightarrow [Pt(dien)py]^{2+} + X^-$$
$$(\text{Rate} = (k_1 + k_2\,[py])[Pt(dien)X]^+)$$

X^-	k_2 ($M^{-1}\,s^{-1}$)
NO_3^-	very fast
Cl^-	5.3×10^{-3}
Br^-	3.5×10^{-3}
I^-	1.5×10^{-3}
N_3^-	1.3×10^{-4}
SCN^-	4.8×10^{-5}
NO_2^-	3.8×10^{-6}
CN^-	2.8×10^{-6}

Source: Calculated from data in F. Basolo, H. B. Gray, and R. G. Pearson, *J. Am. Chem. Soc.*, **1960**, *82*, 4200.

[24] Wilkins, *The Study of Kinetics and Mechanism of Reactions of Transition Metal Complexes*, p. 231.

[25] J. D. Atwood, *Inorganic and Organometallic Reaction Mechanisms*, Brooks/Cole, Monterey, CA, 1985, pp. 60–63.

parallels the increase in η_{Pt}. Each of the parameters s and η_{Pt} may change by a factor of 3 from fast reactions to slow reactions, allowing for an overall ratio of 10^6 in the rates.

12.7 THE *TRANS* EFFECT

In 1926, Chernyaev[26] introduced the concept of the *trans* effect in platinum chemistry. In reactions of square-planar Pt(II) compounds, ligands *trans* to chloride are more easily replaced than those *trans* to ammonia; chloride is said to have a stronger *trans* effect than ammonia. When coupled with the observation that chloride itself is more easily replaced than ammonia, this *trans* effect allows the formation of isomeric Pt compounds, as shown in the reactions of Figure 12.12. In reaction (a), after the first

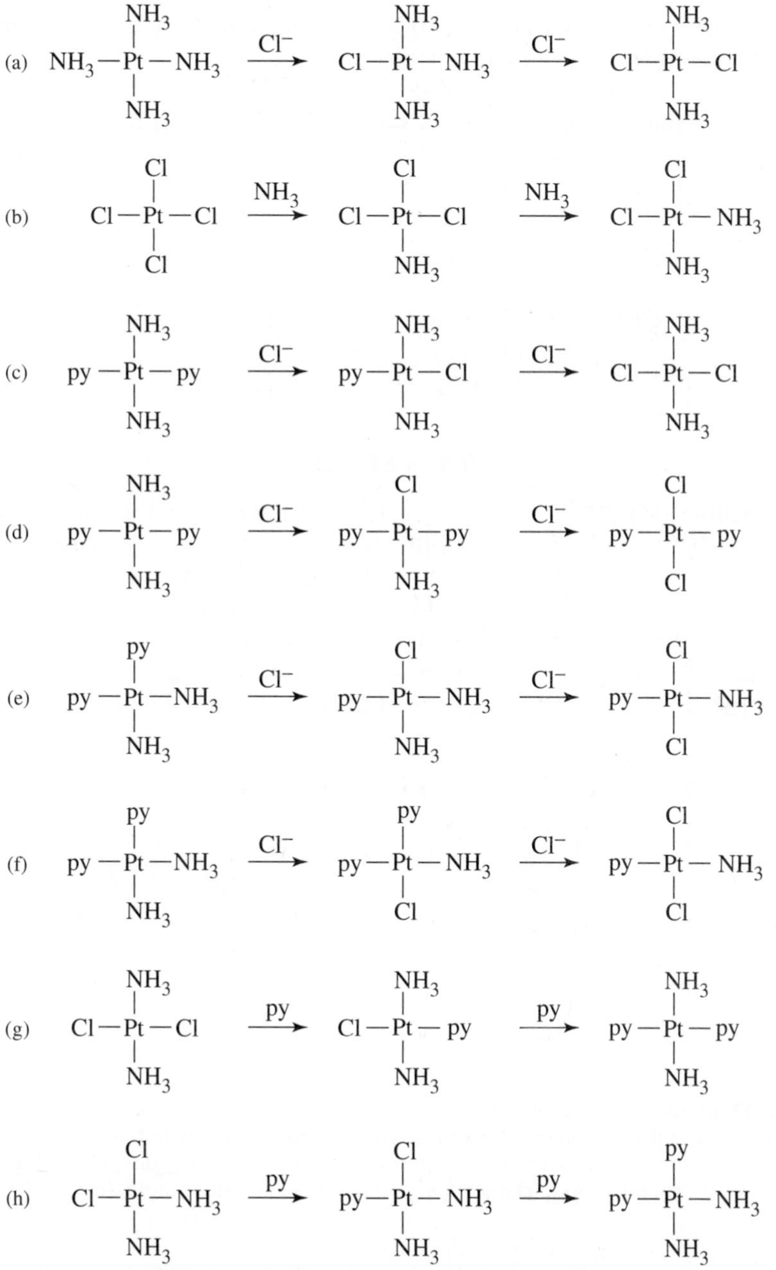

(a)	NH_3 NH_3—Pt—NH_3 NH_3	$\xrightarrow{Cl^-}$	NH_3 Cl—Pt—NH_3 NH_3	$\xrightarrow{Cl^-}$	NH_3 Cl—Pt—Cl NH_3
(b)	Cl Cl—Pt—Cl Cl	$\xrightarrow{NH_3}$	Cl Cl—Pt—Cl NH_3	$\xrightarrow{NH_3}$	Cl Cl—Pt—NH_3 NH_3
(c)	NH_3 py—Pt—py NH_3	$\xrightarrow{Cl^-}$	NH_3 py—Pt—Cl NH_3	$\xrightarrow{Cl^-}$	NH_3 Cl—Pt—Cl NH_3
(d)	NH_3 py—Pt—py NH_3	$\xrightarrow{Cl^-}$	Cl py—Pt—py NH_3	$\xrightarrow{Cl^-}$	Cl py—Pt—py Cl
(e)	py py—Pt—NH_3 NH_3	$\xrightarrow{Cl^-}$	Cl py—Pt—NH_3 NH_3	$\xrightarrow{Cl^-}$	Cl py—Pt—NH_3 Cl
(f)	py py—Pt—NH_3 NH_3	$\xrightarrow{Cl^-}$	py py—Pt—NH_3 Cl	$\xrightarrow{Cl^-}$	Cl py—Pt—NH_3 Cl
(g)	NH_3 Cl—Pt—Cl NH_3	$\xrightarrow{py}$	NH_3 Cl—Pt—py NH_3	$\xrightarrow{py}$	NH_3 py—Pt—py NH_3
(h)	Cl Cl—Pt—NH_3 NH_3	$\xrightarrow{py}$	Cl py—Pt—NH_3 NH_3	$\xrightarrow{py}$	py py—Pt—NH_3 NH_3

FIGURE 12.12 Stereochemistry and the *trans* Effect in Pt(II) Reactions. Charges have been omitted for clarity. In (a) through (f), the first substitution can be at any position, with the second controlled by the *trans* effect. In (g) and (h), both substitutions are controlled by the lability of chloride.

[26] I. I. Chernyaev, *Ann. Inst. Platine USSR.*, **1926**, *4*, 261.

ammonia is replaced, the second replacement is *trans* to the first Cl^-. In reaction (b), the second replacement is *trans* to Cl^- (replacement of ammonia in the second reaction is possible, but then the reactant and product are identical). The first steps in reactions (c) through (f) are the possible replacements, with nearly equal probabilities for replacement of ammonia or pyridine in any position. The second steps of (c) through (f) depend on the *trans* effect of Cl^-. Both steps of (g) and (h) depend on the greater lability of chloride. By using reactions such as these, it is possible to prepare specific isomers with different ligands. Chernyaev and coworkers did much of this, preparing a wide variety of compounds and establishing the order of *trans*-effect ligands:

$$CN^- \sim CO \sim C_2H_4 > PH_3 \sim SH_2 > NO_2^- > I^- > Br^- > Cl^- > NH_3 \sim py > OH^- > H_2O$$

▶ **Exercise 12.3** Predict the products of these reactions (there may be more than one product when there are conflicting preferences).

$$[PtCl_4]^{2-} + NO_2^- \longrightarrow (a) \qquad (a) + NH_3 \longrightarrow (b)$$

$$[PtCl_3NH_3]^- + NO_2^- \longrightarrow (c) \qquad (c) + NO_2^- \longrightarrow (d)$$

$$[PtCl(NH_3)_3]^+ + NO_2^- \longrightarrow (e) \qquad (e) + NO_2^- \longrightarrow (f)$$

$$[PtCl_4]^{2-} + I^- \longrightarrow (g) \qquad (g) + I^- \longrightarrow (h)$$

$$[PtI_4]^- + Cl^- \longrightarrow (i) \qquad (i) + Cl^- \longrightarrow (j)$$

12.7.1 Explanations of the *Trans* Effect[27]

SIGMA-BONDING EFFECTS Two factors dominate the explanations of the *trans* effect, weakening of the Pt—X bond and stabilization of the presumed 5-coordinate transition state. The energy relationships are shown in Figure 12.13, with the activation energy the difference between the reactant ground state and the first transition state.

The Pt—X bond is influenced by the Pt—T bond, because both use the Pt p_x and $d_{x^2-y^2}$ orbitals. When the Pt—T σ bond is strong, it uses a larger part of these orbitals and leaves less for the Pt—X bond (Figure 12.14). As a result, the Pt—X bond is weaker,

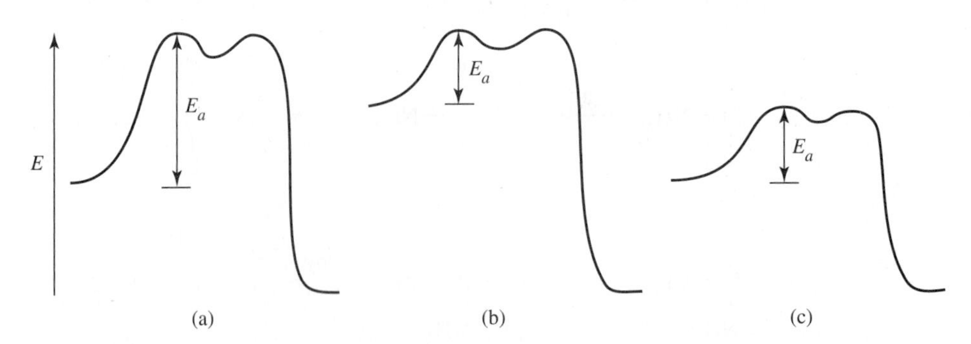

FIGURE 12.13 Activation Energy and the *trans* Effect. The depth of the energy curve for the intermediate and the relative heights of the two maxima will vary with the specific reaction. (a) Poor *trans* effect: low ground state, high transition state. (b) σ-Bonding effect: higher ground state (*trans* influence). (c) π-Bonding effect: lower transition state, (*trans* effect).

[27] Atwood, *Inorganic and Organometallic Reactions Mechanisms*, p. 54; Basolo and Pearson, *Mechanisms of Inorganic Reactions*, p. 355.

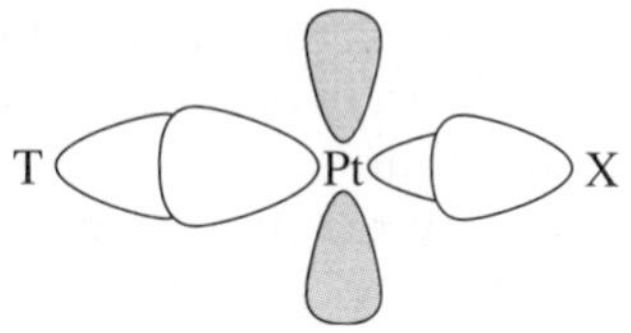

FIGURE 12.14 Sigma-Bonding Effect. A strong σ bond between Pt and T weakens the Pt—X bond.

and its ground state (sigma-bonding orbital) is higher in energy, leading to a smaller activation energy for the breaking of this bond as in Figure 12.13(b). This ground-state effect is sometimes called the ***trans* influence** and applies primarily to the leaving group. It is a thermodynamic effect, contributing to the overall kinetic result by changing the reactant ground state. This part of the explanation predicts the order for the *trans* effect based on the relative σ-donor properties of the ligands:

$$H^- > PR_3 > SCN^- > I^- \sim CH_3^- \sim CO \sim CN^- > Br^- > Cl^- > NH_3 > OH^-$$

The order given here is not quite correct for the *trans* effect, particularly for CO and CN^-, which have strong *trans* effects.

PI-BONDING EFFECTS The additional factor needed is π bonding in the Pt—T bond. When the T ligand forms a strong π-acceptor bond with Pt, charge is removed from Pt and the entrance of another ligand to form a 5-coordinate species is more likely. In addition to the charge effect, the $d_{x^2-y^2}$ orbital, which is involved in σ bonding in the square-planar geometry, and both the d_{xz} and d_{yz} orbitals can contribute to π bonding in the trigonal-bipyramidal transition state. Here, the effect on the ground state of the reactant is small, but the energy of the transition state is lowered, again reducing the activation energy as in Figure 12.13(c). The order of π-acceptor ability of the ligands is

$$C_2H_4 \sim CO > CN^- > NO_2^- > SCN^- > I^- > Br^- > Cl^- > NH_3 > OH^-$$

The expanded overall *trans* effect list is then the result of the combination of the two effects:

$$CO \sim CN^- \sim C_2H_4 > PR_3 \sim H^- > CH_3^- \sim SC(NH_2)_2 > C_6H_5^- >$$
$$NO_2^- \sim SCN^- \sim I^- > Br^- > Cl^- > py, NH_3 \sim OH^- \sim H_2O$$

Ligands highest in the series are strong π acceptors, followed by strong σ donors. Ligands at the low end of the series have neither strong σ-donor nor π-acceptor abilities. The *trans* effect can be very large; rates may differ as much as 10^6 between complexes with strong *trans* effect ligands and those with weak *trans* effect ligands.

▶ Exercise 12.4 It is possible to prepare different isomers of Pt(II) complexes with four different ligands. Predict the products expected if 1 mole of $[PtCl_4]^{2-}$ is reacted successively with the following reagents (e.g., the product of reaction **a** is used in reaction **b**):

a. 2 moles of ammonia
b. 2 moles of pyridine (see Reactions **g** and **h** in Figure 12.12)
c. 2 moles of chloride
d. 1 mole of nitrite, NO_2^-

12.8 OXIDATION-REDUCTION REACTIONS

Oxidation-reduction reactions of transition-metal complexes, like all redox reactions, involve the transfer of an electron from one species to another—in this case, from one complex to another. The two molecules may be connected by a common ligand through which the electron is transferred, in which case the reaction is called a *bridging* or **inner-sphere reaction**, or the exchange may occur between two separate coordination spheres in a *nonbridging* or **outer-sphere reaction**.

The rates have been studied by many different methods, including chemical analysis of the products, stopped-flow spectrophotometry, and the use of radioactive and stable isotope tracers. Taube's research group has been responsible for a large amount of the data, and its reviews cover the field.[28]

The rate of reaction for electron transfer depends on many factors, including the rate of substitution in the coordination sphere of the reactants, the match of energy levels of the reactants, solvation of reactants, and the nature of the ligands.

12.8.1 Inner-Sphere and Outer-Sphere Reactions

When the ligands of both reactants are tightly held, and there is no change in the coordination sphere on reaction, the reaction proceeds by outer-sphere electron transfer. Examples of these reactions are given in Table 12.13 with their rate constants.

The rates show very large differences. Characteristically, the rates depend on the ability of the electrons to tunnel through the ligands. This is a quantum mechanical property whereby electrons can pass through potential barriers that are too high to permit ordinary transfer. Ligands with π or p electrons or orbitals that can be used in bonding (as described in Chapter 10 for π-donor and π-acceptor ligands) provide good pathways for tunneling; those like NH_3, with no extra lone pairs and no low-lying antibonding orbitals, do not.

In outer-sphere reactions, where the ligands in the coordination sphere do not change, the primary change on electron transfer is a change in bond distance. A higher oxidation state on the metal leads to shorter σ bonds, with the extent of change depending on the electronic structure. The changes in bond distance are larger when e_g electrons

TABLE 12.13 Rate Constants for Outer-Sphere Electron Transfer Reactions[a]

	Reductants	
Oxidant	**$[Cr(bipy)_3]^{2+}$**	**$[Ru(NH_3)_6]^{2+}$**
$[Co(NH_3)_5(NH_3)]^{3+}$	6.9×10^2	1.1×10^{-2}
$[Co(NH_3)_5(F)]^{2+}$	1.8×10^3	
$[Co(NH_3)_5(OH)]^{2+}$	3×10^4	4×10^{-2}
$[Co(NH_3)_5(NO_3)]^{2+}$		3.4×10^1
$[Co(NH_3)_5(H_2O)]^{3+}$	5×10^4	3.0
$[Co(NH_3)_5(Cl)]^{2+}$	8×10^5	2.6×10^2
$[Co(NH_3)_5(Br)]^{2+}$	5×10^6	1.6×10^3
$[Co(NH_3)_5(I)]^{2+}$		6.7×10^3

Source: $[Cr(bipy)_3]^{2+}$ data from J. P. Candlin, J. Halpern, and D. L. Trimm, *J. Am. Chem. Soc.*, **1964**, *86*, 1019. $[Ru(NH_3)_6]^{2+}$ data from J. F. Endicott and H. Taube, *J. Am. Chem. Soc.*, **1964**, *86*, 1686.
NOTE: [a] Second-order rate constants in $M^{-1}\ s^{-1}$ at 25 °C.

[28] T. J. Meyer and H. Taube, "Electron Transfer Reactions," in G. Wilkinson, R. D. Gillard, and J. A. McCleverty, eds., Pergamon, *Comprehensive Coordination Chemistry*, Vol. 1, London, 1987, pp. 331–384; H. Taube, *Electron Transfer Reactions of Complex Ions in Solution*, Academic Press, New York, 1970; *Chem. Rev.*, **1952**, *50*, 69; *J. Chem. Educ.*, **1968**, *45*, 452.

are involved, as in the change from high-spin Co(II) ($t_{2g}^5 e_g^2$) to low-spin Co(III) (t_{2g}^6). Because the e_g orbitals are antibonding, removal of electrons from these orbitals results in a more stable complex and shorter bond distances. A larger ligand field stabilization energy makes oxidation easier. Comparing water and ammonia as ligands, we can see that the stronger field of ammonia makes oxidation of Co(II) relatively easy. $[Co(NH_3)_6]^{3+}$ is a very weak oxidizing agent. The aqueous Co(III) ion, on the other hand, has a large enough potential to oxidize water:

$$[Co(NH_3)_6]^{3+} + e^- \rightleftharpoons [Co(NH_3)_6]^{2+} \qquad \mathcal{E}^\circ = +0.108\text{ V}$$

$$Co^{3+}(aq) + e^- \rightleftharpoons Co^{2+}(aq) \qquad \mathcal{E}^\circ = +1.808\text{ V}$$

Inner-sphere reactions also use the tunneling phenomenon, but in this case, a single ligand is the conduit. The reactions proceed in three steps: (1) a substitution reaction that leaves the oxidant and reductant linked by the bridging ligand; (2) the actual transfer of the electron, frequently accompanied by transfer of the ligand; and (3) separation of the products:[29]

$$\underset{\text{Co(III) oxidant}}{[Co(NH_3)_5(Cl)]^{2+}} + \underset{\text{Cr(II) reductant}}{[Cr(H_2O)_6]^{2+}} \longrightarrow \underset{\text{Co(III)}\quad\text{Cr(II)}}{[(NH_3)_5Co(Cl)Cr(H_2O)_5]^{4+}} + H_2O \qquad (1)$$

$$\underset{\text{Co(III)}\quad\text{Cr(II)}}{[(NH_3)_5Co(Cl)Cr(H_2O)_5]^{4+}} \longrightarrow \underset{\text{Co(II)}\quad\text{Cr(III)}}{[(NH_3)_5Co(Cl)Cr(H_2O)_5]^{4+}} \qquad (2)$$

$$[(NH_3)_5Co(Cl)Cr(H_2O)_5]^{4+} + H_2O \longrightarrow [(NH_3)_5Co(H_2O)]^{2+} + [(Cl)Cr(H_2O)_5]^{2+} \qquad (3)$$

In this case, these are followed by a reaction made possible by the labile nature of Co(II):

$$[(NH_3)_5Co(H_2O)]^{2+} + 5\,H_2O \longrightarrow [Co(H_2O)_6]^{2+} + 5\,NH_3$$

The transfer of chloride to the chromium in these reactions is easy to follow experimentally, because Cr(III) is substitutionally inert, and the products can be separated by ion exchange techniques. When this is done, all the Cr(III) appears as $[(Cl)Cr(H_2O)_5]^{2+}$. The $[Cr(H_2O)_6]^{2+}$-$[Cr(H_2O)_5Cl]^{2+}$ exchange reaction, which results in no net change, has also been studied, using radioactive ^{51}Cr as a tracer.[30] All the chloride in the product came from the reactant, with none entering from excess Cl^- in the solution. The rate of the reaction could also be determined by following the amount of radioactivity found in the $[(Cl)Cr(H_2O)_5]^{2+}$ at different times during the reaction.

In many cases, the choice between inner- and outer-sphere mechanisms is difficult. In the examples of Table 12.13, the outer-sphere mechanism is required by the reducing agent. $[Ru(NH_3)_6]^{2+}$ is an inert species and does not allow formation of bridging species fast enough for the rate constants observed. Although $[Cr(bipy)_3]^{2+}$ is labile, the parallels in the rate constants of the two species strongly suggest that its redox reactions are also outer-sphere. In other cases, the oxidant may dictate an outer-sphere mechanism. In Table 12.14, $[Co(NH_3)_6]^{3+}$ and $[Co(en)_3]^{3+}$ have outer-sphere mechanisms, because their ligands have no lone pairs with which to form bonds to the reductant. The other reactions are less certain, although $Cr^{2+}(aq)$ is usually assumed to react by inner-sphere mechanisms in all cases in which bridging is possible.

$V^{2+}(aq)$ reactions appear to be similar to those of $Cr^{2+}(aq)$, although the range of rate constants is smaller than that for Cr^{2+}. This seems to indicate that the ligands are less important and makes an outer-sphere mechanism more likely. This is reinforced by comparison of the rate constants for the reactions of $[Cr(bipy)_3]^{2+}$ (outer-sphere,

[29] J. P. Candlin and J. Halpern, *Inorg. Chem.*, **1965**, *4*, 766.
[30] D. L. Ball and E. L. King, *J. Am. Chem. Soc.*, **1958**, *80*, 1091.

TABLE 12.14 Rate Constants for Aquated Reductants[a]

	Cr^{2+}	Eu^{2+}	V^{2+}
$[Co(en)_3]^{3+}$	$\sim 2 \times 10^{-5}$	$\sim 5 \times 10^{-3}$	$\sim 2 \times 10^{-4}$
$[Co(NH_3)_6]^{3+}$	8.9×10^{-5}	2×10^{-2}	3.7×10^{-2}
$[Co(NH_3)_5(H_2O)]^{3+}$	5×10^{-1}	1.5×10^{-1}	$\sim 5 \times 10^{-1}$
$[Co(NH_3)_5(NO_3)]^{2+}$	$\sim 9 \times 10^{-1}$	$\sim 1 \times 10^{2}$	
$[Co(NH_3)_5(Cl)]^{2+}$	6×10^{5}	3.9×10^{2}	~ 5
$[Co(NH_3)_5(Br)]^{2+}$	1.4×10^{6}	2.5×10^{2}	2.5×10^{1}
$[Co(NH_3)_5(I)]^{2+}$	3×10^{6}	1.2×10^{2}	1.2×10^{2}

Source: Data from J. P. Candlin, J. Halpern, and D. L. Trimm, *J. Am. Chem. Soc.*, **1964**, *86*, 1019; data for Cr^{2+} reactions with halide complexes from J. P. Candlin and J. Halpern, *Inorg. Chem.*, **1965**, *4*, 756; data for $[Co(NH_3)_6]^{3+}$ reactions with Cr^{2+} and V^{2+} from A. Zwickel and H. Taube, *J. Am. Chem. Soc.*, **1961**, *83*, 793.
NOTE: [a] Rate constants in $M^{-1}\ s^{-1}$.

Table 12.13) and V^{2+} (Table 12.14) with the same oxidants. V^{2+} may have different mechanisms for different oxidants, just as Cr^{2+} does.

Eu^{2+} *(aq)* is an unusual case. The rate constants do not parallel those of either the more common inner- or outer-sphere reactants, and the halide data are in reverse order from any others. The explanation offered for these rate constants is that the thermodynamic stability of the EuX^+ species helps drive the reaction faster for F^-, with slower rates and stabilities as we go down the series. Because of the smaller range of rate constants, Eu^{2+} reactions are usually classed as outer-sphere reactions.

When $[Co(CN)_5]^{3-}$ reacts with Co(III) oxidants ($[Co(NH_3)_5X]^{2+}$) that have potentially bridging ligands, the product is $[Co(CN)_5X]^{2+}$, evidence for an inner-sphere mechanism. Rate constants for a number of these reactions are given in Table 12.15. The reaction with hexamminecobalt(III) must be outer-sphere, but has a rate constant similar to the others. The reactions with thiocyanate or nitrite as bridging groups also show interesting behavior. With N-bonded $[(NH_3)_5CoNCS]^{2+}$, it

TABLE 12.15 Rate Constants for Reactions with $[Co(CN)_5]^{3-}$

Oxidant	$k\ (M^{-1}\ s^{-1})$
$[Co(NH_3)_5(F)]^{2+}$	1.8×10^{3}
$[Co(NH_3)_5(OH)]^{2+}$	9.3×10^{4}
$[Co(NH_3)_5(NH_3)]^{3+}$	8×10^{4}[a]
$[Co(NH_3)_5(NCS)]^{2+}$	1.1×10^{6}
$[Co(NH_3)_5(N_3)]^{2+}$	1.6×10^{6}
$[Co(NH_3)_5(Cl)]^{2+}$	$\sim 5 \times 10^{7}$

Source: Data from J. P. Candlin, J. Halpern, and S. Nakamura, *J. Am. Chem. Soc.*, **1963**, *85*, 2517.
NOTE: [a] Outer-sphere mechanism caused by the oxidant. Complexes with other potential bridging groups (PO_4^{3-}, SO_4^{2-}, CO_3^{2-}, and several carboxylic acids) also react by an outer-sphere mechanism, with constants ranging from 5×10^2 to 4×10^4.

reacts by bonding to the free S end of the ligand, because the cyanides soften the normally hard Co^{2+} ion. With S-bonded $[(NH_3)_5CoSCN]^{2+}$, it reacts initially by bonding to the free N end of the ligand and then rearranges rapidly to the more stable S-bonded form. In a similar fashion, a transient O-bonded intermediate is detected[31] in reactions of $[(NH_3)_5Co(NO_2)]^{2+}$ with $[Co(CN)_5]^{3-}$.

Other reactions that follow an inner-sphere mechanism have been studied to determine which ligands bridge best. The overall rate of reaction usually depends on the first two steps (substitution and transfer of electron), and in some cases, it is possible to draw conclusions about the rates of the individual steps. For example, ligands that are reducible provide better pathways, and their complexes are more quickly reduced.[32]

A useful comparison can be made using the ligands benzoate (difficult to reduce) and 4-carboxy-N-methylpyridine (easier to reduce). The rate constants for the reaction of the corresponding pentammine Co(III) complexes of these two ligands with Cr(II) differ by a factor of 10, although both have similar structures and transition states (Table 12.16). For both ligands, the mechanism is inner-sphere, with transfer of the ligand to chromium, indicating that coordination to the Cr(II) is through the carbonyl oxygen. The substitution reactions should have similar rates, so the difference in overall rates is a result of the transfer of electrons through the ligand. The data of Table 12.16 show these effects and extend the data to the ligands glyoxylate and glycolate, which are still more easily reduced. The transfer of an electron through such ligands is very fast when compared to similar reactions with ligands that are not reducible.

Remote attack on ligands with two potentially bonding groups is also found. Isonicotinamide bonded through the pyridine nitrogen can react with Cr^{2+} through the

TABLE 12.16 Ligand Reducibility and Electron Transfer

Rate Constants for the Reaction

$$[(NH_3)_5CoL]^{2+} + [Cr(H_2O)_6]^{2+} \longrightarrow Co^{2+} + 5\,NH_3 + [Cr(H_2O)_5L]^{2+} + H_2O$$

L	$k_2\ (M^{-1}\,s^{-1})$	Comments
$C_6H_5C(=O)-O^-$	0.15	Benzoate is difficult to reduce
$CH_3C(=O)-O^-$	0.34	Acetate is difficult to reduce
$CH_3NC_5H_4C(=O)-O^-$	1.3	N-methyl-4-carboxypyridine is more reducible
$O{=}CHC(=O)-O^-$	3.1	Glyoxylate is easy to reduce
$HOCH_2C(=O)-O^-$	7×10^3	Glycolate is very easy to reduce

Source: H. Taube, *Electron Transfer Reactions of Complex Ions in Solution*, Academic Press, New York, 1970, pp. 64–66.

[31] J. Halpern and S. Nakamura, *J. Am. Chem. Soc.*, **1965**, *87*, 3002, J. L. Burmeister, *Inorg. Chem.*, **1964**, *3*, 919.
[32] Taube, *Electron Transfer Reactions of Complex Ions in Solution*, pp. 64–66; E. S. Gould and H. Taube, *J. Am. Chem. Soc.*, **1964**, *86*, 1318.

TABLE 12.17 Rate Constants for Reduction of Isonicotinamide (4-Pyridine Carboxylic Acid Amide) Complexes by $[Cr(H_2O)_6]^{2+}$

Oxidant	k_2 $(M^{-1}\,s^{-1})$
$[(NH_2C(=O)-C_5H_4N)Cr(H_2O)_5]^{3+}$	1.8
$[(NH_2C(=O)-C_5H_4N)Co(NH_3)_5]^{3+}$	17.6
$[(NH_2C(=O)-C_5H_4N)Ru(NH_3)_5]^{3+}$	5×10^5

Source: H. Taube, *Electron Transfer Reactions of Complex Ions in Solution*, Academic Press, New York, 1970, pp. 66–68.

carbonyl oxygen on the other end of the molecule, transferring the ligand to the chromium and an electron through the ligand from the chromium to the other metal. The rate constants for different metals are shown in Table 12.17. The rate constants for the cobalt pentaammine and the chromium pentaaqua complexes are much closer than usual. The rate for Co compounds with other bridging ligands is frequently as much as 10^5 faster than the rate for corresponding Cr compounds, primarily because of the greater oxidizing power of Co(III). With isonicotinamide compounds, the rate seems to depend more on the rate of electron transfer from Cr^{2+} to the bridging ligand, and the readily reducible isonicotinamide makes the two reactions more nearly equal in rate. The much faster rate found for the ruthenium pentammine has been explained as the result of the transfer of an electron through the pi system of the ligand into the t_{2g} levels of Ru(III)—low-spin Ru(III) has a vacancy in the t_{2g} level. A similar electron transfer to Co(III) or Cr(III) places the incoming electron in the e_g levels, which have σ symmetry.[33]

12.8.2 Conditions for High and Low Oxidation Numbers

The overall stability of complexes with different charges on the metal ion depends on many factors, including LFSE, bonding energy of ligands, and redox properties of the ligands. When other factors are more or less equal, the hard and soft character of the ligands also has an effect. For example, all the very high oxidation numbers for the transition metals are found in combination with hard ligands, such as fluoride and oxide. Examples include MnO_4^-, CrO_4^{2-}, and FeO_4^{2-} with oxide, and AgF_2, RuF_5, PtF_6, and OsF_6 with fluoride. At the other extreme, the lowest oxidation states are found with soft ligands, with carbon monoxide being one of the most common. Zero is a common formal oxidation state for carbonyls; $V(CO)_6$, $Cr(CO)_6$, $Fe(CO)_5$, $Co_2(CO)_8$, and $Ni(CO)_4$ are examples, and examples with negative oxidation states such as $[Co(CO)_4]^-$ and $[Fe(CO)_4]^{2-}$ are also known. Carbonyl complexes are discussed further in Chapters 13 and 14.

Reactions of copper complexes show these ligand effects. Table 12.18 lists some of these reactions and their electrode potentials. If the reactions of the aquated Cu(II) and Cu(I) are taken as the basis for comparison, it can be seen that complexing Cu(II) with the hard ligand ammonia reduces the potential, stabilizing the higher oxidation state as compared with either Cu(I) or Cu(0). On the other hand, the soft ligand cyanide favors Cu(I), as do the halides (increased potentials). The halide cases are complicated

[33] H. Taube and E. S. Gould, *Acc. Chem. Res.*, **1969**, 2, 321.

TABLE 12.18 Electrode Potentials of Cobalt and Copper Species in Aqueous Solution

Cu(II)-Cu(I) Reactions	$\mathscr{E}°$ **(V)**
$Cu^{2+} + 2\,CN^- + e^- \rightleftharpoons [Cu(CN)_2]^-$	+1.103
$Cu^{2+} + I^- + e^- \rightleftharpoons CuI(s)$	+0.86
$Cu^{2+} + Cl^- + e^- \rightleftharpoons CuCl(s)$	+0.538
$Cu^{2+} + e^- \rightleftharpoons Cu^+$	+0.153
$[Cu(NH_3)_4]^{2+} + e^- \rightleftharpoons [Cu(NH_3)_2]^+ + 2\,NH_3$	−0.01
Cu(II)-Cu(0) Reactions	$\mathscr{E}°$ **(V)**
$Cu^{2+} + 2\,e^- \rightleftharpoons Cu(s)$	+0.337
$[Cu(NH_3)_4]^{2+} + 2\,e^- \rightleftharpoons Cu(s) + 4\,NH_4$	−0.05
Co(III)-Co(II) Reactions	$\mathscr{E}°$ **(V)**
$Co^{3+} + e^- \rightleftharpoons Co^{2+}$	+1.808
$[Co(NH_3)_6]^{3+} + e^- \rightleftharpoons [Co(NH_3)_6]^{2+}$	+0.108
$[Co(CN)_6]^{3-} + e^- \rightleftharpoons [Co(CN)_6]^{4-}$	−0.83

Source: T. Moeller, *Inorganic Chemistry*, Wiley InterScience, New York, 1982, p. 742.

by precipitation but still show the effects, and they also show that the soft iodide ligand makes Cu(I) more stable than the harder chloride.

In other cases, almost any ligand can serve to stabilize a particular species, and competing effects will have different results. Perhaps the most obvious example is the Co(III)–Co(II) couple, mentioned earlier in Section 12.8.1. As the hydrated ion, or aqua complex, Co(III) is a very strong oxidizing agent, reacting readily with water to form oxygen and Co(II). However, when coordinated with any ligand other than water or fluoride, Co(III) is kinetically stable, and almost stable in the thermodynamic sense as well. Part of the explanation is that Δ_o is quite large with any ligand, leading to an easy change from the high-spin Co(II) configuration $t_{2g}{}^5\, e_g{}^2$ to the low-spin Co(III) configuration $t_{2g}{}^6$. This means that the reverse reduction is much less favorable, and the complex ions have little tendency to oxidize other species. The reduction potentials (Table 12.18) for Co(III)–Co(II) with different ligands are in the order $H_2O > NH_3 > CN^-$, the order of increasing Δ_o and decreasing hardness. The increasing LFSE change is strong enough to overcome the usual effect of softer ligands stabilizing lower oxidation states.

12.9 REACTIONS OF COORDINATED LIGANDS

The reactions described to this point are either substitution reactions or oxidation-reduction reactions. Other reactions are primarily those of the ligands; in these reactions, coordination to the metal changes the ligand properties sufficiently to change the rate of a reaction or to make possible a reaction that would otherwise not take place. Such reactions are important for many different types of compounds and under many different conditions. Chapter 14 describes such reactions for organometallic compounds. In this chapter, we describe only a few examples of these reactions; the interested reader can find many more examples in the references cited.

Organic chemists have long used inorganic compounds as reagents. For example, Lewis acids such as $AlCl_3$, $FeCl_3$, $SnCl_4$, $ZnCl_2$, and $SbCl_5$ are used in Friedel–Crafts electrophilic substitutions. The labile complexes formed by acyl or alkyl halides and these Lewis acids create positively charged carbon atoms that can react readily with aromatic compounds. The reactions are generally the same as without the metal salts, but their use speeds the reactions and makes them much more useful.

As usual, it is easier to study reactions of inert compounds, such as those of Co(III), Cr(III), Pt(II), and Pt(IV), in which the products remain complexed to the metal and can be isolated for more complete study. However, useful catalysis requires that the products be easily separated from the catalyst, so relatively rapid dissociation from the metal is a desirable feature. Although many of the reactions described here do not have this capability, those with biological significance do, and chemists studying ligand reactions for synthetic purposes try to incorporate it into their reactions.

12.9.1 Hydrolysis of Esters, Amides, and Peptides

Amino acid esters, amides, and peptides can be hydrolyzed in basic solution, and the addition of many different metal ions speeds the reactions. Labile complexes of Cu(II), Co(II), Ni(II), Mn(II), Ca(II), and Mg(II), as well as other metal ions, promote these reactions. Whether the mechanism is through bidentate coordination of the α-amino group and the carbonyl, or only through the amine, is uncertain, but seems to depend on the relative concentrations. Because the reactions depend on complex formation and hydrolysis as separate steps, their temperature dependence is complex, and interpretation of all the effects is difficult.[34]

Co(III) complexes promote similar reactions. When four of the six octahedral positions are occupied by amine ligands, and two *cis* positions are available for further reactions, it is possible to study not only the hydrolysis itself but the steric preferences of the complexes. In general, these compounds catalyze the hydrolysis of N-terminal amino acids from peptides, and the amino acid that is removed remains as part of the complex. The reactions apparently proceed by coordination of the free amine to cobalt, followed either by coordination of the carbonyl to cobalt and subsequent reaction with OH^- or H_2O from the solution (path A in Figure 12.15) or reaction of the carbonyl carbon with coordinated hydroxide (path B).[35] As a result, the N-terminal amino acid is removed from the peptide and left as part of the cobalt complex in which the α-amino nitrogen and the carbonyl oxygen are bonded to the cobalt. Esters and amides are also hydrolyzed by the same mechanism, with the relative importance of the two pathways dependent on the specific compounds used.

Other compounds such as phosphate esters, pyrophosphates, and amides of phosphoric acid are hydrolyzed in similar reactions. Coordination may be through only one oxygen of these phosphate compounds, but the overall effect is similar.

12.9.2 Template Reactions

Template reactions are those in which formation of a complex places the ligands in the correct geometry for reaction. One of the earliest was for the formation of phthalocyanines, shown in Figure 12.16. Although the compounds were known earlier, their study really began in 1928, after discovery of a dark blue impurity in phthalimide prepared by reaction of phthalic anhydride with ammonia in an enameled vessel. This impurity was later discovered to be the iron phthalocyanine complex, created from iron released into the mixture by a break in the enamel surface. A similar reaction takes place with

---

[34] M. M. Jones, *Ligand Reactivity and Catalysis*, Academic Press, New York, 1968. Chapter III summarizes the arguments and mechanisms.

[35] J. P. Collman and D. A. Buckingham, *J. Am. Chem. Soc.*, **1963**, *85*, 3039; D. A. Buckingham, J. P. Collman, D. A. R. Hopper, and L. G. Marzelli, *J. Am. Chem. Soc.*, **1967**, *89*, 1082.

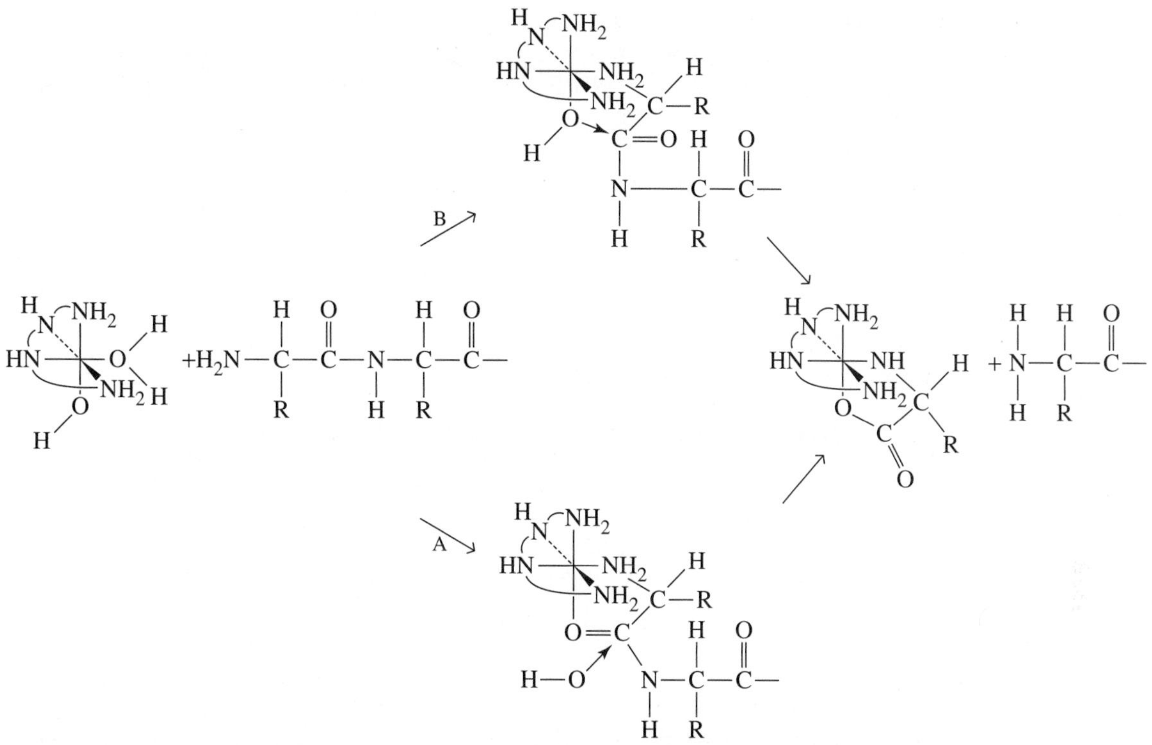

FIGURE 12.15 Peptide Hydrolysis by $[Co(trien)(H_2O)(OH)]^{2+}$.
(Data from D. A. Buckingham, J. P. Collman, D. A. R. Hopper, and L. G. Marzilli, *J. Am. Chem. Soc.*, **1967**, *89*, 1082).

copper, which forms more useful pigments; intermediates isolated from this reaction are shown in Figure 12.16. Phthalic acid and ammonia first form phthalimide, then 1-keto-3-iminoisoindoline, and then 1-amino-3-iminoisoindolenine. The cyclization reaction then occurs, probably with the assistance of the metal ion, which holds the chelated reactants in position. This is confirmed by the lack of cyclization in the absence of the metals.[36] Other reagents can be used for this synthesis, but the essential feature of all these reactions is the formation of the cyclic compound by coordination to a metal ion.

Similar reactions have been used extensively in the formation of macrocyclic compounds. Imine or Schiff base complexes ($R_1\,N{=}CHR_2$) have been extensively studied. In this case, the compounds can be formed without complexation, but the reaction is much faster in the presence of metal ions. An example is shown in Figure 12.17. In the absence of copper, benzothiazoline is favored in the final step, rather than the imine; very little of the Schiff base is present at equilibrium.

A major feature of template reactions is geometric: formation of the complex brings the reactants into close proximity with the proper orientation for reaction. In addition, complexation may change the electronic structure sufficiently to promote the reaction. Both are common to all coordinated ligand reactions, but the geometric factor is more obvious in these; the final product has a structure determined by the coordination geometry. Template reactions have been reviewed, and a large number of reactions and products have been described.[37]

[36] R. Price, "Dyes and Pigments," in G. Wilkinson, R. D. Gillard, and J. A. McCleverty, eds., *Comprehensive Coordination Chemistry*, Vol. 6, Pergamon Press, Oxford, 1987, pp. 88–89.
[37] D. St. C. Black, "Stoichiometric Reactions of Coordinated Ligands," in Wilkinson, Gillard, and McCleverty, *Comprehensive Coordination Chemistry*, pp. 155–226.

FIGURE 12.16 Phthalocyanine Synthesis.

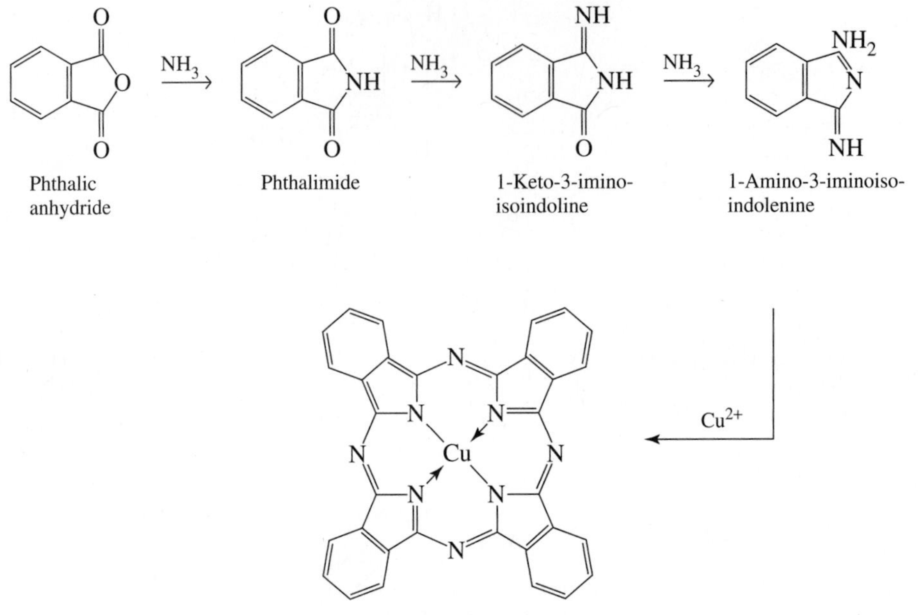

Phthalic anhydride $\xrightarrow{NH_3}$ Phthalimide $\xrightarrow{NH_3}$ 1-Keto-3-imino-isoindoline $\xrightarrow{NH_3}$ 1-Amino-3-iminoiso-indolenine $\xrightarrow{Cu^{2+}}$ Cu(II) phthalocyanine

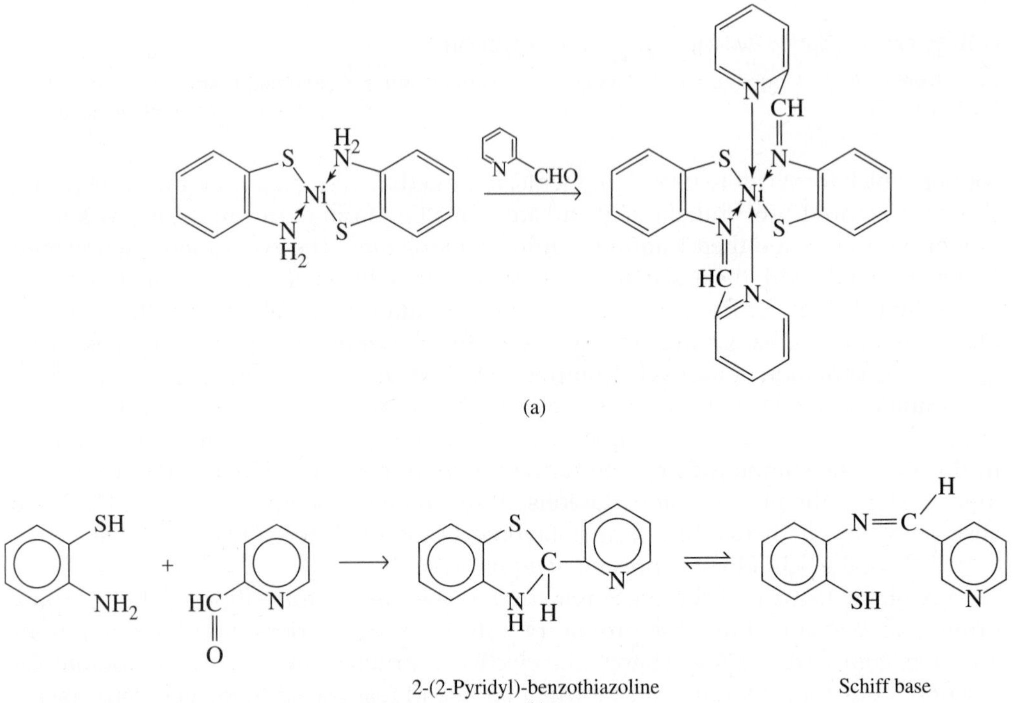

(a)

2-(2-Pyridyl)-benzothiazoline

Schiff base

(b)

FIGURE 12.17 Schiff Base Template Reaction. (a) The Ni(II)-*O*-aminothiophenol complex reacts with pyridine-1-carboxaldehyde to form the schiff base complex. (b) In the absence of the metal ion, the product is benzothiazoline; very little of the schiff base is formed.

(From L. F. Lindoy and S. E. Livingstone, *Inorg. Chem.*, **1968**, *7*, 1149.)

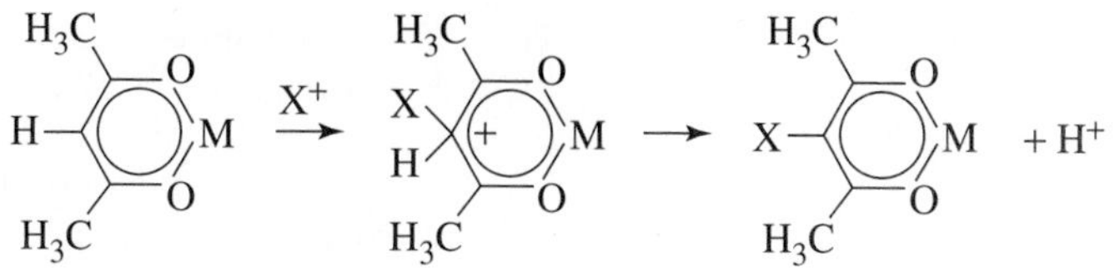

FIGURE 12.18 Electrophilic Substitution on Acetylacetone Complexes. X = Cl, Br, SCN, SAr, SCl, NO_2, CH_2Cl, $CH_2N(CH_3)_2$, COR, CHO.

12.9.3 Electrophilic Substitution

Acetylacetone complexes are known to undergo a wide variety of reactions that are at least superficially similar to aromatic electrophilic substitutions. Bromination, nitration, and similar reactions have been studied.[38] In all cases, coordination forces the ligand into an enol form and promotes reaction at the center carbon by preventing reaction at the oxygens and concentrating negative charge on carbon 3. Figure 12.18 shows the reactions and a possible mechanism.

General References

The general principles of kinetics and mechanisms have been described by J. W. Moore and R. G. Pearson, *Kinetics and Mechanism*, 3rd ed., Wiley InterScience, New York, 1981, and in F. Wilkinson, *Chemical Kinetics and Reaction Mechanisms*, Van Nostrand-Reinhold, New York, 1980). The classic for coordination compounds is F. Basolo and R. G. Pearson, *Mechanisms of Inorganic Reactions*, 2nd ed., John Wiley & Sons, New York, 1967. More recent books are by J. D. Atwood, *Inorganic and Organometallic Reaction Mechanisms*, Brooks/Cole, Monterey, CA, 1985, and D. Katakis and G. Gordon, *Mechanisms of Inorganic Reactions*, Wiley InterScience, New York, 1987. The reviews in G. Wilkinson, R. D. Gillard, and J. A. McCleverty, editors, *Comprehensive Coordination Chemistry*, Pergamon Press, Elmsford New York, 1987, provide a more comprehensive collection and discussion of the data. Volume 1, *Theory and Background*, covers substitution and redox reactions, and Volume 6, *Applications*, is particularly rich in data on ligand reactions.

Problems

12.1 The high-spin d^4 complex $[Cr(H_2O)_6]^{2+}$ is *labile*, but the low-spin d^4 complex ion $[Cr(CN)_6]^{4-}$ is *inert*. Explain.

12.2 Why is the existence of a series of entering groups with different rate constants evidence for an associative mechanism (A or I_a)?

12.3 Predict whether these complexes would be labile or inert and explain your choices. The magnetic moment is given in Bohr magnetons (μ_B) after each complex.

Ammonium oxopentachlorochromate(V)	1.82
Potassium hexaiodomanganate(IV)	3.82
Potassium hexacyanoferrate(III)	2.40
Hexaammineiron(II) chloride	5.45

12.4 The yellow "prussiate of soda" $Na_4[Fe(CN)_6]$ has been added to table salt as an anticaking agent. Why have there been no apparent toxic effects, even though this compound contains cyano ligands?

12.5 Consider the half-lives of substitution reactions of the pairs of complexes:

Half-Lives Shorter than 1 Minute	Half-Lives Longer than 1 Day
$[Cr(CN)_6]^{4-}$	$[Cr(CN)_6]^{3-}$
$[Fe(H_2O)_6]^{3+}$	$[Fe(CN)_6]^{4-}$
$[Co(H_2O)_6]^{2+}$	$[Co(NH_3)_5(H_2O)]^{3+}$ (H_2O exchange)

[38] J. P. Collman, *Angew. Chem., Int. Ed.*, **1965**, 4, 132.

Interpret the differences in half-lives in terms of the electronic structures of each pair.

12.6 The general rate law for substitution in square-planar Pt(II) complexes is valid for the reaction

$$[Pt(NH_3)_4]^{2+} + Cl^- \longrightarrow [Pt(NH_3)_3Cl]^+ + NH_3$$

Design the experiments needed to verify this and to determine the rate constants. What experimental data are needed, and how are the data to be treated?

12.7 The graph shows plots of k_{obs} versus $[X^-]$ for the anation reactions $[Co(en)_2(NO_2)(DMSO)]^{2+} + X^- \longrightarrow [Co(en)_2(NO_2)X]^+ + DMSO$, with △ = NO_2^-, ○ = Cl^-, and • = SCN^-. The broken line shows the rate of the DMSO-exchange reaction. (Reproduced with permission from W. R. Muir and C. H. Langford, *Inorg. Chem.*, **1968**, *7*, 1032.)

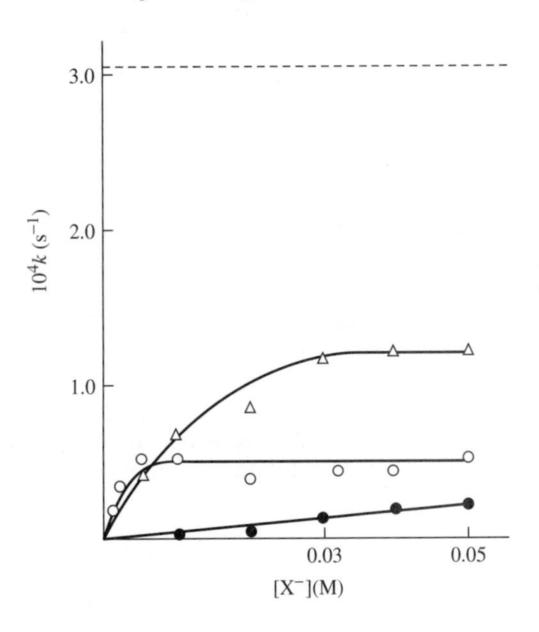

All three reactions are presumed to have the same mechanism.

a. Why is the DMSO exchange so much faster than the other reactions?

b. Why are the curves shaped as they are?

c. Explain what the limiting rate constants (at high concentration) are in terms of the rate laws for D and I_d mechanisms.

d. The limiting rate constants are $0.5 \times 10^{-4}\ s^{-1}$ and $1.2 \times 10^{-4}\ s^{-1}$ for Cl^- and NO_2^- respectively. For SCN^-, the limiting rate constant can be estimated as $1 \times 10^{-4}\ s^{-1}$. Do these values constitute evidence for an I_d mechanism?

12.8 **a.** The CO exchange reaction $Cr(^{12}CO)_6 + {}^{13}CO \longrightarrow Cr(^{12}CO)_5(^{13}CO) + {}^{12}CO$ has a rate that is first order in the concentration of $Cr(^{12}CO)_6$ but independent of the concentration of ^{13}CO. What does this imply about the mechanism of this reaction?

b. The reaction $Cr(CO)_6 + PR_3 \longrightarrow Cr(CO)_5PR_3 + CO$ [$R = P(n\text{-}C_4H_9)_3$] has the rate law of rate = $k_1[Cr(CO)_6] + k_2[Cr(CO)_6][PR_3]$. Why does this rate law have two terms?

c. For the general reaction in part **b**, will bulkier ligands tend to favor the first order or second order pathway? Explain briefly.

12.9 Account for the observation that two separate water exchange rates are found for $[Cu(H_2O)_6]^{2+}$ in aqueous solution.

12.10 Data for the reaction

$$Co(NO)(CO)_3 + As(C_6H_5)_3 \longrightarrow$$
$$Co(NO)(CO)_2[As(C_6H_5)_3] + CO$$

in toluene at 45 °C are given in the table. In all cases, the reaction is pseudo–first order in $Co(NO)(CO)_3$. Determine the rate constant(s) and discuss their probable significance (See E. M. Thorsteinson and F. Basolo, *J. Am. Chem. Soc.*, **1966**, *88*, 3929).

$[As(C_6H_5)_3]$(M)	$k(10^{-5}\ s^{-1})$
0.014	2.3
0.098	3.9
0.525	12
1.02	23

12.11 Shown on the following page is the log of the rate constant for substitution of CO on $Co(NO)(CO)_3$ by phosphorus and nitrogen ligands, plotted against the half-neutralization potential (ΔHNP) of the ligands. ΔHNP is a measure of the basicity of the compounds. Explain the linearity of such a plot and why there are two different lines. Incoming nucleophilic ligands: (1) $P(C_2H_5)_3$, (2) $P(n\text{-}C_4H_9)_3$, (3) $P(C_6H_6)(C_2H_5)_2$, (4) $P(C_6H_5)(C_2H_5)_2$, (5) $P(C_6H_5)_2\,(n\text{-}C_4H_9)$, (6) $P(p\text{-}CH_3O_6H_4)_3$, (7) $P(O\text{-}n\text{-}C_4H_9)_3$, (8) $P(C_6H_5)_3$, (9) $P(OCH_3)_3$, (10) $P(OCH_2)_3\,CCH_3$, (11) $P(OC_6H_5)_3$, (12) 4-picoline, (13) pyridine, (14) 3-chloropyridine. (Reproduced with permission from E. M. Thorsteinson and F. Basolo, *J. Am. Chem. Soc.*, **1966**, *88*, 3929. © 1966 American Chemical Society.)

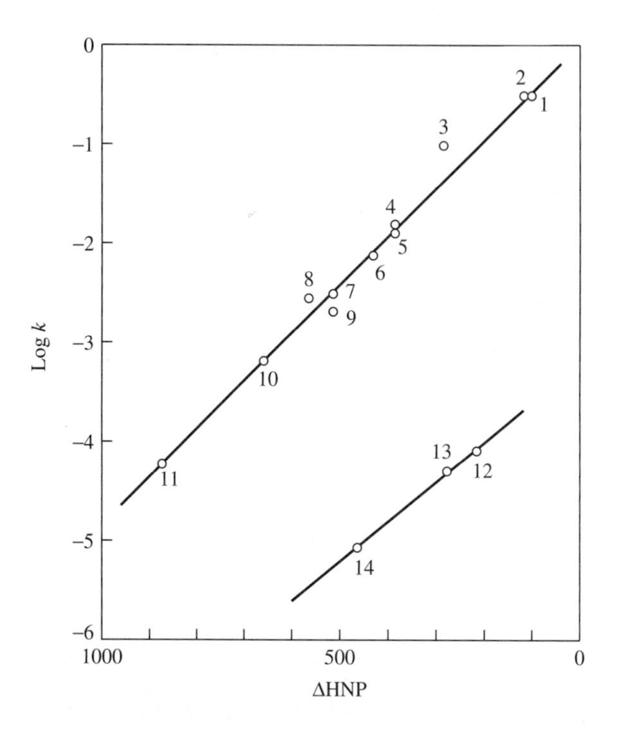

12.12 *cis*-$PtCl_2(PEt_3)_2$ is stable in benzene solution. However, small amounts of free triethylphosphine catalyze establishment of an equilibrium with the *trans* isomer:

$$cis\text{-}PtCl_2(PEt_3)_2 \rightleftharpoons trans\text{-}PtCl_2(PEt_3)_2$$

For the conversion of *cis* to *trans* in benzene at 25 °C, $\Delta H° = 10.3\ \text{kJ mol}^{-1}$ and $\Delta S° = 55.6\ \text{J mol}^{-1}\ \text{K}^{-1}$.

a. Calculate the free energy change, $\Delta G°$, and the equilibrium constant for this isomerization.

b. Which isomer has the higher bond energy? Is this answer consistent with what you would expect on the basis of π bonding in the two isomers? Explain briefly.

c. Why is free triethylphosphine necessary to catalyze the isomerization?

12.13 This table shows the effect of changing ligands on the dissociation rates of CO *cis* to those ligands. Explain the effect of these ligands on the rates of dissociation. Include the effect of these ligands on Cr—CO bonding and on the transition state, presumed to be square-pyramidal. (See J. D. Atwood and T. L. Brown, *J. Am. Chem. Soc.*, **1976**, *98*, 3160).

Compound	k (s^{-1}) for CO Dissociation
$Cr(CO)_6$	1×10^{-12}
$Cr(CO)_5(PPh_3)$	3.0×10^{-10}
$[Cr(CO)_5I]^-$	$< 10^{-5}$
$[Cr(CO)_5Br]^-$	2×10^{-5}
$[Cr(CO)_5Cl]^-$	1.5×10^{-4}

12.14 When the two isomers of $Pt(NH_3)_2Cl_2$ react with thiourea [tu = S=C$(NH_2)_2$], one product is $[Pt(tu)_4]^{2+}$ and the other is $[Pt(NH_3)_2(tu)_2]^{2+}$. Identify the initial isomers and explain the results.

12.15 Predict the products (equimolar mixtures of reactants):

a. $[Pt(CO)Cl_3]^- + NH_3 \longrightarrow$

b. $[Pt(NH_3)Br_3]^- + NH_3 \longrightarrow$

c. $[(C_2H_4)PtCl_3]^- + NH_3 \longrightarrow$

12.16 **a.** Design a sequence of reactions, beginning with $[PtCl_4]^{2-}$, that will result in platinum (II) complexes with four different ligands—py, NH_3, NO_2^-, and CH_3NH_2—with two different sets of *trans* ligands. (CH_3NH_2 is similar to NH_3 in its *trans* effect.)

b. Pt(II) can be oxidized to Pt(IV) by Cl_2 with no change in configuration (chloride ions are added above and below the plane of the Pt(II) complex). Predict the results if the two compounds from (a) are reacted with Cl_2 and then with 1 mole of Br^- for each mole of the Pt compound.

12.17 The rates of exchange of CO on *cis* square-planar Ir complexes have been observed for the following reaction at 298 K:

$$[IrX_2(CO)_2]^- + 2\,{}^*CO \rightleftharpoons [IrX_2({}^*CO)_2]^- + 2\,CO \quad ({}^*C = {}^{13}C)$$

The observed rate constants were:	X	k (L/mol s)
	Cl	1,080
	Br	12,700
	I	98,900

All three reactions have large negative values of entropy of activation, $\Delta S^‡$.

a. Is this reaction associative or dissociative?

b. On the basis of the data, which halide ligand exerts the strongest *trans* effect? (See R. Churlaud, U. Frey, F. Metz, and A. E. Merbach, *Inorg. Chem.*, **2000**, *39*, 304.)

12.18 The rate constant for electron exchange between V^{2+} (*aq*) and V^{3+} (*aq*) is observed to depend on the hydrogen ion concentration:

$$k = a + b / [H^+]$$

Propose a mechanism, and express *a* and *b* in terms of the rate constants of the mechanism. (Hint: V^{3+} (*aq*) hydrolyzes more easily than V^{2+} (*aq*).)

12.19 Is the reaction $[Co(NH_3)_6]^{3+} + [Cr(H_2O)_6]^{2+}$ likely to proceed by an inner-sphere or outer-sphere mechanism? Explain your answer.

12.20 The rate constants for the exchange reaction

$$CrX^{2+} + {}^*Cr^{2+} \longrightarrow {}^*CrX^{2+} + Cr^{2+}$$

where *Cr is radioactive ^{51}Cr are given in the table for reactions at 0 °C and 1 M $HClO_4$. Explain the differences in the rate constants in terms of the probable mechanism of the reaction.

X^-	k ($M^{-1}s^{-1}$)
F^-	1.2×10^{-3}
Cl^-	11
Br^-	60
NCS^-	1.2×10^{-4} (at 24 °C)
N_3^-	>1.2

12.21 The first complex of the ligand NSe (selenonitrosyl), $TpOs(NSe)Cl_2$ [Tp = hydrotris(1-pyrazolyl)borate], is shown. The osmium–nitrogen distances are:

Os–N(1): 210.1(7) pm
Os–N(3): 206.6(8) pm
Os–N(5): 206.9(7) pm

a. Which ligand, Cl or NSe, has the larger *trans* influence? Explain briefly.

b. The nitrogen–selenium distance in this compound is among the shortest N—Se distances known. Why is this distance so short? (See T. J. Crevier, S. Lovell, J. M. Mayer, R. L. Rheingold, and I. A. Guzei, *J. Am. Chem. Soc.*, **1998**, *120*, 6607)

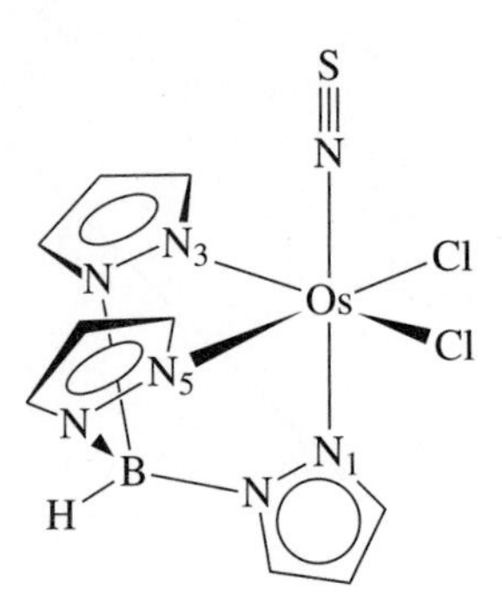

12.22 Exchange of an H_2O ligand on $[(CO)_3Mn(H_2O)_3]^+$ is much more rapid than on the analogous $[(CO)_3Re(H_2O)_3]^+$. The activation volume (change in volume on formation of the activated complex) is $-4.5 \pm 0.4\ cm^3mol^{-1}$. The Mn complex has infrared bands at 2051 and 1944 cm^{-1} that can be attributed to C—O stretching vibrations. (See U. Prinz, A. E. Merbach, O. Maas, and K. Hegetschweiler, *Inorg. Chem.*, **2004**, *43*, 2387.)

a. Suggest why the Mn complex reacts more rapidly than the analogous Re complex.

b. Is the activation volume more consistent with an A (or I_a) or a D (or I_d) mechanism? Explain.

c. On the basis of the IR spectrum, is the reactant more likely a *fac* or *mer* isomer?

13

Organometallic Chemistry

Organometallic chemistry, the chemistry of compounds that contain metal–carbon bonds, has developed enormously as a field of study during the past half century. It encompasses a wide variety of chemical compounds and their reactions, including numerous ligands that can interact in sigma and pi fashions with metal atoms and ions; many cluster compounds, containing one or more metal–metal bonds; and molecules of structural types unusual or unknown in organic chemistry. Some reactions of organometallic compounds are similar to known organic reactions, but in other cases, they are dramatically different. In addition to their intrinsically interesting nature, many organometallic compounds form useful catalysts and consequently are of high industrial interest. In this chapter, we describe a variety of types of organometallic compounds, focusing on the ligands and how they interact with metal atoms. Chapter 14 presents an outline of major types of reactions of organometallic compounds and how these reactions are important in catalytic cycles. Chapter 15 focuses on parallels between organometallic chemistry and main group chemistry.

Some organometallic compounds bear similarities to the types of coordination compounds already discussed in this text. $Cr(CO)_6$ and $[Ni(H_2O)_6]^{2+}$, for example, are both octahedral. Both CO and H_2O are σ-donor ligands; in addition, CO is a strong π acceptor. Other ligands that can exhibit both σ-donor and π-acceptor behaviors include CN^-, PPh_3, SCN^-, and many organic ligands. The metal–ligand bonding and electronic spectra of compounds containing these ligands can be described using concepts discussed in Chapters 10 and 11. However, many organometallic molecules are strikingly different from any we have considered previously. For example, cyclic organic ligands containing delocalized pi systems can team up with metal atoms to form **sandwich compounds**, such as those shown in Figure 13.1.

A characteristic of metal atoms bonded to organic ligands, especially CO, is that they often exhibit the capability to form covalent bonds to other metal atoms to form **cluster compounds**.[1] These clusters may contain only two or three metal atoms or many dozens; there is no limit to their size or variety. They may contain single, double, triple, or quadruple bonds—even molecules that may have quintuple bonds have recently been reported. In some cases cluster compounds have ligands that bridge two or more metal atoms. Examples of metal cluster compounds containing organic ligands are shown in Figure 13.2; clusters will be discussed further in Chapter 15.

Carbon itself may play a distinctly different role than in organic chemistry. Certain metal clusters encapsulate carbon atoms; the resulting carbon-centered clusters, frequently called **carbide clusters**, may contain carbon bonded to five, six, or more surrounding metals. The traditional notion of carbon

[1] Some cluster compounds are also known that contain no organic ligands.

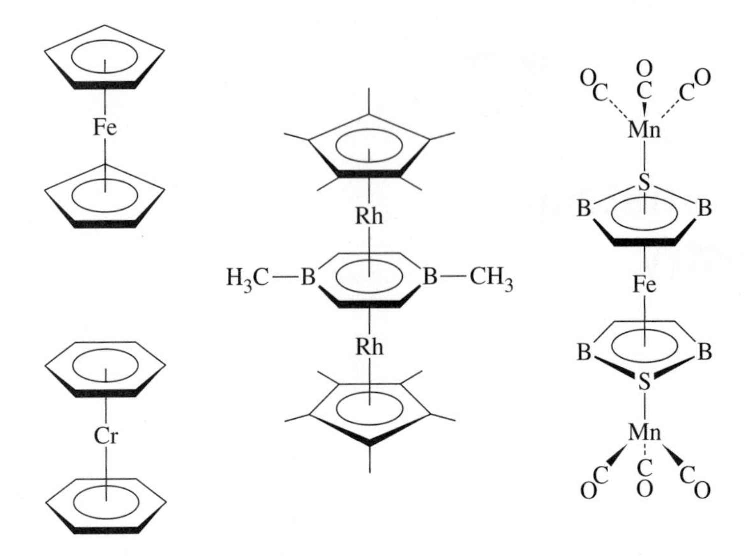

FIGURE 13.1 Examples of Sandwich Compounds.

forming bonds to at most four additional atoms must be reconsidered.[2] Two examples of carbide clusters are included in Figure 13.2.

Many other types of organometallic compounds have interesting structures and chemical properties. Figure 13.3 shows additional examples of the variety of structures that occur in this field.

Strictly speaking, the only compounds classified as organometallic should be ones that have metal–carbon bonds. In practice, however, complexes containing several other ligands similar to CO in their bonding, such as NO and N_2, are often included. (Cyanide also forms complexes in a manner similar to CO but is usually considered a classic, nonorganic ligand.) Other π-acceptor ligands, such as phosphines, often occur in organometallic complexes, and it is useful to consider them also in connection with organometallic chemistry. Even dihydrogen, H_2, now occurs in a new guise, as a donor–acceptor ligand, and it plays an important role in organometallic chemistry, for example in catalytic processes. We will include examples of these and other nonorganic ligands as appropriate in our discussion.

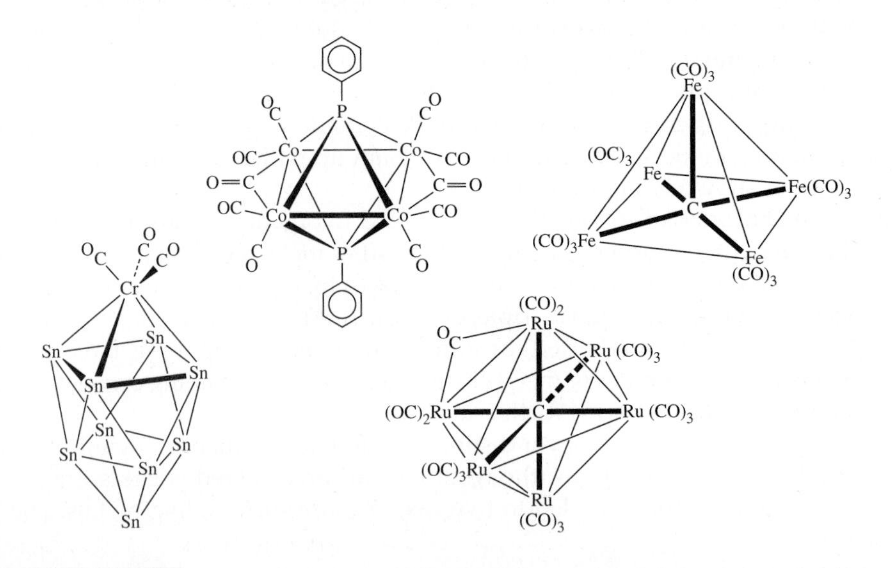

FIGURE 13.2 Examples of Cluster Compounds.

[2] A few examples of carbon bonded to more than four atoms are also known in organic chemistry. See, for example, G. A. Olah and G. Rasul, *Acc. Chem. Res.*, **1997**, *30*, 245.

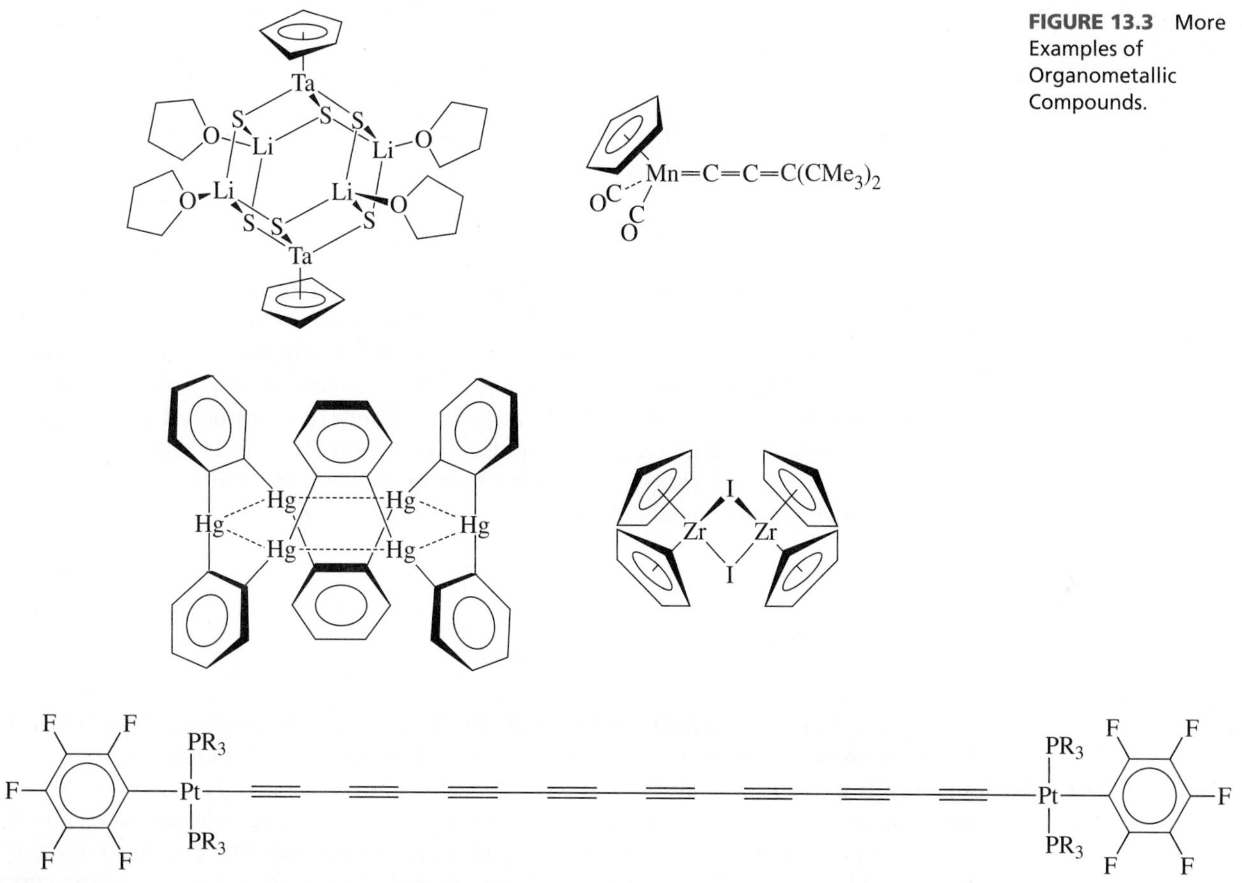

FIGURE 13.3 More Examples of Organometallic Compounds.

13.1 HISTORICAL BACKGROUND

The first organometallic compound to be reported was synthesized in 1827 by Zeise, who obtained yellow needle-like crystals after refluxing a mixture of $PtCl_4$ and $PtCl_2$ in ethanol, followed by addition of KCl solution.[3] Zeise correctly asserted that this yellow product, subsequently dubbed *Zeise's salt*, contained an ethylene group. This assertion was questioned by other chemists, most notably Liebig, and it was not verified conclusively until experiments performed by Birnbaum in 1868. However, the structure of the compound proved elusive and was not determined until more than 100 years later![4] Zeise's salt was the first compound identified as containing an organic molecule attached to a metal using the pi electrons of the organic molecule. It is an ionic compound of formula $K[Pt(C_2H_4)Cl_3] \cdot H_2O$; the structure of the anion, shown in Figure 13.4, is based on a square plane, with three chloro ligands occupying corners of the square and the ethylene occupying the fourth corner, but perpendicular to the plane.

The first compound to be synthesized that contained carbon monoxide as a ligand was another platinum chloride complex, reported in 1867. In 1890, Mond reported the preparation of $Ni(CO)_4$, a compound that became commercially useful for the purification of nickel.[5] Other metal CO (carbonyl) complexes were soon obtained.

---

[3] W. C. Zeise, *Ann. Phys. Chem.*, **1831**, *21*, 497–541. A translation of excerpts from this paper can be found in G. B. Kauffman, ed., *Classics in Coordination Chemistry*, Part 2, Dover, New York, 1976, pp. 21–37. A review of the history of the anion of Zeise's salt, including some earlier references, has been published: D. Seyferth, *Organometallics*, **2001**, *20*, 2.

[4] R. A. Love, T. F. Koetzle, G. J. B. Williams, L. C. Andrews, and R. Bau, *Inorg. Chem.*, **1975**, *14*, 2653.

[5] L. Mond, *J. Chem. Soc.*, **1890**, *57*, 749.

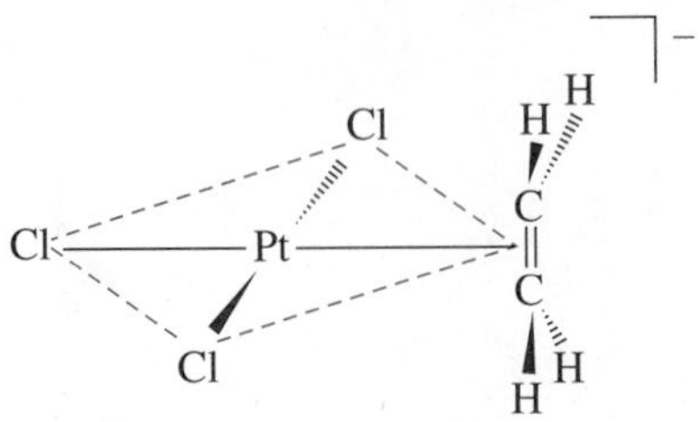

FIGURE 13.4 Anion of Zeise's Salt.

Reactions between magnesium and alkyl halides, performed by Barbier in 1898 and 1899 and subsequently by Grignard,[6] led to the synthesis of alkyl magnesium complexes now known as *Grignard reagents*. These complexes often have a complicated structure and contain magnesium-carbon sigma bonds. Their synthetic utility was recognized early; by 1905, more than 200 research papers had appeared on the topic. Grignard reagents and other reagents containing metal-alkyl σ bonds, such as organozinc and organocadmium reagents, have been of immense importance in the development of synthetic organic chemistry.

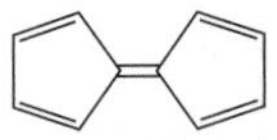

Fulvalene

Organometallic chemistry developed slowly from the discovery of Zeise's salt in 1827 until around 1950. Some organometallic compounds, such as Grignard reagents, found utility in organic synthesis, but there was little systematic study of compounds containing metal–carbon bonds. In 1951, in an attempt to synthesize fulvalene, shown above, from cyclopentadienyl bromide, Kealy and Pauson reacted the Grignard reagent *cyclo*-C_5H_5MgBr with $FeCl_3$, using anhydrous diethyl ether as the solvent.[7] This reaction did not yield the desired fulvalene but rather an orange solid having the formula formula $(C_5H_5)_2Fe$, ferrocene:

$$cyclo\text{-}C_5H_5MgBr + FeCl_3 \longrightarrow \underset{\text{ferrocene}}{(C_5H_5)_2Fe}$$

The product was surprisingly stable; it could be sublimed in air without decomposing and was resistant to catalytic hydrogenation and Diels–Alder reactions. In 1956, X-ray diffraction showed the structure to consist of an iron atom sandwiched between two parallel C_5H_5 rings,[8] but the details of the structure proved controversial.[9] The initial study indicated that the rings were in a staggered conformation (D_{5d} symmetry). Electron diffraction studies of gas-phase ferrocene, on the other hand, showed the rings to be eclipsed (D_{5h}), or very nearly so. More recent X-ray diffraction studies of solid ferrocene have identified several crystalline phases, with an eclipsed conformation at 98 K and with conformations having the rings slightly twisted (D_5) in higher-temperature crystalline modifications (Figure 13.5).[10]

The discovery of the prototype sandwich compound ferrocene rapidly led to the synthesis of other sandwich compounds, of other compounds containing metal atoms

[6] V. Grignard, *Ann. Chim.*, **1901**, *24*, 433. An English translation of most of this paper is in P. R. Jones and E. Southwick, *J. Chem. Ed.*, **1970**, *47*, 290.
[7] T. J. Kealy and P. L. Pauson, *Nature*, **1951**, *168*, 1039.
[8] J. D. Dunitz, L. E. Orgel, and R. A. Rich, *Acta Crystallogr.*, **1956**, *9*, 373.
[9] For an interesting article on the discovery of ferrocene's structure, see P. Laszlo and R. Hoffmann, *Angew. Chem., Int. Ed.*, **2000**, *39*, 123.
[10] E. A. V. Ebsworth, D. W. H. Rankin, and S. Cradock, *Structural Methods in Inorganic Chemistry*, 2nd ed., Blackwell Scientific, Oxford UK, 1991.

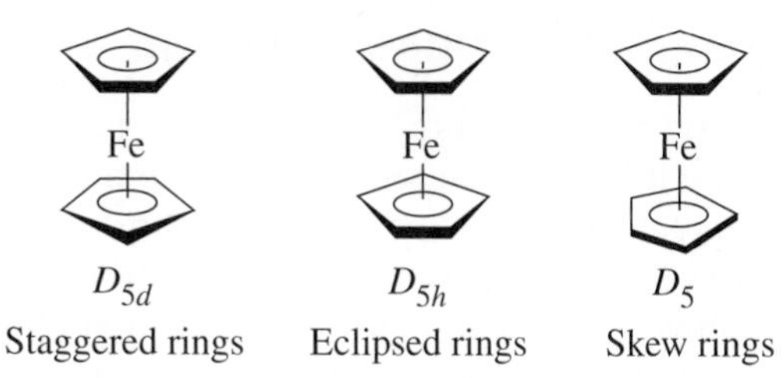

FIGURE 13.5
Conformations of Ferrocene.

bonded to the C_5H_5 ring in a similar fashion, and to a vast array of other compounds containing other organic ligands. Therefore, it is often stated, and with justification, that the discovery of ferrocene began the era of modern organometallic chemistry, an area that continues to grow rapidly.[11]

Finally, an introductory history of organometallic chemistry would be incomplete without mentioning what surely is the oldest organometallic compound known, vitamin B_{12} coenzyme. This naturally occurring cobalt complex, whose structure is illustrated in Figure 13.6, contains a cobalt–carbon sigma bond. It is a cofactor in a number of enzymes that catalyze 1,2 shifts in biochemical systems:

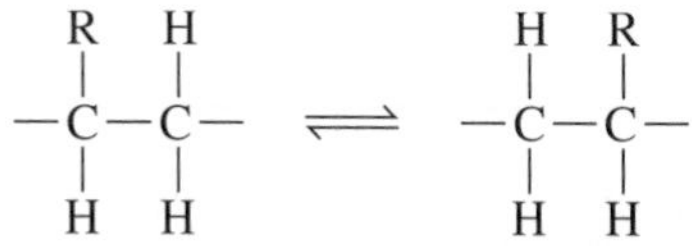

$$-\overset{\displaystyle R}{\underset{\displaystyle H}{C}}-\overset{\displaystyle H}{\underset{\displaystyle H}{C}}- \rightleftharpoons -\overset{\displaystyle H}{\underset{\displaystyle H}{C}}-\overset{\displaystyle R}{\underset{\displaystyle H}{C}}-$$

The chemistry of vitamin B_{12} is described briefly in Chapter 16.

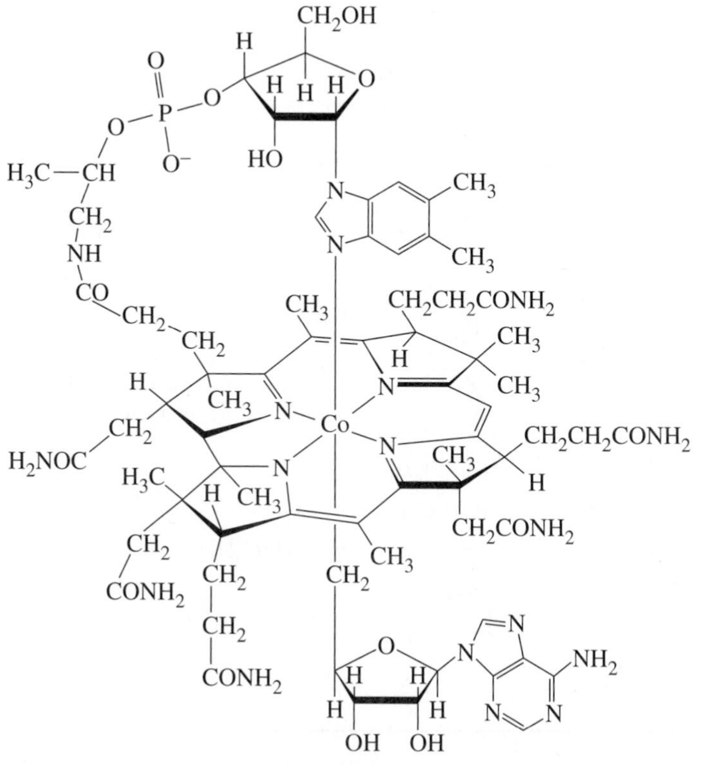

FIGURE 13.6
Vitamin B_{12} Coenzyme.

[11] A special issue of the *Journal of Organometallic Chemistry* (**2002**, *637*, 1) has been devoted to ferrocene, including recollections of some involved in its discovery; a brief summary of some of these recollections appeared in *Chem. Eng. News*, December 3, 2001, p. 37.

13.2 ORGANIC LIGANDS AND NOMENCLATURE

Some of the most common organic ligands are shown in Figure 13.7.

Special nomenclature has been devised to designate the manner in which some of these ligands bond to metal atoms; several of the ligands in Figure 13.7 may bond through different numbers of atoms, depending on the molecule in question. The number of atoms through which a ligand bonds is indicated by the Greek letter η (eta) followed by a superscript indicating the number of ligand atoms attached to the metal. For example, because the cyclopentadienyl ligands in ferrocene bond through all five atoms, they are designated η^5-C_5H_5. The formula of ferrocene may therefore be written $(\eta^5\text{-}C_5H_5)_2Fe$ (in general we will write hydrocarbon ligands before the metal). In written or spoken form, the η^5-C_5H_5 ligand is designated the pentahaptocyclopentadienyl ligand. *Hapto* comes from the Greek word for fasten; therefore, *pentahapto* means "fastened in five places." C_5H_5, probably the second most frequently encountered ligand in organometallic chemistry (after CO), most commonly bonds to metals through five positions, but under certain circumstances, it may bond through only one or three positions. As a ligand, C_5H_5 is commonly abbreviated Cp.

The corresponding formulas and names are designated according to this system as follows:[12]

Number of Bonding Positions	Formula	Name	
1	η^1-C_5H_5	Monohaptocyclopentadienyl	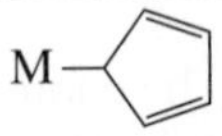
3	η^3-C_5H_5	Trihaptocyclopentadienyl	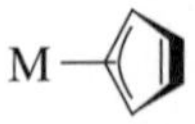
5	η^5-C_5H_5	Pentahaptocyclopentadienyl	

Ligand	Name	Ligand	Name
CO	Carbonyl	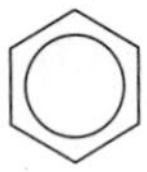	Benzene
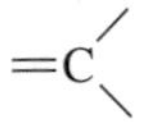	Carbene (alkylidene)	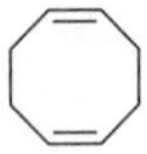	1,5-cyclooctadiene (1,5-COD) (1,3-cyclooctadiene complexes are also known)
$\equiv$C—	Carbyne (alkylidyne)		
	Cyclopropenyl (*cyclo*-C_3H_3)	$H_2C{=}CH_2$	Ethylene
		$HC{\equiv}CH$	Acetylene
			π-Allyl (C_3H_5)
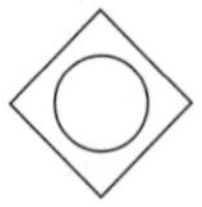	Cyclobutadiene (*cyclo*-C_4H_4)	—CR_3	Alkyl
	Cyclopentadienyl (*cyclo*-C_5H_5)(Cp)	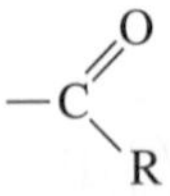	Acyl

FIGURE 13.7 Common Organic Ligands.

[12] For ligands having all carbons bonded to a metal, sometimes the superscript is omitted. Ferrocene may therefore be written $(\eta\text{-}C_5H_5)_2Fe$ and dibenzenechromium $(\eta\text{-}C_6H_6)_2Cr$. Similarly, π with no superscript may occasionally be used to designate that all atoms in the pi system are bonded to the metal; (for example, $(\pi\text{-}C_5H_5)_2Fe$).

As in the case of other coordination compounds, bridging ligands, which are very common in organometallic chemistry, are designated by the prefix μ, followed by a subscript indicating the number of metal atoms bridged. Bridging carbonyl ligands, for example, are designated as follows:

Number of Atoms Bridged	Formula
None (terminal)	CO
2	μ_2-CO
3	μ_3-CO

13.3 THE 18-ELECTRON RULE

In main group chemistry, we have encountered the octet rule, in which the electronic structures of many main group compounds can be rationalized on the basis of a valence shell requirement of 8 electrons. Similarly, in organometallic chemistry, the electronic structures of many compounds are based on a total valence electron count of 18 on the central metal atom. As in the case of the octet rule, there are many exceptions to the 18-electron rule,[13] but the rule nevertheless provides some useful guidelines to the chemistry of many organometallic complexes, especially those containing strong π-acceptor ligands.

13.3.1 Counting Electrons

Several schemes exist for counting electrons in organometallic compounds. We will describe two of these. First, here are two examples of electron counting in 18-electron species:

EXAMPLES

$Cr(CO)_6$

A Cr atom has 6 electrons outside its noble gas core. Each CO is considered to act as a donor of 2 electrons. The total electron count is therefore:

Cr			6 electrons
6(CO)	6 × 2 electrons	=	12 electrons
	Total	=	18 electrons

$Cr(CO)_6$ is therefore considered an 18-electron complex. It is thermally stable; for example, it can be sublimed without decomposition. On the other hand, $Cr(CO)_5$, a 16-electron species, and $Cr(CO)_7$, a 20-electron species, are much less stable and are known only as transient species. Likewise, the 17-electron $[Cr(CO)_6]^+$ and 19-electron $[Cr(CO)_6]^-$ are far less stable than the neutral, 18-electron $Cr(CO)_6$.

The bonding in $Cr(CO)_6$, which provides a rationale for the special stability of many 18-electron systems, will be discussed in Section 13.3.2.

[13] A variation on the 18-electron rule, often called the *effective atomic number (EAN) rule*, is based on electron counts relative to the total number of electrons in noble gases. The EAN rule gives the same results as the 18-electron rule and will not be considered further in this text.

(η^5-C_5H_5)Fe(CO)$_2$Cl

Electrons in this complex may be counted in two ways:

Method A: Donor Pair Method

This method considers ligands to donate electron pairs to the metal. To determine the total electron count, we must take into account the charge on each ligand and determine the formal oxidation state of the metal.

Pentahapto-C_5H_5 is considered by this method as $C_5H_5^-$, a donor of 3 electron pairs; it is a 6-electron donor. As in the first example, CO is counted as a 2-electron donor. Chloride is considered Cl^-, a donor of two electrons. Therefore, (η^5-C_5H_5)Fe(CO)$_2$Cl is formally an iron(II) complex. Iron(II) has 6 electrons beyond its noble gas core. Therefore, the electron count is

Fe(II)	6 electrons
η^5-$C_5H_5^-$	6 electrons
2 (CO)	4 electrons
Cl^-	2 electrons
Total =	18 electrons

Method B: Neutral-Ligand Method

This method uses the number of electrons that would be donated by ligands *if they were neutral*. For simple inorganic ligands, this usually means that ligands are considered to donate the number of electrons equal to their negative charge as free ions. For example,

Cl is a 1-electron donor (charge on free ion $= -1$)

O is a 2-electron donor (charge on free ion $= -2$)

N is a 3-electron donor (charge on free ion $= -3$)

We do not need to determine the oxidation state of the metal to determine the total electron count by this method.

For (η^5-C_5H_5)Fe(CO)$_2$Cl, an iron *atom* has 8 electrons beyond its noble gas core, and η^5-C_5H_5 is now considered a neutral ligand (a 5-electron pi system), in which case it would contribute 5 electrons. CO is a 2-electron donor, and Cl (counted as if it were a neutral species) is a 1-electron donor. The electron count is

Fe atom	8 electrons
η^5-C_5H_5	5 electrons
2 (CO)	4 electrons
Cl	1 electrons
Total =	18 electrons

Many organometallic complexes are charged species, and this charge must be included in determining the total electron count. The reader may wish to verify, by either method of electron counting, that $[Mn(CO)_6]^+$ and $[(\eta^5\text{-}C_5H_5)Fe(CO)_2]^-$ are both 18-electron ions.

In addition, metal–metal bonds must be counted. A metal–metal single bond counts as one electron per metal, a double bond counts as two electrons per metal, and so forth. For example, in the dimeric complex $(CO)_5Mn—Mn(CO)_5$ the electron count per manganese atom is, by either method:

Mn		7 electrons
5 (CO)		10 electrons
Mn—Mn bond		1 electrons
	Total =	18 electrons

Electron counts for common ligands according to both schemes are given in Table 13.1.

TABLE 13.1 Electron Counting Schemes for Common Ligands

Ligand	Method A	Method B
H	2 (H^-)	1
Cl, Br, I	2 (X^-)	1
OH, OR	2 (OH^-, OR^-)	1
CN	2 (CN^-)	1
CH_3, CR_3	2 ($CH_3{}^-$, $CR_3{}^-$)	1
NO (bent M—N—O)	2 (NO^-)	1
NO (linear M—N—O)	2 (NO^+)	3
CO, PR_3	2	2
NH_3, H_2O	2	2
=CRR′ (Carbene)	2	2
$H_2C{=}CH_2$ (Ethylene)	2	2
CNR	2	2
=O, =S	4 (O^{2-}, S^{2-})	2
η^3-C_3H_5 (π-allyl)	2 ($C_3H_5{}^+$)	3
≡CR (Carbyne)	3	3
≡N	6 (N^{3-})	3
Ethylenediamine (en)	4 (2 per nitrogen)	4
Bipyridine (bipy)	4 (2 per nitrogen)	4
Butadiene	4	4
η^5-C_5H_5 (Cyclopentadienyl)	6 ($C_5H_5{}^-$)	5
η^6-C_6H_6 (Benzene)	6	6
η^7-C_7H_7 (Cycloheptatrienyl)	6 ($C_7H_7{}^+$)	7

EXAMPLES

Both methods of electron counting are illustrated for the following complexes.

	Method A		Method B	
$ClMn(CO)_5$	Mn(I)	6 e^-	Mn	7 e^-
	Cl^-	2 e^-	Cl	1 e^-
	5 CO	10 e^-	5 CO	10 e^-
		18 e^-		18 e^-
$(\eta^5\text{-}C_5H_5)_2Fe$	Fe(II)	6 e^-	Fe	8 e^-
(Ferrocene)	2 $\eta^5\text{-}C_5H_5^-$	12 e^-	2 $\eta^5\text{-}C_5H_5$	10 e^-
		18 e^-		18 e^-
$[Re(CO)_5(PF_3)]^+$	Re(I)	6 e^-	Re	7 e^-
	5 CO	10 e^-	5 CO	10 e^-
	PF_3	2 e^-	PF_3	2 e^-
	+ charge	*	+ charge	−1 e^-
		18 e^-		18 e^-

* Charge on ion is accounted for in assignment of oxidation state to Re.

The electron-counting method of choice is a matter of individual preference. Method A includes the formal oxdation state of the metal; Method B does not. Method B may be simpler to use for ligands having extended pi systems; for example, η^5 ligands have an electron count of 5, η^3 ligands an electron count of 3, and so on. Because neither description describes the bonding in any real sense, these methods should, like the Lewis electron-dot approach in main group chemistry, be considered primarily electron bookkeeping tools. Physical measurements are necessary to provide evidence about the actual electron distribution in molecules. Other electron-counting schemes have also been developed. It is generally best to select one method and use it consistently.

In ligands such as CO that can interact with metal atoms in several ways, the number of electrons counted is usually based on σ donation. For example, although CO is a π acceptor and (weak) π donor, its electron-donating count of 2 is based on its σ donor ability alone. However, the π-acceptor and π-donor abilities of ligands have significant effects on the degree to which the 18-electron rule is likely to be obeyed. Linear and cyclic organic pi systems interact with metals in more complicated ways, discussed later in this chapter.

▶ **Exercise 13.1** Determine the valence electron counts for the transition metals in the following complexes:

a. $[Fe(CO)_4]^{2-}$
b. $[(\eta^5\text{-}C_5H_5)_2Co]^+$
c. $(\eta^3\text{-}C_5H_5)(\eta^5\text{-}C_5H_5)Fe(CO)$
d. $Co_2(CO)_8$ (has a single Co—Co bond)

▶ **Exercise 13.2** Identify the first-row transition metal for the following 18-electron species:

a. $[M(CO)_3(PPh_3)]^-$
b. $HM(CO)_5$
c. $(\eta^4\text{-}C_8H_8)M(CO)_3$
d. $[(\eta^5\text{-}C_5H_5)M(CO)_3]_2$ (assume single M—M bond)

13.3.2 Why 18 Electrons?

An oversimplified rationale for the special significance of 18 electrons can be made by analogy with the octet rule in main group chemistry. If the octet represents a complete valence electron shell configuration ($s^2\,p^6$), then the number 18 represents a filled valence shell for a transition metal ($s^2\,p^6\,d^{10}$). Although perhaps a useful way to relate electron configurations to the idea of valence shells of electrons for atoms, this analogy does not provide an explanation for why so many complexes violate the 18-electron rule. In particular, the valence-shell rationale does not distinguish between types of ligands (e.g., σ donors, π acceptors); this distinction is an important consideration in determining which complexes obey and which violate the rule.

A good example of a complex that adheres to the 18-electron rule is $Cr(CO)_6$. The molecular orbitals of interest in this molecule are those that result primarily from interactions between the *d* orbitals of Cr and the σ-donor (HOMO) and π-acceptor orbitals (LUMO) of the six CO ligands. The relative energies of molecular orbitals resulting from these interactions are shown in Figure 13.8.

Chromium(0) has 6 electrons outside its noble gas core. Each CO contributes a pair of electrons to give a total electron count of 18. In the molecular orbital diagram, these 18 electrons appear as the 12σ electrons—the σ electrons of the CO ligands, stabilized by their interaction with the metal orbitals—and the 6 t_{2g} electrons. Addition of one or more electrons to $Cr(CO)_6$ would populate the e_g orbitals, which are antibonding; the consequence would be destabilization of the molecule. Removal of electrons from $Cr(CO)_6$ would depopulate the t_{2g} orbitals, which are bonding as a consequence of the strong π-acceptor ability of the CO ligands; a decrease in electron density in these orbitals would also tend to destabilize the complex. The result is that the 18-electron configuration for this molecule is the most stable.

By considering 6-coordinate molecules of octahedral geometry, we can gain some insight as to when the 18-electron rule can be expected to be most valid. $Cr(CO)_6$ obeys the rule because of two factors: the strong σ-donor ability of CO raises the e_g orbitals in energy, making them considerably antibonding (and raising the energy of electrons in excess of 18); and the strong π-acceptor ability of CO lowers the t_{2g} orbitals in energy, making them bonding (and lowering the energies of electrons 13 through 18). Ligands that are both strong σ donors and π acceptors should therefore be the most effective at forcing adherence to the 18-electron rule. Other ligands, including some organic ligands, do not have these features, and consequently their compounds may or may not adhere to the rule.

Examples of exceptions may be noted. $[Zn(en)_3]^{2+}$ is a 22-electron species; it has both the t_{2g} and $e_g{}^*$ orbitals filled. Although en (ethylenediamine) is a good σ donor, it is not as strong a donor as CO. As a result, electrons in the e_g orbitals are not sufficiently antibonding to cause significant destabilization of the complex, and the 22-electron species, with 4 electrons in e_g orbitals, is stable. An example of a 12-electron species is $TiF_6{}^{2-}$. In this case, the fluoride ligand is a π donor as well as a σ donor. The π-donor ability of F^- destabilizes the t_{2g} orbitals of the complex, making them slightly antibonding. The species $TiF_6{}^{2-}$ has 12 electrons in the bonding σ orbitals and no electrons in the antibonding t_{2g} or

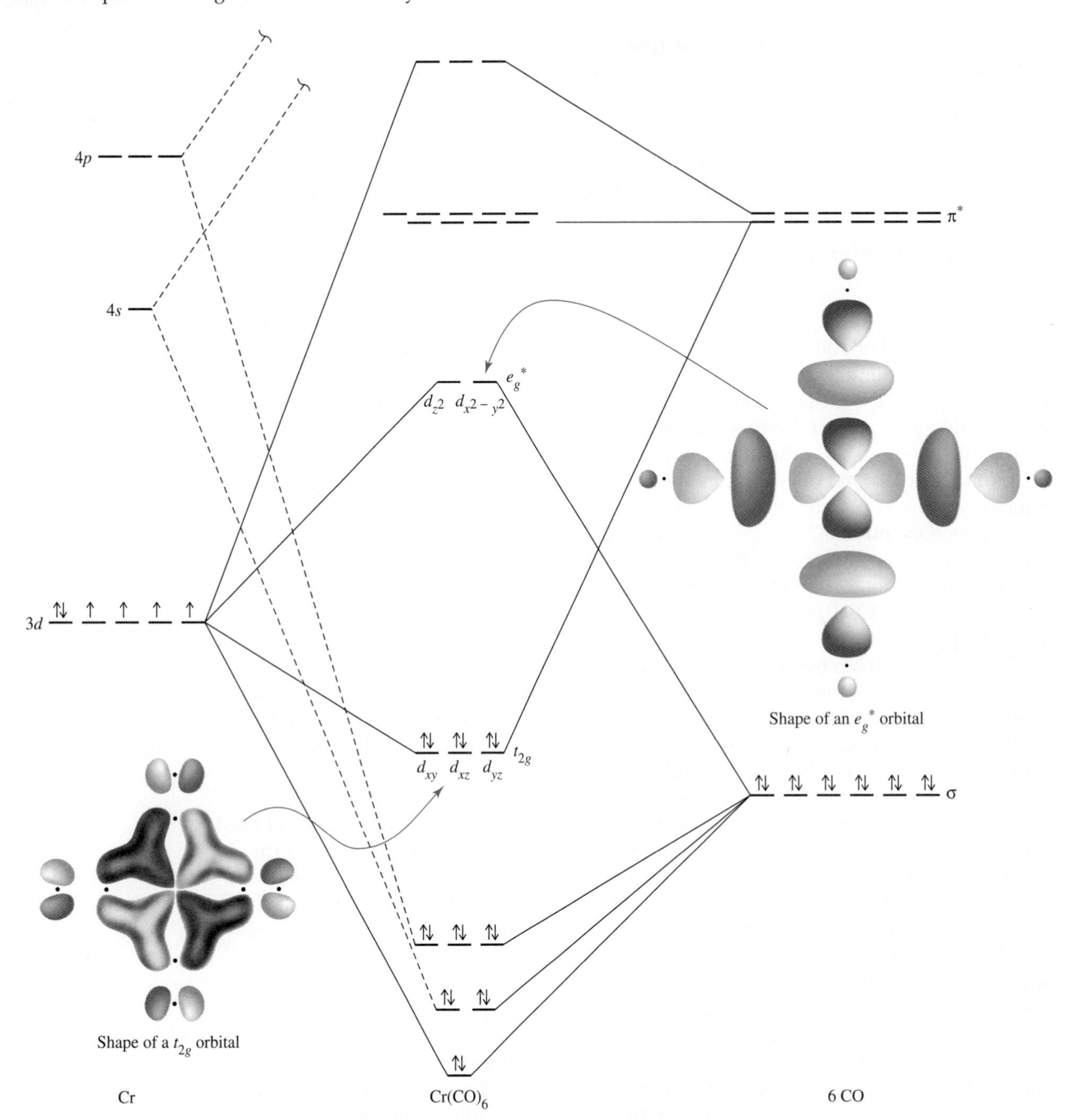

FIGURE 13.8 Molecular Orbital Energy Levels of $Cr(CO)_6$. (Adapted with permission from G. O. Spessard and G. L. Miessler, *Organometallic Chemistry*, Prentice Hall, Upper Saddle River, NJ, 1997, pp. 53–54, Figs. 3.2 and 3.3.)

e_g* orbitals. These examples of exceptions to the 18-electron rule are shown schematically in Figure 13.9.[14]

The same type of argument can be made for complexes of other geometries; in most but not all cases, there is an 18-electron configuration of special stability for complexes of strongly π-accepting ligands. Examples include trigonal-bipyramidal geometry, such as

[14] P. R. Mitchell and R. V. Parish, *J. Chem. Ed.*, **1969**, *46*, 311. See also W. B. Jensen, *J. Chem. Ed.*, **2005**, *82*, 28.

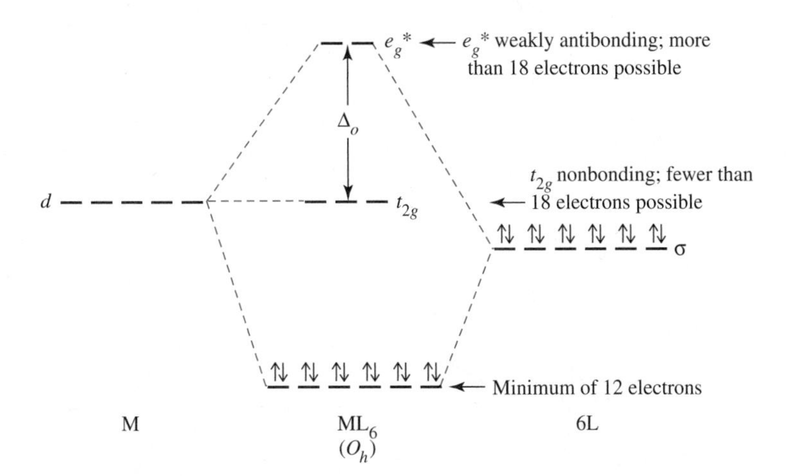

FIGURE 13.9 Exceptions to the 18-Electron Rule.

$Fe(CO)_5$ and tetrahedral geometry, like $Ni(CO)_4$. The most common exception is square-planar geometry, in which a 16-electron configuration may be the most stable, especially for complexes of d^8 metals.

13.3.3 Square-Planar Complexes

Examples of square-planar complexes include the d^8 16-electron complexes shown in Figure 13.10. To understand why 16-electron square-planar complexes might be especially stable, it is necessary to examine the molecular orbitals of such a complex. An energy diagram for the molecular orbitals of a square-planar molecule of formula ML_4, where L is a ligand that can function as both σ donor and π acceptor, is shown in Figure 13.11.[15]

The four lowest energy molecular orbitals in this diagram result from bonding interactions between the σ-donor orbitals of the ligands and the $d_{x^2-y^2}$, d_{z^2}, p_x, and p_y orbitals of the metal. These molecular orbitals are filled by 8 electrons from the ligands. The next four orbitals are either slightly bonding, nonbonding, or slightly antibonding, derived primarily from the d_{xz}, d_{yz}, d_{xy}, and d_{z^2} orbital of the metal.[16] These orbitals are occupied by a maximum of 8 electrons from the metal.[17] Additional electrons would occupy an orbital derived from the antibonding interaction of a metal $d_{x^2-y^2}$ orbital with the σ-donor orbitals of the ligands (the $d_{x^2-y^2}$ orbital points directly toward the ligands; its antibonding interaction is therefore the strongest). Consequently, for square-planar complexes of ligands having both σ-donor and π-acceptor characteristics, a 16-electron configuration is more stable than an 18-electron configuration. Sixteen-electron square-planar

$[Ni(CN)_4]^{2-}$ $PtCl_2(H_2NCH_2CH_2NH_2)$ $RhCl(PPh_3)_3$ (Wilkinson's complex) $IrCl(CO)(PPh_3)_2$ (Vaska's complex)

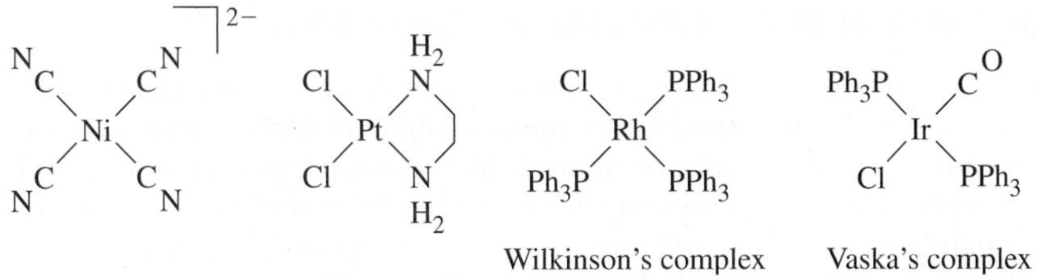

FIGURE 13.10 Examples of Square-Planar d^8 Complexes.

[15] Figure 10.15 shows a more complete diagram.

[16] The d_{z^2} orbital has A_{1g} symmetry and interacts with an A_{1g} group orbital. If this were the only metal orbital of this symmetry, the molecular orbital labeled d_{z^2} in Figure 13.11 would be antibonding. However, the next higher energy s orbital of the metal also has A_{1g} symmetry; the greater the degree to which this orbital is involved, the lower the energy of the molecular orbital.

[17] The relative energies of all four of these orbitals depend on the nature of the specific ligands and metal involved; in some cases, as shown in Figure 10.15, the ability of ligands to π donate can cause the order of energy levels to be different than those shown in Figure 13.11.

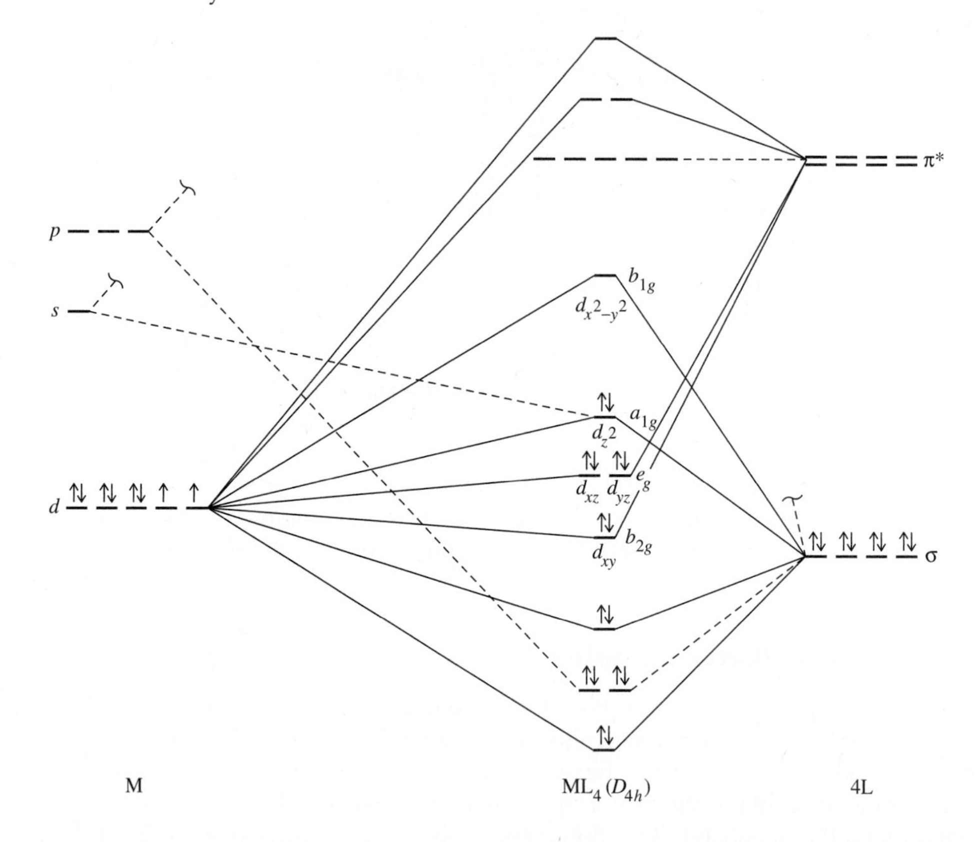

FIGURE 13.11 Molecular Orbital Energy Levels for a Square-Planar Complex.

complexes may also be able to accept one or two ligands at the vacant coordination sites (along the z axis) and thereby achieve an 18-electron configuration. As will be shown in the next chapter, this is a common reaction of 16-electron square-planar complexes.

▶ **Exercise 13.3** Verify that the complexes in Figure 13.10 are 16-electron species.

Sixteen-electron square-planar species are most commonly encountered for d^8 metals, particularly for metals having formal oxidation states of 2+(Ni^{2+}, Pd^{2+}, and Pt^{2+}) and 1+(Rh^+, Ir^+). Square-planar geometry is also more common for second- and third-row transition-metal complexes than for first-row complexes. Some square-planar complexes have important catalytic behavior. Two examples of square-planar d^8 complexes that are used as catalysts are Wilkinson's complex and Vaska's complex, shown in Figure 13.10.

13.4 LIGANDS IN ORGANOMETALLIC CHEMISTRY

Hundreds of ligands are known to bond to metal atoms through carbon. Carbon monoxide forms a very large number of metal complexes and deserves special mention, along with several similar diatomic ligands. Many organic molecules containing linear or cyclic pi systems also form numerous organometallic complexes. Complexes containing such ligands will be discussed next, following a brief review of the pi systems in the ligands themselves. Finally, special attention will be paid to two types of organometallic compounds that are especially important: carbene complexes, containing metal–carbon double bonds, and carbyne complexes, containing metal–carbon triple bonds.

13.4.1 Carbonyl (CO) Complexes

Carbon monoxide is the most common ligand in organometallic chemistry. It serves as the only ligand in binary carbonyls such as $Ni(CO)_4$, $W(CO)_6$, and $Fe_2(CO)_9$ or, more commonly, in combination with other ligands, both organic and inorganic. CO may

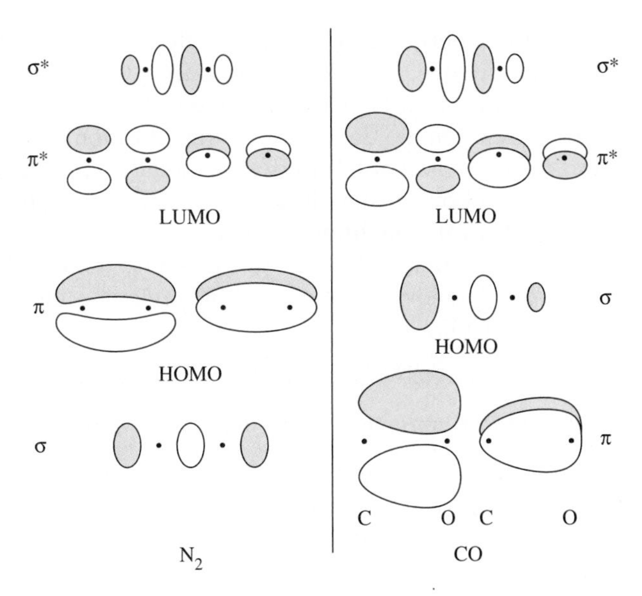

FIGURE 13.12 Selected Molecular Orbitals of N_2 and CO.

bond to a single metal, or it may serve as a bridge between two or more metals. In this section, we will consider the bonding between metals and CO, the synthesis and some reactions of CO complexes, and examples of the various types of CO complexes.

BONDING It is useful to review the bonding in CO. The molecular orbital picture of CO shown in Figure 5.14 is similar to that of N_2. Sketches of the molecular orbitals derived primarily from the 2*p* atomic orbitals of these molecules are shown in Figure 13.12.

Two features of the molecular orbitals of CO deserve attention. First, the highest-energy occupied orbital (the HOMO) has its largest lobe on carbon. It is through this orbital, occupied by an electron pair, that CO exerts its σ-donor function, donating electron density directly toward an appropriate metal orbital, such as an unfilled *d* or hybrid orbital. Carbon monoxide also has two empty π^* orbitals (the lowest unoccupied, or LUMO); these also have larger lobes on carbon than on oxygen. A metal atom having electrons in a *d* orbital of suitable symmetry can donate electron density to these π^* orbitals. These σ-donor and π-acceptor interactions are illustrated in Figure 13.13.

The overall effect is synergistic. CO can donate electron density via a σ orbital to a metal atom; the greater the electron density on the metal, the more effectively it can return electron density to the π^* orbitals of CO. The net effect can be strong bonding between the metal and CO; however, as will be described later, the strength of this bonding depends on several factors, including the charge on the complex and the ligand environment of the metal.

▶ **Exercise 13.4** N_2 has molecular orbitals rather similar to those of CO, as shown in Figure 13.12. Would you expect N_2 to be a stronger or weaker π acceptor than CO?

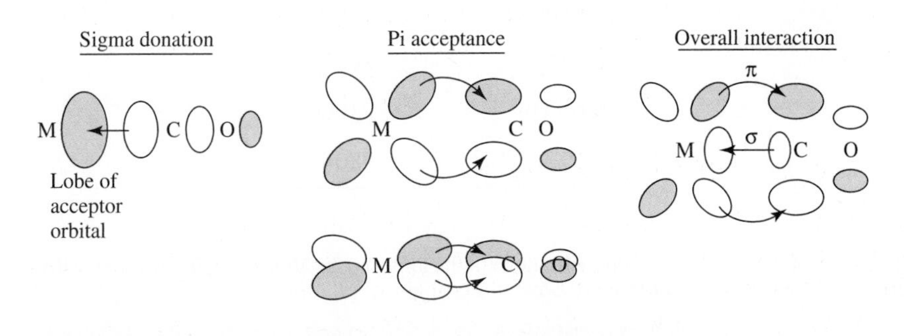

FIGURE 13.13 Sigma and Pi Interactions Between CO and a Metal Atom.

If this picture of bonding between CO and metal atoms is correct, it should be supported by experimental evidence. Two sources of such evidence are infrared spectroscopy and X-ray crystallography. First, any change in the bonding between carbon and oxygen should be reflected in the C—O stretching vibration as observed by IR. As in organic compounds, the C—O stretch in organometallic complexes is often very intense (stretching the C—O bond results in a substantial change in dipole moment), and its energy often provides valuable information about the molecular structure. Free carbon monoxide has a C—O stretch at 2143 cm^{-1}. $Cr(CO)_6$, on the other hand, has its C—O stretch at 2000 cm^{-1}. The lower energy for the stretching mode means that the C—O bond is weaker in $Cr(CO)_6$.

The energy necessary to stretch a bond is proportional to $\sqrt{\frac{k}{\mu}}$, where k = force constant and μ = reduced mass; for atoms of mass m_1 and m_2, the reduced mass is given by

$$\mu = \frac{m_1 m_2}{m_1 + m_2}$$

The stronger the bond between two atoms, the larger the force constant; consequently, the greater the energy necessary to stretch the bond and the higher the energy of the corresponding band (the higher the wavenumber, in cm^{-1}) in the infrared spectrum. Similarly, the more massive the atoms involved in the bond, as reflected in a higher reduced mass, the less energy necessary to stretch the bond, and the lower the energy of the absorption in the infrared spectrum.

Both σ donation (which donates electron density from a bonding orbital on CO) and π acceptance (which places electron density in C—O antibonding orbitals) would be expected to weaken the C—O bond and to decrease the energy necessary to stretch that bond.

Additional evidence is provided by X-ray crystallography. In carbon monoxide, the C—O distance has been measured at 112.8 pm. Weakening of the C—O bond by the factors described above would be expected to cause this distance to increase. Such an increase in bond length is found in complexes containing CO, with C—O distances approximately 115 pm for many carbonyls. Although such measurements provide definitive measures of bond distances, in practice it is far more convenient to use infrared spectra to obtain data on the strength of C—O bonds.

The charge on a carbonyl complex is also reflected in its infrared spectrum. Five isoelectronic hexacarbonyls have the following C—O stretching bands (compare with $\nu(CO) = 2143\ cm^{-1}$ for free CO):[18]

Complex	ν(CO), cm^{-1}
$[Ti(CO)_6]^{2-}$	1748
$[V(CO)_6]^{-}$	1859
$Cr(CO)_6$	2000
$[Mn(CO)_6]^{+}$	2100
$[Fe(CO)_6]^{2+}$	2204

[18] The positions of the C—O stretching vibrations in the ions may be affected by interactions with solvents or counterions, and solid and solution spectra may differ slightly.

Of these five, $[Ti(CO)_6]^{2-}$ has the metal with the smallest nuclear charge; this means that titanium has the weakest ability to attract electrons and the greatest tendency to back-donate electron density to CO. Alternatively, the formal charges on the metals increase from –2 for $[Ti(CO)_6]^{2-}$ to +2 for $[Fe(CO)_6]^{2+}$. The titanium in $[Ti(CO)_6]^{2-}$, with the most negative formal charge, has the strongest tendency to donate to CO. The consequence is strong population of the π^* orbitals of CO in $[Ti(CO)_6]^{2-}$ and reduction of the strength of the C—O bond. In general, the more negative the charge on the organometallic species, the greater the tendency of the metal to donate electrons to the π^* orbitals of CO, and the lower the energy of the C—O stretching vibrations.[19]

▶ **Exercise 13.5** Predict which of the complexes $[V(CO)_6]^-$, $Cr(CO)_6$, or $[Mn(CO)_6]^+$ has the shortest C—O bond.

How is it possible for cationic carbonyl complexes such as $[Fe(CO)_6]^{2+}$ to have C—O stretching bands even higher in energy than those in free CO? It has been argued that in such complexes, the CO ligand does not have π-acceptor activity and that the HOMO of CO, a σ orbital that is slightly antibonding with respect to the carbon–oxygen bond, acts as a donor to the metal. If this orbital were to act as a donor, there would be a decrease in the population of the HOMO and a consequent strengthening of the carbon–oxygen bond. However, calculations have demonstrated that it is much more likely that donation from the HOMO to the metal in cationic complexes is insignificant in comparison with the polarization effect caused by the metal cation.[20]

In free CO, the electrons are polarized toward the more electronegative oxygen. For example, the electrons in the π orbitals are concentrated nearer to the oxygen atom than to the carbon. The presence of a transition metal cation tends to reduce the polarization in the C—O bond by attracting the bonding electrons:

$$\overset{\delta+}{C}\equiv\overset{\delta-}{O} \qquad M^{n+} \longleftarrow \overset{\delta+}{C}\equiv\overset{\delta-}{O}$$

The consequence is that the electrons in the positively charged complex are more equally shared by the carbon and the oxygen, giving rise to a stronger bond and a higher-energy C—O stretch.

BRIDGING MODES OF CO Although CO is most commonly found as a terminal ligand attached to a single metal atom, many cases are known in which CO forms bridges between two or more metals. Many such bridging modes are known; the most common are shown in Table 13.2.

The bridging mode is strongly correlated with the position of the C—O stretching band. In cases in which CO bridges two metal atoms, both metals can contribute electron density into π^* orbitals of CO to weaken the C—O bond and lower the energy of the stretch. Consequently, the C—O stretch for doubly bridging CO is at a much lower energy than for terminal COs. An example is shown in Figure 13.14. Interaction of three metal atoms with a triply bridging CO further weakens the C—O bond; the infrared band for the C—O stretch is still lower than in the doubly bridging case. (For comparison, carbonyl stretches in organic molecules are typically in the range 1700 to 1850 cm^{-1}, with many alkyl ketones near 1700 cm^{-1}.)

[19] For reviews of metal carbonyl anions and other complexes containing metals in negative oxidation states, see J. E. Ellis, *Organometallics*, **2003**, *22*, 3322 and *Inorg. Chem.*, **2006**, *45*, 3167.
[20] A. S. Goldman and K. Krogh-Jespersen, *J. Am. Chem. Soc.*, **1996**, *118*, 12159.

TABLE 13.2 Bridging Modes of CO

Type of CO	Approximate Range for ν (CO) in Neutral Complexes (cm^{-1})
Free CO	2143
Terminal M—CO	1850–2120
Symmetric[a] μ_2—CO	1700–1860
Symmetric[a] μ_3—CO	1600–1700
Asymmetric μ_4—CO	< 1700 (few examples)

NOTE: *Asymmetrically bridging μ_2- and μ_3-CO are also known.

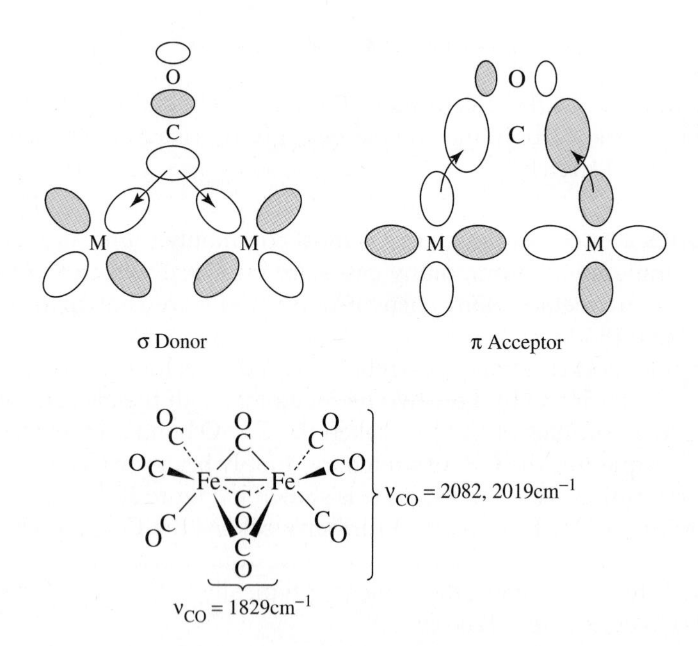

FIGURE 13.14 Bridging CO.

Ordinarily, terminal and bridging carbonyl ligands can be considered 2-electron donors, with the donated electrons shared by the metal atoms in the bridging cases. For example, in the complex

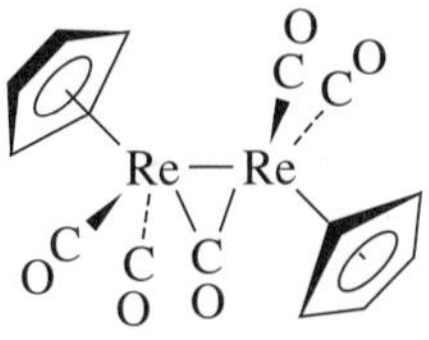

the bridging CO is a 2-electron donor overall, with a single electron donated to each metal. The electron count for each Re atom according to method B is

Re	7 e^-
η^5-C_5H_5	5 e^-
2 (CO) (terminal)	4 e^-
$\frac{1}{2}(\mu_2\text{-CO})$	1 e^-
M—M bond	1 e^-
Total =	18 e^-

A particularly interesting situation is that of nearly linear bridging carbonyls, such as in $[(\eta^5\text{-}C_5H_5)Mo(CO)_2]_2$. When a sample of $[(\eta^5\text{-}C_5H_5)Mo(CO)_3]_2$ is heated, some carbon monoxide is driven off; the product, $[(\eta^5\text{-}C_5H_5)Mo(CO)_2]_2$, reacts readily with CO to reverse this reaction:[21]

$$\underset{1960,\ 1915\ \text{cm}^{-1}}{[(\eta^5\text{-}C_5H_5)Mo(CO)_3]_2} \overset{\Delta}{\rightleftharpoons} \underset{1889,\ 1859\ \text{cm}^{-1}}{[(\eta^5\text{-}C_5H_5)Mo(CO)_2]_2} + 2\,CO$$

This reaction is accompanied by changes in the infrared spectrum in the CO region, as listed above. The Mo—Mo bond distance also shortens by approximately 100 pm, consistent with an increase in the metal–metal bond order from 1 to 3. Although it was originally proposed that the "linear" CO ligands may donate some electron density to the neighboring metal from π orbitals, subsequent calculations have indicated that a more important interaction is donation from a metal d orbital to the π^* orbital of CO, as shown in Figure 13.15.[22] Such donation weakens the carbon–oxygen bond in the ligand and results in the observed shift of the C—O stretching bands to lower energies.

Additional information on infrared spectra of carbonyl complexes is included in Section 13.7 at the end of this chapter.

BINARY CARBONYL COMPLEXES Binary carbonyls, containing only metal atoms and CO, are numerous. Some representative binary carbonyl complexes are shown in Figure 13.16. Most of these complexes obey the 18-electron rule. The cluster compounds $Co_6(CO)_{16}$ and $Rh_6(CO)_{16}$ do not obey the rule, however. More detailed analysis of the bonding in cluster compounds is necessary to satisfactorily account for the electron counting in these and other cluster compounds. This will be discussed in Chapter 15.

[21] D. S. Ginley and M. S. Wrighton, *J. Am. Chem. Soc.*, **1975**, *97*, 3533; R. J. Klingler, W. Butler, and M. D. Curtis, *J. Am. Chem. Soc.*, **1975**, *97*, 3535.
[22] A. L. Sargent and M. B. Hall, *J. Am. Chem. Soc.*, **1989**, *111*, 1563 and references therein.

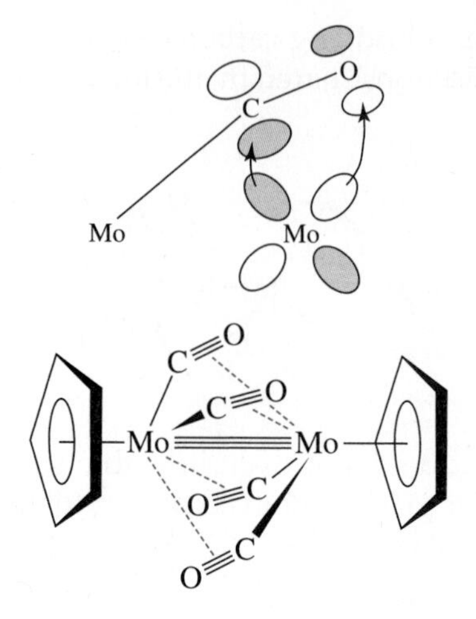

FIGURE 13.15 Bridging CO in $[(\eta^5\text{-}C_5H_5)Mo(CO)_2]_2$.

One other binary carbonyl does not obey the rule, the 17-electron $V(CO)_6$. This complex is one of a few cases in which strong π-acceptor ligands do not succeed in requiring an 18-electron configuration. In $V(CO)_6$, the vanadium is apparently too small to permit a seventh coordination site; hence, no metal–metal bonded dimer, which would give an 18-electron configuration, is possible. However, $V(CO)_6$ is easily reduced to $[V(CO)_6]^-$, a well-studied 18-electron complex.

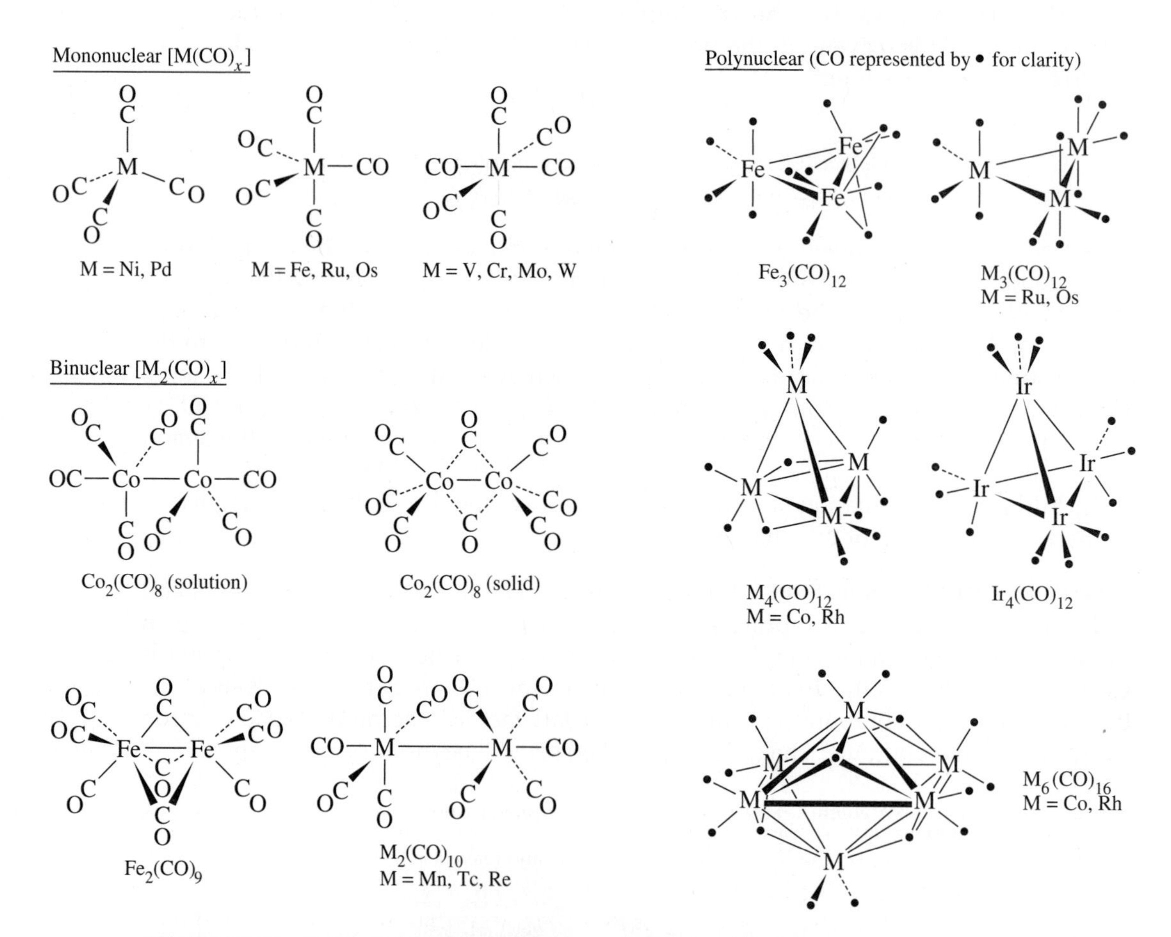

FIGURE 13.16 Binary Carbonyl Complexes.

▶ **Exercise 13.6** Verify the 18-electron rule for five of the binary carbonyls—other than $V(CO)_6$, $Co_6(CO)_{16}$, and $Rh_6(CO)_{16}$—shown in Figure 13.16.

An interesting feature of the structures of binary carbonyl complexes is that the tendency of CO to bridge transition metals decreases going down the periodic table. For example, in $Fe_2(CO)_9$ there are three bridging carbonyls; but in $Ru_2(CO)_9$ and $Os_2(CO)_9$, there is a single bridging CO. A possible explanation is that the orbitals of bridging CO are less able to interact effectively with transition-metal atoms as the size of the metals increases.

Binary carbonyl complexes can be synthesized in many ways. Several of the most common methods are as follows:

1. *Direct reaction of a transition metal with CO.* The most facile of these reactions involves nickel, which reacts with CO at ambient temperature and 1 atm:

$$Ni + 4\,CO \longrightarrow Ni(CO)_4$$

$Ni(CO)_4$ is a volatile, extremely toxic liquid that must be handled with great caution. It was first observed in Mond's study of the reaction of CO with nickel valves.[23] Because the reaction can be reversed at high temperature, coupling of the forward and reverse reactions has been used commercially in the Mond process for obtaining purified nickel from ores. Other binary carbonyls can be obtained from direct reaction of metal powders with CO, but elevated temperatures and pressures are necessary.

2. *Reductive carbonylation:* reduction of a metal compound in the presence of CO and an appropriate reducing agent. Examples are

$$CrCl_3 + 6\,CO + Al \longrightarrow Cr(CO)_6 + AlCl_3$$

$$Re_2O_7 + 17\,CO \longrightarrow Re_2(CO)_{10} + 7\,CO_2$$

(CO acts as a reducing agent in the second reaction; high temperature and pressure are required.)

3. *Thermal or photochemical reaction of other binary carbonyls.* Examples are

$$2\,Fe(CO)_5 \xrightarrow{h\nu} Fe_2(CO)_9 + CO$$

$$3\,Fe(CO)_5 \xrightarrow{\Delta} Fe_3(CO)_{12} + 3\,CO$$

The most common reaction of carbonyl complexes is CO dissociation. This reaction, which may be initiated thermally or by absorption of ultraviolet light, characteristically involves loss of CO from an 18-electron complex to give a 16-electron intermediate, which may react in a variety of ways, depending on the nature of the complex and its environment. A common reaction is replacement of the lost CO by another ligand to form a new 18-electron species as product. For example,

$$Cr(CO)_6 + PPh_3 \xrightarrow[\text{or } h\nu]{\Delta} Cr(CO)_5(PPh_3) + CO$$

$$Re(CO)_5Br + en \xrightarrow{\Delta} fac\text{-}Re(CO)_3(en)Br + 2\,CO$$

O C N Re C O N C O Br

This type of reaction therefore provides a pathway in which CO complexes can be used as precursors for a variety of complexes of other ligands. Additional aspects of CO dissociation reactions will be discussed in Chapter 14.

[23] L. Mond, C. Langer, and F. Quincke, *J. Chem. Soc. (London)*, **1890**, **57**, 749; reprinted in *J. Organomet. Chem.*, **1990**, *383*, 1.

OXYGEN-BONDED CARBONYLS This section would not be complete without mentioning one additional aspect of CO as a ligand: it can sometimes bond through oxygen as well as carbon. This phenomenon was first noted in the ability of the oxygen of a metal-carbonyl complex to act as a donor toward Lewis acids such as $AlCl_3$, with the overall function of CO serving as a bridge between the two metals. Many examples are now known in which CO bonds through its oxygen to transition metal atoms, with the C—O—metal arrangement generally bent. Attachment of a Lewis acid to the oxygen results in significant weakening and lengthening of the C—O bond and a corresponding shift of the C—O stretching vibration to lower energy in the infrared. This shift is typically between 100 and 200 cm^{-1}. Examples of O-bonded carbonyls, sometimes called *isocarbonyls*, are shown in Figure 13.17. The physical and chemical properties of oxygen-bonded carbonyls have been reviewed.[24]

FIGURE 13.17 Oxygen-Bonded Carbonyls.

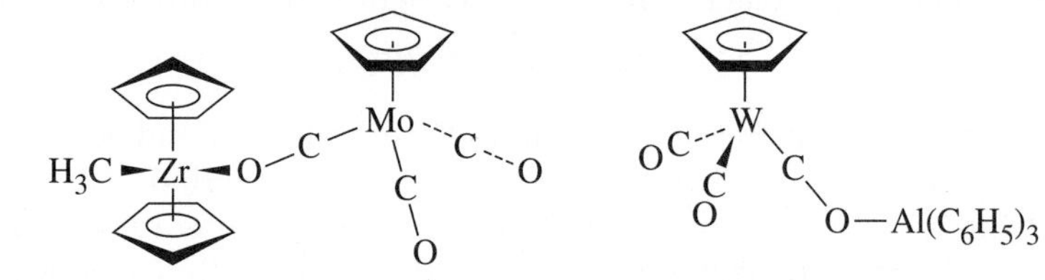

13.4.2 Ligands Similar to CO

Several diatomic ligands similar to CO are worth brief mention. Two of these, CS (thiocarbonyl) and CSe (selenocarbonyl), are of interest in part for purposes of comparison with CO. In most cases, synthesis of CS and CSe complexes is somewhat more difficult than for analogous CO complexes, because CS and CSe do not exist as stable, free molecules and do not, therefore, provide a ready ligand source.[25] Therefore, the comparatively small number of such complexes should not be viewed as an indication of their stability. Thiocarbonyl complexes are also of interest as possible intermediates in certain sulfur-transfer reactions in the removal of sulfur from natural fuels. In recent years, the chemistry of complexes containing these ligands has developed more rapidly as avenues for their synthesis have been devised.

CS and CSe are similar to CO in their bonding modes in that they behave as both σ donors and π acceptors and can bond to metals in terminal or bridging modes. Of these two ligands, CS has been studied more closely. It usually functions as a stronger σ donor and π acceptor than CO.[26]

Several other common ligands are isoelectronic with CO and, not surprisingly, exhibit structural and chemical parallels with CO. Two examples are CN^- and N_2. Complexes of CN^- have been known even longer than carbonyl complexes. For example, blue complexes (Prussian blue and Turnbull's blue) containing the ion $[Fe(CN)_6]^{3-}$ have been used as pigments in paints and inks for approximately three centuries. Cyanide is a stronger σ donor and a substantially weaker π acceptor than CO; overall, it is close to CO in the spectrochemical series.[27] Unlike most organic ligands, which bond to metals in low formal oxidation states, cyanide bonds readily to metals having higher oxidation states. As a good σ donor, CN^- interacts strongly with positively charged metal ions; as a weaker π acceptor than CO, partly a consequence of the negative charge of CN^- and the high energy of its π^* orbitals, cyanide is not as able to stabilize metals in

[24] C. P. Horwitz and D. F. Shriver, *Adv. Organomet. Chem.*, **1984**, *23*, 219.
[25] E. J. Moltzen and K. J. Klabunde, *Chem. Rev.*, **1988**, *88*, 391, provides a detailed review of CS chemistry.
[26] P. V. Broadhurst, *Polyhedron*, **1985**, *4*, 1801.
[27] A comparison of these ligands in mixed ligand complexes of Fe is provided in C. Loschen and G. Frenking, *Inorg. Chem.*, **2004**, *43*, 778.

low oxidation states. Therefore, its compounds are often studied in the context of classic coordination chemistry rather than organometallic chemistry.

The recent discovery that hydrogenase enzymes contain both CO and CN^- bound to iron has stimulated interest in complexes containing both ligands. Remarkably, only two iron complexes containing both CO and CN^- and a single iron atom, $[Fe(CO)(CN)_5]^{3-}$ (reported in 1887) and $[Fe(CO)_4(CN)]^-$ (reported in 1974), were known before 2001. Both the *cis* and *trans* isomers of $[Fe(CO)_2(CN)_4]^{2-}$ and *fac*-$[Fe(CO)_3(CN)_3]^-$ have been prepared by simple pathways. Two of the mixed ligand complexes can be made using $Fe(CO)_4I_2$ as starting material:[28]

$$Fe(CO)_4I_2 \xrightarrow{3\,CN^-} fac\text{-}[Fe(CO)_3(CN)_3]^- \xrightarrow{CN^-} cis\text{-}[Fe(CO)_2(CN)_4]^{2-}$$

The complex *trans*-$[Fe(CO)_2(CN)_4]^{2-}$ can be made simply by the addition of cyanide to a solution of $FeCl_2$ under an atmosphere of CO:[29]

$$FeCl_2(aq \text{ or } CH_3CN) + 4\,CN^- \xrightarrow{CO} trans\text{-}[Fe(CO)_2(CN)_4]^{2-}$$

Dinitrogen is a weaker donor and acceptor than CO. However, N_2 complexes are of great interest, especially as possible intermediates in reactions that may simulate natural processes of nitrogen fixation.

NO COMPLEXES Although not an organic ligand, the NO (nitrosyl) ligand deserves discussion here because of its similarities to CO. Like CO, it is both a σ donor and π acceptor and can serve as a terminal or bridging ligand; useful information can be obtained about its compounds by analysis of its infrared spectra. Unlike CO, however, terminal NO has two common coordination modes, linear (like CO) and bent. Examples of NO complexes are shown in Figure 13.18.

A formal analogy is often drawn between the linear bonding modes of both ligands. NO^+ is isoelectronic with CO; therefore, in its bonding to metals, linear NO is considered by electron-counting scheme A as NO^+, a 2-electron donor. By the neutral ligand method (B), linear NO is counted as a 3-electron donor (it has one more electron than the 2-electron donor CO).

The bent coordination mode of NO can be considered to arise formally from NO^-, with the bent geometry suggesting sp^2 hybridization at the nitrogen. By electron-counting scheme A, therefore, bent NO is considered the 2-electron donor NO^-; by the neutral ligand model, it is considered a 1-electron donor.

Although these electron-counting methods in NO complexes are useful, they do not describe how NO actually bonds to metals. The use of NO^+, NO, or NO^- does not necessarily imply degrees of ionic or covalent character in coordinated NO; these labels are simply convenient means of counting electrons.

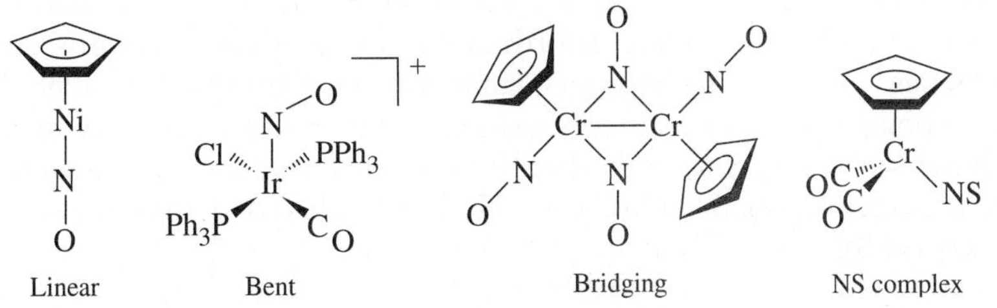

Linear | Bent | Bridging | NS complex

FIGURE 13.18 Examples of NO and NS Complexes.

[28] J. Jiang and S. A. Koch, *Inorg. Chem.*, **2002**, *41*, 158.
[29] J. Jiang and S. A. Koch, *Angew. Chem., Int. Ed.*, **2001**, *40*, 2629; T. B. Rauchfuss, S. M. Contakes, S. C. N. Hsu, M. A. Reynolds, and S. R. Wilson, *J. Am. Chem. Soc.*, **2001**, *123*, 6933; S. M. Contakes, S. C. N. Hsu, T. B. Rauchfuss, and S. R. Wilson, *Inorg. Chem.*, **2002**, *41*, 1670.

Useful information about the linear and bent bonding modes of NO is summarized in Figure 13.19. Many complexes containing each mode are known, and examples are also known in which both linear and bent NO occur in the same complex. Although linear coordination usually gives rise to N—O stretching vibrations at a higher energy than the bent mode, there is enough overlap in the ranges of these bands that infrared spectra alone may not be sufficient to distinguish between the two. Furthermore, the manner of packing in crystals may bend the metal—N—O bond considerably from 180° in the linear coordination mode.

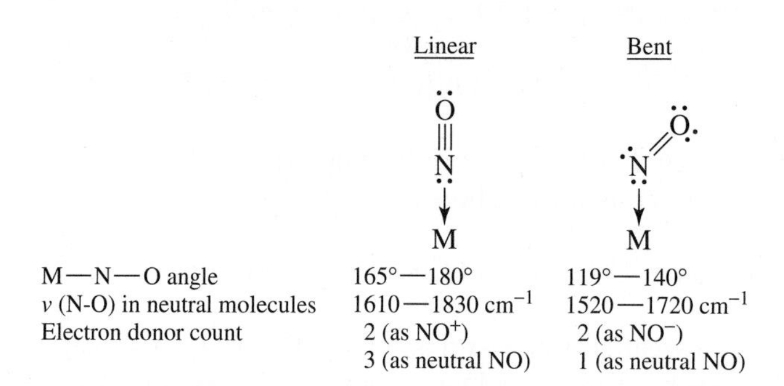

Linear
Bent
O
N
M
O
N
M
M—N—O angle
165°—180°
119°—140°
ν (N-O) in neutral molecules
1610—1830 cm⁻¹
1520—1720 cm⁻¹
Electron donor count
2 (as NO⁺)
3 (as neutral NO)
2 (as NO⁻)
1 (as neutral NO)

FIGURE 13.19 Linear and Bent Bonding Modes of NO.

One compound containing only a metal and NO ligands is known, $Cr(NO)_4$, a tetrahedral molecule that is isoelectronic with $Ni(CO)_4$.[30] Complexes containing bridging nitrosyl ligands are also known, with the neutral bridging ligand formally considered a 3-electron donor. One NO complex, the nitroprusside ion, $[Fe(CN)_5(NO)]^{2-}$, has been widely used as a vasodilator in the treatment of high blood pressure. Its therapeutic effect is a consequence of its ability to release its NO ligand; the NO itself acts as the vasodilating agent.

In recent years, several dozen compounds containing the isoelectronic NS (thionitrosyl) ligand have been synthesized; one of these is shown in Figure 13.18. Infrared data have indicated that, like NO, NS can function in linear, bent, and bridging modes. In general, NS is similar to NO in its ability to act as a π-acceptor ligand; the relative π-acceptor abilities of NO and NS depend on the electronic environment of the compounds being compared.[31]

13.4.3 Hydride and Dihydrogen Complexes

The simplest of all possible ligands is the hydrogen atom; similarly, the simplest possible diatomic ligand is H_2. It is perhaps not surprising that these ligands have gained attention, by virtue of their apparent simplicity, as models for bonding schemes in coordination compounds. Moreover, both ligands have played important roles in the development of applications of organometallic chemistry to organic synthesis and especially to catalytic processes.

HYDRIDE COMPLEXES Although hydrogen atoms form bonds with nearly every element, we will specifically consider coordination compounds containing H bonded to transition metals.[32] Because the hydrogen atom only has a 1*s* orbital of suitable energy for bonding, the bond between H and a transition metal must by necessity be a σ interaction, involving metal *s*, *p*, and/or *d* orbitals (or a hybrid orbital). As a ligand, H may be considered a 2-electron donor as hydride ($:H^-$, method A) or a 1-electron neutral donor (H atom, method B).

---

[30] Compounds containing only a single ligand, such as NO in $Cr(NO)_4$ and CO in $Mo(CO)_6$, are called *homoleptic compounds*.

[31] H. W. Roesky and K. K. Pandey, *Adv. Inorg. Chem. Radiochem.*, **1983**, *26*, 337.

[32] G. J. Kubas, *Comments Inorg. Chem.*, **1988**, *7*, 17; R. H. Crabtree, *Acc. Chem. Res.*, **1990**, *23*, 95; G. J. Kubas, *Acc. Chem. Res.*, **1988**, *21*, 120.

Although some transition-metal complexes containing only the hydride ligand are known—an example of some structural interest is the 9-coordinate $[ReH_9]^{2-}$ ion (Figure 9.33), the classic example of a tricapped trigonal prism[33]—we are principally concerned with complexes containing H in combination with other ligands. Such complexes may be made in a variety of ways. Probably the most common synthesis is by reaction of a transition metal complex with H_2. For example,

$$Co_2(CO)_8 + H_2 \longrightarrow 2\,HCo(CO)_4$$

$$trans\text{-}Ir(CO)Cl(PEt_3)_2 + H_2 \longrightarrow Ir(CO)Cl(H)_2(PEt_3)_2$$

Carbonyl hydride complexes can also be formed by the reduction of carbonyl complexes, followed by the addition of acid. For example,

$$Co_2(CO)_8 + 2\,Na \longrightarrow 2\,Na^+[Co(CO)_4]^-$$

$$[Co(CO)_4]^- + H^+ \longrightarrow HCo(CO)_4$$

One of the most interesting aspects of transition-metal hydride chemistry is the relationship between this ligand and the rapidly developing chemistry of the dihydrogen ligand, H_2.

DIHYDROGEN COMPLEXES Although complexes containing H_2 molecules coordinated to transition metals had been proposed for many years, the first structural characterization of a dihydrogen complex did not occur until 1984, when Kubas and coworkers synthesized $M(CO)_3(PR_3)_2(H_2)$, where M = Mo or W and R = cyclohexyl or isoproplyl.[34] Subsequently, many H_2 complexes have been identified, and the chemistry of this ligand has developed rapidly.[35]

The bonding between dihydrogen and a transition metal can be described as shown in Figure 13.20. The σ electrons in H_2 can be donated to a suitable empty orbital on the metal (such as a *d* orbital or hybrid orbital), and the empty σ^* orbital of the ligand can accept electron density from an occupied *d* orbital of the metal. The result is an overall weakening and lengthening of the H—H bond in comparison with free H_2 Typical H—H distances in complexes containing coordinated dihydrogen are in the range of 82 to 90 pm, in comparison with 74.14 pm in free H_2.

This bonding scheme leads to interesting ramifications that are distinctive from other donor–acceptor ligands such as CO. If the metal is electron rich and donates strongly to the σ^* of H_2 the H—H bond in the ligand can rupture, giving separate H atoms. Consequently, the search for stable H_2 complexes has centered on metals likely to be relatively poor donors, such as those in high oxidation states or surrounded by ligands that function as strong electron acceptors. In particular, good π acceptors, such CO and NO, can be effective at stabilizing the dihydrogen ligand.

▶ **Exercise 13.7** Explain why $Mo(PMe)_5H_2$ is a dihydride (contains two separate H ligands), but $Mo(CO)_3(PR_3)_2(H_2)$ contains the dihydrogen ligand (Me = methyl, *R* = isopropyl).

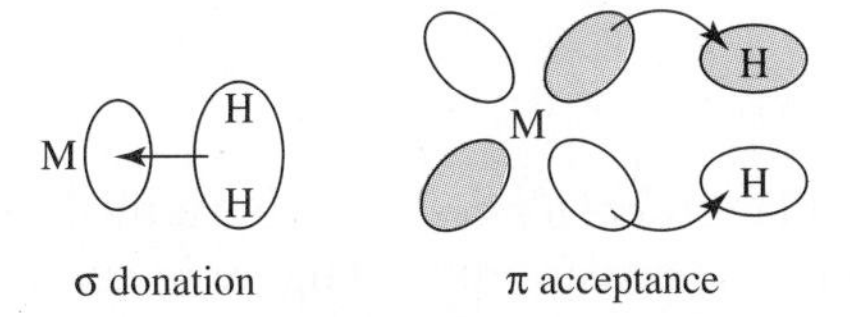

FIGURE 13.20 Bonding in Dihydrogen Complexes.

---

[33] S. C. Abrahams, A. P. Ginsberg, and K. Knox, *Inorg. Chem.*, **1964**, *3*, 558.
[34] G. J. Kubas, R. R. Ryan, B. I. Swanson, P. J. Vergamini, and H. J. Wasserman, *J. Am. Chem. Soc.*, **1984**, *106*, 451.
[35] J. K. Burdett, O. Eisenstein, and S. A. Jackson, "Transition Metal Dihydrogen Complexes: Theoretical Studies," in A. Dedieu, ed., *Transition Metal Hydrides*, VCH, New York, 1992, pp. 149–184.

Dihydrogen complexes have frequently been suggested as possible intermediates in a variety of reactions of hydrogen at metal centers. Some of these reactions are steps in catalytic processes of significant commercial interest. As this ligand becomes more completely understood, the applications of its chemistry are likely to become extremely important.

13.4.4 Ligands Having Extended Pi Systems

Although it is relatively simple to describe pictorially how ligands such as CO and PPh_3 bond to metals, explaining bonding between metals and organic ligands having extended π systems can be more complex. For example, how are the C_5H_5 rings attached to Fe in ferrocene, and how can 1,3-butadiene bond to metals? To understand the bonding between metals and π systems, we must first consider the π bonding within the ligands themselves. In the following discussion, we will first describe linear and then cyclic π systems, after which we will consider how molecules containing such systems can bond to metals.

LINEAR PI SYSTEMS[36] The simplest case of an organic molecule having a linear π system is ethylene, which has a single π bond resulting from the interactions of two $2p$ orbitals on its carbon atoms. Interactions of these p orbitals result in one bonding and one antibonding π orbital, as shown:

$H_2C{=}CH_2$ *p orbitals interacting* *Relative energy*

π*

↑↓ π

The antibonding interaction has a nodal plane perpendicular to the internuclear axis, but the bonding interaction has no such nodal plane.

Next is the three-atom π system, the π-allyl radical, C_3H_5. In this case, there are three $2p$ orbitals to be considered, one from each of the carbon atoms participating in the π system. The possible interactions are as follows:

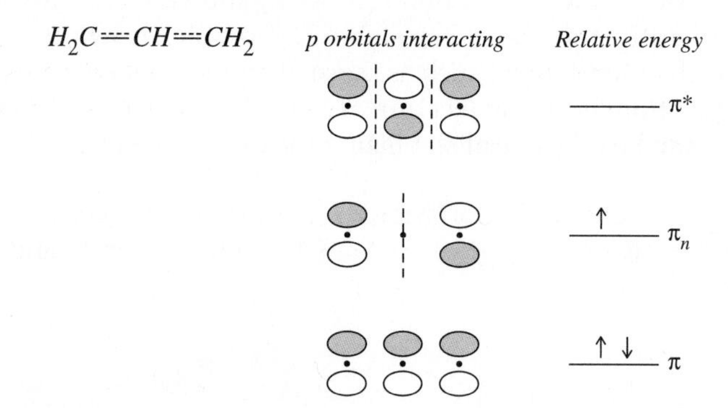

The lowest energy π molecular orbital for this system has all three p orbitals interacting constructively, to give a bonding molecular orbital. Higher in energy is the nonbonding orbital (π_n), in which a nodal plane bisects the molecule, cutting through the central carbon atom.

[36] In this section the term "linear" is used broadly to include not only ligands that have carbons in a straight line but acyclic ligands that are bent at inner sp^2 carbons. For simplicity, the accompanying diagrams are also shown as straight line arrangements.

In this case, the p orbital on the central carbon does not participate in the molecular orbital; a nodal plane passes through the center of this π orbital and thereby cancels it from participation. Highest in energy is the antibonding π^* orbital, in which there is an antibonding interaction between each neighboring pair of carbon p orbitals.

The number of nodes perpendicular to the carbon chain increases in going from lower-energy to higher-energy orbitals; for example, in the π-allyl system, the number of nodes increases from zero to one to two from the lowest to the highest energy orbital. This is a trend that will also appear in the following examples.

One more example should suffice to illustrate this procedure. 1,3-Butadiene may exist in *cis* or *trans* forms. For our purposes, we will treat both as linear systems; the nodal behavior of the molecular orbitals is the same in each case as in a linear π system of four atoms. The $2p$ orbitals of the carbon atoms in the chain may interact in four ways, with the lowest energy π molecular orbital having all constructive interactions between neighboring p orbitals, and the energy of the other π orbitals increasing with the number of nodes between the atoms.

$H_2C{=}CH{-}CH{=}CH_2$ *p orbitals interacting*   *Relative energy*

↑↓

↑↓

Similar patterns can be obtained for longer π systems; two more examples are included in Figure 13.21. As in the other examples, the number of π molecular orbitals is equal to the number of carbons in the π system.

CYCLIC PI SYSTEMS The procedure for obtaining a pictorial representation of the orbitals of cyclic π systems of hydrocarbons is similar to the procedure for the linear systems described above. The smallest such cyclic hydrocarbon is *cyclo*-C_3H_3. The lowest energy π molecular orbital for this system is the one resulting from constructive interaction between each of the $2p$ orbitals in the ring:

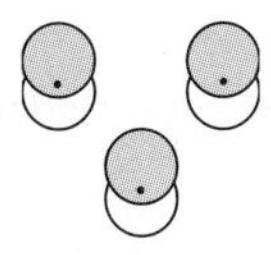

Because the number of molecular orbitals must equal the number of atomic orbitals used, two additional π molecular orbitals are needed. Each of these has a single nodal plane that is perpendicular to the plane of the molecule and bisects the molecule; the nodes for these two molecular orbitals are perpendicular to each other:

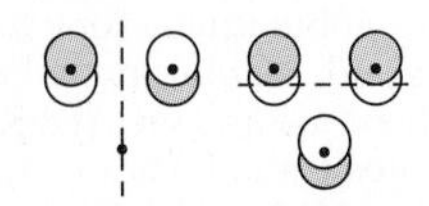

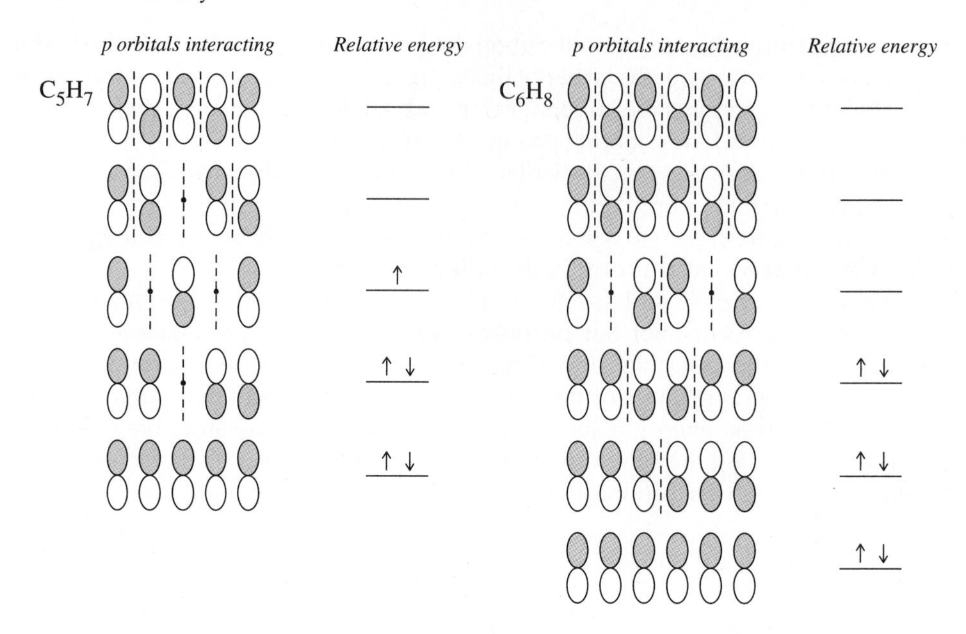

FIGURE 13.21 π Orbitals for Linear Systems.

These molecular orbitals have the same energy; π molecular orbitals having the same number of nodes in cyclic π systems of hydrocarbons are degenerate (have the same energy). The total π molecular orbital diagram for *cyclo*-C_3H_3 can therefore be summarized as follows:

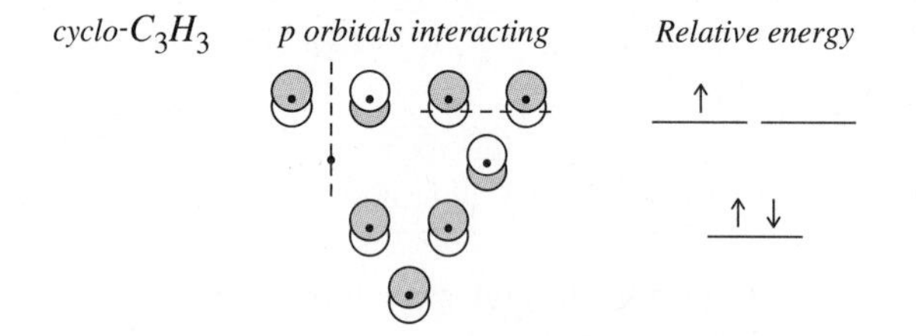

A simple way to determine the *p* orbital interactions and the relative energies of the cyclic π systems that are regular polygons is to draw the polygon with one vertex pointed down. Each vertex then corresponds to the relative energy of a molecular orbital. Furthermore, the number of nodal planes perpendicular to the plane of the molecule increases as one goes to higher energy, with the bottom orbital having zero nodes, the next pair of orbitals a single node, and so on. For example, this scheme predicts that the next cyclic π system, *cyclo*-C_4H_4 (cyclobutadiene), would have molecular orbitals as follows:[37]

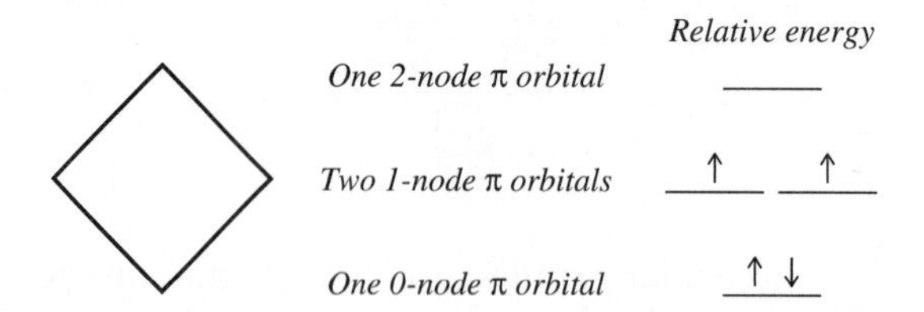

Similar results are obtained for other cyclic π systems; two of these are shown in Figure 13.22. In these diagrams, nodal planes are disposed symmetrically. For example,

[37] This approach predicts a diradical for cyclobutadiene (one electron in each 1-node orbital). Although cyclobutadiene itself is very reactive (P. Reeves, T. Devon, and R. Pettit, *J. Am. Chem. Soc.*, **1969**, *91*, 5890), complexes containing derivatives of cyclobutadiene are known. At 8 K, cyclobutadiene itself has been isolated in an argon matrix (O. L. Chapman, C. L. McIntosh, and J. Pacansky, *J. Am. Chem. Soc.*, **1973**, *95*, 614; A. Krantz, C. Y. Lin, and M. D. Newton, *J. Am. Chem. Soc.*, **1973**, *95*, 2746).

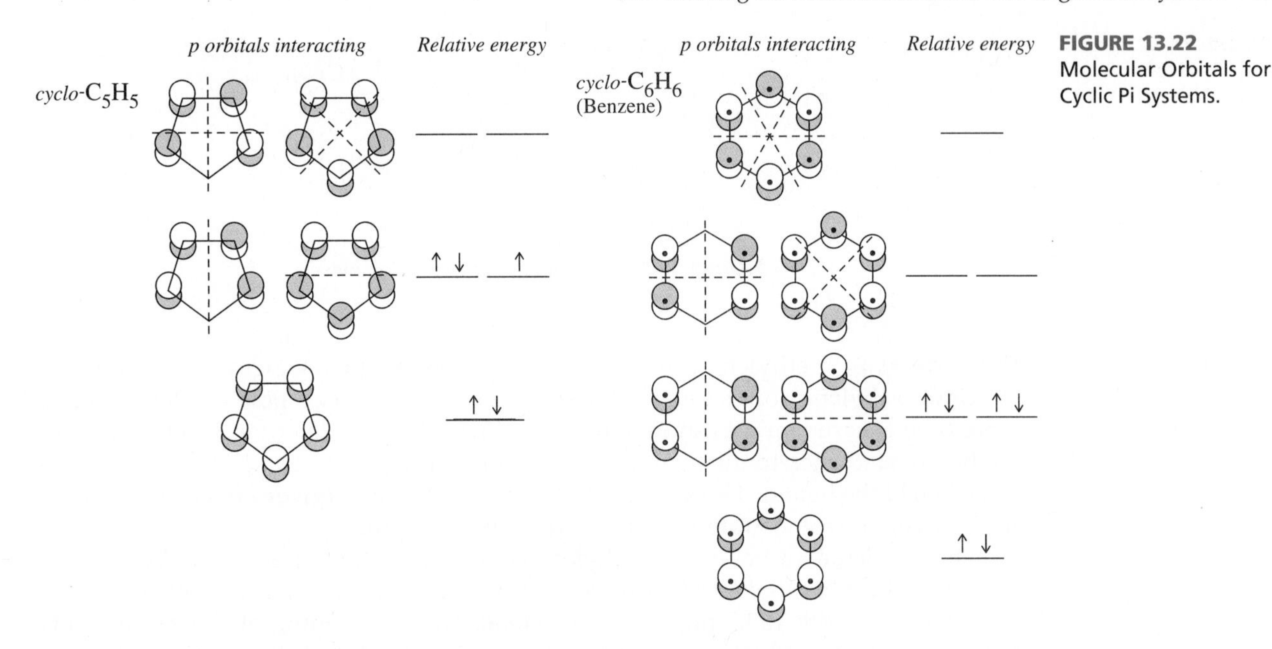

FIGURE 13.22 Molecular Orbitals for Cyclic Pi Systems.

in *cyclo*-C_4H_4 the single-node molecular orbitals bisect the molecule through opposite sides; the nodal planes are oriented perpendicularly to each other. The 2-node orbital for this molecule also has perpendicular nodal planes.

This method may seem oversimplified, but the nodal behavior and relative energies are the same as those obtained from molecular orbital calculations. The method for obtaining equations for the molecular orbitals of cyclic hydrocarbons of formula C_nH_n ($n = 3$ to 8) is given by Cotton.[38] Throughout this discussion, we have shown not the actual shapes of the π molecular orbitals, but rather the *p* orbitals used. The nodal behavior of both sets (the π orbitals and the *p* orbitals used) is identical and therefore sufficient for the discussion of bonding with metals that follows.[39]

13.5 BONDING BETWEEN METAL ATOMS AND ORGANIC PI SYSTEMS

We are now ready to consider metal–ligand interactions involving such systems. We will begin with the simplest of the linear systems, ethylene, and conclude with the classic example of ferrocene.

13.5.1 Linear Pi Systems

PI-ETHYLENE COMPLEXES Many complexes involve ethylene, C_2H_4, as a ligand, including the anion of Zeise's salt, $[Pt(\eta^2\text{-}C_2H_4)Cl_3]^-$, one of the earliest organometallic complexes. In such complexes, ethylene most commonly acts as a sidebound ligand with the following geometry with respect to the metal:

$$\mathrm{Pt}\ \ \begin{matrix} \mathrm{H}\ \mathrm{H} \\ \mathrm{C} \\ \| \\ \mathrm{C} \\ \mathrm{H}\ \mathrm{H} \end{matrix}$$

[38] F. A. Cotton, *Chemical Applications of Group Theory*, 3rd ed., Wiley-Interscience, 1990, pp. 142–159.

[39] Diagrams of many molecular orbitals for linear and cyclic π systems can be found in W. L. Jorgenson and L. Salem, *The Organic Chemist's Book of Orbitals*, Academic Press, New York, 1973.

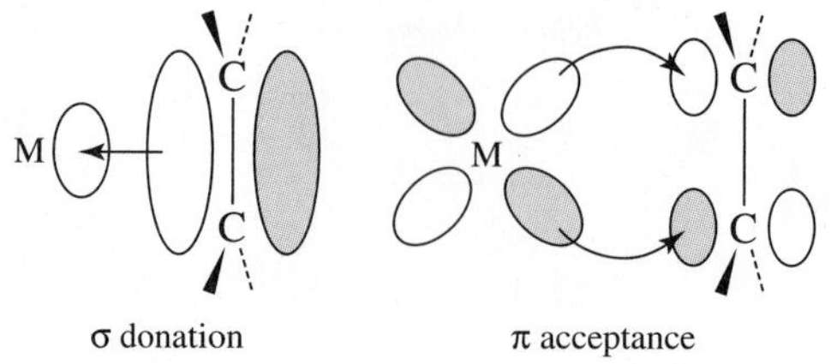

σ donation π acceptance

FIGURE 13.23 Bonding in Ethylene Complexes.

The hydrogens in ethylene complexes are typically bent back away from the metal, as shown. Ethylene donates electron density to the metal in a sigma fashion, using its π-bonding electron pair, as shown in Figure 13.23. At the same time, electron density can be donated back to the ligand in a pi fashion from a metal *d* orbital to the empty π^* orbital of the ligand. This is another example of the synergistic effect of σ donation and π acceptance encountered earlier with the CO ligand.

If this picture of bonding in ethylene complexes is correct, it should be in agreement with the measured C—C distance. The C—C distance in Zeise's salt is 137.5 pm in comparison with 133.7 pm in free ethylene. The lengthening of this bond can be explained by a combination of the two factors involved in the synergistic σ-donor, π-acceptor nature of the ligand: donation of electron density to the metal in a sigma fashion reduces the π-bonding electron density within the ligand, weakening the C—C bond. Furthermore, the back-donation of electron density from the metal to the π^* orbital of the ligand also reduces the C—C bond strength by populating the antibonding orbital. The net effect weakens and lengthens the C—C bond in the C_2H_4 ligand. In addition, vibrational frequencies of coordinated ethylene are at lower energy than in free ethylene; for example, the C=C stretch in the anion of Zeise's salt is at 1516 cm^{-1}, compared to 1623 cm^{-1} in free ethylene.

PI-ALLYL COMPLEXES The allyl group most commonly functions as a trihapto ligand, using delocalized π orbitals as described previously, or as a monohapto ligand, primarily σ bonded to a metal. Examples of these types of coordination are shown in Figure 13.24.

Bonding between η^3-C_3H_5 and a metal atom is shown schematically in Figure 13.25. The lowest energy π orbital can donate electron density in a sigma fashion to a suitable orbital on the metal. The next orbital, nonbonding in free allyl, can act as a donor or acceptor, depending on the electron distribution between the metal and the ligand. The highest energy π orbital acts as an acceptor; thus, there can be synergistic sigma and pi interactions between allyl and the metal. The C—C—C angle within the ligand is generally near 120°, consistent with sp^2 hybridization.

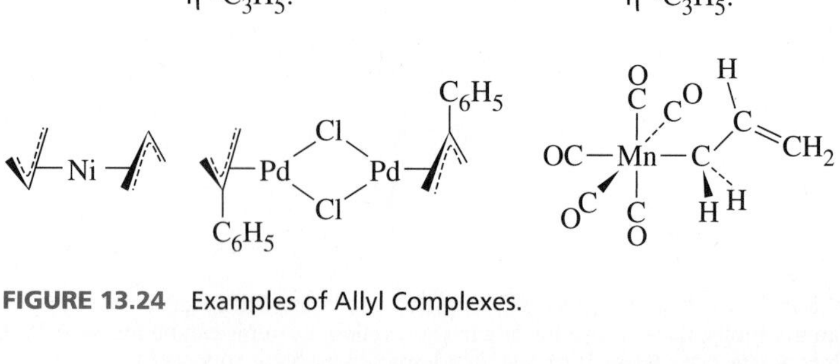

η^3-C_3H_5: η^1-C_3H_5:

FIGURE 13.24 Examples of Allyl Complexes.

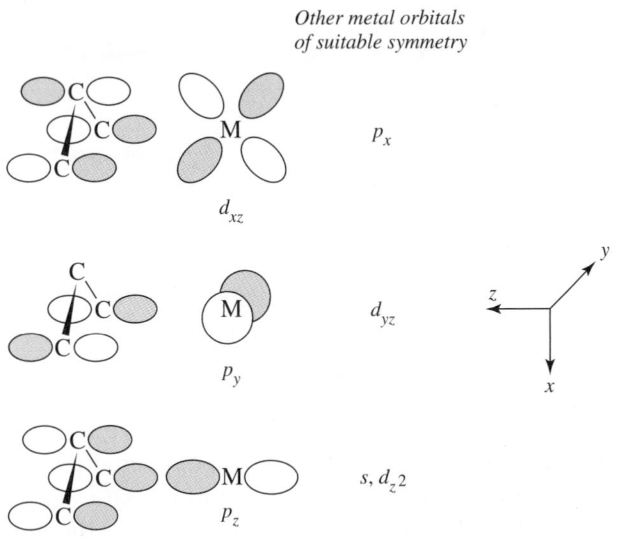

FIGURE 13.25 Bonding in η^3-Allyl Complexes.

Allyl complexes (or complexes of substituted allyls) are intermediates in many reactions, some of which take advantage of the capability of this ligand to function in both a η^3 and η^1 fashion. Loss of CO from carbonyl complexes containing η^1-allyl ligands often results in conversion of η^1- to η^3-allyl. For example,

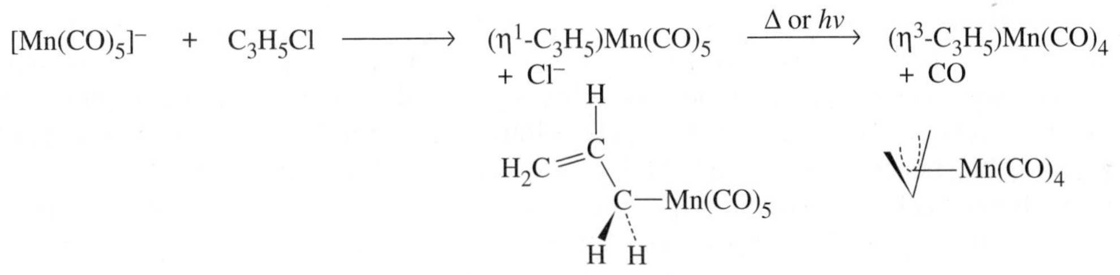

$$[Mn(CO)_5]^- + C_3H_5Cl \longrightarrow (\eta^1\text{-}C_3H_5)Mn(CO)_5 + Cl^- \xrightarrow{\Delta \text{ or } h\nu} (\eta^3\text{-}C_3H_5)Mn(CO)_4 + CO$$

The $[Mn(CO)_5]^-$ ion displaces Cl^- from allyl chloride to give an 18-electron product containing η^1-C_3H_5. The allyl ligand switches to trihapto when a CO is lost, preserving the 18-electron count.

OTHER LINEAR PI SYSTEMS Many other such systems are known; several examples of organic ligands having longer π systems are shown in Figure 13.26. Butadiene and longer conjugated π systems have the possibility of isomeric ligand forms (*cis* and *trans* for butadiene). Larger cyclic ligands may have a π system extending through part of the ring. An example is cyclooctadiene (COD); the 1,3-isomer has a 4-atom π system comparable to butadiene; 1,5-cyclooctadiene has two isolated double bonds, one or both of which may interact with a metal in a manner similar to ethylene.

▶ **Exercise 13.8** Identify the transition metal in the following 18-electron complexes:

a. $(\eta^5\text{-}C_5H_5)(cis\text{-}\eta^4\text{-}C_4H_6)M(PMe_3)_2(H)$ (M = second-row transition metal)
b. $(\eta^5\text{-}C_5H_5)M(C_2H_4)_2$ (M = first-row transition metal)

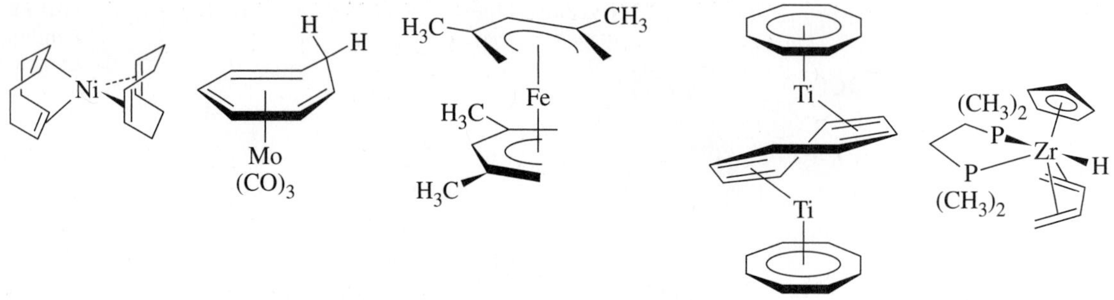

FIGURE 13.26 Examples of Molecules Containing Linear Pi Systems.

13.5.2 Cyclic Pi Systems

CYCLOPENTADIENYL (Cp) COMPLEXES The cyclopentadienyl group, C_5H_5, may bond to metals in a variety of ways, with many examples known of the η^1-, η^3-, and η^5-bonding modes. As described previously in this chapter, the discovery of the first cyclopentadienyl complex, ferrocene, was a landmark in the development of organometallic chemistry and stimulated the search for other compounds containing π-bonded organic ligands. Substituted cyclopentadienyl ligands are also known, such as $C_5(CH_3)_5$, often abbreviated Cp*, and C_5(benzyl)$_5$.

Ferrocene and other cyclopentadienyl complexes can be prepared by reacting metal salts with $C_5H_5^-$.[40]

$$FeCl_2 + 2\,NaC_5H_5 \longrightarrow (\eta^5\text{-}C_5H_5)_2\,Fe + 2\,NaCl$$

Ferrocene, $(\eta^5\text{-}C_5H_5)_2Fe$. Ferrocene is the prototype of a series of sandwich compounds, the metallocenes, with the formula $(C_5H_5)_2M$. Electron counting in ferrocene can be viewed in two ways. One possibility is to consider it an iron(II) complex with two 6-electron cyclopentadienide ($C_5H_5^-$) ions, another to view it as iron(0) coordinated by two neutral, 5-electron C_5H_5 ligands. The actual bonding situation in ferrocene is much more complicated and requires an analysis of the various metal–ligand interactions in this molecule. As usual, we expect orbitals on the central Fe and on the two C_5H_5 rings to interact if they have appropriate symmetry; furthermore, we expect interactions to be strongest if they are between orbitals of similar energy.

For the purposes of our analysis of this molecule, it will be useful to refer to Figure 13.22 for diagrams of the π molecular orbitals of a C_5H_5 ring. Two of these rings are arranged in a parallel fashion in ferrocene to "sandwich in" the metal atom. Our discussion will be based on the eclipsed D_{5h} conformation of ferrocene, the conformation consistent with gas-phase and low-temperature data on this molecule.[41,42] The same approach using the staggered conformation would yield a similar molecular orbital picture. Descriptions of the bonding in ferrocene based on D_{5d} symmetry are common in the chemical literature, because this was once believed to be the molecule's most stable conformation.[43]

---

[40] Solutions of NaC_5H_5 in tetrahydrofuran are available commercially. Alternatively, NaC_5H_5 can be prepared in the laboratory by cracking of dicyclopentadiene, followed by reduction:

$$C_{10}H_{12}\text{ (dicyclopentadiene)} \longrightarrow 2\,C_5H_6\text{ (cyclopentadiene)}$$
$$2\,Na + 2\,C_5H_6 \longrightarrow 2\,NaC_5H_5 + H_2$$

[41] A. Haaland and J. E. Nilsson, *Acta Chem. Scand.*, **1968**, *22*, 2653; A. Haaland, *Acc. Chem. Res.*, **1979**, *12*, 415.

[42] P. Seiler and J. Dunitz, *Acta Crystallogr., Sect. B*, **1982**, *38*, 1741.

[43] The $C_5(CH_3)_5$ and C_5(benzyl)$_5$ analogues of ferrocene have staggered D_{5d} symmetry, as do several other metallocenes. See M.D. Rausch, W-M. Tsai, J. W. Chambers, R. D. Rogers, and H. G. Alt, *Organometallics*, **1989**, *8*, 816. For calculations that compare the energies of the conformations of ferrocene, see S. Coriani, A. Haaland, T. Helgaker, and P. Jørgensen, *ChemPhysChem*, **2007**, *7*, 245.

In developing the group orbitals for a pair of C_5H_5 rings, we pair up molecular orbitals of the same energy and same number of nodes; for example, we pair the zero-node orbital of one ring with the zero-node orbital of the other.[44] We also must pair up the molecular orbitals in such a way that the nodal planes are coincident. Furthermore, in each pairing there are two possible orientations of the ring molecular orbitals: one in which lobes of like sign are pointed toward each other, and one in which lobes of opposite sign are pointed toward each other. For example, the zero-node orbitals of the C_5H_5 rings may be paired in the following two ways:

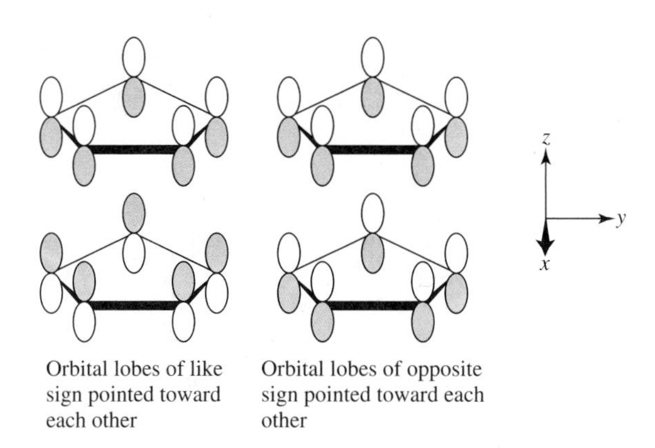

The 10 group orbitals arising from the C_5H_5 ligands are shown in Figure 13.27.

The process of developing the molecular orbital picture of ferrocene now becomes one of matching the group orbitals with the *s*, *p*, and *d* orbitals of appropriate symmetry on Fe.

We will illustrate one of these interactions, between the d_{yz} orbital of Fe and its appropriate group orbital (one of the 1-node group orbitals shown in Figure 13.27). This interaction can occur in a bonding and an antibonding fashion:

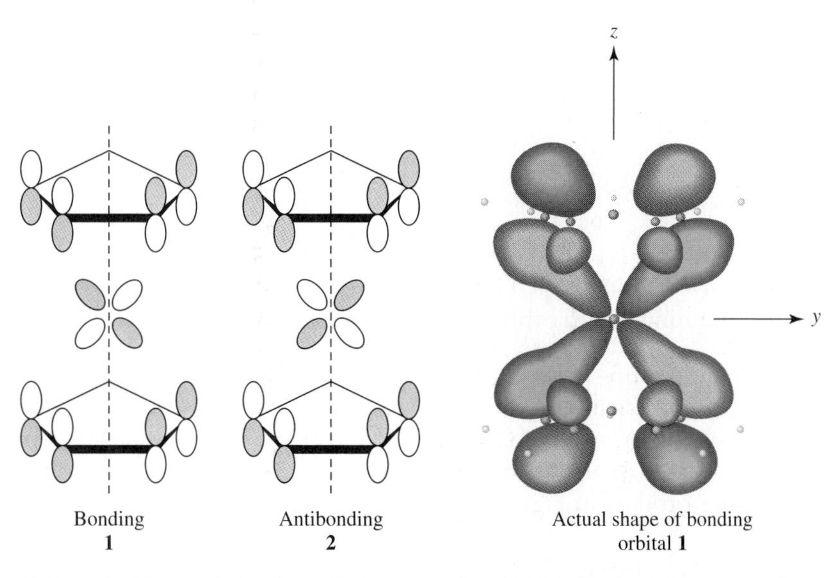

(Adapted with permission from G. O. Spessard and G. L. Miessler, *Organometallic Chemistry*, Prentice Hall, Upper Saddle River, NJ, 1997, p. 93, Fig. 5.7.)

[44] Not counting the nodal planes that are coplanar with the C_5H_5 rings.

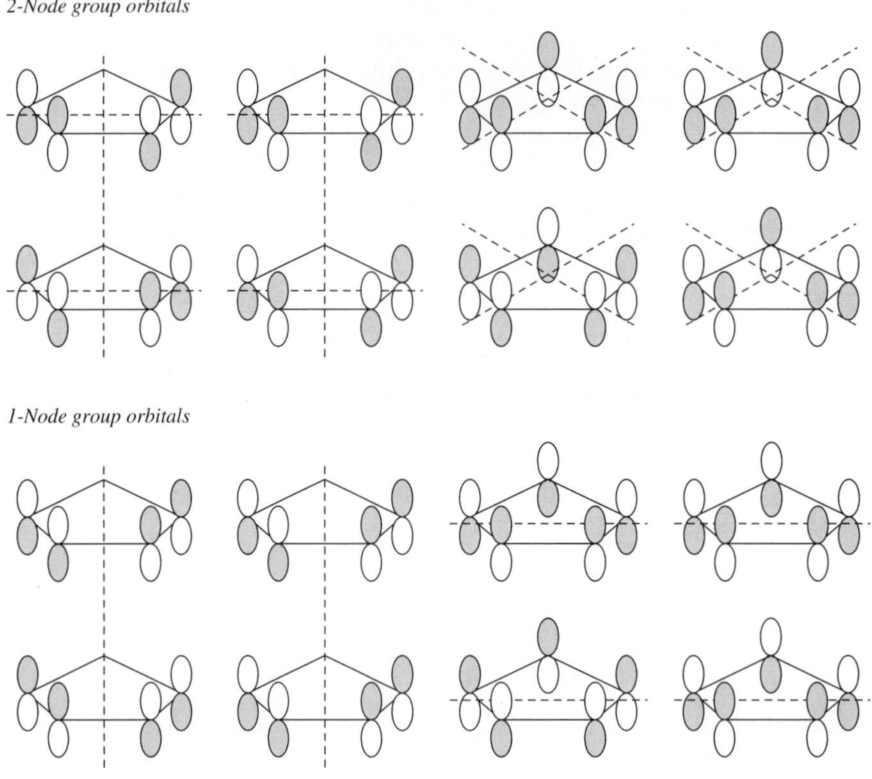

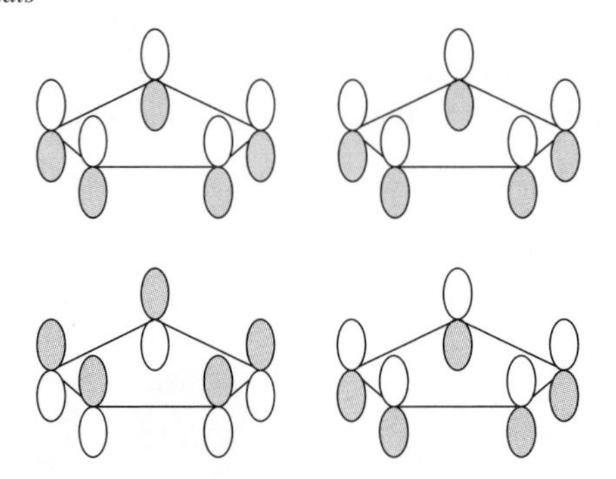

FIGURE 13.27 Group Orbitals for C_5H_5 Ligands of Ferrocene.

▶ **Exercise 13.9** Determine which orbitals on Fe are appropriate for interaction with each of the remaining group orbitals in Figure 13.27.

The complete energy-level diagram for the molecular orbitals of ferrocene is shown in Figure 13.28. The molecular orbital resulting from the d_{yz} bonding interaction, labeled **1** in the MO diagram, contains a pair of electrons. Its antibonding counterpart, **2**, is empty. It is a useful exercise to match the other group orbitals from Figure 13.27 with the molecular orbitals in Figure 13.28 to verify the types of metal–ligand interactions that occur.

The orbitals of ferrocene that are of most interest are those having the greatest *d*-orbital character; these are also the highest occupied and lowest unoccupied orbitals

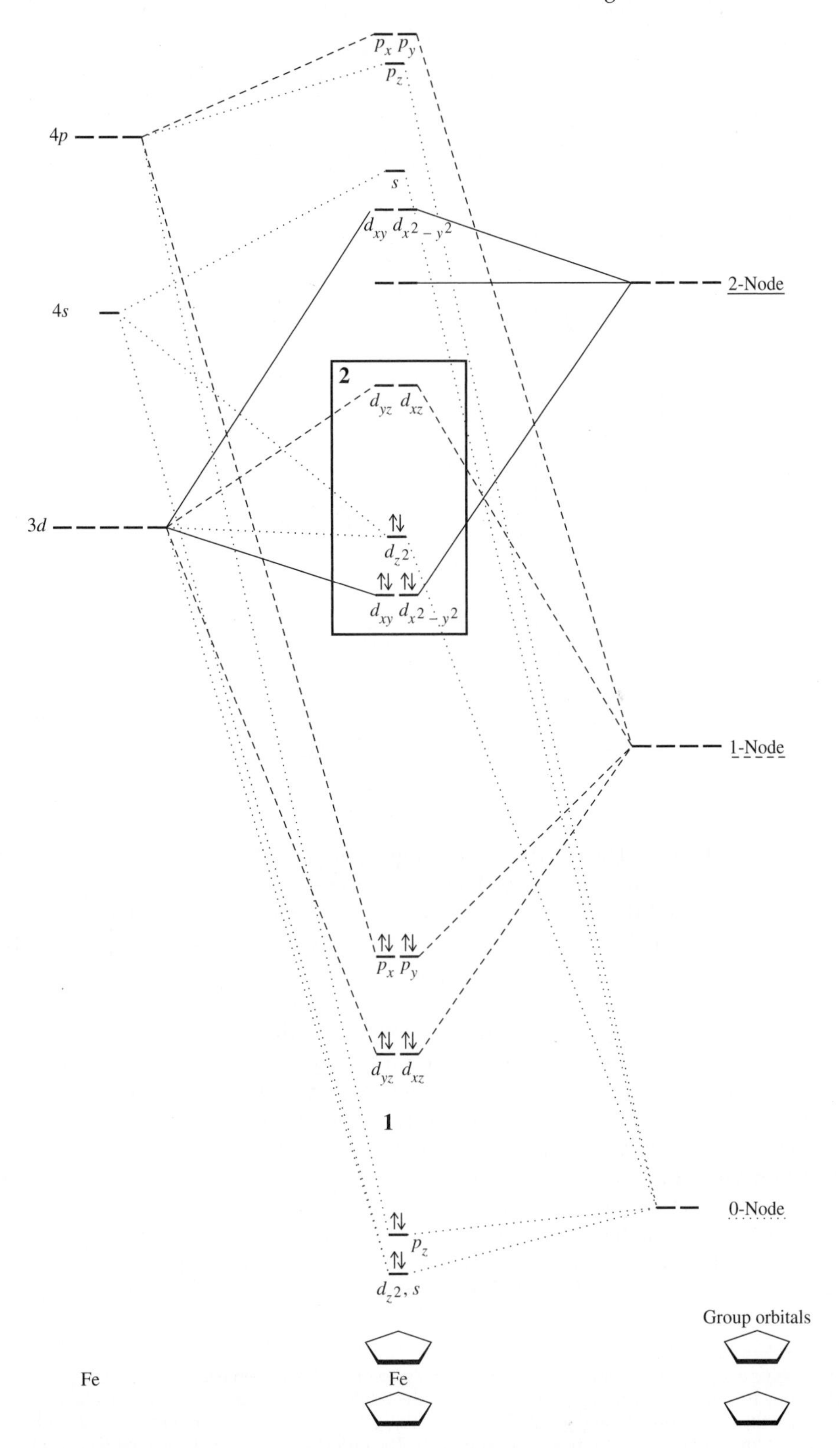

FIGURE 13.28 Molecular Orbital Energy Levels of Ferrocene.

FIGURE 13.29 Molecular Orbitals of Ferrocene Having Greatest *d* Character.

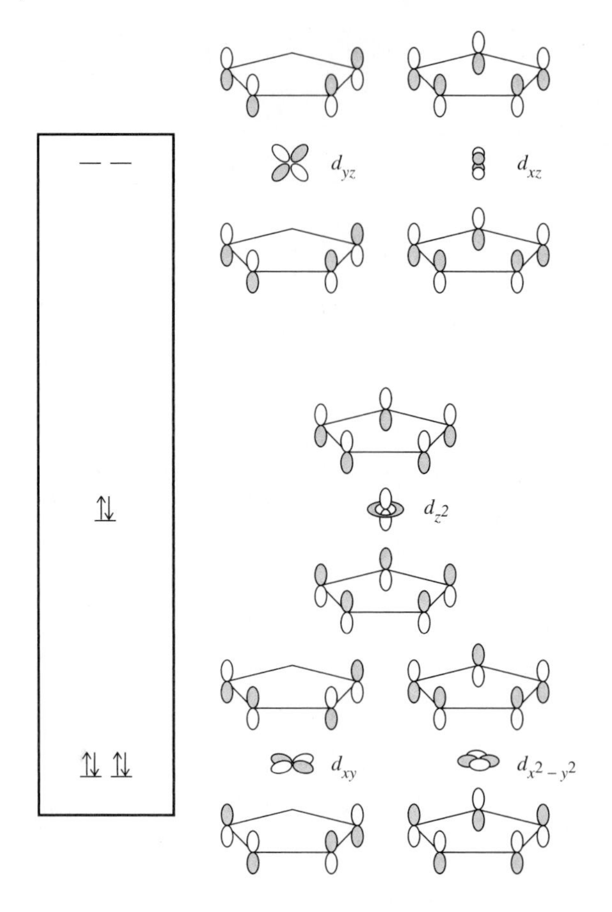

(HOMO and LUMO). These orbitals are highlighted in the box in Figure 13.28. Two orbitals, the degenerate pair having largely d_{xy} and $d_{x^2-y^2}$ character, are weakly bonding and are occupied by electron pairs; one, having largely d_{z^2} character, is essentially nonbonding and is also occupied by an electron pair; and two, having primarily d_{xy} and d_{yz} character, are empty. The relative energies of these orbitals and their *d* orbital–group orbital interactions are shown in Figure 13.29.[45,46]

The overall bonding in ferrocene can now be summarized. The occupied orbitals of the η^5-C_5H_5 ligands are stabilized by their interactions with iron. Note especially the stabilization in energy of 0-node and 1-node group orbitals that have bonding interactions with the metal, forming molecular orbitals that are primarily ligand in nature (these are the orbitals, labeled from lowest to highest energy d_{z^2}, s, p_z, d_{yz}, d_{xz}, p_x, and p_y).

The orbitals next highest in energy are largely derived from iron *d* orbitals; they are populated by 6 electrons as we would expect from iron(II), a d^6 metal ion. These molecular orbitals also have some ligand character, with the exception of the molecular orbital derived from d_{z^2}. The molecular orbital derived from d_{z^2} has almost no ligand character, because its cone-shaped nodal surface points almost directly toward the lobes

[45] The relative energies of the lowest three orbitals in Figure 13.29 have been controversial. UV photoelectron spectroscopy is consistent with the order shown, with the orbital having largely d_{z^2} character slightly higher in energy than the pair having d_{xy} and $d_{x^2-y^2}$ character. However, a relatively recent report places the orbital with d_{z^2} character lower in energy than this degenerate pair. The order of these orbitals may be reversed for some metallocenes. See A. Haaland, *Acc. Chem. Res.*, **1979**, *12*, 415, and Z. Xu, Y. Xie, W. Feng, and H. F. Schaefer III, *J. Phys. Chem. A*, **2003**, *107*, 2716. The 2003 paper also discusses the orbital energies for the metallocenes $(\eta^5\text{-}C_5H_5)_2V$ through $(\eta^5\text{-}C_5H_5)_2Ni$.

[46] J. C. Giordan, J. H. Moore, and J. A. Tossell, *Acc. Chem. Res.*, **1986**, *19*, 281; E. Rühl and A. P. Hitchcock, *J. Am. Chem. Soc.*, **1989**, *111*, 5069.

TABLE 13.3 Comparative Data for Selected Metallocenes

Complex	Electron Count	M—C Distance (pm)	ΔH for M^{2+}-$C_5H_5^-$ Dissociation (kJ / mol)
$(\eta^5\text{-}C_5H_5)_2Fe$	18	206.4	1470
$(\eta^5\text{-}C_5H_5)_2Co$	19	211.9	1400
$(\eta^5\text{-}C_5H_5)_2Ni$	20	219.6	1320

of the matching group orbital, making overlap slight and giving an essentially nonbonding orbital localized on the iron. The molecular orbital description of ferrocene fits the 18-electron rule.

OTHER METALLOCENES AND RELATED COMPLEXES Other metallocenes have similar structures but do not necessarily obey the rule. For example, cobaltocene and nickelocene are structurally similar 19- and 20-electron species. The extra electrons have important chemical and physical consequences, as can be seen from comparative data in Table 13.3. Electrons 19 and 20 of the metallocenes occupy slightly antibonding orbitals (largely d_{yz} and d_{xz} in character); as a consequence, the metal–ligand distance increases, and ΔH for metal–ligand dissociation decreases. Ferrocene itself shows much more chemical stability than cobaltocene and nickelocene; many of the chemical reactions of the latter are characterized by a tendency to yield 18-electron products. For example, ferrocene is unreactive toward iodine and rarely participates in reactions in which other ligands substitute for the cyclopentadienyl ligand. However, cobaltocene and nickelocene undergo the following reactions to give 18-electron products:

$$\underset{19\ e^-}{2(\eta^5\text{-}C_5H_5)_2Co} + I_2 \longrightarrow \underset{\substack{18\ e^- \\ \text{cobalticinium ion}}}{2\,[(\eta^5\text{-}C_5H_5)_2Co]^+} + 2\,I^-$$

$$\underset{20\ e^-}{(\eta^5\text{-}C_5H_5)_2Ni} + PF_3 \longrightarrow \underset{18\ e^-}{Ni(PF_3)_4} + \text{organic products}$$

Cobalticinium reacts with hydride to give a neutral, 18-electron sandwich compound in which one cyclopentadienyl ligand has been modified into $\eta^4\text{-}C_5H_6$, as shown in Figure 13.30.

Ferrocene, however, is by no means chemically inert. It undergoes a variety of reactions, including many on the cyclopentadienyl rings. A good example is that of electrophilic acyl substitution (Figure 13.31), a reaction paralleling that of benzene and its derivatives. In general, electrophilic aromatic substitution reactions are much more rapid for ferrocene than for benzene, an indication of greater concentration of electron density in the rings of the sandwich compound.

Among the most interesting ferrocene-containing compounds is a molecule that was sought for many years before its synthesis in 2006, hexaferrocenylbenzene, a type of molecular "Ferris Wheel," shown in Figure 13.32. This compound, originally obtained in only 4 percent yield from the reaction of hexaiodobenzene and diferrocenylzinc, has six

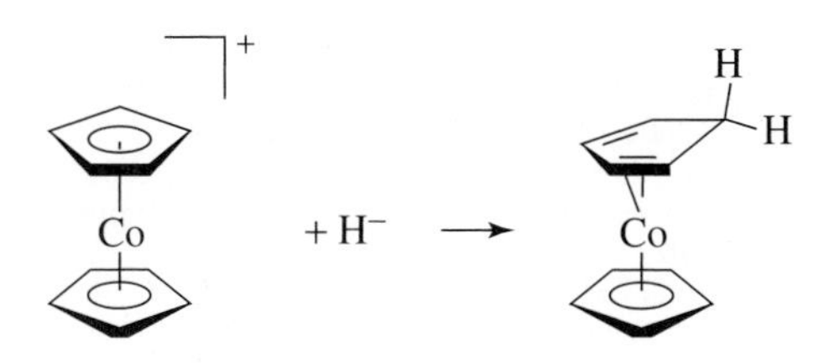

FIGURE 13.30 Reaction of Cobalticinium with Hydride.

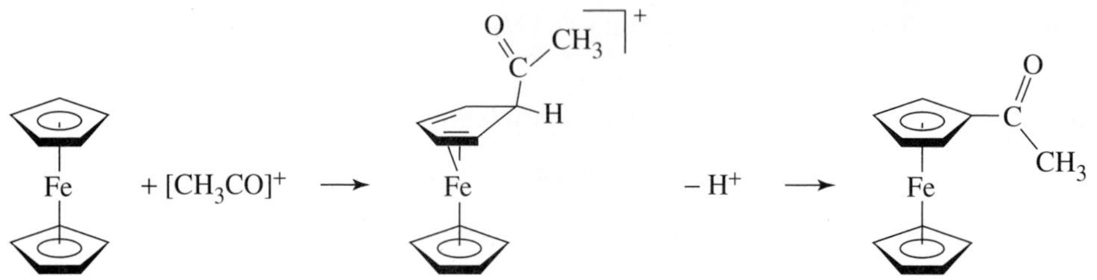

FIGURE 13.31 Electrophilic Acyl Substitution in Ferrocene.

ferrocenyl groups as substituents on a benzene ring.[47] The highly crowded nature of hexaferrocenylbenzene is illustated by the alternating up/down arrangement of ferrocenes around the benzene ring, with the benzene itself adopting a chair conformation with alternating C—C distances of 142.7 and 141.1 pm.

Binuclear metallocenes—with two atoms, rather than one, in the center of a sandwich structure—are also known. Perhaps the best known of these metallocenes is decamethyldizincocene, $(\eta^5\text{-}C_5Me_5)_2Zn_2$, shown in Figure 13.33, which was prepared from decamethylzincocene, $(\eta^5\text{-}C_5Me_5)_2Zn$, and diethylzinc.[48] Particularly notable is $(\eta^5\text{-}C_5Me_5)_2Zn_2$, the first example of a stable molecule with a zinc–zinc bond; moreover, its zinc atoms are in the exceptionally rare +1 oxidation state. Among Group 12 elements, mercury by far exhibits this oxidation state most often, with the best known example the Hg_2^{2+} ion[49]; cadmium and zinc compounds almost always have these metals in the +2 oxidation state. The metallocene $(\eta^5\text{-}C_5Me_5)_2Zn_2$ has parallel C_5Me_5 rings and a zinc–zinc distance of 230.5 pm, consistent with a single bond. Since this compound was first reported, renewed interest in molecules containing zinc–zinc bonds has yielded a variety of interesting such compounds.[50]

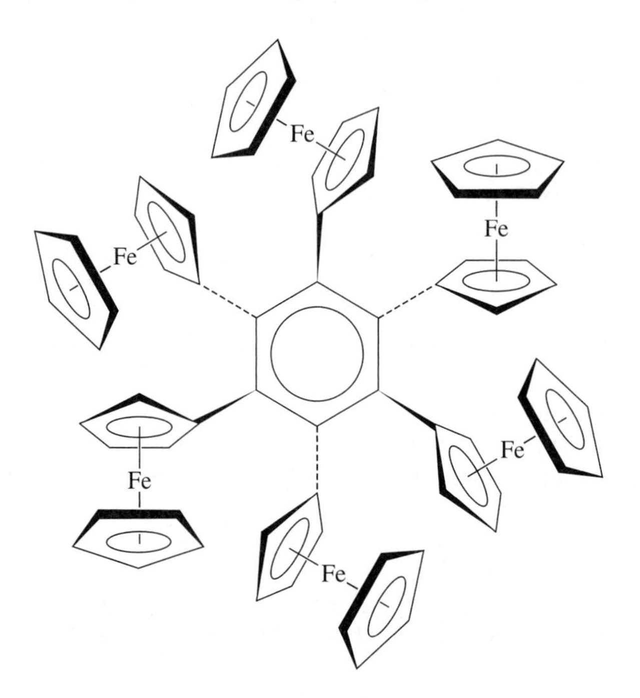

FIGURE 13.32 Hexaferrocenylbenzene.

[47] Y. Yu, A. D. Bond, P. W. Leonard, U. J. Lorenz, T. V. Timofeeva, K. P. C. Vollhardt, G. D. Whitener, and A. A. Yakovenko, *Chem. Commun.*, **2006**, 2572.

[48] I. Resa, E. Carmona, E. Gutierrez-Puebla, and A. Monge, *Science*, **2004**, *305*, 1136; A. Grirrane, I. Resa, A. Rodriguez, E. Carmona, E. Alvarez, E. Gutierrez-Puebla, A. Monge, A. Galindo, D. del Rio, and R. A. Anderson, *J. Am. Chem. Soc.*, **2007**, *129*, 693.

[49] The classic mercury(I) or "mercurous" ion.

[50] D. L. Kays and S. Aldridge, *Angew. Chem., Int. Ed.*, **2009**, *48*, 4109.

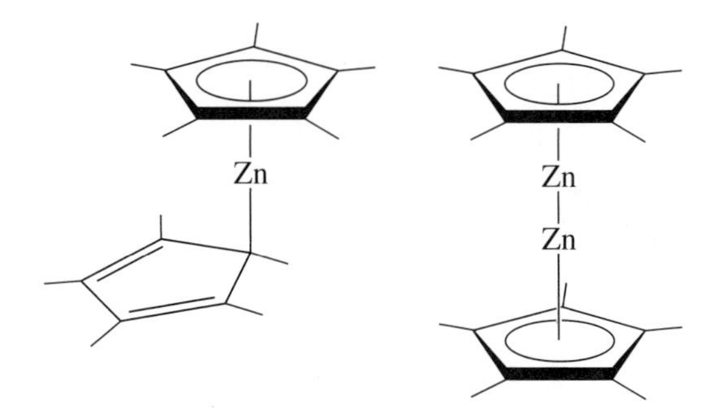

FIGURE 13.33 Decamethylzincocene and Decamethyldizincocene.

A variation on the theme of metallocenes and related sandwich compounds is provided by the "inverse" sandwich shown in Figure 13.34, with calcium(I) ions on the outside and the cyclic pi ligand 1,3,5-triphenylbenzene in between. This compound was most efficiently prepared by reacting 1,3,5-triphenylbenzene with activated calcium in THF solvent using catalytic amounts of 1-bromo-2,4,6-triphenylbenzene.[51] Although the product of this reaction is highly sensitive to moisture and air and is pyrophoric, it represents a rare example of a +1 oxidation state among the alkaline earths.

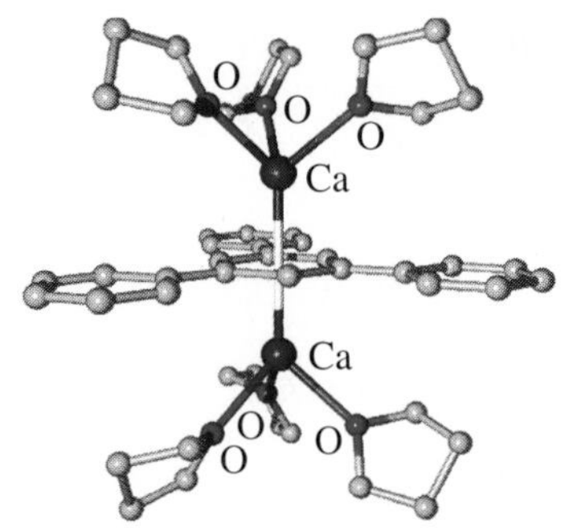

$[(thf)_3Ca\{\mu\text{-}C_6H_3\text{-}1,3,5\text{-}Ph_3\}Ca(thf)_3]$

FIGURE 13.34 An Inverse Sandwich Compound.

COMPLEXES CONTAINING CYCLOPENTADIENYL AND CO LIGANDS Not surprisingly, many complexes are known containing both Cp and CO ligands. These include "half-sandwich" compounds such as $(\eta^5\text{-}C_5H_5)Mn(CO)_3$ and dimeric and larger cluster molecules. Examples are shown in Figure 13.35. As for the binary CO complexes, complexes of the second- and third-row transition metals show a decreasing tendency of CO to act as a bridging ligand.

Many other linear and cyclic pi ligands are known. Examples of complexes containing some of these ligands are shown in Figure 13.36.[52] Depending on the ligand and the electron requirements of the metal (or metals), these ligands may be capable of bonding in a monohapto or polyhapto fashion, and they may bridge two or more metals. Particularly interesting are the cases in which cyclic ligands can bridge metals to give "triple-decker" and even higher order sandwich compounds. (See Figure 13.1.)

---

[51] S. Krieck, H. Görls, L. Yu, M. Reiher, and M. Westerhausen, *J. Am. Chem. Soc.*, **2009**, *131*, 2977.

[52] For interesting historical accounts of the discovery of the first two of these molecules, see the following articles by D. Seyferth: on uranocene: *Organometallics*, **2004**, *23*, 3562; on dibenzenechromium: *Organometallics*, **2002**, *21*, 1520 and **2002**, *21*, 2800.

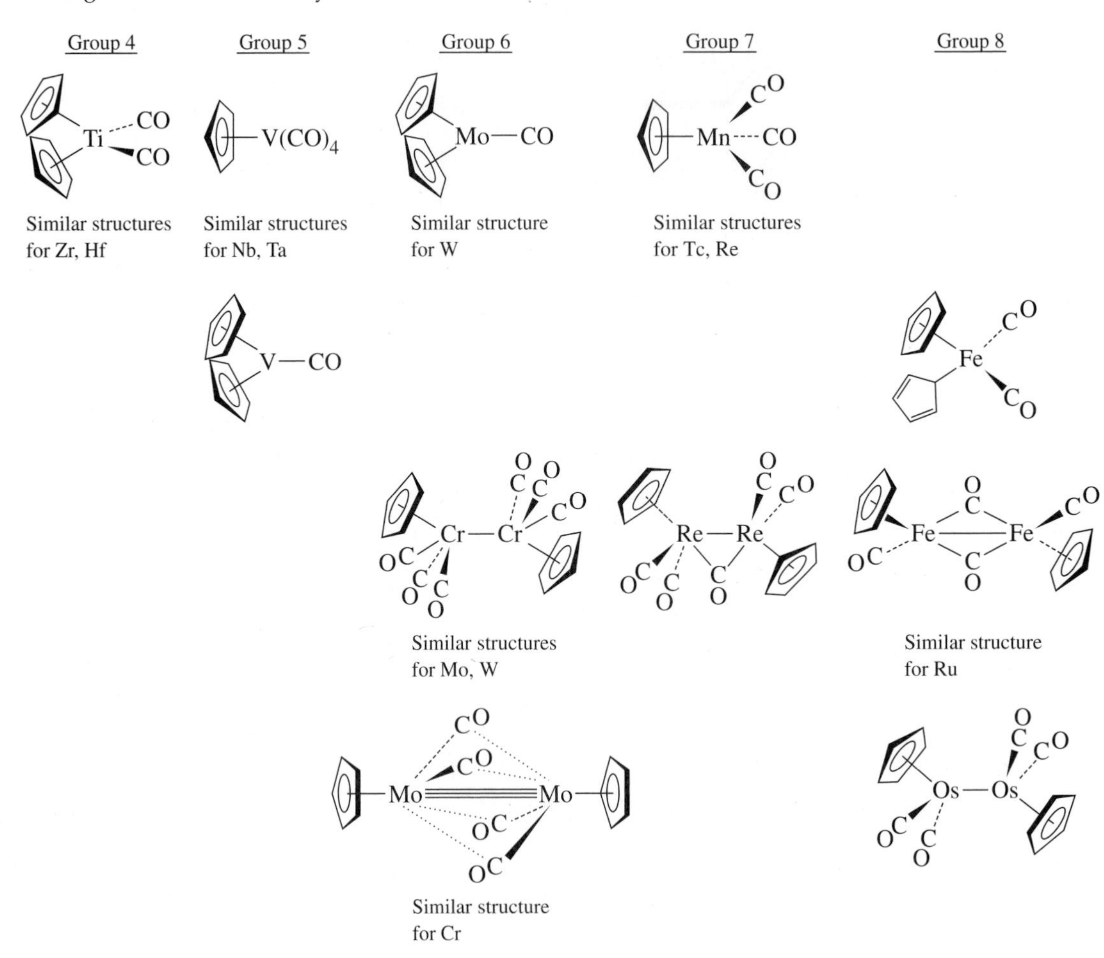

FIGURE 13.35 Complexes Containing C_5H_5 and CO.

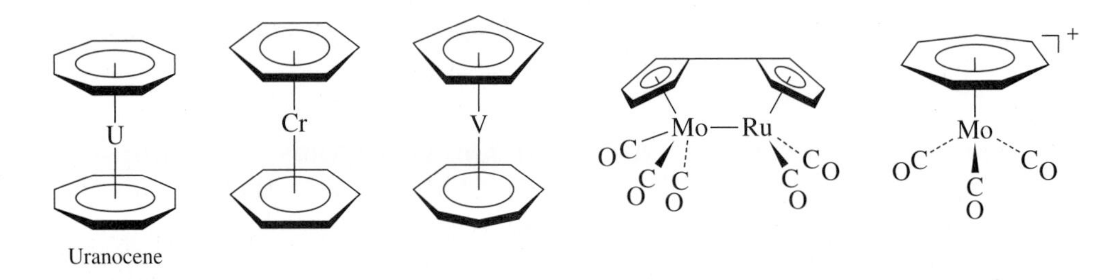

FIGURE 13.36 Examples of Molecules Containing Cyclic Pi Systems.

13.5.3 Fullerene Complexes

As immense pi systems, fullerenes were recognized early as candidates to serve as ligands to transition metals. Fullerene-metal compounds[53] have now been prepared for a variety of metals. These compounds fall into several structural types:

- *Adducts to the oxygens of osmium tetroxide.*[54]
  Example: $C_{60}(OsO_4)$(4-*t*-butylpyridine)$_2$

[53] For a review of metal complexes of C_{60} through 1991, see P. J. Fagan, J. C. Calabrese, and B. Malone, *Acc. Chem. Res.*, **1992**, *25*, 134.
[54] J. M. Hawkins, A. Meyer, T. A. Lewis, S. D. Loren, and F. J. Hollander, *Science*, **1991**, *252*, 312.

- *Complexes in which the fullerene itself behaves as a ligand.*[55]
  Examples: $Fe(CO)_4(\eta^2\text{-}C_{60})$, $Mo(\eta^5\text{-}C_5H_5)_2(\eta^2\text{-}C_{60})$, $[(C_6H_5)_3P]_2Pt(\eta^2\text{-}C_{60})$
- *Fullerenes containing encapsulated (incarcerated) atoms, called* **incarfullerenes**. These may contain one, two, three, or four atoms, sometimes as small molecules, inside the fullerene structure.[56] Although examples of encapsulated nonmetals are known, most incarfullerenes contain metals.
  Examples: UC_{60}, LaC_{82}, Sc_2C_{74}, Sc_3C_{82}
- *Intercalation compounds of alkali metals.*[57] These contain alkali metal ions occupying interstitial sites between fullerene clusters.
  Examples: NaC_{60}, RbC_{60}, KC_{70}, K_3C_{60}

These are conductive, and in some cases superconductive, materials—such as K_3C_{60} and Rb_3C_{60}—that are of great interest in the field of materials science. These are principally ionic, rather than covalent, compounds. The interested reader is encouraged to consult the reference here[58] for additional information about these compounds.

ADDUCTS TO OXYGENS OF OSMIUM TETROXIDE[59] The first pure fullerene derivative to be prepared was $C_{60}(OsO_4)$(4-*t*-butylpyridine)$_2$. The X-ray crystal structure of this compound provided the first direct evidence that the proposed structure for C_{60} was, in fact, correct. Osmium tetroxide, a powerful oxidizing agent, can add across the double bonds of many compounds, including polycyclic aromatic hydrocarbons. When OsO_4 was reacted with C_{60} and 4-*tert*-butylpyridine, 1:1 and 2:1 adducts were formed, products parallel to those anticipated in classic organic chemistry. The 1:1 adduct has been characterized by X-ray crystallography; its structure is shown in Figure 13.37.

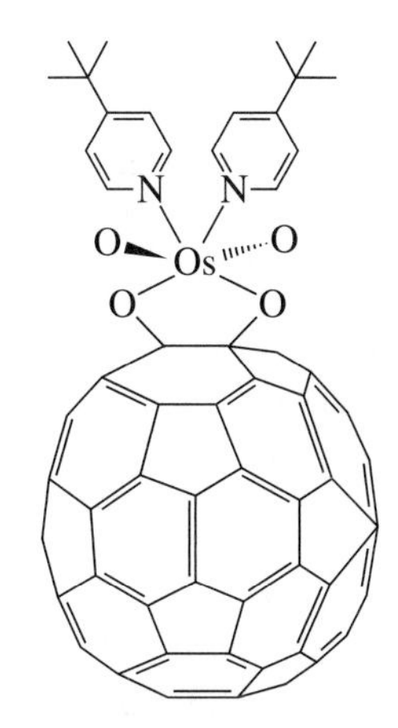

FIGURE 13.37 Structure of $C_{60}(OsO_4)$ (4-*t*-butylpyridine)$_2$.

FULLERENES AS LIGANDS[60] As a ligand, C_{60} behaves primarily as an electron-deficient alkene (or arene) and it bonds to metals in a dihapto fashion through a C—C bond at the fusion of two 6-membered rings, as shown in Figure 13.38. However, there are also instances in which C_{60} bonds in a pentahapto or hexahapto fashion.

Dihapto bonding was observed in the first complex to be synthesized, in which C_{60} acts as a ligand toward a metal, $[(C_6H_5)_3P]_2Pt(\eta^2\text{-}C_{60})$,[61] also shown in Figure 13.38.

A common route to the synthesis of complexes involving fullerenes as ligands is by displacement of other ligands, typically those weakly coordinated to metals. For example, the platinum complex shown in Figure 13.38 can be formed by the displacement of ethylene:

$$[(C_6H_5)_3P]_2Pt(\eta^2\text{-}C_2H_4) + C_{60} \longrightarrow [(C_6H_5)_3P]_2Pt(\eta^2\text{-}C_{60})$$

The *d* electron density of the metal can donate to an empty antibonding orbital of a fullerene. This pulls the two carbons involved slightly away from the C_{60} surface. In

[55] P. J. Fagan, J. C. Calabrese, and B. Malone, "The Chemical Nature of C_{60} as Revealed by the Synthesis of Metal Complexes," in G. S. Hammond and V. J. Kuck, eds., *Fullerenes*, ACS Symposium Series 481, American Chemical Society, Washington, DC, 1992, pp. 177–186; R. E. Douthwaite, M. L. H. Green, A. H. H. Stephens, and J. F. C. Turner, *Chem. Commun. (Cambridge)*, **1993**, 1522; P. J. Fagan, J. C. Calabrese, and B. Malone, *Science*, **1991**, *252*, 1160.
[56] J. R. Heath, S. C. O'Brien, Q. Zhang, Y. Liu, R. F. Curl, H. W. Kroto, and R. E. Smalley, *J. Am. Chem. Soc.*, **1985**, *107*, 7779; H. Shinohara, H. Yamaguchi, N. Hayashi, H. Sato, M. Ohkohchi, Y. Ando, and Y. Saito, *J. Phys. Chem.*, **1993**, *97*, 4259.
[57] R. C. Haddon, A. F. Hebard, M. J. Rosseinsky, D. W. Murphy, S. H. Glarum, T. T. M. Palstra, A. P. Ramirez, S. J. Duclos, R. M. Fleming, T. Siegrist, and R. Tycko, "Conductivity and Superconductivity in Alkali Metal Doped C_{60}," in Hammond and Kuck, *Fullerenes*, pp. 71–89.
[58] R. C. Haddon, *Acc. Chem. Res.*, **1992**, *25*, 127.
[59] J. M. Hawkins, *Acc. Chem. Res.*, **1992**, *25*, 150 and references therein.
[60] P. J. Fagan, J. C. Calabrese, and B. Malone, *Acc. Chem. Res.*, **1992**, *25*, 134.
[61] P. J. Fagan, J. C. Calabrese, and B. Malone, *Science*, **1991**, *252*, 1160.

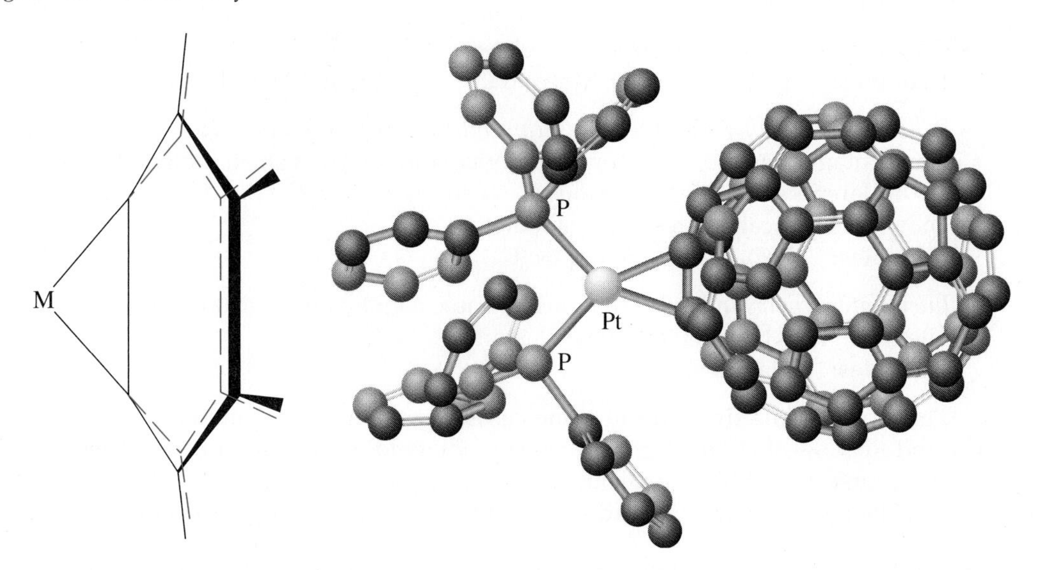

FIGURE 13.38 Bonding of C_{60} to Metal. (Adapted with permission from G. O. Spessard and G. L. Miessler, *Organometallic Chemistry*, Prentice Hall, Upper Saddle River, NJ, 1997, p. 509, Fig. 13.11.)

addition, the distance between these carbons is elongated slightly as a consequence of this interaction, which populates an orbital that is antibonding with respect to the C—C bond. This increase in C—C bond distance is analogous to the elongation that occurs when ethylene and other alkenes bond to metals, as discussed in Section 13.5.1.

In some cases, more than one metal can become attached to a fullerene surface. A spectacular example is $[(Et_3P)_2Pt]_6C_{60}$,[62] shown in Figure 13.39. In this structure, the six $(Et_3P)_2Pt$ units are arranged octahedrally around the C_{60}.

Although complexes of C_{60} have been studied most extensively, some complexes of other fullerenes have also been prepared. An example is $(\eta^2\text{-}C_{70})Ir(CO)Cl(PPh_3)_2$, shown in Figure 13.40. As in the case of the known C_{60} complexes, bonding to the metal occurs at the fusion of two 6-membered rings.

C_{60} bonds to transition metals primarily in a dihapto fashion, but at least one example of a hexahapto structure has been reported. The coordination mode of the C_{60} in the

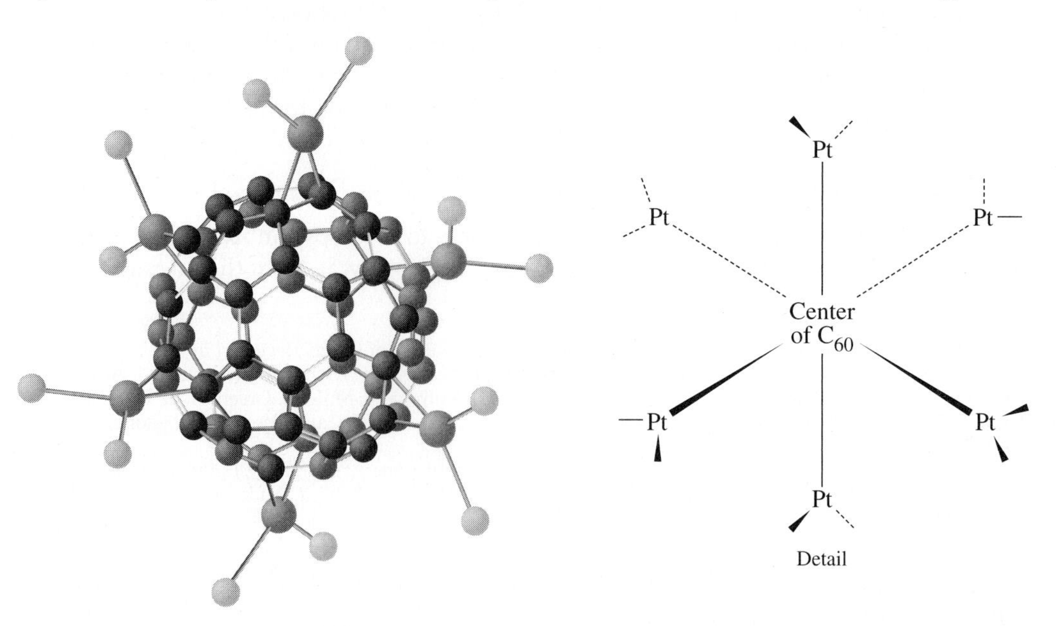

FIGURE 13.39 Structure of $[(Et_3P)_2Pt]_6C_{60}$. (Adapted with permission from G. O. Spessard and G. L. Miessler, *Organometallic Chemistry*, Prentice Hall, Upper Saddle River, NJ, 1997, p. 511, Fig. 13.13).

[62] P. J. Fagan, J. C. Calabrese, and B. Malone, *J. Am. Chem. Soc.*, **1991**, *113*, 9408. See also P. V. Broadhurst, *Polyhedron*, **1985**, *4*, 1801.

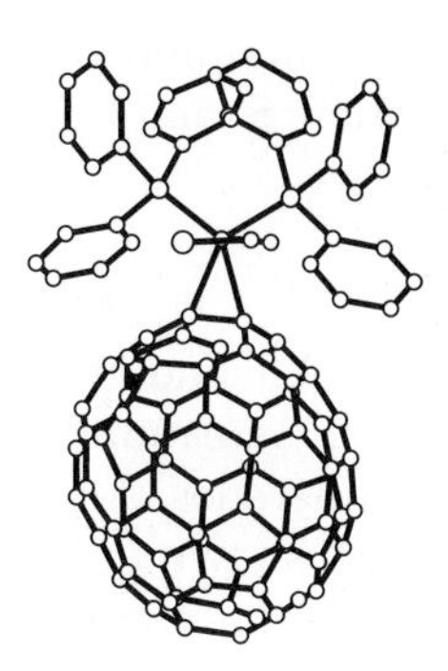

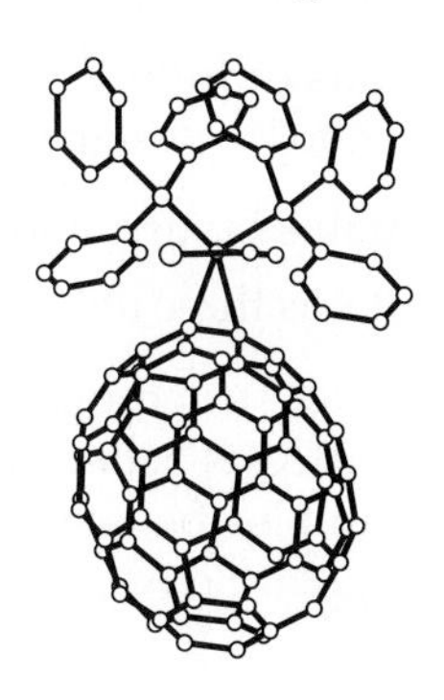

FIGURE 13.40 Stereoscopic View of $(\eta^2\text{-}C_{70})Ir(CO)Cl(PPh_3)_2$.

(Reproduced with permission from A. L. Balch, V. J. Catalano, J. W. Lee, M. M. Olmstead, and S. R. Parkin, *J. Am. Chem. Soc.*, **1991**, *113*, 8953, © 1991 American Chemical Society.)

triruthenium cluster in Figure 13.41(a) is perhaps best described as $\eta^2, \eta^2, \eta^2\text{-}C_{60}$, rather than $\eta^2\text{-}C_{60}$, because the C—C bonds bridged by the ruthenium atoms are slightly shorter than the other C—C bonds in the 6-membered ring.

Hybrids of a fullerene and a ferrocene have been reported in which an ion is sandwiched between a $\eta^5\text{-}C_5H_5$ ring and a η^5-fullerene as shown in Figure 13.41(b). The fullerenes used, $C_{60}(CH_3)_5$ and $C_{70}(CH_3)_3$, have methyl groups that apparently help stabilize these compounds. The methyl groups are bonded to carbons adjacent to the 5-membered ring to which the iron bonds. This pentamethylfullerene has also been used to form pentahapto complexes with a variety of transition metals; one example is shown in Figure 13.41(c).[63]

COMPLEXES WITH ENCAPSULATED METALS[64] These complexes are structural examples of "cage" organometallic complexes in which the metal is completely surrounded by the fullerene. Typically, complexes containing encapsulated metals are prepared by laser-induced vapor phase reactions between carbon and the metals. These compounds contain central metal cations surrounded by a fulleride, a fullerene that has been reduced.

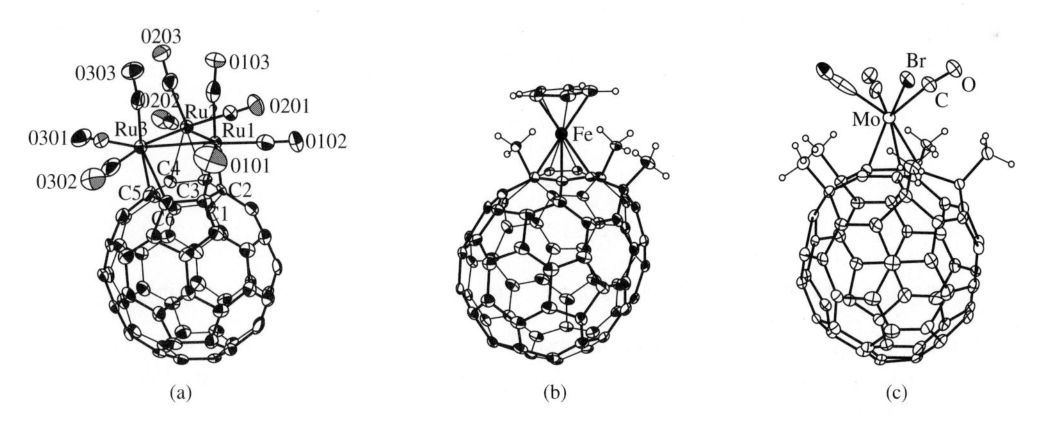

FIGURE 13.41 (a) $Ru_3(CO)_9(\mu_3\text{-}\eta^2, \eta^2, \eta^2\text{-}C_{60})$ (b) $Fe(\eta^5\text{-}C_5H_5)(\eta^5\text{-}C_{70}(CH_3)_3)$, and (c) $MoBr(CO)_3(\eta^5\text{-}C_{60}Me_5)$.

(Reproduced with permission from H.-F. Hsu and J. R. Shapley, *J. Am. Chem. Soc.*, **1996**, *118*, 9192, M. Sawamura, Y. Kuninobu, M. Toganoh, Y. Matsuo, M. Yamanaka, and E. Nakamura, *J. Am. Chem. Soc.*, **2002**, *124*, 9354, and Y. Matsuo, A. Iwashita, and E. Nakamura, *Organometallics*, **2008**, *27*, 4611. © 1996, 2002, 2008, American Chemical Society.)

[63] Y. Matsuo, A. Iwashita, and E. Nakamura, *Organometallics*, **2008**, 27, 4611, also a source of references for a variety of complexes of $\eta^5\text{-}C_{60}R_5$ fullerenes.

[64] A recent listing of fullerenes and their encapsulated atoms and molecules, with references, can be found in F. Langa and J.-F. Nierengarten, *Fullerenes: Principles and Applications*, RSC Publishing, Cambridge, UK, 2007, pp. 8–9.

Chemical formulas of fullerene compounds containing encapsulated metals are written with the @ symbol to designate encapsulation: Examples are

$U@C_{60}$	Contains U surrounded by C_{60}
$Sc_3@C_{82}$	Contains three atoms of Sc surrounded[65] by C_{82}

This designation indicates structure only and does not include charges on ions that may occur. For example, $La@C_{82}$ is believed to contain La^{3+} surrounded by the ${C_{82}}^{3-}$. Small molecules and ions can also be encapsulated in fullerenes. An example is $Sc_3N@C_{78}$, which contains a triangular Sc_3N inside the C_{78} cage, shown in Figure 13.42.

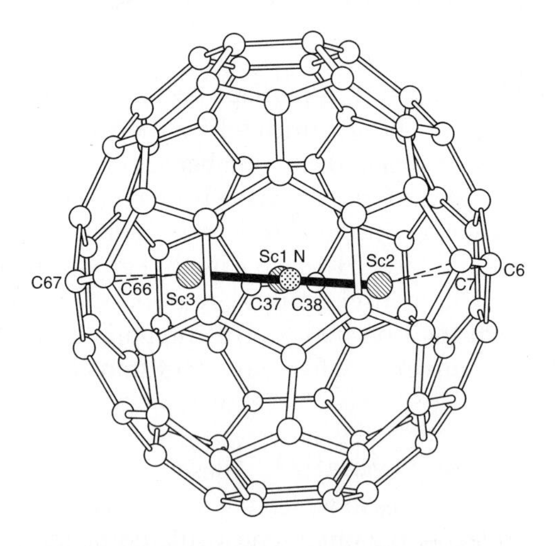

FIGURE 13.42 $Sc_3N@C_{78}$. At the low temperature used for the X-ray study, the Sc_3N is planar with angles of 130.3°, 113.8°, and 115.9°, and each Sc bonds loosely to a C—C bond that is part of two six-membered rings; however, at higher temperatures, the Sc_3N cluster moves freely inside the cage.

(Reproduced with permission from M. M. Olmstead, A. de Bettencort-Dias, J. C. Duchamp, S. Stevenson, D. Marciu, H. C. Dorn, and A. L Balch, *Angew. Chem., Int. Ed.*, **2001**, *40*, 1223.)

13.6 COMPLEXES CONTAINING M—C, M=C, AND M≡C BONDS

Complexes containing direct metal–carbon single, double, and triple bonds have been studied extensively. Table 13.4 gives examples of the most important types of ligands in these complexes.

13.6.1 Alkyl and Related Complexes

Some of the earliest known organometallic complexes were those having σ bonds between main group metal atoms and alkyl groups. Examples include Grignard reagents, having magnesium–alkyl bonds (Figure 8.10), and alkyl complexes with alkali metals, such as methyllithium.

The first stable transition-metal alkyls were synthesized in the first decade of the twentieth century; many such complexes are now known. The metal–ligand bonding in

[65] H. Shinohara, H. Yamaguchi, N. Mayashi, H. Sato, M. Ohkohchi, Y. Ando, and Y. Saito, *J. Phys. Chem.*, **1993**, *97*, 4259.

TABLE 13.4 Complexes Containing M—C, M=C, and M≡C Bonds

Ligand	Formula	Example
Alkyl	$—CR_3$	$W(CH_3)_6$
Carbene (alkylidene)	$=CR_2$	$(OC)_5Cr=C(OCH_3)(C_6H_5)$
Carbyne (alkylidyne)	$\equiv CR$	$X—Cr(CO)_4\equiv C—C_6H_5$
Carbide (carbon)	$\equiv C$	$Cl_2(PR_3)_2Ru\equiv C$
Cumulene	$=C(=C)_nRR'$	$Cl—Ir(P(CH_3)_3)_2=C=C=C=C(C_6H_5)_2$

these complexes may be viewed as primarily involving covalent sharing of electrons between the metal and the carbon in a sigma fashion, as shown here:

M ← CR_3 (R=H, alkyl, aryl)

sp^3 orbital

In terms of electron counting, the alkyl ligand may be considered a 2-electron donor $:CR_3^-$ (method A) or a 1-electron donor $\cdot CR_3$ (method B). Significant ionic contribution to the bonding may occur in complexes of highly electropositive elements, such as the alkali metals and alkaline earths.

Many synthetic routes to transition-metal alkyl complexes have been developed. Two of the most important of these methods are as follows:

1. Reaction of a transition metal halide with organolithium, organomagnesium, or organoaluminum reagent

   Example: $ZrCl_4 + 4\ PhCH_2MgCl \longrightarrow Zr(CH_2Ph)_4\ (Ph = \text{phenyl}) + 4\ MgCl_2$

2. Reaction of a metal carbonyl anion with alkyl halide

   Example: $Na[Mn(CO)_5]^- + CH_3I \longrightarrow CH_3Mn(CO)_5 + NaI$

Although many complexes contain alkyl ligands, transition-metal complexes that contain alkyl groups as the only ligands are relatively rare. Examples include $Ti(CH_3)_4$, $W(CH_3)_6$ and $Cr[CH_2Si(CH_3)_3]_4$. Alkyl complexes have a tendency to be kinetically unstable and difficult to isolate;[66] their stability is enhanced by structural crowding, which protects the coordination sites of the metal by blocking pathways to

[66] An interesting historical perspective on alkyl complexes is in G. Wilkinson, *Science*, **1974**, *185*, 109.

TABLE 13.5 Other Ligands Forming Sigma Bonds to Metals

Ligand	Formula	Example
Aryl	$-C_6H_5$	$(\eta^5\text{-}C_5H_5)_2TaH_2(C_6H_5)$
Alkenyl (vinyl)	$>C=C<$	$Cl-Pt(PR_3)_2-C(H)=CH_2$
Alkynyl	$-C\equiv C-$	$Cl-Pt(PR_3)_2-C\equiv CPh$

decomposition. For example, the 6-coordinate $W(CH_3)_6$ can be melted at 30 °C without decomposition, whereas the 4-coordinate $Ti(CH_3)_4$ is subject to decomposition at approximately –40 °C.[67] In an interesting and unusual use of alkyls, diethylzinc has been used to treat books and documents for their long-term preservation by neutralizing the acid in the paper. Many alkyl complexes are important in catalytic processes; examples of reactions of these complexes will be considered in Chapter 14.

Several other important ligands have direct metal–carbon σ bonds. Examples are given in Table 13.5. In addition, there are many examples of **metallacycles**, complexes in which organic ligands attach to metals at two positions, thereby incorporating the metals into organic rings.[68] The following reaction provides an example of a metallacycle synthesis:

$$PtCl_2(PR_3)_2 + Li(CH_2)_4Li \longrightarrow (R_3P)_2Pt(CH_2)_4 + 2\,LiCl$$

Metallacyclopentane

In addition to being interesting in their own right, metallacycles are important intermediates in a variety of catalytic processes. Examples will be cited in Chapter 14.

13.6.2 Carbene Complexes

Carbene complexes contain metal–carbon double bonds.[69] First synthesized in 1964 by Fischer,[70] carbene complexes are now known for the majority of transition metals and for a wide range of ligands, including the prototype carbene, $:CH_2$. The

[67] A. J. Shortland and G. Wilkinson, *J. Chem. Soc., Dalton Trans.*, **1973**, 872.

[68] B. Blom, H. Clayton, M. Kilkenny, and J. R. Moss, *Adv. Organomet. Chem.*, **2006**, *54*, 149.

[69] IUPAC has recommended that the term "alkylidene" be used to describe all complexes containing metal–carbon double bonds and that "carbene" be restricted to free $:CR_2$. For a detailed description of the distinction between these two terms—and between "carbyne" and "alkylidyne," discussed later in this chapter—see W. A. Nugent and J. M. Mayer, *Metal–Ligand Multiple Bonds*, Wiley InterScience, New York, 1988, pp. 11–16.

[70] E. O. Fischer and A. Maasbol, *Angew. Chem., Int. Ed.*, **1964**, *3*, 580.

TABLE 13.6 Fischer- and Schrock-type Carbene Complexes

Characteristic	Fischer-Type Carbene Complex	Schrock-Type Carbene Complex
Typical metal [oxidation state]	Middle to late transition metal [Fe(0), Mo(0), Cr(0)]	Early transition metal [Ti(IV), Ta(V)]
Substituents attached to $C_{carbene}$	At least one highly electronegative heteroatom (such as O, N, or S)	H or alkyl
Typical other ligands in complex	Good π acceptors	Good σ or π donors
Electron count	18	10–18

majority of such complexes, including those first synthesized by Fischer, contain one or two highly electronegative heteroatoms—such as O, N, or S—directly attached to the carbene carbon. These are commonly designated as *Fischer-type carbene complexes*, and they have been studied extensively. Other carbene complexes contain only carbon and/or hydrogen attached to the carbene carbon. First synthesized several years after the initial Fischer carbene complexes,[71] these have been studied extensively by Schrock's research group and several others. They are sometimes designated as *Schrock-type carbene complexes*, commonly referred to as *alkylidenes*. The distinctions between Fischer- and Schrock-type carbene complexes are summarized in Table 13.6. In this text, we will focus primarily on Fischer-type carbene complexes.

The formal double bond in carbene complexes may be compared with the double bond in alkenes. In the case of a carbene complex, the metal must use a *d* orbital (rather than a *p* orbital) to form the π bond with carbon, as illustrated in Figure 13.43.

Another aspect of bonding of importance to carbene complexes is that complexes having a highly electronegative atom—such as O, N, or S—attached to the carbene carbon tend to be more stable than complexes lacking such an atom. For example, $Cr(CO)_5[C(OCH_3)C_6H_5]$, with an oxygen on the carbene carbon, is much more stable than $Cr(CO)_5[C(H)C_6H_5]$. The stability of the complex is enhanced if the highly electronegative atom can participate in the π bonding, with the result a delocalized, 3-atom π system involving a *d* orbital on the metal and *p* orbitals on the carbon and on the electronegative atom. Such a delocalized 3-atom system provides more stability to the bonding π electron pair than would a simple metal-to-carbon π bond. An example of such a π system is shown in Figure 13.44.

The methoxycarbene complex $Cr(CO)_5[C(OCH_3)C_6H_5]$ illustrates the bonding just described and some important related chemistry.[72] To synthesize this complex, we can begin with the hexacarbonyl, $Cr(CO)_6$. As in organic chemistry, highly nucleophilic reagents can attack the carbonyl carbon. For example, phenyllithium can react with

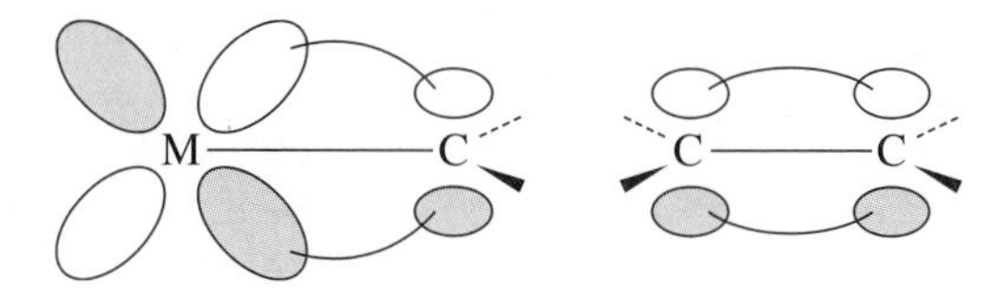

FIGURE 13.43 Bonding in Carbene Complexes and in Alkenes.

[71] R. R. Schrock, *J. Am. Chem. Soc.*, **1974**, *96*, 6796.
[72] E. O. Fischer, *Adv. Organomet. Chem.*, **1976**, *14*, 1.

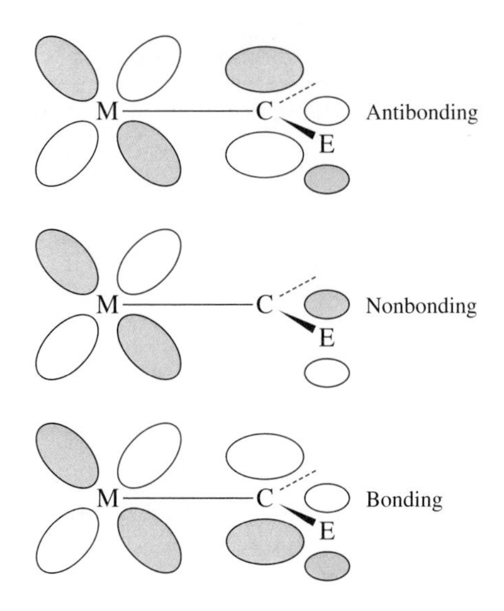

FIGURE 13.44 Delocalized Pi Bonding in Carbene Complexes. E designates a highly electronegative heteroatom, such as O, N, or S.

$Cr(CO)_6$ to give the anion $[C_6H_5C(O)Cr(CO)_5]^-$, which has two important resonance structures, as shown here:

$$Li^+{:}C_6H_5^- + O{\equiv}C{-}Cr(CO)_5 \longrightarrow \underbrace{C_6H_5{-}\overset{\overset{\displaystyle O}{\|}}{C}{-}Cr^-(CO)_5 \longleftrightarrow C_6H_5{-}\overset{\overset{\displaystyle O^-}{|}}{C}{=}Cr(CO)_5}_{C_6H_5-\overset{O}{C}\text{---}Cr(CO)_5} + Li^+$$

Alkylation by a source of CH_3^+, such as $[(CH_3)_3O][BF_4]$ or CH_3I, gives the methoxycarbene complex:

$$C_6H_5{-}\overset{O}{C}\text{---}Cr(CO)_5 + [(CH_3)_3O][BF_4] \longrightarrow C_6H_5{-}\overset{\overset{\displaystyle OCH_3}{|}}{C}{=}Cr(CO)_5 + BF_4^- + (CH_3)_2O$$

Evidence for double bonding between chromium and carbon is provided by X-ray crystallography, which measures this distance at 204 pm, compared with a typical Cr—C single-bond distance of approximately 220 pm.

One very interesting aspect of this complex is that it exhibits a proton NMR spectrum that is temperature dependent. At room temperature, a single resonance is found for the methyl protons; however, as the temperature is lowered, this peak first broadens and then splits into two peaks. How can this behavior be explained?

A single proton resonance, corresponding to a single magnetic environment, is expected for the carbene complex as illustrated, with a double bond between chromium and carbon, and a single bond (permitting rapid rotation about the bond) between carbon and oxygen. The room-temperature NMR is therefore as expected. However, the splitting of this peak at lower temperature into two peaks suggests two different proton environments.[73] Two environments are possible if rotation is hindered about the C—O bond. A resonance structure for the complex can be drawn showing the possibility of some double bonding between C and O; were such double bonding significant, *cis* and *trans* isomers, as shown in Figure 13.45, might be observable at low temperatures.

[73] C. G. Kreiter and E. O. Fischer, *Angew. Chem., Int. Ed.*, **1969**, *8*, 761.

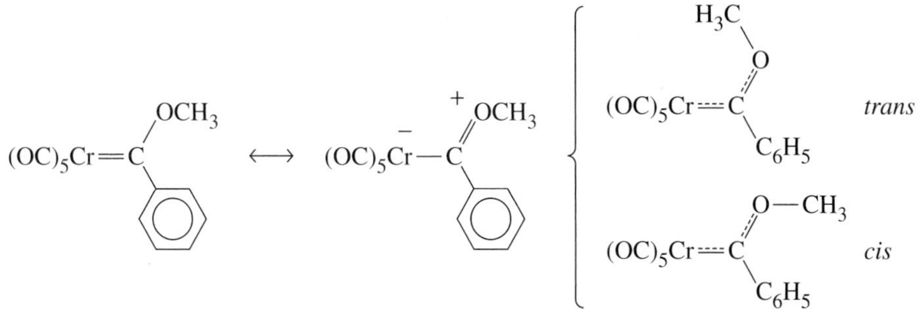

FIGURE 13.45 Resonance Structures and *cis* and *trans* Isomers for $Cr(CO)_5[C(OCH_3)C_6H_5]$.

Evidence for double-bond character in the C—O bond is also provided by crystal structure data, which show a C—O bond distance of 133 pm, compared with a typical C—O single-bond distance of 143 pm.[74] The double bonding between C and O, although weak (typical C=O bonds are much shorter, approximately 116 pm), is sufficient to slow down rotation about the bond so that, at low temperatures, proton NMR detects the *cis* and *trans* methyl protons separately. At higher temperature, there is sufficient energy to cause rapid rotation about the C—O bond, so that the NMR sees only an average signal, which is observed as a single peak.

X-ray crystallographic data, as mentioned, show double-bond character in both the Cr—C and C—O bonds. This supports the statement made at the beginning of this section that π bonding in complexes of this type (containing a highly electronegative atom, in this case oxygen) may be considered delocalized over three atoms. Although not absolutely essential for all carbene complexes, the delocalization of π electron density over three (or more) atoms provides an additional measure of stability to many of these complexes.[75]

Carbene complexes appear to be important intermediates in olefin metathesis reactions, which are of significant industrial interest; these reactions are discussed in Chapter 14.

13.6.3 Carbyne (Alkylidyne) Complexes

Carbyne complexes have metal–carbon triple bonds; they are formally analogous to alkynes.[76] Many carbyne complexes are now known; examples of carbyne ligands include the following:

$$\mathbf{M{\equiv}C{-}R}$$

where R = aryl, alkyl, H, $SiMe_3$, NEt_2, PMe_3, SPh, or Cl. Carbyne complexes were first synthesized fortuitously in 1973 as products of the reactions of carbene complexes with Lewis acids.[77] For example, the methoxycarbene complex $Cr(CO)_5[C(OCH_3)C_6H_5]$ was found to react with the Lewis acids BX_3 (X = Cl, Br, or I).

First, the Lewis acid attacks the oxygen, the basic site on the carbene:

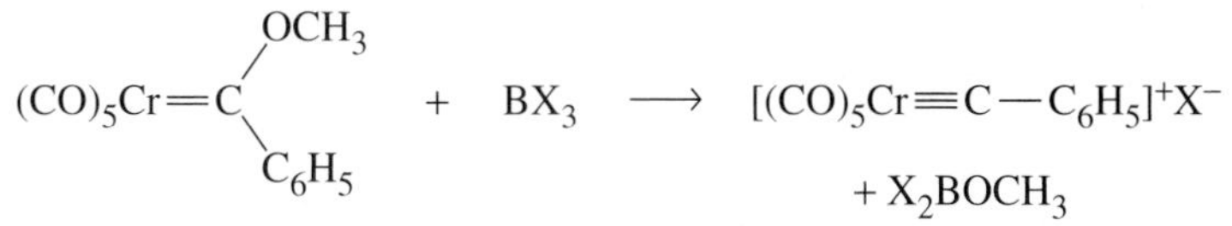

$$(CO)_5Cr{=}C(OCH_3)(C_6H_5) + BX_3 \longrightarrow [(CO)_5Cr{\equiv}C{-}C_6H_5]^+X^- + X_2BOCH_3$$

---

[74] O. S. Mills and A. D. Redhouse, *J. Chem. Soc. A*, **1968**, 642.

[75] K. H. Dötz, H. Fischer, P. Hoffmann, F. R. Kreissl, U. Schubert, and K. Weiss, *Transition Metal Carbene Complexes*, Verlag Chemie, Weinheim, Germany, 1983, pp. 120–122.

[76] IUPAC has recommended that "alkylidyne" be used to designate complexes containing metal–carbon triple bonds.

[77] E. O. Fischer, G. Kreis, C. G. Kreiter, J. Müller, G. Huttner, and H. Lorentz, *Angew. Chem., Int. Ed.*, **1973**, *12*, 564.

Subsequently, the intermediate loses CO, with the halogen coordinating in a position *trans* to the carbyne:

$$[(CO)_5Cr\equiv C-C_6H_5]^+X^- \longrightarrow X-Cr(CO)_4\equiv C-C_6H_5 + CO$$

The best evidence for the carbyne nature of the complex is provided by X-ray crystallography, which gives a Cr—C bond distance of 168 pm (for X = Cl), considerably shorter than the 204 pm for the parent carbene complex. The Cr≡C—C angle is, as expected, 180° for this complex; however, slight deviations from linearity are observed for many complexes in crystalline form, in part a consequence of the manner of packing in the crystal.

Bonding in carbyne complexes may be viewed as a combination of a σ bond plus two π bonds, as illustrated in Figure 13.46.

The carbyne ligand has a lone pair of electrons in an *sp* hybrid on carbon; this lone pair can donate to a suitable orbital on Cr to form a σ bond. In addition, the carbon has two *p* orbitals that can accept electron density from *d* orbitals on Cr to form π bonds. Thus, the overall function of the carbyne ligand is as both a σ donor and π acceptor. (For electron counting purposes, a $:CR^+$ ligand can be considered a 2-electron donor; it is usually more convenient to count neutral CR as a 3-electron donor.)

Carbyne complexes can be synthesized in a variety of ways in addition to Lewis acid attack on carbene complexes. Synthetic routes for carbyne complexes and the reactions of these complexes have been reviewed.[77]

In some cases, molecules have been synthesized containing two or three of the types of ligands discussed in this section (alkyl, carbene, and carbyne). Such molecules provide an opportunity to make direct comparisons of lengths of metal–carbon single, double, and triple bonds, as shown in Figure 13.47.

▶ **Exercise 13.10** Are the compounds shown in Figure 13.47 18-electron species?

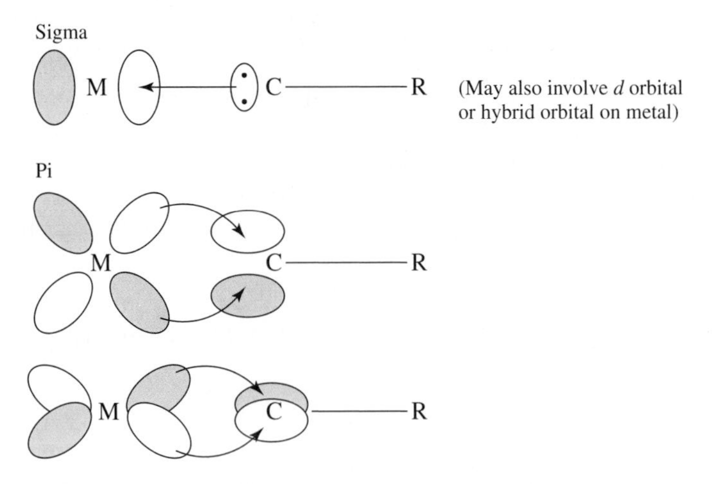

FIGURE 13.46
Bonding in Carbyne Complexes.

[77] H. P. Kim and R. J. Angelici, "Transition Metal Complexes with Terminal Carbyne Ligands," in *Adv. Organomet. Chem.*, **1987**, *27*, 51; H. Fischer, P. Hoffmann, F. R. Kreissl, R. R. Schrock, U. Schubert, and K. Weiss, *Carbyne Complexes*, VCH, Weinheim, Germany, 1988.

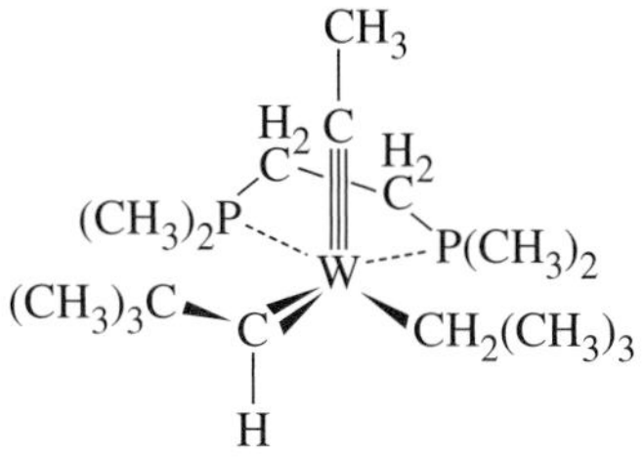

W—C 225.8 pm
W=C 194.2 pm
W≡C 178.5 pm

(a)

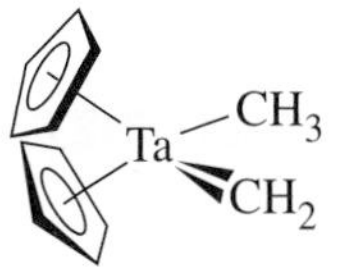

Ta—C 224.6 pm
Ta=C 202.6 pm

(b)

FIGURE 13.47 Complexes Containing Alkyl, Carbene, and Carbyne Ligands. (a) M. R. Churchill and W. J. Young, *Inorg. Chem.*, **1979**, *18*, 2454. (b) L. J. Guggenberger and R. R. Schrock, *J. Am. Chem. Soc.*, **1975**, *97*, 6578.

13.6.4 Carbide and Cumulene Complexes

The simplest possible carbon-containing ligand, a single carbon atom, is known, although examples of such **carbide** or **carbon** ligands remain rare. Although it is tempting, by extending the series —CR_3, =CR_2, ≡CR, to assign a quadruple bond to carbon in complexes containing a metal–carbon ligand, evidence suggests that such metal–ligand bonding is better described as M≡C^-. The first neutral carbide complex was the trigonal-bipyramidal ruthenium complex shown in Figure 13.48.[78] The Ru—C distance in this complex is perhaps longer than might be expected, 165.0 pm—only slightly shorter than the comparable distance in the structurally similar ruthenium carbyne complex, also shown.

Calculations have indicated that bonds between transition metals and terminal carbon atoms are quite strong, with bond dissociation enthalpies comparable to those of transition-metal complexes with M≡N and M = O bonds.[79] In addition, the frontier orbitals of the carbide complex shown in Figure 13.48 (with R = methyl) have many similarities to those of CO, suggesting that such complexes may potentially show similar coordination chemistry to the carbonyl ligand.[80]

Ligands with chains of carbon atoms that have cumulated (consecutive) double bonds, designated *cumulenylidene* ligands, are also known. Such **metallacumulene** complexes have drawn interest because of possible applications as 1-dimensional molecular wires and for use in nanoscale optical devices.[81] In recent years, complexes with 2- and 3-carbon chains have also been developed as effective catalysts. Metallacumulene complexes with carbon chains containing three or more atoms have recently been reviewed.[82]

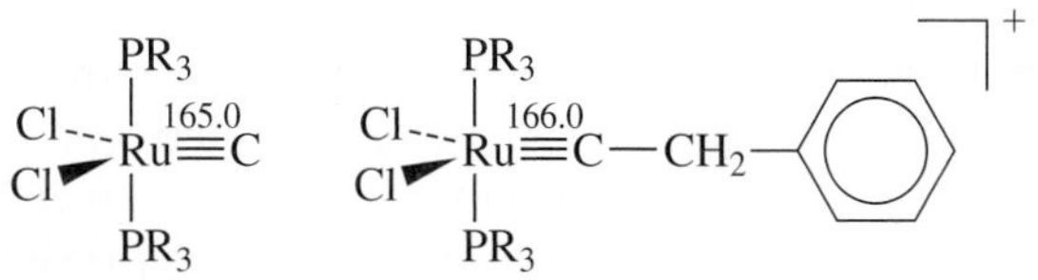

R = cyclohexyl   R = isoprophyl

FIGURE 13.48 Carbide and Carbyne Complexes.

[78] R. G. Carlson, M. A. Gile, J. A. Heppert, M. H. Mason, D. R. Powell, D. Vander Velde, and J. M. Villain, *J. Am. Chem. Soc.*, **2002**, *124*, 1580.
[79] J. B. Gary, C. Buda, M. J. A. Johnson, and B. D. Dunietz, *Organometallics*, **2008**, *27*, 814.
[80] A. Krapp and G. Frenking, *J. Am. Chem. Soc.*, **2008**, *130*, 16646.
[81] M. I. Bruce, *Coord. Chem. Rev.*, **2004**, *248*, 1603.
[82] V. Cadierno and J. Gimeno, *Chem. Rev.*, **2009**, *109*, 3512.

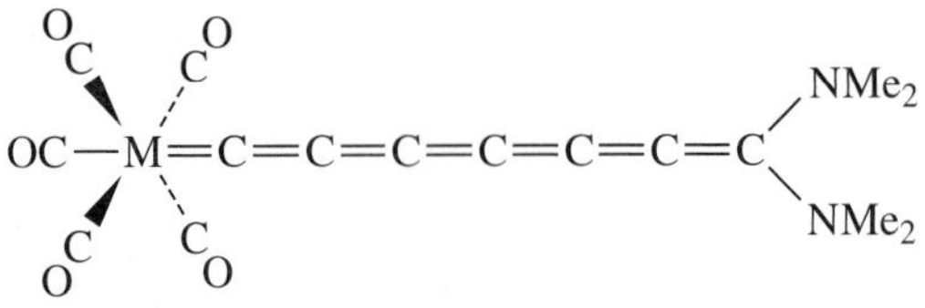

FIGURE 13.49 A Metallacumulene Complex.

$$(OC)_5M{=}C{=}C{=}C{=}C{=}C{=}C{=}C(NMe_2)_2$$

M = Cr, W

The longest cumulenylidene ligand reported to date is the heptahexaenylidene complex shown in Figure 13.49.[83]

13.7 SPECTRAL ANALYSIS AND CHARACTERIZATION OF ORGANOMETALLIC COMPLEXES

One of the most challenging, and sometimes most frustrating, aspects of organometallic research is the characterization of new reaction products. Assuming that specific products can be isolated by chromatographic procedures, recrystallization, or other techniques, determining the structure can present an interesting challenge. Many complexes can be crystallized and characterized structurally by X-ray crystallography; however, not all organometallic complexes can be crystallized, and not all that crystallize lend themselves to structural solution by X-ray techniques. Furthermore, it is frequently desirable to be able to use more convenient techniques than X-ray crystallography (although, in some cases, an X-ray structural determination is the only way to identify a compound conclusively—and may therefore be the most rapid and inexpensive technique). Infrared spectroscopy and NMR spectrometry are often the most useful. In addition, mass spectrometry, elemental analysis, conductivity measurements, and other methods may be valuable in characterizing products of organometallic reactions. We will consider primarily IR and NMR as techniques used in the characterization of organometallic complexes.

13.7.1 Infrared Spectra

IR can be useful in two respects. The number of IR bands, as discussed in Chapter 4, depends on molecular symmetry; consequently, by determining the number of such bands for a particular ligand (such as CO), we may be able to decide among several alternative geometries for a compound or at least reduce the number of possibilities. In addition, the position of the IR band can indicate the function of a ligand (e.g., terminal vs. bridging modes) and, in the case of π-acceptor ligands, can describe the electron environment of the metal.

NUMBER OF INFRARED BANDS In Section 4.4.2, a method was described for using molecular symmetry to determine the number of IR-active stretching vibrations. The basis for this method is that vibrational modes, to be IR active, must result in a change in the dipole moment of the molecule. In symmetry terms, the equivalent statement is that IR-active vibrational modes must have irreducible representations of the same symmetry as the Cartesian coordinates x, y, or z (or a linear combination of these coordinates). The procedure developed in Chapter 4 is used in the following examples. It is suggested as an exercise that the reader verify some of these results using the method described in Chapter 4.

[83] M. Dede, M. Drexler, and H. Fischer, *Organometallics*, **2007**, *26*, 4294.

Our examples will be carbonyl complexes. Identical reasoning applies to other linear monodentate ligands, such as CN^- and NO. We will begin by considering several simple cases.

Monocarbonyl complexes. These complexes have a single possible C—O stretching mode and consequently show a single band in the IR.

Dicarbonyl complexes. Two geometries, linear and bent, must be considered:

O—C—M—C—O

In the case of two CO ligands arranged linearly, only an antisymmetric vibration of the ligands is IR active; a symmetric vibrational mode produces no change in dipole moment and hence is inactive. However, if two CO ligands are oriented in a nonlinear fashion, both symmetric and antisymmetric vibrations result in changes in dipole moment, and both are IR active:

Symmetric Stretch

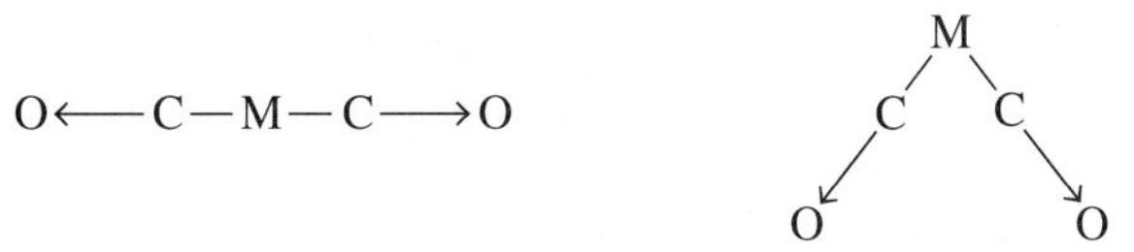

O⟵C—M—C⟶O

No change in dipole moment:
IR inactive

Change in dipole moment:
IR active

Antisymmetric Stretch

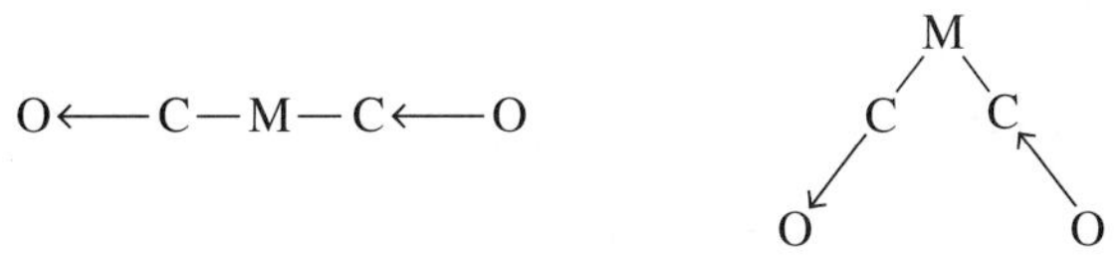

O⟵C—M—C⟵O

Change in dipole moment:
IR active

Change in dipole moment:
IR active

Therefore, an IR spectrum can be a convenient tool for determining structure for molecules known to have exactly two CO ligands: a single band indicates linear orientation of the CO ligands, and two bands indicate nonlinear orientation.

For molecules containing exactly two CO ligands on the same metal atom, the relative intensities of the IR bands can be used to determine the approximate angle between the COs, using the equation

$$\frac{I_{symmetric}}{I_{antisymmetric}} = cotan^2\left(\frac{\phi}{2}\right)$$

where the angle between the ligands is ϕ. For example, for two CO ligands at 90°, $\cot an^2(45°) = 1$. For this angle, two IR bands of equal intensity would be observed. For an angle greater than 90°, the ratio is less than 1; the IR band due to symmetric stretching is less intense than the band due to antisymmetric stretching. If ϕ is less than 90°, the IR band for symmetric stretching is the more intense. (For C—O stretching vibrations, the symmetric band occurs at higher energy than the corresponding antisymmetric band.) In general, this calculation is approximate and requires integrated values of intensities of absorption bands (rather than the more easily determined intensity at the wavelength of maximum absorption).

Complexes containing three or more carbonyls. Here, the predictions are not quite so simple. The exact number of carbonyl bands can be determined according to the symmetry approach of Chapter 4. For convenient reference, the numbers of bands expected for a variety of CO complexes are given in Table 13.7.

Several additional points relating to the number of IR bands are worth noting. First, although we can predict the number of IR-active bands by the methods of group theory, fewer bands may sometimes be observed. In some cases, bands may overlap to such a degree as to be indistinguishable; alternatively, one or more bands may be of very low intensity and not readily observed. In some cases, isomers may be present in the same sample, and it may be difficult to determine which IR absorptions belong to which compound.

In carbonyl complexes, the number of C—O stretching bands cannot exceed the number of CO ligands. The alternative is possible in some cases (more CO groups than IR bands), when vibrational modes are not IR active (do not cause a change in dipole moment). Examples are given in Table 13.7. Because of their symmetry, carbonyl complexes of T_d and O_h symmetry have a single carbonyl band in the IR spectrum.

▶ **Exercise 13.11** The complex $Mo(CO)_3(NCC_2H_5)_3$ has the infrared spectrum shown below. Is this complex more likely the *fac* or *mer* isomer?

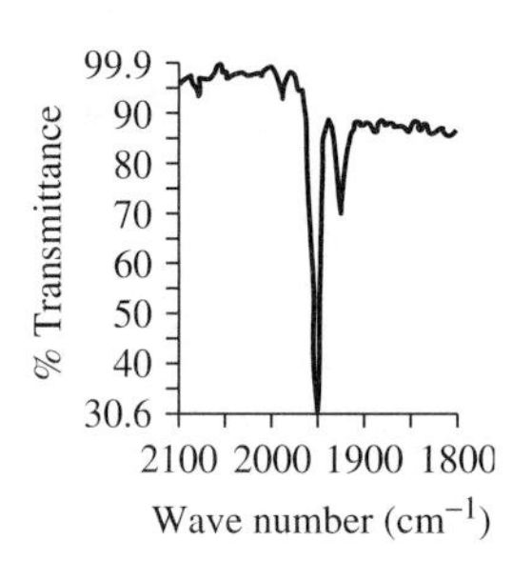

POSITIONS OF IR BANDS We have already encountered in this chapter two examples in which the position of the carbonyl stretching band provides useful information. In the case of the isoelectronic species $[Mn(CO)_6]^+$, $Cr(CO)_6$, and $[V(CO)_6]^-$, an increase in negative charge on the complex causes a significant reduction in the energy of the C—O band as a consequence of additional π back-bonding from the metal to the ligands (Section 13.4.1). The bonding mode is also reflected in the infrared spectrum, with energy decreasing in the order

terminal CO > doubly bridging CO > triply bridging CO

The positions of infrared bands are also a function of other ligands present. For example, consider the data in Tables 13.7 and 13.8.

Going down the series in Table 13.8, the σ-donor ability of the phosphine ligands increases and the π-acceptor ability decreases. PF_3 is the weakest donor (as a consequence of the highly electronegative fluorines) and the strongest acceptor; conversely, PMe_3 is the strongest donor and the weakest acceptor. As a result, the molybdenum in $Mo(CO)_3(PMe_3)_3$ carries the greatest electron density; it is the most able to donate electron density to the π^* orbitals of the CO ligands. Consequently, the CO ligands in $Mo(CO)_3(PMe_3)_3$ have the weakest C—O bonds and the lowest energy stretching bands. Many comparable series are known.

TABLE 13.7 Carbonyl Stretching Bands

	Coordination Number		
Number of CO Ligands	**4**	**5**	**6**
3	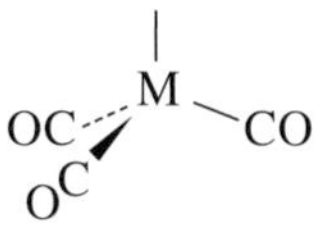 M OC CO OC	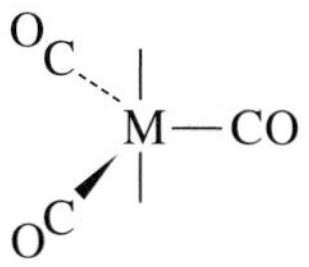 OC M—CO OC	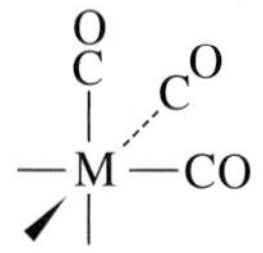 OC CO —M—CO
IR bands:	2	1	2
		OC M—CO CO	CO —M—CO OC
IR bands:		3	3
		OC M—CO OC	
IR bands:		3	
4	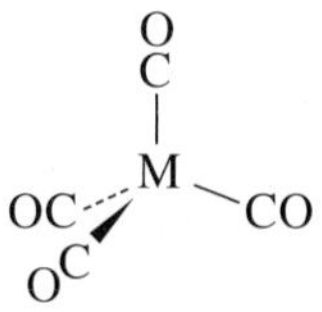 OC M OC CO OC	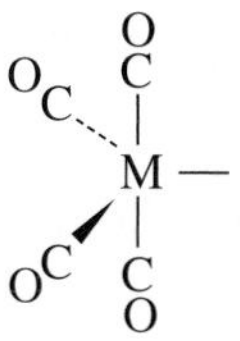 OC OC M— OC CO	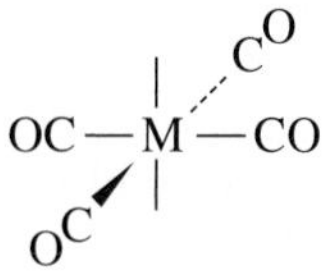 CO OC—M—CO OC
IR bands:	1	4	1
		OC M—CO OC CO	CO OC—M— OC CO
IR bands:		3	4
5 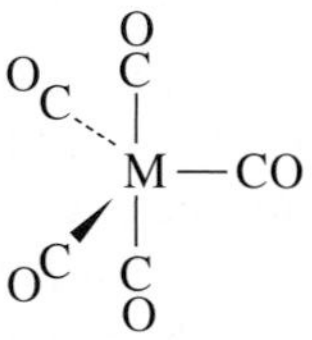 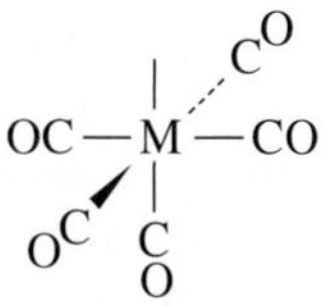		OC OC M—CO OC CO	CO OC—M—CO OC CO
IR bands:		2	3
6 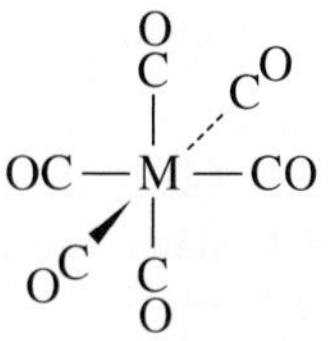			OC CO OC—M—CO OC CO
IR bands:			1

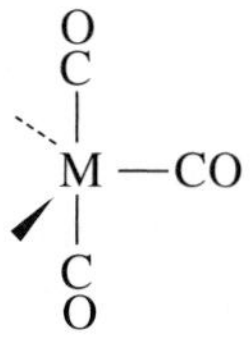

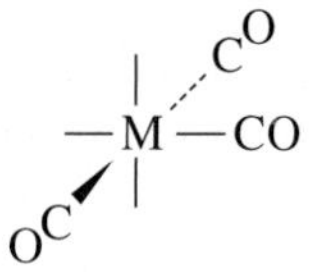

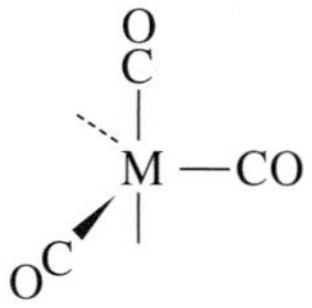

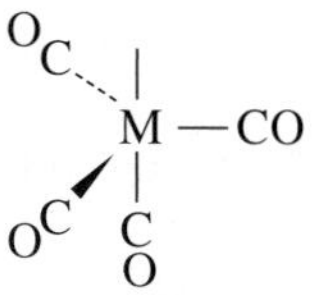

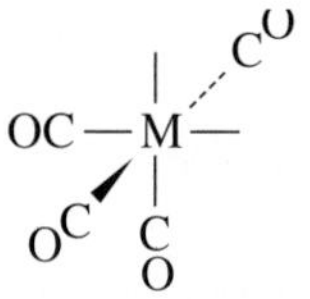

TABLE 13.8 Examples of Carbonyl Stretching Bands: Molybdenum Complexes

Complex	ν (CO), cm^{-1}
fac-$Mo(CO)_3(PF_3)_3$	2090, 2055
fac-$Mo(CO)_3(PCl_3)_3$	2040, 1991
fac-$Mo(CO)_3(PClPh_2)_3$	1977, 1885
fac-$Mo(CO)_3(PMe_3)_3$	1945, 1854

Source: F. A. Cotton, *Inorg. Chem.*, **1964**, *3*, 702.

The important point is that the position of the carbonyl bands can provide important clues to the electronic environment of the metal. The greater the electron density on the metal (and the greater the negative charge), the greater the back-bonding to CO, and the lower the energy of the carbonyl stretching vibrations. Similar correlations between the metal environment and IR spectra can be shown for a variety of other ligands, both organic and inorganic. NO, for example, has an IR spectrum that is strongly correlated with the environment in a manner similar to that of CO. In combination with information on the number of IR bands, the positions of such bands for CO and other ligands can therefore be extremely useful in characterizing organometallic compounds.

13.7.2 NMR Spectra

NMR is also a valuable tool in characterizing organometallic complexes. The advent of high-field NMR instruments using superconducting magnets has in many ways revolutionized the study of these compounds. Convenient NMR spectra can now be taken using many metal nuclei as well as the more traditional nuclei such as ^{1}H, ^{13}C, ^{19}F, and ^{31}P; the combined spectral data of several nuclei make it possible to identify many compounds by their NMR spectra alone.

As in organic chemistry, chemical shifts, splitting patterns, and coupling constants are useful in characterizing the environments of individual atoms in organometallic compounds. The reader may find it useful to review the basic theory of NMR as presented in an organic chemistry text. More detailed discussions of NMR, especially relating to ^{13}C, have been presented elsewhere.[84]

^{13}C NMR Carbon-13 NMR has become increasingly useful with the advent of modern instrumentation. Although the isotope ^{13}C has a low natural abundance (approximately 1.1%) and low sensitivity for the NMR experiment (about 1.6% as sensitive as ^{1}H), Fourier transform techniques now make it possible to obtain useful ^{13}C spectra for most organometallic species of reasonable stability. Nevertheless, the time necessary to obtain a ^{13}C spectrum may still be an experimental difficulty for compounds present in very small amounts or of low solubility. Rapid reactions may also be inaccessible by this technique. Some useful features of ^{13}C spectra include the following:

1. An opportunity to observe organic ligands that do not contain hydrogen, such as CO and $F_3C—C{\equiv}C—CF_3$.

[84] B. E. Mann, " ^{13}C NMR Chemical Shifts and Coupling Constants of Organometallic Compounds," in *Adv. Organomet. Chem.*, **1974**, *12*, 135; P. W. Jolly and R. Mynott, "The Application of ^{13}C NMR Spectroscopy to Organo-Transition Metal Complexes," in *Adv. Organomet. Chem.*, **1981**, *19*, 257; E. Breitmaier and W. Voelter, *Carbon 13 NMR Spectroscopy*, VCH, New York, 1987; E. A. V. Ebsworth, D. W. H. Rankin, and S. Cradock, *Structural Methods in Inorganic Chemistry*, 2nd ed., Blackwell, Oxford, 1991, pp. 414–425.

2. Direct observation of the carbon skeleton of organic ligands.
3. ^{13}C chemical shifts are more widely dispersed than ^{1}H shifts. This often makes it easy to distinguish between ligands in compounds containing several different organic ligands.

^{13}C NMR is also a valuable tool for observing rapid intramolecular-rearrangement processes.[85]

Approximate ranges of chemical shifts for ^{13}C spectra of some categories of organometallic complexes are listed in Table 13.9.

Several features of the data in this table are worth noting.

1. Terminal carbonyl peaks are frequently in the range δ 195 to 225 ppm, a range sufficiently distinctive that the CO groups are usually easy to distinguish from other ligands.
2. The ^{13}C chemical shift is correlated with the strength of the C—O bond; in general, the stronger the C—O bond, the lower the chemical shift.[86]
3. Bridging carbonyls have slightly greater chemical shifts than terminal carbonyls and consequently may lend themselves to easy identification. (However, IR is usually a better tool than NMR for distinguishing between bridging and terminal carbonyls.)
4. Cyclopentadienyl ligands have a wide range of chemical shifts in paramagnetic compounds and a much narrower range in diamagnetic compounds. Other organic ligands may also have fairly wide ranges in ^{13}C chemical shifts.[87]

^{1}H NMR The ^{1}H spectra of organometallic compounds containing hydrogens can also provide useful structural information. For example, protons bonded directly to metals (in hydride complexes, discussed in Section 13.4.3) are very strongly shielded, with chemical shifts commonly in the approximate range of –5 to –20 ppm relative to

TABLE 13.9 ^{13}C Chemical Shifts for Organometallic Compounds

Ligand	^{13}C Chemical Shift (Range)[a]			
M—CH_3	–28.9 to 23.5			
M=C<	190 to 400			
M≡C—	235 to 401			
M—CO	177 to 275			
Neutral binary CO	183 to 223			
M—(η^5-C_5H_5)	68.2 to 121.3 (diamagnetic)	–790 to 1430 (paramagnetic)		
Fe(η^5-C_5H_5)$_2$	69.2			
M—(η^3-C_3H_5)	C_2: 91 to 129	C_1 and C_3: 46 to 79		
M—C_6H_5	M—C: 130 to 193	*ortho*: 132 to 141	*meta*: 127 to 130	*para*: 121 to 131

NOTE: [a]Parts per million (ppm) relative to $Si(CH_3)_4$.

[85] Breitmaier and Voelter, *Carbon 13 NMR Spectroscopy*. pp. 127–133, 166–167, 172–178.
[86] P. C. Lauterbur and R. B. King, *J. Am. Chem. Soc.*, **1965**, *87*, 3266.
[87] Extensive tables of chemical shifts and coupling constants can be found in B. E. Mann, "^{13}C NMR Chemical Shifts and Coupling Constants of Organometallic Compounds," in *Adv. Organomet. Chem.*, **1974**, *12*, 135.

$Si(CH_3)_4$. Such protons are typically easy to detect, because few other protons commonly appear in this region.

Protons in methyl complexes ($M—CH_3$) typically have chemical shifts between 1 and 4 ppm, similar to their positions in organic molecules. Cyclic π ligands, such as η^5-C_5H_5 and η^6-C_6H_6, most commonly have ^1H chemical shifts between 4 and 7 ppm and, because of the relatively large number of protons involved, may lend themselves to easy identification.[88] Protons in other types of organic ligands also have characteristic chemical shifts; examples are given in Table 13.10.

As in organic chemistry, integration of NMR peaks of organometallic complexes can provide the ratio of atoms in different environments. For example, the area of a ^1H peak is usually proportional to the number of nuclei giving rise to that peak. However, for ^{13}C, this calculation is less reliable. Relaxation times of different carbon atoms in organometallic complexes vary widely; this may lead to inaccuracy in correlating peak area with the number of atoms (the correlation between area and number of atoms is dependent on rapid relaxation). Adding paramagnetic reagents may speed up relaxation and thereby improve the validity of integration data. One paramagnetic compound often used is $Cr(acac)_3$ [acac = acetylacetonate = $H_3CC(O)CHC(O)CH_3{}^-$].[89]

TABLE 13.10 Examples of ^1H Chemical Shifts for Organometallic Compounds

Complex	^1H Chemical Shift^a
$Mn(CO)_5\mathbf{H}$	–7.5
$W(C\mathbf{H}_3)_6$	1.80
$Ni(\eta^2\text{-}C_2\mathbf{H}_4)_3$	3.06
$(\eta^5\text{-}C_5\mathbf{H}_5)_2Fe$	4.04
$(\eta^6\text{-}C_6\mathbf{H}_6)_2Cr$	4.12
$(\eta^5\text{-}C_5H_5)_2Ta(CH_3)({=}C\mathbf{H}_2)$	10.22

NOTE: ^{a}Parts per million relative to $Si(CH_3)_4$.

MOLECULAR REARRANGEMENT PROCESSES The compound $(C_5H_5)_2Fe(CO)_2$ has interesting NMR behavior. This compound contains both η^1- and η^5-C_5H_5 ligands (and consequently obeys the 18-electron rule). The ^1H NMR spectrum at room temperature shows two singlets of equal area. A singlet would be expected for the five equivalent protons of the η^5-C_5H_5 ring but is surprising for the η^1-C_5H_5 ring, because the protons are not all equivalent. At lower temperatures, the peak at 4.5 ppm (η^5-C_5H_5) remains constant, but the other peak at 5.7 ppm spreads and then splits into new peaks near 3.5 and between 5.9 and 6.4 ppm—all consistent with a η^1-C_5H_5 ligand. A "ring whizzer" mechanism,[90] Figure 13.50, has been proposed by which the five ring positions of the monohapto ring interchange via 1,2-metal shifts so rapidly at 30 °C that the NMR

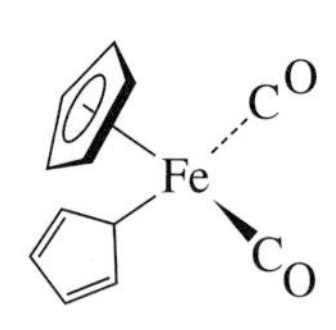

[88] These are ranges for diamagnetic complexes. Paramagnetic complexes may have much larger chemical shifts, sometimes several hundred parts per million relative to tetramethylsilane.
[89] For a discussion of the problems associated with integration in ^{13}C NMR, see J. K. M. Saunders and B. K. Hunter, *Modern NMR Spectroscopy*, W. B. Saunders, New York, 1992.
[90] C. H. Campbell and M. L. H. Green, *J. Chem. Soc., A*, **1970**, 1318.

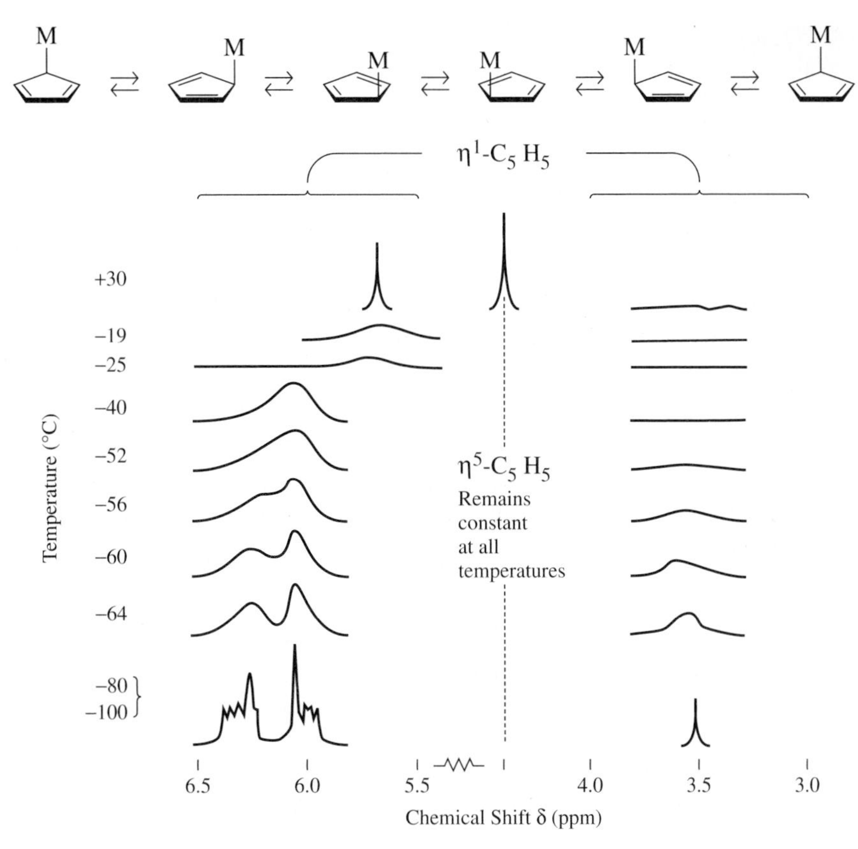

FIGURE 13.50 Ring-Whizzer Mechanism and Variable Temperature NMR Spectra of $(C_5H_5)_2Fe(CO)_2$. The central peak at 4.5 ppm due to the η^5-C_5H_5 ligand, remains constant throughout; it is not shown except in the highest temperature spectrum to simplify the figure. (NMR spectra reproduced with permission from M. H. Bennett, Jr., F. A. Cotton, A. Davison, J. W. Faller, S. J. Lippard, and S. M. Morehouse, *J. Am. Chem. Soc.*, **1966**, *88*, 4371. © 1966 American Chemical Society.)

spectrometer can see only the average signal for the ring.[91] At lower temperatures, this process is slower, and the different resonances for the protons of η^1-C_5H_5 become apparent, also shown in Figure 13.50.

More detailed discussions of NMR spectra of organometallic compounds, including nuclei not mentioned here, have been given by Elschenbroich and Salzer.[92]

13.7.3 Examples of Characterization

In this chapter, we have considered just a few types of reactions of organometallic compounds, principally the replacement of CO by other ligands and the reactions involved in syntheses of carbene and carbyne complexes. Additional types of reactions will be discussed in Chapter 14. We conclude this chapter with two examples of how spectral data may be used in the characterization of organometallic compounds. Further examples can be found in the problems at the end of this chapter and in Chapter 14.

---

[91] M. J. Bennett, Jr., F. A. Cotton, A. Davison, J. W. Faller, S. J. Lippard, and S. M. Morehouse, *J. Am. Chem. Soc.*, **1966**, *88*, 4371. For a summary of early developments in the use of NMR to observe molecular rearrangement processes, see F. A. Cotton, *Inorg. Chem.*, **2002**, *41*, 643.

[92] C. Elschenbroich and A. Salzer, *Organometallics*, 2nd ed., VCH, New York, 1992.

EXAMPLE

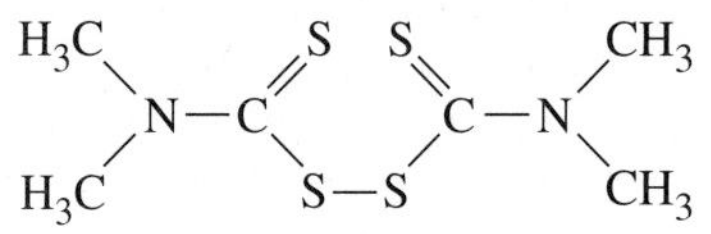

$[(C_5H_5)Mo(CO)_3]_2$ reacts with tetramethylthiuramdisulfide (tds) in refluxing toluene to give a molybdenum-containing product having the following characteristics:

> ^{1}H NMR: Two singlets, at δ 5.48 > (relative area = 5) and δ 3.18 (relative area = 6). (For comparison, $[(C_5H_5)Mo(CO)_3]_2$ has a single ^{1}H NMR peak at δ 5.30.)
>
> IR: Strong bands at 1950 and 1860 cm^{-1}.
>
> Mass spectrum: A pattern similar to the Mo isotope pattern with the most intense peak at m/e = 339. (The most abundant Mo isotope is ^{98}Mo.)

What is the most likely identity of this product?

The ^{1}H NMR singlet at δ 5.48 suggests retention of the C_5H_5 ligand (the chemical shift is a close match for the starting material). The peak at δ 3.18 is most likely due to CH_3 groups originating from the tds. The 5:6 ratio of hydrogens suggests a 1:2 ratio of C_5H_5 ligands to CH_3 groups.

IR shows two bands in the carbonyl region, indicating at least two COs in the product.

The mass spectrum makes it possible to pin down the molecular formula. Subtracting the molecular fragments believed to be present from the total mass:

Total mass:	339
Mass of Mo (from mass spectrum pattern)	–98
Mass of C_5H_5	–65
Mass of two COs	–56
Remaining mass	120

120 is exactly half the mass of tds; it corresponds to the mass of $S_2CN(CH_3)_2$, the dimethyldithiocarbamate ligand, which we have encountered in previous chapters. Therefore, the likely formula of the product is $(C_5H_5)Mo(CO)_2[S_2CN(CH_3)_2]$. This formula has the necessary 5:6 ratio of protons in two magnetic environments and should give rise to two C—O stretching vibrations (because the carbonyls would not be expected to be oriented at 180° angles with respect to each other in such a molecule).

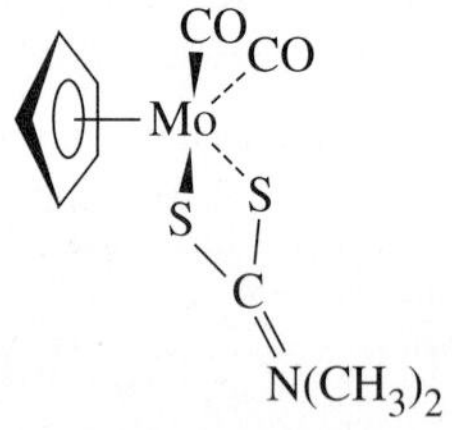

In practice, additional information is likely to be available to help characterize reaction products. For example, additional examination of the infrared spectrum in this case shows a moderately intense band at 1526 cm^{-1}, a common location for C—N stretching bands in dithiocarbamate complexes. Analysis of the fragmentation pattern of mass spectra may also provide useful information on molecular fragments.

EXAMPLE

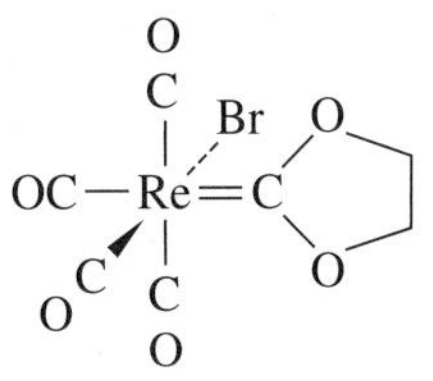

I

When a toluene solution containing **I** and excess triphenylphosphine is heated to reflux, first compound **II** is formed, and then compound **III**. **II** has infrared bands at 2038, 1958, and **III** at 1944 and 1860 cm^{-1}. ^{1}H and ^{13}C NMR data [δ values (relative area)] are as follows:

	I	*II*	*III*
^{1}H:	4.83 singlet	7.62, 7.41 multiplets (15)	7.70, 7.32 multiplets (15)
		4.19 multiplet (4)	3.39 singlet (2)
^{13}C:	224.31	231.02	237.19
	187.21	194.98	201.85
	185.39	189.92	193.83
	184.01	188.98	127.75–134.08 (several peaks)
	73.33	129.03–134.71 (several peaks)	68.80
		72.26	

Additional useful information: the ^{13}C signal of **I** at δ 224.31 is similar to the chemical shift of carbene carbons in similar compounds; the peaks between δ 184 and 202 correspond to carbonyls; and the peak at δ 73.33 is typical of CH_2CH_2 bridges in dioxycarbene complexes.

Identify **II** and **III**.

This is a good example of the utility of ^{13}C NMR. Both **II** and **III** have peaks with similar chemical shifts to the peak at δ 224.31 for **I**, suggesting that the carbene ligand is retained in the reaction. Similarly, **II** and **III** have peaks near δ 73.33, a further indication that the carbene ligand remains intact.

The ^{13}C peaks in the range δ 184 to 202 can be assigned to carbonyl groups. **II** and **III** show new peaks in the range δ 129 to 135. The most likely explanation is that the chemical reaction involves replacement of carbonyls by triphenylphosphines and that the new peaks in the 129 to 135 range are due to the phenyl carbons of the phosphines.

^{1}H NMR data are consistent with replacement of COs by phosphines. In both **II** and **III**, integration of the $-CH_2CH_2-$ peaks (δ 4.19 and 3.39, respectively) and the phenyl peaks (δ 7.32 to 7.70) give the expected ratios for replacement of one and two COs.

Finally, IR data are in agreement with these conclusions. In **II**, the three bands in the carbonyl region are consistent with the presence of three COs, either in a *mer* or a *fac* arrangement.[93] In **III**, the two C—O stretches correspond to two carbonyls *cis* to each other.

The chemical formulas of these products can now be written as follows:

$$\textbf{II}: ReBr(CO)_3(\overline{COCH_2CH_2O})(PPh_3)$$

$$\textbf{III}: \textit{cis}\text{-}ReBr(CO)_2(\overline{COCH_2CH_2O})(PPh_3)_2$$

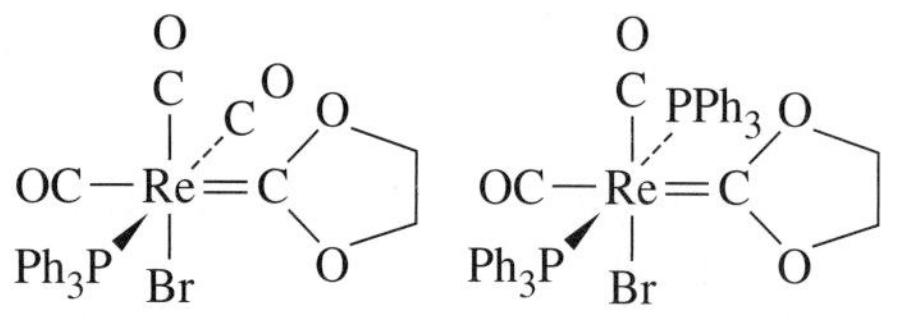

II **III**

▶ **Exercise 13.12** Using ^{13}C NMR data, determine whether **II** is more likely the *fac* or *mer* isomer.[94]

General References

Much information on organometallic compounds is included in two general inorganic references, N. N. Greenwood and A. Earnshaw, *Chemistry of the Elements*, 2nd ed., Butterworth Heinemann, Oxford, 1997, and F. A. Cotton, G. Wilkinson, C. A. Murillo, and M. Bochman, *Advanced Inorganic Chemistry*, 6th ed., Wiley InterScience, New York, 1999. G. O. Spessard and G. L. Miessler, *Organometallic Chemistry*, Oxford University Press, New York, 2010, C. Elschenbroich, *Organometallics*, 3rd ed., Wiley-VCH, Wiesbaden, Germany, 2005, and J. P. Collman, L. S. Hegedus, J. R. Norton, and R. G. Finke, *Principles and Applications of Organotransition Metal Chemistry*, University Science Books, Mill Valley, CA, 1987, provide extensive discussion, with numerous references, of many additional types of organometallic compounds in addition to those discussed in this chapter. The most comprehensive references on organometallic chemistry are the multiple-volume sets edited by G. Wilkinson and F. G. A. Stone, *Comprehensive Organometallic Chemistry*, Pergamon Press, Oxford, 1982, and by E. W. Abel, F. G. A. Stone, and G. Wilkinson, *Comprehensive Organometallic Chemistry II*, Pergamon Press, Oxford, 1995. Each of these sets has an extensive listing of references on organometallic compounds that have been structurally characterized by X-ray, electron, or neutron diffraction. A useful reference to literature sources on the synthesis, properties, and reactions of specific organometallic compounds is J. Buckingham and J. E. Macintyre, editors, *Dictionary of Organometallic Compounds*, Chapman and Hall, London, 1984, to which supplementary volumes have also been published. The series *Advances in Organometallic Chemistry*, Academic Press, San Diego, provides valuable review articles on a variety of organometallic topics.

Problems

13.1 Which of the following obey the 18-electron rule?

a. $Fe(CO)_5$
b. $[Rh(bipy)_2Cl]^+$
c. $(\eta^5\text{-}Cp^*)Re(=O)_3$, where $Cp^* = C_5(CH_3)_5$
d. $Re(PPh_3)_2Cl_2N$
e. $Os(CO)(\equiv CPh)(PPh_3)_2Cl$

[93] In an octahedral complex of formula *fac*-$ML_3(CO)_3$ (having C_{3v} symmetry), only two carbonyl stretching bands are expected if all ligands L are identical. However, in this case, there are three different ligands in addition to CO; the point group is C_1, and three bands are expected.

[94] G. L. Miessler, S. Kim, R. A. Jacobson, and R. A. Angelici, *Inorg. Chem.*, **1987**, *26*, 1690.

13.2 Which of the following square-planar complexes have 16-electron valence configurations?

a. $Ir(CO)Cl(PPh_3)_2$
b. $RhCl(PPh_3)_3$
c. $[Ni(CN)_4]^{2-}$
d. *cis*-$PtCl_2(NH_3)_2$

13.3 On the basis of the 18-electron rule, identify the first-row transition metal for each of the following:

a. $[M(CO)_7]^+$
b. $H_3CM(CO)_5$
c. $M(CO)_2(CS)(PPh_3)Br$
d. $[(\eta^3\text{-}C_3H_3)(\eta^5\text{-}C_5H_5)M(CO)]^-$
e. $(OC)_5M{=}C(OCH_3)(C_6H_5)$
f. $[(\eta^4\text{-}C_4H_4)(\eta^5\text{-}C_5H_5)M]^+$
g. $(\eta^3\text{-}C_3H_5)(\eta^5\text{-}C_5H_5)M(CH_3)(NO)$
h. $[M(CO)_4I(\text{diphos})]^-$
(diphos = 1,2-bis(diphenylphosphino)ethane)

13.4 Determine the metal–metal bond order consistent with the 18-electron rule for the following:

a. $[(\eta^5\text{-}C_5H_5)Fe(CO)_2]_2$
b. $[(\eta^5\text{-}C_5H_5)Mo(CO)_2]^{2-}$

13.5 Identify the most likely second-row transition metal for each of the following:

a. $[M(CO)_3(NO)]^-$
b. $[M(PF_3)_2(NO)_2]^+$ (contains linear M—N—O)
c. $[M(CO)_4(\mu_2\text{-}H)]_3$
d. $M(CO)(PMe_3)_2Cl$ (square-planar complex)

13.6 On the basis of the 18-electron rule, determine the expected charge on the following:

a. $[Co(CO)_3]^z$
b. $[Ni(CO)_3(NO)]^z$ (contains linear M—N—O)
c. $[Ru(CO)_4(GeMe_3)]^z$
d. $[(\eta^3\text{-}C_3H_5)V(CNCH_3)_5]^z$
e. $[(\eta^5\text{-}C_5H_5)Fe(CO)_3]^z$
f. $[(\eta^5\text{-}C_5H_5)_3Ni_3(\mu_3\text{-}CO)_2]^z$

13.7 Determine the unknown quantity:

a. $[(\eta^5\text{-}C_5H_5)W(CO)_x]_2$ (has W—W single bond)
b. $Br(CO)_xRe{=}C$ (ring with two O) $3]$
c. $[(CO)_3Ni{-}Co(CO)_3]^z$
d. $[Ni(NO)_3(SiMe_3)]^z$ (contains linear M—N—O)
e. $[(\eta^5\text{-}C_5H_5)Mn(CO)_x]_2$ (has Mn=Mn bond)

13.8 Determine the unknown quantity:

a. The hapticity of the lower ring in the mixed superphane (See S. Gath, R. Gleiter, F. Rominger, and C. Bleiholder, *Organometallics*, **2007**, *26*, 644.)

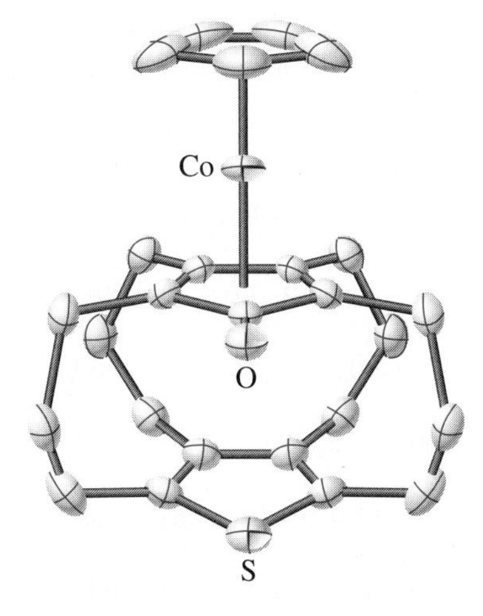

b. The *third* row 16-electron transition metal **M** on the left and the *first* row 16-electron transition metal **M′** on the right (See M. Tamm, A. Kunst, T. Bannenberg, S. Randoll, and P. G. Jones, *Organometallics* **2007**, *26*, 417.)

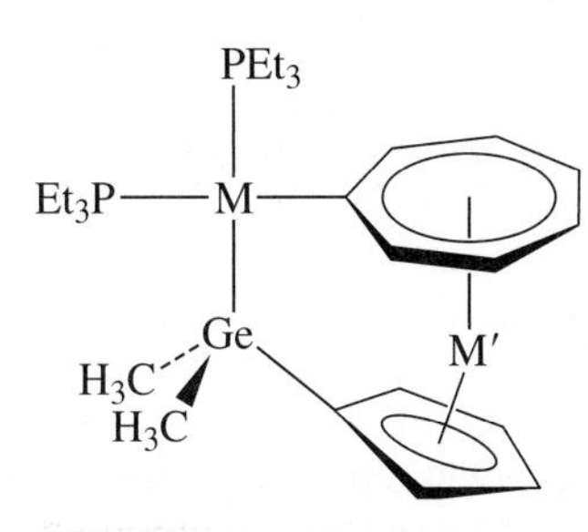

13.9 Nickel tetracarbonyl, $Ni(CO)_4$, is an 18-electron species. Using a qualitative molecular orbital diagram, explain the stability of this 18-valence electron molecule. (See A. W. Ehlers, S. Dapprich, S. V. Vyboishchikov, and G. Frenking, *Organometallics*, **1996**, *15*, 105.)

13.10 The Re—O stretching vibration in $Re(^{16}O)I(HC{\equiv}CH)_2$ is at 975 cm^{-1}. Predict the position of the Re—O stretching band in $Re(^{18}O)I(HC{\equiv}CH)_2$. (See J. M. Mayer, D. L. Thorn, and T. H. Tulip, *J. Am. Chem. Soc.*, **1985**, *107*, 7454.)

13.11 The compound $W(O)Cl_2(CO)(PMePh_2)_2$ has ν (CO) at 2006 cm^{-1}. Would you predict ν (CO) for $W(S)Cl_2(CO)(PMePh_2)_2$ to be at higher or lower energy? Explain briefly. (See J. C. Bryan, S. J. Geib, A. L. Rheingold, and J. M. Mayer, *J. Am. Chem. Soc.*, **1987**, *109*, 2826.)

13.12 The vanadium–carbon distance in $V(CO)_6$ is 200 pm but only 193 pm in $[V(CO)_6]^-$. Explain.

13.13 Describe, using sketches, how the following ligands can act as both σ donors and π acceptors:
a. CN^-
b. $P(CH_3)_3$
c. SCN^-

13.14 **a.** Account for the following trend in IR frequencies:

$[Cr(CN)_5(NO)]^{4-}$	ν (NO) = 1515 cm^{-1}
$[Mn(CN)_5(NO)]^{3-}$	ν (NO) = 1725 cm^{-1}
$[Fe(CN)_5(NO)]^{2-}$	ν (NO) = 1939 cm^{-1}

b. The ion $[RuCl(NO)_2(PPh_3)_2]^+$ has N—O stretching bands at 1687 and 1845 cm^{-1}. The C—O stretching bands of dicarbonyl complexes typically are much closer in energy. Explain.

13.15 Sketch the π molecular orbitals for the following:
a. CO_2
b. 1,3,5-Hexatriene
c. Cyclobutadiene, C_4H_4
d. *Cyclo*-C_7H_7

13.16 For the hypothetical molecule $(\eta^4\text{-}C_4H_4)Mo(CO)_4$:
a. Assuming C_{4v} geometry, predict the number of infrared-active C—O bands.
b. Sketch the π molecular orbitals of cyclobutadiene. For each, indicate which *s*, *p*, and *d* orbitals of Mo are of suitable symmetry for interaction. (Hint: Assign the *z* axis to be collinear with the C_4 axis.)

13.17 Using the D_{5h} character table in Appendix C:
a. Assign symmetry labels (labels of irreducible representations) for the group orbitals shown in Figure 13.27.
b. Assign symmetry labels for the atomic orbitals of Fe in a D_{5h} environment.
c. Verify that the orbital interactions for ferrocene shown in Figure 13.28 are between atomic orbitals of Fe and group orbitals of matching symmetry.

13.18 Dibenzenechromium, $(\eta^4\text{-}C_6H_6)_2Cr$, is a sandwich compound having two parallel benzene rings in an eclipsed conformation. For this molecule,
a. Sketch the π orbitals of benzene.
b. Sketch the group orbitals, using π orbitals of the two benzene rings.
c. For each of the 12 group orbitals, identify the Cr orbital(s) of suitable symmetry for interaction.
d. Sketch an energy level diagram of the molecular orbitals.

13.19 Predict the number of infrared-active C—O stretching vibrations for $W(CO)_3(\eta^6\text{-}C_6H_6)$ assuming C_{3v} geometry.

13.20 In this chapter, the assertion was made that highly symmetric binary carbonyls of T_d and O_h symmetry should show only a single C—O stretching band in the infrared. Check this assertion by analyzing the C—O vibrations of $Ni(CO)_4$ and $Cr(CO)_6$ by the symmetry method described in Chapter 4.

13.21 $Mn_2(CO)_{10}$ and $Re_2(CO)_{10}$ have D_{4d} symmetry. How many IR-active carbonyl stretching bands would you predict for these compounds?

13.22 When molybdenum hexacarbonyl is refluxed in butyronitrile, C_3H_7CN, product **X** is formed first. Continued reflux converts **X** to **Y**, and very long reflux (several days) converts **Y** to **Z**. However, even refluxing for several weeks does not convert compound **Z** into another product. In addition, in each reaction step, a colorless gas is liberated.
The following infrared bands are observed (cm^{-1}):

X:		Y:		Z:	
	2077		2107		1910
	1975		1898		1792
	1938		1842		

a. Propose structures of **X**, **Y**, and **Z**. Where more than one isomer is possible, determine the isomer that is the best match for the infrared data. (Note: a weak band in **Y** may be obscured by other bands.)
b. Account for the trend in the position of the infrared bands in the sequence **X** → **Y** → **Z**.
c. Suggest why **Z** does not react further when refluxed in butyronitrile. (See G. J. Kubas, *Inorg. Chem.* **1983**, *22*, 692.)

13.23 The infrared spectra of *trans*- and *cis*-$[Fe(CO)_2(CN)_4]^{2-}$ and of $[Fe(CO)(CN)_5]^{3-}$ are shown in the figure below.

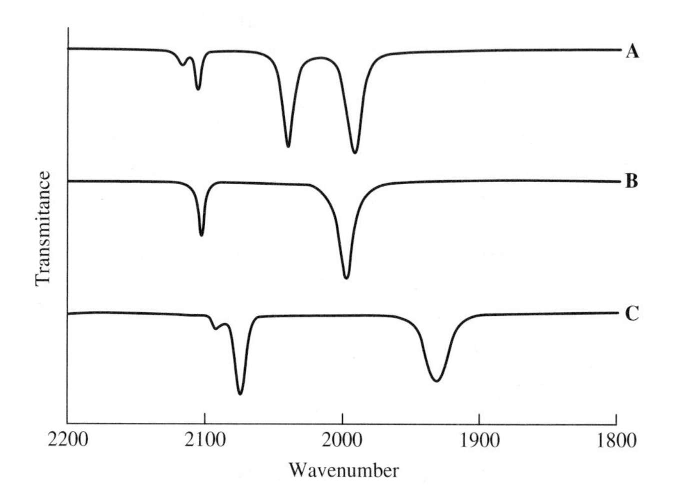

a. Which stretching bands occur at lower energy, those for the CO ligand or those for the CN^- ligand? Explain.
b. How many C—O and C—N stretching bands would you predict for each of these complexes on the basis of their symmetry? Match the complexes with their spectra. (Reproduced with permission from S. M. Contakes, S. C. N. Hsu, T. B. Rauchfuss, and S. R. Wilson, *Inorg. Chem.*, **2002**, *41*, 1670. © American Chemical Society.)

13.24 Samples of $Fe(CO)(PF_3)_4$ show two carbonyl stretching bands, at 2038 and 2009 cm^{-1}.

a. How is it possible for this compound to exhibit two carbonyl bands?

b. $Fe(CO)_5$ has carbonyl bands at 2025 and 2000 cm^{-1}. Would you place PF_3 above or below CO in the spectrochemical series? Explain briefly. (See H. Mahnke, R. J. Clark, R. Rosanske, and R. K. Sheline, *J. Chem. Phys.*, **1974**, *60*, 2997.)

13.25 Evidence has been reported for two isomers of $Ru(CO)_2(PEt_2)_3$, one with the carbonyls occupying axial positions in a trigonal-bipyramidal structure and the other with the carbonyls occupying equatorial positions. Could infrared spectroscopy distinguish between these isomers? How many carbon–oxygen stretching vibrations would be expected for each? (See M. Ogasawara, F. Maseras, N. Gallego-Planas, W. E. Streib, O. Eisenstein, and K. G. Caulton, *Inorg. Chem.*, **1996**, *35*, 7468.)

13.26 Account for the observation that $[Co(CO)_3(PPh_3)_2]^+$ has only a single carbonyl stretching frequency.

13.27 In addition to the hexacarbonyl complexes shown in Section 13.4.1, the ion $[Ir(CO)_6]^{3+}$ has been reported. Predict the position of the carbonyl stretching vibration in this complex. (See C. Bach, H. Willner, C. Wang, S. J. Rettig, J. Trotter, and F. Aubke, *Angew. Chem., Int. Ed.*, **1996**, *35*, 1974.)

13.28 Pathways to a variety of homoleptic transition-metal carbonyl cations have now been developed.

a. Three such cations are $[Hg(CO)_2]^{2+}$, $[Pt(CO)_4]^{2+}$, and $[Os(CO)_6]^{2+}$. Predict which of these has the lowest energy carbon–oxygen stretching vibration in the infrared. (See H. Willner and F. Aubke, *Angew. Chem., Int. Ed.*, **1997**, *36*, 2402.)

b. The cation $[\{Pt(CO)_3\}_2]^{2+}$ is believed to have the structure shown, with D_{2d} symmetry. Predict the number of carbon–oxygen stretching bands observable in the infrared for this ion, and predict the approximate region in the spectrum where these bands might be observed.

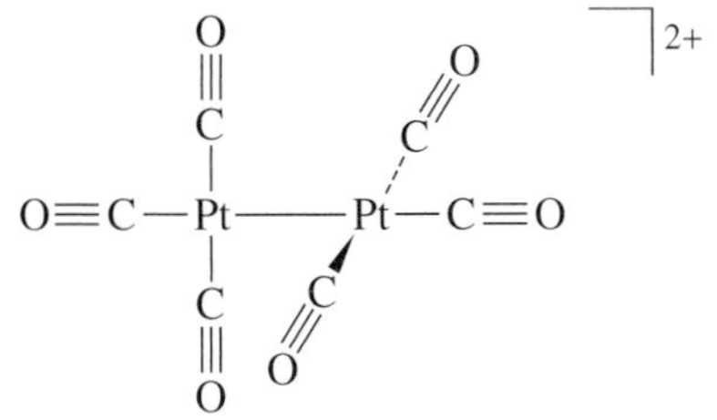

13.29 The Raman spectrum of $Mo(CO)_6$ in pyridine has bands at 2119 and 2015 cm^{-1}. Microwave irradiation of this solution yielded three products with the following Raman bands:

Compound **J**: 2071, 1981 cm^{-1}
Compound **K**: 1892 cm^{-1}
Compound **L**: 1600 cm^{-1}

a. Determine the irreducible representations matching the Raman-active bands of $Mo(CO)_6$.

b. On the basis of the Raman data, propose structures of **J**, **K**, and **L**.
(See T. M. Barnard and N. E. Leadbeater, *Chem. Commun.*, **2006**, 3615.)

13.30 One of the first thionitrosyl complexes to be reported was $(\eta^5\text{-}C_5H_5)Cr(CO)_2(NS)$. This compound has carbonyl bands at 1962 and 2033 cm^{-1}. The corresponding bands for $(\eta^5\text{-}C_5H_5)Cr(CO)_2(NO)$ are at 1955 and 2028 cm^{-1}. On the basis of the IR evidence, is NS behaving as a stronger or weaker pi acceptor in these compounds? Explain briefly. (See T. J. Greenough, B. W. S. Kolthammer, P. Legzdins, and J. Trotter, *Chem. Commun. (Cambridge)*, **1978**, 1036.)

13.31 The ^{14}N and ^{15}N derivatives of $TpOs(NS)Cl_2$ [Tp = hydrotris(1-pyrazolyl)borate, a tridentate ligand] have been prepared. The ^{14}N derivative has a nitrogen–sulfur stretch at 1284 cm^{-1}. Predict the N–S stretch for the ^{15}N derivative. (See T. J. Crevier, S. Lovell, J. M. Mayer, A. L. Rheingold, and I. A. Guzei, *J. Am. Chem. Soc.*, **1998**, *120*, 6607.)

13.32 The compound $[Ru(CO)_6][Sb_2F_{11}]_2$ has a strong infrared band at 2199 cm^{-1}. The spectrum of $[Ru(^{13}CO)_6][Sb_2F_{11}]_2$ has also been reported. Would you expect the band observed at 2199 cm^{-1} for the ^{12}C compound to be shifted to higher or lower energy for the analogous ^{13}C compound? (See C. Wang, B. Bley, G. Balzer-Jöllenbeck, A. R. Lewis, S. C. Siu, H. Willner, and F. Aubke, *Chem. Commun. (Cambridge)*, **1995**, 2071.)

13.33 Predict the products of the following reactions:

a. $Mo(CO)_6 + Ph_2P\text{—}CH_2\text{—}PPh_2 \xrightarrow{\Delta}$

b. $(\eta^5\text{-}C_5H_5)(\eta^1\text{-}C_3H_5)Fe(CO)_2 \xrightarrow{h\nu}$

c. $(\eta^5\text{-}C_5Me_5)Rh(CO)_2 \xrightarrow{\Delta}$ (dimeric product, contains one CO per metal)

d. $V(CO)_6 + NO \longrightarrow$

e. $W(CO)_5[C(C_5H_5)(OC_2H_5)] + BF_3 \longrightarrow$

f. $[(\eta^5\text{-}C_5H_5)Fe(CO)_2]_2 + Al(C_2H_5)_3 \longrightarrow$

13.34 Complexes of formula $Rh(CO)(phosphine)_2Cl$ have the C—O stretching bands shown below. Match the infrared bands with the appropriate phosphine. Phosphines: $P(p\text{-}C_6H_4F)_3$, $P(p\text{-}C_6H_4Me)_3$, $P(t\text{-}C_4H_9)_3$, $P(C_6F_5)_3$
$\upsilon(CO)$, cm^{-1}: 1923, 1965, 1984, 2004

13.35 For each of the following sets, which complex would be expected to have the highest C—O stretching frequency?

a.	$Fe(CO)_5$	$Fe(CO)_4(PF_3)$	$Fe(CO)_4(PCl_3)$	$Fe(CO)_4(PMe_3)$
b.	$[Re(CO)_6]^+$	$W(CO)_6$	$[Ta(CO)_6]^-$	
c.	$Mo(CO)_3(PCl_3Ph)_3$	$Mo(CO)_3(PCl_2Ph)_3$	$Mo(CO)_3(PPh_3)_3$	$Mo(CO)_3\ py_3$ (py = pyridine)

13.36 Arrange the following complexes in order of the expected frequency of their $\upsilon(CO)$ bands. (See M. F. Ernst and D. M. Roddick, *Inorg. Chem.*, **1989**, *28*, 1624.)

$$Mo(CO)_4(F_2PCH_2CH_2PF_2)$$

$$Mo(CO)_4[(C_6F_5)_2PCH_2CH_2P(C_6F_5)_2]$$

$$Mo(CO)_4(Et_2PCH_2CH_2PEt_2)\ (Et = C_2H_5)$$

$$Mo(CO)_4(Ph_2PCH_2CH_2PPh_2)\ (Ph = C_6H_5)$$

$$Mo(CO)_4[(C_2F_5)PCH_2CH_2P(C_2H_5)]$$

13.37 Free N_2 has a stretching vibration (not observable by IR; why?) at 2331 cm^{-1}. Would you expect the stretching vibration for coordinated N_2 to be at higher or lower energy? Explain briefly.

13.38 The ^1H NMR spectrum of the carbene complex shown below shows two peaks of equal intensity at 40 °C. At –40 °C, the NMR shows four peaks, two of a lower intensity and two of a higher intensity. The solution may be warmed and cooled repeatedly without changing the NMR properties at these temperatures. Account for this NMR behavior.

$(OC)_5Cr{=}C(OCH_3)(CH_3)$

13.39 The ^1H NMR spectrum of $(C_5H_2)Fe(CO)_2$ shows two peaks of equal area at room temperature but has four resonances of relative intensity 5:2:2:1 at low temperatures. Explain. (See C. H. Campbell and M. L. H. Green, *J. Chem. Soc., A*, **1970**, 1318.)

13.40 Of the compounds $Cr(CO)_5(PF_3)$ and $Cr(CO)_5(PCl_3)$, which would you expect to have
a. The shorter C—O bonds?
b. The higher energy Cr—C stretching bands in the infrared spectrum?

13.41 Select the best choice for each of the following:
a. Higher N—O stretching frequency:
$[Fe(NO)(mnt)_2]^-$
$[Fe(NO)(mnt)_2]^{2-}$
$mnt^{2-} = [(S)(S)C{=}C(CN)(CN)]^{2-}$
b. Longest N—N bond:
N_2
$(CO)_5\ Cr{:}N{\equiv}N$
$(CO)_5\ Cr{:}N{\equiv}N{:}Cr(CO)_5$
c. Shorter Ta—C distance in $(\eta^5\text{-}C_5H_5)_2Ta(CH_2)(CH_3)$:
Ta—CH_2
Ta—CH_3
d. Shortest Cr—C distance:
$Cr(CO)_6$
Cr—CO in *trans*-$Cr(CO)_4I(CCH_3)$
Cr—CCH_3 in *trans*-$Cr(CO)_4I(CCH_3)$
e. Lowest C—O stretching frequency:
$Ni(CO)_4$
$[Co(CO)_4]^-$
$[Fe(CO)_4]^{2-}$

13.42 A solution of blue $Mo(CO)_2(PEt_3)_2Br_2$ was treated with a tenfold excess of 2-butyne to give **X**, a dark green product. **X** had bands in the ^1H NMR at δ 0.90 (relative area = 3), 1.63 (2), and 3.16 (1). The peak at 3.16 was a singlet at room temperature but split into two peaks at temperatures below –20 °C. ^{31}P NMR showed only a single resonance. IR showed a single strong band at 1950 cm^{-1}. Molecular weight determinations suggest that **X** has a molecular weight of 580±15. Suggest a structure for **X**, and account for as much of the data as possible. (See P. B. Winston, S. J. Nieter Burgmayer, and J. L. Templeton, *Organometallics*, **1983**, *2*, 168.)

13.43 Photolysis at –78 °C of $[(\eta^5\text{-}C_5H_5)Fe(CO)_2]_2$ results in the loss of a colorless gas and the formation of an iron-containing product having a single carbonyl band at 1785 cm^{-1} and containing 14.7 percent oxygen by mass. Suggest a structure for the product.

13.44 Nickel carbonyl reacts with cyclopentadiene to produce a red diamagnetic compound of formula $NiC_{10}H_{12}$. The ^1H NMR spectrum of this compound shows four different types of hydrogen; integration gives relative areas of 5:4:2:1, with the most intense peak in the aromatic region. Suggest a structure of $NiC_{10}H_{12}$ that is consistent with this NMR spectrum.

13.45 The carbonyl carbon—molybdenum—carbon angle in $Cp(CO)_2Mo[\mu\text{-}S_2C_2(CF_3)_2]_2MoCp$ ($Cp = \eta^5\text{-}C_5H_5$) is 76.05°. Calculate the ratio of intensities $I_{symmetric}/I_{antisymmetric}$ expected for the C—O stretching bands of this compound. (See K. Roesselet, K. E. Doan, S. D. Johnson, P. Nicholls, G. L. Miessler, R. Kroeker, and S. H. Wheeler, *Organometallics*, **1987**, *6*, 480.)

13.46 The complex (π-C_4BNH_6)$Cr(CO)_3$ has recently been reported, the first example of the 1,2-dihydro-1,2-azaborine ligand. It has strong absorptions at 1898 and 1975 cm^{-1}, in comparison with 1892 and 1972 cm^{-1} for (π-C_6H_6)$Cr(CO)_3$. The C_4BNH_6 complex has carbon–carbon distances of 1.393, 1.421, and 1.374 Å, in order, in the ring.

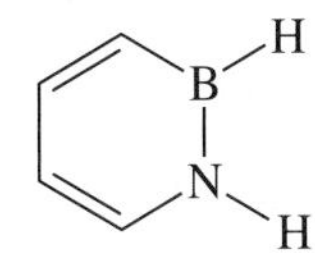

1,2-dihydro-1,2-azaborine

a. Is the C_4BNH_6 ligand a (slightly) stronger or weaker acceptor than C_6H_6? Explain.

b. Account for the differences in the C—C distances in the ring. (See A. J. V. Marwitz, M. H. Matus, L. N. Zakharov, D. A. Dixon, and S.-Y. Liu, *Angew. Chem., Int. Ed.*, **2009**, *48*, 973.)

13.47 Reaction of Ir complex **A** with C_{60} gave a black solid residue **B** with the following spectral characteristics: mass spectrum: $M^+ = 1056$; ^1H NMR: δ 7.65 ppm (multiplet, 2H), 7.48 (multiplet, 2H), 6.89 (triplet, 1H), and 5.97 (doublet, 2H); IR: $\nu(CO) = 1998\ cm^{-1}$.

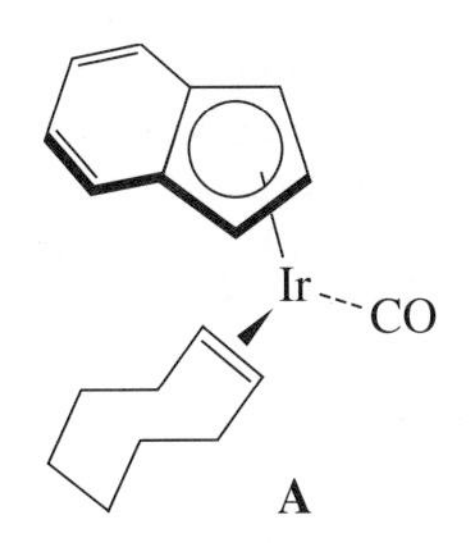

a. Propose a structure for **B**.

b. The carbonyl stretch of **A** was reported at 1954 cm^{-1}. How does the electron density at Ir change in going from **A** to **B**?

c. When **B** was treated with PPh_3, a new complex **C** formed rapidly, along with some C_{60}. What is a likely structure of **C**? (See R. S. Koefod, M. F. Hudgens, and J. R. Shapley, *J. Am. Chem. Soc.*, **1991**, *113*, 8957.)

13.48 Reaction of $Mo(CO)_3(CH_3CN)_3$ with $NaCp^N$ (shown) in THF, followed by reaction with the tetrameric complex $[Cu(PPh_3)Cl]_4$, yields a yellow product having the following characteristics: IR (in THF): strong absorbances at 1906, 1808, 1773 cm^{-1}; ^1H NMR: δ 7.50–7.27 (multiplet, 15 H), 4.64 (apparent triplet, 2H), 4.52 (apparent triplet, 2H), 2.44– 2.39 (multiplet, 4H), 2.19 (singlet, 6H); Elemental analysis: 60.29 % C, 4.82 % H, 2.40 % N by mass.

$NaCp^N$: Na^+ (−) NMe_2

Propose a structure of this product. (See P. J. Fischer, A. P. Heerboth, Z. R. Herm, and B. E. Kucera, *Organometallics*, **2007**, *26*, 6669.)

13.49 Although it is a simple molecule, diisocyanomethane, $H_2C(NC)_2$, was not isolated and characterized by X-ray crystallography until rather recently. This compound has been used in organometallic synthesis as follows: (η^5-C_5H_5)$Mn(CO)_3$ was dissolved in tetrahydrofuran and photolyzed, liberating some CO and forming compound **Q**. At –40 °C, a solution of $H_2C(NC)_2$ in dichloromethane was added to a solution of **Q**. Column chromatography of the resulting solution led to the isolation of compound **R**, which had the following characteristics: ^1H NMR (in CD_2Cl_2): δ 4.71 (relative area 5) 5.01, (2); ^{13}C NMR: δ 50.1, 83.4, 162.0, 210.5, 228.1; IR: 2147, 2086, 2010, 1903 cm^{-1}. Suggest structures for **Q** and **R**. (See J. Buschmann, R. Bartolmäs, D. Lentz, P. Luger, I. Neubert, and M. Röttger, *Angew. Chem., Intl. Ed.*, **1997**, *36*, 2372.)

13.50 In solution, the cyclopentadienyl tricarbonyl dimers $[CpMo(CO)_3]_2$ and $[CpW(CO)_3]_2$ react to form the heterobimetallic complex $Cp(CO)_3Mo{-}W(CO)_3Cp$. However, the reaction does not go to completion; a mixture results in which the $[CpMo(CO)_3]_2$ and $[CpW(CO)_3]_2$ are in equilibrium with the mixed metal compound. The abundance of the three organometallic complexes is governed statistically by the number of $CpMo(CO)_3$ and $CpW(CO)_3$ fragments present. If 0.00100 mmol of $[CpMo(CO)_3]_2$ and 0.00200 mmol of $[CpW(CO)_3]_2$ are dissolved in toluene until equilibrium is achieved, calculate the amounts of the three organometallic complexes in the equilibrium solution. (See T. Madach and H. Vahrenkamp, *Z. Naturforsch.*, **1979**, *34b*, 573.)

13.51 Addition of $BH_3\cdot THF$ to a slurry of $[K([15]crown\text{-}5)_2]_2[Ti(CO)_6]$ in tetrahydrofuran at –60 °C yielded an air-sensitive red solution from which an anion **Z** was isolated. **Z** had strong infrared peaks at 1945 and 1811 cm^{-1}. (Note: $[Ti(CO)_4(\eta^5\text{-}C_5H_5)]^-$ has bands at 1921 and 1779 cm^{-1}.) Other peaks are observed at 2495, 2132, and 2058 cm^{-1}; these are in the spectral region where B—H stretches commonly occur. The ^1H NMR spectrum at –95 °C showed broad singlets of relative intensity 3:1; these peaks became more complex at higher temperature. Propose a formula and structure for **Z**. (See P. J. Fischer, V. G. Young, Jr., and J. E. Ellis, *Angew. Chem., Int. Ed.*, **2000**, *39*, 189.)

13.52 When a solution containing $(C_5Me_5)_2Os_2Br_4$ and $LiAlH_4$ is stirred at low temperature in diethyl ether, followed by addition of a small amount of ethanol, a white product is formed that can be isolated by sublimation. Analysis of the white product, an 18-electron complex, provided the following information:

Elemental analysis: 36.6 % C, 6.07 % H

^{1}H NMR: δ 2.02 (s, relative area-3), –11.00 (s, relative area-1)

^{13}C NMR: δ 2.02 (s), 94.2 (s)

IR: the region where Os–H stretches most likely occur is shown.

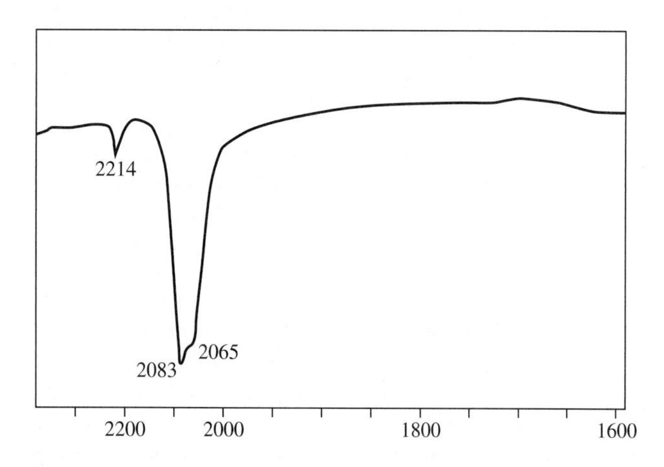

Propose the structure of the product, and account for as much of the data as possible. (See C. L. Gross and G. S. Girolami, *Organometallics*, **2007**, *26*, 160.)

Use molecular modeling software for the following problems:

13.53 **a.** Generate and display the π and π^* orbitals of the cyclopentadienyl group, C_5H_5. Compare the results with the diagrams in Figure 13.22. Identify the nodes that cut through the plane of the atoms.

b. Generate and display the molecular orbitals of ferrocene. Identify the molecular orbitals that result from interactions of π orbitals of the C_5H_5 ligands with iron. Locate orbitals that show interactions of metal d_{xz} and d_{yz} with the ligands, and compare with the diagrams preceding Figure 13.27.

c. Compare the relative energies of the molecular orbitals with the molecular orbitals shown in Figure 13.28. How do the relative energies of the orbitals in the box in Figure 13.28 compare with the results of your calculations?

d. The d_{z^2} orbital is often described as essentially nonbonding in ferrocene. Do your results support this designation?

13.54 Generate and display the molecular orbitals of the anion of Zeise's salt, shown in Figure 13.4. Identify the molecular orbitals that show how the π and π^* orbitals of ethylene interact with orbitals on platinum.

13.55 Generate and display the molecular orbitals of $Cr(CO)_6$. Identify the t_{2g} and $e_g{}^*$ orbitals (Figure 13.8). Verify that there are three equivalent and degenerate t_{2g} orbitals and that there are two degenerate $e_g{}^*$ orbitals. Why do the $e_g{}^*$ orbitals have different shapes?

13.56 Suppose a transition metal is bonded to a *cyclo*-C_3H_3 ligand as shown below.

a. Sketch each of the π molecular orbitals of η^3-C_3H_3.

b. For each of these orbitals, determine which *s*, *p*, and *d* orbitals of the metal are of appropriate symmetry for interaction.

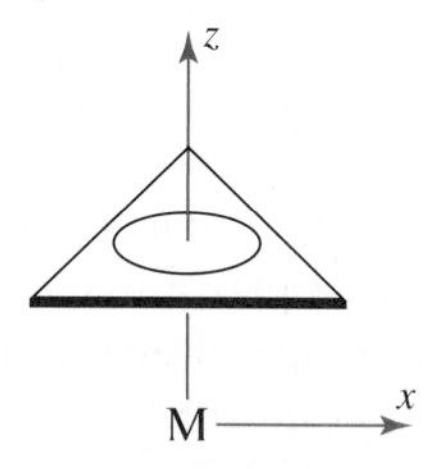

c. Using molecular modeling software, generate the molecular orbitals of *cyclo*-C_3H_3, and compare the π orbitals with your sketches in **a**.

d. Generate and display the molecular orbitals of the theoretical sandwich complex $[(\eta^3\text{-}C_3H_3)_2Ni]^{2-}$. Observe the orbitals showing overlap between lobes of the rings and the central Ni, and compare these with the orbitals you indicated for metal–ring interactions in **b**.

13.57 $(\eta^4\text{-}C_4H_4)_2Ni$ has apparently not been synthesized. However, calculations predict reasonable stability and eclipsed geometry for this molecule, with the rings parallel. (See Q. Li and J. Guan, *J. Phys. Chem. A* **2003**, *107*, 8584.)

a. Sketch the group orbitals, using the π orbitals of the cyclobutadiene rings.

b. For each of the group orbitals, identify the Ni orbitals of suitable symmetry for interaction.

c. Using the *p* orbitals of the ligands as a basis, construct a (reducible) representation. Reduce it to its irreducible components, and match these with appropriate orbitals on the central Ni.

d. Generate and display the molecular orbitals of $(\eta^4\text{-}C_4H_4)_2Ni$. Observe the orbitals showing overlap between lobes of the rings and the central Ni, and compare with your results from **c**.

14

Organometallic Reactions and Catalysis

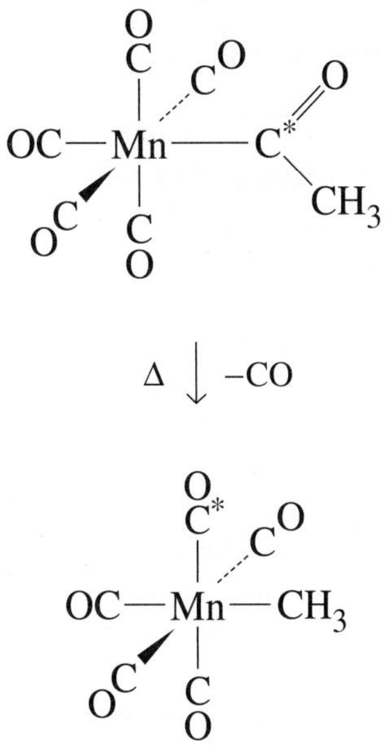

Organometallic compounds undergo a rich variety of reactions, comparable in diversity to the reactions of organic molecules. These may involve loss or gain of ligands (or both), molecular rearrangement, formation or breaking of metal–metal bonds, or reactions at the ligands themselves. Often, reaction mechanisms involve multiple steps and, frequently, reactions yield not one but a variety of products. Sequences of reactions may be combined into catalytic cycles that may be useful, in some cases commercially. In this chapter, we will not attempt to cover all possible types of organometallic reactions but will concentrate on those that have proved most common and useful, particularly for synthetic and catalytic processes. We will discuss organometallic reactions according to the following outline:

I. Reactions involving gain or loss of ligands
   A. Ligand dissociation and substitution
   B. Oxidative addition
   C. Reductive elimination
   D. Nucleophilic displacement

II. Reactions involving modification of ligands
   A. Insertion
   B. Carbonyl insertion (alkyl migration)
   C. Hydride elimination
   D. Abstraction

14.1 REACTIONS INVOLVING GAIN OR LOSS OF LIGANDS

Some of the most important reactions of organometallic compounds involve a change in coordination number of the metal by a gain or loss of ligands. If the formal oxidation state of the metal is retained, these reactions are considered addition or dissociation reactions; if the formal oxidation state is changed, they are termed oxidative additions or reductive eliminations.

Type of Reaction	Change in Coordination Number	Change in Formal Oxidation State of Metal
Addition	Increase	None
Dissociation	Decrease	None
Oxidative addition	Increase	Increase
Reductive elimination	Decrease	Decrease

In classifying these reactions, it will frequently be necessary to determine formal oxidation states of the metals in organometallic compounds. In general, method A (the donor pair method) described in Chapter 13 can be used in assigning oxidation states. Examples will be given later in this chapter in the discussion of oxidative addition reactions.

We will first consider ligand dissociation reactions. When coupled with addition reactions, dissociation reactions can be useful synthetically, providing an avenue to replace ligands such as carbon monoxide and phosphines by other ligands.

14.1.1 Ligand Dissociation and Substitution

CO DISSOCIATION Chapter 13 gave a brief introduction to carbonyl dissociation reactions, in which CO may be lost thermally or photochemically. Such a reaction may result in rearrangement of the remaining molecule or replacement of CO by another ligand, as shown here:

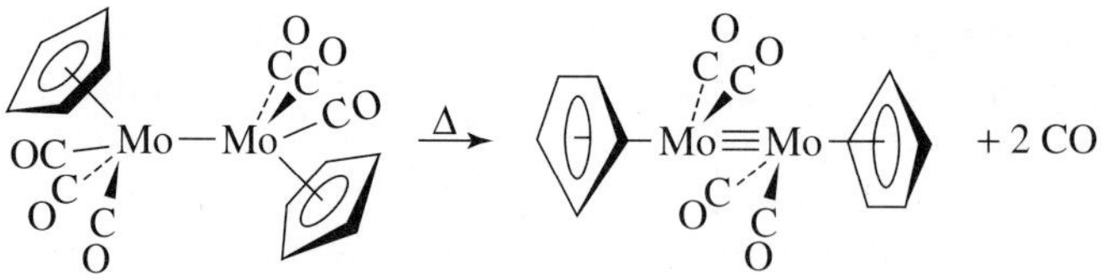

$$Fe(CO)_5 + P(CH_3)_3 \xrightarrow{\Delta} Fe(CO)_4(P(CH_3)_3) + CO$$

The second type of reaction, involving ligand replacement, is an important way to introduce new ligands into complexes and deserves further discussion.

Most thermal reactions involving replacement of CO by another ligand, L, have rates that are independent of the concentration of L; they are first order with respect to the metal complex. This behavior is consistent with a **dissociative** mechanism involving slow loss of CO, followed by rapid reaction with L:

$$\underset{18\,e^-}{Ni(CO)_4} \longrightarrow \underset{16\,e^-}{Ni(CO)_3} + CO \quad \text{(slow)} \quad \text{loss of CO from 18-electron complex}$$

$$\underset{16\,e^-}{Ni(CO)_3} + L \longrightarrow \underset{18\,e^-}{Ni(CO)_3L} \quad \text{(fast)} \quad \text{addition of L to 16-electron intermediate}$$

Loss of CO from the stable, 18-electron $Ni(CO)_4$ is slow relative to the addition of L to the more reactive, 16-electron $Ni(CO)_3$. Consequently, the first step is rate limiting, and this mechanism has the following rate law:

$$\text{Rate} = k_1[Ni(CO)_4]$$

Some reactions show more complicated kinetics. For example, study of the reaction

$$Mo(CO)_6 + L \xrightarrow{\Delta} Mo(CO)_5L + CO \ (L = \text{phosphine})$$

has shown that, for some phosphine ligands, the rate law has the following form:

$$\text{Rate} = k_1[\text{Mo(CO)}_6] + k_2[\text{Mo(CO)}_6][\text{L}]$$

The two terms in the rate law imply parallel pathways for the formation of $Mo(CO)_5L$. The first term is again consistent with a dissociative mechanism:

$$\text{Mo(CO)}_6 \xrightarrow{k_1} \text{Mo(CO)}_5 + \text{CO} \qquad \text{(slow)}$$

$$\text{Mo(CO)}_5 + \text{L} \longrightarrow \text{Mo(CO)}_5\text{L} \qquad \text{(fast)}$$

$$\text{Rate}_1 = k_1[\text{Mo(CO)}_6]$$

The second term in the rate law is consistent with an **associative** process, involving a bimolecular reaction of $Mo(CO)_6$ and L to form a transition state that then loses CO:

$$\text{Mo(CO)}_6 + \text{L} \xrightarrow{k_2} [\text{Mo(CO)}_6\text{----L}] \qquad \text{association of Mo(CO)}_6 \text{ and L}$$

$$[\text{Mo(CO)}_6\text{----L}] \longrightarrow \text{Mo(CO)}_5\text{L} + \text{CO} \qquad \text{loss of CO from transition state}$$

Formation of the transition state is the rate-limiting step in this mechanism; the rate law for this pathway is therefore

$$\text{Rate}_2 = k_2[\text{Mo(CO)}_6][\text{L}]$$

There is also strong evidence that solvent is involved in the first-order mechanism for the replacement of CO; however, because the solvent is in great excess, it does not appear in the rate law, and the observed rate law obtained in this case is the same as that shown above.[1]

Because of the two pathways, the overall rate of formation of $Mo(CO)_5L$ is the sum of the rates of the unimolecular and bimolecular mechanisms, $\text{Rate}_1 + \text{Rate}_2$.

Although most CO substitution reactions proceed primarily by a dissociative mechanism, an associative path is more likely for complexes of large metals (providing favorable sites for incoming ligands to attack) and for reactions involving highly nucleophilic ligands.

As pointed out in the introduction to this section, even though ligand dissociation and association involve changes in coordination number, they do not involve changes in the oxidation state of the metal.[2]

DISSOCIATION OF PHOSPHINE Carbon monoxide is by no means the only ligand that can undergo dissociation from metal complexes. Many other ligands can dissociate, with the ease of dissociation a function of the strength of metal–ligand bonding and, in some cases, the degree of crowding of ligands around the metal. These steric effects have been investigated for a variety of ligands, especially phosphines and similar ligands.

To describe steric effects, Tolman has defined the **cone angle** as the apex angle, θ, of a cone that encompasses the van der Waals radii of the outermost atoms of a ligand, as shown in Figure 14.1.[3] Values of cone angles of selected ligands are given in Table 14.1.

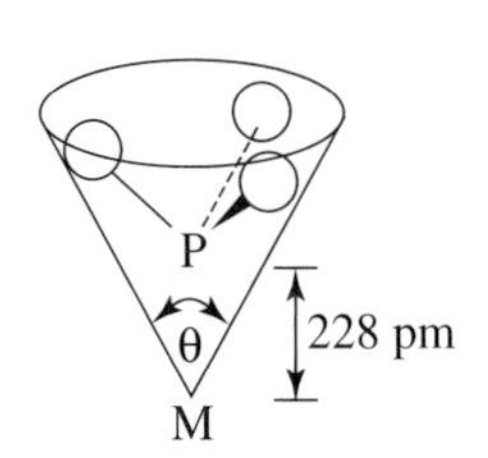

FIGURE 14.1
Ligand Cone Angle.

As might be expected, the presence of bulky ligands, having large cone angles, can lead to more rapid ligand dissociation as a consequence of crowding around the metal. For example, the rate of the reaction

$$cis\text{-Mo(CO)}_4\text{L}_2 + \text{CO} \longrightarrow \text{Mo(CO)}_5\text{L} + \text{L} \qquad (\text{L} = \text{phosphine or phosphite})$$

[1] W. D. Covey and T. L. Brown, *Inorg. Chem.*, **1973**, *12*, 2820.
[2] Assuming that no oxidation–reduction reaction occurs between the ligand and the metal.
[3] C. A. Tolman, *J. Am. Chem. Soc.*, **1970**, *92*, 2953; *Chem. Rev.*, **1977**, *77*, 313. See also K. A. Bunten, L. Chen, A. L. Fernandez, and A. J. Poë, *Coord. Chem. Rev.*, **2002**, *233-234*, 41.

TABLE 14.1 Ligand Cone Angles

Ligand	Cone Angle θ	Ligand	Cone Angle θ
PH_3	87°	$P(CH_3)(C_6H_5)_2$	136°
PF_3	104°	$P(CF_3)_3$	137°
$P(OCH_3)_3$	107°	$P(C_6H_5)_3$	145°
$P(OC_2H_5)_3$	109°	$P(\textit{cyclo}\text{-}C_6H_{11})_3$	170°
$P(CH_3)_3$	118°	$P(\textit{t}\text{-}C_4H_9)_3$	182°
PCl_3	124°	$P(C_6F_5)_3$	184°
PBr_3	131°	$P(\textit{o}\text{-}C_6H_4CH_3)_3$	194°
$P(C_2H_5)_3$	132°	P(Me, Me, Me-substituted benzene ring)$_3$	212°

which is first order in *cis*-$Mo(CO)_4L_2$, increases with increasing ligand bulk, as shown in Figure 14.2; the larger the cone angle, the more rapidly the phosphine or phosphite is lost.[4] The overall effect is substantial; for example, the rate for the most bulky ligand shown is more than four orders of magnitude greater than that for the least bulky ligand.

Many other examples of the effect of ligand bulk on the dissociation of ligands have been reported in the chemical literature.[5] For many dissociation reactions, the effect of ligand crowding may be more important than electronic effects in determining reaction rates.

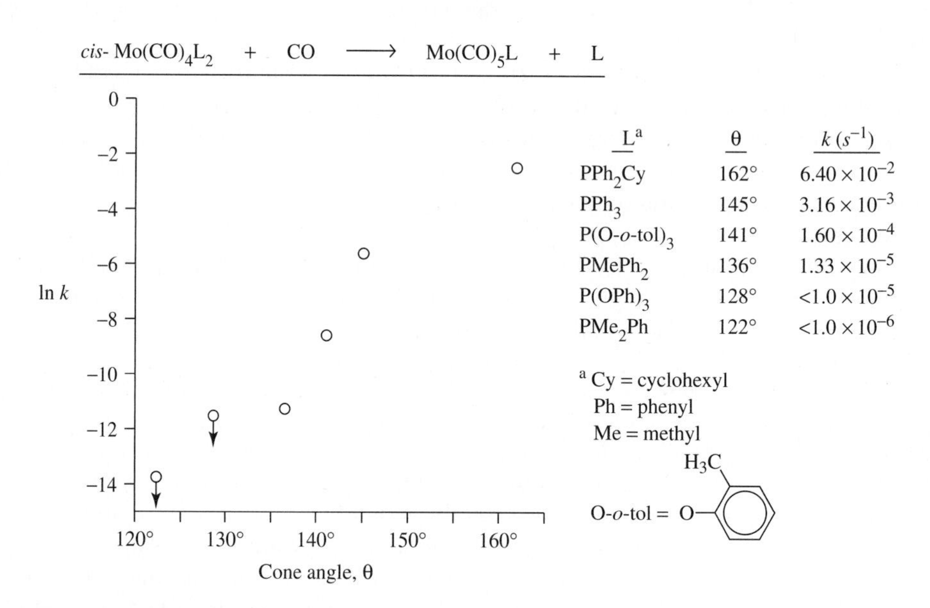

FIGURE 14.2 Reaction Rate Constant versus Cone Angle for Phosphine Dissociation.

14.1.2 Oxidative Addition

These reactions, as the name suggests, involve an increase in both the formal oxidation state and the coordination number of the metal. *Oxidative addition (OA)* reactions are among the most important of organometallic reactions and are essential steps in many catalytic processes. The reverse type of reaction, designated *reductive elimination*

[4] D. J. Darensbourg and A. H. Graves, *Inorg. Chem.*, **1979**, *18*, 1257.

[5] For example, M. J. Wovkulich and J. D. Atwood, *Organometallics*, **1982**, *1*, 1316; J. D. Atwood, M. J. Wovkulich, and D. C. Sonnenberger, *Acc. Chem. Res.*, **1983**, *16*, 350.

(RE), is also very important. These reactions can be described schematically by the following equation:

$$L_nM + X{-}Y \underset{RE}{\overset{OA}{\rightleftharpoons}} L_nM\begin{matrix} X \\ Y \end{matrix}$$

For example, heating $Fe(CO)_5$ in the presence of I_2 leads to formation of *cis*-$I_2Fe(CO)_4$. The reaction has two steps:

$$Fe(CO)_5 + I_2 \xrightarrow{\Delta} Fe(CO)_4 + CO$$

$$\underset{16\,e^-}{Fe(CO)_4} + I_2 \longrightarrow \underset{18e^-}{cis{-}I_2Fe(CO)_4} \quad \text{(oxidative addition)}$$

The first step involves dissociation of CO to give a 4-coordinate iron(0) intermediate. In the second step, iron is formally oxidized to iron(II), and the coordination number is expanded by the addition of two iodo ligands. This second step is an example of oxidative addition. Like most oxidative additions, this step involves an increase by 2 in both the oxidation state and coordination number of the metal.

It may be useful at this point to review briefly the assignment of oxidation states. Coordinated ligands are generally assigned the charges of the free ligand (e.g., zero for neutral ligands such as CO, 1– for Cl^-, CN^-). Hydrogen atom ligands and organic radicals are treated as anions:

H^-	CH_3^-	$C_6H_5^-$	$C_5H_5^-$
hydride	methyl	phenyl	η^5-C_5H_5

(The assigned charges on these ligands may have little chemical significance. For example, in methyl complexes, the carbon–metal bond is largely covalent, and such complexes should not be viewed as containing the free ion CH_3^-. The assignment of these charges is a formalism, another electron-counting scheme.)

OA reactions of square-planar d^8 complexes have special chemical significance, and we will therefore use one such complex, *trans*-$Ir(CO)Cl(PEt_3)_2$, to illustrate these reactions (Figure 14.3).

In each of the examples shown, the formal oxidation state of iridium increases from (I) to (III), and its coordination number increases from 4 to 6. The new ligands may add in a *cis* or *trans* fashion, with their orientation a function of the mechanistic pathway involved. An important feature of such reactions is that, in the expansion of the

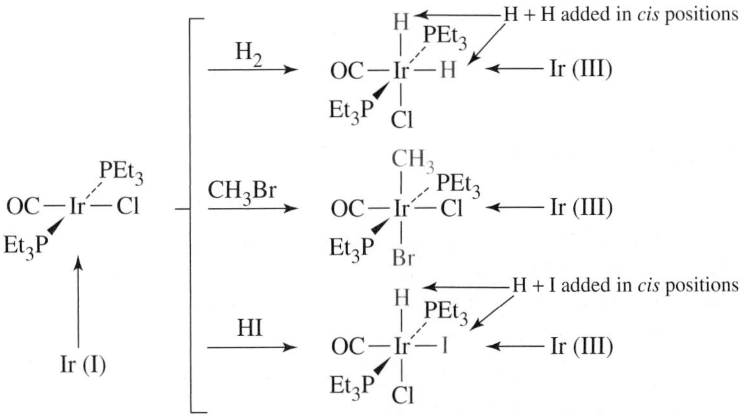

FIGURE 14.3 Examples of Oxidative Addition Reactions.

coordination number of the metal, the newly added ligands are brought into close proximity to the original ligands; this may enable chemical reactions to occur between ligands. Such reactions, encountered frequently in the mechanisms of catalytic cycles involving organometallic compounds, will be discussed later in this chapter.

FIGURE 14.4 Cyclometallation Reactions.

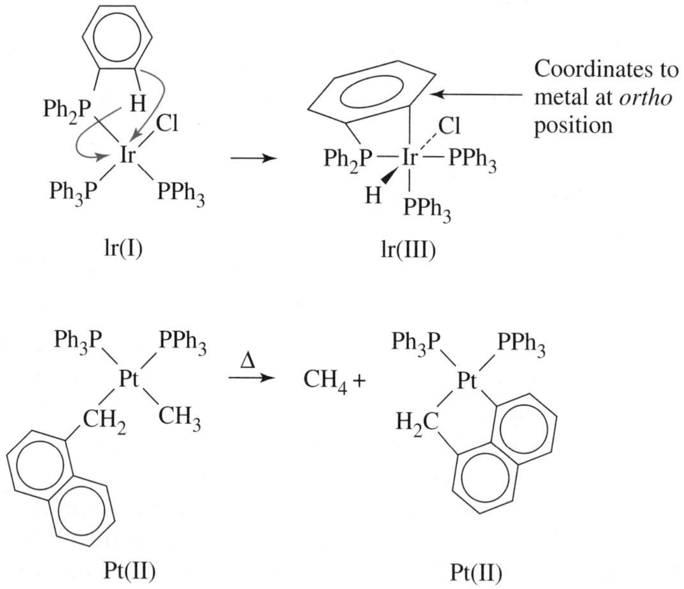

CYCLOMETALLATIONS These are reactions that incorporate metals into organic rings. The most common of these are *orthometallations*, oxidative additions in which the *ortho* position of an aromatic ring becomes attached to the metal. The first example in Figure 14.4 is an OA in which an *ortho* carbon and the hydrogen originally in the *ortho* position add to iridium.

Not all cyclometallation reactions are OAs; the second example in Figure 14.4 shows a cyclometallation that is not an OA overall (although one step in the mechanism may be OA).

14.1.3 Reductive Elimination

Reductive elimination is the reverse of oxidative addition. To illustrate this distinction, consider the following equilibrium:

$$\underset{\text{Ta(III)}}{(\eta^5\text{-}C_5H_5)_2TaH} + H_2 \underset{RE}{\overset{OA}{\rightleftharpoons}} \underset{\text{Ta(V)}}{(\eta^5\text{-}C_5H_5)_2TaH_3}$$

The forward reaction involves formal oxidation of the metal, accompanied by an increase in coordination number; it is an OA. The reverse reaction is an example of RE, which involves a decrease in both oxidation number and coordination number.

RE reactions often involve elimination of molecules such as

$$R{-}H \quad R{-}R' \quad R{-}X \quad H{-}H \quad (R, R' = \text{alkyl, aryl; } X = \text{halogen})$$

The products eliminated by these reactions may be important and useful organic compounds (R—H, R—R′, R—X). In some cases, the organic fragments (R, R′) undergo rearrangement or other reactions while coordinated to the metal. Examples of this phenomenon will be discussed later in this chapter.

As might be expected, the rates of RE reactions are also affected by ligand bulk. An example of this effect is shown in Table 14.2. The three *cis*-dimethyl complexes

TABLE 14.2 Relative Rates of Reductive Elimination

Complex	Rate Constant (s^{-1})	T(°C)
$(Ph_3P)_2Pd(CH_3)_2$	1.04×10^{-3}	60
$(MePh_2P)_2Pd(CH_3)_2$	9.62×10^{-5}	60
$(Ph_2PCH_2CH_2PPh_2)Pd(CH_3)_2$	4.78×10^{-7}	80

shown undergo RE following replacement of a phosphine ligand by a solvent molecule (solv):

$$L_2Pd(CH_3)_2 + \text{solv} \xrightarrow[-L]{} L(\text{solv})Pd(CH_3)_2 \xrightarrow{RE} LPd(\text{solv}) + H_3C{-}CH_3$$

RE yields ethane in each case. The most crowded complex, $Pd(CH_3)_2(PPh_3)_2$, undergoes reductive elimination the most rapidly.[6]

14.1.4 Nucleophilic Displacement

Ligand displacement reactions may be described as nucleophilic substitutions, involving incoming ligands as nucleophiles. Organometallic complexes, especially those carrying negative charges, may themselves behave as nucleophiles in displacement reactions. For example, the anion $[(\eta^5\text{-}C_5H_5)Mo(CO)_3]^-$ can displace iodide from methyl iodide:

$$[(\eta^5\text{-}C_5H_5)Mo(CO)_3]^- + CH_3I \longrightarrow [(\eta^5\text{-}C_5H_5)(CH_3)Mo(CO)_3] + I^-$$

$[Fe(CO)_4]^{2-}$ is a versatile organometallic nucleophile. Cooke and Collman developed the synthesis for the parent compound of this nucleophile, $Na_2Fe(CO)_4$, commonly known as *Collman's reagent*, by reacting sodium with $Fe(CO)_5$ in dioxane:[7]

$$2\,Na + Fe(CO)_5 \xrightarrow{\text{dioxane}} Na_2Fe(CO)_4 \cdot 1.5\text{ dioxane} + CO$$

The product of this reaction can be used to synthesize a variety of organic compounds. For example, the nucleophilic attack of $[Fe(CO)_4]^{2-}$ on an organic halide RX yields $[RFe(CO)_4]^-$, which can subsequently be converted to alkanes, ketones, carboxylic acids, aldehydes, acid halides, or other organic products. These reactions are outlined

[6] A. Gillie and J. K. Stille, *J. Am. Chem. Soc.*, **1980**, *102*, 4933. Rates of other RE reactions are also reported in this reference.

[7] M. P. Cooke, *J. Am. Chem. Soc.*, **1970**, *92*, 6080; J. P. Collman, *Acc. Chem. Res.*, **1975**, *8*, 342; R. G. Finke and T. N. Sorrell, *Org. Synth.*, **1979**, *59*, 102.

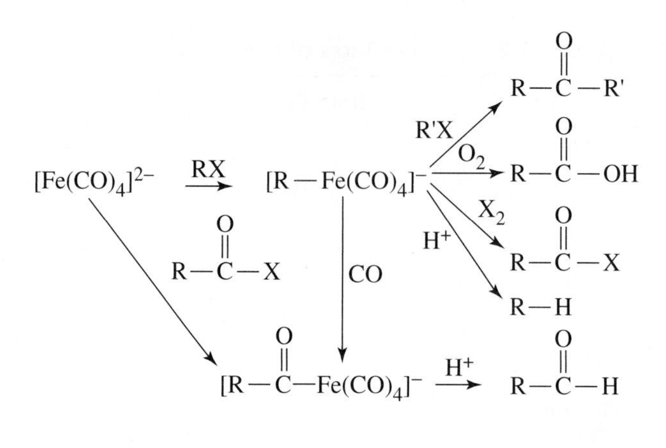

FIGURE 14.5 Synthetic Pathways Using $[Fe(CO)_4]^{2-}$.

in Figure 14.5. $[RFe(CO)_4]^-$ also undergoes other types of reactions; for example, insertion of CO, as shown in the figure. Additional details of these reactions can be found in the literature.[8]

Another useful anionic nucleophile is $[Co(CO)_4]^-$, whose chemistry has been developed by Heck.[9] A rather mild nucleophile, $[Co(CO)_4]^-$ can be synthesized by the reduction of $Co_2(CO)_8$ by sodium; it reacts with organic halides to generate alkyl complexes:

$$[Co(CO)_4]^- + RX \longrightarrow RCo(CO)_4 + X^-$$

The alkyl complex reacts with carbon monoxide to apparently insert CO into the cobalt–alkyl bond (insertion reactions will be discussed in the following section) to give an acyl complex (containing a —C(=O)R ligand):

$$RCo(CO)_4 + CO \longrightarrow RC(=O)Co(CO)_4$$

The acyl complex can then react with alcohols to generate esters:

$$RC(=O)Co(CO)_4 + R'OH \longrightarrow RC(=O)OR' + HCo(CO)_4$$

Reaction of $HCo(CO)_4$, a strong acid, with a base can regenerate the $[Co(CO)_4]^-$ to repeat the reaction cycle.

Many other nucleophilic anionic organometallic complexes have been studied, and the relative nucleophilicities of various carbonyl anions have been reported.[10] Parallels between these anions and anions of main group elements will be discussed in Chapter 15.

14.2 REACTIONS INVOLVING MODIFICATION OF LIGANDS

Many cases are known in which a ligand or molecular fragment appears to insert itself into a metal–ligand bond. Two examples have already been shown in the preceding section, in which CO becomes placed between a metal and an alkyl ligand. Although

[8] J. P. Collman, R. G. Finke, J. N. Cawse, and J. I. Brauman, *J. Am. Chem. Soc.*, **1977**, *99*, 2515; *J. Am. Chem. Soc.*, **1978**, *100*, 4766.

[9] R. F. Heck, in I. Wender and P. Pino, eds., *Organic Synthesis via Metal Carbonyls*, Vol. 1, Wiley, New York, 1968, pp. 373–404.

[10] R. E. Dessy, R. L. Pohl, and R. B. King, *J. Am. Chem. Soc.*, **1966**, *88*, 5121.

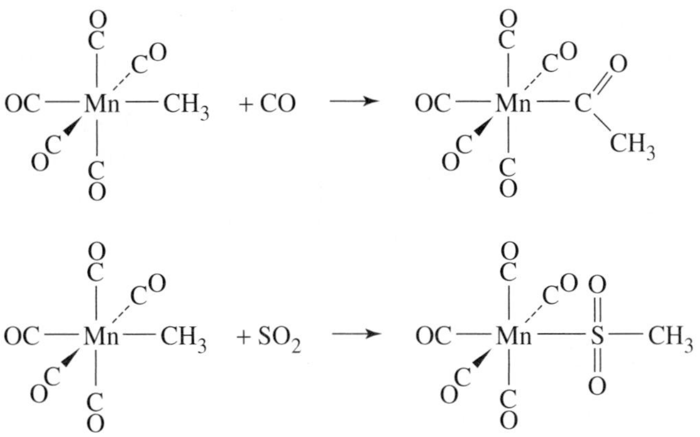

FIGURE 14.6
Examples of
1,1 Insertion Reactions.

some of these reactions are believed to occur by direct, single-step insertion, many "insertion" reactions are much more complicated and do not involve a direct insertion step at all. The most studied of these reactions are the carbonyl insertions; these will be discussed following a brief introduction to some common insertion reactions.

14.2.1 Insertion

The reactions in Figure 14.6 may be designated as 1,1 insertions, indicating that both bonds to the inserted molecule are made to the same atom in that molecule. For example, in the second reaction, both the Mn and CH_3 are bonded to the sulfur of the inserted SO_2.

In 1,2 insertions, bonds to the inserted molecule are made to adjacent atoms in that molecule. For example, in the reaction of $HCo(CO)_4$ with tetrafluoroethylene, as shown in Figure 14.7, the product has the $Co(CO)_4$ group attached to one carbon and the H attached to the neighboring carbon.

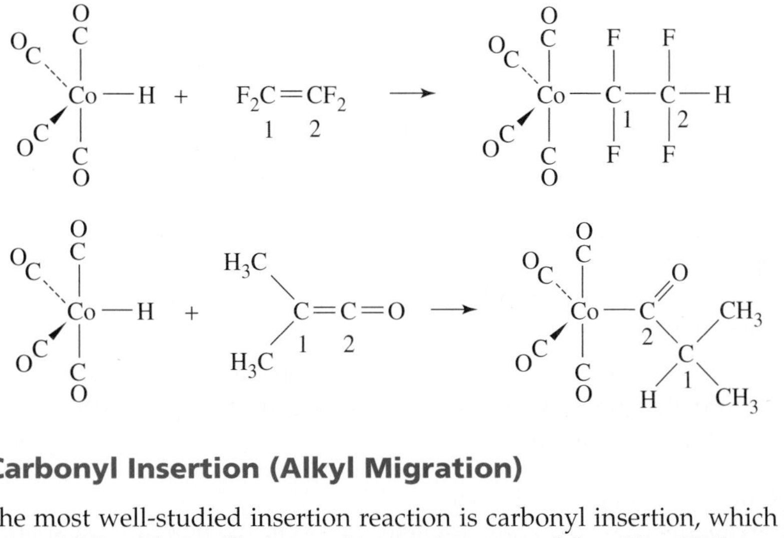

FIGURE 14.7
Examples of
1,2 Insertion Reactions.

14.2.2 Carbonyl Insertion (Alkyl Migration)

Perhaps the most well-studied insertion reaction is carbonyl insertion, which involves the reaction of CO with an alkyl complex to give an acyl [—C(=O)R] product. For example, the reaction of $CH_3Mn(CO)_5$ with CO has the following stoichiometry:

$$H_3C-Mn(CO)_5 + CO \longrightarrow CH_3C(=O)-Mn(CO)_5$$

The insertion of CO into a metal–carbon bond in alkyl complexes is of particular interest for its potential applications to organic synthesis and catalysis (examples will be discussed in Section 14.3), and its mechanism deserves careful consideration.

From the net equation, we might expect that the CO inserts directly into the $Mn—CH_3$ bond. However, other mechanisms are possible that would give the overall reaction stoichiometry while involving steps other than the insertion of an incoming CO. Three plausible mechanisms have been suggested for this reaction:

Mechanism 1: CO Insertion

Direct insertion of CO into a metal–carbon bond.

Mechanism 2: CO Migration

Migration of CO to give intramolecular CO insertion. This would yield a 5-coordinate intermediate, with a vacant site available for attachment of an incoming CO.

Mechanism 3: Alkyl Migration

In this case, the alkyl group would migrate, rather than the CO, and attach itself to a CO *cis* to the alkyl. This would also give a 5-coordinate intermediate with a vacant site available for an incoming CO.

These mechanisms are described schematically in Figure 14.8. In both mechanisms 2 and 3, the intramolecular migration is considered to occur to one of the migrating group's nearest neighbors, located in *cis* positions.

Experimental evidence that may be used to evaluate these mechanisms includes the following:[11]

1. Reaction of $CH_3Mn(CO)_5$ with ^{13}CO gives a product with the labeled CO in carbonyl ligands only; *none* is found in the acyl position.
2. The reverse reaction

$$CH_3C(=O)—Mn(CO)_5 \longrightarrow H_3C—Mn(CO)_5 + CO$$

occurs readily on heating $CH_3C(=O)Mn(CO)_5$. When this reaction is carried out with ^{13}C in the acyl position, the product $CH_3Mn(CO)_5$ has the labeled CO entirely *cis* to CH_3. No labeled CO is lost in this reaction.
3. The reverse reaction, when carried out with ^{13}C in a carbonyl ligand *cis* to the acyl group, gives a product that has a 2:1 ratio of *cis* to *trans* product (*cis* and *trans* refer to the position of labeled CO relative to CH_3 in the product). Some labeled CO is also lost in this reaction.

The mechanisms can now be evaluated on the basis of these data. First, mechanism 1 is definitely ruled out by the first experiment. Direct insertion of ^{13}CO must result in ^{13}C in the acyl ligand; because none is found, the mechanism cannot be a direct insertion. Mechanisms 2 and 3, on the other hand, are both compatible with the results of this experiment.

The principle of microscopic reversibility requires that any reversible reaction must have identical pathways for the forward and reverse reactions, simply proceeding in opposite directions. (This principle is similar to the idea that the lowest pass

[11] T. C. Flood, J. E. Jensen, and J. A. Statler, *J. Am. Chem. Soc.*, **1981**, *103*, 4410 and references therein.

CO Insertion Reactions

Mechanism 1

Direct insertion

$OC{-}Mn(CO)_4{-}CH_3 + CO \longrightarrow OC{-}Mn(CO)_4{-}C(=O){-}CH_3$

New CO in the acyl group

Mechanism 2

Intramolecular insertion

$OC{-}Mn(CO)_4{-}CH_3 \longrightarrow OC{-}Mn(CO)_3{-}C(=O){-}CH_3 + CO \longrightarrow OC{-}Mn(CO)_4{-}C(=O){-}CH_3$

New CO *cis* to $\overset{O}{\overset{\|}{C}}{-}CH_3$

Mechanism 3

Methyl migration

$OC{-}Mn(CO)_4{-}CH_3 \longrightarrow OC{-}Mn(CO)_3(C(=O)CH_3) + CO \longrightarrow OC{-}Mn(CO)_3(C(=O)CH_3){-}CO$

New CO *cis* to $\overset{O}{\overset{\|}{C}}{-}CH_3$

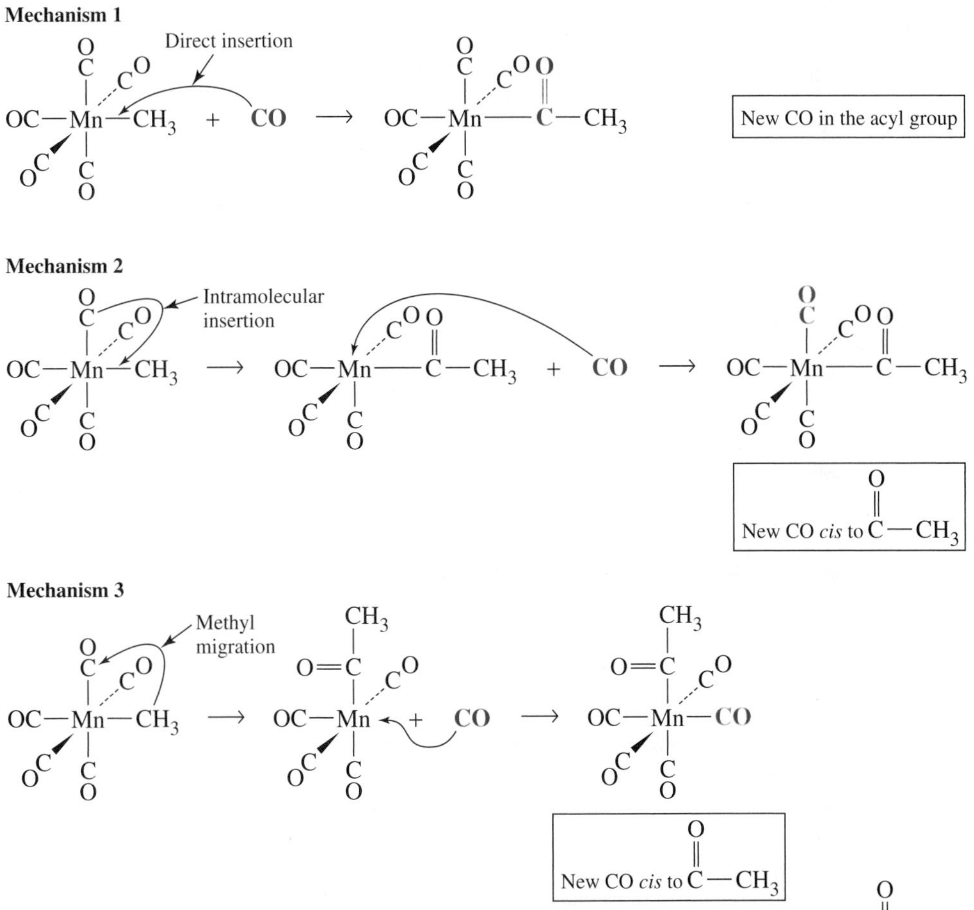

FIGURE 14.8 Possible Mechanisms for CO Insertion Reactions. Acyl groups are shown as $-\overset{O}{\overset{\|}{C}}-CH_3$ for clarity; the actual geometry around acyl carbons is trigonal.

over a mountain chain must be the same regardless of the direction of travel.) If the forward reaction is carbonyl migration (mechanism 2), the reverse reaction must proceed by loss of a CO ligand, followed by migration of CO from the acyl ligand to the empty site. Because this migration is unlikely to occur to a *trans* position, all the product should be *cis*. If the mechanism is alkyl migration (mechanism 3), the reverse reaction must proceed by loss of a CO ligand, followed by migration of the alkyl portion of the acyl ligand to the vacant site. Again, all the product should be *cis*. Both mechanisms 2 and 3 would transfer labeled CO in the acyl group to a *cis* position and are therefore consistent with the experimental data for the second experiment (Figure 14.9).

▶ **Exercise 14.1** Show that heating of $CH_3{-}{}^{13}\overset{O}{\overset{\|}{C}}{-}Mn(CO)_5$ would not be expected to give the *cis* product by mechanism 1.

The third experiment differentiates conclusively between mechanisms 2 and 3. The CO migration of mechanism 2, with ^{13}CO *cis* to the acyl ligand, requires migration of CO from the acyl ligand to the vacant site. As a result, 25% of the product should

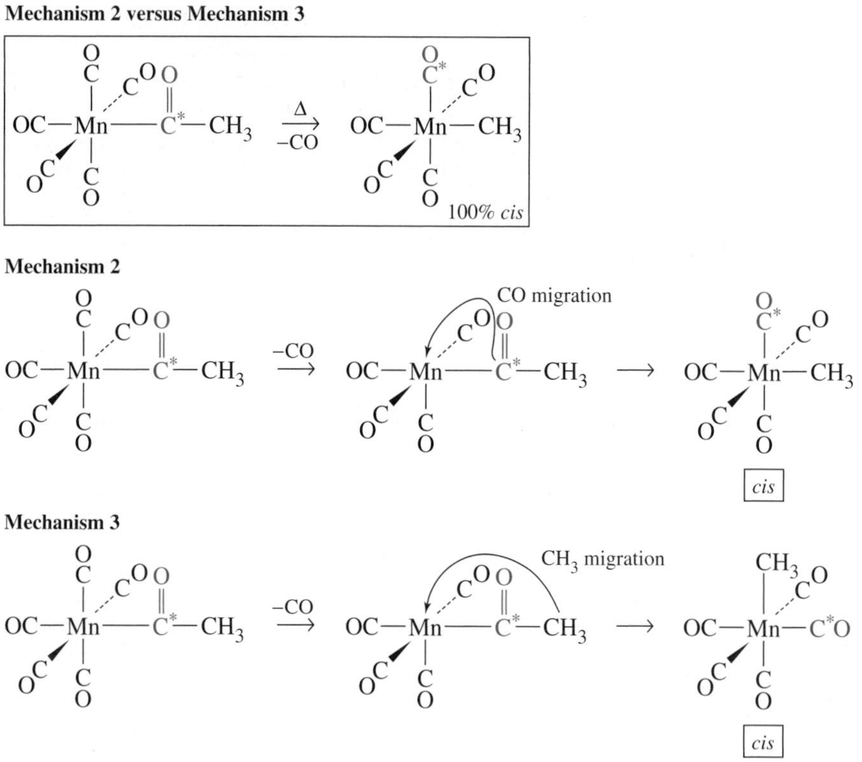

Mechanism 2 versus Mechanism 3

$\xrightarrow[-CO]{\Delta}$ 100% *cis*

Mechanism 2

$-CO$ CO migration *cis*

Mechanism 3

$-CO$ CH_3 migration *cis*

FIGURE 14.9 Mechanisms of Reverse Reactions for CO Migration and Alkyl Insertion (1). C* indicates the location of ^{13}C.

have no ^{13}CO label and 75% should have the labeled CO *cis* to the alkyl, as shown in Figure 14.10. On the other hand, alkyl migration (mechanism 3) should yield 25% with no label, 50% with the label *cis* to the alkyl, and 25% with the label *trans* to the alkyl. Because this is the ratio of *cis* to *trans* found in the experiment, the evidence supports mechanism 3, which is the accepted pathway for this reaction.

The result is that a reaction that initially appears to involve CO insertion, and is often so designated, does not involve CO insertion at all! It is not uncommon, on close study, for reactions to differ substantially from how they might at first appear; the "carbonyl insertion" reaction may in fact be more complicated than described here. In this reaction, as well as in all chemical reactions, it is extremely important for chemists to be willing to undertake mechanistic studies and to keep an open mind about possible alternative mechanisms. No mechanism can be proved; it is always possible to suggest alternatives consistent with the known data.

One final point about the mechanism of these reactions should be made. In the previous discussion of mechanisms 2 and 3, it was assumed that the intermediate was a square pyramid and that no rearrangement to other geometries (such as trigonal-bipyramidal) occurred. Other labeling studies, involving reactions of labeled $CH_3Mn(CO)_5$ with phosphines, have supported a square-pyramidal intermediate.[12]

▶ **Exercise 14.2** Predict the product distribution for the reaction of *cis*-$CH_3Mn(CO)_4$ (^{13}CO) with PR_3 ($R = C_2H_5$).

[12] T. C. Flood, J. E. Jensen, and J. A. Statler, *J. Am. Chem. Soc.*, **1981**, *103*, 4410 and references therein.

Mechanism 2

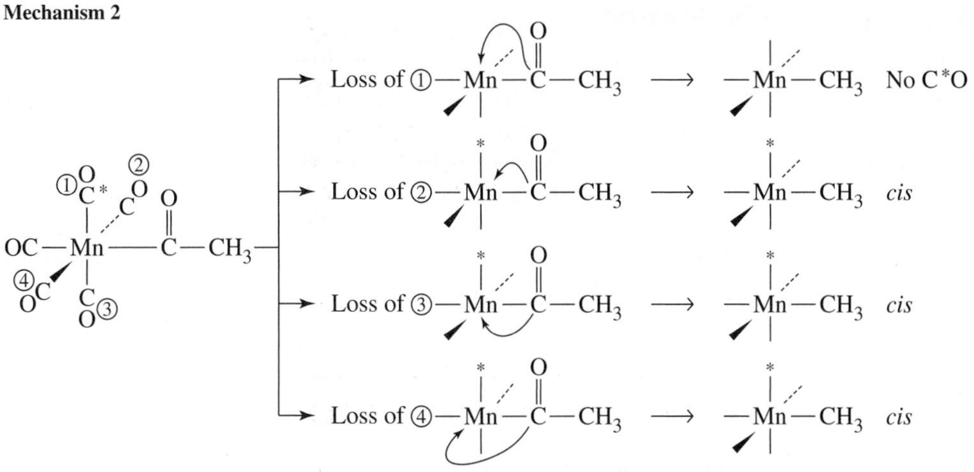

Loss of ① — No C*O
Loss of ② — *cis*
Loss of ③ — *cis*
Loss of ④ — *cis*

Mechanism 3

Loss of ① — No C*O
Loss of ② — *cis*
Loss of ③ — *trans*
Loss of ④ — *cis*

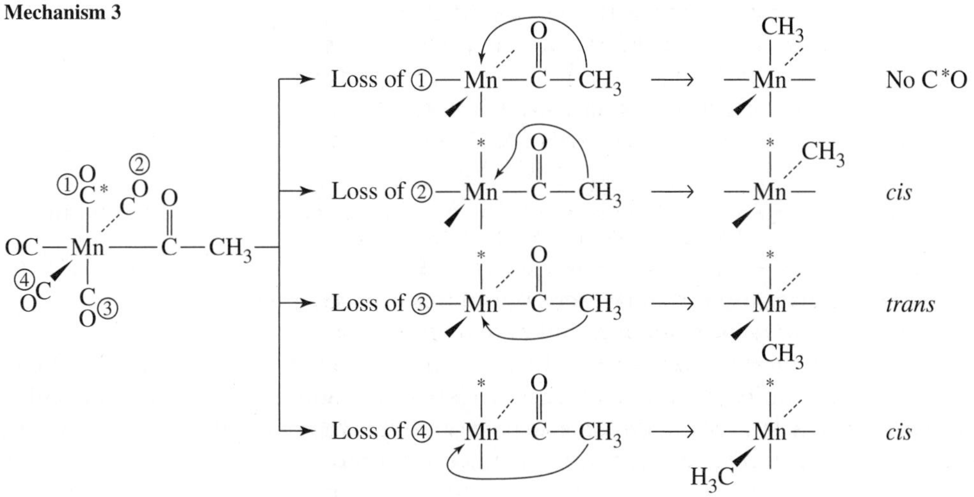

FIGURE 14.10 Mechanisms of Reverse Reactions for CO Migration and Alkyl Insertion (2). C* indicates the location of ^{13}C.

14.2.3 Examples of 1,2 Insertions

Two examples of 1,2 insertions have been shown in Figure 14.7. An important application of 1,2 insertions of alkenes into metal–alkyl bonds is in the formation of polymers. One such process is the Cossee–Arlman mechanism,[13] proposed for the Ziegler–Natta polymerization of alkenes (discussed in Section 14.4.1). According to this mechanism, a polymer chain can grow, as a consequence of repeated 1,2 insertions, into a vacant coordination site, as follows:

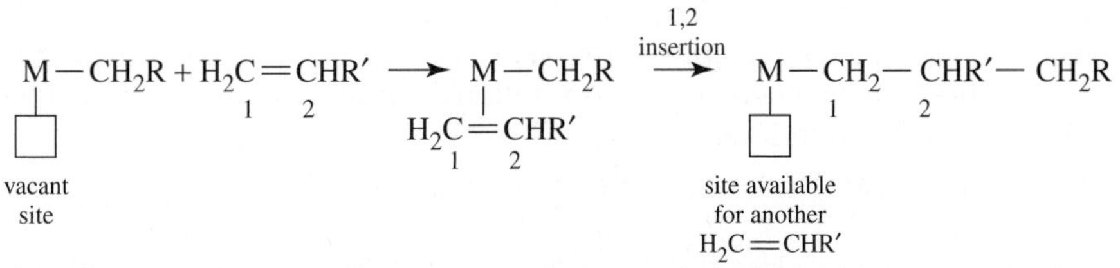

$$M{-}CH_2R + H_2\underset{1}{C}{=}\underset{2}{C}HR' \longrightarrow M{-}CH_2R\,(H_2\underset{1}{C}{=}\underset{2}{C}HR') \xrightarrow{\text{1,2 insertion}} M{-}\underset{1}{C}H_2{-}\underset{2}{C}HR'{-}CH_2R$$

vacant site

site available for another $H_2C{=}CHR'$

[13] P. Cossee, *J. Catal.*, **1964**, *3*, 80; E. J. Arlman and P. Cossee, *J. Catal.*, **1964**, *3*, 99.

14.2.4 Hydride Elimination

Hydride elimination reactions are characterized by the transfer of a hydrogen atom from a ligand to a metal. Effectively, this may be considered an oxidative addition, with both the coordination number and the formal oxidation state of the metal increased (the hydrogen transferred is formally considered as hydride, H^-). The most common type of hydride elimination is ***β* elimination**, with a proton in a β position[14] on an alkyl ligand transferred to the metal by way of an intermediate in which the metal, the α and β carbons, and the hydride are coplanar. An example of β elimination—the reverse of 1,2, insertion—is shown in Figure 14.11.

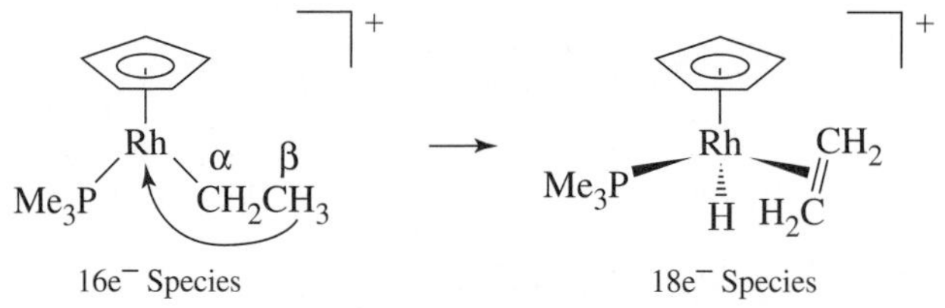

Rh α β Me₃P CH₂CH₃ → Me₃P Rh CH₂ H H₂C

16e⁻ Species 18e⁻ Species

FIGURE 14.11 Example of β Elimination.

▶ Exercise 14.3 Show that the reverse of the reaction shown in Figure 14.11 would be 1,2 insertion.

Beta eliminations, as will be seen later in this chapter, are important in many catalytic processes involving organometallic complexes.

Several general comments can be made about β-elimination reactions. First, because only complexes that have β hydrogens can undergo these reactions, alkyl complexes that lack β hydrogens tend to be more stable thermally than those that have such hydrogens (although the former may undergo other types of reactions). Furthermore, coordinatively saturated complexes—complexes in which all coordination sites are filled—containing β hydrogens are in general more thermally stable than complexes having empty coordination sites; the β-elimination mechanism requires transfer of a hydrogen to an empty coordination site. Finally, other types of elimination reactions are also known, such as the elimination of hydrogen from α and γ positions; the interested reader is referred to other sources for examples of such reactions.[15]

14.2.5 Abstraction

Abstraction reactions are elimination reactions in which the coordination number of the metal does not change. In general, they involve removal of a substituent from a ligand, often by the action of an external reagent, such as a Lewis acid. Two types of abstractions, α and β, are illustrated in Figure 14.12; they involve, respectively, removal of substituents from the α and β positions (with respect to the metal) of coordinating ligands. Alpha abstraction has been encountered previously, in the synthesis of carbyne complexes discussed in Section 13.6.3.

[14] The Greek letter α is used to designate the carbon atom directly attached to the metal, β is used for the next carbon atom, and so forth.

[15] J. D. Fellmann, R. R. Schrock, and D. D. Traficante, *Organometallics*, **1982**, *1*, 481; J. P. Collman, L. S. Hegedus, J. R. Norton, and R. G. Finke, *Principles and Applications of Organotransition Metal Chemistry*, University Science Books, Mill Valley, CA, 1987, and references therein.

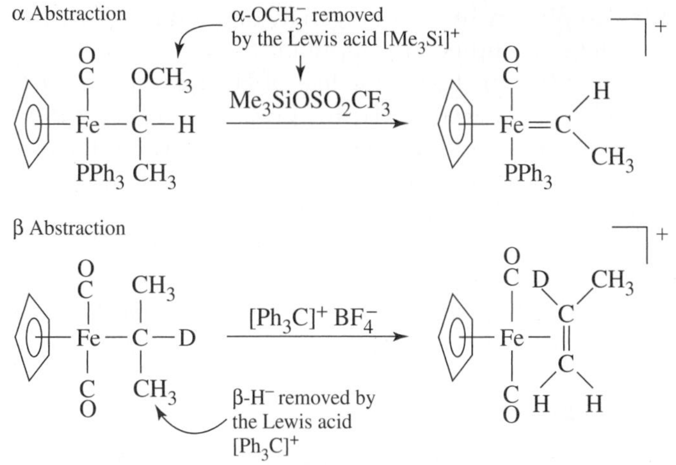

α Abstraction

α-OCH_3^- removed by the Lewis acid $[Me_3Si]^+$

$Me_3SiOSO_2CF_3$

β Abstraction

$[Ph_3C]^+ BF_4^-$

β-H^- removed by the Lewis acid $[Ph_3C]^+$

FIGURE 14.12 Abstraction Reactions.

14.3 ORGANOMETALLIC CATALYSTS

In addition to holding an intrinsic interest for chemists, organometallic reactions are also of great interest industrially, especially in the development of catalysts for reactions of commercial importance. The commercial interest in catalysis has been spurred by the fundamental problem of how to convert relatively inexpensive feedstocks (e.g., coal, petroleum, and water) into molecules of greater commercial value. This frequently involves, as part of the industrial process, conversion of simple molecules into more complex molecules (e.g., ethylene into acetaldehyde, methanol into acetic acid, or organic monomers into polymers), conversion of one molecule into another of the same type (one alkene into another), or a selective reaction at a particular molecular site (e.g., replacement of hydrogen by deuterium, selective hydrogenation of a specific double bond). Historically, many catalysts have been heterogeneous in nature—that is, solid materials having catalytically active sites on their surface, with only the surface in contact with the reactants.

Homogeneous catalysts, soluble in the reaction medium, are molecular species that are easier to study and modify for specific applications than heterogeneous catalysts. Transition-metal complexes have a variety of characteristics that can make them attractive for consideration as homogeneous catalysts: they exhibit a variety of oxidation states; an immense variety of ligands can become attached to them, including ligands that can engage in pi, sigma, and more complex interactions; the metals can change coordination numbers as they gain and lose ligands; and they can have different coordination numbers and geometries. Appropriate design of catalyst molecules may provide high selectivity in the processes catalyzed; it is not surprising that development of highly selective homogeneous catalysts has been of considerable industrial interest. Only a small percent of catalytic cycles, however, are efficient or profitable enough to be commercially feasible.

In the examples of catalysis that follow, the reader will find it useful to identify the catalysts, the species regenerated in each complete reaction cycle. In addition, the individual steps in these cycles will provide examples of the various types of organometallic reactions introduced earlier in this chapter. In each case, the proposed mechanisms presented in this section should be viewed as subject to modification as additional research is conducted.

14.3.1 Example of Catalysis: Catalytic Deuteration

If deuterium gas (D_2) is bubbled through a benzene solution of $(\eta^5\text{-}C_5H_5)_2TaH_3$ at an elevated temperature, the hydrogen atoms of benzene are slowly replaced by deuterium; eventually, perdeuterobenzene, C_6D_6, can be obtained (for use, for example, as an NMR solvent).[16] Replacement of hydrogen by deuterium occurs in a series of alternating RE and OA steps, as outlined in Figure 14.13.

The initial step in this process is the loss of H_2 (formally, reductive elimination) from the 18-electron $(\eta^5\text{-}C_5H_5)_2TaH_3$ to give the 16-electron $(\eta^5\text{-}C_5H_5)_2TaH$. $(\eta^5\text{-}C_5H_5)_2TaH$ can then react with benzene in the second step (oxidative addition) to give an 18-electron species containing a phenyl group σ-bonded to the metal. This species can undergo a second loss of H_2 to give another 16-electron species, $(\eta^5\text{-}C_5H_5)_2TaH{-}C_6H_5$. $(\eta^5\text{-}C_5H_5)_2TaH{-}C_6H_5$ subsequently adds D_2 (another oxidative addition) to form an 18-electron species (Step 4), which in the last step eliminates C_6H_5D. Repetition of this sequence in the presence of excess D_2 eventually leads to C_6D_6. In each subsequent cycle, the catalytic species $(\eta^5\text{-}C_5H_5)_2TaD$ is regenerated.

14.3.2 Hydroformylation

The hydroformylation, or oxo, process was introduced in 1938 and is the oldest homogeneous catalytic process still in commercial use. It is used to convert terminal alkenes into aldehydes and a variety of other organic products, especially those having their carbon chain increased by one. Approximately 10 million tons of hydroformylation products are produced annually. The conversion of an alkene of formula $R_2C{=}CH_2$ into an aldehyde $R_2CH{-}CH_2{-}CHO$ is outlined in Figure 14.14.[17]

Each step of the hydroformylation cycle may be categorized according to its characteristic type of organometallic reaction, as indicated in the figure. The cobalt-containing intermediates in this cycle alternate between 18- and 16-electron species. The 18-electron species react to formally reduce their electron count by 2 (by ligand dissociation, 1,2 insertion of coordinated alkene, alkyl migration, or reductive elimination), whereas the 16-electron species can increase their formal electron count (by coordination of alkene or CO or by oxidative addition). Such a pattern is commonly encountered in catalytic cycles involving organometallic complexes, with the catalytic activity in large part a consequence of the capability of the metal to react by way of a variety of 18- and 16-electron intermediates.

A few steps of the hydroformylation process are worth comment. The first step, involving dissociation of CO from $HCo(CO)_4$, is inhibited by high CO pressure, yet the fourth step requires CO; thus, careful control of this pressure is necessary for optimum yields and rates.[18] The second step is first order in alkene; it is the slow (rate-determining) step. In Step 3, the product shown has a CH_2 group, rather than a CR_2 group, bonded to the metal; such a preference for CH_2 attaching to the metal is enhanced by bulky R groups. However, coordination of the CR_2 can also occur, leading to a branched product. Because linear products are generally more valuable, a challenge in the development of hydroformylation has been to design processes with high linear to branched ratios.

Detailed calculations of the geometries and energies of the steps in hydroformylation have been reported.[19] The calculated geometry of the product of the third reaction step is more interesting than that shown in Figure 14.14; it has an **agostic hydrogen**, a

[16] J. W. Lauher and R. Hoffmann, *J. Am. Chem. Soc.*, **1976**, *98*, 1729 and references therein.
[17] R. F. Heck and D. S. Breslow, *J. Am. Chem. Soc.*, **1961**, *83*, 4023; see also F. Heck, *Adv. Organomet. Chem.*, **1966**, *4*, 243.
[18] For more information on reaction conditions, see G. W. Parshall and S. D. Ittel, *Homogeneous Catalysis*, 2nd ed., John Wiley & Sons, New York, 1992, pp. 106–111.
[19] C-F. Huo, Y.-W. Li, M. Beller, and H. Jiao, *Organometallics*, **2003**, *22*, 4665.

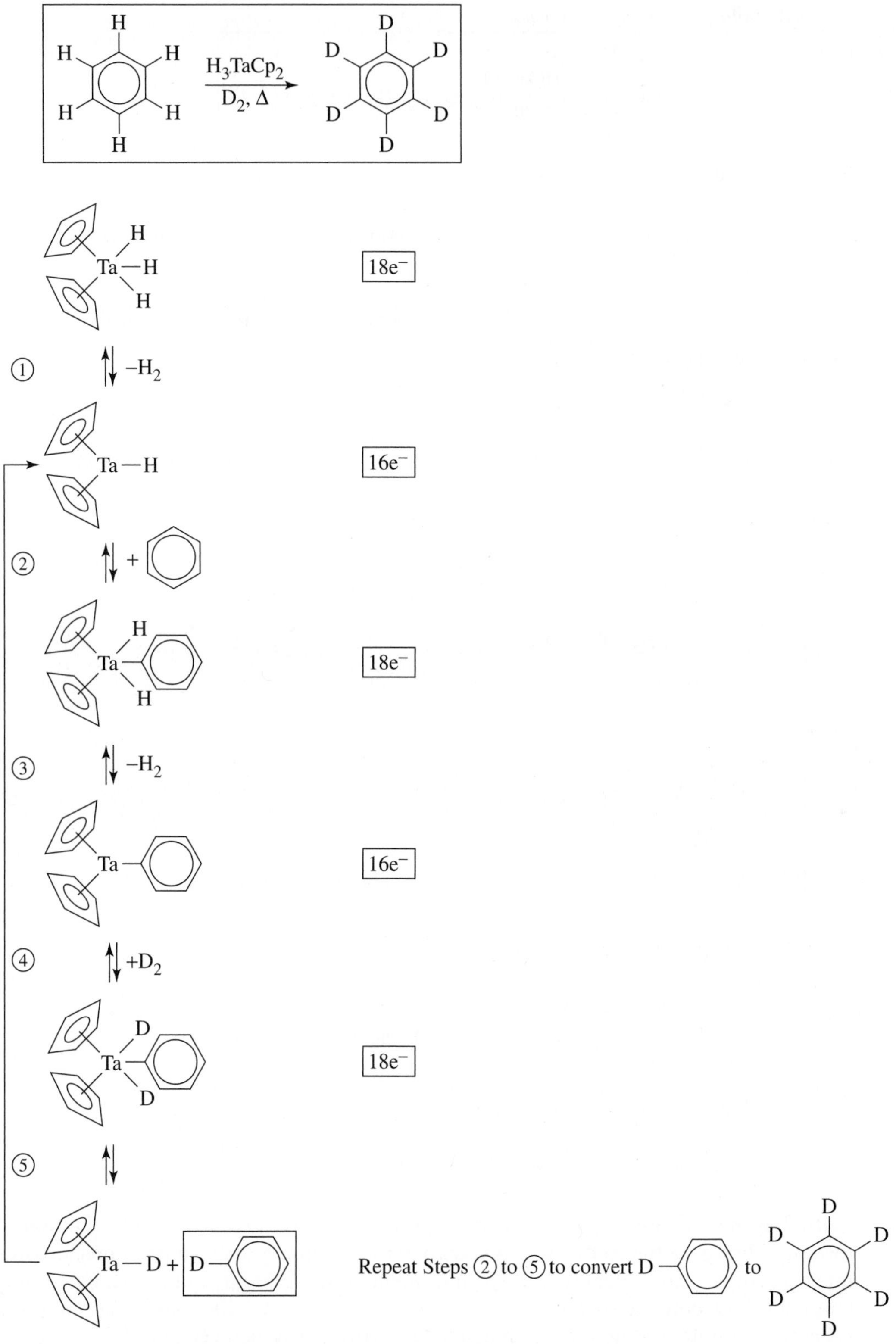

$$C_6H_6 \xrightarrow[D_2,\ \Delta]{H_3TaCp_2} C_6D_6$$

Cp_2TaH_3 — 18e$^-$

① $\rightleftharpoons$ $-H_2$

$Cp_2Ta{-}H$ — 16e$^-$

② $\rightleftharpoons$ + C_6H_6

$Cp_2Ta(H)_2(C_6H_5)$ — 18e$^-$

③ $\rightleftharpoons$ $-H_2$

$Cp_2Ta{-}C_6H_5$ — 16e$^-$

④ $\rightleftharpoons$ $+D_2$

$Cp_2Ta(D)_2(C_6H_5)$ — 18e$^-$

⑤ $\rightleftharpoons$

$Cp_2Ta{-}D + D{-}C_6H_5$

Repeat Steps ② to ⑤ to convert $D{-}C_6H_5$ to C_6D_6

FIGURE 14.13 Catalytic Deuteration.

FIGURE 14.14 Hydroformylation Process.

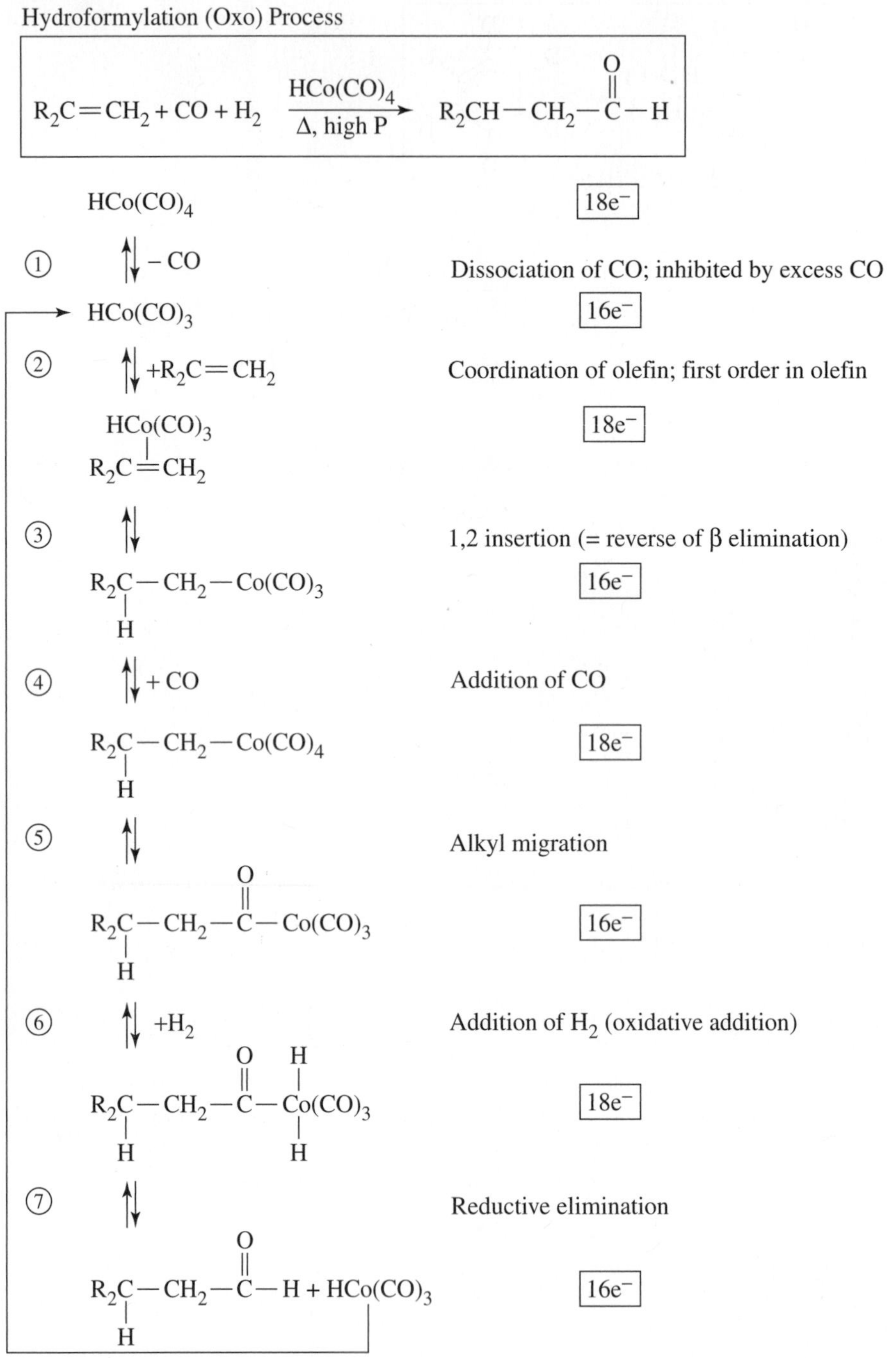

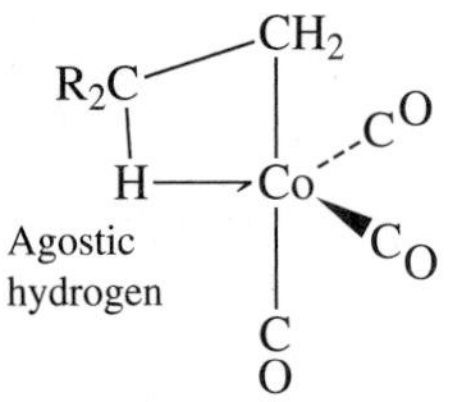

Hydroformylation (Oxo) Process

$$R_2C{=}CH_2 + CO + H_2 \xrightarrow[\Delta,\ \text{high P}]{HCo(CO)_4} R_2CH{-}CH_2{-}\overset{O}{\overset{\|}{C}}{-}H$$

$HCo(CO)_4$ — 18e⁻

① ⇅ $-CO$ — Dissociation of CO; inhibited by excess CO

$HCo(CO)_3$ — 16e⁻

② ⇅ $+R_2C{=}CH_2$ — Coordination of olefin; first order in olefin

$HCo(CO)_3(R_2C{=}CH_2)$ — 18e⁻

③ ⇅ — 1,2 insertion (= reverse of β elimination)

$R_2C(H){-}CH_2{-}Co(CO)_3$ — 16e⁻

④ ⇅ $+CO$ — Addition of CO

$R_2C(H){-}CH_2{-}Co(CO)_4$ — 18e⁻

⑤ ⇅ — Alkyl migration

$R_2C(H){-}CH_2{-}C({=}O){-}Co(CO)_3$ — 16e⁻

⑥ ⇅ $+H_2$ — Addition of H_2 (oxidative addition)

$R_2C(H){-}CH_2{-}C({=}O){-}Co(H)_2(CO)_3$ — 18e⁻

⑦ ⇅ — Reductive elimination

$R_2C(H){-}CH_2{-}C({=}O){-}H + HCo(CO)_3$ — 16e⁻

R_2C, CH_2, H, Co, CO, CO, CO

Agostic hydrogen

hydrogen that is bonded to the α carbon and also is strongly attracted to the metal, forming a bridge between the two, as shown in the margin. Agostic interactions, involving a hydrogen atom weakly bonded to a metal and having a weakened, elongated bond with a carbon have been found in a variety of organometallic complexes and have been proposed in intermediates in reactions of hydrogen at metal centers.[20]

[20] M. Brookhart, M. L. H. Green, and G. Parkin, *Proc. Nat. Acad. Sci. USA*, **2007**, *104*, 6908.

Step 6 involves addition of H_2 (OA); however, high H_2 pressure can lead to addition of H_2 to the 16-electron intermediate from Step 3, which would then eliminate an alkane:

$$\underset{16\ e^-}{R_2CH{-}CH_2{-}Co(CO)_3} + H_2 \longrightarrow \underset{18\ e^-}{R_2CH{-}CH_2{-}Co(H)_2(CO)_3} \quad \text{oxidative addition}$$

$$\underset{18\ e^-}{R_2CH{-}CH_2{-}Co(H)_2(CO)_3} \longrightarrow R_2CH{-}CH_3 + \underset{16\ e^-}{HCo(CO)_3} \quad \text{reductive elimination}$$

Again, careful control of the experimental conditions is necessary to maximize yield of the desired products.[21] The actual catalytic species in this mechanism is the 16-electron $HCo(CO)_3$.

The main industrial application of hydroformylation is in the production of butanal from propene ($CH_3CH{=}CH_2 \longrightarrow CH_3CH_2CH_2CHO$). Subsequent hydrogenation gives butanol, an important industrial solvent. Other aldehydes are also produced industrially by hydroformylation, using either cobalt catalysts such as the one in Figure 14.14 or rhodium-based catalysts.

▶ **Exercise 14.4** Show how $(CH_3)_2CHCH_2CHO$ can be prepared from $(CH_3)_2C{=}CH_2$ by the hydroformylation process.

A shortcoming of the cobalt carbonyl-based hydroformylation process outlined in Figure 14.14 is that it produces only about 80 percent of the much more valuable linear aldehydes, with the remainder having branched chains. Modifying the catalyst by replacing one of the CO ligands of the starting complex by PBu_3 (Bu = *n*-butyl) to give $HCo(CO)_3(PBu_3)$ increases the selectivity of the process to give an approximately 9:1 ratio of linear to branched aldehydes.[22] Finally, replacing the cobalt with rhodium yields far more active catalysts (much less catalyst needs to be present) that can function with higher linear to branched selectivity at significantly lower temperatures and pressures than cobalt-based catalysts.[23] A proposed mechanism for an example of such a catalytic process using $HRh(CO)_2(PPh_3)_2$ is shown in Figure 14.15.[24]

▶ **Exercise 14.5** Classify each step of the mechanism in Figure 14.15 according to its reaction type.

---

[21] For a discussion of additional details, including possible alternative steps in this mechanism, see T. Ziegler and L. Versluis, "The Tricarbonylhydridocobalt-Based Hydroformylation Reaction," in W. R. Moser and D. W. Slocum, eds., *Homogeneous Transition Metal-Catalyzed Reactions*, American Chemical Society, Washington, DC, 1992, pp. 75–93.

[22] L. H. Slaugh and R. D. Mullineaux, *J. Organomet. Chem.*, **1968**, *13*, 469.

[23] J. A. Osborne, J. F. Young, and G. Wilkinson, *Chem. Commun. (Cambridge)*, **1965**, 17; C. K. Brown and G. Wilkinson, *J. Chem. Soc., A*, **1970**, 2753.

[24] For a more detailed outline of the various cobalt- and rhodium-based hydroformylation catalysts, and for additional related references, see G. O. Spessard and G. L. Miessler, *Organometallic Chemistry*, Oxford University Press, New York, 2010, pp. 322–339.

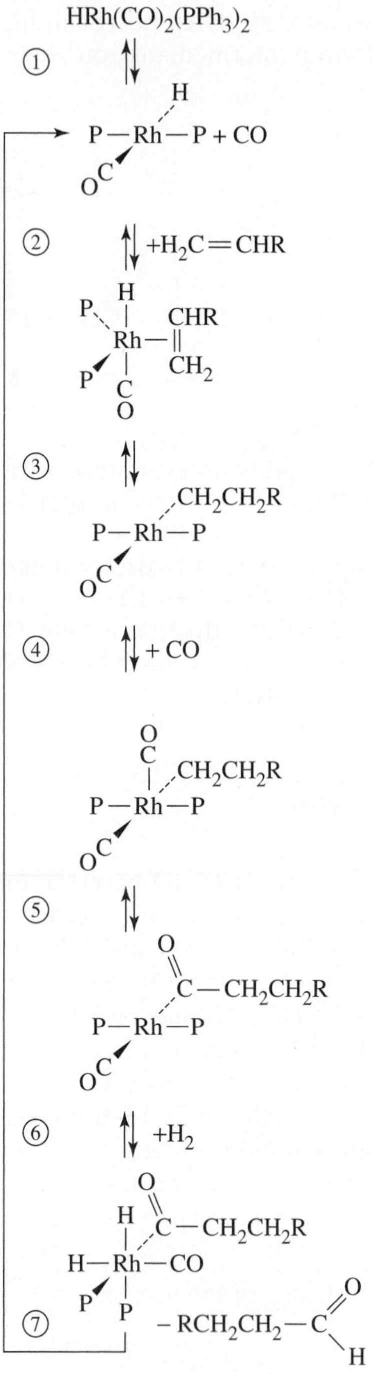

FIGURE 14.15 Hydroformylation using $HRh(CO)_2(PPh_3)_3$. P = PPh_3.

From C. K. Brown and G. Wilkinson, *J. Chem. Soc., A*, **1970**, 2753.

$HRh(CO)_2(PPh_3)_2$

① ⇅

$P{-}Rh(H)(CO){-}P + CO$

② ⇅ $+H_2C{=}CHR$

$HRh(P)_2(CO)(\eta^2\text{-}CH_2{=}CHR)$

③ ⇅

$P{-}Rh(CH_2CH_2R)(CO){-}P$

④ ⇅ $+ CO$

$P{-}Rh(CO)_2(CH_2CH_2R){-}P$

⑤ ⇅

$P{-}Rh(CO)(C(=O){-}CH_2CH_2R){-}P$

⑥ ⇅ $+H_2$

$H{-}Rh(H)(CO)(P)_2(C(=O){-}CH_2CH_2R)$

⑦ $- RCH_2CH_2{-}C(=O)H$

More recently, higher linear to branched ratios for rhodium-catalyzed hydroformylation have been achieved using bidentate ligands, which broaden the range of steric and electronic options available. For example, the BISBI ligand has enabled a linear to branched ratio of 66:1 to be achieved.[25] In a series of oxygen-bridged diphosphine ligands

[25] C.P. Casey, G. T. Whiteker, M. G. Melville, L. M. Petrovich, J. A. Garvey, Jr., and D. R. Powell, *J. Am. Chem. Soc.*, **1992**, *114*, 5535.

such as the one shown below, the linear to branched ratio was found to increase with the **bite angle**, the angle of coordination β_n between the donor atoms and the metal.[26]

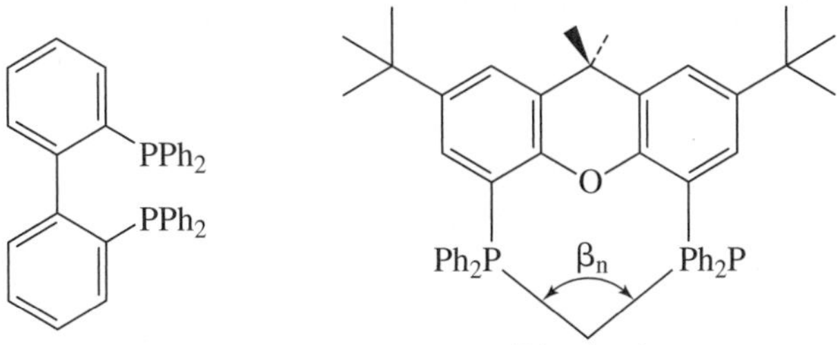

PPh_2
PPh_2
BISBI

Ph_2P β_n Ph_2P
Bite Angle

14.3.3 Monsanto Acetic Acid Process

The synthesis of acetic acid from methanol and CO is a process that has been used with great commercial success by Monsanto since 1971. The mechanism of this process is complex; a proposed outline is shown in Figure 14.16. As in the hydroformylation process, the individual steps of this mechanism are the characteristic types of organometallic reactions described previously in this chapter; the intermediates are 18- or 16-electron species having the capability to lose or gain, respectively, 2 electrons. (Solvent molecules may occupy empty coordination sites in the 4- and 5-coordinate 16-electron intermediates.) The first step, oxidative addition of CH_3I to $[RhI_2(CO)_2]^-$, is rate determining.[27]

The final step involving rhodium is reductive elimination of $IC(=O)CH_3$. Acetic acid is formed by hydrolysis of this compound. The catalytic species, $[Rh(CO)I_2]^-$, which may contain solvent in the empty coordination sites, is regenerated.

In addition to rhodium-based catalysts, iridium-based catalysts have also been developed for carbonylation of methanol. The iridium system, known as the *Cativa process*, follows a cycle similar to the rhodium system in Figure 14.16, beginning with oxidative addition of CH_3I to $[Ir(CO)_2I_2]^-$. The first step in the iridium system is several hundred times faster than in the Monsanto process; the second step, involving alkyl migration, is much slower, and it is rate determining for the Cativa process.[28] In addition to the catalytic cycle involving the anion $[Ir(CO)_2I_2]^-$, an alternative neutral cycle involving $Ir(CO)_3I$ or $Ir(CO)_2I$ has been reported.[29]

14.3.4 Wacker (Smidt) Process

The Wacker or Smidt process, used to synthesize acetaldehyde from ethylene, involves a catalytic cycle that uses $PdCl_4^{2-}$. A brief outline of a cycle proposed for this process is shown in Figure 14.17. The fourth step in this cycle is substantially more complex than that shown in the figure and has been the subject of much study.[30]

An important feature of this process is that it uses the ability of palladium to form complexes with the reactant ethylene, with the important chemistry of ethylene occurring while it is attached to the metal. In other words, the palladium modifies the chemical behavior of ethylene to enable reactions to occur that would not be possible for free

---

[26] L. A. van der Veen, P. H. Keeven, G. C. Schoemaker, J. N. H. Reek, P. C. J. Kramer, P. W. N. M. van Leeuwen, M. Lutz, and A. L. Spek, *Organometallics*, **2000**, *19*, 872 and references cited therein.

[27] A discussion of the mechanism of this reaction can be found in D. Forster and T. W. Deklava, *J. Chem. Ed.*, **1986**, *63*, 204, and references therein.

[28] M. Cheong, R. Schmid, and T. Ziegler, *Organometallics*, **2000**, *19*, 1973 and references therein.

[29] J. Forster, *J. Chem. Soc., Dalton Trans.*, **1979**, 1639; A. Haynes, et al., *J. Am. Chem. Soc.*, **2004**, *126*, 2847.

[30] For example, see J. M. Francis and P. M. Henry, *Organometallics*, **1991**, *10*, 3498; **1992**, *11*, 2832.

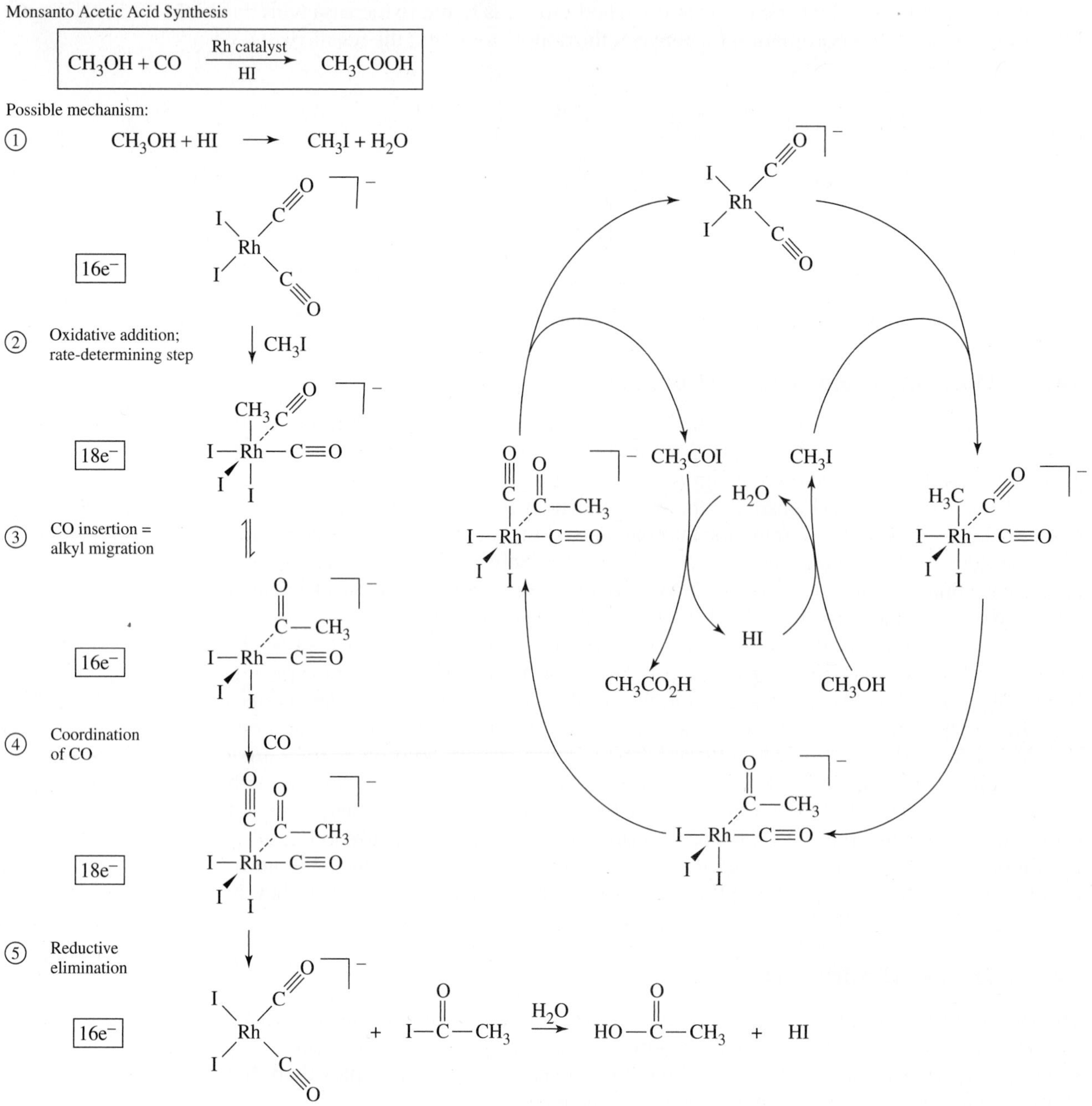

Monsanto Acetic Acid Synthesis

$$CH_3OH + CO \xrightarrow[HI]{\text{Rh catalyst}} CH_3COOH$$

Possible mechanism:

① $CH_3OH + HI \longrightarrow CH_3I + H_2O$

16e⁻ $[RhI_2(CO)_2]^-$

② Oxidative addition; rate-determining step (CH_3I)

18e⁻ $[CH_3RhI_3(CO)_2]^-$

③ CO insertion = alkyl migration

16e⁻ $[(CH_3CO)RhI_3(CO)]^-$

④ Coordination of CO (CO)

18e⁻ $[(CH_3CO)RhI_3(CO)_2]^-$

⑤ Reductive elimination

16e⁻ $[RhI_2(CO)_2]^-$ + $I{-}C({=}O){-}CH_3$ $\xrightarrow{H_2O}$ $HO{-}C({=}O){-}CH_3$ + HI

Catalytic cycle labels: CH_3COI, H_2O, HI, CH_3CO_2H, CH_3I, CH_3OH

FIGURE 14.16 Monsanto Acetic Acid Process.

(From A. Haynes, B. E. Mann, D. J. Gulliver, G. E. Morris, and P. M. Maitlis, *J. Am. Chem. Soc.*, **1991**, *113*, 8567; M. Cheong, R. Schmid, and T. Ziegler, *Organometallics*, **2000**, *19*, 1973.)

ethylene. Incidentally, the first ethylene complex with palladium in Figure 14.17 is isoelectronic with Zeise's complex, $[PtCl_3(\eta^2\text{-}H_2C{=}CH_2)]^-$.

14.3.5 Hydrogenation by Wilkinson's Catalyst

Wilkinson's catalyst, $RhCl(PPh_3)_3$, is not itself an organometallic compound but participates in the same types of reactions as expected for 4-coordinate organometallic compounds; for example, many reactions bear similarities to Vaska's catalyst,

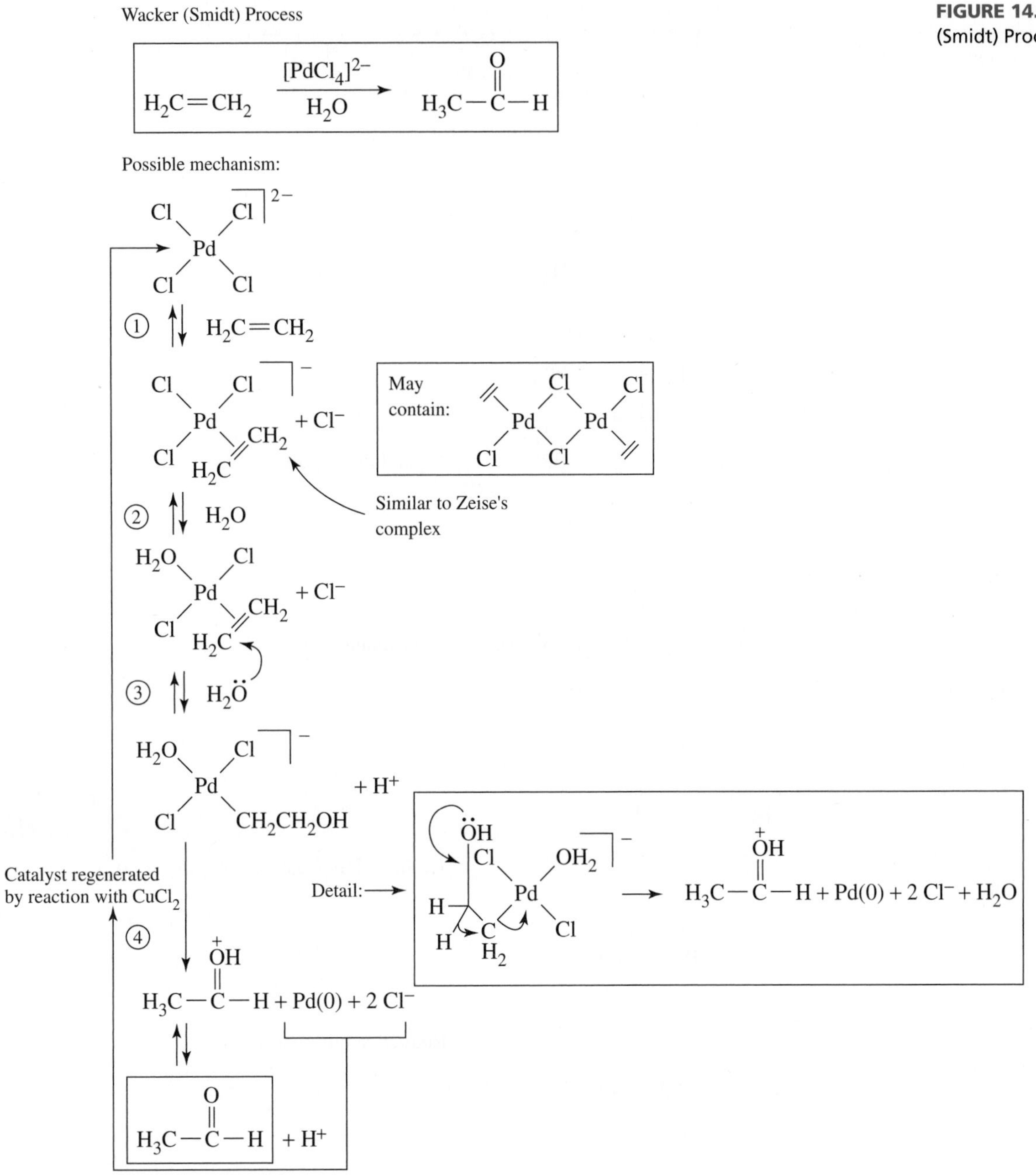

Wacker (Smidt) Process

$$H_2C{=}CH_2 \xrightarrow[H_2O]{[PdCl_4]^{2-}} H_3C{-}C({=}O){-}H$$

Possible mechanism:

FIGURE 14.17 Wacker (Smidt) Process.

trans-$IrCl(CO)(PPh_3)_2$. $RhCl(PPh_3)_3$ participates in a wide variety of catalytic and non-catalytic processes. The bulky phosphine ligands play an important role in making the complex selective—for example, they limit coordination of Rh to unhindered positions on alkenes. One example, involving catalytic hydrogenation of an alkene, is shown in Figure 14.18.[31]

The first two steps in this process give the catalytic species $RhCl(H)_2(PPh_3)_2$, which has a vacant coordination site. A C=C double bond can coordinate to this site, gain the two hydrogens coordinated to Rh, and subsequently leave if the double bond is

[31] B. R. James, *Adv. Organomet. Chem.*, **1979**, *17*, 319; see also J. P. Collman, L. S. Hegedus, J. R. Norton, and R. G. Finke, *Principles and Applications of Organotransition Metal Chemistry*, University Science Books, Mill Valley, CA, 1987, pp. 531–535 and references therein.

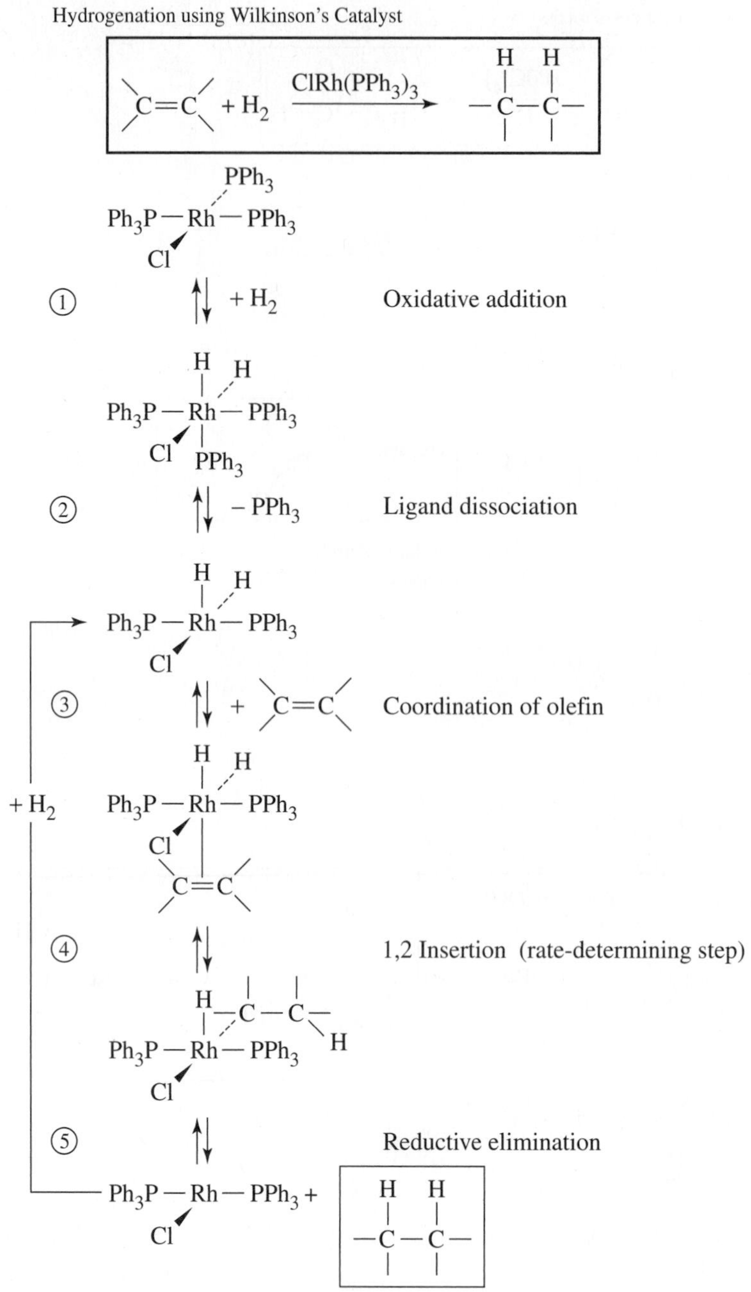

FIGURE 14.18 Catalytic Hydrogenation Involving Wilkinson's Catalyst.

Hydrogenation using Wilkinson's Catalyst

$$\mathrm{>C{=}C< + H_2 \xrightarrow{ClRh(PPh_3)_3} -\overset{H}{\underset{|}{C}}-\overset{H}{\underset{|}{C}}-}$$

$ClRh(PPh_3)_3$

① $+H_2$ Oxidative addition

$H_2RhCl(PPh_3)_3$

② $-PPh_3$ Ligand dissociation

$H_2RhCl(PPh_3)_2$

③ $+ >C{=}C<$ Coordination of olefin

$H_2RhCl(PPh_3)_2(>C{=}C<)$

④ 1,2 Insertion (rate-determining step)

$HRhCl(PPh_3)_2(-\overset{|}{C}-\overset{|}{C}-H)$

⑤ Reductive elimination

$Ph_3P-Rh(Cl)-PPh_3 + -\overset{H}{\underset{|}{C}}-\overset{H}{\underset{|}{C}}-$

$+H_2$

not sterically hindered. This effect is illustrated in Table 14.3, which shows relative rates of hydrogenation using Wilkinson's catalyst.

In molecules containing several double bonds, the least hindered double bonds are reduced. The most hindered positions cannot coordinate effectively to Rh, largely because of the presence of the bulky phosphines, and hence do not react as rapidly. Consequently, Wilkinson's catalyst is useful for selective hydrogenations of C=C bonds that are not sterically hindered. Examples are shown in Figure 14.19.

Because the selectivity of Wilkinson's catalyst is largely a consequence of the bulky triphenylphosphine ligands, the selectivity can be fine-tuned somewhat by using phosphines having different cone angles than PPh_3. Wilkinson's catalyst and similar compounds having different phosphine ligands are useful in a variety of other catalytic cycles.

TABLE 14.3 Relative Rates of Hydrogenation Using Wilkinson's Catalyst at 25 °C

Compound Hydrogenated	Rate Constant × 100 ($L\ mol^{-1}\ s^{-1}$)
(cyclohexene)	31.6
H, H, H_3C, C_2H_5	9.9
H_3C, H, H, C_2H_5	1.8
CH_3 (1-methylcyclohexene)	0.6
H_3C, CH_3, C_2H_5, C_2H_5	< 0.1

Source: A. J. Birch and D. H. Williamson, *Org. React.*, **1976**, *24*, 1.

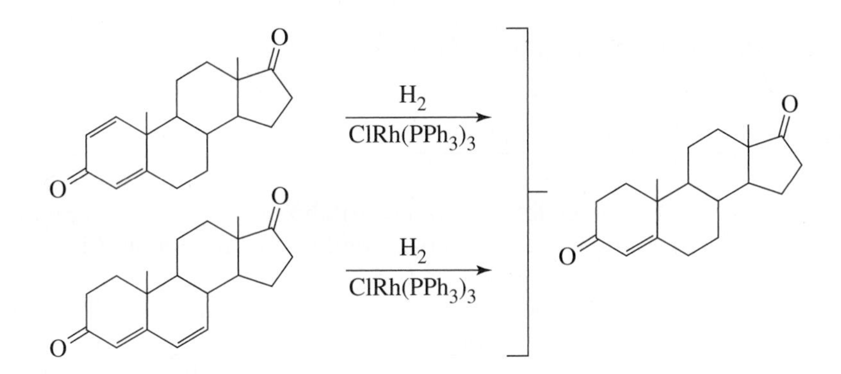

FIGURE 14.19 Selective Hydrogenation by Wilkinson's Catalyst.

14.3.6 Olefin Metathesis

Olefin metathesis, first discovered in the 1950s, involves the formal exchange of $:CR_2$ fragments (R = H or alkyl) between alkenes, also known as olefins. For example, metathesis between molecules of formula $H_2C{=}CH_2$ and $HRC{=}CHR$ would yield two molecules of $H_2C{=}CHR$:

$$\begin{matrix} H_2C & & CHR \\ \| & & \| \\ H_2C & & CHR \end{matrix} \rightleftharpoons \begin{matrix} H_2C{=}CHR \\ H_2C{=}CHR \end{matrix}$$

New double bonds are formed between the top and bottom two carbons in the diagram, and the original double bonds are severed.[32]

[32] Discussions of the history of the metathesis reaction written by two of its discoverers can be found in R. L. Banks, *Chemtech*, **1986**, *16*, 112 and H. Eleuterio, *Chemtech*, **1991**, *21*, 92.

EXAMPLE

Predict the possible products of metathesis of the following olefins. Be sure to consider that two molecules of the same structure can also metathesize (undergo self-metathesis).

a. Between propene and 1-butene

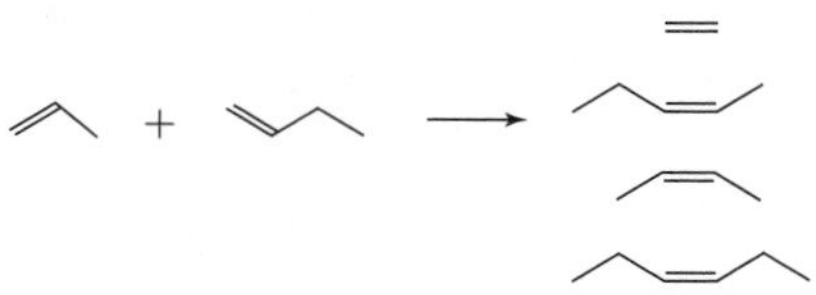

b. Between ethylene and cyclohexene.

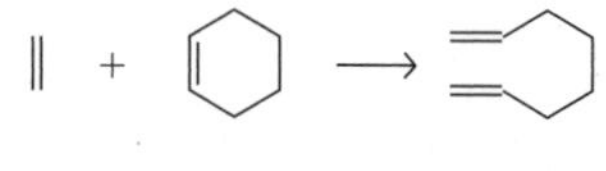

Example **b** is an example of **ring-opening metathesis (ROM)**, in which metathesis opens a ring of a cyclic alkene. The reverse of this process is called, appropriately, **ring-closing metathesis (RCM)**. An example of ring-closing metathesis is shown later, in Figure 14.29.

▶ **Exercise 14.6** Predict the products of metathesis:

a. Between two molecules of propene
b. Between propene and cyclopentene

Metathesis, which is reversible and can be catalyzed by a variety of organometallic complexes, has been the subject of considerable investigation and is now highly important industrially. In recognition of the significance of this area and the fundamental research to shed light on the pathway of metathesis, the 2005 Nobel Prize in Chemistry was awarded to three leaders in its development, Yves Chauvin, Robert Grubbs, and Richard Schrock. Their Nobel lectures provide both an excellent background on the experiments leading to an understanding of metathesis and the flavor of the early work in this field.[33]

In the early stages of metathesis research, several mechanisms were proposed. We will consider three of the most important: an alkyl group transfer mechanism, a diolefin ("pairwise") mechanism, and a mechanism involving carbene complexes ("non-pairwise"). These are shown schematically in Figure 14.20.

AKLYL EXCHANGE The alkyl exchange mechanism received an early test. In 1968 Calderon and coworkers reacted two 2-butenes, $H_3C{-}CH{=}CH{-}CH_3$ and the fully deuterated $D_3C{-}CD{=}CD{-}CD_3$. If aklyl transfer could occur, $-CH_3$ and $-CD_3$ groups would be expected to exchange, giving mixtures of H and D atoms on each side of the double bond ($=CH{-}CD_3$ and $=CD{-}CH_3$). The result, shown in Figure 14.21, was that the halves of the molecule on each side of the double bond exchanged as a whole; there was no evidence for exchange of alkyl groups.[34]

[33] R. H. Grubbs, *Angew. Chem, Int. Ed.*, **2006**, *45*, 3760; R. R. Schrock, *Adv. Synth. Cat.*, **2007**, *349*, 41; Y. Chauvin, *Adv. Synth. Cat.*, **2008**, *349*, 27. These Nobel lectures can also be found on the Nobel Prize Web site at http://nobelprize.org/nobel_prizes/chemistry/laureates/2005/.
[34] N. Calderon, E. A. Ofstead, J. P. Ward, W. A. Judy, and K. W. Scott, *J. Am. Chem. Soc.*, **1968**, *90*, 4133.

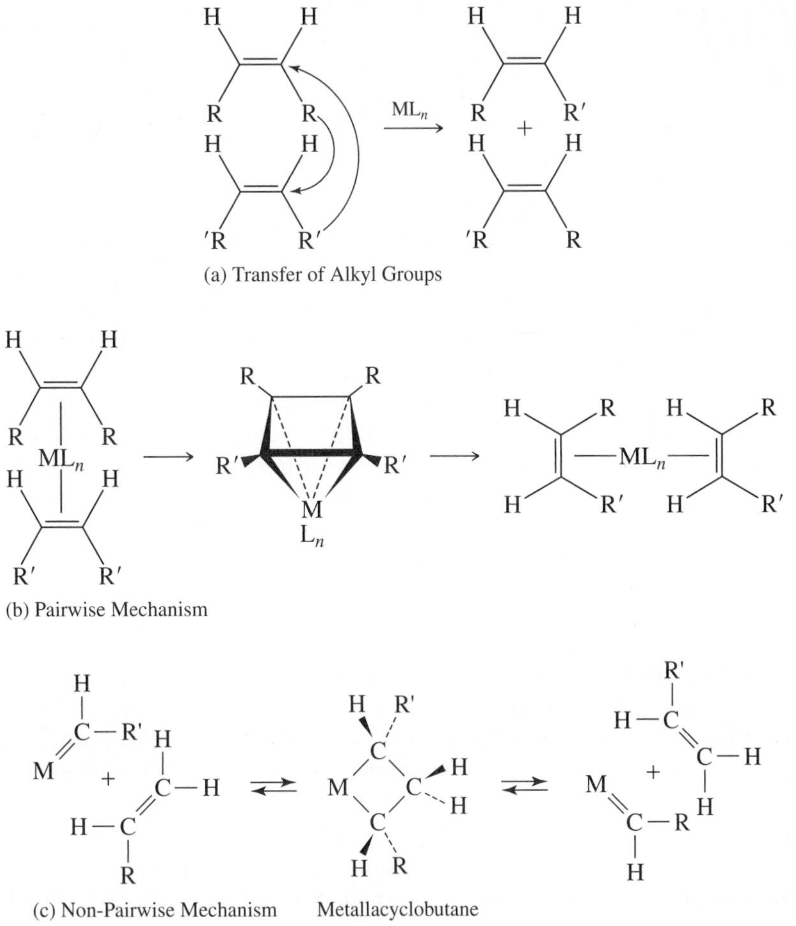

(a) Transfer of Alkyl Groups

(b) Pairwise Mechanism

(c) Non-Pairwise Mechanism Metallacyclobutane

FIGURE 14.20 Proposed Metathesis Mechanisms (a) Transfer of Alkyl Groups (b) Pairwise Mechanism (c) Non-Pairwise Mechanism.

$$\begin{matrix} H_3C{-}CH{=}CH{-}CH_3 \\ + \\ D_3C{-}CD{=}CD{-}CD_3 \end{matrix} \longrightarrow H_3C{-}CH{=}CD{-}CD_3$$

The *only* product

FIGURE 14.21 Experiment to Test the Alkyl Exchange Mechanism.

Disinguishing between the other two mechanisms was somewhat more challanging; each could account for the results in Figure 14.21.

DIOLEFIN (PAIRWISE) MECHANISM In 1967 Bradshaw proposed that "the dismutation of olefins should proceed via a quasicyclobutane intermediate."[35] In this mechanism the two alkenes would first coordinate to a transition metal, forming a quasicyclobutane as shown in Figure 14.22, after which the metal-complex intermediate would break apart to form the new alkenes. Because formation of the intermediate would involve two alkenes attaching to the metal in pairs, this has been called both the *diolefin* and *pairwise mechanism*.

In 1971 Hérisson and Chauvin reported the experiment shown in Figure 14.23. Metathesis of cyclopentene and 2-pentene, according to the diolefin mechanism, would

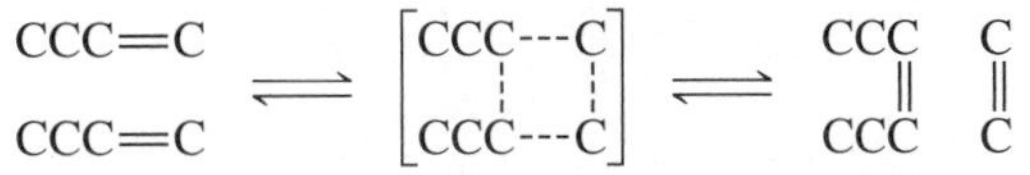

FIGURE 14.22 Quasicyclobutane Intermediate.

[35] C. P. C. Bradshaw, E. J. Howman, and L. Turner, *J. Catal.*, **1967**, *7*, 269.

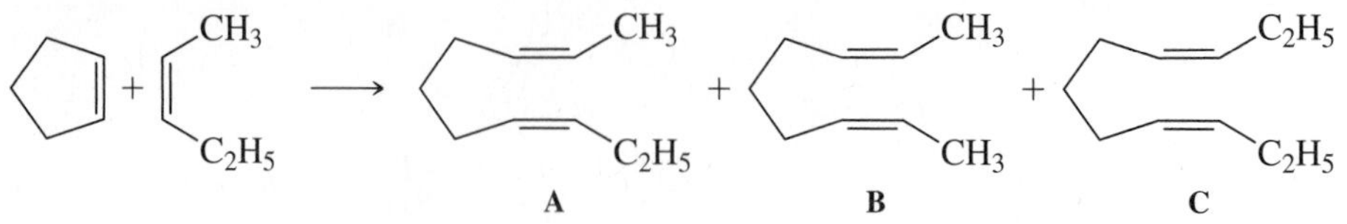

FIGURE 14.23 Hérisson and Chauvin's Experiment. Metathesis of cyclopentene and 2-pentene.

form the first product shown, **A**, with terminal methyl and ethyl groups. This product could undergo further metathesis with $CH_3{-}CH{=}CH{-}C_2H_5$ to form products **B** and **C** (in addition to *trans* isomers and other products). Because the metathesis reactions are equilibria, the ultimate distribution of products would be expected to be in the proportion expected statistically on the basis of equal numbers of terminal methyl and ethyl groups in 2-pentene.[36] In such a case, analysis of products should initially show a higher proportion of the first product, and the statistical distribution of products would come later. In Hérisson and Chauvin's work, however, the statistical distribution of products was found even if the reaction was quenched before equilibrium was achieved, arguing against the pairwise mechanism.

▶ **Exercise 14.7** Show how the pairwise mechanism could form the last two products by metathesis involving the first product.

CARBENE (NON-PAIRWISE) MECHANISM Hérisson and Chauvin proposed that metathesis reactions are catalyzed by carbene (alkylidene) complexes that react with alkenes via the formation of a cyclic intermediate, a metallacyclobutane, as shown in Figure 14.20(c). In this mechanism, a metal carbene complex first reacts with an alkene to form the metallacyclobutane. This intermediate can either revert to reactants or form new products; because all steps in the process are in equilibria, an equilibrium mixture of alkenes results. This non-pairwise mechanism would enable the statistical mixture of products to form from the start by the action of catalytic amounts of the necessary carbene complexes, with both R and R′ groups, as shown in Figure 14.20(c).

Subsequent to the work of Hérisson and Chauvin, a variety of reactions continued to show results consistent with a non-pairwise mechanism involving carbene complexes and a metallacyclobutadiene intermediate. For example, in the reaction shown in Figure 14.24, the gaseous products $H_2C{=}CH_2$, $D_2C{=}CD_2$, and $H_2C{=}CD_2$ formed in the expected statistical ratio early in the reaction, with no indication that the mixed product $H_2C{=}CD_2$, the first product by the pairwise mechanism, forms before the others.[37]

The first direct demonstration that a carbene complex could engage in metathesis involved the carbene complex shown in Figure 14.25, which was able to replace its carbene ligand with a terminal alkene in a reaction consistent with the non-pairwise mechanism.[38]

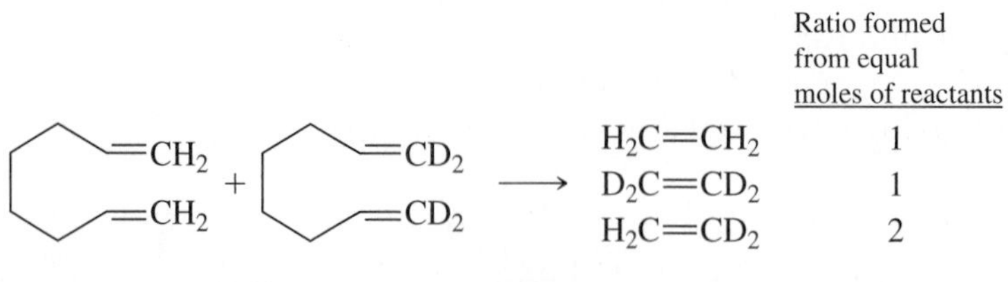

FIGURE 14.24 A Mechanistic Test.

[36] J.-L. Hérisson and Y. Chauvin, *Makromol. Chem.*, **1971**, *141*, 161.
[37] R. H. Grubbs, P. L. Burk, and D. D. Carr, *J. Am. Chem. Soc.*, **1975**, *97*, 3265.
[38] T. J. Katz and J. McGinness, *J. Am. Chem. Soc.*, **1975**, *97*, 1592 and **1977**, *99*, 1903.

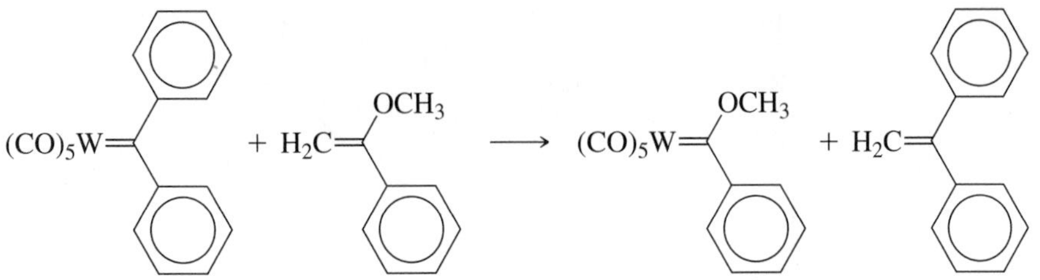

FIGURE 14.25 Metathesis of a Carbene Complex.

Although alternatives to the non-pairwise mechanism were proposed, the preponderance of evidence over many years has strongly supported the role of carbene complexes as catalysts for olefin metathesis. This mechanism, now known as the **Chauvin mechanism**, is believed to be the pathway of the majority of transition metal–catalyzed olefin metathesis reactions.[39]

The most thoroughly studied catalysts that effect alkene metathesis are of two types, shown in Figure 14.26. Schrock metathesis catalysts are the most effective of all metathesis catalysts but in general are highly sensitive to oxygen and water. These catalysts are now available commercially; the catalyst having M = Mo and R = isopropyl is sometimes called Schrock's catalyst. An example of a reaction utilizing this catalyst is the final step of the synthesis of the natural product dactylol, shown in Figure 14.27.[40]

The reaction shown in Figure 14.27 is an example of **ring-closing metathesis (RCM)**, in which the metathesis of two double bonds leads to ring formation. Like ordinary metathesis, ring-closing metathesis is believed to occur by way of a metallacyclobutane intermediate; this intermediate is responsible for joining the originally separate carbons into a ring.

Grubbs metathesis catalysts in general have less catalytic activity than Schrock catalysts but are less sensitive to oxygen and water. They are also substantially less expensive than the molybdenum and tungsten catalysts. The catalyst having R = cyclohexyl, X = Cl, and R′ = phenyl has received particular attention and is marketed as Grubbs's catalyst. One requirement of these catalysts is the presence of bulky phosphine ligands. This bulkiness facilitates phosphine dissociation, a key step in the proposed mechanism involving the Grubbs catalyst, shown in Figure 14.28.[41]

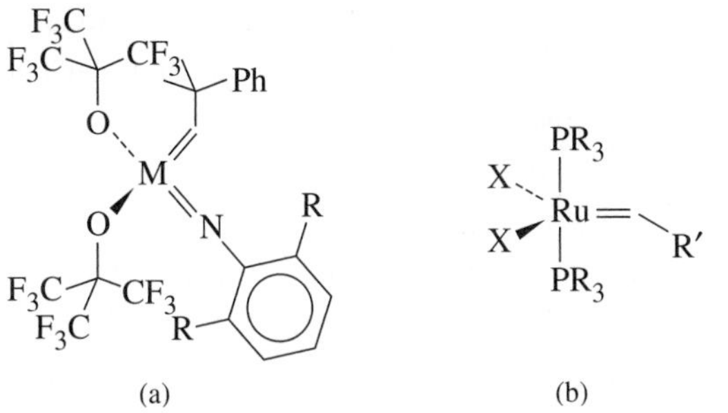

FIGURE 14.26 Metathesis Catalysts. (a) Schrock catalyst (M = Mo, W). (b) Grubbs catalyst (X = Cl, Br).

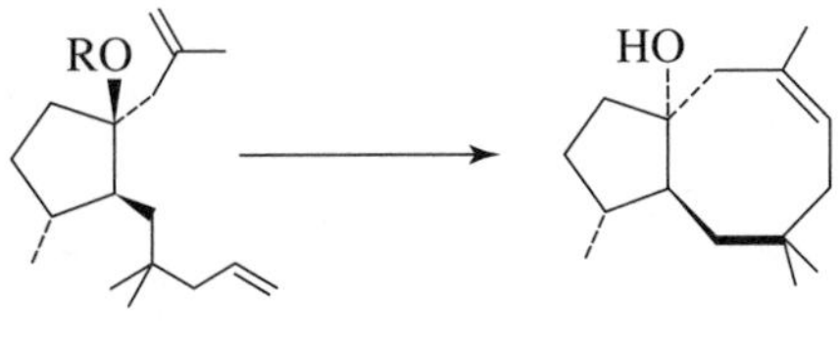

FIGURE 14.27 Ring-Closing Metathesis (RCM).

[39] For a short outline of the development of olefin metathesis, see C. P. Casey, *J. Chem. Educ.*, **2006**, *83*, 192.
[40] A. Fürstner and K. Langemann, *J. Org. Chem.*, **1996**, *61*, 8746.
[41] E. L. Dias, S. T. Nguyen, and R. H. Grubbs, *J. Am. Chem. Soc.*, **1997**, *119*, 3887.

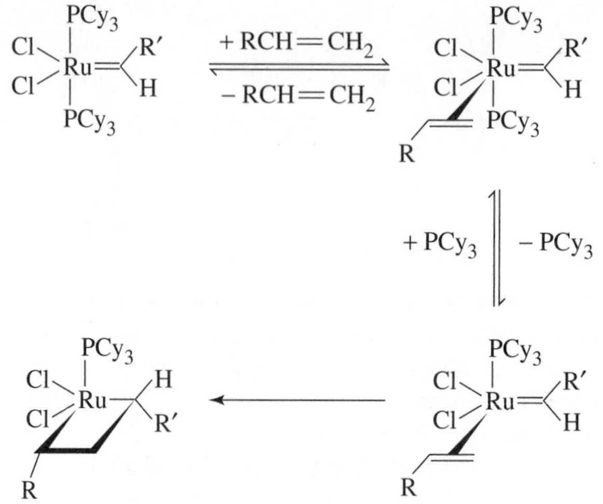

FIGURE 14.28 Proposed Mechanism for Formation of Metallacyclobutane from Ruthenium Catalyst.

Although much research in the field of homogeneous metathesis catalysis has focused on complexes resembling those of Schrock and Grubbs, various other avenues have also been pursued. A promising recent development has been the introduction of catalysts that contain ruthenium and *N*-heterocyclic carbene ligands.[42] These ligands exceed trialkylphosphines in steric requirements and are more strongly electron donating;[43] both features support improved catalytic activity. An example of such a catalyst, and a ring-closing metathesis process that it catalyzes, are shown in Figure 14.29.[44]

Such catalysts compare favorably in activity with Schrock's catalyst and typically are thermally stable with low sensitivity toward oxygen and water. The process shown in Figure 14.29 has also been performed using Schrock's catalyst and Grubbs's catalyst. As shown in Table 14.4, the *N*-heterocyclic catalyst compares favorably with Schrock's catalyst and is far superior to Grubbs's catalyst—at least for this reaction.

An interesting variation on olefin metathesis is the use of carbene complexes to catalyze alkene polymerization, also via a metallacyclobutane intermediate. An example is the use of W(CH—*t*-Bu)(OCD_2—*t*-Bu)$_2Br_2$ as a catalyst in the ring-opening polymerization of norbornene in the presence of $GaBr_3$, as shown in Figure 14.30.[45] Proton and ^{13}C NMR data are consistent with the proposed structure of the metallacyclobutane, as well as the polymer growing off the carbene carbon.

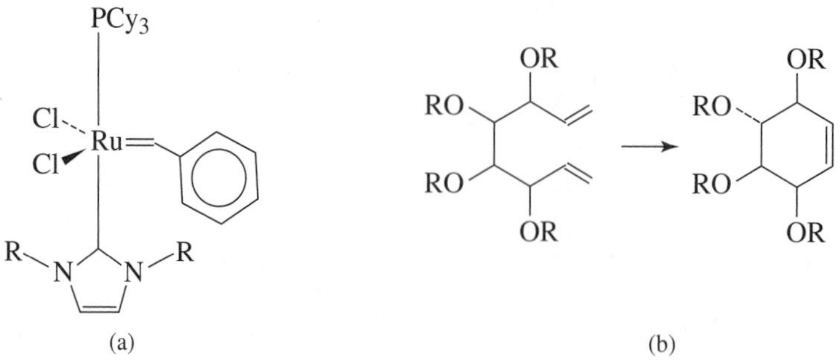

FIGURE 14.29 Ring-Closing Metathesis Catalyzed by the *N*-Heterocyclic Carbene Complex. (a) Catalyst (R = mesityl) (b) Ring-closing reaction (R = benzyl).

[42] M. S. Sanford, J. A. Love, and R. H. Grubbs, *J. Am. Chem. Soc.*, **2001**, *123*, 6543.
[43] J. Huang, H.-J. Schanz, E. D. Stevens, and S. P. Nolan, *Organometallics*, **1999**, *18*, 2370.
[44] L. Ackermann, D. El Tom, and A. Fürstner, *Tetrahedron*, **2000**, *56*, 2195.
[45] J. Kress, J. A. Osborn, R. M. E. Greene, K. J. Ivin, and J. J. Rooney, *J. Am. Chem. Soc.*, **1987**, *109*, 899.

TABLE 14.4 Relative Activity of Metathesis Catalysts

Catalyst	Reaction Time (*h*)	Yield (%)
Schrock's catalyst	1	92
Grubbs's catalyst	60	32
Catalyst in Figure 14.29	2	89

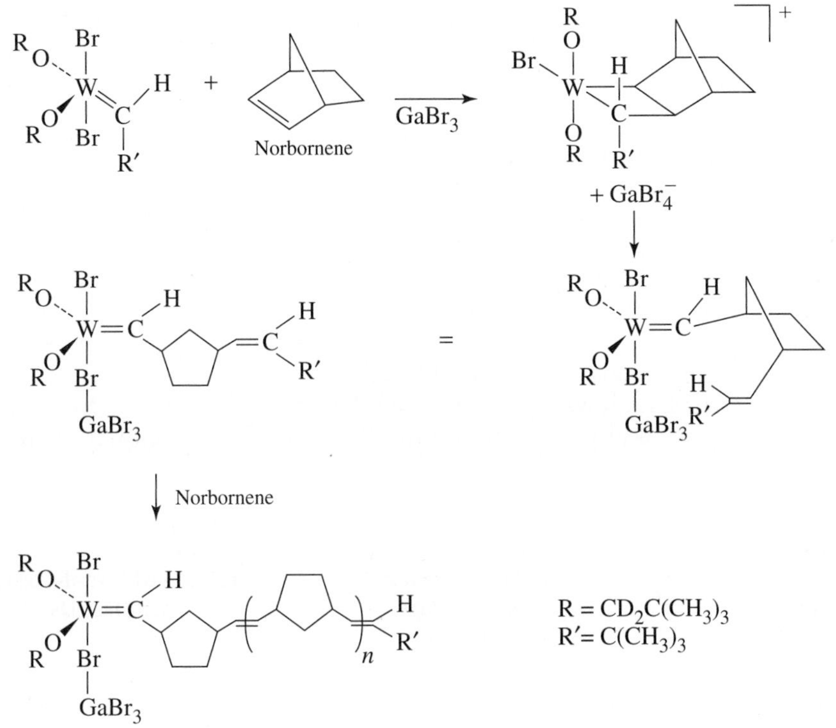

Norbornene

$GaBr_3$

$+ GaBr_4^-$

Norbornene

$R = CD_2C(CH_3)_3$
$R' = C(CH_3)_3$

FIGURE 14.30 Polymerization of Norbornene using a Carbene Catalyst.

Alkynes can also undergo metathesis reactions catalyzed by transition-metal carbyne complexes. The intermediates in these reactions are believed to be metallacyclobutadiene species, formed from the addition of an alkyne across a metal–carbon triple bond of the carbyne (Figure 14.31). The structures of a variety of metallacyclobutadiene complexes have been determined, and some have been shown to catalyze alkyne metathesis.[46]

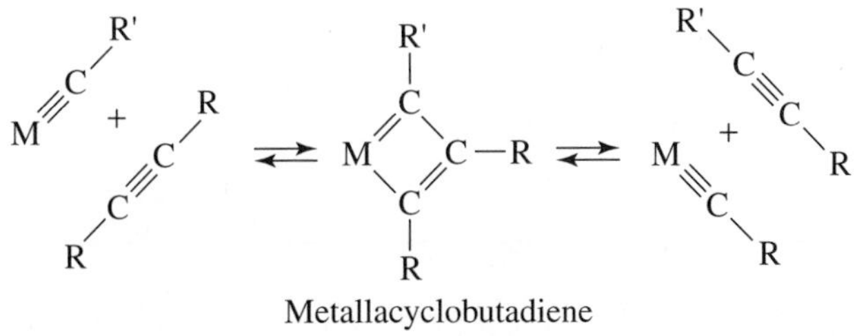

Metallacyclobutadiene

FIGURE 14.31 Alkyne Metathesis.

[46] W. A. Nugent and J. M. Mayer, *Metal–Ligand Multiple Bonds*, Wiley Interscience, New York, 1988, p. 311 and references therein; U. H. W. Bunz and L. Kloppenburg, *Angew. Chem., Int. Ed.*, **1999**, *38*, 478.

14.4 HETEROGENEOUS CATALYSTS

In addition to homogeneous catalytic processes, heterogeneous processes, involving solid catalytic species, are very important, although the exact nature of the reactions occurring on the surface of the catalyst may be extremely difficult to ascertain. The vast majority of the organic chemicals produced in greatest quantities in the United States are produced commercially by processes that involve metal catalysts; most of these processes involve heterogeneous catalysis. Selected examples from 2008 are given in Table 14.5.[47]

In many cases, the methods of preparing the catalysts and information on their function are proprietary, the product of substantial corporate investment. Nevertheless, it is important to mention several of these processes as important practical applications of organometallic reactions.

14.4.1 Ziegler–Natta Polymerizations

In 1955, Ziegler and coworkers reported that solutions of $TiCl_4$ in hydrocarbon solvents in the presence of $Al(C_2H_5)_3$ gave heterogeneous solutions capable of polymerizing ethylene.[48] Subsequently, many other heterogeneous processes were developed for polymerizing alkenes, using aluminum alkyls in combination with transition-metal complexes. An outline of a possible mechanism for the Ziegler–Natta process proposed by Cossee and Arlman is given in Figure 14.32.[49]

First, reaction of $TiCl_4$ with aluminum alkyl gives $TiCl_3$, which on further reaction with the aluminum alkyl gives a titanium alkyl complex, as shown in the figure.

TABLE 14.5 Leading Organic Compounds and Metal Catalysts

Compound	U.S. Production 2008 ($\times 10^9$ kg)	Example of Metal-Containing Catalyst Used
Ethylene	22.55	
Propylene	14.79	$TiCl_3$ or $TiCl_4$ + AlR_3 (R = alkyl)
1,2-Dichloroethane	8.97	$FeCl_3$, $AlCl_3$
Benzene	5.61	Pt on Al_2O_3 support
Urea	5.29	
Ethylbenzene	4.10	$AlCl_3$
Styrene	4.10	ZnO, Cr_2O_3
Cumene	3.39	
Ethylene oxide	2.90	Ag
1,3 Butadiene	1.63	Fe_2O_3 or other metal oxide
Vinyl acetate	1.27	Pd salts
Acrylonitrile	1.02	$BiPMe_{12}O_{40}$

[47] R. Chang and W. Tikkanen, *The Top Fifty Industrial Chemicals*, Random House, New York, 1988; *Chem. Eng. News*, July 6, 2009, p. 53.

[48] K. Ziegler, E. Holzkamp, H. Breiland, and H. Martin, *Angew. Chem.*, **1955**, *67*, 541.

[49] J. Cossee, *J. Catal.*, **1964**, *3*, 80; E. J. Arlman, *J. Catal.*, **1964**, *3*, 89; E. J. Arlman and J. Cossee, *J. Catal.*, **1964**, *3*, 99.

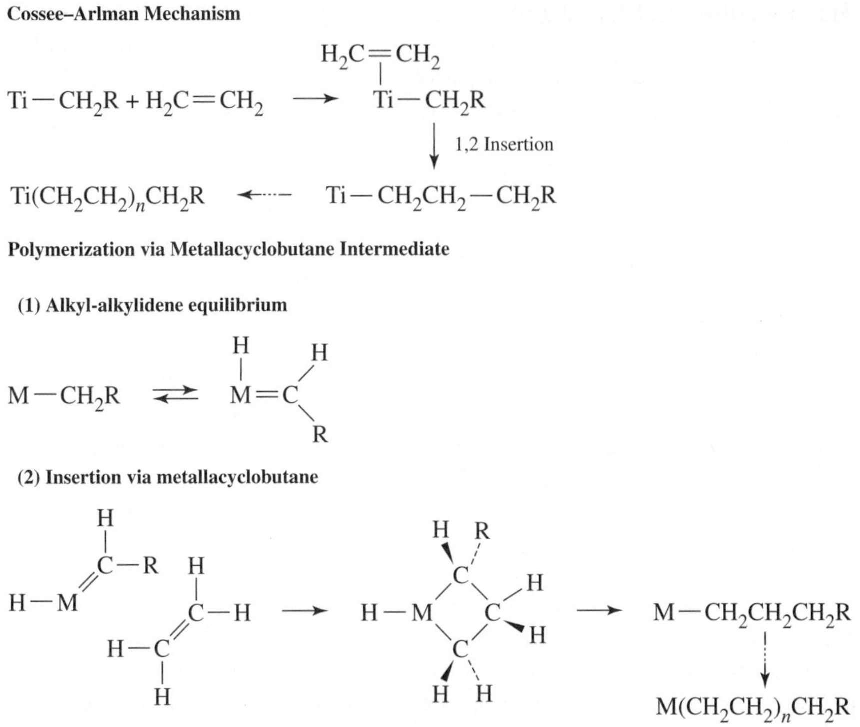

Cossee–Arlman Mechanism

$$Ti-CH_2R + H_2C{=}CH_2 \longrightarrow \underset{Ti-CH_2R}{\overset{H_2C=CH_2}{|}}$$

$$\downarrow \text{1,2 Insertion}$$

$$Ti(CH_2CH_2)_nCH_2R \longleftarrow Ti-CH_2CH_2-CH_2R$$

Polymerization via Metallacyclobutane Intermediate

(1) Alkyl-alkylidene equilibrium

$$M-CH_2R \rightleftharpoons \overset{H}{\overset{|}{M}}{=}C(H)(R)$$

(2) Insertion via metallacyclobutane

H–M(H)=C(H)–R + H(H)C=C(H)–H ⟶ H–M[metallacyclobutane: C(H)(R), C(H)(H), C(H)(H)] ⟶ $M-CH_2CH_2CH_2R$ ⇣ $M(CH_2CH_2)_nCH_2R$

FIGURE 14.32 Ziegler–Natta Polymerization.

Ethylene (or propylene) can then insert into the titanium–carbon bond, forming a longer alkyl. This alkyl is further susceptible to insertion of ethylene to lengthen the chain. Although the mechanism of the Ziegler–Natta process has proved difficult to understand, direct insertions of multiply bonded organics into titanium–carbon bonds have been demonstrated, supporting the Cossee–Arlman mechanism.[50]

However, an alternative mechanism, involving a metallacyclobutane intermediate, has also been proposed.[51] This mechanism, also shown in Figure 14.32, involves the initial formation of alkylidene from a metal alkyl complex, followed by addition of ethylene to give the metallacyclobutane, which then yields a product having ethylene inserted into the original metal–carbon bond. Distinguishing between these mechanisms has been a long and difficult process, but experiments by Grubbs and coworkers have strongly supported the Cossee–Arlman mechanism as the likely pathway for polymerization in most cases.[52] In at least one example, however, there is strong evidence for ethene polymerization involving a metallacycle intermediate.[53]

14.4.2 Water Gas Reaction

This reaction occurs at elevated temperatures and pressures between water (steam) and natural sources of carbon, such as coal or coke:

$$H_2O + C \longrightarrow H_2 + CO$$

[50] J. J. Eisch, A. M. Piotrowski, S. K. Brownstein, E. J. Gabe, and F. L. Lee, *J. Am. Chem. Soc.*, **1985**, *107*, 7219.
[51] K. J. Ivin, J. J. Rooney, C. D. Stewart, and M. L. H. Green, *Chem. Commun. (Cambridge)*, **1978**, 604.
[52] L. Clauson, J. Sato, S. L. Buchwald, M. L. Steigerwald, and R. H. Grubbs, *J. Am. Chem. Soc.*, **1985**, *107*, 3377. For a brief review of experiments used to distinguish between the two mechanisms, see G. O. Spessard and G. L. Miessler, *Organometallic Chemistry*, Prentice Hall, Upper Saddle River, NJ, 1997, pp. 357–369.
[53] W. H. Turner and R. R. Schrock, *J. Am. Chem. Soc.*, **1982**, *104*, 2331.

The products of this reaction, an equimolar mixture of H_2 and CO, called *synthesis gas* or *syn gas* (some CO_2 may be produced as a by-product), can be used with metallic heterogeneous catalysts in the synthesis of a variety of useful organic products. For example, the **Fischer–Tropsch process**, developed by German chemists in the early 1900s, uses transition metal catalysts to prepare hydrocarbons, alcohols, alkenes, and other products from synthesis gas.[54] For example:

$$H_2 + CO \longrightarrow \text{Alkanes} \qquad \text{Co catalyst}$$

$$3\,H_2 + CO \longrightarrow CH_4 + H_2O \qquad \text{Ni catalyst}$$

$$2\,H_2 + CO \longrightarrow CH_3OH \qquad \text{Co or Zn/Cu catalyst}$$

Various heterogeneous catalysts are used industrially—for example, transition metals on Al_2O_3 and mixed transition-metal oxides.

Most of these processes have been conducted under heterogenous conditions. However, there has been considerable interest in developing homogenous systems to catalyze the Fischer–Tropsch conversion.

These processes for obtaining synthetic fuels were used by a number of countries during World War II. They are, however, uneconomical in most cases, because hydrogen and carbon monoxide in sufficient quantities must be obtained from coal or petroleum sources. Currently, South Africa, which has large coal reserves, makes the greatest use of Fischer–Tropsch reactions in the synthesis of fuels in its Sasol plants.

In **steam reforming**, natural gas—consisting chiefly of methane—is mixed with steam at high temperatures and pressures over a heterogeneous catalyst to generate carbon monoxide and hydrogen:

$$CH_4 + H_2O \longrightarrow CO + 3\,H_2 \qquad \text{Ni catalyst, 700 °C to 1000 °C}$$

(Other alkanes also react with steam to give mixtures of CO and H_2.) Steam reforming is the principal industrial source of hydrogen gas. Additional hydrogen can be produced by recycling the CO to react further with steam in the **water gas shift reaction**:

$$CO + H_2O \longrightarrow CO_2 + H_2 \qquad \text{Fe–Cr or Zn–Cu catalyst, 400 °C}$$

This reaction is favored thermodynamically: at 400 °C, $\Delta G° = -14.0$ kJ/mol. Removal of CO_2 by chemical means from the product can yield hydrogen of greater than 99 percent purity. This reaction has been studied extensively with the objective of being able to catalyze formation of H_2 *homo*geneously.[55] An example is shown in Figure 14.33.[56] However, these processes have not yet proved efficient enough for commercial use.

In general, these processes, when performed using heterogeneous catalysts, require significantly elevated temperatures and pressures. Consequently, as in the case of the water gas shift reaction, there has been great interest in developing homogeneous catalysts that can perform the same functions under much milder conditions.

---

[54] E. Fischer and H. Tropsch, *Brennst. Chem.*, **1923**, *4*, 276.
[55] For example, see M. M. Taqui Khan, S. B. Halligudi, and S. Shukla, *Angew. Chem., Int. Ed.*, **1988**, *27*, 1735 and R. Ziessel, *Angew. Chem., Int. Ed.*, **1992**, *30*, 844.
[56] J. P. Collins, R. Ruppert, and J. P. Sauvage, *Nouv. J. Chim.*, **1985**, *9*, 395.

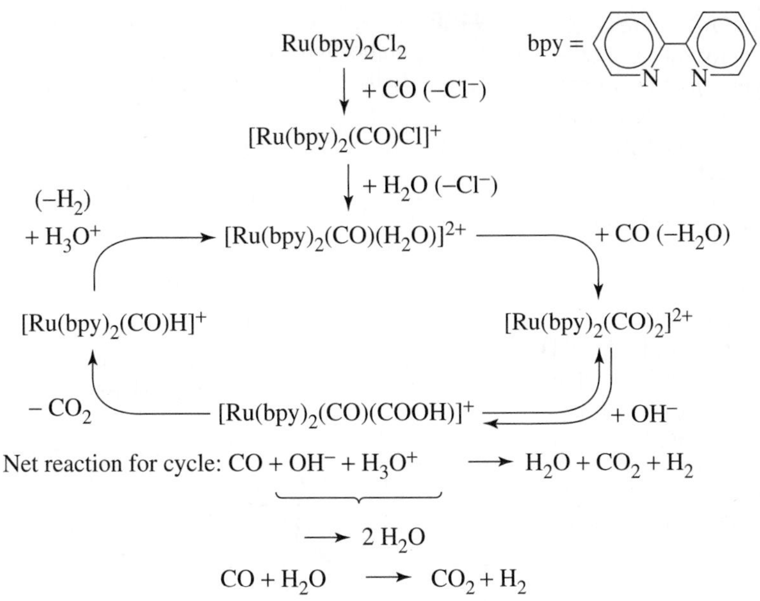

$Ru(bpy)_2Cl_2$ bpy = (2,2′-bipyridine)

$\downarrow + CO\ (-Cl^-)$

$[Ru(bpy)_2(CO)Cl]^+$

$\downarrow + H_2O\ (-Cl^-)$

$(-H_2)$
$+ H_3O^+ \longrightarrow [Ru(bpy)_2(CO)(H_2O)]^{2+} \longrightarrow + CO\ (-H_2O)$

$[Ru(bpy)_2(CO)H]^+$ $[Ru(bpy)_2(CO)_2]^{2+}$

$- CO_2$ $[Ru(bpy)_2(CO)(COOH)]^+ \rightleftharpoons + OH^-$

Net reaction for cycle: $CO + OH^- + H_3O^+ \longrightarrow H_2O + CO_2 + H_2$

$\longrightarrow 2\ H_2O$

$CO + H_2O \longrightarrow CO_2 + H_2$

FIGURE 14.33 Homogeneous Catalysis of Water Gas Shift Reaction.

(Adapted with permission from H. Ishida, K. Tanaka, M. Morimoto, and T. Tanaka, *Organometallics*, **1986**, 5, 724. © 1986 American Chemical Society.)

General References

J.P. Collman, L. S. Hegedus, J. R. Norton, and R. G Finke, *Principles and Applications of Organotransition Metal Chemistry*, University Science Books, Mill Valley, CA, 1987, provides a detailed discussion, with numerous references, of many of the reactions and catalytic processes described in this chapter, as well as a variety of other types of organometallic reactions. In addition to providing extensive information on the structural and bonding properties of organometallic compounds, G. Wilkinson, F. G. A. Stone, and E. W. Abel, editors of *Comprehensive Organometallic Chemistry*, Pergamon Press, Oxford, UK, 1982, and E. W. Abel, F. G. A. Stone, and G. Wilkinson, editors of *Comprehensive Organometallic Chemistry II*, Pergamon Press, Oxford, 1995, give the most comprehensive information on organometallic reactions, with numerous references to the original literature. Two recent sources of general information and recent references on catalytic processes are P. W. N. M. van Leeuwen's, *Homogeneous Catalysis: Understanding the Art*, Kluwer Academic Publishers, Dordrecht, the Netherlands, 2004, and J. Hagen's *Industrial Catalysis: A Practical Approach*, 2nd ed., Wiley-VCH, Weinheim, Germany, 2006. S. T. Oyama and G. A. Somorjai, "Homogeneous, Heterogeneous, and Enzymatic Catalysis" in *J. Chem. Educ.*, **1988**, *65*, 765, gives examples of the types and amounts of catalysts used in a variety of industrial processes. The other references listed at the end of Chapter 13 are also useful in connection with this chapter.

Problems

14.1 Predict the transition metal-containing products of the following reactions:

a. $[Mn(CO)_5]^- + H_2C{=}CH{-}CH_2Cl \longrightarrow$ initial product $\xrightarrow{-CO}$ final product

b. *trans*-$Ir(CO)Cl(PPh_3)_2 + CH_3I \longrightarrow$

c. $Ir(PPh_3)_3Cl \xrightarrow{\Delta}$

d. $(\eta^5\text{-}C_5H_5)Fe(CO)_2(CH_3) + PPh_3 \longrightarrow$

e. $(\eta^5\text{-}C_5H_5)Mo(CO)_3[C({=}O)CH_3] \xrightarrow{\Delta}$

f. $H_3C{-}Mn(CO)_5 + SO_2 \longrightarrow$ (no gases are evolved)

14.2 Predict the transition metal–containing products of the following reactions:

a. $H_3C{-}Mn(CO)_5 + P(CH_3)(C_6H_5)_2 \longrightarrow$ (no gases are evolved)

b. $[Mn(CO)_5]^- + (\eta^5\text{-}C_5H_5)Fe(CO)_2Br \xrightarrow{\Delta}$

c. *trans*-$Ir(CO)Cl(PPh_3)_2 + H_2 \longrightarrow$

d. $W(CO)_6 + C_6H_5Li \longrightarrow$

e. *cis*-$Re(CH_3)(PEt_3)(CO)_4 + {}^{13}CO \longrightarrow$ (show all expected products, percentage of each)

f. *fac*-$Mn(CO)_3(CH_3)(PMe_3)_2 + {}^{13}CO \longrightarrow$ (show all expected products, percentage of each)

14.3 Predict the transition metal–containing products of the following reactions:

a. *cis*-$Mn(CO)_4({}^{13}CO)(COCH_3) \xrightarrow{\Delta}$ (show all expected products, percentage of each)

b. $C_6H_5CH_2{-}Mn(CO)_5 \xrightarrow{h\nu} CO +$

c. $V(CO)_6 + NO \longrightarrow$

d. $Cr(CO)_6 + Na/NH_3 \longrightarrow$

e. $Fe(CO)_5 + NaC_5H_5 \longrightarrow$

f. $[Fe(CO)_4]^{2-} + CH_3I \longrightarrow$

g. $H_3C{-}Rh(PPh_3)_3 \xrightarrow{\Delta} CH_4 +$

14.4 Heating $[(\eta^5\text{-}C_5H_5)Fe(CO)_3]^+$ with NaH in solution gives **A**, which has the empirical formula $C_7H_6O_2Fe$. **A** reacts rapidly at room temperature to eliminate a colorless gas **B**, forming a purple-brown solid **C** having the empirical formula $C_7H_5O_2Fe$. Treatment of **C** with iodine generates a brown solid **D** with the empirical formula $C_7H_5O_2FeI$, which on treatment with TlC_5H_5 gives a solid **E** with the formula $C_{12}H_{10}O_2Fe$. **E**, on heating, gives off a colorless gas, leaving an orange solid **F** with the formula $C_{10}H_{10}Fe$. Propose structural formulas for **A** through **F**.

14.5 $Na[\eta^5\text{-}C_5H_5)Fe(CO)_2]$ reacts with $ClCH_2CH_2SCH_3$ to give **A**, a monomeric and diamagnetic substance of stoichiometry $C_{10}H_{12}FeO_2S$ having two strong IR bands at 1980 and 1940 cm^{-1}. Heating of **A** gives **B**, a monomeric, diamagnetic substance having strong IR bands at 1920 and 1630 cm^{-1}. Identify **A** and **B**.

14.6 The reaction of $V(CO)_5(NO)$ with $P(OCH_3)_3$ to give $V(CO)_4\,[P(OCH_3)_3](NO)$ has the rate law

$$\frac{-d[V(CO)_5(NO)]}{dt} = k_1\,[V(CO)_5\,(NO)] + k_2[P(OCH_3)_3][V(CO)_5(NO)]$$

a. Suggest a mechanism for this reaction consistent with the rate law.

b. One possible mechanism consistent with the last term in the rate includes a transition state of formula $V(CO)_5[P(OCH_3)_3](NO)$. Would this necessarily be a 20-electron species? Explain.

14.7 The rate law for the reaction $H_2 + Co_2(CO)_8 \longrightarrow 2\,HCo(CO)_4$ is

$$\text{Rate} = \frac{k[Co_2(CO)_8][H_2]}{[CO]}$$

Propose a mechanism consistent with this rate law.

14.8 Which of the following *trans* complexes would you expect to react most rapidly with CO? Which would you expect to react least rapidly? Briefly explain your choices.

$Cr(CO)_4(PPh_3)_2$

$Cr(CO)_4(PPh_3)(PBu_3)$ (Bu = *n*-butyl)

$Cr(CO)_4(PPh_3)[P(OMe)_3]$

$Cr(CO)_4(PPh_3)[P(OPh)_3]$

(See M. J. Wovkulich and J. D. Atwood, *Organometallics*, **1982**, *1*, 1316.)

14.9 The equilibrium constants for the ligand dissociation reaction $NiL_4 \rightleftharpoons NiL_3 + L$ have been determined for a variety of phosphines. (See C. A. Tolman, W. C. Seidel, and L. W. Gosser, *J. Am. Chem. Soc.*, **1974**, *96*, 53). For L = PMe_3, PEt_3, $PMePh_2$, and PPh_3, arrange these equilibria in order of the expected magnitudes of their equilibrium constants (from largest *K* to smallest).

14.10 The complex shown below loses carbon monoxide on heating. Would you expect this carbon monoxide to be ^{12}CO, ^{13}CO, or a mixture of both? Why?

O=^{13}C(CH$_3$)–Mn(CO)$_3$(PPh$_3$)$_2$ [structure: acetyl (^{13}C=O, CH$_3$) on Mn; OC—Mn—PPh$_3$; CO; OC; PPh$_3$]

14.11 **a.** Predict the products of the following reaction, showing clearly the structure of each:

[structure: Mn bearing CH$_3$, three CO (OC—Mn, C=O above, OC below) and two PMe$_3$] $+\ ^{13}CO \longrightarrow$

b. Each product of this reaction has a new, rather strong IR band that is distinctly different in energy from any bands in the reactants. Account for this band, and predict its approximate location (in cm^{-1}) in the IR spectrum.

14.12 Give structural formulas for **A** through **D**:

$$(\eta^5\text{-}C_5H_5)_2Fe_2(CO)_4 \xrightarrow{Na/Hg} \mathbf{A} \xrightarrow{Br_2}$$

$$\mathbf{B} \xrightarrow{LiAlH_4} \mathbf{C} \xrightarrow{PhNa} \mathbf{A} + \mathbf{D}\ \text{(a hydrocarbon)}$$

ν_{CO} = 1961, 1942, 1790 cm^{-1} for $(\eta^5\text{-}C_5H_5)_2Fe_2(CO)_4$

A has strong IR bands at 1880 and 1830 cm^{-1}; **C** has a ^1H NMR spectrum consisting of two singlets of relative intensity 1:5 at approximately $\delta\,-12$ ppm and δ 5 ppm, respectively. (Hint: Metal hydrides often have protons with negative chemical shifts.)

14.13 $Re(CO)_5Br$ reacts with the ion $Br{-}CH_2CH_2{-}O^-$ to give compound **Y** + Br^-.

a. What is the most likely site of attack of this ion on $Re(CO)_5Br$? (Hint: Consider the hardness (see Chapter 6) of the Lewis base.)

b. Using the following information, propose a structural formula of **Y** and account for each of the following:

Y obeys the 18-electron rule.
No gas is evolved in the reaction.
^{13}C NMR indicates that there are five distinct magnetic environments for carbon in **Y**.
Addition of a solution of Ag^+ to a solution of **Y** gives a white precipitate.

(See M.M. Singh and R.J. Angelici, *Inorg. Chem.*, **1984**, *23*, 2699.)

14.14 The carbene complex **I** shown below undergoes the following reactions. Propose structural formulas for the reaction products.

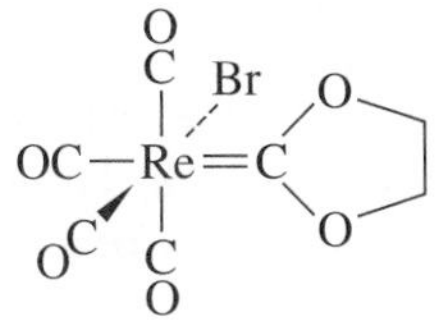

a. When a toluene solution containing **I** and excess triphenylphosphine is heated to reflux, compound **II** is formed first, and then compound **III**. **II** has infrared bands at 2038, 1958, and 1906 cm^{-1} and **III** at 1944 and 1860 cm^{-1}. ^1H NMR data δ values (relative area) are as follows:

II: 7.62, 7.41 multiplets (15)
4.19 multiplet (4)

III: 7.70, 7.32 multiplets (15)
3.39 singlet (2)

b. When a solution of **I** in toluene is heated to reflux with 1,1-bis(diphenylphosphino)methane, a colorless product **IV** is formed that has the following properties:

IR: 2036, 1959, 1914 cm^{-1}
Elemental analysis (accurate to ±0.3%): 35.87% C, 2.73% H

c. I reacts rapidly with the dimethyldithiocarbamate ion $S_2CN(CH_3)_2^-$ in solution to form $Re(CO)_5Br$ + **V**, a product that does not contain a metal atom. This product has no infrared bands between 1700 and 2300 cm^{-1}. However, it does show moderately intense bands at 1500 and 977 cm^{-1}. The ^1H NMR spectrum of **V** shows bands at δ 3.91 (triplet), 3.60 (triplet), 3.57 (singlet), and 3.41 (singlet).

(See G. L. Miessler, S. Kim, R. A. Jacobson, and R. J. Angelici, *Inorg. Chem.*, **1987**, *26*, 1690.)

14.15 The complex **I** in the preceding problem can be synthesized from $Re(CO)_5Br$ and 2-bromoethanol in ethylene oxide solution with solid NaBr present. Suggest a mechanism for the formation of the carbene ligand.

14.16 $BrCH_2CH_2CH_2Mn(CO)_5$ is formed by reaction of $[Mn(CO)_5]^-$ with 1,3-dibromopropane. However, the reaction does not stop here; the product reacts with additional $[Mn(CO)_5]^-$ to yield a carbene complex. Propose a structure for this complex and suggest a mechanism for its formation.

14.17 An acyl metal carbonyl ($R{-}C({=}O)M(CO)_x$) is generally easier to protonate than either a metal carbonyl or an organic ketone, such as acetone. Suggest an explanation.

14.18 Show how transition-metal complexes could be used to effect the following syntheses:

a. Acetaldehyde from ethylene
b. $CH_3CH_2COOCH_3$ from CH_3CH_2Cl
c. $CH_3CH_2CH_2CH_2CHO$ from $CH_3CH_2CH{=}CH_2$
d. $PhCH_2CH_2CH_2CHO$ from an alkene (Ph = phenyl)
e.

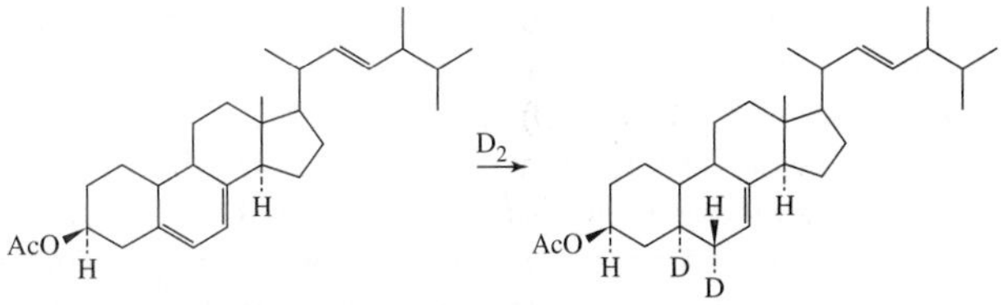

f. $C_6D_5CH_3$ from toluene, $C_6H_5CH_3$.

14.19 The complex $Rh(H)(CO)_2(PPh_3)_2$ can be used in the catalytic synthesis of *n*-pentanal from an alkene having one less carbon. Propose a mechanism for this process. Give an appropriate designation for each type of reaction step (such as oxidative addition or alkyl migration), and identify the catalytic species.

14.20 It is possible to synthesize the following aldehyde from an appropriate 5-carbon alkene by using an appropriate transition-metal catalyst:

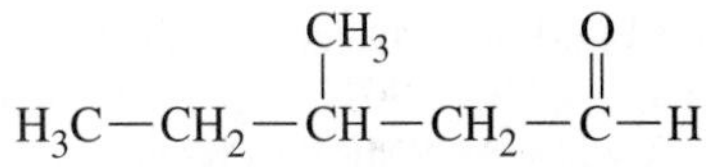

$$H_3C{-}CH_2{-}CH(CH_3){-}CH_2{-}C({=}O){-}H$$

Show how this synthesis could be effected catalytically. Identify the catalytic species.

14.21 Predict the products if the following compounds undergo metathesis:

a.

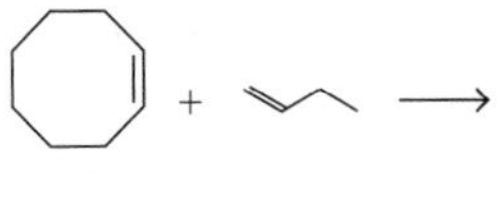

b. + C_2H_5 CH_3

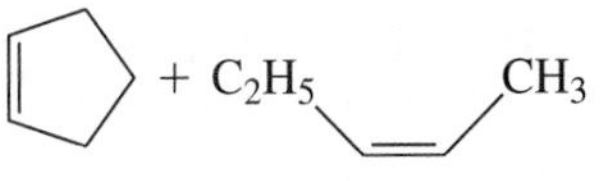

c. 1-butene + 2-butene
d. 1,7-octadiene

14.22 One of the classical experiments in the development of olefin metathesis was the "double cross" metathesis in which a mixture of cyclooctene, 2-butene, and 4-octene underwent metathesis.

a. What products would be expected from this metathesis?
b. How would the formation of these products differ in the pairwise and non-pairwise mechanisms? (See T. J. Katz and J. McGinness, *J. Am. Chem. Soc.*, **1975**, *97*, 1592.)

14.23 The complex $(\eta^5\text{-}C_5H_5)_2Zr(CH_3)_2$ reacts with the highly electrophilic borane $HB(C_6F_5)_2$ to form a product having stoichiometry $(CH_2)[HB(C_6F_5)_2]_2[(C_5H_5)_2Zr]$; the product is a rare example of pentacoordinate carbon.

a. Suggest a structure for this product.

b. An isomer of this product, $[(C_5H_5)_2\,ZrH]^+$ $[CH_2\,\{B(C_6F_5)_2\}_2\,(\mu - H)]^-$, has been proposed as a potential Ziegler–Natta catalyst. Suggest a mechanism by which this isomer might serve as such a catalyst.

(See R. E. von H. Spence, D. J. Parks, W. E. Piers, M. MacDonald, M. J. Zaworotko, and S. J. Rettig, *Angew. Chem., Int. Ed.*, **1995**, *34*, 1230.)

14.24 At low temperature and pressure, a gas-phase reaction can occur between iron atoms and toluene. The product, a rather unstable sandwich compound, reacts with ethylene to give compound **X**. Compound **X** decomposes at room temperature to liberate ethylene; at –20 °C it reacts with $P(OCH_3)_3$ to give $Fe(toluene)[P(OCH_3)_3]_2$. Suggest a structure for compound **X**. (See U. Zenneck and W. Frank, *Angew. Chem., Int. Ed.*, **1986**, *25*, 831.)

14.25 The reaction of $RhCl_3 \cdot 3\,H_2O$ with tri-*o*-tolylphosphine in ethanol at 25 °C gives a blue-green complex **I** $(C_{42}\,H_{42}\,P_2\,Cl_2\,Rh)$ that has $\nu\,(Rh—Cl)$ at 351 cm^{-1} and μ_{eff} = 2.3 BM. At a higher temperature, a diamagnetic yellow complex **II** that has an Rh:Cl ratio of 1:1 is formed that has an intense band near 920 cm^{-1}. Addition of NaSCN to **II** replaces Cl with SCN to give a product **III** having the following ^1H NMR spectrum:

Chemical Shift	Relative Area	Type
6.9–7.5	12	Aromatic
3.50	1	Doublet of 1:2:1 triplets
2.84	3	Singlet
2.40	3	Singlet

Treatment of **II** with NaCN gives a phosphine ligand **IV** with the empirical formula $C_{21}H_{19}P$ and a molecular weight of 604. **IV** has an absorption band at 965 cm^{-1} and the following ^1H NMR spectrum:

Chemical Shift	Relative Area	Type
7.64	1	Singlet
6.9–7.5	12	Aromatic
2.37	6	Singlet

Determine the structural formulas of compounds **I** through **IV**, and account for as much of the data as possible. (See M. A. Bennett and P. A. Longstaff, *J. Am. Chem. Soc.*, **1969**, *91*, 6266.)

14.26 Ring-closing metathesis (RCM) is not restricted to alkenes; similar reactions are also known for alkynes. The tungsten alkylidyne complex $W(\equiv CCMe_3)(OCMe_3)_3$ has been used to catalyze such reactions. Predict the structures of the cyclic products for metathesis of

a. $[MeC\equiv C(CH_2)_2OOC(CH_2)]_2$

b. $MeC\equiv C(CH_2)_8COO(CH_2)_9\,C\equiv CMe$

(See A Fürstner and G. Seidel, *Angew. Chem., Int. Ed.*, **1998**, *37*, 1734.)

14.27 More than two decades elapsed between the development of a homogeneous catalytic process for producing acetic acid in the late 1960s and convincing experimental evidence for the key intermediate $[CH_3Rh(CO)_2I_3]^-$ (Figure 14.16). Describe the evidence presented for this intermediate. A. Haynes, B. E. Mann, D. J. Gulliver, G. E. Morris, and P. M. Maitlis, *J. Am. Chem. Soc.*, **1991**, *113*, 8567.

14.28 The compound $Fe(CO)_4I_2$ reacts with cyanide in methanol solution to form complex **A**, which has intense IR bands at 2096 and 2121 cm^{-1} and less intense bands at 2140 and 2162 cm^{-1}. Reaction of **A** with additional cyanide yields **B**. Ion **B** also has two pairs of infrared bands, a more intense pair at 1967 and 2022 cm^{-1} and a less intense pair at 2080 and 2106 cm^{-1}. Neither **A** nor **B** contains iodine. Propose structures of **A** and **B**. (See J. Jiang and S. A. Koch, *Inorg. Chem.*, **2002**, *41*, 158.)

14.29 Cation **1** reacts with the ion $HB(sec\text{-}C_4H_9)_3^-$, a potential source of hydride, to form **2**. The following data are reported for **2**:

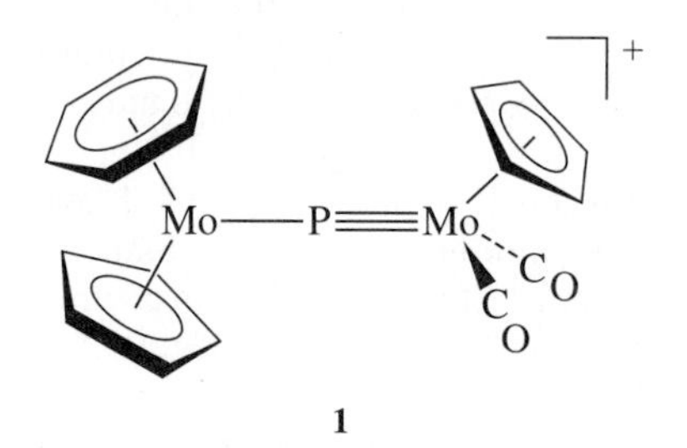

IR: strong bands at 1920, 1857 cm^{-1}.

^1H NMR: chemical shift (relative area):

5.46 (2)
5.28 (5)
5.15 (3)
4.22 (2)
1.31 (27)

In addition, a small peak is believed to be hidden under other peaks.

^{13}C NMR: resonance at 236.9 ppm, seven additional peaks and clusters of peaks between 32.4 and 115.7 ppm.

Propose a structure for **2**, and account for as much of the experimental data as possible. (See I. Amor, D. García-Vivó, M. E. García, M. A. Ruiz, D. Sáez,

H. Hamidov, and J. C. Jeffery, *Organometallics*, **2007**, *26*, 466.)

14.30 Sometimes an organometallic complex can act as the intermediate in the synthesis of organic compounds. The heterocyclic compound shown at right was reacted with $LiBH(C_2H_5)_3$ to form **X**, a halogen-free heterocycle. When **X** was treated with $Cr(CO)_3(CH_3CN)_3$, followed by reaction with HF · pyridine, transition-metal complex **Y**, which had strong absorptions at 1898 and 1975 cm^{-1}, formed. Addition of triphenylphosphine gave a new transition-metal complex plus heterocycle **Z**. The ^1H NMR spectrum of **Z** showed four single peaks of equal intensity between δ 6.4 and 7.8 ppm, a quartet at 4.9 ppm, and a triplet at 8.44 ppm. Propose structures for **X**, **Y**, and **Z**. (See A. J. V. Marwitz, M. H. Matus, L. N. Zakharov, D. A. Dixon, and S-H. Liu, *Angew. Chem, Int. Ed.*, **2009**, *48*, 973.)

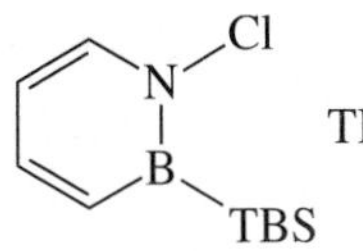

Cl
N
B
TBS

TBS = *tert*-butyldimethylsilylallylamine

14.31 When hexaiodobenzene, C_6I_6, reacts with diferrocenylzinc, $[(\eta^5\text{-}C_5H_5)FeC_5H_4]_2Zn$, one of the products has a C to Fe atom ratio exactly 10 percent higher than in ferrocene and an H to Fe atom ratio 10 percent lower than in ferrocene. The product contains no elements other than those that occur in ferrocene. Suggest a structure of this product. (See Y. Yu, A. D. Bond, P. W. Leonard, U. J. Lorenz, T. V. Timofeeva, K. P. C. Vollhardt, G. D. Whitener, and A. A. Yakovenko, *Chem. Commun.*, **2006**, 2572.)

15

Parallels Between Main Group and Organometallic Chemistry

It is common to treat organic and inorganic chemistry as separate topics and, within inorganic chemistry, to consider separately the chemistry of main group compounds and organometallic compounds, as we have generally done so far in this text. However, valuable insights can be gained by examining parallels between these different classifications of compounds. Such an examination may lead to a more thorough understanding of the different types of compounds being compared and may suggest new chemical compounds or new types of reactions. The objective of this chapter is to consider several of these parallels, especially between main group and organometallic compounds.

15.1 MAIN GROUP PARALLELS WITH BINARY CARBONYL COMPLEXES

Comparisons within main group chemistry have already been discussed in earlier chapters. These included the similarities and differences between borazine and benzene, the relative instability of silanes in comparison with alkanes, and differences in bonding in homonuclear and heteronuclear diatomic species, such as the isoelectronic N_2 and CO. In general, these parallels have centered around isoelectronic species. Similarities also occur between main group and transition-metal species that are electronically equivalent, species that require the same number of electrons to achieve a filled valence configuration.[1] For example, a halogen atom, one electron short of a valence shell octet, may be considered electronically equivalent to $Mn(CO)_5$, a 17-electron species one electron short of an 18-electron configuration. In this section, we will discuss briefly some parallels between main group atoms and ions and electronically equivalent binary carbonyl complexes.

Much chemistry of main group and metal carbonyl species can be rationalized from the way in which these species can achieve closed-shell (octet or 18-electron) configurations. These methods of achieving more stable configurations will be illustrated for the following electronically equivalent species:

Electrons Short of a Filled Shell	Examples of Electronically Equivalent Species	
	Main Group	Metal Carbonyl
1	Cl, Br, I	$Mn(CO)_5$, $Co(CO)_4$
2	S	$Fe(CO)_4$, $Os(CO)_4$
3	P	$Co(CO)_3$, $Ir(CO)_3$

[1] J. E. Ellis, *J. Chem. Educ.*, **1976**, *53*, 2.

TABLE 15.1 Parallels Between Cl and $Co(CO)_4$

Characteristic	Examples	Examples
Ion of 1− charge	Cl^-	$[Co(CO)_4]^-$
Neutral dimeric species	Cl_2	$[Co(CO)_4]_2$
Hydrohalic acid	HCl (strong acid in aqueous solution)	$HCo(CO)_4$ (strong acid in aqueous solution)[a]
Formation of interhalogen compounds	$Br_2 + Cl_2 \rightleftharpoons 2\ BrCl$	$I_2 + [Co(CO)_4]_2 \longrightarrow 2\ ICo(CO)_4$
Formation of heavy-metal salts of low solubility in water	AgCl	$AgCo(CO)_4$
Addition to unsaturated species	$Cl_2 + H_2C{=}CH_2 \longrightarrow H{-}CHCl{-}CHCl{-}H$	$[Co(CO)_4]_2 + F_2C{=}CF_2 \longrightarrow (CO)_4Co{-}CF_2{-}CF_2{-}Co(CO)_4$
Disproportionation by Lewis bases	$Cl_2 + N(CH_3)_3 \longrightarrow [ClN(CH_3)_3]Cl$	$[Co(CO)_4]_2 + C_5H_{10}NH$ (Piperidine) $\longrightarrow [(CO)_4Co(C_5H_{10}NH)][Co(CO)_4]$

NOTE: [a] However, $HCo(CO)_4$ is only slightly soluble in water.

Halogen atoms, one electron short of a valence shell octet, exhibit chemical similarities with 17-electron organometallic species; some of the most striking are the parallels between halogen atoms and $Co(CO)_4$, as summarized in Table 15.1. Both can reach filled-shell electron configurations by acquiring an electron or by dimerization. The neutral dimers are capable of adding across multiple carbon–carbon bonds and can undergo disproportionation by Lewis bases. Anions of both electronically equivalent species have a 1− charge and can combine with H^+ to form acids; both HX (X = Cl, Br, or I) and $HCo(CO)_4$ are strong acids in aqueous solution. Both types of anions form precipitates with heavy metal ions such as Ag^+ in aqueous solution. The parallels between 7-electron halogen atoms and 17-electron binary carbonyl species are sufficiently strong to justify extending the label *pseudohalogen* (see Chapter 8) to these carbonyls.

Similarly, 6-electron main group species show chemical similarities with 16-electron organometallic species. As for the halogens and 17-electron organometallic complexes, many of these similarities can be accounted for on the basis of ways in which the species can acquire or share electrons to achieve filled-shell configurations. Some similarities between sulfur and the electronically equivalent $Fe(CO)_4$ are listed in Table 15.2.

The concept of electronically equivalent groups can also be extended to 5-electron main group elements [Group 15 (VA)] and 15-electron organometallic species. For example, phosphorus and $Ir(CO)_3$ both form tetrahedral tetramers, as shown in Figure 15.1. The 15-electron $Co(CO)_3$, which is isoelectronic with $Ir(CO)_3$, can replace one or more phosphorus atoms in the P_4 tetrahedron, as also shown in this figure.

The parallels between electronically equivalent main group and organometallic species are interesting and summarize a considerable amount of their chemistry. The limitations of these parallels should also be recognized, however. For example, main group compounds having expanded shells (central atoms exceeding an electron count of 8) may not have organometallic analogues; organometallic analogues of such compounds as IF_7 and XeF_4 are not known. Organometallic complexes of ligands significantly weaker than CO in the spectrochemical series may not follow the 18-electron rule and may consequently behave quite differently from electronically equivalent main

TABLE 15.2 Parallels Between Sulfur and $Fe(CO)_4$

Characteristic	Examples	
Ion of 2– charge	S^{2-}	$[Fe(CO)_4]^{2-}$
Neutral compound	S_8	$Fe_2(CO)_9$, $[Fe(CO)_4]_3$
Hydride	H_2S: $pK_1 = 7.24$, $pK_2 = 14.92$[a]	$H_2Fe(CO)_4$: $pK_1 = 4.44$, $pK_2 = 14$[a]
Phosphine adduct	Ph_3PS	$Ph_3PFe(CO)_4$
Polymeric mercury compound	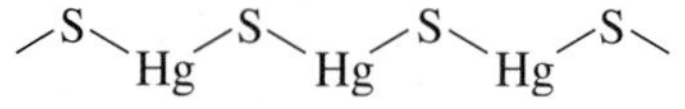 –S–Hg–S–Hg–S–Hg–S–	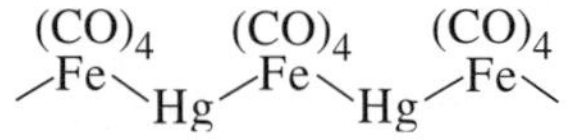 –$Fe(CO)_4$–Hg–$Fe(CO)_4$–Hg–$Fe(CO)_4$–
Compound with ethylene	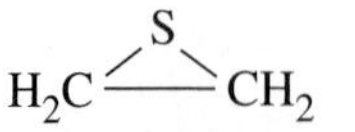 H_2C–S–CH_2 (ring) Ethylene sulfide	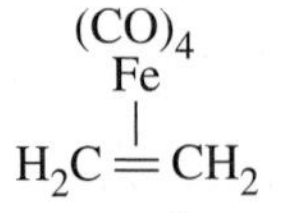 $(CO)_4Fe$–($H_2C{=}CH_2$) π complex

NOTE: [a] *pK* values in aqueous solution at 25 °C.

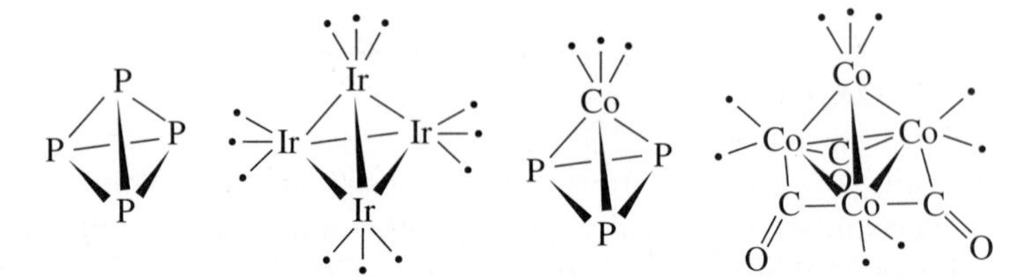

FIGURE 15.1 P_4, $[Ir(CO)_3]_4$, $P_3[Co(CO)_3]$, and $Co_4(CO)_{12}$.

• = terminal CO

group species. In addition, the reaction chemistry of organometallic compounds may be very different from main group chemistry. For example, loss of ligands such as CO is far more common in organometallic chemistry than in main group chemistry. Therefore, as in any scheme based on as simple a framework as electron counting, the concept of electronically equivalent groups, although useful, has its limitations. It serves as valuable background, however, for a potentially more versatile way to seek parallels between main group and organometallic chemistry: the concept of isolobal groups.

15.2 THE ISOLOBAL ANALOGY

An important contribution to the understanding of parallels between organic and inorganic chemistry has been the concept of isolobal molecular fragments, described most elaborately by Roald Hoffmann in his 1981 Nobel lecture.[2] Hoffmann defined molecular fragments to be isolobal

> if the number, symmetry properties, approximate energy and shape of the frontier orbitals and the number of electrons in them are similar—not identical, but similar.

To illustrate this definition, we will compare fragments of methane with fragments of an octahedrally coordinated transition-metal complex, ML_6. For simplicity, we will consider only σ bonding between the metal and the ligands in this complex.[3] The fragments to be discussed are shown in Figure 15.2.

[2] R. Hoffmann, *Angew. Chem., Int. Ed.* **1982**, *21*, 711; see also H-J. Krause, *Z. Chem.*, **1988**, *28*, 129.

[3] The model can be refined further to include pi interactions between d_{xy}, d_{xz}, and d_{yz} orbitals with ligands having suitable donor and/or acceptor orbitals.

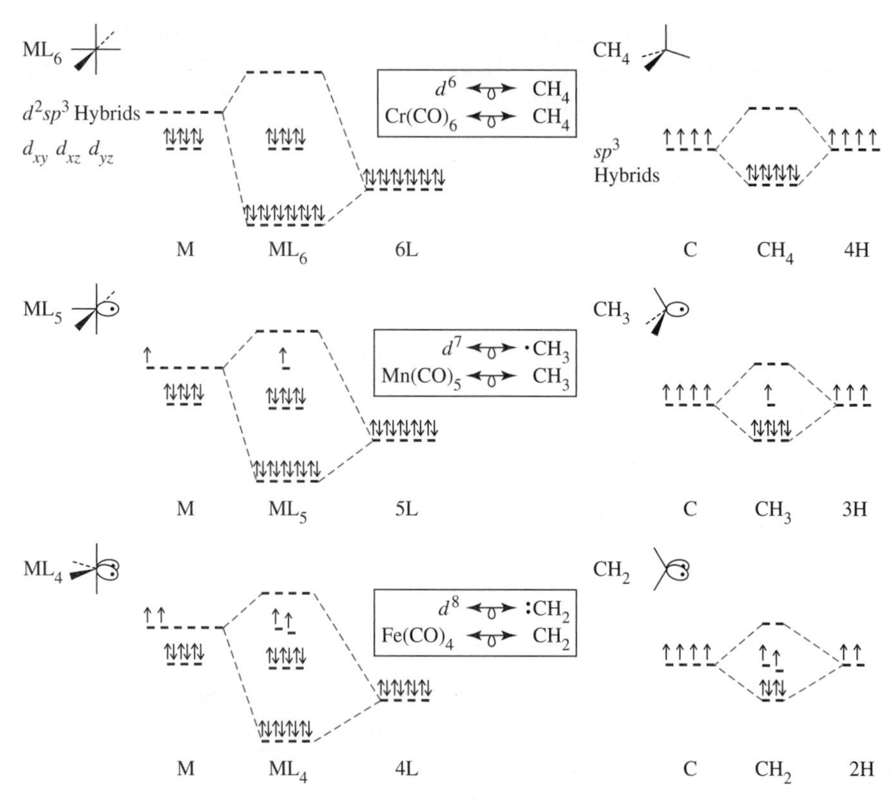

FIGURE 15.2 Orbitals of Octahedral and Tetrahedral Fragments.

The parent compounds have filled valence shell electron configurations, an octet for CH_4, and 18 electrons for ML_6 [$Cr(CO)_6$ is an example of such an ML_6 compound]. Methane may be considered to use sp^3 hybrid orbitals in bonding, with 8 electrons occupying bonding pairs formed from interactions between the hybrids and 1*s* orbitals on hydrogen. The metal in ML_6, by similar reasoning, uses d^2sp^3 hybrids in bonding to the ligands, with 12 electrons occupying bonding orbitals and 6 essentially nonbonding electrons occupying d_{xy}, d_{xz}, and d_{yz} orbitals.

Molecular fragments containing fewer ligands than the parent polyhedra can now be described. For the purpose of the analogy, these fragments will be assumed to preserve the geometry of the remaining ligands.

In the 7-electron fragment CH_3, three of the sp^3 orbitals of carbon are involved in σ bonding with the hydrogens. The fourth hybrid is singly occupied and at higher energy than the σ-bonding pairs of CH_3, as shown in Figure 15.2. This situation is similar to the 17-electron fragment $Mn(CO)_5$. The sigma interactions between the ligands and Mn in this fragment may be considered to involve five of the metal's d^2sp^3 hybrid orbitals. The sixth hybrid is singly occupied and at higher energy than the five σ-bonding orbitals.

As Figure 15.2 shows, each of these fragments has a single electron in a hybrid orbital at the vacant site of the parent polyhedron. These orbitals are sufficiently similar to meet Hoffmann's isolobal definition. Using Hoffmann's symbol $\leftarrow\!\!\circ\!\!\rightarrow$ to designate groups as isolobal, we may write

$$CH_3 \leftarrow\!\!\circ\!\!\rightarrow ML_5$$

H, H, H, C ←o→ L, L, L, L, L, M

Similarly, 6-electron CH_2 and 16-electron ML_4 are isolobal. Both CH_2 and ML_4 are 2 electrons short of a filled-shell octet or 18-electron configuration, so they are electronically equivalent; each has two single electrons occupying hybrid orbitals at otherwise vacant sites. Absence of a third ligand similarly gives a pair of isolobal fragments, CH and ML_3.

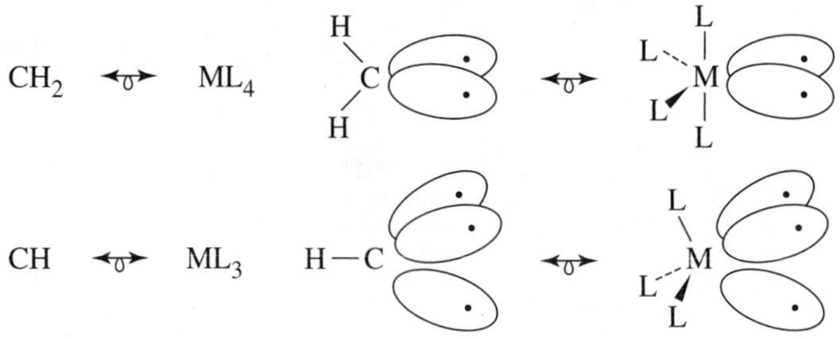

$CH_2 \leftrightarrow ML_4$

$CH \leftrightarrow ML_3$

To summarize:

	Organic	Inorganic	Organo-metallic Example	Vertices Missing from Parent Polyhedron	Electrons Short of Filled Shell
Parent	**CH_4**	**ML_6**	**$Cr(CO)_6$**	**0**	**0**
Fragments	CH_3	ML_5	$Mn(CO)_5$	1	1
	CH_2	ML_4	$Fe(CO)_4$	2	2
	CH	ML_3	$Co(CO)_3$	3	3

These fragments can be combined into molecules. For example, two CH_3 fragments form ethane, and two $Mn(CO)_5$ fragments form $(OC)_5Mn—Mn(CO)_5$. Furthermore, these organic and organometallic fragments can be combined into $H_3C—Mn(CO)_5$, which is also a known compound.

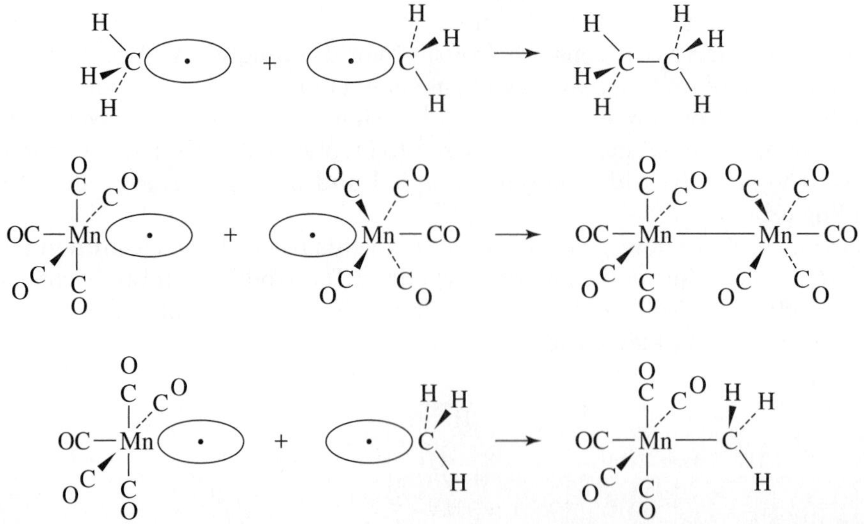

The organic and organometallic parallels are not always this complete. For example, although two 6-electron CH_2 fragments form ethylene, $H_2C{=}CH_2$, the dimer of the isolobal $Fe(CO)_4$ is not nearly as stable; it is known as a transient species, obtained photochemically[4] from $Fe_2(CO)_9$. However, both CH_2 and $Fe(CO)_4$ form three-membered rings, cyclopropane and $Fe_3(CO)_{12}$. Although cyclopropane is a trimer of three CH_2 fragments, $Fe_3(CO)_{12}$ has two bridging carbonyls and is therefore not a perfect trimer of $Fe(CO)_4$. The isoelectronic $Os_3(CO)_{12}$, on the other hand, is a trimeric combination of three $Os(CO)_4$ fragments, which are isolobal with both $Fe(CO)_4$ and CH_2 and can correctly be described as $[Os(CO)_4]_3$.

C_3H_6 $\quad$ $Fe_3(CO)_{12}$ $\quad$ $Os_3(CO)_{12}$

• = terminal carbonyl

The isolobal species $Ir(CO)_3$, $Co(CO)_3$, CR, and P may also be combined in several different ways. As mentioned previously, $Ir(CO)_3$, a 15-electron fragment, forms $[Ir(CO)_3]_4$, which has T_d symmetry. The isoelectronic complex $Co_4(CO)_{12}$ has a nearly tetrahedral array of cobalt atoms but has three bridging carbonyls and hence C_{3v} symmetry. Compounds are also known that have a central tetrahedral structure, with one or more $Co(CO)_3$ fragments replaced by the isolobal CR fragment, as shown in Figure 15.3. This is similar to the replacement of phosphorus atoms in the P_4 tetrahedron by $Co(CO)_3$ fragments; P may also be described as isolobal with CR.

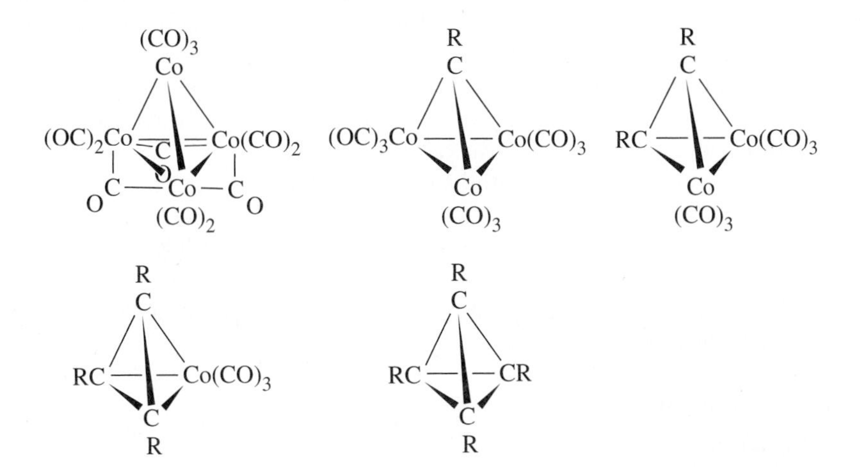

FIGURE 15.3 Structures Resulting from Combinations of Isolobal $Co(CO)_3$ and CR.

[4] M. Poliakoff and J. J. Turner, *J. Chem. Soc., A*, **1971**, 2403.

15.2.1 Extensions of the Analogy

The concept of isolobal fragments can be extended beyond the examples given so far to include charged species, a variety of ligands other than CO, and organometallic fragments based on structures other than octahedral. Some of the ways of extending the isolobal parallels can be summarized as follows:

1. The isolobal definition may be extended to isoelectronic fragments having the same coordination number. For example, because

$$\underset{\text{(17-electron fragment)}}{Mn(CO)_5} \underset{o}{\longleftrightarrow} \underset{\text{(7-electron fragment)}}{CH_3}, \quad \underset{\text{(17-electron fragment)}}{\begin{matrix} Re(CO)_5 \\ [Fe(CO)_5]^+ \\ [Cr(CO)_5]^- \end{matrix}} \underset{o}{\longleftrightarrow} \underset{\text{(7-electron fragment)}}{CH_3}$$

2. Gain or loss of electrons from two isolobal fragments yields isolobal fragments. For example, because

$$\underset{\text{(17-electron fragment)}}{Mn(CO)_5} \underset{o}{\longleftrightarrow} \underset{\text{(7-electron fragment)}}{CH_3}, \quad \underset{\text{(16-electron fragment)}}{\begin{matrix} Cr(CO)_5 \\ Mo(CO)_5 \\ W(CO)_5 \end{matrix}} \underset{o}{\longleftrightarrow} \underset{\text{(6-electron fragment)}}{CH_3^+}$$

$$\underset{\text{(18-electron fragment)}}{\begin{matrix} Fe(CO)_5 \\ Ru(CO)_5 \\ Os(CO)_5 \end{matrix}} \underset{o}{\longleftrightarrow} \underset{\text{(8-electron fragment)}}{CH_3^-}$$

Note that all the examples shown above are one ligand short of the parent complex. $Fe(CO)_5$ is isolobal with CH_3^-, for example, because both have filled electron shells and both are one vertex short of the parent polyhedron. By contrast, $Fe(CO)_5$ and CH_4 are not isolobal. Both have filled electron shells (18 and 8 electrons, respectively), but CH_4 has all vertices of the tetrahedron occupied, whereas $Fe(CO)_5$ has an empty vertex in the octahedron.

3. Other 2-electron donors are treated similarly to CO:[5]

$$Mn(CO)_5 \underset{o}{\longleftrightarrow} Mn(PR_3)_5 \underset{o}{\longleftrightarrow} [MnCl_5]^{5-} \underset{o}{\longleftrightarrow} Mn(NCR)_5 \underset{o}{\longleftrightarrow} CH_3$$

4. Ligands η^5-C_5H_5 and η^6-C_6H_6 are considered to occupy three coordination sites and to be 6-electron donors:[6]

$$\begin{matrix} (\eta^5\text{-}C_5H_5)Fe(CO)_2 \\ (\eta^6\text{-}C_6H_6)Mn(CO)_2 \end{matrix} \underset{o}{\longleftrightarrow} [Fe(CO)_5]^+ \underset{o}{\longleftrightarrow} Mn(CO)_5 \quad \text{(17-electron fragments)}$$

$$\begin{matrix} (\eta^5\text{-}C_5H_5)Mn(CO)_2 \\ (\eta^6\text{-}C_6H_6)Cr(CO)_2 \end{matrix} \underset{o}{\longleftrightarrow} [Mn(CO)_5]^+ \underset{o}{\longleftrightarrow} Cr(CO)_5 \quad \text{(16-electron fragments)}$$

---

[5] Hoffmann uses electron-counting method A, in which chloride is considered a negatively charged, 2-electron donor.

[6] η^5-C_5H_5 is considered the 6-electron donor $C_5H_5^-$.

5. Octahedral fragments of formula ML_n (where M has a d^x configuration) are isolobal with square-planar fragments of formula ML_{n-2} (where M has a d^{x+2} configuration and L is a 2-electron donor):

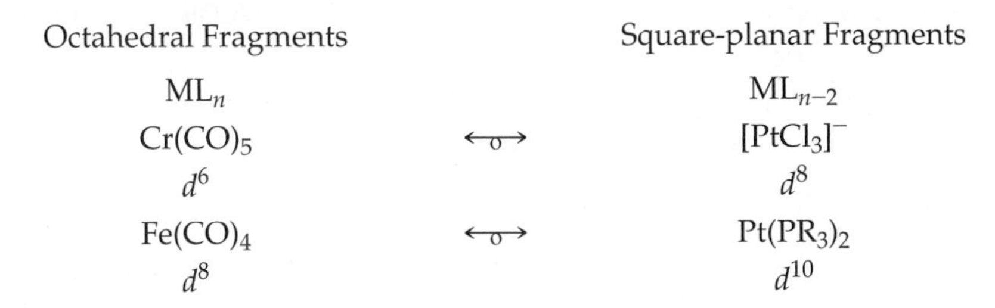

Octahedral Fragments
Square-planar Fragments
ML_n
ML_{n-2}
$Cr(CO)_5$
$[PtCl_3]^-$
d^6
d^8
$Fe(CO)_4$
$Pt(PR_3)_2$
d^8
d^{10}

The fifth of these extensions of the isolobal analogy is less obvious than the others and deserves explanation. We will consider two examples, the parallels between d^6 ML_5 (octahedral) and d^8 ML_3 (square-planar) fragments and the parallels between d^8 ML_4 (octahedral) and d^{10} ML_2 (square-planar) fragments. The ML_3 and ML_2 fragments of a square-planar parent structure are shown in Figure 15.4. They will be compared with the fragments of an octahedral ML_6 molecule, shown in Figure 15.2.

A square-planar d^8 ML_3 fragment, such as $[PtCl_3]^-$, has an empty lobe of a nonbonding hybrid orbital as its LUMO. This is comparable to the LUMO of a d^6 ML_5 fragment of an octahedron (e.g., $Cr(CO)_5$):[7]

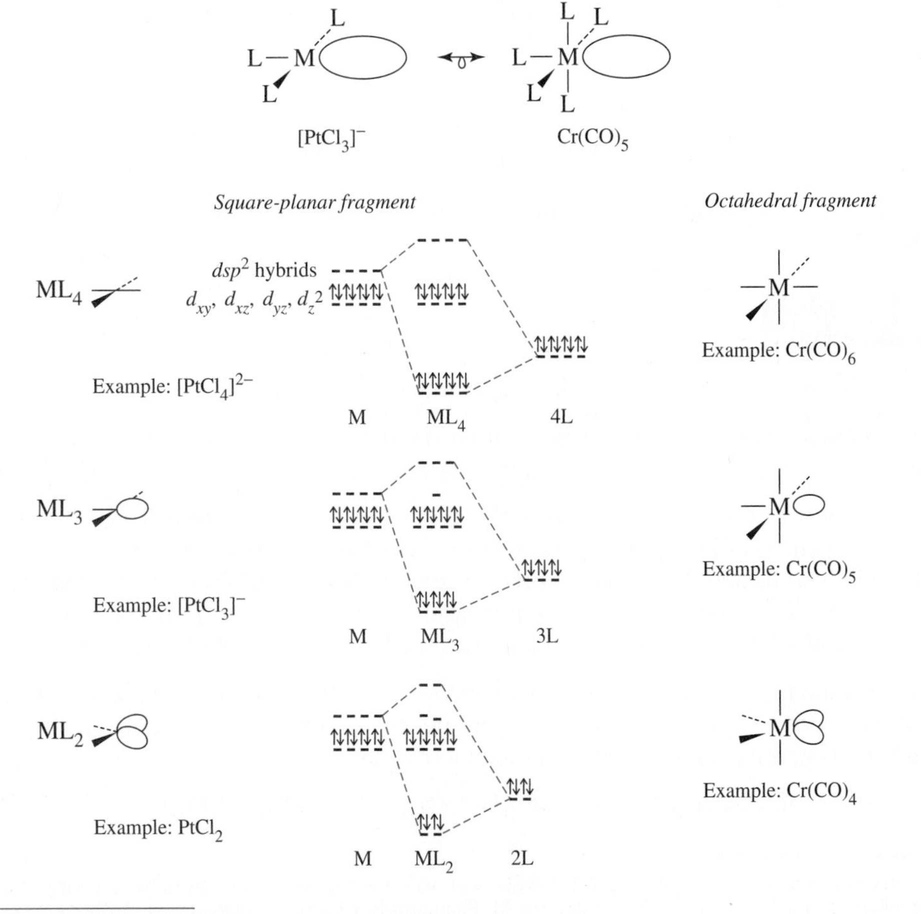

FIGURE 15.4 Comparison of Square-Planar Fragments with Octahedral Fragments.

[7] Such a fragment would have one less electron than is shown for $Mn(CO)_5$ in Figure 15.2.

TABLE 15.3 Examples of Isolobal Fragments

Neutral hydrocarbons	CH_4	CH_3	CH_2	CH	C
Isolobal organometallic fragments (Cp = η^5-C_5H_5)	$Cr(CO)_6$ $[Mn(CO)_6]^+$ $CpMn(CO)_3$	$Mn(CO)_5$ $[Fe(CO)_5]^+$ $CpFe(CO)_2$	$Fe(CO)_4$ $[Co(CO)_4]^+$ $CpCo(CO)$	$Co(CO)_3$ $[Ni(CO)_3]^+$ CpNi	$Ni(CO)_2$ $[Cu(CO)_2]^+$
Anionic hydrocarbon fragments obtained by loss of H^+	CH_3^-	CH_2^-	CH^-		
Isolobal organometallic fragments	$Fe(CO)_5$	$Co(CO)_4$	$Ni(CO)_3$		
Cationic hydrocarbon fragments obtained by gain of H^+		CH_4^+	CH_3^+	CH_2^+	CH^+
Isolobal organometallic fragments		$V(CO)_6$	$Cr(CO)_5$	$Mn(CO)_4$	$Fe(CO)_3$

A d^8 fragment such as $[PtCl_3]^-$ would therefore be isolobal with $Cr(CO)_5$ and other ML_5 fragments, provided the empty lobe in each case had suitable energy.[8]

A d^{10} ML_2 fragment such as $Pt(PR_3)_2$ would have two valence electrons more than the example of $PtCl_2$ shown in Figure 15.4. These electrons are considered to occupy two nonbonding hybrid orbitals. This situation is very comparable to the $Fe(CO)_4$ fragment (Figure 15.2); each complex has two singly occupied lobes:

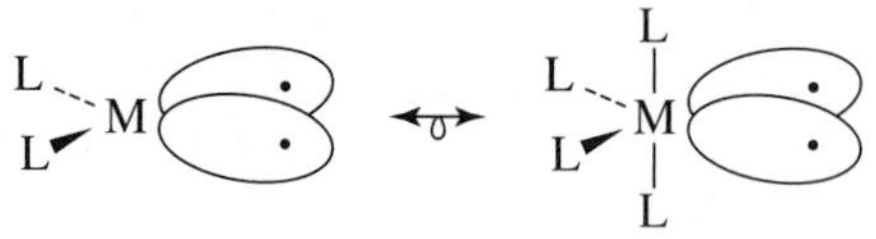

L, L, M, L, L, M, L, L

$Pt(PR_3)_2$ $\quad$ $Fe(CO)_4$

Examples of isolobal fragments containing CO and η^5-C_5H_5 ligands are given in Table 15.3.

EXAMPLE

Propose examples of organometallic fragments isolobal with CH_2^+.

For the purpose of this example, we will limit ourselves to the ligand CO and first-row transition metals. Other ligands and other metals may be used with equally valid results.

CH_2^+ is two ligands and 3 electrons short of its parent compound (CH_4). The corresponding octahedral fragment will therefore be a 15-electron species with the formula ML_4, two ligands and 3 electrons short of its parent, $M(CO)_6$. If L = CO, the four carbon monoxides contribute 8 electrons, requiring that the metal contribute the remaining 7. The first-row d^7 metal is Mn. The overall result is $CH_2^+ \leftarrow\!\!\!\!\!\!\!\!\;\circ\!\!\!\!\to Mn(CO)_4$.

Other octahedral isolobal fragments can be found by changing the metal and the charge on the complex. A positive charge compensates for a metal with one more electron, and a negative charge compensates for a metal with one less electron:

d^7: $Mn(CO)_4$ $\qquad$ d^8: $[Fe(CO)_4]^+$ $\qquad$ d^6: $[Cr(CO)_4]^-$

[8] The highest occupied orbitals of ML_3 have similar energies. For a detailed analysis of the energies and symmetries of ML_5, ML_3, and other fragments, see M. Elian and R. Hoffmann, *Inorg. Chem.*, **1975**, *14*, 1058, and T. A. Albright, R. Hoffmann, J. C. Thibeault, and D. L. Thorn, *J. Am. Chem. Soc.*, **1979**, *101*, 3801.

▶ **Exercise 15.1** For the following, propose examples of isolobal organometallic fragments other than those just cited and in Table 15.3:

a. A fragment isolobal with CH_2^+.
b. A fragment isolobal with CH^-.
c. Three fragments isolobal with CH_3.

▶ **Exercise 15.2** Find organic fragments isolobal with each of the following:

a. $Ni(\eta^5\text{-}C_5H_5)$
b. $Cr(CO)_2(\eta^6\text{-}C_6H_6)$
c. $[Fe(CO)_2(PPh_3)]^-$

Analogies are by no means limited to octahedral and square-planar organometallic fragments; similar arguments can be used to derive fragments of different polyhedra. For example, $Co(CO)_4$, a 17-electron fragment of a trigonal bipyramid, is isolobal with $Mn(CO)_5$, a 17-electron fragment of an octahedron:

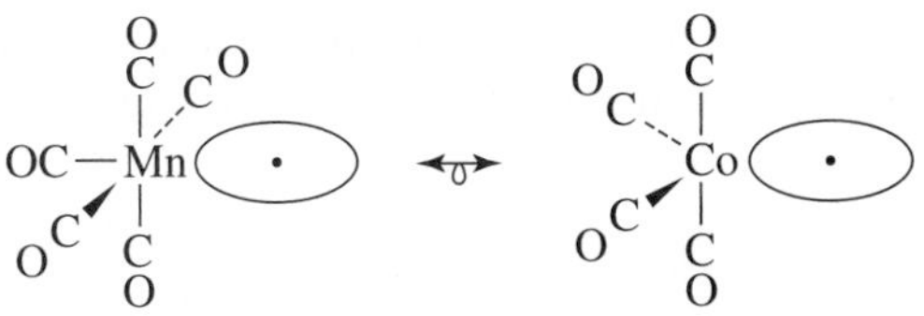

Examples of electron configurations of isolobal fragments of polyhedra having five through nine vertices are given in Table 15.4.

The interested reader is encouraged to refer to Hoffmann's Nobel lecture for further information on how the isolobal analogy can be extended to include other ligands and geometries.

15.2.2 Examples of Applications of the Analogy

The isolobal analogy can be extended to any molecular fragment having frontier orbitals of suitable size, shape, symmetry, and energy. For example, the 5-electron fragment CH is isolobal with P and other Group 15 atoms. A potential application of this relationship is to seek phosphorus-containing analogues to organometallic complexes containing cyclic π ligands such as C_5H_5 and C_6H_6. Most of the examples developed to date have been with metallocenes, $[(C_5H_5)_2M]^n$. Not only can P_5^-, the analogue to the cyclopentadienide ion $C_5H_5^-$, be prepared in solution,[9] but sandwich compounds containing P_5 rings, such as those shown in Figure 15.5, have been synthesized. The first of

TABLE 15.4 Isolobal Relationships for Fragments of Polyhedra

Organic Fragment	Coordination Number of Transition Metal for Parent Polyhedron					Valence Electrons of Fragment
	5	6	7	8	9	
CH_3	d^9-ML_4	d^7-ML_5	d^5-ML_6	d^3-ML_7	d^1-ML_8	17
CH_2	d^{10}-ML_3	d^8-ML_4	d^6-ML_5	d^4-ML_6	d^2-ML_7	16
CH		d^9-ML_3	d^7-ML_4	d^5-ML_5	d^3-ML_6	15

---

[9] M. Baudler, S. Akpapoglou, D. Ouzounis, F. Wasgestian, B. Meinigke, H. Budzikiewicz, and H. Münster, *Angew. Chem., Int. Ed.* **1988**, 27, 280.

FIGURE 15.5 Metallocenes Containing P_5 Rings.

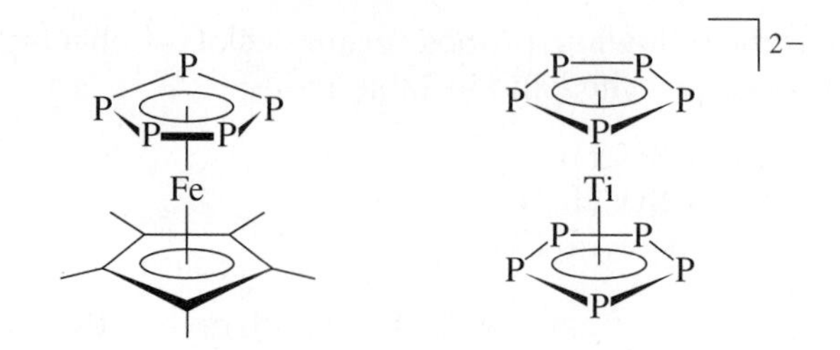

these, $(\eta^5\text{-}C_5Me_5)Fe(\eta^5\text{-}P_5)$, was prepared not directly from P_5^-, but rather from the reaction of $[(\eta^5\text{-}C_5Me_5)Fe(CO)_2]_2$ with white phosphorus (P_4).[10]

Perhaps the most interesting of all the phosphorus analogues of metallocenes is the first carbon-free metallocene, $[(\eta^5\text{-}P_5)_2Ti]^{2-}$. This complex, prepared by the reaction of $[Ti(naphthalene)_2]^{2-}$ with P_4, contains parallel, eclipsed P_5 rings.[11] The P_5 ligand in this and other complexes functions as a weaker donor, but substantially stronger acceptor, than the cyclopentadienyl ligand.

Another example, $Au(PPh_3)$, a 13-electron fragment, has a single electron in a hybrid orbital pointing away from the phosphine.[12] This electron is in an orbital of similar symmetry but of somewhat higher energy than the singly occupied hybrid in the $Mn(CO)_5$ fragment.

Nevertheless, $Au(PPh_3)$ can combine with the isolobal $Mn(CO)_5$ and CH_3 to form $(OC)_5\,Mn{-}Au(PPh_3)$ and $H_3C{-}Au(PPh_3)$.

Even a hydrogen atom, with a single electron in its 1*s* orbital, can in some cases be viewed as a fragment isolobal with such species as CH_3, $Mn(CO)_5$, and $Au(PPh_3)$. Hydrides of the first two are well known, and $Au(PPh_3)$ and H in some cases show surprisingly similar behavior, such as their ability to bridge the following triosmium clusters.[13,14]

[10] O. J. Scherer and T. Brück, *Angew. Chem., Int. Ed.*, **1987**, *26*, 59.

[11] E. Urnezius, W. W. Brennessel, C. J. Cramer, J. E. Ellis, and P. von Ragué Schleyer, *Science*, **2002**, *295*, 832. For a discussion of the development of complexes containing ligands composed of group 15 atoms, see M. Scheer, *Dalton Trans.*, **2008**, 4372.

[12] D. G. Evans and D. M. P. Mingos, *J. Organomet. Chem.*, **1982**, *232*, 171.

[13] A. G. Orpen, A. V. Rivera, E. G. Bryan, D. Pippard, G. Sheldrick, and K. D. Rouse, *Chem. Commun. (Cambridge)*, **1978**, 723.

[14] B. F. G. Johnson, D. A. Kaner, J. Lewis, and P. R. Raithby, *J. Organomet. Chem.*, **1981**, *215*, C33.

Potentially, the greatest practical use of isolobal analogies is in the suggested syntheses of new compounds. For example, CH_2 is isolobal with 16-electron $Cu(\eta^5\text{-}C_5Me_5)$ (extension 4 of the analogy) and 14-electron PtL_2 ($L = PR_3$, CO; extension 5). Recognition of these fragments as isolobal has been exploited in the syntheses of organometallic compounds composed of fragments isolobal with fragments of known compounds.[15] Some of the compounds obtained in these studies are shown in Figure 15.6.

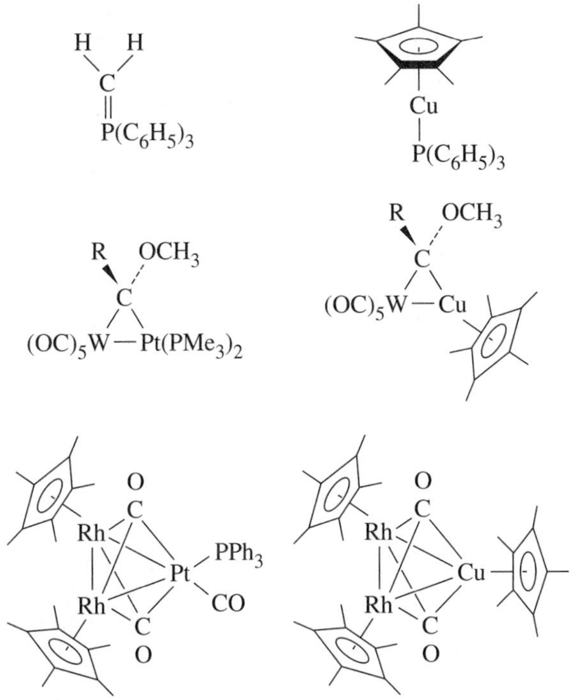

FIGURE 15.6 Compounds Composed of Isolobal Fragments.

15.3 METAL–METAL BONDS

The isolobal approach was used in the previous section to describe the formation of metal–metal bonds. These bonds differ from others only in the use of *d* orbitals on both atoms. In addition to the usual σ and π bonds, δ bonds are possible in transition metal compounds. Furthermore, bridging by ligands and the ability to form cluster compounds make for great variety in structures containing metal–metal bonds. Examples of compounds with carbon–carbon, other main group, and metal–metal single, double, and triple bonds, together with a metal–metal quadruple bond, are shown in Figure 15.7.

For approximately a century, compounds containing two or more metal atoms have been known. The first of these compounds to be correctly identified, by Werner, were held together by bridging ligands shared by the metals involved; X-ray crytallographic studies eventually showed that the metal atoms were too far apart to be likely participants in direct metal–metal orbital interactions.

Not until 1935 did X-ray crystallography demonstrate direct metal–metal bonding. In that year, Brosset reported the structure of $K_3W_2Cl_9$, which contained the

[15] G. A. Carriedo, J. A. K. Howard, and F. G. A. Stone, *J. Organomet. Chem.*, **1983**, *250*, C28.

FIGURE 15.7 Single, Double, Triple, and Quadruple Bonds.

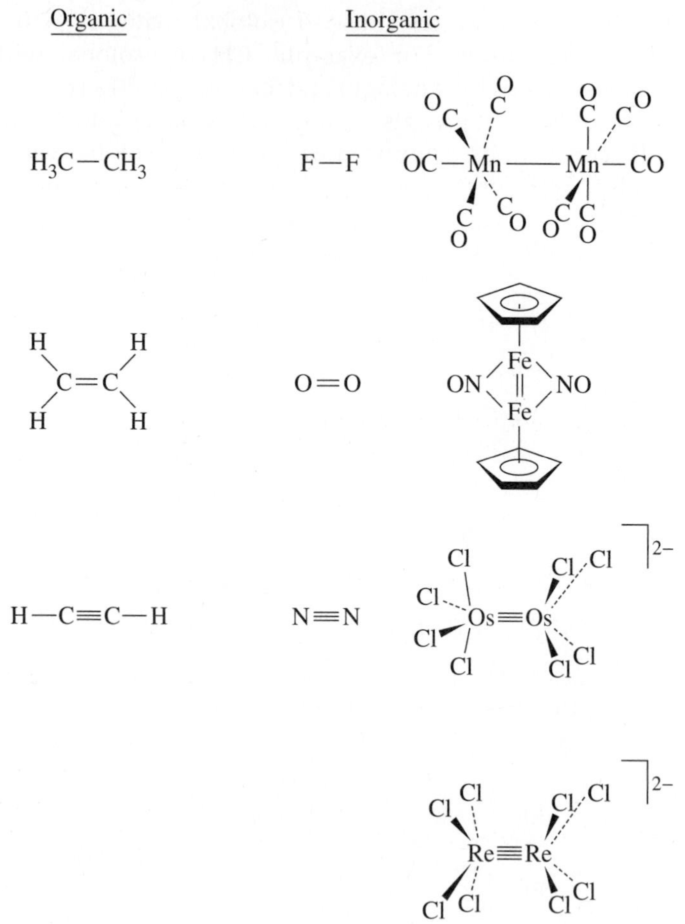

Organic	Inorganic	
H_3C-CH_3	F—F	$(OC)_5Mn-Mn(CO)_5$
$H_2C=CH_2$	O=O	$[CpFe(NO)]_2$
H—C≡C—H	N≡N	$[Cl_4Os≡OsCl_4]^{2-}$
		$[Cl_4Re≣ReCl_4]^{2-}$

$[W_2Cl_9]^{3-}$ ion. In this ion, the tungsten–tungsten distance (240 pm) was found to be substantially shorter than the interatomic distance in tungsten metal (275 pm):

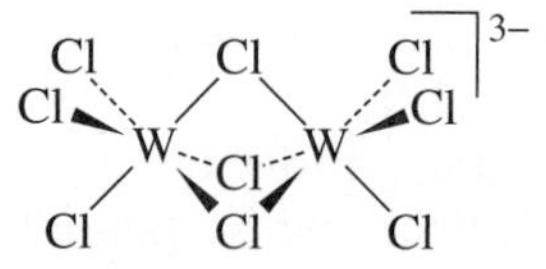

$$[Cl_3W(\mu\text{-}Cl)_3WCl_3]^{3-}$$

The short distance between the metal atoms in this ion raised for the first time the serious possibility of direct bonding interactions between metal orbitals. However, little attention was paid to this interesting question for many years, even though several additional compounds having very short metal–metal distances were synthesized.

The modern development of the chemistry of metal–metal-bonded species was spurred by the crystal structures of $[Re_3Cl_{12}]^{3-}$ and $[Re_2Cl_8]^{2-}$.[16] $[Re_3Cl_{12}]^{3-}$, originally believed to be monomeric $ReCl_4^-$, was shown in 1963 to be a trimeric cyclic ion having very short rhenium–rhenium distances (248 pm). In the following year, during a study on the synthesis of triruthenium complexes, the dimeric $[Re_2Cl_8]^{2-}$

[16] F. A. Cotton, *Chem. Soc. Rev.*, **1975**, *4*, 27.

was synthesized. This ion had a remarkably short metal–metal distance (224 pm) and was the first complex found to have a quadruple bond:

$[Re_3Cl_{12}]^{3-}$ $[Re_2Cl_8]^{2-}$

During the succeeding decades, many thousands of cluster compounds of transition metals have been synthesized, including hundreds containing quadruple bonds and some that may contain quintuple bonds. Therefore, we need to consider briefly how metal atoms can bond to each other and, in particular, how high order bonds between metals are possible.

15.3.1 Multiple Metal–Metal Bonds

QUADRUPLE BONDS Transition metals may form single, double, triple, or quadruple bonds (or bonds of fractional order) with other metal atoms. How are quadruple bonds possible? In main group chemistry, atomic orbitals in general can interact in a sigma or pi fashion, with the highest possible bond order of 3 a combination of one σ bond and two π bonds. When two transition metal atoms interact, the most important interactions are between their outermost d orbitals. These d orbitals can combine to form not only σ and π orbitals, but also δ (delta) orbitals, as shown in Figure 15.8. If the z axis is

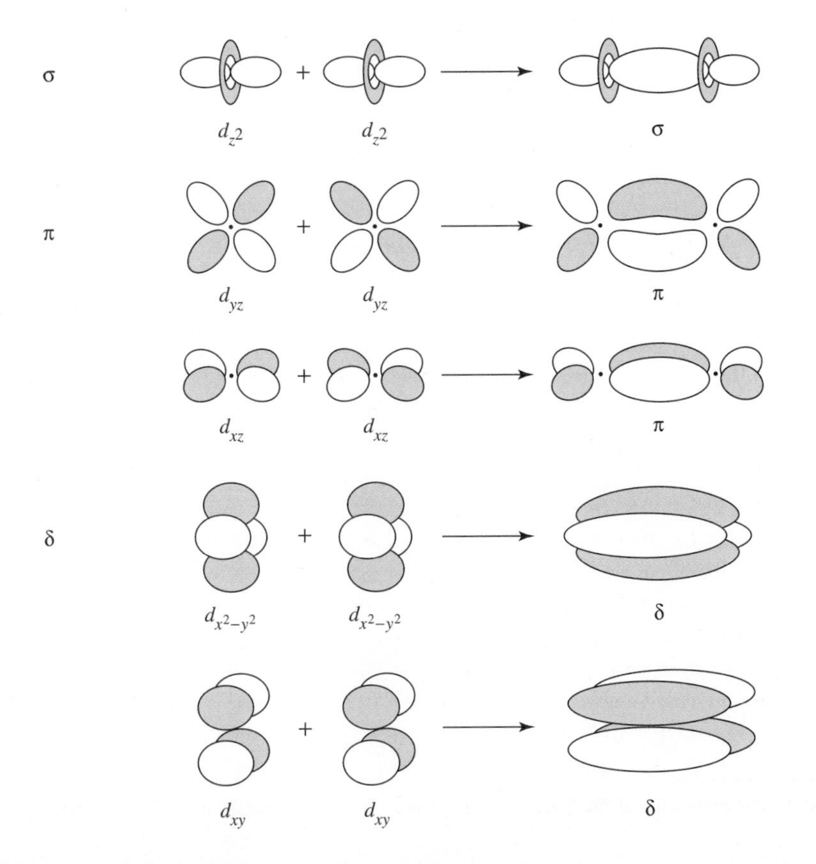

FIGURE 15.8 Bonding Interactions Between Metal d Orbitals.

chosen as the internuclear axis, the strongest interaction (involving greatest overlap) is the sigma interaction between the d_{z^2} orbitals. Next in effectiveness of overlap are the d_{xz} and d_{yz} orbitals, which form π orbitals as a result of interactions in two regions in space. The last, and weakest, of these interactions are between the d_{xy} and $d_{x^2-y^2}$ orbitals, which interact in four regions in the formation of δ molecular orbitals.

The relative energies of the resulting molecular orbitals are shown schematically in Figure 15.9. In the absence of ligands, an M_2 fragment would have five bonding orbitals resulting from *d–d* interactions, with molecular orbitals increasing in energy in the order σ, π, δ, δ^*, π^*, σ^*, as shown. In $[Re_2Cl_8]^{2-}$, our example of quadruple bonding, the configuration is eclipsed (D_{4h} symmetry). For convenience, we can choose the Re—Cl bonds to be oriented in the *xz* and *yz* planes. The ligand orbitals interact most strongly with the metal orbitals pointing toward them, in this case the δ and δ^* orbitals originating primarily from the $d_{x^2-y^2}$ atomic orbitals.[17] The consequence of these interactions is that new molecular orbitals are formed, as shown on the right side of Figure 15.9. The relative energies of these orbitals depend on the strength of the metal–ligand interactions and therefore vary for different complexes.

In $[Re_2Cl_8]^{2-}$, each rhenium is formally Re(III) and has four *d* electrons. If the eight *d* electrons for this ion are placed into the four lowest energy orbitals shown in Figure 15.9 (not including the low-energy orbital arising from the $d_{x^2-y^2}$ interactions, occupied by ligand electrons), the total bond order is 4, corresponding to—in increasing

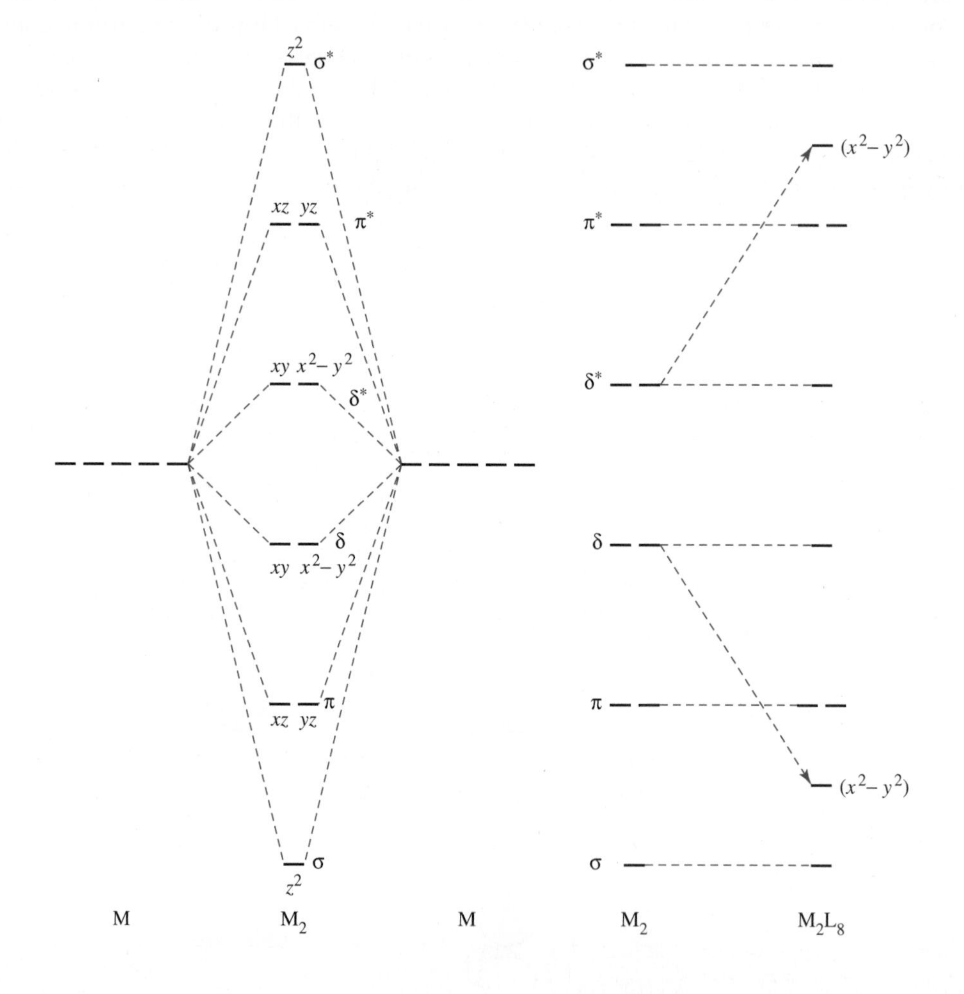

FIGURE 15.9 Relative Energies of Orbitals Formed from *d*-Orbital Interactions.

[17] Analysis of the symmetry of this ion shows that the *s*, p_x, and p_y orbitals are also involved.

energy—one σ bond, two π bonds, and one δ bond.[18] The δ bond is weakest; however, it is strong enough to maintain this ion in its eclipsed conformation. The weakness of the δ bond is illustrated by the small separation in energy of the δ and δ^* orbitals. This energy difference typically corresponds to the energy of visible light, with the consequence that most quadruply bonded complexes are vividly colored. For example, $[Re_2Cl_8]^{2-}$ is royal blue, and $[Mo_2Cl_8]^{4-}$ is bright red. By comparison, main group compounds having filled π and empty π^* orbitals are often colorless (e.g., N_2 and CO), because the energy difference between these orbitals is commonly in the ultraviolet part of the spectrum.

Additional electrons populate δ^* orbitals and reduce the bond order. For example, $[Os_2Cl_8]^{2-}$, an osmium(III) species with a total of 10 *d* electrons, has a triple bond. The δ bond order in this ion is zero; in the absence of such a bond, the eclipsed geometry—as found in quadruply bonded complexes, such as $[Re_2Cl_8]^{2-}$—is absent. X-ray crystallographic analysis has shown $[Os_2Cl_8]^{2-}$ to be very nearly staggered (D_{4d} geometry), as would be expected from VSEPR considerations.

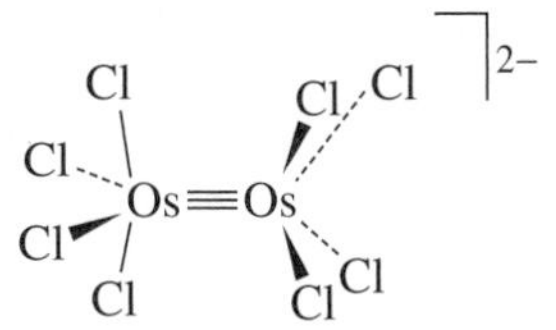

$[Os_2Cl_8]^{2-}$

Similarly, fewer than 8 valence electrons would also give a bond order less than 4. Examples of such complexes are shown in Figure 15.10.

Metal–metal multiple bonding can have dramatic effects on bond distances, as measured by X-ray crystallography. One way of describing the shortening of interatomic distances by multiple bonds is by comparing the bond distances in multiple bonds to the distances for single bonds. The ratios of these distances is sometimes called the **formal shortness ratio**. Values of this ratio are compared below for main group triple bonds and for some of the shortest of the measured transition-metal quadruple bonds:

Multiple Bond Distance/Single Bond Distance			
Bond	**Ratio**	**Bond**	**Ratio**
C≡C	0.783	Cr ≣ Cr	0.771
N≡N	0.786	Mo ≣ Mo	0.807
		Re ≣ Re	0.848

The ratios found for several quadruply bonded chromium complexes are the smallest ratios found to date for any compounds. Considerable variation in bond distances has been observed. Mo—Mo quadruple bonds, for example, have been found in the range 203.7 to 230.2 pm.[19]

[18] More elaborate calculations have yielded an "effective" bond order of 3.2 for $[Re_2Cl_8]^{2-}$, recommended as an alternative label to a "weak" quadruple bond. See L. Gagliardi and B. O. Roos, *Inorg. Chem.*, **2003**, *42*, 1599.
[19] F. A. Cotton and R. A. Walton, *Multiple Bonds Between Metal Atoms*, John Wiley & Sons, New York, 1982, pp. 161–165.

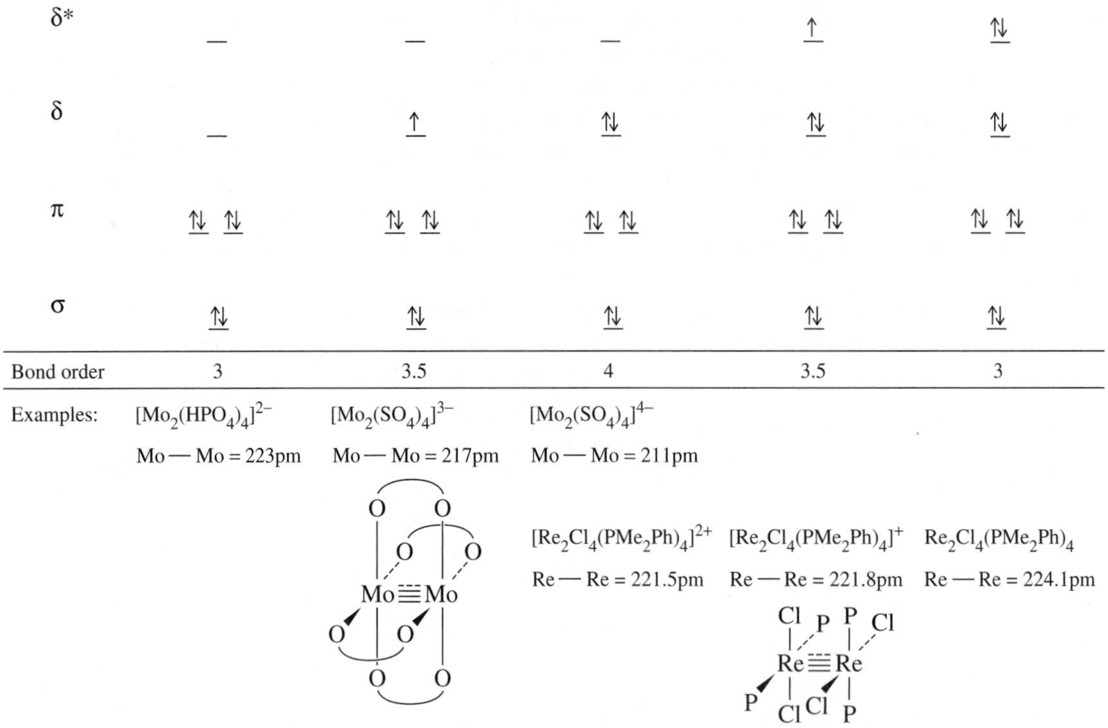

δ^*	—	—	—	↑	⇅
δ	—	↑	⇅	⇅	⇅
π	⇅ ⇅	⇅ ⇅	⇅ ⇅	⇅ ⇅	⇅ ⇅
σ	⇅	⇅	⇅	⇅	⇅
Bond order	3	3.5	4	3.5	3

Examples: $[Mo_2(HPO_4)_4]^{2-}$ Mo — Mo = 223pm; $[Mo_2(SO_4)_4]^{3-}$ Mo — Mo = 217pm; $[Mo_2(SO_4)_4]^{4-}$ Mo — Mo = 211pm

$[Re_2Cl_4(PMe_2Ph)_4]^{2+}$ Re — Re = 221.5pm; $[Re_2Cl_4(PMe_2Ph)_4]^{+}$ Re — Re = 221.8pm; $Re_2Cl_4(PMe_2Ph)_4$ Re — Re = 224.1pm

FIGURE 15.10 Bond Order and Electron Count in Dimetal Clusters. (From A. Bino and F. A. Cotton, *Inorg. Chem.*, **1979**, *18*, 3562; and F. A. Cotton, *Chem. Soc. Rev.*, **1983**, *12*, 35.)

The effect of population of δ and δ^* orbitals on bond distances can sometimes be surprisingly small. For example, removal of δ^* electrons on oxidation of $Re_2Cl_4(PMe_2Ph_4)_4$ gives only very slight shortening of the Re—Re distances, as shown in Table 15.5.[20]

A possible explanation for the small change in bond distance is that, with increasing oxidation state of the metal, the *d* orbitals contract. This contraction may cause overlap of *d* orbitals in π bonding to become less effective. Thus, as δ^* electrons are removed, the pi interactions become weaker; the two factors—increase in bond order and increase in oxidation state of Re—very nearly offset each other.

TABLE 15.5 Effect of Oxidation on Re—Re Bond Distances in Re_2 Complexes

Complex	Number of *d* Electrons	Formal Re—Re Bond Order	Formal Oxidation State of Re	Re—Re Distance (pm)
$Re_2Cl_4(PMe_2Ph_4)_4$	10	3	2	224.1
$[Re_2Cl_4(PMe_2Ph_4)_4]^+$	9	3.5	2.5	221.8
$[Re_2Cl_4(PMe_2Ph_4)_4]^{2+}$	8	4	3	221.5

QUINTUPLE BONDS Can there be such a thing as a quintuple bond? Figure 15.8 shows five possible interactions between *d* orbitals, including two δ interactions—so it is reasonable to propose a compound having bonding electron pairs in orbitals arising from all five. In 2005 Power and colleagues reported a compound with "fivefold" bonding.[21] This dimeric

[20] F. A. Cotton, *Chem. Soc. Rev.*, **1983**, *12*, 35.
[21] T. Nguyen, A. D. Sutton, M. Brynda, J. C. Fettinger, G. J. Long, and P. P. Power, *Science*, **2005**, *310*, 844.

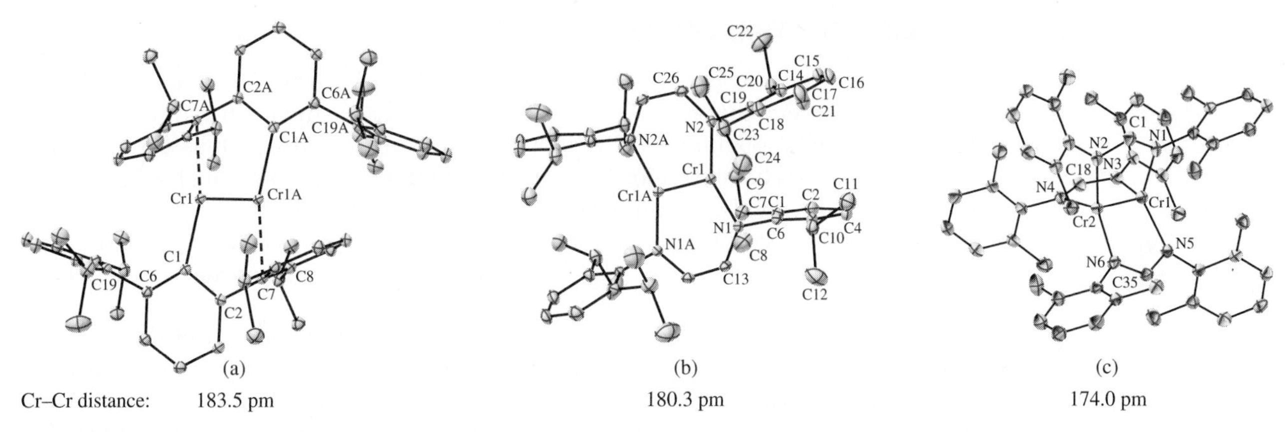

FIGURE 15.11 Chromium(I) Complexes with Extremely Short Metal–Metal Bonds.

((a) From T. Nguyen, A. D. Sutton, M. Brynda, J. C. Fettinger, G. J. Long, and P. P. Power, *Science*, **2005**, *310*, 844. Reprinted with permission from AAAS. (b) Reprinted with permission from K. Kreisel, G. P. A. Yap, O. Dmitrenko, C. R. Landis, and K. H. Theopold, *Journal of the American Chemical Society*, **2007**, *129*, 14162. Copyright 2007 American Chemical Society. (c) Y.-C. Tsai, C.-W. Hsu, J.-S. K. Yu, G.-H. Lee, Y. Wang, and T.-S. Kuo. Remarkably Short Metal–Metal Bonds: A Lantern-Type Quintuply Bonded Dichromium(I) Complex. *Angew. Chem. Int. Ed.*, **2008**, *47*, 7251. Copyright Wiley-VCH Verlag GmbH & Co. KGaA. Reproduced with permission.)

chromium(I) complex, shown in Figure 15.11(a) and on a larger scale in Figure 1.3, has ligands that are sufficiently bulky to protect the metals from reactions that might reduce the bond order. In addition, the low oxidation state of chromium in this complex leaves the metals each with the five electrons necessary to fully occupy molecular orbitals from the five sets of *d* orbital interactions. The Cr—Cr distance of 183.5 pm gives a formal shortness ratio of 0.774. However, this is not even as short as the lowest ratio (0.771) reported for quadruple bonds. While Power's complex may in fact have one σ, two π, and two δ interactions, this does not imply bonding that has the full strength of the five interactions in Figure 15.8. The *trans*-bent geometry and apparent interactions between the chromium atoms and the aromatic rings make the bonding in this compound more complex; nevertheless, calculations support that the σ, π, and δ interactions between *d* orbitals do occur.[22]

Since the initial report of fivefold bonding in 2005, the record for shortest Cr—Cr bond has been reduced to 180.3 pm in 2007[23] and to 174.0 pm in 2008.[24] These structures are also shown in Figure 15.11. The formal shortness ratio of the last of these compounds, reported by Tsai and coworkers, is 0.733, the smallest such value reported to date; the authors make the case for very strong Cr—Cr bonding with electron pairs occupying one σ, two π, and two δ orbitals. In contrast to Power's complex, Tsai's complex does not have the complication of *trans*-bent geometry, and more direct *d* orbital interactions may be one factor in making the bonding between the metals more efficient.

The bond order in these and other complexes with very short metal–metal bonds has been controversial. Various calculations have given bond orders as low as 3.3 for "fivefold" bonded chromium(I) complexes. Factors such as the influence of bridging ligands on metal–metal distances, nonlinearity in Power's original complex that may reduce orbital interactions, and more complex interactions than shown in this chapter are among the factors that merit consideration. Discussion of these factors is beyond the scope of this text, but the interested reader is encouraged to consult the sources cited[25] for discussions of the numerous factors involved in Cr—Cr multiple bonds in these dimers.

---

[22] G. Frenking, *Science*, **2005**, *310*, 796; U. Radius and F. Breher, *Angew. Chem., Int. Ed.*, **2006**, *45*, 3006.
[23] K. A. Kreisel, G. P. A. Yap, O. Dmitrenko, C. R. Landis, and K. H. Theopold, *J. Am. Chem. Soc.*, **2007**, *129*, 14162.
[24] Y.-C. Tsai, C.-W. Hsu, J.-S. K. Yu, G.-H. Lee, Y. Wang, and T.-S. Kuo, *Angew. Chem., Int. Ed.*, **2008**, *47*, 7250.
[25] G. Merino, K. J. Donald, J. S. D'Acchioli, and R. Hoffmann, *J. Am. Chem. Soc.*, **2007**, *129*, 15295; M. Brynda, L. Gagliardi, and B. O. Roos, *Chem. Phys. Lett.*, **2009**, *471*, 1; D. B. DuPré, *J. Chem. Phys. A*, **2009**, *113*, 1559.

15.4 CLUSTER COMPOUNDS

Examples of cluster compounds have been given in previous sections of this chapter and in several earlier chapters. Transition-metal cluster chemistry has developed rapidly since the 1980s. Beginning with simple dimeric molecules, such as $Co_2(CO)_8$ and $Fe_2(CO)_9$,[26] chemists have developed syntheses of far more complex clusters, some with interesting and unusual structures and chemical properties. Large clusters have been studied with the objective of developing catalysts that may duplicate or improve on the properties of heterogeneous catalysts; the surface of a large cluster may in these cases mimic the behavior of the surface of a solid catalyst.

Before discussing transition-metal clusters in more detail, it is useful to consider compounds of boron, which has a highly detailed cluster chemistry. As mentioned in Chapter 8, boron forms numerous hydrides (boranes) with interesting structures. Some of these compounds exhibit similarities in their bonding and structures to transition-metal clusters.

15.4.1 Boranes

Boron and hydrogen form many neutral and ionic species, far too many to describe in this text. For the purposes of illustrating parallels between these species and transition-metal clusters, we will first consider one category of boranes, *closo* (Greek, "cagelike") boranes that have the formula $B_nH_n^{2-}$. These boranes consist of closed polyhedra with n corners and all triangular faces (triangulated polyhedra). Each corner is occupied by a BH group.

Molecular orbital calculations have shown that *closo* boranes have $2n+1$ bonding molecular orbitals, including n B—H σ bonding orbitals and $n + 1$ bonding orbitals in the central core, described as **framework** or **skeletal** bonding orbitals.[27] A useful example is $B_6H_6^{2-}$, which has O_h symmetry. In this ion, each boron has four valence orbitals that can participate in bonding, giving a total of 24 boron orbitals for the cluster. These orbitals can be classified into two sets. If the z axis of each boron atom is chosen to point toward the center of the octahedron (Figure 15.12), the p_z and s orbitals are a set of suitable symmetry to bond with the hydrogen atoms. A second set of orbitals, consisting of the p_x and p_y orbitals of the borons, is then available for boron–boron bonding.

The p_z and s orbitals of the borons collectively have the same symmetry (which reduces to the irreducible representations $A_{1g} + E_g + T_{1u}$; an analysis of the orbitals in terms of symmetry is left as an exercise in Problem 15.19). Therefore, they may be considered to form *sp* hybrid orbitals. These hybrid orbitals, two on each boron, point out toward the hydrogen atoms and in toward the center of the cluster, as shown for A_{1g} and T_{1u} in Figure 15.13.

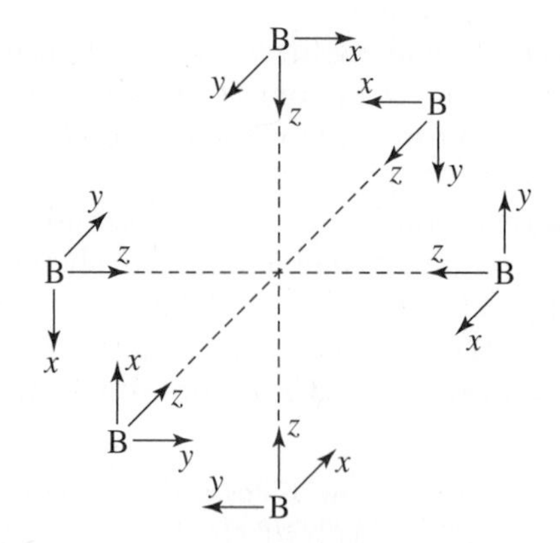

FIGURE 15.12 Coordinate System for Bonding in $B_6H_6^{2-}$.

[26] Some chemists define clusters as having at least three metal atoms.
[27] K. Wade, *Electron Deficient Compounds*, Thomas Nelson & Sons, London, 1971.

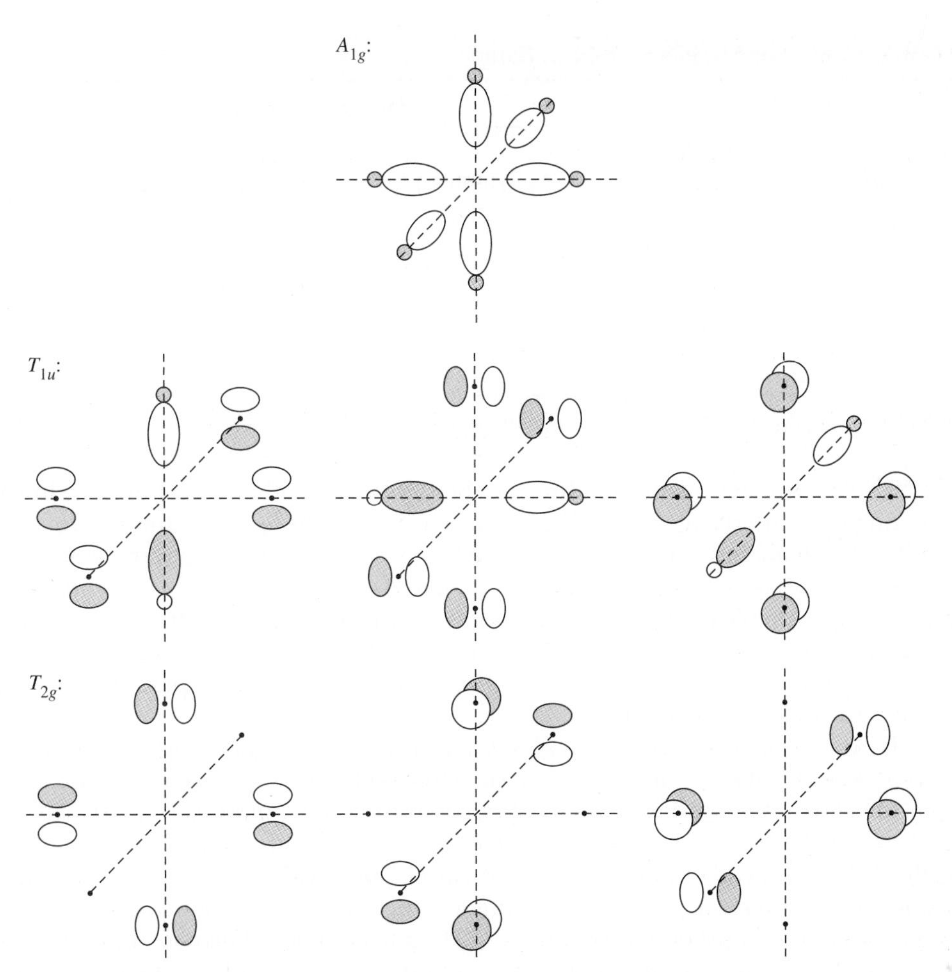

FIGURE 15.13
Bonding in $B_6H_6^{2-}$.

Six of the hybrids form bonds with the 1*s* orbitals of the hydrogens. The six remaining hybrids and the unhybridized 2*p* orbitals of the borons remain to participate in bonding within the B_6 core. Seven orbital combinations lead to bonding interactions; these are also shown in Figure 15.13. Constructive overlap of all six hybrid orbitals at the center of the octahedron yields a framework bonding orbital of A_{1g} symmetry; as its symmetry label indicates, this orbital is completely symmetric with respect to all symmetry operations of the O_h point group. Additional bonding interactions are of two types: overlap of two *sp* hybrid orbitals with parallel *p* orbitals on the remaining four boron atoms (three such interactions, collectively of T_{1u} symmetry) and overlap of *p* orbitals on four boron atoms within the same plane (three interactions, T_{2g} symmetry). The remaining orbital interactions lead to nonbonding or antibonding molecular orbitals. To summarize, from the 24 valence atomic orbitals of boron are formed:

13 bonding orbitals (= $2n + 1$), consisting of
- 7 framework molecular orbitals (= $n + 1$), consisting of
  - 1 bonding orbital (A_{1g}) from overlap of *sp* hybrid orbitals
  - 6 bonding orbitals from overlap of *p* orbitals of boron with *sp* hybrid orbitals (T_{1u}) or with other boron *p* orbitals (T_{2g})
- 6 boron–hydrogen bonding orbitals (= n)

11 nonbonding or antibonding orbitals

TABLE 15.6 Bonding Pairs for *Closo* Boranes

		Framework Bonding Pairs		
Formula	Total Valence Electron Pairs	A_1 Symmetry[a]	Other Symmetry	B—H Bonding pairs
$B_6H_6^{2-}$	13	1	6	6
$B_7H_7^{2-}$	15	1	7	7
$B_8H_8^{2-}$	17	1	8	8
$B_nH_n^{2-}$	$2n + 1$	1	n	n

NOTE: [a]Symmetry designation depends on the point group (such as A_{1g} for O_h symmetry).

Similar descriptions of bonding can be derived for other *closo* boranes. In each case, one particularly useful similarity can be found: there is one more framework bonding pair than the number of corners in the polyhedron. The extra framework bonding pair is in a totally symmetric orbital (like the A_{1g} orbital in $B_6H_6^{2-}$), resulting from the overlap of atomic (or hybrid) orbitals at the center of the polyhedron. In addition, a significant gap in energy exists between the highest bonding orbital (HOMO) and the lowest nonbonding orbital (LUMO).[28] The numbers of bonding pairs for common geometries are shown in Table 15.6.

Together, the *closo* structures make up only a very small fraction of all known borane species. Additional structural types can be obtained by removing one or more corners from the *closo* framework. Removal of one corner yields a ***nido*** (nestlike) structure, removal of two corners an ***arachno*** (weblike) structure, removal of three corners a ***hypho*** (netlike) structure, and removal of four corners a ***klado*** (branched) structure.[29] Examples of three related *closo*, *nido*, and *arachno* borane structures are shown in Figure 15.14, and the structures for these boranes having 6 to 12 boron atoms are shown in Figure 15.15.

The classification of structural types can often be done more conveniently on the basis of valence electron counts. Various schemes for relating electron counts to structures have been proposed, with most proposals based on the set of rules formulated by

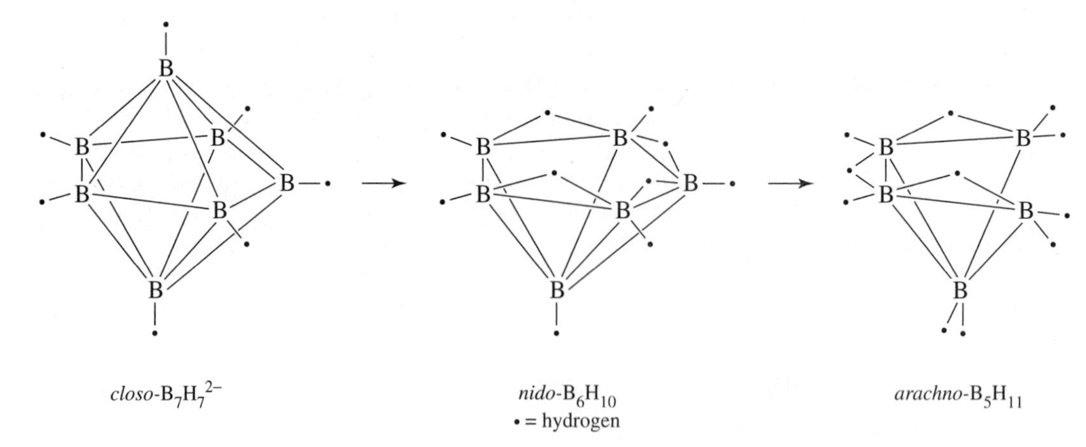

FIGURE 15.14 *Closo*, *Nido*, and *Arachno* Borane Structures.

[28] K. Wade, "Some Bonding Considerations," in B. F. G. Johnson, ed., *Transition Metal Clusters*, John Wiley & Sons, New York, 1980, p. 217.

[29] *Hypho-* and *klado-* structures appear to be known only as derivatives. Additional details on naming boron hydrides and related compounds can be found in the IUPAC publication, G. J. Leigh, ed., *Nomenclature of Inorganic Chemistry: Recommendations 1990*, Blackwell Scientific Publications, Cambridge, MA, 1990, pp. 207–237.

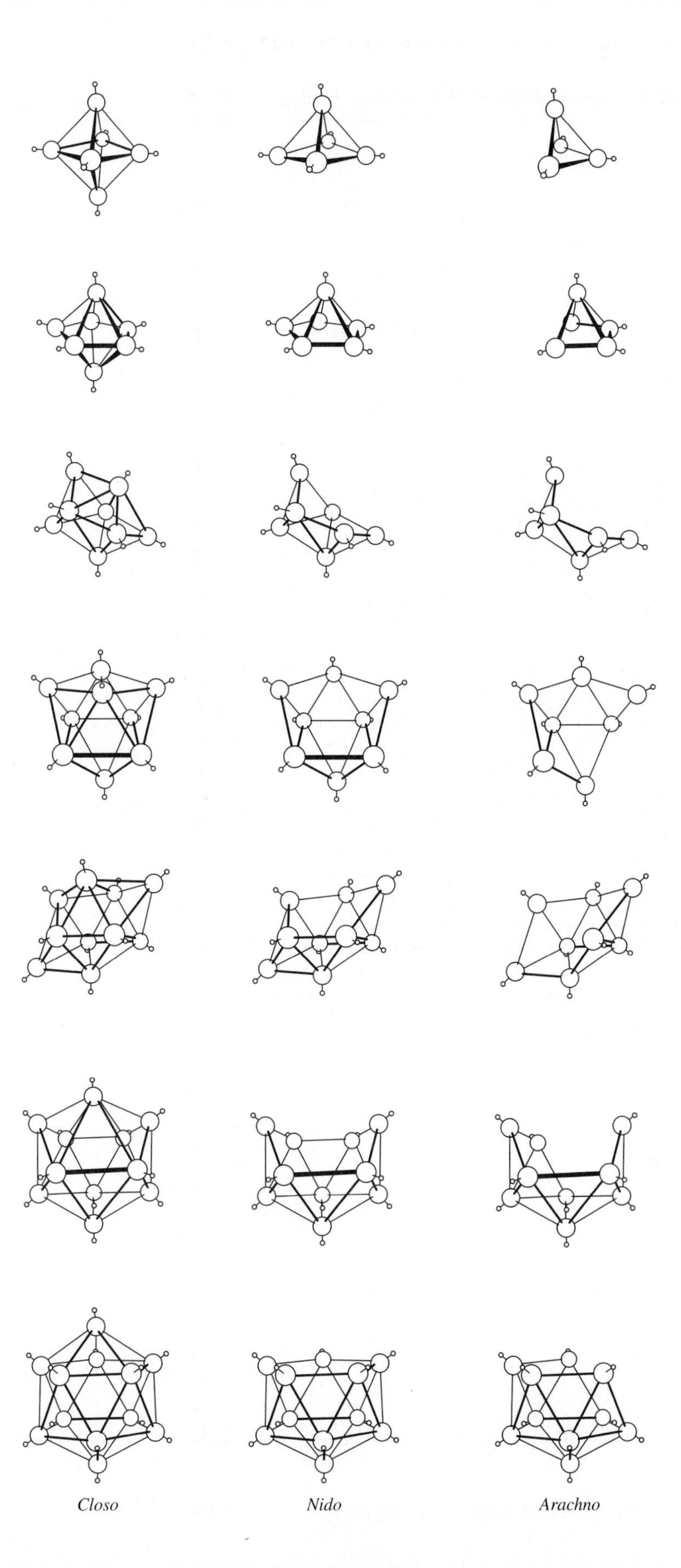

FIGURE 15.15 Structures of *Closo*, *Nido*, and *Arachno* Boranes having 6 to 12 Borons.

(Reproduced and adapted with permission from R. W. Rudolph, *Acc. Chem. Res.*, **1976**, *9*, 446. © 1976 American Chemical Society.)

TABLE 15.7 Classification of Cluster Structures

Structure Type	Corners Occupied	Pairs of Framework Bonding Electrons	Empty Corners
Closo	n corners of n-cornered polyhedron	$n + 1$	0
Nido	$(n - 1)$ corners of n-cornered polyhedron	$n + 1$	1
Arachno	$(n - 2)$ corners of n-cornered polyhedron	$n + 1$	2
Hypho	$(n - 3)$ corners of n-cornered polyhedron	$n + 1$	3
Klado	$(n - 4)$ corners of n-cornered polyhedron	$n + 1$	4

Wade.[30] The classification scheme based on these rules is summarized in Table 15.7. In this table, the number of pairs of framework bonding electrons is determined by subtracting one B—H bonding pair per boron; the $n + 1$ remaining framework electron pairs may be used in boron–boron bonding or in bonds between boron and other hydrogen atoms.

In addition, it is sometimes useful to relate the total valence electron count in boranes to the structural type. In *closo* boranes, the total number of valence electron pairs is equal to the sum of the number of vertices in the polyhedron (each vertex has a boron–hydrogen bonding pair) and the number of framework bond pairs. For example, in $B_6H_6^{2-}$ there are 26 valence electrons, or 13 pairs (= $2n + 1$, as mentioned previously). Six of these pairs are involved in bonding to the hydrogens (one per boron), and seven pairs are involved in framework bonding. The polyhedron of the *closo* structure is the parent polyhedron for the other structural types. Table 15.8 summarizes electron counts and classifications for several examples of boranes.

TABLE 15.8 Examples of Electron Counting in Boranes

Vertices in Parent Polyhedron	Classification	Boron Atoms in Cluster	Valence Electrons	Framework Electron Pairs	Examples	Formally Derived From
6	*Closo*	6	26	7	$B_6H_6^{2-}$	$B_6H_6^{2-}$
	Nido	5	24	7	B_5H_9	$B_5H_5^{4-}$
	Arachno	4	22	7	B_4H_{10}	$B_4H_4^{6-}$
7	*Closo*	7	30	8	$B_7H_7^{2-}$	$B_7H_7^{2-}$
	Nido	6	28	8	B_6H_{10}	$B_6H_6^{4-}$
	Arachno	5	26	8	B_5H_{11}	$B_5H_5^{6-}$
12	*Closo*	12	50	13	$B_{12}H_{12}^{2-}$	$B_{12}H_{12}^{2-}$
	Nido	11	48	13	$B_{11}H_{13}^{2-}$	$B_{11}H_{11}^{4-}$
	Arachno	10	46	13	$B_{10}H_{15}^{2-}$	$B_{10}H_{10}^{6-}$

[30] K. Wade, *Adv. Inorg. Chem. Radiochem.*, **1976**, *18*, 1–66.

A METHOD FOR CLASSIFYING BORANES Boranes can conveniently be classified by considering

closo boranes to have the formula $B_nH_n^{2-}$
nido boranes to be derived from $B_nH_n^{4-}$ ions
arachno boranes to be derived from $B_nH_n^{6-}$ ions
hypho boranes to be derived from $B_nH_n^{8-}$ ions
klado boranes to be derived from $B_nH_n^{10-}$ ions

The formulas of boranes can be related to these formulas by formally subtracting H^+ ions from the formula to make the number of B and H atoms equal. For example, to classify $B_9H_{14}^-$, we can formally consider it to be derived from $B_9H_9^{6-}$:

$$B_9H_{14}^- - 5\,H^+ = B_9H_9^{6-}$$

The classification for this borane is therefore *arachno*.

EXAMPLES

Classify the following boranes by structural type.

$B_{10}H_{14}$

$B_{10}H_{14} - 4\,H^+ = B_{10}H_{10}^{4-}$ The classification is *nido*.

$B_2H_7^-$

$B_2H_7^- - 5\,H^+ = B_2H_2^{6-}$ The classification is *arachno*.

B_8H_{16}

$B_8H_{16} - 8\,H^+ = B_8H_8^{8-}$ The classification is *hypho*.

▶ **Exercise 15.3** Classify the following boranes by structural type:

a. $B_{11}H_{13}^{2-}$ **b.** $B_5H_8^-$ **c.** $B_7H_7^{2-}$ **d.** $B_{10}H_{18}$

15.4.2 Heteroboranes

The electron-counting schemes can be extended to isoelectronic species such as the carboranes, also known as *carbaboranes*. The CH^+ unit is isoelectronic with BH; many compounds are known in which one or more BH groups have been replaced by CH^+ (or by C, which also has the same number of electrons as BH). For example, replacement of two BH groups by CH^+ in *closo*-$B_6H_6^{2-}$ yields *closo*-$C_2B_4H_6$, a neutral compound. *Closo*, *nido*, and *arachno* carboranes are all known, most commonly containing two carbon atoms; examples are shown in Figure 15.16. Chemical formulas corresponding to these designations are given in Table 15.9.

Carboranes may be classified by structural type using the same method described previously for boranes. Because a carbon atom has the same number of valence electrons as a boron atom plus a hydrogen atom, formally each C should be converted to BH in the classification scheme. For example, for a carborane having the formula $C_2B_8H_{10}$,

$$C_2B_8H_{10} \longrightarrow B_{10}H_{12}$$
$$B_{10}H_{12} - 2\,H^+ = B_{10}H_{10}^{2-}$$

the classification of the carborane $C_2B_8H_{10}$ is therefore *closo*.

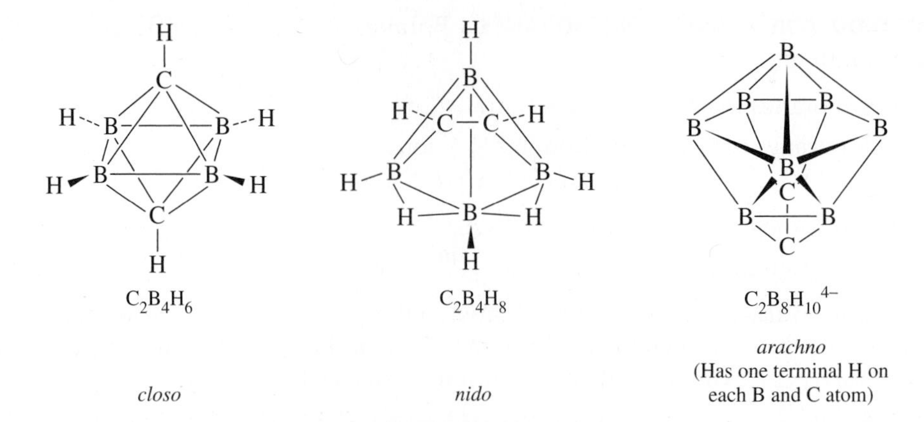

FIGURE 15.16 Examples of Carboranes.

TABLE 15.9 Examples of Formulas of Boranes and Carboranes

Type	Borane	Example	Carborane	Example
Closo	$B_nH_n^{2-}$	$B_{12}H_{12}^{2-}$	$C_2B_{n-2}H_n$	$C_2B_{10}H_{12}$
Nido	B_nH_{n+4}[a]	$B_{10}H_{14}$	$C_2B_{n-2}H_{n+2}$	$C_2B_8H_{12}$
Arachno	B_nH_{n+6}[a]	B_9H_{15}	$C_2B_{n-2}H_{n+4}$	$C_2B_7H_{13}$

NOTE: [a]*Nido* boranes may also have the formulas $B_nH_{n+3}^{-}$ and $B_nH_{n+2}^{2-}$; *arachno* boranes may also have the formulas $B_nH_{n+5}^{-}$ and $B_nH_{n+4}^{2-}$.

EXAMPLES

Classify the following carboranes by structural type:

$C_2B_9H_{12}^{-}$

$C_2B_9H_{12}^{-} \longrightarrow B_{11}H_{14}^{-}$

$B_{11}H_{14}^{-} - 3H^{+} = B_{11}H_{11}^{4-}$ The classification is *nido*.

$C_2B_7H_{13}$

$C_2B_7H_{13} \longrightarrow B_9H_{15}$

$B_9H_{15} - 6H^{+} = B_9H_9^{6-}$ The classification is *arachno*.

$C_4B_2H_6$

$C_4B_2H_6 \longrightarrow B_6H_{10}$

$B_6H_{10} - 4H^{+} = B_6H_6^{4-}$ The classification is *nido*.

▶ **Exercise 15.4** Classify the following carboranes by structural type:

a. $C_3B_3H_7$ **b.** $C_2B_5H_7$ **c.** $C_2B_7H_{12}^{-}$

Many derivatives of boranes containing other main group atoms, designated *heteroatoms*, are also known. These *heteroboranes* may be classified by formally converting the heteroatom to a BH_x group having the same number of valence electrons, and then proceeding as in previous examples. For some of the more common heteroatoms, the substitutions are

Heteroatom	Replace with
C, Si, Ge, Sn	BH
N, P, As	BH_2
S, Se	BH_3

EXAMPLES

Classify the following heteroboranes by structural type:

SB_9H_{11}

$SB_9H_{11} \longrightarrow B_{10}H_{14}$

$B_{10}H_{14} - 4\,H^+ = B_{10}H_{10}{}^{4-}$ The classification is *nido*.

$CPB_{10}H_{11}$

$CPB_{10}H_{11} \longrightarrow PB_{11}H_{12} \longrightarrow B_{12}H_{14}$

$B_{12}H_{14} - 2H^+ = B_{12}H_{12}{}^{2-}$ The classification is *closo*.

▶ **Exercise 15.5** Classify the following heteroboranes by structural type:

a. SB_9H_9 **b.** $GeC_2B_9H_{11}$ **c.** $SB_9H_{12}{}^-$

Although it may not be surprising that the same set of electron-counting rules can be used to describe satisfactorily such similar compounds as boranes and carboranes, we should examine how far the comparison can be extended. Can Wade's rules, for example, be used effectively on compounds containing metals bonded to boranes or carboranes? Can the rules be extended even further to describe the bonding in polyhedral metal clusters?

15.4.3 Metallaboranes and Metallacarboranes

The CH group of a carborane is isolobal with 15-electron fragments of an octahedron such as $Co(CO)_3$. Similarly, BH, which has 4 valence electrons, is isolobal with 14-electron fragments such as $Fe(CO)_3$ and $CO(\eta^5\text{-}C_5H_5)$. These organometallic fragments have been found in substituted boranes and carboranes in which the organometallic fragments substitute for the isolobal main group fragments. For example, the organometallic derivatives of B_5H_9 shown in Figure 15.17 have been synthesized. Theoretical calculations on the iron derivative have supported the view that $Fe(CO)_3$ in this compound bonds in a fashion isolobal with BH.[31] In both fragments, the orbitals involved in

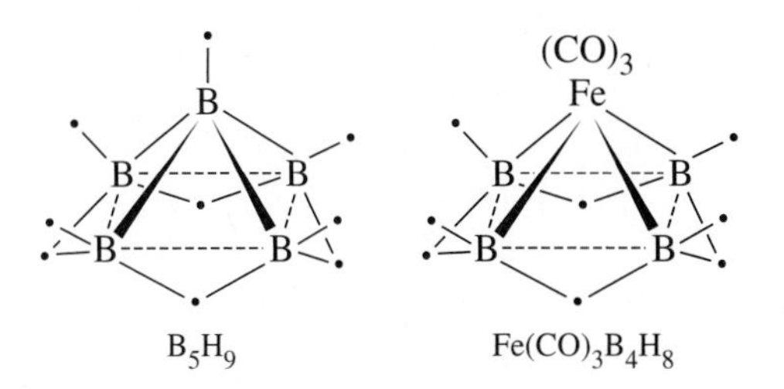

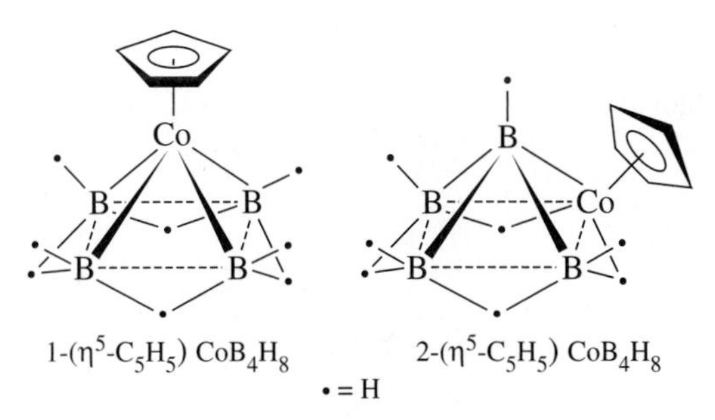

FIGURE 15.17 Organometallic Derivatives of B_5H_9.

[31] R. L. DeKock and T. P. Fehlner, *Polyhedron*, **1982**, *1*, 521.

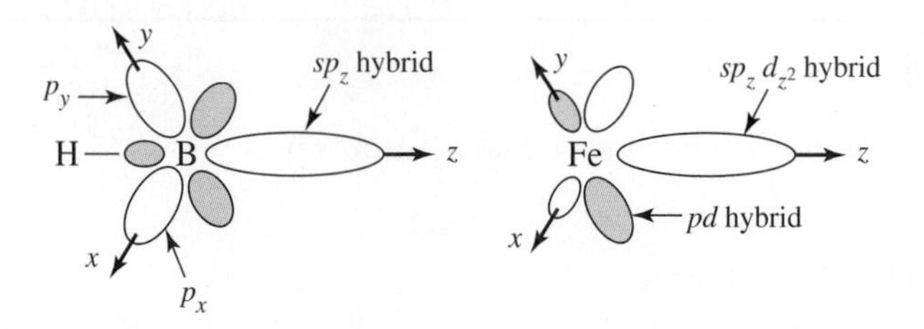

FIGURE 15.18 Orbitals of Isolobal Fragments BH and $Fe(CO)_3$.

framework bonding within the cluster are similar (Figure 15.18). In BH, the orbitals participating in framework bonding are an sp_z hybrid pointing toward the center of the polyhedron (similar to the orbitals participating in the bonding of A_{1g} symmetry in $B_6H_6^{2-}$ in Figure 15.13) and p_x and p_y orbitals tangential to the surface of the cluster. In $Fe(CO)_3$, an $sp_zd_{z^2}$ hybrid points toward the center, and *pd* hybrid orbitals are oriented tangentially to the cluster surface.

There are many metallaboranes and metallacarboranes. Selected examples with *closo* structures are given in Table 15.10.

TABLE 15.10 Metallaboranes and Metallacarboranes with *Closo* Structures

Number of Skeletal Atoms	Shape	Examples	
6	Octahedron	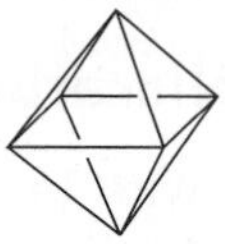 $B_4H_6(CoCp)_2$	$C_2B_3H_5Fe(CO)_3$
7	Pentagonal bipyramid	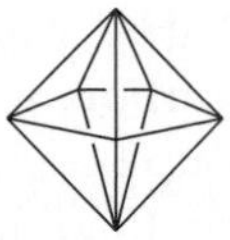 $C_2B_4H_6Ni(PPh_3)_2$	$C_2B_3H_5(CoCp)_2$
8	Dodecahedron	 $C_2B_4H_4[(CH_3)_2Sn]CoCp$	
9	Capped square antiprism	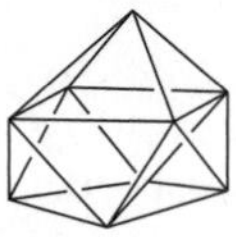 $C_2B_6H_8Pt(PMe_3)_2$	$C_2B_5H_7(CoCp)_2$
10	Bicapped square antiprism	 $[B_9H_9NiCp]^-$	$CB_7H_8(CoCp)(NiCp)$
11	Octadecahedron	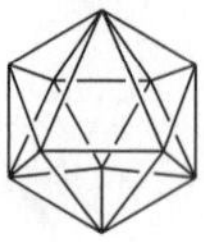 $[CB_9H_{10}CoCp]^-$	$C_2B_8H_{10}IrH(PPh_3)_2$
12	Icosahedron	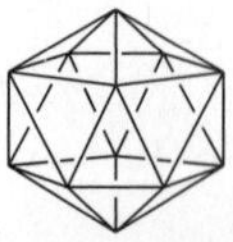 $C_2B_7H_9(CoCp)_3$	$C_2B_9H_{11}Ru(CO)_3$

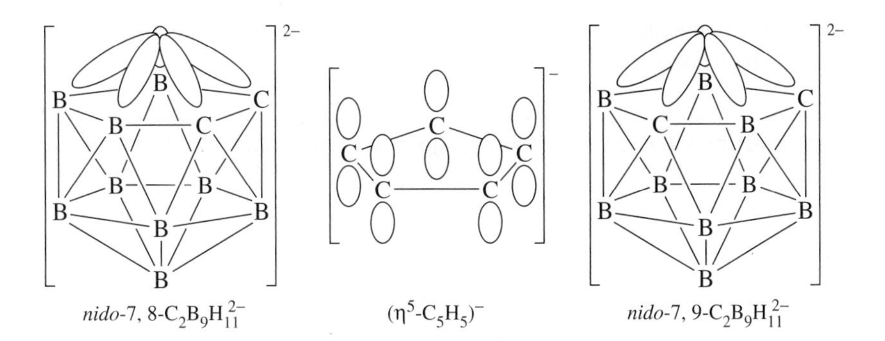

FIGURE 15.19 Comparison of $C_2B_9H_{11}{}^{2-}$ with $C_5H_5{}^-$.

(Adapted with permission from N. N. Greenwood and A. Earnshaw, *Chemistry of the Elements*, Pergamon Press, Oxford, 1984, p. 210. © 1984, Pergamon Press PLC.)

Anionic boranes and carboranes can also act as ligands toward metals in a manner resembling that of cyclic organic ligands. For example, *nido* carboranes of formula $C_2B_9H_{11}{}^{2-}$ have *p* orbital lobes pointing toward the "missing" site of the icosahedron (remember that the *nido* structure corresponds to a *closo* structure, in this case the 12-vertex icosahedron, with one vertex missing). This arrangement of *p* orbitals can be compared with the *p* orbitals of the cyclopentadienyl ring, as shown in Figure 15.19.

Although the comparison between these ligands is not exact, the similarity is sufficient that $C_2B_9H_{11}{}^{2-}$ can bond to iron to form a carborane analogue of ferrocene, $[Fe(\eta^5-C_2B_9H_{11})_2]^{2-}$. A mixed-ligand sandwich compound containing one carborane and one cyclopentadienyl ligand, $[Fe(\eta^5-C_2B_9H_{11})(\eta^5-C_5H_5)]$, has also been made (Figure 15.20).[32] Many other examples of boranes and carboranes serving as ligands to transition metals are also known.[33]

Metallaboranes and metallacarboranes can be classified structurally by using a procedure similar to the method described previously for boranes and their main group derivatives. To classify borane derivatives with transition metal–containing fragments, it is convenient to determine how many electrons the metal-containing fragment needs to satisfy the requirement of the 18-electron rule. This fragment can be considered equivalent to a BH_x fragment needing the same number of electrons to satisfy the octet rule. For example, a 14-electron fragment such as $Co(\eta^5-C_5H_5)$ is 4 electrons short of 18; this fragment may be considered the equivalent of the 4-electron fragment BH,

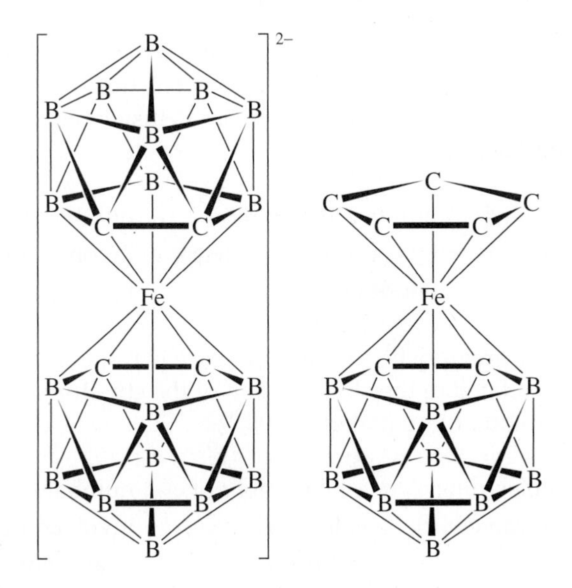

FIGURE 15.20 Carborane Analogs of Ferrocene.

(Adapted with permission from N. N. Greenwood and A. Earnshaw, *Chemistry of the Elements*, Pergamon Press, Oxford, 1984, pp. 211–212. © 1984, Pergamon Press PLC.)

[32] M. F. Hawthorne, D. C. Young, and P. A. Wegner, *J. Am. Chem. Soc.*, **1965**, *87*, 1818.
[33] K. P. Callahan and M. F. Hawthorne, *Adv. Organomet. Chem.*, **1976**, *14*, 145.

which is four electrons short of an octet. Shown here are examples of organometallic fragments and their corresponding BH_x fragments:

Valence Electrons in Organometallic Fragment	Example	Replace with
13	$Mn(CO)_3$	B
14	CoCp	BH
15	$Co(CO)_3$	BH_2
16	$Fe(CO)_4$	BH_3

EXAMPLES

Classify the following metallaboranes by structural type:

$B_4H_6(CoCp)_2$

$$B_4H_6(CoCp)_2 \longrightarrow B_4H_6(BH)_2 = B_6H_8$$
$$B_6H_8 - 2\,H^+ = B_6H_6^{\,2-}$$

The classification is *closo*.

$B_3H_7[Fe(CO)_3]_2$

$$B_3H_7[Fe(CO)_3]_2 \longrightarrow B_3H_7[BH]_2 = B_5H_9$$
$$B_5H_9 - 4\,H^+ = B_5H_5^{\,4-}$$

The classification is *nido*.

▶ **Exercise 15.6** Classify the following metallaboranes by structural type:

a. $C_2B_7H_9(CoCp)_3$
b. $C_2B_4H_6Ni(PPh_3)_2$

15.4.4 Carbonyl Clusters

The structures of several carbonyl cluster compounds were shown in Chapter 13. Many carbonyl clusters have structures similar to boranes; it is therefore of interest to determine to what extent the approach used to describe bonding in boranes may also be applicable to bonding in carbonyl clusters and other clusters.

According to Wade, the valence electrons in a cluster can be assigned to framework and metal–ligand bonding.[34]

$$\text{Total number of valence electrons in cluster} = \text{number of electrons involved in framework bonding} + \text{number of electrons involved in metal–ligand bonding}$$

As we have seen previously, the number of electrons involved in framework bonding in boranes is related to the classification of the structure as *closo*, *nido*, *arachno*, *hypho*, or *klado*. Rearranging this equation gives

$$\text{Number of electrons involved in framework bonding} = \text{total number of valence electrons in cluster} - \text{number of electrons involved in metal–ligand bonding}$$

[34] K. Wade, *Adv. Inorg. Chem. Radiochem.*, **1980**, *18*, 1.

For a borane, one electron pair is assigned to one boron–hydrogen bond on each boron. The remaining valence electron pairs are regarded as framework bonding pairs.[35] For a transition metal carbonyl complex, on the other hand, Wade suggests that 6 electron pairs per metal are involved either in metal–carbonyl bonding (to all carbonyls on a metal) or are nonbonding and therefore unavailable for participation in framework bonding. A metal–carbonyl cluster has 5 more electron pairs per framework atom, or 10 more electrons, than the corresponding borane. A metal–carbonyl analogue of *closo*-$B_6H_6^{2-}$, which has 26 valence electrons, would therefore need a total of 86 valence electrons to adopt a *closo* structure. An 86-electron cluster that satisfies this requirement is $Co_6(CO)_{16}$, which has an octahedral framework similar to $B_6H_6^{2-}$. As in the case of boranes, *nido* structures correspond to *closo* geometries from which one vertex is empty, *arachno* structures lack two vertices, and so on.

A simpler way to compare electron counts in boranes and transition-metal clusters is to consider the different numbers of valence orbitals available to the framework atoms. Transition metals, with nine valence orbitals (one *s*, three *p*, and five *d* orbitals), have five more orbitals available for bonding than boron, which has only four valence orbitals; these five extra orbitals, when filled as a consequence of bonding within the framework and with surrounding ligands, give an increased electron count of 10 electrons per framework atom. Consequently, a useful rule of thumb is to increase the electron requirement of the cluster by 10 per framework atom when replacing a boron with a transition-metal atom. In the example cited previously, replacing the six borons in *closo*-$B_6H_6^{2-}$ with six cobalts should, therefore, increase the electron count from 26 to 86 for a comparable *closo* cobalt cluster. $Co_6(CO)_{16}$, an 86-electron cluster, meets this requirement.

The valence electron counts corresponding to the various structural classifications for main group and transition-metal clusters are summarized in Table 15.11.[36] In this table, *n* designates the number of framework atoms.

TABLE 15.11 Electron Counting in Main Group and Transition Metal Clusters

Structure Type	Main Group Cluster	Transition Metal Cluster
Closo	$4n + 2$	$14n + 2$
Nido	$4n + 4$	$14n + 4$
Arachno	$4n + 6$	$14n + 6$
Hypho	$4n + 8$	$14n + 8$

Examples of *closo, nido,* and *arachno* borane and transition-metal clusters are given in Table 15.12. Transition-metal clusters formally containing seven metal–metal framework bonding pairs are among the most common; examples illustrating the structural diversity of these clusters are given in Table 15.13 and Figure 15.21.

The predictions of structures of transition metal–carbonyl complexes, using Wade's rules are often, but not always, accurate. For example, the clusters $M_4(CO)_{12}$ (M = Co, Rh, Ir) have 60 valence electrons and are predicted to be *nido* complexes ($14n + 4$ valence electrons). A *nido* structure would correspond to a trigonal bipyramid (the parent structure) with one position vacant. X-ray crystallographic studies, however, have shown these complexes to have tetrahedral metal cores.

[35] For structures involving bridging hydrogen atoms, the bridging hydrogens are considered to be involved in framework bonding.

[36] D. M. P. Mingos, *Acc. Chem. Res.*, **1984**, *17*, 311.

TABLE 15.12 ***Closo*, *Nido*, and *Arachno* Borane and Transition-Metal Clusters**

Atoms in Cluster	Vertices in Parent Polyhedron	Framework Electron Pairs	Valence Electrons (Boranes)				Valence Electrons (Transition Metal Clusters)			
			Closo	*Nido*	*Arachno*	Examples	*Closo*	*Nido*	*Arachno*	Examples
4	4	5	18				58			
	5	6		20		$B_4H_7^-$		60		$Co_4(CO)_{12}$
	6	7			22	B_4H_{10}			62	$[Fe_4C(CO)_{12}]^{2-}$
5	5	6	22			$C_2B_3H_5$	72			$Os_5(CO)_{16}$
	6	7		24		B_5H_9		74		$Os_5C(CO)_{15}$
	7	8			26	B_5H_{11}			76	$[Ni_5(CO)_{12}]^{2-}$
6	6	7	26			$B_6H_6^{2-}$	86			$Co_6(CO)_{16}$
	7	8		28		B_6H_{10}		88		$Os_6(CO)_{17}[P(OMe)_3]_3$
	8	9			30	B_6H_{12}			90	

TABLE 15.13 Clusters Formally Containing Seven Metal–Metal Framework Bond Pairs

Number of Framework Atoms	Cluster Type	Shape	Examples
7	Capped *closo*[a]	Capped octahedron	$[Rh_7(CO)_{16}]^{3-}$
			$Os_7(CO)_{21}$
6	*Closo*	Octahedron	$Rh_6(CO)_{16}$
			$Ru_6C(CO)_{17}$
6	Capped *nido*[a]	Capped square pyramid	$H_2Os_6(CO)_{18}$
5	*Nido*	Square pyramid	$Ru_5C(CO)_{15}$
4	*Arachno*	Butterfly	$[Fe_4(CO)_{13}H]^{-}$[b]

Source: K. Wade, "Some Bonding Considerations," in B. F. G. Johnson, ed., *Transition Metal Clusters*, John Wiley & Sons, 1980, p. 232.

NOTE: [a]A capped *closo* cluster has a valence electron count equivalent to neutral B_nH_n. A capped *nido* cluster has the same electron count as a *closo* cluster.

[b]This complex has an electron count matching a *nido* structure, but it adapts the butterfly structure expected for *arachno*. This is one of the many examples in which the structure of metal clusters is not predicted accurately by Wade's rules. Limitations of Wade's rules are discussed in R. N. Grimes, "Metallacarboranes and Metalloboranes," in G. Wilkinson, F. G. A. Stone, and W. Abel, eds., *Comprehensive Organometallic Chemistry*, Vol. 1, Pergamon Press, Elmsford, NY, 1982, p. 473.

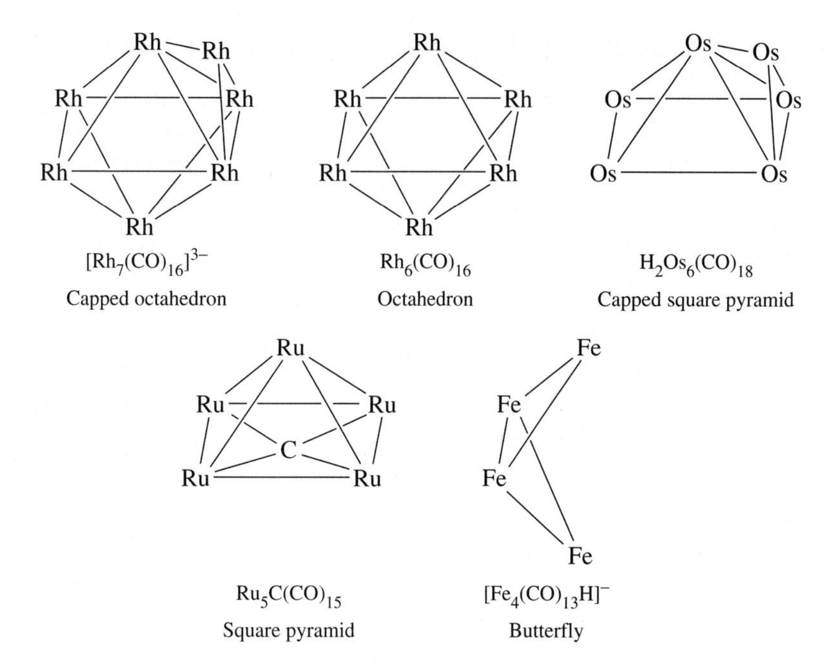

FIGURE 15.21 Metal Cores for Clusters Containing Seven Skeletal Bond Pairs.

Ionic clusters of main group elements can also be classified by a similar approach to that used for other clusters. Many such clusters are known;[37] they are sometimes called **Zintl ions**. Examples are shown in Figure 15.22.

EXAMPLES

Classify the following main group clusters:

a. Pb_5^{2-}: Total valence-electron count = 22 (including each of the 4 valence electrons per Pb, plus 2 electrons for the charge). Because $n = 5$, the total electron count = $4n + 2$; the classification is *closo*. (See Table 15.11.)

b. Sn_9^{4-}: Total number of valence electrons = 40. For $n = 9$, the electron count = $4n + 4$; the classification is *nido*. The structure, as shown in Figure 15.22, has one missing vertex.

c. Sb_4^{2-}: Total number of valence electrons = 22 = $4n + 6$. The classification is *arachno*. The square structure of this ion (Figure 15.22) corresponds to an octahedron with two vertices missing.

▶ **Exercise 15.7** Classify the following main group clusters:

a. Ge_9^{2-}
b. Bi_5^{3+}

[37] J. D. Corbett, *Angew. Chem., Int. Ed.*, **2000**, *39*, 670.

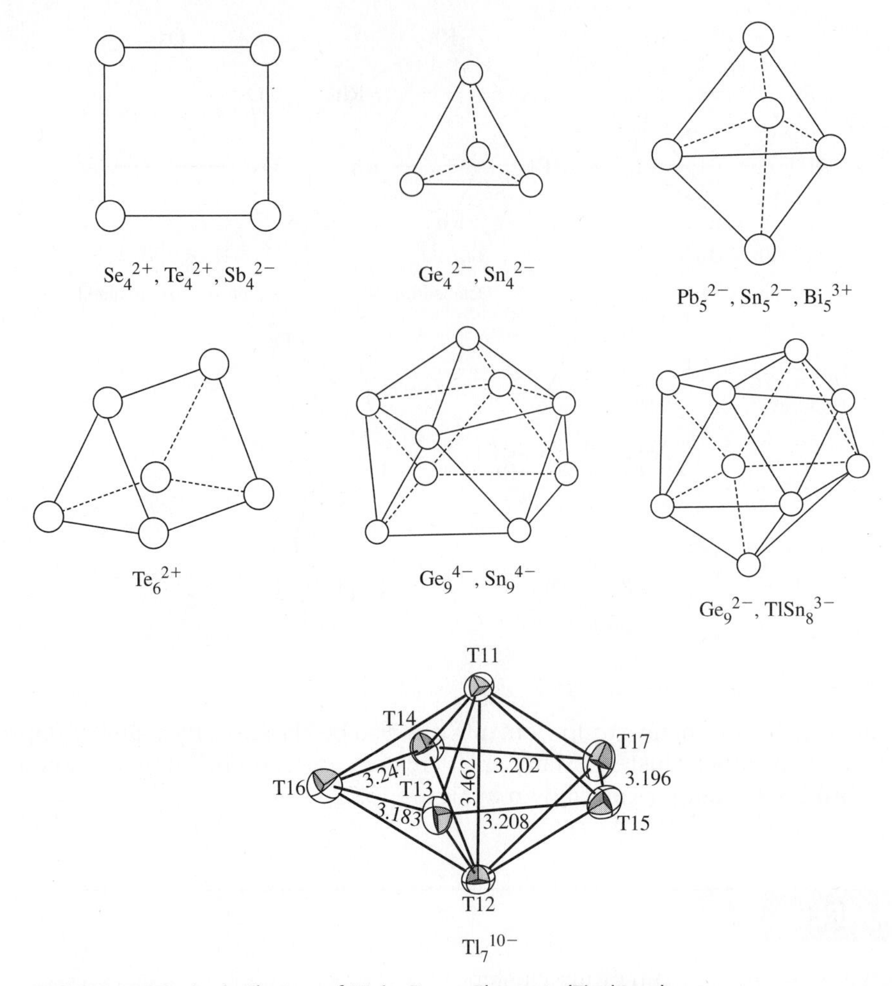

FIGURE 15.22 Ionic Clusters of Main Group Elements (Zintl Ions). (Tl_7^{10-} diagram reprinted with permission from S. Kaskel and J. D. Corbett, *Inorg. Chem.* **2000**, *39*, 778. © 2000, American Chemical Society.)

An extension of Wade's rules has been described for electron counting in boranes, heteroboranes, metallaboranes, other clusters, and even metallocenes.[38] This approach, called the ***mno* rule**, states that for a closed cluster structure to be stable, there must be $m + n + o$ skeletal electron pairs, where

m = number of condensed (linked) polyhedra

n = total number of vertices

o = number of single-atom bridges between two polyhedra

A fourth term, p, must be added for structures with missing vertices:

p = number of missing vertices (e.g., $p = 1$ for *nido*, $p = 2$ for *arachno*.)

This approach has been particularly developed for application to macropolyhedral structures, clusters involving linked polyhedra, and many examples have been described in detail.[39]

[38] E. D. Jemmis, M. M. Balakrishnarajan, and P. D. Pancharatna, *J. Am. Chem. Soc.*, **2001**, *123*, 4313.
[39] E. D. Jemmis, M. M. Balakrishnarajan, and P. D. Pancharatna, *Chem. Rev.*, **2002**, *102*, 93.

This approach is best illustrated by some examples.

EXAMPLES

Determine the number of skeletal electron pairs predicted by the *mno* rule for the following:

$B_{12}H_{12}{}^{2-}$ (see Figure 15.15)

m: This structure consists of a single polyhedron.	$m = 1$
n: Each boron atom in the polyhedron is a vertex.	$n = 12$
o: There are no bridges between polyhedra.	$o = 0$
p: The structure is *closo*, so $p = 0$.	
	$m + n + o =$ **13** electron pairs

$(\eta^5\text{-}C_2B_9H_{11})_2Fe^{2-}$ (see Figure 15.20)

m: This structure has two linked polyhedra.	$m = 2$
n: All carbons, borons, and the Fe are vertices.	$n = 23$
o: The Fe atom serves to bridge the polyhedra.	$o = 1$
p: The structure is *closo*, so $p = 0$.	
	$m + n + o =$ **26** electron pairs

Ferrocene, $(\eta^5\text{-}C_5H_5)_2Fe$ (see Figure 13.5)

m: The structure may be viewed as two linked polyhedra (pentagonal pyramids).	$m = 2$
n: Each atom in the structure is a vertex.	$n = 11$
o: The iron atom bridges the polyhedra.	$o = 1$
p: The structure is *not* closo; the top or bottom may be viewed as a pentagonal bipyramid lacking one vertex; the classification is *nido*.	$p = 2$ (one open face per polyhedron)
	$m + n + o + p =$ **16** electron pairs

▶ **Exercise 15.8** Determine the number of skeletal electron pairs predicted by the *mno* rule for the following:

a. $(\eta^5\text{-}C_5H_5)Fe(\eta^5\text{-}C_2B_9H_{11})$ (see Figure 15.20)
b. *nido*-7,8-$C_2B_9H_{11}{}^{2-}$ (see Figure 15.19)

15.4.5 Carbide Clusters

Many compounds have been synthesized, often fortuitously, in which one or more atoms have been partially or completely encapsulated within metal clusters. The most common of these cases have been the carbide clusters, with carbon exhibiting

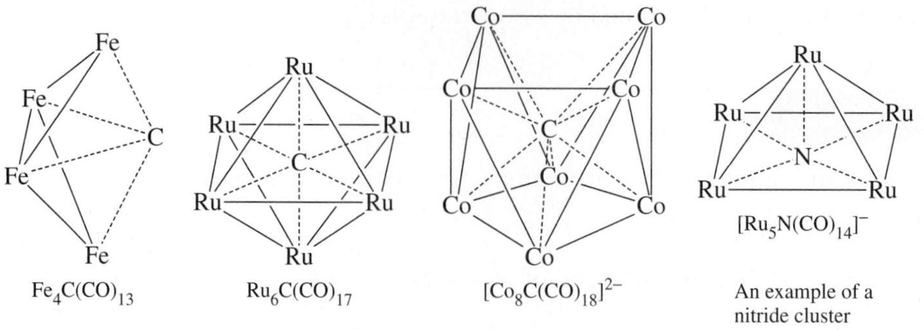

$Fe_4C(CO)_{13}$ $Ru_6C(CO)_{17}$ $[Co_8C(CO)_{18}]^{2-}$ $[Ru_5N(CO)_{14}]^-$

An example of a nitride cluster

FIGURE 15.23 Carbide Clusters. CO ligands have been omitted for clarity.

coordination numbers and geometries not found in classic organic structures. Examples of these unusual coordination geometries are shown in Figure 15.23.

Encapsulated atoms contribute their valence electrons to the total electron count. For example, carbon contributes its 4 valence electrons in $Ru_6C(CO)_{17}$ to give a total of 86 electrons, corresponding to a *closo* electron count (Table 15.12).

How can carbon, with only four valence orbitals, form bonds to more than four surrounding transition metal atoms? $Ru_6C(CO)_{17}$, with a central core of O_h symmetry, is a useful example. The 2*s* orbital of carbon has A_{1g} symmetry and the 2*p* orbitals have T_{1u} symmetry in the O_h point group. The octahedral Ru_6 core has framework bonding orbitals of the same symmetry as in $B_6H_6^{\,2-}$ described earlier in this chapter (see Figure 15.13): a centrally directed A_{1g} group orbital and two sets of orbitals, oriented tangentially to the core, of T_{1u} and T_{2g} symmetry. Therefore, there are two ways in which the symmetry match is correct for interactions between the carbon and the Ru_6 core: the interactions of A_{1g} and T_{1u} symmetry shown in Figure 15.24 (the T_{2g} orbitals participate in Ru—Ru bonding but not in bonding with the central carbon). The net

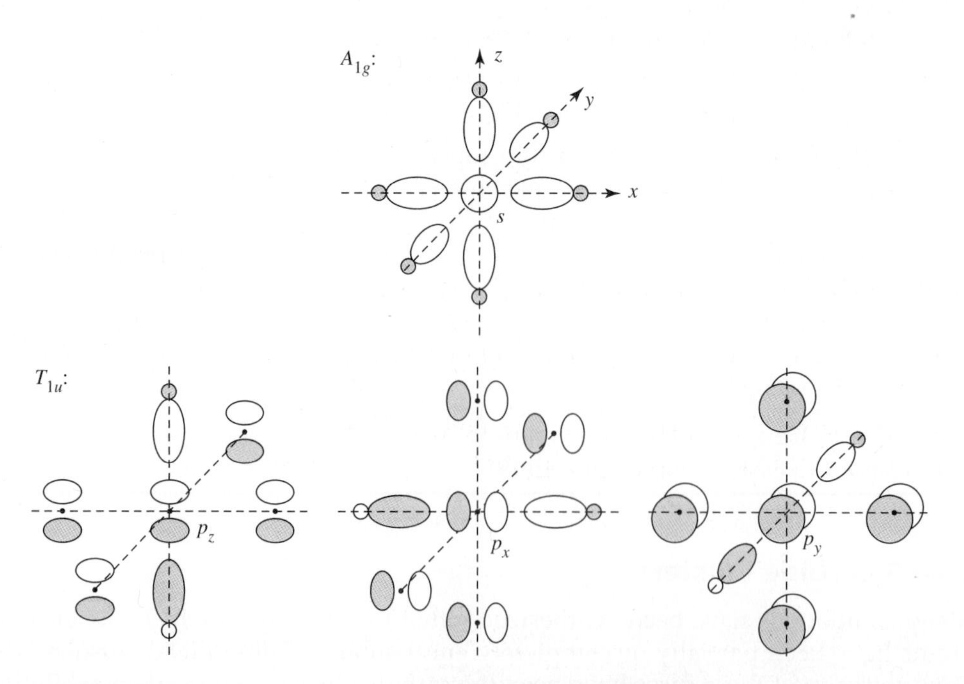

FIGURE 15.24 Bonding Interactions Between Central Carbon and Octahedral Ru_6.

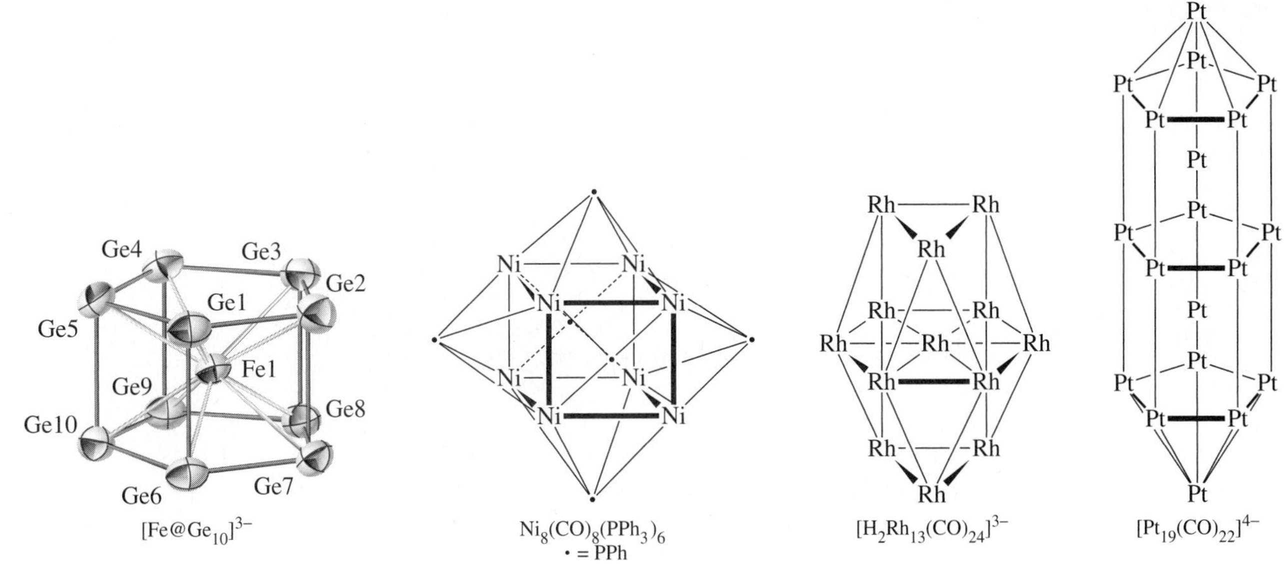

FIGURE 15.25 Examples of Large Clusters. CO and hydride ligands have been omitted to show the metal–metal bonding more clearly.

result is the formation of four C—Ru bonding orbitals, occupied by electron pairs in the cluster, and four unoccupied antibonding orbitals.[40]

▶ **Exercise 15.9** Classify the following clusters by structural type:

a. $[Re_7C(CO)_{21}]^{3-}$
b. $[Fe_4N(CO)_{12}]^{-}$

A particularly interesting recent example of encapsulation is that of an iron atom inside a pentagonal prismatic Zintl ion, $[Fe@Ge_{10}]^{3-}$, shown in Figure 15.25. This ion is formed from the potassium salt of the Zintl species Ge_9^{4-} and $Fe(2,6\text{-}Mes_2C_6H_3)_2$ (mes = mesityl) in the presence of 2,2,2-crypt; the crypt and the solvent, ethylenediamine, are also present in the crystal. $[Fe@Ge_{10}]^{3-}$ has been compared with the $[Ti(\eta^5\text{-}P_5)_2]^{2-}$ ion (see Section 15.2.2), with significant differences noted in structure and bonding despite their similarity in appearance.[41]

15.4.6 Additional Comments on Clusters

As we have seen, transition-metal clusters can adopt a wide variety of geometries and can involve metal–metal bonds of order as high as 5. Clusters may also include much larger polyhedra than shown so far in this chapter; polyhedra linked through vertices, edges, or faces; and extended three-dimensional arrays. Examples of these types of clusters are given in Figure 15.25. Even an example of a hydride-centered cluster, with a hydride ion within a cage of eight lithium ions, has been reported (Figure 15.26).

---

[40] G. A. Olah, G. K. S. Prakash, R. E. Williams, L. D. Field, and K. Wade, *Hypercarbon Chemistry*, John Wiley & Sons, New York, 1987, pp. 123–133.

[41] B. Zhou, M. S. Denning, D. L. Kays, and J. M. Goicoechea, *J. Am. Chem. Soc.*, **2009**, *131*, 2802.

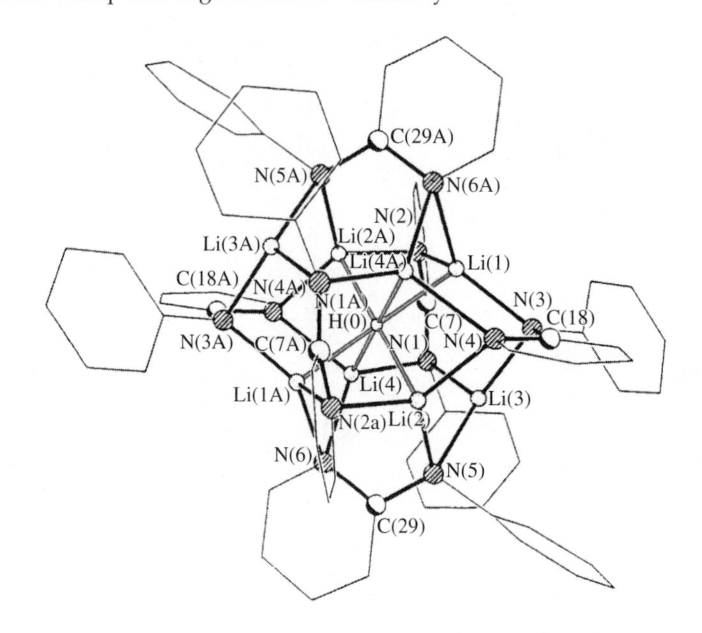

FIGURE 15.26 A Hydride Ion in a Cage of Eight Lithium Ions.

(Reproduced with permission from D. R. Armstrong, W. Clegg, R. P. Davies, S. T. Liddle, D. J. Linton, P. R. Raithby, R. Snaith, and A. E. H. Wheatley, *Angew. Chem., Int. Ed.*, **1999**, *38*, 3367. © 1999, Wiley-VCH and A. E. H. Wheatley.)

General References

A classic reference on parallels between main group and organometallic chemistry is Roald Hoffmann's 1982 Nobel lecture, "Building Bridges between Inorganic and Organic Chemistry," in *Angew. Chem., Int. Ed.*, **1982**, *21*, 711–724, which describes in detail the isolobal analogy. Another very useful paper is John Ellis's "The Teaching of Organometallic Chemistry to Undergraduates," in *J. Chem. Educ.*, **1976**, *53*, 2–6. K. Wade's, *Electron Deficient Compounds*, Thomas Nelson, New York, 1971, provides detailed descriptions of bonding in boranes and related compounds. Metallacarboranes have been reviewed extensively by R. N. Grimes in E. W. Abel, F. G. A. Stone, and G. Wilkinson, editors, *Comprehensive Organometallic Chemistry II*, Pergamon Press, Oxford, 1995, Vol. 1, Chapter 9, pp. 373–430. Topics related to multiple bonds between metal atoms are discussed in detail in F. A. Cotton and R. A. Walton's, *Multiple Bonds Between Metal Atoms*, John Wiley & Sons, New York, 1982. Two articles in *Chemical and Engineering News* are recommended for further discussion of applications of cluster chemistry: E. L. Muetterties' "Metal Clusters," Aug. 20, 1982, pp. 28–41, and F. A. Cotton and M. H. Chisholm's, "Bonds between Metal Atoms," June 28, 1982, pp. 40–46.

Problems

15.1 Predict the following products:

a. $Mn_2(CO)_{10} + Br_2 \longrightarrow$

b. $HCCl_3$ + excess $[Co(CO)_4]^- \longrightarrow$

c. $Co_2(CO)_8 + (SCN)_2 \longrightarrow$

d. $Co_2(CO)_8 + C_6H_5{-}C{\equiv}C{-}C_6H_5$ (product has a single Co—Co bond)

e. $Mn_2(CO)_{10} + [(\eta^5\text{-}C_5H_5)Fe(CO)_2]_2 \longrightarrow$

15.2 Propose organic fragments that are isolobal with

a. $Tc(CO)_5$

b. $[Re(CO)_4]^-$

c. $[Co(CN)_5]^{3-}$

d. $[CpFe(\eta^6\text{-}C_6H_6)]^+$

e. $[Mn(CO)_5]^+$

f. $Os_2(CO)_8$ (Find an organic molecule isolobal with this dimeric molecule.)

15.3 Propose two organometallic fragments not mentioned in this chapter that are isolobal with

a. CH_3

b. CH

c. CH_3^+

d. CH_3^-

e. $(\eta^5\text{-}C_5H_5)Fe(CO)_2$

f. $Sn(CH_3)_2$

15.4 Propose an organometallic molecule that is isolobal with each of the following:

a. Ethylene

b. P_4

c. Cyclobutane

d. S_8

15.5 Hydrides such as $NaBH_4$ and $LiAlH_4$ have been reacted with the complexes $[(C_5Me_5)Fe(C_6H_6)]^+$, $[(C_5H_5)Fe(CO)_3]^+$, and $[(C_5H_5)Fe(CO)_2(PPh_3)]^+$. (See P. Michaud, C. Lapinte, and D. Astruc, *Ann. N. Y. Acad. Sci.*, **1983**, *415*, 97).

a. Show that these complexes are isolobal.

b. Predict the products of the reactions of these complexes with hydride reagents.

15.6 Hoffmann has described the following molecules to be composed of isolobal fragments. Subdivide the molecules into fragments, and show that the fragments are isolobal.

a.

b.

c.

15.7 Verify that the following compounds are composed of isolobal fragments:

(See G. A. Carriedo, J. A. K. Howard, and F. G. A. Stone, *J. Organomet. Chem.*, **1983**, *250*, C28.)

15.8 Calculations reported on the fragments $Mn(CO)_5$, $Mn(CO)_3$, $Cu(PH_3)$, and $Au(PH_3)$ have shown that the energies of their singly occupied hybrid orbitals are in the order $Au(PH_3) > Cu(PH_3) > Mn(CO)_3 > Mn(CO)_5$.

a. In the compound $(OC)_5Mn—Au(PH_3)$, would you expect the electrons in the Mn—Au bond to be polarized toward Mn or Au? Why? (Hint: Sketch an energy-level diagram for the molecular orbital formed between Mn and Au.)

b. The $Cu(PPh_3)$ fragment bonds to C_5H_5 in a manner similar to the isolobal $Mn(CO)_3$ fragment. However, the geometry of the corresponding $Au(PPh_3)$ complex is significantly different:

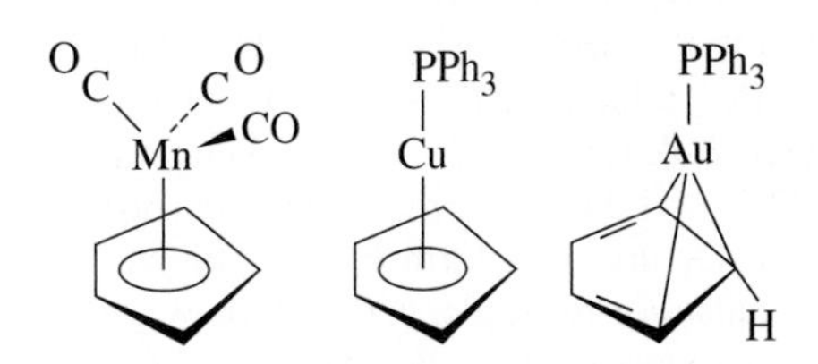

Suggest an explanation. (See D. G. Evans and D. M. P. Mingos, *J. Organomet. Chem.*, **1982**, *232*, 171.)

15.9 **a.** A tin atom can bridge two $Fe_2(CO)_8$ groups in a structure similar to that of spiropentane. Show that these two molecules are composed of isolobal fragments.

b. Tin can also bridge two $Mn(CO)_2(\eta^5\text{-}C_5Me_5)$ fragments; the compound formed has Mn—Sn—Mn arranged linearly. Explain this linear arrangement. (Hint: Find a hydrocarbon isolobal with this compound. See W. A. Herrmann, *Angew. Chem., Int. Ed.*, **1986**, *25*, 56.)

15.10 The $AuPPh_3$ fragment is isolobal with the hydrogen atom. Furthermore, analogues of the unstable CH_5^+, CH_6^{2+}, and CH_7^{3+} ions can be prepared using $AuPPh_3$ instead of H. Predict the structures of the $AuPPh_3$ analogues of these ions, and suggest a reason for their stability. (Hint: See G. A. Olah and G. Rasul, *Acc. Chem. Res.*, **1997**, *25*, 56, and references therein.)

15.11 The isolobal fragments $Fe(CO)_3$ and CpCo are predicted to form a variety of analogous carbonyl complexes, for example:

$$Fe_2(CO)_6(\mu_2\text{-}CO)_3 \text{ and } Cp_2Co_2(\mu_2\text{-}CO)_3$$

$$Fe_2(CO)_6(\mu_2\text{-}CO) \text{ and } Cp_2Co_2(\mu_2\text{-}CO)$$

In each case the cobalt compound is predicted to have a substantially shorter metal–metal bond than the analogous iron compound, even though the bond orders of the related compounds are the same. Suggest two reasons for this phenomenon. (See H. Wang, Y. Xie, R. B. King, and H. F. Schaefer III, *J. Am. Chem. Soc.*, **2005**, *127*, 11646.)

15.12 The molecular orbitals of the known complex $[Ti(\eta^5\text{-}P_5)_2]^{2-}$, the fragment $[Ti(\eta^5\text{-}P_5)]^-$, and *cyclo*-$P_5^-$ have been calculated.

a. Compare the orbital interactions between *d* orbitals of Ti and P_5^- orbitals with the corresponding interactions between *d* orbitals of Fe and cyclopentadiene (Figure 13.28).

b. Of the species $[Ti(\eta^5\text{-}P_5)_2]^{2-}$, $[Ti(\eta^5\text{-}P_5)]^-$, and P_5^-, which has the longest P—P distance? Why? (See Z-Z. Liu, W-Q. Tian, J-K. Feng, G. Zhang, and W-Q. Li, *J. Phys. Chem. A*, **2005**, *109*, 5645.)

15.13 The C_2 unit can form bridges between transition metals. For example, $(CO)_5Mn(\mu_2\text{-}C_2)Mn(CO)_5$ has been reported. (See P. Balanzoni, N. Re, A. Sgamellotti, and C. Floriani, *J. Chem. Soc., Dalton Trans.*, **1997**, 4773.)

a. Show how two $Mn(CO)_5$ fragments can interact in a sigma fashion with a bridging C_2 ligand.

b. Show how pi interactions can occur between C_2 and the $Mn(CO)_5$ fragments, with the $Mn(CO)_5$, using orbitals *not* involved in the sigma interactions.

c. Compare your results with those reported in the literature.

15.14 The complex $Mo_2(NMe_2)_6$ (Me = methyl) contains a metal–metal triple bond. Would you predict this molecule to be more likely eclipsed or staggered? Explain.

15.15 In $[Re_2Cl_8]^{2-}$, the $d_{x^2-y^2}$ orbitals of rhenium interact strongly with the ligands. For one $ReCl_4$ unit in this ion, sketch the four group orbitals of the chloride ligands (assume one σ donor orbital per Cl). Identify the group orbital of suitable symetry to interact with the $d_{x^2-y^2}$ orbital of rhenium.

15.16 $[Tc_2Cl_8]^{2-}$ has a higher bond order than $[Tc_2Cl_8]^{3-}$; however, the Tc—Tc bond distance in $[Tc_2Cl_8]^{2-}$ is longer. Suggest an explanation. (Hint: See F. A. Cotton, *Chem. Soc. Rev.*, **1983**, *122*, 35, or F. A. Cotton and R. A Walton, *Multiple Bonds Between Metal Atoms*, Clarendon Press, Oxford, 1993, pp. 122–123.)

15.17 Formal bond orders can sometimes be misleading. For example, $[Re_2Cl_9]^-$, which has a metal–metal bond order of 3.0, has a longer Re—Re bond (270.4 pm) than $[Re_2Cl_9]^{2-}$ (247.3 pm), which has a bond order of 2.5. Account for the shorter bond in $[Re_2Cl_9]^{2-}$. (Hint: See G. A. Heath, J. E. McGrady, R. G. Raptis, and A. C. Willis, *Inorg. Chem.*, **1996**, *35*, 6838.)

15.18 The molybdenum compounds $[Mo_2(DTolF)_3]_2(\mu\text{-}OH)_2$, **1**, and $[Mo_2(DTolF)_3]_2(\mu\text{-}O)_2$, **2**, have similar core structures, as shown (DTolF = $[(p\text{-tolyl})NC(H)N(p\text{-tolyl})]^-$). The molybdenum–molybdenum distance in **2** (214.0 pm) is slightly greater than the comparable distance in **1** (210.7 pm). Suggest an explanation for this difference. Hydrogens and the toluene rings are omitted to show the bonding to the Mo more clearly. (Reproduced with permission from F. A. Cotton, L. M. Daniels, I. Guimet, R. W. Henning, G. T. Jordan IV, C. Lin, C. A. Murillo; and A. J. Schultz, *J. Am. Chem. Soc.*, **1998**, *120*, 12531. © 1998, American Chemical Society.)

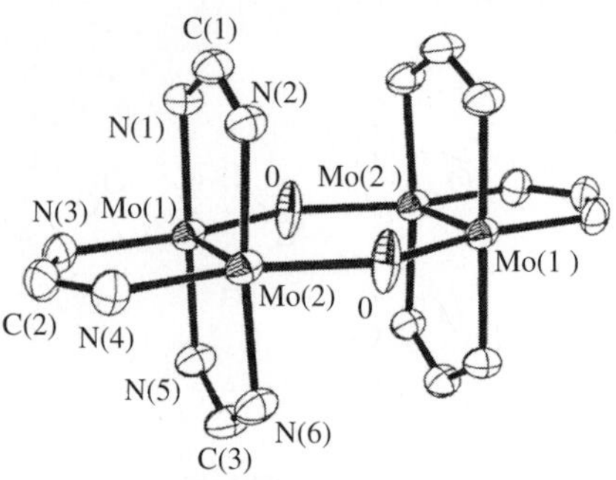

1 $[Mo_2(DTolF)_3]_2(\mu\text{-}OH)_2$

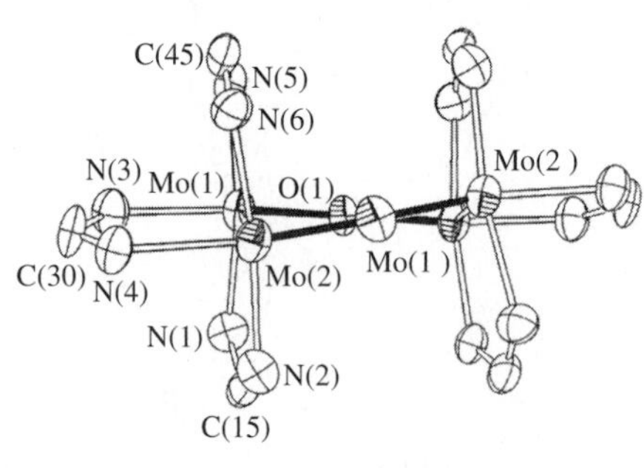

2 $[Mo_2(DTolF)_3]_2(\mu\text{-}O)_2$

15.19 Using the coordinate system of Figure 15.12, for $B_6H_6^{2-}$:

a. Show that the p_z orbitals of the borons collectively have the same symmetry as the *s* orbitals (generate a representation based on the six p_z orbitals of the borons, and do the same for the six *s* orbitals).

b. Show that these representations reduce to $A_{1g} + E_g + T_{1u}$.

c. Show that the p_x and p_y orbitals of the borons form molecular orbitals of T_{2g} and T_{1u} symmetry.

15.20 For the *closo* cluster $B_7H_7^{2-}$, which has D_{5h} symmetry, verify that there are 8 framework bonding electron pairs.

15.21 Classify the following as *closo*, *nido*, or *arachno*:

a. $C_2B_3H_7$ **b.** B_6H_{12}
c. $B_{11}H_{11}^{2-}$ **d.** $C_3B_5H_7$
e. $CB_{10}H_{13}^-$ **f.** $B_{10}H_{14}^{2-}$

15.22 Classify the following as *closo*, *nido*, or *arachno*:

a. $SB_{10}H_{10}^{2-}$ **b.** $NCB_{10}H_{11}$
c. $SiC_2B_4H_{10}$ **d.** $As_2C_2B_7H_9$
e. $PCB_9H_{11}^-$

15.23 Classify the following as *closo* or *nido*:

a. $B_3H_8Mn(CO)_3$ **b.** $B_4H_6(CoCp)_2$
c. $C_2B_7H_{11}CoCp$ **d.** $B_5H_{10}FeCp$
e. $C_2B_9H_{11}Ru(CO)_3$

15.24 Classify the following as *closo*, *nido*, or *arachno*:

a. Ge_9^{4-} **b.** $InBi_3^{2-}$ **c.** Bi_8^{2+}

15.25 Determine the number of skeletal electron pairs predicted by the *mno* rule for the following:
- **a.** *arachno*-B_5H_{11}
- **b.** 1-(η^5-C_5H_5)CoB_4H_{10} (see Figure 15.17)
- **c.** (η^5-C_5H_5)Fe(η^5-$C_2B_9H_{11}$) (see Figure 15.20)

15.26 Assign the point groups of the following structures from this chapter:
- **a.** The isolobal symbol
- **b.** The metallocenes with P_5 rings shown in Figure 15.5
- **c.** $[Re_2Cl_8]^{2-}$ and $[Os_2Cl_8]^{2-}$
- **d.** A T_{2g} orbital of $B_6H_6{}^{2-}$ shown in Figure 15.13
- **e.** The *nido* borane in the bottom center of Figure 15.15
- **f.** The carborane analogs of ferrocene shown in Figure 15.20
- **g.** $Te_6{}^{2+}$ and $Ge_9{}^{4-}$ in Figure 15.22

15.27 Assign the point groups:
- **a.** The core of the iron(III) cluster $[Fe_8O_4(sao)_8(py)_4]\cdot 4py$ (sao = salicylaldoxime; py = pyridine), which has an Fe_4O_4 cubane structure inside an Fe_4 tetrahedron. (See I. A. Gass, C. J. Milios, A. G. Whittaker, F. P. A. Fabiani, S. Parsons, M. Murrie, S. P. Perlepes, and E. K. Brechin, *Inorg. Chem.*, **2006**, *45*, 5281.)

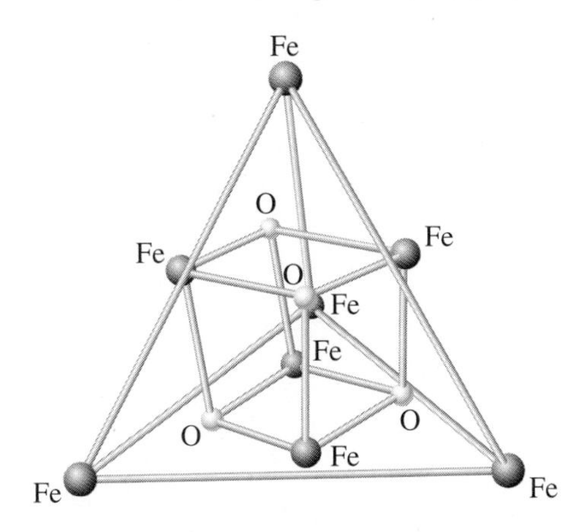

- **b.** The $[As@Ni_{12}@As_{20}]^{3-}$ ion, which contains an As atom at the center of a Ni_{12} icosahedron, which in turn is surrounded by an As_{20} dodecahedron. (See B. W. Eichhorn, *Science*, **2003**, *300*,)

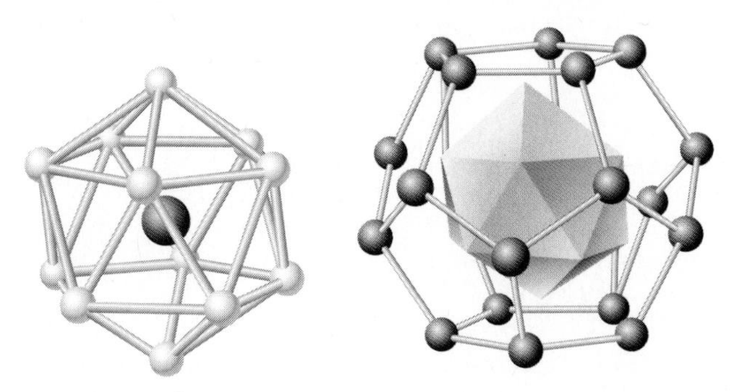

The following problems require the use of molecular modeling software.

15.28 The ligand $P_5{}^-$ has been reported to act as a stronger acceptor than η^5-$C_5H_5{}^-$.
- **a.** Using the group orbital approach as for ferrocene (see Figure 13.27), sketch the group orbitals arising from the P_5 rings. Then show how the group orbitals can interact with appropriate orbitals on a central transition metal.
- **b.** On the basis of your diagram, suggest why the $P_5{}^-$ ligand might act as a stronger acceptor than η^5-$C_5H_5{}^-$.
- **c.** Using molecular modeling software, calculate and display the molecular orbitals of $[(\eta^5\text{-}P_5)_2Ti]^{2-}$. Compare your results with those described in the literature for this complex. (See E. Urnezius, W. W. Brennessel, C. J. Cramer, J. E. Ellis, and P. von Ragué Schleyer, *Science*, **2002**, *295*, 832; a comparison of the molecular orbitals of η^5-$C_5H_5{}^-$ and $P_5{}^-$ can be found in H-J. Zhai, L-S. Wang, A. E. Kuznetsov, and A. I. Boldyrev, *J. Phys. Chem. A*, **2002**, *106*, 5600.)

15.29 Construct the ions $Re_2Cl_8{}^{2-}$ and $Os_2Cl_8{}^{2-}$ (see Figure 15.7), and calculate and display their molecular orbitals. Compare these orbitals with Figure 15.8, focusing on the metal–metal bonds and on *d*-orbital interactions with the ligands, and classify orbitals involved in metal–metal bonding as σ, π, or δ. How do your results compare with the quadruple and triple bonds reported respectively for these ions? (Depending on the software used, you may need to fix atoms in position before doing orbital calculations, especially for $Re_2Cl_8{}^{2-}$. Structural information on this ion can be found in F. A. Cotton and C. B. Harris, *Inorg. Chem.*, **1965**, *4*, 330.)

15.30 The maximum bond order for a diatomic chemical species has been calculated to be six, for example in Cr_2. What types of orbital interactions would be possible in this molecule? Calculate and display the molecular orbitals for Cr_2, and classify them as σ, π, or δ and as bonding or antibonding. Identify the atomic orbitals used in each molecular orbital. Are your results consistent with a bond order of six? (See G. Frenking and R. Tonner, *Nature*, **2007**, *446*, 276 and references cited therein.)

15.31 Draw *closo*-$B_6H_6{}^{2-}$ (see Figure 15.12), and calculate its molecular orbitals. Identify and display the following:
- **a.** The orbital having A_{1g} symmetry resulting from overlap of *p* or hybrid orbitals pointing toward the center of the octahedron (see Figure 15.13).

b. Three orbitals having T_{1u} symmetry involving pi interactions between sets of four boron *p* orbitals.

c. Three orbitals having T_{2g} symmetry involving overlap of sets of four boron *p* orbitals within the same plane.

15.32 Draw the core carbon-centered cluster of $Ru_6C(CO)_{17}$ (see Figure 15.23), and calculate its molecular orbitals. Identify and display the following:

a. The orbital having A_{1g} symmetry resulting from overlap of *p* or hybrid orbitals pointing toward the center of the octahedron with the 2*s* orbital of carbon (see Figure 15.24).

b. Three orbitals having T_{1u} symmetry involving pi interactions between sets of four ruthenium *p* or *d* orbitals with 2*p* orbitals on carbon.

c. Three orbitals having T_{2g} symmetry involving overlap of sets of four boron *p* or *d* orbitals within the same plane.

16

Bioinorganic and Environmental Chemistry

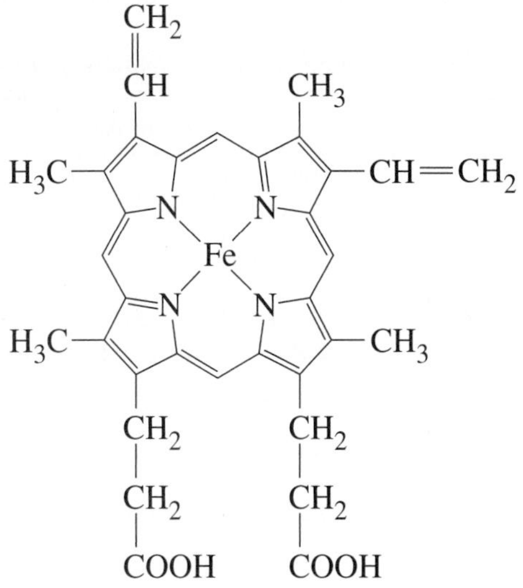

Inorganic compounds of many types have biological action—for example, as toxins or medicines when ingested, as part of the body's normal functioning, or by enabling essential processes in plants. The list of such compounds is far too long to cover with any degree of completeness in a short chapter. The approach here is to give only a few representative examples of bioinorganic compounds and their actions, along with some examples of the environmental effects of both metals and nonmetals.

Many biochemical reactions depend on the presence of metal ions. These ions may be present in specific coordination complexes, or they may act to facilitate or inhibit reactions in solution. In the first part of this chapter, we describe a few of these compounds and reactions, together with the biochemistry of NO, which has many functions that have only recently been discovered.

Many metals are essential to plant and animal life, although in many cases their role is uncertain. The list includes all the first-row transition metals except scandium and titanium, but only molybdenum and perhaps tungsten from the heavier transition metals.[1] Table 16.1 lists several that are important in mammalian biochemistry. The importance of iron is obvious from the number of roles it plays, from oxygen carrier in hemoglobin and myoglobin to electron carrier in the cytochromes to detoxifying agent in catalase and peroxidase.

How do inorganic compounds and ions help cause biochemical reactions? A partial list of their actions is given here, most related to metal ion complexes.

1. Promotion of reactions by providing appropriate geometry for breaking or forming bonds. Although many coordination sites in bioinorganic molecules are approximately tetrahedral, octahedral, or square-planar, they have subtle variations that provide for unusual reactions. An organic ligand may provide a pocket that is slightly too large or small for a particular reactant, or it may have angles that make other sites on the metal more reactive. The binding of small molecules can also create reactive species by forcing them to adopt unusual angles or bond distances.
2. Changes in acid–base activity. Water bound to a metal ion frequently is more acidic than free water, and coordination to proteins enhances this effect still more. This results in M—OH species that can then react with other substrates. Mg^{2+} and Zn^{2+} are common metal ions that serve this function.
3. Changes in redox potentials. Coordination by different ligands changes redox potentials, making some reactions easier and some more difficult, and it provides pathways for electron transfer.
4. Some ions (Na^+, K^+, Ca^{2+}, Cl^-) act as specific charge carriers, with concentration gradients maintained and modified by membrane ion pumps and trigger mechanisms. Sudden changes in these concentration gradients are signals for nerve or muscle action.

[1] E. Frieden, *J. Chem. Educ.*, **1985**, *65*, 917.

TABLE 16.1 Metal-Containing Enzymes and Proteins

Metal	Compounds and Actions
Fe (heme)	Hemoglobin, peroxidase, catalase, cytochrome P-450, tryptophan dioxygenase, cytochrome c, nitrite reductase
Fe (non-heme)	Pyrocatechase, ferredoxin, hemerythrin, transferrin, aconitase, nitrogenase
Cu	Tyrosinase, amine oxidases, laccase, ascorbate oxidase, ceruloplasmin, superoxide dismutase, plastocyanin, nitrite reductase
Co (B_{12} coenzyme)	Glutamate mutase, dioldehydrase, methionine synthetase
Co(II) (non-corrin)	Dipeptidase
Zn(II)	Carbonic anhydrase, carboxypeptidase, alcohol dehydrogenase, DNA polymerase
Mg(II)	Activates phosphotransferases and phosphohydrases, DNA polymerase
K(I)	Activates pyruvate phosphokinase and K-specific ATPase
Na(I)	Activates Na-specific ATPase
Mo	Nitrogenase, nitrate reductase, xanthine oxidase, formate dehydrogenase, sulfite oxidase, DMSO reductase
W	Aldehyde ferredoxin oxidoreductase

5. Organometallic reactions can create species that are otherwise not attainable. Cobalamin enzymes are particular examples of these catalysts.
6. Inorganic ions, both cationic and anionic, are used as structural units to form bone and other hard structures. Maintenance of cell membranes and DNA structure also depends on the presence of cations to balance charges in the organic portions.
7. A few small molecules have specific effects that do not fit easily into any of the categories above. Perhaps the most obvious is NO, which has many functions, primarily related to control of blood flow, neurotransmission, learning, memory, and, at higher concentrations, as a defensive cytotoxin against tumor cells and pathogens.

A relatively new feature of the study of bioinorganic molecules is the use of molecular orbital calculations to guide research into their mechanism of action. This is similar to calculations of minimum energies and transition states for other reactions, but it requires either careful design of models to include the essential features of the large protein and nucleic acid molecules, or their inclusion in the calculation, at considerable cost in complexity and time. The results frequently depend on the design of the model, the complexity of the data sets used, and the computation methods; the environment around the active site may also be very important to the results.

16.1 PORPHYRINS AND RELATED COMPLEXES

One of the most important groups of compounds is the ***porphyrins***, in which a metal ion is surrounded by the four nitrogens of a porphine ring in a square-planar geometry, and the axial sites are available for other ligands. Different side chains, metal ions, and surrounding species result in very different reactions and roles for these compounds. The parent porphine ring and some specific porphyrin compounds are shown in Figure 16.1.

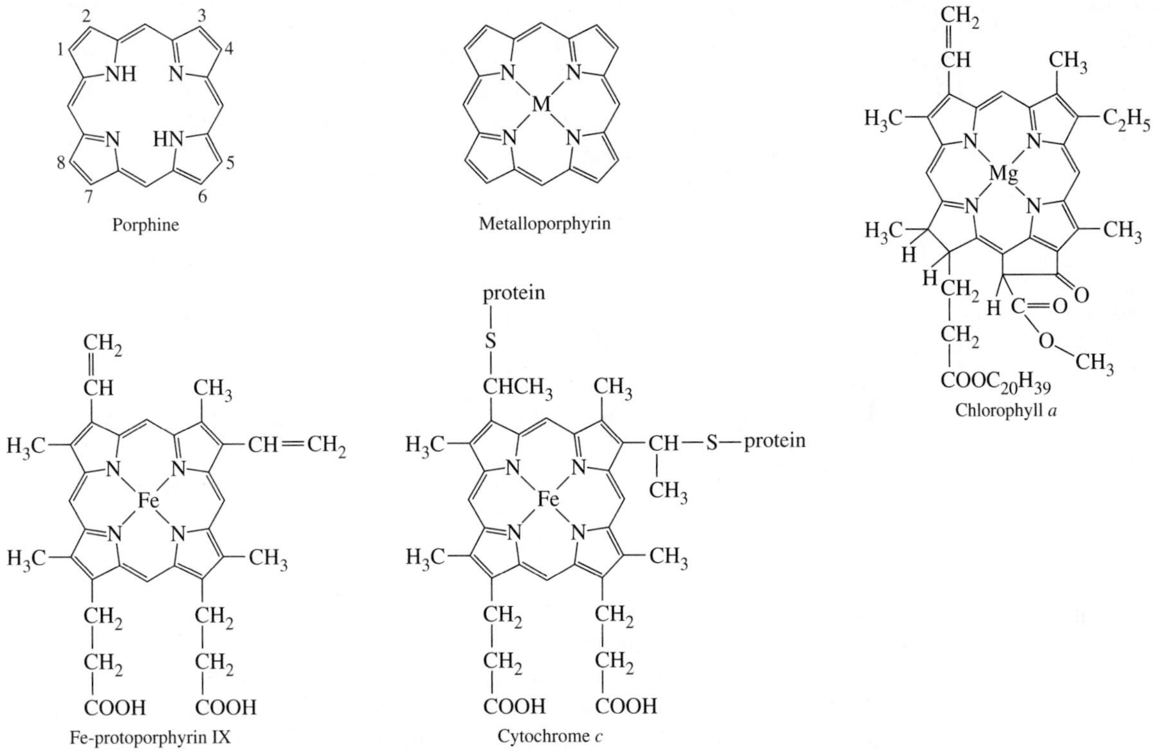

Porphine

Metalloporphyrin

Chlorophyll *a*

Fe-protoporphyrin IX

Cytochrome *c*

FIGURE 16.1 Porphine, Porphyrin, and Related Compounds.

16.1.1 Iron Porphyrins

HEMOGLOBIN AND MYOGLOBIN The best known iron porphyrin compounds are hemoglobin and myoglobin, oxygen transfer and storage agents in the blood and muscle tissue, respectively. Each of us has nearly 1 kg of hemoglobin in our body, to pick up molecular oxygen in the lungs and deliver it to the rest of the body. Each hemoglobin molecule is made up of four globin protein subunits, two α and two β. In each of these, the protein molecule partially encloses the heme group, bonding to one of the axial positions through an imidazole nitrogen, as shown in Figure 16.2. The other axial position is vacant or has water bound to it (the imidazole ring from histidine E7 is too far from the iron atom to bond). When dissolved oxygen is present, it can occupy this position, and subtle changes in the conformation of the proteins result. As one iron binds an oxygen molecule, the molecular shape changes to make binding of additional oxygen molecules easier. The four irons can each carry one O_2, with generally increasing equilibrium constants:

$$Hb + O_2 \rightleftharpoons HbO_2 \qquad K_1 = 5 \text{ to } 60$$

$$HbO_2 + O_2 \rightleftharpoons Hb(O_2)_2$$

$$Hb(O_2)_2 + O_2 \rightleftharpoons Hb(O_2)_3$$

$$Hb(O_2)_3 + O_2 \rightleftharpoons Hb(O_2)_4 \qquad K_4 = 3000 \text{ to } 6000$$

The equilibrium constants increase, with the fourth constant many times larger than the first, depending on the biological species from which the hemoglobin came. In the absence of the structural changes, K_4 would be much smaller than K_1. As a result, as soon as some oxygen has been bound to the molecule, all four irons are readily oxygenated. In a similar fashion, initial removal of oxygen triggers the release of the

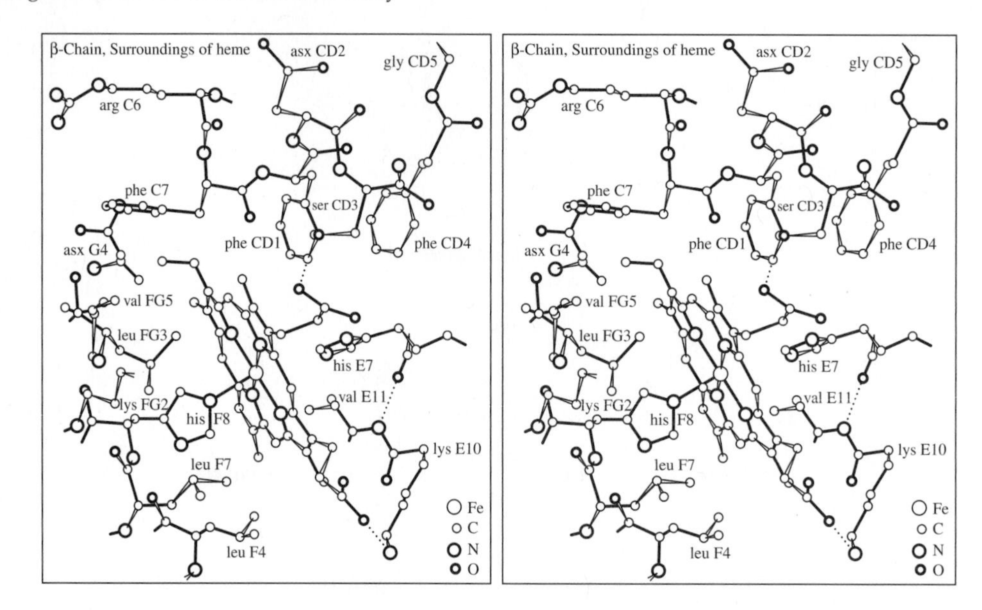

FIGURE 16.2 Heme Group Binding in Hemoglobin. Illustrated here is a stereo drawing of the surroundings of the heme in the β chain of hemoglobin. Broken lines indicate hydrogen bonds.

(Reprinted by permission from J. F. Perutz and H. Lehmann, *Nature*, **1968**, *219*, 902. © 1968 Macmillan Magazines Limited.)

remainder, and the entire load of oxygen is delivered at the required site. The structural changes accompanying oxygenation have been described thoroughly by Baldwin and Chothia[2] and by Dickerson and Geis.[3] This effect is also favored by pH changes caused by increased CO_2 concentration in the capillaries. As the concentration of CO_2 increases, formation of bicarbonate ($2\,H_2O + CO_2 \rightleftharpoons HCO_3^- + H_3O^+$) causes the pH to decrease, and the increased acidity favors release of O_2 from the oxyhemoglobin. This is known as the *Bohr effect*.

Myoglobin has only one heme group per molecule and serves as an oxygen storage molecule in the muscles. The myoglobin molecule is similar to a single subunit of hemoglobin. Bonding between the iron and the oxygen molecule is similar to that in hemoglobin, but the equilibrium is simpler, because only one oxygen molecule is bound:

$$\mathrm{Mb} + \mathrm{O_2} \rightleftharpoons \mathrm{MbO_2}$$

When hemoglobin releases oxygen to the muscle tissue, myoglobin picks it up and stores it until it is needed. The Bohr effect and the cooperation of the four hemoglobin binding sites make the transfer more complete, when the oxygen concentration is low and the carbon dioxide concentration is high; the opposite conditions in the lungs promote the transfer of oxygen to hemoglobin and the transfer of CO_2 to the gas phase in the lungs. As shown in Figure 16.3, myoglobin binds O_2 more strongly than the first O_2 of hemoglobin. However, the fourth equilibrium constant of hemoglobin is larger than that for myoglobin by a factor of about 50.

In hemoglobin, the iron is formally Fe(II), and bonding to oxygen does not oxidize it to Fe(III). However, when the heme group is removed from the protein, exposure to oxygen oxidizes the iron quickly to a μ-oxo dimer containing two Fe(III) ions. The presence of hydrophilic protein around the heme seems to prevent oxidation of Fe(II) in hemoglobin, but the presence of water alone allows oxidation of the free heme. In a test of this hypothesis, Wang[4] embedded a heme derivative saturated with CO in a polystyrene matrix and studied its equilibrium with CO and O_2. He was able to cycle the material between the oxygenated form, the CO-bound form, and the free heme with no oxidation to Fe(III). From this evidence, he concluded that a nonaqueous environment

---

[2] J. Baldwin and C. Chothia, *J. Mol. Biol.*, **1979**, *129*, 175.
[3] R. E. Dickerson and I. Geis, *Hemoglobin*, Benjamin/Cummings, Menlo Park, CA, 1983.
[4] J. H. Wang, *J. Am. Chem. Soc.*, **1958**, *80*, 3168.

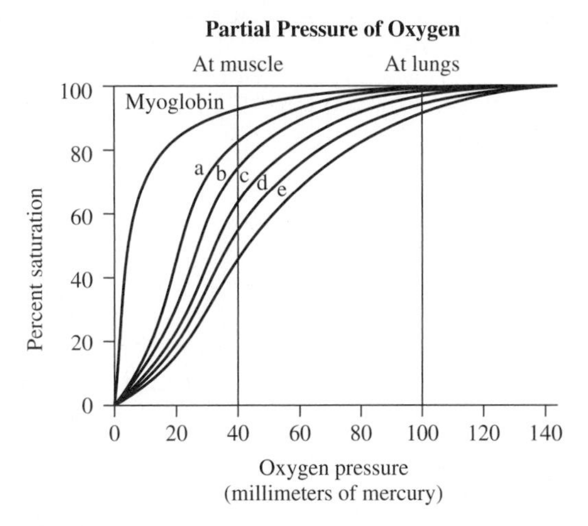

FIGURE 16.3 Myoglobin and Hemoglobin Binding Curves. Myoglobin, and hemoglobin at five different pH values: (a) 7.6, (b) 7.4, (c) 7.2, (d) 7.0, (e) 6.8.

(Reproduced with permission from R. E. Dickerson and I. Geis, *Hemoglobin*, Benjamin/Cummings, Menlo Park, CA, 1983, p. 24. Irving Geis rights owned by Howard Hughes Medical Institute. Not to be reproduced without permission.)

is required for reversible O_2 or CO binding. In hemoglobin, the protein surrounding the heme groups provides this nonaqueous environment and prevents oxidation. Others have argued that oxidation results from one oxygen molecule simultaneously bonding to two hemes, which is effectively prevented by Wang's polystyrene matrix or the globin of native hemoglobin.

One method of studying hemoglobin, and many other complex biological systems, is through model compounds such as those used by Wang. Many heme derivatives have been synthesized and tested for oxygen binding, with a more complete understanding of the process as a goal.[5] These compounds have been designed to protect the heme from the approach of another heme to prevent oxidation of the iron(II) by a cooperative reaction between two heme irons bridged by O_2. In addition, some of the model compounds have an imidazole or pyridine nitrogen linked to the heme to hold it in a convenient location for binding to the iron.

For more than a century, efforts have been made to develop blood substitutes, primarily based on human and bovine hemoglobin. Producing such a substitute that does not have averse effects on many patients has been truly challenging, and the goal of developing a safe blood substitute remains elusive.[6]

In hemoglobin, the Fe(II) is about 70 pm out of the plane of the porphyrin nitrogens in the direction of the imidazole nitrogen bonding to the axial position (Figure 16.2) and is a typical high-spin d^6 ion. When oxygen or carbon monoxide bond to the sixth position, the iron becomes coplanar with the porphyrin, and the resulting compound is diamagnetic. Carbon monoxide is a strong enough ligand to force spin pairing, and the resulting pi back-bonding stabilizes the complex. Oxygen bonds at an angle of approximately 130°, also with considerable pi back-bonding. Some have described the bonding as nearly that of Fe(III)—O_2^-, with enough metal-to-ligand electron transfer to result in a simple double bond between the oxygens. A structure,

[5] K. S. Suslick and T. J. Reinert, *J. Chem. Ed.*, **1985**, *62*, 974.
[6] J. E. Squires, *Science*, **2002**, *295*, 1004; J.-Y. Chen, M. Scerbo, and G. Kramer, *Clinics*, **2009**, *64*, 803.

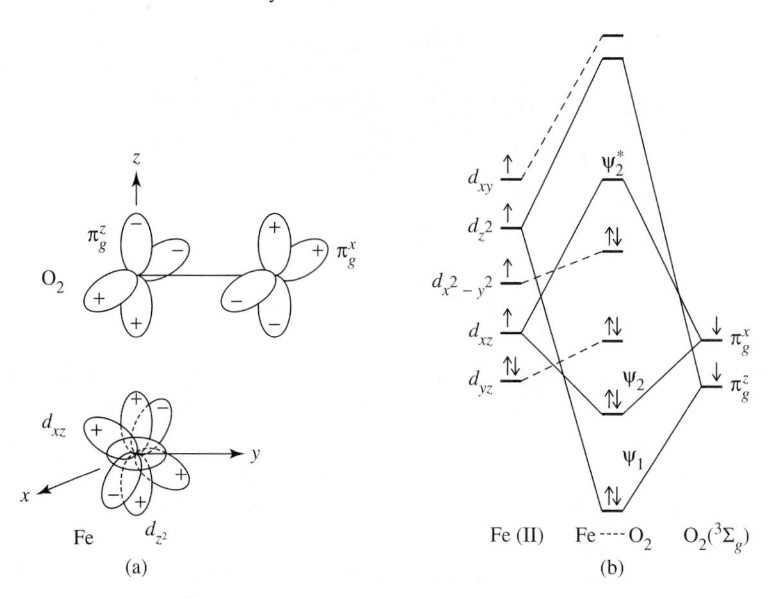

FIGURE 16.4 Electronic Structure of Oxyhemoglobin. (a) The most likely interaction between O_2 in the ground state ($^3\Sigma_g$) and Fe(II)-heme in the high-spin state; *x* and *y* axes bisect the angle N—Fe—N. The signs on the oxygen orbitals are appropriate for the π^* orbitals. (b) The interaction between O_2 and Fe(II)-heme expressed in an energy level diagram. Fe in the high-spin state is located a little out of the porphyrin plane and, as the reaction proceeds, it is thought to move to the center in the plane. This effect is shown by the broken lines.

(Reprinted with permission from E.-I. Ochiai, *J. Inorg. Nucl. Chem.*, **1974**, *36*, 2129. © 1974, Pergamon Press PLC.)

shown in Figure 16.4, has been proposed in which the triplet O_2 and the high-spin Fe(II) combine to form a spin-paired compound. The stronger sigma interaction is between the d_{z^2} and $\pi_g{}^z$ (antibonding π^*) orbitals. The weaker pi interaction is between d_{xz} and $\pi_g{}^x$ (antibonding π^*). The increased ligand field results in pairing of the electrons and a weakened O—O bond. In hemoglobin, CO also forms bent bonds to Fe, probably because surrounding groups in the hemoglobin force it out of the linear form. This reduces the formation constant for CO—Hb. Without this reduction, normal amounts of CO in the body would be enough to interfere with oxygen transport.

CYTOCHROMES, PEROXIDASES, AND CATALASES Other heme compounds are also active biochemically. Cytochrome P-450 catalyzes oxidation reactions in the liver and adrenal cortex, helping to detoxify some substances by adding hydroxyl groups that make the compounds more water soluble and more susceptible to further reactions. Unfortunately, at times this process has the reverse effect because some relatively safe molecules are converted into potent carcinogens. Peroxidases and catalases are Fe(III)-heme compounds that decompose hydrogen peroxide and organic peroxides. The reactions seem to proceed through Fe(IV) compounds with another unpaired electron on the porphyrin, which becomes a radical cation. Similar intermediates are also known in simpler porphyrin molecules.[7]

A model compound that decomposes hydrogen peroxide rapidly has been made from Fe(III) and triethylenetetramine (trien).[8] Although the rate is not as high as that for catalase (Table 16.2), it is many times faster than that for hydrated iron oxide, which

[7] D. L. Hickman, A. Nanthakumar, and H. M. Goff, *J. Am. Chem. Soc.*, **1988**, *110*, 6384.

[8] J. H. Wang, *J. Am. Chem. Soc.*, **1955**, *77*, 822, 4715; *Acc. Chem. Res.*, **1970**, *3*, 90; R. C. Jarnagin and J. H. Wang, *J. Am. Chem. Soc.*, **1958**, *80*, 786.

TABLE 16.2 Rates of Hydrogen Peroxide Decomposition

Catalyst	Relative Rate
Catalase	10^8
$[Fe(trien)]^{3+}$	10^4
Methemoglobin [Fe(III) Hb]	1

seems to have a large surface effect. The proposed mechanism for the $[Fe(trien)]^{3+}$ reaction is shown in Figure 16.5. Tracer studies using ^{18}O-labeled water have shown that the reaction produces oxygen gas in which all of the oxygen atoms come from the peroxide; as a result, the steps forming O_2 must involve removal of hydrogen from H_2O_2. Formation of water as the other product requires breakage of the oxygen–oxygen bond.

A group of cytochromes—labeled *a*, *b*, and *c*, depending on their spectra—serve as oxidation-reduction agents, converting the energy of the oxidation process into the synthesis of adenosine triphosphate (ATP), which makes the energy more available to other reactions. Copper is also involved in these reactions. The copper cycles between Cu(II) and Cu(I), and the iron cycles between Fe(III) and Fe(II) during the reactions. Details of the reactions are available in other sources.[9, 10]

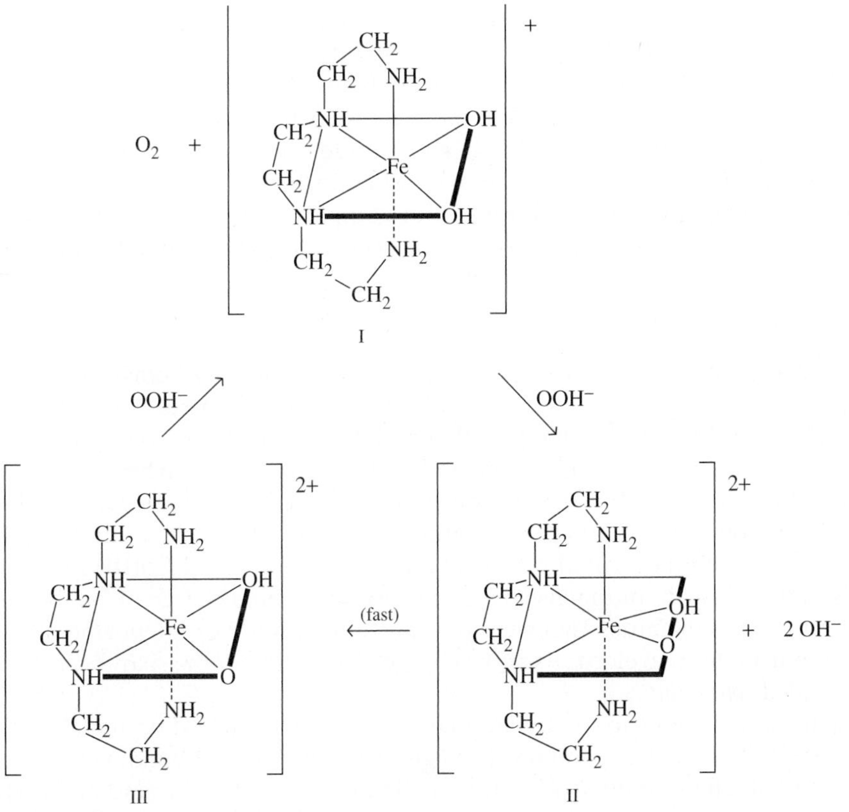

FIGURE 16.5 Mechanism of the $[Fe(trien)]^{3+} - H_2O_2$ Reaction.

(Reproduced with permission from J. H. Wang, *J. Am. Chem. Soc.*, **1955**, *77*, 4715. © 1955 American Chemical Society.)

[9] J. T. Groves, *J. Chem. Educ.*, **1988**, *11*, 928.

[10] E.-I. Ochiai, *Bioinorganic Chemistry*, Allyn and Bacon, Boston, 1977, pp. 150–165; T. E. Meyer and M. D. Kamen, *Adv. Protein Chem.*, **1982**, *35*, 105; G. R. Moore, C. G. S. Eley, and G. Williams, *Adv. Inorg. Bioinorg. Mech.*, **1984**, *3*, 1.

16.1.2 Similar Ring Compounds

CHLOROPHYLLS A porphine ring with one double bond reduced is called a *chlorin*. The chlorophylls (Figure 16.1) are examples of compounds containing this ring. They are green pigments found in plants, contain magnesium, and start the process of photosynthesis. They absorb light at the red end of the visible spectrum, transfer an electron to adjacent compounds, and, by a series of complex reactions, finally transfer the energy of the light to the metabolic processes of the plant. The overall process can be summarized in the two reactions

$$2\,H_2O \longrightarrow O_2 + 4\,H^+ + 4\,e^-$$

$$CO_2 + 4\,H^+ + 4\,e^- \longrightarrow [CH_2O] + H_2O$$

where $[CH_2O]$ represents sugars, carbohydrates, and cellulose synthesized in the plant. In effect this process also reverses the oxidation process that produces the energy for animal life, in which the $[CH_2O]$ compounds are converted back to water and carbon dioxide. The entire process is very complicated and is far from being completely understood, but it includes a vital role for manganese in the first reaction.

Other compounds containing metal ions, such as the *ferredoxins*,[11] are involved in electron-transfer reactions, part of the photosynthetic pathway in plants, and in electron-transfer chains linked to hydroxylation and other reactions in mammals and bacteria. Ferredoxins are iron-sulfur proteins with active sites involving iron, sulfur, and sulfur-containing amino acids, for example $Fe_2S_2(cys)_4$ (cys = cysteine, $HSCH_2CH(NH_2)COOH$), shown in Figure 16.6(a). They typically have Fe(II) and Fe(III) in tetrahedral sites bridged by sulfide ions and bound into the protein by Fe—S bonds to the amino acid. There are also other more complex ferredoxins that contain Fe_4S_4 or Fe_3S_4 units, again with tetrahedral iron and sulfide bridges. Structures for the Fe—S active sites of three ferredoxins are shown in Figure 16.6. The clostridial ferredoxin in this figure contains two identical iron-sulfur complexes, each containing four sulfurs and four irons in a distorted cube, sometimes called a *cubane* structure, with four cysteine sulfurs also coordinated to the iron.

COENZYME B_{12} A vitamin known as coenzyme B_{12} is the only known organometallic compound in nature. It incorporates cobalt into a corrin ring structure, which has one less =CH— bridge between the pyrrole rings than the porphyrins (Figure 16.7). This compound is known to prevent anemia and also has been found to have many catalytic properties. During isolation of this compound from natural sources, the adenosine group is usually replaced by cyanide, and it is in this cyanocobalamin (vitamin B_{12}) form that it is used medicinally. The cobalt can be counted as Co(III) in these compounds; the four corrin nitrogens contribute electrons and a charge of 2–, the benzimidazole nitrogen contributes two electrons, and the cyanide or adenosine in the sixth position contributes two electrons and a charge of 1–. Without the sixth ligand, the molecule is called *cobalamin*.

Methylcobalamin can methylate many compounds, including metals. The reactions of alkylcobalamins depend on cleavage of the alkyl–cobalt bond, which can result in Co(I) and an alkyl cation, Co(II) and an alkyl radical, or Co(III) and an alkyl anion,

---

[11] C. R. Crossnoe, J. P. Germanas, P. LeMagueres, B. Mustata, and K. L. Krause, *J. Mol. Biol.*, **2002**, *318*, 503; R. Morales, M. Frey, and J.-M. Mouesca, *J. Am. Chem. Soc.*, **2002**, *124*, 6714.

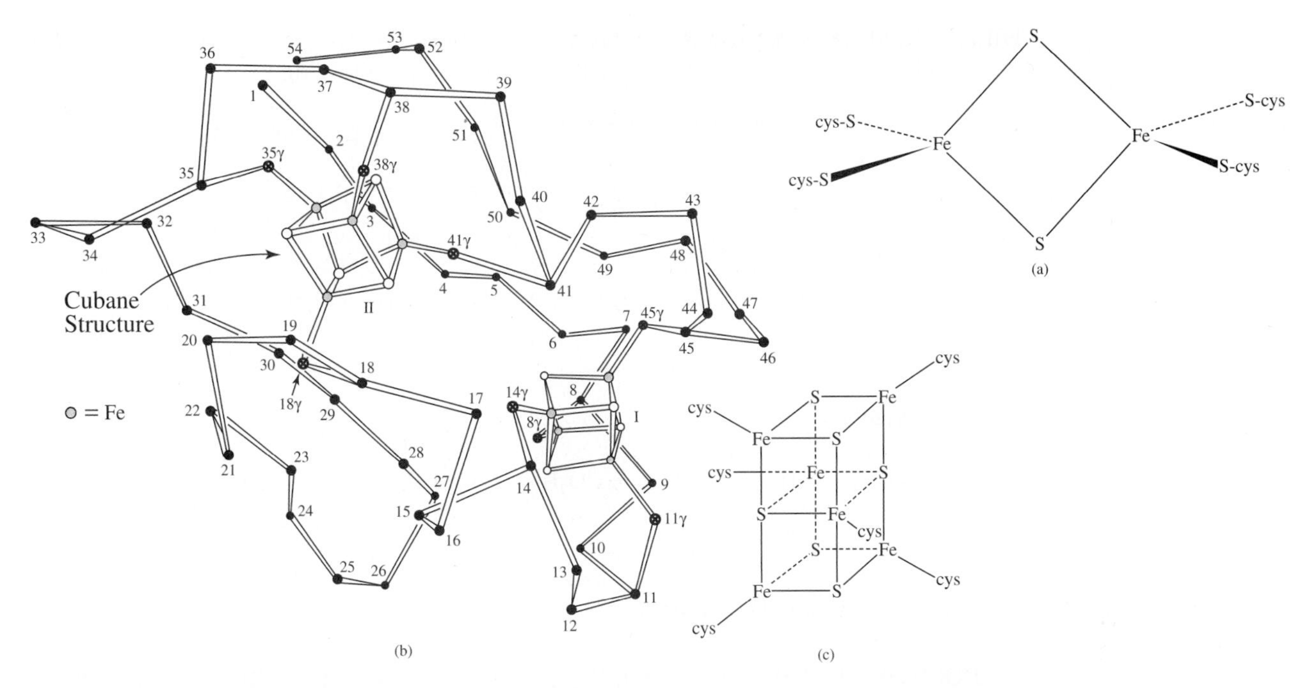

FIGURE 16.6 Structures of Fe—S Protein-Active Sites. (a) Ferredoxin. (From E.-I. Ochiai, *Bioinorganic Chemistry*, Allyn and Bacon, Boston, 1977, p. 184.) (b) Clostridial ferredoxin. (Reproduced with permission from E. T. Adman, L. C. Sieker, and L. H. Jensen, *J. Biol. Chem.*, **1973**, *248*, 3987.) (c) A model for the structure of the Fe_6S_6 active unit.

(Reproduced with permission from E.-I. Ochiai, *Bioinorganic Chemistry*, Allyn and Bacon, Boston, 1977, p. 192.)

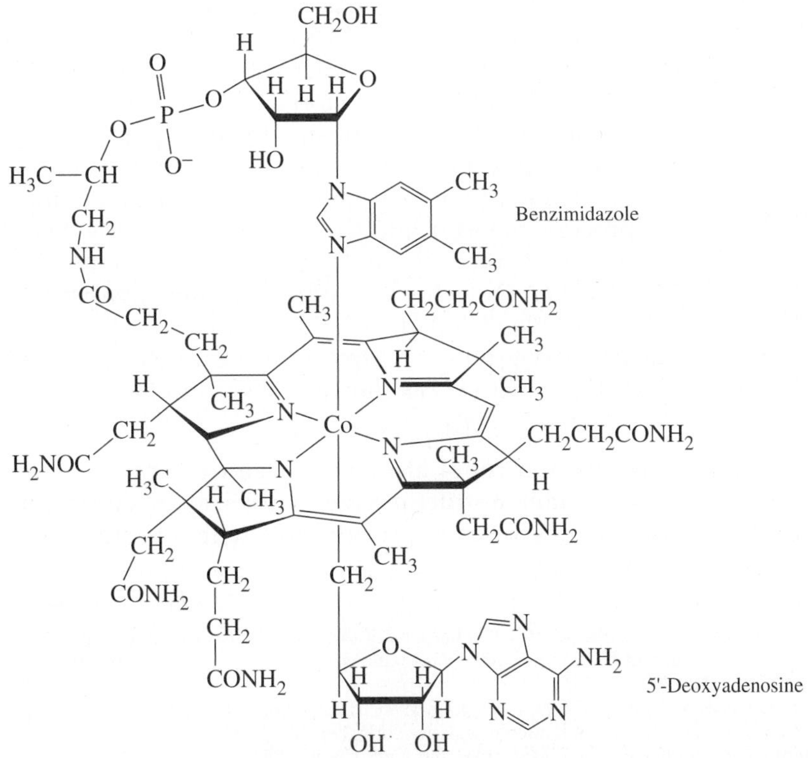

Benzimidazole

5'-Deoxyadenosine

FIGURE 16.7 Coenzyme B_{12}.

with the radical mechanism being the most common. The alkyl products can then react in a number of ways. Some of the reactions include the following:[12]

Methylation or Hydroxymethylation

$$HO_2CCH(NH_2)CH_2CH_2SH \longrightarrow HO_2CCH(NH_2)CH_2CH_2SCH_3$$

$$H_2N{-}CH_2{-}CO_2H \longrightarrow H_2N{-}CH(CH_2OH){-}CO_2H$$

Isomerization

$$HO_2C{-}CH(NH_2){-}CH_2{-}CH_2{-}CO_2H \longrightarrow HO_2C{-}CH(NH_2){-}CH(CH_3){-}CO_2H$$

Isomerization and Dehydration

$$HOCH_2{-}CH(OH){-}CH_3 \longrightarrow HOCH(OH){-}CH_2{-}CH_3 \longrightarrow HC({=}O){-}CH_2{-}CH_3 + H_2O$$

16.2 OTHER IRON COMPOUNDS

FERRITIN AND TRANSFERRIN Iron is stored in both plant and animal organisms in combination with a protein called *apoferritin*. The resulting ferritin contains a micelle of ferric hydroxide–oxide–phosphate surrounded by the protein and is present mainly in the spleen, liver, and bone marrow in mammals. Individual subunits of the apoferritin have a molecular weight of about 18,500; and 24 of these subunits combine to form the complex, with a protein molecular weight of about 445,000 and up to 4,300 atoms of Fe in the iron core, stored as ferrihydrite phosphate, $[(Fe(O)OH)_8(FeOPO_3H_2) \cdot xH_2PO_4]$. The mechanisms for incorporation of iron into this complex and removal for use in the body are uncertain, but it appears that reduction to Fe(II) and chelation of the Fe(II) are required to remove iron from the core, and the reverse process moves it into the storage core of the complex. It is known from tracer experiments that all the oxygen atoms in the ferrihydrite are derived from water, rather than O_2. Other iron proteins, called *transferrins*, serve to transport iron as Fe(III) in the blood and other fluids. One of these has iron bound as Fe(III) by two tyrosine phenoxy groups, an aspartic acid carboxyl group, a histidine imidazole, and either HCO_3^- or CO_3^{2-}, as shown in Figure 16.8.[13]

SIDEROCHROMES Bacteria and fungi also synthesize iron transfer compounds, called *siderochromes*.[14] The common structures are complex hydroxamates, also called ferrichromes or ferrioxamines, or complex catechols (enterobactin), all shown in

---

[12] R. H. Abeles, "Current Status of the Mechanism of Action of B_{12}-Coenzyme" in A. W. Addison, W. R. Cullen, D. Dolphin, and B. R. James, eds., *Biological Aspects of Inorganic Chemistry*, Wiley InterScience, New York, 1977, pp. 245–260.

[13] R. E. Feeney and S. K. Komatsu, *Struct. Bonding*, **1966**, *1*, 149–206; E E. Hazen, cited in B. L. Vallee and W. E. C. Wacker, *Metalloproteins*, Academic Press, New York, 1969, p. 89.

[14] K. N. Raymond, G. Müller, and B. F. Matzanke, *Top. Curr. Chem.*, **1984**, *123*, 49.

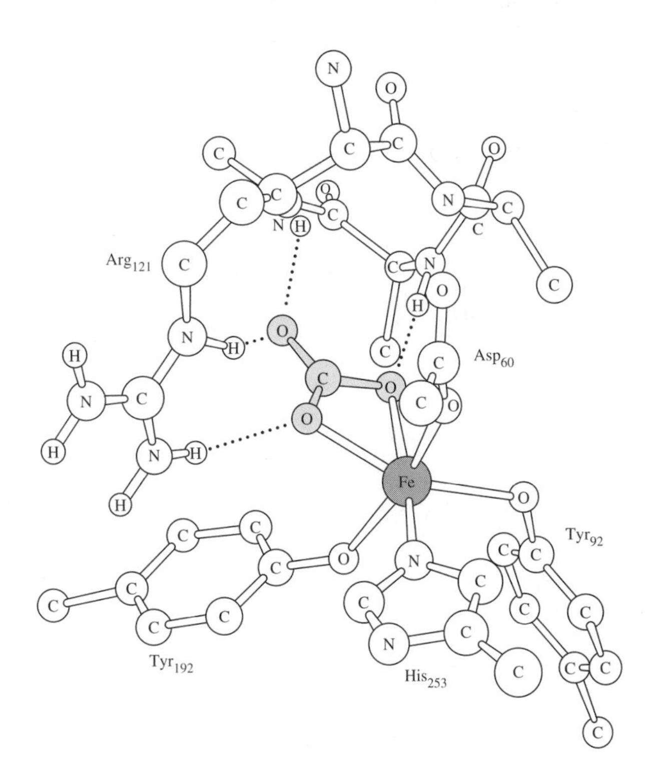

FIGURE 16.8 Lactotransferrin.

(Reproduced with permission from S. J. Lippard and J. M. Berg, *Principles of Bioinorganic Chemistry*, University Science Books, Mill Valley, CA, 1994, p. 144.)

Figure 16.9. They have peptide backbones and are very strong chelating agents ($K_f \approx 10^{30}$ to 10^{50}), allowing the organism to extract iron from surroundings that contain very little iron or are basic enough that the iron is present as insoluble hydroxides or oxides. Some of these compounds act as growth factors for bacteria, and others act as antibiotics. There are also examples in which the iron is bound by a mixture of phenolic hydroxyl, hydroxamate, amine, and alcoholic hydroxyl groups.

16.3 ZINC AND COPPER ENZYMES

Zinc is found in more than 80 enzymes. Two of these, carboxypeptidase and carbonic anhydrase, will be discussed here.[15] Copper is also a common metal in enzymes and is present in four different forms. Two of the copper enzymes will also be described.

CARBOXYPEPTIDASE Carboxypeptidase is a pancreatic enzyme that catalyzes the hydrolysis of the peptide bond at the carboxyl end of proteins and peptides, with a strong preference for amino acids with an aromatic or branched aliphatic side chain. The zinc ion is bound in a 5-coordinate site by two histidine nitrogens, both oxygens from a glutamic acid carboxyl group, and a water molecule. A pocket in the protein structure accommodates the side chain of the substrate. Evidence indicates that the negative carboxyl group of the substrate hydrogen bonds to an arginine on the enzyme, while the zinc bonds to the oxygen of the peptide carbonyl, as shown in Figure 16.10. A Zn—OH or Zn—OH_2 combination seems to be the group that reacts with the carbonyl carbon, with assistance of a glutamic acid carboxyl group from the enzyme that assists in the transfer of H^+ from the bound water to the amino acid product.[16] An artificial peptidase model compound has been made with a Cu(II) bound by four nitrogens in a

[15] I. Bertini, C. Luchinat, and R. Monnanni, *J. Chem. Educ.*, **1985**, *62*, 924.
[16] D. W. Christianson and W. N. Lipscomb, *Acc. Chem. Res.*, **1989**, *22*, 62.

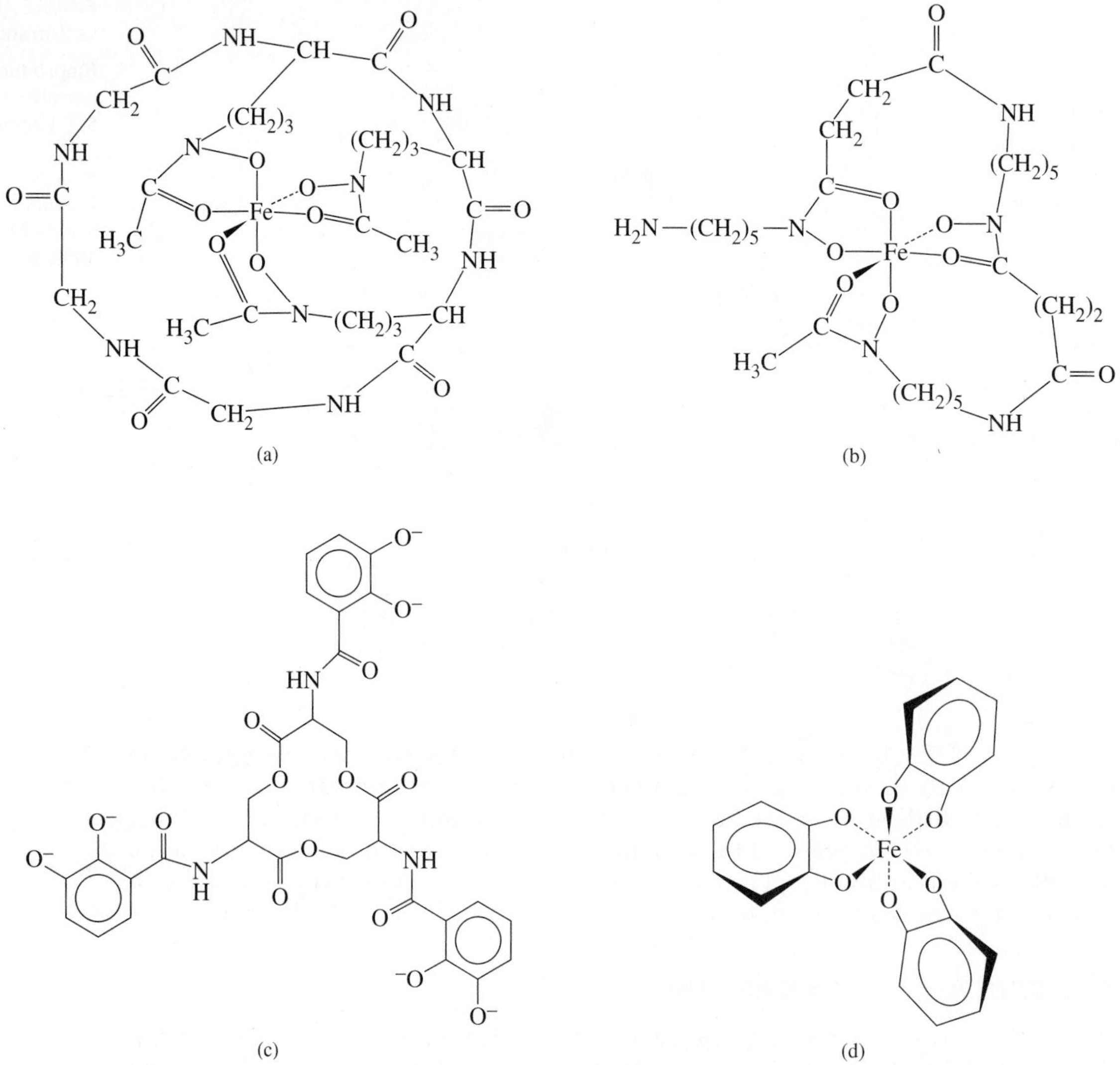

(a) (b) (c) (d)

FIGURE 16.9 Ferrichromes, Ferrioxamines, and Catechol Siderochromes. (a) Ferrichrome A. (b) Ferrioxamine B. (c) Enterobactin, a catechol siderochrome. (d) The catechol complex of enterobactin with Fe(III). The trilactone ring (omitted in the drawing) has all *S* conformation in the chiral atoms, which in turn results in a Δ conformation when the six catechol oxygen atoms complex with Fe(III).

chain that ends in a guanidinium ion, all attached to a cross-linked polystyrene.[17] The catalytic activity is high for hydrolysis of amides with carboxyl groups attached, similar to a carboxypeptidase activity. The H^+ on the guanidinium group can hydrogen bond to the carboxyl group, holding the substrate in position near the Cu, which is the active site.

CARBONIC ANHYDRASE In a few cases, theoretical calculations of transition state and intermediate energies and geometries provide confirmation of experimental studies of the mechanism of enzyme reactions and suggest directions for further study. One of these is the hydration of carbon dioxide catalyzed by carbonic anhydrase, a zinc enzyme. Below pH 7, the uncatalyzed reaction $HCO_3^- + H^+ \rightleftharpoons H_2O + CO_2$ is favored. Above pH 7, the reaction is $HCO_3^- \rightleftharpoons OH^- + CO_2$, catalyzed by carbonic

[17] J. Suh and S.-J. Moon, *Inorg. Chem.*, **2001**, *40*, 4890.

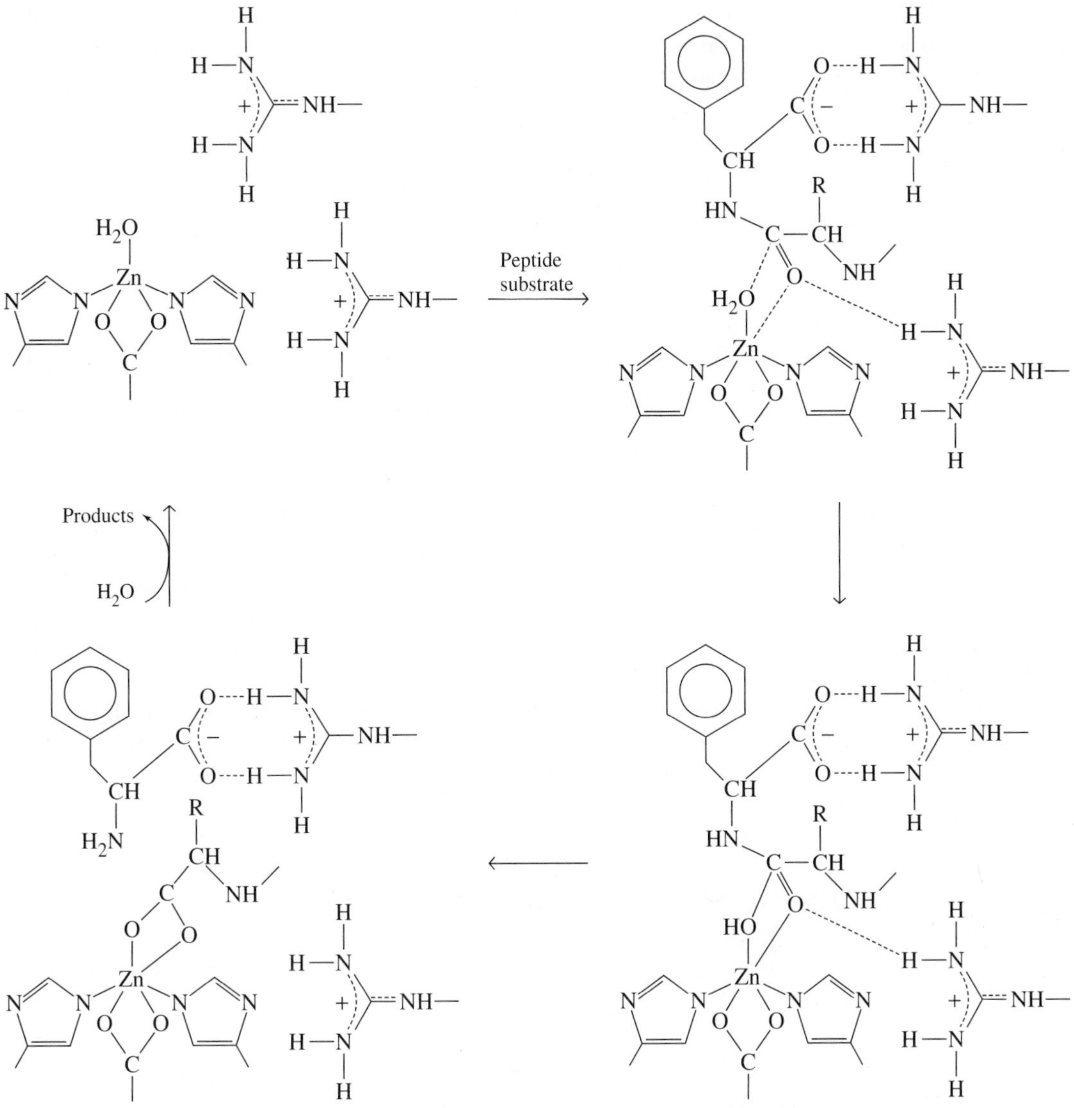

FIGURE 16.10 Proposed Mechanism of Carboxypeptidase Action. Transfer of several hydrogen ions is not shown.

anhydrases I, II, and III; II is particularly active, with a rate enhancement of 10^6 or more, approaching a diffusion-controlled rate.[18]

The active site has a zinc(II) ion bonded to three histidine imidazole groups and a fourth site occupied by water or hydroxide ion, $L_3Zn^{2+}OH^-$, where L = imidazole N from histidine. Experimentally, the enzyme loses activity below pH 7, indicating that an ionizable group with $pK = 7$ is part of the active site. In addition, it is known that the product of the hydration of CO_2 is HCO_3^-, as would be expected in neutral or basic solution.

Calculations have shown that water bound to the zinc ion can lose a proton readily, but imidazole bound to the zinc ion cannot. There is still an unsettled question about the first ionization from the zinc-bound water molecule. This reaction seems to be much

[18] S. Lindskog, *Adv. Bioinorg. Chem.*, **1982**, *4*, 116; P. J. Stein, S. P. Merrill, and R. W. Henkens, *J. Am. Chem. Soc.*, **1977**, *99*, 3194; *Biochemistry*, **1985**, *24*, 2459; S. Lindskog in T. G. Spiro, ed., *Zinc Enzymes*, Wiley, New York, 1983, p. 77.

too fast and is dependent on buffer concentration. The role of the buffer is still unknown, but in some fashion it assists in the reaction. The sequence of reactions usually used to describe the reaction is as follows:

$$L_3Zn^{2+}OH^- + CO_2 \longrightarrow L_3Zn^{2+}OH^- \cdot CO_2 \quad (1)$$

$$L_3Zn^{2+}OH^- \cdot CO_2 \longrightarrow L_3Zn^{2+}HOCO_2{}^- \quad (2)$$

$$L_3Zn^{2+}HOCO_2{}^- \longrightarrow L_3Zn^{2+}OCOOH^- \quad (3)$$

$$L_3Zn^{2+}OCOOH^- + H_2O \longrightarrow L_3Zn^{2+}(OCOOH^-)(H_2O) \quad (4)$$

$$L_3Zn^{2+}(OCOOH^-)(H_2O) \longrightarrow L_3Zn^{2+}OH_2 + HCO_3{}^- \quad (5)$$

$$L_3Zn^{2+}OH_2 \longrightarrow L_3Zn^{2+}OH^- + H^+ \text{ (which may be on a histidine N)} \quad (6)$$

The complex formed in (1) is loosely bound, moving to the more tightly bound product of (2). The transition state for reaction (3) may be a bidentate hydrogen carbonate,[19] or there may be a proton transfer between the bound oxygen atom and one of the unbound oxygen atoms.[20] In either case, the result is probably a bound hydrogen carbonate ion that has the OH group at as great a distance from the Zn as possible. Whether reaction (5) has a 5-coordinate Zn with addition of the water molecule is uncertain; it may just be part of the transition state. There have been several attempts to determine the mechanism and the transition states by theoretical calculations,[21] but the details remain uncertain. Some possibilities are shown in Figure 16.11. Future calculations that incorporate of the protein structure surrounding the active site promise more reliable mechanistic information.

A study of spinach carbonic anhydrase showed very similar kinetic behavior but also showed that the Zn is bound to a sulfur atom.[22] It was concluded that the two enzymes are convergently evolved, with different structures, but have equivalent functions.

CERULOPLASMIN AND SUPEROXIDE DISMUTASE Copper is present in mammals in ceruloplasmin and superoxide dismutase, and it is also part of a number of enzymes in plants and other organisms, including laccase, ascorbate oxidase, and plastocyanin. In these compounds, it is present in four different forms, listed in Table 16.3.

FIGURE 16.11 Proposed Mechanism for Carbonic Anhydrase.

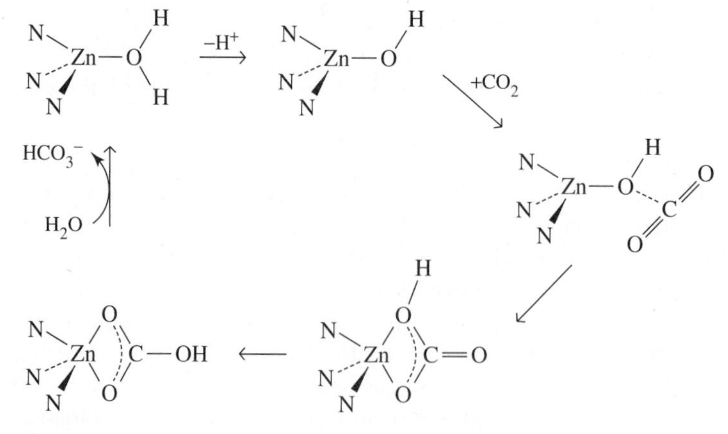

---

[19] S. Lindskog, *Ann. N.Y. Acad. Sci.*, **1984**, *429*, 61; D. N. Silverman and S. Lindskog, *Acc. Chem. Res.*, **1988**, *21*, 30.

[20] W. N. Lipscomb, *Ann. Rev. Biochem.*, **1983**, *429*, 17.

[21] J.-Y. Liang and W. N. Lipscomb, *Biochemistry*, **1987**, *26*, 5293; K. M. Merz, Jr., R. Hoffman, and M. J. S. Dewar, *J. Am. Chem. Soc.*, **1989**, *111*, 5636; Y.-J. Zheung and K. M. Merz, Jr., *J. Am. Chem. Soc.*, **1992**, *114*, 10498; M. Solà, A. Lledós, M. Duran, and J. Bertrán, *J. Am. Chem. Soc.*, **1992**, *114*, 869.

[22] R. S. Rowlett, M. R. Chance, M. D. Wirt, D. E. Sidelinger, J. R. Royal, M. Woodroffe, Y.-F. A. Wang, R. P. Saha, and M. G. Lam, *Biochemistry*, **1994**, *33*, 13967.

TABLE 16.3 Forms of Copper in Proteins

	Absorption Maximum (nm)	Extinction Coefficients ($L\ mol^{-1}\ cm^{-1}$)	Comments
Type 1	600 nm	1000–4000	Responsible for the blue color of blue oxidases and electron–transfer proteins, L $\longrightarrow$ M charge transfer spectrum of Cu—S bond
Type 2	Near 600 nm	300	Similar to ordinary tetragonal Cu(II) complexes but more intensely colored
Type 3	330 nm	3000–5000	Paired Cu(II) ions, diamagnetic, associated with redox reactions of O_2, where it undergoes a 2-electron change, bypassing superoxide
Cu(I)			Colorless, diamagnetic, no epr spectrum (no unpaired electrons)

Ceruloplasmin[23] is an intensely blue glycoprotein of the α_2-globulin fraction of mammalian blood, which acts as a copper transfer protein and probably has a role in iron storage. The structure is known;[24] it contains three Type 1 (T1) sites, one of which seems to be inactive, and a Type 2/Type 3 (T2/T3) trinuclear cluster. It is believed to be part of the process of oxidizing Fe(II) to Fe(III) in the transfer of iron from ferritin to transferrin. Reduction of two T1 sites and the T3 pair is fast, but reduction of the T2 Cu site is slow; the pathways of electron transfer between the sites have been investigated, but the complete mechanism is still unknown.[25]

Bovine superoxide dismutase[26] contains one atom of Cu(II) and one atom of Zn(II) in each of two subunits, with a molecular weight of about 16,000. The copper is in a distorted square-pyramidal site, bound to four histidine nitrogens and a water; the zinc is bound to three histidines, including a bridging imidazole ring bound to both metal ions, and an aspartate carboxyl oxygen in a distorted tetrahedral structure, as shown in Figure 16.12. Cu is the more essential metal, which cannot be replaced while retaining activity. On the other hand, the Zn can be replaced by other divalent metals with retention of most of the catalytic activity. The major role of zinc may be to provide structural stability, as evidenced by the stability of the enzyme at high temperatures, but the enzyme with Cu in both sites is still active in the presence of SCN^-, which breaks the histidine bridge between the Cu atoms.[27] The superoxide ion, O_2^-—which can be formed by dissociation of oxygen from heme proteins, leaving behind Fe(III)—is relatively unreactive; but one of its products, HO_2, is very reactive, so the superoxide must be removed quickly. O_2^- is found in several metabolic processes and appears to be essential for a few (e.g., tumor necrosis factor, antibacterial effect of myeloperoxidases), but large amounts form in some pathological conditions and cause serious damage. One pathway of reaction is the formation of OH radicals and singlet oxygen, both of which are toxic.

---

[23] S. H. Lawrie and E. S. Mohammed, *Coord. Chem. Rev.*, **1980**, *33*, 279.

[24] I. Zaitseva, V. Zaitsev, G. Card, K. Moshov, B. Bax, A. Ralph, and P. Lindley, *J. Biol. Inorg. Chem.*, **1996**, *1*, 15; P. F. Lindley, G. Card, I. Zaitseva, V. Zaitsev, B. Reinhammar, E. Selin-Lindgren, and K. Yoshida, *J. Biol. Inorg. Chem.*, **1997**, *2*, 454; V. N. Zaitsev, I. Zaitseva, M. Papiz, and P. F. Lindley, *J. Biol. Inorg. Chem.*, **1999**, *4*, 579.

[25] T. E. Machonkin and E. I. Solomon, *J. Am. Chem. Soc.*, **2000**, *122*, 12547.

[26] I. Fridovich, *Adv. Inorg. Biochem.* **1979**, *1*, 67; J. S. Valentine and D. M. de Freitas, *J. Chem. Educ.*, **1985**, *62*, 728.

[27] K. G. Strothcamp and S. J. Lippard, *Biochemistry*, **1981**, *20*, 7488.

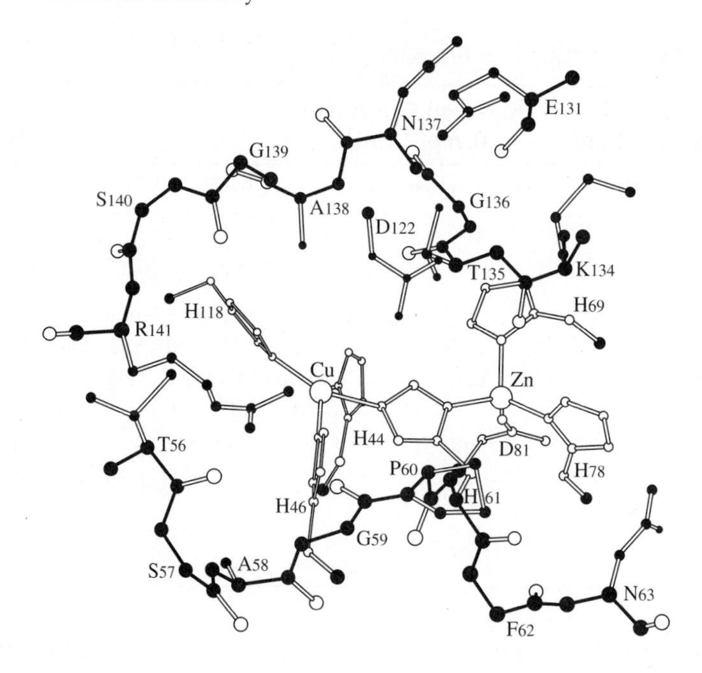

FIGURE 16.12 Active Site of Bovine Superoxide Dismutase. Shown is a drawing of the active site channel as viewed from the solvent. The main chain is shown in black, the ligand side chains as open circles and bonds, and the other side chains as solid atoms and open bonds.

(Reproduced with permission from J. A. Tainter, E. D. Getzoff, J. S. Richardson, and D. C. Richardson, *Nature*, **1983**, *306*, 284. © 1983 Macmillan Magazines Limited.)

Some reactions of superoxide follow:

$$2\,O_2^- + 2\,H^+ \longrightarrow H_2O_2 + O_2$$

$$O_2^- + HO_2 \longrightarrow HO_2^- + O_2$$

$$2\,HO_2 \longrightarrow O_2 + H_2O_2$$

$$H_2O_2 + O_2^- + H^+ \longrightarrow O_2 + H_2O + \cdot OH$$

$$HO_2 + O_2^- + H^+ \longrightarrow O_2 + H_2O_2$$

The reactions catalyzed by superoxide dismutases

$$2\,O_2^- + 2\,H^+ \rightleftharpoons H_2O_2 + O_2$$

and

$$2\,O_2^- + H_2O \rightleftharpoons HO_2^- + O_2 + OH^-$$

have large equilibrium constants and can proceed by the reactions

$$HA + O_2^- \longrightarrow A^- + HO_2$$

and

$$H_2O + HO_2 + O_2^- \longrightarrow H_2O_2 + O_2 + OH^-$$

or

$$HO_2 + HO_2 \longrightarrow H_2O_2 + O_2$$

Reactions seem to involve a Cu(II)–Cu(I) cycle, with H^+ replacing the Cu(I) on the bridging histidine at one stage. The rate-limiting step is the approach and/or bonding of O_2^- to Cu. In saturated conditions, H^+ transfer may be rate-limiting.

In a simple model of reaction, O_2^- and H^+ react to form Cu(I) and H^+–histidine–Zn and O_2, then O_2^- reacts with the enzyme to reform the Cu(II)–histidine–Zn and H_2O_2:

$$(His)_3Cu^{II}—N—C—N—Zn(His)_2(Asp) + O_2^- + H^+ \longrightarrow (His)_3Cu^I + H^+N—C—N—Zn(His)_2(Asp) + O_2$$

$$(His)_3Cu^I + H^+N—C—N—Zn(His)_2(Asp) + O_2^- \longrightarrow (His)_3Cu^{II}—N—C—N—Zn(His)_2(Asp) + HO_2^-$$

A more detailed model[28] has O_2^- replacing H_2O as the fifth ligand on Cu, with hydrogen bonding to an arginine guanidinium group, transfer of H^+ from water to the histidine, and release of O_2, forming Cu(I) and H^+–histidine–Zn. These react with O_2^- and H^+ to form the arginine–H^+–O_2–Cu species, with hydrogen bonding to the H^+ histidine and water. Release of H_2O_2 reforms the native enzyme. It has also been suggested[29] that the Zn–histidine role is to promote the release of HO_2^- from the copper in this final step by forcing the HO_2^- into the axial position, where it would be more weakly bound.

The roles of the copper enzymes in electron transport, oxygen transport, and oxidation reactions have guaranteed continued interest in their study. In addition to studies of the natural compounds, there have been many attempts to design model structures of these enzymes, particularly of the binuclear species. Many of these include both nitrogen and oxygen donors built into macrocyclic ligands, although sulfur has been used as well.[30]

16.4 NITROGEN FIXATION

A very important sequence of reactions converts nitrogen from the atmosphere into ammonia:

$$N_2 + 6\,H^+ + 6\,e^- \longrightarrow 2\,NH_3$$

The NH_3 can then be further converted into nitrate or nitrite or directly used in the synthesis of amino acids and other essential compounds. This reaction takes place at 0.8 atm N_2 pressure and ambient temperatures in *Rhizobium* bacteria in nodules on the roots of legumes such as peas and beans, as well as in other independent bacteria. In contrast to these mild conditions, industrial synthesis of ammonia requires high temperatures and pressures with iron oxide catalysts, and even then yields only 15% to 20% conversion of the nitrogen to ammonia. Intensive efforts to determine the bacterial mechanism and to improve the efficiency of the industrial process have so far been only moderately successful; the goal of approaching enzymatic efficiency on an industrial scale motivates significant current research.

The nitrogenase enzymes responsible for nitrogen fixation contain two proteins. The iron-molybdenum protein contains two metal centers. One, called the *FeMo-cofactor*, contains molybdenum, iron, and sulfur (Figure 16.13). This may be the site of nitrogen reduction; there are open binding sites on some of the iron atoms in the middle and a pocket large enough for substrate binding. The other site, called the *P-cluster*, contains eight iron atoms and eight sulfur atoms in two subunits that are nearly cubic.[31] The P-cluster is believed to assist the reaction by transfer of electrons, but little more is known about the mechanism of the reaction. The second protein contains two identical subunits with a single 4Fe:4S cluster and an adenosine diphosphate (ADP) molecule

[28] R. Osman and H. Basch, *J. Am. Chem. Soc.*, **1984**, *106*, 5710.
[29] L. S. Ellerby, D. E. Cabelli, J. A. Graden, and J. S. Valentine, *J. Am. Chem. Soc.*, **1996**, *118*, 6556.
[30] K. D. Karlin and Y. Gultneh, *J. Chem. Educ.*, **1985**, *62*, 983; K. G. Strothcamp and S. J. Lippard, *Acc. Chem. Res.*, **1982**, *15*, 318.
[31] M. M. Georgiadis, H. Komiya, P. Chakrabarti, D. Woo, J. J. Kornuc, and D. C. Rees, *Science*, **1992**, *257*, pp. 1653, 1677.

FIGURE 16.13 The FeMo-Cofactor Site of Nitrogenase.

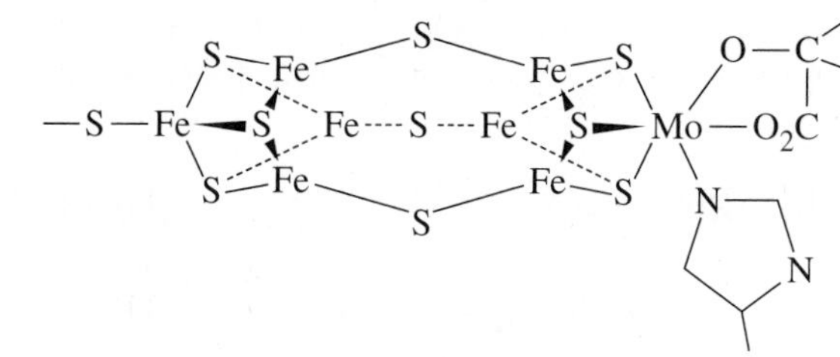

bound between the two subunits. In some fashion, this protein is reduced and transfers single electrons to the FeMo-cofactor protein, where the reaction with nitrogen takes place. Eight electrons are required for N_2 conversion to 2 NH_3 by the enzymes, because the reaction also forms H_2.[32] In addition, 16 molecules of MgATP are converted to MgADP and inorganic phosphate.

$$N_2 + 8\,H^+ + 8\,e^- \longrightarrow 2\,NH_3 + H_2$$

The nitrogenase reaction seems to begin with a series of four electron transfers, leading to a reduced form of the enzyme that then can bind four H^+ ions. After these changes, N_2 can be bound as H_2 is released and, finally, the N_2 is reduced to NH_3 and released from the complex.[33] Two alternative sites for the first proton to bind have been suggested, one in the middle of the Fe_6S_3 cluster in the MoFe active site and the other at an alkoxy oxygen of homocitrate bound to Mo at one end of the cluster.[34] Calculations show the possibility of N_2 bonding asymmetrically with one of the N atoms near the center of the four Fe atoms, approximately at the corners of a square on the front face shown in Figure 16.13 and the other sticking out toward the top front.[35] This more distant N is positioned to accept H atoms from the nearby S atoms, which could release NH_3. The remaining N can then accept H atoms in a similar fashion from S atoms to form the second NH_3.

In addition to the nitrogenases whose cofactor contains the structure $MoFe_7S_9$, others contain clusters with no Mo. Theoretical calculations[36] based on an Fe_2S_5 cluster—a portion of the larger cofactor, with one S bridging two Fe(II) atoms—have shown that adding an H atom to the bridging S is required to allow formation of an N_2 bridge between the iron atoms. Once this occurs, addition of H to the N_2 can proceed through exothermic formation of N_2H, N_2H_2, N_2H_3, and N_2H_4. Subsequent steps are less easily predicted, but the suggestion is that the next H atom combines with the H from the bridging sulfur to form H_2, known to be one of the products of the reaction. The next H atom could add to the N_2H_4, forming NH_3 and NH_2, each bound to an iron atom, and a final H creates the second NH_3, with a large exothermic value. Preliminary calculations based on an $Fe_8S_9{}^{2-}$ cluster, of the same structure as the cofactor but with more symmetry because Fe replaces Mo, gives similar results, in which addition of H atoms to the bridging sulfur atoms is required to open the structure sufficiently for addition of N_2.

There have been many attempts to make model compounds for ammonia production, but none have been successful. How the enzyme manages to carry out the reaction at ambient temperature and less than 1 atm pressure of N_2 is still an unanswered question.

NITRIFICATION AND DENITRIFICATION Oxidation of ammonia to nitrite, $NO_2{}^-$, and nitrate, $NO_3{}^-$, is called *nitrification*; the reverse reaction is ammonification. Reduction from nitrite to nitrogen is called *denitrification*. All these reactions and more occur in enzyme systems, many of which include transition metals. A molybdenum enzyme,

[32] F. B. Simpson .and R. H. Burris, *Science*, **1984**, *224*, 1095.
[33] R. N. F. Thornley and D. Lowe, in T. G. Spiro, ed., *Molybdenum Enzymes*, Wiley InterScience, New York, 1985.
[34] T. Lovell, J. Li, D. A. Case, and L. Noodleman, *J. Am. Chem. Soc.*, **2002**, *124*, 4546.
[35] I. Dance, *Chem. Commun. (Cambridge)*, **1997**, 165.
[36] P. E. M. Siegbahn, J. Westerberg, M. Svensson, and R. H. Crabtree, *J. Phys. Chem.*, **1998**, *102*, 1615.

nitrate reductase, reduces nitrate to nitrite. Further reduction to ammonia seems to proceed by 2-electron steps through an uncertain intermediate with a +1 oxidation state (possibly hyponitrite, $N_2O_2^{2-}$) and hydroxylamine:

$$NO_2^- \longrightarrow N_2O_2^{2-} \longrightarrow NH_2OH \longrightarrow NH_3$$

Some nitrite reductases contain iron and copper; other enzymes active in these reactions contain manganese. Reactions catalyzed by copper and iron enzymes with NO, N_2O, and N_2 as products have also been reported.

Nitrite reductase from *Alcaligenes xylosoxidans* is made up of three idential subunits, each with an embedded Cu (Type I) and a Cu (Type II) bound by residues from two of the subunits. Catalysis[37] seems to proceed by binding of NO_2^- to the Type II Cu, with carboxylate from an aspartate residue hydrogen–bonded to one of the NO_2^- oxygens, followed by transfer of an electron from the Type I Cu, transfer of H^+ to the same oxygen from a histidine residue, and release of NO to regenerate the active site hydrogen bonded to the aspartate through a water molecule, as shown in Figure 16.14.

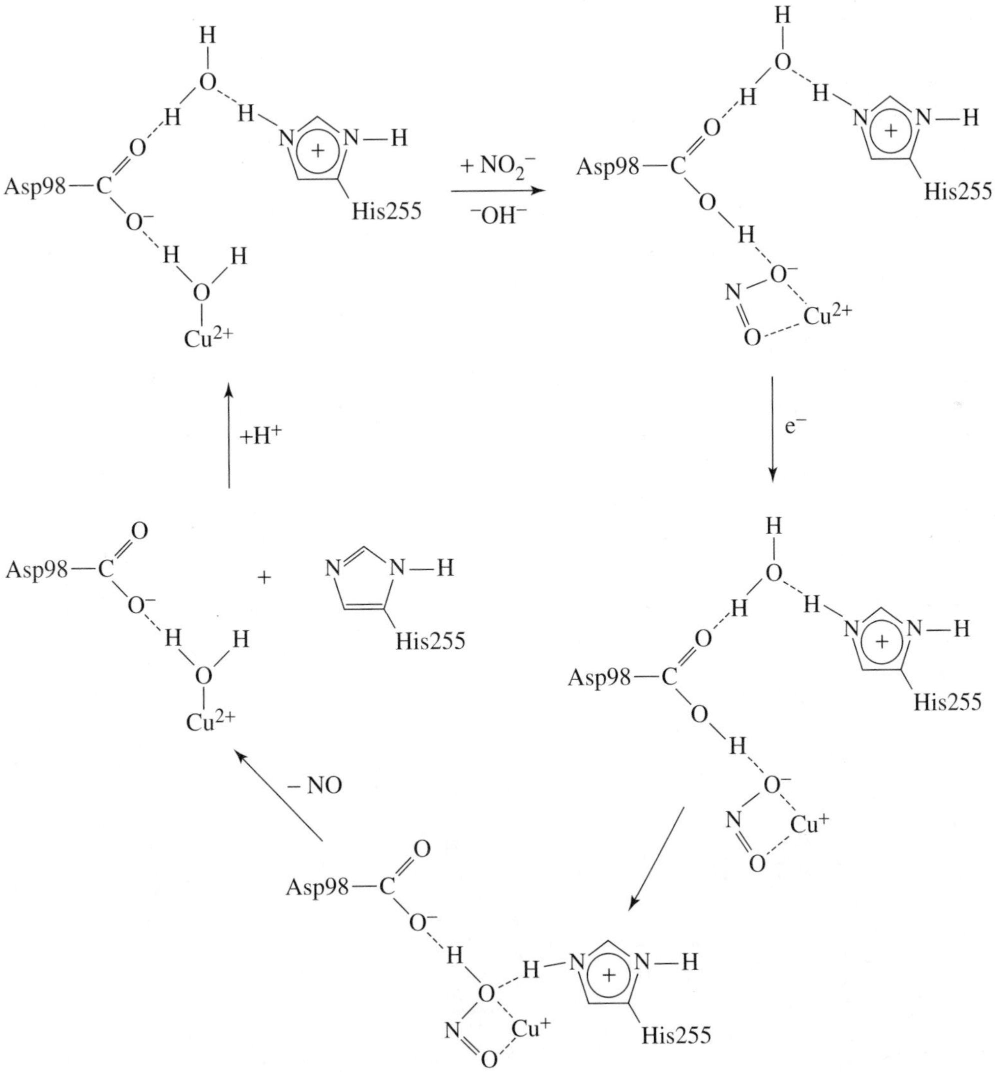

FIGURE 16.14 Proposed Mechanism of Nitrite Reductase.

(Redrawn from M. J. Boulanger, M. Kukimoto, M. Nishiyama, S. Horinouchi, and M. E. P. Murphy, *J. Biol. Chem.*, **2000**, *275*, 23957.)

[37] S. Suzuki, K. Kataoka, and K. Yamaguchi, *Acc. Chem. Res.*, **2000**, 33, 728.

This mechanism is supported by study of mutants in which the aspartate and histidine were modified, which reduced the activity of the enzyme.[38]

Cytochrome cd_1 nitrite reductase from *Paracoceus pantotrophus* has a different mechanism[39] with two identical subunits, each with domains containing a *c*-type cytochrome heme and a d_1-type cytochrome heme. Electrons from external donors enter through the *c* heme; the d_1 heme is the site of nitrite reduction to NO and oxygen reduction to water. One of the puzzles of the mechanism is how the NO can escape from the heme, for which it has a strong affinity. As in the case of the copper nitrite reductases, protonated nitrogen atoms on histidine residues play an important part. As shown in Figure 16.15,

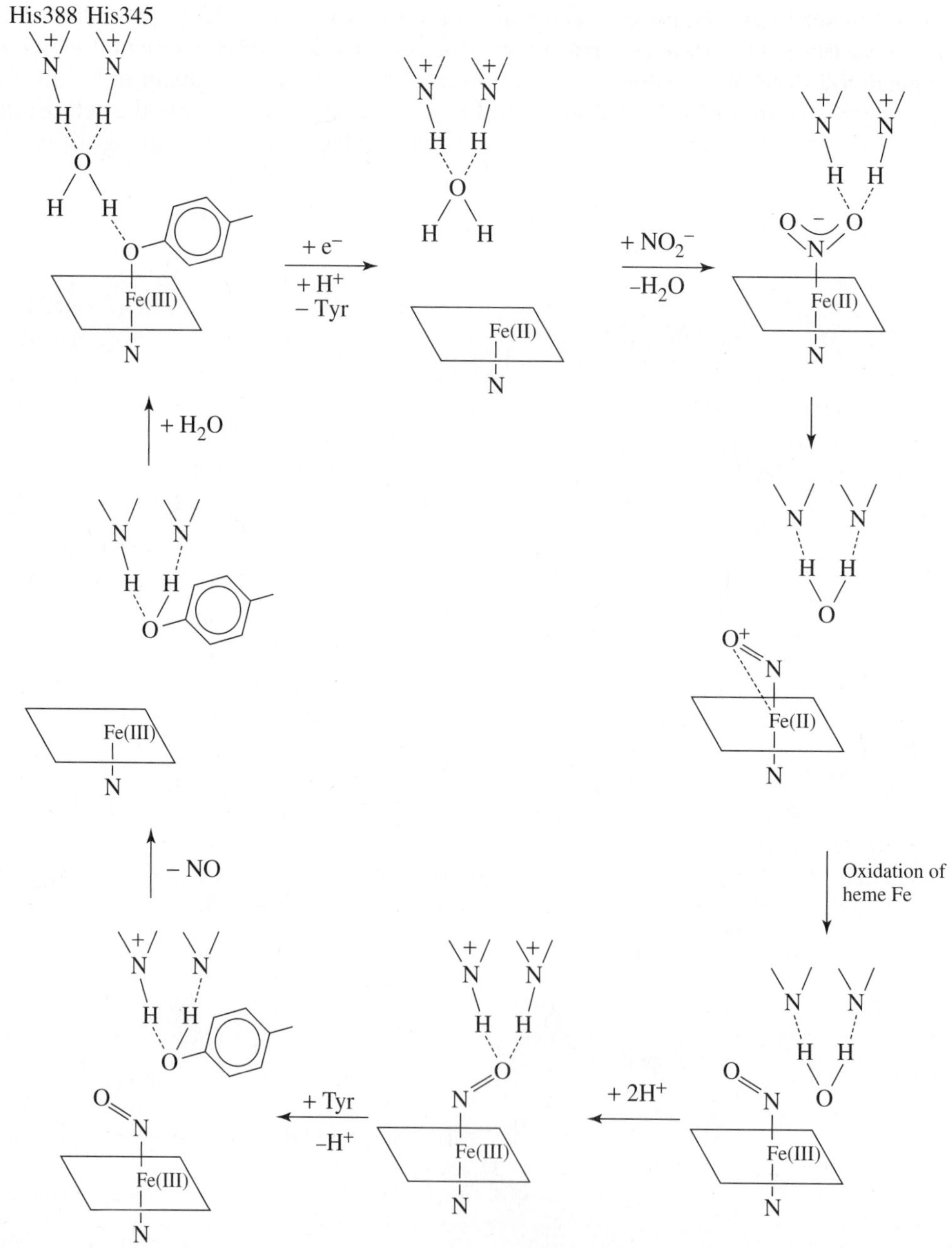

FIGURE 16.15 Possible Routes for Nitrite Reduction by Cytochrome cd_1 Nitrite Reductase.

(Redrawn from G. Ranghino, E. Scorza, T. Sjögren, P A. Williams, M. Ricci, and J. Hajdu, *Biochemistry*, **2000**, *39*, 10958.)

[38] M. J. Boulanger, M. Kukimoto, M. Nishiyama, S. Horinouchi, and M. E. P. Murphy, *J. Biol. Chem.*, **2000**, *275*, 23957.
[39] G. Ranghino, E. Scorza, T. Sjögren, P. A. Williams, M. Ricci, and J. Hajdu, *Biochemistry*, **2000**, *39*, 10958.

crystallographic evidence points to an oxidized enzyme with the heme bound by a histidine—Fe bond on the bottom of the heme ring and a tyrosine oxygen on the top. The oxygen is hydrogen bonded to a water molecule that in turn is hydrogen bonded to two histidines. On reduction, the tyrosine bond and the water molecule are lost, leaving room for nitrite to enter and bond through the nitrogen atom to the iron, its entry perhaps assisted by the positive charges on the protonated histidines. The two histidines apparently participate in removal of one of the nitrite oxygens; hydrogen bonds to the oxygen become stronger as the N—O bond weakens, and the bent Fe—N—O is thought to change from Fe(II)—NO^+ to Fe(III)—NO, which remains bent in contrast to other Fe(III)—NO structures. The two protonated histidine nitrogens may be involved in hydrogen bonding to the NO before it is released, and the cycle can begin again. The tyrosine oxygen can replace the NO on the Fe(III), with the release of over 330 kJ/mol of energy.

The proposed mechanism is as follows:

1. His345 and His388 are protonated in the active site of the unliganded reduced form of cytochrome cd_1.
2. Nitrite binds to this doubly protonated enzyme form.
3. Nitrite reduction starts with proton transfer from the histidines to the bound nitrite ion. This process cleaves off a water molecule from the substrate and leaves an NO^+ cation on the still-reduced d_1 heme.
4. Electron transfer from the d_1 heme to the bound NO^+ cation creates the more stable [Fe(III)–NO] product complex. The orientation and stability of nitric oxide in this complex depend on possible hydrogen bonding interactions with His345 and His388 in the active site.
5. Release of nitric oxide from the d_1 heme can happen in more than one way.

Delocalization is not continuous in the d_1 heme and, as a consequence, the four nitrogen ligands surrounding the iron in the heme plane are not equivalent. Tyr25 or another ligand that can bond temporarily between the histidines may facilitate NO release, influenced by their protonation (or lack thereof).

Still another nitrite reductase, cytochrome *c* NIR, contains five heme groups, only one of which functions as the active site.[40] A combination of calculations and crystallographic studies has suggested a mechanism in which nitrite replaces a water molecule on one side of the Fe(II) heme (a lysine N is on the opposite side), one of the oxygens of NO_2^- is protonated, and the N—O bond is broken with loss of H_2O, leaving a linear Fe(III)—NO species with a low-spin Fe(III). Addition of two electrons and H^+ leads to $Fe^{II}HNO$, which is then reduced to $Fe^{II}H_2NOH$. Yet another electron and another H^+ allow release of H_2O and formation of an $Fe^{III}NH_3$ complex. Release of ammonia and a final electron and water addition complete the cycle. Overall, six electrons and seven hydrogen ions react with the nitrite:

$$6\,e^- + 7\,H^+ + NO_2^- \longrightarrow NH_3 + 2\,H_2O$$

With this enzyme, NO is not released, and NO added to the enzyme is only about 1% reduced.

FeS clusters have been known in nitrogenases and other enzymes for some time and have been studied as less complicated species, partly in the hope of elucidating the nitrogenase mechanism and partly because of the large number of possibilities and their interesting nature. The Fe_4S_4 cubane structure is present in high-potential Fe proteins and ferredoxins; nitrogenase contains a Fe—S—Mo region (shown in Figure 16.13) at the presumed active site. The clusters show a wide variety of structures

---

[40] O. Einsle, A. Messerschmidt, R. Huber, P. M. H. Kroneck, and F. Neese, *J. Am. Chem. Soc.*, **2002**, *124*, 11737.

and reactions, which makes them difficult to categorize. A review describes a large number of abiological iron-sulfur clusters and their reactions.[41]

16.5 NITRIC OXIDE

The importance of NO in biochemistry has only been recognized since the middle of the 1980s, but it was named Molecule of the Year in 1992 by *Science*.[42] Before that time, it was known primarily as a very reactive gas that is formed during combustion and reacts with oxygen in the air to form NO_2. These two gases, together with tiny amounts of other oxides of nitrogen, are known as NO_x in environmental chemistry, where they are the starting compounds for many reactions. It is now known that another large set of reactions is possible in the body, and the effects of NO are still being discovered.[43] For example, overproduction of NO is linked to immune-type diabetes, inflammatory bowel disease, rheumatoid arthritis, carcinogenesis, septic shock, multiple sclerosis, transplant rejection, and stroke. Insufficient NO production is linked to hypertension, impotence, arteriosclerosis, and susceptibility to infection.[44]

NO is synthesized in the body by a number of enzymes; some produce small amounts for nerve transmission and blood flow regulation, and some produce large amounts for defense against tumor cells. When large amounts are produced, NO can also have negative effects, such as large blood pressure drops and destruction of tissue, leading to inflammatory disease and degeneration of nerve and brain tissue. The structure of the active site of one of these enzymes, inducible nitric oxide synthase oxygenase, has been determined.[45] It contains a heme group in a large pocket of the protein, with one side of the iron atom bound to a cysteine sulfur atom and the other side available for substrate binding. It functions by oxidizing arginine in what is believed to be the following two-step reaction:

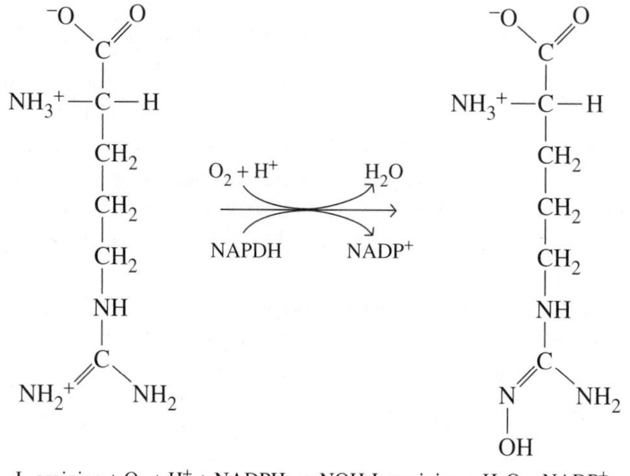

$$\text{L-arginine} + O_2 + H^+ + \text{NADPH} \rightarrow \text{NOH-L-arginine} + H_2O + \text{NADP}^+$$

[41] H. Ogino, S. Inomata, and H. Tobita, *Chem. Rev.*, **1998**, *98*, 2093.

[42] D. E. Koshland, Jr., *Science*, **1992**, *258*, 1861.

[43] P. J. Feldman, O. W. Griffith, and D. J. Stuehr, *Chem. Eng. News*, Dec. 20, 1993, p. 26.

[44] S. Moncada and A. Higgs, *N. Engl. J. Med.*, **1993**, *329*, 2002; C. Nathan and Q. Xie, *Cell*, **1994**, *78*, 915; H. H. Schmidt and U. Walter, *Cell*, **1994**, *78*, 919; O. W. Griffith and D. J. Stuehr, *Ann. Rev. Physiol.*, **1995**, *57*, 707; O. W. Griffith and C. Szabo, *Biochem. Pharmacol.*, **1996**, *51*, 383.

[45] B. R. Crane, A. S. Arvai, R. Gachhui, C. Wu, D. K. Ghosh, E. D. Getzoff, D. J. Stuehr, and J. A. Tainer, *Science*, **1997**, *278*, 425.

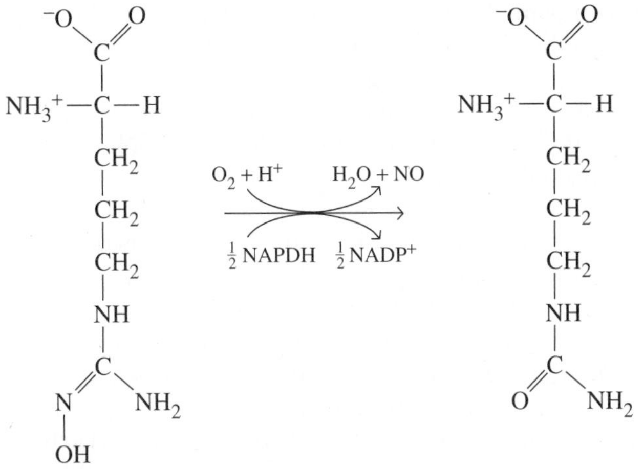

$$\text{NOH-L-arginine} + O_2 + \tfrac{1}{2}(\text{NADPH} + H^+) \rightarrow \text{L-citrulline} + \text{NO} + H_2O + \tfrac{1}{2}\,\text{NADP}^+$$

The energy for these reactions comes from the oxidation of nicotine-adenine dinucleotide phosphate (NADPH) to $NADP^+$ and from the conversion of molecular oxygen to water.

Synthesis of this enzyme is triggered by external stimuli, such as cytokines, released by cancer cells. Once synthesized, the enzyme produces large quantities of NO, which then diffuses into the tumor cells, disrupting DNA synthesis and inhibiting cell growth. The other NO synthases are present at all times but are activated in a sequence of steps dependent on Ca^{2+} concentration. An activated neuron releases a chemical messenger that opens calcium channels in the next neuron. As Ca^{2+} enters the nerve cell, it binds with calmodulin and the NO synthase to activate it. The reactions described earlier for formation of NO take place, and the NO then activates another enzyme, guanylyl cyclase. From this point on, the effects are uncertain but may include diffusion back to the first cell and reinforcement of the stimulus. One of the end results seems to be relaxation of smooth muscle related to the effect seen in blood vessels.

In blood vessels, a similar NO synthase is also activated by Ca^{2+} and calmodulin binding. Increased Ca^{2+} concentration in the endothelial cells of the blood vessels is controlled by calcium channels that can be opened in response to the action of a number of hormones and drugs or by increased pressure in the blood vessel. Again, this activates the enzyme, and the NO formed diffuses into the next layer of smooth muscle cells, where it activates guanylyl cyclase to form cyclic guanosine monophosphate (GMP). This compound, in turn, causes a decrease in free Ca^{2+}. Because Ca^{2+} is required for muscle contraction, the net result is muscle relaxation, dilation of the blood vessels, and lowering of the blood pressure. A similar, nonenzymatic effect can be achieved by nitroglycerin, a common heart medicine. It releases NO directly and dilates the blood vessels, thereby increasing blood flow to the heart and other parts of the body. Maintenance of proper blood pressure appears to require continual synthesis of NO at low levels, because the lifetime of NO in the blood or in cells is very short; the half-life is a few seconds, depending on the surroundings. NO can also diffuse into the blood, where it decreases clotting ability. In red blood cells, NO is rapidly converted into nitrate by reaction with oxyhemoglobin, in which the Fe(II) is simultaneously converted to the inactive Fe(III) form, or methemoglobin. Other enzyme reactions reduce the Fe(III) back to Fe(II) and restore the activity.

In a different organism, the effect of pH on NO bound to a heme group in the protein nitrophorin 1 helps the bloodsucking insect *Rhodnius prolixus* obtain a meal.[46] In the saliva of the insect, the pH is about 5 and the complex is stable. When the complex is injected with the saliva into the blood of a victim, the pH rises to about 7, and the NO is released. The vasodilator and anticoagulant action of the NO make it easier for the insect to draw blood from the victim.

The chemistry of transition-metal nitrosyls has been reviewed,[47] with spectra of many types used to study the electronic structure. Bonding, as described in Chapter 13, can be thought of as a linear complex of NO^+, isoelectronic with CO and with NO stretching frequencies of 1700 to 2000 cm^{-1}, or a bent complex of NO^-, isoelectronic with O_2 and with NO stretching frequencies of 1500 to 1700 cm^{-1}. The number of electrons on the metal ion and the influence of the other ligands on the metal provide for changes from one to the other during reactions.

NO has a half-life on the order of seconds and is converted to many other products, including NO^+, NO^-, and $ONOO^-$, which rapidly decomposes to $\cdot OH + NO_2$ or isomerizes to $NO_3^- + H^+$ after protonation. Each of these undergoes further reactions, with $\cdot OH$ and $ONOO^-$ in particular causing many reactions with adverse effects.

16.6 INORGANIC MEDICINAL COMPOUNDS

Historically, a number of metallic compounds have been used in medicine, including arsenic compounds for the treatment of syphilis and mercury compounds as antiseptics and diuretics. The general toxicity of these compounds has prompted their replacement, but others have been developed for other diseases: lithium has activity in the brain and is used to treat hyperactivity; gold compounds are used in arthritis treatment; antimony compounds are used for the treatment of schistosomiasis; barium sulfate is used in gastrointestinal X-rays as an imaging agent. Although barium is toxic, the extremely low solubility of the sulfate prevents negative effects. There are other examples in ordinary use: antacids, fluoride as a tooth decay preventative, and other drugs using copper, zinc, and tin. We will describe only three groups of these compounds: the anticancer platinum complexes, gold compounds used in arthritis treatment, and vanadium compounds used in diabetes and cancer treatment.

16.6.1 Cisplatin and Related Complexes

One compound that is currently being used for the treatment of certain cancers is *cis*-diamminedichloroplatinum(II), or cisplatin. This compound shares the common action of chemotherapeutic agents by preventing cell growth and proliferation. It also shares the common trait of affecting normal cells as well as cancerous cells but of having a larger effect on the cancerous cells because of their rapid growth rate.

H_3N Cl
Pt
H_3N Cl

Cisplatin

[46] J. M. C. Ribeiro, J. M. Hazzare, R. H. Suxxenzveig, D. E. Champagne, and F. A. Walker, *Science*, **1993**, *260*, 539.
[47] B. L. Westcott and J. H. Enemark, "Transition Metal Nitrosyls," in E. I. Solomon and A. B. P. Lever, eds., *Inorganic Electronic Structure and Spectroscopy*, John Wiley & Sons, New York, 1999, pp. 403–450.

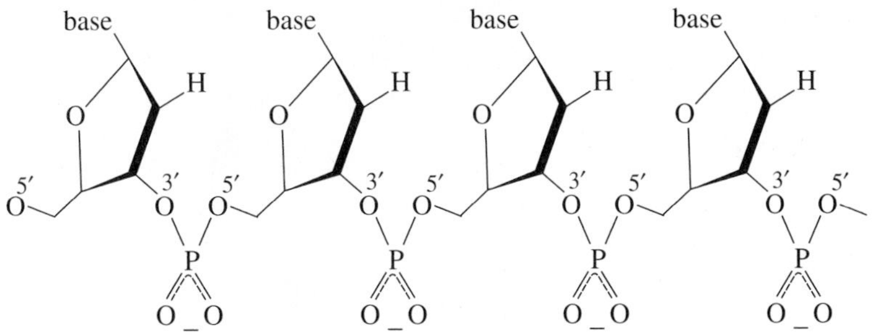

FIGURE 16.16 DNA Backbone Structure. The bases are cytosine, guanine, thymine, and adenine (CGTA).

Cisplatin's effect on cell growth was discovered by B. Rosenberg,[48] when *E. coli* bacteria placed in an electric field stopped dividing and grew into long filaments, similar to their action when treated with antitumor agents. It was found that the ammonium chloride buffer and the platinum electrode were forming compounds, including cisplatin. Cisplatin acts on the deoxyribonucleic acid (DNA) of the cells, disturbing the usual helical structure and thus preventing duplication.

Deoxyribonucleic acids are chains of 5-membered deoxyribose sugar rings connected by phosphate links between the 3′ and 5′ oxygen atoms, with each sugar connected to one of four bases (cytosine, guanine, thymine, and adenine, abbreviated C, G, T, and A) shown in Figures 16.16 and 16.17.

DNA usually adopts the double-chain twisted ladder structure, the double helix, shown in Figure 16.18, with complementary base sequences allowing hydrogen bonding between the two chains, as in the figure shown here, with the 3′ end of one chain opposite the 5′ end of the other:

```
3′ A C G T T G G C C A T A 5′
   | | | | | | | | | | | |
5′ T G C A A C C G G T A T 3′
```

In this structure, the planar rings of the bases are stacked in parallel planes on the inside of the helix with the negative phosphate groups on the outside. Ribonucleic acids (RNAs) have a similar backbone, but the sugars have OH instead of H in the 2′ positions,

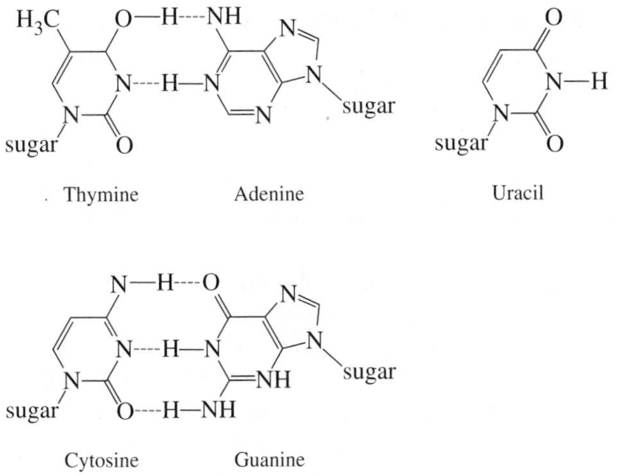

Thymine Adenine Uracil

Cytosine Guanine

FIGURE 16.17 Purine and Pyrimidine Bases of DNA and RNA. Adenine and guanine are purines, and cytosine, thymine, and uracil are pyrimidines. The hydrogen-bonding combinations of complementary base pairs are shown.

[48] *Chem. Eng. News*, June 21, 1999, p. 9.

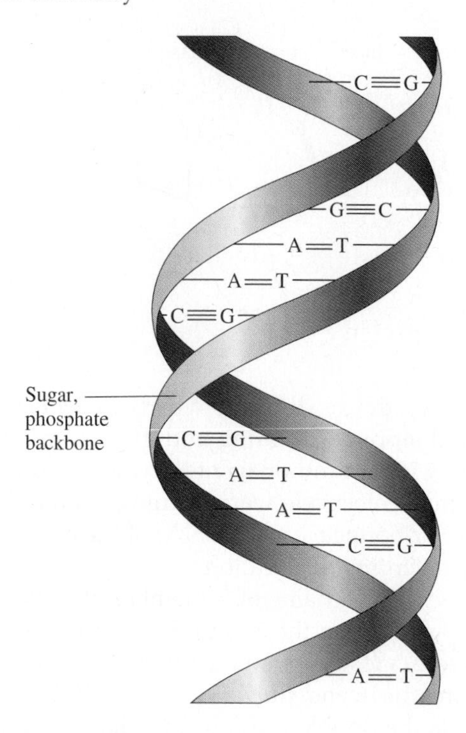

FIGURE 16.18 The DNA Double Helix.

and uracil replaces thymine. RNAs have more varied structures and generally do not form the double helix that is common in DNAs.

DNA carries the genetic code that dictates the amino acid sequence during the synthesis of proteins, which in turn dictates the form of life and the details of structure and action. RNA was until recently thought to be primarily a messenger, carrying information from the DNA to the synthetic site. More recently, however, RNA has been found to have enzymatic activity of its own.

During growth, the DNA molecule "unzips," and new partner molecules are formed on each of the chains, resulting in two molecules where one existed before. Many cancer treatments depend on interrupting this process to prevent the rampant growth characteristic of cancer.

Cisplatin hydrolyzes to the diaqua complex, which then reacts with the nitrogen atoms of guanine in the DNA to form a crosslink between adjacent guanine bases, usually within the same strand or occasionally between strands. The result is a kink in the DNA helix with angles up to 34°. This change in shape is enough to interfere with the self-replication of the DNA and slows growth of the cancer. In fact, this treatment actually results in shrinkage of cancers, although the mechanism for this is not yet clear. The structure of cisplatin bound to a short segment of double-stranded DNA is shown in Figure 16.19.

The structure of a protein believed to be involved in anticancer activity when combined with a cisplatin-modified DNA complex[49] shows a larger kink (61°) in the DNA and intercalation of a phenylalanine ring from the protein into the resulting notch, where the phenylalanine ring is stacked between two base layers. Binding such as this might prevent removal of the cisplatin and other repair reactions of the DNA.

Cisplatin must be administered intravenously rather than orally, and it is effective against only a limited number of cancers; it is highly toxic, especially to the kidneys, and has a variety of other adverse side effects. The high toxicity of cisplatin has been attributed in part to the lability of its chloro ligands; when they are lost and replaced by aqua

[49] U.-M. Ohndorf, M. A. Rould, Q. He, C. O. Pabo, and S. J. Lippard, *Nature*, **1999**, *399*, 708.

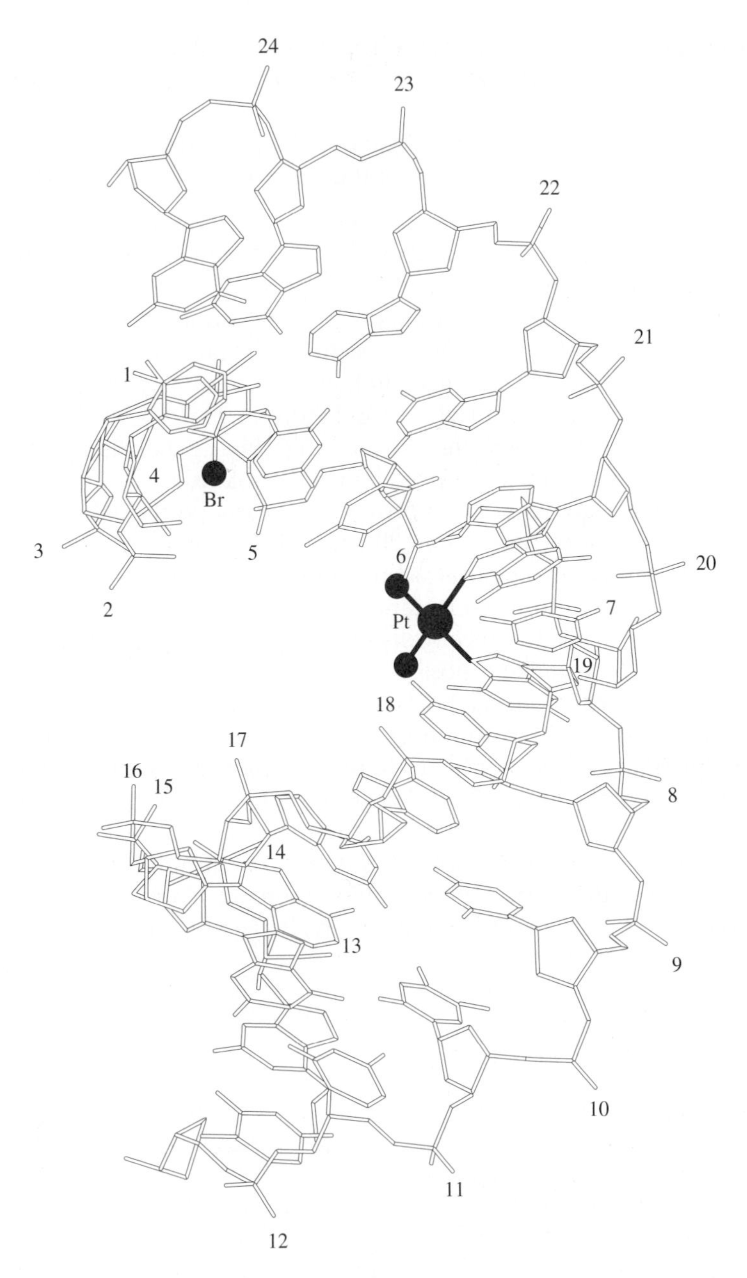

FIGURE 16.19
Structure of a Cisplatin-DNA Complex. Shown is a 26° bend imposed by the GG bonding.

(Reproduced with permission from P. M. Takahara, C. A. Frederick, and S. J. Lippard, *J. Am. Chem. Soc.*, **1996**, *118*, 12309.)

ligands, the remaining cationic platinum complex may bind to proteins in blood plasma before reaching its target. Intensive efforts have been devoted to developing other platinum-based reagents that do not have these drawbacks. In addition, some tumors have been found to be inherently cisplatin resistant or to develop such resistance.

Various structural requirements for effective platinum-containing antitumor compounds have been suggested, including:[50]

1. A pair of hard *cis*-anionic ligands (chloride or ligands with oxygen donor atoms) that are subject to substitution by DNA nitrogen bases

---

[50] D. B. Brown, A. R. Khokhar, M. P. Hacker, J. J. MacCormack, and R. A. Newman, "Synthesis and Biological Studies of a New Class of Antitumor Platinum Complexes," in S. J. Lippard, ed., *Platinum, Gold, and Other Metal Chemotherapeutic Agents*, American Chemical Society, Washington, DC, 1983, pp. 265–277; T. W. Hambley, *Coord. Chem. Rev.*, **1997**, *166*, 181 and references cited therein.

2. Water solubility and ability to pass through cell membranes
3. Unreactive ligands on the other sites that are primary or secondary amines in *cis* geometry
4. At least one proton on each amine ligand and few alkyl substituents on the amine; the fewer the substituents, the greater the activity

In addition, the activity and toxicity of the antitumor compound are affected by the ease with which the leaving groups (the ligands in item 1) are lost. Such ligands should be only moderately easy to remove.

Although a broad range of potential platinum-based antitumor compounds have been studied, only a small number have reached the stage of clinical trials, and few have been approved for use. Examples that have either been approved for clinical use or are in advanced clinical trials are shown in Figure 16.20.[51] These all have ammine or amine ligands in *cis* positions and, *trans* to these ligands, ligands that can dissociate (or partly dissociate) as anions, to be replaced by water molecules.

The first of these agents to be approved was carboplatin, which has now been used in cancer treatment for more than 20 years. The bidentate cyclobutadienedicarboxylate ligand forms a more stable attachment to platinum than chloro ligands, allowing more time for the drug to reach its target without affecting healthy cells. Although carboplatin is also less toxic to cancer cells than cisplatin, its lower nephrotoxicity and overall milder side effects make it possible to administer much higher doses of carboplatin. Carboplatin is also more easily used in combination with other drugs in cancer treatment. Like cisplatin, carboplatin must be administered intravenously.

The first platinum antitumor agent to reach clinical trials for oral administration was satraplatin, also called *JM 216*, whose acetato ligands increase its solubility. Satraplatin has been found to be effective against some tumor cell lines that have acquired resistance to cisplatin. Picoplatin, the most structurally similar of these molecules to cisplatin, was developed specifically to target cisplatin-resistant tumors. Like satraplatin, picoplatin can be administered orally; its ligands protect it from reaction in

FIGURE 16.20 Platinum Antitumor Agents.

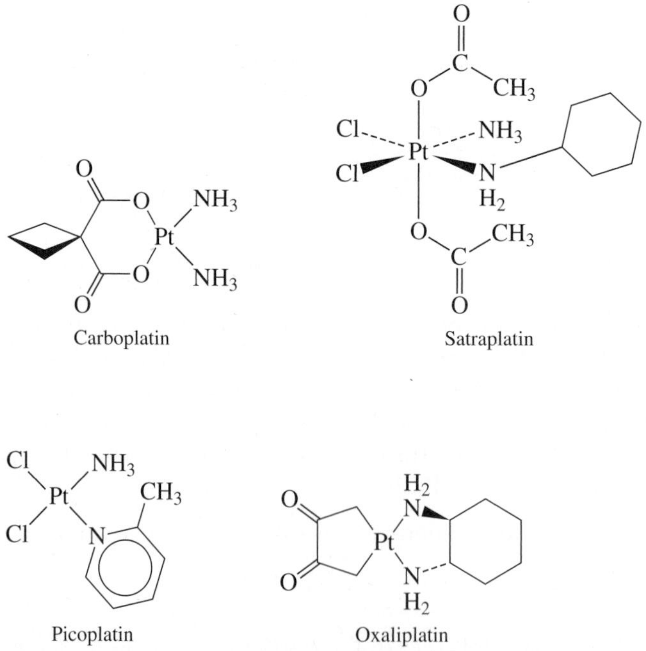

Carboplatin

Satraplatin

Picoplatin

Oxaliplatin

[51] S. P. Fricker, *Dalton Trans.*, **2007**, 4903; R. A. Alderden, M. D. Hall, and T. W. Hambley, *J. Chem. Ed.*, **2006**, *83*, 728.

the digestive system and allow it to be absorbed into the bloodstream. Oxaliplatin, approved for use in 2002, is also effective against some cisplatin-resistant tumors; it has become a leading anticancer drug. The mechanisms of cellular uptake and processing of cisplatin and related drugs have recently been reviewed.[52]

Coordination complexes of a variety of other metals, most notably ruthenium, have also been explored as anticancer agents.[53] In addition, new methods of delivering platinum-based drugs have recently been reported. Two notable examples are to adapt apoferritin, known to be internalized in some tumor tissues, to encapsulate cisplatin and carboplatin,[54] and to use a single-walled carbon nanotube bioconjugate to target cisplatin delivery to cancer cells.[55]

16.6.2 Auranofin and Arthritis Treatment

Gold in many forms has been used medicinally for hundreds of years with relatively few proven benefits and many examples of toxicity. More recently, gold complexes of thiols, Figure 16.21(a) and (b), have been used for treatment of arthritis but have the major disadvantage that they must be administered by injection into the site of inflammation.

More recently the compound auranofin, (2,3,4,6-tetra-*O*-acetyl-1-thio-β-glucopyranosato-S-)(triethylphosphine)gold(I), Figure 16.21(c), has been developed. It has the advantage that it can be administered orally and still be effective.

The mechanism of action of these compounds is still not known. One possibility is that they act through the formation of gold-sulfur complexes, which can inhibit the formation of disulfide bonds. Because much of the biochemistry of arthritis is still uncertain, the design of drugs for specific action is difficult.

16.6.3 Vanadium Complexes in Medicine

Several vanadium (IV) compounds have been found to have insulin-like activity. However, their toxicity prevents medical use. (Dipicolinato)oxovanadate(V), $[VO_2\ dipic]^-$, is effective as an oral agent in animals,[56] and has less toxicity. The acid–base properties of the compound make it likely that it is absorbed in the acidic environment of the stomach or the first part of the small intestine; it protonates at pH ~1. Several V(IV) compounds, shown in Figure 16.22, also have anticancer activity.[57]

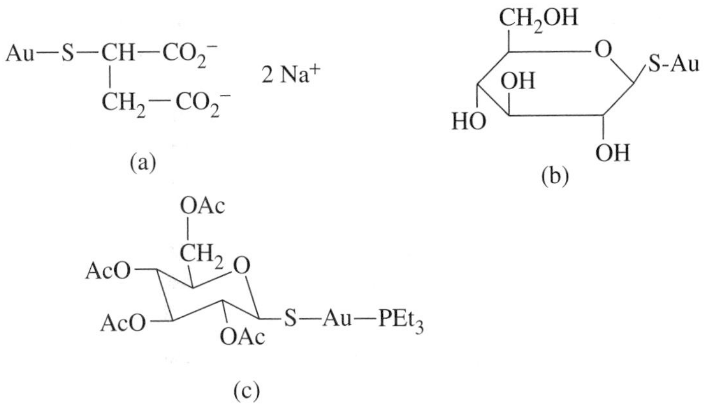

FIGURE 16.21 Gold Antiarthritic Drugs. (a) Sodium aurothiomalate. (b) Aurothioglucose. (c) Auranofin.

[52] F. Arnesano and G. Natile, *Coord. Chem. Rev.*, **2009**, *253*, 2070.
[53] P. C. A. Bruijnincx and P. J. Sadler, *Curr. Opin. Chem. Bio.*, **2008**, *12*, 197.
[54] Z. Yang, X. Wang, H. Diao, J. Zhang, H. Li, H. Sun, and Z. Guo, *Chem. Commun.*, **2007**, 3453.
[55] A. A. Bhirde, V. Patel, J. Gavard, G. Zhang, A. A. Sousa, A. Masedunskas, R. D. Leapman, R. Weigert, J. S. Gutkind, and J. F. Rusling, *ACS Nano*, **2009**, *3*, 307.
[56] D. C. Crans, L. Yang, T. Jakusch, and T. Kiss, *Inorg. Chem.*, **2000**, *39*, 4409 and references therein.
[57] F. M. Uckun, Y. Dong, and P. Gosh, U.S. Patent 6,245,808 (2002).

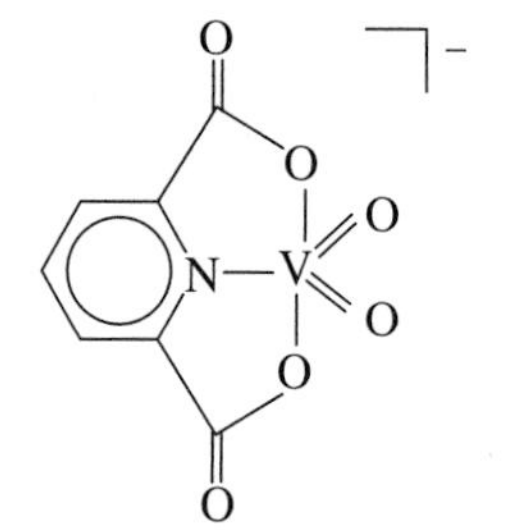

(Dipicolinato)oxovanadate(V)

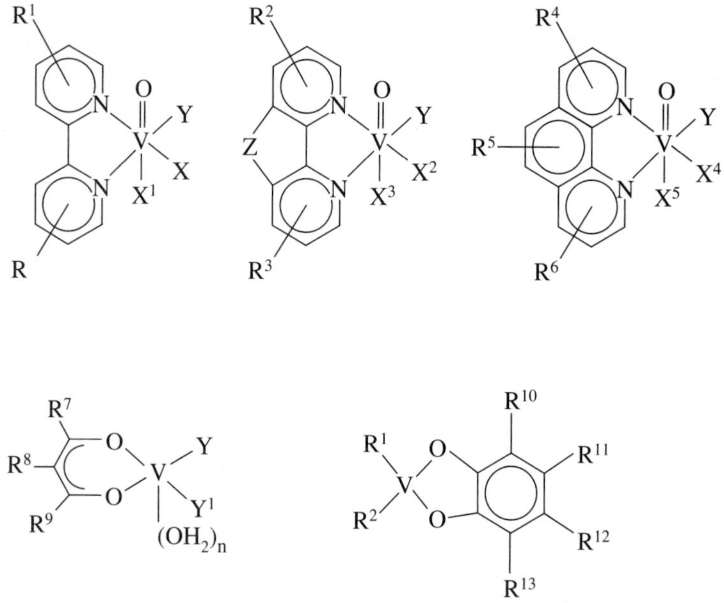

FIGURE 16.22 (Dipicolinato)oxovanadate(V) and Anticancer V(IV) Compounds. R–R^9 are H, alkyl, alkoxy, halogenated alkyl, alkanoyloxy, or NO_2; R^{10}–R^{13} are H, halogen, or C_1–C_6 alkyl; and X^4, X^5, Y, and Y^1 are monodentate or bidentate ligands.

16.7 STUDY OF DNA USING INORGANIC AGENTS

DNA polymerases work by stitching together two nucleotides, one at the 3′ end of a DNA chain and the other a deoxynucleic acid triphosphate. The two are held in position near each other by hydrogen bonding to the template DNA chain; these are joined by reaction of the first phosphate of the triphosphate with the OH of the saccharide ring of the other base, releasing diphosphate. Mg^{2+} ions separated by about 390 pm are bound on each side of the phosphorus in the transition state, as in Figure 16.23.

Information on the three-dimensional structure of RNA can be obtained by tethering cleavage agents to known positions and then studying the fragments produced by the cleavage reactions. This has been used to study *Escherichia coli* ribosomal RNA.[58] Fe(II) as a cleavage agent was attached to the 5′ terminus of the 3′ fragment and to two other specific sites as 1-(*p*-bromoacetamidobenzyl)-EDTA Fe(II). Addition of H_2O_2 and ascorbic acid generated hydroxyl radicals by the following reaction:

$$Fe^{2+} + H_2O_2 \longrightarrow Fe^{3+} + OH^- + \cdot OH$$

These free radicals can diffuse to nearby sites and react with the saccharide ring by abstracting H from the 1′ position. Several reactions follow, with the net result that the RNA chain is cleaved, with formation of a small organic molecule as the result of destruction of the saccharide ring. Some of the cleavage points are near the tethering point in the same chain; others are distant in terms of chain position but close in terms of three-dimensional folding. As a result of these experiments, several parts of the chain

[58] L. F. Newcomb and H. F. Noller, *Biochemistry*, **1999**, *38*, 945.

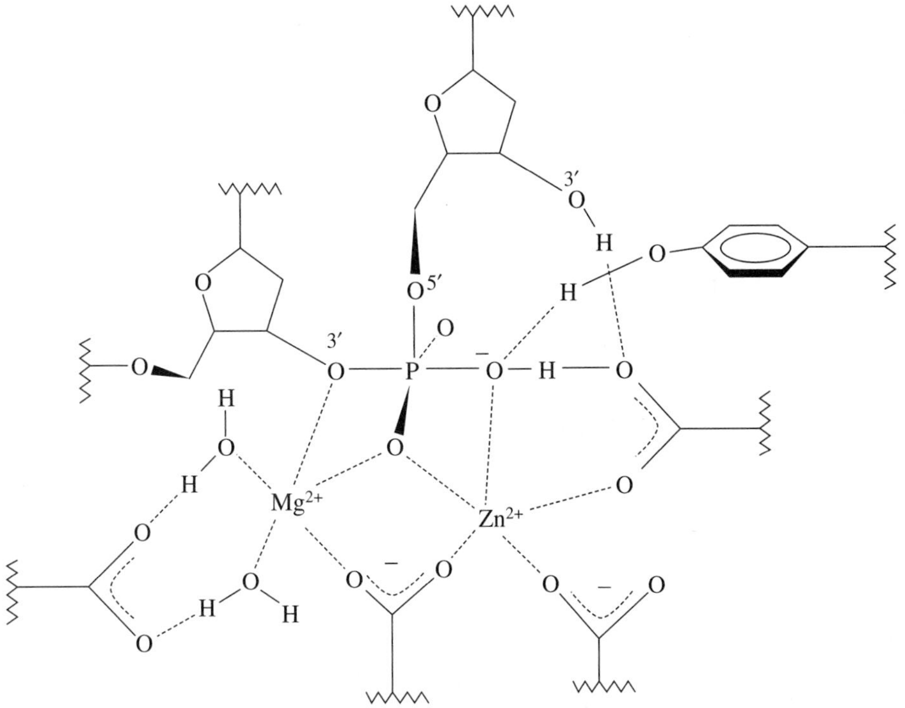

FIGURE 16.23 DNA Polymerase Transition State.
(Redrawn from T. A. Steitz and J. A. Steitz, *Proc. Natl. Acad. Sci. U.S.A.*, **1993**, *90*, 6498.)

are now known to be near each other in the folded structure. Similar studies offer the promise of further elucidation of the complete structure. Methidium-EDTA-Fe(II) is another tethered cleavage reagent; it generates superoxide from O_2 and hydroxyl radical from peroxide and is reduced back to Fe(II) by dithiothreitol.[59]

Hydroxyl radicals can also be generated by the reaction

$$[Cu(phen)_2]^+ + H_2O_2 \longrightarrow [Cu(phen)_2]^{2+} + OH^- + \cdot OH$$

after which thiols or ascorbic acid can reduce Cu(II) back to Cu(I). $[Cu(phen)_2]^+$ intercalates (fitting between the parallel rings of the bases) in the minor groove of right-handed double-helix DNA.[60] Intercalation unwinds the DNA by about 11°, binding to two base pairs. The reaction shows only a slight GC preference. Tethered Cu-phen complexes are also possible; these complexes and their cleavage reactions have been reviewed.[61]

$[Ru(en)_2phi]^{3+}$ (phi = phenanthrenequinone diimine) intercalates with B-DNA. The Δ isomer prefers 5′-GC sites in the major groove, and the Λ isomer is site neutral.[62] On photoactivation, the complex abstracts H3′ from the deoxyribose ring, and the chain is cleaved, leaving 3′ and 5′ phosphates, propenoic acid, and 3′-phosphoglycaldehyde consistent with a reaction between O_2 and the 3′ carbon, as shown in Figure 16.24.

[59] R. P. Hertzberg and P. B. Dervan, *J. Am. Chem. Soc.*, **1982**, *104*, 313.
[60] L. E. Marshall, D. R. Graham, K. A. Reich, and D. S. Sigman, *Biochemistry*, **1981**, *20*, 244; C. Yoon, M. D. Kubawara, A. Spassky, and D. S. Sigman, *Biochemistry*, **1990**, *29*, 2116.
[61] D. S. Sigman, T. C. Bruice, A. Mazunder, C. L. Sutton, *Acc. Chem. Res.*, **1993**, *26*, 98.
[62] T. P. Schields and J. K. Barton, *Biochemistry*, **1995**, *34*, 15037.

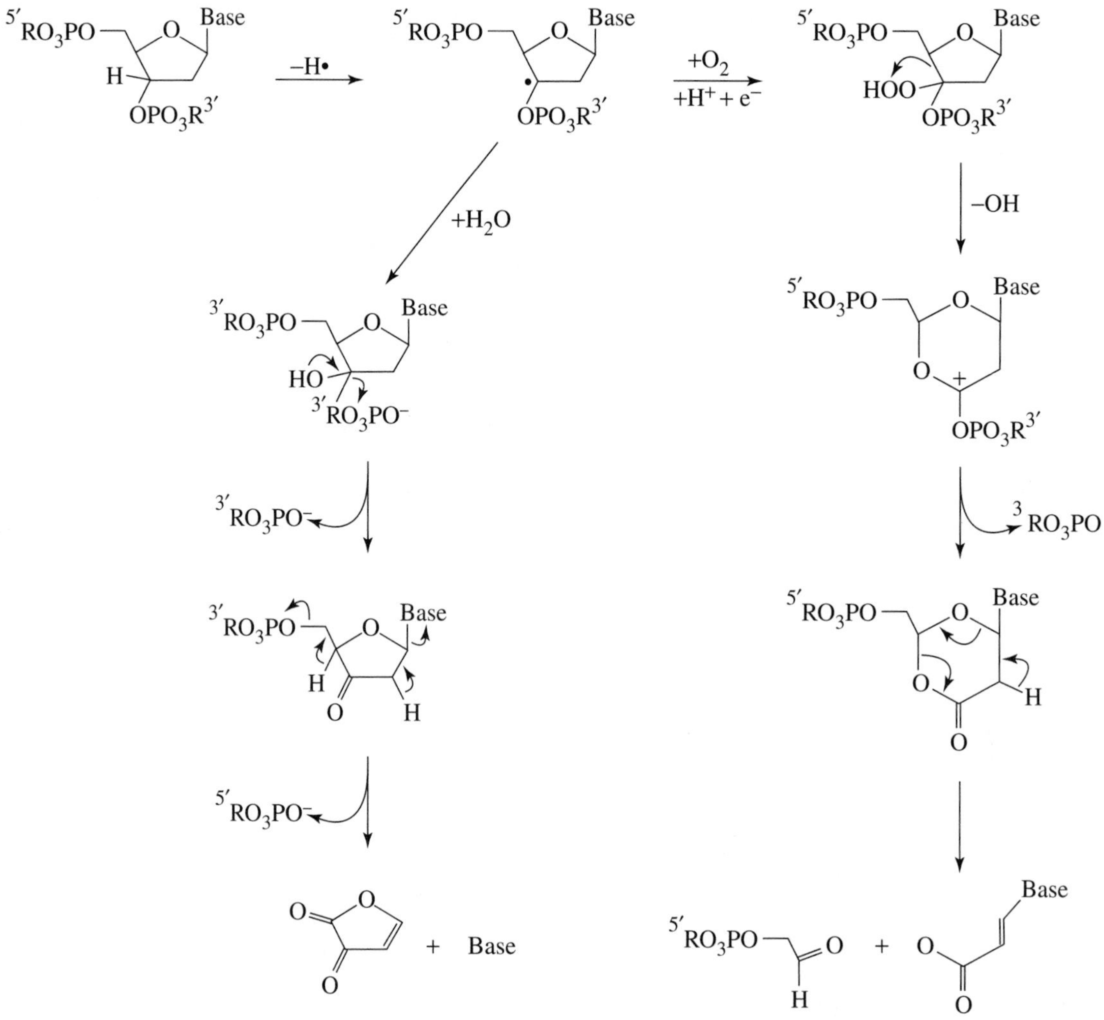

FIGURE 16.24 Proposed Mechanisms of DNA Cleavage Initiated by $[Ru(en)_2(phi)]^{3+}$.
(Redrawn from A. Sitlani, E. C. Long, A. M. Pyle, and J. K. Barton, *J. Am. Chem. Soc.*, **1992**, *114*, 2303.

16.8 ENVIRONMENTAL CHEMISTRY

16.8.1 Metals

Mercury and lead are two of the most prominent metallic environmental contaminants today. Although there have been continued efforts to prevent distribution of these metals and to clean up sources of contamination, they are still serious problems. Other metals and semimetals, such as arsenic, also cause significant health effects. Some of them are described here.

MERCURY Because mercury has a significant vapor pressure, the pure metal can be as serious a problem as its compounds. Although the problem is usually less severe in laboratories today, mercury contamination and poisoning have been a problem in chemistry and physics laboratories for many years. Large amounts of the liquid have been used in manometers, Toeppler pumps, and mercury diffusion pumps on vacuum lines. Because liquid mercury breaks into tiny drops, cleanup is extremely difficult, and contamination can remain even after strenuous efforts to remove it. As a result, a low level of mercury vapor is present in many laboratories and can result in toxic reactions. Mercury interferes with nerve action, causing both physical and psychological symptoms. Whether the behavior of the Mad Hatter in Lewis Carroll's *Alice in Wonderland* had an origin in fact is uncertain, but mercury compounds were used in felt making, and some hatters were victims of mercury poisoning as a result.

Several industrial processes use mercury in large amounts, and the resulting potential for spills and damage to the environment is great. One of the largest is the chloralkali industry, in which mercury is used as an electrode for the electrolysis of brine to form chlorine gas and sodium hydroxide:

$$2\,H_2O + 2\,e^- \longrightarrow H_2 + 2\,OH^-$$

and

$$2\,Cl^- \longrightarrow Cl_2 + 2\,e^-$$

In one tragic incident, an entire community on Minamata Bay in Japan was poisoned, resulting in extremely serious birth defects, very painful reactions, mental disorders, and many deaths. Only after lengthy research was the cause determined to be mercury compounds discarded into a river by a plastics factory. Whether it was inorganic salts or methylmercury compounds seems uncertain, but the contamination was immense, and methylmercury compounds were found in the silt and in animals and humans. The methylmercury was readily taken up by the organisms living in the bay. Because the people of the community depended on fish and other seafood from the bay for much of their diet, the entire community was poisoned.

This incident showed the concentrating effect of the food chain and the need for extreme caution in predicting the outcome of dumping any material into the environment. Even though the concentration of methylmercury was low, it was readily taken up by the plants and microorganisms in the water. As these organisms were eaten by larger ones, each organism in the food chain retained the mercury, and the concentration of mercury in these predators increased, leading to harmful concentrations in the larger fish and in other organisms eaten by people living nearby.

During the research on mercury reactions in the environment, it was also discovered that insoluble metallic mercury can be converted to soluble methylmercury by bacterial action involving methylcobalamin. Earlier, it had been thought that elemental mercury was unreactive in lakes and rivers; now it is known to be dangerous. As a result, there are now many more toxic metal sources than had once been recognized. Historically, large amounts of metallic mercury were discharged into the Great Lakes and other bodies of water in the belief that it was harmless.[63] Cleanup of these sites seems impossible, so the problem will remain with us for the foreseeable future.

Although concern about mercury contamination is now more visible, and industries have reduced its release into the environment, the increasing use of mercury in small batteries and other products results in a greater distribution of mercury into the environment as a whole. As a result, the problem is changing from one of a few large sources of contamination to many small ones; and techniques for dealing with the problem must change as well. Concerns are being expressed about heavy metal contamination of the atmosphere by incinerators burning municipal garbage and trash, and it is likely that removal of these materials from the trash before burning, or scrubbing of the flue gases to remove the volatile products, will be needed. Another source of atmospheric mercury and other elements is the burning of coal for electric power. In one plant burning 6×10^8 tons of coal per year, 60 tons of mercury, 12,000 tons of lead, 240 tons of cadmium, 3,000 tons of arsenic, 3,000 tons of selenium, 2,400 tons of

---

[63] A. T. Schwartz, D. M. Bance, R. G. Silberman, C. L. Stanitsti, W. J. Stratton, and A. P. Zipp, *Chemistry in Context*, 2nd ed., American Chemical Society, Washington, DC, Wm. C. Brown, Dubuque, IA, 1997. Chapter 7, describes the effects on Onondaga Lake in New York.

antimony, 15,000 tons of vanadium, and 120,000 tons of zinc were released as particulates or gases.[64] It is now believed that the majority of mercury in many lakes comes from the atmosphere.

Dimethylmercury has been found in the gases from landfills in Florida.[65] Landfills may be a source of the methylated mercury species that appear in rain. The dimethylmercury quickly breaks down to monomethylmercury, CH_3Hg^+, a species that is water soluble and is commonly found in fish and other aquatic organisms.

A Canadian study designed to track the cycles of mercury in the atmosphere and lakes has used three stable mercury isotopes—^{198}Hg for wetlands, ^{200}Hg for uplands, and ^{202}Hg for lakes,[66]—that are common in nature (10%, 23%, and 30%, respectively). These isotopes were added to the area over a period of 3 years in amounts similar to those received by the most atmospherically polluted lakes in eastern North America, and their fate is tracked by mass spectroscopy to determine the sources of mercury. The isotopes were applied to the uplands and wetlands by plane during rainstorms, and collectors were placed under the forest canopy to determine the amounts reaching the ground. Preliminary results showed different reactions for new mercury initially, but after a few days or weeks, the compounds and reactions were the same as for the mercury already in the soil and vegetation.

Mercury in the Arctic cycles with the seasons between the atmosphere and snow on the ground.[67] In the spring, as the sun reappears after the winter darkness, mercury levels in the troposphere decline for about 3 months. At the same time, the level of mercury in the snow increases significantly, both as methylmercury and as inorganic compounds of mercury. Later in the year, as the snow melts and the levels in the snow drop, mercury reappears in the troposphere. The elemental mercury in the atmosphere is converted to particulates or reactive species, parallelling a decrease in atmospheric ozone, and is then deposited in the snow. Later in the summer, the mercury levels in the atmosphere increase, probably due to temperature- or sunlight-induced emission of volatile mercury species from the surface.

LEAD Lead is another metal that is widespread in the environment, principally as a result of human activities. Two of the largest sources for environmental lead were paint pigments and leaded gasoline, both now much reduced in importance. White lead [basic lead carbonate, $2\ PbCO_3 \cdot Pb(OH)_2$] was used as a paint pigment for many years, and older buildings still have lead-containing paint, frequently under layers of more modern paint. If children living in these buildings eat paint chips, they are likely to ingest significant amounts of lead. In fact, in some cities, lead poisoning of children is a very common problem.[68] As is the case with mercury, lead can affect nerve action and can cause retardation and other mental problems, as well as cause acute illness. Unfortunately, the only cure is complete removal of the paint, a very time-consuming and expensive process.

Although heavy metal glazes are prohibited in commercial manufacture of ceramics in many countries, there are still reports of lead and other toxic heavy metals showing up in dishes imported from countries without similar controls or in ceramic items made by individuals who do not take the appropriate precautions. Because the glaze seems permanent and impervious to water and ordinary foods, it might seem that such materials would not be a hazard, but acidic solutions can

[64] N. E. Bolton, J. A. Carter, J. F. Emery, C. Feldman, W. Fulkerson, L. D. Hulett, and W. S. Lyon, in S. P. Babu, ed., *Trace Elements in Fuel*, American Chemical Society, Washington, DC, 1975, p. 175.
[65] S. E. Lindberg, *Atmos. Environ.*, **2001**, *35*, 4011.
[66] *Chem. Eng. News*, Sept. 24, 2001, pp. 35–38.
[67] W. H. Schroeder, K. G. Anlauf, L. A. Barrie, J. Y. Lu, A. Steffen, D. R. Schneeberger, and T. Berg, *Nature*, **1998**, *394*, 331.
[68] M. W. Oberle, *Science*, **1969**, *165*, 991; P. Mushak and A. F. Crocetti, *Environ. Res.*, **1989**, *50*, 210.

extract significant amounts of the heavy metals and result in chronic low-level lead poisoning.

Lead in gasoline is being phased out in industrialized countries, but it is a continuing problem in developing countries. Tetraethyl lead, $Pb(C_2H_5)_4$, has been used as an antiknock compound in gasoline for many years. When this compound is present, a low grade of gasoline burns as efficiently in automobile engines as a higher grade without the lead. Unfortunately, the lead from the gasoline has been distributed throughout the environment. Some studies have found increased lead levels in roadside plants and soil, and the population in general has been exposed to higher levels of lead as a result of this use. Laws requiring the use of nonleaded gasoline in newer cars have required other changes in the engines and in the refining of gasoline to compensate. The history of tetraethyllead—its synthesis and development, use in fuels, and decline as its consequenes to public health and the environment became evident—has been described in interesting detail by Seyferth.[69]

CATALYTIC CONVERTERS The use of catalytic converters to reduce the amount of unburned hydrocarbons in exhaust gases is an additional example of the use of metals. Reactions of these unburned hydrocarbons in the atmosphere are described later in the section on photochemical smog. The catalyst currently used is a cordierite or alumina support treated with an Al_2O_3 wash coat containing rare earth oxides and 0.10% to 0.15% Pt, Pd, and/or Rh, which catalyzes the combustion of hydrocarbons in the exhaust gases to carbon dioxide and water. Platinum, palladium, and nickel are among the most reactive, and most widely used, catalytic materials. They are used in many different, specific compounds and physical forms for reactions of surprising specificity in the petroleum and chemical industries.

In another of the many interactions between problems and their solutions, catalysts in catalytic converters are poisoned by lead. For this reason, cars with catalytic converters are required to use only unleaded gasoline. One negative side effect of the use of catalytic converters is an increase in N_2O emission. The converters reduce NO and NO_2 to N_2O, which has less immediate effects but has a greenhouse effect (described later in this chapter).

Still another recently discovered side effect is the deposition of platinum, palladium, and rhodium along roadsides as a result of catalyst breakdown.[70] The amounts are small (maximum was 70 ng/g of Pt), but it has been suggested that the amounts approach those that would make recovery economically feasible, because the material could be easily collected in comparison with usual mining operations.

ARSENIC Efforts to remove toxic materials from industrial sites, homes, and farms have unearthed other problems. For example, during the 1930s, farmers fought grasshopper infestations with bran poisoned with arsenic compounds. Fifty or more years later, burlap bags of arsenic-laced bran were found in barns and storage sheds, where they were potentially serious hazards. Several states have begun programs to locate and remove these poisons for safe disposal, but because there is no way to detoxify a heavy metal, the material will remain toxic forever. The only possible way to alleviate the problem is to seal the material in a toxic waste dump and take every possible means to prevent leaching, or other ways of spreading the material, or to find some other use for the heavy metal compounds that is profitable enough to make reprocessing feasible. So far, such uses have been very rare.

Arsenic is also present in groundwater whenever it percolates through minerals containing arsenic compounds. One area where this is particularly common is

[69] D. Seyferth, *Organometallics*, **2003**, *22*, 2346 and **2003**, *22*, 5154.

[70] J. C. Ely, C. R. Neal, C. F. Kulpa, M. A. Schneegurt, J. A. Seidler, and J. C. Jain, *Environ. Sci. Technol.*, **2002**, *35*, 3816.

Bangladesh. In an attempt to reduce waterborne illness, international agencies have been helping drill wells to provide water uncontaminated by bacteria and other surface contaminants. However, increased incidence of birth defects and other health problems began developing at the same time. It was finally discovered that the problem was the high levels of arsenic in the water, 50 ppm or higher. Drinking water standards in the United States require levels less than 50 ppb; a new standard required reduction in public water supplies to 10 ppb by 2005.[71] There is now evidence that As(III) disrupts endocrine function even at very low concentrations, and it interferes with DNA repair capacity.[72] As further research uncovers the detailed mechanisms of these actions, even more stringent limits may be indicated, in spite of their high costs.

Other heavy metals are also toxic but are fortunately less widespread and are present in smaller amounts. Mine tailings (waste rock remaining after the valuable minerals have been removed) and waste material from processing plants are major sources of such metals. Many major rivers and lakes have sources of metal contamination from industries whose processes were developed and whose facilities were built before control of waste was recognized as a major problem.

RADIOACTIVE WASTE Disposal of radioactive waste is a continuing controversial topic. Some argue that the technical problems have been solved and that only politics remain in the way of efficient permanent storage of such wastes, primarily in the Yucca Mountain site now being constructed in Nevada. Others maintain that the technical problems are far from being solved and that the long half-lives of some of the isotopes will require protection of the disposal sites for hundreds or even thousands of years. At this time, it is impossible to predict the outcome, beyond noting that no location is perfect, either geologically or politically. Reports of contamination of water and land around processing sites have led to even more suspicion of any reported solution, which has made the choices even more difficult. An additional concern is the need to transport the wastes to whatever storage site is selected. Although some argue that the containers and transportation modes, either by truck or rail, have been developed and adequately tested, others argue that public safety demands even more than has been done. One factor that has been neglected in much of the discussion is that new fuel rods have been delivered to nuclear reactors for years with few incidents. Again, the question is what degree of safety is required and how it can be guaranteed.

As in the case of the heavy metals described earlier, the problem is the permanent nature of the atoms. Even though they are undergoing radioactive decay, the process is one that will leave some radioactive materials for thousands of years, and the radiation will be dangerous for that length of time. A related problem is the wide variety of elements in much of the radioactive waste. Spent fuel rods from nuclear reactors contain ^{238}U in large amounts, ^{235}U in small amounts (largely depleted by the chain reaction), ^{239}Pu, fission and other decay products of a bewildering variety, and the metal cladding material that has become radioactive because of the intense neutron flux of the reactor. Structural materials from decommissioned reactors and by-products from ore processing and isotope enrichment plants are other examples of relatively high-level wastes. Low-level wastes from laboratories and hospitals present different technical difficulties because of their relatively large volume but low radioactive level. For some purposes, concentration of such wastes would be desirable, but loss to the environment during processing is an additional problem. As a relatively small but very important part of the overall problem of waste disposal, the disposal of radioactive waste will long be the subject of many fiercely fought battles.

---

[71] Details can be found at http://www.epa.gov/safewater/arsenic.html.

[72] A. S. Andrew, M. R. Karagas, and J. W. Hamilton, *Int. J. Cancer*, **2003**, *104*, 263; R. C. Kaltreider, A. M. Davis, J. P. Lariviere, and J. W. Hamilton, *Env. Health Persp.*, **2001**, *109*, 225.

16.8.2 Nonmetals

SULFUR Mine tailings are a source of both metal and nonmetal contamination. A common material in coal mines is iron pyrite, FeS_2. As a contaminant of coal, this compound and similar compounds contribute to the production of sulfur oxides in flue gases when coal is burned. As a material in mine tailings, it contributes both iron and sulfur to water pollution when the sulfide is oxidized in a series of reactions to sulfate, and the Fe(II) is oxidized to Fe(III):

$$4\,FeS_2 + 15\,O_2 + 6\,H_2O \longrightarrow 4\,[Fe(OH)]^{2+} + 8\,HSO_4^{\,-}$$

Because Fe(III) is a strongly acidic cation, the net result is a dilute solution of sulfuric acid containing Fe(II), Fe(III), and other heavy metal ions dissolved in the acidic solution (pH values of 2 to 3.5 have been measured). In areas with played-out mines, such solutions are common in the streams and rivers, effectively killing most plant and animal life in the water.

When coal containing sulfur compounds is burned, the resulting sulfur dioxide and sulfur trioxide can result in atmospheric contamination. There is much worldwide controversy regarding such contamination, because it travels across political and natural boundaries, and those who generate the contamination are rarely those who suffer its direct consequences. The sulfur oxides and nitrogen oxides from high-temperature combustion are readily dissolved in water droplets in the atmosphere and returned to the earth as acid rain. Although the evidence is still being debated, there seems little doubt that such acid rain has damaged forests and lakes globally, as well as attacking building materials and artistic works. Studies of the damage to limestone statues and building materials show an accelerating rate of destruction, and many carvings and sculptures have become completely unrecognizable over a relatively short time.

Although the amount of sulfur released by smelting is only about 10% of the total released into the atmosphere, the dramatic effects of sulfur oxides can be seen locally around smelting industries, where nickel or copper are mined and purified. The major ores of these metals are sulfides, and the method of extracting the metal begins with roasting the ore in air to convert it to the oxide:

$$MS + \tfrac{3}{2}\,O_2 \longrightarrow MO + SO_2\ (M = Cu, Ni)$$

When compared with the United States, a larger fraction of the sulfur dioxide generated in Canada is caused by smelting operations, because more of Canada's power generation is hydroelectric, and the total amount of power generated is smaller. Two sites that have been studied thoroughly are in Trail, British Columbia, and Sudbury, Ontario. When the area around Trail was studied from 1929 through 1936, after 30 to 40 years of smelter operation, no conifers were found within 12 miles, and damage to vegetation could be seen as far as 39 miles from the source.[73] Similar effects could also be seen around Sudbury, with evidence of acidified lakes up to 40 miles away. Efforts to control the emission of SO_2 and SO_3 have reduced the contamination, but recovery of the environment is a very slow process.

One advantage of the recovery of sulfur oxides from smelting is that the amounts are large enough to be economically useful; in most cases, the concentration of sulfur dioxide and sulfur trioxide found in power plant flue gases is so small that it is simply an added expense to remove them. Two techniques are used: removal of the sulfur compounds from the coal before burning and scrubbing of the stack gases to remove the oxides. Because FeS_2 is much more dense than coal, much of it can be removed by reducing the coal to a powder and separating the two by gravitational techniques.

---

[73] C. G. Down and J. Stocks, *Environmental Impact of Mining*, Wiley, New York, 1977, p. 63.

Leaching with sodium hydroxide also removes much of the sulfide contaminant, but scrubbing of the stack gases with a substance such as an aqueous slurry of $CaCO_3$ is still required for complete removal. The resulting $CaSO_3$ and $CaSO_4$ must also be disposed of or used in some way. Other techniques require gasification of the coal (partial combustion in steam to CO and H_2) and scrubbing of the gas to remove the resulting H_2S; combustion of a fluidized bed of finely pulverized coal and limestone; or complete conversion of SO_2 to SO_3 on a V_2O_5 catalyst and removal of SO_3 as H_2SO_4.

NITROGEN OXIDES AND PHOTOCHEMICAL SMOG Nitrogen oxides are also major contaminants, primarily from automobiles. The combustion process in automotive engines takes place at a high enough temperature that NO and NO_2 are formed. In the air, NO is rapidly converted to NO_2, and both can react with the hydrocarbons that are also released by cars. The resulting compounds are among the primary causes of smog seen in urban areas, particularly those where geography prevents easy mixing of the atmosphere and removal of contaminants. Although improvements have been made, there are still serious problems. The nitrogen oxides can also form nitric acid, which can contribute to acid rain:

$$3\,NO_2 + H_2O \longrightarrow 2\,HNO_3 + NO\cdot$$

Photochemical smog can form whenever air heavily laden with exhaust gases is trapped by atmospheric and topographic conditions and exposed to sunlight. Ozone and formaldehyde formed in the atmosphere from nitrogen oxides and hydrocarbons are also major contributors to smog. Some major reactions in this sequence are shown here.[74]

Reactions during combustion of gasoline include the following:

$$N_2 + O_2 \longrightarrow 2\,NO\cdot$$

$$C_nH_m + O_2 \longrightarrow CO_2 + CO + H_2O$$

Traces of ozone can be photolyzed, with hydroxyl radical an important product:

$$O_3 + h\nu \longrightarrow O + O_2$$

$$O + H_2O \longrightarrow 2\cdot OH$$

Another important species, the hydroperoxyl radical, is formed by the photolysis of formaldehyde:

$$HCHO + h\nu \longrightarrow H\cdot + HCO\cdot$$

$$H\cdot + O_2 + M \longrightarrow HO_2\cdot + M$$

(M is an unreactive molecule or larger particle that removes kinetic energy from the products after this exothermic reaction.)

$$HCO\cdot + O_2 \longrightarrow HO_2\cdot + CO$$

Oxidation of $NO\cdot$ at high concentration yields $NO_2\cdot$

$$2\,NO\cdot + O_2 \longrightarrow 2\,NO_2\cdot$$

and oxidation of NO by $HO_2\cdot$ at low NO concentrations, which is more common, also yields NO_2:

$$NO\cdot + HO_2\cdot \longrightarrow NO_2\cdot + \cdot OH$$

Photolysis of NO_2 forms oxygen atoms:

$$NO_2\cdot + h\nu \longrightarrow NO\cdot + O$$

---

[74] B. J. Finlayson-Pitts and J. N. Pitts, Jr., *Atmospheric Chemistry: Fundamentals and Experimental Techniques*, Wiley, New York, 1986, pp. 29–37.

(This requires light with $\lambda < 395$ nm, at the ultraviolet edge of the visible region.) Finally, production of ozone occurs:

$$O + O_2 + M \longrightarrow O_3 + M$$

Oxygen atoms and ozone react with NO and NO_2 to form NO_2, NO_3, and N_2O_5. These products then react with water to form HNO_2 and HNO_3. They also react with hydrocarbons to form aldehydes, oxygen-containing free radical species, and finally alkyl nitrites and nitrates, all of which are very reactive and contribute to eye and lung irritation and the damaging effects on vegetation, rubber, and plastics. One of the most reactive is peroxyacetyl nitrate, formed by the reaction of aldehydes with hydroxyl radical and NO_2:

$$CH_3\overset{\overset{O}{\|}}{C}H + \bullet OH \longrightarrow CH_3\overset{\overset{O}{\|}}{C}\bullet + H_2O$$

$$CH_3\overset{\overset{O}{\|}}{C}\bullet + O_2 \longrightarrow CH_3\overset{\overset{O}{\|}}{C}OO\bullet$$

$$CH_3\overset{\overset{O}{\|}}{C}OO\bullet + NO_2\bullet \longrightarrow CH_3\overset{\overset{O}{\|}}{C}OONO_2$$

Photochemical reactions of the aldehydes and alkyl nitrites generate more radicals and continue the chain of reactions.

THE OZONE LAYER Although it is an injurious pollutant in the lower atmosphere, ozone is an essential protective agent in the stratosphere. It is formed by photochemical dissociation of oxygen.

$$O_2 + h\nu \longrightarrow O + O^*$$

This requires light of $\lambda < 242$ nm, in the far UV. The activated oxygen atoms, O^*, react with molecular oxygen to form ozone:

$$O^* + O_2 + M \longrightarrow O_3 + M$$

The ozone formed in this way absorbs ultraviolet radiation with $\lambda < 340$ nm, regenerating molecular oxygen:

$$O_3 + h\nu \longrightarrow O_2 + O$$

followed by

$$O + O_3 \longrightarrow 2\,O_2$$

This mechanism filters out much of the sun's ultraviolet radiation, protecting plant and animal life on the surface of the Earth from other damaging photochemical reactions. This natural equilibrium is affected by compounds added to the atmosphere by humans. The most well known of these compounds are the chlorofluorocarbons, especially CF_2Cl_2 and CCl_3F, known as CFC 12 and 11, respectively. The names can be deciphered by adding 90 to the numbers; the resulting sequence of numbers gives the number of carbon, hydrogen, and fluorine atoms, and the number of chlorine atoms can be deduced from this information. These compounds were once widely used as refrigerants, blowing agents for the manufacture of plastic foams, and propellants in aerosol cans. Because their damaging effects have been demonstrated conclusively, substitutes for chlorofluorocarbons have been found, and nonessential uses are now restricted.

The destruction of ozone by these compounds is caused, paradoxically, by their extreme stability and lack of reaction under ordinary conditions. Because they are so stable, they remain in the atmosphere indefinitely and finally diffuse to the stratosphere. The intense high-energy ultraviolet radiation in the stratosphere causes dissociation and forms chlorine atoms, which then undergo a series of reactions that destroy ozone:[75]

$$CCl_2F_2 + h\nu \longrightarrow Cl\cdot + \cdot CClF_2$$ (This requires ultraviolet radiation with $\lambda \approx 200$ nm.)

$$Cl\cdot + O_3 \longrightarrow ClO\cdot + O_2$$

$$ClO\cdot + O \longrightarrow Cl\cdot + O_2$$

(These two reactions remove O_3 and oxygen atoms without reducing the number of chlorine atoms, $Cl\cdot$.)

Other compounds, such as NO and NO_2, also contribute to the chain of events:

$$NO\cdot + ClO\cdot \longrightarrow \cdot Cl + NO_2\cdot$$

$$NO_2\cdot + O_3 \longrightarrow NO_3\cdot + O_2$$

$$NO\cdot + O_3 \longrightarrow NO_2\cdot + O_2$$

$$NO_2\cdot + NO_3\cdot + M \longrightarrow N_2O_5 + M$$

The chains are terminated by reactions such as

$$\cdot Cl + CH_4 \longrightarrow HCl + \cdot CH_3 \quad \text{and} \quad \cdot Cl + H_2 \longrightarrow HCl + H\cdot$$

that are followed by combinations of the new radicals to form stable molecules such as CH_4, H_2, and C_2H_6 and by a reaction that ties up the chlorine:

$$ClO\cdot + NO_2\cdot + M \longrightarrow ClONO_2 + M$$

Although Rowland and Molina had predicted depletion of ozone concentrations by these reactions, there were many who doubted their conclusions. The phenomenon that finally brought the problem to the attention of the world was the discovery of the ozone "hole" over the Antarctic in 1985.[76] During the winter, a combination of air flow pattern and low temperature create stratospheric clouds of ice particles. The surface of these particles is an ideal location for reaction of NO_2, OCI, and O_3. These clouds contain nitric acid hydrate, formed by

$$N_2O_5 + H_2O \longrightarrow 2\,HNO_3$$

and

$$ClONO_2 + H_2O \longrightarrow HOCl + HNO_3$$

These reactions, plus

$$HCl + HOCl \longrightarrow Cl_2 + H_2O \quad \text{and} \quad HCl + ClONO_2 \longrightarrow Cl_2 + HNO_3$$

remove chlorine from the air and generate Cl_2 on the surface of the ice crystals. In the spring, increased sunlight splits these molecules into chlorine atoms, and the decomposition of ozone proceeds at a much higher rate. The reaction with NO_2 that would remove ClO from the air is prevented, because the NO_2 is mostly tied up as HNO_3 in the ice. The polar vortex prevents mixing with air containing a higher concentration of ozone, and the result is a reduced concentration of ozone over the Antarctic. As the air warms in the summer, the circulation changes, the clouds dissipate, and the level of ozone returns to a more nearly normal level. Within 2 years of the discovery of the ozone hole, the international community had accepted this as evidence of a global problem, and the Montreal

[75] M. J. Molina and F. S. Rowland, *Nature*, **1974**, *249*, 810; F. S. Rowland, *Am. Sci.*, **1989**, *77*, 36.
[76] J. C. Farman, B. G. Gardiner, and J. D. Shanklin, *Nature*, **1985**, *315*, 207.

Protocol on Substances that Deplete the Ozone Layer was signed in 1987, which set a schedule for decreasing use and production of CFCs and eventually for their complete ban.

After the signing of the Montreal Protocol, the size of the ozone hole continued to increase for several years. For about the past 15 years, the maximum size of the ozone hole has fluctuated from year to year, with the largest size observed in 2006. It is uncertain how long it will take for ozone levels to return to their previous amounts—or if in fact they will do so; many decades may well be necessary. Current and historical information on the ozone hole is available from NASA.[77]

Whether reduction in use of these chlorofluorocarbons will be sufficient to prevent serious worldwide results caused by destruction of the ozone layer remains to be seen. Predictions based on the materials already in the atmosphere indicate that the damage will be significant, even if production could be stopped immediately, but such predictions are based on untested computer models and are subject to considerable error. Production has stopped or declined drastically in most countries, but the compounds proposed as substitutes are primarily those containing C, H, Cl, and F with lower stability. Whether they really reduce the effects is still uncertain, and complete replacement will require many years to achieve.[78] Methods for recycling CFCs from air conditioners and refrigeration units have been developed, but there are still large amounts of CFCs in use that will eventually make their way into the atmosphere.

More recent observations have detected a similar ozone hole in the Arctic, but it is smaller and much more variable, largely because the temperatures vary more there.[79] Volcanic activity that injects sulfur dioxide into the atmosphere also has an effect that depends on temperature and on the height of the SO_2 injection. The SO_2 reacts with air to form SO_3, which then reacts with water to form sulfuric acid aerosols. These volcanic aerosols, particularly at polar temperatures, reduce the nitrogen oxide concentration of the air and activate chlorine species that destroy ozone, as do the polar stratospheric clouds described earlier. Because these aerosols are stable at warmer temperatures $(\sim 200\ \text{K})$ than the natural stratospheric clouds, and because they can exist at lower altitudes, they can have significant effects. Until the level of chlorine is reduced to preindustrial levels, low temperatures and volcanic activity are likely to create Arctic ozone holes each spring as a result of reactions during the winter.

THE GREENHOUSE EFFECT Another atmospheric problem is the greenhouse effect. The major cause in this case is carbon dioxide released by combustion and decomposition of organic matter. Other gases, including methane and chlorofluorocarbons, also contribute to the problem. In the greenhouse effect, visible and ultraviolet radiation from the sun that is not absorbed in the stratosphere and upper atmosphere reaches the surface of the Earth and is absorbed and converted to heat. This heat, in the form of infrared radiation, is transmitted out from the Earth through the atmosphere. Molecules such as CO_2 and CH_4, which have low-energy vibrational energy levels, absorb this radiation and reradiate the energy, much of it toward the Earth. As a result, the energy cannot escape, and the Earth's surface and the atmosphere are warmed.

The main contributor to the greenhouse effect, CO_2, has received considerable public attention. Since the beginning of the industrial revolution, the atmospheric concentration has increased considerably, from approximately 280 ppm to more than 385 ppm;[80] it continues to increase annually and, like many other greenhouse gases, is monitored at a variety of sites around the globe.[81]

---

[77] See http://ozonewatch.gsfc.nasa.gov/
[78] L. E. Manaer, *Science*, **1990**, *249*, 31.
[79] A. Tabazadeh, K. Drdla, M. R. Schoeberl, P. Hajill, and O. B. Toon, *Proc. Natl. Acad. Sci. U.S.A.*, **2002**, *99*, 2609.
[80] Information on the concentrations of many greenhouse gases is available from the Carbon Dioxide Information Analysis Center at http://cdiac.esd.ornl.gov/pns/current_ghg.html.
[81] For example, see the longest continuous record, at Mauna Loa in Hawaii at http://www.esrl.noaa.gov/gmd/ccgg/trends/.

Like carbon dioxide, many other greenhouse gases have increased in abundance at least in part as a consequence of human activity. In recent years, several relatively new inorganic atmospheric components have been observed that have unusually high global warming potential (GWP), a measure of the relative radiative effect of gases in comparison with an equal mass of CO_2, which is assigned a GWP of 1. The global warming potential is calculated over a specific time frame, for example 100 years. The larger the value of GWP, the greater the ability of a particular mass of a gas to radiate thermal energy and, therefore, to contribute to atmospheric warming.[80]

Information on selected inorganic greenhouse gases, together with methane, is provided in Table 16.4.

Although the quantities of the last four gases in the table are comparatively low, they are of concern because of their extremely high global warming potentials and because of their rate of increase in the atmosphere. Sulfur hexafluoride is a nonflammable and nontoxic gas used widely as an insulating gas for high-voltage generators and switches and as an inert blanketing gas in magnesium metal casting. Because of its low reactivity, it has an exceptionally long lifetime in the atmosphere. The sources of SF_5CF_3 are less clear, because it is not produced commercially in significant amounts. It has been proposed that SF_5CF_3 may be produced in the atmosphere from the reaction of SF_6 with the trace atmospheric gases CHF_3 and CH_2F_2; this reaction has been reported under laboratory conditions.[82] Sulfuryl fluoride, SO_2F_2, has been used to replace the fumigant methyl bromide, which is being phased out under the Montreal Protocol because of its large ozone depletion potential. It has also been used to replace SF_6 as a blanketing gas in magnesium manufacture in response to the recognition of the high global warming potential of SF_6.[83] Nitrogen trifluoride, although toxic, has been used extensively in recent years by the electronics industry for etching silicon, for production of liquid crystal displays, and for other uses. Although the current concentration of NF_3 is very low, the rapid growth of its use, high global warming potential, and high atmospheric lifetime make this gas of potential long term concern.[84]

TABLE 16.4 Selected Greenhouse Gases

Gas	Pre-1750 Concentration	Current Concentration	GWP (100 yr)	Atmospheric Lifetime (yr)
CO_2	280 ppm	384.8 ppm	1	100
CH_4	700 ppb	1800 ppb[a]	25	12
N_2O	270 ppb	320 ppb	298	114
SF_6	0	6.2 ppt[a]	22,800	3200
SO_2F_2[b]	0	1.5 ppt	4800	36
SF_5CF_3[c]	0	0.12 ppt	17,500	800
NF_3[b, d]	0	0.6 ppt	17,200	740

Sources: Except where noted, data from Carbon Dioxide Information Analysis Center at http://cdiac.ornl.gov, updated July 2009.

NOTES: [a]Average of Northern and Southern Hemisphere values. [b]P. Fraser, "Nitrogen Trifluoride and Sulfuryl Fluoride," Centre for Australian Weather and Climate Research, Greenhouse 2009, Perth, Australia. [c]*World Meteorological Organization, Scientific Assessment of Ozone Depletion: 2002*, Geneva, Switzerland, 2002. [d]Intergovernmental Panel on Climate Change (IPCC), *Climate Change 2007*, IPCC, Geneva, Switzerland, 2007.

[82] L. Huang, L. Zhu, X. Pan, J. Zhang, B. Ouyang, and H. Hou, *Atmos. Environ.*, **2005**, *39*, 1641.
[83] See J. Mühle, et al., *J. Geophys. Res.*, **2009**, *114*, D05306 for additional data and references on SO_2F_2 in the atmosphere.
[84] W.-T. Tsai, *J. Hazard. Mat.*, **2008**, *159*, 257.

A milestone in addressing the greenhouse effect was the international conference in Kyoto, Japan, in 1997 that reached preliminary agreement on reduction of greenhouse gases. The United States officially objected to the proposed reductions in CO_2 production. The largest ever United Nations meeting on climate change was held in Copenhagen in December 2009. Disputes between rich and poor countries and between the world's biggest CO_2 polluters, China and the United States, dominated the conference, which resulted in a modest political compromise that sets the stage for more talks.

If unabated, the increase in atmospheric greenhouse gases is likely to have dramatic consequences, which have been described through many avenues. Addressing these consequences will be one of humanity's great challenges throughout the rest of this century.

General References

A number of bioinorganic chemistry books are available, including S. J. Lippard and J. M. Berg's, *Principles of Bioinorganic Chemistry*, University Science Books, Mill Valley, CA, 1994; J. A. Cowan's, *Inorganic Biochemistry*, VCH, New York, 1993; W. Kaim and B. Schwederski, *Bioinorganic Chemistry: Inorganic Elements in the Chemistry of Life*, John Wiley & Sons, Chichester, England, 1994; and I. Bertini, H. B. Gray, S. J. Lippard, and J. S. Valentine, editors, *Bioinorganic Chemistry*, University Science Books, Sausalito, CA, 1994.

J. E. Fergusson's, *Inorganic Chemistry and the Earth*, Pergamon Press, Elmsford, NY, 1982, includes several chapters on environmental chemistry, and J. O'M. Bockris, editors, *Environmental Chemistry*, Plenum Press, New York, 1977, offers the viewpoints of many different authors. R. A. Bailey, H. M. Clarke, J. P. Farris, S. Krause, and R. L. Strong's, *Chemistry of the Environment*, Academic Press, New York, 1978, covers a very broad range of environmental topics at an easily accessible level. S. E. Manahan's, *Environmental Chemistry*, 7th ed., CRC Press, Boca Raton, FL, and P. O'Neill's, *Environmental Chemistry*, 2nd ed., Chapman & Hall, London, are two comprehensive texts on the subject. B. J. Finlayson-Pitts and J. N. Pitts, Jr.'s, *Atmospheric Chemistry: Fundamentals and Experimental Techniques*, John Wiley & Sons, New York, 1986, offers complete coverage of both the laboratory and field studies of all kinds of chemicals and their reactions. A more specific report on the greenhouse effect is a National Academy of Sciences report by the Carbon Dioxide Assessment Committee, *Changing Climate*, National Academy Press, Washington, DC, 1983. One reference handbook on ozone is by D. E. Newton, *The Ozone Dilemma*, Instructional Horizons and ABC-CLIO, Santa Barbara, CA, 1995. Finally, two of the standard references used throughout this book must be mentioned again. F. A. Cotton, G. Wilkinson, C. A. Murillo, and M. Bochman's, *Advanced Inorganic Chemistry*, 6th ed., Wiley InterScience, New York, 1999, includes a good review of bioinorganic chemistry, as does G. Wilkinson, R. D. Gillard, and J. A. McCleverty's *Comprehensive Coordination Chemistry*, Vol. 6, *Applications*, Pergamon Press, Oxford, 1987.

Problems

16.1 Describe possible mechanisms for the vitamin B_{12}-catalyzed reactions given in the text.

16.2 The curves in the graph that follows show the degree of saturation of hemoglobin and myoglobin as a function of the pressure of oxygen. Explain how these two compounds are each best suited to their specific roles, with hemoglobin transferring oxygen from the lungs to the blood and myoglobin transferring oxygen from the blood to the other tissues. (Graph from J. M. Rifkin "Hemoglobin and Myoglobin," in G. L. Eichhorn, ed., *Inorganic Biochemistry*, Vol. 2, Elsevier, New York, 1973, p. 853. © Elsevier Science Publishing Co., Inc. Reprinted by permission of the publisher.)

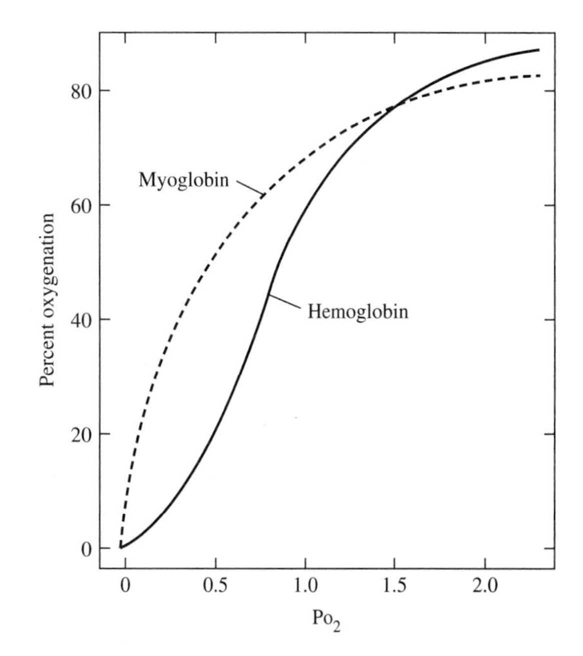

16.3 Acetohydroxamic acid (AcHA) is $CH_3C(=O)NHOH$. Sketch the structure expected for a complex of AcHA and Fe^{3+}, including the possible resonance structures.

16.4 Oxygen can bind to metals in several ways. Proposals have been made for linear, bent, and side-on bonding in hemoglobin and similar compounds. Review the arguments for each of these and the evidence for the structure used in this chapter for HbO_2. (See W. Kaim and B. Schwederski, *Bioinorganic Chemistry: Inorganic Elements in the Chemistry of Life*, John Wiley & Sons, Chichester, England, 1994, pp. 92–96, or a similar source.)

16.5 Dissociation of NO_2 to NO and O requires light with a wavelength less than 395 nm. Calculate the dissociation energy of NO_2, assuming all the energy is concentrated in this reaction. Dissociation of O_2 requires $\lambda < 242$ nm. Calculate the dissociation energy of O_2 with the same assumption. Do the results of these two calculations match the bonding in these molecules described in Chapters 3 and 5?

16.6 Methyl cobalamin is usually described as a Co(II) compound, which changes to Co(III) on dissociation of CH_3^-. Describe the probable electronic structure (splitting of *d* levels and number of unpaired electrons) of the cobalt in both cases.

16.7 Lead can accumulate in the bones and other body tissues unless removed soon after ingestion. In some cases, treatment with chelating agents such as EDTA has been used to remove lead, mercury, or other heavy metals from the body. Discuss the advantages and disadvantages of such treatment. Include both thermodynamic and kinetic arguments in your answer.

16.8 Some of the reactions of NO in the blood do not cause problems at low concentrations but can upset the normal reactions of hemoglobin if there is a large concentration of NO. The same reactions can cause trouble if a synthetic blood substitute contains a molecule similar to hemoglobin but does not contain all the other enzymes normally contained in red blood cells. Explain what this problem is and how it arises.

16.9 There are three families of enzymes having molybdenum at their active sites: the sulfide oxidase, xanthine oxidase, and DMSO reductase families.

a. Crystal structures of these enzymes show that in each case molybdenum is coordinated by a complex dithiolene ligand. What is the structure of this ligand, and what is it called?

b. Using sketches, show how the coordination environment of Mo is different in these three enzyme families.

c. What types of reactions do these enzymes catalyze? (See J. McMaster and J. H. Enemark, *Curr. Opin. Chem. Bio.*, **1998**, *2*, 201; J. H. Enemark, J. J. A. Cooney, J.-J. Wang, and R. H. Holm, *Chem. Rev.*, **2004**, *104*, 1175.)

16.10 There are two families of enzymes having tungsten at their active sites: the formate dehydrogenase and aldehyde oxidoreductase families.

a. Using sketches, show how the coordination environment of W is different in these families. Which family is structurally similar to a molybdenum enzyme family?

b. What types of reactions are catalyzed by these enzymes?

c. Sometimes tungsten is considered an "old" element in the evolutionary context of its enzymes, in contrast to the "newer" molybdenum. Why?

(See L. E. Bevers, P.-L. Hagedoorn, and W. R. Hagen, *Coord. Chem. Rev.*, **2009**, *253*, 269; J. H. Enemark, J. J. A. Cooney, J.-J. Wang, and R. H. Holm, *Chem. Rev.*, **2004**, *104*, 1175.)

16.11 Single-walled carbon nanotubes have been used to assist in the delivery of cisplatin to cancer cells (A. A. Bhirde, et al., *ACS Nano*, **2009**, *3*, 307). Briefly describe how this delivery was effected, the types of cancer cells studied, and the importance of this research. What role did quantum dots play in this study?

16.12 Medicinal applications of organometallic chemistry are developing rapidly. Ferrocifen is an example that has shown effectiveness against specific cancer cell lines. What is ferrocifen, against what type of cancer does it show activity, and, briefly, how is this activity explained? (See U. Schatzschneider and N. Metzler-Nolte, *Angew. Chem., Int. Ed. Engl.*, **2006**, *45*, 1504 and E. Hillard, A. Vessieres, L. Thouin, G. Jaouen, and C. Amatore, *Angew. Chem., Int. Ed. Engl.*, **2006**, *45*, 285.)

16.13 The longest continuous measurement of atmospheric concentration of carbon dioxide has been obtained at Mauna Loa, Hawaii, beginning in 1959; data are available at http://www.esrl.noaa.gov/gmd/ccgg/trends.

a. Monthly data show annual maxima and minima in carbon dioxide concentrations. How can these fluctuations be explained?

b. Create a graph using annual data for every 3 to 5 years, ending with the most recent year for which data are available. Do the data suggest that the rate of carbon dioxide growth is increasing, staying the same, or decreasing? Explain.

c. Use a mathematical function to fit the data and to predict when the concentration of carbon dioxide will reach 450 ppm. Comment on how valid your prediction is likely to be. What factors are important in predicting future greenhouse gas concentrations?

16.14 The World Data Centre for Greenhouse gases provides data from a wide variety of sampling sites for many greenhouse gases, including inorganic gases such as CO_2, N_2O, and SF_6 at http://gaw.kishou.go.jp/wdcgg. Other sources provide data on minor atmospheric components such as SF_5CF_3 and NF_3. Select an inorganic greenhouse gas and find information on its current atmospheric concentration, whether and how rapidly its concentration is increasing, its sources, its atmospheric lifetime and proposed mechanisms by which it is removed from the atmosphere, and its potential impact on global warming. Cite all sources used.

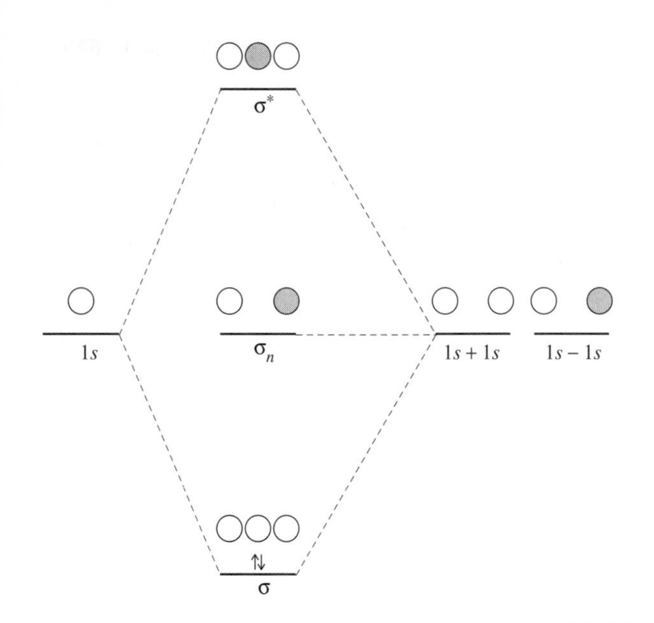

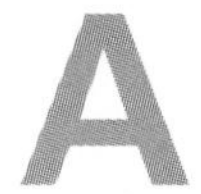

Answers to Exercises

CHAPTER 2

2.1 $E = R_H\left(\frac{1}{2^2} - \frac{1}{3^2}\right) = R_H\left(\frac{5}{36}\right) = 2.179 \times 10^{-18}\,\text{J}\left(\frac{5}{36}\right) = 3.026 \times 10^{-19}\ \text{J}$

$= 1.097 \times 10^7\ \text{m}^{-1}\left(\frac{5}{36}\right) = 1.524 \times 10^6\ \text{m}^{-1} \times \frac{\text{m}}{100\ \text{cm}} = 1.524 \times 10^4\,\text{cm}^{-1}$

2.2 The nodal surfaces require $2z^2 - x^2 - y^2 = 0$, so the angular nodal surface for a d_{z^2} orbital is the conical surface where $2z^2 = x^2 + y^2$.

2.3 The angular nodal surfaces for a d_{xz} orbital are the planes where $xz = 0$, which means that either x or z must be zero. The yz and xy planes satisfy this requirement.

2.4 If the 3p electrons all have the same spin, as in $\underline{\uparrow_1}$ $\underline{\uparrow_2}$ $\underline{\uparrow_3}$, there are three exchange possibilities (1 and 2, 1 and 3, or 2 and 3) and no pairs. Overall, the total energy is $3\Pi_e$.

If there is one unpaired electron, as in $\underline{\uparrow\downarrow}$ $\underline{\uparrow}$ $\underline{\quad}$, there is one electron with $\downarrow$ spin, and no possibility of exchange; two electrons with $\uparrow$ spin, with one exchange possibility; and one pair. Overall, the total energy is $\Pi_e + \Pi_c$.

Because Π_e is negative and Π_c is positive, the configuration with three unpaired electrons has a much lower energy.

2.5

Tin	Total	5p	5s	4d
Z	50	50	50	50
($1s^2$)	2	2	2	2
($2s^22p^6$)	8	8	8	8
($3s^23p^6$)	8	8	8	8
($3d^{10}$)	10	10	10	10
($4s^24p^6$)	8	8 × 0.85	8 × 0.85	8
($4d^{10}$)	10	10 × 0.85	10 × 0.85	9 × 0.35
($5s^25p^2$)	4	3 × 0.35	3 × 0.35	
Z^*		5.65	5.65	10.85

2.6

Uranium	Total	$7s$	$5f$	$6d$
Z	**92**	**92**	**92**	**92**
$(1s^2)$	2	2	2	2
$(2s^2 2p^6)$	8	8	8	8
$(3s^2 3p^6)$	8	8	8	8
$(3d^{10})$	10	10	10	10
$(4s^2 4p^6)$	8	8	8	8
$(4d^{10})$	10	10	10	10
$(4f^{14})$	14	14	14	14
$(5s^2 5p^6)$	8	8	8	8
$(5d^{10})$	10	10	10	10
$(5f^3)$	3	3	2×0.35	3
$(6s^2 6p^6)$	8	8×0.85		8
$(6d^1)$	1	1×0.85		
$(7s^2)$	2	1×0.35		
Z^*		3.00	13.30	3.00

CHAPTER 3

3.1 POF_3: The octet rule results in single P—F and P—O bonds; formal charge arguments result in a double bond for P=O. The actual distance is 143 pm, considerably shorter than a regular P—O bond (164 pm).

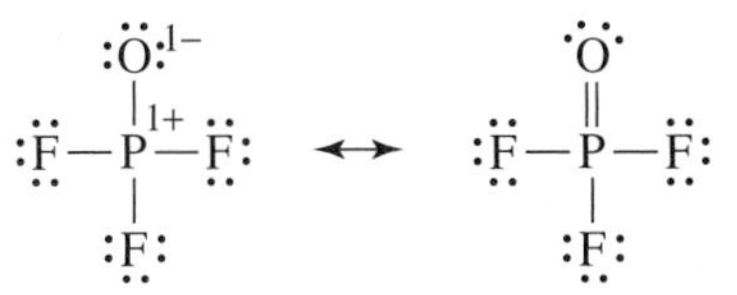

SOF_4: This is a distorted trigonal bipyramidal structure, with an S=O double bond and S—F single bonds required by formal charge arguments. The short S=O bond length of 140 pm is in agreement.

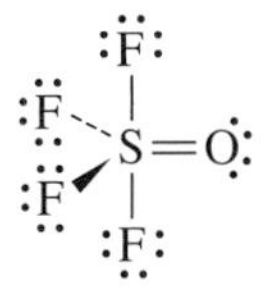

SO_3F^-: This is basically a tetrahedral structure, with two double bonds to oxygen atoms and single bonds to fluorine and the third oxygen. The S—O bond order is then 1.67 and the bond length is 143 pm, shorter than the 149 pm of SO_4^{2-}, which has a bond order of 1.5.

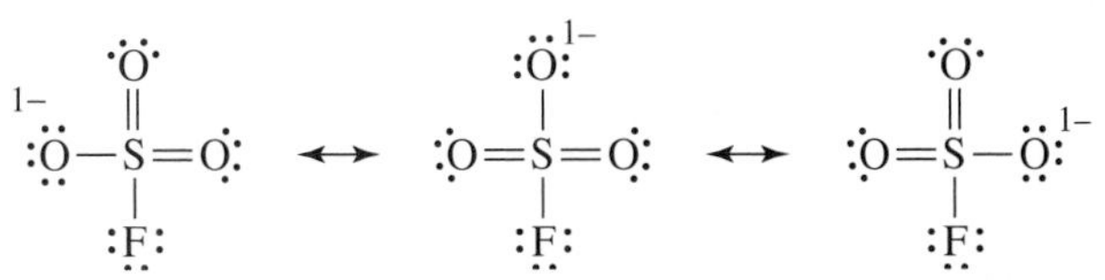

3.2

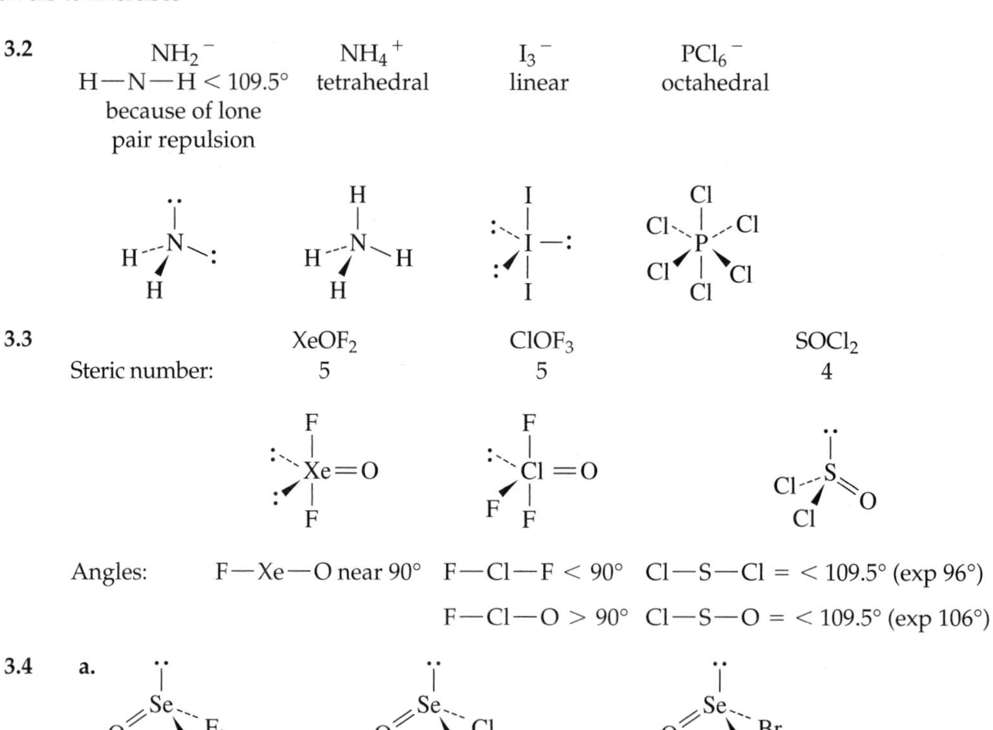

In $OSeF_2$, the fluorines exhibit the strongest attraction for electrons in the Se—F bonds. This reduces electron-electron repulsions near the Se atom, enhancing the ability of the lone pair and double bond to squeeze the rest of the molecule together.

b.

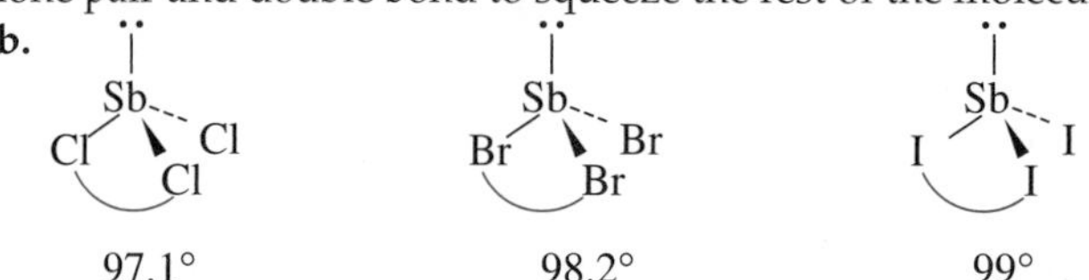

Cl, the most electronegative of the halogens in this series, pulls shared electrons the most strongly away from Sb, reducing electron density near Sb. The consequence is that the lone pair exerts the strongest influence on shape in $SbCl_3$.

c.

PI_3	AsI_3	SbI_3
102°	100.2°	99°

Phosphorus is the most electronegative of the central atoms. Consequently, it exerts the strongest pull on shared electrons, concentrating these electrons near P and increasing bonding pair–bonding pair repulsions, hence the largest angle in PI_3. Sb, the least electronegative central atom, has the opposite effect: Shared electrons are attracted away from Sb, reducing repulsions between the Sb—I bonds. The consequence is that the effect of the lone pair is greatest in SbI_3, which has the smallest angle.

Atomic size arguments can also be used for these species. Larger outer atoms result in larger angles; larger central atoms result in smaller angles.

3.5 **a.** Because fluorine is more electronegative than chlorine, electrons in N—F bonds are pulled more strongly away from nitrogen than electrons in N—Cl bonds. As a result, in NF_3 the bonding pair–bonding pair repulsions near the nitrogen are weaker. Repulsion by the lone pair then leads to smaller F—N—F angles than Cl—N—Cl angles.

b. In SOF_4 the bonding pairs to axial fluorine atoms are repelled at approximately 90° angles by two bonding pairs and the double bond to oxygen. By contrast, equatorial bonding pairs engage in 90° interactions only with the axial bonding pairs. Because of greater overall repulsions, the bonds to axial positions are longer.

c. Because the CH_3 groups are less electronegative than iodine atoms, the Te—C bonds are more polarized toward Te. As a result, the methyl groups occupy the equatorial positions, where there is less crowding and the methyl groups are farther from the lone pair.

3.6 According to the LCP model, the Cl · · · Cl distance in BCl_4^- should be approximately the same as in BCl_3. By analogy with the preceding example, and using the tetrahedral bond angle of 109.5° for BCl_4^-:

$$x = \text{B—Cl bond distance} = \frac{150.5 \text{ pm}}{\sin 54.75°} = 184 \text{ pm}$$

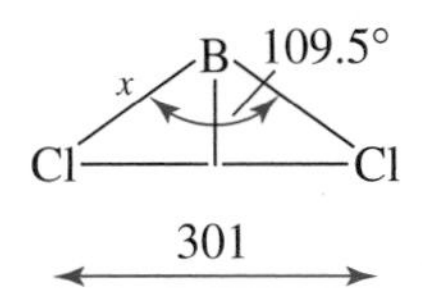

CHAPTER 4

4.1 S_2 is made up of C_2 followed by $\sigma_\perp$, which is shown in the figure below to be the same as i.

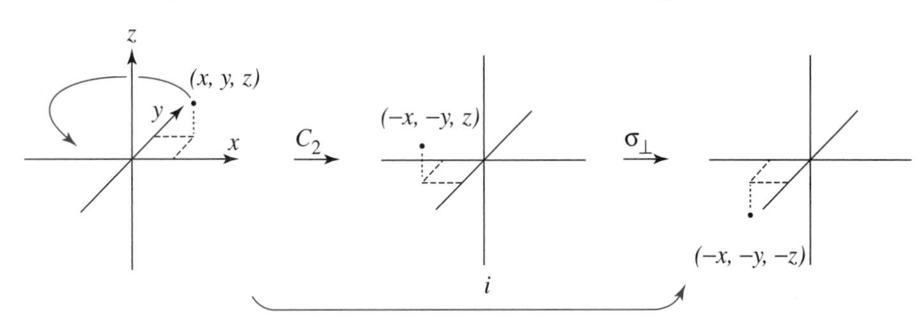

S_1 is made up of C_1 followed by $\sigma_\perp$, which is shown in the figure below to be the same as σ.

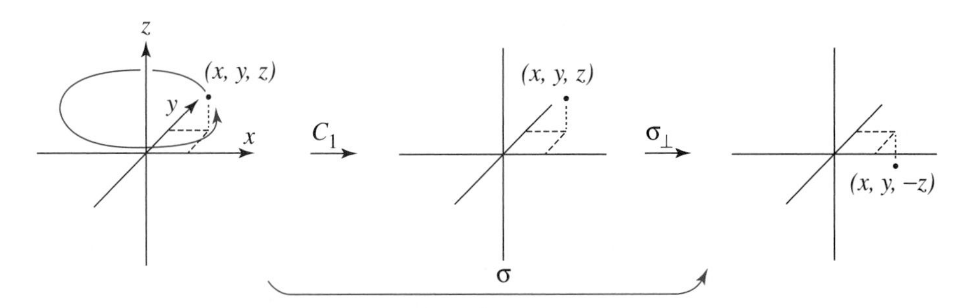

4.2 NH_3 has a threefold axis through the N perpendicular to the plane of the three hydrogen atoms and three mirror planes, each including the N and one H. $2C_3$, $3\sigma_v$

C_3

N, H, H, H

Cyclohexane in the boat conformation has a C_2 axis perpendicular to the plane of the lower four carbon atoms and two mirror planes that include this axis and are perpendicular to each other. C_2, $2\sigma_v$

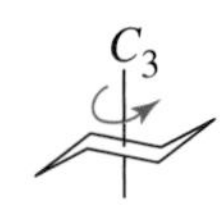

Cyclohexane in the chair conformation has a C_3 axis perpendicular to the average plane of the ring, three perpendicular C_2 axes passing between carbon atoms, and three mirror planes passing through opposite carbon atoms and perpendicular to the average plane of the ring. It also contains a center of inversion and an S_6 axis collinear with the C_3 axis. A model is very useful in analysis of this molecule. $2C_3$, $3C_2$, $3\sigma_d$, i, $2S_6$

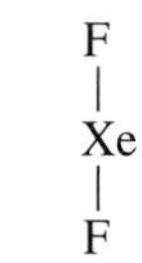

XeF_2 is a linear molecule, with a C_∞ axis through the three nuclei, an infinite number of perpendicular C_2 axes, a horizontal mirror plane (which is also an inversion center), and an infinite number of mirror planes that include the C_∞ axis. C_∞, $\infty\ C_2$, $i = \sigma_h$, $\infty\ \sigma_v$.

4.3 Several of the molecules below have symmetry elements in addition to those used to assign the point group. See the character tables in Appendix C for the complete list for each point group.

N_2F_2 has a mirror plane through all the atoms, which is the σ_h plane, perpendicular to the C_2 axis through the N=N bond. There are no other symmetry elements, so it is C_{2h}.

$B(OH)_3$ also has a σ_h mirror plane, the plane of the molecule, perpendicular to the C_3 axis through the B atom. Again, there are no others, so it is C_{3h}.

H_2O has a C_2 axis in the plane of the drawing, through the O atom and between the two H atoms. It also has two mirror planes, one in the plane of the drawing and the other perpendicular to it; overall, C_{2v}.

PCl_3 has a C_3 axis through the P atom and equidistant from the three Cl atoms. Like NH_3, it also has three σ_v planes, each through the P atom and one of the Cl atoms; overall, C_{3v}.

BrF_5 has one C_4 axis through the Br atom and the F atom in the plane of the drawing, two σ_v planes (each through the Br atom, the F atom in the plane of the drawing, and two of the other F atoms), and two σ_d planes between the equatorial F atoms; overall, C_{4v}.

HF, CO, and HCN all are linear, with the infinite rotation axis through the center of all the atoms. There are also an infinite number of σ_v planes, all of which contain the C_∞ axis; overall, $C_{\infty v}$.

N_2H_4 has a C_2 axis perpendicular to the N—N bond and splitting the angle between the two lone pairs. There are no other symmetry elements, so it is C_2.

$P(C_6H_5)_3$ has only a C_3 axis, much like that in NH_3 or $B(OH)_3$. The twist of the phenyl rings prevents any other symmetry; C_3.

BF_3 has a C_3 axis perpendicular to the σ_h plane of the molecule and three C_2 axes, each through the B atom and an F atom; overall, D_{3h}.

$PtCl_4{}^{2-}$ has a C_4 axis perpendicular to the σ_h plane of the molecule. It also has four C_2 axes in the plane of the molecule, two through opposite Cl atoms and two splitting the Cl—Pt—Cl angles, thus making it D_{4h}.

$Os(C_5H_5)_2$ has a C_5 axis through the center of the two cyclopentadienyl rings and the Os, five C_2 axes parallel to the rings and through the Os atom, and a σ_h plane parallel to the rings through the Os atom, for a D_{5h} assignment.

Benzene has a C_6 axis perpendicular to the σ_h plane of the ring and six C_2 axes in the plane of the ring, three through two C atoms each and three between the atoms. These are sufficient to make it D_{6h}.

F_2, N_2, and H—C≡C—H are all linear, each with a C_∞ axis through the atoms. There are also an infinite number of C_2 axes perpendicular to the C_∞ axes and a σ_h plane perpendicular to the C_∞ axes, sufficient to make them $D_{\infty h}$.

Allene, $H_2C{=}C{=}CH_2$, has a C_2 axis through the three carbon atoms and two C_2 axes perpendicular to the line of the carbon atoms, both at 45° angles to the planes of the H atoms. Two σ_d mirror planes through each H—C—H combination complete the assignment of D_{2d}.

$Ni(C_4H_4)_2$ has a C_4 axis through the centers of the C_4H_4 rings and the Ni, four C_2 axes perpendicular to the C_4 through the Ni, and four σ_d planes, each including two opposite carbon atoms of the same ring and the Ni; overall, D_{4d}.

$Fe(C_5H_5)_2$ has a C_5 axis through the centers of the rings and the Fe, five C_2 axes perpendicular to the carbon atoms and through the Fe, and five σ_d planes that include the C_5 axis; overall, D_{5d}.

$[Ru(en)_3]^{2+}$ has a C_3 axis perpendicular to the drawing through the Ru and three C_2 axes in the plane of the paper, each intersecting an en ring at the midpoint and passing through the Ru; overall, D_3.

4.4 **a.** $\begin{bmatrix} 5 & 1 & 3 \\ 4 & 2 & 2 \\ 1 & 2 & 3 \end{bmatrix} \times \begin{bmatrix} 2 & 1 & 1 \\ 1 & 2 & 3 \\ 5 & 4 & 3 \end{bmatrix}$

$$= \begin{bmatrix} (5\times 2)+(1\times 1)+(3\times 5) & (5\times 1)+(1\times 2)+(3\times 4) & (5\times 1)+(1\times 3)+(3\times 3) \\ (4\times 2)+(2\times 1)+(2\times 5) & (4\times 1)+(2\times 2)+(2\times 4) & (4\times 1)+(2\times 3)+(2\times 3) \\ (1\times 2)+(2\times 1)+(3\times 5) & (1\times 1)+(2\times 2)+(3\times 4) & (1\times 1)+(2\times 3)+(3\times 3) \end{bmatrix}$$

$$= \begin{bmatrix} 26 & 19 & 17 \\ 20 & 16 & 16 \\ 19 & 17 & 16 \end{bmatrix}$$

b. $\begin{bmatrix} 1 & -1 & -2 \\ 0 & 1 & -1 \\ 1 & 0 & 0 \end{bmatrix} \times \begin{bmatrix} 2 \\ 1 \\ 3 \end{bmatrix} = \begin{bmatrix} (1\times 2)-(1\times 1)-(2\times 3) \\ (0\times 2)+(1\times 1)-(1\times 3) \\ (1\times 2)+(0\times 1)+(0\times 3) \end{bmatrix} = \begin{bmatrix} -5 \\ -2 \\ 2 \end{bmatrix}$

c. $[1 \;\; 2 \;\; 3] \times \begin{bmatrix} 1 & -1 & -2 \\ 2 & 1 & -1 \\ 3 & 2 & 1 \end{bmatrix}$

$$= [(1\times 1)+(2\times 2)+(3\times 3) \quad 1\times(-1)+(2\times 1)+(3\times 2) \quad 1\times(-2)+2\times(-1)+(3\times 1)]$$

$$= [14 \;\; 7 \;\; -1]$$

4.5 E: The new coordinates are

$$\begin{aligned} x' &= \text{new } x = x \\ y' &= \text{new } y = y \\ z' &= \text{new } z = z \end{aligned} \qquad \begin{bmatrix} 1 & 0 & 0 \\ 0 & 1 & 0 \\ 0 & 0 & 1 \end{bmatrix} = \text{transformation matrix for } E$$

In matrix notation,

$$\begin{bmatrix} x' \\ y' \\ z' \end{bmatrix} = \begin{bmatrix} 1 & 0 & 0 \\ 0 & 1 & 0 \\ 0 & 0 & 1 \end{bmatrix} \begin{bmatrix} x \\ y \\ z \end{bmatrix} = \begin{bmatrix} x \\ y \\ z \end{bmatrix} \quad \text{or} \quad \begin{bmatrix} x' \\ y' \\ z' \end{bmatrix} = \begin{bmatrix} x \\ y \\ z \end{bmatrix}$$

$\sigma_v'(yz)$: Reflect a point with coordinates (x, y, z) through the yz plane.

$$\begin{aligned} x' &= \text{new } x = -x \\ y' &= \text{new } y = y \\ z' &= \text{new } z = z \end{aligned} \qquad \begin{bmatrix} -1 & 0 & 0 \\ 0 & 1 & 0 \\ 0 & 0 & 1 \end{bmatrix} = \text{transformation matrix for } \sigma_v'(yz)$$

In matrix notation,

$$\begin{bmatrix} x' \\ y' \\ z' \end{bmatrix} = \begin{bmatrix} -1 & 0 & 0 \\ 0 & 1 & 0 \\ 0 & 0 & 1 \end{bmatrix} \begin{bmatrix} x \\ y \\ z \end{bmatrix} = \begin{bmatrix} -x \\ y \\ z \end{bmatrix} \quad \text{or} \quad \begin{bmatrix} x' \\ y' \\ z' \end{bmatrix} = \begin{bmatrix} -x \\ y \\ z \end{bmatrix}$$

4.6 Representation Flow Chart: N_2F_2 (C_{2h})

Symmetry Operations

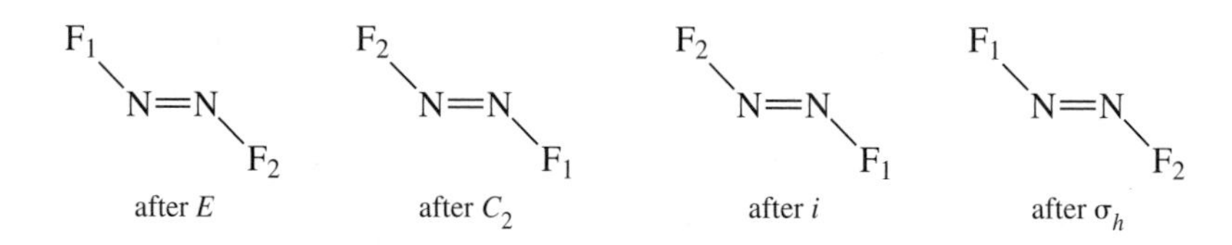

Matrix Representations (Reducible)

$$E: \begin{bmatrix} 1 & 0 & 0 \\ 0 & 1 & 0 \\ 0 & 0 & 1 \end{bmatrix} \quad C_2: \begin{bmatrix} -1 & 0 & 0 \\ 0 & -1 & 0 \\ 0 & 0 & 1 \end{bmatrix} \quad i: \begin{bmatrix} -1 & 0 & 0 \\ 0 & -1 & 0 \\ 0 & 0 & -1 \end{bmatrix} \quad \sigma_h: \begin{bmatrix} 1 & 0 & 0 \\ 0 & 1 & 0 \\ 0 & 0 & -1 \end{bmatrix}$$

Characters of Matrix Representations

3   –1   –3   1

Block Diagonalized Matrices

$$\begin{bmatrix} [1] & 0 & 0 \\ 0 & [1] & 0 \\ 0 & 0 & [1] \end{bmatrix} \quad \begin{bmatrix} [-1] & 0 & 0 \\ 0 & [-1] & 0 \\ 0 & 0 & [1] \end{bmatrix} \quad \begin{bmatrix} [-1] & 0 & 0 \\ 0 & [-1] & 0 \\ 0 & 0 & [-1] \end{bmatrix} \quad \begin{bmatrix} [1] & 0 & 0 \\ 0 & [1] & 0 \\ 0 & 0 & [-1] \end{bmatrix}$$

Irreducible Representations

	E	C_2	i	σ_h	Coordinate Used
	1	–1	–1	1	x
	1	–1	–1	1	y (x and y give the same irreducible representation)
	1	1	–1	–1	z
Γ	3	–1	–3	1	

4.7 Chiral molecules may have only proper rotations. The C_1, C_n, and D_n groups—along with the rare T, O, and I groups—meet this condition.

4.8

$$\sigma_{xz}: \begin{bmatrix} 1 & 0 & 0 & 0 & 0 & 0 & 0 & 0 & 0 \\ 0 & -1 & 0 & 0 & 0 & 0 & 0 & 0 & 0 \\ 0 & 0 & 1 & 0 & 0 & 0 & 0 & 0 & 0 \\ 0 & 0 & 0 & 1 & 0 & 0 & 0 & 0 & 0 \\ 0 & 0 & 0 & 0 & -1 & 0 & 0 & 0 & 0 \\ 0 & 0 & 0 & 0 & 0 & 1 & 0 & 0 & 0 \\ 0 & 0 & 0 & 0 & 0 & 0 & 1 & 0 & 0 \\ 0 & 0 & 0 & 0 & 0 & 0 & 0 & -1 & 0 \\ 0 & 0 & 0 & 0 & 0 & 0 & 0 & 0 & 1 \end{bmatrix} \quad \sigma_{yz}: \begin{bmatrix} -1 & 0 & 0 & 0 & 0 & 0 & 0 & 0 & 0 \\ 0 & 1 & 0 & 0 & 0 & 0 & 0 & 0 & 0 \\ 0 & 0 & 1 & 0 & 0 & 0 & 0 & 0 & 0 \\ 0 & 0 & 0 & 0 & 0 & 0 & -1 & 0 & 0 \\ 0 & 0 & 0 & 0 & 0 & 0 & 0 & 1 & 0 \\ 0 & 0 & 0 & 0 & 0 & 0 & 0 & 0 & 1 \\ 0 & 0 & 0 & -1 & 0 & 0 & 0 & 0 & 0 \\ 0 & 0 & 0 & 0 & 1 & 0 & 0 & 0 & 0 \\ 0 & 0 & 0 & 0 & 0 & 1 & 0 & 0 & 0 \end{bmatrix}$$

4.9 The reducible representation (D_{2h} symmetry) is:

D_{2h}	E	$C_2(z)$	$C_2(y)$	$C_2(x)$	i	$\sigma(xy)$	$\sigma(xz)$	$\sigma(yz)$
Γ	18	0	0	−2	0	6	2	0

This reduces to: $3\,A_g + 3\,B_{1g} + 2\,B_{2g} + B_{3g} + A_u + 2\,B_{1u} + 3\,B_{2u} + 3\,B_{3u}$

Translational modes (matching x, y, and z): $B_{1u} + B_{2u} + B_{3u}$

Rotational modes (matching R_x, R_y, and R_z): $B_{1g} + B_{2g} + B_{3g}$

Vibrational modes (all that remain): $3\,A_g + 2\,B_{1g} + B_{2g} + A_u + B_{1u} + 2\,B_{2u} + 2\,B_{3u}$

4.10 **a.** $\Gamma_1 = A_1 + T_2$:

T_d	E	$8C_3$	$3C_2$	$6S_4$	$6\sigma_d$
Γ_1	4	1	0	0	2
A_1	1	1	1	1	1
A_2	1	1	1	−1	−1
E	2	−1	2	0	0
T_1	3	0	−1	1	−1
T_2	3	0	−1	−1	1

For A_1: $\frac{1}{24} \times [(4 \times 1) + 8(1 \times 1) + 3(0 \times 1) + 6(0 \times 1) + 6(2 \times 1)] = 1$

For A_2: $\frac{1}{24} \times [(4 \times 1) + 8(1 \times 1) + 3(0 \times 1) + 6(0 \times (-1)) + 6(2 \times (-1))] = 0$

For E: $\frac{1}{24} \times [(4 \times 2) + 8(1 \times (-1)) + 3(0 \times 2) + 6(0 \times 0) + 6(2 \times 0)] = 0$

For T_1: $\frac{1}{24} \times [(4 \times 3) + 8(1 \times 0) + 3(0 \times (-1)) + 6(0 \times 1) + 6(2 \times (-1))] = 0$

For T_2: $\frac{1}{24} \times [(4 \times 3) + 8(1 \times 0) + 3(0 \times (-1)) + 6(0 \times (-1)) + 6(2 \times 1)] = 1$

Adding the characters of A_1 and T_2 for each operation confirms the result.

b. $\Gamma_2 = A_1 + B_1 + E$:

D_{2d}	E	$2S_4$	C_2	$2C_2'$	$2\sigma_d$
Γ_2	4	0	0	2	0
A_1	1	1	1	1	1
A_2	1	1	1	−1	−1
B_1	1	−1	1	1	−1
B_2	1	−1	1	−1	1
E	2	0	−2	0	0

For A_1: $\frac{1}{8} \times [(4 \times 1) + 2(0 \times 1) + (0 \times 1) + 2(2 \times 1) + 2(0 \times 1)] = 1$

For A_2: $\frac{1}{8} \times [(4 \times 1) + 2(0 \times 1) + (0 \times 1) + 2(2 \times (-1)) + 2(0 \times (-1))] = 0$

For B_1: $\frac{1}{8} \times [(4 \times 1) + 2(0 \times (-1)) + (0 \times 1) + 2(2 \times 1) + 2(0 \times (-1))] = 1$

For B_2: $\frac{1}{8} \times [(4 \times 1) + 2(0 \times (-1)) + (0 \times 1) + 2(2 \times (-1)) + 2(0 \times 1)] = 0$

For E: $\frac{1}{8} \times [(4 \times 2) + 2(0 \times 0) + (0 \times (-2)) + 2(2 \times 0) + 2(0 \times 0)] = 1$

Adding the characters of A_1, B_1, and E for each operation confirms the result.

c. $\Gamma_3 = A_2 + B_1 + B_2 + 2E$:

C_{4v}	E	$2C_4$	C_2	$2\sigma_v$	$2\sigma_d$
Γ_3	7	−1	−1	−1	−1
A_1	1	1	1	1	1
A_2	1	1	1	−1	−1
B_1	1	−1	1	1	−1
B_2	1	−1	1	−1	1
E	2	0	−2	0	0

For A_1: $\frac{1}{8} \times [(7 \times 1) + 2((-1) \times 1) + ((-1) \times 1) + 2((-1) \times 1) + 2((-1) \times 1)] = 0$

For A_2: $\frac{1}{8} \times [(7 \times 1) + 2((-1) \times 1) + ((-1) \times 1) + 2((-1) \times (-1)) + 2((-1) \times (-1))] = 1$

For B_1: $\frac{1}{8} \times [(7 \times 1) + 2((-1) \times (-1)) + ((-1) \times 1) + 2((-1) \times 1) + 2((-1) \times (-1))] = 1$

For B_2: $\frac{1}{8} \times [(7 \times 1) + 2((-1) \times (-1)) + ((-1) \times 1) + 2((-1) \times (-1)) + 2((-1) \times 1)] = 1$

For E: $\frac{1}{8} \times [(7 \times 2) + 2((-1) \times 0) + ((-1) \times (-2)) + 2((-1) \times 0) + 2((-1) \times 0)] = 2$

4.11 The A_{2u} (matching symmetry of z) and both E_u (matching symmetry of x and y together) vibrational modes are IR active.

4.12 Vibrational analysis for NH_3:

C_{3v}	E	$2C_3$	$3\sigma_v$	
Γ	12	0	2	
A_1	1	1	1	z
A_2	1	1	−1	R_z
E	2	−1	0	(x, y) (R_x, R_y)

a. A_1: $\frac{1}{6}[(12 \times 1) + 2(0 \times 1) + 3(2 \times 1)] = 3$

A_2: $\frac{1}{6}[(12 \times 1) + 2(0 \times 1) + 3(2 \times (-1))] = 1$

E: $\frac{1}{6}[(12 \times 2) + 2(0 \times (-1)) + 3(2 \times 0)] = 4$

$\Gamma = 3A_1 + A_2 + 4E$

b. Translation: $A_1 + E$, based on the x, y, and z entries in the table
Rotation: $A_2 + E$, based on the R_x, R_y, and R_z entries in the table
Vibration: $2A_1 + 2E$ remaining from the total; the A_1 vibrations are symmetric stretch and symmetric bend. The E vibrations are asymmetric.

c. There are three translational modes, three rotational modes, and six vibrational modes, for a total of 12. With 4 atoms in the molecule, $3N = 12$, so there are $3N$ degrees of freedom in the ammonia molecule.

d. All the vibrational modes are IR active (all have x, y, or z symmetry).

4.13 Taking only the C—O stretching modes for $Mn(CO)_5Cl$ (only the vectors between the C and O atoms):

C_{4v}	E	$2C_4$	C_2	$2\sigma_v$	$2\sigma_d$	
Γ	5	1	1	3	1	
A_1	1	1	1	1	1	z
A_2	1	1	1	−1	−1	R_z
B_1	1	−1	1	1	−1	
B_2	1	−1	1	−1	1	
E	2	0	−2	0	0	(x, y) (R_x, R_y)

$\Gamma = 2A_1 + B_1 + E$

$Mn(CO)_5Cl$ should have four IR-active stretching modes, two from A_1 and two from E. The E modes are a degenerate pair; they give rise to a single infrared band. The B_1 mode is IR inactive.

4.14 Using as basis the I═O bonds, the following representation is obtained in the D_{5h} point group:

D_{5h}	E	$2C_5$	$2C_5^2$	$5C_2$	σ_h	$2S_5$	$2S_5^2$	$5\sigma_v$
Γ	2	2	2	0	0	0	0	2

This reduces to:

A_1'	1	1	1	1	1	1	1	1	$x^2 + y^2, z^2$
A_2''	1	1	1	−1	−1	−1	−1	1	z

The A_1' vibration matches the symmetry of $x^2 + y^2$ and z^2; it is Raman active. The A_2'' does not match xy, xz, yz, or a squared term; it matches z and is therefore active in the IR but not in the Raman spectrum. Therefore, the single Raman band is consistent with *trans* orientation.

CHAPTER 5

5.1 p_x and d_{xz} $\qquad$ p_z and d_{z^2} $\qquad$ s and $d_{x^2-y^2}$

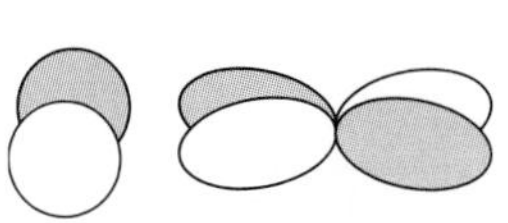

π interaction

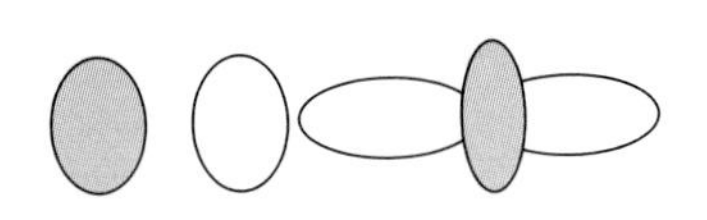

σ interaction

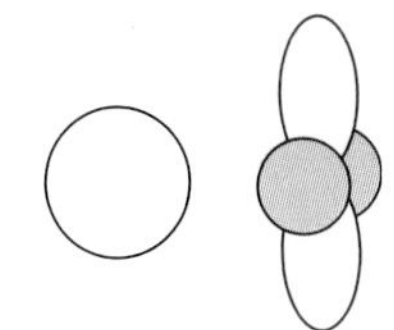

no interaction

5.2 In Figure 5.3, (a), σ^* is σ_u, σ is σ_g, π^* is π_g, π is π_u, δ^* is δ_u, and δ is δ_g.

5.3 Bonding in the OH^- ion.

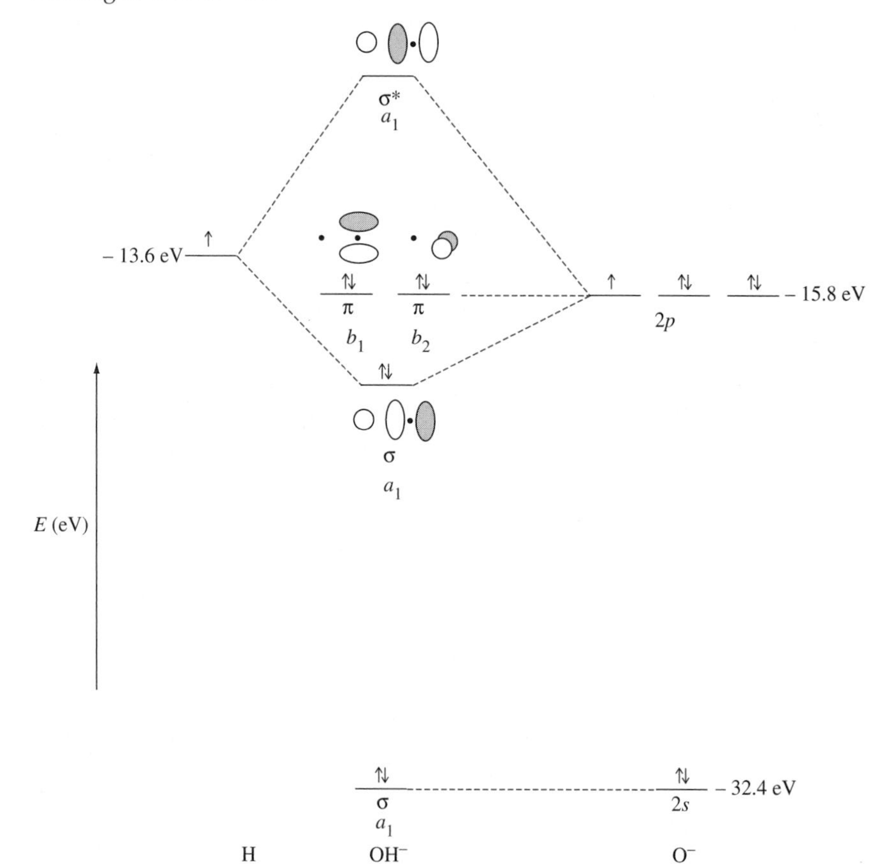

The energy match of the H 1*s* and the O 2*p* orbitals is fairly good, but that of the H 1*s* with the O 2*s* is poor. Therefore, molecular orbitals are formed between the H 1*s* and the O $2p_z$, as shown above (*z* is the axis through the nuclei). All the other O orbitals are nonbonding, either because of poor energy match or lack of useful overlap.

5.4 H_3^+ energy levels.

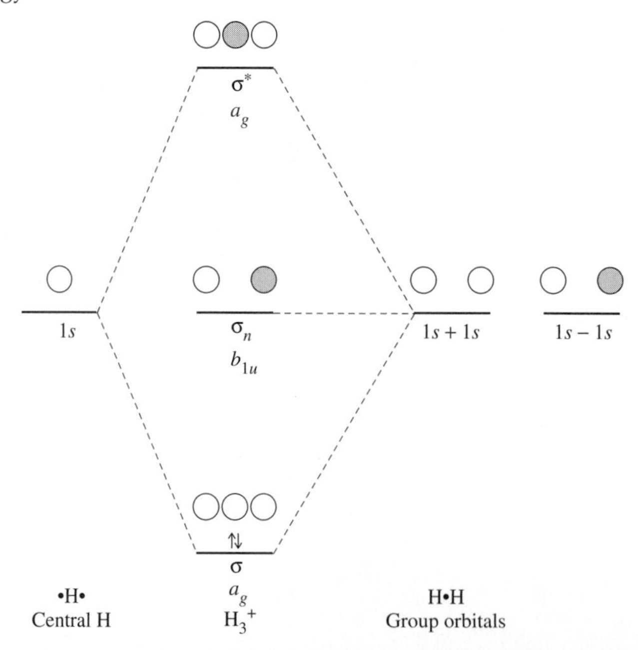

5.5 Group orbital 1: Every operation in the D_{2h} point group transforms the orbital into one identical to the original, so the character for each operation is 1, matching the top (A_g) row in the character table.

Group orbital 2: Each operation that transforms this orbital into one identical to the original (· → ·) has a character of 1, and each operation that reverses the signs of the lobes (· → ·) has a character of –1. When this is carried out for all eight operations, the result matches the B_{1u} row of the table.

Group orbital 3 has identical symmetry properties as group orbital 1, so it is also classified A_g.

Group orbital 4 has identical symmetry properties as group orbital 2, so it is also classified B_{1u}.

Group orbitals 5 through 8: As described for group orbital 2, operations that result in orbital lobes identical to those of the original group orbital give a character of 1, and operations that reverse the signs of the lobes give a character of –1. The results for these four group orbitals match the labels in Figure 5.19.

5.6 Group orbital 2 is made up of oxygen 2*s* orbitals, with an orbital potential energy of –32.4 eV. Group orbital 4 is made up of oxygen $2p_z$ orbitals, with an orbital potential energy of –15.9 eV. The carbon $2p_z$ orbital has an orbital potential energy of –10.7 eV, a much better match for the energy of group orbital 4. In general, energy differences greater than about 12 eV are too large for effective combination into molecular orbitals.

5.7 The molecular orbitals of N_3^- differ from those of CO_2 described in Section 5.4.2 because all the atoms have the same initial orbital energies. Therefore, the best orbitals are formed by combinations of the three 2*s* orbitals or three 2*p* orbitals of the same type (*x*, *y*, or *z*). The resulting pattern of orbitals is shown in the diagram. Note that the σ_n orbital is slightly higher in energy than the π_n orbitals as a consequence of an antibonding interaction with the 2*s* orbital of the central nitrogen. For symmetry labels on the orbitals, see Figure 5.26.

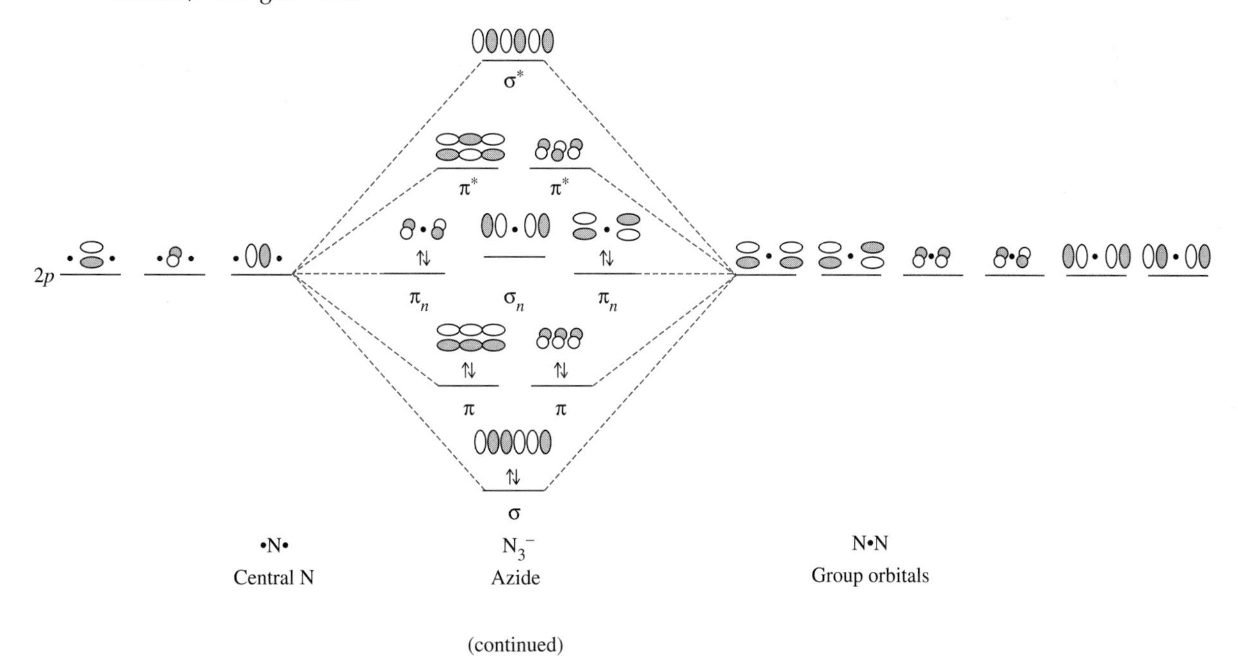

(continued)

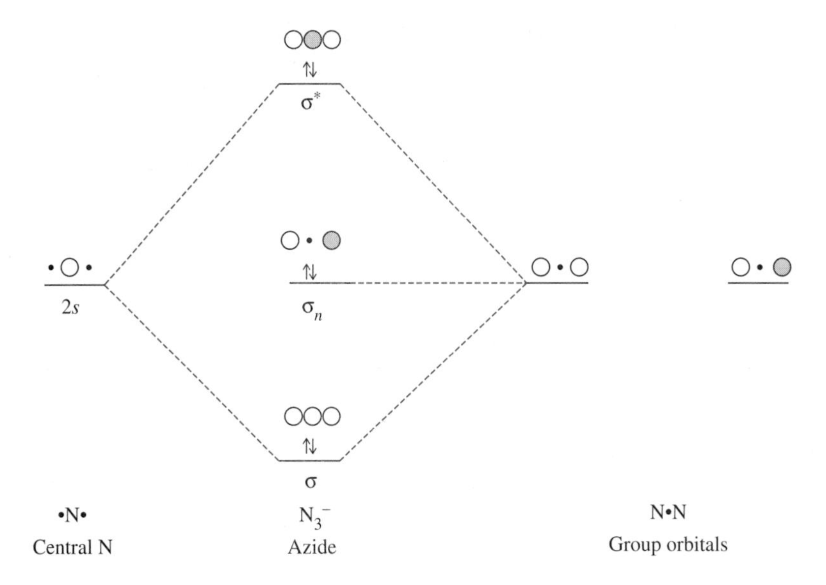

5.8 BeH_2 molecular orbitals.

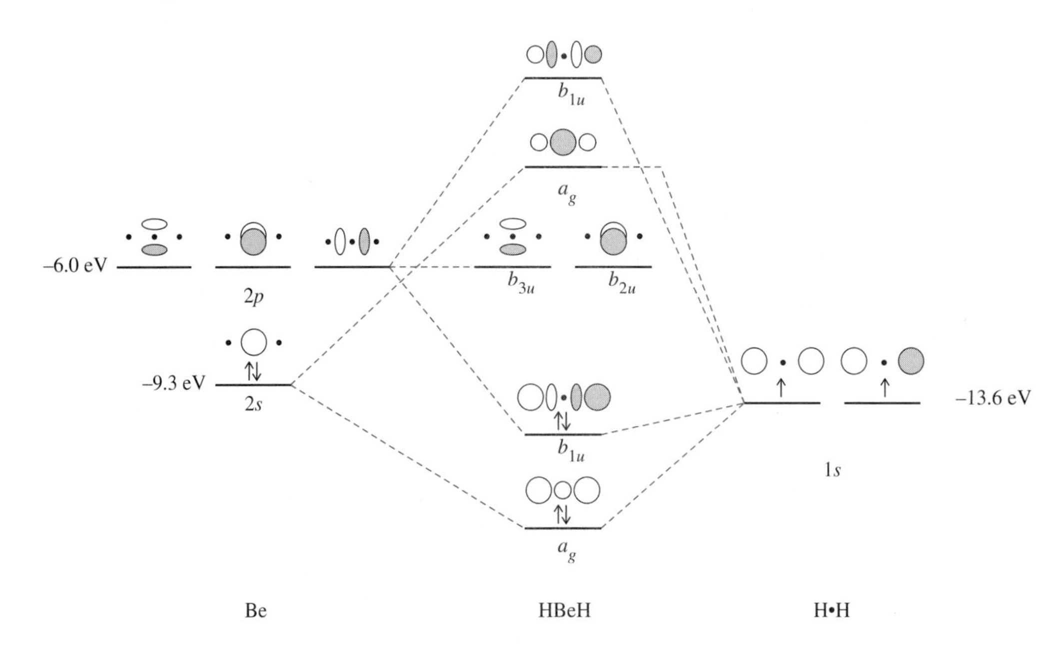

5.9 **a.** PF_5 has a trigonal bipyramidal geometry with D_{3h} symmetry. The group orbitals for the five fluorine 2*s* or $2p_y$ orbitals (the *y* axes are directed toward the P) are found from the reducible representation.

D_{3h}	E	$2C_3$	$3C_2$	σ_h	$2S_3$	$3\sigma_v$		
Γ	5	2	1	3	0	3		
A_1'	1	1	1	1	1	1		$x^2 + y^2, z^2$
A_2'	1	1	–1	1	1	–1	R_z	
E'	2	–1	0	2	–1	0	(x, y)	$(x^2 - y^2, xy)$
A_1''	1	1	1	–1	–1	–1		
A_2''	1	1	–1	–1	–1	1	z	
E''	2	–1	0	–2	1	0	(R_x, R_y)	(xz, yz)

The reduction to $\Gamma = 2A_1' + E' + A_2''$ can be verified by the usual procedures. The P orbitals that match are then $3s$, $3d_{z^2}$, $3p_x$, $3p_y$, and $3p_z$, for dsp^3 hybrids.

b. $[PtCl_4]^{2-}$ has a square-planar geometry and a D_{4h} point group. The four group orbitals can be found from the reducible representation, where $\Gamma = A_{1g} + B_{1g} + E_u$.

D_{4h}	E	$2C_4$	C_2	$2C_2'$	$2C_2''$	i	$2S_4$	σ_h	$2\sigma_v$	$2\sigma_d$		
Γ	4	0	0	2	0	0	0	4	2	0		
A_{1g}	1	1	1	1	1	1	1	1	1	1		$x^2 + y^2, z^2$
B_{1g}	1	−1	1	1	−1	1	−1	1	1	−1		$x^2 - y^2$
E_u	2	0	−2	0	0	−2	0	2	0	0	(x, y)	

The Pt orbitals used in bonding are then the s, d_{z^2} (both A_{1g}), $d_{x^2-y^2}(B_{1g})$, p_x, and p_y (E_u). Symmetry allows two sets of hybrids, dsp^2 or d^2p^2.

5.10 $SOCl_2$ has only a mirror plane and belongs to group C_s. Using s orbitals on O and the two Cls, we can obtain the reducible representation and its irreducible components shown in the table.

C_s	E	σ_h		
Γ	3	1		
A'	1	1	x, y, R_z	x^2, y^2, z^2, xy
A''	1	−1	z, R_x, R_y	yz, xz

$\Gamma = 2A' + A''$

The sulfur orbitals used in σ bonding are the $3p_x$, $3p_y$, and $3p_z$. The $3s$ could be involved, but the nonplanar shape of the molecule requires that all three p orbitals be used.

CHAPTER 6

6.1 $2\,IF_5 \rightleftharpoons IF_4^+ + IF_6^-$

$IF_5 + SbF_5 \rightleftharpoons IF_4^+ + SbF_6^-$ increases the concentration of IF_4^+, the acid in this solvent.

$IF_5 + F^- \rightleftharpoons IF_6^-$ results in an increased concentration of IF_6^-, the base in this solvent.

6.2 If SO_2 is labeled with ^{18}O, it becomes $SO^{18}O$. Reaction with another molecule of SO_2 then gives the results

$$SO^{18}O + SO_2 \rightleftharpoons S^{18}O^{2+} + SO_3^{2-}$$

or

$$SO^{18}O + SO_2 \rightleftharpoons SO^{2+} + SO_2{}^{18}O^{2-}$$

If the $S^{18}O^{2+}$ product reacts with Cl^-, the result is labeled $SOCl_2$:

$$S^{18}O^{2+} + 2\,Cl^- \rightleftharpoons S^{18}OCl_2.$$

If $SOCl_2$ is labeled, it again forms $S^{18}O^{2+}$ when it dissociates:

$$S^{18}OCl_2 \rightleftharpoons S^{18}O^{2+} + 2\,Cl^-$$

If this product reacts with SO_3^{2-} from SO_2 dissociation, the result is labeled SO_2:

$$S^{18}O^{2+} + SO_3^{2-} \rightleftharpoons SO^{18}O + SO_2$$

6.3 **a.** Cu^{2+} is a borderline soft acid and will react more readily with NH_3 than with the harder OH^-, with the product $[Cu(NH_3)_4]^{2+}$ in a solution containing significant amounts of both NH_3 and OH^-. In the same fashion, it will react more readily with S^{2-} than with O^{2-}, forming CuS in basic solutions of sulfide.

b. On the other hand, Fe^{3+} is a hard acid and will react more readily with OH^- and O^{2-}. The product in basic ammonia is $Fe(OH)_3$ (approximately; the product is an ill-defined hydrated Fe(III) oxide and hydroxide mixture). In basic sulfide solution, the same $Fe(OH)_3$ product is formed. (There may also be some reduction of Fe(III) to Fe(II) and precipitation of FeS.)

c. Silver ion is a soft acid and is likely to combine more readily with PH_3 than with NH_3.

d. CO is a relatively soft base, Fe^{3+} is a very hard acid, Fe^{2+} is a borderline acid, and Fe is a soft acid. Therefore, CO is more likely to combine effectively with Fe(0) than with Fe(II) or Fe(III).

6.4 **a.** Al^{3+} has $I = 119.99$ and $A = 28.45$. Therefore,

$$\chi = \frac{119.99 + 28.45}{2} = 74.22 \qquad \eta = \frac{119.99 - 28.45}{2} = 45.77$$

Fe^{3+} has $I = 54.8$ and $A = 30.65$. Therefore,

$$\chi = \frac{54.8 + 30.65}{2} = 42.7 \qquad \eta = \frac{54.8 - 30.655}{2} = 12.1$$

Co^{3+} has $I = 51.3$ and $A = 33.50$. Therefore,

$$\chi = \frac{51.3 + 33.50}{2} = 42.4 \qquad \eta = \frac{51.3 - 33.50}{2} = 8.9$$

b. OH^- has $I = 13.17$ and $A = 1.83$. Therefore,

$$\chi = \frac{13.17 + 1.83}{2} = 7.50 \qquad \eta = \frac{13.17 - 1.83}{2} = 5.67$$

Cl^- has $I = 13.01$ and $A = 3.62$. Therefore,

$$\chi = \frac{13.01 + 3.62}{2} = 8.31 \qquad \eta = \frac{13.01 - 3.62}{2} = 4.70$$

NO_2^- has $I > 10.1$ and $A = 2.30$. Therefore,

$$\chi = \frac{> 10.1 + 2.3}{2} > 6.2 \qquad \eta = \frac{> 10.1 - 2.3}{2} > 3.9$$

c. H_2O has $I = 12.6$ and $A = -6.4$. Therefore,

$$\chi = \frac{12.6 + (-6.4)}{2} = 3.1 \qquad \eta = \frac{12.6 - (-6.4)}{2} = 9.5$$

NH_3 has $I = 10.7$ and $A = -5.6$. Therefore,

$$\chi = \frac{10.7 + (-5.6)}{2} = 2.6 \qquad \eta = \frac{10.7 - (-5.6)}{2} = 8.2$$

PH_3 has $I = 10.0$ and $A = -1.9$. Therefore,

$$\chi = \frac{10.0 + (-1.9)}{2} = 4.0 \qquad \eta = \frac{10.0 - (-1.9)}{2} = 6.0$$

6.5 **a.**

Acid	E_A	C_A	Base	E_B	C_B	ΔH (E)	ΔH (C)	ΔH (total)
BF_3	9.88	1.62	NH_3	1.36	3.46	−13.44	−5.60	−19.04
			CH_3NH_2	1.30	5.88	−12.84	−9.53	−22.37
			$(CH_3)_2NH$	1.09	8.73	−10.77	−14.14	−24.91
			$(CH_3)_3N$	0.808	11.54	−7.98	−18.70	−26.68

b.

Base	E_B	C_B	Acid	E_A	C_A	ΔH (E)	ΔH (C)	ΔH (total)
py	1.17	6.40	Me_3B	6.14	1.70	−7.18	−10.88	−18.06
			Me_3Al	16.9	1.43	−19.77	−9.15	−28.92
			Me_3Ga	13.3	0.881	−15.56	−5.64	−21.20

The amine series shows a steady increase in C_B and decrease in E_B as methyl groups are added. The methyl groups push electrons onto the N, and the lone-pair electrons are then made more available to the acid BF_3. As a result, covalent bonds to BF_3 are more likely with more methyl groups. The lone pairs of the molecules with fewer methyl groups are more tightly held, with more ionic bonding and a larger E_B and ΔH (*E*). The covalent effect is stronger and has a larger change in ΔH, so it determines the order in the total.

The B, Al, Ga series is less regular. Possible arguments for Al being the strongest in E_A include the following:

1. The larger central atom leads to more electrostatic bonding and less covalent bonding. This would make the order B < Al < Ga for electrostatic bonding, rather than the B < Ga < Al that is calculated.
2. The *d* electrons in Ga shield the outer electrons, so they are held less tightly. This results in less electrostatic attraction for the outer electrons and those of the pyridine, which makes them less likely to form either an electrostatic (ionic) or covalent bond.

6.6 Dissociation of acetic acid

a. By Hess's law:

$$\Delta H° = +55.9 - 56.3 = -0.4 \text{ kJ mol}^{-1}$$

$$\Delta S° = -80.4 - 12.0 = -92.4 \text{ J K}^{-1} \text{ mol}^{-1}$$

b. By temperature dependence:

$$\Delta H° = -2.8 \text{ kJ mol}^{-1} \text{ from the slope}$$

$$\Delta S° = -100 \text{ JK}^{-1} \text{ mol}^{-1} \text{ from the intercept}$$

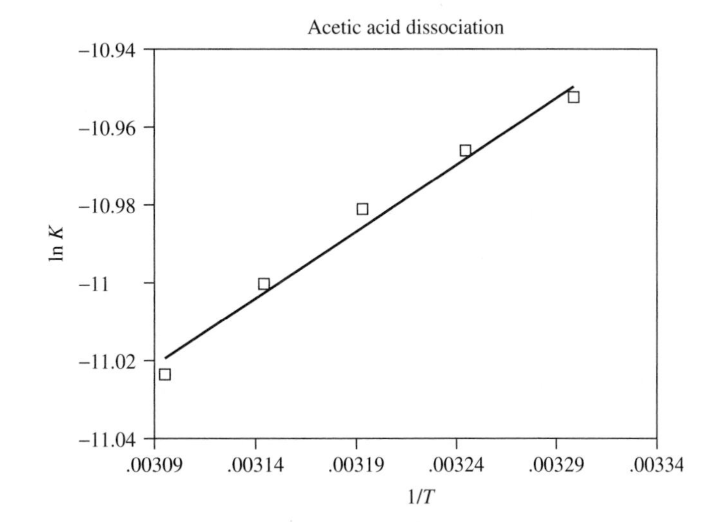

6.7 **a.**

	H_2SO_3	HSO_3^-
pK_a (9 – 7*n*)	2	7
pK_a (8 – 5*n*)	3	8
pK_a (exptl)	1.9	7.2

b.

	H_3PO_3	$H_2PO_3^-$
pK_a (9 – 7*n*)	2	7
pK_a (8 – 5*n*)	3	8
pK_a (exptl)	1.8	6.2

6.8 Boron trifluoride has electron-withdrawing fluorine atoms; trimethyl boron has electron-donating methyl groups. As a result, boron trifluoride should be a stronger acid, with a more positive boron atom, toward ammonia. There are no significant steric interactions with ammonia.

With bulkier bases, the same results should be seen, because BF_3 is less bulky and would have less steric interference, regardless of the structure of the base. In this case, BF_3 is the stronger acid toward both sterically hindered and sterically unhindered bases.

6.9 **a.** Acetic acid in water is only slightly dissociated according to the equation

$$HOAc \rightleftharpoons H^+ + OAc^-$$

Sodium hydroxide is completely dissociated into Na^+ and OH^-. During the titration, the primary reaction is

$$HOAc + OH^- \longrightarrow H_2O + OAc^-$$

At the midpoint, half the original acetic acid is present as HOAc and half as OAc^-. At the end point, all the HOAc has been converted to OAc^-, and the solution contains primarily Na^+ and OAc^-. The next increment of OH^- added does not react but remains as OH^-.

b. Acetic acid acts as a strong base in pyridine, forming pyridinium ion and acetate:

$$HOAc + py \longrightarrow Hpy^+ + OAc^-$$

Tetramethylammonium hydroxide is a strong base, completely dissociated into $(CH_3)_4N^+ + OH^-$. During the titration, the hydroxide reacts with the pyridinium ion:

$$OH^- + Hpy^+ \longrightarrow py + H_2O$$

At the midpoint, half the pyridine is Hpy^+ and half is py. At the end point, all the pyridine is converted to the free base, py, and the remaining ions are $(CH_3)_4N^+$ and OAc^-. Any additional titrant added simply adds $(CH_3)_4N^+$ and OH^-.

CHAPTER 7

7.1 **a.** Atom or ion in the center of the cell = 1
8 atoms or ions at the corners of the cell, each $\frac{1}{8}$ within the cell = 1 Total = 2 atoms or ions per unit cell

b. Eight atoms at the corners of the cell, four of which are $\frac{1}{12}$ $\left(\frac{1}{2} \times \frac{1}{6}\right)$ in the cell and four of which are $\frac{1}{6}$ $\left(\frac{1}{2} \times \frac{1}{3}\right)$ in the cell $= \frac{4}{12} + \frac{4}{6} = 1$ atom per unit cell.

7.2 If the edge of the unit cell has a length of a, the face diagonal is $\sqrt{2}a$ and the body diagonal is $\sqrt{3}a$, from the Pythagorean theorem. The diagonal through a body-centered unit cell also has a length of $4r$, where r is the radius of each of the atoms (r for each of the corner atoms and $2r$ for the body-centered atom). Therefore,

$$4r = \sqrt{3}a, \quad \text{or} \quad a = 2.31r$$

7.3 A simple cubic array of anions with radius r_- has a body diagonal of length $2\sqrt{3}r_-$. With a cation of the ideal size in the body center, this distance is also $2r_+ + 2r_-$. Setting the two equal to each other and solving, $r_+ = 0.732r_-$, or a radius ratio of $r_+/r_- = 0.732$. Any ratio between 0.732 and 1.00 (the ideal for CN = 12) should fit the fluorite structure. $CaCl_2$ has r_+/r_- between 126/167 = 0.754 (CN = 8) and 114/167 = 0.683 (CN = 6). $CaBr_2$ has r_+/r_- between 126/182 = 0.692 (CN = 8) and 114/182 = 0.626 (CN = 6). $CaCl_2$ is on the edge of the CN = 6, CN = 8 boundary, but $CaBr_2$ is clearly in the CN = 6 region. Both crystallize with CN = 6 in structures that are similar to rutile (TiO_2).

7.4 Using the Born Mayer equation from Section 7.2.1:

$$U = \frac{NMZ_+Z_-}{r_0}\left[\frac{e^2}{4\pi\varepsilon_0}\right]\left(1 - \frac{\rho}{r_0}\right)$$

with $r_0 = r_+ + r_- = 116 + 167 = 283$ pm, $M = 1.74756$, $N = 6.02 \times 10^{23}$, $Z_+ = Z_- = 1$, $\rho = 30$ pm, and $\frac{e^2}{4\pi\varepsilon_0} = 2.3071 \times 10^{-28}$ J m. Changing r_0 to meters in the first fraction, the result is 766 kJ mol^{-1}.

7.5

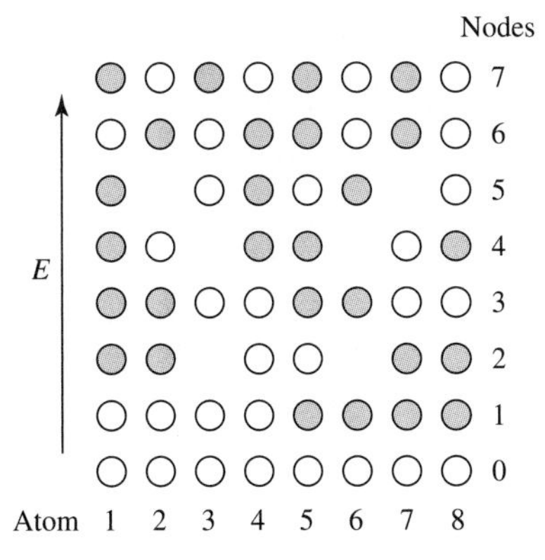

The diagram is approximate because some of the nodes are near, but not exactly over, an atomic nucleus. Using a much larger number of atoms would show the alternating signs more clearly.

7.6 Consider the bottom tetrahedron to have four oxygen atoms, or SiO_4. Proceeding clockwise, the next tetrahedron duplicates one oxygen (shared with the first tetrahedron), adding SiO_3. The last tetrahedron shares oxygens (one each) with the first two tetrahedra, adding SiO_2, for a total of three silicons and nine oxygens. Using the charges Si^{+4} and O^{2-} gives the formula $Si_3O_9{}^{6-}$.

CHAPTER 8

8.1 The fraction remaining for first-order decay = 0.5^n, where n = number of half-lives.

$$3.5 \times 10^{-2} = 0.5^n \qquad n \log 0.5 = \log(3.5 \times 10^{-2})$$

$$n = \frac{\log(3.5 \times 10^{-2})}{\log 0.5} = 4.836 \text{ half-lives}$$

$$\text{Age} = n(5730 \text{ years}) = 2.77 \times 10^4 \text{ years}$$

8.2

$$\begin{aligned}
\mathrm{NO} &\longrightarrow \mathrm{N_2} & \mathscr{E}^\circ &= 1.68\ \mathrm{V} \\
\mathrm{N_2} &\longrightarrow \mathrm{N_2O} & \mathscr{E}^\circ &= -1.77\ \mathrm{V} \\
2\,\mathrm{H^+} + 2\,\mathrm{NO} + 4\,\mathrm{e^-} &\longrightarrow \mathrm{N_2} + 2\,\mathrm{H_2O} & \Delta G^\circ &= -n\mathscr{F}\mathscr{E}^\circ = -(4)(1.68\ \mathrm{V})\mathscr{F} = -6.72\ \mathrm{V}\mathscr{F} \\
\mathrm{H_2O} + \mathrm{N_2} &\longrightarrow \mathrm{N_2O} + 2\,\mathrm{e^-} & \Delta G^\circ &= -n\mathscr{F}\mathscr{E}^\circ = -(2)(-1.77\ \mathrm{V})\mathscr{F} = 3.54\ \mathrm{V}\mathscr{F} \\
\hline
2\,\mathrm{H^+} + 2\,\mathrm{NO} + 2\,\mathrm{e^-} &\longrightarrow \mathrm{N_2O} + \mathrm{H_2O} & \Delta G^\circ &= -n\mathscr{F}\mathscr{E}^\circ = -(2)\mathscr{F}\mathscr{E}^\circ = -3.18\ \mathrm{V}\mathscr{F} \\
 & & \mathscr{E}^\circ &= 3.18\ \mathrm{V}/2 = 1.59\ \mathrm{V}
\end{aligned}$$

8.3 In NH_4^+, N has oxidation state –3; in NO_3^-, N has oxidation state +5. The average is +1, which fits N_2O, so the reaction is $NH_4NO_3 \longrightarrow 2\,N_2O + 2\,H_2O$.

$$\begin{aligned}
\mathrm{H_2O} + 2\,\mathrm{NH_4^+} &\longrightarrow \mathrm{N_2O} + 10\,\mathrm{H^+} + 8\,\mathrm{e^-} & \mathscr{E}^\circ &= -1.30\ \mathrm{V} \\
8\,\mathrm{e^-} + 10\,\mathrm{H^+} + 2\,\mathrm{NO_3^-} &\longrightarrow \mathrm{N_2O} + 5\,\mathrm{H_2O} & \mathscr{E}^\circ &= 2.34\ \mathrm{V} \\
\hline
2\,\mathrm{NH_4^+} + 2\,\mathrm{NO_3^-} &\longrightarrow 2\,\mathrm{N_2O} + 4\,\mathrm{H_2O} & \mathscr{E}^\circ &= 1.04\ \mathrm{V}
\end{aligned}$$

With a positive net potential, the reaction will go spontaneously. With heating, this reaction can be used to prepare small amounts of N_2O but is hazardous, because it can become explosive.

CHAPTER 9

9.1
a. Triamminetrichlorochromium(III)
b. Dichloroethylenediamineplatinum(II)
c. Bis(oxalato)platinate(II) or bis(oxalato) platinate (2 –)
d. Pentaaquabromochromium(III) or pentaaquabromochromium (2 +)
e. Tetrachloroethylenediaminecuprate(II) or tetrachloroethylenediaminecuprate (2 –)
f. Tetrahydroxoferrate(III) or tetrahydroxoferrate (1 –)

9.2
a. Tris(acetylacetonato) iron(III)

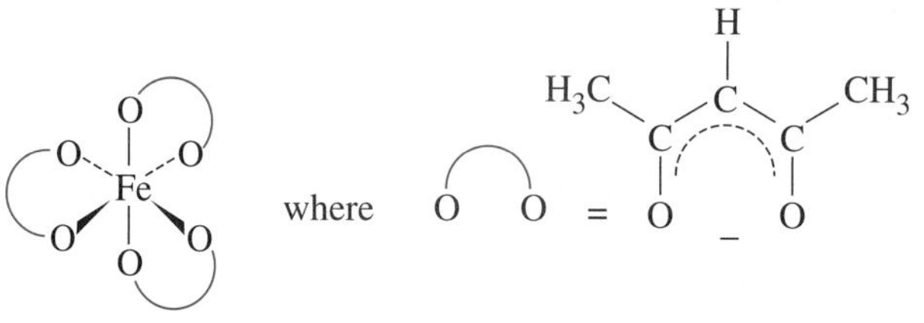

where

Δ isomer; Λ is also possible (Section 9.3.5)

b. Hexabromoplatinate (2 –)

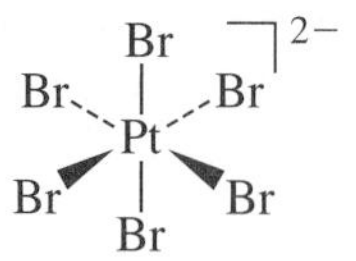

c. Diamminetetrabromocobaltate(III)

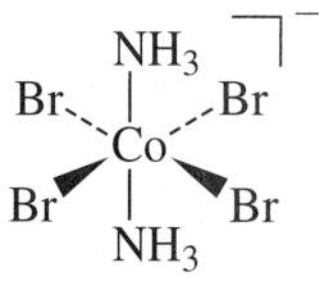

or

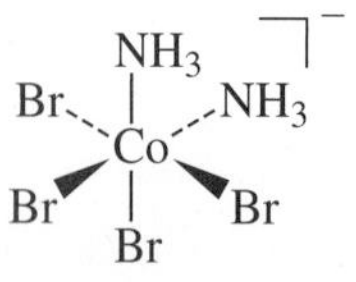

trans *cis*

d. Tris(ethylenediamine)copper(II)

N N N 2+
Cu
N N N

where N N = $H_2NCH_2CH_2NH_2$

Δ isomer; Λ is also possible (Section 9.3.5)

e. Hexacarbonylmanganese(I)

O
C +
OC CO
Mn
OC CO
C
O

f. Tetrachlororuthenate (1 –)

Cl –
Cl---Ru
Cl
Cl

9.3 Ma_2b_2cd has eight isomers, including two pairs of enantiomers, according to the method of Section 9.3.4.

M<aa><bb><cd> M<aa><bc><bd> M<ac><ad><bb> M<ab><ab><cd>

d a b / b a c   b a b / d a c   d b a / c b a   b c a / b d a

M<ab><ad><bc> M<ab><ac><bd>

b a b / a d c   b a b / c d a   c b a / b d a   a b c / a d b

9.4 M(AA)bcde has six geometric isomers, each with enantiomers, for a total of 12 isomers.

M<Ab><Ac><de>	M<Ab><Ad><ce>	M<Ab><Ae><cd>
M<Ac><Ad><be>	M<Ac><Ae><bd>	M<Ad><Ae><bc>

9.5 This is a Λ configuration:

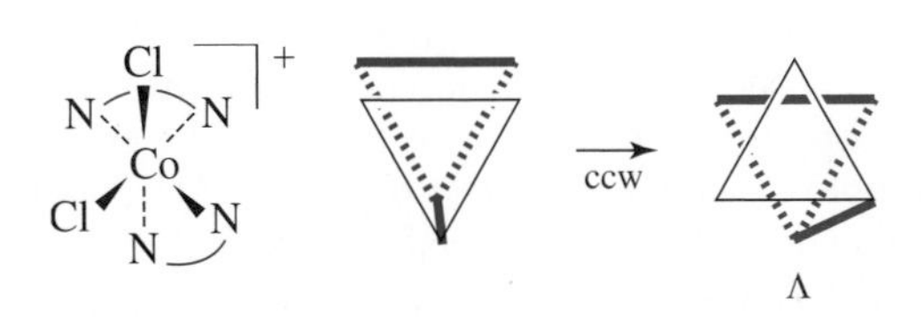

9.6 This can be analyzed more easily if it is flipped over, rotating about a horizontal axis in the plane of the paper. The result on the right can be checked for the two rings on the lower front as they relate to the ring across the back. One is Λ and one is Δ:

Rotating the complex to bring the other upper ring (top to back right) into the horizontal position, we find that one of the front rings is in the same plane, and the other is Δ. Overall, $\Delta\Delta\Lambda$.

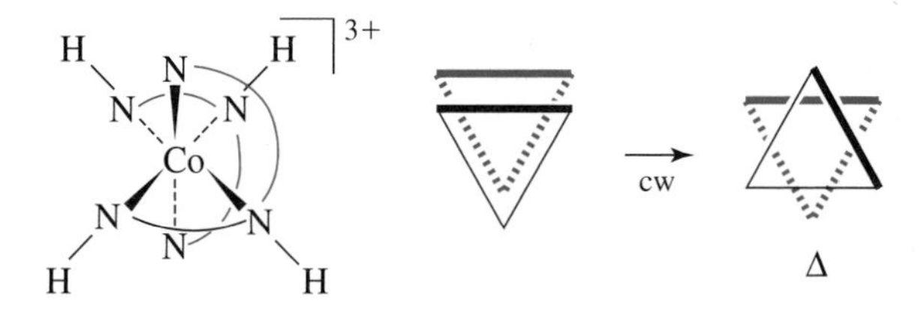

The other isomer is a meridional form that has two Λ and two Λ combinations. The first set has rings as originally shown. The second set is shown after a clockwise C_4 rotation about the axis, through the top front and bottom rear nitrogens of the vertical diethylenetriamine.

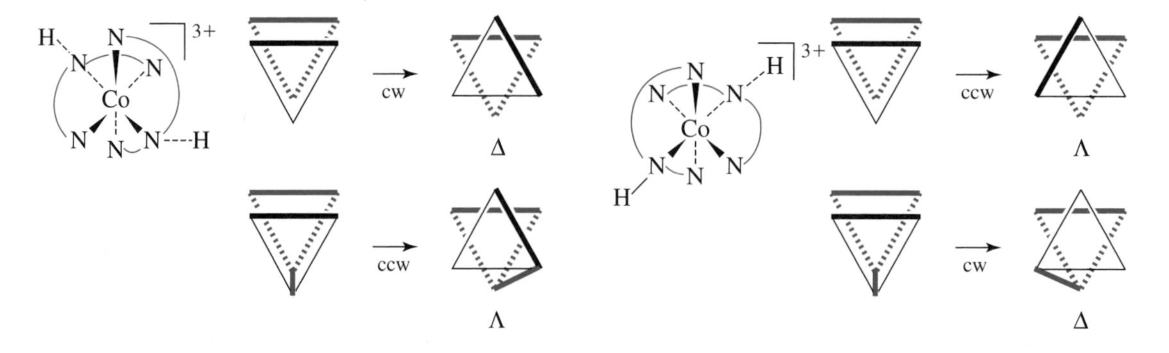

9.7 In the thiocyanate ligand, the harder, less polarizable terminal atom is nitrogen. This end of the ligand tends to interact more strongly with more polar, harder solvents, leading to stronger solvation. Consequently, formation of a bond between the metal and the less solvated sulfur at the opposite end of SCN^- is favored. The more polarizable, softer sulfur tends to interact more effectively with less polar solvents, thereby favoring bond formation between the metal and the nitrogen end of SCN^-.

CHAPTER 10

10.1 Nitrogen has three electrons in the $2p$ levels, with $m_l = -1, 0, +1$ and all with $m_s = +\frac{1}{2}$.

$$M_S = \frac{1}{2} + \frac{1}{2} + \frac{1}{2} = \frac{3}{2}, \quad M_L = -1 + 0 + 1 = 0, \quad \text{so} \quad S = \frac{3}{2} \text{ and } L = 0.$$

10.2 $S = n/2$, with n the number of unpaired electrons.

$$4\,S(S+1) = 4(n/2)(n/2+1) = n^2 + 2n = n(n+2), \quad \text{so} \quad \sqrt{4S(S+1)} = \sqrt{n(n+2)}$$

10.3 Fe has the electron configuration $4s^2\,3d^6$, with four unpaired electrons. From the equations in Exercise 10.2,

$$\mu = \sqrt{4(4+2)} = 4.9\,\mu_B \quad \text{(Bohr magnetons)}$$

Fe^{2+} has the configuration $3d^6$ (the $4s$ electrons are lost first) with four unpaired electrons:

$$\mu = \sqrt{4(4+2)} = 4.9\,\mu_B$$

Cr has the electron configuration $4s^1\,3d^5$ with six unpaired electrons.

$$\mu = \sqrt{6(6+2)} = 6.9\,\mu_B$$

Cr^{3+} has the electron configuration $3d^3$ with three unpaired electrons.

$$\mu = \sqrt{3(3+2)} = 3.9\,\mu_B$$

Cu has the electron configuration $4s^1\,3d^{10}$ with one unpaired electron.

$$\mu = \sqrt{1(1+2)} = 1.7\,\mu_B$$

Cu^{2+} has the electron configuration $3d^9$ with one unpaired electron.

$$\mu = \sqrt{1(1+2)} = 1.7\,\mu_B$$

10.4 E: All six vectors remain unchanged; character = 6.

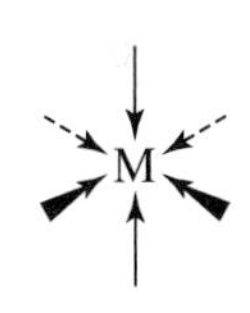

$8C_3$ and $6C_2$: All vectors are rotated into another position; character = 0.

$6C_4$: Rotation leaves only the vectors along the C_4 axis unchanged; character = 2.

$3C_2(= C_4{}^2)$: Rotation leaves only the vectors along the C_2 axis unchanged; character = 2.

i: All vectors are switched to opposite sides; character = 0.

$6S_4$: Rotation moves four vectors, reflection switches the other two; character = 0.

$8S_6$: All vectors are moved; character = 0.

$3\sigma_h$: Reflection keeps all four vectors in the plane of reflection unchanged, switches the other two; character = 4.

$6\sigma_d$: This type of reflection has only two atoms in the reflection plane, and all others are moved; character = 2.

By the method of Section 4.4.2, the representation reduces to $A_{1g} + T_{1u} + E_g$. It can easily be verified—by summing the columns for these three representations, as provided in the chapter—that the total in each column matches the characters shown above.

10.5 The representations of the octahedral π orbitals can be found by using the x and z coordinates of the ligands in Figure 10.7. The characters for the symmetry operations are in the row labeled Γ_π in the table that precedes this exercise. In the C_2, C_4, and σ operations, some of the vectors do not move, but they are balanced by vectors whose direction is reversed. Only the E and $C_2 = C_4^{\,2}$ operations have nonzero totals, so they are the only ones used in finding the irreducible representations.

O_h	E	$8C_3$	$6C_2$	$6C_4$	$3C_2(= C_4^{\,2})$	i	$6S_4$	$8S_6$	$3\sigma_h$	$6\sigma_d$	
Γ_π	12	0	0	0	−4	0	0	0	0	0	
T_{1g}	3	0	−1	1	−1	3	1	0	−1	−1	
T_{2g}	3	0	1	−1	−1	3	−1	0	−1	1	(d_{xy}, d_{xz}, d_{yz})
T_{1u}	3	0	−1	1	−1	−3	−1	0	1	1	(p_x, p_y, p_z)
T_{2u}	3	0	1	−1	−1	−3	1	0	1	−1	

T_{1g}, T_{2g}, T_{1u}, and T_{2u} each total $= \frac{1}{48}[(12 \times 3) + 3(-4)(-1)] = 1.$

All other representations total 0.

10.6 A high-spin d^5 ion has three exchange possibilities in the t_{2g} level (1-2, 1-3, 2-3), and one exchange possibility in the e_g level (4-5), for a total exchange energy of $4\Pi_e$.

A low-spin d^5 ion has one unpaired electron. The set of three with the same spin has three exchange possibilities (1-2, 1-3, 2-3) and the set of two has one exchange possibility (4-5), for a total of four and a total exchange energy of $4\Pi_e$, the same as in the high spin case.

10.7 A low-spin d^6 ion has all six electrons in the lower t_{2g} levels, each at $-2/5\Delta_o$, so the total LFSE $= 6(-2/5\Delta_o) = -12/5\Delta_o$.

A high-spin d^6 ion has four electrons in the t_{2g} levels at $-2/5\Delta_o$ and two in the e_g levels at $3/5\Delta$. LFSE $= 4(-2/5\Delta_o) + 2(3/5\Delta_o) = -2/5\Delta_o$.

10.8 Using p_y orbitals of the four ligands, the reducible representation has four unchanged vectors for the E and σ_h operations and two for the C_2' and σ_v operations (the vectors along the C_2' axis and contained in the σ_v plane). All other operations result in changed positions and a character of 0. The p_x and p_z orbitals are similar, except that the p_x changes direction with the C_2' and σ_v operations, and the p_z changes direction with the C_2' and σ_v operations.

D_{4h}	E	$2C_4$	C_2	$2C_2'$	$2C_2''$	i	$2S_4$	σ_h	$2\sigma_v$	$2\sigma_d$	
Γ_{p_y}	4	0	0	2	0	0	0	4	2	0	σ
Γ_{p_x}	4	0	0	−2	0	0	0	4	−2	0	$\pi_{\parallel}$
Γ_{p_z}	4	0	0	−2	0	0	0	−4	2	0	$\pi_{\perp}$

Only the nonzero operations need to be used in finding the irreducible representations.

For Γ_{p_y}: A_{1g}, and B_{1g} each total $\frac{1}{16}[(4 \times 1) + 2(2)(1) + 1(4)(1) + 2(2)(1)] = 1$

E_u totals $\frac{1}{16}[(4 \times 2) + 2(2)(0) + 1(4)(2) + 2(2)(0)] = 1$

For Γ_{p_x}: A_{2g}, B_{2g} each total $\frac{1}{16}[(4 \times 1) + 2(-1)(-2) + 1(1)(4) + 2(-1)(-2)] = 1$

E_u totals $\frac{1}{16}[(4 \times 2) + 2(-2)(0) + 1(4)(2) + 2(-2)(0)] = 1$

For Γ_{p_z}: A_{2u}, B_{2u} each total $\frac{1}{16}[(4 \times 1) + 2(-1)(-2) + 1(-1)(-4) + 2(1)(2)] = 1$

E_g totals $\frac{1}{16}[(4 \times 2) + 2(-2)(0) + 1(-4)(-2) + 2(2)(0)] = 1$

All others total 0.

10.9 Energy changes:

d_{xy} total for 7, 8, 9, 10 $= 1.33e_{\sigma}$

d_{xz} total for 7, 8, 9, 10 $= 1.33e_{\sigma}$

d_{yz} total for 7, 8, 9, 10 $= 1.33e_{\sigma}$

d_{z^2} total for 7, 8, 9, 10 $= 0$

$\Delta_0 = 3e_{\sigma},\ \Delta_t = 1.33e_{\sigma},\ \Delta_t = \frac{4}{9}\Delta_0$

$d_{x^2-y^2}$ total for 7, 8, 9, 10 = 0

Ligands are lowered by e_{σ} each.

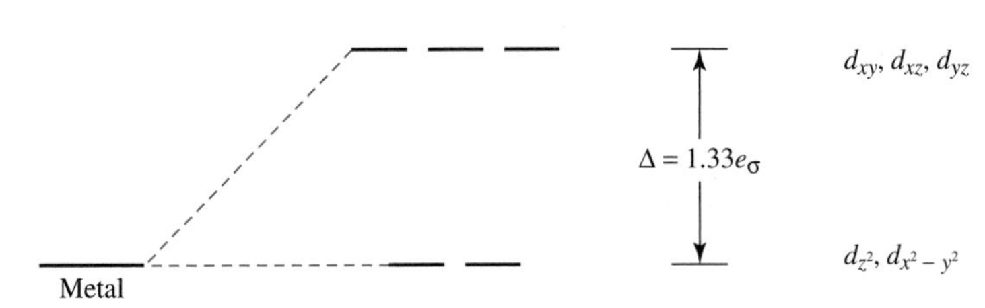

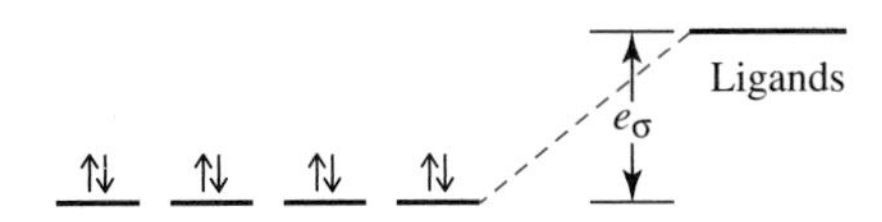

10.10 Adding π bonding to the results of Exercise 10.9 results in these changes in energy:

d_{xy}, d_{xz}, d_{yz} each total $0.89e_{\pi}$ for 7, 8, 9, 10

d_{z^2} and $d_{x^2-y^2}$ each total $2.67e_{\pi}$ for 7, 8, 9, 10

Ligands decrease by $2e_{\pi}$ each.

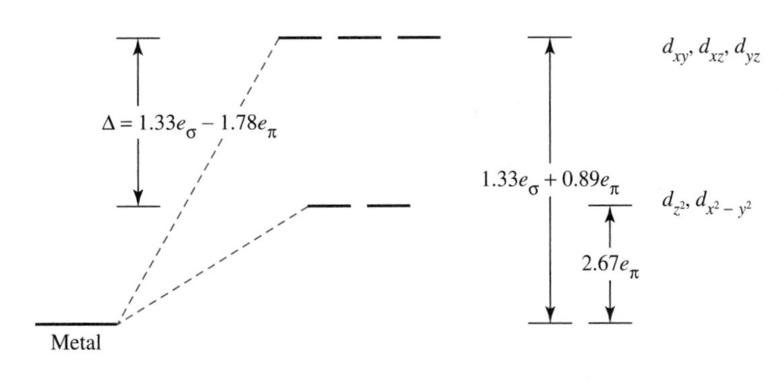

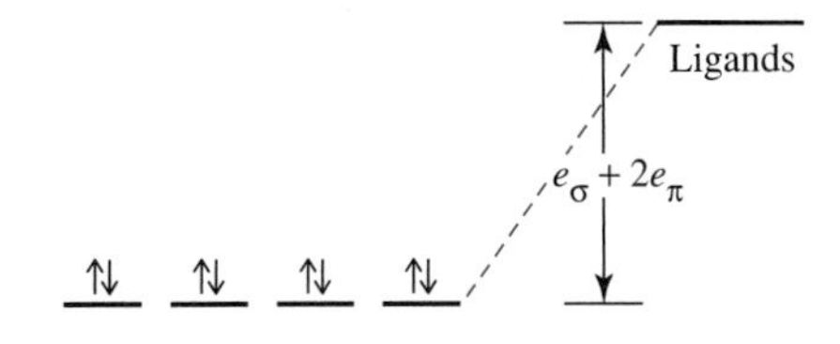

10.11 Square planar (positions 2, 3, 4, 5)

a. Energy changes for σ only:

d_{z^2} total for 2, 3, 4, 5 $= e_{\sigma}$

$d_{x^2-y^2}$ total for 2, 3, 4, 5 $= 3e_{\sigma}$

d_{xy}, d_{xz}, d_{yz} total for 2, 3, 4, 5 = 0

Ligands decrease by e_σ each.

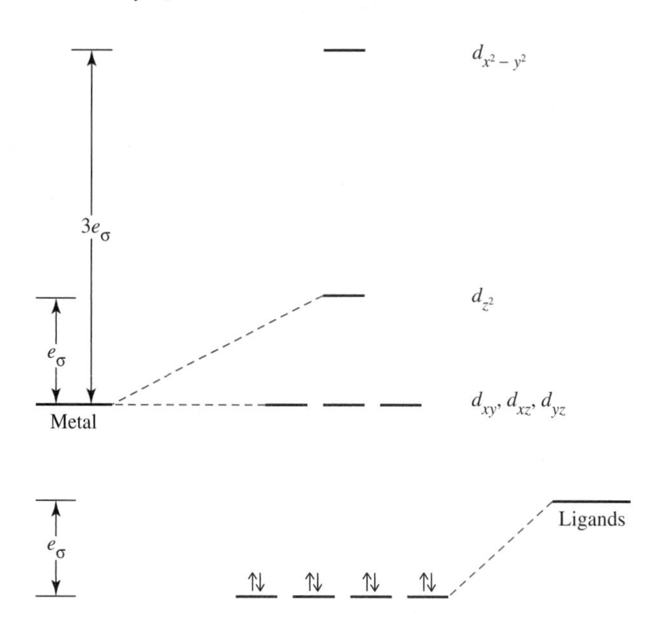

b. Adding π:

$d_{z^2}, d_{x^2-y^2}$ total for 2, 3, 4, 5 = 0

d_{xz}, d_{yz} total for 2, 3, 4, 5 = $2e_\pi$

d_{xy} total for 2, 3, 4, 5 = $4e_\pi$

Ligand π^* orbitals increase by $2e_\pi$ each.

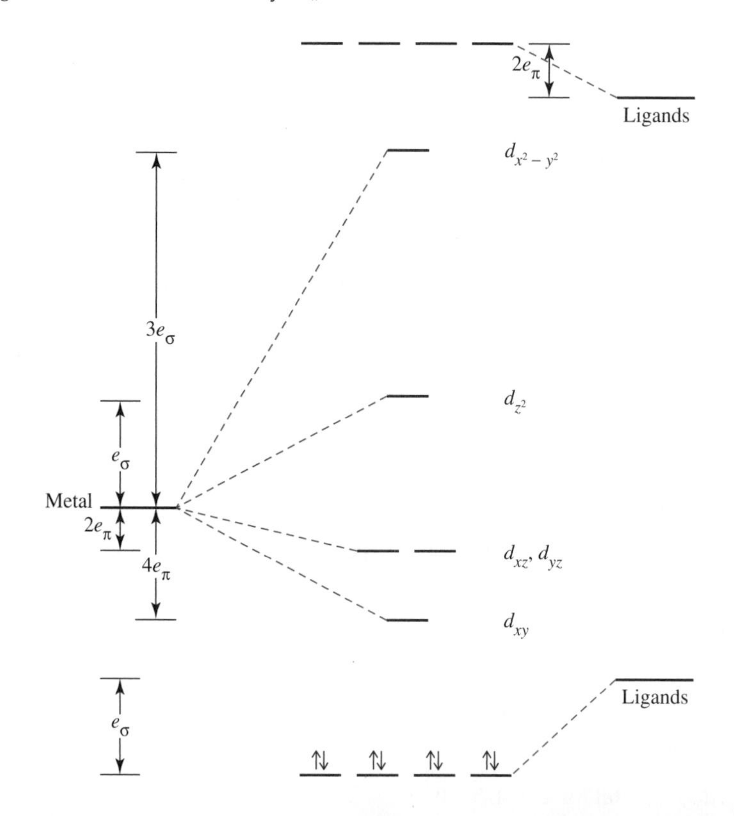

10.12 See Table 10.5 for the complete high-spin and low-spin configurations.

Number of electrons		1	2	3	4	5	6	7	8	9	10
High spin	e_g	0	0	0	1	2	2	2	2	3	4
	Jahn-Teller	w	w		s		w	w		s	
	t_{2g}	1	2	3	3	3	4	5	6	6	6
Low spin	e_g	0	0	0	0	0	0	1	2	3	4
	Jahn-Teller	w	w		w	w		s		s	
	t_{2g}	1	2	3	4	5	6	6	6	6	6

The weak Jahn–Teller cases have unequal occupation of t_{2g} orbitals; the strong Jahn–Teller cases have unequal occupation of e_g orbitals.

10.13 Ligands are in positions 1, 2, 6, 11, and 12 as shown in Table 10.10. The angular overlap parameters for these positions are as follows:

Position	z^2	$x^2 - y^2$	xy	xz	yz
1	1	0	0	0	0
2	$\frac{1}{4}$	$\frac{3}{4}$	0	0	0
6	1	0	0	0	0
11	$\frac{1}{4}$	$\frac{3}{16}$	$\frac{9}{16}$	0	0
12	$\frac{1}{4}$	$\frac{3}{16}$	$\frac{9}{16}$	0	0
Total	$2\frac{3}{4}$	$1\frac{1}{8}$	$1\frac{1}{8}$	0	0

The results are consistent with Figure 10.31. The highest energy molecular orbital involves the d_{z2} orbital and has symmetry label A_1' in the D_{3h} point group; it is predicted by the angular overlap approach to have energy 2.75 e_σ. The degenerate pair involving the $d_{x^2-y^2}$ and d_{xy} orbitals (symmetry label E') has energy 1.125 e_σ, and the lowest energy degenerate pair (symmetry label E''), consisting of the d_{xz} and d_{yz} orbitals, does not interact with the ligand orbitals; its energy is 0 e_σ.

CHAPTER 11

11.1 Microstate table for d^2.

M_S (columns), M_L (rows)

M_L \ M_S	−1	0	+1
+4		$2^+\ 2^-$	
+3	$2^-\ 1^-$	$2^+\ 1^-$ $2^-\ 1^+$	$2^+\ 1^+$
+2	$2^-\ 0^-$	$2^+\ 0^-$ $2^-\ 0^+$ $1^+\ 1^-$	$2^+\ 0^+$
+1	$2^-\ {-1}^-$ $1^-\ 0^-$	$2^+\ {-1}^-$ $2^-\ {-1}^+$ $1^+\ 0^-$ $1^-\ 0^+$	$2^+\ {-1}^+$ $1^+\ 0^+$
0	$-2^-\ 2^-$ $-1^-\ 1^-$	$-2^+\ 2^-$ $-1^+\ 1^-$ $0^+\ 0^-$ $-1^-\ 1^+$ $-2^-\ 2^+$	$-2^+\ 2^+$ $-1^+\ 1^+$
−1	$-1^-\ 0^-$ $-2^-\ 1^-$	$-1^+\ 0^-$ $-1^-\ 0^+$ $-2^-\ 1^+$ $-2^+\ 1^-$	$-1^+\ 0^+$ $-2^+\ 1^+$
−2	$-2^-\ 0^-$	$-1^+\ {-1}^-$ $-2^+\ 0^-$ $-2^-\ 0^+$	$-2^+\ 0^+$
−3	$-2^-\ {-1}^-$	$-2^+\ {-1}^-$ $-2^-\ {-1}^+$	$-2^+\ {-1}^+$
−4		$-2^+\ {-2}^-$	

11.2 ^2D $L = 2, S = \frac{1}{2}$
$M_L = -2, -1, 0, 1, 2$
$M_S = -\frac{1}{2}, \frac{1}{2}$

M_L \ M_S	$-\frac{1}{2}$	$+\frac{1}{2}$
+2	x	x
+1	x	x
0	x	x
−1	x	x
−2	x	x

^1P $L = 1, S = 0$
$M_L = -1, 0, 1$
$M_S = 0$

M_L \ M_S	0
+1	x
0	x
−1	x

^2S $L = 0, S = \frac{1}{2}$
$M_L = 0$
$M_S = -\frac{1}{2}, \frac{1}{2}$

M_L \ M_S	$-\frac{1}{2}$	$+\frac{1}{2}$
0	x	x

11.3
$L = 4, S = 0, J = 4$ ^1G
$L = 3, S = 1, J = 4, 3, 2$ ^3F
$L = 2, S = 0, J = 2$ ^1D
$L = 1, S = 1, J = 2, 1, 0$ ^3P
$L = 0, S = 0, J = 0$ ^1S

Following Hund's rules:

1. The highest spin (S) is 1, so the ground state is ^3F or ^3P.
2. The highest L in Step 1 is $L = 3$, so ^3F is the ground state.

11.4 The J values for each term are shown in the solution to Exercise 11.3. The full set of term symbols for a d^2 configuration is $^1G_4$, $^3F_4$, $^3F_3$, $^3F_2$, $^1D_2$, $^3P_2$, $^3P_1$, $^3P_0$, $^1S_0$. Hund's third rule predicts which J value corresponds to the lowest energy state. The d orbitals are less than half-filled, so the minimum J value for ^3F, $J = 2$, is the ground state. Overall, it is $^3F_2$.

11.5 High-spin d^6

1. ↑ ↑
   ↑↓ ↑ ↑
2. Spin multiplicity $= 4 + 1 = 5$, $S = 2$
3. Maximum possible value of $M_L = 2 + 2 + 1 + 0 - 1 - 2 = 2$, therefore, D term, $L = 2$.
4. ^5D

Low-spin d^6

1. —— ——
   ↑↓ ↑↓ ↑↓
2. Spin multiplicity $= 0 + 1 = 1$, $S = 0$
3. Maximum value of $M_L = 2 + 2 + 1 + 1 + 0 + 0 = 6$; therefore, I term, $L = 6$.
4. ^1I

11.6 **a.** The e_g level is asymmetrically occupied, so this is an E state.
b. The e_g and t_{2g} levels are symmetrically occupied, making this an A state.
c. The t_{2g} levels are asymmetrically occupied; this is a T state.

11.7 $[Fe(H_2O)_6]^{2+}$ is a high-spin d^6 complex. The weak field (left) part of the Tanabe–Sugano diagram for d^6 shows that the only excited state with the same spin multiplicity (5) as the ground state is the ^5E. The transition is therefore $^5T_2 \longrightarrow {}^5E$. The excited state $t_{2g}{}^3 e_g{}^3$ is subject to Jahn–Teller distortion; consequently, as in the d^1 complex $[Ti(H_2O)_6]^{3+}$, the absorption band is split.

11.8 First assigning transitions, which are to the left of the crossover point of $^4A_2$ and $^4T_1$:

$$^4T_1 \longrightarrow {}^4T_2 \qquad \text{not seen in this range } \nu_1$$

$$^4T_1 \longrightarrow {}^4A_2 \qquad 16{,}000 \text{ cm}^{-1} = \nu_2$$

$$^4T_1 \longrightarrow {}^4T_1 \qquad 20{,}000 \text{ cm}^{-1} = \nu_3 \qquad \nu_3/\nu_2 = 1.25$$

From the Tanabe-Sugano diagram, $\nu_3/\nu_2 = 1.25$ at $\Delta/B = 10$; $\nu_3 = 25$ and $\nu_2 = 20$; and $\nu_3/\nu_2 = 1.25$ B and Δ can then be calculated:

$$\nu_2 = 20 \qquad B = E/\nu_2 = 16{,}000/20 = 800 \text{ cm}^{-1}$$

$$\nu_3 = 25 \qquad B = E/\nu_2 = 20{,}000/25 = 800 \text{ cm}^{-1}$$

$$\Delta/B = 10, \Delta = 10 \times 800 = 8{,}000 \text{ cm}^{-1}$$

11.9

$VO_4{}^{3-}$, vanadate	$CrO_4{}^{2-}$, chromate	$MnO_4{}^{-}$, permanganate
colorless	yellow	purple

As the charge on the nucleus increases, the vacant metal d orbitals are pulled to lower energies. The difference between the oxygen donor orbitals and the metal d orbitals grows smaller, and less energy is required for the CTTM transition. Permanganate, which absorbs yellow (the complement of purple) requires the least energy.

CHAPTER 12

12.1

$$ML_5X + Y \underset{k_{-1}}{\overset{k_1}{\rightleftharpoons}} ML_5 + XY$$

$$ML_5XY \xrightarrow{k_2} ML_5Y + X$$

Applying the stationary-state approach to ML_5XY,

$$\frac{d\,[ML_5XY]}{dt} = k_1\,[ML_5X][Y] - k_{-1}[ML_5XY] - k_2[ML_5XY] = 0$$

$$\text{and } [ML_5XY] = \frac{k_1[ML_5X][Y]}{k_{-1} + k_2}$$

From the second equation,

$$\frac{d\,[ML_5Y]}{dt} = k_2\,[ML_5XY]$$

Combining the two,

$$\frac{d[ML_5Y]}{dt} = \frac{k_1k_2[ML_5X][Y]}{k_{-1} + k_2} = k[ML_5X][Y]$$

12.2

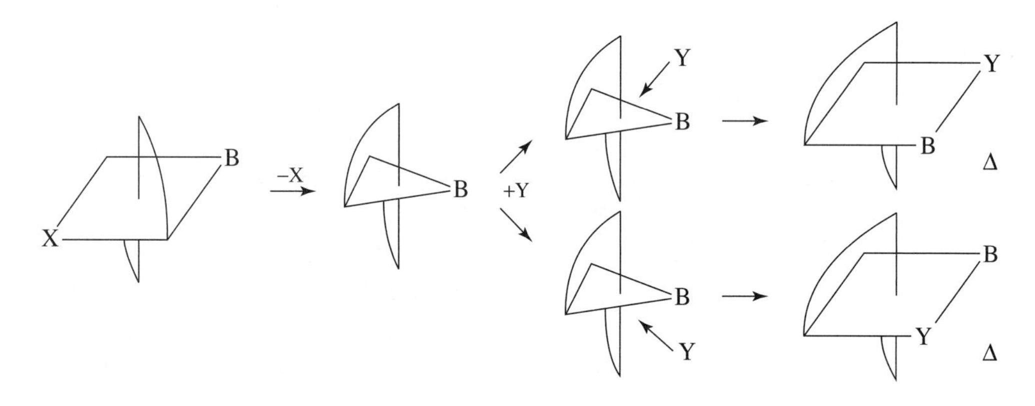

12.3 $[PtCl_4]^{2-} + NO_2^- \longrightarrow$ (a) (a) + $NH_3 \longrightarrow$ (b)

(a) = $[PtCl_3(NO_2)]^{2-}$ (b) = *trans*-$[PtCl_2(NO_2)(NH_3)]^-$

NO_2^- is a better *trans* director than Cl^-.

$[PtCl_3(NH_3)]^- + NO_2^- \longrightarrow$ (c) (c) + $NO_2^- \longrightarrow$ (d)

(c) = *cis*-$[PtCl_2(NO_2)(NH_3)]^-$ (d) = *trans*-$[PtCl(NO_2)_2(NH_3)]^-$

Cl^- has a larger *trans* effect than NH_3, and NO_2^- has a larger *trans* effect than either Cl^- or NH_3. In the first step, Cl^- is a better leaving group than NH_3.

$[PtCl(NH_3)_3]^+ + NO_2^- \longrightarrow$ (e) (e) + $NO_2^- \longrightarrow$ (f)

(e) = *trans*-$[PtCl(NO_2)(NH_3)_2]$ (f) = *trans*-$[Pt(NO_2)_2(NH_3)_2]$

Cl^- has a larger *trans* effect than NH_3, and NO_2^- has a larger *trans* effect than either Cl^- or NH_3.

$[PtCl_4]^{2-} + I^- \longrightarrow$ (g) (g) + $I^- \longrightarrow$ (h)

(g) = $[PtCl_3I]^{2-}$ (h) = *trans*-$[PtCl_2I_2]^{2-}$

I^- has a larger *trans* effect than Cl^-.

$[PtI_4]^{2-} + Cl^- \longrightarrow$ (i) (i) + $Cl^- \longrightarrow$ (j)

(i) = $[PtClI_3]^{2-}$ (j) = *cis*-$[PtCl_2I_2]^{2-}$

I^- has a larger *trans* effect than Cl^- and replacement of Cl^- in the second step would give no net change.

12.4 **a.** $[PtCl_4]^{2-} + 2\,NH_3 \longrightarrow$ *cis*-$PtCl_2(NH_3)_2 + 2\,Cl^-$
Because chloride is both a stronger *trans* director and a better leaving group, the major product is the *cis* isomer.

b. *cis*-$PtCl_2(NH_3)_2 + 2$ py $\longrightarrow$ *cis*-$[Ptpy_2(NH_3)_2]^{2+} + 2\,Cl^-$
As shown in (h) of Figure 12.12, the chloride is again a better leaving group, so the major product is the *cis* isomer.

c. *cis*-$[Ptpy_2(NH_3)_2]^{2+} + 2\,Cl^- \longrightarrow$ *trans*-$PtCl_2(NH_3)$py + py + NH_3
The major product is the *trans* isomer, as shown in (e) and (f) of Figure 12.12. Chloride is the strongest *trans* director, so the *trans* isomer is formed regardless of whether py or NH_3 is the first ligand replaced.

d. *trans*-$PtCl_2(NH_3)py + NO_2^- \longrightarrow Pt < Cl(NO_2)> < (NH_3)py> + Cl^-$ (Cl^- is *trans* to NO_2^-; NH_3 is *trans* to py.)
Chloride is the strongest *trans* director (and the best leaving group), so one of them is replaced.

This sequence has been used to form the complex with four different groups. With some modification, it has been used with oxidation by Cl_2 and another substitution to form complexes with six different groups, with the geometry predictable to the last stage.

CHAPTER 13

13.1

	Method A		Method B	
a. $[Fe(CO)_4]^{2-}$	Fe^{2-}	10	Fe	8
	4 CO	8	4 CO	8
			2–	2
		18		18
b. $[(\eta^5\text{-}C_5H_5)_2\,Co]^+$	Co^{3+}	6	Co	9
	2 Cp^-	12	2 Cp	10
			1 +	−1
		18		18
c. $(\eta^3\text{-}C_5H_5)(\eta^5\text{-}C_5H_5)Fe(CO)$	Fe	8	Fe	8
	$\eta^3\text{-}Cp^+$	2	η^3-Cp	3
	$\eta^5\text{-}Cp^-$	6	η^5-Cp	5
	CO	2	CO	2
		18		18
d. $Co_2(CO)_8$	Co	9	Co	9
	4 CO	8	4 CO	8
	bridging CO	1	bridging CO	1
		18		18

13.2

	Method A		Method B	
a. $[M(CO)_3PPh_3]^-$	3 CO	6	3 CO	6
	PPh_3	2	PPh_3	2
			1–	1
		8		9
Need 10 electrons for M^- or 9 electrons for M, so the metal is Co.				
b. $HM(CO)_5$	5 CO	10	5 CO	10
	H^-	2	H	1
		12		11
Need 6 electrons for M^+ or 7 electrons for M, so the metal is Mn.				

Continued

	Method A		Method B	
c. $(\eta^4\text{-}C_8H_8)M(CO)_3$	3 CO	6	3 CO	6
	$\eta^4\text{-}C_8H_8$	4	$\eta^4\text{-}C_8H_8$	4
		10		10
Need 8 electrons for M, so the metal is Fe.				
d. $[(\eta^5\text{-}C_5H_5)M(CO)_3]_2$	3 CO	6	3 CO	6
	$\eta^5\text{-}C_5H_5^-$	6	$\eta^5\text{-}C_5H_5$	5
	M—M	1	M—M	1
		13		12
Need 5 electrons for M^+ or 6 electrons for M, so the metal is Cr.				

13.3

	Method A		Method B	
$[Ni(CN)_4]^{2-}$	Ni(II)	8	Ni	10
	4 CN^-	8	4 CN	4
			2–	2
		16		16
$PtCl_2en$	Pt(II)	8	Pt	10
	2 Cl^-	4	2 Cl	2
	en	4	en	4
		16		16
$RhCl(PPh_3)_3$	Rh(I)	8	Rh	9
	Cl^-	2	Cl	1
	3 PPh_3	6	3 PPh_3	6
		16		16
$IrCl(CO)(PPh_3)_2$	Ir(I)	8	Ir	9
	Cl^-	2	Cl	1
	CO	2	CO	2
	2 PPh_3	4	2 PPh_3	4
		16		16

13.4 The N_2 sigma and pi levels are very close together in energy (see Chapter 5) and they are distributed evenly over both atoms. The CO levels are farther apart in energy and concentrated on C. Therefore, the geometric overlap for CO is better, and CO has better σ–donor and π–acceptor qualities than N_2.

13.5 The greater the negative charge on the complex, the greater the degree of pi acceptance by the CO ligands. This increases the population of the π^* orbitals of CO, weakening the carbon-oxygen bond. Consequently, $[V(CO)_6]^-$ has the longest C—O bond and $[Mn(CO)_6]^+$ has the shortest.

13.6 The two methods of electron counting are equivalent for these examples.

$M(CO)_4$	(M = Ni, Pd)	M	10
		4 CO	8
			18
$M(CO)_5$	(M = Fe, Ru, Os)	M	8
		5 CO	10
			18

$M(CO)_6$	(M = Cr, Mo, W)	M	6
		6 CO	12
			18
$Co_2(CO)_8$ (solution)		Co	9
(for each Co)		4 CO	8
		Co—Co	1
			18
$Co_2(CO)_8$ (solid)		Co	9
(for each Co)		3 CO	6
		2 μ_2-CO	2
		Co—Co	1
			18
$Fe_2(CO)_9$		Fe	8
(for each Fe)		3 CO	6
		3 μ_2-CO	3
		Fe—Fe	1
			18
$M_2(CO)_{10}$	(M = Mn, Te, Re)	M	7
(for each M)		5 CO	10
		M—M	1
			18
$Fe_3(CO)_{12}$	Fe on left	Fe	8
		4 CO	8
		2 Fe—Fe	2
			18
	Other Fe	Fe	8
		3 CO	6
		2 μ_2-CO	2
		2 Fe—Fe	2
			18
$M_3(CO)_{12}$	(Ru, Os)	M	8
(for each M)		4 CO	8
		2 M—M	2
			18
$M_4(CO)_{12}$	(M = Co, Rh), M on top	M	9
		3 CO	6
		3 M—M	3
			18
	Other M	M	9
		2 CO	4
		2 μ_2-CO	2
		3 M—M	3
			18
$Ir_4(CO)_{12}$		Ir	9
(for each Ir)		3 CO	6
		3 Ir—Ir	3
			18

13.7 PMe_3 is a stronger σ donor and weaker π acceptor than CO. Therefore, the Mo in $Mo(PMe_3)_5H_2$ has a greater concentration of electrons and a greater tendency to back-bond to the hydrogens by donating to the σ^* orbital of H_2. This donation is strong enough to rupture the H—H bond, converting H_2 into two hydride ligands.

13.8

	Method A		Method B	
a. $(\eta^5\text{-}C_5H_5)(cis\text{-}\eta^4\text{-}C_4H_6)M(PMe_3)_2(H)$	$\eta^5\text{-}C_5H_5^-$	6	$\eta^5\text{-}C_5H_5$	5
	$\eta^4\text{-}C_4H_6$	4	$\eta^4\text{-}C_4H_6$	4
	PMe_3	2	PMe_3	2
	H^-	2	H	1
		14		12
M^{2+} needs 4 electrons, M needs 6. Mo fits.				
b. $(\eta^5\text{-}C_5H_5)M(C_2H_4)_2$	$\eta^5\text{-}C_5H_5^-$	6	$\eta^5\text{-}C_5H_5$	5
	$2\ C_2H_4$	4	$2\ C_2H_4$	4
		10		9
M^+ needs 8 electrons, M needs 9; Co fits.				

13.9 2-Node group orbitals.

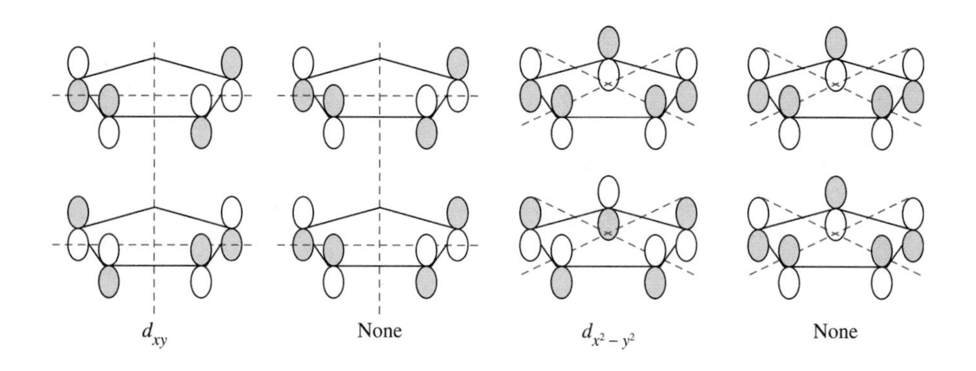

1-Node group orbitals.

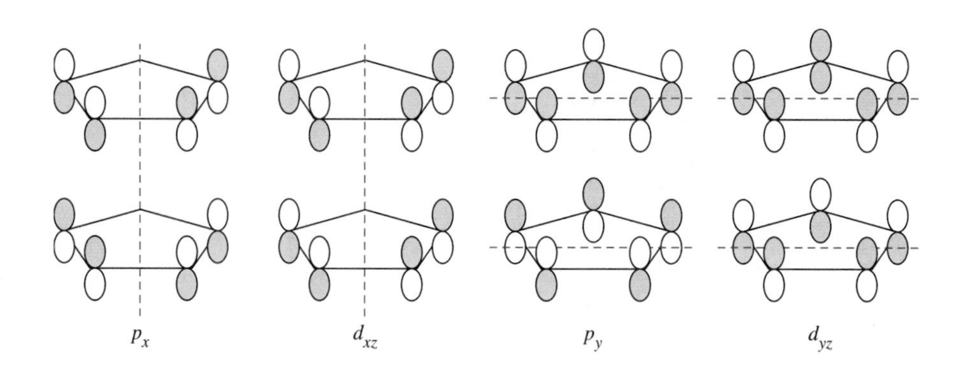

13.10

	Method A		Method B	
a. $[((CH_3)_3CCH)((CH_3)_3CCH_2)(CCH_3)(C_2H_4(P(CH_3)_2)_2)W]$				
	W^+	5	W	6
	$(CH_3)_3)CCH$	2	$(CH_3)_3CCH$	2
	$(CH_3)_3CCH_2^-$	2	$(CH_3)_3CCH_2$	1
	CCH_3	3	CCH_3	3
	$C_2H_4(P(CH_3)_2)_2$	4	$C_2H_4(P(CH_3)_2)_2$	4
		16		16
b. $Ta(Cp)_2(CH_3)(CH_2)$	Ta^{3+}	2	Ta	5
	$2\ Cp^-$	12	2 Cp	10
	CH_3^-	2	CH_3	1
	CH_2	2	CH_2	2
		18		18

13.11 With just two bands, this is more likely the *fac* isomer. The *mer* isomer should show three bands; it would show two only if two bands coincide in energy.

13.12 II has three separate resonances in the CO range (at 194.98, 189.92, and 188.98) and is more likely the *fac* isomer. The *mer* isomer would be expected to have two carbonyls of magnetically equivalent environments and one that is different.

CHAPTER 14

14.1 The *cis* product is one with the labeled CO *cis* to CH_3. The reverse of Mechanism 1 removes the acetyl ^{13}CO from the molecule completely, which means that the product should have no ^{13}CO label at all.

14.2 The product distribution for the reaction of *cis*-$CH_3Mn(CO)_4(^{13}CO)$ with PR_3 ($R = C_2H_5$, $^*C = {}^{13}C$):

25% has ^{13}C in the CH_3CO.

25% has ^{13}CO *trans* to the CH_3CO.

Continued

50% has ^{13}CO *cis* to the CH_3CO.

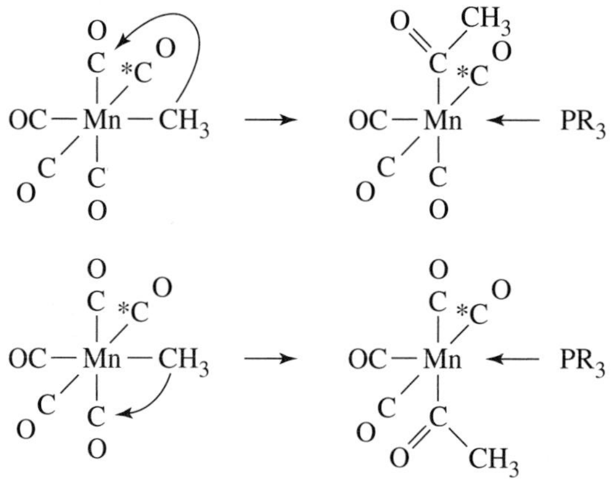

All the products have PEt_3 *cis* to the CH_3CO.

14.3 The reverse of the reaction has a pi-bonded ethylene and a hydride bonded to Rh rearranging, with Rh going to carbon 1 of the ethylene and the hydrogen going to carbon 2 of the ethylene, a 1,2 insertion of Rh and H into the double bond.

14.4 The hydroformylation process for the preparation of $(CH_3)_3CH{-}CH_2{-}CHO$ from $(CH_3)_2C = CH_2$ is exactly that of Figure 14.14, with $R = CH_3$.

14.5
1. Ligand dissociation
2. Coordination of olefin
3. 1,2 insertion
4. Ligand coordination
5. Alkyl migration (carbonyl insertion)
6. Oxidative addition
7. Reductive elimination

14.6 **a.** Metathesis between two molecules of propene, $H_2C{=}CHCH_3$:

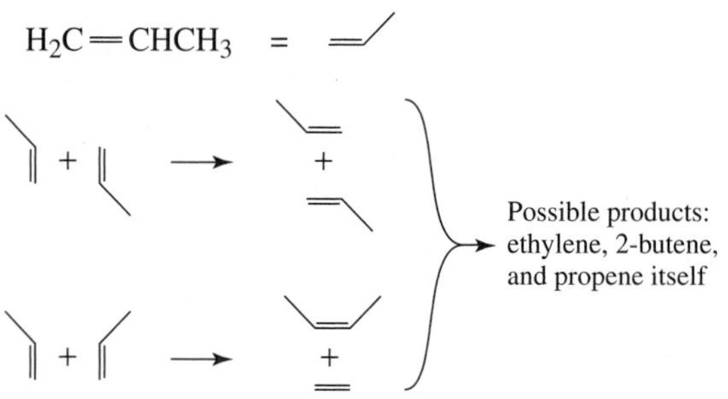

$H_2C{=}CHCH_3$ =

Possible products: ethylene, 2-butene, and propene itself

b. Metathesis between propene and cyclopentene:

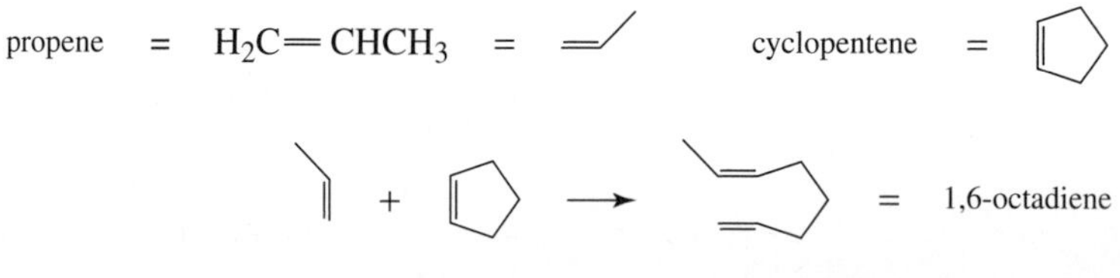

propene = $H_2C{=}CHCH_3$ =   cyclopentene =

= 1,6-octadiene

1,6-octadiene can participate in further metathesis reactions. For example,

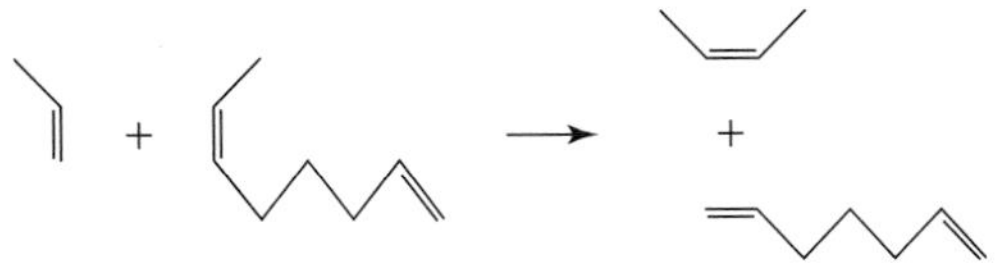

14.7

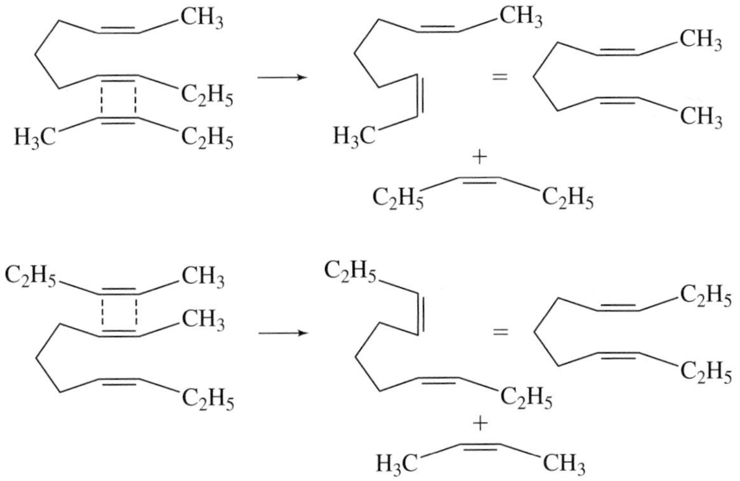

CH_3 C_2H_5 C_2H_5 H_3C → CH_3 H_3C = CH_3 CH_3 + C_2H_5 C_2H_5

C_2H_5 CH_3 CH_3 C_2H_5 → C_2H_5 C_2H_5 = C_2H_5 C_2H_5 + H_3C CH_3

CHAPTER 15

15.1 There are many possible answers. Examples include the following:

a. $Re(CO)_4$ $(\eta^5\text{-}C_5H_5)Fe(CO)$

b. $Pt(CO)_3$ $[(\eta^5\text{-}C_5H_5)Co]^{2-}$

c. $Re(CO)_5$ $[(\eta^5\text{-}C_5H_5)Mn(CO)_2]^-$ $(\eta^6\text{-}C_6H_6)Mn(CO)_2$

15.2 **a.** This is a 15-electron species with three vacant positions, isolobal with CH.
b. This is a 16-electron species with one vacant position, isolobal with CH_3^+.
c. This is a 15-electron species with three vacant positions, isolobal with CH.

15.3 **a.** $B_{11}H_{13}^{2-}$ is derived from $B_{11}H_{11}^{4-}$, a *nido* species.
b. $B_5H_8^-$ is derived from $B_5H_5^{4-}$, a *nido* species.
c. $B_7H_7^{2-}$ is a *closo* species.
d. $B_{10}H_{18}$ is derived from $B_{10}H_{10}^{8-}$, a *hypho* species.

15.4 **a.** $C_3B_3H_7$ is equivalent to B_6H_{10}, derived from $B_6H_6^{4-}$, a *nido* species.
b. $C_2B_5H_7$ is equivalent to B_7H_9, derived from $B_7H_7^{2-}$, a *closo* species.
c. $C_2B_7H_{12}^-$ is equivalent to $B_9H_{14}^-$, derived from $B_9H_9^{6-}$, an *arachno* species.

15.5 **a.** SB_9H_9 is equivalent to $B_{10}H_{12}$, derived from $B_{10}H_{10}^{2-}$, a *closo* species.
b. $GeC_2B_9H_{11}$ is equivalent to $B_{12}H_{14}$, derived from $B_{12}H_{12}^{2-}$, a *closo* species.
c. $SB_9H_{12}^-$ is equivalent to $B_{10}H_{15}^-$, derived from $B_{10}H_{10}^{6-}$, an *arachno* species.

15.6 **a.** $C_2B_7H_9(CoCp)_3$ is equivalent to $B_9H_{11}(CoCp)_3$ or $B_{12}H_{14}$, derived from $B_{12}H_{12}^{2-}$, a *closo* species.
b. $C_2B_4H_6Ni(PPh_3)_2$ is equivalent to $B_6H_8Ni(PPh_3)_2$ or B_7H_9, derived from $B_7H_7^{2-}$, a *closo* species.

15.7 **a.** Ge_9^{2-} The total valence electron count is 38, four for each Ge plus two for the charge; $n = 9$, so the electron count is $4n + 2$, and the structure is *closo*.
b. Bi_5^{3+} The valence electron count is 22, five for each Bi minus three for the charge. Because $n = 5$, the total count again is $4n + 2$, and the structure is *closo*.

15.8 **a.** $(\eta^5\text{-}C_5H_5)(\eta^5\text{-}C_2B_9H_{11})Fe$

m = 2, two linked polyhedra
n = 17; all Fe, B, and C atoms are vertices
o = 1; one atom, Fe, bridges the polyhedra
p = 1; the top polyhedron is not complete
$m + n + o + p$ = 21 electron pairs

b. *nido* – 7,8 – $C_2B_9H_{11}{}^{2-}$

m = 1 single polyhedron
n = 11; each B and C atom is a vertex
o = 0; no bridges between polyhedra
p = 1; *nido* classification
$m + n + o + p$ = 13 electron pairs

15.9 **a.** $[Re_7C(CO)_{21}]^{3-}$

7 Re	49
C	4
21 Co	42
3–	3
Total	98

A 98-electron, seven-metal cluster is two electrons short of a *closo* configuration. The structure would be expected to be a capped *closo*, such as the 98-electron species $[Rh_7(CO)_{16}]^{3-}$ and $Os_7(CO)_{21}$ in Table 15.13.

b. $[Fe_4N(CO)_{12}]^-$

4 Fe	32
N	5
12 CO	24
1–	1
Total	62

A 62-electron, four-metal cluster is classified as *arachno*.

B

Useful Data

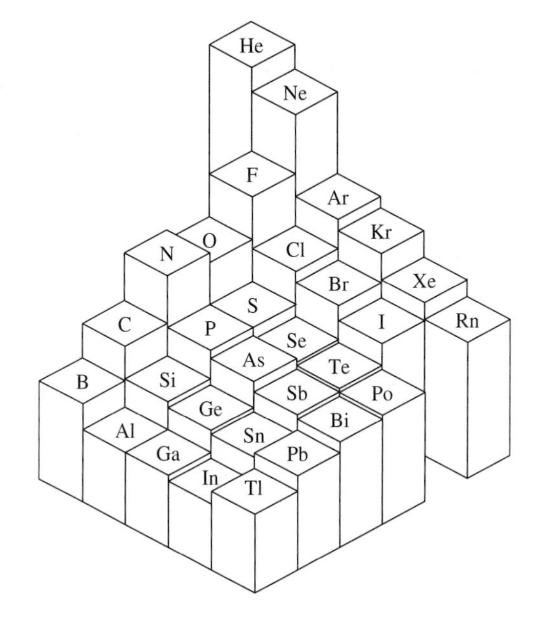

APPENDIX B.1: IONIC RADII

The values given are the crystal radii of Shannon, calculated using electron density maps and internuclear distances from X-ray data. Some of the trends that can be seen in these radii are the following:

1. Increase in size with increasing coordination number
2. Increase in size for a given coordination number with increasing Z within a periodic group
3. Decreasing size with increasing nuclear charge for isoelectronic ions
4. Decreasing size with increasing ionic charge for the same Z
5. Irregular, slowly decreasing size with increasing Z for transition metal, lanthanide, or actinide ions of the same charge
6. Larger size for high-spin ions than for low-spin ions of the same species and charge

Not shown in the table, but another apparent factor, is the decrease in anion size with increasing cation field strength, determined by the charge and size of the cation in the crystal. (See O. Johnson, *Inorg. Chem.*, **1973**, *12*, 780, for details.)

Ionic radii are given in picometers.

		Coordination Number						
Z		**2**	**4**	**6**	**8**	**10**	**12**	**14**
1	H	−4						
2	(He)							
3	Li^+		73	90	106			
4	Be^{2+}		41	59				
5	B^{3+}		25					
6	C^{4+}		29					
7	N^{3-}		132					
8	O^{2-}	121	124	126	128			
	OH^-	118	121	123				
9	F^-	115	117	119				
10	(Ne)							
11	Na^+		113	116	132		153	
12	Mg^{2+}		71	86	103			
13	Al^{3+}		53	68				
14	Si^{4+}		40	54				

Continued

Z		Coordination Number						
		2	4	6	8	10	12	14
15	P^{3+}			58				
16	S^{2-}			170				
17	Cl^{-}			167				
18	(Ar)							
19	K^{+}		151	152	165	173	178	
20	Ca^{2+}			114	126	137	148	
21	Sc^{3+}			89	101			
22	Ti^{2+}			100				
	Ti^{3+}			81				
	Ti^{4+}		56	75	88			
23	V^{2+}			93				
	V^{3+}			78				
24	Cr^{2+}			hs 94				
	Cr^{2+}			ls 87				
	Cr^{3+}			76				
25	Mn^{2+}		hs 80	hs 97				
	Mn^{2+}			ls 81				
	Mn^{3+}			hs 79				
	Mn^{3+}			ls 72				
26	Fe^{2+}		hs 77	hs 92				
	Fe^{2+}			ls 75				
	Fe^{3+}		hs 63	hs 79				
	Fe^{3+}			ls 69				
27	Co^{2+}		hs 72	hs 89				
	Co^{2+}			ls 79				
	Co^{3+}			hs 75				
	Co^{3+}			ls 69				
28	Ni^{2+}		69	83				
	Ni^{2+}		sq 63					
	Ni^{3+}			hs 74				
	Ni^{3+}			ls 70				
29	Cu^{+}	60	74	91				
	Cu^{2+}		71	87				
30	Zn^{2+}		74	88	104			
31	Ga^{3+}		61	76				
32	Ge^{4+}		53	67				
33	As^{3+}			72				
	As^{5+}		48	60				
34	Se^{2-}			184				
35	Br^{-}			182				
36	(Kr)							
37	Rb^{+}			166	175	180	186	197
38	Sr^{2+}			132	140	150	158	
39	Y^{3+}			104				
40	Zr^{4+}		73	86	98			
41	Nb^{3+}			86				
	Nb^{4+}			82	93			
42	Mo^{3+}			83				
	Mo^{4+}			79				
43	Tc^{4+}			79				
44	Ru^{3+}			82				
	Ru^{4+}			76				
45	Rh^{3+}			81				
	Rh^{4+}			74				

Continued

Z		Coordination Number						
		2	4	6	8	10	12	14
46	Pd^{2+}		sq 78	100				
47	Ag^{+}	81	114	129	142			
	Ag^{+}		sq 116					
48	Cd^{2+}		92	109	124		145	
49	In^{3+}		76	94	106			
50	Sn^{4+}		69	83	95			
51	Sb^{3+}			90				
52	Te^{2-}			207				
53	I^{-}			206				
54	(Xe)							
55	Cs^{+}			181	188	195	202	
56	Ba^{2+}			149	156	166	175	
57	La^{3+}			117	130	141	150	
58	Ce^{3+}			115	128	139	148	
59	Pr^{3+}			113	127			
60	Nd^{3+}			112	125		141	
61	Pm^{3+}			111	123			
62	Sm^{3+}			110	122		138	
63	Eu^{3+}			109	121			
64	Gd^{3+}			108	119			
65	Tb^{3+}			106	118			
66	Dy^{3+}			105	117			
67	Ho^{3+}			104	116	126		
68	Er^{3+}			103	114			
69	Tm^{3+}			102	113			
70	Yb^{3+}			101	113			
71	Lu^{3+}			100	112			
72	Hf^{4+}		72	85	97			
73	Ta^{3+}			86				
	Ta^{4+}			82				
74	W^{4+}			80				
75	Re^{4+}			77				
76	Os^{4+}			77				
77	Ir^{3+}			82				
	Ir^{4+}			77				
78	Pt^{2+}		sq 74	94				
	Pt^{4+}			77				
79	Au^{+}			151				
	Au^{3+}		sq 82	99				
80	Hg^{2+}	83	110	116	128			
81	Tl^{3+}		89	103	112			
82	Pb^{2+}		112	133	143	154	163	
	Pb^{4+}		79	92	108			
83	Bi^{3+}			117	131			
84	Po^{4+}			108	122			
85	At^{7+}			76				
86	(Rn)							
87	Fr^{+}			194				
88	Ra^{2+}				162		184	
89	Ac^{3+}			126				
90	Th^{4+}			108	119	127	135	

Source: R. D. Shannon, *Acta Crystallogr.*, **1976**, *A32*, 751.

NOTE: hs = high spin, ls = low spin, sq = square planar; values for CN = 4 are for tetrahedral geometry unless designated square planar.

APPENDIX B.2: IONIZATION ENERGY

Atomic No.	Element	eV	kJ mol^{-1}	Atomic No.	Element	eV	kJ mol^{-1}
1	H	13.598	1,312.0	52	Te	9.009	869.2
2	He	24.587	2,372.8	53	I	10.451	1,008.4
3	Li	5.392	520.2	54	Xe	12.130	1,170.4
4	Be	9.322	899.4	55	Cs	3.894	375.7
5	B	8.298	800.6	56	Ba	5.212	502.9
6	C	11.260	1,086.5	57	La	5.577	538.1
7	N	14.534	1,402.3	58	Ce	5.47	528
8	O	13.618	1,314.0	59	Pr	5.42	523
9	F	17.422	1,681.0	60	Nd	5.49	530
10	Ne	21.564	2,080.6	61	Pm	5.55	535
11	Na	5.139	495.8	62	Sm	5.63	543
12	Mg	7.646	737.8	63	Eu	5.67	547
13	Al	5.986	577.6	64	Gd	6.14	592
14	Si	8.151	786.5	65	Tb	5.85	564
15	P	10.486	1,011.7	66	Dy	5.93	572
16	S	10.360	999.6	67	Ho	6.02	581
17	Cl	12.967	1,251.1	68	Er	6.10	589
18	Ar	15.759	1,520.5	69	Tm	6.18	596
19	K	4.341	418.8	70	Yb	6.254	603.4
20	Ca	6.113	589.8	71	Lu	5.426	523.5
21	Sc	6.54	631	72	Hf	7.0	675
22	Ti	6.82	658	73	Ta	7.89	761
23	V	6.74	650	74	W	7.98	770
24	Cr	6.766	652.8	75	Re	7.88	760
25	Mn	7.435	717.4	76	Os	8.7	839
26	Fe	7.870	759.3	77	Ir	9.1	878
27	Co	7.86	758	78	Pt	9.0	868
28	Ni	7.635	736.7	79	Au	9.225	890.1
29	Cu	7.726	745.5	80	Hg	10.437	1,007.0
30	Zn	9.394	906.4	81	Tl	6.108	589.3
31	Ga	5.999	578.8	82	Pb	7.416	715.5
32	Ge	7.899	762.1	83	Bi	7.289	703.3
33	As	9.81	947	84	Po	8.42	812
34	Se	9.752	940.9	85	At	7.289	703.3
35	Br	11.814	1,139.9	86	Rn	10.748	1,037.1
36	Kr	13.999	1,350.7	87	Fr	4	400
37	Rb	4.177	403.0	88	Ra	5.279	509.3
38	Sr	5.695	549.5	89	Ac	6.9	666
39	Y	6.38	616	90	Th	6.1	590
40	Zr	6.84	660	91	Pa	5.9	570
41	Nb	6.88	664	92	U	6.1	590
42	Mo	7.099	684.9	93	Np	6.2	600
43	Tc	7.28	702	94	Pu	6.06	585
44	Ru	7.37	711	95	Am	5.99	578
45	Rh	7.46	720	96	Cm	6.02	581
46	Pd	8.34	805	97	Bk	6.23	601
47	Ag	7.576	731.0	98	Cf	6.30	608
48	Cd	8.993	867.7	99	Es	6.42	619
49	In	5.786	558.3	100	Fm	6.50	627
50	Sn	7.344	708.6	101	Md	6.58	635
51	Sb	8.641	833.7	102	No	6.65	642

Source: C. E. Moore, *Ionization Potentials and Limits Derived from the Analyses of Optical Spectra*, NSRDS-NBS 34, National Bureau of Standards, Washington, DC, 1970; W. C. Martin, L. Hagan, J. Reador and J. Sugar, *J. Phys. Chem. Ref. Data*, **1974**, *3*, 771; and J. Sugar, *J. Opt. Soc. Am.*, **1975**, *65*, 1366.

NOTE: 1 eV = 96.4853 kJ mol^{-1}.

Updated values of ionization energies can be found online at http://physics.nist.gov/PhysRefData/IonEnergy/tblNew.html

APPENDIX B.3: ELECTRON AFFINITY

Atomic No.	Element	eV	kJ mol^{-1}	Atomic No.	Element	eV	kJ mol^{-1}
1	H	0.754	72.8	45	Rh	1.137	109.7
2	He	- 0.5*	- 50	46	Pd	0.557	53.7
3	Li	0.618	59.6	47	Ag	1.302	125.6
4	Be	- 0.5*	- 50	48	Cd	- 0.7*	- 68
5	B	0.277	26.7	49	In	0.3	29
6	C	1.263	121.9	50	Sn	1.2	116
7	N	- 0.07	- 7	51	Sb	1.07	103
8	O	1.461	141.0	52	Te	1.971	190.2
9	F	3.399	328.0	53	I	3.059	295.2
10	Ne	- 1.2*	- 116	54	Xe	- 0.8*	- 77
11	Na	0.548	52.9	55	Cs	0.472	45.5
12	Mg	- 0.4*	- 39	56	Ba	- 0.3*	- 29
13	Al	0.441	42.6	57	La	0.5	48
14	Si	1.385	133.6	58	Ce	<0.5[a]	<48
15	P	0.747	72.0	59	Pr	<0.5[a]	<48
16	S	2.077	200.4	60	Nd	<0.5[a]	<48
17	Cl	3.617	349.0	61	Pm	<0.5[a]	<48
18	Ar	- 1.0*	- 97	62	Sm	<0.5[a]	<48
19	K	0.501	48.4	63	Eu	<0.5[a]	<48
20	Ca	- 0.3*	- 29	64	Gd	<0.5[a]	<48
21	Sc	0.188	18.1	65	Tb	<0.5[a]	<48
22	Ti	0.079	7.6	66	Dy	<0.5[a]	<48
23	V	0.525	50.7	67	Ho	<0.5[a]	<48
24	Cr	0.666	64.3	68	Er	<0.5[a]	<48
25	Mn	<0	< 0.0	69	Tm	<0.5[a]	<48
26	Fe	0.163	15.7	70	Yb	<0.5[a]	<48
27	Co	0.661	63.8	71	Lu	<0.5[a]	<48
28	Ni	1.156	111.5	72	Hf	~0	~0
29	Cu	1.228	118.5	73	Ta	0.322	31.1
30	Zn	- 0.6*	- 58	74	W	0.815	78.6
31	Ga	0.3	29	75	Re	0.15	14.5
32	Ge	1.2	115.8	76	Os	1.1	106.1
33	As	0.81	78	77	Ir	1.565	151.0
34	Se	2.021	195.0	78	Pt	2.128	205.3
35	Br	3.365	324.7	79	Au	2.309	222.8
36	Kr	- 1.0*	- 97	80	Hg	- 0.5*	- 48
37	Rb	0.486	46.9	81	Tl	0.2	19
38	Sr	- 0.3*	- 29	82	Pb	0.364	35.1
39	Y	0.307	29.6	83	Bi	0.946	91.3
40	Zr	0.426	41.1	84	Po	1.9	183
41	Nb	0.893	86.2	85	At	2.8	270
42	Mo	0.746	72.0	86	Rn	- 0.7*	- 68
43	Tc	0.55	53.1	87	Fr	0.6*	58
44	Ru	1.05	101.3	88	Ra	- 0.3*	- 29

Source: All data from W. Hotop and W. C. Lineberger, *J. Phys. Chem. Ref. Data*, **1985**, *14*, 731, except those marked*, which are from S. G. Bratsch and J. J. Lagowski, *Polyhedron*, **1986**, 5, 1763.

NOTE: Many of these data are known to greater accuracy than that shown in the table, some to 10 significant figures. Updated electron affinities can be found online at http://webbook.nist.gov.

[a] Estimated values

APPENDIX B.4: ELECTRONEGATIVITY[a]

1	2	3	4	5	6	7	8	9	10	11	12	13	14	15	16	17	18
H 2.300																	He 4.160
Li 0.912	Be 1.576											B 2.051	C 2.544	N 3.066	O 3.610	F 4.193	Ne 4.787
Na 0.869	Mg 1.293											Al 1.613	Si 1.916	P 2.253	S 2.589	Cl 2.869	Ar 3.242
K 0.734	Ca 1.034	Sc 1.19	Ti 1.38	V 1.53	Cr 1.65	Mn 1.75	Fe 1.80	Co 1.84	Ni 1.88	Cu 1.85	Zn 1.588	Ga 1.756	Ge 1.994	As 2.211	Se 2.424	Br 2.685	Kr 2.966
Rb 0.706	Sr 0.963	Y 1.12	Zr 1.32	Nb 1.41	Mo 1.47	Tc 1.51	Ru 1.54	Rh 1.56	Pd 1.58	Ag 1.87	Cd 1.521	In 1.656	Sn 1.824	Sb 1.984	Te 2.158	I 2.359	Xe 2.582
Cs 0.659	Ba 0.881	Lu 1.09	Hf 1.16	Ta 1.34	W 1.47	Re 1.60	Os 1.65	Ir 1.68	Pt 1.72	Au 1.92	Hg 1.765	Tl 1.789	Pb 1.854	Bi (2.01)	Po (2.19)	At (2.39)	Rn (2.60)

Source: J. B. Mann, T. L. Meek, and L. C. Allen, *J. Am. Chem. Soc.*, **2000**, *122*, 2780, and J. B. Mann, T. L. Meek, E. T. Knight, J. F. Capitani, and L. C. Allen, *J. Am. Chem. Soc.*, **2000**, *122*, 5132.

NOTE:[a] The shaded elements are metalloids, based on their electronegativities.

APPENDIX B.5: ABSOLUTE HARDNESS PARAMETERS

Hardness Parameters (in eV)					
	Ion or Molecule	**I**	**A**	χ	η
Cations	B^{3+}	259.37	37.93	148.65	110.72
	Be^{2+}	153.89	18.21	86.05	67.84
	Al^{3+}	119.99	28.45	74.22	45.77
	Li^{+}	75.64	5.39	40.52	35.12
	Mg^{2+}	80.14	15.04	47.59	32.55
	Na^{+}	47.29	5.14	26.21	21.08
	Ca^{2+}	50.91	11.87	31.39	19.52
	Sr^{2+}	43.6	11.03	27.3	16.3
	K^{+}	31.63	4.34	17.99	13.64
	Fe^{3+}	54.8	30.65	42.73	12.08
	Rb^{+}	27.28	4.18	15.77	11.55
	Rh^{3+}	53.4	31.1	42.4	11.2
	Zn^{2+}	39.72	17.96	28.84	10.88
	Cs^{+}	25.1	3.89	14.5	10.6
	Cd^{2+}	37.48	16.91	27.20	10.29
	Cr^{3+}	49.1	30.96	40.0	9.1
	Mn^{2+}	33.67	15.64	24.66	9.02
	Mn^{3+}	51.2	33.67	42.4	8.8
	Co^{3+}	51.3	33.50	42.4	8.9

Continued

Hardness Parameters (in eV)

	Ion or Molecule	I	A	χ	η
	V^{3+}	46.71	29.31	38.01	8.70
	Ni^{2+}	35.17	18.17	26.67	8.50
	Pb^{2+}	31.94	15.03	23.49	8.46
	Au^{3+}	54.1	37.4	45.8	8.4
	Cu^{2+}	36.83	20.29	28.56	8.27
	Co^{2+}	33.50	17.06	25.28	8.22
	Pt^{2+}	35.2	19.2	27.2	8.0
	Sn^{2+}	30.50	14.63	22.57	7.94
	Ir^{3+}	45.3	29.5	37.4	7.9
	Hg^{2+}	34.2	18.76	26.5	7.7
	V^{2+}	29.31	14.65	21.98	7.33
	Fe^{2+}	30.65	16.18	23.42	7.24
	Cr^{2+}	30.96	16.50	23.73	7.23
	Ag^{+}	21.49	7.58	14.53	6.96
	Ti^{2+}	27.49	13.58	20.54	6.96
	Pd^{2+}	32.93	19.43	26.18	6.75
	Rh^{2+}	31.06	18.08	24.57	6.49
	Cu^{+}	20.29	7.73	14.01	6.28
	Sc^{2+}	24.76	12.80	18.78	5.98
	Ru^{2+}	28.47	16.76	22.62	5.86
	Au^{+}	20.5	9.23	14.90	5.6
Molecules	BF_3	15.81	− 3.5	6.2	9.7
	H_2O	12.6	− 6.4	3.1	9.5
	N_2	15.58	− 2.2	6.70	8.9
	NH_3	10.7	− 5.6	2.6	8.2
	CH_3CN	12.2	− 2.8	4.7	7.5
	C_2H_2	11.4	− 2.6	4.4	7.0
	PF_3	12.3	− 1.0	5.7	6.7
	$(CH_3)_3N$	7.8	− 4.8	1.5	6.3
	C_2H_4	10.5	− 1.8	4.4	6.2
	PH_3	10.0	− 1.9	4.1	6.0
	O_2	12.2	0.4	6.3	5.9
	$(CH_3)_3P$	8.6	− 3.1	2.8	5.9
	$(CH_3)_3As$	8.7	− 2.7	3.0	5.7
	SO_2	12.3	1.1	6.7	5.6
	SO_3	12.7	1.7	7.2	5.5
	C_6H_6	9.3	− 1.2	4.1	5.3
	C_5H_5N	9.3	− 0.6	4.4	5.0
	Butadiene	9.1	− 0.6	4.3	4.9
	PCl_3	10.2	0.8	5.5	4.7
	PBr_3	9.9	1.6	5.6	4.2

Hardness Parameters for Atoms and Radicals (in eV)[a]

Atom or Radical	I	A	χ	η
F	17.42	3.40	10.41	7.01
H	13.60	0.75	7.18	6.43
OH	13.17	1.83	7.50	5.67
NH_2	11.40	0.74	6.07	5.33
CN	14.02	3.82	8.92	5.10
CH_3	9.82	0.08	4.96	4.87
Cl	13.01	3.62	8.31	4.70
C_2H_5	8.38	− 0.39	4.00	4.39
Br	11.81	3.36	7.60	4.24
C_6H_5	9.20	1.1	5.2	4.1
NO_2	> 10.1	2.30	> 6.2	> 3.9
I	10.45	3.06	6.76	3.70
SiH_3	8.14	1.41	4.78	3.37
C_6H_5O	8.85	2.35	5.60	3.25
$Mn(CO)_5$	8.44	2.0	5.2	3.2
CH_3S	8.06	1.9	5.0	3.1
C_6H_5S	8.63	2.47	5.50	3.08

Source: R. G. Pearson, *Inorg. Chem.*, **1988**, *27*, 734.

[a] The hardness values approximate those of the corresponding anions.

APPENDIX B.6: C_A, E_A, C_B, AND E_B VALUES

Acid	C_A	E_A
Trimethylboron, $B(CH_3)_3$	1.70	6.14
Boron trifluoride (gas), BF_3	1.62	9.88
Trimethylaluminum, $Al(CH_3)_3$	1.43	16.9
Iodine (standard), I_2	1.00[a]	1.00[a]
Trimethylgallium, $Ga(CH_3)_3$	0.881	13.3
Iodine monochloride, ICl	0.830	5.10
Sulfur dioxide, SO_2	0.808	0.920
Phenol, C_6H_5OH	0.442	4.33
tert-Butyl alcohol, C_4H_9OH	0.300	2.04
Pyrrole, C_4H_4NH	0.295	2.54
Chloroform, $CHCl_3$	0.159	3.02

Continued

Base	C_B	E_B
1-Azabicyclo[2.2.2] octane, (quinuclidine)	13.2	0.704
Trimethylamine, $(CH_3)_3N$	11.54	0.808
Triethylamine, $(C_2H_5)_3N$	11.09	0.991
Dimethylamine, $(CH_3)_2NH$	8.73	1.09
Diethyl sulfide, $(C_2H_5)_2S$	7.40[a]	0.339
Pyridine, C_5H_5N	6.40	1.17
Methylamine, CH_3NH_2	5.88	1.30
Pyridine-N-oxide, C_5H_5NO	4.52	1.34
Tetrahydrofuran, C_4H_8O	4.27	0.978
7-Oxabicyclo[2.2.1] heptane, $C_6H_{10}O$	3.76	1.08
Ammonia, NH_3	3.46	1.36
Diethyl ether, $(C_2H_5)_2O$	3.25	0.963
Dimethyl sulfoxide, $(CH_3)_2SO$	2.85	1.34
N,N-dimethylacetamide, $(CH_3)_2NCOCH_3$	2.58	1.32[a]
p-Dioxane, $O(C_2H_4)_2O$	2.38	1.09
Acetone, CH_3COCH_3	2.33	0.987
Acetonitrile, CH_3CN	1.34	0.886
Benzene, C_6H_6	0.681	0.525

Source: R. S. Drago, *J. Chem. Educ.*, **1974**, *51*, 300.
NOTE: [a] Reference values.

APPENDIX B.7: LATIMER DIAGRAMS FOR SELECTED ELEMENTS[1]

ACIDIC SOLUTION

$$H^+ \xrightarrow{0} H_2 \xrightarrow{-2.25} H^-$$

Group 1

$$Li^+ \xrightarrow{-3.04} Li \quad Na^+ \xrightarrow{-2.71} Na \quad K^+ \xrightarrow{-2.92} K \quad Rb^+ \xrightarrow{-2.92} Rb \quad Cs^+ \xrightarrow{-2.92} Cs$$

Group 2

$$Be^{2+} \xrightarrow{-1.97} Be \quad Mg^{2+} \xrightarrow{-2.36} Mg \quad Ca^{2+} \xrightarrow{-2.84} Ca \quad Sr^{2+} \xrightarrow{-2.89} Sr \quad Ba^{2+} \xrightarrow{-2.92} Ba$$

[1] Data from A. J. Bard, R. Parsons, and J. Jordan, eds., *Standard Potentials in Aqueous Solution*, Marcel Dekker, New York, 1985; A. Kaczmarcyzk, W. C. Nichols, W. H. Stockmayer, and T. B. Ames, *Inorg. Chem.*, **1968**, *7*, 1057; M. Pourbaix, *Atlas of Electrochemical Equilibria in Aqueous Solution*, 2d ed., translated by J. A. Franklin, National Association of Corrosion Engineers, Houston, TX, 1974.

Group 3

$$Sc^{3+} \xrightarrow{-2.03} Sc \quad Y^{3+} \xrightarrow{-2.37} Y \quad La^{3+} \xrightarrow{-2.38} La \quad Ac^{3+} \xrightarrow{-2.13} Ac$$

Group 4

$$TiO^{2+} \xrightarrow{0.10} Ti^{3+} \xrightarrow{-0.37} Ti^{2+} \xrightarrow{-1.63} Ti$$

Group 5

$$V(OH)_4{}^{+} \xrightarrow{1.00} VO^{2+} \xrightarrow{0.34} V^{3+} \xrightarrow{-0.26} V^{2+} \xrightarrow{-1.13} V$$

Group 6

$$Cr_2O_7{}^{2-} \xrightarrow{1.38} Cr^{3+} \xrightarrow{-0.42} Cr^{2+} \xrightarrow{-0.90} Cr$$

Group 7

$$MnO_4{}^{-} \xrightarrow{0.564} MnO_4{}^{2-} \xrightarrow{0.274} MnO_4{}^{3-} \xrightarrow{4.27} MnO_2 \xrightarrow{0.95} Mn^{3+} \xrightarrow{1.51} Mn^{2+} \xrightarrow{-1.18} Mn$$

$$MnO_4{}^{-} \xrightarrow{1.507} Mn^{2+}$$

$$MnO_4{}^{-} \xrightarrow{1.70} MnO_2$$

Group 8

$$FeO_4{}^{2-} \xrightarrow{2.20} Fe^{3+} \xrightarrow{0.771} Fe^{2+} \xrightarrow{-0.44} Fe$$

Group 9

$$CoO_2 \xrightarrow{1.416} Co^{3+} \xrightarrow{1.92} Co^{2+} \xrightarrow{-0.277} Co$$

Group 10

$$NiO_4{}^{2-} \xrightarrow{>1.8} NiO_2 \xrightarrow{1.59} Ni^{2+} \xrightarrow{-0.257} Ni$$

$$NiO_4{}^{2-} \xrightarrow{>1.6} Ni$$

Group 11

$$CuO^{2+} \xrightarrow{1.8} Cu^{2+} \xrightarrow{0.159} Cu^{+} \xrightarrow{0.521} Cu$$

$$Cu^{2+} \xrightarrow{0.340} Cu$$

Group 12

$$Zn^{2+} \xrightarrow{-0.763} Zn \quad Cd^{2+} \xrightarrow{-0.403} Cd \quad Hg^{2+} \xrightarrow{0.911} Hg_2{}^{2+} \xrightarrow{0.796} Hg$$

Group 13

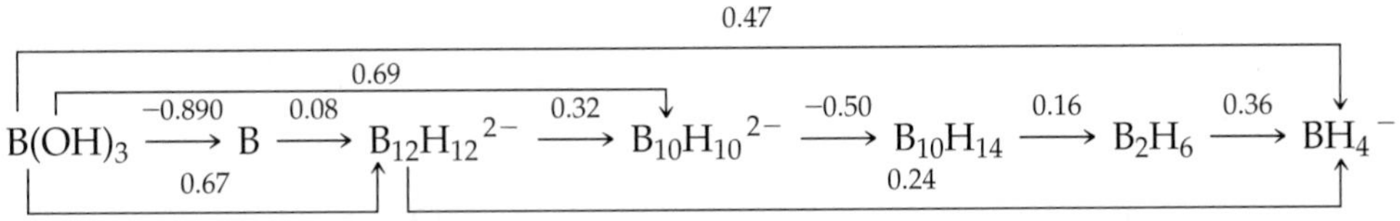

$$B(OH)_3 \xrightarrow{-0.890} B \xrightarrow{0.08} B_{12}H_{12}^{2-} \xrightarrow{0.32} B_{10}H_{10}^{2-} \xrightarrow{-0.50} B_{10}H_{14} \xrightarrow{0.16} B_2H_6 \xrightarrow{0.36} BH_4^-$$

Connecting potentials: $B(OH)_3 \rightarrow BH_4^-$: 0.47; $B(OH)_3 \rightarrow B_{10}H_{10}^{2-}$: 0.69; $B(OH)_3 \rightarrow B_{12}H_{12}^{2-}$: 0.67; $B_{12}H_{12}^{2-} \rightarrow BH_4^-$: 0.24

$$Al^{3+} \xrightarrow{-1.68} Al \qquad Ga^{3+} \xrightarrow{-0.65} Ga^{2+} \xrightarrow{-0.45} Ga \qquad In^{3+} \xrightarrow{-0.49} In^{2+} \xrightarrow{-0.40} In^{+} \xrightarrow{-0.126} In$$

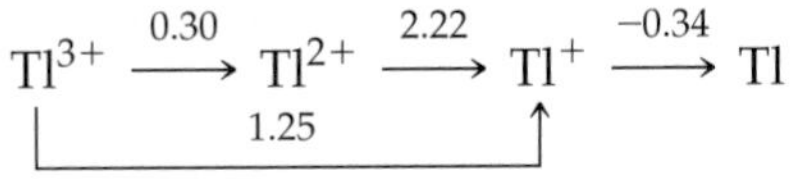

$$Tl^{3+} \xrightarrow{0.30} Tl^{2+} \xrightarrow{2.22} Tl^{+} \xrightarrow{-0.34} Tl$$

Connecting potential: $Tl^{3+} \rightarrow Tl^{+}$: 1.25

Group 14

$$CO_2 \xrightarrow{-0.106} CO \xrightarrow{0.517} C \xrightarrow{0.132} CH_4$$

$$SiO_2 \xrightarrow{-0.848} Si \xrightarrow{-0.143} SiH_4$$

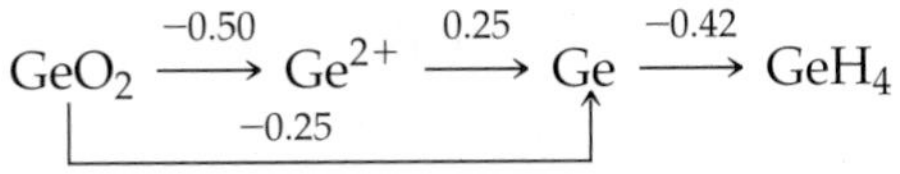

$$GeO_2 \xrightarrow{-0.50} Ge^{2+} \xrightarrow{0.25} Ge \xrightarrow{-0.42} GeH_4$$

Connecting potential: $GeO_2 \rightarrow Ge$: −0.25

$$Sn^{4+} \xrightarrow{0.15} Sn^{2+} \xrightarrow{-0.137} Sn$$

$$PbO_2 \xrightarrow{1.46} Pb^{2+} \xrightarrow{-0.125} Pb$$

Group 15

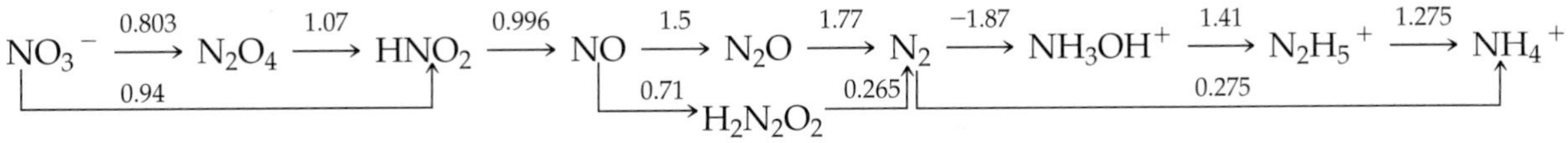

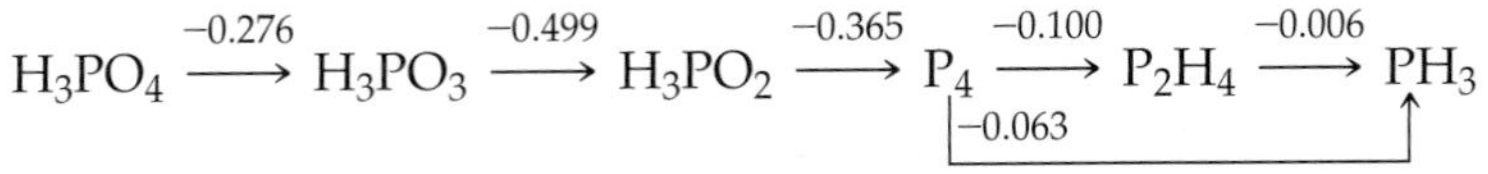

$$NO_3^- \xrightarrow{0.803} N_2O_4 \xrightarrow{1.07} HNO_2 \xrightarrow{0.996} NO \xrightarrow{1.5} N_2O \xrightarrow{1.77} N_2 \xrightarrow{-1.87} NH_3OH^+ \xrightarrow{1.41} N_2H_5^+ \xrightarrow{1.275} NH_4^+$$

Connecting potentials: $NO_3^- \rightarrow HNO_2$: 0.94; $NO \rightarrow H_2N_2O_2$: 0.71; $H_2N_2O_2 \rightarrow N_2$: 0.265; $N_2 \rightarrow NH_4^+$: 0.275

$$H_3PO_4 \xrightarrow{-0.276} H_3PO_3 \xrightarrow{-0.499} H_3PO_2 \xrightarrow{-0.365} P_4 \xrightarrow{-0.100} P_2H_4 \xrightarrow{-0.006} PH_3$$

Connecting potential: $P_4 \rightarrow PH_3$: −0.063

$$H_3AsO_4 \xrightarrow{0.560} H_3AsO_3 \xrightarrow{0.240} As \xrightarrow{-0.225} AsH_3$$

$$Sb_2O_5 \xrightarrow{0.605} SbO^+ \xrightarrow{0.204} Sb \xrightarrow{-0.510} SbH_3$$

$$Bi_2O_5 \xrightarrow{1.60} BiO^+ \xrightarrow{0.317} Bi \xrightarrow{-0.97} BiH_3$$

Group 16

$$O_3 \xrightarrow{2.075} O_2 \xrightarrow{-0.125} HO_2 \xrightarrow{1.51} H_2O_2 \xrightarrow{1.763} H_2O$$

$O_2 \xrightarrow{0.695} H_2O_2$; $O_2 \xrightarrow{1.229} H_2O$

$$SO_4^{2-} \xrightarrow{-0.25} S_2O_6^{2-} \xrightarrow{0.57} SO_2 \xrightarrow{0.51} S_4O_6^{2-} \xrightarrow{0.07} S_2O_3^{2-} \xrightarrow{0.60} S \xrightarrow{0.14} H_2S$$

$SO_4^{2-} \xrightarrow{0.16} SO_2$; $SO_2 \xrightarrow{0.45} S$

$$SeO_4^{2-} \xrightarrow{1.1} H_2SeO_3 \xrightarrow{0.74} Se \xrightarrow{-0.11} H_2Se$$

$$H_2TeO_4 \xrightarrow{1.00} TeO_2(s) \xrightarrow{0.53} Te \xrightarrow{-0.64} H_2Te$$

Group 17

$$F_2 \xrightarrow{2.87} F^-$$

$$ClO_4^- \xrightarrow{1.201} ClO_3^- \xrightarrow{1.18} HClO_2 \xrightarrow{1.70} HClO \xrightarrow{1.63} Cl_2 \xrightarrow{1.358} Cl^-$$

$$BrO_4^- \xrightarrow{1.85} BrO_3^- \xrightarrow{1.45} HBrO \xrightarrow{1.60} Br_2(l) \xrightarrow{1.065} Br^-$$

$$H_5IO_6 \xrightarrow{1.60} IO_3^- \xrightarrow{1.13} HIO \xrightarrow{1.44} I_2(s) \xrightarrow{0.535} I^-$$

BASIC SOLUTION

Group 1

$$H_2O \xrightarrow{-0.828} H_2 \xrightarrow{-2.25} H^-$$

Group 2

$$Be_2O_3^{2-} \xrightarrow{2.62} Be \quad Mg(OH)_2 \xrightarrow{-2.69} Mg \quad Ca(OH)_2 \xrightarrow{-3.03} Ca \quad Sr(OH)_2 \xrightarrow{-2.88} Sr \quad Ba(OH)_2 \xrightarrow{-2.81} Ba$$

Group 3

$$Sc(OH)_3 \xrightarrow{-2.6} Sc \qquad Y(OH)_3 \xrightarrow{-2.85} Y \qquad La(OH)_3 \xrightarrow{-2.80} La$$

Group 4

$$TiO_2 \xrightarrow{-1.90} Ti \quad H_3ZrO_4 \xrightarrow{-2.36} Zr \quad HfO(OH)_2 \xrightarrow{-2.50} Hf$$

Group 5

$$VO_4^{3-} \xrightarrow{0.120} V$$

Group 6

$$CrO_4^{2-} \xrightarrow{-0.11} Cr(OH)_3 \xrightarrow{-1.1} Cr(OH)_2 \xrightarrow{-1.4} Cr$$

$$CrO_4^{2-} \xrightarrow{-0.72} Cr(OH)_4^- \xrightarrow{-1.33} Cr$$

Group 7

$$MnO_4^- \xrightarrow{0.564} MnO_4^{2-} \xrightarrow{0.27} MnO_4^{3-} \xrightarrow{0.96} MnO_2 \xrightarrow{0.15} Mn_2O_3 \xrightarrow{-0.25} Mn(OH)_2 \xrightarrow{-1.56} Mn$$

Group 8

$$FeO_4^{2-} \xrightarrow{0.72} Fe(OH)_3 \xrightarrow{-0.56} Fe(OH)_2 \xrightarrow{-0.887} Fe$$

Group 9

$$CoO_2 \xrightarrow{0.7} Co(OH)_3 \xrightarrow{0.17} Co(OH)_2 \xrightarrow{-0.73} Co$$

Group 10

$$NiO_2^{2-} \xrightarrow{>0.4} NiO_2 \xrightarrow{0.49} Ni(OH)_2 \xrightarrow{-0.72} Ni$$

Group 11

$$CuO \xrightarrow{-0.29} Cu_2O \xrightarrow{-0.365} Cu$$

Group 12

$$ZnO_2^{2-} \xrightarrow{-1.285} Zn$$

Group 13

$$B_4O_7^{2-} \xrightarrow{-0.76} B \xrightarrow{-1.04} B_{12}H_{12}^{2-} \xrightarrow{0.32} B_{10}H_{10}^{2-} \xrightarrow{1.15} B_{10}H_{14}(s) \xrightarrow{0.98} B_2H_6 \xrightarrow{0.78} BH_4^-$$

$$H_2AlO_3^- \xrightarrow{-2.31} Al$$

$$H_2GaO_3^- \xrightarrow{-1.22} Ga$$

$$In(OH)_3 \xrightarrow{1.0} In$$

$$Tl(OH)_3 \xrightarrow{-0.05} Tl(OH) \xrightarrow{-0.34} Tl$$

Group 14

$$CO_3^{2-} \xrightarrow{-1.01} HCO_2^- \xrightarrow{-0.52} C \xrightarrow{-0.70} CH_4$$

$$SiO_3^{2-} \xrightarrow{-1.69} Si \xrightarrow{-0.93} SiH_4$$

$$HGeO_3^- \xrightarrow{-0.89} Ge \xrightarrow{<-1.1} GeH_4$$

$$Sn(OH)_6^{2-} \xrightarrow{-0.90} HSnO_2^- \xrightarrow{-0.91} Sn$$

$$PbO_2 \xrightarrow{0.25} PbO \xrightarrow{-0.58} Pb$$

Group 15

$$NO_3^- \xrightarrow{-0.86} N_2O_4 \xrightarrow{0.867} NO_2^- \xrightarrow{-0.46} NO \xrightarrow{0.76} N_2O \xrightarrow{0.94} N_2 \xrightarrow{-3.04} NH_2OH \xrightarrow{0.73} N_2H_4 \xrightarrow{0.1} NH_3$$

$$PO_4^{3-} \xrightarrow{-1.12} HPO_3^{2-} \xrightarrow{-1.57} H_2PO_2^- \xrightarrow{-2.05} P_4 \xrightarrow{-0.89} PH_3$$

$$AsO_4^{3-} \xrightarrow{-0.67} H_2AsO_3^- \xrightarrow{-0.68} As \xrightarrow{-1.37} AsH_3$$

$$Sb(OH)_6^- \xrightarrow{-0.465} SbO_2^- \xrightarrow{-0.639} Sb \xrightarrow{-1.34} SbH_3$$

$$Bi_2O_5 \xrightarrow{0.78} Bi_2O_4 \xrightarrow{0.56} Bi_2O_3 \xrightarrow{-0.46} Bi \xrightarrow{<-1.6} BiH_3$$

Group 16

$$O_3 \xrightarrow{2.075} O_2 \xrightarrow{-0.069} HO_2^- \xrightarrow{0.867} OH^-$$

$$SO_4^{2-} \xrightarrow{-0.94} SO_3^{2-} \xrightarrow{0.79} S_4O_6^{2-} \xrightarrow{0.08} S_2O_3^{2-} \xrightarrow{-0.74} S \xrightarrow{-0.45} S^{2-}$$

$$SO_3^{2-} \xrightarrow{-0.66} S$$

$$SeO_4^{2-} \xrightarrow{0.03} SeO_3^{2-} \xrightarrow{-0.36} Se \xrightarrow{-0.67} Se^{2-}$$

$$TeO_2(OH)_4^{2-} \xrightarrow{0.07} TeO_3^{2-} \xrightarrow{-0.42} Te \xrightarrow{-1.14} Te^{2-}$$

Group 17

$$F_2 \xrightarrow{2.87} F^-$$

$$ClO_4^- \xrightarrow{0.374} ClO_3^- \xrightarrow{-0.48} ClO_2 \xrightarrow{1.07} ClO_2^- \xrightarrow{0.681} ClO^- \xrightarrow{0.421} Cl_2 \xrightarrow{1.358} Cl^-$$

$$ClO_3^- \xrightarrow{0.295} ClO_2^-$$

$$BrO_4^- \xrightarrow{1.025} BrO_3^- \xrightarrow{0.49} BrO^- \xrightarrow{0.455} Br_2(l) \xrightarrow{1.065} Br^-$$

$$H_3IO_6^{2-} \xrightarrow{0.65} IO_3^- \xrightarrow{0.15} IO^- \xrightarrow{0.42} I_2(s) \xrightarrow{0.535} I^-$$

Group 18

$$HXeO_6^{3-} \xrightarrow{0.94} HXeO_4^- \xrightarrow{1.24} Xe$$

APPENDIX B.8: ANGULAR FUNCTIONS FOR HYDROGEN ATOM *f* ORBITALS

There is no unique set of angular functions for *f* orbitals. The most commonly used sets are provided here.

CUBIC SET:

Orbital	Angular Function $\Theta\Phi$ (x, y, z)
f_{x^3}	$\frac{1}{4}\sqrt{\frac{7}{\pi}}\,x(5x^2 - 3r^2)/r^3$
f_{y^3}	$\frac{1}{4}\sqrt{\frac{7}{\pi}}\,y(5y^2 - 3r^2)/r^3$
f_{z^3}	$\frac{1}{4}\sqrt{\frac{7}{\pi}}\,z(5z^2 - 3r^2)/r^3$
$f_{x(z^2-y^2)}$	$\frac{1}{4}\sqrt{\frac{105}{\pi}}\,x(z^2 - y^2)/r^3$
$f_{y(z^2-x^2)}$	$\frac{1}{4}\sqrt{\frac{105}{\pi}}\,y(z^2 - x^2)/r^3$
$f_{z(x^2-y^2)}$	$\frac{1}{4}\sqrt{\frac{105}{\pi}}\,z(x^2 - y^2)/r^3$
f_{xyz}	$\frac{1}{4}\sqrt{\frac{105}{\pi}}\,xyz/r^3$

GENERAL SET:

Orbital	Angular Function $\Theta\Phi$ (x, y, z)
f_{z^3}	$\frac{1}{4}\sqrt{\frac{7}{\pi}}\,z(5z^2 - 3r^2)/r^3$
f_{xz^2}	$\frac{1}{8}\sqrt{\frac{42}{\pi}}\,x(5z^2 - r^2)/r^3$
f_{yz^2}	$\frac{1}{8}\sqrt{\frac{42}{\pi}}\,y(5z^2 - r^2)/r^3$
$f_{y(3x^2-y^2)}$	$\frac{1}{8}\sqrt{\frac{70}{\pi}}\,y(3x^2 - y^2)/r^3$
$f_{x(x^2-3y^2)}$	$\frac{1}{8}\sqrt{\frac{70}{\pi}}\,x(x^2 - 3y^2)/r^3$
f_{xyz}	$\frac{1}{4}\sqrt{\frac{105}{\pi}}\,xyz/r^3$
$f_{z(x^2-y^2)}$	$\frac{1}{4}\sqrt{\frac{105}{\pi}}\,z(x^2 - y^2)/r^3$

APPENDIX B.9: ORBITAL POTENTIAL ENERGIES

Atomic Number	Element	Potential Energy (eV)	
		1*s*	
1	H	−13.61	
2	He	−24.59	
		2*s*	**2*p***
3	Li	−5.39	
4	Be	−9.32	
5	B	−14.05	−8.30
6	C	−19.43	−10.66
7	N	−25.56	−13.18
8	O	−32.38	−15.85
9	F	−40.17	−18.65
10	Ne	−48.47	−21.59
		3*s*	**3*p***
11	Na	−5.14	
12	Mg	−7.65	
13	Al	−11.32	−5.98
14	Si	−15.89	−7.78
15	P	−18.84	−9.65
16	S	−22.71	−11.62
17	Cl	−25.23	−13.67
18	Ar	−29.24	−15.82
		4*s*	**4*p***
19	K	−4.34	
20	Ca	−6.11	
30	Zn	−9.39	
31	Ga	−12.61	−5.93
32	Ge	−16.05	−7.54
33	As	−18.94	−9.17
34	Se	−21.37	−10.82
35	Br	−24.37	−12.49
36	Kr	−27.51	−14.22

Continued

Atomic Number	Element	Potential Energy (eV)	
		5*s*	**5*p***
37	Rb	–4.18	
38	Sr	–5.70	
48	Cd	–8.99	
49	In	–11.89	–5.60
50	Sn	–14.56	–7.01
51	Sb	–16.74	–8.41
52	Te	–18.71	–9.79
53	I	–20.89	–11.18
54	Xe	–23.40	–12.56
		6*s*	**6*p***
55	Cs	–3.90	
56	Ba	–5.21	
	Hg	–10.44	
81	Tl	–13.14	–5.47
82	Pb	–15.12	–6.81
83	Bi	(–17.52)	–8.15
84	Po	(–20.05)	(–9.42)
85	At	(–22.69)	(–10.71)
86	Rn	(–25.47)	–12.03

Source: J. B. Mann, T. L. Meek, and L. C. Allen, *J. Am. Chem. Soc.*, **2000**, *122*, 2780.

C

Character Tables

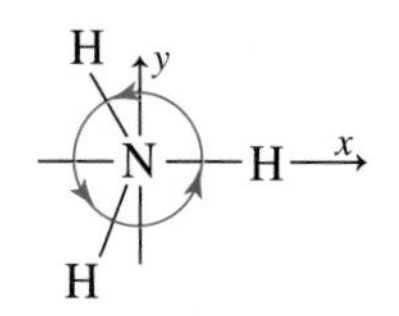

C_{3v}	E	$2C_3$	$3\sigma_v$	
A_1	1	1	1	z
A_2	1	1	−1	R_z
E	2	−1	0	$(x, y), (R_x, R_y)$

NOTE: $i = \sqrt{-1}$; $\varepsilon^* = \varepsilon$ with $-i$ substituted for i.

1. GROUPS OF LOW SYMMETRY

C_1	E
A	1

C_s	E	σ_h		
A'	1	1	x, y, R_z	x^2, y^2, z^2, xy
A''	1	−1	z, R_x, R_y	yz, xz

C_i	E	i		
A_g	1	1	R_x, R_y, R_z	$x^2, y^2, z^2, xy, xz, yz$
A_u	1	−1	x, y, z	

2. C_n, C_{nv}, AND C_{nh} GROUPS

C_n Groups

C_2	E	C_2		
A	1	1	z, R_z	x^2, y^2, z^2, xy
B	1	−1	x, y, R_x, R_y	yz, xz

C_3	E	C_3	C_3^2		
A	1	1	1	z, R_z	$x^2 + y^2, z^2$
E	$\begin{Bmatrix} 1 \\ 1 \end{Bmatrix}$	$\begin{matrix} \varepsilon \\ \varepsilon^* \end{matrix}$	$\begin{matrix} \varepsilon^* \\ \varepsilon \end{matrix}$	$(x, y), (R_x, R_y)$	$(x^2 - y^2, xy), (yz, xz)$

$\varepsilon = e^{(2\pi i)/3}$

C_4	E	C_4	C_2	C_4^3		
A	1	1	1	1	z, R_z	$x^2 + y^2, z^2$
B	1	−1	1	−1		$x^2 - y^2, xy$
E	$\begin{Bmatrix} 1 \\ 1 \end{Bmatrix}$	$\begin{matrix} i \\ -i \end{matrix}$	$\begin{matrix} -1 \\ -1 \end{matrix}$	$\begin{matrix} -i \\ i \end{matrix}$	$(x, y), (R_x, R_y)$	(yz, xz)

C_5	E	C_5	C_5^2	C_5^3	C_5^4		
A	1	1	1	1	1	z, R_z	$x^2 + y^2, z^2$
E_1	1	ε	ε^2	ε^{2*}	ε^*	$(x, y), (R_x, R_y)$	(yz, xz)
	1	ε^*	ε^{2*}	ε^2	ε		
E_2	1	ε^2	ε^*	ε	ε^{2*}		$(x^2 - y^2, xy)$
	1	ε^{2*}	ε	ε^*	ε^2		

$\varepsilon = e^{(2\pi i)/5}$

C_6	E	C_6	C_3	C_2	C_3^2	C_6^5		
A	1	1	1	1	1	1	z, R_z	$x^2 + y^2, z^2$
B	1	−1	1	−1	1	−1		
E_1	1	ε	$-\varepsilon^*$	−1	$-\varepsilon$	ε^*	$(x, y), (R_x, R_y)$	(yz, xz)
	1	ε^*	$-\varepsilon$	−1	$-\varepsilon^*$	ε		
E_2	1	$-\varepsilon^*$	$-\varepsilon$	1	$-\varepsilon^*$	$-\varepsilon$		$(x^2 - y^2, xy)$
	1	$-\varepsilon$	$-\varepsilon^*$	1	$-\varepsilon$	$-\varepsilon^*$		

$\varepsilon = e^{(\pi i)/3}$

C_7	E	C_7	C_7^2	C_7^3	C_7^4	C_7^5	C_7^6		
A	1	1	1	1	1	1	1	z, R_z	$x^2 + y^2, z^2$
E_1	1	ε	ε^2	ε^3	ε^{3*}	ε^{2*}	ε^*	(x, y),	(yz, xz)
	1	ε^*	ε^{2*}	ε^{3*}	ε^3	ε^2	ε	(R_x, R_y)	
E_2	1	ε^2	ε^{3*}	ε^*	ε	ε^3	ε^{2*}		$(x^2 - y^2, xy)$
	1	ε^{2*}	ε^3	ε	ε^*	ε^{3*}	ε^2		
E_3	1	ε^3	ε^*	ε^2	ε^{2*}	ε	ε^{3*}		
	1	ε^{3*}	ε	ε^{2*}	ε^2	ε^*	ε^3		

$\varepsilon = e^{(2\pi i)/7}$

C_8	E	C_8	C_4	C_2	C_4^3	C_8^3	C_8^5	C_8^7		
A	1	1	1	1	1	1	1	1	z, R_z	$x^2 + y^2, z^2$
B	1	−1	1	1	1	−1	−1	−1		
E_1	1	ε	i	−1	$-i$	$-\varepsilon^*$	$-\varepsilon$	ε^*	(x, y),	(yz, xz)
	1	ε^*	$-i$	−1	i	$-\varepsilon$	$-\varepsilon^*$	ε	(R_x, R_y)	
E_2	1	i	−1	1	−1	$-i$	i	$-i$		$(x^2 - y^2, xy)$
	1	$-i$	−1	1	−1	i	$-i$	i		
E_3	1	$-\varepsilon$	i	−1	$-i$	ε^*	ε	$-\varepsilon^*$		
	1	$-\varepsilon^*$	$-i$	−1	i	ε	ε^*	$-\varepsilon$		

$\varepsilon = e^{(\pi i)/4}$

C_{nv} Groups

C_{2v}	E	C_2	$\sigma_v(xz)$	$\sigma_v'(yz)$		
A_1	1	1	1	1	z	x^2, y^2, z^2
A_2	1	1	−1	−1	R_z	xy
B_1	1	−1	1	−1	x, R_y	xz
B_2	1	−1	−1	1	y, R_x	yz

C_{3v}	E	$2C_3$	$3\sigma_v$		
A_1	1	1	1	z	$x^2 + y^2, z^2$
A_2	1	1	−1	R_z	
E	2	−1	0	$(x, y), (R_x, R_y)$	$(x^2 - y^2, xy), (xz, yz)$

C_{4v}	E	$2C_4$	C_2	$2\sigma_v$	$2\sigma_d$		
A_1	1	1	1	1	1	z	$x^2 + y^2, z^2$
A_2	1	1	1	−1	−1	R_z	
B_1	1	−1	1	1	−1		$x^2 - y^2$
B_2	1	−1	1	−1	1		xy
E	2	0	−2	0	0	$(x, y), (R_x, R_y)$	(xz, yz)

C_{5v}	E	$2C_5$	$2C_5^2$	$5\sigma_v$		
A_1	1	1	1	1	z	$x^2 + y^2, z^2$
A_2	1	1	1	−1	R_z	
E_1	2	2 cos 72°	2 cos 144°	0	$(x, y), (R_x, R_y)$	(xz, yz)
E_2	2	2 cos 144°	2 cos 72°	0		$(x^2 - y^2, xy)$

C_{6v}	E	$2C_6$	$2C_3$	C_2	$3\sigma_v$	$3\sigma_d$		
A_1	1	1	1	1	1	1	z	$x^2 + y^2, z^2$
A_2	1	1	1	1	−1	−1	R_z	
B_1	1	−1	1	−1	1	−1		
B_2	1	−1	1	−1	−1	1		
E_1	2	1	−1	−2	0	0	$(x, y), (R_x, R_y)$	(xz, yz)
E_2	2	−1	−1	2	0	0		$(x^2 - y^2, xy)$

C_{nh} Groups

C_{2h}	E	C_2	i	σ_h		
A_g	1	1	1	1	R_z	x^2, y^2, z^2, xy
B_g	1	−1	1	−1	R_x, R_y	xz, yz
A_u	1	1	−1	−1	z	
B_u	1	−1	−1	1	x, y	

C_{3h}	E	C_3	C_3^2	σ_h	S_3	S_3^5		
A'	1	1	1	1	1	1	R_z	$x^2 + y^2, z^2$
E'	$\begin{cases}1\\1\end{cases}$	ε ε^*	ε^* ε	1 1	ε ε^*	ε^* ε	(x, y)	$(x^2 - y^2, xy)$
A''	1	1	1	−1	−1	−1	z	
E''	$\begin{cases}1\\1\end{cases}$	ε ε^*	ε^* ε	−1 −1	$-\varepsilon$ $-\varepsilon^*$	$-\varepsilon^*$ $-\varepsilon$	(R_x, R_y)	(xz, yz)

$\varepsilon = e^{(2\pi i)/3}$

C_{4h}	E	C_4	C_2	C_4^3	i	S_4^3	σ_h	S_4		
A_g	1	1	1	1	1	1	1	1	R_z	$x^2 + y^2, z^2$
B_g	1	−1	1	−1	1	−1	1	−1		$x^2 - y^2, xy$
E_g	$\begin{cases}1\\1\end{cases}$	i $-i$	−1 −1	$-i$ i	1 1	i $-i$	−1 −1	$-i$ i	(R_x, R_y)	(xz, yz)
A_u	1	1	1	1	−1	−1	−1	−1	z	
B_u	1	−1	1	−1	−1	1	−1	1		
E_u	$\begin{cases}1\\1\end{cases}$	i $-i$	−1 −1	$-i$ i	−1 −1	$-i$ i	1 1	i $-i$	(x, y)	

C_{5h}	E	C_5	C_5^2	C_5^3	C_5^4	σ_h	S_5	S_5^7	S_5^3	S_5^9		
A'	1	1	1	1	1	1	1	1	1	1	R_z	$x^2 + y^2, z^2$
E_1'	1	ε	ε^2	ε^{2*}	ε^*	1	ε	ε^2	ε^{2*}	ε^*	(x, y)	
	1	ε^*	ε^{2*}	ε^2	ε	1	ε^*	ε^{2*}	ε^2	ε		
E_2'	1	ε^2	ε^*	ε	ε^{2*}	1	ε^2	ε^*	ε	ε^{2*}		$(x^2 - y^2, xy)$
	1	ε^{2*}	ε	ε^*	ε^2	1	ε^{2*}	ε	ε^*	ε^2		
A''	1	1	1	1	1	−1	−1	−1	−1	−1	z	
E_1''	1	ε	ε^2	ε^{2*}	ε^*	−1	$-\varepsilon$	$-\varepsilon^2$	$-\varepsilon^{2*}$	$-\varepsilon^*$	(R_x, R_y)	(xz, yz)
	1	ε^*	ε^{2*}	ε^2	ε	−1	$-\varepsilon^*$	$-\varepsilon^{2*}$	$-\varepsilon^2$	$-\varepsilon$		
E_2''	1	ε^2	ε^*	ε	ε^{2*}	−1	$-\varepsilon^2$	$-\varepsilon^*$	$-\varepsilon$	$-\varepsilon^{2*}$		
	1	ε^{2*}	ε	ε^*	ε^2	−1	$-\varepsilon^{2*}$	$-\varepsilon$	$-\varepsilon^*$	$-\varepsilon^2$		

$\varepsilon = e^{(2\pi i)/5}$

C_{6h}	E	C_6	C_3	C_2	C_3^2	C_6^5	i	S_3^5	S_6^5	σ_h	S_6	S_3		
A_g	1	1	1	1	1	1	1	1	1	1	1	1	R_z	$x^2 + y^2, z^2$
B_g	1	−1	1	−1	1	−1	1	−1	1	−1	1	−1		
E_{1g}	1	ε	$-\varepsilon^*$	−1	$-\varepsilon$	ε^*	1	ε	$-\varepsilon^*$	−1	$-\varepsilon$	ε^*	(R_x, R_y)	(xz, yz)
	1	ε^*	$-\varepsilon$	−1	$-\varepsilon^*$	ε	1	ε^*	$-\varepsilon$	−1	$-\varepsilon^*$	ε		
E_{2g}	1	$-\varepsilon^*$	$-\varepsilon$	1	$-\varepsilon^*$	$-\varepsilon$	1	$-\varepsilon^*$	$-\varepsilon$	1	$-\varepsilon^*$	$-\varepsilon$		$(x^2 - y^2, xy)$
	1	$-\varepsilon$	$-\varepsilon^*$	1	$-\varepsilon$	$-\varepsilon^*$	1	$-\varepsilon$	$-\varepsilon^*$	1	$-\varepsilon$	$-\varepsilon^*$		
A_u	1	1	1	1	1	1	−1	−1	−1	−1	−1	−1	z	
B_u	1	−1	1	−1	1	−1	−1	1	−1	1	−1	1		
E_{1u}	1	ε	$-\varepsilon^*$	−1	$-\varepsilon$	ε^*	−1	$-\varepsilon$	ε^*	1	ε	$-\varepsilon^*$	(x, y)	
	1	ε^*	$-\varepsilon$	−1	$-\varepsilon^*$	ε	−1	$-\varepsilon^*$	ε	1	ε^*	$-\varepsilon$		
E_{2u}	1	$-\varepsilon^*$	$-\varepsilon$	1	$-\varepsilon^*$	$-\varepsilon$	−1	ε^*	ε	−1	ε^*	ε		
	1	$-\varepsilon$	$-\varepsilon^*$	1	$-\varepsilon$	$-\varepsilon^*$	−1	ε	ε^*	−1	ε	ε^*		

$\varepsilon = e^{(\pi i)/3}$

3. D_n, D_{nd}, AND D_{nh} GROUPS

D_n Groups

D_2	E	$C_2(z)$	$C_2(y)$	$C_2(x)$		
A	1	1	1	1		x^2, y^2, z^2
B_1	1	1	−1	−1	z, R_z	xy
B_2	1	−1	1	−1	y, R_y	xz
B_3	1	−1	−1	1	x, R_x	yz

D_3	E	$2C_3$	$3C_2$		
A_1	1	1	1		$x^2 + y^2, z^2$
A_2	1	1	−1	z, R_z	
E	2	−1	0	$(x, y), (R_x, R_y)$	$(x^2 - y^2, xy), (xz, yz)$

D_4	E	$2C_4$	$C_2 (=C_4^2)$	$2C_2'$	$2C_2''$		
A_1	1	1	1	1	1		$x^2 + y^2, z^2$
A_2	1	1	1	−1	−1	z, R_z	
B_1	1	−1	1	1	−1		$x^2 - y^2$
B_2	1	−1	1	−1	1		xy
E	2	0	−2	0	0	$(x, y), (R_x, R_y)$	(xz, yz)

D_5	E	$2C_5$	$2C_5^2$	$5C_2$		
A_1	1	1	1	1		$x^2 + y^2, z^2$
A_2	1	1	1	−1	z, R_z	
E_1	2	2 cos 72°	2 cos 144°	0	$(x, y), (R_x, R_y)$	(xz, yz)
E_2	2	2 cos 144°	2 cos 72°	0		$(x^2 - y^2, xy)$

D_6	E	$2C_6$	$2C_3$	C_2	$3C_2'$	$3C_2''$		
A_1	1	1	1	1	1	1		$x^2 + y^2, z^2$
A_2	1	1	1	1	−1	−1	z, R_z	
B_1	1	−1	1	−1	1	−1		
B_2	1	−1	1	−1	−1	1		
E_1	2	1	−1	−2	0	0	$(x, y), (R_x, R_y)$	(xz, yz)
E_2	2	−1	−1	2	0	0		$(x^2 - y^2, xy)$

D_{nd} Groups

D_{2d}	E	$2S_4$	C_2	$2C_2'$	$2\sigma_d$		
A_1	1	1	1	1	1		$x^2 + y^2, z^2$
A_2	1	1	1	−1	−1	R_z	
B_1	1	−1	1	1	−1		$x^2 - y^2$
B_2	1	−1	1	−1	1	z	xy
E	2	0	−2	0	0	$(x, y), (R_x, R_y)$	(xz, yz)

D_{3d}	E	$2C_3$	$3C_2$	i	$2S_6$	$3\sigma_d$		
A_{1g}	1	1	1	1	1	1		$x^2 + y^2, z^2$
A_{2g}	1	1	−1	1	1	−1	R_z	
E_g	2	−1	0	2	−1	0	(R_x, R_y)	$(x^2 - y^2, xy), (xz, yz)$
A_{1u}	1	1	1	−1	−1	−1		
A_{2u}	1	1	−1	−1	−1	1	z	
E_u	2	−1	0	−2	1	0	(x, y)	

D_{4d}	E	$2S_8$	$2C_4$	$2S_8^3$	C_2	$4C_2'$	$4\sigma_d$		
A_1	1	1	1	1	1	1	1		$x^2 + y^2, z^2$
A_2	1	1	1	1	1	−1	−1	R_z	
B_1	1	−1	1	−1	1	1	−1		
B_2	1	−1	1	−1	1	−1	1	z	
E_1	2	$\sqrt{2}$	0	$-\sqrt{2}$	−2	0	0	(x, y)	
E_2	2	0	−2	0	2	0	0		$(x^2 - y^2, xy)$
E_3	2	$-\sqrt{2}$	0	$\sqrt{2}$	−2	0	0	(R_x, R_y)	(xz, yz)

D_{5d}	E	$2C_5$	$2C_5^2$	$5C_2$	i	$2S_{10}^3$	$2S_{10}$	$5\sigma_d$		
A_{1g}	1	1	1	1	1	1	1	1		$x^2 + y^2, z^2$
A_{2g}	1	1	1	−1	1	1	1	−1	R_z	
E_{1g}	2	2 cos 72°	2 cos 144°	0	2	2 cos 72°	2 cos 144°	0	(R_x, R_y)	(xz, yz)
E_{2g}	2	2 cos 144°	2 cos 72°	0	2	2 cos 144°	2 cos 72°	0		$(x^2 - y^2, xy)$
A_{1u}	1	1	1	1	−1	−1	−1	−1		
A_{2u}	1	1	1	−1	−1	−1	−1	1	z	
E_{1u}	2	2 cos 72°	2 cos 144°	0	−2	−2 cos 72°	−2 cos 144°	0	(x, y)	
E_{2u}	2	2 cos 144°	2 cos 72°	0	−2	−2 cos 144°	−2 cos 72°	0		

D_{6d}	E	$2S_{12}$	$2C_6$	$2S_4$	$2C_3$	$2S_{12}^5$	C_2	$6C_2'$	$6\sigma_d$		
A_1	1	1	1	1	1	1	1	1	1		$x^2 + y^2, z^2$
A_2	1	1	1	1	1	1	1	−1	−1	R_z	
B_1	1	−1	1	−1	1	−1	1	1	−1		
B_2	1	−1	1	−1	1	−1	1	−1	1	z	
E_1	2	$\sqrt{3}$	1	0	−1	$-\sqrt{3}$	−2	0	0	(x, y)	
E_2	2	1	−1	−2	−1	1	2	0	0		$(x^2 - y^2, xy)$
E_3	2	0	−2	0	2	0	−2	0	0		
E_4	2	−1	−1	2	−1	−1	2	0	0		
E_5	2	$-\sqrt{3}$	1	0	−1	$\sqrt{3}$	−2	0	0	(R_x, R_y)	(xz, yz)

D_{nh} Groups

D_{2h}	E	$C_2(z)$	$C_2(y)$	$C_2(x)$	i	$\sigma(xy)$	$\sigma(xz)$	$\sigma(yz)$		
A_g	1	1	1	1	1	1	1	1		x^2, y^2, z^2
B_{1g}	1	1	−1	−1	1	1	−1	−1	R_z	xy
B_{2g}	1	−1	1	−1	1	−1	1	−1	R_y	xz
B_{3g}	1	−1	−1	1	1	−1	−1	1	R_x	yz
A_u	1	1	1	1	−1	−1	−1	−1		
B_{1u}	1	1	−1	−1	−1	−1	1	1	z	
B_{2u}	1	−1	1	−1	−1	1	−1	1	y	
B_{3u}	1	−1	−1	1	−1	1	1	−1	x	

D_{3h}	E	$2C_3$	$3C_2$	σ_h	$2S_3$	$3\sigma_v$		
A_1'	1	1	1	1	1	1		$x^2 + y^2, z^2$
A_2'	1	1	−1	1	1	−1	R_z	
E'	2	−1	0	2	−1	0	(x, y)	$(x^2 - y^2, xy)$
A_1''	1	1	1	−1	−1	−1		
A_2''	1	1	−1	−1	−1	1	z	
E''	2	−1	0	−2	1	0	(R_x, R_y)	(xz, yz)

D_{4h}	E	$2C_4$	C_2	$2C_2'$	$2C_2''$	i	$2S_4$	σ_h	$2\sigma_v$	$2\sigma_d$		
A_{1g}	1	1	1	1	1	1	1	1	1	1		$x^2 + y^2, z^2$
A_{2g}	1	1	1	−1	−1	1	1	1	−1	−1	R_z	
B_{1g}	1	−1	1	1	−1	1	−1	1	1	−1		$x^2 - y^2$
B_{2g}	1	−1	1	−1	1	1	−1	1	−1	1		xy
E_g	2	0	−2	0	0	2	0	−2	0	0	(R_x, R_y)	(xz, yz)
A_{1u}	1	1	1	1	1	−1	−1	−1	−1	−1		
A_{2u}	1	1	1	−1	−1	−1	−1	−1	1	1	z	
B_{1u}	1	−1	1	1	−1	−1	1	−1	−1	1		
B_{2u}	1	−1	1	−1	1	−1	1	−1	1	−1		
E_u	2	0	−2	0	0	−2	0	2	0	0	(x, y)	

D_{5h}	E	$2C_5$	$2C_5^2$	$5C_2$	σ_h	$2S_5$	$2S_5^3$	$5\sigma_v$		
A_1'	1	1	1	1	1	1	1	1		$x^2 + y^2, z^2$
A_2'	1	1	1	−1	1	1	1	−1	R_z	
E_1'	2	2 cos 72°	2 cos 144°	0	2	2 cos 72°	2 cos 144°	0	(x, y)	
E_2'	2	2 cos 144°	2 cos 72°	0	2	2 cos 144°	2 cos 72°	0		$(x^2 - y^2, xy)$
A_1''	1	1	1	1	−1	−1	−1	−1		
A_2''	1	1	1	−1	−1	−1	−1	1	z	
E_1''	2	2 cos 72°	2 cos 144°	0	−2	−2 cos 72°	−2 cos 144°	0	(R_x, R_y)	(xz, yz)
E_2''	2	2 cos 144°	2 cos 72°	0	−2	−2 cos 144°	−2 cos 72°	0		

D_{6h}	E	$2C_6$	$2C_3$	C_2	$3C_2'$	$3C_2''$	i	$2S_3$	$2S_6$	σ_h	$3\sigma_d$	$3\sigma_v$		
A_{1g}	1	1	1	1	1	1	1	1	1	1	1	1		$x^2 + y^2, z^2$
A_{2g}	1	1	1	1	−1	−1	1	1	1	1	−1	−1	R_z	
B_{1g}	1	−1	1	−1	1	−1	1	−1	1	−1	1	−1		
B_{2g}	1	−1	1	−1	−1	1	1	−1	1	−1	−1	1		
E_{1g}	2	1	−1	−2	0	0	2	1	−1	−2	0	0	(R_x, R_y)	(xz, yz)
E_{2g}	2	−1	−1	2	0	0	2	−1	−1	2	0	0		$(x^2 - y^2, xy)$
A_{1u}	1	1	1	1	1	1	−1	−1	−1	−1	−1	−1		
A_{2u}	1	1	1	1	−1	−1	−1	−1	−1	−1	1	1	z	
B_{1u}	1	−1	1	−1	1	−1	−1	1	−1	1	−1	1		
B_{2u}	1	−1	1	−1	−1	1	−1	1	−1	1	1	−1		
E_{1u}	2	1	−1	−2	0	0	−2	−1	1	2	0	0	(x, y)	
E_{2u}	2	−1	−1	2	0	0	−2	1	1	−2	0	0		

D_{8h}	E	$2C_8$	$2C_8^3$	$2C_4$	C_2	$4C_2'$	$4C_2''$	i	$2S_8$	$2S_8^3$	$2S_4$	σ_h	$4\sigma_d$	$4\sigma_v$		
A_{1g}	1	1	1	1	1	1	1	1	1	1	1	1	1	1		$x^2 + y^2, z^2$
A_{2g}	1	1	1	1	1	−1	−1	1	1	1	1	1	−1	−1	R_z	
B_{1g}	1	−1	−1	1	1	1	−1	1	−1	−1	1	1	1	−1		
B_{2g}	1	−1	−1	1	1	−1	1	1	−1	−1	1	1	−1	1		
E_{1g}	2	$\sqrt{2}$	$-\sqrt{2}$	0	−2	0	0	2	$\sqrt{2}$	$-\sqrt{2}$	0	−2	0	0	(R_x, R_y)	(xz, yz)
E_{2g}	2	0	0	−2	2	0	0	2	0	0	−2	2	0	0		$(x^2 - y^2, xy)$
E_{3g}	2	$-\sqrt{2}$	$\sqrt{2}$	0	−2	0	0	2	$-\sqrt{2}$	$\sqrt{2}$	0	−2	0	0		
A_{1u}	1	1	1	1	1	1	1	−1	−1	−1	−1	−1	−1	−1		
A_{2u}	1	1	1	1	1	−1	−1	−1	−1	−1	−1	−1	1	1	z	
B_{1u}	1	−1	−1	1	1	1	−1	−1	1	1	−1	−1	−1	1		
B_{2u}	1	−1	−1	1	1	−1	1	−1	1	1	−1	−1	1	−1		
E_{1u}	2	$\sqrt{2}$	$-\sqrt{2}$	0	−2	0	0	−2	$-\sqrt{2}$	$\sqrt{2}$	0	2	0	0	(x, y)	
E_{2u}	2	0	0	−2	2	0	0	−2	0	0	2	−2	0	0		
E_{3u}	2	$-\sqrt{2}$	$\sqrt{2}$	0	−2	0	0	−2	$\sqrt{2}$	$-\sqrt{2}$	0	2	0	0		

4. LINEAR GROUPS

$C_{\infty v}$	E	$2C_\infty^\phi$	...	$\infty\sigma_v$		
$A_1 \equiv \Sigma^+$	1	1	...	1	z	x^2+y^2, z^2
$A_2 \equiv \Sigma^-$	1	1	...	−1	R_z	
$E_1 \equiv \Pi$	2	$2\cos\phi$	...	0	$(x, y), (R_x, R_y)$	(xz, yz)
$E_2 \equiv \Delta$	2	$2\cos 2\phi$	...	0		(x^2-y^2, xy)
$E_3 \equiv \Phi$	2	$2\cos 3\phi$	...	0		
...	...	...	...	...		

$D_{\infty h}$	E	$2C_\infty^\phi$	...	$\infty\sigma_v$	i	$2S_\infty^\phi$	...	∞C_2		
$A_{1g} \equiv \Sigma_g^+$	1	1	...	1	1	1	...	1		x^2+y^2, z^2
$A_{2g} \equiv \Sigma_g^-$	1	1	...	−1	1	1	...	−1	R_z	
$E_{1g} \equiv \Pi_g$	2	$2\cos\phi$	...	0	2	$-2\cos\phi$	...	0	(R_x, R_y)	(xz, yz)
$E_{2g} \equiv \Delta_g$	2	$2\cos 2\phi$	...	0	2	$2\cos 2\phi$	...	0		(x^2-y^2, xy)
...	...	...	...	...	...	...	...	...		
$A_{1u} \equiv \Sigma_u^+$	1	1	...	1	−1	−1	...	−1	z	
$A_{2u} \equiv \Sigma_u^-$	1	1	...	−1	−1	−1	...	1		
$E_{1u} \equiv \Pi_u$	2	$2\cos\phi$	...	0	−2	$2\cos\phi$	...	0	(x, y)	
$E_{2u} \equiv \Delta_u$	2	$2\cos 2\phi$	...	0	−2	$-2\cos 2\phi$	...	0		
...	...	...	...	...	...	...	...	...		

5. S_{2n} GROUPS

S_4	E	S_4	C_2	S_4^3		
A	1	1	1	1	R_z	x^2+y^2, z^2
B	1	−1	1	−1	z	x^2-y^2, xy
E	1	i	−1	$-i$	$(x, y), (R_x, R_y)$	(xz, yz)
	1	$-i$	−1	i		

S_6	E	C_3	C_3^2	i	S_6^5	S_6		
A_g	1	1	1	1	1	1	R_z	x^2+y^2, z^2
E_g	1	ε	ε^*	1	ε	ε^*	(R_x, R_y)	(x^2-y^2, xy), (xz, yz)
	1	ε^*	ε	1	ε^*	ε		
A_u	1	1	1	−1	−1	−1	z	
E_u	1	ε	ε^*	−1	$-\varepsilon$	$-\varepsilon^*$	(x, y)	
	1	ε^*	ε	−1	$-\varepsilon^*$	$-\varepsilon$		

$\varepsilon = e^{(2\pi i)/3}$

S_8	E	S_8	C_4	S_8^3	C_2	S_8^5	C_4^3	S_8^7		
A	1	1	1	1	1	1	1	1	R_z	x^2+y^2, z^2
B	1	−1	1	−1	1	−1	1	−1	z	
E_1	1	ε	i	$-\varepsilon^*$	−1	$-\varepsilon$	$-i$	ε^*	(x, y), (R_x, R_y)	
	1	ε^*	$-i$	$-\varepsilon$	−1	$-\varepsilon^*$	i	ε		
E_2	1	i	−1	$-i$	1	i	−1	$-i$		(x^2-y^2, xy)
	1	$-i$	−1	i	1	$-i$	−1	i		
E_3	1	$-\varepsilon^*$	$-i$	ε	−1	ε^*	i	$-\varepsilon$		(xz, yz)
	1	$-\varepsilon$	i	ε^*	−1	ε	$-i$	$-\varepsilon^*$		

$\varepsilon = e^{(\pi i)/4}$

6. TETRAHEDRAL, OCTAHEDRAL, AND ICOSAHEDRAL GROUPS

T	E	$4C_3$	$4C_3^2$	$3C_2$		
A	1	1	1	1		$x^2 + y^2 + z^2$
E	$\left\{\begin{matrix} 1 \\ 1 \end{matrix}\right.$	$\begin{matrix} \varepsilon \\ \varepsilon^* \end{matrix}$	$\begin{matrix} \varepsilon^* \\ \varepsilon \end{matrix}$	$\left.\begin{matrix} 1 \\ 1 \end{matrix}\right\}$		$(2z^2 - x^2 - y^2, x^2 - y^2)$
T	3	0	0	−1	$(R_x, R_y, R_z), (x, y, z)$	(xy, xz, yz)

$\varepsilon = e^{(2\pi i)/3}$

T_d	E	$8C_3$	$3C_2$	$6S_4$	$6\sigma_d$		
A_1	1	1	1	1	1		$x^2 + y^2 + z^2$
A_2	1	1	1	−1	−1		
E	2	−1	2	0	0		$(2z^2 - x^2 - y^2, x^2 - y^2)$
T_1	3	0	−1	1	−1	(R_x, R_y, R_z)	
T_2	3	0	−1	−1	1	(x, y, z)	(xy, xz, yz)

T_h	E	$4C_3$	$4C_3^2$	$3C_2$	i	$4S_6$	$4S_6^5$	$3\sigma_h$		
A_g	1	1	1	1	1	1	1	1		$x^2 + y^2 + z^2$
A_u	1	1	1	1	−1	−1	−1	−1		
E_g	$\left\{\begin{matrix} 1 \\ 1 \end{matrix}\right.$	$\begin{matrix} \varepsilon \\ \varepsilon^* \end{matrix}$	$\begin{matrix} \varepsilon^* \\ \varepsilon \end{matrix}$	$\begin{matrix} 1 \\ 1 \end{matrix}$	$\begin{matrix} 1 \\ 1 \end{matrix}$	$\begin{matrix} \varepsilon \\ \varepsilon^* \end{matrix}$	$\begin{matrix} \varepsilon^* \\ \varepsilon \end{matrix}$	$\left.\begin{matrix} 1 \\ 1 \end{matrix}\right\}$		$(2z^2 - x^2 - y^2, x^2 - y^2)$
E_u	$\left\{\begin{matrix} 1 \\ 1 \end{matrix}\right.$	$\begin{matrix} \varepsilon \\ \varepsilon^* \end{matrix}$	$\begin{matrix} \varepsilon^* \\ \varepsilon \end{matrix}$	$\begin{matrix} 1 \\ 1 \end{matrix}$	$\begin{matrix} -1 \\ -1 \end{matrix}$	$\begin{matrix} -\varepsilon \\ -\varepsilon^* \end{matrix}$	$\begin{matrix} -\varepsilon^* \\ -\varepsilon \end{matrix}$	$\left.\begin{matrix} -1 \\ -1 \end{matrix}\right\}$		
T_g	3	0	0	−1	3	0	0	−1	(R_x, R_y, R_z)	(xy, xz, yz)
T_u	3	0	0	−1	−3	0	0	1	(x, y, z)	

$\varepsilon = e^{(2\pi i)/3}$

O	E	$6C_4$	$3C_2(=C_4^2)$	$8C_3$	$6C_2$		
A_1	1	1	1	1	1		$x^2 + y^2 + z^2$
A_2	1	−1	1	1	−1		
E	2	0	2	−1	0		$(2z^2 - x^2 - y^2, x^2 - y^2)$
T_1	3	1	−1	0	−1	$(R_x, R_y, R_z), (x, y, z)$	
T_2	3	−1	−1	0	1		(xy, xz, yz)

O_h	E	$8C_3$	$6C_2$	$6C_4$	$3C_2(=C_4^2)$	i	$6S_4$	$8S_6$	$3\sigma_h$	$6\sigma_d$		
A_{1g}	1	1	1	1	1	1	1	1	1	1		$x^2+y^2+z^2$
A_{2g}	1	1	−1	−1	1	1	−1	1	1	−1		
E_g	2	−1	0	0	2	2	0	−1	2	0		$(2z^2-x^2-y^2, x^2-y^2)$
T_{1g}	3	0	−1	1	−1	3	1	0	−1	−1	(R_x, R_y, R_z)	
T_{2g}	3	0	1	−1	−1	3	−1	0	−1	1		(xy, xz, yz)
A_{1u}	1	1	1	1	1	−1	−1	−1	−1	−1		
A_{2u}	1	1	−1	−1	1	−1	1	−1	−1	1		
E_u	2	−1	0	0	2	−2	0	1	−2	0		
T_{1u}	3	0	−1	1	−1	−3	−1	0	1	1	(x, y, z)	
T_{2u}	3	0	1	−1	−1	−3	1	0	1	−1		

I	E	$12C_5$	$12C_5^2$	$20C_3$	$15C_2$		
A	1	1	1	1	1		$x^2+y^2+z^2$
T_1	3	$\frac{1}{2}(1+\sqrt{5})$	$\frac{1}{2}(1-\sqrt{5})$	0	−1	$(x, y, z), (R_x, R_y, R_z)$	
T_2	3	$\frac{1}{2}(1-\sqrt{5})$	$\frac{1}{2}(1+\sqrt{5})$	0	−1		
G	4	−1	−1	1	0		
H	5	0	0	−1	1		$(xy, xz, yz, x^2-y^2, 2z^2-x^2-y^2)$

I_h	E	$12C_5$	$12C_5^2$	$20C_3$	$15C_2$	i	$12S_{10}$	$12S_{10}^3$	$20S_6$	15σ		
A_g	1	1	1	1	1	1	1	1	1	1		$x^2+y^2+z^2$
T_{1g}	3	$\frac{1}{2}(1+\sqrt{5})$	$\frac{1}{2}(1-\sqrt{5})$	0	−1	3	$\frac{1}{2}(1-\sqrt{5})$	$\frac{1}{2}(1+\sqrt{5})$	0	−1	(R_x, R_y, R_z)	
T_{2g}	3	$\frac{1}{2}(1-\sqrt{5})$	$\frac{1}{2}(1+\sqrt{5})$	0	−1	3	$\frac{1}{2}(1+\sqrt{5})$	$\frac{1}{2}(1-\sqrt{5})$	0	−1		
G_g	4	−1	−1	1	0	4	−1	−1	1	0		
H_g	5	0	0	−1	1	5	0	0	−1	1		$(2z^2-x^2-y^2, x^2-y^2, xy, xz, yz)$
A_u	1	1	1	1	1	−1	−1	−1	−1	−1		
T_{1u}	3	$\frac{1}{2}(1+\sqrt{5})$	$\frac{1}{2}(1-\sqrt{5})$	0	−1	−3	$-\frac{1}{2}(1-\sqrt{5})$	$-\frac{1}{2}(1+\sqrt{5})$	0	1	(x, y, z)	
T_{2u}	3	$\frac{1}{2}(1-\sqrt{5})$	$\frac{1}{2}(1+\sqrt{5})$	0	−1	−3	$-\frac{1}{2}(1+\sqrt{5})$	$-\frac{1}{2}(1-\sqrt{5})$	0	1		
G_u	4	−1	−1	1	0	−4	1	1	−1	0		
H_u	5	0	0	−1	1	−5	0	0	1	−1		

INDEX

F

G

H

M

Q

R